UK AIM
The UK Aeronautical Information Manual

The UK Aeronautical Information Manual
(UK AIM) 2006

CAMBER

1st Edition 2001
2nd Edition 2002
3rd Edition 2005
4th Edition 2006

© Copyright 2001, 2002, 2005 & 2006 Camber Publishing Limited
Crown copyright

Crown copyright legislation and explanatory notes are reproduced under the terms of Crown Copyright Policy Guidance issued by HMSC.

The assistance of the Aeronautical Information Service is gratefully acknowledged.

Aeronautical Information Circulars are reproduced in the interests of flight safety for information, guidance and necessary action.

Copyright in the typographical *arrangement, design and layout vests in* the *publisher. No part of this publication may be reproduced, stored in a retrieval system, or transmitted in any form or by any means, electronic or otherwise, without the prior permission of the copyright holder.*

The UK Aeronautical Information Manual (UK AIM)

Effective Information Date 16 March 2006

ISBN 1 874783 49 7

Published by
Camber Publishing Limited
1a Ringway Trading Estate,
Shadowmoss Road,
Manchester M22 5LH
Tel: 0161 499 0023 **Fax:** 0161 499 0298

Distributed by:

Airplan Flight Equipment Limited
1a Ringway Trading Estate,
Shadowmoss Road,
Manchester M22 5LH
Tel: 0161 499 0023 **Fax:** 0161 499 0298
www.afeonline.com

Editorial information:

Compiler:	Louise Southern
Editor:	Louise Southern
Contributors:	David Cockburn
	Civil Aviation Authority
	Jeremy M Pratt
	Louise Southern
	Dr John Wray
	Chris Walsh
Design:	Robert Taylor, GDi studio

EFFECTIVE INFORMATION DATE:
16 March 2006

IMPORTANT
The UK Aeronautical Information Manual is **a guide only** and is not to be taken as an authoritative document. Whilst reasonable efforts have been made to check the accuracy of this publication against the official source documents, the publisher and the editorial team will not be liable in any way whatsoever for any errors, inaccuracies or omissions. The information on which this guide is based can, and does, change frequently. The user is strongly recommended to check information against the official sources such as the UK AIP, current AICs and other information published by the Authority and the Aeronautical Information Service.

How to use the UK Aeronautical Information Manual

The UK Aeronautical Information Manual (UK AIM) is based on official information sources, including Acts of Parliament, Aeronautical information Service (AIS) publications and Aeronautical Information Circulars (AICs). Whilst reasonable efforts have been made to check the accuracy of this guide against the relevant official information source documents, the UK AIM is still to be treated as a guide only, not an authoritative document. Only the official documentation (e.g. Parliamentary legislation, Aeronautical Information Publication, Aeronautical Information Circulars etc.), as amended and revised, can be treated as authoritative. If you have not done so already, please read the important note on the copyright information page.

For convenience of use, this guide uses the same numbering systems and order as the original source documents, and is broadly arranged into the equivalents of the Aeronautical Information Publication (AIP), Air Navigation Order (ANO) and selected Aeronautical Information Circulars (AICs). The contents list shows the main subject headings of each section, and there is a subject index at the back of this book referring to the relevant sub-sections.

For those studying for a UK JAR PPL or an IMC rating there are lists of recommended study sections in an appendix at the back of this publication.

The source documentation is subject to frequent amendment and revision. This guide has been compiled to the effective information date shown on the copyright information page. There is no formal revision process for this publication at present, other than the fact that it will be republished annually.

The aim of this publication is to give pilots and others involved in the aviation industry an inexpensive, userfriendly, practical and portable guide to the legislation, rules and procedures that govern them. All suggestions and comments will be gratefully received by the publisher and the editorial team via our website at: www.afeonline.com

Contents

AERONAUTICAL INFORMATION PUBLICATION

GENERAL

AIP GEN 0 1	PREFACE	21
AIP GEN 1	NATIONAL REGULATIONS AND REQUIREMENTS	23
AIP GEN 1.1	Designated authorities	23
AIP GEN 1.2	Entry, transit and departure of aircraft	24
AIP GEN 1.3	Entry, transit and departure of passengers and crew	34
AIP GEN 1.4	Hazardous Cargo Regulations	36
AIP GEN 1.5	Aircraft instruments, equipment and flight documents	37
AIP GEN 1.7	Differences from ICAO standards, recommended practices and procedures	42
AIP GEN 2	TABLES AND CODES	74
AIP GEN 2.1	Measuring system, aircraft markings, holidays	74
AIP GEN 2.2	Abbreviations used in AIS publications	76
AIP GEN 2.3	Chart symbols	86
AIP GEN 2.4	Location indicators encode/decode	91
AIP GEN 2.5	List of radio navigation aids encode/decode	96
AIP GEN 2.6	Conversion tables	103
AIP GEN 3	SERVICES	104
AIP GEN 3.1	Aeronautical information services	104
AIP GEN 3.2	Aeronautical charts	109
AIP GEN 3.3	Air traffic services	116
AIP GEN 3.4	Communication services	122
AIP GEN 3.5	Meteorological services	131
AIP GEN 3.6	Search and rescue	162
AIP GEN 4	CHARGES FOR AERODROMES/HELIPORTS AND AIR NAVIGATION SERVICES	171
AIP GEN 4.1	Aerodrome/heliport charges	171
AIP GEN 4.2	Air navigation services charges	172

ENROUTE

AIP ENR 1	GENERAL RULES AND PROCEDURES	176
AIP ENR 1.1	General rules	176
AIP ENR 1.2	Visual flight rules	201
AIP ENR 1.3	Instrument flight rules	202
AIP ENR 1.4	ATS airspace classification	203
AIP ENR 1.5	Holding, approach and departure procedures	211
AIP ENR 1.6	Radar services and procedures	215
AIP ENR 1.7	Flight level graph	230
AIP ENR 1.8	Regional supplementary procedures (Doc 7030)	235
AIP ENR 1.10	Flight planning	235
AIP ENR 1.11	Addressing of flight plan messages	248
AIP ENR 1.12	Interception of civil aircraft	249
AIP ENR 1.14	Air traffic incidents	253
AIP ENR 1.15	Off-shore operations	255
AIP ENR 2.2	Other regulated airspace, military aerodrome traffic zones	265
AIP ENR 5.5	Aerial sporting and recreational activities	268
AIP ENR 5.6	Bird migration and areas with sensitive fauna	287
AIP ENR 6	EN ROUTE CHARTS	289
	Lower airspace radar service	289
	Military middle airspace radar service	290
	Warton radar advisory service area	291
	UK altimeter setting and flight information regions	292
	London ACC – FIS sectors	294
	Scottish ACC – FIS sectors	295
	Military aerodrome traffic zones	296
	Class B gliding areas south of 5520N	297
	Class B gliding areas north of 5520N	298
	Bird concentration areas March – July	299
	Bird concentration areas Oct – March	300

Contents

AERODROMES
AIP AD 1	AERODROME/HELIPORTS INTRODUCTION	303
AIP AD 1.1	Aerodrome/heliport availability	303
AIP AD 1.2	Rescue and fire fighting services and snow plan	317
AIP AD 1.4	Grouping of aerodromes/heliports	321

AIR NAVIGATION ORDER 323

ARTICLES TO THE AIR NAVIGATION ORDER
CITATION, COMMENCEMENT AND REVOCATION 329
Article 1	Citation and commencement	329
Article 2	Revocation	329

PART 1 — REGISTRATION AND MARKING OF AIRCRAFT 329
Article 3	Aircraft to be registered	329
Article 4	Registration of aircraft in the UK	329
Article 5	Nationality and registration marks	330

PART 2 — AIR OPERATORS' CERTIFICATES 331
Article 6	Grant of air operators' certificates	331
Article 7	Grant of police air operators' certificates	331

PART 3 — AIRWORTHINESS AND EQUIPMENT OF AIRCRAFT 331
Article 8	Certificate of airworthiness to be in force	331
Article 9	Issue, renewal, etc., of certificates of airworthiness	331
Article 10	Validity of certificate of airworthiness	332
Article 11	Issue, renewal etc; of permits to fly	332
Article 12	Issue of EASA permits fly	333
Article 13	Issue etc; certificates of validation of permits to fly or equivalent document	333
Article 14	Certificate of maintenance review	333
Article 15	Technical log	334
Article 16	Requirement for a certificate of release to service	334
Article 17	Requirement for a certificate of release to service under Part 145	335
Article 18	Licensing of maintenance engineers	335
Article 19	Equipment of aircraft	336
Article 20	Radio equipment of aircraft	336
Article 21	Minimum equipment requirements	337
Article 22	Aircraft, engine and propeller log books	337
Article 23	Aircraft weight schedule	337
Article 24	Access and inspection for airworthiness purposes	337

PART 4 — AIRCRAFT CREW AND LICENSING 338
Article 25	Composition of crew of aircraft	338
Article 26	Members of flight crew – requirement for licence	339
Article 27	Grant, renewal and effect of flight crew licences	340
Article 28	Maintenance of privileges of aircraft ratings in United Kingdom licences	341
Article 29	Maintenance of privileges of aircraft ratings in JAR-FCL licences, United Kingdom licences for which there are JAR-FCL equivalents, United Kingdom Basic Commercial Pilot's licences and United Kingdom Flight Engineer's Licences	342
Article 30	Maintenance of privileges of aircraft ratings in National Private Pilot's Licences	342
Article 31	Maintenance of privileges of other ratings	342
Article 32	Medical requirements	342
Article 33	Miscellaneous licensing provisions	343
Article 34	Validation of licences	343
Article 35	Personal flying logbook	343
Article 36	Instruction in flying	343
Article 37	Glider pilot – minimum age	344

PART 5 — OPERATION OF AIRCRAFT 344
Article 38	Operations Manual	344
Article 39	Police operations manual	344
Article 40	Training Manual	345
Article 41	Flight data monitoring, accident prevention and flight safety programme	345

Contents

Article 42	Public transport – operator's responsibilities	345
Article 43	Loading – public transport aircraft and suspended loads	345
Article 44	Public transport – aeroplanes -operating conditions and performance requirements	346
Article 45	Public transport – helicopters - operating conditions and performance requirements	347
Article 46	Public transport operations at night or in instrument meteorological conditions aeroplanes with one power unit which are registered elsewhere than in the United Kingdom	348
Article 47	Public transport aircraft registered in the United Kingdom – aerodrome operating minima	348
Article 48	Public transport aircraft registered elsewhere than in the United Kingdom – aerodrome operating minima	349
Article 49	Non-public transport aircraft – aerodrome operating minima	349
Article 50	Pilots to remain at controls	350
Article 51	Wearing of survival suits by crew	350
Article 52	Pre-flight action by commander of aircraft	350
Article 53	Passenger briefing by commander	350
Article 54	Public transport of passengers – additional duties of commander	351
Article 55	Operation of radio in aircraft	351
Article 56	Minimum navigation performance	352
Article 57	Height keeping performance – aircraft registered in the United Kingdom	352
Article 58	Height keeping performance – aircraft registered elsewhere than in the United Kingdom	352
Article 59	Area navigation and required navigation performance capabilities – aircraft registered in the United Kingdom	352
Article 60	Area navigation and required navigation performance capabilities – aircraft registered elsewhere than in the United Kingdom	353
Article 61	Use of airborne collision avoidance system	353
Article 62	Use of flight recording systems and preservation of records	353
Article 63	Towing of gliders	353
Article 64	Operation of self-sustaining gliders	354
Article 65	Towing, picking up and raising of persons and articles	354
Article 66	Dropping of articles and animals	354
Article 67	Dropping of persons and grant of parachuting permissions	354
Article 68	Grant of aerial application certificates	355
Article 69	Carriage of weapons and of munitions of war	355
Article 70	Carriage of dangerous goods	356
Article 71	Method of carriage of persons	356
Article 72	Exits and break-in markings	356
Article 73	Endangering safety of an aircraft	357
Article 74	Endangering safety of any person or property	357
Article 75	Drunkenness in aircraft	357
Article 76	Smoking in aircraft	357
Article 77	Authority of commander of an aircraft	357
Article 78	Acting in a disruptive manner	357
Article 79	Stowaways	357
Article 80	Flying displays	357
PART 6	**FATIGUE OF CREW AND PROTECTION OF CREW FROM COSMIC RADIATION**	358
Article 81	Application and interpretation of Part 6	358
Article 82	Fatigue of crew – operator's responsibilities	358
Article 83	Fatigue of crew – responsibilities of crew	359
Article 84	Flight times - responsibilities of flight crew	359
Article 85	Protection of aircrew from cosmic radiation	359
PART 7	**DOCUMENTS AND RECORDS**	359
Article 86	Documents to be carried	359
Article 87	Keeping and production of records of exposure to cosmic radiation	359
Article 88	Production of documents and records	360
Article 89	Production of air traffic service equipment documents and records	360
Article 90	Power to inspect and copy documents and records	360
Article 91	Preservation of documents, etc.	360
Article 92	Revocation, suspension and variation of certificates, licences and other documents	360
Article 93	Revocation, suspension and variation of permissions, etc. granted under article 138 or article 140	361
Article 94	Offences in relation to documents and records	361

Contents

PART 8	**MOVEMENT OF AIRCRAFT**	362
Article 95	Rules of the air	362
Article 96	Power to prohibit or restrict flying	362
Article 97	Balloons, kites, airships, gliders and parascending parachutes	362
Article 98	Regulation of small aircraft	363
Article 99	Regulation of rockets	363
PART 9	**AIR TRAFFIC SERVICES**	364
Article 100	Requirement for an air traffic control approval	364
Article 101	Duty of person in charge to satisfy himself as to competence of controllers	364
Article 102	Manual of air traffic services	364
Article 103	Provision of air traffic services	364
Article 104	Making of an air traffic direction in the interests of safety	364
Article 105	Making of a direction for airspace policy purposes	365
Article 106	Use of radio call signs at aerodromes	365
PART 10	**LICENSING OF AIR TRAFFIC CONTROLLERS**	365
Article 107	Prohibition of unlicensed air traffic controllers and student air traffic controllers	365
Article 108	Grant and renewal of air traffic controller's and student air traffic controller's licences	365
Article 109	Privileges of an air traffic controller licence or a student air traffic controller licence	366
Article 110	Maintenance of validity of ratings and endorsements	366
Article 111	Obligation to notify rating ceasing to be valid and change of unit	366
Article 112	Requirement for medical certificate	366
Article 113	Appropriate licence	366
Article 114	Incapacity of air traffic controllers	366
Article 115	Fatigue of air traffic controllers – air traffic controller's responsibilities	366
Article 116	Prohibition of drunkenness etc. of controllers	366
Article 117	Failing exams	366
Article 118	Use of simulators	367
Article 119	Approval of courses and persons	367
Article 120	Acting as an air traffic controller and a student air traffic controller	367
PART 11	**FLIGHT INFORMATION SERVICES AND LICENSING OF FLIGHT INFORMATION SERVICE OFFICERS**	367
Article 121	Prohibition of unlicensed flight information service officers	367
Article 122	Licensing of flight information service officers	367
Article 123	Flight information service manual	367
PART 12	**AIR TRAFFIC SERVICE EQUIPMENT**	368
Article 124	Air traffic service equipment	368
Article 125	Air traffic service equipment records	368
PART 13	**AERODROMES, AERONAUTICAL LIGHTS AND DANGEROUS LIGHTS**	369
Article 126	Aerodromes – public transport of passengers and instruction in flying	369
Article 127	Use of Government aerodromes	369
Article 128	Licensing of aerodromes	369
Article 129	Charges at aerodromes licensed for public use	370
Article 130	Use of aerodromes by aircraft of Contracting States and of the Commonwealth	370
Article 131	Noise and vibration caused by aircraft on aerodromes	370
Article 132	Aeronautical lights	370
Article 133	Lighting of en-route obstacles	371
Article 134	Lighting if wind turbine generators in United Kingdom territorial waters	371
Article 135	Dangerous lights	372
Article 136	Customs and Excise aerodromes	372
Article 137	Aviation fuel at aerodromes	372
PART 14	**GENERAL**	372
Article 138	Restriction with respect to carriage for valuable consideration in aircraft registered elsewhere than the UK	372
Article 139	Filing and approval of tariffs	373
Article 140	Restriction on aerial photography, aerial survey and aerial work in aircraft registered elsewhere than in the UK	373

Contents

Article 141	Flights over any foreign country	373
Article 142	Mandatory reporting of occurrences	373
Article 143	Mandatory reporting of birdstrikes	374
Article 144	Power to prevent aircraft flying	375
Article 145	Right of access to aerodromes and other places	375
Article 146	Obstruction of persons	375
Article 147	Directions	375
Article 148	Penalties	375
Article 149	Extra-territorial effect of the Order	376
Article 150	Aircraft in transit over certain United Kingdom territorial waters	376
Article 151	Application of Order to British -controlled aircraft registered elsewhere than in the United Kingdom	376
Article 152	Application of Order to the Crown and visiting forces, etc.	376
Article 153	Exemption from Order	376
Article 154	Appeal to County Court or Sheriff Court	377
Article 155	Interpretation	377
Article 156	Meaning of aerodrome traffic zone	385
Article 157	Public transport and aerial work – general rules	385
Article 158	Public transport and aerial work – exceptions – flying displays etc.	386
Article 159	Public transport and aerial work – exceptions – charity flights	386
Article 160	Public transport and aerial work – exceptions – cost sharing	386
Article 161	Public transport and aerial work – exceptions – recovery of direct costs	387
Article 162	Public transport and aerial work – exceptions – jointly owned aircraft	387
Article 163	Public transport and aerial work – exceptions – parachuting	387
Article 164	Exceptions from application of provisions of the order for certain classes of aircraft	387
Article 165	Approval of persons to furnish reports	387
Article 166	Certificates, authorisations, approvals and permissions	388
Article 167	Competent authority	388
Article 168	Saving	388

SCHEDULES TO THE AIR NAVIGATION ORDER		389
Schedule 1	Orders revoked	389
Schedule 3	A and B Conditions and categories of certificate of airworthiness	389
	Part A A and B Conditions	392
	Part B Categories of certificate of airworthiness and purposes for which aircraft may fly	392
Schedule 4	Aircraft equipment	393
Schedule 5	Radio communication and radio navigation equipment to be carried in aircraft	410
Schedule 6	Aircraft, engine and propeller log books	414
Schedule 7	Areas specified in connection with the carriage of flight navigators as members of the flight crews or suitable navigational equipment on public transport aircraft	415
Schedule 8	Flight crew of aircraft – licences, ratings, qualification and maintenance of licence privileges	416
	Part A Flight crew licences	416
	Part B Ratings and qualifications	425
	Part C Maintenance of licence privileges	427
Schedule 9	Public Transport – operational requirements	431
	Part A Operations Manual	431
Schedule 10	Part B Training Manual	431
	Part C Crew training and tests	432
Schedule 11	Air traffic controllers – licences, ratings, endorsements and maintenance of licence privileges	
	Part A Air traffic controller licences	434
	Part B Ratings, rating endorsements and licence endorsements	435
Schedule 12	Air traffic service equipment – records required and matters to which the CAA may have regard	436
	Part A Records to be kept in accordance with article 125(1)	436
	Part B Records required in accordance with article 125(4)©	436
	Part C Matters to which the CAA may have regard in granting an approval of apparatus under article 125(5)	436
Schedule 13	Aerodrome manual	437
Schedule 14	Penalties	437
	Part A Provisions referred to in article 148(5)	437
	Part B Provisions referred to in article 148(6)	440
	Part C Provisions referred to in article 148(7)	440
Schedule 15	Parts of straits specified in connection with the flight of aircraft in transit over the United Kingdom territorial waters	441

Contents

THE RULES OF THE AIR		443
Section I	**INTERPRETATION**	444
Rule 1	Interpretation	444
Section II	**GENERAL**	444
Rule 2	Application of Rules to aircraft	444
Rule 3	Misuse of signals and markings	444
Rule 4	Reporting hazardous conditions	444
Rule 5	Low flying	444
Rule 6	Simulated instrument flight	446
Rule 7	Practice instrument approaches	446
Section III	**LIGHTS AND OTHER SIGNALS TO BE SHOWN OR MADE BY AIRCRAFT**	446
Rule 8	General	446
Rule 9	Display of lights by aircraft	446
Rule 10	Failure of navigation and anti-collision lights	446
Rule 11	Flying machines	447
Rule 12	Gliders	447
Rule 13	Free balloons	447
Rule 14	Captive balloons and kites	447
Rule 15	Airships	448
Section IV	**GENERAL FLIGHT RULES**	448
Rule 16	Weather reports and forecasts	448
Rule 17	Rules for avoiding aerial collisions	448
Rule 18	Aerobatic manoeuvres	449
Rule 19	Right-hand traffic rule	449
Rule 20	Notification of arrival and departure	449
Rule 21	Flight in Class A airspace	450
Rule 22	Choice of VFR or IFR	450
Rule 23	Speed limitation	450
Section V	**VISUAL FLIGHT RULES**	450
Rule 24	Visual flight and reported visibility	450
Rule 25	Flight within controlled airspace	450
Rule 26	Flight outside controlled airspace	451
Rule 27	VFR flight plan and air traffic control clearance	451
Section VI	**INSTRUMENT FLIGHT RULES**	452
Rule 28	Instrument Flight Rules	452
Rule 29	Minimum height	452
Rule 30	Quadrantal rule and semi-circular rule	452
Rule 31	Flight plan and air traffic control clearance	453
Rule 32	Position reports	453
Section VII	**AERODROME TRAFFIC RULES**	453
Rule 33	Application of aerodrome traffic rules	453
Rule 34	Visual signals	453
Rule 35	Movement of aircraft on aerodromes	453
Rule 36	Access to and movement of persons and vehicles on the aerodrome	453
Rule 37	Right of way on the ground	453
Rule 38	Launching, picking up and dropping of tow ropes, etc.	454
Rule 39	Flight within aerodrome traffic zones	454
Section VIII	**SPECIAL RULES**	454
Rule 40	Use of radio navigation aids	454
Section IX	**AERODROME SIGNALS AND MARKINGS – VISUAL AND AURAL SIGNALS**	454
Rule 41	General	454
Rule 42	Signals in the signals area	455
Rule 43	Markings for paved runways and taxiways	456
Rule 44	Markings on unpaved manoeuvring areas	457
Rule 45	Signals visible from the ground	458
Rule 46	Lights and pyrotechnic signals for control of aerodrome traffic	459
Rule 47	Marshalling signals (from marshaller to an aircraft)	460

Contents

Rule 48	Marshalling signals (from a pilot of an aircraft to a marshaller)	462
Rule 49	Distress, urgency and safety signals	462

THE AIR NAVIGATION GENERAL REGULATIONS 2005 463

PART 2	LOAD SHEETS	464
Regulation 4	Particulars and weighing requirements	464
PART 3	AIRCRAFT PERFORMANCE	465
Regulation 5	Aeroplanes to which article 44(5) applies	465
Regulation 6	Helicopters to which article 45(1) applies	465
Regulation 7	Weight and performance: general provisions	466
PART 4	NOISE AND VIBRATION, MAINTENANCE AND AERODROME FACILITIES	466
Regulation 8	Noise and vibration caused by aircraft on aerodromes	466
Regulation 9	Pilots maintenance – prescribed repairs or replacements	466
Regulation 10	Aeroplanes flying for the purpose of public transport of passengers – aerodrome facilities for approach to landing and landing	467
PART 5	MANDATORY REPORTING	467
Regulation 11	Reportable occurrences, time and manner of reporting and information to be reported	467
Regulation 12	Mandatory reporting of bird strikes – time and manner of reporting and information to be reported	468
PART 6	NAVIGATION PERFORMANCE AND EQUIPMENT	468
Regulation 13	Minimum navigation performance and height keeping equipment	468
Regulation 14	Airborne Collision Avoidance System	469
Regulation 15	Mode S Transponder	469

SCHEDULE 2	AEROPLANE PERFORMANCE	469
1	Weight and performance of public transport aeroplanes designated as aeroplanes of performance group A or performance group B	469
2	Weight and performance of public transport aeroplanes designated as aeroplanes of performance group C	471
3	Weight and performance of public transport aeroplanes designated as aeroplanes of performance group D	473
4	Weight and performance of public transport aeroplanes designated as aeroplanes of performance group E	474
5	Weight and performance of public transport aeroplanes designated as aeroplanes of performance group F	475
	Weight and performance of public transport aeroplanes designated as aeroplanes of performance group X	476
	Weight and performance of public transport aeroplanes designated as aeroplanes of performance group Z	477

SCHEDULE 3	HELICOPTER PERFORMANCE	480
1	Weight and performance of public transport helicopters carrying out Performance Class 1 operation	480
2	Weight and performance of public transport helicopters carrying out Performance Class 2 operation	480
3	Weight and performance of public transport helicopters carrying out Performance Class 3 operation	481

THE AIR NAVIGATION (RESTRICTION OF FLYING) (SCOTTISH HIGHLANDS) REGULATIONS 1981...............482

THE AIR NAVIGATION (RESTRICTION OF FLYING) (NUCLEAR INSTALLATIONS) REGULATIONS 2002 483

THE AIR NAVIGATION (RESTRICTION OF FLYING) (PRISONS) REGULATIONS 2001...............................484

THE AIR NAVIGATION (RESTRICTION OF FLYING (HIGHGROVE HOUSE) REGULATIONS 1991...................486

THE AIR NAVIGATION (RESTRICTION OF FLYING) (SPECIFIED AREA) REGULATIONS 2005 486

THE AIR NAVIGATION (RESTRICTION OF FLYING) (HYDE PARK) REGULATIONS 2004................................487

THE AIR NAVIGATION (RESTRICTION OF FLYING) (ISLE OF DOGS) REGULATIONS 2004............................488

Contents

THE AIR NAVIGATION (RESTRICTION OF FLYING) (CITY OF LONDON) REGULATIONS 2004..................488

DANGEROUS GOODS REGULATIONS 2002..489

PART I	**PRELIMARY**..489	
Article 1	Citation and commencement..489	
Article 2	Revocation..489	
Article 3	Interpretation..489	
PART II	**REQUIREMENTS FOR CARRIAGE OF DANGEROUS GOODS**........................490	
Article 4	Requirement for approval of operator..490	
Article 5	Prohibition of carriage of dangerous goods...490	
PART III	**OPERATORS OBLIGATIONS**...491	
Article 6	Provision of information by the operator to crew etc....................................491	
Article 7	Acceptance of dangerous goods by the operator..491	
Article 8	Method of loading by operator...492	
Article 9	Inspections by the operator for damage, leakage or contamination..............492	
Article 10	Removal of contamination by the operator..492	
PART IV	**SHIPPERS RESPONSIBILITIES**...492	
Article 11	Shipper's responsibilities..492	
PART V	**COMMANDERS OBLIGATIONS**..493	
Article 12	Commander's duty to inform air traffic services..493	
PART VI	**TRAINING**...493	
Article 13	Provision of training..493	
PART VII	**PROVISION OF INFORMAITON TO PASSENGERS AND IN RESPECT OF CARGO**................493	
Article 14	Provision of information to passengers...493	
Article 15	Provision of information in respect of cargo...494	
PART VIII	**DOCUMENTS AND RCORDS, ENFORCEMENR POWERS AND GENERAL**........................494	
Article 16	Keeping of documents and records...494	
Article 17	Production of documents and records..494	
Article 18	Powers in relation to enforcement of the Regulations..................................494	
Article 19	Occurrence reporting...495	
Article 20	Dropping articles for agricultural, horticultural, forestry or pollution control purposes	
Article 21	Police aircraft...495	

CIVIL AVIATION REGULATIONS 1991..496

PART 1	**GENERAL**..496	
Article 1	Citation and commencement...496	
Article 2	Revocation..496	
Article 3	Interpretation..496	
Article 4	Service of documents..497	
Article 5	Publication by the Authority...498	
PART II	**FUNCTIONS CONFERRED ON THE AUTHORITY BY OR UNDER**........................498	
	AIR NAVIGATION ORDERS	
Article 6	Regulation of the conduct of the Authority..498	
Article 7	Reasons for decisions..499	
Article 8	Inspection of aircraft register...499	
Article 9	Dissemination of reports of reportable occurrences.....................................499	
Article 10-14	Substitution of a public use aerodrome licence for an ordinary aerodrome licence for a public use aerodrome licence	...499

Contents

PART III	AIR TRANSPORT LICENCING	501
Article 15	Regulation of the conduct of the Authority	501
Article 16	Application for the grant, revocation, suspension or variation of licences	501
Article 17	Revocation, suspension or variation of licences without application being made	502
Article 18	Variation of schedules of terms	502
Article 19	Environmental cases	503
Article 20	Objections and representations	503
Article 21	Consultation by the Authority	503
Article 22	Furnishing of information by the Authority	504
Article 23	Preliminary hearings	504
Article 24	Preliminary hearings of allegations of behaviour damaging to a competitor	504
Article 25	Hearings in connection with licences	505
Article 26	Procedure at hearings	505
Article 27	Appeals to the Secretary of State	506
Article 28	Appeal from decisions after preliminary hearings of allegations of behaviour damaging to a competitor	507
Article 29	Decisions of appeals	507
Article 30	Transfer of licences	508
Article 31	Surrender of licences	508
PART IIIA	REFERENCES IN RESPECT OF AN AIR TRAFFIC SERVICES LICENCE	508
Article 31 A	Determination by the authority	508
Article 31 B	Representations	508
Article 31 C	Hearings in connection with licences	509
Article 31 D	Procedure at hearings	509
Article 31 E	Determination by Authority and Appeal to the Secretary of State	509
Article 31 F	Decision by Secretary of State on appeal	510
PART IV	OTHER FUNCTIONS OF THE AUTHORITY	510
Article 32	Participation in civil proceedings	510
GENERAL		513
	Aviation organisations	515
	CAA contact details	527
	AM Radio stations broadcasting in the UK	531
	Medical Examiners	532
	CAA medical department	549
	ICAO Aircraft designators	553
	Aircraft registration and country codes	562
	ICAO code, aircraft registration and country codes	568
	ICAO document listing	575
	ICAO wake turbulence classifications	577
	Aerodrome fire/crash protection categories	578
	HF Volmet broadcasts	579
	VOR Tacan channel pairings	581
	World zones & map	584
	Worldwide SR-SS tables	586
	AERAD ENC chart legend	589
	CAA Chart legend	600
	Interception procedures	602
	RTF standard words and phrases	604
	Fluids	605
	Conversion factors	608
	UK Pressure setting chart	609
	UK Call signs	610

Contents

CAA SAFETY SENSE LEAFLETS 619
Leaflet 1	Good Airmanship Guide	621
Leaflet 2	Care of Passengers	630
Leaflet 3	Winter Flying	635
Leaflet 4	Use of MOGAS	642
Leaflet 5	VFR Navigation	647
Leaflet 6	Aerodrome Sense	653
Leaflet 7	Aeroplane performance	661
Leaflet 8	Air Traffic Services Outside Controlled Airspace	667
Leaflet 9	Weight and Balance	673
Leaflet 10	Bird Avoidance	679
Leaflet 11	Interception Procedures	684
Leaflet 12	Strip Sense	689
Leaflet 13	Collision Avoidance	695
Leaflet 14	Piston Engine Icing	703
Leaflet 15	Wake Vortex	709
Leaflet 16	Balloon Airmanship Guide	715
Leaflet 17	Helicopter Airmanship	724
Leaflet 18	Military Low Flying	735
Leaflet 19	Aerobatics	741
Leaflet 20	VFR Flight Plans	746
Leaflet 21	Ditching	753
Leaflet 22	Radiotelephony for General Aviation	763
Leaflet 23	Pilots: Its your decision	776
Leaflet 24	Pilot Health	782
Leaflet 25	Use of GPS	787
Leaflet 26	Visiting Military Aerodromes	794

AERONAUTICAL INFORMATION CIRCULARS 803
AIC 19/2002	Low altitude windshear	805
AIC 67/2002	Take-off, climb and landing performance of light aeroplanes	816
AIC 6/2003	Flight over and in the vicinity of high ground	824
AIC 81/2004	The effect of thunderstorms and associated turbulence on aircraft operations	830
AIC 93/2004	VHF Radio telephony emergency communications	844
AIC 106/2004	Frost, ice and snow on aircraft	846
AIC 17/1999	Wake turbulence	851
AIC 52/1999	Guidance to training captains – simulation of engine failure in aeroplanes	863
AIC 61/1999	Risks and factors associated with operations on runways affected by snow, slush or water	868

Appendix 1		875
Appendix 2		877
Index		878

Notes

Notes

Effective date March 2006

Aeronautical Information Publication
UK AIM

The UK Aeronautical Information Manual

AIP
GEN section

PART 1 GENERAL (GEN)
GEN 0
GEN 0 PREFACE

1 Aeronautical Authority
1.1 The United Kingdom AIP is published by authority of the Civil Aviation Authority (CAA)

2 Applicable ICAO Documents
2.1 The AIP is prepared in accordance with the Standards and Recommended Practices (SARP) of Annex 15 to the Convention on International Civil Aviation and with the Aeronautical Information Services Manual (ICAO Doc 8126). Charts contained in the AIP are produced in accordance with Annex 4 to the Convention on International Civil Aviation and with the Aeronautical Chart Manual (ICAO Doc 8697). Differences from ICAO Standards and Recommended Practices are given at GEN 1.7.

3 The AIP structure and amendment interval
3.1 The AIP Structure
3.1.1 The AIP forms part of the Integrated Aeronautical Information Package, details of which are given at GEN 3.1. The principal AIP structure is shown in graphic form on Page GEN 0.1.3. The AIP is made up of three Parts, General (GEN), En-Route (ENR), and Aerodromes (AD), each divided into sections and sub-sections as applicable, containing various types of information subjects.

Part 1 – General (GEN)
GEN consists of five sections containing information briefly described hereafter.

GEN 0 Preface; Record of AIP Amendments; Record of AIP Supplements; Checklist of AIP pages; List of hand amendments to Part 1; Table of Contents to Part 1.

GEN 1 National regulations and requirements – Designated authorities; Entry, transit and departure of aircraft; Entry, transit and departure of passengers and crew; Entry, transit and departure of cargo; Aircraft instruments, equipment and flight documents; Summary of national regulations and international agreements/conventions; Differences from ICAO Standards and Recommended Practices.

GEN 2 Tables and Codes – Measuring system, aircraft markings, holidays; Abbreviations used in AIS publications; Chart symbols; Location indicators; List of radio navigation aids; Conversion tables; Sunrise/Sunset Tables; Rate of Climb Table.

GEN 3 Services – Aeronautical Information Services; Aeronautical charts; Air Traffic Services; Communications services; Meteorological services; Search and Rescue.

GEN 4 Charges for aerodrome/heliport and air navigation services – Aerodrome/heliport charges; Air navigation service charges.

Part 2 – En-route (ENR)
ENR consists of seven sections containing information briefly described hereafter.

ENR 0 Preface; List of hand amendments to Part 2; Table of Contents to Part 2.

ENR 1 General rules and procedures – General rules; Visual flight rules; Instrument flight rules; ATS airspace classification; Holding, approach and departure procedures; Radar services and procedures; Altimeter setting procedures; Regional supplementary procedures; Air traffic flow management; Flight planning; Addressing of flight plan messages; Interception of civil aircraft; Unlawful interference; Air traffic incidents; Off-shore operations.

ENR 2 Air traffic services airspace – Detailed description of Flight Information Regions (FIR); Upper Flight Information Regions (UIR); Terminal Control Areas (TMA); Other regulated airspace.

ENR 3 ATS routes – Detailed description of Lower ATS routes; Upper ATS routes; Area navigation routes; Helicopter routes; Other routes; En-route holding.

ENR 4 Radio Navigation aids/systems – Radio navigation aids – en-route; Special navigation systems; Name-code designators for significant points; Aeronautical ground lights – en-route.

ENR 5 Navigation warnings – Prohibited, restricted and danger areas; Military exercise and training areas; Other activities of a dangerous nature; Air navigation obstacles – en-route; Aerial sporting and recreational activities; Bird migration and areas of sensitive fauna.

ENR 6 En-route charts – En-route Chart-ICAO and index charts.

Part 3 – Aerodromes (AD)
AD consists of four sections containing information as briefly described hereafter.

AD 0 Preface; List of hand amendments to Part 3; Table of Contents to Part 3.

AD 1 Aerodromes/Heliports – Introduction – Aerodrome/heliport availability; Rescue and fire fighting services and Snow plan; Index to aerodromes and heliports; Grouping of aerodromes/heliports.

AD 2 Aerodromes – Detailed information about aerodromes (including helicopter landing areas if located at the aerodromes listed under 24 sub-sections.

AD 3 Heliports – Detailed information about heliports (not located at the aerodromes), listed under 23 sub-sections.

3.2 Amendment Interval
3.2.1 Regular amendments to the AIP will be issued every 28 days.

4 Service to contact

4.1 The information in UK AIS publications is collected from a number of varied sources and is considered to be as reliable as possible at the time of publication. The Civil Aviation Authority, while exercising great care in the compilation of this information, will not be responsible for the accuracy of the contents of the UK AIP and other AIS publications, omissions therein, the adequacy or the receipt of Amendments or Supplements.

4.2 Any inaccuracies or omissions should be notified immediately to AIS at the address given below, giving as much detail as possible, including the source and date of the information.

Aeronautical Information Service
UK Publications Office
National Air Traffic Services Ltd
Control Tower Building
London Heathrow Airport
Hounslow
Middlesex
TW6 1JJ

4.3 Enquiries regarding the content of the following publications should be addressed as follows:

(a) UK AIC and UK SUP — to the 'Content' telephone number given at the top of each AIC or SUP.
(b) Pre-Flight Information Bulletins — Central Pre-Flight Information Bulletin (PIB) Section Tel: 020 8745 3464.
(c) All other UK AIS Publications — AIS UK Publications Office Fax: 020 8745 3456.

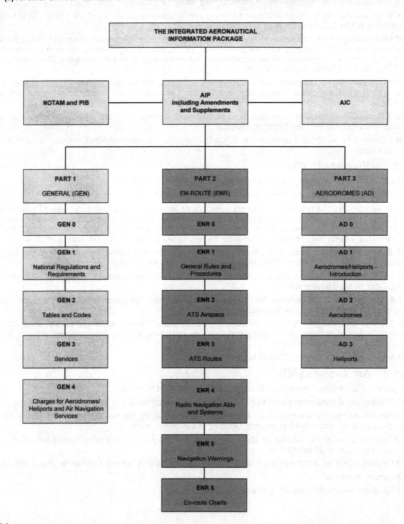

GEN 1 NATIONAL REGULATIONS AND REQUIREMENTS

GEN 1.1 DESIGNATED AUTHORITIES

1 Designated Authorities

1.1 The addresses of the designated authorities concerned with facilitation of international air navigation are:

(a) Civil Aviation
Department of Transport, Local Government and the Regions
Great Minster House
76 Marsham Street
London
SW1P 4DR
Tel: 020 7944 8300 (Main Switchboard) Office Hours: 0900-1730
Tel: 020 7944 5999 (Out of Office Hours)
Fax: 020 7944 9622
Telex: 22221 DOE MAR G
AFS: EGGCYAYX

Civil Aviation Authority
Headquarters
CAA House
45-59 Kingsway
London
WC2B 6TE
Tel: 020 7379 7311
Telex: 883092
AFS: EGGAYAYX

Civil Aviation Authority
Safety Regulation Group
Aviation House
Gatwick Airport South
West Sussex
RH6 0YR
Tel: 01293 567171
Fax: 01293 573999
Telex: 878753
AFS: EGGRYAYX

Civil Aviation Office for Scotland
7 Melville Terrace
Stirling
FK8 2ND
Tel: 01786 431410
Fax: 01786 448030

(b) Meteorology
The Civil Aviation Authority is the Meteorological Authority for the United Kingdom. Meteorological forecasting and climatological services for civil aviation are provided by the Meteorological Office as the agent for the Civil Aviation Authority.
See GEN 3.5.1

(c) Customs and Excise
HM Customs and Excise
New Kings Beam House
22 Upper Ground
London
SE1 9PJ
Tel: 020 7620 1313
Fax: 020 7865 4744

(d) Immigration
Under Secretary of State
Immigration and Nationality Directorate
Home Office
Lunar House
40 Wellesley Road
Croydon
CR9 2BY
Tel: 0870 606 7766 (Central Telephone Enquiry Bureau)
Calls are held and answered in sequence, if not answered immediately please hold.
Telex: 946606
E-mail: indpublicenquiries@ind.homeoffice.gsi.gov.uk

(e) Health
Department of Health
International Relations Unit
Richmond House
79 Whitehall
London
SW1A 2SN
Tel: 020 7972 2000
Fax: 020 7210 4884/0

(f) En Route and Aerodrome/Heliport Charges
See GEN 4

(g) Agricultural Quarantine
Department for Environment, Food and Rural Affairs
Nobel House
17 Smith Square
London
SW1P 3JR
Tel: 020 7238 6000

Department of Agriculture and Rural Development for Northern Ireland
Dundonald House
Upper Newtownards Road
Belfast
BT4 3SB
Tel: 028 9052 4999
Fax: 028 9052 5546
Telex: 74578

(h) Aircraft Accident Investigation
Department of Transport
Air Accidents Investigation Branch
Berkshire Copse Road
Aldershot
Hants
GU11 2HH
Tel: 01252 510300
Tel: 01252 512299 Accident Reports (H24)
Fax: 01252 376999
Telex: 858119 ACCINV-G
AFS: EGGCYLYX

23

GEN 1.2 ENTRY, TRANSIT AND DEPARTURE OF AIRCRAFT

GEN 1.2.1 NATIONAL REGULATIONS

1 Crossing UK Boundaries

1.1 Aircraft flying from or to places abroad may cross the coastline or the land frontier at any point, subject, however, to the requirements of regulations in force including those contained in the UK AIP ENR 1.12.

2 Restrictions on Use

2.1 All operators are reminded of the need to comply with local flying restrictions, and noise characteristics and noise abatement procedures in respect of jet aircraft at United Kingdom airports, details of which are shown on the relevant pages of the AD Section. Care must be taken to ensure that advance arrangements have been made for the ground handling of the aircraft and that, unless special arrangements have been made with the airport authorities concerned, arrivals are scheduled during the airport's normal hours of watch.

3 Aerodrome Operating Minima

3.1 Articles 39 and 40 of the Air Navigation Order 2000 state that public and non-public transport aircraft registered in a country other than the United Kingdom, when making a descent to an aerodrome, shall not descend from a height of 1000ft or more above the aerodrome to a height less than 1000ft above the aerodrome if the relevant runway visual range at the aerodrome is at the time less than the specified minimum for landing. In addition the aircraft referred to above when making a descent to an aerodrome shall not:

(a) Continue an approach to landing at any aerodrome by flying below the relevant specified decision height; or

(b) Descend below the relevant specified minimum descent height, unless in either case from such height the specified visual reference for landing is established and is maintained.

3.2 An application to conduct Category II and/or III operations, or low visibility take offs in less than 150m (200m for Category D aeroplanes) runway visual range, must be accompanied by a copy of the relevant Permission or Certificate of Competence issued to the operator by the aviation authority for the State of Registry of the aircraft. This Permission or Certificate of Competence must state the operator's name, aircraft type and the specific operating minima approved.

3.3 Applications should be addressed to:

Civil Aviation Authority
Flight Operations Department
Aviation House
Gatwick Airport South Area
West Sussex
RH6 0YR

4 Jet Aircraft

4.1 For flights to and from London Gatwick, London Heathrow and London Stansted, operators are reminded that information appertaining to noise characteristics and noise abatement procedures in respect of their jet aircraft must have been submitted to and approved by the Department of Transport, Local Government and the Regions (DTLR), AED 2, Great Minster House, 76 Marsham Street, London, SW1P 4DR as prescribed in the relevant pages of the AD Section. Permits will not be granted by the Secretary of State until this requirement has been satisfied. Moreover, the number of jet aircraft movements at night is subject to limitation at London Gatwick, London Heathrow and London Stansted. Details of the limitations are published by Supplement to the AIP, and are also available from AED 2, address as above Tel: 020 7944 4089.

5 Arrival and Departure of Civil Aircraft on Flights between Great Britain, Republic of Ireland, Northern Ireland, Isle of Man or the Channel Islands

5.1 The Terrorism Act 2000 came into operation on 19th February 2001, replacing the Prevention of Terrorism Act 1989.

5.2 The owners or agents of aircraft employed to carry passengers for reward, coming to Great Britain from the Republic of Ireland, Northern Ireland or any of the Islands or going from Great Britain to any other of those places shall not, without the approval of an examining officer, arrange for the aircraft to call at an airport other than a designated airport for the purpose of disembarking or embarking passengers.

5.3 The captain of an aircraft not employed to carry passengers for reward and coming to Great Britain from the Republic of Ireland, Northern Ireland or any of the Islands or going from Great Britain to any other of those places shall not, without the approval of an examining officer, permit the aircraft to call or leave an airport in Great Britain other than a designated airport.

5.4 The purpose of these requirements is to enable the police to exercise their powers under the Act to carry out security checks and pilots must comply with the requirements of the examining officer in respect of any examination of the captain, crew or passengers where carried. Arrangements in respect of scheduled flights are well established.

5.5 If a pilot or an aircraft operator wishes to make a flight to or from a non-designated airport without an intermediate landing at a designated airport, he must seek prior permission from the Chief Officer of Police in whose area the non-designated airport is located. In practice this is usually via the force Special Branch and should be done as far in advance as possible of the flight being made. (Most Police forces require at least 24 hours notice either directly or via the airport operator).

5.6 Although Filton is not a Designated Airport under the 2000 Act, pilots wishing to land at or depart from Filton Aerodrome on a flight to or from the Republic of Ireland, Northern Ireland, the Isle of Man, or the Channel Islands, without an intermediate landing at an airport designated in the Act, should apply direct to Filton Aerodrome, during office hours and at least 24 hours prior to the flight, for permission to use it for this purpose.

GEN 1.2   GEN 1.2.1.2 Arrival in the United Kingdom from another EU Country

5.7. The following Acts continue to remain in force. Prevention of Terrorism (Jersey) Law 1996. Prevention of Terrorism (Bailwick of Guernsey) Law 1990, soon to be amended by, Crime and Terrorism and Crime (Bailliwick of Guernsey) Law 2002. The Isle of Man Prevention of Terrorism 1990.

5.8 Airports now available and designated under the 2000 Act are listed at GEN 1.2, Appendix A.

5.9 Addresses of Chief Constables are given at GEN 1.2, Appendix B.

6 Notification of Civil Helicopter Flights to Northern Ireland

6.1 Under the present arrangements contained in the Prevention of Terrorism (Temporary Provisions) Act 1989, the Secretary of State at the Northern Ireland Office, requires pilots in command of civil helicopters flying into Northern Ireland to notify the Royal Ulster Constabulary Force Control and Information Centre – Tel: 028 906 50222 Ex 33607, of the point and time of arrival in Northern Ireland. In addition the usual security arrangements for the destination should be followed.

6.2 Any amendment to the arrival point/time must be advised to Belfast Aldergrove Approach Control on RTF 120.900 MHz who will advise the Royal Ulster Constabulary for the pilot.

7 Flights between the Isle of Man and Great Britain or Northern Ireland

7.1 All aircraft flying to or from the Isle of Man must land at Isle of Man Airport. All aircraft carrying from the Isle of Man any goods on which United Kingdom or European Union duties have not been paid must land at one of the Customs and Excise airports listed at Appendix C (paragraph 9.1 refers).

8 Forced Landings

8.1 If an aircraft while on a flight to which the restrictions as to place of landing set out in GEN 1.2.1.3 paragraph 1 apply, is required under or by virtue of any enactment relating to air navigation, or is compelled by accident, stress of weather or other unavoidable cause, to land at a place other than a Customs and Excise airport the commander of that aircraft must as soon as possible report the landing to an officer of Customs and Excise or a police constable and an immigration officer. The commander must comply with any directions given by a customs or immigration officer with respect to any passengers, crew or goods on board the aircraft. In all forced landing situations, any action taken by officers of Customs and Excise and the Immigration Service will take full account of the health and safety needs of passengers and crew.

9 Customs and Excise, Immigration and Health Airports

9.1 The aerodromes in the United Kingdom which have been designated as Customs and Excise Airports by the DTLR and as designated ports of entry for the purposes of the Immigration Act 1971 are listed at Appendix C. The appendix also gives information on those aerodromes designated as Sanitary Airports under International Sanitary Regulations.

9.2 Only designated Customs and Excise Airports listed in Appendix C which are also ports of entry for immigration purposes may handle unlimited traffic to and from places outside the European Union (EU) (see GEN 1.2.1.2 paragraph 1 for details of EU countries) although certain non-designated aerodromes handle a limited range of traffic to and from such places. For detailed arrangements at specific aerodromes contact the aerodrome manager.

9.3 There are no Customs or Immigration restrictions on the places of landing and take-off of aircraft on flights to and from other EU countries but see the notification procedures detailed in GEN 1.2.1.2, paragraph 2.

9.4 Charges are payable for the attendance of Customs and Excise staff outside certain authorised hours for any duties other than the acceptance of inward report of the aircraft, and the clearance of passengers and their baggage. Details of such charges are published in Customs and Excise Notice No. 112 or they may be obtained from any Customs and Excise office. A charge may also be levied if attendance by Immigration staff is required; contact the local Immigration Office for further information.

9.5 For information about the International traffic authorised to be handled by individual aerodromes, and about the hours of attendance and arrangements for immigration clearance contact the aerodrome operator or the relevant office (see GEN 1.2 Appendices D and E).

GEN 1.2.1.2 FLIGHTS FROM/TO COUNTRIES WITHIN THE EU

1 As a result of the completion of the European Single Market on 1 January 1993, the Customs and Excise requirements referred to in GEN 1.2.1.1 paragraph 1, GEN 1.2.1.2 paragraphs 2 and 3; GEN 1.2.1.3 paragraphs 1, 2 and 3; GEN 1.3.1 paragraph 1, 2 and 3; and GEN 1.4.1 paragraph 1, now vary according to whether or not the place of departure or the destination is within the European Union (EU). For the purposes of these requirements, the Countries of the EU (also known as Member States) are: Austria, Belgium, Denmark, Finland, France, Germany, Greece, Ireland, Italy, Luxembourg, Netherlands, Sweden, Portugal (including the Azores and Madeira), Spain (including the Balearic Islands), United Kingdom (including the Isle of Man).

Note: The Channel Islands and the Canary Islands are regarded as places outside the EU for the purposes of the arrival and departure procedures set out in GEN 1.2.1.3 paragraph 1.

2 Arrival in the United Kingdom from another EU Country

2.1 Except in the circumstances referred to in GEN 1.2.1.3 paragraph 1, there is no Customs restriction on the places of landing for aircraft arriving in the United Kingdom from another EU Country.

2.2 For aircraft landing at:

(i) any place other than one of the Customs and Excise Airports listed at Appendix C, Flight Plan details and the names and nationalities of passengers should be advised to the local Customs contact point (see list at GEN 1.2 Appendix D).

(ii) any place listed at Appendix C which is not an approved Port of Entry under the Immigration Act and any other place information as at (i) above should be passed to the local Immigration Office (see GEN 1.2 Appendix E) unless Immigration advise that this is not required.

2.2.1 This information may be supplied either by the aircraft commander or the operator of the aerodrome at which the aircraft is due to land. The actual amount of advance notice needed should be checked beforehand.

2.3 At Customs and Excise Airports, there is no requirement for the arrival of intra EC services to be reported to Customs, but report to the Immigration Service is required if non EEA nationals are on board. For details, contact the airport concerned. There is no need for the aircraft commander to bring the aircraft to the examination station unless requested to do so by the Customs officer. If the aircraft has stores on board, they must be declared to Customs in accordance with the procedure in operation at the airport concerned (generally using Form C209).

2.4 Liability of Aircraft to Customs Duty and VAT

(a) Duty: There is no customs duty on EU registered civil aircraft entering the United Kingdom from another EU country. For other civil aircraft see GEN 1.2.1.3 paragraph 1.3.2.

(b) VAT: For aircraft liable to VAT, there are no formalities provided that VAT on the aircraft has been accounted for in one of the EU member states.

3 Departure from the United Kingdom to another EU Country

3.1 With the exception of the stopover flights referred to in GEN 1.2.1.3 paragraph 1.3.3, there is no Customs restriction on the places of take-off for flights to other EU Countries, nor is there any need for details of the flight to be notified to Customs. However, if the aircraft is carrying stores, these together with a copy of the stores list on form C208 must be made available for customs examination if required.

GEN 1.2.1.3 FLIGHTS FROM/TO COUNTRIES OUTSIDE THE EU

1 Arrival in the United Kingdom from outside the EU

1.1 Place of arrival. Unless special permission has been obtained from Customs and Immigration the commander of an aircraft entering the United Kingdom from a place outside the EU may not land at any place other than one of the airports listed at Appendix C which is approved both for Customs and Immigration purposes.

(a) For the first time after its arrival in the United Kingdom; or

(b) at any time while it is carrying passengers or goods brought in that aircraft from a place outside the EU and not yet cleared.

1.2 Timely requests for the special permission referred to should be made by or on behalf of the commander of an aircraft through the operator of the aerodrome to the appropriate local offices of Customs and Excise and the Immigration Service. It should be noted that the grant of such permission in no way affects the need for compliance with the relevant Article of the Air Navigation Order which requires particular aircraft to use a licensed aerodrome when engaged in the public transport of passengers.

1.3 When an aircraft lands at a Customs and Excise airport, the aircraft commander must bring it to the examination station and report its arrival to Customs. The Customs office for the airport of arrival should be contacted for advice on the procedure for making report there (including any documentary requirements), as such procedures may vary from place to place depending on local circumstances.

1.3.1 If the aircraft has stores on board, they must be declared to Customs in accordance with the procedure in operation at the airport concerned (generally using Form C209).

1.3.2 Liability of Aircraft to Customs Duty and VAT

(a) Duty: All civil aircraft are eligible for relief of customs duty on importation into the EU under the end-use arrangements (see HM Customs and Excise Notice 770 (The European Community: Imported Goods: End-use relief));

(b) VAT: Aircraft having a maximum take-off weight of less than 8000 kg are liable to VAT at the standard rate. VAT may be suspended on such aircraft if they are temporarily imported into the EU, provided that the importer:

(i) is a non-EU resident; and

(ii) intends to re-export the aircraft when he next leaves the EU, or when it has been in the EU for a total of 6 months in any 12 month period, whichever comes first.

If these conditions are not met liability to VAT will arise.

1.3.3 The provisions of this paragraph apply both to aircraft which have entered the United Kingdom direct from a non-EU Country and to those which have made a stopover at an airport in another EU Country.

1.4 Departure from the United Kingdom for a destination outside the EU

1.4.1 Unless special permission has been obtained from the Commissioners of Customs and Excise the commander of an aircraft may not depart from the United Kingdom to any place or area outside the EU except from a Customs and Excise airport. Timely requests for the special permission referred to should be made by or on behalf of the commander of an aircraft through the operator of the aerodrome to the appropriate local office of Customs and Excise.

1.4.2 It should be noted that the grant of such permission in no way affects the need for compliance with the relevant Article of the Air Navigation Order which requires particular aircraft to use a licensed aerodrome when engaged in the public transport of passengers.

1.4.3 Any stores on board or loaded onto an aircraft, together with the stores list on form C208 must be made available for customs examination if required. The Customs office will advise details of any other documentary requirements which may vary from place to place depending on local circumstances.

1.4.4 The provisions of this paragraph apply both to direct flights to non-EU destinations and to those where the aircraft makes a stopover at an airport in another EU country.

GEN 1.2.3.2 DOCUMENTARY REQUIREMENTS

1 Flights to and from the United Kingdom – Non-scheduled Flights – Commercial

1.1 Documentary Requirements

1.2 In addition to operational requirements as set out in other parts of the UK AIP operators are required to provide copies of the following:

(a) Certificate of Competency and/or Air Operator's Certificate and/or an Operating Licence issued by the aeronautical authority of the country of origin.

(b) Certificate of Airworthiness issued by the aeronautical authority of the country of origin for each aircraft to be used on services to the UK.

(c) Certificate of liability insurance for passenger and third party risks in respect of each aircraft to be used on services to the UK. For cargo flights, certificate of third party liability. The level of insurance must be compliant with the obligations of insurance contained in ECAC Resolution 25-1 on minimum levels of insurance cover for passenger and third party liability.

(d) Dangerous Goods confirmation as set out in GEN 1.2.3.1 paragraph 9.5.2.

(e) Air Travel Organisers Licence (ATOL) as set out in 1.2.3.1 paragraph 7 (not for cargo services).

(f) If the airline has not previously filed minima with the Civil Aviation Authority for the type of aircraft to be used, a statement that flight crews Non Precision and Category 1 Aerodrome Operating Minima comply with Article 39(3) of the Air Navigation Order 2000.

Note: The notified method for calculating UK minima is set out in the UK AIP Aerodrome (AD Volume 1).

(g) Statement that the aircraft operator is aware of the UK Approach Ban requirements set out in Article 39(6) of the Air Navigation Order 2000 and that flight crews will be issued with written instructions with regard to Article 39(6) prior to operating the flights into the UK.

Note: Statements about Aerodrome Minima and Approach Ban are to be made to the DTLR, but any enquiries should be made to the Flight Operations Policy Dept, CAA, Aviation House, London Gatwick Airport West Sussex. RH6 0YR. Tel: 01293 573414, Fax: 01293 573991, Telex: 878753.

(h) Noise certificate for each aircraft to be used.

2 Leased Aircraft

2.1 The Department will also require the above documents in regard to any foreign registered aircraft leased from another carrier which the applicant carrier proposes to use to the UK together with the following information:

(a) conformation that the lease has been approved by the lessee's aeronautical authorities.

(b) conformation of which airline's operations and flight manuals will be used.

(c) contact details of the lessor airline (name, address, telephone, fax, telex).

Note: Where possible, documents should be provided in the English language. If they are submitted in a foreign language, DTLR may request an English translation to be provided.

GEN 1.2.4 PRIVATE FLIGHTS

1 Advanced Notification – Non-Scheduled Flights – Non-Commercial

1.1 Advance Notice of Permit Requirements

1.1.1 Except for airport purposes (GEN 1.2.1.1) prior permission is not required for private flights by aircraft registered in States which are parties to the Chicago Convention for overflying the United Kingdom or for making non-traffic stops in United Kingdom territory. Permission for such flights by aircraft registered in States not parties to the Chicago Convention should be sought in accordance with the procedure as set out in GEN 1.2.3.1 paragraph 8 for commercial flights.

1.1.2 Pilots of non-scheduled non-commercial flights, have the obligation in respect of passport-control requirements set out in GEN 1.3.2 paragraph 1 and are to present their passengers on arrival and departure to the Immigration Officer or to the Customs Officer acting as Immigration Officer, in accordance with the arrangements approved by the Home Office and notified to the airport manager.

1.1.3 The exemption for aircrew detailed in GEN 1.3.2 paragraph 1.1 is not applicable to non-commercial crew and the provisions of GEN 1.3.2 paragraph 2.1 apply.

GEN 1.2 APPENDIX A – AIRPORTS DESIGNATED UNDER PREVENTION OF TERRORISM LEGISLATION

APPENDIX A – AIRPORTS DESIGNATED UNDER PREVENTION OF TERRORISM LEGISLATION (GEN 1.2.1.1 paragraph 5 refers)

Designated Airports (Great Britain)

Aberdeen	Exeter	Manchester	**Designated Airports**
Biggin Hill	Glasgow	Manston	**(Northern Ireland)**
Birmingham	Gloucestershire	Newcastle	Belfast Aldergrove
Blackpool	Humberside	Norwich	Belfast City
Bournemouth	Leeds Bradford	Nottingham East	Londonderry Eglington (City of
Bristol	Liverpool	Midlands	Derry)
Cambridge	London City	Plymouth	**Designated Airports**
Cardiff	London Gatwick	Prestwick	**(Isle of Man and Channel Islands)**
Carlisle	London Heathrow	Sheffield City	Alderney
Coventry	London Luton	Southampton	Isle of Man
Durham Tees Valley	London Stansted	Southend	Guernsey
Edinburgh	Lydd		Jersey

APPENDIX B – UNITED KINGDOM POLICE FORCES (GEN 1.2.1.1 paragraph 5 refers)

Constabulary	Addresses (unless 'Police' in title)	Telephone Number
England and Wales		
Avon & Somerset	PO Box 37, Valley Road, Portishead, Bristol, Avon, BS20 8QJ	0845 456 7000
Bedfordshire Police	Woburn Road, Kempston, Bedford, MK43 9AX	01234 841212
Cambridgeshire	Hinchingbrooke Park, Huntingdon, PE29 6NP	01480 456111
Cheshire Police	Clemonds Hey, Oaksmere Road, Winsford, CW7 2UA	01244 350000
Cleveland Police	Ladgate lane, Middlesborough, Cleveland, TS8 9EH	01642 326326
Cumbria	Carleton Hall, Penrith, Cumbria, CA10 2AU	01768 891999
Derbyshire	Butterley Hall, Ripley, Derbyshire, DE5 3RS	0845 1233333
Devon & Cornwall	Middlemoor, Exeter, Devon, EX2 7HQ	0845 2777444
Dorset Police	Winfrith, Dorchester, Dorset, DT2 8DZ	01305 251212
Durham	Aykley Heads, Durham, DH1 5TT	0191 386 4929
Dyfed/Powys Police	PO Box 99, Carmarthen, SA31 2PF	01267 232000
Essex Police	PO Box 2, Springfield, Chelmsford, Essex, CM2 6DA	01245 491491
Gloucestershire	Holland House, Lansdown Road, Cheltenham, GL51 6QH	0845 0901234
Greater Manchester Police	PO Box 22, Manchester, M16 0RE	0161 872 5050
Gwent	Croesyceiliog, Cwmbran, Gwent, NP44 2XJ	01633 838111
Hampshire	West Hill, Winchester, Hants, SO22 5DB	0845 0454545
Hertfordshire	Stanborough Road, Welwyn Garden City, Herts, AL8 6XF	01707 354000
Humberside Police	Priory Police Station, Kingston upon Hull, HU5 5SF	01482 326111
Kent Police	Sutton Road, Maidstone, Kent, ME15 9BZ	01622 690690
Lancashire	PO Box 77, Hutton, Preston, PR4 5SB	0845 1253545
Leicestershire	St John's, Enderby, Leicester, LE19 2BX	0116 2222222
Lincolnshire Police	PO Box 999, Lincoln, LN5 7PH	01522 532222
London Metropolitan Police	New Scotland Yard, Broadway, London, SW1H 0BG	020 7230 1212
London, City of, Police	37 Wood Street, London, EC2P 2NQ	020 7601 2455
Merseyside Police	Canning Place, Liverpool, L69 1JD	0151 7096010
Norfolk	Falconers Chase, Wymondham, Norfolk, NR18 0WW	01953 424242
Northamptonshire Police	Wootton Hall, Northampton, NN4 0JQ	01604 700700
Northumbria Police	North Road, Ponteland, Newcastle upon Tyne, NE20 0BL	01661 872555
North Wales Police	Glan-Y-Don, Colwyn Bay, Conway, LL29 8AW	0845 6071002
North Yorkshire Police	Newby Wiske, Northallerton, DL7 9HA	0845 6060247
Nottinghamshire Police	Sherwood Lodge, Arnold, Nottingham, NG5 8PP	0115 9670999
South Wales Police	Cowbridge Road, Bridgend, CF31 3SU	01656 655555
South Yorkshire Police	Snig Hill, Sheffield, S3 8LY	0114 2202020
Staffordshire Police	Cannock Road, Stafford, ST17 0QG	01785 257717
Suffolk	Martlesham Heath, Ipswich, Suffolk, IP5 3QS	01473 613500
Surrey Police	Mount Browne, Sandy Lane, Guildford, Surrey, GU3 1HG	0845 1252222

GEN 1.2   APPENDIX C – DESIGNATED AIRPORTS – CUSTOMS, IMMIGRATION AND HEALTH (GEN 1.2.1.2 paragraph 2 refers)

Constabulary	Addresses (unless 'Police' in title)	Telephone Number
Sussex Police	Malling House, Lewes, East Sussex, BN7 2DZ	0845 6070999
Thames Valley Police	Oxford Road, Kidlington, Oxon, OX5 2NX	0845 8505505
Warwickshire Police	PO Box 4, Leek Wootton, Warwick, CV35 7QB	01926 415000
West Mercia	Hindlip Hall, Hindlip, PO Box 55, Worcester, WR3 8SP	01905 723000
West Midlands Police	Lloyd House, Colmore Circus, Birmingham, B4 6NQ	0845 1135000
West Yorkshire Police	PO Box 9, Wakefield, WF1 3QP	0845 6060606
Wiltshire	London Road, Devizes, Wiltshire, SN10 2DN	01380 722341
Scotland		
Central Scotland Police	Randolphfield, Stirling, FK8 2HD	01786 456000
Dumfries & Galloway	Cornwall Mount, Dumfries, DG1 1PZ	01387 252112
Fife	Detroit Road, Glenrothes, Fife, KY6 2RJ	01592 418888
Grampian Police	Queen Street, Aberdeen, AB10 1ZA	0800 371553
Lothian & Borders Police	Fettes Avenue, Edinburgh, EH4 1RB	0131 311 3131
Northern	Perth Road, Inverness, IV2 3SY	01463 715555
Strathclyde Police	173 Pitt Street, Glasgow, G2 4JS	0141 532 2000
Tayside Police	PO Box 59, West Bell Street, Dundee, DD1 9JU	01382 223200
Northern Ireland		
Police Service of Northern Ireland	'Brooklyn', 65 Knock Road, Belfast, BT5 6LE	028 9065 0222
Isle of Man and Channel Islands		
Isle of Man	Glencrutchery Road, Douglas, IM2 4RG	01624 631212
Jersey Police	PO Box 789, Jersey, Channel Islands, JE2 3ZA	01534 612612
Guernsey Police	Hospital Lane, St Peter Port, Guernsey, Channel Islands. GY1 2QN	01481 725111

Note: Alderney, Channel Islands, is within the jurisdiction of Guernsey and communications in respect of the Island should be addressed to the Guernsey Police.

APPENDIX C – DESIGNATED AIRPORTS – CUSTOMS, IMMIGRATION AND HEALTH (GEN 1.2.1.2 paragraph 2 refers)

1 The Aerodromes listed below have been designated as Customs and Excise Airports by the DTLR.
2 All are also designated points of entry for the purposes of the Immigration Act 1971 except those indicated by footnote 1.
3 For those aerodromes which are also designated as Sanitary Airports are also identified under International Sanitary Regulations (see footnote 2).

Aberdeen
Belfast Aldergrove (2)
Biggin Hill (1)
Birmingham
Blackpool (1)
Bournemouth
Bristol
Cambridge (1)
Cardiff
East Midlands
Edinburgh
Exeter (1)
Glasgow (2)
Humberside (1)
Isle of Man (1)

Leeds Bradford
Coventry (1)
London City (1)
London Gatwick (2)
London Heathrow (2)
London Luton
London Stansted
Lydd (1)
Liverpool
Manchester (2)
Manston (1)
Newcastle
Norwich
Plymouth (1)
Prestwick (2)

Sheffield (1)
Shoreham (1)
Southampton
Southend
Sumburgh (1)
Teesside

Footnotes:
(1) These aerodromes are NOT Ports of Entry under the Immigration Act 1971.
(2) These aerodromes are Sanitary Airports under International Sanitary Regulations.

APPENDIX D – CUSTOMS CONTACT POINTS
(GEN 1.2.1.2 paragraph 2 refers)

Customs Contact Points for giving Advance Notice of Arrival of Intra EU Flights at Places other than Customs Designated Airports

Collection	Address	Telephone Facsimile
Anglia	HM Customs and Excise, CCU, Haven House, 17 Lower Brook Street, Ipswich, Suffolk, IP4 1DN	01473 235704 01473 255001
Central England	HM Customs and Excise, Passenger Terminal, Birmingham International Airport, Birmingham, B26 3QZ	0121 782 6655 0121 782 1282
Eastern England	HM Customs and Excise, CCU, Bowman House, 100-102 Talbot Street, Nottingham, NG1 5NF	0115 971 2377 0115 971 2291
London Airports	HM Customs and Excise, PSD5, Custom House, Heathrow Airport, Hounslow, Middlesex, TW6 2LA	020 8910 3721 or 020 8910 3743 020 8910 3800 (if no response)
London Central	HM Customs and Excise, CCU, Towergate, 163 Tower Bridge Road, London, SE1 3LT	020 7865 354 4020 7865 3575
Northern England	HM Customs and Excise, CCU, Custom House, Hedon Road, Hull, HU9 5PW	01482 782107 01482 702413
Northern Ireland	HM Customs and Excise, Carne House, 20 Corry Place, Belfast, BT3 9HY	028 9035 8255 028 9074 3332 56/57/58/59/60
North West	HM Customs and Excise, Intelligence Co-ordination Unit, Building 302, Cargo Centre, Manchester International Airport, Manchester, M90 5XX	0161 912 6977/78 0161 912 6986
Scotland	HM Customs and Excise, CCU, Falcon House, Inchinnan Road, Paisley, PA3 2RE	0141 887 9369 0141 848 1639
South East England	HM Customs and Excise, CCU, Priory Court, St John's Road, Dover, CT17 9SH	0800 373229 01304 210017 (Freephone) or 01304 215639
Southern England	HM Customs and Excise, CCU, Jubilee House, Western Docks, Southampton, SO15 1AU.	01703 701811 01703 702966
South London & Thames	HM Customs and Excise, CCU, The Boathouse, Thames Custom House, The Terrace, Gravesend, DA12 2BW	01474 537115 01474 335069 029 2076 7000
Wales, the West & Borders	HM Customs and Excise, CCU, Room G310, The Podium, Phase 2 Government Buildings, Llanishen, Cardiff. CF14 5FT	029 2076 7001

Note: CCU = Collection Co-ordination Unit

APPENDIX E – UK IMMIGRATION SERVICE OFFICES

District (*) and Local offices for Flight Notification and Approval Requests

District	Address	Telephone/Fax Nos	Area Served
East Anglia	(*) Parkeston Quay, Harwich, CO12 4SX	Tel: 01255 509770 Fax: 01255 509718	Essex (east of A12/A130/A131/A604)
	Dock Police Station, Dock Road, Felixstowe, Suffolk, IP11 8SE	Tel: 01394 674915 Fax: 01394 673260	Suffolk
	Norwich Airport, Norwich, NR6 6EP	Tel: 01603 408859 Fax: 01603 417358	Norfolk
London Airports District 1	(*) Terminal 2, Heathrow Airport, Hounslow, Middx, TW6 1BN	Tel: 020 8745 6852 Fax: 020 8745 6867	Buckinghamshire (North of M25)
	Terminal 1, Heathrow Airport, Hounslow, Middx, TW6 1BN	Tel: 020 8745 6800 Fax: 020 8745 6828	Berkshire (East of A34), Hampshire (north of A31/A287/A339), Surrey (Woking/Addlestone Police Divisions only)
	Terminal Building, RAF Brize Norton, Oxon, OX18 3LX	Tel/Fax: 01993 845992	Oxfordshire, Wiltshire (east of A429/A350 and north of A4), Gloucestershire (east of A441/A429),and Berkshire (west of A34)
London Airports District 2	(*) South Terminal, Gatwick Airport, West Sussex, RH6 0NP	Tel: 01293 502019 Fax: 01293 553643	Sussex (excluding SW section), Surrey (south of M25 excluding Woking/Addlestone Police Divisions), Kent (north and west of M25).
	Biggin Hill Airport, Westerham, Kent, TN16 3BN	Tel/Fax: 01959 572225	
London Airports District 3	(*) Terminal 3, Heathrow Airport, Hounslow, Middlesex, TW6 1ND	Tel: 020 8745 6941 Fax: 020 8745 6943	Buckinghamshire (south of M25), Hertfordshire (south of M25)

APPENDIX E – UK IMMIGRATION SERVICE OFFICES

District	Address	Telephone/Fax Nos	Area Served
Midlands	(*) Main Terminal, Birmingham Airport, Birmingham. B26 3QZ	Tel: 0121 606 7350 Fax: 0121 782 0006	West Midlands, Warickshire, Staffordshire, Shropshire, Worcestershire, Herefordshire
	East Midlands Airport, Castle Donnington, Derby, DE74 2SA	Tel: 01332 812000 Fax: 01332 811569	Derbyshire, Leicestershire, Nottinghamshire
	Luton Airport, Luton, LU2 9LU	Tel: 01582 421891 Fax: 01582 405215	Bedfordshire, Hertfordshire (North of M25), Northamptonshire
North East	(*) Leeds/Bradford Airport, Yeadon, Leeds, LS19 7TZ	Tel: 0113 250 1196 Fax: 0113 250 5716	South Yorkshire, West Yorkshire and southern section of North Yorkshire
	Sheraton House, 2 Skinner Street, Stockton-on-Tees, TS18 1BR	Tel: 01642 614280 Fax: 01642 618567	Cleveland and northern section of North Yorkshire
	Unit 4, Anchor Court, Francis Street, Hull, HU2 8DT	Tel: 01482 223017 Fax: 01482 219034	East Yorkshire
	Humberside Airport, Kirmington, South Humberside, DN39 6YH	Tel: 01652 688584 Fax: 01652 688551	Lincolnshire
	Newcastle Airport, Woolsington, Newcastle-upon-Tyne, NE13 8BZ	Tel: 0191 286 9469 Fax: 0191 214 0143	Durham, Northumberland
North West	(*) 6th Floor, Tower Block, Manchester Airport, Manchester, M90 2AA	Tel: 0161 489 2659 Fax: 0161 489 2346	
	Ringway House, Preston, PR1 3HQ	Tel: 01772 254212 Fax: 01772 204993	Cumbria, Lancashire
	Graeme House, Derby Square, Liverpool, L2 7SF	Tel: 0151 236 8974 Fax: 0151 236 4656	Merseyside, Cheshire, and North Wales Police District
Scotland and Northern Ireland	Glasgow Airport, Paisley, PA3 2TD	Tel: 0141 887 4115 Fax: 0141 887 1566	Strathclyde Central, Dumfries and Galloway Police Authority Regions
	Aberdeen Airport, Dyce, AB2 0DY	Tel: 01224 722890 Fax: 01224 724095	Grampian and Northern Police Authority Regions

APPENDIX E – UK IMMIGRATION SERVICE OFFICES

District	Address	Telephone/Fax Nos	Area Served
	Edinburgh Airport, Edinburgh, EH12 9DN	Tel: 0131 333 1041 Fax: 0131 335 3197	Tayside, Fife, Lothian and Borders Police Authority Regions
	Olive Tree House, Fountain Street, Belfast, BT1 5EA	Tel: 028 9032 2547 Fax: 028 9024 4939	Northern Ireland
South East	(*) No 1 Control Building, Eastern Docks, Dover, CT16 1JD	Tel: 01304 244900 Fax: 01304 213594	
	SEPST, Dover Hoverport, Dover, CT17 9TF	Tel: 01304 216405 Fax: 01304 216303	Kent (south of M25/M26/M20/A249)
	Ground Floor, Apex House, Northfleet, DA11 9PD	Tel: 01474 534731 Fax: 01474 531731	Kent (north of M26/M20/A249, east of M25)
Southern	(*)Terminal Building, Continental Ferry Port, Wharf Road, Portsmouth, PO2 8QN	Tel: 023 9285 2700 Fax: 023-9286 4149	Hampshire (east of A32/A339) and south of A31/A287/A339), southwest Sussex
	Southampton Airport, Wide Lane, Southampton, SO18 2HG	Tel: 023 8062 7107 Fax: 023 8062 7262	Hampshire (west of A32/A339), Wiltshire (south of A4 and east of A350/A429)
	Car Ferry Freight Terminal, New Harbour Road, Poole, BH15 4AJ	Tel: 01202 673658 Fax: 01202 680004	Dorset
South West	(*) 2nd Floor, Phase 1, Government Site, Ty Glas, Cardiff, CF4 5UN	Tel: 029 2076 4474 Fax: 029 2076 4014	
	Greystoke Business Centre, High Street, Portishead, Bristol, BS20 9PY	Tel: 01275 818688 Fax: 01275 818680	Somerset, Gloucestershire (west of A44/A424/A429)
	Ballard House, West Hoe Street, Plymouth, PL1 3BJ	Tel: 01752 261547 Fax: 01752 226175	Cornwall, Devon and Isles of Scilly
	Poplar House, Tawe Business Village, Phoenix Way, Swansea. SA7 9LA	Tel: 01792 700944 Fax: 01792 700954	West Glamorgan, Dyfed-Powys Police District
	Cardiff Airport, Rhoose, Barry, CF6 9BD	Tel: 01446 710485 Fax: 01446 710606	South Glamorgan, Mid Glamorgan and Gwent
Stansted	(*) Stansted Passenger Terminal, Bassingbourn Road, Stansted, CM24 1RW	Tel: 01279 680118 Fax: 01279 680145	Essex (west of A12, A130/A131 and A604), Cambridgeshire

GEN 1.3 ENTRY, TRANSIT AND DEPARTURE OF PASSENGERS AND CREW

GEN 1.3.1 CUSTOMS REQUIREMENTS

1 Aircrew

1.1 Arriving on Flights from other EU Countries

1.1.1 The duty/tax free allowances do not apply to intra EU crew. No declaration is required to be made. Also, there is no Customs restriction on crew members' exit route from the airport (although in practice at most larger airports their movements are constrained by security measures).

1.2 Arriving on Flights from Non-EU Countries

1.2.1 The Customs office at the airport of arrival will be able to advise on the arrangements in operation there for clearance of aircrew. Normally, this will involve the crew members in either:

(a) Making a declaration on Form C909; or

(b) making an oral declaration in the Red Channel or at the Red Point, if they are carrying goods in excess of the customs allowances for aircrew.

1.2.2 The terms of this paragraph apply not only to crew who have arrived on a direct flight from outside the EU, but also to crew whose aircraft has made a stopover at another EU airport.

1.3 Departing on Flights to Non-EU Destinations

1.3.1 It is not normally necessary for crews' effects to be made available for Customs, except when refund of VAT is being claimed under the Retail Export Scheme. VAT leaflet 704/1/93 explains the conditions under which aircrew are eligible for the Scheme, and the procedures to be followed.

2 Passengers

2.1 Arriving on Domestic Flights

2.1.1 The hold baggage of passengers who arrived in the United Kingdom from a non-EU Country and have transferred to a domestic flight will be subject to Customs control at the destination airport, if it has not been cleared at the airport of arrival in the United Kingdom. The Customs office at the destination airport should be contacted for details of the arrangements.

2.2 Arriving on Flights from other EU Countries

2.2.1 Passengers on direct flights from other EU Countries are not normally required to make any declaration and at most airports proceed through a separate EU exit.

2.2.2 The only exception is for passengers who commenced their journey outside the EU and have transferred to a flight to the United Kingdom after arriving in another EU Country. After reclaiming their hold baggage, such passengers must make an oral declaration in the Red Channel or at the Red Point, if they are carrying goods in excess of Customs allowances. Passengers with nothing to declare should proceed through the Green Channel.

2.3 Arriving on Flights from Non-EU Countries

2.3.1 After disembarkation, passengers completing their journey at a United Kingdom airport reclaim any hold baggage, and if they are carrying goods in excess of Customs allowances they must make an oral declaration in the Red Channel or at the Red Point. Passengers with nothing to declare should proceed through the Green Channel. These procedures apply not only to passengers who have arrived on a direct flight from outside the EU, but also to passengers whose aircraft has made a stopover at another EU airport.

2.3.2 Passengers who are transferring to a flight to another EU Country do not reclaim their hold baggage, but must declare any goods in their cabin baggage which are in excess of customs allowances.

2.3.3 Passengers who are transferring to a flight to a non-EU Country are not required to make any declaration to Customs.

2.3.4 The Customs office for the airport of arrival should be contacted for advice on the arrangements there passengers transferring to a flight to another United Kingdom airport, as these may vary depending on local circumstances.

2.4 Departing on Flights to Non-EU Destinations

2.4.1 It is not normally necessary for passengers baggage to be made available to Customs, except when refund of VAT is being claimed under the Retail Export Scheme. VAT leaflet 704/1/93 explains the conditions under which passengers are eligible for the Scheme and the procedures to be followed.

3 Further Information

3.1 Further information on the Customs requirements for international travellers, including details of Customs allowances is in HM Customs and Excise Notice 1.

4 Customs and Excise Forms

4.1 Copies of Forms (eg Cargo Manifest, Stores Lists, etc), Customs Notices and leaflets may be obtained free of charge from the Customs and Excise officer at any Customs and Excise airport.

GEN 1.3.2 IMMIGRATION REQUIREMENTS

1 Aircrew

1.1 In the case of an aircrew member arriving or departing as such, a valid crew licence or crew member certificate which includes a certification that the holder may at all times re-enter the state of issuance, is acceptable as a document of identity. Aircrew travelling as passengers are required to comply with the provisions of paragraph 2.1.

1.2 When a person subject to immigration control arrives as a member of the crew of an aircraft and is under an engagement requiring him to leave within seven days as a member of the crew of that or another aircraft, he may enter the United Kingdom without leave and remain until the departure of the aircraft on which he is required by his engagement to leave unless either:

(a) There is in force a deportation order against him; or

(b) He has at any time been refused leave to enter the United Kingdom and has not since been given leave to enter or remain in the United Kingdom; or

(c) An Immigration Officer requires him to submit to examination.

1.3 An aircrew member who lawfully enters the United Kingdom without leave by virtue of the foregoing provision, should seek the permission of the Immigration Officer if he wishes to stay longer than seven days.

2 Passengers

2.1 For immigration purposes the United Kingdom, the Channel Islands, the Isle of Man and the Republic of Ireland collectively form a common travel area. A person arriving in the United Kingdom direct from, or departing from the United Kingdom direct to, a place outside the common travel area, is liable to be examined by an Immigration Officer and must produce to him (if required to do so) a valid passport or some other acceptable document satisfactorily establishing identity and nationality or citizenship, endorsed when necessary with a current United Kingdom visa or entry clearance. He must furnish the Immigration Officer with such information as may be required for the purpose of deciding whether he requires leave to enter and, if so, whether and on what terms leave should be given. Information on visa or entry clearance requirements may be obtained from a British Government representative overseas, the Home Office or the local Immigration Office.

2.2 For the purposes of these requirements the term European Economic Area (EEA) is used to refer to the group of countries listed at GEN 1.2.1.2 paragraph 1 but also includes Iceland, Norway and Liechtenstein.

2.3 Powers and Obligations of Captains and Owners or Agents of Aircraft Under the Immigration Act 1971

(a) Except with the prior approval of the Secretary of State, to embark and disembark passengers only at an airport designated as a port of entry in the GEN 1.2.1.1 Appendix C unless there is reasonable cause to believe all of them to be British citizens.

(b) to ensure that persons who have arrived in the United Kingdom do not disembark until examined by an Immigration Officer or other than in accordance with arrangements approved by an Immigration Officer.

(c) to ensure that passengers are presented for examination in an orderly manner.

(d) to provide a list of crew and passengers if required to do so.

(e) to provide landing cards for all passengers who are not British citizens or other nationals of Member States of the EEA.

(f) to remove or make arrangements for the removal from the United Kingdom of a passenger arriving in the United Kingdom who is refused leave to enter.

(g) to detain in custody persons placed on board an aircraft under the authority of an Immigration Officer.

(h) to pay the Secretary of State on demand any expenses incurred by the latter in respect of the custody, accommodation or maintenance of a person at any time after his arrival while he was detained or liable to be detained under the authority of an Immigration Officer unless on arrival that person held a certificate of entitlement, an entry clearance or work permit, or was subsequently given leave to enter before removal directions had been carried out.

(i) to remove a person against whom a deportation order is in force, if so directed by the Secretary of State.

3 Immigration (Carriers' Liability) Act 1987

3.1 Under the Immigration (Carriers' Liability) Act a charge will be levied on carriers who bring to the United Kingdom passengers without proper documentation. The charge will be £2000 for each passenger. The charge will be enforceable by civil action and would arise where a person requiring leave to enter (ie not a British Citizen or other national of the EEA) arrives at the Immigration Control without:

(a) A valid passport or document satisfactorily establishing identity and nationality or citizenship; and

(b) a valid visa where one is required under the Immigration Rules.

3.2 Information on visas and documentation can be obtained from British Consulates abroad or from Immigration Service offices in the United Kingdom; the principal air travel information-systems also carry up-dated information on visas.

4 Landing Cards

4.1 In addition to the general requirements described in paragraph 2 passengers who are not British citizens or other nationals of the EEA will normally be required to produce completed landing cards to the Immigration Officer.

GEN 1.3.3 PUBLIC HEALTH

1 Public Health Requirements

1.1 The Airport Medical Officer may examine the following persons and take any necessary measures for preventing danger to public health:

(a) A person entering the United Kingdom who is suspected of suffering from, or to have been exposed to infection from, an infectious disease or suspected of being verminous.

(b) a person who proposes to depart from the United Kingdom if there are reasonable grounds for believing him to be suffering from a (quarantinable) disease subject to the International Health Regulations.

1.2 Subject to certain exceptions, foreign nationals and Commonwealth citizens arriving in the United Kingdom may be subjected to medical examinations under the Immigration Act 1971.

GEN 1.3.4 FLYING LICENCES AND RATINGS

1 Experienced Holders of Non-UK Professional Pilot's Licences

1.1 Where the holder of a non-UK professional pilot's licence, who has flown in UK controlled airspace in circumstances governed by the requirements of Annex 6 (Operation of Aircraft) Part 1 (International Commercial Air Transport) to the ICAO Convention, wishes to fly a UK registered aircraft in a private capacity and exercise the privileges of an instrument rating, the CAA will be prepared to consider granting a certificate of validation of the foreign licence to permit this to be done. Such pilots should contact Personnel Licensing Department at the address given in paragraph 2.2. No similar arrangements exist for issuing certificates of validation to the pilots wishing to give flying instruction.

2 Visiting Pilots – Instrument and Flying Instructor Ratings

2.1 Article 21(4) of the Air Navigation Order 2000 allows, unless the CAA in the particular case gives a direction to the contrary, the holder of any valid pilot's licence (other than a student pilot's licence) granted under the law of a Contracting State of ICAO or relevant Overseas Territory to exercise the privileges of that licence, excluding the privileges of an Instrument Rating or Flying Instructor's Rating, in United Kingdom registered aircraft as long as flights are not made for public transport, aerial work or remuneration. In effect, this means that the holders of a valid pilots licence issued by other States may, within the limitations noted above, exercise private privileges on UK registered aircraft.

2.2 Pilots holding valid licences, including an instrument and/or flying instructor's rating issued by ICAO Contracting States, need to obtain at least a UK Private Pilot's Licence (PPL) and the appropriate rating if they wish to fly UK registered aircraft in controlled airspace in circumstances requiring compliance with the Instrument Flight Rules or to give flying instruction. Full details of the requirements to obtain a UK PPL and the associated Flying Instructor Ratings and Instrument Ratings are available from:

Civil Aviation Authority
Personnel Licensing Department
Aviation House
London Gatwick Airport
West Sussex
RH6 0YR
Tel: 01293 573700

3 JAA Licences issued by Joint Aviation Authorities Member States

3.1 Article 21(4)(b) of the Air Navigation Order 2000 allows, unless the CAA in the particular case gives a direction to the contrary, the holder of a JAA licence to exercise all the privileges of that licence in United Kingdom registered aircraft.

3.2 For the purposes of Article 21(4)(b), a JAA licence means a licence granted in accordance with Joint Aviation Requirements – Flight Crew Licensing (JAR-FCL) by a JAA member state that has been recommended for mutual recognition by Central JAA.

3.3 A list of JAA Member States that have implemented JAR-FCL and been recommended for mutual recognition in respect of JAR-FCL can be found on the CAA website at – www.caa.co.uk/srg/licensing

Flight crew licences issued in accordance with JAR-FCL by these countries are automatically recognised by UK CAA.

GEN 1.4.2 AGRICULTURAL QUARANTINE REQUIREMENTS

To be developed

GEN 1.4.3 HAZARDOUS CARGO REQUIREMENTS

1 Carriage of Munitions of War

1.1 Article 59 of the Air Navigation Order 2000, prohibits the carriage of munitions of war (ie any weapon, with or without ammunition, ammunition or article containing an explosive or any noxious liquid, gas or other thing which is designed or made for use in warfare or against the person including parts and accessories). No flight on which these articles are to be carried may be operated unless permission has been granted by the Civil Aviation Authority.

1.2 Application must be made at least 10 working days before the proposed date of the flight and should state precisely what munitions of war are involved, the manufacturer, the import/export licence number and its expiry date, air waybill number, the names and addresses of both consignor and consignee, the destination, the airports of departure and arrival and the date of the flight and flight number.

1.3 If the consignment contains dangerous goods the United Nations number, hazard class or division, compatibility group (where applicable) and net explosive content (for explosives) should be stated, together with information on the method of packing.

1.4 The above is the minimum information that is required. On occasions further information may be requested. (See also Carriage of Dangerous Goods, paragraph 2).

2 Carriage of Dangerous Goods

2.1 Regulation 4(1)(a) of the Air Navigation (Dangerous Goods) Regulations 1994 made pursuant to Article 60 of the Air Navigation Order 2000, prohibits the carriage of dangerous goods unless carried with the written permission of the Civil Aviation Authority. No flight on which dangerous goods are to be carried may be ordered unless prior permission for their carriage has been obtained.

2.2 Application must be made at least 10 working days before the proposed date of the flight and should include full details of the following: United Nations number, hazard class or division, compatibility group (when applicable), net explosive content (for explosives), full packing details, weight, air waybill number, the names and addresses of both consignor and consignee, the destination, the airports of arrival and departure and the date of the flight and flight number.

2.3 The above is the minimum information that is required. On occasions further information may be requested.

3 Applications

3.1 Application and enquiries for such permissions should be made in writing and addressed to:

Civil Aviation Authority
FO(PT)S Dangerous Goods Office
Floor 1W
Aviation House
South Area
London Gatwick Airport
West Sussex
RH6 0YR
Telex: 878753/4
Fax: 01293 573991
AFS: EGGRYAYF

GEN 1.5 AIRCRAFT INSTRUMENTS, EQUIPMENT AND FLIGHT DOCUMENTS

GEN 1.5.1 GENERAL

1 Introduction

1.1 Sub-sections GEN 1.5.1 and GEN 1.5.2 have yet to be developed by the United Kingdom.

1.2 The following agreed entries are published in accordance with European RVSM implementation.

1.3 Details of other UK General and Special Equipment requirements may be added once determined.

2 RVSM

2.1 Except for designated airspace where RVSM transition tasks are carried out, only RVSM approved aircraft and non-RVSM approved State aircraft shall be permitted to operate within the EUR RVSM airspace.

2.2 RVSM approved aircraft are those aircraft for which the operator has obtained RVSM approval, either from the State in which the aircraft is based, or from the State in which the aircraft is registered.

2.3 Guidance material on the airworthiness, continued airworthiness and the operational practices and procedures for the EUR RVSM airspace is provided in the Joint Aviation Authorities (JAA) Temporary Guidance Leaflet (TGL) Number 6, Revision 1, and the ICAO EUR Regional Supplementary Procedures (Doc 7030/4-EUR).

2.4 Except for State aircraft, RVSM approval is required for aircraft to operate in the RVSM airspace within the London and Scottish UIR(s), as described in ENR 2.1

2.5 Exceptionally, aircraft without RVSM approval operating in the North Atlantic Region airspace will be permitted to climb/descend through RVSM airspace in the London and Scottish UIR(s) under specified conditions or as directed by ATC.

Note: The provisions applicable to non-RVSM approved civil operations in EUR RVSM airspace where RVSM transition tasks are carried out as specified in the ICAO EUR Regional Supplementary Procedures (Doc 7030/4-EUR).

3 RNAV

3.1 AIP Text to be developed.

GEN 1.5.2 – SPECIAL EQUIPMENT TO BE CARRIED

To be delveloped

GEN 1.5.3 – EQUIPMENT TO BE CARRIED

1 Carriage of Radio and Radio Navigation Equipment

1.1 The requirements for the carriage of radio, radio navigation and area navigation equipment are contained in Articles 15, 50, 51 and Schedule 5 to the Air Navigation Order 2000. In the case of the Shanwick Oceanic Control Area, additionally in Article 47 of the Air Navigation Order 2000 and Regulation 18 of the Air Navigation (General) Regulations 1993, as amended.

1.2 The requirements for all aircraft flying in United Kingdom Airspace, in specified circumstances, together with the requirements for flight in the Shanwick Oceanic Control Area are tabulated in paragraphs 1.2.1 to 1.2.4 and in the case of area navigation equipment in paragraph 1.2.5. SSR transponder equipment requirements are detailed in paragraph 1.3 Additional requirements for Public Transport aircraft are detailed in Schedule 5 to the Air Navigation Order 2000.

Paragraph	Type of Flight	Equipment
1.2.1	Flight in Controlled Airspace	(a) **Communications Equipment** VHF RTF for 760 channel operation
1.2.2	Flight under IFR in Controlled Airspace below FL 245	(a) **Communications Equipment** VHF RTF for 760 channel operation (b) **Navigation Equipment** (†) VOR receiver, DME and automatic DF (c) **Approach Aid Equipment** For landing at certain aerodromes within control zones, ILS (†) Aircraft with special VFR clearance are not required to carry the navigation equipment specified in paragraph 1.2.2 (b)
1.2.3	Flight in the Upper Airspace (except Gliders)	(a) **Communications Equipment** VHF RTF with appropriate frequencies available (b) **Navigation Equipment** VOR receiver, DME and automatic DF
1.2.4	Flight in Shanwick Oceanic Control Area	(a) **Communications Equipment** HF RTF with appropriate frequencies available (b) **Navigation Equipment** Suitable long-range navigation equipment **Note:** Flights operating between FL 285 and FL 420 are required to meet the minimum navigation performance specifications (MNPS)
1.2.5	Flight on the ATS Route Structure 1&3 above FL 952 (with the exception of TMA Airspace)	(a) **Communications Equipment** VHF RTF for 760 channel operation (b) **Navigation Equipment** State approved equipment which meets the RNAV RNP5 requirement contained in ICAO Doc 7030 Regional Supplementary Procedures (EUR RAC Section 16)

1 The term ATS Route Structure is used to mean variously:
Upper ATS Route; Airway; Advisory Route (participating aircraft only) and Arrival and Departure Route.

2 Certain ATS Routes eg Airway N862 are not aligned with VOR and require area navigation equipment at all route levels. In this case the requirement for area navigation equipment is denoted by use of one of the following route designators: L, M, N, P, Q, T, Y or Z.

3 Inbound aircraft from Oceanic Airspace are permitted to navigate the MNPS Airspace navigational accuracy requirements but are to be RNP5 compliant on reaching Landfall Fix

1.2.6 Exemptions

1.2.6.1 State aircraft are exempt from the requirement for carriage of RNAV equipment. No other general exemption from these requirements will be granted, but, in very special circumstances, relaxation of the requirements, for a single flight, may be approved by the appropriate ATC Unit (see paragraph 1.3.3 for exemptions from the carriage of SSR for limited periods).

GEN 1.5   GEN 1.5.3.1 Carriage of Radio and Radio Navigation Equipment

1.3 Carriage of SSR Transponders

1.3.1 The requirements for the carriage of radar equipment are hereby notified for the purpose of Article 20 and Schedule 5 to the Air Navigation Order 2005. With the exceptions listed at paragraph 1.3.2, all aircraft are required to carry SSR transponder equipment with the following capability:

(a) Outside the lateral and vertical bounds of the London TMA: (1) The whole of the United Kingdom Airspace at and above FL 100 (2) United Kingdom Controlled Airspace below FL 100 when operating under IFR (3) The Scottish TMA between 6000ft ALT and FL 100 (4) Sumburgh CTR/CTA (**Note:** Requirement expected to continue until March 2006)	Mode A and Mode C with altitude reporting
(b) Inside the lateral and vertical bounds of the London TMA with Effect from 31 March 2005: (1) When operating with a valid exemption from the Mode S carriage requirements at sub-paragraphs 1.3.1 (c) and 1.3.1 (d) below issued at the EUROCONTROL Exemption Co-ordination Cell or the Civil Aviation Authority.	
(c) Inside the lateral and vertical bounds of the London TMA with Effect from 31 March 2005: (1) Fixed-wing aeroplanes having a maximum take-off mass in excess of 5700kg (2) Fixed-wing aeroplanes having a maximum cruising true airspeed capability in excess of 250kt.	Mode S Enhanced Surveillance functionality
(d) Inside the lateral and vertical bounds of the London TMA with Effect from 31 March 2005: (1) Fixed-wing aeroplanes having a maximum take-off mass not exceeding 5700kg and a maximum cruising true airspeed capability not exceeding 250kt. (2) Helicopters	Mode S Enhanced Surveillance functionality

1.3.2 Exceptions

1.3.2.1 The requirement will not apply to:

(a) Gliders.

(b) aircraft below FL 100 in Controlled Airspace outside of the London TMA receiving an approved crossing service.

1.3.3 Exemptions

1.3.3.1 Mode A and Mode C Requirements. Exemptions from the requirement for Mode A and Mode C with altitude reporting will not normally be issued. However, if it is considered that exceptional circumstances exist, an application must be made to the Airspace Utilisation Section (AUS), Directorate of Airspace Policy, K1, CAA House, 45-59 Kingsway, London, WC213 6TE (Tel: +44 (0)20 7453 6599, Fax +44 (0)20 7453 6593). Applications should state details of the inability to meet the requirement, including aircraft type, registration and, where applicable, the forecast date by which installation of equipment will be complete.

1.3.3.2 Mode S Requirements. A Transition Period of 2 years between 31 March 2005 and 31 March 2007 has been agreed, during which exemptions from the requirement for Mode S in the London TMA will be issued. In order to co-ordinate and process Mode S exemption applications on behalf of the European Mode S implementing States, an Exemption Co-ordination Cell (ECC) has been set up within the EUROCONTROL Agency in Brussels. Operators of aircraft requiring access to notified Mode S airspace in the UK and elsewhere will need to register with the EUROCONTROL ECC. Registration will include the requirement for a declaration of compliance or a request for exemption and can be completed manually or electronically. Registration forms, together with full details of the exemption application process, are available on the Mode S section of the EUROCONTROL website (http://www.eurocontrol.int/mode_S). After 31 March 2007, exemptions from the requirement for Mode S will not normally be issued. However, if exceptional circumstances exist, an appropriate application will need to be submitted to the ECC for consideration by the Civil Aviation Authority and the ATS regulators of the other Mode S implementing States.

1.3.3.3 GA Aircraft. GA aircraft wishing to enter the London TMA after 31 March 2007 are to be Mode S equipped, as per the weight/speed criteria in AIC 4912026 (Yellow 171). After 31 March 2007, access to the London TMA by Mode S non-compliant GA aircraft will be denied.

1.3.3.4 In specific cases, where short-notice exemption from the carriage and operation of SSR transponder equipment is required by an operator, entry may be permitted on an individual flight basis provided that the approval of the ATC Unit responsible for the airspace has been obtained and that the pilot complies with the appropriate Rules of the Air and ATC, which apply to the airspace concerned. Operators should note that this does not apply in respect of the London TMA where, for safety and practical purposes, short-notice exemptions cannot be accommodated and the procedure outlined in paragraph 1.3.3.2 must be followed. The controlling ATC Unit is to be contacted prior to departure or, in exceptional circumstances, on the appropriate ATC frequency prior to entry to the specified airspace. Entry under an exemption is not guaranteed in normal circumstances and will only be permitted at the discretion of ATC where it does not impinge on the safe operation within the airspace involved. Short-potice exemptions will only be granted on a case-by-case basis and will not be given to operators who require regular access to airspace to which mandatory carriage and operation of SSR regulations apply. Such applications should be submitted either to AUS in accordance with paragraph 1.3.3.1 or the EUROCONTROL ECC in accordance with paragraph 1.3.3.2, as appropriate. Cases in respect of transponder failure are to be dealt with in accordance with the SSR Operating Procedures as promulgated in ENR 1.6.2, paragraph 4. An aircraft in an emergency situation will be afforded the appropriate level of priority, which shall include implicit exemptions from the appropriate legislation for the purpose of saving life.

1.3.4 Channel Islands Control Zone

1.3.4.1 Requirements for the Channel Islands Control Zone are given at EGJJ AD 2.22, paragraph 1.

1.4 Carriage of Airborne Collision Avoidance Systems (ACAS) in the United Kingdom FIR and UIR.

1.4.1 The requirements of the carriage of Airborne Collision Avoidance Systems (ACAS) are contained in Articles 20 and 61 and Schedule 5 to the Air Navigation Order 2005. The Traffic Alert and Collision Avoidance System (TCAS) II is accepted as a suitable ACAS system provided its installation is certified by the State of Registry, and that its operation by flight crew is in accordance with instructions for the use of this equipment specified in their company's manual.

1.4.2 With the exception of those circumstances at paragraph 1.4.3, all aeroplanes powered by one or more turbine jets or turbine propeller engines and either having a maximum take-off weight exceeding 5700kg or a maximum approved passenger seating configuration of more than 19 passengers are to be fitted with, and operate, TCAS II software Version 7.0 (RTCA DO-185A) with a Mode S transponder compliant with Annex 10 Mode S SARPs Amendment 73) within UK Airspace. Use of ACAS in the United Kingdom FIR and UIR is detailed at ENR 1.1.3 General Flight Procedures Section 4. This includes operation of aircraft when ACAS II is unserviceable.

1.4.3 Exemptions

1.4.3.1 A General Exemption from the requirements of Schedule 5 Scale J of the Air Navigation Order 2000 concerning the carriage of ACAS II in UK airspace has been granted for aeroplanes operating under certain conditions. Two classes of flights are affected:

(a) Delivery Flights. Aeroplanes newly manufactured within European Civil Aviation Conference (ECAC) member states, which are not fitted with ACAS II. These will be permitted to transit direct flights only, out of the airspace of ECAC member states to regions where the carriage and operation of ACAS II is not required. ECAC membership consists of the following States:

Albania, Armenia, Austria, Azerbaijan, Belgium, Bosnia and Herzegovina, Bulgaria, Croatia, Cyprus, Czech Republic, Denmark, Estonia, Finland, France, Germany, Greece, Hungary, Iceland, Ireland, Italy, Latvia, Lithuania, Luxembourg, Malta, Moldova, Monaco, Netherlands, Norway, Poland, Portugal, Romania, Serbia and Montenegro, Slovakia, Slovenia, Spain, Sweden, Switzerland, The former Yugoslav Republic of Macedonia, Turkey, Ukraine and the United Kingdom.

(b) Maintenance Flights. Direct flights by aeroplanes, which are not fitted with ACAS II, from outside ECAC member states, for the purpose of maintenance and engineering at facilities located within the ECAC member states. Following notification approval of an ACAS II exemption for the flight, the aircraft operator should indicate on the Flight Plan that the flight is being operated under the provisions of the ACAS II Delivery and Maintenance Flight Exemption provisions, by inserting, in Field 18, the information:

RMK/Delivery flight ACAS II exemption approved', or

RMK/Maintenance flight ACAS II exemption approved'.

1.4.3.2 Flights operated under the provisions of these exemptions must be nonrevenue flights. An ACAS II delivery or maintenance flight exemption is not available for those flights seeking only to transit through the airspace of ECAC member states.

1.4.3.3 The following conditions apply:

(a) Where agreed Regulations and Procedures exist, these shall be maintained.

(b) An ICAO compliant altitude reporting transponder must be fitted and serviceable **before** departure.

(c) An ACAS II exemption approval will be valid for a 3 day period from estimated departure date, and solely for the purpose for which it has been issued. If the flight is subsequently delayed beyond the maximum 3 day exemption period a fresh application must be submitted; this may take a further 3 working days to process.

(d) An ACAS II Exemption Letter issued by the ASU, must be carried onboard the aircraft.

(e) Conditions may be imposed by one or more ECAC Member States: such as operating within certain restrictive hours, or via specific routes, or at stated flight levels (for safety reasons or otherwise).

(f) The flight must be conducted along the most direct (or permissible) route to the delivery or maintenance destination airport. It must be noted that the onus is on the aircraft operator to ensure compliance with the above conditions and that the exempted flight is in accordance with the operator's originally stated intentions, and that it must comply with any conditions laid down by the ASU and subsequently by the ATC authorities.

1.4.3.4 In addition, test flights are to be subject to established national regulations, procedures and authorisation. Carriage of ACAS II equipment will be addressed under the current provisions for avionics equipment required for these flights.

1.4.3.5 Operators of aeroplanes intended to be operated under the provisions of these exemptions must apply for an exemption on an individual case by case basis, to the EUROCONTROL ACAS Support Unit (ASU) at least 3 working days before a flight is due to depart for or from one of the ECAC States. The ASU, on behalf of all ECAC member states, will then assess, process and notify to the aircraft operator, the ACAS II Delivery or Maintenance flight exemption, coordinating as appropriate with the national civil aviation authorities of the ECAC member states through whose airspace the aircraft is planned to fly.

1.4.3.6 Applications for an ACAS II exemption under the provisions of this procedure should be made to the ASU, at the following address:
ACAS Support Unit
Tel: +32 2 729 3133/3170/3113
Fax: +32 2 729 3719
SITA: BRUAC7X
email: acas@eurocontrol.int
web site: www.eurocontrol.int/acas

1.4.4 A further General Exemption from the requirements of Schedule 5 Scale J of the Air Navigation Order 2005 concerning the carriage of ACAS II in UK airspace has been granted for certain historical and ex military aeroplanes. These may be considered to be those types to which Article 4(1) to Regulation (EC) No 1592/2002 does not apply, namely those:

GEN 1.5 GEN 1.5.3.1 Carriage of Radio and Radio Navigation Equipment

(a) Aircraft having a clear historical relevance, related to:
(i) participation in a noteworthy historic event; or
(ii) a major step in the development of aviation; or
(iii) a major role played in the armed forces of a Member State; and meeting one or more of the following criteria:
(iv) its initial design is established as being more than 40 years old;
(v) its production stopped at least 25 years ago;
(vi) fewer than 50 aircraft of the same basic design are still registered in the Member States;
(b) Aircraft specifically designed or modified for research, experimental or scientific purposes, and likely to be produced in very limited numbers;
(c) Aircraft whose initial design was intended for military purposes only.

1.4.5 Operators of affected historical and ex military aeroplanes intended to be operated under the provisions of this exemption must apply for exemptions, on an individual case by case basis, to the Terminal Airspace section of the Directorate of Airspace Policy **no later than 31 January 2005** and thereafter upon the renewal or award of a Permit to Fly. The Directorate of Airspace Policy will then assess, process and notify to the aircraft operator the ACAS II 'historical and exmilitary' exemption. Co ordination of such flights through the airspace of other ECAC member states must be actioned by the aircraft operator through the ASU.

1.4.6 Following notification of an ACAS II exemption approval for a historical or ex military aeroplane, the aircraft operator should indicate in Field 18 of the Flight Plan that the flight is being operated under the provisions of the ACAS II exemption provisions by inserting the information '**RMK/Historic aircraft ACAS II exemption approved**'.

1.4.7 The following conditions apply:

(a) An ACAS II Exemption Letter issued by the CAA must be carried on board the aircraft.

(b) Proof of approval by the ASU for flight through the airspace of other ECAC member states.

1.4.8 Aircraft operators should note that the 'historical and exmilitary' exemption arrangements described above are temporary in nature and that standing arrangements will be developed in due course.

1.4.9 Further information and advice concerning the operation and carriage of ACAS equipment in UK airspace can be obtained from:

Manager Terminal Airspace
Directorate of Airspace Policy
CAA House
45-59 Kingsway
London
WC213 6TE
Tel: +(0)20 7453 6510
Fax: +(0)20 7453 6565
email: terminal.airspace@dap.caa.co.uk

1.5 8.33 kHz Channel Spacing in the VHF Radio Communications Band

1.5.1 As required by ICAO Regional Supplementary Procedures Doc 703014 EUR/RAC-4 and further to the delayed decision agreed by Eurocontrol AC13 (23 July 1998), the carriage and operation of 8.33 kHz channel spacing radio equipment is mandatory throughout the ICAO EUR Region for flights above FL 245.

1.5.2 Non-equipped flights which are flight planned to enter any FIR/UIR in the EUR Region where no exemptions have been published, except for those applicable to UHF equipped State flights (refer to the AIP/Supplement of the State covering the FIR/UIR concerned), must flight plan to operate below FL 245 throughout the entire EUR Region.

1.5.3 8.33 kHz Exemptions

1.5.3.1 With the exception of State aircraft, there are no exemptions.

1.5.4 State Aircraft

1.5.4.1 Those State aircraft which are infrequent users of the FIR/UIR are permanently exempted from the above carriage requirements, provided that they are able to communicate on UHF, where available. Where UHF is not available, State (Military) aircraft not equipped with 8.33 kHz channel spacing equipment shall be excluded from 8.33 kHz Airspace.

1.5.4.2 Provision for State aircraft exempted from the carriage of 8.33 kHz channel spaced communications equipment will be made on a tactical basis through the provision of an alternative UHF channel, the details of which will be given at time of use.

1.6 Use of GPS for North Sea Operations

1.6.1 UK AOC Holders intending to use GPS for enroute navigation for North Sea flight operations are to use GPS equipment that meets or exceeds CAA Specification 22. AOC holders requiring further information should contact their assigned flight operations Inspector. Non UK AOC holders are recommended to operate to at least the CAA Specification 22 standard.

GEN 1.7 DIFFERENCES FROM ICAO STANDARDS, RECOMMENDED PRACTICE AND PROCEDURES

Annex 1	Personnel Licensing (9th Edition)
Reference	Difference
Chapter 1	Definitions and General Rules Concerning Licences
1.1	**Flight-time.** United Kingdom legislation continues to define a piloted aircraft as being in flight: 'from the moment when, after the embarkation of its crew for the purpose of taking off, it first moves under its own power until the moment when it next comes to rest after landing.'
	Pilot-in-command. In United Kingdom legislation, 'Pilot-in-command' in relation to an aircraft means a person who for the time being is in charge of the piloting of the aircraft without being under the direction of any other pilot in the aircraft. The flight crew member charged with the safe conduct of a flight is called the 'Commander'.
1.2.2.1	Provision is made within United Kingdom legislation which renders valid a flight crew licence of Contracting States for flight for private purposes unless the United Kingdom Civil Aviation Authority gives a direction to the contrary. The licence holder cannot receive remuneration or exercise the privileges of an Instrument Rating or Flying Instructor Rating included in the license. No certificate of validation is issued.
	Flight crew licenses and aircraft maintenance engineer licences issued by other Member States of the joint Aviation Authorities in accordance with the Joint Aviation requirements are automatically rendered valid unless the United Kingdom Civil Aviation Authority gives a direction to the contrary. No certificate of validation is issued. The United Kingdom does not otherwise validate aircraft maintenance engineer licences.
	The United Kingdom does not validate air traffic controller licences issued by any other state.
1.2.4.2	Not in compliance for private pilot licences. See under below for differences relating to medicals for private pilot licences.
1.2.5.2	The standards for Annex 1 compliant PPLs for aeroplanes and helicopters adopted with effect from 1 July 1999 are:
	60 months until age 30;
	24 months until age 50;
	12 months until age 65;
	6 months thereafter.
	The United Kingdom minimum medical standard for a private pilot licence – free balloon is a declaration of Health. The United Kingdom does not issue glider pilot licences.
1.2.5.2.2	See 1.2.5.2.
Chapter 2	Licences and Ratings for Pilots
1.1.1.1	United Kingdom legislation also provides for the issue of privat pilot licences for gyroplanes (autogyros) and private and commercial pilot licences for airships. The United Kingdom also issues a recreational private pilot licence called a 'National Private Pilots Licence (Aeroplanes) which may include class ratings for microlight aeroplanes, self launching motor gliders (powered sailplanes) and for other simple single engine aeroplanes not exceeding 2000Kg take of weight. These licences have no equivalent in Annex 1.
	Glider pilot licences are not issued although there is provision for a commercial pilot licence for gliders. All gliding in the United Kingdom is recreational in nature and the British Gliding Association issues certificates to pilots under the auspices of the Federation Aeronautique Internationale.
1.1.1.1	The United Kingdom grants separate licences for each category.
2.1.2.3	The United Kingdom grants separate PPL, CPL and ATPL licences.
2.1.2.4	The United Kingdom grants separate licences for each category.
2.1.3.1.1	A type rating is required for each helicopter type which the licence holder flies.
2.1.7	An Instrument Meteorology Conditions Rating which is purely a National rating, is also issued for use within UK airspace boundaries to allow private flight under IFR outside controlled airspace and in class D, E and F controlled airspace. It has no equivalent under ICAO as it is not a full instrument rating.
2.1.9.2	Licence holders may be fully credited with co-pilot flight time towards the total time required for a higher grade of pilot licence.
2.1.10.1	United Kingdom legislation provides for public transport operations up to the age of 65 provided that it is a multi-crew operation and the other pilot is under the age of 60.
2.1.10.2	United Kingdom legislation provides for public transport operations up to the age of 65 provided that it is a multi-crew operation and the other pilot is under the age of 60.

Reference	Difference
2.2.3	The United Kingdom minimum medical standard for a student for a private pilot licence – free balloon is a declaration of health.
2.3.1.3.1	As applicant shall have completed not less than 45 hours of flight time as pilot of aeroplanes for a private pilot licence – aeroplane issued in compliance with Annex 1.
	The United Kingdom also issues a recreational private pilot licence called a 'National Private Pilot's Licence' which mau include class rating for microlight aeroplanes, self launching motor gliders (powered sailplanes) and for other simple single engine aeroplanes not exceeding 2000kg take off weight. The hours of flight time that must be completed vary according to the class rating to be included in the licence. This licence has no equivalent in Annex 1.
2.3.1.6	For the United Kingdom 'National Private Pilot's Licence (Aeroplanes)' different medical assessment requirements are applied. This licence has no equivalent in Annex 1.
2.3.2.1	A holder of a private pilot licence which includes a flight instructor rating valid for microlights or self-launching motor gliders (powered sailplanes) may be paid for giving instruction or conducting flight tests on microlights or self-launching motor gliders when doing so as and with members of the same flying club.
2.5.1.3.1	The holder of an aeroplane pilot licence who is also the holder of a Flight Engineer licence may be credited with 50% of the time spent undertaking the duties of a Flight Engineer up to a maximum of 250 hours towards the 1500 hours requirement.
2.6.1.2.2 (b)	See 2.6.1.3.1
2.6.1.3.1	Flight instruction for a single engine instrument rating shall comprise at least 50 hours dual instruction of which up to 35 hours may be instrument ground time in a flight simulator. This training shall be approved training.
	Flight instruction for a multi engine Instrument Rating shall comprise at least 55 hours dual instruction of which up to 40 hours may be instrument ground time in a flight simulator. This training shall be approved training.
2.6.1.5.2	United Kingdom private pilot licence holders with Instrument Rating are not required to meet the full ICAO class 1 medical assessment requirements. A hearing test to class 1 standards is required.
2.7.1.3.1	An applicant shall have completed not less than 45 hours of flight time as pilot of helicopters for a private pilot licence – helicopter.
2.7.2.1	The holder of a private pilot licence (helicopters) issued before 1 July 2000 which includes a flight instructor rating valid for helicopters may be paid for giving instruction or conducting flight tests on helicopters when doing so as and with members of the same flying club.
2.8.1.3.1	An applicant for a Commercial Pilot Licence for helicopters shall have completed not less than 185 hours of flight time or 135 hours if completed during a course of approved training.
2.9.1.3.1.1	An applicant shall have completed in helicopters not less than: 70 hours of instrument time of which not more than 30 hours may be instrument ground time: & 100 hours of night flight as pilot in command or co pilot.
2.10.1.2.2 (b)	See 2.10.1.3
2.10.1.3	Flight instruction for a single engine Instrument Rating shall comprise at least 50 hours dual instruction of which up to 35 hours may be instrument ground time in a flight simulator. This training shall be approved.
	Flight instruction for a multi engine Instrument Rating shall comprise at least 55 hours dual instruction of which up to 40 hours may be instrument ground time in a flight simulator. This training shall be approved training.
2.10.1.5.2	United Kingdom private pilot licence holders with Instrument Rating are not required to meet the full ICAO class 1 medical assessment requirements. A hearing test to class 1 standards is required.
2.10.3	The United Kingdom does not issue a combined Instrument Rating for Helicopters and Aeroplanes.
2.11.1.3	An applicant for a Flight Instructor Rating – Aeroplane must complete approved training comprising not less than:
	125 hours theoretical knowledge instruction:
	30 hours flight instruction in aeroplanes.
2.11.2	Flight instruction for ratings in licences may be given only by a licence holder who holds an appropriate instructor rating. In addition to Flight Instructor Ratings, the United Kingdom grants:
	Single Pilot Aeroplane Class Rating Instructor Rating,
	Multi Pilot Aeroplane Type Rating Instructor Ratings;
	Helicopter Type Rating Instructor Ratings;
	Aeroplane Instrument Rating Instructor Ratings;
	Helicopter Instrument Rating Instructor Ratings;

Reference	Difference
2.12	Provision is made in the United Kingdom legislation for the issue of the commercial pilot licence (gliders) only. Private and club glider flying is regulated by the British Gliding Association whose certificates are issued under the auspices of the Federation Aeronautique Internationale.
2.13	The United Kingdom issues both private and commercial pilot licences for free balloons.
2.13.1.1	An applicant for a private pilot licence shall be not less than 17 years of age. An applicant for a commercial pilot licence shall be not less than 18 years of age.
2.13.1.2.1	The United Kingdom issues separate Flight Radiotelephony Operators Licences.
2.13.1.3	For a private pilot licence only 6 launches and ascents are required.
	For a commercial pilot licence for aerial work, an applicant shall have completed not less than:
	35 hours flight time as pilot of balloons including:
	15 hours of instruction in flying;
	20 hours as pilot in command.
	Public transport operations (commercial air transportation) may not be undertaken until the commercial licence holder has completed at least 75 hours flight time as pilot in balloons of which 60 hours must be as pilot in command.
2.13.1.5	The United Kingdom minimum medical standard for a private pilot licence – free balloon is a declaration of health.

Chapter 3	Licences for Flight Crew Members other than Licences for Pilots
3	The United Kingdom continues to issue separate Flight Radiotelephony Operators Licences.
3.2.1.1	The United Kingdom does not currently issue Flight Navigator's Licences although there is provision in the legislation.
3.3.1.1	The applicant shall be not less than 21 years of age.
3.3.1.2.1	The UK does not require this knowledge for Flight Engineers Licence issue.

Chapter 4	Licences and Ratings for Personnel other than Flight Crew Members
4.2.1.4	Training – The United Kingdom does not require the completion of a course of training for basic licence issue or for certain aircraft types prior to type endorsement.
4.2.1.5	Skill – The United Kingdom does not require a demonstration of practical ability. The candidates knowledge of maintenance procedures, use of tools and troubleshooting/decision making is examined.
4.3.1.1	The minimum age is 20 for an ATC licence.
4.3.1.4	The United Kingdom requires air traffic controllers to hold United Kingdom class 1 medical certificates except for aerodromes controllers who require, as a minimum, to hold a United Kingdom class 2 certificate. See also 6.1.1.
4.4.1.1 (d)	There is no separate precision approach radar rating in the United Kingdom, this rating having been subsumed into the control surveillance rating.
4.4.1.1 (e) & (f)	See comments under 4.4.1.1 (d) above.
4.4.2.2.1 (b) & (c)	Under the harmonised ATC licensing scheme, all UK units will be required to have approved unit training plans that setail training requirements.
4.4.2.2.2	Operational constraints do not always allow for completion within 6 months. However, the United Kingdom has developed Unit Training Plans which ensure operational requirements are met before a rating is issued.
4.4.3	The privileges differ due to the revised rating system in use. See 4.4.1.1 above.
4.5	The United Kingdom does not issue Flight Operations Officer/Flight Dispatcher Licences. The activity is controlled as part of the approval of the Air Operators Certificate.
4.6.1.4.1	The United Kingdom does not regulate the operational use of an Aeronautical Station Operator Licence and has no method to establish the experience of an applicant.

Chapter 6	Medical Provisions for Licensing
6.1.1	The United Kingdom complies with ICAO class 1 and 2 medical assessments and certificates for Airline Transport Pilot Licence, Commercial Pilot Licence and Private Pilot Licence holders.
	The United Kingdom requires Air Traffic Control Officers to hold a United Kingdom class 1 medical certificate which meets or exceeds the ICAO class 3 standards. These certificates continue to be issued to Air Traffic Control Officers to avoid confusion within the United Kingdom by changing certificate designations.
	The CAA does not use the same definitions as ICAO for classes of medical assessment for holders of the United Kingdom 'National Private Pilots Licence (Aeroplanes)', pilots of balloons, airships or gyroplanes.

Reference	Difference
6.2.3.1	The level of illumination recommended for use in visual acuity tests is 80 cd/m2. The difference between 30-60 cd/m2 and 80 cd/m2 will not result in any important practical differences in the test results.
6.2.4.3.1	No recommendations are made on the colour of sunglasses for air traffic controllers.
6.3.2.8.1	Unless clinically required radiography examinations of the chest are not repeated following initial assessment.
6.4.1.2	Private pilots require a medical examination only every 5 years when under 30 years of age.
6.4.2.8.1	Not required for the initial or repeat medical assessments unless clinically indicated.
6.5.1.1	Class 3 medical certificates for Air Traffic Controllers are not issued. A United Kingdom class 1 medical certificate is issued for ATCO duties.
6.5.2.8.1	Unless clinically required radiography examinations of the chest are not repeated following initial assessment.
6.5.3.2	Air Traffic Control Offices currently have no requirements to have their binocular visual acuity measured, as the previous Annex 1 SARPS did not require this.
6.5.3.2.2	Large refractive errors (greater than 5 dioptres) are not currently permitted under UK ATCO requirements.
6.5.3.2.3	Large refractive errors (greater than 5 dioptres, equivalent to approximately 6/60) are not currently permitted.
6.5.3.4.1	ATCO's who require vision correction are not currently required to carry a spare set of correcting spectacles for near vision, as the previous Annex 1 SARPS did not require this.

Annex 2 Rules of the Air (9th Edition) (AMDT 37)

Chapter 1 Definitions

Acrobatic Flight: The UK uses the term 'Aerobatic Manoeurves'.

Aerodrome Control Service: For aircraft in the air the service is limited to aircraft flying in or in the vicinity of the aerodrome traffic zone by visual reference to the surface.

Approach Control Service: For aircraft in the air the service is limited to aircraft flying in or in the vicinity of the aerodrome traffic zone by visual reference to the surface.

Cloud Ceiling: In relation to an aerodrome means the vertical distance from the elevation of the aerodrome to the lowest part of any cloud visible from the aerodrome which is sufficient to obscure more than one half of the sky so visible.

Controlled Aerodrome: The UK does not use this term but lists in the AIP those aerodromes at which air traffic control service is provided.

Flight Crew: Those members of the crew of the aircraft who respectively undertake to act as pilot, flight navigator, flight engineer and flight radiotelephony operator of the aircraft.

Manoeuvring Area: The part of an aerodrome provided for the take-off and landing of aircraft and for the movement of aircraft on the surface, excluding the apron and any part of the aerodrome provided for the maintenance of aircraft.

Pilot in Command: In relation to an aircraft means a person who for the time being is in charge of the piloting of the aircraft without being under the direction of any other pilot in the aircraft.

Runway: An area, whether or not paved, which is provided for the take-off or landing run of aircraft.

Special VFR Flight: A flight made at any time in a Control Zone which is Class A airspace, or in any other control zone in Instrument Meteorological Conditions or at night, in respect of which the appropriate air traffic control unit has given permission for the flight to be made in accordance with special instructions given by that unit instead of in accordance with the Instrument Flight Rules and in the course of which the aircraft complies with any instructions given by that unit and remains clear of cloud and in sight of the surface.

Chapter 3 General Rules

3.3.1.2 (b)	Flight plans are not required for aircraft flying within Advisory Airspace unless they intend to participate in the Advisory Service.
3.3.1.2 (c)	For a flight of more than 10 miles from the coast or over sparsely populated or mountainous areas, particularly if the aircraft is not equipped with radio, it is advisable to file a flight plan to facilitate the provision of Alerting and Search and Rescue. A flight plan may be filed for any flight.
3.3.5.3 & 3.3.5.4	The UK requires a pilot flying to a destination without ATS or an AFS facility, prior to departure, to notify a responsible person at the destination of his ETA. The responsible person will inform the Parent ATSU if the aircraft fails to arrive within 30 minutes of the ETA. In the event of a Pilot unable to find a responsible person at his destination he may request his Parent ATSU to act in this capacity. Should this occur, the Pilot is required to inform the Parent ATSU within 30 minutes of arrival at destination.

45

3.6.5.2.2		Additional procedures appropriate to specific circumstances are detailed in the UK AIP ENR 1.1.3 General Flight Procedures Section. Further detailed procedures for individual major aerodromes may differ from the basic procedure and are notified in the UK AIP Aerodrome (AD2) Sections.
3.9		**1. Class A Airspace:** The UK has not yet notified VMC minima for class A airspace as adopted in Annex 2 on 4 November 1999. However, comparable VMC minima are specified for certain applications in Class A airspace (UK AIP ENR 1-4-1).

2. Class B Airspace: The UK has not yet implemented the distance from cloud requirements introduced in Annex 2 on 4 November 1999.

3. Class C, D and E Airspace: In Addition to the minima specified in Table 3-1, the VFR flight is allowed by aircraft, other than helicopters, at or below 300ft amsl at a speed of 140kt or less, which remain clear of cloud and in sight of the surface and in a flight visibility of at least 5km. Helicopters may fly under VFR in Class C, D or E Airspace at or below 3000ft amsl provided that they remain clear of cloud and in sight of the surface.

4. Class F and G Airspace: The VMC minima at and below FL100 applies down to the surface (instead of down to 3000ft amsl) with the minima at and below 3000ft as an alternative. The proviso 'or 300m above terrain whichever is higher' does not apply in the UK.

5. For the purposes of an aeroplane taking off from or approaching to land at an aerodrome within Class B, C or D airspace, the visibility, if any communicated to the commander of an aeroplane by the appropriate air traffic control unit shall be taken to be flight visibillity for the time being.

Chapter 4   Visual Flight Rules

4.1	See details above for para 3.9
4.2	The UK does not permit VFR flights in certain Control Zones which are notified in the UK AIP as class A airspace.
4.3	VFR flight is not permitted at night. (Night as defined in the UK legislation).
4.4 (a)	VFR flight is permitted above FL200, except in certain areas notified in the UK AIP as Class A airspace.
4.5	VFR flight by General Air Traffic (as defined in UK AIP GEN 1-7-56 Table 1.7.2) is not permitted at and above FL290. VFR flight by Operational Air Traffic (as defined in UK AIP GEN 1-7-56 Table 1.7.2) is permitted and will be provided with 200-ft vertical separation.
4.6	In the UK:

(a) Minimum height over congested areas is 1500ft.

(b) There is no minimum height above the surface, but aircraft must maintain a minimum distance of 500ft from persons, vessels, vehicles and structures.

The minimum heights apply to all flights whether under VFR or IFR and in all meteorological conditions.

4.7	It is not mandatory in the UK for VFR flights to adopt any particular cruising level system. However, when operating above Transition Altitude they are recommended to conform to the cruising level system prescribed in the UK for IFR flights. In those parts of controlled airspace where VFR flight is permitted such flights are not required to adopt any particular cruising level system.

In those parts of controlled airspace where VFR flight is permitted, such flights are not required to adopt any particular cruising level system.

Chapter 5   Instrument Flight Rules

5.1.2   Minimum Flight Altitude.

(a) The UK has no statutory requirements relating specifically to minimum IFR altitude when operating over high terrain or mountainous areas.

(b) The UK regulations require that an aircraft operating under IFR shall not fly at a height less than 1000ft above the highest fixed obstacle within a distance of 5nm of the aircraft unless the aircraft is flying on a route so notified or is operating at or below 3000ft amsl and remains clear and in sight of the surface.

In addition to the minimum height requirements in respect of obstacles, the minimum height over congested areas is 1500ft.

Appendix 3   Cruising Levels for IFR Flight

IFR flights operating in level cruising flight above 3000ft amsl outside controlled airspace or above the appropriate transition altitude in the UK will use Table I if flying below 24500ft or Table II if flying above 24500ft. The altimeter shall be set to a pressure setting of 1013.2 hectopascals.

These levels do not apply to flying in conformity with ATC instructions or in accordance with notified holding procedures in relation to an aerodrome.

TABLE I

Flights at levels below 24500ft

Magnetic Track	Cruising Level
Less than 90°	Odd thousands of ft
090° but less than 180°	Odd thousands of ft, plus 500ft
180° but less than 270°	Even thousands of ft
270° but less than 360°	Even thousands of ft, plus 500ft

TABLE II

Flights at levels above 24500ft

Magnetic Track	Cruising Level
Less than 180°	25000ft and above at intervals of 2000ft to 29000ft; 33000ft and above at intervals of 4000ft.
180° but less than 360°	26000ft and above at intervals of 2000ft to 31000ft. 35000ft and above at intervals of 4000ft.

Appendix 1 Signals

Section 3 Visual signals used to warn an unauthorized aircraft flying in, or about to enter, a Restricted, Prohibited or Danger Area

These visual warning signals are not utilised by the UK.

Section 4 Signals for aerodrome traffic

4.2 At land aerodromes ground signals may be displayed for the guidance of Air Traffic. Such signals will normally be displayed in the Signals Area or on the Signals Mast, and as near as possible to the Control Tower. The signals which may be displayed and the interpretation of the signals are shown in Rules 42 to 46 of the Schedule to the Rules of the Air Regulations 1996. The signals are in accordance with ICAO Annex 2, Appendix A with the exception of those described below which either differ from or supplement those in Annex 2.

4.2.5.1 **Directions for Landing or Take-off.** A white disc displayed alongside the cross arm of the T and in line with the shaft of the T signifies that the direction of landing and take-off do not necessarily coincide. This may also be indicated by a black ball suspended from a mast.

4.2.5.2 Black numerals in two-figure groups, and where parallel runways are provided the letter or letters L (left), LC (left centre), C (centre), RC (right centre) and R (right), placed against a yellow background, indicate the direction for take-off or the runway-in-use.

4.2.6 **Right-Hand Circuit.** A red and yellow striped arrow. This may also be indicated by a rectangular green flag flown from a mast.

4.2.8 **Glider flights in operation.** In addition to the double white cross, two red balls suspended from a mast one above the other signify that glider flying is in progress at the aerodrome. A yellow cross indicates the tow-rope dropping area.

The following additional signals may be used in the UK:

Aerodrome Control in Operation. A checkered flag or board containing 12 equal squares, coloured red and yellow alternately, signifies that aircraft may move on the manoeuvring area and apron only in accordance with the permission of the Air Traffic Control Unit at the aerodrome.

Landing area for light aircraft. A white letter L indicates a part of the manoeuvring area which shall be used only for the taking-off and landing of light aircraft. A red letter L displayed on the standard dumb-bell signifies that light aircraft are permitted to take-off and land either on a runway or on the area designated above.

Helicopter operations. When helicopters are required to take-off and land only within a designated area a white letter H is displayed in the signals area and a white letter H indicates the area to be used by helicopters.

Boundary Markers

(a) Unserviceable portions of paved runway taxiway or apron: Markers with alternate orange and white stripes.

(b) Unserviceable portions of unpaved manoeuvring area: Orange and white markers alternating with flags coloured orange and white. (One or more white crosses indicate the area that is unserviceable.)

(c) Aerodrome boundary, where not otherwise evident: Orange and white markers.

(d) Boundary of an unpaved runway or of a stop way where not otherwise evident: White flat rectangular markers.

Additional Ground Signals. The following ground signals, not provided for in air navigation legislation, may be displayed at military aerodromes and at other aerodromes not normally available for civil aircraft in general.

Unserviceable areas. A yellow and black solid of triangular section will be displayed on areas which are unserviceable owing to bad ground or to the presence of stationary vehicles, working parties or other obstacles.

Landing dangerous. A white cross displayed at the end of a runway shall indicate that runway is non-usable. The aerodrome may be used for storage purposes.

Emergency use only. A white cross and a single white bar displayed at the end of the runway at a disused aerodrome indicates that the runway is fit for emergency use. Runways so marked are not safeguarded and may be temporarily obstructed.

Land in emergency only. Two vertical yellow bars on a red square on the Signals Area indicate that the landing areas are serviceable but the normal safety facilities are not available. Aircraft should land in emergency only.

Variable circuit. If the direction of the circuit is variable, a red flag will be flown from the Signals Mast when a left-hand circuit is in operation and a green flag when a right-hand circuit is in operation.

Light aircraft. A red L shall indicate that light aircraft may land on a special grass area which is delimited by white corner markings; taxiing of light aircraft on grass is permitted.

Appendix 3	**Tables of Cruising Levels**	
	See entry under para 5.3.1	
Appendix 4	**Unmanned Free Balloons**	
	The UK requires permission to be obtained for operations of unmanned balloons and details restrictions on the release of large numbers of small balloons, but not to the extent of Appendix 4.	
Attachment A	**Interception of Civil Aircraft**	
2.3 (f)	Not all UK interception aircraft and intercept control units have the capability to communicate on 121.500 MHz. Where an intercept control unit does not have such a capability use would be made of direct communications between that unit and another ATC unit which does have a 121.500 MHz capability. This would ensure that the establishment of communication on 121.500 MHz is not jeopardised.	
Annex 3	**Meteorological Service for International Air Navigation (15th Edition) (AMDT 73)**	
Chapter 4	**Meteorological Observations and Reports**	
4.1.2	The UK has established a number of meteorological stations on off-shore structures that produce fully automated observations. However, these observations do not meet the requirements of ICAO Annex 3 with respect to the measurement of present weather and cloud.	
4.1.5	Not all UK aerodromes with precision approach runways intended for Category II operations have automated equipment for the measurement if visibility installed. At these aerodromes human observed visibility shall be reported. Such aerodromes will not have fully observed automatic systems for acquisition, processing, dissemination and display in real time of the meteorological parameters affecting landing and take-off operations.	
4.1.6	Not all UK aerodromes with precision approach runways intended for Category I operations have automated equipment for the measurement of visibility and runway visual installed. At these aerodromes human observed visibility shall be reported. Such aerodromes will not have fully integrated automatic systems for acquisition, processing, dissemination/and display in real time of the meteorological parameters affecting landing and take-off operations.	
4.4.2	UK aerodromes issue half hourly METAR for flight planning and local special reports for arriving and departing aircraft SPECI reports are not issued.	
4.6.2.2	In local routine and special reports in the UK, the visibility reported is the prevailing visibility, supplemented by runway visual range measurements, where appropriate.	
Chapter 6	**Forecasts**	
6.6.2	Abbreviated plain language area forecasts issued in the UK to cover the layer between the ground and flight level 100 are not prepared as GAMET area forecasts. UK forecasts provide full details of the weather conditions to be expected in the areas concerned, employing approved ICAO abbreviations and numerical values.	
Appendix 2	**Technical specifications for Local Routine Reports, Local Special Reports and Reports in the Metar/Speci Code Forms**	
1.3.4	WAFC London will continue to issue amendments to forecasts of significant weather using abbreviated plain language messages, but will not issue amended BUFR files.	

Appendix 3	Critearia for reporting Meterological and related parameters in automated Air-Reports
2.3.2	2.3.2(e) Additional visibility threshold of 2000m, 5000m (irrespective of numbers of VFR flights) and 10km will be used. Additional visibility thresholds of 150m, 350m and 600m will be used where an RVR is not available, by local arrangement at each aerodrome.
	2.3.2(f) Special reports for RVR will only be prepared by local arrangement at each aerodrome.
	2.3.2(i) Additional cloud base thresholds of 300ft, 700ft, 1500ft (irrespective of numbers of VFR flights) and 2000ft for cloud layers of BKN or OVC extent will be used.
	2.3.2(k) Observations of vertical visibility will not be made in the UK.
4.1.4.2	Variations from the mean wind direction during the past 10 minutes are reported in the UK in (b) (1) when the wind speed is more than 3 knots and in (b) (2) when the wind speed is 3 knots or less.
4.2.4.2	In local routine and special reports, the visibility reported is the prevailing visibility, supplemented by runway visual range measurements, where appropriate.
4.3.2.2	Runway visual range on some runways intended for Category 1 instrument approach and landing operations my be assessed by human observer. Further information is given in UK AIP GEN 3.5
4.3.6.6	UK aerodromes do not report RVR trends and significant variations.
4.5.1	Sensors to determine cloud amount and height of base cloud for local routine and special reports in the UK are sited to give suitable indications of the height of cloud base and cloud amount at the threshold.
4.5.4	The abbreviation NSC (no significant cloud) will be used if there are no clouds of operational significance, no TCU or CB and the abbreviations CAVOK and SCK are not appropriate. Observations of vertical visibility are not made in the UK.
4.8.1.3	Information on wind shear is not added to METAR reports in the UK.
4.8.1.4	10 minute automated reports from equipment located on some offshore structures contain information on the state of the sea (specifically mean wave height, maximum wave height and wave period). This information is available in the associated meteorological forecast office.
4.9.1.2	The abbreviation 'NDV' will be introduced in automated METAR reports in the UK from 6 October 2005.
4.9.1.3	The abbreviation 'UP' will be introduced in automated METAR reports in the UK from 6 October 2005.
4.9.1.4	The reporting of missing cloud types using'///' and the abbreviation 'NCD' will be introduced in automated METAR reports in the UK from 6 October 2005.
Table A3-2	The use of the term 'METAR COR' to indicate the type of report will be introduced in the UK from 6 October 2005.
Appendix 5	Technical Specifications for SIGMET and AIRMET messages and special Air-Reports
1.3.1	1.3.1 (b) Additional visibility thresholds of 5000m (irrespective of numbers of VFR flights) and 10km will be used in the UK. Visibility thresholds of 150m, 600m and 3000m will not be used. However visibility thresholds of 3000m and 7km will be used additionally for eleven UK civil aerodromes serving offshore helicopter operations.
	1.3.1 (e) Additional cloud base thresholds of 1500ft and 5000ft for cloud layers of BKN or OVC extent will be used in the UK. An additional threshold of 700ft will be used for eleven UK civil aerodromes serving offshore helicopters operations.
	1.3.1 (h) Forecasts of vertical visibility are not made in the UK.
3.3.3	The visibility threshold of 5000m is used in the UK, irrespective of numbers of VFR flights.
2.2.5	The cloud base threshold of 1500ft will be used in the UK, irrespective of numbers of VFR flights, for cloud layers of BKN or OVC extent. Ad additional cloud base threshold of 300ft will be used. An additional threshold of 700ft will be used for eleven UK civil aerodromes serving offshore helicopter operations.
2.2.6	Forecasts of vertical visibility are not made in the UK.
Appendix 6	Technical specification related to SIGMET and AIRMET information aerodrome warnings and windshear warnings
5.1.2	Aerodrome warnings in the UK are not for tropical cyclone, dust storm, sand storm, rising sand or dust.

Annex 4	Aeronautical Charts (9th Edition) (AMDT 53)
Chapter 1	Definitions, Applicability and Availability
1.2.2	Some charts produced do not conform to all relevant standards.
1.2.2.1	Some charts produced do not conform to all relevant recommendations.
Chapter 2	General Specifications
2.1.8	The basic sheet size is 279mm x 210mm.
2.2	ICAO should be included in the title if conformity with all SARPS.
2.18.2.2	Geiod undulation not shown.

Chapter 3	Aerodrome Obstacle Chart – ICAO Type A (Operating Limitations)
3.4.2	A scale of 1:20 000 is sometimes used.
3.7	Magnetic variation is shown to one tenth of a degree.
3.8.1.1	1 per cent slope is used in place of 1.2 per cent. At 9000m from the point of origin, the surface plane changes from a 1 per cent slope to horizontal.
3.8.1.2	For the first 900m of the take-off flight path area, the shadow planes are horizontal and beyond this point to 9000m such planes have an upward slope of 1 per cent. A further safety factor is applied. Obstacles within the outer 25m edges of the take-off flight path do not eliminate other obstacles, other than those also found in the 25m zone.
3.8.2.1(b)	The width at the point of origin is 180m and this width increases at a rate of 0.25D to a maximum of 3930m where D is the distance from the point of origin.
3.8.2.2	The take-off flight path area specified in 3.8.2.1(c) remains at 15 km. The slope of the plane surface shall remain as specified in 3.8.1.1 and 3.8.1.2.

Chapter 4	Aerodrome Obstacle chart ICAO type B
	Chart not produced.

Chapter 5	Aerodrome Obstacle chart ICAO type C
	Chart not produced.

Chapter 7	En Route Chart – ICAO
	Chart not produced.

Chapter 8	Area Chart – ICAO
	Chart not produced.

Chapter 9	Standard Departure Chart – Instrument (SID) – ICAO
9.3.2	Charts are not drawn to scale.
9.3.3	No scale shown.
9.4.1	Projection not shown.
9.4.2	Parallels and meridians not shown.
9.4.3	Graduation marks not shown.
9.6.1	Topography and culture not shown.
9.6.2	Relief and smoothed contours not shown.
9.7	Magnetic variation not shown.
9.8.2	True and Grid north is not used as geo reference system.
9.8.3	Reference to tracks and radials as True or Grid north is not shown.
9.9.4.1.1	Minimum flight altitudes along the route or route segments are not shown. Radar vectoring procedures not shown.
9.9.4.1.1.2 (e)	Elevation and channel of the DME is not shown
9.9.4.1.1.3	Bearings are shown to the nearest degree and distance to the nearest nautical mile.
9.9.4.1.6	Terrain information is not shown.
9.9.4.2	Communication failure procedure is not shown.

Chapter 10	Standard Arrival Chart – Instrument (STAR) – ICAO
10.3.2	Charts are not drawn to scale.
10.3.3	Scale bar not shown.
10.4.1	Projection not shown.
10.4.2	Parallels and meridians not shown.
10.6.1	Topography and culture not shown.
10.6.2	Relief and smoothed contours not shown.
10.6.3	Area Minimum Altitudes not shown.
10.7	Magnetic variation is not shown.
10.8.2	True and Grid north is not used as GEO reference system.
10.8.3	Reference to tracks and radials as True or Grid North is not shown.
10.9.1.4.1.1.2 (e)	Elevation and channel of the DME is not shown.
10.9.1.4.1.1 (f)	Minimum flight altitudes along the route or route segments are not shown, Radar Vectoring procedures not shown.
10.9.1.4.1.1.3	Bearings are shown to the nearest degree and distance to the nearest nautical mile.
10.9.1.4.1.1.6	Terrain information is not shown.
10.9.4.2	Communication failure procedure is not shown.

Chapter 11	Instrument Approach Chart – ICAO
11.2.4	Chart is not separated by aircraft category.
11.4	Sheet size is 297mm x 210mm.
11.7.2	Chart shows contour intervals of 300ft with the first interval being at 300ft above aerodrome elevation rounded down to the nearest 50ft.
11.8.2	Magnetic variation shown only for non VOR procedures.
11.10.2.2	Obstacles determining an obstacle clearance altitude/height may not be identified.
11.10.2.4	Only the AMSL height of an obstacle is shown.
11.10.2.7	Precision Approach procedures without obstacle free zones are not described or indicated as such.
11.10.4.3	The FAF/FAP geographical coordinates are not shown.
11.10.4.5	Only radio communication frequencies are shown, not callsigns.
11.10.6.3 (f)	Transition Altitude is situated above the plan view, not within the profile area.
11.10.6.5	The ground profile depicts the highest elevations of relief occurring along the extended runway centre line only.
1.1.1.1	Aerodrome operating minima is not shown.
11.10.7.2	Basic CAT D OCA(H) are shown only.
11.10.9	Associated instrument approach procedure data is not shown.

Chapter 12	Visual Approach Charts – ICAO
	Not produced.

Chapter 13	Aerodrome/Heliport Chart – ICAO
13.5	Annual change of magnetic variation is not shown.
13.6.1 (b/c)	Geoid undulation and highest elevation/threshold of the touchdown zone of a precision approach is not shown.
13.6.1 (d)	Runway bearing strengths not shown. Clearways and runway markings are not shown.
13.6.1 (e)	Apron lighting, markings and their visual guidance and control aids including docking systems and bearing strengths not shown.
13.6.1 (f)	Geographical coordinates not shown.
13.6.1 (g)	Taxiway surface, width and nearing strengths not shown.
13.6.1 (l)	Routes for taxing aircraft and designators not shown.
13.6.2 (a)	Heliport type is not shown.
13.6.2 (b)	TLOF slope and bearing strength not shown.
13.6.2 (c)	True bearing not shown, magnetic bearing is shown to the nearest degree. Slope not shown.
13.6.2 (d)	Clearway not shown.
13.6.2 (h)	Declared distances are not shown.

Chapter 14	Aerodrome Ground Movement Chart – ICAO
	Charts not produced in the uK.

Chapter 15	Aircraft Parking/Docking Chart – ICAO
15.5.2	Annual change of magnetic variation is not shown.
15.6 (b)	Apron lighting, markings and other visual guidance and control aids including docking systems and bearing strengths not shown.
15.6 (c)	Geographical coordinates not shown.
15.6 (e)	Geographical coordinates not shown.

Chapter 16	World Aeronautical Chart – ICAO 1 100 000
	Not produced in the UK.

Chapter 17	Aeronautical Chart – ICAO 1:500 000
17.3.2	A conversion scale is not shown.
17.4.3	All charts are sold flat.
17.4.4	The 1:1 000 000 chart is not produced.
17.7.11	Escarpments are not shown.
17.9.2.2	Lighting available is not shown apart from aerodrome light beacons. Surface and length of longest runway are not shown.

GEN 1.7   Annex 4 Aeronautical Charts (9th Edition) (AMDT 53) Chapter 18

Chapter 18	Aeronautical Navigation Chart – ICAO small scale
	Not produced.
Chapter 19	Plotting Chart – ICAO
	Not produced.
Chapter 20	Electronic Chart Display – ICAO
	Chart not produced in the UK.

Annex 5   Units of Measurement to be used in Air and Ground Operations (4th Edition) (AMDT 16)

Chapter 3   Standard Application of Units of Measurement

Measurement of	Units
Distance used in navigation, position report etc – generally in excess of 2 or 3 nautical miles	*Nautical miles and tenths
Relatively short distances such as those relating to aerodromes (eg runway lengths)	Metres
Altitudes, elevations and heights	Feet
Horizontal speed including wind speed	Knots
Vertical speed	Feet per minute
Wind direction for landing and taking off	Degrees Magnetic
Wind direction except for landing and taking off	Degrees True
Visibility including runway visual range	Kilometres or metres
Altimeter setting	Millibars (Hectopascals)
Temperature	Degrees Celsius (Centigrade)
Weight	Metric tonnes or kilogrammes
Date/Time	Year, Month, Day, Hour and minute the day of 24 hours beginning at midnight Co-ordinated Universal Time

*International nautical miles, for which conversion into metres is given by 1 international nautical mile = 1852m.

3.1   There is no sharp dividing line between the usage of nautical miles or metres for the two types of horizontal distances referred to in the table above. There is no hard and fast rule, but distances having a navigational or position reporting aspect are given in nautical miles even if they are less than 2 nautical miles (eg ranges from touchdown during a precision approach). These are given in nautical miles and fractions. Distances on the aerodrome (eg runway lengths, etc.) are given in metres. Distances from obstacles in the vicinity of aerodromes and ILS marker distances are generally shown in nautical miles and tenths, except when it is necessary for greater accuracy for some middle markers to be shown in metres.

3.2   Nautical miles or feet, as appropriate, will be used in designating horizontal distances at military aerodromes, joint-user aerodromes controlled by Service personnel and in RAF Sections at Air Traffic Control Centres. Horizontal distances will, however, be given to civil pilots in metric units, on request. Similarly, at civil aerodromes, joint-user aerodromes under civil control and in civil sections at Air Traffic Control Centres, Service pilots will be given distances in nautical miles or feet, as appropriate, on request.

Annex 6   Operation of Aircraft Part 1 (International Commercial Air Transport – Aeroplanes) (8th Edition) (AMDT 28)

Chapter 1   Definitions

Approach and Landing Operations using Instrument Approach Procedures. Subpart E of JAR-OPS 1 (Aeroplanes) – now the sole UK code for aerodrome operating minima policy – specifies a minimum RVR that is 50m less than the 350m specified in the ICAO Category II definition.

Crew Member. The UK definition is based upon the functions that crew members undertake. Although different, the UK definition is more precise than that of ICAO.

GEN 1.7 Annex 6 Operation of Aircraft Part 1 (International Commercial Air Transport – Aeroplanes) (8th Edition) (AMDT 28) Chapter 6

Flight Crew Member. The UK definition is based upon the functions that flight crew members undertake. Although different, it is more precise than the ICAO definition. Effectively, both definitions achieve the same result.

Pilot-in-Command. In United Kingdom legislation, "Pilot in command' in relation to an aircraft means a person who for the time being is in charge of the piloting of the aircraft without being under the direction of any other pilot in the aircraft.

Chapter 4 Flight Operations

4.1.2	The UK does not explicitly require operators to specify in their operations manuals this instruction on reporting without delay any inadequacy of facilities that may be observed.
4.2.5	The UK does not explicitly require operators to ensure that the design and uitilisation of checklists shall observe Human Factors principles.
4.2.7.4	The UK allows meteorology visibility to be converted to RVR. No limiting visibility is prescribed: if a reported RVR is not available, then an approach may be made if the conversion results in an RVR-equivalent value that is not less than the relevant aerodrome operating minima.
4.2.8	The UK does not explicitly require operators to establish operational procedures designed to ensure that an aeroplane being used to conduct precision approaches crosses the threshold by a safe margin.
4.3.6	The UK does not distinguish between turbo-jet/turbo-prop powered aeroplanes and reciprocating engine powered aeroplanes.
4.3.7.2	The UK does not at the moment require operators to specify in their operations manuals this instruction on establishing two-way communication by means of the aeroplane's intercommunication system or other suitable means between ground crew supervising the refuelling and the qualified personnel on board the aeroplane.
4.4.7, 4.6.1, 4.6.2 & 10.1	The UK does not explicitly require operators to specify in their operations manuals these instructions on the duties and training associated with the employment of flight operations officers / flight despatchers.
4.5.1	The UK prescribes duties for the pilot designated by the operator as commander of the aircraft, covering essentially the same requirements as in the ICAO text, but described in a different and more precise manner.

Chapter 6 Aeroplane Instruments, Equipment and Flight Documents

6.1.2	The UK does not require the operator to include an MEL in the Operations Manual. However, where an operator has an MEL this must be included in the Operations Manual.
6.1.3	The UK does not require operators to observe Human Factors principles in the design of the aircraft operating manual.
6.3.1.4	The UK does not accept the use of analogue Flight Data Recorders using FM on aircraft new or second hand first brought onto the UK register after 1 July 1981.
6.3.1.4.1	The UK does not prohibit this type of Flight Data Recorder.
6.3.1.5, 6.3.1.5.1 & 6.3.1.5.2	The UK has not promulgated any requirements for compliance with these Standards (which apply from 1 January 2005) on recording digital communications.
6.3.1.6	This type of recording is only allowed up to 5,700kg.
6.3.1.8 (6.3.1.8.1 to 6.3.1.8.5 inc)	Not all of the parameters listed are required by the UK CAA.
6.3.6	
6.3.7.2	The UK has the same requirement but only for aircraft with C of A issued after 31/5/90.
6.3.9.2	The UK requires at least the last 30 minutes preceding removal of electrical power from the equipment.
6.3.9.3	The UK requires at least the last 30 minutes preceding removal of electrical power from the equipment.
6.9.1 (i)	The UK does not require public transport aeroplanes of maximum total weight not exceeding 5,700kg to provide a means of indicating outside air temperature.
6.15.3	The UK prescribes an implementation date of 1 January 2005 for all turbine-engined aeroplanes of a maximum certificated take-off mass in excess of 15,000kg or authorised to carry more than 30 passengers that the Standard states shall be equipped by 1 January 2003 with a ground proximity warning system which has a predictive terrain warning function.
6.15.6	The UK does not require compliance with this recommendation relating to carriage of GPWS in turbine-engined aeroplanes.
6.15.7	The UK does not require compliance with this recommendation relating to carriage of GPWS in piston-engined aeroplanes.
6.17.2, 6.17.3, 6.17.5 & 6.17.6	The UK does not require carriage of automatically activated Emergency Locator Transmitters in public transport aeroplanes. (Over specified surfaces/locations).

GEN 1.7 Annex 6 Operation of Aircraft Part 1 (International Commercial Air Transport – Aeroplanes) (8th Edition) (AMDT 28) Chapter 6

6.17.7	The UK does not require carriage of automatically activated Emergency Locator Transmitters in all public transport aeroplanes. (Over all surfaces/locations).
6.18.2	The UK does not require compliance with this Standard that will require from 1 January 2005 all turbine-engined aeroplanes with a maximum certificated take-off mass in excess of 5,700kg or authorised to carry more than 19 passengers to be equipped with an Airborne Collision Avoidance System II.
6.18.3	The UK does not require carriage of an Airborne Collision Avoidance System II in all public transport aeroplanes.
6.20	The UK proscribes the use of hand-held microphones below Flight Level 150 in controlled airspace.
6.21.1 & 6.21.2	The UK does not prescribe this recommendation for carriage of a forward looking windshear warning system.

Chapter 7 Aeroplane Communication and Navigation Equipment

7.2.2	The UK does not prescribe a requirement that track deviations be displayed when flying in MNPS airspace.

Chapter 8 Aeroplane Maintenance

8.3.1	United Kingdom does not require operators to observe Human Factors principles in the design and application of the maintenance programme.
8.7.5.4	The United Kingdom does not require maintenance organisations to ensure maintenance personnel receive training in knowledge and skill related to human performance.

Chapter 10 Flight Operations Officer/Flight Dispatcher

10.2, 10.3 & 10	The UK does not explicitly require operators to specify in their operations manuals these instructions on the duties and training associated with the employment of flight operations officers/flight dispatchers.

Chapter 13 Security

13.2.3	(a) The UK will require the door to be locked from engine start until the engine shut down.
	(b) The UK will not require that the 'entire' door area be monitored.
13.2.4	The UK has no intention of mandating this recommendation.
13.2.5	Not applicable as the UK does not intend to implement Recommendation 13.2.4.
13.5.1	The UK does not currently prescribe that specialised means of attenuating and directing the blast should be provided for for use in the least-risk bomb location.

Annex 6 Operation of Aircraft Part II (International General Aviation – Aeroplanes) (6th Edition) (AMDT 23)

Chapter 1 Definitions

Approach and landing operations using Instrument Approach Procedures. Subpart E of JAR-OPS 1 (Aeroplanes) – now the sole UK code for aerodrome operating minima policy – specifies a minimum RVR that is 50m less than the 350m specified in the ICAO Category II definition.

Pilot-in-Command. In United Kingdom legislation, "Pilot-in-command" in relation to an aircraft means a person who for the time being is in charge of the piloting of the aircraft without being under the direction of any other pilot in the aircraft.

Chapter 4 Flight Preparation and In-Flight Procedures

4.6.2.2	The commander of the aircraft must be satisfied before flight that the flight can be safely made, taking into account the weather reports and forecasts and any alternative course of action in case the flight cannot be completed as planned.
4.6.3.1	The requirement to discontinue the flight towards the destination is not mandated.
4.9 & 4.10	Oxygen requirements are not mandated.

Chapter 6 Aeroplane Instruments and Equipment

6.1.1	The following are not required to be approved: maps, charts and codes; first aid equipment; timepieces; torches; whistles; sea anchors; rocket signals; equipment for mooring, anchoring and manoeuvring on water; paddles; food and water; stoves, cooking utensils, snow shovels, ice saws, sleeping bags, arctic suits; megaphones.
6.1.3.1.1 (a), (b) & (c) (1)	The requirements for first aid kit, fire extinguisher and seats are not mandated for all types of flight.
6.1.4.1 & 6.1.4.2	The method of marking break-in areas may differ.
6.3.1	Seaplane special equipment is not mandated.
6.3.3	The UK currently has no legislation covering this requirement. The UK currently relies on the provision of guidance material.

GEN 1.7   Annex 6 Operation of Aircraft Part III
                              (International Operations – Helicopters) (5th Edition) Section II Chapter 2

6.4	Signalling and life saving equipment is not mandated for areas where search and rescue would be especially difficult.
6.7 (c), (e) & (f)	Landing light, passenger compartment lights and torches for each crew member station are not mandated.
6.9.6	TAWS Class B parameters are not specified in CAA regulatory material.
6.10.1.5	The requirements for recording, correlation and duration of data link communications are not mandated for new aeroplanes from 1 January 2005.
6.10.1.5.1	The requirements for recording, correlation and duration of data link communications are not mandated for all aeroplanes from 1 January 2007.
6.10.1.5.2	The requirement to record the content and time of data link messages is not mandated.
6.10.1.7	The parameters for Type IA flight data recorders are not specified in UK regulatory material.
6.10.3.1	The requirement for Type I flight data recorders over 27,000kg is not mandated for all C of A categories.
6.10.4.1	The requirement for Type I A flight data recorders in new aeroplanes over 5,700kg from 1 January 2005 is not mandated.
6.10.5.1	The requirements for cockpit voice recorders over 27,000kg are not mandated for all C of A categories.
6.10.6.3	The requirements for the cockpit voice recorder in new aeroplanes over 5,700kg from 1 January 2003 to retain the last two hours of information is not mandated. The UK requires at least the last 30 minutes preceding removal of electrical power from the equipment.
6.12.1	The ELT requirements applicable until 1 January 2005 for extended flights over water and the designated land areas (6.4) are not mandated.
6.12.2	The automatic ELT requirements for extended flights over water and the designated land areas (6.4) are not mandated for new aeroplanes from 1 January 2002.
6.12.3	The automatic ELT requirements for extended flights over water and the designated land areas (6.4) are not mandated for all aeroplanes from 1 January 2005.
6.13.1	Partially implemented. UK requires pressure-altitude reporting transponder for flight in designated airspace.
6.13.2	The recommendation that all aeroplanes should be equipped with a pressure-altitude reporting transponder is not implemented.

Chapter 7   Aeroplane Communication and Navigation Equipment

7.1.1   UK requires radio communication equipment for IFR flight in controlled airspace and notified airspace. Gliders are excepted.

Annex 6   Operation of Aircraft Part III
           (International Operations – Helicopters) (5th Edition)

Section 1   General

Chapter 1   Definitions

Cabin Crew Member. The UK definition is based upon the functions that crew members undertake. Although different, the UK definition is more precise than that of ICAO.

Flight Crew Member. The UK definition is based upon the functions that flight crew members undertake. Although different, it is more precise than the ICAO definition. Effectively, both definitions achieve the same result.

Flight Time – Helicopters. The UK definition differs only in minor detail, with flight time beginning when the helicopter first moves under its own power, and ending when the rotors are next stopped.

Instrument Approach and Landing Operations. Subpart E of JAR-OPS 3 (Helicopters) – now the sole UK code for aerodrome operating minima policy

1 specifies a minimum RVR that is;

(i) 50m less than the 550m specified in the ICAO Category I definition;

(ii) 50m less than the 350m specified in the ICAO Category II definition;

2 does not permit CAT III A, B or C Operations.

Pilot-in-Command. In United Kingdom legislation, "Pilot-in-command" in relation to an aircraft means a person who for the time being is in charge of the piloting of the aircraft without being under the direction of any other pilot in the aircraft.

Section II   International Commercial Air Transport

Chapter 2   Flight Operations

2.1.2   The UK does not explicitly require operators to specify in their operations manuals this instruction on reporting without delay any inadequacy of facilities that may be observed.

55

GEN 1.7 Annex 6 Operation of Aircraft Part III
(International Operations – Helicopters) (5th Edition) Section II Chapter 2

2.2.3.2	The UK does not require operators to specify this prohibition (on turning rotors under power without a qualified pilot at the controls) in operations manuals.
2.2.5	The UK does not explicitly require operators to ensure that the design and utilisation of checklists shall observe Human Factors principles.
2.2.7.4	The UK allows meteorology visibility to be converted to RVR. No limiting visibility is prescribed: if a reported RVR is not available, then an approach may be made if the conversion results in an RVR-equivalent value that is not less than the relevant aerodrome operating minima.
2.2.11	The UK does not require all helicopters operated over water to be certified for ditching, but makes provision for floatation by other means.
2.3.4.2	The UK does not require operators to specify in their operations manuals these instructions on the use of suitable off-shore alternates.
2.3.4.3	The UK does not require operators to specify in their operations manuals the recommendation on the carriage of fuel rather than payload in adverse weather conditions so as to then use an on-shore alternate.
2.4.7, 2.6.1 & 2.6.2	The UK does not explicitly require operators to specify in their operations manuals these instructions on the duties and training associated with the employment of flight operations officers/flight dispatchers.
2.5.1	The UK prescribes duties for the pilot designated by the operator as commander of the aircraft, covering essentially the same requirements as in the ICAO text, but described in a different and more precise manner.

Chapter 3	Helicopter Performance Operating Limitations
3.1.5	The UK does not require operators to specify in their operations manual this prohibition on operating Performance Class 3 helicopters from elevated heli decks.

Chapter 4	Helicopter Instruments, Equipment and Flight Documents
4.3.1.4	The UK does not accept the use of analogue Flight Data Recorders using FM on aircraft new or second hand first brought onto the UK register after 1 July 1981.
4.3.1.4.1	The UK does not prohibit this type of equipment.
4.3.1.4.1	The UK does not prohibit this type of equipment.
4.3.1.5, 4.3.1.5.1 & 4.3.1.5.2	The UK has not promulgated any requirements for compliance with these Standards (which apply from 1 January 2005) on recording digital communications.
4.3.1.7	Not all parameters listed are required by the UK CAA.
4.3.1.7.1	Not all parameters listed are required by the UK CAA.
4.3.1.7.2	Not all parameters listed are required by the UK CAA.
4.3.1.7.3	Not all parameters listed are required by the UK CAA.
4.3.1.7.4	Not all parameters listed are required by the UK CAA.
4.3.1.7.5	Not all parameters listed are required by the UK CAA.
4.3.4.1	Not all parameters listed are required by the UK CAA.
4.3.7.2	The UK already requires that the last 30 minutes of operation is retained.
4.3.7.3	The UK currently requires the last 30 minutes of operation to be retained.
4.7.5 & 4.7.6	The UK does not require carriage of automatically activated Emergency Locator Transmitters in public transport helicopters over designated land areas.
4.7.7	The UK does not require carriage of automatically activated Emergency Locator Transmitters in all public transport helicopters.
4.10.1 (i)	The UK does not require public transport helicopters of a maximum total weight not exceeding 5,700kg to provide a means of indicating outside air temperature.
4.16	The UK proscribes the use of hand-held microphones below Flight Level 150 in controlled airspace.

Chapter 6	Helicopter Maintenance
6.3	The United Kingdom does not require maintenance organisations to ensure maintenance personnel receive training in knowledge and skill related to human performance.

Chapter 8	Flight Operations Officer/Flight Dispatcher
8.1, 8.2, 8.3 & 8.4	The UK does not explicitly require operators to specify in their operations manuals these instructions on the duties and training associated with the employment of flight operations officers/flight dispatchers.

Section III	International General Aviation
Chapter 2	Flight Operations
2.6.2.2	The commander of the aircraft must be satisfied before the flight that the flight can be safely made, taking into account the weather reports and forecasts and any alternative course of action in case the flight cannot be completed as planned.
1.1.1.1	The requirement to discontinue flight towards the destination is not mandated.

2.9.1, 2.9.2 & 2.10	Oxygen requirements are not mandated.	
2.19	The requirement for helicopters on over water flights in 4.3.1 to be certificated for ditching is not mandated.	

Chapter 4 Helicopter Instruments, Equipment and Flight Documents

4.1.1	The following are not required to be approved: maps, charts and codes; first aid equipment; timepieces; torches; whistles; sea anchors; rocket signals; equipment for mooring, anchoring and manoeuvring on water; paddles; food and water; stoves, cooking utensils, snow shovels, ice saws, sleeping bags, arctic suits; megaphones.
4.1.3.1 (a), (b) & (c)1	The requirements for first aid kit, fire extinguisher and seats are not mandated for all types of flight.
4.1.4.1 & 4.1.4.2	The method of marking break-in areas may differ.
4.3.1	The floatation equipment requirements for helicopters on over water flights (more than 10 minutes/beyond forced landing distance from land) is not mandated.
4.3.2.1	UK relies on the provision of guidance material. UK recommendations on life jackets and rafts provide a higher level of safety.
4.3.2.2, 4.3.2.3 & 4.3.2.4	The UK relies on the provision of guidance material on carriage and use of life jackets and rafts.
4.4	Signalling and life-saving equipment is not mandated for areas where search and rescue would be especially difficult.
4.5.1	Oxygen requirements are not mandated.
4.9.1.5	The requirements for recording, correlation and duration of data link communications are not mandated for new helicopters from 1 January 2005.
4.9.1.5.1	The requirements for recording, correlation and duration of data link communications are not mandated for all helicopters from 1 January 2007.
4.9.1.5.2	The requirement to record the content and time of data link messages is not mandated.
4.9.1.7	The parameters for Type IV A flight data recorders are not specified in UK regulatory material.
4.9.4.1	The requirement for Type IV A flight data recorders in new helicopters over 3,180kg from 1 January 2005 is not mandated.
4.9.5.1	The requirement for cockpit voice recorders for helicopters having maximum certificated take-off mass over 7,000kg are not mandated for all C of A categories.
4.9.6.3	The requirement for the cockpit voice recorder in new helicopters form 1 January 2003 to retain the last two hours of information is not mandated. The UK requires at least the last 30 minutes preceding removal of electrical power from the equipment.
4.10.1	The ELT requirements applicable until 1 January 2005 for over water flights (as 4.3.1) are not mandated.
4.10.2	The automatic ELT/survival ELT requirements for over water flights (as 4.3.1) are not mandated for new helicopters from 1 January 2002.
4.10.3	The automatic ELT/survival ELT requirements for over water flights (as 4.3.1) are not mandated for all helicopters from 1 January 2005.
4.10.4	The ELT requirements applicable until 1 January 2005 for flights over designated land areas (as 4.4) are not mandated.
4.10.5	The automatic ELT requirements for flights over designated land areas (as 4.4) are not mandated for new helicopters from 1 January 2002.
4.10.6	The automatic ELT requirements for flights over designated land areas (as 4.4) are not mandated for all helicopters from 1 January 2005.
4.11.1	Partially implemented. UK requires pressure-altitude reporting transponder for flight in designated airspace.
4.11.2	The recommendation that all helicopters should be equipped with a pressure-altitude reporting transponder is not implemented.

Chapter 5 Helicopter Communication and Navigation Equipment

5.1.1	The UK requires radio communication equipment for IFR flight in controlled and notified airspace.

Annex 7 Aircraft Nationality and Registration Marks (5th Edition) (AMDT 5)

Chapter 3 Location of Nationality, Common and registration Marks

3.2.5	Balloons of not more than 2metres in any linear dimension at any stage of its flight, including any basket or other equipment attached to the balloon are exempt from registration and also from the need to carry a fireproof identification plate.

GEN 1.7 Annex 7 Aircraft Nationality and Registration Marks
(5th Edition) (AMDT 5) Chapter 6

Chapter 6		Register of Nationality, Common and Registration Marks
6		Balloons of not more than 2 metres in any linear dimension at any stage of its flight, including any basket or other equipment attached to the balloon are exempt from registration. For those unmanned balloons that are registered the UK Register of Civil Aircraft does not contain the time, date and location of the release of the balloon.
Chapter 8		Identification Plate
8		Balloons of not more than 2metres in any linear dimension at any stage of its flight, including any basket or other equipment attached to the balloon are exempt from registration and also from the need to carry a fireproof identification plate.
Annex 8		**Airworthiness of Aircraft (9th Edition) (AMDT 98)**
Part I		Definitions
		Performance class 1, 2 and 3 helicopters. The United Kingdom classifies Helicopters as either Category A or B for certification.
Part II		Large Aeroplanes
Part IIIA 2.2.3		The United Kingdom complies except that it does not require the scheduling of landing distance with runway slope.
Part IIIA 2.2.3		The United Kingdom complies except that performance is not scheduled for variations in water surface conditions, density of water and strength of current.
Part IIIA 2.3.4.1		In the United Kingdom stall testing with one engine **inoperative is not required.**
Part IIIA 4.1, Part IIIB D1.1		For design of the flight deck the United Kingdom has a means of compliance other than legislation for Human Factor Principles. For the design of other parts of the aeroplane the United Kingdom has no requirement or guidance material for Human Factor Principles.
Part III		Part of these provisions implement ICAO's initative to incorporate security into aircraft design. At this time, the UK has not implemented these requirements
Part IIIB d.2(b)(g)3 (h) (I)		Differences are associated with explosives and incendiary devices being the casual factor.
Part IIIB F1.1		Not implemented
Part IIIB G2.5		
Part IIIA 9.3.5		These provisions implement ICAO's initiative to incorporate security into aircraft design. At this time, the UK has not implemented these requirements.
Part IIIA 11.1, 11.3 &		These provisions implement ICAO's initiative to incorporate security into aircraft design. At this time, the UK has not implemented these requirements.
Part IIIA 11.2 &		These provisions implement ICAO's initiative to incorporate security into aircraft design. At this time,
Part IIIB K2		the UK has not implemented these requirements.
Part IIIB B2.7(b)		The United Kingdom does not require accelerate-stop distance to be determined with worn brakes for commuter category aeroplanes.
Part IIIB B2.7(e)		The UK does not require landing distance to be determined with fully worn brakes. However, the UK requires the landings to be measured over six landings using the same tyres, wheels and brakes so some brake wear is accounted for. Additionally, factors on landing distance are applied by operational rules, where appropriate.
Part III Part IIIB D2(a)		The prevention of mis-assembly is not implemented in the UK.
Part IIIB I1		The United Kingdom does not require account to be taken of developments in the subject of crashworthiness in the design of aeroplanes
Part IV 2.2.3.1		Helicopters
		In the United Kingdom for Category B helicopters, only take-off distance is required to be included in the performance data while take-off distance, path and rejected take-off distance information is required for Category A helicopters.
Part IV 2.2.3.2(b)		In the United Kingdom en-route performance is based on climb performance both for all engines operating and one engine inoperative situations. The case of the two critical power units inoperative for helicopters having three or more engines is not addressed.
Part IV 4.1.6(f)		There are no requirements in the United Kingdom for design precautions to be taken to protect against instances of cabin depressurisation.
		Unpressurised cabins and compliance with JAR 27/29.831 ensures compliance with the standard relating to incapacitation from 'smoke or other toxic gases'.
Part IV 6.7		There is no comparable requirements for Category B helicopters. (CS/JAR 27 only complies for Category A helicopters).

Part IV 2.2.2.1	Performance class 1, 2 and helicopters. The United Kingdom
2.2.2.2	classifies helicopters as either Category A or B
2.2.3.1	certification.
2.2.3.1.1	
2.2.3.1.2	
2.2.3.1.3	
2.2.3.1.4	
2.2.3.3.1	
6.8.1	

Annex 9 Facilitation (12th Edition) (AMDT 19)

Chapter 2 Entry and Departure of Aircraft

2.10.1	In certain circumstances particulars of members of crew may be required.
2.12	In certain circumstances carriers may be required to provide a passenger list showing the names and nationalities of passengers.
2.19	General customs supervision should at all times be possible: such supervision may include a documents check (Article 12 of the European Community Customs code refers).

Chapter 3 Entry and Departure of Persons and their Baggage

3.26	Disembarkation cards must normally be completed by all passengers except nationals of Member Sates of the European Economic Area.
3.29	Disembarkation cards must be provided by the carrier at his expense and distributed to all passengers who need to complete them.
3.38	The UK retains the right to introduce export controls in certain circumstances.
3.44	An operator remains liable for the care and custody of inadmissible persons, including associated costs, in certain circumstances.
3.41.1	Where the UK imposes a requirement to provide API, this requirement will apply regardless of whether the information in the passenger's travel document is available in machine readable form.
3.47.2	UK legislation provides for a charge of (currently) £2000 to be levied on operators who bring to the UK persons who are inadequately documented irrespective of the introduction of API. Inadequately documented persons are those who fail to produce a valid national passport or other travel document satisfactorily establishing their identity or nationality, a valid UK visa if required or produce a forged/counterfeit national passport or other document where the forgery is reasonably apparent.
3.52	Where required UK visa and entry clearances should be obtained prior to travel and a person will normally be refused entry in the absence of the necessary clearance. The Immigration Officer has discretion to waive the requirement for an entry clearance in exceptional circumstances.
3.60	The UK permits transit without visas for passengers who normally require visas, provided that the passenger has: (a) entry facilities for the countries en-route for the final destination; (b) a firm booking for travel by air within 24 hours; (c) no purpose in entering the UK other than to pass through in transit; NB. This information is regularly updated in the Travel Information Manual.
3.67, 3.68, 3.68.1	Crew member certificates are not issued by the UK public authorities to crew members of UK airlines, whether or not they are required to be licensed. Identification documents bearing photographs of the holders are issued to UK aircrew members, licensed and unlicensed, by UK airlines and by airport authorities on their behalf, the validity of which may be checked by contacting the issuing authorities. UK flight crew licences conform to the specification for personnel licences set forth in paragraph 5.1.1 of Annex 1. The date of birth is also included, Following the introduction of computerised licence issues a photograph of the holder is no longer required, neither is the place of birth nor a statement of the right of re-entry to the State of issue – these items are part of the Annex 9 Appendix 7 crew member certificate but are not called for in paragraph 5.1.1 of Annex 1.
3.71.1, 3.72, 3.73	The UK requires aircrew who arrive as passengers, or who are supernumerary to be in possession of a valid passport or other satisfactory document establishing identity and nationality and, where applicable, of a valid visa. The UK visa requirement is waived in respect of vosa nationals who arrive and leave as aircrew within seven days.

Chapter 4 Entry and Departure of Cargo and Other Articles

4.2	The European Community Customs Code does not foresee a guarantee waiver for transport by road (including airfreight by road); however, provisions exist to authorise a reduction of the guarantee level.

4.3	Under European Community Custom legislation consultation with operators and other parties concerned is not compulsory in every case. Close co-operation and consultation with the operators is however generally sought in order to improve the quality and effectiveness of new regulations and of amendments to existing rules.
4.13	In the UK this provision applies to customs matters for which the 'declarant' is the relevant person. With regard to other policies (such as phyto-sanitary measures etc) the person responsible for the information concerned may be a person other than the declarant.
4.20	In the European Community a wide range of simplified customs procedures are in practice available for operators as regards export (for example incomplete declarations, simplified declarations, local clearance procedure – Article 76 of the European Community Customs Code refers). Some of these procedures are subject to prior authorisation from the customs authorities. As an authorised operator, the exporter is allowed to carry out any number of operations. The authorisation is based on general criteria, for example the ability to ensure that effective checks can be undertaken. Depending on the simplified procedure used, the declarant must make available to the Customs authorities all of the required documents required for application of the provisions governing the export of goods.
4.22	This standard and in particular the words 'at any customs office', does not conform to Article 161 ß 5 of the European Community Customs Code which provides that the export declaration must be lodged where the goods are packed or where the exporter is established.
4.24	The Recommended Practice would seriously frustrate control by public authorities over goods loaded on departing aircraft. Furthermore, the return of certain goods after their departure would not be guaranteed despite the lodging of a security.
4.27	Currently, no European Community provision determines in which cases the use of simplified arrangements is obligatory or must be granted to the operators. In the European Community a wide range of simplified customs procedures is in practice available for operators as regards export (for example incomplete declarations simplified declarations, local clearance procedure – Article 76 of the European Community Customs Code refers). Some of these procedures are subject to prior authorisation from the customs authorities. As an authorised operator, the exporter is allowed to carry out any number of operations.
4.29	While Customs clearance is expedited as far as possible, there may be other agencies involved in the clearance procedure. Customs cannot therefore undertake to release all goods within three hours of their arrival. One of the objectives of customs is nevertheless to perform checks and release goods within the shortest possible time.
4.30	This Recommended Practice is acceptable in as far as the Contracting States have a common interpretation of the term 'part consignment'. According to Article 73(2) of the European Community Customs Code, all the goods covered by the same declaration shall be released at the same time on the understanding that, where a declaration form covers two or more items, the particulars relating to each item shall be deemed to constitute a separate declaration.
4.34	UK and European Community provisions concerning export and transit licences remain applicable, in certain cases, when the goods are redirected to another destination (for example weapons, dual use goods, precursors, etc).

Chapter 5 Inadmissible Persons and Deportees

5.4	An operator is required to remove an inadmissable person in accordance with the directions given by the Immigration Officer.
5.9.1	Under UK legislation, where a passenger is refused entry, the operator will normally be responsible for any detention costs up to a maximum of 14 days unless the passenger is in possession of a current entry clearance/visa.
5.11	UK legislation requires an operator to remove an inadmissible person to a country of which he is a national or citizen, a country or territory in which he has obtained a passport or other document of identity, a country or territory in which he embarked for the UK or a country or territory to which there is reason to believe that he will be admitted.
5.14	Under UK carrier liability legislation a change may be imposed on the operator if a person arrives without the required documents. However, the operator is not liable if: (i) It can show that the required documents were produced when the passenger embarked for the UK; (ii) a false document is produced or the passenger impersonates the rightful holder of a documents unless the falsity f the document or the impersonation is reasonably apparent; In addition, an operator may apply for Approved Gate Check status at individual ports of embarkation. If the operator satisfies the UK authorities that it meets the published criteria, which include an audited high standard of document checking and security procedures, the UK will normally waive charges relating to persons who arriving with no documents from the station.
5.26	The UK will co-operate fully with the requesting State to investigate and validate the persons claim to be a British citizen and to resolve the claim quickly, within 30 days if possible.
5.27	This provision only applies where the person concerned is admissible or is to be expelled by the authorities.

GEN 1.7 Annex 10 Aeronautical Telecommunications Vol II (Communications Procedures including those with PANS status) (6th Edition) (AMDT 79) Chapter 3

Chapter 6	International Airports – Facilities and Services for Traffic
6.37	The UK does not offer duty-free goods for sale to inbound passengers.
6.49	In the UK permission to remove goods to off-airport facilities is subject to prior authorisation by Customs.
6.58	UK law and practice, which applies to air and other means of transportation, requires that the parties responsible for handling the traffic shall provide and maintain such facilities as may be necessary for the proper control and examination of goods and passengers.
6.60	UK law, which applies to air and other means of transportation, allows for a change to be made for immigration clearance requested by operators additional to the basic service at ports of entry in the UK.
Chapter 8	Other Facilitation Provisions
8.19, 8.20, 8.21	The UK strongly supports close co-ordination between civil aviation security and facilitation programmes. It has established a Facilitation Stakeholders Forum which meets regularly under Department of Transport chairmanship. The Government itself does not establish facilitation committees at airports. There are, however, national consultative bodies for particular subjects, and ad hoc meetings are arranged when necessary to discuss particular topics. UK law allows the Government to require that adequate facilities for consultation be established at airports. Consultation arrangements have been established under these powers at 50 airports.

Annex 10	Aeronautical Telecommunications Vol I (Radio Navigation Aids) (5th Edition) (AMDT 79)
Chapter 2	General Provisions for Radio Navigation Aids
2.1.1.2	The UK has no specific criteria published that prescribes the operational duration of established non-visual aids.
2.1.1.3	The UK does not prescribe the category of performance to be provided by an airport.
2.1.4 & 2.1.4.1	PAR's are not used for civil aviation in the UK.
2.2.1.1	No requirement exists to meet this recommendation.
2.7.1	Whereas the UK is compliant with this requirement for ILS, ILS associated DME, En-Route DME, VOR and NDB's it does not require regular flight testing of non ILS associated Airports DME's.
Chapter 3	Specifications for Radio Navigation Aids
3.1.3.3.2	Some localisers are promulgated in AIP as having specific areas where signals do not meet specifications.
3.1.3.3.5	This is only applied to new CAT III localisers installed since the date this recommendation was introduced.
3.1.3.5.3.6	Several old CAT I and uncategorised systems do not meet this recommendation.
3.1.3.6.2	CAP 670 Inspection limits for CAT II are ± 17.5 feet.
3.1.3.7.3	CAT II Inspection limits are ± 17%.
3.1.5.1.5	Some CAT I systems have reference datum heights between 40 and 50 feet.
3.1.5.3.1	The UK accepts that some G/P have restricted coverage – this is published in AIP's for each specific system.
3.1.7.3.1	UK still uses fly through time corrected to 96 knots as in the original Annex 10. The rounding up and down of distances by ICAO means that UK limits are not precisely those now in Annex 10.
3.1.7.6.6	The UK permits DME as an alternative to makers regardless of whether provision is impracticable.
3.1.7.7.2	A few older beacons may not meet this recommendation.
3.3.4.1	The UK does not promulgate a specific requirement for coverage.
3.3.6.1	The UK does not promulgate a specification for radiation polarisation.
3.4.6.4	The UK allows a fall of up to 0.5 dB.

Annex 10	Aeronautical Telecommunications Vol II (Communications Procedures including those with PANS status) (6th Edition) (AMDT 79)
Chapter 3	General procedures for the International Aeronautical Telecommunication Service
3.3.6	UK complies only at ATC units and recommends compliance at certain AFIS units.
3.3.6.1	UK has no requirement to record telephone system calls although they are recorded at many ATC units. The only aeronautical stations with speaker systems which are required to record messages are ATC units.

GEN 1.7 Annex 10 Aeronautical Telecommunications Vol II (Communications Procedures including those with PANS status) (6th Edition) (AMDT 79) Chapter 3

3.5.1.1 & 3.5.1.1.1	UK complies only at ATC units and recommends compliance at certain AFIS units.
3.5.1.1.2	Aircraft logs are not required by the UK.
3.5.1.6	The UK has some requirements applicable to certain aeronautical stations.
Chapter 5	**Aeronautical Mobile Service – Voice Communications**
5.2.1.3.1.1	On exception. Flight levels ending in hundreds are transmitted as 'HUNDRED' eg, "FLIGHT LEVEL ONE HUNDRED" in order to differentiate from flight level one one zero.
5.2.1.5.8	CONTACT shall have the meaning "Establish communications with… (your details have been passed)".
	Additional word – FREECALL shall have the meaning "Call (unit…. (Your details have not been passed)". Mainly used by military ATC.
	GO AHEAD not used. Different phrase – PASS YOUR MESSAGE used instead.
	Additional phrase – PASS YOUR MESSAGE shall have the meaning "proceed with your message".
	RECLEARED Not used.
5.2.1.7.3.2.3	Under certain circumstances the answering ground station may omit its call sign.
5.2.1.7.3.2.6	Air to air comms on frequency 123.450 MHz are not permitted.
5.2.1.7.3.2.6.1	
5.2.1.9.2.3	This method of acknowledging receipt is not used in UK.
5.2.1.9.2.3.1	This method of acknowledging position reports is not used in UK.
5.2.1.9.3	This method of ending conversations is not used in UK.
5.2.1.9.4.7	If an aircraft read back of clearance or instruction is incorrect, the controller shall transmit the word 'NEGATIVE' followed by the correct version.
5.2.2.1.1.4	Not applicable.
5.2.2.1.3	Not applied in UK.

Annex 10	**Aeronautical Telecommunications Vol III Part 1 (Digital Data Communications Systems) and Part 2 (Voice Communication Systems) (1st Edition) (AMDT 79)**
Part II	
Chapter 2	**Aeronautical Mobile Service**
2.2.1.2	The UK interprets 'On a high percentage of occasions' to be the 95 percentile value and thus requires the effective radiated power to be such as to provide a field strength of at least 188 microvolts per metre (minus 101 dBW/m2).
2.2.2.2	The UK specifies receiver sensitivity in terms of the minimum level of input signal (dBm), modulated 30% by a sinewave of 1 kHz, applied to the receiver which is required to produce a SINAD ratio of 12dB at the audio output measured with a psophometric filter.
2.3.1.2	The UK does not specify the effective radiated power, but provides for classes of transmitter grouped into two classifications of 16 watts and 4 watts minimum output power, having an estimated radio-line-of-sight distances of 200nm and 100nm respectively. A recommendation that the output power be limited to 25 watts to reduce interference is also made.
2.3.1.3	The UK does not specify the adjacent channel power but defines a spectral mask for the transmitter occupied spectrum.
2.3.1.4	The UK specifies the modulation as 'not less than 70%' when modulated by a 1000 Hz audio frequency signal.
2.3.2.1	The UK does not define the frequency stability of receiver.
2.3.2.2.1	The UK specifies the sensitivity in terms of a radio frequency input signal not exceeding 10 microvolts (-93 dBm), with 30% modulation at 1000 Hz to produce a signal plus noise to noise ratio of 6 dBm with an output power not less than 10 dB below the declared output power.
2.3.2.3 & 2.3.2.4	The UK does not state the acceptance bandwidth but defines the effective bandwidth relative to the selected channel frequency of the receiver at the 6 dB and 60 dB points.
2.3.2.5	The specification the UK applies only states that adjacent channel rejection for 8.33 KHz channel spacing.
	For 8.33 kHz channel spacing an adjacent channel rejection of 45 dB is specified at the first upper and lower adjacent channels for defined desired and interfering signals.
2.3.2.6	The UK does not specify the adjacent channel rejection for 25 kHz, 50 kHz or 100 kHz channel spacing.
2.3.2.8.1	Not yet implemented.
2.3.2.8.2	
2.3.2.8.4 & 2.3.2.8.4.1	

GEN 1.7   Annex 11 Air Traffic Services (Air Traffic Service, Flight Information Service and Alerting Service) (13th Edition) (AMDT 42 Chapter 6

2.3.3.1, 2.3.3.2	The UK requires that for aircraft (including helicopters) of 5,700 kg MTWA or less non-immune VFR comm receivers may be permitted and the aircraft permitted to operate under IFR provided that crews are alerted to potential sources of interference.

Annex 10   Aeronautical Telecommunications Vol IV (Surveillance Radar and Collision Avoidance Systems) (3nd Edition) (AMDT 79)

Chapter 2	**General**
2.1.3.2.1 & 2.1.3.2.4	Carriage and operation of SSR is mandated in UK by in designated airspace only.
Chapter 4	**Utilization of Frequencies above 30 MHz**
Para 4.1.1	VHF communications frequencies are planned in accordance with planning agreements reached within Europe. These agreements do not respect the table of allocations given in Annex 10.
4.1.3.1.1	The UK encourage the use of practice PAN calls on 121.5 MHz in contradiction with the Annex 10 requirement for the frequency to only be used in genuine emergencies.
4.1.8.1.3	Within Europe the bands 131.400 – 132.000 & 136.875 MHz inclusive are designated for operational control communications. This has been agreed at a European regional level and hence frequencies to meet aircraft operating obligations under Annex 6 may not be assigned in the band 128.825-132.025 MHz.

Annex 11   Air Traffic Services (Air Traffic Service, Flight Information Service and Alerting Service) (13th Edition) (AMDT 42)

Chapter 1	**Definitions**
	Traffic Information. The UK includes alerting a controller and deleted 'and to help a pilot avoid a collision' in CAP 493.
Chapter 2	**General**
2.1.2	UK has arranged for services to be provided in accordance with the practices and procedures established for their territorial airspace.
2.5.2.2.1.1	The UK does not implement control zones in all portions of the airspace where ATC service is provided.
2.5.2.3	The UK does not use the term 'controlled aerodrome' but lists in the AIP those aerodromes at which ATC service is provided.
2.6	In certain notified portions of Class A airspace gliders are permitted to operate without reference to ATC in accordance with specified conditions and neither separation nor traffic information is provided in respect of such flights.
2.9.3.2.2	**Lower limit of CTA.** UK does not necessarily apply VFR cruising levels as the lower limit. A level is chosen appropriate to the circumstances.
2.9.3.3	**Upper limit of CTA.** UK does not apply VFR cruising levels.
5.5.5.5	**Upper limit of CTR.** UK does not necessarily use VFR cruising levels.
2.2.2	UK implementing ESARR 4 which comes into place in May 2004.
2.26.4	UK already specifies SMS for ATC units. Paragraphs (a) and (b) will be met through application of ESARR 3 and ESARR 4, the later to be implemented by May 2004.
Chapter 3	**Air Traffic Control Service**
3.1	In certain notified portions of Class A airspace gliders are permitted to operate without reference to ATC.
3.3.1	In certain notified portions of Class A airspace gliders are permitted to operate without reference to ATC.
3.3.4	Vertical separation of aircraft. The UK uses the Quadrantal system of cruising levels for flights below 24500ft.
Chapter 6	**Air Traffic Services Requirements for Communications**
7.7.7.7.7	Automatic recording is not available in each and every case
7.7.7.7.8	in the UK.
6.2.3.1.2 & 6.2.3.1.3	
6.2.3.4	Automatic recording is not available in each and every case in the UK.

GEN 1.7   Annex 11 Air Traffic Services (Air Traffic Service, Flight Information Service and Alerting Service) (13th Edition) (AMDT 42 Appendix 3

Appendix 3		Principals Governing the Identification of Standard Departure and Arrival Routes and Associated Procedures
2.1.2		In the UK the basic indicator for Standard Arrival Routes is the name or name code of the holding facility or fix where the arrival route terminates.
Appendix 4		ATS Airspace Classes – Services Provided and Flight Requirements
		The UK complies with the requirements of the table at Appendix 4 except in the following areas:
		(a) In certain notified portions of Class A airspace gliders are permitted to operate without reference to air traffic control in accordance with specified conditions and neither separation nor traffic information is provided in respect of such flights.
		(b) Class A airspace, VMC minima for the purposes of:
		(i) Climbs and descents maintaining VMC;
		(ii) Powered aircraft – Airways crossings and
		(iii) Powered aircraft – other penetrations of Airways Class A airspace, the VMC minima are to be:
		At or above FL 100: 8km flight visibility, 1500m horizontal and 1000ft vertical distance from cloud.
		Below FL 100: 5km flight visibility, 1500m horizontal and 1000ft vertical distance from cloud.
		(c) Class C, D and E airspace, VMC minima – Additionally in Class C, D and E airspace VFR flight is allowed by aircraft, other than helicopters, at or below 3000ft amsl, at a speed of 140kt or less which remain clear of cloud and in sight of the surface and in a flight visibility of at least 5km. Helicopters may fly VFR in Class C, D and E airspace at and below 3000ft amsl provided that they remain clear of cloud and in sight of the surface.
		(d) Class F and G Airspace the VMC minima at and below FL 100 applies down to the surface (instead of down to 3000ft) with the minima at and below 3000ft as an alternative. The proviso 'or 300m above terrain whichever is the higher' does not apply in the UK.
		(e) There is no mandatory requirement for continuous two-way radio communications in F and G Airspace under IFR.
Appendix 5		Aeronautical Data Quality Requirements
(Table 4 Bearing)		All types are calculated to the required accuracy. However, they are not published in the AIP to this accuracy. They are all published as rounded values to the nearest whole degree.
(Table 5 Length/Distance/Dimension)		All types are calculated to the required accuracy. However, they are not published in the AIP to this accuracy. They are published as rounded values to the nearest whole degree.

Annex 12	Search and Rescue (7th Edition) (AMDT 17)
Chapter 1	Definitions
	Pilot-in-Command. In United Kingdom legislation, 'Pilot-in-command' in relation to an aircraft means a person who for the time being is in charge of the piloting of the aircraft without being under the direction of any other pilot in the aircraft

Annex 13	Aircraft Accident and Incident Investigation (9th Edition) (AMDT 10)
	Nil

Annex 14	Aerodromes: Vol I (Aerodrome Design and Operations) (3rd Edition) (AMDT 6)
Chapter 1	General
1.2.2	There is no requirement for Annex 14 to apply to 'Government' Aerodromes open to public use.
1.4	There is no requirement for the certification of 'Government' Aerodromes.
1.6.1	UK determines code number in accordance with characteristics of the aerodrome. UK uses the greater of TODA/ASDA to determine the reference code number.
1.6.2	Column (2) ARFL is replaced by greater of TODA or ASDA.
1.6.3	Column (2) ARFL is replaced by greater of TODA or ASDA.
Chapter 3	Physical Characteristics
3.6.3	UK allows for the width to be not less than that of the visual strip width for that runway.
3.6.4	UK allows 2% up slope where the codes are 1 and 2.
3.9.7	Difference to Table 3.1

Column	(10)	(11)
Code A	21.0	13.5
Code B	31.5	19.5

GEN 1.7 Annex 14 Aerodromes: Vol I (Aerodrome Design and Operations)
(3rd Edition) (AMDT 6) Chapter 9

| 3.12.3 | UK takes into consideration only the interference with radio aids, the penetration of the OFZ by a holding aeroplane and the holding aeroplane being accountable in the calculation of OCA/H. |

Chapter 5   Visual Aids for Navigation

| 5.2.5.5 | UK uses a different style as shown below: |

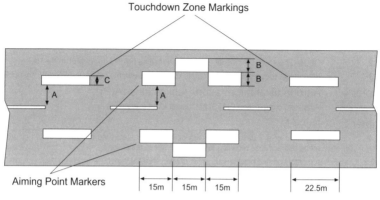

Runway Width (m)	Distance A Rwy C/L to marker (m)	Marker Width B (m)	Marker Width C (m)
45	9	5.5	3
30	3	5	3
23	5	2.5	1.5
18	3	2.5	1.5

5.2.6.4	UK uses different style of TDZ marking.
5.3.5.42	UK uses a plane 1 degree below the lower boundary of an on slope signal originating 90m from units where the LDA is 1200m or greater, 60m where the LDA is 800m, 1199m and 30m where the LDA is less than 800m, diverging at 15 degrees from the runway edge at the threshold out to 15nm.
5.3.16.1	Centre line lights are not required between the taxiway centre line and the stop position on the stand.
5.3.16.7	UK uses amber/green both ways within the ILS protected area.
5.3.16.12	UK uses 30m spacing in RVR >200m and 12m spacing in RVR <200m.
5.3.16.14	UK uses 15m spacing in RVR 200m and above, and 7.5m spacing in RVR <200m.
5.3.16.18	UK uses 15m spacing in RVR 200m and above, and 7.5m spacing in RVR <200m. UK uses systems not aligned for use by both pilots.
5.3.22.1	Required in RVR <1200m.
5.3.22.2	Required in RVR <1200m.
5.3.24.11	UK uses systems not aligned for use by both pilots.
5.3.24.14	UK permits systems where the pilot has to turn head.
5.3.24.16	UK uses systems not aligned for use by both pilots.
5.3.26.1	Required in RVR <1200m.
5.3.26.2	Required in RVR <1200m.
5.3.26.4	(b) flashing amber used instead of red.
5.4.3.4	Not used in UK.
5.5.3.1	Stopway edges marked with lights in UK.
5.5.3.2	Not used in UK.
Table 5-4	UK uses visibility to determine one of two sizes. Location varies according to runway coding.

Chapter 9   Equipment and Installations

| 9.2.11 | At all aerodromes, up to a maximum of 50% of the complementary extinguishing agents may be replaced by water for the production of a foam meeting performance level B.
For the purposes of substitution the following rates will apply:
1kg dry chemical powder or halogenated hydrocarbon = 1 litre water.
2kg carbon dioxide = 1 litre water. |

Annex 14	Aerodromes: Vol II (Heliports) (2nd Edition) (AMDT 3)
	Nil

Annex 15	Aeronautical Information Services (11th Edition) (AMDT 33)
Chapter 3	General
3.2.9	Co-ordinate information to required standard published on CD ROM version of the AIP and as separate database.
3.2.10	Only co-ordinate information is CRC wrapped.
3.7.2.2	OSGM02 is the Geiod model used for determining heights above MSL in the UK.
3.7.2.3	Parameters required for height transformation between the model and EGM-96 not published.
Chapter 4	Aeronautical Information Publications (AIP)
4.2.5	Producing organisation (NATS) not identified on every page.
Chapter 5	NOTAM
5.1.1.4	Unable to give 7 days notice of the activation of established danger, restricted or prohibited areas and activities requiring temporary airspace restrictions.
5.2.13.3	A monthly printed plain language list of valid NOTAM is not used.
Chapter 10	Electronic Terrain and Obstacle Data
10.1.1	Electronic terrain and obstacle databases not provided.
10.2.1	Electronic terrain and obstacle databases not provided.
10.2.2	Electronic terrain and obstacle databases not provided.
10.2.3	Electronic terrain and obstacle databases not provided.
10.2.4	Electronic terrain and obstacle databases not provided.
10.2.5	Electronic terrain and obstacle databases not provided.
10.3.1	Electronic terrain database not provided.
10.3.2	Electronic terrain database not provided.
10.3.3	Electronic terrain database not provided.
10.4.1	Electronic terrain database not provided.
10.4.2	Electronic terrain database not provided.
10.5.1	Electronic terrain and obstacle databases not provided.
10.5.2	Electronic terrain and obstacle databases not provided.
10.5.3	Electronic terrain database not provided.
10.5.4	Electronic terrain database not provided.
10.5.5	Electronic terrain database not provided.
10.5.6	Electronic terrain and obstacle databases not provided.
10.5.7	Electronic terrain database not provided.
10.5.8	Electronic terrain and obstacle databases not provided.
10.5.9	Electronic terrain and obstacle databases not provided.
10.6.1	Electronic terrain and obstacle databases not provided.
10.6.1.1	Electronic terrain and obstacle databases not provided.
10.6.1.2	Electronic terrain and obstacle databases not provided.
10.6.1.3	Electronic terrain and obstacle databases not provided.
2.2.2.2	Electronic terrain and obstacle databases not provided.
2.2.2.3	Electronic terrain and obstacle databases not provided.
Appendix 1	
GEN 1.5	General and special equipment section not populated.
GEN 2.1.4	Parameters required for height transformation between the model used and EGM-96 not published.
GEN 2.6	Item 3 Conversion Tables
	Decimals minutes of arc/seconds of arc.
GEN 3.1.6	Information not yet available.
GEN 3.3.5	Section blank, decoder to be developed.
ENR 1.4	Airspace classification not depicted as per Annex 11, Appendix 4.

GEN 1.7 Doc 4444 Procedures for Air Navigation Services Rules of the Air and Air Traffic Services (14th Edition)

ENR 1.6.1	Radar services & procedures.
	Section 3 not used.
	Section 4 contained in ENR 6.
ENR 1.6.2	Graphic portrayal of SSR coverage is contained in ENR 6-1-6-1 and ENR 6-1-6-2, not in ENR 1.6.2.
ENR 1.13	Information not published.
ENR 3.3	RNAV routes.
ENR 3.4	Section not established.
ENR 3.6	Item 7 'Controlling Unit' not published.
ENR 4.3	GNSS.
ENR 4.4	Name code designators.
ENR 4.5	En-route lights.
ENR 5.4	Air navigation obstacles.
ENR 6	ICAO En-route chart is not published. Chart index published in GEN 3.2.
AD 2.10	AD obstacles – use of Areas 1, 2, 3.
AD 2.12	Bearings given to nearest tenth of a degree runway end co-ords not declared TDZ high point not published.
AD 2.19	Declination not published.
AD 3.10	Heliport obstacles, new areas not yet applied.
AD 3.18	Declination not published.

Annex 16 Environmental Protection Vol I (Aircraft Noise) (3rd Edition) (AMDT 7)

Nil

Annex 16 Vol II (Aircraft Engine Emissions) (2nd Edition) (AMDT 4)

Nil

Annex 17 Security (6th Edition) (AMDT 10)

Nil

Annex 18 The Safe Transport of Dangerous Goods by Air (3nd Edition) (AMDT 7)

Chapter 1 Definitions

Pilot-in-Command. In United Kingdom legislation, 'Pilot-in-command' in relation to an aircraft means a person who for the time being is in charge of the piloting of the aircraft without being under the direction of any other pilot in the aircraft.

Chapter 2 Applicability

2.5.2 Legislation not appropriate.

Chapter 9 Provision of Information

9.6.1 & 9.6.2 Although Reg 6(3) of Air Navigation (Dangerous Goods) Regulations 2002 requires CAA to be told, it is not considered appropriate to make specific reference only to emergency services in UK law.

9.4 Partial compliance only; Air Navigation (Dangerous Goods Regulations 2002 require operators, shippers and others to train staff according to the requirements in the Technical Instructions and these include emergency procedures. However, the regulations only require operators to give instructions for emergencies. It is not considered appropriate in UK law to require shippers and others to give instructions in emergencies.

Chapter 10 Establishment of Training Programmes

Partial compliance only; the provisions of chapter 4 of the Technical Instructions, which is the amplification of Chapter 10 in Annex 18, are fully met with the exception of requiring training for shippers in emergency procedures as it is not considered appropriate in UK law to require shippers and others to be so trained.

Doc 4444 Procedures for Air Navigation Services Rules of the Air and Air Traffic Services (14th Edition)

The UK differences against the current version of ICAO Doc 4444 are under review by the UK CAA and will be published in due course.

GEN 1.7   Doc 8168 Procedures for Air Navigation Services Aircraft Operations Vol 1
(Flight Procedures) (4th Edition) Part II Departure Procedures Chapter 6

Doc 8168	Procedures for Air Navigation Services Aircraft Operations Vol 1 (Flight Procedures) (4th Edition)
Part II	Departure Procedures
Chapter 6	Use of FMS/RNAV equipment to follow conventional departure procedures:
6.1	Additional requirements:
	(i) the conventional procedure must have been inserted into the FMS from a recognised database and cannot be manually loaded or modified by the crew other than to follow ATC instructions;
	(ii) after the procedure has been loaded into the FMS as above, it must be cross-checked against the published conventional procedure before any attempt is made to follow the procedure using the FMS.
Part III	Approach Procedures
Chapter 6	Use of FMS/RNAV equipment to follow conventional non-precision approach procedures:
6.1	Additional requirements:
	(i) the conventional procedure must have been inserted into the FMS from a recognised database and cannot be manually loaded or modified by the crew other than to follow ATC instructions;
	(ii) after the procedure has been loaded into the FMS as above, it must be cross-checked against the published conventional procedure before any attempt is made to follow the procedure using the FMS.
Part V	Noise Abatement Procedures
Chapter 2	Noise Preferential Runways and Routes
2.2.3	In general, where turns are required shortly after take-off for noise abatement or other operational purposes, the nominal track has not been designed in accordance with the criteria in Volume II Part 2 Chapter 3 para 3.3. However, no turns are to be commenced below a height of 500ft aal. Airport Operators may specify the criteria used to determine individual Noise Preferential Routes. These criteria are for guidance only and aircraft operators should adhere to the routes to the maximum extent practicable commensurate with the safe operation of the aircraft.
Chapter 3	Aeroplane Operating Procedures
3.6	Unless otherwise stated, the upper limit for noise abatement procedures is 3000ft alt. However, aircraft operators are expected to operate their aircraft at all times in a manner calculated to cause the least noise disturbance on the ground.

Doc 8168	Procedures for Air Navigation Services – Aircraft Operations Volume II (Construction of Visual and Instrument Flight Procedures) (4th Edition)
Part II	Procedure Construction and Obstacle Clearance Criteria for Departure Procedures
Chapter 3	Departure Routes
3.3	In general, where turns are required shortly after take-off for noise abatement or other operational purposes, the nominal track has not been designed in accordance with these criteria. However, no turns are to be commenced below a height of 500ft aal. Primary and secondary areas for obstacle clearance on Standard Instrument Departure Procedures, where published, are determined along the nominal ground track of the Noise Preferential Route as specified by the Airport Operator. Obstacle clearance is not assessed for any routes other than published standard Instrument Departure Procedures.
Chapter 4	Omni-Directional Departures
	Obstacle clearance is not assessed for omni-directional departures and procedures are not published. Obstacles in the vicinity of the aerodrome are detailed in AD 2 item AD 2.10 and aircraft operators are responsible for ensuring that adequate terrain/obstacle clearance exists relevant to the stage of flight. No turns are to be commenced below 500ft aal.
Chapter 5	Published Information
Para C	Procedures for omni-directional departures are not published.

GEN 1.7 Doc 8168 Procedures for Air Navigation Services – Aircraft Operations Volume II (Construction of Visual and Instrument Flight Procedures) (4th Edition) Chapter 7

Part III	**Frontispiece Procedure Construction and Obstacle Clearance Criteria for Approach Procedures**
3 (i) (Note)	**Applicability.** The procedures in this part are applicable to all aircraft capable of complying with the speeds specified in Table III 1-2. For this purpose helicopters, when operating as aeroplanes, are classified as Category A aircraft.
Chapter 1	**General**
1.9.2	Procedure speed restrictions. Unless otherwise stated, procedures are speed restricted to a maximum IAS of 185kt.
Chapter 4	**Initial Approach Segment**
4.4.5.1 & 4.5.4.1	Requirement for separate instrument approach charts. In certain procedures different outbound tracks and/or timings may be specified for Category A/B and Category C/D aircraft. These tracks/ timings will normally be published on a common instrument approach chart. Separate charts will normally be published whenever Category A/B and Category C/D aircraft have different procedure altitudes or different missed approach points.
4.6.7 (Note)	**Reduction in width of secondary areas.** There is no reduction in width for secondary areas in race-track/reversal procedures.
Chapter 5	**Intermediate Approach Segment**
5.3	**Intermediate approach segment based on a straight track alignment.** Where the turn at the FAF is greater than 10°, the final approach area is widened on the outer side of the turn as in paragraph 7.3.5.3.2. This adjustment is based on the fastest 'final approach IAS' from Table III 1-2 appropriate to the procedure corrected to TAS for aerodrome elevation at ISA + 15, the standard ICAO wind for the height, and a bank angle of 25° (or the angle giving a rate of turn of 3°/second, whichever is the lesser).
5.4.1	**Length.** The length of the intermediate segment should conform to the standard given in paragraph 5.4.1 whenever possible. However, when an operational advantage may be gained the minimum length of the intermediate segment may be reduced to 5.5km (3nm).
Chapter 7	**Missed Approach Segment**
7.3.2	**Turn parameters – IAS.** Unless otherwise specified, all missed approach turns are limited to 185kt IAS maximum.
7.3.4.2	(last sentence) If the TO has to be located before the SOC calculated for the final approach and straight missed approach, then either the MAPt shall be moved back and, if necessary, the OCA/H increased or, when essential, the turn must be specified as a 'turn as soon as practicable' (see paragraph 7.3.4.7).
7.3.4.4.2	**Turn of 15° or less.** Where a turn of 15° or less is specified the criteria above apply except that the MOC in the primary area and the clearance above obstacles in the turn initiation area shall be 30m (98ft).
7.3.4.7	**Turn as soon as practicable.** (Paragraph 7.3.4.7 – UK additions to PANS-OPS.)
7.3.4.7.1	**General.** A turn as soon as practicable is prescribed in non-precision procedures when it is essential to locate the TO before the SOC associated with a normal turn at an altitude or at a fix, and when it is not convenient to move the MAPt. When specified, the missed approach procedures shall be annotated 'turn left (or right) as soon as practicable'. The criteria are the same as those for turn at a designated altitude, modified in accordance with paragraph 7.3.4.7.2 to paragraph 7.3.4.7.4.
7.3.4.7.2	**Turn altitude/height.** The turn altitude/height is also the OCA/H for the procedure. The TO is plotted at distance c after the latest limit of the MAPt tolerance area.
7.3.4.7.3 & 7.3.4.7.3.1	**Areas.** **Turn initiation area.** The turn initiation area is bounded by the edges of the MAPt tolerance area, starting at the earliest MAPt and extended beyond the latest MAPt to the TO.
7.3.4.7.3.2	**Turn area.** The inner and outer boundaries of the turn area are constructed as specified in paragraph 7.3.3 with the following exceptions: (a) The boundaries are based on the intermediate missed approach speed of the appropriate aircraft category; (b) the outer boundary starts at the range of the TO (distance c has already been included in the turn initiation area).
7.3.4.7.4	**Obstacle clearance.** The obstacle clearance in the turn initiation and turn areas is adjusted to preserve the normal MOC associated with the transitional tolerance X into the turn area as follows: (a) Obstacle clearance in the turn initiation area. Obstacle elevation/height in the turn initiation area shall be less than: OCA/H – MOC approach segment. (b) Obstacle clearance in the turn area. Obstacle elevation/height in the turn area and subsequently shall be less than:

GEN 1.7 Doc 8168 Procedures for Air Navigation Services – Aircraft Operations Volume II (Construction of Visual and Instrument Flight Procedures) (4th Edition) Chapter 7

OCA/H – MOC missed approach + (d° – X) tan Z with the additional provision that obstacle height need not be less than (OCH – MOC approach segment). where:

d° is measured from the obstacle to the nearest point on the turn initiation area boundary.

MOC approach is the primary area MOC associated with the final approach segment.

MOC missed approach is the MOC applicable to the missed approach; 50m (164ft) for turns exceeding 15° and 30m (98ft) for turns of 15° or less, reduced if appropriate for obstacles within any secondary areas.

Chapter 9 Minimum Sector Altitudes (MSA)

9.4 **Combining sectors for adjacent facilities:** Where more than one facility provides instrument approaches to an aerodrome, and unless otherwise specified, the minimum sector altitude for each sector is the highest of those calculated for that specific sector for every facility serving the aerodrome, regardless of the distance between the facilities.

Chapter 21 ILS

21.1.3 **OCA/H values for non-standard aircraft – ILS.** The ILS OCA/H values published are based on the 'standard dimensions' specified in paragraph 21.1.3. When the OCA/H value calculated for an aircraft of larger than 'standard dimensions' exceeds that published for the 'standard dimensions' the procedure will be annotated 'increased OCA/H for aircraft exceeding standard dimensions' and the appropriate value will be published separately.

Chapter 24 Radar

24.2.1 **Surveillance Radar – General.** See paragraph 24.3 for separate criteria for approved 'high resolution' equipment with a termination range of 0.5nm or less.

24.2.2.3 Additionally, within a specified area aligned with an Instrument Runway, when an aircraft is being vectored to an Instrument Approach, minimum obstacle clearance may be reduced to 150m (500ft). The specified area is shown on the Radar Vectoring Area Chart and is of the following dimensions:

A line 2.5nm long, centred on the runway centre-line, 1.5nm from the threshold in the approach and a line 5nm long, centered on the runway centreline, 9.5nm from the threshold in the approach, joined at the ends to form a quadrangle.

24.2.4.3 **Area.** The area to be considered for obstacle clearance begins at the FAF and ends at the MAPt.

5.5.5 **Termination Range.** A Surveillance Radar Approach shall be terminated 2nm before touchdown except where a termination range of 1nm has been specifically approved. See paragraph 24.3 for separate criteria for approved 'high resolution' equipment with a termination range of 0.5nm or less.

The Missed Approach Point (MAPt) is located at the point where the radar approach terminates. However, where operationally advantageous, the MAPt for 2nm SRAs may be designated as 1nm before touchdown.

5.6 **Surveillance Radar (high resolution)** – UK addition to PANS-OPS.

General. Certain approved Surveillance Radar equipments can provide final approach guidance of better quality than that provided for in paragraph 24.2. The criteria for procedures using these radars are the same as those contained in paragraph 24.2 except for the final approach and missed approach areas and obstacle clearance described below:

Note: Approval of 'high resolution' SRE procedures is based on an operational and technical evaluation of the equipment. In all cases:

(a) There is a continuous talkdown, on a discrete frequency, from 4nm with ranges and advisory heights being given every 0.5nm;

(b) The approach controller providing final approach guidance is allocated full time to the task;

(c) The display system incorporates a centre-line with associated reflectors to conform centre-line accuracy;

(d) The accuracy, resolution, antenna rotation rate, low level cover, and extent of permanent echoes are assessed as capable of giving a high probability of a successful approach with a termination range of 0.5nm or less.

Area. The area to be considered for obstacle clearance begins at the FAF and ends at the MAPt and is centred on the Final Approach Track. The minimum length of the Final Approach Track shall be 3nm. The length shall be established by taking account of the permissible descent gradient (see paragraph 24.2.4.5). The maximum length should not exceed 6nm. Where a turn is required over the FAF, Table 26-1 applies (paragraph 26.4.2). The width of the area is proportional to the distance from the radar antenna, according to the following formulae:

$W/2 = 1.9 + 0.1\ D$ km, for D greater than 10km.

$W/2 = 0.3 + 0.26\ D$ km, for D equal or less than 10km.

Where:

W = total area width in km.

D = distance from antenna to track in km.

The maximum value for D is 37km (20nm) subject to the accuracy of the radar equipment as determined by the Authority.

A secondary area comprising 25% of the total width lies on each side of the primary area, which comprises 50% of the total width.

Obstacle Clearance. The MOC is 75m (246ft) in the primary area, reducing to zero at the outer edges of the secondary areas.

Missed Approach Secondary Areas. Secondary areas are established on each side of the primary area, with width equal to 25% of the total area width at the MAPt, reducing to zero width at the SOC.

GEN 1.7 – DIFFERENCES FROM 1CAO STANDARDS, RECOMMENDED PRACTICES AND PROCEDURES

1 Definitions

The definitions listed in Chapter 1 of ICAO Annexes 2 and 11 and/or part 1 of ICAO Doc 4444 and/or ICAO Doc 8168 apply throughout the UK AIP except for the terms given in Table 1.7.1 where the UK has an interpretation which differs from the ICAO definition.

2 Table 1.7.2 gives definitions of terms which are not defined in ICAO Annex 2 and/or ICAO Doc 4444 and/or ICAO Doc 8168 but which the UK has found it necessary to clarify.

Table 1.7.1 – UK interpretation Differences

Term	Defined in ICAO Doc	UK Definition
Aerodrome Traffic Zone	Annex 2	In relation to an aerodrome at which: The length of the longest runway is notified as 1850m or less, the airspace extending from the surface to a height of 2000ft above the level of the aerodrome within the area bounded by a circle centred on the notified mid-point of the longest runway and having a radius of 2nm. Provided that where such an Aerodrome Traffic Zone would extend less than 1.5nm beyond the end of any runway at the aerodrome and this proviso is notified as being applicable, sub-paragraph (b) hereof shall apply as though the length of the longest runway is notified as greater than 1850m; The length of the longest runway is notified as greater then 1850m, the airspace extending from the surface to a height of 2000ft above the level of the aerodrome within the area bounded by a circle centred on the notified mid-point of the longest runway and having a radius of 2.5nm: Except any part of that airspace which is within the Aerodrome Traffic Zone of another aerodrome which is notified as being the controlling aerodrome.
Apron	Annex 2 Doc 4444	This part of an aerodrome provided for the stationing of aircraft for the embarkation and disembarkation of passengers, the loading and unloading of cargo, refuelling and for parking.
Decision Height	Doc 4444	(1) The minimum height specified by the operation, or ascertainable by reference to, the operations manual as being the minimum height to which an approach to landing can safely be made by that aircraft at that aerodrome without visual reference to the ground. (2) A specified height at which a missed approach must be iniated if the required visual reference to continue the approach to land has not been established.
Holding Point	Doc 4444	Add second meaning: On the manoeuvring area of an aerodrome, a location at which an aircraft carries out an engine run-up or is held before entering a runway for take off.
Instrument Meteorological Conditions (IMC)	Annex 2 Annex 11 Doc 4444	Meteorological conditions expressed in terms of visibility, horizontal and vertical distance from cloudless than the minima specified for visual meteorological conditions.
Manoeuvring Area	Annex 2 Annex 11 Doc 4444	This part of an aerodrome provided for the take-off and landing of aircraft and for the movement of aircraft on the surface excluding the apron and any part of the aerodrome provided for the maintenance of aircraft.
Secondary Surveillance Radar (SSR)	Doc 4444	A system of radar using ground interrogators and airbourne transponders to determine the position of aircraft in range and azimuth and, when the agreed modes and codes are used, height and identity as well.
Special VFR Flight	Annex 2 Annex 11 Doc 4444	In the UK this means a flight at any time in a Control Zone which is Class A Airspace or in any other Control Zone in IMC or at night in respect of which the appropriate Air Traffic Control Unit has given permission for the flight to be made in accordance with special instructions given by that Unit instead of in accordance with the Instrument Flight Rules and in the course of which flight the aircraft complies with any instructions given by that unit and remains clear of cloud and in sight of the surface.

Table 1.7.2 – UK DEFINITIONS FOR TERMS USED BY ICAO

Term	UK Definition (no definition in either Annex 2 or Doc 4444)
Aerodrome Approach	That part of an Instrument Approach Procedure commencing at the designated height over the radio aid to be used and ending when the aircraft has broken cloud.
AIRPORX	A situation in which, in the opinion of a pilot or a controller, the distance between aircraft as well as their relative positions and speed have been such that the safety of the aircraft involved was or may have been comprised.
Cloud Ceiling	In relation to an aerodrome, the distance measured vertically from the notified elevation of the aerodrome to the lowest part of any cloud visible form the aerodrome which is sufficient to obscure more than one half of the sky so visible.
Competent ATS Authority	Means in relation to the UK, the Civil Aviation Authority, and in relation to any other country the authority responsible under the law of that country for promoting for safety of civil aviation.
Continuous Descent Approach (CDA)	A noise abatement technique for arriving aircraft in which the pilot, when given descent clearance below Transition Altitude by ATC, will descent at the rate he judges will be best suited to the achievement of continuous descent, whilst meeting the ATC speed control requirements, the objective being to join the glide-path at the appropriate height for the distance without recourse to level flight.
Dropping Zone/Drop Zone	Means the notified portion of airspace within which parachute descents are made.
General Air Traffic (GAT)	Flights conducted in accordance with the Regulations and Procedures for flight promulgates by the State Civil Aviation Authorities and operating under the control or authority of the Civil ATS organisation.
Known Traffic	Traffic the current flight details and intentions of which are known to the controller concerned through direct communications or co-ordination.
Low Power/Low Drag Approach (LP/LD)	A noise abatement technique for arriving aircraft in which the pilot delays the extension of wing flaps and undercarriage until the final stages of the approach, subject to compliance with ATC speed control requirements and the safe operation of the aircraft.
Night	The time between half an hour after sunset and half an hour before sunrise, sunset and sunrise being determined at surface level.
Operational Air Traffic (OAT)	Flights conducted under the control or authority of the military ATS organisation.
Quadrantal Cruising Levels	Specified cruising levels determined in relation to magnetic track within quadrants of the compass.
Parachute Landing Area	Means the terrain onto which parachute descents are made.
Radar Clear Range	A range surveyed by radar within which the operating authority accepts responsibility for with holding fire is an aircraft is within the area into which, and through which, missiles are liable to fall.
Radar Departure	The control of a departing aircraft by the use of surveillance radar to assist it to leave the vicinity of an aerodrome safely and expeditiously.
Radar Handover	Transfer of responsibility for the control of an aircraft between two controllers using radar. Following identification of the aircraft by both controllers.
Radar Surveillance	Observance of the movements of aircraft on a radar display and the passing of advice and information to identified aircraft and, where appropriate to other ATS units.
Radar Vectoring Area	A defined area in the vicinity of an aerodrome, in which the minimum safe levels allocated by a radar controller vectoring IFR flights have been pre-determined.
Radial	A magnetic bearing extending from a VOR/VORTAC/TACAN.
Stack Departure Time	The time at which an aircraft is required to leave the holding facility to commence its approach.
Straight Ahead	When used in departure clearances means: 'track extended runway centre-line' And when given in Missed Approach Procedures means; 'maintain final approach track'.
Upper ATS Route	A designated route within the Upper Airspace CTA.

GEN 2 – TABLES AND CODES

GEN 2.1 MEASURING SYSTEM, AIRCRAFT MARKINGS, HOLIDAYS

1 Dimensional Units

1.1 The table of dimensional units used in the United Kingdom is that which is published in ICAO Annex 5 (Fourth Edition) Chapter 3 tables 3.3 and 3.4 under Non-SI alternative unit.

Table 2.1.1 Units of Measurement used in the UK

Measurement of	Units
Distance used in navigation, position report etc – generally in excess of 2 or 3nm	* nautical miles and tenths
Relatively short distances such as those relating to aerodromes (eg runway lengths)	Metres
Altitudes, elevations and heights	Feet and Flight Levels
Horizontal speed including wind speed	Knots
Vertical speed	Feet per minute
Wind direction for landing and taking off	Degrees Magnetic
Wind direction except for landing and taking off	Degrees True
Visibility <5000 metres (including RVR)	Metres
Visibility >5000 metres	Kilometres
Altimeter setting	Millibars
Temperature	Degrees Celsius (Centigrade)
Weight	Metric tonnes or Kilogrammes
	Date/Time Year, Month, Day, Hour and minute, the day of 34 hours beginning at midnight co-ordinated Universal Time

* International nautical miles, for which conversion into metres is given 1 international nautical mile = 1852m.

1.2 There is no sharp dividing line between the usage of nautical miles or metres for the two types of horizontal distances referred to in the table above. There is no hard and fast rule, but distances having a navigational or position reporting aspect are given in nautical miles even if they are less than 2 nautical miles (eg ranges from touchdown during a precision approach). These are given in nautical miles and fractions. Distances on the aerodrome (eg runway lengths, etc.) are given in metres. Distances from obstacles in the vicinity of aerodromes and ILS marker distances are generally shown in nautical miles and tenths, except when it is necessary for greater accuracy for some middle markers to be shown in metres.

1.3 Nautical miles or feet, as appropriate, will be used in designating horizontal distances at military aerodromes, joint-user aerodromes controlled by Service personnel and in RAF Sections at Air Traffic Control Centres. Horizontal distances will, however, be given to civil pilots in metric units, on request. Similarly, at civil aerodromes, joint-user aerodromes under civil control and in civil sections at Air Traffic Control Centres, Service pilots will be given distances in nautical miles or feet, as appropriate, on request.

2 Time System

2.1 Time in the case of the following services and publications is expressed in terms of Co-ordinated Universal Time (UTC):

(a) Air Traffic Services
(b) Aeronautical Telecommunications Service
(c) UK AIP
(d) Supplements to the UK AIP (unless otherwise stated)
(e) UK AIC (unless otherwise stated)
(f) UK NOTAM (unless otherwise stated)

2.1.1 In the case of Air Traffic Services, time is given in position reporting and estimates is expressed to the nearest minute. Time checks given to aircraft by Air Traffic Services Units may be expressed in terms of seconds or fractions of a minute dependent upon the type and accuracy of clocks available.

2.1.2 In the UK, local time during the summer will be 1 hour in advance of UTC (UTC + 1 hour), in winter it will be UTC.

2.1.3 Summer time will commence at 0100 on 26 March 2006, and end at 0100 on 29 October 2006.

3 Geodetic Reference Data

3.1. The geographical co-ordinates indicating Latitude and Longitude are expressed in terms of the World Geodetic Survey of 1984 (WGS84). geodetic reference datum.

3.2 Area of Application

3.2.1 The area of application for the published geographical co-ordinates coincides with the Area of Responsibility for UK Air Traffic Services, ie London and Scottish Flight Information Regions; Scottish and Shanwick Oceanic Control Areas. In particular the airspace types listed in Table 2.1.2

Table 2.1.2 – Air navigation Co-ordinates

Area/En-route Co-ordinates	Aerodrome Co-ordinates
FIR Boundaries and Crossing points	Reference Points
ATS, RNAV and Conditional Routes	Thresholds
Holding Points	Radio Nav Aids – Precision and Non-Precision
Radio Nav Aids (off Aerodrome)	Runway centre-line
CTA, TMA, CTR	Extended Runway centre-line FAFs
Restricted, Prohibited and Danger Areas	Stands and Nav Checkpoints
Obstacles – En-route	
Other significant points and areas	

3.1.3 The application of WGS84 will be achieved either by survey or mathematical conversion of co-ordinates. Where the information has been transformed mathematically into WGS84 co-ordinates they are published accompanied by an asterisk indicating that the information is of low integrity.

4 Aircraft Nationality and Registration Marks

4.1 The nationality mark for British Civil aircraft is the letter 'G'. This nationality mark is followed by a hyphen and a registration mark consisting of four letters. Example: G-BAAA.

5 Public Holidays in the UK

5.1 The following dates are notified as Public Holidays during 2005:

Public Holiday	England & Wales	Scotland	Northern Ireland
New Years Day	2 Jan	2 Jan	2 Jan
Bank Holiday	–	3 Jan	-
Bank Holiday – St Patrick's Day	–	–	17 Mar
Good Friday	14 Apr	14 Apr	14 Apr
Easter Monday	17 Apr	–	17 Apr
Early May Bank Holiday	1 May	1 May	1 May
Spring Bank Holiday	29 May	29 May	29 May
Bank Holiday	–	–	12 Jul
Summer Bank Holiday	–	7 Aug	-
Summer Bank Holiday	28 Aug	–	28 Aug
Christmas Day	25 Dec	25 Dec	25 Dec
Boxing Day	26 Dec	–	26 Dec

GEN 2.2 – ABBREVIATIONS USED IN AIS PUBLICATIONS

The abbreviations used in this AIP and in the general dissemination of information are extracted from ICAO DOC 8400. Abreviations which differ from the ICAO abbreviations are shown in italics.

† When radiotelephony is used, the abbreviations and terms are transmitted as spoken words.

‡ When radiotelephony is used, the abbreviations and terms are transmitted using the individual letters in non-phonetic form.

A

A	Amber
AAA	(or AAB, AAC.etc, in sequence) Amended meteorological message (message type designator)
A/A	Air-to-air
AAL	Above Aerodrome Level
AAR	*Air to Air Refuelling*
AARA	*Air to Air Refuelling Areas*
ABM	Abeam
ABN	Aerodrome Beacon
ABT	About
ABV	Above
AC	Altocumulus
ACARS	Aircraft Communications Addressing and Reporting System
ACAS†	Airborne Collision Avoidance Systems
ACC‡	Area Control Centre OR Area Control
ACCID	Notification of an Aircraft Accident
ACFT	Aircraft
ACK	Acknowledge
ACL	Altimeter Check Location
ACN	Aircraft Classification Number
ACN	*Airspace Co-ordination Notification (CAN in ICAO is aircraft class number)*
ACP	Acceptance (message type designator)
ACPT	Accept OR Accepted
ACT	Active OR Activated OR Activity
AD	Aerodrome
ADA	Advisory Area
ADDN	Addition OR Additional
ADF‡	Automatic Direction-Finding Equipment
ADIZ†	(to be pronounced 'AY-DIZ') Air Defence Identification Code
ADJ	Adjacent
ADR	Advisory Route
ADS	Automatic Dependent Surveillance
ADSU	Automatic Dependent Surveillance Unit
ADVS	Advisory Service
ADZ	Advise
AES	Aircraft Earth Station
AEW	*Airborne Early Warning*
AFIL	Flight Plan Filed in the Air
AFIS	Aerodrome Flight Information Service
AFM	Yes OR Affirm OR Affirmative OR That is Correct
AFS	Aeronautical Fixed Service
AFT	After (time or place)
AFTN‡	Aeronautical Fixed Telecommunication Network
A/G	Air-to-Ground
AGA	Aerodromes, Air Routes and Ground Aids
AGL	Above Ground Level
AGN	Again
AGNIS	Azimuth Guidance for Nose-In Stand
AIAA	*Area of Intense Air Activity*
AIC	Aeronautical Information Circular
AIM	ATFM Information Message
AIP	Aeronautical Information Publication
AIRAC	Aeronautical Information Regulation and Control
AIREP†	Air-Report
AIREX	*Air Exercise*
AIS	Aeronautical Information Services
ALA	Alighting Area
ALERFA†	Alert Phase
ALR	Alerting (message type designator)
ALRS	Alerting Service
ALS	Approach Lighting System
ALT	Altitude
ALTN	Alternate OR Alternating (Light alternates in colour)
ALTN	Alternate (Aerodrome)
AMA	Area Minimum Altitude
AMC	Airspace Management Cell
AMD	Amend OR Amended (used to indicate amended meteorological message; message type designator)
AMDT	Amendment (AIP Amendment)
AMS	Aeronautical Mobile Service
AMSL	Above Mean Sea Level
AMSS	Aeronautical Mobile Satellite Service
ANM	*ATFM Notification Message*
ANO	*Air Navigation Order*
ANS	Answer
AO	Aircraft Operators
AOC	Aerodrome Obstacle Chart
AOC	Air Operator Certificate
AOM	*Aerodrome Operating Minima*
AP	Airport
APAPI	Abbreviated Precision Approach Path Indicator
APCH	Approach
APIS	*Aircraft Positioning and Information System*
APP	Approach Control Office OR Approach Control OR Approach Control Service
APR	April
APRX	Approximate OR Approximately
APSG	After Passing
APV	Approve OR Approved OR Approval
ARFOR	Area Forecast (In aeronautical Meteorological Code)
ARNG	Arrange
ARO	Air Traffic Services Reporting Office
ARP	Aerodrome Reference Point
ARP	Air-Report (message type designator)
ARQ	Automatic Error Correction
ARR	Arrive OR Arrival
ARR	Arrival (message type designator)
ARS	Special Air-Report (message type designator)
ARST	Arresting (Specify (part of) Aircraft Arresting Equipment)
AS	Altostratus
ASC	Ascent to OR Ascending to

GEN 2.2 – ABBREVIATIONS USED IN AIS PUBLICATIONS

ASDA	Accelerate-Stop Distance Available	BS	Commercial Broadcasting Station
ASPH	Asphalt	BTL	Between Layers
ASR	*Altimeter Setting Region*	BTN	Between
ASR	*Airspace Revervation*	**C**	
AT	At (followed by time at which weather change is forecast to occur)	C	Centre (runway identification)
		C	Degrees Celsius (Centigrade)
ATA‡	Actual Time of Arrival	*CAA*	*Civil Aviation Authority*
ATA	*Aerial Tactics Area*	*CANP*	*Civil Aircraft Notification Procedure*
ATC‡	Air Traffic Control (in general)	*CAP*	*Civil Aviation Publication*
ATD‡	Actual Time of Departure	CAS	Controlled Airspace
ATFM	Air Traffic Flow Management	CAT	Category
ATIS†	Automatic Terminal Information Service	CAT	Clear Air Turbulence
ATM	Air Traffic Management	CATZ	Combined Aerodrome Traffic Zone
ATN	Aeronautical Telecommunication Network	CAVOK†	(To be pronounced 'KAV-OH-KAY') Visibility, Cloud and present weather better than prescribed values or Conditions
ATOTN	Air Traffic Operational Telephone Network		
ATP	At (time or place)		
ATS	Air Traffic Service	CB‡	(To be pronounced 'CEE BEE') Cumulonimbus
ATSOCAS	*ATS Outside Controlled Airspace*		
ATSU	*Air Traffic Service Unit*	CC	Counter Clockwise
ATTN	Attention	CC	Cirrocumulus
ATZ	Aerodrome Traffic Zone	CCA	(Or CCB, CCC Corrected meteorological message etc, in sequence) (message type designator)
AUG	August		
AUS	*Airspace Utilisation Section*	CD	Candela
AUTH	Authorised OR Authorisation	CDN	Co-ordination (message type designator)
AUW	All Up Weight	*CDR*	*Conditional Route*
AUX	Auxiliary	CF	Change frequency to
AVASIS	Abbreviated Visual Approach Slope Indicator System	*CFMU*	*Central Flow Management Unit (Europe)*
		CGL	Circling Guidance Light(s)
AVBL	Available OR Availability	*CHAPI*	*Compact Helicopter Approach Path Indicator*
AVG	Average		
AVGAS†	Aviation Gasoline	CH	Channel
AVTUR	*Aviation Turbine Fuel*	CHG	Modification (message type designator)
AWY	Airway	CI	Cirrus
AZM	Azimuth	CIDIN†	Common ICAO Data Interchange Network
B		CIT	Near OR over large towns
B	Blue	CIV	Civil
BA	Braking Action	CK	Check
BAA	*British Airports Authority plc*	CL	Centre-Line
BASE	Cloud Base	CLA	Clear Type of Ice Formation
BBMF	*Battle of Britain memorial Flight*	CLBR	Calibration
BCFG	Fog Patches	CLD	Cloud
BCN	Beacon (Aeronautical ground light)	CLG	Calling
BCST	Broadcast	CLR	Clear(s) OR Cleared to OR Clearance
BDRY	Boundary	CLSD	Close OR Closed OR Closing
BECMG	Becoming	CM	Centimetre
BFR	Before	*CMATZ*	*Combined Military Aerodrome Traffic Zone*
BKN	Broken	CMB	Climb to OR Climbing to
BL	Blowing (followed by DU = Dust, SA = Sand or SN = Snow)	CMPL	Completion OR Completed OR Complete
		CNL	Cancel OR Cancelled
BLCP	*Base Level Change Point*	CNL	Flight Plan Cancellation (message type designator)
BLDG	Building		
BLO	Below Clouds	CNS	Communications, Navigation and Surveillance
BLW	Below		
BOMB	Bombing	*COL*	*Column (in tables and text)*
BOTA	*Brest Oceanic Transition Area*	COM	Communications
BR	Mist	CONC	Concrete
BRF	Short (Used to indicate the type of approach desired or required)	COND	Condition
		CONS	Continuous
BRG	Bearing	CONST	Construction OR Constructed
BRKG	Braking	CONT	Continue(s) OR Continued
B-RNAV†	Basic – (To be pronounced 'AR-NAV') Area Navigation	COOR	Co-ordinate OR Co-ordination
		CO-ORD	*Geographical Co-ordinates*

77

GEN 2.2 – ABBREVIATIONS USED IN AIS PUBLICATIONS

COP	Change-Over Point	DPT	Depth
COR	Correct OR Correction OR Corrected (Used to indicate Corrected meteorological message; message type designator)	DR	Dead Reckoning
		DR	Low Drifting (followed by DU = Dust, SA = Sand or SN = Snow)
COT	At the Coast	DRG	During
COV	Cover OR Covered OR Covering	DS	Duststorm
CPDLC	Controller to Pilot Data Link	DSB	Double Sideband
CPL	Current Flight Plan (message type designator)	DTAM	Descend to and Maintain
		DTG	Date-Time Group
CRAM	*Conditional Route Availability Message*	DTRT	Deteriorate OR Deteriorating
CRZ	Cruise	DTW	Dual Tandem Wheels
CS	Cirrostratus	DU	Dust
CTA	Control Area	DUC	Dense Upper Cloud
CTAM	Climb to and Maintain	DUR	Duration
CTC	Contact	DVOR	Doppler VOR
CTL	Control	DW	Dual Wheels
CTN	Caution	DZ	Drizzle
CTOT	*Calculated Take-off Time*	**E**	
CTR	Control Zone	E	East OR Eastern Longitude
CU	Cumulus	EAT	Expected Approach Time
CUF	Cumuliform	EB	Eastbound
CUST	Customs	ECA	*Emergency Controlling Authority*
CW	Continuous Wave	*ECAC*	*European Civil Aviation Conference*
CWY	Clearway	EDT	Estimated Departure Time
D		EET	Estimated Elapsed Time
D...	DME Range (prefix used in graphics)	EFC	Expected Further Clearance
D...	Danger Area (Followed by Identification)	*EFIS*	*Electronic Flight Instrument System*
D	Downward (tendency in RVR During previous 10 minutes)	EGNOS	European Geostationary Navigation Overlay Service
DA	Decision Altitude	EHF	Extremely High Frequency (30000 to 300000MHz)
DAAIS	*Danger Area Activity Information Service*		
DACS	*Danger Area Crossing Service*	ELBA†	Emergency Location Beacon – Aircraft
DATIS	Data Link Automatic Terminal Information Service	ELEV	Elevation
		ELR	Extra Long Range
DCD	Double Channel Duplex	ELT	Emergency Locator Transmitter (GEN 3.6.6)
DCKG	Docking		
DCS	Double Channel Simplex	EM	Emission
DCT	Direct (In relation to flight path clearances and type of approach)	EMBD	Embedded in a Layer (To indicate cumulonimbus Embedded in layers of other clouds)
DEC	December		
DEG	Degrees	EMERG	Emergency
DEMO	*Demonstration*	END	Stop-end (related to RVR)
DEP	Depart OR Departure	ENE	East North East
DEP	Departure (message type Designator)	ENG	Engine
DES	Descend to OR Descending to	ENRT	En-Route
DEST	Destination	EOBT	Estimated Off-Block Time
DETRESFA†	Distress Phase	EQPT	Equipment
DEV	Deviation OR Deviating	ER	Here...
DF	*Direction Finding*	ESE	East South East
DfT	*Department of Transport*	EST	Estimate OR Estimated OR Estimate (message type designator)
DFTI	Distance from Touchdown Indicator		
DH	Decision Height	ETA‡	Estimated Time of Arrival OR Estimating Arrival
DIF	Diffuse		
DIST	Distance	ETD‡	Estimated Time of Departure OR Estimating Departure
DIV	Divert OR Diverting		
DLA	Delay (message type Designator)	*ETFMS*	*Enhanced Tactical Flight Management System*
DLA	Delay OR Delayed		
DLY	Daily	ETO	Estimated Time Over Significant Point
DME‡	Distance Measuring Equipment	*ETOPS*	*Extended Twin-jet Operations*
DNG	Danger OR Dangerous	EV	Every
DOC	*Designated Operational Coverage*	EXC	Except
DOM	Domestic	EXER	Exercises OR Exercising OR To Exercise
DP	Dew Point Temperature	EXP	Expect OR Expected OR Expecting
		EXT	Extension

GEN 2.2 – ABBREVIATIONS USED IN AIS PUBLICATIONS

EXTD	Extend OR Extending	FZDZ	Freezing Drizzle
F		FZFG	Freezing Fog
F	Fixed	FZRA	Freezing Rain
FAC	Facilities	**G**	
FAF	Final Approach Fix	G	Green
FAL	Facilitation of International Air Transport	G/A	Ground-to-Air
FAM	*Flight Activation Monitoring*	G/A/G	Ground-to-Air and Air-to-Ground
FAP	Final Approach Point	GA	*General Aviation*
FAT	*Final Approach Track*	GAGAN	GPS & Geostationary Earth Orbit Augmented Navigation
FATO	Final Approach and Take-off Area	*GAT*	*General Air Traffic*
FAX	Facsimile Transmission	GBAS	Ground Based Augmentation System
FBL	Light (Used to indicate the intensity of weather phenomena, interference or static reports, eg FBL RA = Light rain)	GCA‡	Ground Controlled Approach System OR Ground Controlled Approach
FBU	*Flight Briefing Unit*	GEN	General
FC	Funnel Cloud (tornado or water spout)	GEO	Geographic OR True
FCST	Forecast	GES	Ground Earth Station
FCT	Friction Coefficient	GLD	Glider
FDOD	*Flight Data Operations Division*	GLONASS	Global Orbiting Navigation Satellite System
FDPS	Flight Data Processing System	*GMC*	*Ground Movement Control*
FEB	February	*GMR*	*Ground Movement Radar*
FG	Fog	GND	Ground
FIC	Flight Information Centre	GNDCK	Ground Check
FIR‡	Flight Information Region	GNSS	Global Navigation Satellite System
FIS	Flight Information Service	GP	Glide Path
FISO	*Flight Information Service Officer*	GPS	Global Positioning System
FISA	Automated Flight Information Service	*GPWS*	*Ground Proximity Warning System*
FL	Flight Level	GR	Hail
FLAS	*Flight Level Allocation Scheme*	GRAS	Ground Based Regional Augmentation System
FLD	Field		
FLG	Flashing	GRASS	Grass Landing Area
FLR	Flares	GRVL	Gravel
FLT	Flight	GS	Ground Speed
FLTCK	Flight Check	GS	Small hail and/or snow pellets
FLUC	Fluctuating OR Fluctuation OR Fluctuated	GUND	Geoid Undulation
FLW	Follow(s) OR Following	*GVS*	*Gas Venting Site*
FLY	Fly OR Flying	**H**	
FM	From	H24	Continuous Day and Night Service
FM...	From (followed by time weather change is forecast to begin)	HAPI	Helicopter Approach Path Indicator
		HBN	Hazard Beacon
FMC	*Flight Management Computer*	HDF	High Frequency Direction-Finding Station
FMD	*Flow Management Division*	HDG	Heading
FMP	*Flow Management Position*	HEL	Helicopter
FMS	Flight Management System	*HEMS*	*Helicopter Emergency Medical Service*
FMU	Flow Management Unit	HF‡	High Frequency (3000 to 30000 kHz)
FNA	Final Approach	HGT	Height OR Height Above
FPL	Filed Flight Plan (message type designator)	*HIAL*	*Highlands and Islands Airports Ltd*
FPM	Feet Per Minute	HI	High Intensity directional lights
FPR	Flight Plan Route	*HIRTA*	*High Intensity Radio Transmission Area*
FR	Fuel Remaining	HJ	Sunrise to sunset
FREQ	Frequency	HLDG	Holding
FRI	Friday	*HMR*	*Helicopter Main Routes*
FRNG	Firing	HN	Sunset to Sunrise
FRONT†	Front (Relating to Weather)	HO	Service available to meet operational requirements
FRQ	Frequent		
FSL	Full Stop Landing	HOL	Holiday
FSS	Flight Service Station	*HOPA*	*Helicopter Operational Area*
FST	First	HOSP	Hospital Aircraft
FT	Feet (Dimensional Unit)	HPA	Hectopascal
FTT	Flight Technical Tolerance	HR/HRS	Hours
FU	Smoke	HS	Service Available During Hours of Scheduled Operations
FZ	Freezing		

79

GEN 2.2 – ABBREVIATIONS USED IN AIS PUBLICATIONS

HT	*High Tension (power)*		JTST	Jet Stream
HTA	*Helicopter Training Area*		JUL	July
HURCN	Hurricane		JUN	June
HVDF	High and Very High Frequency Direction Finding Stations (At the Same Location)		**K**	
			KG	Kilogrammes
HVY	Heavy		KHz	Kilohertz
HVY	Heavy (used to indicate the intensity of weather phenomena, eg HVY RA = Heavy rain)		KM	Kilometres
			KMH	Kilometres per Hour
			KPA	Kilopascal
HX	No Specific Working Hours		KT	Knots
HYR	Higher		KW	Kilowatts
HZ	Dust Haze		**L**	
Hz	Hertz (Cycle Per Second)		L	Left (Runway Identification)
I			L	Locator (NDB with published approach procedure. See LM, LO)
IAC	Instrument Approach Chart			
IAF	Initial Approach Fix		*LACC*	*London Area Control Centre*
IAO	In and Out of Clouds		LAM	Logical Acknowledgement (message type designator)
IAP	*Instrument Approach Procedure*			
IAR	Intersection of Air Routes		LAN	Inland
IAS	Indicated Air Speed		*LARS*	*Lower Airspace Radar Advisory Service*
IBN	Identification Beacon		LAT	Latitude
IC	Diamond Dust (very small Ice crystals in suspension)		*LATCC (Mil)*	*London Air Traffic Control Centre (Mil)*
			LDA	Landing Distance Available
ICE	Icing		LDAH	Landing Distance Available, Helicopter
ID	Identifier OR Identify		LDG	Landing
IDENT†	Identification		LDI	Landing Direction Indicator
IF	Intermediate Approach Fix		LEN	Length
IFF	Identification Friend/Foe		LF	Low Frequency (30 to 300kHz)
IFPS	*Integrated Flight Planning System*		*LFA*	*Low Flying Area*
IFR‡	Instrument Flight Rules		*LFZ*	*Low Flying Zone*
IGA	International General Aviation		LGT	Light OR Lighting
ILS‡	Instrument Landing System		LGTD	Lighted
IM	Inner marker		*LHA*	*Lowest Holding Altitude*
IMC‡	Instrument Meteorological Condition		*LHS*	*Left hand side*
IMG	Immigration		LI	Low Intensity omni-directional lights
IMPR	Improve OR Improving		LIH	Light Intensity High
IMT	Immediate OR Immediately		LIL	Light Intensity Low
INA	Initial Approach		LIM	Light Intensity Medium
INBD	Inbound		*LITAS*	*Low Intensity Two Colour Approach Slope Indicators at… and… metres from threshold bracketing approach angle of degrees*
INC	In Cloud			
INCERFA†	Uncertainty Phase			
INFO	Information		LLZ	Localizer
INOP	Inoperative		LM	Locator, Middle
INP	If Not Possible		LMT	Local Mean Time
INPR	In Progress		LNG	Long (Used to indicate the type of approach desired or required)
INS	Inertial Navigation System			
INSTL	Install OR Installed OR Installation		LO	Locator, outer
INSTR	Instrument		LOC	Local OR Locally OR Location OR Located
INT	Intersection		LONG	Longitude
INTL	International		LORAN†	(Long Range Air Navigation System)
INTRG	Interrogator		LRG	Long Range
INTRP	Interrupt OR Interruption OR Interrupted		LSQ	Line Squall
INTSF	Intensify OR Intensifying		*LTCC*	*London Terminal Control Centre*
INTST	Intensity		LTD	Limited
IR	Ice on Runway		LTT	Landline teletypewriter
IRVR	*Instrumented Runway Visual Range*		LV	Light and Variable (Relating to Wind)
ISA	International Standard Atmosphere		LVE	Leave OR Leaving
ISB	Independent Sideband		LVL	Level
ISOL	Isolated		LVP	*Low Visibility Procedures*
J			LYR	Layer OR Layered
JAN	January		**M**	
JMC	Joint Maritime Course		M	Mach Number (Followed by figures)

GEN 2.2 — ABBREVIATIONS USED IN AIS PUBLICATIONS

M	Metres (Preceded by figures)	MRG	Medium Range
MAA	Maximum Authorised Altitude	MRP	ATS/MET Reporting Point
MAG	Magnetic	*MRSA*	*Military Mandatory Radar Service Area*
MAINT	Maintenance	MS	Minus
MAP	Aeronautical Maps and charts	MSA	Minimum Sector Altitude
MAPt	Missed Approach Point	MSG	Message
MAR	At sea	MSL	Mean Sea Level
MAR	March	MT	Mountain
MAS	Manual A1 Simplex	*MTA*	*Military Training Area*
MASA	Multifunctional Transport Satellite (MTSAT) Satellite Based Augmentation System	*MTOW*	*Maximum Take-off Weight*
		MTRA	*Military Temporary Reserved Airspace*
MATZ	*Military Aerodrome Traffic Zone*	MTU	Metric Units
MAX	Maximum	MTW	Mountain Waves
MAY	May	*MTWA*	*Maximum Total Weight Authorised*
MB	*Millibars*	MVDF	Medium and Very High Frequency Direction Finding Stations (At the same location)
MCA	Minimum Crossing Altitude		
MCW	Modulated Continuous Wave	MWARA	Major World Air Route Area
MDA	Minimum Descent Altitude	MWO	Meteorological Watch Office
MDA	*Managed Danger Area*	MX	Mixed type of ice formation (white and clear)
MDF	Medium frequency Direction Finding Station		
MDH	Minimum Descent Height	**N**	
MEA	Minimum En-route Altitude	N	North OR Northern latitude
MEDA	*Military Emergency Diversion Aerodrome*	N	No distinct tendency (in RVR during previous 10 minutes)
MEHT	Minimum Eye Height over Threshold (For VASIS and PAPI)	NAT	North Atlantic
MET†	Meteorological OR Meteorology	NAV	Navigation
METAR†	Aviation routine weather report (In aeronautical meteorological code)	NB	Northbound
		NBFR	Not Before
MF	Medium Frequency (300 to 3000 kHz)	NC	No Change
MHDF	Medium and High Frequency Direction Finding Stations (At the same location)	NDB	Non-Directional Radio Beacon
		NDS	*Non-deviating Status*
MHVDF	Medium, High and Very High Frequency Direction Finding Stations (At the same location)	NE	North East
		NEB	North Eastbound
		NEG	No OR Negative OR Permission not granted OR That is not correct
MHz	Megahertz		
MID	Mid-point (related to RVR)	*NERS*	*North Atlantic European Routing System*
MIFG	Shallow fog	NGT	Night
MIL	Military	NIL*†	None OR I Have Nothing to send to you
MIN	Minutes	NM	Nautical Miles
MKR	Marker radio beacon	NML	Normal
MLS‡	Microwave Landing System	NNE	North North East
MM	Middle Marker	NNW	North North West
MNM	Minimum	NOF	International NOTAM Office
MNPS	Minimum Navigation Performance Specifications	NOSIG†	No Significant Change (Used in trend-type landing forecasts)
MNPSA	*Minimum Navigation Performance Specification Airspace*	NOTAM†	A notice containing information concerning the establishment, condition or change in any aeronautical facility, service, procedure or hazard, the timely knowledge of which is essential to personnel concerned with flight operations
MNT	Monitor OR Monitoring OR Monitored		
MNTN	Maintain		
MOA	Military Operating Area		
MOC	Minimum Obstacle Clearance (required)		
MOD	Moderate (Used to indicate the intensity of weather phenomena, interference or static reports, eg MOD RA = Moderate rain)	*NOTA*	*Northern Oceanic Transition Area*
		NOV	November
		NPR	*Noise Preferential Routeing*
MOGAS	*Motor Gasoline*	NR	Number
MON	Above Mountains	NRH	No Reply Heard
MON	Monday	NS	Nimbostratus
MOTNE	Meteorological Operational Telecommunications Network Europe	NSC	Nil Significant Cloud
		NSF	*Non Standard Flights*
MOV	Move OR Moving OR Movement	NSW	Nil Significant Weather
MPH	Statute Miles Per Hour	*NVG*	*Night Vision Goggles*
MPS	Metres Per Second	NW	North West
MRA	Minimum Reception Altitude	NWB	North Westbound

GEN 2.2 – ABBREVIATIONS USED IN AIS PUBLICATIONS

NXT	Next		PNR	Point of No Return
O			PO	Dust Devils
OAC	Oceanic Area Control Centre		POB	Persons On Board
OAS	Obstacle Assessment Surface		POSS	Possible
OAT	*Operational Air Traffic*		PPI	Plan Position Indicator
OBS	Observe OR Observed OR Observation		PPR	Prior Permission Required
OBSC	Obscure OR Obscured OR Obscuring		PPSN	Present Position
OBST	Obstacle		PRI	Primary
OCA	Obstacle Clearance Altitude		PRKG	Parking
OCA	Oceanic Control Area		PROB†	Probability
OCC	Occulting (light)		PROC	Procedure
OCH	Obstacle Clearance Height		PROV	Provisional
OCNL	Occasional OR Occasionally		PS	Plus
OCS	Obstacle Clearance Surface		PSG	Passing
OCT	October		PSN	Position
OFZ	Obstacle Free Zone		PSP	Pierced Steel Plank
OHD	Overhead		PTN	Procedure Turn
OLDI	On-Line Data Interchange		PTS	Polar Track Structure
OM	Outer Marker		PWR	Power
OPA	Opaque, white type of ice formation		Q	
OPC	The control indicated is operational control		QDM‡	Magnetic Heading (zero wind)
OPMET†	Operational Meteorological (information)		QDR	Magnetic Bearing
OPN	Open OR Opening OR Opened		QFE‡	Atmospheric pressure at aerodrome elevation (OR at runway threshold)
OPR	Operator OR Operate OR Operative OR Operating OR Operational		QFU	Magnetic orientation of runway
OPS†	Operations		QNH‡	Altimeter sub-scale setting to obtain elevation when on the ground
O/R	On Request			
ORCA	*Oceanic Route Clearance Authorisation System*		QTE	True bearing
			QUAD	Quadrant
ORD	Indication of an order		R	
OSV	Ocean Station Vessel		R	Red
OTLK	Outlook (used in SIGMET messages for volcanic ash and tropical cyclones)		R...	Restricted Area (followed by identification)
			R...	*Radial (prefix for use in graphics)*
OTP	On Top		R	Right (runway identification)
OTS	Organised Track System		RA	Rain
OUBD	Outbound		*RA*	*Resolution Advisory/Advisories (ACAS)*
OVC	Overcast		RAC	Rules of the Air and Air Traffic Services
P			RAD	Radar Approach Aid
P...	Prohibited area (Followed by identification)		*RAD*	*Radius*
PALS	Precision Approach Lighting System (Specify category)		*RAD*	*Route Availability Document*
			RAF	Royal Air Force
PANS	Procedures for Air Navigation Services		RAFC	Regional Area Forecast Centre
PAOAS	Parallel Approach Obstacle Assessment Surfaces		*RAFAT*	*Royal Air Force Aerobatic Team*
			RAFCT	*Royal Air Force Combined Training*
PAPA	*Parallax Aircraft Parking Aid*		RAG	Ragged
PAPI†	Precision Approach Path Indicator		RAG	Runway Arresting Gear
PAR‡	Precision Approach Radar		RAI	Runway Alignment Indicator
PARL	Parallel		*RAS*	*Radar Advisory Service*
PAX	Passenger(s)		*RASA*	*Radar Advisory Service Area*
PCD	Proceed OR Proceeding		RASC	Regional AIS System Centre
PCN	Pavement Classification Number		RB	Rescue boat
PDG	Procedure Design Gradient		RCA	Reach Cruising Altitude
PER	Performance		RCC	Rescue Co-ordination Centre
PERM	Permanent		RCF	Radio Communication Failure (message type designator)
PH	*Public Holiday*			
PIB	Pre-flight Information Bulletin		RCH	Reach OR Reaching
PJE	Parachute Jumping Exercise		RCL	Runway Centre Line
PLA	Practice Low Approach		RCLL	Runway Centre Line Light(s)
PLN	Flight Plan		RCLR	Recleared
PLVL	Present Level		RDH	Reference Datum Height (For ILS)
PN	Prior Notice required		RDL	Radial
PNdB	*Perceived Noise Decibels*		RDO	Radio

GEN 2.2 – ABBREVIATIONS USED IN AIS PUBLICATIONS

RE…	Recent (Used to qualify weather phenomena, eg RERA = recent rain)	RTODAH	Rejected Take-off Distance Available, Helicopter
REC	Receive OR Receiver	*RTR*	*Radar Termination Range*
REDL	Runway Edge Light(s)	RTS	Return To Service
REF	Reference to… OR Refer to…	RTT	Radio teletypewriter
REG	Registration	RTZL	Runway Touchdown Zone Light(s)
REQ	Request OR Requested	RUT	Standard Regional Route transmitting frequencies
RERTE	Re-route		
RESA	Runway End Safety Area	RV	Rescue Vessel
RET	*Rapid Exit Taxiway*	*RVA*	*Radar Vectoring Area*
RET	*Rapid Exit Taxiway Indicator Lights*	*RVP*	*Rendezvous Point*
RFF	Fire and Rescue Equipment	RVR‡	Runway Visual Range
RG	Range (lights)	RVSM	Reduced Vertical Separation Minimum
RHAG	*Rotary Hydraulic Arrester Gear*	RWY	Runway
RHS	*Right hand side*	**S**	
RIF	Reclearance In Flight	S	South OR Southern Latitude
RIS	*Radar Information Service*	SA	Sand
RITE	Right (Direction of Turn)	SALS	Simple Approach Lighting System
RL	Report Leaving	SAN	Sanitary
RLA	Relay to	SAP	As Soon as possible
RLCE	Request Level Change En-route	SAR	Search and Rescue
RLLC	*Royal Low Level Corridor*	SARPS	Standards and Recommended Practices (ICAO)
RLLS	Runway Lead-in Lighting System		
RLNA	Requested Level Not Available	*SARSAT*	*Search and Rescue Satellite Aided Tracking System*
RMK	Remark		
RN	*Royal Navy*	SAT	Saturday
RNAV†	(To be pronounced 'AR-NAV') Area Navigation	SATCOM†	Satellite Communication
		SB	Southbound
RNG	Radio Range	SC	Stratocumulus
RNHF	*Royal Navy Historical Flight*	ScACC	Scottish Area Control Centre
RNP	Required Navigation Performance	ScATCC	Scottish Area and Terminal Control Centre
ROBEX†	Regional OPMET Bulletin Exchange (Scheme)	SCT	Scattered
		SDBY	Stand by
ROC	Rate Of Climb	*SDF*	*Step Down Fix*
ROD	Rate Of Descent	SE	South East
ROFOR	Route Forecast (In aeronautical meteorological code)	SEB	South Eastbound
		SEC	Seconds
RON	Receiving Only	SECT	Sector
RPL	Repetitive flight plan	SEG	Stand Entry Guidance
RPLC	Replace OR Replaced	SELCAL†	Selective calling System
RPS	Radar Position Symbol	SEP	September
RQMNTS	Requirements	SER	Service OR Servicing OR Served
RQP	Request flight plan (message type designator)	SEV	Severe (Used eg to qualify icing and turbulence reports)
RQS	Request supplementary flight plan (message type designator)	SFC	Surface
		SG	Snow Grains
RR	Report Reaching	SGL	Signal
RRA	(OR RRB, RRC… etc, in sequence) Delayed meteorological message (message type designator)	SH…	Showers (followed by RA = Rain, SN = Snow, PL = Ice pellets, GR = Hail, GS = Small hail and/or snow pellets or combinations thereof, eg SHRASN = Showers of rain and Snow)
RSC	Rescue Sub-Centre		
RSCD	Runway Surface Condition		
RSP	Responder beacon	SHF	Super High Frequency (3000 to 30000 MHz)
RSR	En-Route Surveillance Radar		
RTD	Delayed (used to indicate delayed meteorological message; message type designator)	*SI*	*Statutory Instruments*
		SID†	Standard Instrument Departure
		SIF	Selective Identification Feature
RTE	Route	SIGMET†	Information concerning en-route weather phenomena which may affect the Safety of aircraft operations
RTF	Radiotelephone		
RTG	Radiotelegraph		
RTHL	Runway threshold light(s)	SIGWX	Significant weather
RTN	Return OR Returned OR Returning	SIMUL	Simultaneous OR Simultaneously
RTOAA	*Rejected Take-off Area Available*	SIWL	Single Isolated Wheel Load

GEN 2.2 – ABBREVIATIONS USED IN AIS PUBLICATIONS

SKC	Sky Clear		TACAN†	UHF Tactical Air Navigation Aid
SKED	Schedule OR Scheduled		TAF†	Aerodrome forecast
SLP	Speed Limiting Point		TAIL†	Tail wind
SLW	Slow		TAR	Terminal Area Surveillance Radar
SMB	*Side Marker Boards*		TAS	True Airspeed
SMC	Surface Movement Control		TAX	Taxiing OR Taxi
SMR	Surface Movement Radar		TC	Tropical Cyclone
SN	Snow		*TCAS*	*Traffic Alert and Collision Avoidance System*
SNOCLO	Aerdrome closed due to snow (used in METAR/SPECI)		TCU	Towering Cumulus
			TDA	Temporary Danger Area
SNOWTAM†	A Special Series NOTAM notifying the presence or removal of hazardous conditions due to snow, ice, Slush or Standing water associated with Snow, Slush and ice on the movement area, by means of a Specific format		TDO	Tornado
			TDZ	Touch Down Zone
			TECR	Technical Reason
			TEL	Telephone
			TEMPO†	Temporary OR Temporarily
SPECI†	Aviation Selected Special weather report (In aeronautical meteorological code)		TFC	Traffic
			TGL	Touch-and-Go Landing
SOTA	*Shannon Oceanic Transition Area*		TGS	Taxiing Guidance System
SPECIAL†	Special meteorological report (In abbreviated plain language)		THR	Threshold
			THRU	Through
SPL	Supplementary flight plan (message type designator)		THU	Thursday
			TIL†	Until
SPOT†	Spot wind		TIP	Until past… (place)
SQ	Squall		TKOF	Take-off
SR	Sunrise		TL	Till (followed by Time by which weather change is forecast to end)
SRA	Surveillance Radar Approach			
SRD	*Standard Route Document*		TLOF	Touchdown and Lift-off Area
SRE	Surveillance Radar Element of precision approach radar System		TLP	Tactical Leadership Programme
			TMA‡	Terminal Control Area
SRG	Short range		TNA	Turn Altitude
SRR	Search and Rescue Region		TNH	Turn Height
SRY	Secondary		TO	To… (place)
SS	Sandstorm		TOC	Top Of Climb
SS	Sunset		TODA	Take-off Distance Available
SSB	Single Sideband		TODAH	Take-off Distance Available, Helicopter
SSE	South South East		TOP†	Cloud Top
SSR‡	Secondary Surveillance Radar		TORA	Take-off Run Available
SST	Supersonic transport		TP	Turning Point
SSW	South South West		TR	Track
ST	Stratus		TRA	Temporary Reserved Airspace
STA	Straight in approach		*TRA*	*Temporary Restricted Area*
STAR	Standard instrument arrival		TRANS	Transmits OR Transmitter
STD	Standard		TRL	Transition Level
STF	Stratiform		TROP	Tropopause TS Thunderstorm (in aerodrome reports and forecasts TS used alone means Thunder heard but no precipitation at the aerodrome)
STN	Station			
STNR	Stationary			
STOL	Short Take-Off and Landing			
STS	Status			
STWL	Stopway light(s)		TS…	Thunderstorm (followed by RA = Rain, SN = Snow, PL = Ice pellets, GR = Hail, GS = Small hail and/or snow pellets or combinations Thereof, eg TSRASN = Thunderstorm with rain and snow)
SUBJ	Subject to			
SUN	Sunday			
SUP	Supplement (AIP Supplement)			
SUPPS	Regional Supplementary procedures			
SVC	Service message			
SVCBL	Serviceable		TT	Teletypewriter
SVFR	*Special Visual Flight Rules*		TUE	Tuesday
SW	South West		TURB	Turbulence
SWB	South Westbound		TVOR	Terminal VOR
SWY	Stopway		TWR	Aerodrome control Tower OR aerodrome control
T				
T	Temperature		TWY	Taxiway
TA	Transition Altitude		TWYL	Taxiway-Link

GEN 2.2 – ABBREVIATIONS USED IN AIS PUBLICATIONS

TYP	Type of Aircraft	VOT	VOR airborne equipment test facility
TYPH	Typhoon	VRB	Variable
U		*VRP*	*Visual Reference Point*
U	Upward (tendency in RVR during previous 10 minutes)	VSA	By visual reference to the ground
		VSP	Vertical speed
UAA	*Unusual Aerial Activity*	VSTOL	Very Short Take-Off and Landing
UAB	Until Advised By…	VTOL	Vertical Take-Off and Landing
UAC	Upper Area Control Centre	**W**	
UAR	Upper Air Route	W	West or Western longitude
UAV	*Unmanned Aerial Vehicle*	W	White
UDF	Ultra High Frequency Direction Finding Station	WAAS	Wide Area Augmentation System
		WAC	World Aeronautical Chart ICAO 1:1,000,000
UFN	Until Further Notice		
UHDT	Unable Higher Due Traffic	WAFC	World Area Forecast Centre
UHF‡	Ultra High Frequency (300 to 3000 MHz)	WB	Westbound
UIC	Upper Information Centre	WBAR	Wing bar lights
UIR‡	Upper Flight Information Region	WDI	Wind Direction Indicator
UK	*United Kingdom*	WDSPR	Widespread
ULR	Ultra Long Range	WED	Wednesday
UNA	Unable	WEF	With Effect From OR Effective From
UNAP	Unable to Approve	WGS	World Geodetic System
UNL	Unlimited	WI	Within
UNREL	Unreliable	WID	Width
U/S	Unserviceable	WIE	With Immediate Effect OR Effective Immediately
UTA	Upper Control Area		
UTC‡	Co-ordinated Universal Time	WILCO†	Will comply
V		WINTEM	Forecast upper Wind and temperature for aviation
VA	Volcanic Ash		
VAC	Visual Approach Chart	WIP	Work In Progress
VAL	In Valleys	WKN	Weaken OR Weakening
VAN	Runway Control Van	WNW	West North West
VAR	Magnetic Variation	WO	Without
VAR	Visual-aural radio range	WPT	Way-point
VASIS†	Visual Approach Slope Indicator System	WRNG	Warning
VC…	Vicinity of aerodrome (followed by FG = Fog, FC = Funnel cloud, SH = Showers, PO = Dust/sand whirls, BLDU = Blowing dust, BLSA = Blowing sand or BLSN = Blowing snow, eg VC FG = Vicinity fog)	WS	Windshear
		WSW	West South West
		WT	Weight
		WTSPT	Waterspout
		WX	Weather
VCY	Vicinity	**X**	
VDF	Very High Frequency Direction Finding Station	X	Cross
		XBAR	Crossbar (of approach lighting system)
VER	Vertical	XNG	Crossing
VFR‡	Visual Flight Rules	XS	Atmospheres
VHF‡	Very High Frequency (30 to 300 MHz)	**Y**	
VIP‡	Very Important Person	Y	Yellow
VIS	Visibility	YCZ	Yellow caution zone (runway lighting)
VLF	Very Low Frequency (3 to 30 KHz)	YR	Your
VLR	Very Long Range	**Z**	
VMC‡	Visual Meteorological Conditions	Z	Co-ordinated Universal Time (in meteorological messages)
VM(C)	*Visual Manoeuvring (Circling)*		
VOLMET†	Meteorological information for aircraft in flight		
VOR‡	Very High Frequency Omnidirectional Radio Range		
VORTAC†	VOR and TACAN combination		

85

GEN 2.3 – CAA CHART

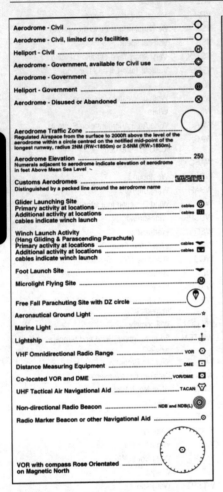

GEN 2.3 – CAA CHART

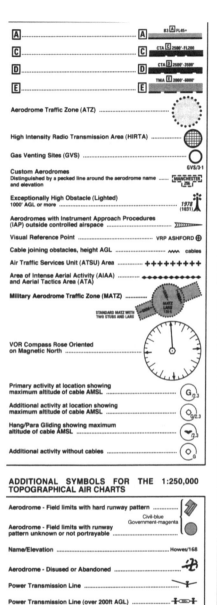

87

GEN 2.3 – CAA CHART

Symbol	Meaning	Symbol	Meaning
⌐⩘	Thunderstorm	″ ″ / ″ ″	Rain
ⓎTropical	Tropical cyclone	✶	Snow
⤨	Severe squall line	✛	Widespread blowing snow
△	Hail	▽	Shower
⌢	Moderate turbulence	ʂ	Severe sand or dust haze
⌢⤒	Severe turbulence	ʂ̧	Widespread sandstorm or duststorm
◯	Marked mountain waves	∞	Widespread haze
⋃	Light aircraft icing	═	Widespread mist
⋃⫯	Moderate aircraft icing	≡	Widespread fog
⋃⫯⫯	Severe aircraft icing	≢	Freezing fog
⌒⌣	Freezing precipitation	⌒	Widespread smoke
•	Drizzle	⏣	Volcanic eruption

Note: Altitudes between which phenomena and any associated cloud are expected are indicated by flight levels, top over base or top followed by base. 'XXX' means the phenomenon is expected to continue above and/or below the vertical coverage of the chart. Phenomena of relatively lesser significance, for example light aircraft icing or drizzle, are not usually shown on charts even when the phenomenon is expected. The thunderstorm symbol implies hail, moderate or severe icing and/or turbulence.

Symbol	Meaning
400	Tropopause spot altitude (eg FL400)
H 440	High point or maximum in tropopause topography (eg FL440)
340 L	Low point or minimum in tropopause topography (eg FL340)
0°:100	Freezing level

Symbol	Meaning
〰〰〰	Boundary of area of significant weather
– – – –	Boundary of area of clear air turbulence. The CAT area may be marked by a numeral inside a square and a legend describing the numbered CAT area may be entered in the margin
🚩 10	State of sea (wave height in metres)
ⓘ 18	Sea surface temperature (°C)

2 Fronts and Convergence Zones

Symbol	Meaning	Symbol	Meaning
▲▲▲	Cold front at the surface	―――	Axis of trough
●●●	Warm front at the surface	∿∿∿	Axis of ridge
▲●▲●	Occluded front at the surface	→⤳→	Convergence line
●▲●▲	Quasi-stationary front at the surface	⫴▭	Inter-tropical convergence zone

Note: An arrow with associated figures indicates the direction and the speed of the movement of the front (knots). Dots inserted at intervals along the line of a front indicate it is a developing feature (frontogenesis), while bars indicate it is a weakening feature (frontolysis).

GEN 2.3

GEN 2.3 – CAA CHART

3 Cloud Abbreviations

3.1 Type

CI = Cirrus
CC = Cirrocumulus
CS = Cirrostratus
AC = Altocumulus
AS = Altostratus
NS = Nimbostratus
SC = Stratocumulus
ST = Strratus
CU = Cumulus
CB = Cumulonimbus (its insertion implies hail moderate or severe icing and/or turbulence)
LYR = Layer or layered (instead of the cloud type)

3.2 Amount

Clouds except CB
SKC = clear (0 okta)
FEW = few (1/8 or 2/8)
SCT = scattered (3/8 or 4/8)
BKN = broken (5/8 to 7/8)
OVC = overcast (8/8)
CB only
ISOL = individual CB's (isolated)
OCNL = well separated CB's (occasional)
FRQ = CB's with little or no separation (frequent)
EMBD = thunderstorm clouds contained in layers of other clouds (embedded).

4 Example Weather Abbreviations

RA = rain
DZ = drizzle
SN = snow
SH = showers
FZ = freezing
TS = thunderstorms
Other phenomena may be expressed as a combination of abbreviations or written in full. TS implies severe turbulence and icing.

5 Wind Symbols

5.1 Wind/Temperature Chart

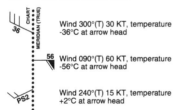

Wind 300°(T) 30 KT, temperature -36°C at arrow head

Wind 090°(T) 60 KT, temperature -56°C at arrow head

Wind 240°(T) 15 KT, temperature +2°C at arrow head

5.2 Significant Weather/Tropopause/ Maximum Wind Chart

FL380
Maximum wind 270°(T), 110 KT at FL380

A double bar marks a speed change of 20 KT, and/or height change of 3000 ft

GEN 2.3

2.3.5 STANDARD INSTRUMENT DEPARTURE (SID) AND ARRIVAL ROUTES (STAR)

2.3.5 STANDARD INSTRUMENT DEPARTURE (SID) AND ARRIVAL ROUTES (STAR)

1 SID procedure charts are located in AD 2. They consist of a textual description of the procedure, a graphical illustration and explanatory notes. Only aeronautical information pertinent to the procedure is shown and these charts should therefore be used together with a suitable En-route chart which gives details of Airspace Reservations, Controlled Airspace and ATS routes.

2 SID charts are arranged by Main Exit Points: the various runway directions which can be used to the relevant Main Exit Point will be found in one chart.

3 The procedure charts ARE NOT DRAWN TO SCALE. Unless otherwise indicated:

(a) Distances are in nautical miles;

(b) Headings, Bearings, Tracks and Radials are in degrees magnetic;

(c) Heights/Altitudes where stated are based on QFE/QNH;

(d) Horizontal datum WGS 84 (CO-ORDS IN DEG MIN SEC).

4 Area Minimum Altitude

The lowest altitude which will provide a minimum obstacle clearance of 1000 feet in a sector of a circle of 25NM (46KM) radius centred on the departure Aerodrome Reference Point. For all SID procedures within the United Kingdom FIRs, Area Minimum Altitudes will be shown for each quadrant of a circle of radius 25nm (46nm) centred on the departure Aerodrome Reference Point, with True North as the quadrant datum. Area Minimum Altitude is given in hundreds of feet AMSL. eg 23 = 2300'.

5 Net Climb Gradient

The climb gradient, expressed as a percentage, that the aircraft is required to achieve to meet standard (ICAO PANS-OPS) obstacle clearance requirements, will be detailed in the textual description of the SID procedure when the required gradient is greater than 3.3% to be achieved. Procedure design gradients are annotated on charts as necessary. A table for conversion of percentage climb gradients to rates of climb for various speeds is given in the GEN 2.8 section.

6 Arrival Charts

STARs or established inbound routes are shown in a similar fashion to SIDs. Tracks terminate at the main inbound holding point from which the Instrument Approach commences.

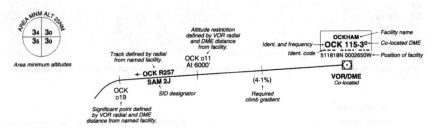

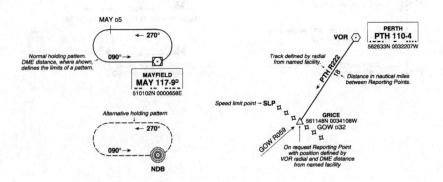

GEN 2.4 – LOCATION INDICATORS ENCODE

Location	Indicator	Location	Indicator
Aberdeen	EGPD	Cranfield	EGTC
Aberporth	EGUC	Cranwell	EGYD
Albourne	EGKD	Croughton (MOD)	EGWR
Alderney	EGJA	Crowfield	EGSO
ALFENS (Mobile) (MOD)	EGDF	Culdrose	EGDR
Andrewsfield	EGSL	Cumbernauld	EGPG
Ascot Racecourse	EGLT	Deanland	EGKL
Aylesbury/Thame	EGTA	Denham	EGLD
Bagby	EGNG	Derby	EGBD
Ballykelly	EGQB	Dishforth	EGXD
Barkston Heath	EGYE	Doncaster Sheffield	EGCN
Barra	EGPR	Donna Nook (MOD)	EGXS
Barrow/Walney Island	EGNL	Dundee	EGPN
Beccles	EGSM	Dunkeswell	EGTU
Bedford/Castle Mill	EGSB	Dunsfold	EGTD
Belfast Aldergrove	EGAA	Durham Tees Valley	EGNV
Belfast City	EGAC	Duxford	EGSU
Bembridge	EGHJ	Eaglescott	EGHU
Benbecula	EGPL	Earls Colne	EGSR
Benbecula (MOD)	EGXM	Eday	EGED
Benson	EGUB	Edinburgh	EGPH
Bentley Priory (MOD)	EGWS	Elmsett	EGST
Beverley/Linley Hill	EGNY	Elstree	EGTR
Bicester	EGDD	Enniskillen/St Angelo	EGAB
Biggin Hill	EGKB	Exeter	EGTE
Birmingham	EGBB	Fairford	EGVA
Blackbushe	EGLK	Fair Isle	EGEF
Blackpool	EGNH	Fairoaks	EGTF
Bodmin	EGLA	Farnborough	EGLF
Boscombe Down	EGDM	Farthing Corner	EGMF
Boulmer (MOD)	EGQM	Faslane (MOD)	EGXO
Bourn	EGSN	Fenland	EGCL
Bournemouth	EGHH	Fife	EGPJ
Brampton (MOD)	EGYB	Forest Moor	EGXF
Bristol	EGGD	Fowlmere	EGMA
Bristol Filton	EGTG	Full Sutton	EGNU
Bristol Weather Centre	EGRD	Garvie Island (MOD)	EGQC
Brize Norton	EGVN	Gatwick CAA SRG (Aviation House)	EGGR
Brooklands	EGLB	Glasgow	EGPF
Brough	EGNB	Glasgow City Heliport	EGEG
Buchan (MOD)	EGQN	Glasgow Prestwick	EGPK
Caernarfon	EGCK	Glasgow Weather Centre	EGRA
Cambridge	EGSC	Gloucestershire	EGBJ
Campbeltown	EGEC	Goodwood Racecourse	EGKG
Cardiff	EGFF	Great Yarmouth/North Denes	EGSD
Cardiff/Tremorfa Foreshore Heliport	EGFC	Guernsey	EGJB
Carlisle	EGNC	Halley Research Station	EGAH
Chalgrove	EGLJ	Halton	EGWN
Challock	EGKE	Haverfordwest	EGFE
Cheltenham Racecourse	EGBC	Hawarden	EGNR
Chichester/Goodwood	EGHR	Henlow	EGWE
Chivenor	EGDC	Henstridge	EGHS
Church Fenton	EGXG	Hereford	EGVH
Clacton	EGSQ	Hethel	EGSK
Colerne	EGUO	High Wycombe (MOD)	EGUH
Coltishall	EGYC	High Wycombe (Met)	EGRH
Compton Abbas	EGHA	HM Ships (All)	EGYY
Coningsby	EGXC	Holbeach (MOD)	EGYH
Cosford	EGWC	Holyhead	EGCH
Cottesmore	EGXJ	Honington	EGXH
Coventry	EGBE	HQ 2 Group (MOD)	EGDH

91

GEN 2.4 – LOCATION INDICATORS ENCODE

Location	Indicator	Location	Indicator
HQ NORIRELAND (MOD)	EGOG	MOD UK Air	EGWB
HQ SPTA	EGDS	MOD UK Navy	EGWI
Hucknall	EGNA	Mount Pleasant	EGYP
Humberside	EGNJ	Mountwise	EGDB
Inverness	EGPE	Neatishead (MOD)	EGUJ
Islay	EGPI	Netheravon	EGDN
Isle of Man	EGNS	Netherthorpe	EGNF
Isle of Wight/Sandown	EGHN	Newcastle	EGNT
Isleworth	EGLI	Newmarket Racecourse	EGSW
Jersey	EGJJ	Newtownards	EGAD
Kemble	EGBP	Northampton/Sywell	EGBK
Kinloss	EGQK	Northolt (AIDU)	EGVC
Kirkwall	EGPA	Northolt	EGWU
Lakenheath	EGUL	Northolt (RN NAIC)	EGXA
Land's End/St Just	EGHC	North Ronaldsay	EGEN
Langford Lodge	EGAL	North Weald	EGSX
Lasham	EGHL	Northwood (MOD)	EGWX
Lashenden/Headcorn	EGKH	Norwich	EGSH
Leconfield	EGXV	Norwich Weather Centre	EGRN
Leeds Bradford	EGNM	Nottingham	EGBN
Leeds Weather Centre	EGRY	Nottingham East Midlands	EGNX
Lee-on-Solent	EGHF	Oaksey Park	EGTW
Leeming	EGXE	Oban	EGEO
Leicester	EGBG	Odiham	EGVO
Lerwick/Tingwall	EGET	Old Buckenham	EGSV
Leuchars	EGQL	Old Sarum	EGLS
Linton-on-Ouse	EGXU	Oxford	EGTK
Lisburn	EGQD	Panshanger	EGLG
Little Gransden	EGMJ	Papa Westray	EGEP
Liverpool	EGGP	Pembrey (AD)	EGFP
Llanbedr	EGOD	Pembrey (MOD)	EGOP
London ACC (Civil)	EGTT	Penzance Heliport	EGHK
London Area	EGGO	Perranporth	EGTP
London City	EGLC	Perth/Scone	EGPT
London (CAA HQ)	EGGA	Peterborough Conington	EGSF
London (DTLR Aviation Directorate)	EGGC	Peterborough Sibson	EGSP
London Gatwick	EGKK	Peterhead/Longside Heliport	EGPS
London Heathrow	EGLL	Plymouth (Mil)	EGVE
London Heliport	EGLW	Plymouth	EGHD
London Luton	EGGW	Popham	EGHP
London Stansted	EGSS	Portland	EGDP
Londonderry/Eglinton	EGAE	Portsmouth/Fleetlands	EGVF
Long Marston	EGBL	RAF Mobiles	EGWW
Lossiemouth	EGQS	Redhill	EGKR
Lydd	EGMD	Retford/Gamston	EGNE
Lyneham	EGDL	Rochester	EGTO
Manchester	EGCC	Rothera Research Station	EGAR
Manchester Barton	EGCB	St Athan	EGDX
Manchester Woodford	EGCD	St Mawgan	EGDG
Manston	EGMH	Sanday	EGES
Marham	EGYM	Sandtoft	EGCF
Marshland Wisbech	EGSI	Scampton	EGXP
Maypole	EGHB	Scatsta	EGPM
MET Office Aberdeen	EGRQ	Scilly Isles/St Mary's	EGHE
MET Office Birmingham	EGRO	Scottish ACC (Civil)	EGPX
MET Office Cardiff	EGRG	Scottish ACC (Mil)	EGQQ
MET Office Exeter	EGRR	Seething	EGSJ
MET Office London	EGRB	Shanwick OACC	EGGX
MET Office Manchester	EGRC	Shawbury	EGOS
Middle Wallop	EGVP	Sheffield City	EGSY
Mildenhall	EGUN	Sherburn-in-Elmet	EGCJ

GEN 2.4 – LOCATION INDICATORS DECODE

Location	Indicator
Shipdham	EGSA
Shobdon	EGBS
Shoreham	EGKA
Silverstone	EGBV
Skegness	EGNI
Sleap	EGCV
Southampton	EGHI
Southampton Weather Centre	EGRI
Southend	EGMC
Southport Birkdale Sands	EGCO
Spadeadam (MOD)	EGOM
Stapleford	EGSG
Stornoway	EGPO
Stronsay	EGER
Strubby Heliport	EGCG
Sturgate	EGCS
Sumburgh	EGPB
Swansea	EGFH
Swanwick ATCC (Mil)	EGVV
Tain (MOD)	EGQA
Tatenhill	EGBM
Ternhill	EGOE
Thorne	EGCP
Thruxton	EGHO
Tilstock	EGCT
Tiree	EGPU
Topcliffe	EGXZ
Tresco Heliport	EGHT
Truro	EGHY
Turweston	EGBT
UK AFTN/CIDIN Centre	EGGG
UK AIRPROX BOARD	EGGF
UK MCC	EGQP
UK MOTNE Centre	EGGY
UK NOTAM Office (NOF)	EGGN
Unst	EGPW
Uxbridge	EGUU
Valley	EGOV
Waddington	EGXW
Wainfleet (MOD)	EGYW
Warton	EGNO
Wattisham	EGUW
Wellesbourne Mountford	EGBW
Welshpool	EGCW
West Drayton ATCC (Mil)	EGWD
West Wales/Aberporth (AD)	EGFA
West Freugh	EGOY
Westray	EGEW
Whalsay	EGEH
White Waltham	EGLM
Wick	EGPC
Wickenby	EGNW
Wittering	EGXT
Wolverhampton	EGBO
Woodvale	EGOW
Wrexham/Boras	EGCE
Wycombe Air Park/Booker	EGTB
Wyton	EGUY
Yeovil	EGHG
Yeovilton	EGDY

GEN 2.4 – LOCATION INDICATORS DECODE

Indicator	Location
EGAA	Belfast Aldergrove
EGAB	Enniskillen/St Angelo
EGAC	Belfast City
EGAD	Newtownards
EGAE	Londonderry/Eglinton
EGAH	Halley Research Station
EGAL	Langford Lodge
EGAR	Rothera Research Station
EGBB	Birmingham
EGBC	Cheltenham Racecourse
EGBD	Derby
EGBE	Coventry
EGBG	Leicester
EGBJ	Gloucestershire
EGBK	Northampton/Sywell
EGBL	Long Marston
EGBM	Tatenhill
EGBN	Nottingham
EGBO	Wolverhampton
EGBP	Kemble
EGBS	Shobdon
EGBT	Turweston
EGBV	Silverstone
EGBW	Wellesbourne Mountford
EGCB	Manchester Barton
EGCC	Manchester
EGCD	Manchester Woodford
EGCE	Wrexham/Boras
EGCF	Sandtoft
EGCG	Strubby Heliport
EGCH	Holyhead
EGCJ	Sherburn-in-Elmet
EGCK	Caernarfon
EGCL	Fenland
EGCO	Southport Birkdale Sands
EGCP	Thorne
EGCS	Sturgate
EGCT	Tilstock
EGCV	Sleap
EGCW	Welshpool
EGDB	Mountwise
EGDC	Chivenor
EGDD	Bicester
EGDF	ALFENS (Mobile)
EGDG	St Mawgan
EGDH	HQ 2 Group (MOD)
EGDL	Lyneham
EGDM	Boscombe Down
EGDN	Netheravon
EGDP	Portland
EGDR	Culdrose
EGDS	HQ SPTA
EGDX	St Athan
EGDT	Yeovilton
EGEC	Campbeltown
EGED	Eday
EGEF	Fair Isle
EGEG	Glasgow City Heliport
EGEH	Whalsay

GEN 2.4 — LOCATION INDICATORS DECODE

Indicator	Location	Indicator	Location
EGEN	North Ronaldsay	EGLG	Panshanger
EGEO	Oban	EGLI	Isleworth
EGEP	Papa Westray	EGLJ	Chalgrove
EGER	Stronsay	EGLK	Blackbushe
EGES	Sanday	EGLL	London Heathrow
EGET	Lerwick/Tingwall	EGLM	White Waltham
EGEW	Westray	EGLS	Old Sarum
EGFA	West Wales/Aberporth	EGLT	Ascot Racecourse
EGFC	Cardiff/Tremorfa Foreshore Heliport	EGLW	London Heliport
EGFE	Haverfordwest	EGMA	Fowlmere
EGFF	Cardiff	EGMC	Southend
EGFH	Swansea	EGMD	Lydd
EGFP	Pembrey (AD)	EGMF	Farthing Corner
EGGA	London (CAA HQ)	EGMH	Manston (Civil)
EGGC	London (DFT Aviation Directorate)	EGMJ	Little Gransden
EGGD	Bristol	EGNA	Hucknall
EGGF	UK AIRPROX BOARD	EGNB	Brough
EGGG	UK AFTN/CIDIN Centre	EGNC	Carlisle
EGGN	UK NOTAM Office (NOF)	EGNE	Retford/Gamston
EGGO	London Area	EGNF	Netherthorpe
EGGP	Liverpool	EGNG	Bagby
EGGR	Gatwick CAA SRG (Aviation House)	EGNH	Blackpool
EGGW	London Luton	EGNI	Skegness
EGGX	Shanwick OACC	EGNJ	Humberside
EGGY	UK MOTNE Centre	EGNL	Barrow/Walney Island
EGHA	Compton Abbas	EGNM	Leeds Bradford
EGHB	Maypole	EGNO	Warton
EGHC	Land's End/St Just	EGNR	Hawarden
EGHD	Plymouth	EGNS	Isle of Man
EGHE	Scilly Isles/St Mary's	EGNT	Newcastle
EGHF	Lee-on-Solent	EGNU	Full Sutton
EGHG	Yeovil	EGNV	Durham Tees Valley
EGHH	Bournemouth	EGNW	Wickenby
EGHI	Southampton	EGNX	Nottingham East Midlands
EGHJ	Bembridge	EGNY	Beverley/Linley Hill
EGHK	Penzance Heliport	EGOD	Llanbedr
EGHL	Lasham	EGOE	Ternhill
EGHN	Isle of Wight/Sandown	EGOG	HQ NORIRELAND (MOD)
EGHO	Thruxton	EGOM	Spadeadam (MOD)
EGHP	Popham	EGOP	Pembrey (MOD)
EGHR	Chichester/Goodwood	EGOS	Shawbury
EGHS	Henstridge	EGOV	Valley
EGHT	Tresco Heliport	EGOW	Woodvale
EGHU	Eaglescott	EGOY	West Freugh
EGHY	Truro	EGPA	Kirkwall
EGJA	Alderney	EGPB	Sumburgh
EGJB	Guernsey	EGPC	Wick
EGJJ	Jersey	EGPD	Aberdeen
EGKA	Shoreham	EGPE	Inverness
EGKB	Biggin Hill	EGPF	Glasgow
EGKD	Albourne	EGPG	Cumbernauld
EGKE	Challock	EGPH	Edinburgh
EGKG	Goodwood Racecourse	EGPI	Islay
EGKH	Lashenden/Headcorn	EGPJ	Fife
EGKK	London Gatwick	EGPK	Glasgow Prestwick
EGKL	Deanland	EGPL	Benbecula
EGKR	Redhill	EGPM	Scatsta
EGLA	Bodmin	EGPN	Dundee
EGLB	Brooklands	EGPO	Stornoway
EGLC	London City	EGPR	Barra
EGLD	Denham	EGPS	Peterhead/Longside Heliport
EGLF	Farnborough	EGPT	Perth/Scone

GEN 2.4 — LOCATION INDICATORS DECODE

Indicator	Location
EGPU	Tiree
EGPW	Unst
EGPX	Scottish ACC (Civil)
EGQA	Tain (MOD)
EGQB	Ballykelly
EGQD	Lisburn
EGQK	Kinloss
EGQL	Leuchars
EGQM	Boulmer (MOD)
EGQN	Buchan (MOD)
EGQP	UK MCC
EGQQ	Scottish ACC (Mil)
EGQS	Lossiemouth
EGRA	Glasgow Weather Centre
EGRB	MET Office London
EGRC	MET Office Manchester
EGRD	MET Office Bristol
EGRG	MET Office Cardiff
EGRH	High Wycombe (Met)
EGRI	Southampton Weather Centre
EGRN	Norwich Weather Centre
EGRO	MET Office Birmingham
EGRQ	MET Office Aberdeen
EGRR	MET Office Exeter
EGRY	Leeds Weather Centre
EGSA	Shipdham
EGSB	Bedford/Castle Mill
EGSC	Cambridge
EGSD	Great Yarmouth/North Denes
EGSF	Peterborough Conington
EGSG	Stapleford
EGSH	Norwich
EGSI	Marshland, Wisbech
EGSJ	Seething
EGSK	Hethel
EGSL	Andrewsfield
EGSM	Beccles
EGSN	Bourn
EGSO	Crowfield
EGSP	Peterborough Sibson
EGSQ	Clacton
EGSR	Earls Colne
EGSS	London Stansted
EGST	Elmsett
EGSU	Duxford
EGSV	Old Buckenham
EGSW	Newmarket Racecourse
EGSX	North Weald
EGSY	Sheffield City
EGTA	Aylesbury/Thame
EGTB	Wycombe Air Park/Booker
EGTC	Cranfield
EGTD	Dunsfold
EGTE	Exeter
EGTF	Fairoaks
EGTG	Bristol Filton
EGTH	Old Warden
EGTK	Oxford
EGTO	Rochester
EGTP	Perranporth
EGTR	Elstree
EGTT	London ACC (Civil)

Indicator	Location
EGTU	Dunkeswell
EGTW	Oaksey Park
EGUB	Benson
EGUC	Aberporth
EGUH	High Wycombe (MOD)
EGUJ	Neatishead (MOD)
EGUL	Lakenheath
EGUN	Mildenhall
EGUO	Colerne
EGUU	Uxbridge
EGUW	Wattisham
EGUY	Wyton
EGVA	Fairford
EGVC	Northolt (AIDU)
EGVE	Plymouth (Mil)
EGVF	Portsmouth/Fleetlands
EGVH	Hereford
EGVN	Brize Norton
EGVO	Odiham
EGVP	Middle Wallop
EGVV	Swanwick ATCC (Mil)
EGWB	MOD UK Air
EGWC	Cosford
EGWD	West Drayton ATCC (Mil)
EGWE	Henlow
EGWI	MOD UK Navy
EGWN	Halton
EGWR	Croughton (MOD)
EGWS	Bentley Priory (MOD)
EGWU	Northolt
EGWW	RAF Mobiles
EGWX	Northwood (MOD)
EGXA	Northolt (RN NAIC)
EGXC	Coningsby
EGXD	Dishforth
EGXE	Leeming
EGXF	Forest Moor
EGXG	Church Fenton
EGXH	Honington
EGXJ	Cottesmore
EGXM	Benbecula (MOD)
EGXO	Faslane (MOD)
EGXP	Scampton
EGXS	Donna Nook (MOD)
EGXT	Wittering
EGXU	Linton-on-Ouse
EGXV	Leconfield
EGXW	Waddington
EGXZ	Topcliffe
EGYB	Brampton (MOD)
EGYC	Coltishall
EGYD	Cranwell
EGYE	Barkston Heath
EGYH	Holbeach (MOD)
EGYM	Marham
EGYP	Mount Pleasant
EGYW	Wainfleet (MOD)
EGYY	HM Ships (All)

GEN 2.5 – LIST OF RADIO NAVIGATION AIDS ENCODE

Station Name	Facility	IDENT	Purpose	Station Name	Facility	IDENT	Purpose
Aberdeen	VOR	AND	AE	Bristol Filton	ILS	I BRF	A
Aberdeen	DME	AND	AE	Bristol Filton	DME	I BRF	A
Aberdeen	NDB	AQ	A	Bristol Filton	ILS	I FB	A
Aberdeen/Dyce	L	ATF	A	Bristol Filton	DME	I FB	A
Aberdeen/Dyce	ILS	I ABD	A	Bristol Filton	L	OF	A
Aberdeen/Dyce	DME	I ABD	A	Brookmans Park	VOR	BPK	E
Aberdeen/Dyce	ILS	I AX	A	Brookmans Park	DME	BPK	E
Aberdeen/Dyce	DME	I AX	A	Brough	NDB	BV	A
Alderney	L	ALD	A	Burnham	NDB	BUR	E
Barkway	VOR	BKY	E	Caernarfon	NDB	CAE	A
Barkway	DME	BKY	E	Cambridge	DME	I CMG	A
Barra	NDB	BRR	A	Cambridge	ILS	I CMG	A
Belfast	VOR	BEL	AE	Cambridge	L	CAM	A
Belfast	DME	BEL	AE	Campbeltown	NDB	CBL	A
Belfast/Aldergrove	ILS	I AG	A	Cardiff	L	CDF	A
Belfast/Aldergrove	DME	I AG	A	Cardiff	DME	I CDF	A
Belfast/Aldergrove	ILS	I FT	A	Cardiff	DME	I CWA	A
Belfast/Aldergrove	DME	I FT	A	Cardiff	ILS	I CDF	A
Belfast/Aldergrove	L	OY	A	Cardiff	ILS	I CWA	A
Belfast/City	L	HB	A	Carlisle	L	CL	A
Belfast/City	ILS	HBD	A	Carlisle	DME	CO	A
Belfast/City	DME	HBD	A	Carnane	NDB	CAR	A
Belfast/City	ILS	I BFH	A	Chiltern	NDB	CHT	E
Belfast/City	DME	I BFH	A	Clacton	VOR	CLN	E
Bembridge	NDB	IW	A	Clacton	DME	CLN	E
Benbecula	L	BRA	A	Compton	VOR	CPT	E
Benbecula	DME	BCL	A	Compton	DME	CPT	E
Benbecula	VOR	BEN	AE	Compton Abbas	NDB	COM	A
Benbecula	DME	BEN	AE	Coventry	L	CT	A
Berry Head	VOR	BHD	E	Coventry	ILS	I CT	A
Berry Head	DME	BHD	E	Coventry	DME	I CT	A
Biggin	VOR	BIG	AE	Coventry	ILS	I CTY	A
Biggin	DME	BIG	AE	Coventry	DME	I CTY	A
Biggin Hill	ILS	I BGH	A	Cranfield	VOR	CFD	AE
Biggin Hill	DME	I BGH	A	Cranfield	L	CIT	A
Birmingham	L	BHX	A	Cranfield	ILS	I CR	A
Birmingham	ILS	I BIR	A	Cumbernauld	NDB	CBN	A
Birmingham	DME	I BIR	A	Cumbernauld	DME	CBN	A
Birmingham	ILS	I BM	A	Daventry	VOR	DTY	E
Birmingham	DME	I BM	A	Daventry	DME	DTY	E
Blackbushe	DME	BLC	A	Dean Cross	VOR	DCS	E
Blackbushe	NDB	BLK	A	Dean Cross	DME	DCS	E
Blackpool	L	BPL	A	Detling	VOR	DET	E
Blackpool	ILS	I BPL	A	Detling	DME	DET	E
Blackpool	DME	I BPL	A	Doncaster Sheffield	DME	I FNL	A
Bourn	NDB	BOU	A	Doncaster Sheffield	ILS	I FNL	A
Bournemouth	L	BIA	A	Doncaster Sheffield	L	FNY	A
Bournemouth	ILS	I BH	A	Dover	VOR	DVR	E
Bournemouth	DME	I BH	A	Dover	DME	DVR	E
Bournemouth	ILS	I BMH	A	Dundee	ILS	I DDE	A
Bournemouth	DME	I BMH	A	Dundee	DME	I DDE	A
Bovingdon	VOR	BNN	E	Dundee	L	DND	A
Bovingdon	DME	BNN	E	Durham Tees Valley	ILS	I TD	A
Brecon	VOR	BCN	E	Durham Tees Valley	DME	I TD	A
Brecon	DME	BCN	E	Durham Tees Valley	ILS	I TSE	A
Bristol	L	BRI	A	Durham Tees Valley	DME	I TSE	A
Bristol	ILS	I BON	A	Durham Tees Valley	L	TD	A
Bristol	DME	I BON	A	Edinburgh	L	EDN	A
Bristol	ILS	I BTS	A	Edinburgh	ILS	I TH	A
Bristol	DME	I BTS	A	Edinburgh	DME	I TH	A

GEN 2..5 — LIST OF RADIO NAVIGATION AIDS ENCODE

Station Name	Facility	IDENT	Purpose	Station Name	Facility	IDENT	Purpose
Edinburgh	ILS	I VG	A	Islay	NDB	LAY	A
Edinburgh	DME	I VG	A	Isle of Man	VOR	IOM	AE
Edinburgh	L	UW	A	Isle of Man	DME	IOM	AE
Enniskillen/St Angelo	DME	ENN	A	Isle of Man	ILS	I RH	A
Enniskillen/St Angelo	NDB	EKN	A	Isle of Man	DME	I RH	A
Epsom	NDB	EPM	E	Isle of Man	ILS	I RY	A
Exeter	L	EX	A	Isle of Man	DME	I RY	A
Exeter	ILS	I ET	A	Jersey	ILS	I DD	A
Exeter	DME	I ET	A	Jersey	DME	I DD	A
Exeter	ILS	I XR	A	Jersey	ILS	I JJ	A
Exeter	DME	I XR	A	Jersey	DME	I JJ	A
Fairoaks	NDB	FOS	A	Jersey	VOR	JSY	AE
Fairoaks	DME	FRK	A	Jersey	DME	JSY	AE
Farnborough	LLZ	I FNB	A	Jersey	L	JW	A
Farnborough	DME	I FNB	A	Kirkwall	ILS	I KIR	A
Farnborough	LLZ	I FRG	A	Kirkwall	DME	I KIR	A
Farnborough	DME	I FRG	A	Kirkwall	ILS	I ORK	A
Fenland	NDB	FNL	A	Kirkwall	DME	I ORK	A
Gamston	VOR	GAM	E	Kirkwall	L	KW	A
Gamston	DME	GAM	E	Kirkwall	VOR	KWL	A
Glasgow	L	GLW	A	Kirkwall	DME	KWL	A
Glasgow	VOR	GOW	AE	Lambourne	VOR	LAM	E
Glasgow	DME	GOW	AE	Lambourne	DME	LAM	E
Glasgow	ILS	I OO	A	Lands End	VOR	LND	E
Glasgow	DME	I OO	A	Lands End	DME	LND	E
Glasgow	ILS	I UU	A	Lashenden/Headcorn	NDB	LSH	A
Glasgow	DME	I UU	A	Lashenden/Headcorn	DME	HLS	A
Gloucestershire	DME	GOS	A	Leeds Bradford	ILS	I LBF	A
Gloucestershire	L	GST	A	Leeds Bradford	NDB	LBA	AE
Goodwood	VOR	GWC	AE	Leeds Bradford	DME	I LF	A
Goodwood	DME	GWC	AE	Leeds Bradford	ILS	I LF	A
Great Yarmouth/N Denes	L	ND	A	Leeds Bradford	NDB	LBA	AE
Guernsey	L	GUY	A	Leicester	NDB	LE	A
Guernsey	VOR	GUR	AE	Lerwick/Tingwall	NDB	TL	A
Guernsey	DME	GUR	AE	Lichfield	NDB	LIC	E
Guernsey	DME	I GH	A	Liverpool	ILS	I LQ	A
Guernsey	DME	I UY	A	Liverpool	DME	I LQ	A
Guernsey	ILS	I GH	A	Liverpool	L	LPL	A
Guernsey	ILS	I UY	A	Liverpool	ILS	I LVR	A
Haverfordwest	NDB	HAV	A	Liverpool	DME	I LVR	A
Haverfordwest	DME	HDW	A	London	VOR	LON	E
Hawarden	L	HAW	A	London	DME	LON	E
Hawarden	ILS	I HDN	A	London Gatwick	NDB	GE	A
Hawarden	DME	I HDN	A	London Gatwick	L	GY	A
Hawarden	ILS	I HWD	A	London Gatwick	ILS	I GG	A
Hawarden	DME	I HWD	A	London Gatwick	DME	I GG	A
Henton	NDB	HEN	E	London Gatwick	ILS	I WW	A
Honiley	VOR	HON	E	London Gatwick	DME	I WW	A
Honiley	DME	HON	E	London Heathrow	ILS	I AA	A
Humberside	ILS	I HS	A	London Heathrow	DME	I AA	A
Humberside	DME	I HS	A	London Heathrow	ILS	I BB	A
Humberside	L	KIM	A	London Heathrow	DME	I BB	A
Inverness	ILS	I DX	A	London Heathrow	ILS	I LL	A
Inverness	DME	I DX	A	London Heathrow	DME	I LL	A
Inverness	ILS	I LN	A	London Heathrow	ILS	I RR	A
Inverness	DME	I LN	A	London Heathrow	DME	I RR	A
Inverness	NDB	IVR	A	London Heathrow	MLS	M HER	A
Inverness	VOR	INS	AE	London Luton	ILS	I LJ	A
Inverness	DME	INS	AE	London Luton	DME	I LJ	A
Islay	DME	ISY	A	London Luton	ILS	I LTN	A

GEN 2.5 – LIST OF RADIO NAVIGATION AIDS ENCODE

Station Name	Facility	IDENT	Purpose	Station Name	Facility	IDENT	Purpose
London Luton	DME	I LTN	A	Nottingham East Midlands	ILS	I EME	A
London Luton	L	LUT	A	Nottingham East Midlands	DME	I EME	A
London Stansted	ILS	I SED	A	Nottingham East Midlands	ILS	I EMW	A
London Stansted	DME	I SED	A	Nottingham East Midlands	DME	I EMW	A
London Stansted	ILS	I SX	A	Ockham	VOR	OCK	E
London Stansted	DME	I SX	A	Ockham	DME	OCK	E
London Stansted	NDB	SSD	A	Ottringham	VOR	OTR	E
London City	NDB	LCY	A	Ottringham	DME	OTR	E
London City	ILS	LSR	A	Oxford/Kidlington	L	OX	A
London City	DME	LSR	A	Oxford/Kidlington	DME	OX	A
London City	ILS	LST	A	Pembrey	L	PMB	A
London City	DME	LST	A	Penzance	NDB	PH	A
Londonderry/Eglinton	L	EGT	A	Perth	VOR	PTH	E
Londonderry/Eglinton	ILS	I EGT	A	Plymouth	ILS	I PLY	A
Londonderry/Eglinton	DME	I EGT	A	Plymouth	DME	I PLY	A
Lydd	DME	LDY	A	Plymouth	L	PY	A
Lydd	VOR	LYD	E	Pole Hill	VOR	POL	E
Lydd	DME	LYD	E	Pole Hill	DME	POL	E
Lydd	NDB	LYX	A	Prestwick	ILS	I KK	A
Machrihanish	VOR	MAC	AE	Prestwick	DME	I KK	A
Machrihanish	DME	MAC	AE	Prestwick	ILS	I PP	A
Manchester	ILS	I MM	A	Prestwick	DME	I PP	A
Manchester	DME	I MM	A	Prestwick	NDB	PIK	A
Manchester	ILS	I NN	A	Prestwick	L	PW	A
Manchester	DME	I NN	A	Redhill	NDB	RDL	A
Manchester	ILS	I MC	A	Rochester	NDB	RCH	A
Manchester	DME	I MC	A	Ronaldsway	L	RWY	A
Manchester	L	MCH	A	Scatsta	L	SS	A
Manchester	VOR	MCT	AE	Scilly Isles/St Marys	L	STM	A
Manchester	DME	MCT	AE	Scotstown Head	NDB	SHD	E
Manchester Woodford	ILS	IWU	A	Seaford	VOR	SFD	E
Manchester Woodford	DME	IWU	A	Seaford	DME	SFD	E
Manchester Woodford	L	WFD	A	Sheffield City	DME	I SFH	A
Manston	ILS	I MSN	A	Sheffield City	NDB	SMF	A
Manston	DME	I MSN	A	Sherburn-in-Elmet	NDB	SBL	A
Manston	L	MTN	A	Shipdham	NDB	SDM	A
Manston	LLZ	MOZ	A	Shobdon	NDB	SH	A
Manston	DME	MOZ	A	Shoreham	L	SHM	A
Mayfield	VOR	MAY	E	Shoreham	DME	SRH	A
Mayfield	DME	MAY	E	Sleap	NDB	SLP	A
Midhurst	VOR	MID	E	Southampton	L	EAS	A
Midhurst	DME	MID	E	Southampton	DME	I SN	A
New Galloway	NDB	NGY	E	Southampton	ILS	I SN	A
Newcastle	ILS	I NC	A	Southampton	VOR	SAM	AE
Newcastle	DME	I NC	A	Southampton	DME	SAM	AE
Newcastle	ILS	I NWC	A	Southend	ILS	I ND	A
Newcastle	DME	I NWC	A	Southend	DME	I ND	A
Newcastle	L	NT	A	Southend	L	SND	A
Newcastle	VOR	NEW	AE	St Abbs	VOR	SAB	E
Newcastle	DME	NEW	AE	St Abbs	DME	SAB	E
Northampton/Sywell	NDB	NN	A	Stornoway	L	SAY	A
Northolt	ILS	I NHT	A	Stornoway	LLZ	SOY	A
Northolt	DME	I NHT	A	Stornoway	DME	SOY	A
Norwich	DME	I NH	A	Stornoway	LLZ	STW	A
Norwich	ILS	I NH	A	Stornoway	DME	STW	A
Norwich	L	NH	A	Stornoway	VOR	STN	AE
Norwich	L	NWI	A	Stornoway	DME	STN	AE
Nottingham	NDB	NOT	A	Strumble	VOR	STU	E
Nottingham East Midlands	L	EME	A	Strumble	DME	STU	E
Nottingham East Midlands	L	EMW	A	Sumburgh	ILS	I SG	A

GEN 2.5 – LIST OF RADIO NAVIGATION AIDS DECODE

Station Name	Facility	IDENT	Purpose
Sumburgh	DME	I SG	A
Sumburgh	L	SBH	A
Sumburgh	ILS	SUB	A
Sumburgh	DME	SUB	A
Sumburgh	VOR	SUM	AE
Sumburgh	DME	SUM	AE
Swansea	L	SWN	A
Swansea	DME	SWZ	A
Talla	VOR	TLA	E
Talla	DME	TLA	E
Tatenhill	NDB	TNL	A
Test (on)	(Various)	TST	-
Tiree	VOR	TIR	AE
Tiree	DME	TIR	AE
Trent	VOR	TNT	E
Trent	DME	TNT	E
Turnberry	VOR	TRN	E
Turnberry	DME	TRN	E
Wallasey	VOR	WAL	E
Wallasey	DME	WAL	E
Warton	DME	WQ	A
Warton	NDB	WTN	A
Welshpool	NDB	WPL	A
Welshpool	DME	WPL	A
Westcott	NDB	WCO	E
West Wales/Aberporth	NDB	AP	E
Whitegate	NDB	WHI	E
Wick	L	WIK	A
Wick	VOR	WIK	AE
Wick	DME	WIK	AE
Wolverhampton	L	WBA	A
Wolverhampton	DME	WOL	A
Woodley	NDB	WOD	E
Yeovil/Westland	L	YVL	A
Yeovil/Westland	DME	YVL	A

GEN 2.5 – LIST OF RADIO NAVIGATION AIDS DECODE

IDENT	Station Name	Facility	Purpose
AND	Aberdeen	VOR	AE
AND	Aberdeen	DME	AE
ALD	Alderney	L	A
AP	West Wales/Aberporth	NDB	A
AQ	Aberdeen	NDB	A
ATF	Aberdeen/Dyce	L	A
BBA	Benbecula	L	A
BCL	Benbecula	DME	A
BCN	Brecon	VOR	E
BCN	Brecon	DME	E
BEL	Belfast	VOR	AE
BEL	Belfast	DME	AE
BEN	Benbecula	VOR	AE
BEN	Benbecula	DME	AE
BHD	Berry Head	VOR	E
BHD	Berry Head	DME	E
BHX	Birmingham	L	A
BIA	Bournemouth	L	A
BIG	Biggin	VOR	AE
BIG	Biggin	DME	AE
BKY	Barkway	VOR	E
BKY	Barkway	DME	E
BLC	Blackbushe	DME	A
BLK	Blackbushe	NDB	A
BNN	Bovingdon	VOR	E
BNN	Bovingdon	DME	E
BOU	Bourn	NDB	A
BPK	Brookmans Park	VOR	E
BPK	Brookmans Park	DME	E
BPL	Blackpool	L	A
BRI	Bristol	L	A
BRR	Barra	NDB	A
BUR	Burnham	NDB	E
BV	Brough	NDB	A
CAE	Caernarfon	NDB	A
CAM	Cambridge	L	A
CAR	Carnane	L	A
CBL	Campbeltown	NDB	A
CBN	Cumbernauld	NDB	A
CBN	Cumbernauld	DME	A
CDF	Cardiff	L	A
CFD	Cranfield	VOR	AE
CHT	Chiltern	NDB	E
CIT	Cranfield	L	A
CL	Carlisle	L	A
CLN	Clacton	VOR	E
CLN	Clacton	DME	E
CO	Carlisle	DME	A
COM	Compton Abbas	NDB	A
CPT	Compton	VOR	E
CPT	Compton	DME	E
CT	Coventry	L	A
DCS	Dean Cross	VOR	E
DCS	Dean Cross	DME	E
DET	Detling	VOR	E
DET	Detling	DME	E
DND	Dundee	L	A

GEN 2.5 – LIST OF RADIO NAVIGATION AIDS DECODE

IDENT	Station Name	Facility	Purpose	IDENT	Station Name	Facility	Purpose
DTY	Daventry	VOR	E	IBM	Birmingham	DME	A
DTY	Daventry	DME	E	I BMH	Bournemouth	ILS	A
DVR	Dover	VOR	E	I BMH	Bournemouth	DME	A
DVR	Dover	DME	E	I BON	Bristol	ILS	A
EAS	Southampton	L	A	I BON	Bristol	DME	A
EDN	Edinburgh	L	A	I BPL	Blackpool	ILS	A
EGT	Londonderry/Eglinton	L	A	I BPL	Blackpool	DME	A
EKN	Enniskillen/St Angelo	NDB	A	I BRF	Bristol Filton	ILS	A
EME	Nottingham East Midlands	L	A	I BRF	Bristol Filton	DME	A
EMW	Nottingham East Midlands	L	A	I BTS	Bristol	ILS	A
ENN	Enniskillen/St Angelo	DME	A	I BTS	Bristol	DME	A
EPM	Epsom	NDB	E	I CDF	Cardiff	ILS	A
EX	Exeter	L	A	I CDF	Cardiff	DME	A
FNL	Fenland	NDB	A	I CMG	Cambridge	ILS	A
FNY	Doncaster Sheffield	L	A	I CMG	Cambridge	DME	A
FOS	Fairoaks	NDB	A	ICR	Cranfield	ILS	A
FRK	Fairoaks	DME	A	ICT	Coventry	ILS	A
GAM	Gamston	VOR	E	ICT	Coventry	DME	A
GAM	Gamston	DME	E	I CTY	Coventry	ILS	A
GE	London Gatwick	NDB	A	I CTY	Coventry	DME	A
GLG	Glasgow	L	A	I CWA	Cardiff	ILS	A
GOS	Gloucestershire	DME	A	I CWA	Cardiff	DME	A
GOW	Glasgow	VOR	AE	IDD	Jersey	DME	A
GOW	Glasgow	DME	AE	IDD	Jersey	ILS	A
GST	Gloucestershire	L	A	I DDE	Dundee	ILS	A
GUY	Guernsey	L	A	I DDE	Dundee	DME	A
GUR	Guernsey	VOR	A	I DX	Inverness	ILS	A
GUR	Guernsey	DME	A	I DX	Inverness	DME	A
GWC	Goodwood	VOR	AE	I EGT	Londonderry/Eglinton	DME	A
GWC	Goodwood	DME	AE	I EGT	Londonderry/Eglinton	ILS	A
GY	London Gatwick	NDB	A	I EME	Nottingham East Midlands	DME	A
HAV	Haverfordwest	NDB	A	I EME	Nottingham East Midlands	ILS	A
HAW	Hawarden	L	A	I EMW	Nottingham East Midlands	DME	A
HB	Belfast City	L	A	I EMW	Nottingham East Midlands	ILS	A
HBD	Belfast City	ILS	A	IET	Exeter	DME	A
HBD	Belfast City	DME	A	IET	Exeter	ILS	A
HDW	Haverfordwest	DME	A	IFB	Bristol Filton	DME	A
HEN	Henton	NDB	E	IFB	Bristol Filton	ILS	A
HON	Honiley	VOR	E	I FNB	Farnborough	LLZ	A
HON	Honiley	DME	E	I FNB	Farnborough	DME	A
HLS	Lashenden/Headcorn	DME	A	I FNL	Doncaster Sheffield	DME	A
IAA	London Heathrow	ILS	A	I FNL	Doncaster Sheffield	ILS	A
IAA	London Heathrow	DME	A	I FRG	Farnborough	LLZ	A
I ABD	Aberdeen/Dyce	ILS	A	I FRG	Farnborough	DME	A
I ABD	Aberdeen/Dyce	DME	A	IFT	Belfast Aldergrove	DME	A
IAG	Belfast/Aldergrove	ILS	A	IFT	Belfast Aldergrove	ILS	A
IAG	Belfast/Aldergrove	DME	A	IGG	London Gatwick	DME	A
IAX	Aberdeen/Dyce	ILS	A	IGG	London Gatwick	ILS	A
IAX	Aberdeen/Dyce	DME	A	IGH	Guernsey	ILS	A
IBB	London Heathrow	ILS	A	IGH	Guernsey	DME	A
IBB	London Heathrow	DME	A	I HDN	Hawarden	DME	A
I BFH	Belfast City	ILS	A	I HDN	Hawarden	ILS	A
I BFH	Belfast City	DME	A	I HS	Humberside	DME	A
I BGH	Biggin Hill	ILS	A	I HS	Humberside	ILS	A
I BGH	Biggin Hill	DME	A	I HWD	Hawarden	ILS	A
IBH	Bournemouth	ILS	A	I HWD	Hawarden	DME	A
IBH	Bournemouth	DME	A	IJJ	Jersey	DME	A
I BIR	Birmingham	ILS	A	IJJ	Jersey	ILS	A
I BIR	Birmingham	DME	A	I KIR	Kirkwall	ILS	A
IBM	Birmingham	ILS	A	I KIR	Kirkwall	DME	A

GEN 2.5 — LIST OF RADIO NAVIGATION AIDS DECODE

IDENT	Station Name	Facility	Purpose	IDENT	Station Name	Facility	Purpose
IKK	Prestwick	DME	A	I TH	Edinburgh	ILS	A
IKK	Prestwick	ILS	A	I TH	Edinburgh	DME	A
I LBF	Leeds Bradford	ILS	A	I TSE	Teesside	ILS	A
I LBF	Leeds Bradford	DME	A	I TSE	Teesside	DME	A
ILF	Leeds Bradford	ILS	A	IUU	Glasgow	ILS	A
ILF	Leeds Bradford	DME	A	IUU	Glasgow	DME	A
ILJ	London Luton	ILS	A	IUY	Guernsey	ILS	A
ILJ	London Luton	DME	A	IUY	Guernsey	DME	A
ILL	London Heathrow	ILS	A	IVG	Edinburgh	ILS	A
ILL	London Heathrow	DME	A	IVG	Edinburgh	DME	A
ILN	Inverness	ILS	A	IWU	Manchester Woodford	ILS	A
ILN	Inverness	DME	A	IWU	Manchester Woodford	DME	A
ILQ	Liverpool	ILS	A	IWW	London Gatwick	ILS	A
ILQ	Liverpool	DME	A	IWW	London Gatwick	DME	A
I LTN	London Luton	ILS	A	IXR	Exeter	ILS	A
I LTN	London Luton	DME	A	IXR	Exeter	DME	A
I LVR	Liverpool	ILS	A	INS	Inverness	VOR	AE
I LVR	Liverpool	DME	A	INS	Inverness	DME	AE
IMC	Manchester	ILS	A	IOM	Isle of Man	VOR	AE
IMC	Manchester	DME	A	IOM	Isle of Man	DME	AE
IMM	Manchester	ILS	A	ISY	Islay	DME	A
IMM	Manchester	DME	A	IVR	Inverness	NDB	A
I MSN	Manston	ILS	A	IW	Bembridge	NDB	A
I MSN	Manston	DME	A	JSY	Jersey	VOR	AE
INC	Newcastle	ILS	A	JSY	Jersey	DME	AE
INC	Newcastle	DME	A	JW	Jersey	L	A
IND	Southend	ILS	A	KIM	Humberside	L	A
IND	Southend	DME	A	KW	Kirkwall	L	A
INH	Norwich	DME	A	KWL	Kirkwall	VOR	A
INH	Norwich	ILS	A	KWL	Kirkwall	DME	A
I NHT	Northolt	ILS	A	LAM	Lambourne	VOR	E
I NHT	Northolt	DME	A	LAM	Lambourne	DME	E
INN	Manchester	ILS	A	LAY	Islay	NDB	A
INN	Manchester	DME	A	LBA	Leeds Bradford	NDB	AE
I NWC	Newcastle	ILS	A	LCY	London City	NDB	A
I NWC	Newcastle	DME	A	LDY	Lydd	DME	A
IOO	Glasgow	ILS	A	LE	Leicester	NDB	A
IOO	Glasgow	DME	A	LIC	Lichfield	NDB	E
I ORK	Kirkwall	ILS	A	LND	Lands End	VOR	E
I ORK	Kirkwall	DME	A	LND	Lands End	DME	E
I PLY	Plymouth	ILS	A	LON	London	VOR	E
I PLY	Plymouth	DME	A	LON	London	DME	E
IPP	Prestwick	ILS	A	LPL	Liverpool	L	A
IPP	Prestwick	DME	A	LSH	Lashenden/Headcorn	NDB	A
IRH	Isle of Man	ILS	A	LSR	London City	ILS	A
IRH	Isle of Man	DME	A	LSR	London City	DME	A
IRR	London Heathrow	ILS	A	LST	London City	ILS	A
IRR	London Heathrow	DME	A	LST	London City	DME	A
IRY	Isle of Man	ILS	A	LUT	London Luton	L	A
IRY	Isle of Man	DME	A	LYD	Lydd	VOR	E
I SED	London Stansted	ILS	A	LYD	Lydd	DME	E
I SED	London Stansted	DME	A	LYX	Lydd	NDB	A
ISG	Sumburgh	ILS	A	M HER	London Heathrow	MLS	A
ISG	Sumburgh	DME	A	MAC	Machrihanish	VOR	AE
ISN	Southampton	DME	A	MAC	Machrihanish	DME	AE
ISN	Southampton	ILS	A	MAY	Mayfield	VOR	E
ISX	London Stansted	ILS	A	MAY	Mayfield	DME	E
ISX	London Stansted	DME	A	MCH	Manchester	L	A
ITD	Durham Tees Valley	ILS	A	MCT	Manchester	VOR	AE
ITD	Durham Tees Valley	DME	A	MCT	Manchester	DME	AE

101

GEN 2.5 – LIST OF RADIO NAVIGATION AIDS DECODE

IDENT	Station Name	Facility	Purpose
MID	Midhurst	VOR	E
MID	Midhurst	DME	E
MOZ	Manston	LLZ	A
MOZ	Manston	DME	A
MTN	Manston	L	A
ND	Great Yarmouth/N.Denes	L	A
NEW	Newcastle	VOR	AE
NEW	Newcastle	DME	AE
NGY	New Galloway	NDB	E
NH	Norwich	L	A
NN	Northampton/Sywell	NDB	A
NOT	Nottingham	NDB	A
NT	Newcastle	L	A
NWI	Norwich	L	A
OCK	Ockham	VOR	E
OCK	Ockham	DME	E
OF	Bristol Filton	L	A
OTR	Ottringham	VOR	E
OTR	Ottringham	DME	E
OX	Oxford/Kidlington	L	A
OX	Oxford/Kidlington	DME	A
OY	Belfast/Aldergrove	L	A
PH	Penzance	NDB	A
PIK	Prestwick	NDB	A
PMB	Pembrey	L	A
PLY	Plymouth	DME	A
POL	Pole Hill	VOR	E
POL	Pole Hill	DME	E
PTH	Perth	VOR	E
PW	Prestwick	L	A
PY	Plymouth	L	A
RCH	Rochester	NDB	A
RDL	Redhill	NDB	A
RWY	Ronaldsway	L	A
SAB	St Abbs	VOR	E
SAB	St Abbs	DME	E
SAM	Southampton	VOR	AE
SAM	Southampton	DME	AE
SAY	Stornoway	L	A
SBH	Sumburgh	L	A
SBL	Sherburn-in-Elmet	NDB	A
SDM	Shipdham	NDB	A
SFD	Seaford	VOR	E
SFD	Seaford	DME	E
SFH	Sheffield City	DME	A
SH	Shobdon	NDB	A
SHD	Scotstown Head	NDB	E
SHM	Shoreham	L	A
SLP	Sleap	NDB	A
SMF	Sheffield/City	NDB	A
SND	Southend	L	A
SOY	Stornoway	LLZ	A
SOY	Stornoway	DME	A
SRH	Shoreham	DME	A
SS	Scatsta	L	A
SSD	London Stansted	NDB	A
STM	Scilly Isles/St Marys	L	A
STN	Stornoway	VOR	AE
STN	Stornoway	DME	AE
STU	Strumble	VOR	E

IDENT	Station Name	Facility	Purpose
STU	Strumble	DME	E
STW	Stornoway	LLZ	A
STW	Stornoway	DME	A
SUB	Sumburgh	ILS	A
SUB	Sumburgh	DME	A
SUM	Sumburgh	VOR	AE
SUM	Sumburgh	DME	AE
SWN	Swansea	L	A
SWZ	Swansea	DME	A
TD	Durham Tees Valley	L	A
TIR	Tiree	VOR	AE
TIR	Tiree	DME	AE
TL	Lerwick/Tingwall	NDB	A
TLA	Talla	VOR	E
TLA	Talla	DME	E
TNL	Tatenhill	NDB	A
TNT	Trent	VOR	E
TNT	Trent	DME	E
TRN	Turnberry	VOR	E
TRN	Turnberry	DME	E
TST	Test (on)	(Various)	-
UW	Edinburgh	L	A
WAL	Wallasey	VOR	E
WAL	Wallasey	DME	E
WBA	Wolverhampton	L	A
WCK	Wick	L	A
WCO	Westcott	NDB	E
WFD	Manchester Woodford	L	A
WHI	Whitegate	NDB	E
WIK	Wick	L	A
WIK	Wick	VOR	AE
WIK	Wick	DME	AE
WOD	Woodley	NDB	E
WOL	Wolverhampton	DME	A
WPL	Welshpool	NDB	A
WPL	Welshpool	DME	A
WQ	Warton	DME	A
WTN	Warton	NDB	A
YVL	Yeovil/Westland	L	A
YVL	Yeovil/Westland	DME	A

GEN 2.6 – CONVERSION TABLES

1 Conversions

1.1 Conversion Factors:

KM to NM		NM to KM		LBS to KG		KG to LBS	
1KM = 0.54 NM		1NM = 1.8520KM		1LB = 0.4536 KG		1Kg = 2.2046 LBS	
1KM = 0.6214 Statute Miles		1NM = 1.1508 Statute Miles		1	0.454	1	2.205
				2	0.907	2	4.409
1	0.54	1	1.85	3	1.361	3	6.614
2	1.08	2	3.43	4	1.814	4	8.818
3	1.62	3	5.56	5	2.268	5	11.023
4	2.16	4	7.41	6	2.722	6	13.228
5	2.70	5	9.26	7	3.175	7	15.432
6	3.24	6	11.11	8	3.629	8	17.637
7	3.78	7	12.96	9	4.082	9	19.842
8	4.32	8	14.82	10	4.536	10	22.046
9	4.86	9	16.67	20	9.072	20	44.092
10	5.40	10	18.52	30	13.608	30	66.139
20	10.80	20	37.04	40	18.144	40	88.185
30	16.20	30	55.56	50	22.680	50	110.231
40	21.60	40	74.08	60	27.216	60	132.277
50	27.00	50	92.60	70	31.751	70	154.324
60	32.40	60	111.12	80	36.287	80	176.370
70	37.80	70	129.64	90	40.823	90	198.416
80	43.20	80	148.16	100	45.359	100	220.462
90	48.60	90	166.68	1 metric tonnes = 1000Kgs		1000kgs = 1 metric tonne	
100	54.00	100	185.20				
200	108.00	200	370.40				
300	162.00	300	555.60				
400	216.00	400	740.80				
500	270.00	500	926.00				
600	324.00	600	1111.20				
700	378.00	700	1296.40				
800	432.00	800	1481.60				
900	486.00	900	1666.80				
1000	540.00	1000	1852.00				

Feet to Metres		Metres to Feet	
1 Foot = 0.3048 Metres		1 Metre = 3.2808 Feet	
0.305		1	3.281
2	0.610	2	6.562
3	0.914	3	9.843
4	1.219	4	13.123
5	1.524	5	16.404
6	1.829	6	19.685
7	2.134	7	22.966
8	2.438	8	26.247
9	2.743	9	29.528
10	3.048	10	32.808
20	6.096	20	65.617
30	9.144	30	98.425
40	12.192	40	131.234
50	15.240	50	164.042
60	18.288	60	196.850
70	21.336	70	229.659
80	24.384	80	262.467
90	27.432	90	295.276
100	30.48	100	328.084
200	60.96	200	656.168
300	91.44	300	984.252
400	121.92	400	1312.336
500	152.40	500	1640.420
1000	304.80	1000	3280.840
2000	609.60	2000	6561.680
3000	914.40	3000	9842.520
4000	1219.20	4000	13123.360
5000	1524.00	5000	16404.199

GEN 3 – SERVICES

GEN 3.1 AERONAUTICAL INFORMATION SERVICES

1 Organisation of the UK AIS

1.1 The United Kingdom Aeronautical Information Service is operated by National Air Traffic Services (NATS) Ltd on behalf of the UK Civil Aviation Authority.

1.2 UK AIS is centrally located at London Heathrow Airport:

UK Aeronautical Information Service
National Air Traffic Services Ltd
Control Tower Building
London Heathrow Airport
Hounslow
TW6 1JJ
Tel: See paragraph 1.3
Fax (General): +44 (0)20 8745 3453
AFS: EGGNYNYX
Internet Web site: http://www.ais.org.uk

1.3 Principal UK AIS sections and service hrs are as follows:

UK NOTAM Office (NOF)	Daily H24	+44 (0)20 8745 3450/3451
Fax (NOTAM Proposals)	Daily H24	+44 (0)20 8557 0054
UK AIS Publications Section (AIP/SUP/AIC etc.)	Monday to Friday (office hrs)	+44 (0)20 8745 3456
UK AIS Library	Monday to Friday	+44 020 8745 3473

Enquiries & self-briefing visits
(All visits subject to prior arrangement)

1.4 Telephone calls to AIS may be recorded.

2 Area of Responsibility

2.1 The UK Aeronautical Information Service is responsible for the collection and dissemination of information/data necessary for the safety, regularity and efficiency of air navigation throughout the entire territory and airspace of the UK and the airspace over the high seas under the jurisdiction of the UK for Air Traffic Control purposes.

3 Aeronautical Publications

3.1 Integrated Aeronautical Information Package

3.1.1 The UK AIS conforms to ICAO Annex 15 requirements and publishes aeronautical information as an integrated package.

3.1.2 The Integrated Aeronautical Information Package consists of the following elements:

(a) The Aeronautical Information Publication (AIP) and Amendment service (see paragraph 3.2 and 3.3);

(b) AIP Supplements (SUP) (see paragraph 3.4);

(c) Aeronautical Information Circulars (AIC) (see paragraph 3.5);

(d) NOTAM (see paragraph 3.6);

(e) Pre-Flight Information Bulletins (PIB) (see paragraph 3.10);

(f) Check Lists.

3.1.3 Documents may be obtained from Documedia Solutions Ltd, who publish and distribute AIS Publications on behalf of NATS Ltd (see Paragraph 3.11).

3.2 The United Kingdom Aeronautical Information Publication (AIP)

3.2.1 The UK AIP is published in accordance with the provisions of Annex 15 to the Convention on International Civil Aviation and is, in addition, the official document used to publish Notifications required by the UK Air Navigation Order.

3.2.2. In order to register any permanent changes to information contained in the AIP, information providers are required to complete CAA/AIS Form 933 and forward it to both the CAA and AIS. Guidance is available from the AIS Publications Section on the procedure for submitting such information. This process is also required for AIC's and AIP Supplements (SUP) (see paragraph 3.4). CAA/AIS Form 933 is available on the AIS web site.

Note: Changes of short term/temporary nature should follow the processes required for submission of NOTAM (see paragraph 3.6).

3.2.3 Civil Aviation Legislation

3.2.3.1 UK legislation makes specific provision in a number of cases for Notification. The term 'notified' is defined as set forth in a document published by the Civil Aviation Authority and entitled 'United Kingdom AIP'. A list of principal civil aviation legislation and air navigation regulations is shown at GEN 1.6 and references to the legislation under which Notification is made appear, as appropriate, within the UK AIP.

3.2.3.2 Notices relating to the limitation of noise at London Gatwick, London Heathrow and London Stansted aerodromes (Section 78 of the Civil Aviation Act 1982 refers) are required to be published in a format that may not be compatible with the host AIS document. Notices requiring operators of aircraft to secure that specified requirements for limiting or mitigating the effect of noise and vibration are complied with by aircraft landing or taking-off at those designated aerodromes appear in the UK AIP under each of the aerodromes' item AD 2.2.1. Notices specifying, for those designated aerodromes, a maximum number of take-offs or landings which may be permitted

GEN 3.1   GEN 3.1.3 Aeronautical Publications

during specific periods and determining the operators who are entitled to land and take off during these periods and the number of occasions on which their aircraft may take off or land during those periods will appear in UK AIP Supplements.

3.3 UK AIP Amendment service (AMDT and AMDT AIRAC)

3.3.1 The UK AIP is available in both paper and electronic format (CD-ROM and Internet). AIP Amendments may contain both AIRAC and Non-AIRAC changes (paragraphs 3.3.2, 3.3.3 and 4 refer). Each paper Amendment will be accompanied by a pink coloured cover sheet which will identify the nature and status of each change. The electronic version will be updated to a specified AIRAC date and the CD-ROM will also present a preview of the following AIRAC changes.

3.3.2 The ICAO AIRAC system is used to provide advance notice of the introduction of permanent operationally significant changes on an internationally recognised AIRAC effective date. AIRAC Amendment pages are identified by the footnote AMDT AIRAC and do not replace the existing AIP pages until the AIRAC effective date on which the changes take place. (See also paragraph 3.6.1.2).

3.3.3 Non-AIRAC AIRAC amendments (AMDT) to the AIP comprise permanent operationally significant changes that have received previous notification by NOTAM and other permanent information that is not required to be announced by NOTAM. Non-AIRAC changes to the AIP are published together with AIRAC changes but may be considered to be effective on or before receipt, unless otherwise indicated. To remain consistent with the CD-ROM version, AIP pages (AIRAC and Non-AIRAC) should not be replaced before the stated AIRAC date or AIP insertion date.

3.3.4 The AIP Amendment cover sheet, will indicate any NOTAM or permanent AIP Supplements that have been incorporated. On each replacement page, changes are either annotated or identified by a vertical line or arrow in the outer margin of the page by a vertical line or arrow adjacent to the change/addition/deletion.

3.3.5 Each AIP page is dated to reflect the Amendment's AIRAC effective date or AIP insertion date and a complete checklist of AIP pages, relating page reference to date, is reissued with each amendment as AIP section GEN 0.4.

3.3.6 Each combined AIP amendment is allocated and AIRAC Cycle serial number that is consecutive and based on the calendar year. The year, indicated by two digits, is a part of the serial number of the amendment, e.g. AIRAC Cycle 1/03. When necessary to provide additional advance notice of AIRAC changes, the Amendment will be issued in several parts, each relating to a common effective date. These Amendments will be identified by a part number suffix, e.g. AIRAC Cycle 2/03 Part 1.

3.3.7 Further explanation of AIRAC changes may, when appropriate, be promulgated in an AIC.

3.4 Supplements to the AIP (AIP SUP)

3.4.1 UK AIP Supplements will normally contain items of a temporary nature only. To be included in an AIP Supplement, information must be of operational significance and contain comprehensive text and/or graphics (e.g. major air exercises or aerodrome work programmes) that preclude 'complete' promulgation by NOTAM. See also paragraph 3.6.1.2.

3.4.2 UK AIP Supplements are are available in both paper format (coloured yellow) and in electronic format (CD-ROM and Internet) and are normally issued in batches every 28 days.

3.4.3 AIP Supplement are to be kept in the AIP binder for as long as all or some of their contents remain valid. The period of validity of the information contained in the AIP Supplement will normally be given in the Supplement itself. Alternatively, NOTAM may be used to indicate changes to the period of validity or to advise of the cancellation of an AIP Supplement.

3.4.4 A checklist of UK AIP SUP currently in force is normally included within each issued batch of Supplements.

3.5 Aeronautical Information Circulars (AIC)

3.5.1 As a general rule, AICs refer to subjects of an administrative rather than of an operational nature. They are, however, also used to publish advance warnings of impending operational changes and to add explanation or emphasis on matters of safety or operational significance. Aeronautical chart issues are also notified through the medium of the AIC.

3.5.2 UK AIC are available in both paper format (paragraph 3.5.3 refers) and in an electronic format (CD-ROM and Internet) and are normally issued every 28 days.

3.5.3 UK AIC's are subject to a dual referencing system. The standard year-based reference is supplemented by a colour group reference; eg AIC 31/2003 (Yellow 103). AIC are colour coded according to their subject matter as follows:

    White   Administrative matters (e.g. licence examination dates, services and publications).
    Yellow   Operational matters (including ATS facilities and requirements).
    Pink   Safety related matters.
    Mauve   UK Airspace Restrictions imposed under the Restriction of Flying Regulations.
    Green   Maps and charts.

3.6 NOTAM (Notices to Airmen)

3.6.1 General

3.6.1.1 All operationally significant information not covered by AIP Amendment or AIP Supplement will be issued as a NOTAM (via the Aeronautical Fixed Service AFS).

3.6.1.2 All operationally significant information issued as AIRAC UK AIP Amendments or UK AIP Supplements will be additionally announced by NOTAM. The NOTAM will give an abbreviated description of the change, condition or activity together with the effective date(s) and the Amendment or Supplement reference number. These 'trigger' NOTAM ensure that brief entries appear in the appropriate Pre-flight Information Bulletins (PIB).

3.6.1.3 'Trigger' NOTAM will remain valid for 15 days after the effective date of a permanent change and for the complete duration of any temporary change, condition or activity.

105

3.6.1.4 NOTAM are available on the UK AIS Web site (www.ais.org.uk), in the form of Pre-flight Information Bulletins (PIB).

3.6.2 NOTAM Construction

3.6.2.1 To assist the automated processing of data, ICAO format NOTAM are issued with a 'qualifier line' (identified by the letter 'Q') which is composed by the issuing NOF.

3.6.2.2 Three NOTAM types are issued and are identified as follows:
NOTAMN contains new information;
NOTAMR replaces a previous NOTAM;
NOTAMC cancels a previous NOTAM.

3.6.3 UK NOTAM Series

3.6.3.1 See Table 3.1.1 for details of the content of each individual series.

3.6.4 NOTAM Handling

3.6.4.1 ICAO NOTAM format and conditions require that:

(a) Each NOTAM deals only with one subject and one condition concerning that subject;

(b) NOTAM text is both precise and concise, using plain language and commonly used ICAO abbreviations.

(c) An activity or condition applying 'with immediate effect' (WIE) is indicated by a ten figure group (Year/Month/Day/Time) reflecting the actual date/time of NOTAM issue.

(d) All temporary NOTAM must include an expiry date/time. The term 'until further notice' (UFN) will not be used. If estimated, then a ten figure group will be suffixed by 'EST' (e.g. 0104032100 EST). NOTAM with such an estimated expiry date/time remain in force until cancelled by a NOTAMC or replaced by a NOTAMR. Information providers are required to monitor their estimates and to advise the UK NOTAM Office of any change or cancellation. The original NOTAM must be replaced if the estimated date/time is changed.

(e) If information is permanent then the abbreviation 'PERM' will appear in the NOTAM.

3.6.5 Requests for NOTAM Issue

3.6.5.1 Guidance on the submission of requests for NOTAM action is available from the UK AIS NOF (see paragraph 1.3).

3.6.5.2 Requests may be submitted by AFS (EGGNYNYX) or by using fax +44 (0)20 8557 0054 in system NOTAM format. In the event of being unable to submit NOTAM requests via either of these two methods, then the AIS general fax no +44 (0)20 8745 3453 may be used.

3.6.6 Requests for NOTAM Reception

3.6.6.1 Requests from Foreign NOF and/or UK aeronautical organisations to be included in AFS address lists for the distribution of UK NOTAM should be submitted in writing to UK AIS (see paragraph 1.2), or via AFS (EGGNYNYX).

3.6.6.2 States with whom NOTAM are exchanged are listed in ICAO Doc 7383.

3.6.6.3 Automatic Query/Response – UK International NOTAM Database. UK NOTAM and NOTAM from member States that distribute their NOTAM to UK are available by automatic query/response to UK users. Limited non-UK NOTAM information is available by query/response via AFS to International users. These users will predominantly be the International NOTAM offices from member states.

Examples:
Request single NOTAM:
GG EGGNYNYX
040131 EBBRYNYX
RQN EGGN A0054/04
Request consecutive NOTAM:
GG EGGNYNYX
040131 EBBRYNYX
RQN EGGN B0054/04-B0075/04
Request several non-consecutive NOTAM: (maximum of 4 NOTAM in ascending order)
GG EGGNYNYX
040131 EBBRYNYX
RQN EGGN A0054/04 A0057/04 A0060/04 A0072/04
Request a list of current NOTAM: (for a particular series)
GG EGGNYNYX
040131 EBBRRYNYX
RQL EGGN A
All replies will be transmitted to the recipient in RQR format.

3.7 Evacuation of UK NOTAM Office (NOF)

3.7.1 In the event of an evacuation of the UK NOF (EGGNYNYX) of more than 1 hour and less than 12 hours, the Netherlands NOF (EHAMYNYX) will originate civil UK NOTAM in accordance with a bilateral agreement. If the evacuation lasts longer than 12 hours, UK AIS will establish a temporary NOF in the UK. Details of the location, AFS indicator, facsimile and telephone numbers, etc. would then be notified by NOTAM.

3.7.2 During an evacuation, only NOTAM of urgent operational significance will be originated (Checklists of current NOTAM will not be issued). NOTAM will be issued in a special series 'X' as follows:

(a) All NOTAM will have a maximum validity of 12 hours regardless of whether or not the known duration is longer. 'EST' will not be used after the date/time group in Field 'C';

(b) NOTAM will continue to be issued to all recipients within the UK international address lists;

(c) When the UK NOF has resumed normal operation, all current NOTAM series 'X' issued by the Netherlands NOF will be reissued as new NOTAM (NOTAMN) in an appropriate 'normal' UK Series.

3.8 SNOWTAM

3.8.1 SNOWTAM messages must be submitted by the Aerodrome Operator/representative or Controlling Authority, using a strict format for automatic processing through the AFS system. Information on message submission, format and transmission protocols is contained in an AIC issued on a seasonal basis. The role of UK AIS regarding SNOWTAM is limited to monitoring the automatic AFS handling system.

3.9 Foreign NOTAM AFS Addressing

3.9.1 When addressing NOTAM to the UK, foreign NOTAM Offices (NOF) should use the collective address indicator EGZZN followed by the first three letters of the location indicator of the NOF of origin.

Example: EGZZNEBB for a NOTAM from Belgium (EBBR)

3.10 Pre-flight Information Bulletins (PIB)

3.10.1 NOTAM are available on the AIS web site in the form of Pre-flight Information Bulletins (PIB). Users are required to register onto the web site in order to obtain a briefing. The AIS web site offers several methods by which users can obtain PIB. Both UK and International NOTAM are available using the dynamic database facility whereby the user inputs selectable parameters in order to obtain Narrow Route/Route/Aerodrome and Area briefings. Alternatively pre-prepared PIB for the London and Scottish FIRs/Aerodromes are obtainable in electronic form or via a poll fax facility. Full details as to the complete range of PIB products are available from the AIS web site using the help pages and frequently asked questions (FAQs).

3.11 Supply of UK AIS Documents

3.11.1 Most UK AIS publications are available in both paper and electronic format. The UK AIP, AIP SUPs and AICs may be obtained from the UK AIS CD-ROM and, in addition, these documents feature on the UK AIS Web site (www.ais.org.uk).

3.11.2 All purchase requests and enquiries regarding the supply of UK AIS publications (including the CD-ROM) should be addressed to Documedia Solutions Ltd, 37 Windsor Street, Cheltenham, GL52 20G (Tel: 0870 8871410; Fax: 0870 8871411).

3.11.3 Details and prices of UK AIS publications are published annually in an AIC.

4 AIRAC System

4.1 AIRAC AIP Amendments are originated and distributed by AIS with the objective of reaching chart producers and data handlers at least 28 days in advance of the effective date. Information providers should note that strict adherence to both the AIRAC publication and effective dates is essential if the information is to be incorporated in flight-deck documentation and flight management systems by the effective date of the selected AIRAC Cycle.

4.2 The information listed below is required by ICAO to be published and brought into effect in accordance with the AIRAC System.

(a) The establishment and withdrawal of, and predetermined significant changes (including operational trials) to: Horizontal and vertical limits, regulations and procedures applicable to FIR/UIR, CTA, CTR. Advisory Areas and ATS Routes; Permanent danger, prohibited and restricted areas (including type and periods of activity when known) and ADIZ, Permanent areas or routes or portions thereof where the possibility of interception exists;

(b) Positions, frequencies, call signs, known irregularities and maintenance periods of radio navigational aids and communications facilities;

(c) Holding and approach to land procedures, arrival and departure procedures, noise abatement procedures and any other pertinent ATS procedures;

(d) Meteorological facilities (including broadcasts) and procedures;

(e) Runways and stopways.

4.3 In addition, the establishment and withdrawal of, and predetermined significant changes to the information listed below may be published and brought into effect in accordance with the AIRAC System.

(a) Position, height and lighting of navigational obstacles;

(b) Taxiways and aprons;

(c) Hours of service: Aerodromes, facilities and services;

(d) Customs, immigration and health services;

(e) Temporary danger, prohibited and restricted areas and navigational hazards, military exercises and mass movements of aircraft;

(f) Temporary areas or routes or portions thereof where the possibility of interception exists.

4.4 When operationally necessary, the ICAO AIRAC System permits major changes to be promulgated two Cycles (56 days) in advance. Similarly, additional notice is required if the introduction of an intended change cannot be planned to take place on an AIRAC effective date. Publication would then be required no later than the AIRAC Cycle within which the actual effective date falls.

4.5 Information providers should consult the UK AIS Publications Section (see paragraph 1.3) for guidance and details of the promulgation schedules.

4.6 Schedule of AIRAC effective dates:

2004	2005	2006	2007	2008	2009	2010
Published	20 January	19 January	18 January	17 January	15 January	14 January
Published	17 February	16 February	15 February	14 February	12 February	11 February
Published	17 March	16 March	15 March	13 March	12 March	11 March
Published	14 April	13 April	12 April	10 April	9 April	8 April
Published	12 May	11 May	10 May	8 May	7 May	6 May
10 June	9 June	8 June	7 June	5 June	4 June	3 June
8 July	7 July	6 July	5 July	3 July	2 July	1 July
5 August	4 August	3 August	2 August	31 July	30 July	29 July
2 September	1 September	31 August	30 August	28 August	27 August	26 August
30 September	29 September	28 September	27 September	25 September	24 September	23 September
28 October	27 October	26 October	25 October	23 October	22 October	21 October
25 November	24 November	23 November	22 November	20 November	19 November	18 November
23 December	22 December	21 December	20 December	18 December	17 December	16 December

UK NOTAM Series

Series	Content
A	Aerodromes: Aberdeen/Dyce, Belfast Aldergrove, Belfast City, Edinburgh, Glasgow, Inverness, London Gatwick, London Heathrow, London Luton, London Stansted, Manchester International, Prestwick and Sumburgh
B	En-route Airspace London & Scottish FIR/UIR: Regulations & Procedures, En-Route Navigation Aids (inc facilities used as Approach Aids) ATS and Air/Ground Communications.
C	Aerodromes: Alderney, Biggin Hill, Birmingham, Blackpool, Bournemouth, Bristol, Cambridge, Cardiff, Coventry, Doncaster Sheffield, Durham Tees Valley, Exeter, Farnborough, Guernsey, Humberside, Isle of Man, Jersey, Leeds Bradford, Liverpool, Clondon City, Lydd, manston, Newcastle, Norwich, Nottingham East Midlands, Plymouth, Shoreham, Southampton and Southend.
D	Not used
E	Not used
F	Not used
G	En-route Airspace Shanwick FIR/UIR: Regulations & Procedures, ATS, Air/Ground Communications, Airspace Reservations, Navigation Warnings and Notifiable Danger Area Activity
H	Navigation warnings (except those catered for in Series G).
J	Temporary Danger Areas, Restrictions of Flying, Prohibited, Restricted and Notifiable Danger Area Activity (except those catered for in Series G).
K	Not used.
L	Aerodromes listed in UK AIP, not covered in any of the above series.
M	Not used.
N	Unserviceabilities of lighting on en-route Air Navigation.
O & P	Not used.
Q	Military Services (Sovereign Bases).
R	Not used.
U	Military Services (UK).
V & W	Reserved for Military use.
X	NOF Evacuation Series (For use by Netherlands NOF).
Y & Z	Reserved for Military use.

5 Pre-Flight Briefing

5.1 UK AIS is operated from a single central location from which information is made available via multi-media output.

5.2 Aerodrome operators are responsible for providing briefing, including UK AIP, AIP SUPs, AICs, NOTAM and/or selected Pre-flight Information Bulletins (PIB), is normally held and made available for reference and self briefing. Alternatively information and assistance is available from UK AIS (see paragraph 1.3).

6 Post-Flight Information

6.1 United Kingdom Facilities

6.1.1 Reports on the adequacy or otherwise of navigation facilities or services provided by UK ATS are welcomed by UK AIS, but should first be directed to the organisation responsible for providing the service or facility (see also GEN 3.3 and 3.4).

6.2 Overseas Facilities

6.2.1 Post-flight information, makes an invaluable contributor to flight safety and, therefore, flight crews are requested to complete a 'Post-flight Report' and submit it via a Flight Briefing Unit or ATS Reporting Office.

GEN 3.2 AERONAUTICAL CHARTS
GEN 3.2.1 RESPONSIBLE AUTHORITY

1 General

1.1 The UK Civil Aviation Authority publishes a wide range of aeronautical charts for use by all types of civil aviation:
Civil Aviation Authority,
Aeronautical Charts & Data (AC & D) DAP,
K6 CAA House,
45-59 Kingsway
London
WC2B 6TE
Tel: 020 7453 6572
Fax: 020 7453 6565

2 Applicable ICAO Documents

2.1 The Standards, Recommended Practices and, when applicable, the procedures contained in the following ICAO documents are applied:

Annex 4	Aeronautical Charts
Doc 8168-OPS/611	Aircraft Operations (Holding patterns, OCH and Instrument Approach Procedures diagrams)
Doc 8697 AN/889	Aeronautical Chart Manual

3 ICAO IAP Charts

3.1 Responsibility for the design and approval of Instrument Flight Procedures (IFP) for civil aerodromes in the UK and associated policy matters, lies with the Terminal Airspace Section (TAS), Directorate of Airspace Policy (DAP), Civil Aviation Authority. Instrument Approach Charts are produced and notified by AC & D on behalf of TAS. Enquiries concerned with the design of IAP, as opposed to the charting or communications issues, should be addressed to TAS, DAP.

Civil Aviation Authority,
TAS, DAP,
K6 CAA House,
45-59 Kingsway,
London
WC2B 6TE
Tel: 020 7453 6513
Fax: 020 7453 6565

GEN 3.2.2 MAINTENANCE OF CHARTS

1 Chart Amendment and Revision

1.1 New editions of the CAA charts are published as often as resources permit with priority being given to those charts affected by major changes to aeronautical information. However, due to high production costs some CAA charts – particularly those listed at GEN 3.2.5 paras 1 & 2 have an edition life of at least 1 year. The publication of each new edition will replace the previous edition, which will then become obsolete. Details will be notified by AIC and/or AIP Amendment as appropriate. The aeronautical information validity date will be specified on each edition and topographical and hydrographical information will have been revised where necessary.

1.2 Before using any CAA chart operationally, users must determine what aeronautical changes have taken place since its validity date and amend the chart accordingly. This will require users to consult all UK AIP amendments issued since the charts validity date. If in doubt, they should consult AIS Central Office. A list of amendments for the VFR chart series is available on the CAA web site: www.caa.co.uk/dap/dapcharts

1.3 Items of information found after publication to have been incorrect at the aeronautical information validity date specified on the chart are corrected immediately by NOTAM and AIC. A list of corrections is at GEN 3.2.8.

1.4 Amendments to the Aerodrome Obstacle Charts – ICAO – Type A and to the Chart of United Kingdom Airspace Restrictions and Hazardous Areas may be published as or when necessary, but the Chart of UK Areas of Intense Air Activity, Aerial Tactics Areas and Military Low Flying System is amended by annual re-issue.

GEN 3.2.3 PURCHASE ARRANGEMENTS

1 The charts listed at GEN 3.2.5 paragraphs 1, 2 and 6, may be obtained through Airplan Flight Equipment Ltd. The main chart agent acting on behalf of the CAA.

Airplan Flight Equipment Ltd
Unit 1a Ringway Trading Estate
Shadowmoss Road
Manchester
M22 5LH
Tel: 0161 499 0013/0023
Fax: 0161 499 0298

2 The charts listed at GEN 3.2.5 paragraphs 3, 4, 5 and 6 may be obtained from:

Documedia Solutions Ltd
37 Windsor Street
Cheltenham
Glos
GL52 2DG
Tel: 0870 887 1410
Fax: 0870 887 1411

3 A selection of Royal Air Force aeronautical charts and planning documents is available from:

AIDU (Sales Office)
Royal Air Force Northolt
West End Road
Ruislip
Middlesex
HA4 6NG
Tel: 020 8845 2300 Ex 7209
Fax: 020 8845 2300 Ex 7510

A catalogue and price list are available on request.

4 A series of LORAN C Charts are published by the Hydrographic Department of the Admiralty for all areas where coverage exists (with the exception of the Mediterranean), and can be obtained from Admiralty Chart Agents.

GEN 3.2.4 AERONAUTICAL CHART SERIES AVAILABLE

1 Specifications

1.1 Internationally agreed specifications for charts are set down by the International Civil Aviation Organisation in Annex 4 to the convention on International Civil Aviation. In general, the charts listed below conform to these specifications where they apply.

1.2 Co-ordinate Datum Reference = WGS 84.

2 General Description of CAA Aeronautical Charts

2.1 Aeronautical Charts ICAO Scale 1:500,000 – United Kingdom

2.1.1 This series, covering the United Kingdom, is constructed on a Lambert conic projection and conforms with ICAO specifications for Topography, Culture and Aeronautical Information. A Chart List is at GEN 3.2.5 with an Index at GEN 3.2.6.

2.2 Topographical Air Charts of the United Kingdom – Scale 1:250,000

2.2.1 This series is constructed on a Transverse Mercator projection. The Aeronautical Information generally conforms with the series at paragraph 2.1, though runway patterns of both active and disused aerodromes are shown where reliable information is available. The vertical limit of the series is 5000ft ALT. To assist users, airspace with a base of FL 55 is shown, EXCEPT where a minimum ALT in excess of 5000ft applies. If the QNH is below 1013mb Controlled Airspace not shown on the charts may be below 5000ft ALT and reference must be made to Aeronautical Chart ICAO 1:500,000 to ensure adequate vertical separation. A chart list is at GEN 3.2.5 with an index at GEN 3.2.7. Under normal conditions the revision cycle for sheets 6 and 8 is annual, while the remaining sheets will be bi-annual.

2.3 Procedural and Aerodrome Charts

(a) Instrument Approach and Aerodrome Charts – ICAO

(i) Refer to GEN 3.2.1 paragraph 3 for information on departmental responsibilities within the CAA for Instrument Approach Charts – ICAO.

(ii) These charts are available for all aerodromes where Instrument Approach Procedures have been established and approved by the CAA. They conform to ICAO specifications, and appear in AD 2.

(iii) Separate charts for initial arrival procedures, instrument approach procedures, aerodrome charts and, where necessary, parking/docking charts, are now available for all aerodromes used by international commercial air transport listed in Sections AD 2 and AD 3 of the UK AIP. See GEN 3.2.3 paragraph 2.

(b) Aerodrome Obstacle Charts – ICAO Type 'A'

These charts are available for most aerodromes designated for use by scheduled international air services. Where a Type 'A' Chart is not published, operators are advised to refer to the appropriate 1:50,000 Ordnance Survey Maps.

GEN 3.2   GEN 3.2.3 Instrument Approach and Aerodrome Charts – ICAO, SID/STAR Charts, and Noise Abatement Charts

(c) Precision Approach Terrain Charts – ICAO
These charts are designed to provide detailed terrain profile information within a defined portion of the final approach so as to enable aircraft operating agencies to assess the effect of the terrain on decision height determination by the use of radio altimeters. They are published for aerodromes with precision approach runways Categories II and III.

(d) Standard Instrument Departure, Standard Instrument Arrival, Standard Terminal Arrival Route and Noise Abatement Charts
These charts show designated Departure and Arrival routes for all aerodromes where such procedures are established, and are suitable for operational use. They conform to ICAO specifications. See GEN 3.2.3 paragraph 2.

2.4 Miscellaneous Charts

(a) UK Airspace Restrictions and Hazardous Areas – 1:1,000,000
This chart shows all UK airspace restrictions including danger areas and other areas where a potential hazard to air navigation may exist.

(b) UK Areas of Intense Air Activity, Aerial Tactics Areas and Military Low Flying System – 1:1,000,000
This chart portrays areas of intense military air activity and elements of the military low flying system within the UK FIR.

(c) Helicopter Routes in the London Control Zone – Scale 1:50,000
This chart is based on a modified version of the 1:50,000 Ordnance Survey Landranger series, and shows the London Control Zone, Helicopter Routes, the Specified Area, aerodromes, heliports and reporting points within the Zone.

(d) Miscellaneous AIP Charts
The UK AIP charts are available to the public.

2.5 Refer to GEN 3.2.5 for lists of available charts.

3 Symbol Sheet

3.1 The aeronautical symbols used on charts produced by the Authority are shown at GEN 2.3. They correspond with the internationally agreed symbols contained in ICAO Annex 4, except that some additional symbols are included where appropriate ICAO equivalents do not exist.

GEN 3.2.5 AERONAUTICAL CHARTS AVAILABLE

1 Aeronautical Charts, ICAO, United Kingdom (Scale 1:500,000) (see Index at GEN 3.2.6)

Copies of these charts are obtainable from Airplan Flight Equipment Ltd (See GEN 3.2.3). Price £13.99 per sheet.

Sheet Ref	Chart Title	Edition	Aero Info Date
2150ABCD	Scotland	23	9 Jun 2005
2171AB	Northern England and Northern Ireland	28	12 May 2005
2171CD	Southern England and Wales	31	15 Apr 2005

2 Topographical Air Charts of the United Kingdom (Scale 1:250,000) New Series (see Index at GEN 3.2.7)

Copies of these charts are obtainable from Airplan Flight Equipment Ltd (See GEN 3.2.3). Price £13.99 per sheet.

Sheet Ref	Chart Title	Edition	Aero Info Date
1	Northern Scotland West	3	19 Feb 2004
2	Northern Scotland East	3	18 Mar 2004
3	Northern Ireland	3	7 Jul 2005
4	The Borders	4	30 Oct 2003
5	Central England and Wales	6	14 Apr 2005
6	England East	6	9 Jun 2005
7	The West and South Wales	4	4 Aug 2005
8	England South	9	17 Feb 2005

Note: All the above Charts (listed in para 1 and 2) are available as laminated sheets. (Postage and Packing are extra).

3 Instrument Approach and Aerodrome Charts – ICAO, SID/STAR Charts, and Noise Abatement Charts
Refer to paragraph 6, and to Sections AD 2 and AD 3.

4 Aerodrome Obstacle Charts – ICAO TYPE A – Operating Limitations

(i) Copies of these charts may be obtained from Documedia Solutions Ltd. (See GEN 3.2.2) Price £15.00 per sheet
(ii) Charts are also available from online from the Civil Aviation Authority website: www.caa.uk/typecharts

Chart Title	Date	Chart Title	Date
Aberdeen/Dyce		**Jersey**	
Runway 16/34	Aug 2004	Runway 09/27	Aug 2005
Alderney		**Kirkwall**	
Runway 08/26	Feb 2001	Runway 09/27	Jan 2005
Belfast Aldergrove		**Leeds Bradford**	
Chart No 1 Runway 07/25	Nov 2004	Runway 14/32	May 2005
Chart No 2 Runway 17/35	Nov 2004	**Liverpool**	
Belfast City		Runway 09/27	Oct 2002
Runway 04/22	Oct 2005	**London City**	
Benbecula		Runway 10/28	May 2004
Runway 06/24	Mar 2002	**London Gatwick**	
Biggin Hill		Chart No 1 Runway 08R/26L	Dec 2003
Runway 03/21	Jun 2003	Chart No 2 Runway 08L/26R	Dec 2003
Birmingham		**London Heathrow**	
Chart No 1 Runway 15/33	May 2005	Chart No 1 Runway 09R/27L	Nov 2005
Chart No 2 Runway 06/24	May 2005	Chart No 2 Runway 09L/27R	Nov 2005
Blackpool		**London Luton**	
Runway 10/28	Jun 2003	Runway 08/26	Jan 2003
Bournemouth		**London Stansted**	
Runway 08/26	Oct 2005	Runway 05/23	Feb 2005
Bristol		**Londonderry/Eglinton**	
Runway 09/27	Jul 2004	Runway 08/26	Jul 2004
Bristol Filton (Filton on chart)		**Manchester**	
Runway 09/27	Dec 2001	Chart No 1 Runway 06L24R	Aug 2002
Cambridge		Chart No 2 Runway 06R/24L	Sep 2002
Runway 05/23	May 2003	**Manchester Woodford** (Woodford on chart)	
Cardiff		Runway 07/25	Oct 2001
Runway 12/30	Jul 2004	**Manston**	
Carlisle		Runway 08/26	May 2004
Runway 07/25	Jan 2003	**Newcastle**	
Coventry		Runway 07/25	Sep 2005
Runway 05/23	Aug 2004	**Norwich**	
Cranfield		Runway 09/27	Jul 2005
Runway 22	Sep 2003	**Nottingham East Midlands**	
Doncaster Sheffield		Runway 09/27	Feb 2005
Runway 202/20	Feb 2005	**Plymouth**	
Dundee		Runway 13/31	Mar 2003
Runway 10/28	Apr 2004	**Prestwick**	
Durham Tees Valley		Chart No 1 Runway 13/31	May 2005
Runway 05/23	Apr 2005	Chart No 2 Runway 03/21	May 2005
Edinburgh		**Sheffield City**	
Chart No 1 Runway 06/24	Oct 2004	Runway 10/28	Feb 2001
Chart No 2 Runway 12/30	Oct 2004	**Shoreham**	
Exeter		Runway 03/21	Sep 2001
Chart No 1 Runway 26	Apr 2004	**Southampton**	
Chart No 2 Runway 08	Apr 2004	Chart No 1 Runway 02/20	Apr 2005
Farnborough		Chart No 2 Runway 02/20	May 2005
Runway 06/24	Sep 2005	(RWY 20-15° angled flight path area)	
Glasgow		**Southend**	
Chart No 1 Runway 05	Feb 2003	Runway 06/24	Feb 2003
Chart No 2 Runway 23	Feb 2003	**Stornoway**	
Guernsey		Runway 18/36	Feb 2003
Runway 09/27	Dec 2001	**Sumburgh**	
Hawarden		Runway 15/33	Nov 2003
Chart No 1 Runway 23	Sep 2004	**Warton**	
Chart No 2 Runway 05	Sep 2004	Runway 08/25	Jun 2003
Humberside		**Wick**	
Runway 03/21	Jul 2003	Chart No 1 Runway 13/31	May 2002
Inverness		Chart No 2 Runway 08/26	May 2002
Runway 06/24	Feb 2002	**Wolverhampton**	
Isle of Man		Runway 16/34	Sep 2004
Chart No 1 Runway 08/26	Aug 2002		
Chart No 2 Runway 03/21	Aug 2002		

5 Precision Approach Terrain Charts – ICAO

Copies of these charts may be obtained from Documedia Solutions Ltd (See GEN 3.2.3). (Price £15.00 per sheet, postage extra).

Chart Title	Date
Belfast/Aldergrove	
Runway 17	May 2001
Runway 26	May 2001
Birmingham	
Runway 15	Apr 2002
Runway 33	Apr 2002
Bristol	
Runway 27	Mar 2001
Bristol Filton	
Runway 27	Apr 2001
East Midlands	
Runway 27	Feb 2001
Runway 09	Feb 2001
Edinburgh	
Runway 06	Apr 2002
Runway 24	Apr 2002
Glasgow	
Runway 05	May 2001
Runway 23	May 2001
Isle of Man	
Runway 08	May 2001
Runway 26	May 2001
Leeds Bradford	
Runway 32	Aug 1992

Chart Title	Date
Liverpool	
Runway 27	Jun 2001
London Gatwick	
Runway 08R	Oct 2001
Runway 26L	Oct 2001
London Heathrow	
Runway 09L	Feb 2001
Runway 09R	Feb 2001
Runway 27L	Feb 2001
Runway 27R	Feb 2001
London Luton	
Runway 08	Jan 2003
Runway 26	Jan 2003
London Stansted	
Runway 05	Jul 2002
Runway 23	Jul 2002
Manchester	
Runway 06L	Aug 2002
Runway 24R	Aug 2002
Newcastle	
Runway 25	Apr 2001
Runway 07	Apr 2001

6 Other Charts

6.1 Selected AIP Chart Pages

Copies of these charts may be obtained from Airplan Flight Equipment Ltd or Documedia Solutions Ltd (See GEN 3.2.3).

Chart Title	Edition	UK AIP Page No	Aero Info Date	Price
Selected AIP Chart Pages				
Chart of United Kingdom ATS Airspace Classifications	9	ENR 6-1-4-1	27 Nov 2005	£5.00 (See note)
Chart of United Kingdom Airspace Restrictions and Hazardous Areas	33	ENR 6-5-1-1	27 Nov 2005	£5.00 (See note)
(Scale 1:1,000,000)				
Chart of United Kingdom Areas of Intense Air Activity, Aerial Tactics Areas	25	ENR 6-5-2-1	27 Nov 2005	£5.00 (See note)

Note: A set of the three ENR charts is available priced at £12.95 when purchased together Available from CAA Chart Sales

6.2 Miscellaneous Charts

Chart Title	Edition	UK AIP Page No	Aero Info Date	Price
Chart of Helicopter Routes in the London Control Zone. (Laminated) (Scale 1:50,000)	11	-	25 Nov 04	£13.99

Available from CAA Charts Sales

---

*Charts are also available from online from the Civil Aviation Authority website: www.caa.uk/typecharts

GEN 3.2 GEN 3.2.6 INDEX TO AERONAUTICAL CHARTS – ICAO 1:500,000

GEN 3.2.6 INDEX TO AERONAUTICAL CHARTS – ICAO 1:500,000

INDEX TO AERONAUTICAL CHARTS ICAO 1:500,000
OF THE UNITED KINGDOM

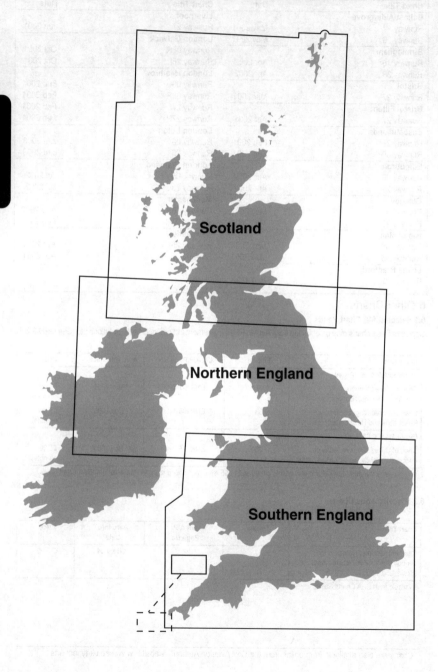

GEN 3.2.7 INDEX TO TOPOGRAPHICAL AIR CHARTS 1:250,000

INDEX TO TOPOGRAPHICAL AIR CHARTS OF THE UK 1:250,000

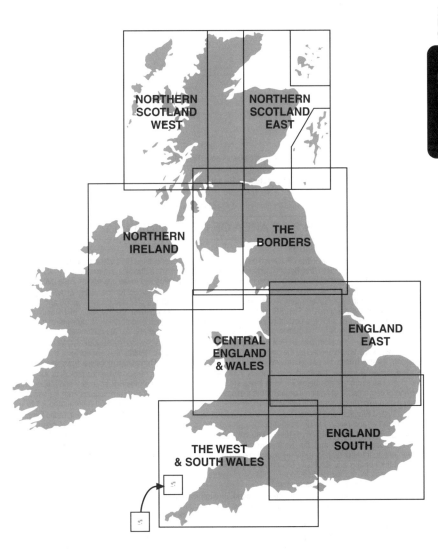

GEN 3.2.8 CHART CORRECTIONS

1 At the compilation date of this amendment there are no known corrections to be made to current charts published by the Authority.

GEN 3.3 AIR TRAFFIC SERVICES
GEN 3.3.1 RESPONSIBLE AUTHORITY

1 Responsibility for the overall administration of the Air Traffic Services in the United Kingdom is vested in the Chief Executive of the National Air Traffic Services Ltd (NATS Ltd) acting under the powers of the Secretaries of State for Transport and for Defence.

Chief Executive
NATS Ltd
Corporate and Technical Centre
4000 Parkway
Whiteley
Fareham
Hants PO15 7FL
Tel: 01489 61001

2 Applicable ICAO Documents

2.1 The Standards, Recommended Practices and, when applicable, the procedures contained in the following ICAO documents are applied:

Annex 2	Rules of the Air;
Annex 11	Air Traffic Services;
Doc 4444	Procedures for Air Navigation Services – Air Traffic Management;
Doc 7030	Regional Supplementary Procedures;
Doc 7754	Air Navigation Plan – European Region;
Doc 8755	Air Navigation Plan – North Atlantic.

2.2 Differences from ICAO Standards Recommended Practices and Procedures are given at GEN 1.7.

GEN 3.3.2 AREA OF RESPONSIBILITY

1 National Responsibilities

1.1 Air Traffic Services, notified in the UK AIP are provided for that Airspace above the UK and the surrounding seas within the London and Scottish FIRs/UIRs. By arrangement with the appropriate State authorities, this responsibility has, in some areas, been delegated to the UK. In other areas, responsibility has been delegated to the appropriate State authority:

(a) Under the terms of a bi-lateral contract between the European Organisation for the safety of Air Navigation (Eurocontrol) and the UK Government, Air Traffic Services above FL 245 are provided by the London and Scottish ACCs throughout the entire UK FIR/UIR and certain portions of Airspace over the Republic of Ireland and its territorial waters, but excluding a portion of the London UIR in the southwest and portions of routes over the North Sea, where arrangements for delegation of control have been made with the appropriate authorities.

(b) By agreement with the North Atlantic Provider States, Air Traffic Services are provided by Scottish ACC for the Airspace over the high seas encompassed by the boundaries of the Shanwick Oceanic Control Area, except for a portion of the Shannon Oceanic Transition Area, where arrangements for delegation of control have been made with the Irish authorities.

(c) The United Kingdom, Denmark and Norway have agreed, to transfer responsibility for providing Air Traffic Services to all aircraft at and below FL 85 in those areas of their FIRs between the FIR boundary and Median Line (the line of demarcation of National areas for the exploration and exploitation of natural resources from the sea bed), to the nation exploiting the natural resources of the area. The areas are shown at ENR 2-2-1-1 paragraph 1.1 to 1.3

(d) The United Kingdom and Iceland have agreed to transfer responsibility to the UK for providing Air Traffic Services to all aircraft at and below FL 85 within a defined area of the Reykjavik FIR/OCA during the hours of operation of Sumburgh ATSU as listed at ENR 1-15-4, paragraph 2.2. The area is shown at ENR 2-2-1-1 paragraph 1.4.

(e) The United Kingdom and the Netherlands have arranged, through the exchange of a bi-lateral Letter of Agreement, to transfer responsibility for providing Air Traffic Services to all aircraft at 3000ft amsl and below in those areas of the London and Scottish FIRs between the FIR boundaries and Median Line (the line of demarcation of National areas for the exploration and exploitation of natural resourcs from the sea bed), to the Netherlands. The area is shown at ENR 2-2-1-1 paragraph 2.1.

(f) The United Kingdom and Denmark have arranged, through the exchange of a bi-lateral Letter of Agreement, to transfer responsibility for providing Air Traffic Services to all aircraft between FL 195 and FL 660 (inclusive) in parts of the London and Scottish FIRs/UIRs to Denmark. The area is shown at ENR 2-2-1-1 paragraph 3.1.

(g) The United Kingdom and the Netherlands have arranged, through the exchange of a bi-lateral Letter of Agreement, to transfer responsibility for providing Air Traffic Services to aircraft between FL 175 and FL 245 (inclusive) in parts of the London FIR to the Netherlands. The areas are shown at ENR 2-2-1-2 paragraphs 4.1.

(h) The United Kingdom and the Netherlands have arranged, through the exchange of a bi-lateral Letter of Agreement, to transfer responsibility for providing Air Traffic Services to all aircraft between FL 235 and FL 660 (inclusive) and between FL 55 and FL 660 (inclusive) in parts of the Amsterdam FIR to the United Kingdom. The areas are shown at ENR 2-2-1-2 paragraphs 5.1 and 6.1.

(i) The United Kingdom, France and the Irish Republic have arranged, through the exchange of a bi-lateral Letter of Agreement, to transfer responsibility for providing Air Traffic Services to all aircraft between FL 245 and FL 660 (inclusive) in parts of the London FIR to France and the Irish Republic. The areas are shown at ENR 2-2-1-1/2 paragraphs 7.1 and 7.2

(j) The United Kingdom and the Irish Republic have arranged through the exchange of a bi-lateral Letter of Arrangement, to transfer responsibility for providing Air Traffic Services to all GAT in a part of the London FIR to the Irish Republic. These areas are shown at ENR 2-2-1-2 paragraphs 8.1 and 8.2

(k) The United Kingdom and the Irish Republic have arranged, through the exchange of a bi-lateral Letter of Agreement, to transfer responsibility for providing Air Traffic Services to all GAT in a part of the Shannon UIR to the United Kingdom. These areas are shown at ENR 2-2-1-3 paragraphs 9.1.

1.2 Flight Information Service and Alerting Service will be provided throughout all the Airspace described above. Whenever part of the FIR or UIR has been further classified, FIS and Alerting Service will be provided by the controller giving other ATS appropriate to the airspace classification. In those parts of the FIR or UIR which have not been further classified, FIS and Alerting Service will be provided by a special Area FIS Officer operating from the appropriate ACC.

2 Provision of Air Traffic Services (ATS)

2.1 The provision of ATS within the United Kingdom's areas of responsibility described in paragraph 1 is shared by several organisations, both Civil and Military.

2.2 National Air Traffic Services (NATS Ltd) provides the principal en-route (ACC/OAC) services together with aerodrome ATS at the airports listed at GEN 3.4.2 paragraph 2.2. Other en-route and aerodrome ATS are provided by civil and military organisations identified within the specific AIP entries.

GEN 3.3.3 TYPES OF SERVICES

1 Area Air Traffic Services

1.1 In addition to the Area Control, Area Flight Information (FIS) and Alerting Services provided by UK ACC/OAC, several other civil and military ATS units provide Radar and Flight Information services which are detailed at ENR 1.6 (LARS/RASA), ENR 1.15 (Off-shore), ENR 2.2 (MATZ) and ENR 5.1 (DACS/DAAIS).

2 Aerodrome Air Traffic Services

2.1 ATC at an aerodrome is responsible for the control of aircraft in the air in the vicinity of the aerodrome and for the control of all traffic on the manoeuvring are subject to prior permission from ATC.

2.2 Control of movements of vehicles and persons on the apron is the responsibility of the aerodrome authority. Movement of aircraft on the apron is subject to prior permission from ATC, who will provide advice and instructions to assist in the prevention of collisions between moving aircraft.

2.3 The total ATC responsibility at an aerodrome is shared between Aerodrome Control and Approach Control. Aerodrome Control is responsible for aircraft on the manoeuvring area except the runways-in-use. The point dividing the responsibilities of Aerodrome Control and of Approach Control for aircraft on the runways-in-use and in the air may vary with different weather conditions or for other considerations, but it is the normal rule that departing aircraft contact Aerodrome Control first and that arriving aircraft contact Approach Control first for ATC instructions.

2.4 Three types of service are used at United Kingdom aerodromes for the control or supervision of aerodrome traffic. Where Air Traffic Control is required an Aerodrome Control Service (TWR) is provided. At other aerodromes, either an Aerodrome Flight Information Service (AFIS) or an Air-Ground Service (A/G) may be provided. Where traffic levels are variable, the available service may be changed at specific times or by arrangement.

2.5 ATC fulfils its functions at an aerodrome by giving aircraft by RTF the instructions and information required for taxiing, take-off or landing.

2.6 At some busy airports to alleviate RTF loading on the operational channels, Automatic Terminal Information Service (ATIS) broadcast messages are used to pass routine arrival/departure information on a discrete RTF frequency or on an appropriate VOR. Pilots of aircraft inbound to these airports are required on first contact with the aerodrome ATS Unit to acknowledge receipt of current information by quoting the code letter of the broadcast. Pilots of outbound aircraft are not normally required to acknowledge receipt of departure ATIS but are requested to ensure that they are in possession of up-to-date information. ATIS is described in ICAO Doc 7030, EUR/RAC paragraph 10.

3 Approach Control Service (APP)

3.1 Approach Control Service is provided at some aerodromes which are within Controlled Airspace and at others which are not. In the latter case, however, there is no legal requirement for pilots flying IMC to comply with the instructions issued by Approach Control unless they are within the Aerodrome Traffic Zone. Nor is there any legal requirement for such pilots to report their presence. It is, therefore, impossible for Approach Control to be sure that they are giving separation from all aircraft in their area and for this reason Approach Control Service at aerodromes outside Controlled Airspace must be regarded as advisory only.

3.2 The more aircraft that are known to Approach Control at an aerodrome outside Controlled Airspace, the better will be the service provided and pilots are therefore strongly recommended either:

(a) To avoid flying under IFR within 10nm radius at less than 3000ft above an aerodrome having Approach Control; or

(b) if it is necessary to fly under IFR in such an airspace, to contact Approach Control when at least 10 minutes flying time away and to comply with any instructions they may give.

3.3 Responsibility of APP at Aerodromes within Controlled Airspace

3.3.1 Approach Control will provide standard separation to IFR flights from the time or place at which:

(a) Inbound aircraft are released by the ACC or Zone Control until they are transferred to Aerodrome Control; and

(b) outbound aircraft are taken over from Aerodrome Control until they are handed over to the ACC or Zone Control;

(c) aircraft inbound from the FIR come under its jurisdiction within the Controlled Airspace until they are transferred to Aerodrome Control.

3.4 Responsibility of APP Control at Aerodromes outside Controlled Airspace

3.4.1 Subject to the reservation in paragraph 3.1, that approach control services outside Controlled Airspace are advisory only, Approach Control will provide separation between aircraft under its jurisdiction from the time and place at which:

(a) Arriving aircraft are released by the ACC until they are transferred to Aerodrome Control;
(b) arriving aircraft first place themselves under Approach Control until they are transferred to Aerodrome Control;
(c) departing aircraft are taken over from Aerodrome Control until they are transferred to the ACC, or they state that they no longer wish to be controlled or they are more than 10 minutes flying time away from the aerodrome, whichever is the sooner;
(d) transit aircraft first place themselves under the control of Approach Control until they are clear of the approach pattern or state they no longer wish to be controlled.

4 Aerodrome Flight Information Service (AFIS)

4.1 The Flight Information Service (FIS) is a service provided at an aerodrome to give information useful for the safe and efficient conduct of flights in the Aerodrome Traffic Zone. From the information received pilots will be able to decide the appropriate course of action to be taken to ensure the safety of flight.

4.2 FIS is available at aerodromes during the hours of operation indicated at AD 2 and AD 3. The service is easily identifiable by the call sign suffix 'INFORMATION'. Only the holder of an FIS officers licence is permitted to use this suffix.

4.3 The Flight Information Service Officer (FISO) is responsible for:
(a) Issuing information to aircraft flying in the Aerodrome Traffic Zone to assist pilots in preventing collisions;
(b) issuing information to aircraft on the manoeuvring area to assist pilots in preventing collisions between aircraft and vehicles/obstructions on the manoeuvring area, or between aircraft moving on the apron;
(c) informing aircraft of essential aerodrome information (ie the state of the aerodrome and its facilities);
(d) alerting the safety services;
(e) initiating overdue action.

5 Services and Procedures for Arriving Flights

5.1 Pilots of arriving aircraft should contact Approach Control when instructed to do so by the ACC. If flying outside Controlled Airspace in IMC the first call should be made to Approach Control at least 10 minutes before ETA at the aerodrome.

5.1.1 As soon as practicable after pilots have made contact with Approach Control they will be given:
(a) Runway-in-use;
(b) Current meteorological information which will include:
(i) Surface wind direction (in degrees magnetic) and speed;
(ii) visibility;
(iii) present weather;
(iv) significant cloud amount and height of base;
(v) QFE or QNH (with height of aerodrome);
(vi) any other relevant information (gusts, icing, etc);
(vii) Runway Visual Range.
Note: This information may be reduced to items (i) and (v) when aircraft are below cloud, flying in VMC and able to continue in VMC to the landing, or at busy aerodromes within Control Zones when visibility and cloud base permit approaches under VFR.
(c) Current runway surface conditions where appropriate;
(d) any changes in the operational status of visual and non-visual aids essential for approach and landing.

5.2 Holding

5.2.1 Holding will normally be carried out in accordance with the procedures set out on IAC. Certain procedures are mandatory and are notified for the purpose of Rule 31(3) of the Rules of the Air Regulations 1996. If none have been published, or if a pilot does not know them, Approach Control should be advised and full instructions will then be given.

5.2.2 If for any reason a pilot is unable to comply with any particular holding instruction he should advise Approach Control and ask permission to follow an alternative procedure. Permission will normally be given if traffic conditions permit.

5.2.3 A pilot will be given a time at which to leave the holding point and commence his approach. This will be given sufficiently in advance for him to arrange his flight so as to arrive at the holding point at the time specified.

5.3 Procedures for Arriving VFR Flights

5.3.1 An aircraft approaching an aerodrome under VFR where an Approach Control Service is available should make initial RTF contact when 15nm or five minutes flying time from the Aerodrome Traffic Zone boundary, whichever is the greater. If the aircraft is not equipped with the Approach frequency, communication on the Aerodrome Control frequency will be acceptable. As well as landing information, ATC will pass information on pertinent known traffic to assist pilots of VFR flights to maintain separation from both IFR and other VFR flights.

5.3.2 If radar sequencing of IFR flights is in progress, ATC will provide VFR flights with information to enable them to fit into the landing sequence.

5.4 Instrument Approaches

5.4.1 Pilots will be expected to be conversant with the correct notified Instrument Approach Procedures detailed in published charts, but on request, in exceptional circumstances, Approach Control will supply the following information:
(a) The aid concerned, aircraft category and Final Approach Track;
(b) arrival level;
(c) type of reversal manoeuvre, including outbound track, length in time or distance, level instructions, and direction of procedure turn where applicable;
(d) intermediate and final approach tracks and fixes, and step down fixes (where applicable), with level instructions;
(e) Obstacle Clearance Height;

(f) Missed Approach Point and Missed approach procedure.

5.4.2 A pilot finding that he can see the ground before he has completed the approach procedure must, nevertheless, carry out the entire procedure, unless he specifically requests and ATC gives permission for him to complete his approach visually. This permission will only be given when:

(a) The pilot can maintain visual reference to the surface; and

(b) the reported cloud ceiling is not below the initial approach level or the pilot reports that visibility will permit a visual approach and he is reasonably confident that a landing can be accomplished. ATC will continue to provide IFR separation from other aircraft.

5.4.3 Providing that it is not at night or in Class A Airspace, a pilot is entitled to ask ATC to cancel his IFR plan during his approach to land providing that he can continue in uninterrupted VMC. In this case, he must accept responsibility for maintaining his own separation from other aircraft.

5.4.4 An arriving aircraft may be cleared, by day only, to descend remaining in VMC and maintaining own separation if reports indicate that this is possible. Essential traffic information will be given.

5.5 Visual Circuit Reporting Procedures

5.5.1 In order that the maximum use may be made of aerodromes for the purpose of landing and taking-off, it is essential that pilots accurately report their positions in the circuit.

5.5.2 Standard Overhead Join

5.5.2.1 An aircraft which has been instructed to complete a standard overhead join will:

(a) Overfly the aerodrome at 2000ft aal;

(b) descend on the 'dead side' to circuit height;

(c) join the circuit by crossing the upwind end of the runway at circuit height;

(d) position downwind.

5.5.3 Position reports are to be made as follows:

(a) **Downwind** – Aircraft are to report 'Downwind' when abeam the upwind end of the runway. Aircraft report 'Late downwind' if they are on downwind leg, have been unable to report 'Downwind' and have passed abeam the downwind end of the runway.

(b) **Base Leg** – Aircraft are to report 'Base Leg', if requested by ATC, immediately on completion of the turn on to base leg.

(c) **Final** – Aircraft are to report 'Final' after the completion of the turn on to final approach and when at a range of not more than 4nm from the approach end of the runway.

(d) **Long Final** – Aircraft flying a final approach of a greater length than 4nm are to report 'Long Final' when beyond that range, and 'Final' when a range of 4nm is reached. Aircraft flying a straight-in approach are to report 'Long Final' at 8nm from the approach end of the runway, and 'Final' when a range of 4nm is reached.

Note 1: At grass aerodromes, the area to be used for landing should be regarded as the runway for the purposes of position reporting.

Note 2: Pilots are reminded that, when operating in the vicinity of aerodromes and not in receipt of Air Traffic Control instructions to the contrary, Rule 17 para 5 (a) of the Rules of the Air Regulations 1996 applies and they should 'conform to the pattern of traffic formed by other aircraft intending to land at that aerodrome'. Flights contemplating carrying out a straight in approach should bear in mind that, on many occasions, this method of joining will not permit compliance with this Rule.

6 Runway Utilisation Procedures

6.1 Runway-in-Use

6.1.1 The runway-in-use is selected by Aerodrome Control as the best for general purposes. If it is unsuitable for a particular operation, the pilot can obtain permission from ATC to use another but must accept that he may thereby incur a delay.

6.2 Clearance for Immediate Take-Off

6.2.1 A pilot receiving the ATC instruction 'cleared for immediate take-off' is required to act as follows:

(a) If waiting clear of the runway, taxi immediately on to it and begin his take-off run without stopping his aircraft;

(b) if already lined up on the runway, take-off without delay;

(c) if unable to comply with the instruction, inform ATC immediately.

6.3 Land after Procedure

6.3.1 Normally, only one aircraft is permitted to land or take-off on the runway-in-use at any one time. However, when the traffic sequence is two successive landing aircraft, the second one may be allowed to land before the first one has cleared the runway-in-use, providing:

(a) The runway is long enough;

(b) it is during daylight hours;

(c) the second aircraft will be able to see the first aircraft clearly and continuously until it is clear of the runway;

(d) the second aircraft has been warned.

ATC will provide this warning by issuing the second aircraft with the instruction 'land after (first aircraft type)' in place of the usual instruction 'Cleared to land'. Responsibility for ensuring adequate separation between the two aircraft rests with the pilot of the second aircraft.

6.4 Special Landing Procedures at London Heathrow, London Gatwick, London Stansted and Manchester Airports

6.4.1 Special landing procedures may be in force at London Heathrow, London Gatwick (except for Runway 08L/26R at London Gatwick), London Stansted (Land after Departure only) and Manchester (Land after Departure only) in conditions shown hereunder, when the use will be as follows:

GEN 3.3.3.6 Runway Utilisation Procedures

(a) When the runway-in-use is temporarily occupied by other traffic, landing clearance will be issued to an arriving aircraft provided that at the time the aircraft crosses the threshold of the runway-in-use the following separation distances will exist:

London Heathrow and London Gatwick

(i) Landing following landing – The preceding landing aircraft will be clear of the runway-in-use or will be at least 2500m from the threshold of the runway-in-use.

(ii) Landing following departure – The departing aircraft will be airborne and at least 2000m from the threshold of the runway-in-use, or if not airborne, will be at least 2500m from the threshold of the runway-in-use.

Manchester

Landing following departure – The departing aircraft will be airborne and at least 2000m from the landing threshold of the runway-in-use, or if not airborne, will be at least 2400m from the landing threshold of the runway-in-use.

London Stansted

Landing following departure – The departing aircraft will be airborne and at least 2000m from the threshold of the runway-in-use, or if not airborne, will be at least 2500m from the threshold of the runway-in-use.

(b) Reduced separation distances as follows will be used where both the preceding and succeeding landing aircraft or both the landing and departing aircraft are propeller driven and have a maximum total weight authorised not exceeding 5700kg:

London Heathrow and London Gatwick

(i) Landing following landing – The preceding aircraft will be clear of the runway-in-use or will be at least 1500m from the threshold of the runway-in-use.

(ii) Landing following departure – The departing aircraft will be airborne or will be at least 1500m from the threshold of the runway-in-use.

Manchester and London Stansted

Landing following departure – The departing aircraft will be airborne and at least 1500m from the landing threshold, or if not airborne, will be at least 1500m from the landing threshold.

Note: The reduced distances do not apply to those jets which are 5700kg MTWA or less.

(c) Conditions of Use. The procedures will be used by DAY only under the following conditions:

London Heathrow and London Stansted

(i) When the reported meteorological conditions are equal to or better than a visibility of 6km and a cloud ceiling of 1000ft and the Air Controller is satisfied that the pilot of the next arriving aircraft will be able to observe continuously the relevant traffic.

(ii) When both the preceding and succeeding aircraft are being operated in the normal manner. (Pilots are responsible for notifying ATC if they are operating their aircraft in other than the normal manner; eg final approach speed greater than 160kt).

(iii) When the runway is dry and free of all precipitants such that there is no evidence that the braking action may be adversely affected.

(iv) When the Air Controller is able to assess the separation either visually or by means of Aerodrome Traffic Monitor.

London Heathrow and London Stansted

(i) When the reported meteorological conditions are equal to or better than a visibility of 6km and a cloud ceiling of 1000ft and the Air Controller is satisfied that the pilot of the next arriving aircrafts will be able to observe continuously the relevant traffic;

(ii) When both the preceding and succeeding aircraft are being operated in the normal manner. (Pilots are responsible for notifying ATC if they are operating their aircraft in other than the normal manner; eg final approach speed greater than 160kt.

(iii) When the runway is dry and free of all precipitants such that there is no evidence that the braking action may be adversely affected;

(iv) When the Air controller is able to asses the separation either visually or by means of Aerodrome Traffic monitor.

London Gatwick

(i) When 26L/08R is in use;

(ii) When the controller is satisfied that the pilot of the next arriving aircraft will be able to observe the relevant traffic clearly and continuously;

(iii) When the pilot of the following aircraft is warned;

(iv) When there is no evidence that the braking action may be adversely affected;

(v) When the controller is able to assess separation visually or by radar derived information.

Manchester

(i) When the reported meteorological conditions are equal to or better than a visibility of 6km and a cloud ceiling of 1000ft.

(ii) When both the preceding and succeeding aircraft are being operated in the normal manner. (Pilots are responsible for notifying ATC if they are operating their aircraft in other than the normal manner; eg final approach speed greater than 160kt).

(iii) When the runway is dry and free of all precipitants such that there is no evidence that the braking action may be adversely affected.

(iv) When the Air Controller is able to assess the separation visually.

(d) When issuing a landing clearance following the application of these procedures ATC will issue the second aircraft with the following instructions:

London Heathrow and London Gatwick
.... (call sign) .... after the landing/departing ....
(Aircraft Type) cleared to land Runway .... (Designator).

Manchester and London Stansted
.... (call sign) .... after the departing .... (Aircraft Type) cleared to land Runway .... (Designator).

7 Departure Clearances

7.1 ATC clearances shall specify some or all of the following as necessary:
(a) Direction of take-off and turn after take-off;
(b) track to be made good before proceeding on desired heading;
(c) level to maintain before continuing climb to assigned cruising level, time, point and/or rate at which level change shall be made;
(d) and any other manoeuvre necessary to maintain separation as appropriate.

7.2 A departing aircraft may be cleared to climb, subject to remaining in VMC and maintaining own separation, until a specified time, place or level if reports indicate that this is possible. Essential traffic information will be given.

7.3 The Pilots of all aircraft flying Instrument Departures should include at least the following items of information on first contact with Approach Control/Departure Radar:
(a) Callsign;
(b) SID Designator where appropriate;
(c) current or passing ALT/FL; PLUS:
(d) cleared ALT/FL. For Standard Instrument Departures (SID) involving stepped climb profiles, state the initial ALT/FL to which the aircraft is climbing.

8 Lamp, Pyrotechnic and Ground Signals

8.1 Non-radio aircraft may be given instructions or information by signal lamp or pyrotechnics from the Control Tower and by ground signals in the Aerodrome Signals Area.

8.2 Lamp and pyrotechnic signals may be made to any aircraft, radio-equipped or otherwise from a subsidiary control point such as a Runway Control Vehicle.

9 Initial Call

9.1 Pilots of aircraft flying Instrument Departures (including those outside controlled airspace) shall include the following information on initial contact with the first en-route ATS Unit:
(a) Callsign;
(b) SID or Standard Departure Route Designator (where appropriate);
(c) Current or passing level; PLUS
(d) Initial climb level (ie the first level at which the aircraft will level off unless otherwise cleared. For example, on a Standard Instrument Departure that involves a stepped climb profile, the initial climb level will be the first level specified in the profile).

9.2 Unless otherwise instructed or where paragraph 9.1 applies, when changing communication channel to an ATC unit (including changes within the same ATS unit), the initial call on the new frequency shall include aircraft identification and level only.

9.2.1 When making such an initial call and the aircraft is in level flight but cleared to another level, the call shall include the aircraft identification followed by the current level and the cleared level.

9.2.2 When making such an initial call and the aircraft is not in level flight, the call shall include the aircraft identification followed by the cleared level only.

9.2.3 When making such an initial call and the aircraft has been assigned a speed, this information shall also be included.

Note: Except as described above, a pilot receiving a Radar Control Service is not required to report leaving a level, passing a level, or reaching a level, unless specifically requested to do so. Pilots receiving a Radar Advisory Service (RAS) must report before changing heading or level, or if receiving a Radar Information Service (RIS), must report before changing level, level band, or route (as described in the UK AIP ENR 1.6.1, Use of Radar in Air Traffic Services).

GEN 3.3.4 CO-ORDINATION BETWEEN THE OPERATOR AND ATS

1 Co-ordination between the Operator and ATS is effected in accordance with ICAO Annex 11 Chapter 2 paragraph 2.15 and ICAO Doc 4444 – Chapter 11 paragraphs 11.2.1.1.4 and 11.2.1.1.5.

GEN 3.3.5 MINIMUM FLIGHT ALTITUDE

TO BE DEVELOPED

GEN 3.3.6 ATS UNITS ADDRESS LIST

Name	Postal Address	Tel No	Fax No	AFS
London TCC	Porters Way West Drayton Middlesex UB7 9AX	01895 445566	01895 423993	EGTTZGZC
London ACC	Sopwith Way Swanwick Southampton Hampshire SO31 7AY	01489 572288	01489 612421	EGTTZRZO
	ATC Watch Supervisor	01489 612440 Or 612420	01489 612421	
	Flight Information Service	01489 611970	01489 612421	
	Flight Plan Processing Section	01489 612423/ 612424/612425	01489 612448	
Manchester ACC	Tower Block Manchester Airport M90 2PL	0161 499 5320/1	0161 499 5501/5504	EGCCYFYG
Scottish ACC	Atlantic House	01292 479800	01292 692872	EGPXZRZX
Shanwick OAC	Sherwood Road Prestwick Ayrshire KA9 2NR			
	ATC Watch Manager (ACC)	01292 692842	01292 692872	
	ATC Watch Manager (OAC)	01292 692663	01292 671048	
	Flight Plan Reception Suite		01292 671048	

Details of other ATS unit contact numbers and addresses appear within the AD and ENR sections

GEN 3.4 COMMUNICATION SERVICES
GEN 3.4.1 RESPONSIBLE AUTHORITY

1 The Civil Aviation Telecommunications Services in the United Kingdom are administered by the Civil Aviation Authority.

Civil Aviation Authority
CAA House
45-59 Kingsway
London
WC2B 6TE
Tel: 020 7379 7311
Telex: 883092
Fax: 020 7240 1153
AFTN: EGGAYAYX

2 Applicable ICAO Documents

2.1 The Standards, Recommended Practices and, when applicable, the procedures contained in the following ICAO documents are applied:

- Annex 10   Aeronautical Telecommunications.
- Doc 7030   Regional Supplementary Procedures.
- Doc 7910   Location Indicators for geographical locations.
- Doc 8400   ICAO Abbreviations and Codes.
- Doc 8585   Designators for Aircraft Operating Agencies, Aeronautical Authorities and Services.

2.2 Differences from ICAO Standards Recommended Practices and Procedures are given at GEN 1.7.

GEN 3.4.2 AREAS OF RESPONSIBILITY FOR PROVIDING AERONAUTICAL TELECOMMUNICATIONS SERVICES

1 En-route Telecommunications Services

1.1 All en-route telecommunications (air-interpreted navigational aids and communications) services, together with those services at aerodromes administered by Highlands and Islands Airports Ltd, except where otherwise identified in the AIP, are provided by the National Air Traffic Services Ltd (NATS Ltd).

National Air Traffic Services Ltd
Director – Infrastructure Services
Spectrum House
Gatwick Road
Gatwick Airport South
West Sussex
RH6 0LG
Tel: 01293 576000
Fax: 01293 576655

2 Communications and Navigational Aids at UK Aerodromes

2.1 Several organisations are approved by the CAA to provide Civil Aviation Telecommunications Services at UK aerodromes.

2.2 These services at the following aerodromes are administered by NATS Ltd: Aberdeen (including Off-shore responsibilities), Belfast Aldergrove, Birmingham, Cardiff, Edinburgh, Farnborough, Glasgow, London City, London Gatwick, London Heathrow (including Thames Radar), London Stansted, Manchester, Southampton and Sumburgh (including Off-shore responsibilities).

2.3 All correspondence relating to the services provided should be addressed as follows.

(a) Aerodromes at which ATS is provided under contract by NATS Ltd:

National Air Traffic Services Ltd
Head of Aerodrome Engineering Support (Head of AES)
NATS – Airports Services
Control Tower Building
London Heathrow Airport
Hounslow
Middlesex
TW6 1JJ
Tel: 020 8745 3703
Fax: 020 8745 3690

(b) Aerodromes operated by Highlands and Islands Airports Ltd (HIAL):
Contact NATS Ltd at Gatwick (paragraph 1.1 refers).

(c) Aerodromes at which ATS is not provided by NATS Ltd and that are not listed in paragraph 2.2 or operated by HIAL:
Contact the individual aerodrome/heliport operators identified at AD 2.2/AD 3.2.

3 AFTN and associated Data Services

National Air Traffic Services Ltd
UK Civil Aviation Communications Centre.
Control Tower Building
London Heathrow Airport
Hounslow
Middlesex
TW6 1JJ
Tel: 020 8745 3356
Fax: 020 8745 3673 or 3437
AFS: EGGGYFYX

4 Enquiries and Complaints

4.1 Enquiries or complaints about the performance of civil aviation telecommunications services should be referred in the first instance to the operational organisation providing the services. Urgent matters should be communicated verbally and supported by written reports.

GEN 3.4.3 TYPES OF SERVICE

1 Radio Navigation Services

1.1 The following types of radio aids to navigation are provided in the UK:

(a) MF Non-Directional Beacon (NDB);
(b) VHF Direction-Finding Station (VDF);
(c) Approach Radar (RAD);
(d) Microwave Landing System (MLS);
(e) Precision Approach Radar (PAR) at certain military aerodromes;
(f) Instrument Landing System (ILS);
(g) VHF Omni-directional Radio range (VOR);

(h) Distance Measuring Equipment (DME).

1.2 MF Non-directional Beacon (NDB)

(a) The range promulgated for UK NDBs is based on a daytime protection ratio between wanted and unwanted signals that limits bearing errors at that distance to ±5 degrees or less. At ranges greater than those promulgated bearing errors will increase. Adverse propagation conditions particularly at night will also increase bearing errors.

(b) NDBs provided for use as approach aids, during the notified hours of ATS, at aerodromes for which instrument approach procedures are published in the AD section are notified in this AIP as locator beacons (L). Most locator beacons continue to transmit a usable signal outside notified hours, these signals are provided for the purposes of navigational aid only. Aerodrome NDBs not notified as locator beacons are provided for navigational use only.

(c) Details of the Maritime Radio beacons (NDB) are published by the Hydrographer of the Navy, Hydrographic Department (MOD), Taunton, Somerset TA1 2DN, and are available from Admiralty Chart Agents.

1.3 VHF Direction-Finding Station (VDF)

(a) VDF bearings are classified as follows;
Class A: accurate to within ±2 degrees;
Class B: accurate to within ±5 degrees;
Class C accurate to within ±10 degrees.

(b) VDF bearing information will only be given when conditions are satisfactory. Normally no better than Class B bearing will be available.

1.4 Approach Radar (RAD)

(a) The VHF communications frequencies for use with Approach Radars are listed in AD 2 item 2.18.

1.5 Instrument Landing System (ILS)

(a) Aircraft overflying the localizer or manoeuvring on or near the runway may disturb the ILS guidance signals. ATC will apply increased separation and such other measures considered necessary to prevent interference during Category II and III operations.

(b) Such measures will also be applied at the discretion of ATS when requested by pilots wishing to use Category II and III landing procedures when meteorological conditions do not necessitate them.

(c) For all civil ILS notified in AD 2 item 2.19 the Localizer usable coverage sector is ±35° about the nominal course line and the Glide Path provides coverage to the minimum ICAO requirement of 10nm unless otherwise stated. Pilots using these instrument landing systems are advised not to attempt to intercept and follow the Glide Path until the aircraft is established on the Localizer centre-line. Due to the presence of false courses on some localizers operating in the UK, pilots are advised not to attempt to use any ILS facility outside ±35° of the front course line. This advice is in addition to the notes promulgated in AD 2 item 2.19 for individual ILS. Use of ILS facilities in the UK is the subject of AIC 34/1997 (Pink 141).

(d) Steep Angle facilities listed in the ILS entries in AD 2 item 2.19 provide a more limited coverage than that described in paragraph 1.5(c), the Localizer usable coverage sector being limited to 10nm at ±35° and to 18nm at ±10°. The glidepath provides coverage to 8nm.

(e) Although all MOD ILS facilities are technically classed as 'uncategorised', they are flight checked to ICAO CAT I standards. However, pilots of aircraft cleared to carry out practice auto-coupled approaches with the appropriate visual references to below CAT I limits are to note the following: Unless specifically promulgated otherwise, ILS facilities at MOD airfields are not capable of providing the required quality of beam structure to enable auto-coupled approaches to be continued below the minimum CAT I Decision Height.

1.6 Instrument Landing System (ILS) and Distance Measuring Equipment (DME)

(a) A DME facility at an aerodrome, which is frequency paired with the ILS, is arranged to give zero range indication with respect to the threshold of the runway with which it is associated and precise ranges will only be indicated when aircraft are in line with the runway on the approach path. As a consequence of this, if used other than in accordance with promulgated procedures indicated range should be taken as an approximate range to the aerodrome.

1.7 VHF Omni-directional Radio Range (VOR) and Distance Measuring Equipment (DME)

(a) The Designated Operational Coverage promulgated for UK VORs and DMEs together with details of any unsatisfactory conditions known to exist, are listed in AD 2 item 2.19 or at ENR 4.1.

(b) Because inaccurate bearing information may be radiated by a VOR during a changeover to the standby transmitter, no identification signal is radiated until the changeover is completed. Pilots are advised to continually monitor the identification signal throughout a VOR approach.

(c) Where a VOR and TACAN are frequency paired, but not within the co-location limit of 600m, the last letter of the TACAN identification will be a 'Z'. Civil Pilots are advised not to make operational use of distance information provided by Military TACAN facilities promulgated as unreliable and/or transmitting a series of dots after the identification code.

(d) Where an en-route VOR or VOR/DME facility has an instrument approach procedure published in the AD section, a note in Column 7 at ENR 4.1 indicates the aerodrome so served. The hours of service as an approach aid are within the notified hours of the Air Traffic Services at the aerodrome served.

1.8 Microwave Landing Systems (MLS)

(a) For all MLS notified in AD 2 item 2.19 the azimuth coverage section is ±40° about the nominal course line. Range is 20nm unless otherwise stated.

1.9 Aerodrome Distance Measuring Equipment (DME)

(a) Aerodrome DME referred to in instrument approach procedures published in the AD section are provided for use as approach aids during the notified hours of ATS only. Most DME continue to transmit outside ATC notified hours for the purpose of navigational aid only.

(b) Aerodrome DME with their zero range offset to occur at specific runway thresholds are identified by comments in Column 7 at AD 2 item 2.19. Any DME range indications observed when between runway thresholds should be ignored. Other aerodrome DME indicate true slant range from aircraft to DME site.

1.10 Radio Navigation Aids – Designated Operational Coverage

1.10.1 Due to the limitations in the availability of spectrum, most NDB, VOR and DME facilities operate on a shared frequency channel. Frequency planning criteria based on ICAO Standards and Recommended Practices are employed to reduce the risk of interference from other facilities operating on the same frequency (co-channel interference).

1.10.2 When interference occurs, it is likely to be manifested in one of the following ways:
(a) Garbled identification;
(b) Bearing/range errors;
(c) Inability to acquire the navigational signal;
(d) Acquiring the undesired signal instead of the desired one.

1.10.3 Under abnormal propagation conditions interference can occur. For VOR and DME this is rare in the European region. However for a NDB, such conditions occur for a number of reasons (see paragraph 1.11).

1.10.4 The attention of pilots and aircraft operators is drawn to the following:
(a) Using a VOR/DME outside the Designated Operational Coverage (DOC) can lead to errors in navigation. Such errors can be dangerous. This is particularly to be noted when using multiple DME in the RNAV configuration where it is difficult, if not impossible, to make a positive identification of the beacon being used.
(b) DOCs are published in the UK AIP AD 2 item 2.19 and ENR 4.1. Where pilot channel selection is made, it is essential that this document be consulted as part of the pre-flight briefing to determine the DOC of every radio navigation aid upon which the safety of the intended flight may depend.

1.11 Limitations of Non-Directional Beacons and Automatic Direction Finding Equipment

1.11.1 Although VHF Omni-Range facilities (VOR) have increasingly replaced Non-Directional Beacons (NDBs) in many parts of the world, NDBs are still in use as navigation and instrument approach aids and are likely to be utilised in these roles for several years to come.

1.11.2 The increasing use of VOR may result in pilots losing sight of the inherent limitations of the NDB and its associated airborne Automatic Direction-Finding (ADF) equipment which if used under certain conditions is capable of producing large and potentially dangerous errors.

1.11.3 The principal factors liable to affect ND13/A13F system performance and integrity are:
Static Interference
Station Interference
Night Effect
Mountain Effect
Coastal Refraction
Absence of failure warning system

1.11.3.1 Very occasionally the Authority becomes aware of other conditions in ADF/NDB systems which give rise to false indications. In all such cases notification is given to the affected aircraft and ground equipment operators.

1.11.3.2 Static Interference – All kinds of precipitation (including failing snow) and thunderstorms can cause static interference of varying intensity to ADF systems. Precipitation static reduces the effective range and accuracy of bearing information and thunderstorms can give rise to bearing errors of considerable magnitude and even to false 'overhead' indications. Indeed it is often said that in an area affected by thunderstorm activity, the ADF bearing pointer is useful only as an indication of the direction of the most active storm cell.

1.11.3.3 Station Interference – Most countries adopt measures to minimise the possibility of interference between transmissions from different stations by spacing frequencies and limiting the power outputs of those which might conflict. However, the LF and MF frequency bands remain inevitably congested and there is a risk that some interference will occasionally occur. When interference is experienced, bearing errors of varying degree will result. By day, the use of a NDB within the promulgated service range (based on daylight conditions) will normally afford protection against interference. Providing the NDB is correctly selected and identified reliable performance can usually be expected. By night, however, it is possible for skywave signals from other (more distant) transmitters to penetrate those areas considered protected during the day, thus giving rise to the possibility of two signals being received and resulting in unreliable bearing indications. Extreme care should therefore be exercised when making use of NDBs during night or twilight hours, even when well within the promulgated service range. A similar degree of care is necessary by day when close to the limit of the promulgated service range. Positive identification of the call sign of the required NDB is essential and is just as important with modern incrementally-tuned crystal controlled ADF sets, as with the earlier designs since frequency references alone cannot guarantee that the required NDB is being unambiguously received. Following initial identification and when ADIF indications are being followed, further checks on reception of the correct call sign and on the accuracy of tuning should also be made at frequent intervals.

1.11.3.4 Night Effect – At night in addition to the interference which can occur between transmissions from different stations (already described in paragraph 1.11.3.3) it is possible for the reception of a ground wave signal from an individual NDB to be contaminated by a sky wave signal from the same transmission source. This will give rise to bearing errors of varying magnitude depending on the heights of the ionised layers and the polarisation of the signals on arrival at the receiver. Night effect is usually most marked during the twilight hours when sky wave contamination can cause 'fading' of signal strength with resultant wandering of the ADIF bearing pointer. Caution should therefore be exercised whenever fluctuations in bearing indications are observed in the circumstances described.

1.11.3.5 Mountain Effect – ADF systems may be subject to errors caused by the reflection and refraction of the transmitted radio waves in mountainous areas. High ground between the aircraft and the NDB may increase the errors especially at low altitude.

1.11.3.6 Coastal Refraction – In coastal areas the differing radio energy absorption properties of land and water result in refraction of NDB transmissions. This error, known as 'Coastal Refraction', is most marked when the transmissions cross the coastline at an oblique angle and when the NDB is located away from the coast. Such bearings should, therefore, be used with caution.

1.11.3.7 Lack of Failure Warning System – Because of the absence of failure warning devices on most ADF instruments, failure in any part of the system (including the NDB) may produce false indications which are not readily detectable. NDB failures in particular could adversely effect both systems of a dual ADF installation in the aircraft. Having selected and identified the NDB, monitoring the audio identification signal and the pointer behaviour is the correct method of assuring normal system operation. This will reduce the risk of a false indication being followed and applies particularly when making an approach toward the NDB, when, in the event of failure, the ADF pointer could indicate that the beacon is ahead of the aircraft even though the beacon has been passed. Particular care should be exercised when an instrument approach procedure is commenced at ADF pointer reversal, in Instrument Meteorological Conditions to below sector safety altitude where no independent cross check is available.

1.11.4 Two methods of modulation are used to transmit the Morse identification signal of a NDB. In the UK all NDB utilise Modulated Continuous Wave (NON A2A) type of modulation but in many other countries Interrupted Continuous Wave (NON A1A) modulation may be employed which requires the use of a Beat Frequency Oscillator (BFO) or tone generator in the ADIF receiver. Pilots must therefore be aware of the type of emission to expect and to pre-select the ADF receiver controls accordingly. In the case of Interrupted Continuous Wave (NON A1A) type emissions, the ADF bearing pointer may wander during the ident period, due to the interruptions in the carrier frequency.

1.11.6 In conclusion it has to be stressed that at the comparatively short distances, ie less than 50 nautical miles, over which NDB are most commonly used, the most potentially dangerous errors are those resulting from all types of precipitation static, thunderstorms and station interference (particularly at night). When these are experienced, the ADF system should be used only when necessary and then with extreme caution; VHF aids are much less affected and these should be used in preference wherever possible. Cross-checking on the accuracy of ADF indications by reference to other available navigational aids is not only a matter of good airmanship but also a most necessary safeguard wherever any difficulty is experienced in the reception or identification of the intended NDB.

2 The Aeronautical Mobile Service

2.1 General

(a) Facilities are provided to meet the Air/Ground communications requirements of the Air Traffic Services described in the AIP and the Emergency Services detailed at paragraph 2.4. These services include coverage over the greater part of the United Kingdom Flight Information Regions above 3000ft. Some limited cover may be possible below this altitude.

(b) United Kingdom Air/Ground facilities will communicate with aircraft on frequencies within the Aeronautical Mobile (R) Service which has been allocated to the band 118 to 136.975MHz.

(c) United Kingdom Airspace is a 25kHz channel spaced environment with 760 communication channels, including some offset carrier systems.

(d) The language to be used when communicating on the United Kingdom Aeronautical Mobile Service is English.

(e) Procedures to follow in the event of Radio failures are contained in the AD and ENR Sections.

(f) At a civil aerodrome the following words in a call sign identify an Air Traffic Control Service: TOWER, APPROACH, GROUND, ZONE, RADAR, DIRECTOR, DELIVERY.

(g) In a call sign, only the word INFORMATION is used to identify an Aerodrome Flight Information Service, Aerodrome Terminal Information Service or Area Flight Information Service.

(h) In a call sign, only the word RADIO is used to identify an aerodrome Air/Ground communication service.

2.2 Use of VHF R/T Channels

(a) Geographical separation between international services using the same or adjacent frequencies is determined so as to ensure as far as possible that aircraft at the limits of height and range to each service do not interfere with one another. In the case of en-route sectors these limits correspond to that of the ATC sector concerned and those for international aerodrome services are appropriate a radius of 25nm up to a height of 4000ft (TWR) or 10000ft (APP).

(b) Except in emergency, or unless otherwise instructed by the Air Traffic Services, pilots should observe these limits. Services other than international services are provided on frequencies which are shared between numerous ground stations and have to operate to a higher utilisation in order to satisfy the demand for frequencies. Pilots using these frequencies should assist in reducing interference by keeping communications to a minimum and by limiting the use of aircraft transmitters to the minimum height and distance from the aerodrome that are operationally necessary. In the case of TWR, AFIS and A/G facilities, communications on these frequencies should be restricted as far as possible to heights up to 1000ft in the immediate vicinity of the aerodrome concerned and in any event within 10nm and 3000ft.

2.3 Common Frequency for Helicopter Departures

(a) At locations having no ground radio facilities a VHF channel is available to assist departing helicopters.

(b) Conditions of use are:

(i) It shall only be used at locations having no radio facilities. If another VHF assignment is valid for that location, it must be used even outside the normal operating hours;

(ii) Transmissions shall occur only when helicopters are below 500ft agl;

(iii) Helicopters approaching a site should monitor the channel. Blind transmissions are not permitted.

(c) Departing helicopters shall state:

(i) 'To all stations';

(ii) The callsign of the aircraft;

(iii) The location either by name or by reference to a readily identifiable feature;

(iv) The direction and height of the intended departure.

(d) The frequency assigned is 122.950MHz and shall be known as 'DEPCOM'.

2.4 Common VHF frequency for Use at Aerodromes having no notified Ground Radio Frequency

(a) At aerodromes having no notified ground radio frequency a VHF frequency is available to assist pilots to avoid potential collisions between arriving and departing aircraft. Pilots may use this frequency to broadcast their intentions for safety purposes.

(b) the frequency assigned is 135.475 MHz and is known as 'SAFETYCOM'.

(c) the conditions of use are:

GEN 3.4   GEN 3.4.3.3 Aeronautical Fixed Services

(i) SAFETYCOM shall only be used at aerodromes having no notified ground radio frequency. If a VHF frequency is notified for a location, that notified frequency must be used even outside the notified operating hours.

(ii) transmissions shall be made only within a maximum range of 10nm of the aerodrome of intended landing, and below 2000 feet above the aerodrome elevation.

(iii) SAFETYCOM shall only be used to transmit information regarding the pilot's intentions, and there should be no response, except where the pilot of another aircraft also needs to transmit his intentions or, exceptionally, has information critical to the safety of an aircraft in a condition of distress or urgency.

(iv) phraseology is to comply with the requirements of CAP 413 (Radiotelephony manual) Chapter 4 Section 6.

(v) SAFETYCOM is not o be used for the conduct of formation flights unless landing at or taking off from an aerodrome for which no other frequency is notified and within the limits specified at sub para (ii).

(vi) Pilots operating at aerodromes without a notified frequency are recommended to use SAFETYCOM, but its use is not mandatory. However, if pilots choose to use it, they must make the transmissions listed in CAP 413 as 'essential'. It must not be assumed that all other pilots in the vicinity are monitoring the frequency and, as at all other times, pilots must maintain a good lookout.

(vii) No air traffic service is associated with SAFETYCOM. Where an aerodrome lies within controlled airspace, pilots must establish contact with the responsible air traffic services unit, and obtain clearance prior to entering controlled airspace.

(viii) information transmitted on SAFETYCOM confers no priority or right of way. Pilots shall comply with the Rules of the Air regulations, including the provisions in relation to avoiding aerial collisions.

(d) unless, specifically approved by the CAA, SAFETYCOM is not to be used for special events as defined in CAP 403 (Flying Displays and Special Events: A Guide to Safety and Administrative Arrangements). Frequencies for special events should continue to be requested through existing channels.

2.5 Emergency Service

(a) An emergency communications and aid service is continuously available on 121.500MHz from two Distress and Diversion (D & D) sections, one located in the London Area and Terminal Control Centre (Mil) (LATCC Mil) at West Drayton and the other in the Scottish Area Control Centre (ScACC) at Prestwick.

(b) Operational control is exercised, south of 55°N, from the LATCC D & D, call sign 'LONDON CENTRE' and north of 55°N, from the ScACC at Prestwick, call sign 'SCOTTISH CENTRE'. The service provides coverage over the greater part of the United Kingdom above 3000ft. In addition, the stations and units listed at GEN 3.6 have the capability of providing an emergency service on 121.500MHz. (For further details see GEN 3.6.6, paragraph 5).

(c) Pilots of aircraft in emergency and using 121.500MHz should broadcast the initial 'MAYDAY' or 'PAN PAN' call; it is not necessary to address the call to any specific Centre or Station. The Air Traffic Controller at West Drayton or Prestwick will answer the call depending on the location of the aircraft, and initiate appropriate action. (For use of SSR in emergency see ENR 1.6).

(d) If the emergency is ended the pilot should inform the controlling authority of the fact and state his intentions before leaving the frequency. This will ensure that any action to alert diversion aerodromes or other assistance will be cancelled.

(e) Details of the ATSUs with Emergency Facilities are shown at GEN 3.6.

2.6 Emergency Satellite Voice Calls from Aircraft

See GEN 3.6.

2.7 Radio Communications between Aerodrome Fire Services and Aircraft during an Emergency

(a) Whenever an emergency has been declared at an aerodrome where this service is notified, aircraft may communicate direct with the Fire Service in attendance with the following conditions:

(i) The service must be used only when the aircraft is on the ground.

(ii) Contact with ATC, on the appropriate frequency must be maintained.

Note: This service is only available by arrangement via ATC and may only be used for the duration of the emergency. The fire service does not normally monitor this service at other times.

(b) The availability of this service is indicated in the AD section at item 2.18. It should be noted that the service provided is not an Air Traffic Service.

2.8 Relay of RTF Communications to the Public

(a) ATS RTF communications may be relayed to departing passengers in certain specified public areas at the following airport(s): London Heathrow (Terminal 4 'Holideck'). This is in compliance with the Wireless Telegraphy Act 1949. The ATS provider reserves the right to disconnect the communication feed during any emergency occurrence without prior notice.

3 Aeronautical Fixed Services

3.1 In the United Kingdom the following Aeronautical Fixed Services are provided:

(a) The Operational Telephone Network for use by ATC and supporting operational services;

(b) the Administrative Telephone Network for use by authorised agencies connected with air traffic operations;

(c) the Aeronautical Fixed Telecommunications Network (AFTN), for the exchange of messages between aeronautical fixed stations within the network.

Note: Messages of authorised categories can be accepted at designated stations for transmission on the AFS. The rules and procedures for handling of communications on the AFS are contained in Annex 10, Vol II Chapters 3-4.

4 Aeronautical Broadcast Service

(a) The Aeronautical Broadcast Service provides broadcasts which contain meteorological, navigation and aerodrome information.

(b) Details of these broadcasts are listed under the name of the controlling aerodrome or Air Traffic Control Unit in the AD section at 2.18/3.17 and in GEN 3.5.7.

GEN 3.4   GEN 3.4.4.1 Approval and Licensing of Aircraft Radio Stations

GEN 3.4.4 REQUIREMENTS AND CONDITIONS

1 Approval and Licensing of Aircraft Radio Stations

1.1 General

1.1.1 The Civil Aviation Authority must approve in writing the design and installation of radio equipment in aircraft and the station must be licensed by the Radiocommunications Agency, Aeronautical Licensing Section before such radio equipment may be operated in an aircraft. The regulations governing the compulsory carriage of radio equipment are contained in Part III of the Air Navigation Order 2000.

1.2 Approval and Licensing Procedure

1.2.1 Full details of the procedure governing the approval and licensing of aircraft radio stations, together with information regarding modifications to such stations, are given in Chapter A3-5 of Section A of the British Civil Airworthiness Requirements Handbook. This publication, or appropriate sections of it, are obtainable from Documedia Solutions Ltd, (see GEN 3.2.3). The approval of an aircraft radio station is based, among other things, upon the results of radio tests in flight; details of the associated procedures are given below.

1.3 Carriage of Radio and Radar Equipment

1.3.1 The requirements for the carriage of radio and radar equipment are contained in the Air Navigation Order 2000 and the Air Navigation (General) Regulations 1993 as amended. The main provisions are published at GEN 1.5.3 paragraph 1.

1.4 Radio Tests in Flight

1.4.1 The CAA only expects Radio Tests in flight to be carried out in exceptional circumstances.

1.4.2 Tests of VHF RTF communication equipment may be carried out with one of the Air Traffic Service Units listed below. Such tests may only be carried out on the frequency stated and when the aircraft is within the Designated Operational Coverage (DOC) of the station as described in the table below.

Air Traffic Service Unit	RTF Call Sign	Frequency (MHz)	DOC (based upon ARP unless otherwise stated)
Aberdeen/Dyce	Aberdeen Approach / Aberdeen Radar	119.500	55nm / FL 250
Belfast Aldergrove	Aldergrove Approach / Aldergrove Radar	125.500	60nm / FL 245
Birmingham	Birmingham Approach / Birmingham Radar	118.050	25nm / FL 100
Bournemouth	Bournemouth Approach / Bournemouth Radar	119.475	50nm / Fl 120
Cambridge	Cambridge Approach	123.600	25nm / FL 100
Cardiff	Cardiff Approach / Cardiff Radar	125.850	50nm / FL 190
Edinburgh	Edinburgh Approach / Edinburgh Radar	121.200	40nm / FL 100
Exeter	Exeter Approach / Exeter Radar	128.975	40nm / FL 100
Glasgow	Glasgow Approach / Glasgow Radar	119.100	25nm / Fl 100
Isle of Man	Ronaldsway Approach / Ronaldsway Radar	120.850	40nm / FL 100
Leeds Bradford	Leeds Bradford Approach	123.750	25nm / FL 100
Liverpool	Liverpool Approach / Liverpool Radar	119.850	40nm / FL 100
London terminal Control	Essex Radar	120.625	Within the operational area of Essex radar
Newcastle	Newcastle Approach / Newcastle Radar	124.375	40nm / FL 150
Norwich	Norwich Approach / Norwich Radar	119.350	40nm / FL 70
Nottingham East Midlands	East Midlands Approach / East Midlands Radar	134.175	60nm / FL 150
Prestwick	Prestwick Approach / Prestwick Radar	120.550	25nm / FL 100
Shoreham	Shoreham Apporach / Shoreham Tower/ Shoreham Radio	123.150	25nm / FL 100
Southend	Southend Approach / Southend Radar	130.775	40nm / FL 100
Sumburgh	Sumburgh Tower	118.250	25nm / FL 40

1.4.3 Please note that not all these aeronautical radio stations operate H24, for hours of operation refer to the relevant aerodrome's AIP entry.

1.4.4 Prior arrangements with these stations are not required, however, when radio traffic conditions are unfavourable it may not be possible for tests to be carried out. Where possible prior arrangements with the ATSU concerned should be made.

1.4.5 Tests of HF Communication equipment may be carried out with any ATSU that is suitably equipped, it should be noted that the UK does not have any ATSU that operate HF for this purpose.

1.4.6 Radio tests in flight of other radio equipment, including the testing of all prototype radio equipment, shall only be carried out with the prior arrangement of the Safety Regulation Group of the Civil Aviation Authority.

1.4.7 Conditions for Tests

1.4.7.1 Except where problems are suspected to have developed during the current flight, RTF equipment tests in flight shall only be made following satisfactory ground testing.

1.4.7.2 VHF RTF test transmission may only be made with the following emission characteristics:
1.4.7.2.1 6K80A3EJN for frequency assignments.

1.4.7.2.2 5K00A3EJN for 8.33 channel assignments.
1.4.7.3 HF RTF test transmissions may only be made, using fixed or tailing antenna, with either 2K70J3EJN or 6K00A3EJN emission characteristics as appropriate to the radio station that the communications is with.
1.4.8 Priority of Messages
1.4.8.1 Communications concerning safety or flight regularity will always be given priority over messages transmitted for test purposes.
1.4.9 Radio Test in Flight Procedure
1.4.9.1 Aircraft must meet minimum airworthiness requirements before commencing any flight for radio test in flight purposes.
1.4.9.2 Aircraft must comply with the Air Traffic Service rules applicable to the area within which they are flying.
1.4.9.3 All radio transmissions for test in flight purposes shall be if the minimum duration necessary for the test and shall not continue for more than 10 seconds. The recurrence of such transmissions shall be kept to the minimum necessary for the test.
1.4.9.4 The nature of the test transmission shall be such that it is identifiable as a test transmission and can not be confused with other communications. To achieve this the following format shall be used:
(a) 'the call sign' of the aeronautical radio station being called, followed by the words 'THIS IS';
(b) 'the aircraft identification';
(c) the words 'RADIO CHECK ON';
(d) 'the frequency (or 8.33 channel)' being used for the test;
(e) 'the aircraft identification'.
1.4.9.5 The operator of the aeronautical radio station being called will assess the transmission and will advise the aircraft making the test transmission in terms of the readability scale below, together with a comment on the nature of any abnormality noted (ie excessive noise) using the following format:
(a) 'the aircraft identification' followed by the words 'THIS IS';
(b) 'the call sign' of the aeronautical radio station replying;
(c) information regarding the readability of the aircraft transmission using the words 'READABILITY x' where 'x' is a number taken from the table below that equates to the assessment of the transmission;
(d) additional concise and unambiguous information with respect to the noted abnormality may be given;
(e) 'the call sign' of the aeronautical radio station replying;
(f) for practical reasons it may be necessary for the operator of an aeronautical radio station to reply with 'THIS IS' followed by 'the call sign' of the aeronautical radio station 'STATION CALLING ON' state 'the frequency (or 8.33 channel) UNREADABLE'.

Quality	Scale
Unreadable	1
Readable now and then	2
Readable but with difficulty	3
Readable	4
Perfectly readable	5

1.4.9.7 The test transmission and reply thereto are recorded at the ATSU.
1.4.9.8 The operator of the airborne station shall complete a 'Flight Test Report' based upon the assessment information conveyed to them, this should be recorded in the aircraft maintenance records. Action should be taken to rectify any identified problems before further test r use.

2 Approval and Licensing of Ground Radio Stations
2.1 General
2.1.1 The Civil Aviation Authority must approve in writing the operation of a ground radio station and the station must be licensed under the terms of the Wireless Telegraphy Act before it may be operated.
2.2 Approval and Licensing Procedure
2.2.1 Full details of the procedures governing the approval and licensing of ground radio stations are obtainable from:
Civil Aviation Authority
ATS Standards Department
Aviation House
Gatwick Road
Gatwick Airport South
West Sussex
RH6 0YR
Tel: 01293 573692
Fax: 01293 573974
AFS: EGGRYAYA
Telex: 878753

2.2.2 Full details of the procedures governing the Wireless Telegraphy Act, aeronautical licensing are obtainable from:

Radio Licensing Section
Surveillance and Spectrum Management
Civil Aviation Authority
K6 G6 CAA House
45-59 Kingsway
London
WC2B 6NN
Tel: 020 7453 6555
Fax: 020 7453 6556
email: radio.licensing@dap.caa.co.uk

3 Malfunctions, Maintenance and Test Transmissions

3.1 During periods of malfunction or maintenance of navigational aids, the promulgated identification signal is suppressed as a means of warning users that the transmission cannot be safely used for navigation purposes. The identification signal will be suppressed in one of two ways:

(a) by complete removal;

(b) by radiating a continuous tone.

3.2 Any transmission using the identification 'TST' is radiating for test purposes only and must not be used for operational purposes.

4 Interference from High Powered Transmitters

4.1 Pilots are advised that interference may be experienced in aircraft flying in the vicinity of high power broadcast stations. If such interference is troublesome or is experienced well beyond the vicinity of the ground transmitter, pilots are requested to file a Ground Fault Report Form CA 647 which should include the following information:

(a) Frequency on which interference occurred;

(b) position and height of aircraft;

(c) aircraft registration letters;

(d) date and time of interference;

(e) description of interfering signal eg music, speech, language, etc.

4.2 Other sources of High Intensity Radio Transmission are listed in the ENR Section. Pilots are warned that within the areas defined, interference or damage to aircraft electronic equipment may occur. Navigation Information from equipment may be unreliable.

4.3 ILS/VOR navigation and VHF communication receivers may still suffer interference from high powered FM broadcast stations in the 88 to 108MHz band.

4.3.1 Since 1 Jan 2001, the use of FM-immune ILS/VOR navigation and VHF communications receivers is required for IFR operations within the UK FIR. Performance standards for FM-immune receivers are specified at sections 3.1.4 and 3.3.8 of ICAO Annex 10, Volume I, and the applicable performance standards and the requirement for carriage of such equipment in UK airspace in CAP 455, Airworthiness Notice, No 84 (as amended), which may be accessed from the Publications page on the website www.srg.caa.co.uk.

4.3.2 The improved performance of these receivers means that interference from licensed FM broadcast stations should no longer occur.

4.3.3 Users of ILS/VOR navigation and VHF communications receivers that do not comply with FM immunity standards may continue to suffer interference from FM broadcast stations. Furthermore, the CAA will no longer conduct an assessment of the potential for interference to such receivers, nor promulgate areas where interference is likely to be suffered.

GEN 3.5 METEOROLOGICAL SERVICES
GEN 3.5.1 RESPONSIBLE AUTHORITY

1 The Civil Aviation Authority is the Meteorological Authority for the United Kingdom (UK). This authority is derived from Directions issued under section 66(1) of the Transport Act 2000 relating to the Civil Aviation Authority's performance of air navigation functions. The policy of the UK Met Authority is to discharge its responsibilities for the provision of meteorological services to UK based national and international civil aviation operations in accordance with ICAO Annex 3 and other national and international requirements as may be promulgates from time to time.

Head of Met Authority
Civil Aviation Authority
Directorate of Airspace Policy
K6 CAA House
45-59 Kingsway
London
WC2B 6TE
Tel: 020 7453 6526
Fax: 020 7453 6565
AFS: EGGAYFYM
e-mail: metauthority@dap.caa.co.uk

2 Meteorological forecasting and climatological services for civil aviation in the United Kingdom are provided by the Met Office acting as the agent for the Civil Aviation Authority.

Head of Civil Aviation
Met Office
Fitzroy Road
Exeter
EX1 3PB
Tel: 01392 886420
Fax: 01392 884156 (Administrative)
AFS: EGRRYTYH (Administrative) or;
AFS: EGRRYMYX (National Meteorological Centre (NMC))

3 Applicable ICAO Documents

3.1 The Standards, Recommended Practices and, when applicable, the procedures contained in the following ICAO documents are applied:

Annex 3	Meteorological Service for International Air Navigation.
Doc 7754	Air Navigation Plan – EUR Region.
Doc 8400	PANS – ICAO Abbreviations and Codes.
Doc 8755	Air Navigation Plan – NAT Region.
Doc 8896	Manual of Aeronautical Meteorological Practice.
Doc 9328	Manual of Runway Visual Range Observing and Reporting Practices.

3.2 The UK Met Authority's objective is to supply operators, flight crew members, ATS units, airport management and other civil aviation users with the meteorological information necessary for the performance of their respective functions, thus contributing towards the safety, regularity and efficiency of air navigation. All ICAO Annex 3 standards and recommended practices, including ICAO definitions listed in Chapter 1 of Annex 3, are applied in the UK unless a difference has been filed with ICAO. UK differences from ICAO standards and recommended practices are listed in GEN 1.7.

GEN 3.5.2 AREAS OF RESPONSIBILITY

1 The United Kingdom provides area Meteorological Watch for the London and Scottish FIR/UIR and for the Shanwick FIR/OCA. The Met Office Operations Centre, Exeter, acts as the Meteorological Watch Office (MWO) for these areas.

2 The UK operates one of the two World Area Forecast Centres (WAFC), responsible for the provision of global forecasts of significant weather and the following global grid point data; wind, temperature, humidity, tropopause height and temperature, maximum wind speed, direction and height. In the event of an interruption of the operation of a WAFC, its functions will be provided by the other WAFC. Additionally, the UK operates a Volcanic Ash Advisory Centre; further information on this service can be found at http://www,icao.int/icao/en/anb/met/IAVW.html

GEN 3.5.3 OBSERVATIONS AND REPORTS

1 Observing Systems and Operating Procedures

1.1 Surface wind sensors on aerodromes are positioned to give the best practical indication of the winds which an aircraft will encounter during take-off and landing within the layer between 6 and 10m above the runway(s). The surface wind reported for take-off and landing by ATS Units at aerodromes supporting operations by aircraft whose maximum total weight authorized is below 5700kg is usually an instantaneous wind measurement with direction referenced to Magnetic North. However, at other designated aerodromes the wind reports for take-off and landing are averaged over the previous 2 minutes. Variations in the wind direction are given when the total variation is 60° or more and the mean speed above 3kt, the directional variations are expressed as the two extreme directions between which the wind has varied in the past 10 minutes. In reports for take-off, surface winds of 3kt or less include a range of wind directions whenever possible if the total variation is 60° or more. Variations from the mean wind speed (gust and lulls) during the past 10 minutes are only reported when the variation from the mean speed has exceeded 10kt. Such variations are expressed as the maximum and minimum speeds attained.

GEN 3.5   GEN 3.5.3.1 Observing Systems and Operating Procedures

1.1.1 At aerodromes which normally report surface wind averaged over the previous 2 minutes, the instantaneous wind velocity is available on request. Where an instantaneous wind velocity has been requested the word 'instant' will be inserted in the report (eg. 'G-CD cleared to land Runway 34 instant surface wind 270 7' or 'G-CD cleared to land Runway 34 instant 270 7'). An indication of the wind velocity normally reported at particular aerodromes is included in Table 3.5.3.2. Aerodromes not featuring in the Table use instant reports with the exception of Wycombe Air Park where an averaged value is normally used.

1.1.2 Surface wind measurements contained in METAR and SPECI reports are referenced to True North and are averaged over the previous 10 minutes, except when during the 10 minute period there is an abrupt and sustained change in wind direction of 30° or more, with a wind speed of at least 10kt both before and after the change, or a change in wind speed of 10kt or more lasting more than 2 minutes. In this case only data occurring since the abrupt change will be used to obtain the mean values. METAR and SPECI reports may give variations in wind direction if during the 10 minute period preceding the time of observation, the total variation in wind direction is 60° or more and the speed greater than 3kt. The maximum speed is only given if it exceeds the mean speed by 10kt or more. At aerodromes with wind sensors at two or more sites, METAR surface wind reports are always obtained from one designated 'aerodrome system' irrespective of the system currently in use by the ATS Unit for take-off and landing reports.

1.2 Information on cloud height is obtained by the use of ceilometers (nodding beam or laser), cloud searchlights and alidades, balloons, pilot reports and observer estimation. At some aerodromes an additional, cloud ceilometer may be installed on the approach. The cloud heights reported from an approach ceilometer are:

(a) The most frequently occurring value during the past 10 minutes if this value is 1000ft or less;

(b) if cloud is being indicated at heights 100ft or more below that indicated at (a) above then the height of the lowest cloud is also reported, prefaced by 'OCNL';

(c) if the most frequently occurring value is above 1000ft but the lowest value is 1000ft or below then only the lowest is reported; for example: most frequent 1200ft, lowest 900ft 'CBR 24 OCNL 900 FEET' (24 refers to the nearest runway).

1.3 Temperature is reported in whole degrees from liquid-in-glass or electrical resistance thermometers located in a ventilated screen.

1.4 Horizontal surface visibility is assessed by human observer, assisted by a visiometer. Visibility is reported in increments of 50m up to 800m, and then increments of 100m up to 5000m and in units of kilometres for 5000m or more.

1.4.1 Pilots are reminded that surface visibility forecast in a TAF, TREND or Area Forecast might be subject to marked deterioration caused by smoke at any time. Such deteriorations in surface visibility will be reported as they occur in routine or special aerodrome meteorological reports, and forecasts might consequently be amended. It is not possible to forecast the onset or cessation of the smoke, or the precise amount of the visibility deterioration. Turbulence and breathing difficulty might also be encountered in the area affected by the smoke.

2 Accuracy of Meteorological Measurement or Observation

The United Nation's World Meteorological Observation (WMO) has assessed the attainable accuracy of meteorological measurement or observation of a number of meteorological parameters, described below. However it should be noted that in most cases this exceeds the requirements for aeronautical meteorological observations specified by ICAO.

Element	Accuracy of Measurement or Observation (1994)
Mean surface wind	Direction ± 5° Speed ± 1kt up to 20kt, ± 5% above 20kt
Variations from the mean surface wind speed	± 1kt
Visibility	± 50m up to 500m ± 10% between 150m and 2000m ± 20% above 2000m up to 10km
RVR	± 25m up to 150m ± 50m between 150m and 500m ± 10% above 500m up to 2000m
Cloud amount	± 1 okta in daylight, worse in darkness and during atmospheric obstruction
Cloud height	± 33ft up to 3300ft ± 100ft above 3300ft up to 10000ft
Air temperature and dew point temperature	± 0.2°C
Pressure value (QNH, QFE)	± 0.3mb

Note: The accuracy stated refers to assessment by instruments (except cloud amount); it is not usually attainable in observations made without the aid of instruments.

3 Details of Meteorological Observations and Reports for UK aerodromes are listed in table 3.5.3.2.

4 Aerodrome Warnings

4.1 Aerodrome Warnings are issued as appropriate when one or more of the following phenomena occurs or is expected to occur:

(a) Gales (when the mean surface wind is expected to exceed 33kt, or if gusts are expected to exceed 42kt) or strong winds according to locally-agreed criteria;

(b) squalls, hail or thunderstorms;

(c) snow, including the expected time of beginning, duration and intensity of fall; the expected depth of accumulated snow, and the time of expected thaw. Amendments or cancellations are issued as necessary;
(d) frost warnings are issued when any of the following conditions are expected:
(i) A ground frost with air temperatures not below freezing point;
(ii) the air temperature above the surface is below freezing point (air frost);
(iii) hoar frost, rime or glaze deposited on parked aircraft.
(e) fog (normally when the visibility is expected to fall below 600m);
(f) rising dust or sand;
(g) freezing precipitation.

4.2 Aerodrome operators requiring notification of the above warnings should apply to the MET Authority (see GEN 3.5.1).
4.3 The normal method of notifying Aerodrome Warnings is by a single AFS, telephone message or Fax by special arrangement to the aerodrome, with local dissemination of the warning being the responsibility of the aerodrome operator.

5 Special Facilities

5.1 Marked Temperature Inversion
5.1.1 At certain aerodromes annotated at Table 3.5.3.2, a Warning of Marked Temperature Inversion is issued whenever a temperature difference of 10°C or more exists between the surface and any point up to 1000ft above the aerodrome. This warning is broadcast on departure and arrival ATIS at aerodromes so equipped, or in the absence of ATIS passed by radio to departing aircraft before take-off, and to arriving aircraft as part of the report of aerodrome meteorological conditions.

5.2 Windshear Alerting Service – London Heathrow Airports
5.2.1 Forecasters for London Heathrow airport regularly review the weather conditions at the airport and monitor aircraft reports of windshear experienced on the approach or climb out. Where a potential low level (below 1600ft) windshear condition exists an Alert is issued, based on one or more of the following criteria:
(a) Mean surface wind speed at least 20kt;
(b) the magnitude of the vector difference between the mean surface wind and the gradient wind (an estimate of the 2000ft wind) at least 40kt;
(c) thunderstorm(s) or heavy shower(s) within approximately 5nm of the airport.
Note: Alerts are also issued based on recent pilot reports of windshear on the approach or climb-out.
5.2.2 The Alert message is given in the arrival and departure ATIS broadcasts at Heathrow in one of three formats:
(a) 'WINDSHEAR FORECAST' (WSF) – when the meteorological conditions indicate that low level windshear on the approach or climb-out (below 2000ft) may be encountered;
(b) 'WINDSHEAR FORECAST AND REPORTED' (WSFR) – as above, supported by a report from at least one aircraft of windshear on the approach or climb-out within the last hour;
(c) 'WINDSHEAR REPORTED' (WSR) – when an aircraft has reported windshear on the approach or climb-out within the last hour, but insufficient meteorological evidence exists for the issue of a forecast of windshear.
5.2.3 Pilot reports of windshear on approach or climb-out can greatly enhance the operational efficiency of this service. In addition, they also serve in the continuous evaluation of the criteria upon which Alerts are forecast. Thus pilots who experience windshear on the approach or climb-out are requested to report the occurrence to ATC, as soon as it is operationally possible to do so, even if an Alert has been issued. Windshear reporting criteria are shown at GEN 3.5 6 paragraph 3.2. Pilots who experience windshear at any UK aerodrome are requested to report it in the same way.

5.3 Runway Visual Range (RVR)
5.3.1 RVR assessment is made by either a Human Observer or an Instrumented RVR system (IRVR). Most IRVR systems have an upper limit of 1500m; the upper limit of the Human Observer system is normally less than this. The system in use at particular aerodromes is indicated in Table 3.5.3.2, and explained in Table 3.5.3.3.
5.3.2 The United Kingdom standard RVR reporting incremental scale is 25m between 0 and 400m, 50m between 400 and 800m , and 100m above 800m. Some IRVR systems are unable to report every incremental point in the scale. Known limitations in IRVR systems are as follows:

  Liverpool (EGGP) IRVR 50 – 1500m (175m not reported)
  London City (EGLC) IRVR 50 – 1100m IRVR remains serviceable if TDZ fails
  London Gatwick (EGKK) RWY 08L IRVR remains serviceable if TDZ fails
           RWY 26R TDZ IRVR is considerably displaced from Start-to-Roll position.

5.3.3 The assessment and reporting of RVR begins whenever the horizontal visibility or the RVR is observed to be less than 1500m. At those aerodromes where IRVR is available, RVR may also be reported when the observed value is at or below the maximum reportable value or when shallow fog is forecast or reported.
5.3.4 RVR is passed to aircraft before take-off and during the approach to landing. Changes in the RVR are passed to aircraft throughout the approach. Additionally, information from pilot reports or ATC observation that the visibility on the runway is worse than that indicated by the RVR report, for example patches of thick fog, are passed.
5.3.5 Table 3.5.3.2 shows which IRVR system is provided at an aerodrome. Aerodromes using AGIVIS systems suppress mid-point and/or stop-end values when:
(a) They are equal to or higher than the touchdown zone value unless they are less than 400m; or
(b) they are 800m or more.
Aerodromes using MET-1 systems suppress mid-point and/or stop-end values unless they are 550m or less.
5.3.6 At those aerodromes having multi-site IRVR the standard UK procedure is that the touchdown zone RVR is always given first, followed by any values for the mid-point RVR and/or stopped RVR which have not been suppressed. With the above in mind, the co-operation of the pilots is sought in avoiding unnecessary radio requests for mid-point and/or stop-end values when they have not been given. When all three values are given they are passed as a series of numbers, for example, 'RVR 600, 500, 550' relates to touchdown zone, mid-point and stopped

133

respectively. If two values are to be passed, they are to be individually identified, eg, 'touchdown 650, stop-end 550'.

5.3.7 If a single transmissometer fails, and the remainder of the IRVR system is still serviceable, RVR readings are not suppressed for the remaining sites and these values are passed to pilots. For example, if the touchdown zone transmissometer is unserviceable, 'RVR: touchdown not available, mid-point 600, stopped 400'. If two transmissometers fail in a three-site IRVR system, the remaining value is passed and identified provided that it is not the stop-end value, in which event the system is considered unserviceable for that runway direction. In a two-site IRVR system, giving touchdown zone and stop-end values, if the touchdown zone transmissometer fails, the system is considered unserviceable for that runway direction. Exceptionally, if the Civil Aviation Authority determines in a particular case that the distance between the two transmissometers is sufficiently small, the system may be considered serviceable for that runway direction.

5.3.8 When RVR information is not available, or when the RVR element of Aerodrome Operating Minima falls outside the range of reportable RVR, pilots should use meteorological visibility in the manner specified in their operations manuals, or at AD 1.1.2 (Aerodrome Operating Minima – Non-public transport flights by aircraft).

5.3.9 Information on specific types and locations of observation systems at particular aerodromes is available on request from the Meteorological Authority or the aerodrome operator.

6 Climatological information for certain UK aerodromes

is available for civil aviation purposes from the Meteorological Office, according to the following criteria: The aerodromes are so classified at Table 3.5.3.2.

Table 3.5.3.1 Climatological Information

A Climatological statistics readily available based on at least 10 years' of three-hourly (usually hourly) data.
B Climatological statistics readily available based on less than 10 years' data and/or some gaps in night-time data.
C Limited climatological statistics available based on available METAR reports starting July 1983 or later.
D No data available or insufficient data available to provide climatological statistics.

The aerodromes are so classified in Table 3.5.3.2

6.1 Climatological statistics for routes and areas in the United kingdom are not available. However, global climatology of upper wind and temperature data is held by the Met Office.

Table 3.5.3.2

Aerodrome/ Location Ind	Observations			Surface Wind	Sites	RVR Eqpt	Obs Hrs	Climatological Data
	Type	Freq	Warnings					
Aberdeen/Dyce EGPD	METAR	h H	AW MTI	Average	16 TDZ/END 34 TDZ/END	AGIVIS AGIVIS	HO+	A
Alderney EGJA	METAR	h	AW	Average	08/26 TDZ	OBS	HO	C
Barra EGPR	METAR‡	l	AW	Average			HO	B
Belfast Aldergrove EGAA	METAR	h	AW MTI Windshear	Average	07 TDZ/MID/END 25 TDZ/MID/END 07 TDZ 17 TDZ 25 TDZ/MID 35 TDZ	AGIVIS AGIVIS OBS OBS OBS OBS	H24	A
Belfast City EGAC	METAR	h	AW MTI	Average	04/22 TDZ	OBS	HO	C
Benbecula EGPL	METAR	h	AW	Average	06/24 TDZ	OBS	HO	A
Biggin Hill EGKB	METAR	h	AW	Average	21 TDZ	OBS	HO	C
Birmingham EGBB	METAR	h	AW MTI	Average	15 TDZ/MID/END 33 TDZ/MID/END	AGIVIS AGIVIS	H24 H24	C
Blackpool EGNH	METAR	h	AW MTI	Average	10/28 TDZ	OBS	HO	A, C
Bournemouth EGHH	METAR	h	AW	Average	08/26 TDZ	OBS	HO	A
Bristol EGGD	METAR	h	AW MTI	Average	09/27 TDZ/MID/END	Vaisala	H24	C
Britstol Filton EGTG	METAR	h	AW	Average			HO	A, C
Cambridge EGSC	METAR	H h	AW MTI	Average	05/23 TDZ	OBS	HO	C
Campletown EGEC	METAR	H	AW	Average			HO	B
Cardiff EGFF	METAR	h	AW MTI	Average	12/30 TDZ/END	AGVIS	H24	C

GEN 3.5

GEN 3.5.3.6 Climatology Information for UK Airports

Aerodrome/ Location Ind	Type	Observations Freq	Warnings	Surface Wind	Sites	RVR Eqpt	Obs Hrs	Climatological Data
Carlisle EGNC	METAR	h	AW	Average	07/25 TDZ	OBS	HO	C
Coventry EGBE	METAR	h	AW	Average	05/23 TDZ	OBS	H24	C
Cranfield EGTC	METAR	H	AW	Average			HO	C
Doncaster Sheffield EGCN	METAR	h	AW			H24		D
Dundee EGPN	METAR	h	AW	Average			HO	D
Durham Tees Valley EGNV	METAR	h	AW	Average	05/23 TDZ		HO	C
Edinburgh EGPH	METAR	h	AW MTI	Average	06 TDZ/MID/END 24 TDZ/MID/END	AGIVIS AGIVIS	H24	A
Exeter EGTE	METAR	h	AW MTI	Average	08/26 TDZ	OBS	HO OBS	A, C
Fairoaks EGTF	METAR‡	l	AW	Instant			HO	D
Farnborough EGLF	METAR	h	AW	Average	06/24 TDZ/END 06/24 TDZ	AGIVIS	HO OBS	B
Glasgow EGPF	METAR	h	AW MTI	Average	05 TDZ/MID/END 23 TDZ/MID/END	AGIVIS AGIVIS	H24	A
Gloucestershire EGBJ	METAR	h	AW	Average	09/27 TDZ	OBS	HO	C
Guernsey EGJB	METAR	h	AW	Average	09/27 TDZ/END	HO	HO	A
Hawarden EGNR	METAR‡	l	AW	Average			HO	D
Humberside EGNJ	METAR	h	AW	Average	03/21 TDZ	OBS	HO	C
Inverness EGPE	METAR	h	AW	Average	05/23 TDZ/END	AGIVIS	HO	D
Islay EGPI	METAR	h	AW	Average			HO	D
Isle of Man EGNS	METAR	h	AW MTI	Average	26 TDZ	OBS	HO+	A
Jersey EGJJ	METAR	h	AW MTI	Average	09/27 TDZ/END	AGIVIS	HO+	A
Kirkwall EGPA	METAR	h	AW	Average			HO+	A
Leeds Bradford EGNM	METAR	h	AW MTI	Average	14 TDZ 32 TDZ/MID 14 TDZ/MID 32 TDZ/MID	OBS OBS MET-1 MET-1	H24	C
Liverpool EGGP	METAR	h	AW MTI	Average	09 TDZ/MID/END 27 TDZ/MID/END	MET-1 MET-1	H24	A, C
London City EGLC	METAR	h	AW MTI	Average	10 TDZ/END 28 TDZ/END	MET-1	HO+ MET-1	C
London Gatwick EGKK	METAR	h	AW MTI	Average	08R TDZ/MID/END 26L TDZ/MID/END 08L TDZ/MID 26R TDZ/END	AGIVIS AGIVIS AGIVIS AGIVIS	H24	A
LondonHeathrow EGLL	METAR	h	AW MTI Windshear	Average	09L TDZ/MID/END 09R TDZ/MID/END 27L TDZ/MID/END 27R TDZ/MID/END	AGIVIS AGIVIS AGIVIS AGIVIS	H24	A
London Luton EGGW	METAR	h	AW MTI	Average	08 TDZ/MID/END 26 TDZ/MID/END	AGIVIS AGIVIS	H24	C
London Stansted EGSS	METAR	h	AW MTI	Average	05 TDZ/MID/END 23 TDZ/MID/END	AGIVIS AGIVIS	H24	A

GEN 3.5.3.6 Climatology Information for UK Airports

Aerodrome/ Location Ind	Type	Observations Freq	Warnings	Surface Wind	Sites	RVR Eqpt	Obs Hrs	Climatological Data
London Heliport	METAR‡	I	AW	Instant			HO	D
Londonderry Eglinton EGAE	METAR	h	AW	Average	08/26 TDZ	OBS	HO	C
Lydd EGMD	METAR	h	AW	Average	03/21 TDZ	OBS	HO	C
Manchester EGCC	METAR	h	AW MTI	Average	06 TDZ/MID/END 24 TDZ/MID/EMD	AGIVIS AGIVIS	H24	A
Manchester Woodford EGCD	METAR‡	H	AW	Average	07/25 TDZ 25 Alt TDZ/DEP	OBS	HO	D
Manston EGMH	METAR	h	AW	Instant			HO	A
Newcastle EGNT	METAR	h	AW MTI	Average	07 TDZ/MID/END 25 TDZ/MID/END 07/25 TDZ/END	Elecma Elecma OBS	H24	C
Norwich EGSH	METAR	h	AW MTI	Average	09/27 TDZ	OBS	HO	C
Nottingham East Midlands EGNX	METAR	h	AW MTI	Average	09 TDZ/MID/END 27 TDZ/MID/END	MET-1 MET-1	H24	C
Oxford EGTK	METAR‡	H	AW	Instant			HO	D
Penzance Heliport EGHK	METAR‡	h	IAW	Average			HO	D
Plymouth EGHD	METAR	h	AW	Average	31 TDZ	OBS	HO	C
Prestwick EGPK	METAR	h	AW MTI	Average	13/31 TDZ	OBS	H24	A H24
Scatsta EGPM	METAR	h	AW	Average	09/27 TDZ	AGIVIS	HO	C
Scilly Isles/St Marys EGHE	METAR	h	AW	Instant			HO	A, C
Shoreham EGKA	METAR	H	AW	Average			HO	C
Southampton EGHI	METAR	h	AW	Average	02/20 TDZ	OBS	HO	C
Southend EGMC	METAR	h	AW MTI	Average	06/24 TDZ	OBS	H24	C
Stornoway EGPO	METAR	h	AW	Average	18/36 TDZ	OBS	HO	A
Sumburgh EGPB	METAR	h	AW MTI	Average	09/27 TDZ	OBS	HO+	A
Swansea EGFH	METAR‡	I	AW	Instant	04/22 TDZ	OBS	HO	C
Tiree EGPU	METAR	h	AW	Average			HO	A
Warton EGNO	METAR‡	h	AW	Average			HO	D
Wick EGPC	METAR	h	AW	Average			HO	A
Yeovil/Westland EGHG	METAR‡	H	AW	Instant			HO	D

Table 3.5.3.3

Observation Type (Column 2)	METAR	Aviation Routine Weather Report (actual)
	MEATR ‡	METAR not distributed routinely
Observation Frequency (column 3)	h	half-hourly
	H	hourly
	I	irregular
Observation Warnings (column 4)	AW	Aerodrome warning
	MTI	Marked Temperature Inversion
Surface		see GEN 3.5.3 paragraphs 1 and 2
RVR (column 6 & 7)	TDZ	touchdown zone
	MID	mid-point
	END	stop end
	OBS	human observer
	AGVIS, MET-1 Elecma & Vaisala	types of IRVR system
Observing Hours (column 8)	HO	available to meet operational requirements (ie during aerodrome opening hours)
	HO+	more than HO but not H24
	H24	24 hours
Climatological data (column 9)	A, B, C, D	see table 3.5.3.1

Note: Table 3.5.3.2 lists only those aerodromes with accredited observers that produce METARs. Observations from other aerodromes not listed shall be regarded as unofficial.

GEN 3.5.4 TYPES OF SERVICE

1 Forecast Offices providing a service to Civil Aviation

1.1 The designated Forecast Office(s) for principal aerodromes are given in AD 2.11 (Meteorological Information Provided), in the Aerodromes section of the AIP. Designated forecast offices operate H24.

1.2 Some Military and Government aerodromes provide a forecast and briefing service for Civil Aviation, but only for departures from those aerodromes.

1.3 The designated Forecast Office for Intercontinental flights departing any UK Civil aerodrome is the Met Office National Met Centre.

2 Pre-flight Briefing

2.1 The primary method of meteorological briefing for flight crew in the UK is by self-briefing, using information and documentation routinely displayed in aerodrome briefing areas. Alternatively, flight crew and operators may obtain information direct by using the Met Office's PC Service MIST (Meteorological Information Self-briefing Terminal), the Fax services jointly provided by the Met Office and the CAA, and the CAA's AIRMET telephone service, see GEN 3.5.9 for fuller details of the automated services available. English is the language used for all UK documentation and forecast clarification. The primary method of briefing does not require prior notification to a Forecast Office.

2.2 Where this primary method is not available, or is inadequate for the intended flight, Special Forecasts, as described at GEN 3.5.5, may be provided.

2.3 When necessary the personal advice of a forecaster, or other meteorological information, can be obtained from the appropriate Forecast Office. Pilots departing from Military or Government aerodromes may find an on-site Forecaster Briefing Service available. Forecaster advice or other information for safety related clarification/amplification will only be given from a Forecast Office on the understanding that full use has already been made of all meteorological briefing material. Forecaster clarification/amplification of conditions en-route is not provided for flights departing from locations outside UK.

2.4 The Met Office operates a centralised telephone enquiry system. Pilots should state their requirements to the operator who may provide the necessary information or who will promptly transfer the call to the most suitable location.

Table 3.2.4.1 – Forecast Offices providing service to Civil Aviation

Forecast Office	Services Available	Telephone	Remarks
Met Office	A, B, C, D, E	0870 900 0100	In case of restricted access to 0870 numbers, use 01392 885680 0870 calls from anywhere in the UK are charged at the national call rate
Isle of Man Airport	A, B, C	01624 821641	Isle of Man services B and C available only to departures from within the Isle of Man
Jersey Airport	A, B, C	0907 1557777	Jersey services B and C available only to departures from the Channel Islands. Calls charges at premium rate. In case restricted access to 090 numbers, telephone 01534 492256 for further information
Jersey AIRMET		09006 650033	Jersey AIRMET available 0600 2100 and only to callers in the Channel Islands and UK. Calls charged at premium rate. In case of restricted access to 090 numbers, telephone 01534 492256 for further information.

Key to Services Available:

A   Provision of TAF, warnings and take-off data for the assigned principal aerodrome. Amplification/clarification of these aerodrome forecasts and warnings only.

B   Dictation of TAFs and METARs unobtainable from the automated services (usually limited to four aerodromes).

C   Amplification/clarification of AIRMET Regional/Area Forecasts and Metforms 214/215, and requests for Special Forecasts, including specialised supplementary information for ballooning/Gliding.

D   Dictation of AIRMET amendments and Regional Forecasts.

E   Amplification/clarification and amendments for Metforms 414/415 and EUR charts.

2.5 Forecast Offices and self-briefing facilities are under no obligation to prepare documentation packages.

2.6 Meteorological observations and forecasts have certain expected tolerances of accuracy. Pilots interpreting observations and forecasts should be aware that information could vary within these tolerances, which are shown at GEN 3.5.3 paragraph 2 and Table 3.5.4.3 respectively. Additionally, observations and forecasts are not normally amended until certain criteria for change are exceeded. These are shown at GEN 3.5.4 paragraph 7.

2.7 The specific value of any of the elements given in a forecast shall be understood to be the most probable value which the element is likely to assume during the period of the forecast. Similarly, when the time occurrence or change of an element is given in a forecast, this time shall be understood to be the most probable time.

2.8 The issue of a new forecast, such as an aerodrome forecast, shall be understood to automatically cancel any forecast of the same type previously issued for the same place and for the same period of validity or part thereof.

3 U K Low Level Weather and Spot Wind Forecast Charts – Metform 215/214

3.1 **The UK Low Level Forecast (Metform 215)** is a forecast of in-flight conditions from the surface to 10000 ft, covering the UK and near Continent. The form comprises:

(a) a fixed time forecast weather chart and text box describing the expected visibility and weather, the cloud and the height of the zero degree isotherm in each separate area of weather highlighted on the chart. An outlook box describes the main weather developments in the 7-hour period beyond the end of the validity period of the forecast.

(b) a separate outlook chart, available on the internet only, shows the expected position of the principle synoptic features, 1 hour after the end of the validity period of the forecast.

3.2 Information on Form. The following sub-paragraphs summarise the contents of Metform 216 (Explanatory Notes for Form 215), available in A4 or larger size on application to the address at GEN 3.5.1 paragraph 2 or from METFAX by dialling 09060 700 505.

3.2.1 Main Forecast Weather Chart and Text

(a) The fixed time weather chart

(i) the weather chart shows the forecast position, direction and speed of movement of surface fronts and pressure centres for the fixed time shown in the chart legend. The position of highs (H) and lows (L), with pressure values in millibars is shown by the symbols 0 and X. The direction and speed of movement (in knots) of fronts and other features is given by arrows and figures. Speeds of less than 5 knots are shown as 'SLOW.

(ii) zones of distinct weather patterns are enclosed by continuous scalloped lines, each zone being identified by a letter within a rectangle. The forecast weather conditions (visibility, weather, cloud and height of the zero degree isotherm) during the period of validity, together with warnings and any remarks are given in the text to the right of the charts, each zone being dealt with separately and completely.

(b) In the text:

(i) surface visibility and weather

(1) surface visibility is expressed in metres (m) or kilometres (km), with the change over at 5000 m;

(2) weather is described using the METAR code form; the full list of terms is available at GEN 3.5. 10 paragraph 6.

(3) warnings and the expected occurrence of icing and turbulence are highlighted, using standard ICAO symbolism and abbreviations where possible (see GEN 2.2);

GEN 3.5   GEN 3.5.4.5 European Medium/High Level Spot Wind/Temperature Forecast Chart – Metform 614

(ii) Cloud
(1) cloud amount is described using the METAR code form, where FEW indicates 1 to 2 oktas, SCT (scattered) indicates 3 to 4 oktas, BI(N (broken) indicates 5 to 7 oktas and OVC (overcast) indicates 8 oktas.
(2) cloud type is given using standard meteorological and ICAO abbreviations (see GEN 2.2).
(3) the height of the cloud base and top is given in the form height of cloud base 1 height of cloud top with all heights in hundreds of feet (ft) above mean sea level;
(iii) The height of the zero degree Celsius isotherm.
(1) the height of the zero degree Celsius isotherm is given with all heights in hundreds of feet (ft) amsl.
(2) the height of any sub-zero layer below the main layer will also be given.
(c) Outlook text
An outlook box describes the main weather developments in the 7-hour period beyond the end of the validity period of the forecast.
(d) Notes
(i) a forecast of thunderstorm (TS) and/or cumulonimbus (C13) implies hail and severe turbulence and icing;
(ii) hill fog will be included as a warning in the text whenever the base of any cloud is forecast to be at the same height or below the height of the highest ground in the zone. Hill fog implies visibility less than 200m;
(iii) single numerical values given for any element represent the most probable mean in a range of values, covering approximately ±25%.

3.2.2 Outlook Chart
A separate outlook chart, available on the internet only, shows the expected position of the principle synoptic features 1 hour after the end of the validity period of the forecast. No weather zones are given on the outlook chart but the pattern of surface isobars and frontal positions are shown.

3.3 The UK Low Level Spot Wind Forecast (Metform 214) is fixed time, and suitable for use for a period three hours before or after the validity time.
(a) The data provided is for Latitude/Longitude positions shown at the top of each box;
(b) Wrid Speed and temperature information is provided for a selected range of attitudes and are shown in thousands of feet above mean sea level and Degrees Celsius.

3.4 Weather Forecast Chart Issues
3.4.1 The routine issue of weather charts is detailed below. The date and time that each Metform 215/415 is issued by the Met Office will be shown at the bottom of the form. A summary of the times of issue and validity times of the charts is given in the table below.

Time when forecast Chart becomes available	Fronts and weather zones forecast time	Period of validity	Outlook to
0300	1200	0800-1700	1800
0900	1800	1400-2300	0000
1500	0000	2000-0500	0600
2100	0600	0200-1100	1200

3.4.2 Amendments
(a) Amendments may appear as complete re-issues of the Metform in which case the validity start time may be different from the routine issue.
(b) An amended Metform 215 is indicated by the word AMENDED at the top of the form, and the element amended written in bold and underlined.

4 Northwest Europe Low Level Weather and Spot Wind Forecast Charts – Metforms 415/414

(a) These charts are similar in format to Metforms 214/215 and extend the low-level flight forecast coverage more into continental Europe.
(b) They are issued daily at the same times as Metform 214/215 and are valid for the same periods.
(c) Amendments will appear as complete re-issues of the Metform in which case the validity start time may be different from the routine issue.

5 European Medium/High Level Spot Wind/Temperature Forecast Chart – Metform 614

(a) This chart is similar in format to Metforms 214 and 414 but extends the coverage to most of Europe and western parts of the Mediterranean and North Africa.
(b) It is available only from METFAX (09060 700 541) and provides a single sheet alternative to part of the area covered by the six standard EUR wind/temperature charts between FL 050 and FL 340 to accompany the EUR significant weather chart.

6 Global upper-air wind and temperature data

(a) Global upper-air wind and temperature data, as well as data on upper-air humidity, tropopause heights and temperatures and maximum wind speed, direction and height are available in grid points in digital form, updated four times per day, from the address given in 3.5.1 paragraph 2.

(b) Specific chart areas representing part or all of the above data may be available from some suppliers, but does not form part of the ICAO World Area Forecast System service provision.

Table 3.5.4.2 – Meteorological Forecast Charts – Coverage and Validity Times

Area	Chart	Levels	Coverage	Projection	Issue Times	Validity Times
UK (F215)	Weather*3	SFC-10000ft amsl	British Isles		1500*2	2000 to 0500
UK (F214)	Spot Wind/Temperature	24000, 18000, 10000,	And near		2100*2	0200 to 1100
		5000, 2000, 1000ft*4 amsl	Continent		0300	0800 to 1700
					0900	1400 to 2300
Europe (EUR)	Weather/Tropopause	FL100-FL450	N53 E065	Polar	1800	0000
	Max wind		N25 E034	Stereographic	0000	0600
			N26 W018		0600	1200
			N54 W050		1200	1800
North Atlantic	Weather/Tropopause/	FL250-FL630	N24 E056	Polar	1100*2	0000
(NAT)	Max Wind		N02 W004	Stereographic	1700*2	0600
			N03 W083		2300*2	1200
			N28 W148		0500	1800
	Isobaric and Frontal	Surface	N37 E050	Polar	0200	0000
	Analysis (ASXX)		N68 W105	Stereographic	0800	0600
			N34 W055		1400	1200
			N20 E010		2000	1800
	Isobaric and frontal	Surface			0400*2	0000
	prognosis (FSXX)				1000*2	0600
					1600*2	1200
					2200*2	1800
Mid/Far East	Weather/Tropopause	FL250-FL630	N23 E150	Polar	1100*2	0000
(MID)	Max Wind		S06 E102	Stereographic	1700*2	0600
			S03 E033		2300*2	1200
			N30 W020		0500	1800
Africa	Weather/Tropopause/	FL250 – FL630	N70 W032	Mercator	1100*2	0000
(AFI)	Max Wind		N70 E065		1700*2	0600
			S38 W083		2300*2	1200
			S38 W032		0500	1800
Caribbean/	Weather/Tropopause/	FL250 – FL630	N23 W113	Tilted	1100*2	0000
South America	Max Wind		N72 W003	Mercator	1700*2	0600
(CARSAM)			N33 E059		2300*2	1200
			S48 W052		0500	1800

Note *1 Charts cover the period within 3 hours either side of the quoted fixed time, except 215 charts and surface isobaric charts, which are valid for the time secified.
Note *2 Previous day
Note *3 This chart includes an outlook to the end of the next forecast period
Note *4 Where terrain permits

Table 3.5.4.3 – Accuracy of Meteorological Forecasts
The percentages in this table are ICAO minimum standards.

Element	Operationally desirable accuracy of forecast	Minimum percentage of cases within range
Aerodrome Forecast (TAF)		
Wind direction	± 30°	80
Wind speed	± 5kt up to 25kt	80
	± 20% above 25kt	
Visibility	± 200m up to 700m	80
	± 30% between 700m and 10km	
Precipitation	Occurrence or non-occurrence	80
Cloud amount	± 2 oktas	70
Cloud height	± 100ft up to 400ft	70
	± 30% between 400ft and 10000ft	
Air temperature (if forecast)	± 1C∫	70
Landing Forecast (TREND)		
Wind direction	± 30°	90
Wind speed	± 5kt up to 25kt	90
	± 20% above 25kt	
Visibility	± 200m up to 700m	90
	± 30% between 700m and 10km	
Precipitation	occurrence or non-occurrence	90
Cloud amount	± 2 oktas	90
Cloud height	± 100ft up to 400ft	90
	± 30% between 400ft and 10000ft	
Take-off Forecast		
Wind direction	± 30°	90
Wind speed	± 5kt up to 25kt	90
	± 20% above 25kt	
Air Temperature	± 1°C	90
Pressure Valve (QNH)	± 1mb	90
Area, Flight and Route Forecast		
Upper air temperature	± 3°C (mean for 500m)	90
Upper wind	± 15kt up to FL250	90
	± 20kt above FL250	
	(modulus of vector difference for 500m)	
Significant en-route	Occurrence or non-occurrence	80
WX phenomena	Location ± 60nm	70
and cloud	Vertical extent ± 2000ft	70

6 Aerodrome Forecasts (TAF)

6.1 The Aerodrome Forecast (TAF) is the primary method of providing the forecast weather information that pilots require about an airfield in an abbreviated format. The TAF consists of a concise statement of the mean or average meteorological conditions expected at an aerodrome or heliport during the specified period of validity.

6.2 UK civil TAFs are prepared to cover the notified hours of operation of those principal civil aerodromes that have accredited meteorological observers, who produce regular aerodrome weather reports. Being site-specific, to provide an aerodrome forecast in TAF form requires the forecaster to be confident in the knowledge of the weather conditions prevailing at that aerodrome. In the interests of flight safety, continuity of regular reports, and ideally special reports when significant changes occur (particularly if the deterioration or improvement has not been forecast or is mis-timed), are essential for the routine updates and an adequate amendment service to be provided by the forecast office.

6.3 Therefore, where an aerodrome is not openH24, the issue of a TAF will be delayed until at least two consecutive METARs have been received and accepted by the forecaster at the forecast office responsible for its operation.

6.4 The METARs will be produced by an accredited observer and separated by an interval of not less than 20 minutes and not more than 1 hour. In practice, when METARs are prepared every 30 minutes, a TAF will be drafted by the forecaster once the first METAR has been seen, and when the second METAR is received 30 minutes later and confirms the prevailing weather over the aerodrome the forecaster will issue the TAF.

6.5 however, in the event that an automatic or semi-automatic observing system located on the aerodrome regularly issues information on wind speed and direction, visibility, cloud amount and height, temperature and dew point to the forecast office when an aerodrome is closed, the forecaster will draft the TAF on the basis of the automatic observation. The TAF will not be issued until a METAR has been produced by an accredited observer and received in the forecast office, in order to confirm the weather conditions.

GEN 3.5   GEN 3.5.4.6 Aerodrome Forecasts (TAF)

6.6 If a gap of more than one hour between METAR reports occurs, or if a key element is missing from the report, the TAF will be withdrawn. The TAF will not be re-issued until two complete METARs have been received.

6.7 Accredited observers at some H24 aerodromes take a duty break overnight, of maximum two hours duration. A supply of automatic observations will be provided to the forecast office during this period, but amendments to the TAF will be made only at the discretion of the forecaster. If the duty observer has not recommenced observations after two hours, the TAF will be withdrawn.

6.8 if a TAF needs to be amended due to a deterioration or improvement that has not been forecast or is mis-timed, such amendments shall be issued within 15 minutes of receipt of he observation at the forecast office.

7 Criteria for Special Meteorological Reports and Forecasts

7.1 The following are the criteria for the issue of Special Aerodrome Meteorological Reports, TRENDS, TAF Variants/Amendments and Amended Route/Area Forecasts:

(a) Special Report

(i) Surface Wind. Issued only when no serviceable wind indicator in ATC; criteria to be agreed locally, based on changes of operational significance at aerodrome; for example:

(1) A change in mean direction of 60° or more, mean speed before or after change being 10kt or more; and/or a change of 30°, the speed 20kt or more;

(2) A change in mean speed of 10kt or more;

(3) A change in gust speed of 10kt or more, the mean speed before or after the change being 15kt or more.

(ii) Visibility

(1) A change in the prevailing visibility from one of the following ranges to another:
10km or more
5000m to 9km
3000 to 4900m
2000m to 2900m
1500m to 1900m
800m to 1400m
750m or less

(2) At the onset or cessation of the requirement to report minimum visibility, ie When the minimum visibility in one or more directions is less than 50% of the prevailing visibility.

(3) If the minimum visibility is being reported, when the minimum visibility changes from one of the ranges, given in (1) above, to another.

(4) Additional change groups of 100m or less, 150 to 300m, 350 to 550m and 600 to 750m are used where an RVR is not available, either permanently or during temporary unserviceability. These criteria will apply by local arrangement.

(5) Additional change groups of 3000 to 3900m and 4000 to 4900m apply at Aberdeen/Dyce airport.

(iii) Runway Visual Range (RVR)

(1) A change from one of the following ranges to another:
800m or more
600m to 750m
350m to 550m
150m to 325m
125m or less

(2) Note that special reports for RVR are only made by local arrangements.

(iv) Weather

(1) The onset or cessation of:
– Moderate or heavy: precipitation, including showers;
– Freezing fog and freezing precipitation;
– Thunderstorm, squall, funnel cloud;
– Low drifting or blowing: snow, sand or dust.

(2) A change in the intensity of the precipitation and blowing snow from slight to moderate/heavy and visa versa.

Note: The phenomenon associated with a significant change in visibility or cloud shall be reported, whatever the intensity.

(v) Cloud

(1) When the base or the lowest cloud of over 4 oktas (BKN or OVC) changes from one of the following ranges to another:
2000ft or more
1500ft to 1900ft
1000ft to 1400ft
700ft to 900ft
500ft to 600ft
300ft to 400ft
200ft
100ft
Less than 100ft*

(2) * this includes state of sky obscured.

(3) When the amount of cloud below 1500ft changes from 4 oktas or less (nil, FEW, SCT) to more than 4 oktas (BKN or OVC), and visa versa.

GEN 3.5 GEN 3.5.4.7 Criteria for Special Meteorological Reports and Forecasts

(vi) QFE/QNH
When the QNH or QFE changes by 1.0mb or more.
(vii) Severe Icing/Turbulence.
After confirmation by the duty forecaster, pilot reports of severe icing or severe turbulence, either on the approach to, or climb out from, the aerodrome.
(b) TREND
(i) Surface Wind
(1) A change in mean direction of 30° or more, the mean speed before or after the change being 20kt or more; a change in mean direction of 60° or more, the mean speed before or after the change being 10kt or more.
(2) A change in mean speed of 10kt or more.
(3) A change in gust speed of 10kt or more, the mean speed before or after the change being 15kt or more.
(ii) Surface Visibility.
(1) A change in the prevailing visibility from one of the following ranges to another:
5000m or more
3000m to 4900m
1500m to 2900m
800m to 1400m
600m to 750m
350m to 550m
150m to 300m
100m or less
(iii) Weather.
(1) Onset, cessation or change in intensity of:
– freezing precipitation;
– freezing fog;
– moderate or heavy: precipitation, including showers;
– low drifting: sand, dust or snow; blowing: sand, dust or snow;
– thunderstorm;
– squall, funnel cloud;
– other phenomena if associated with a significant change in visibility or cloud, whatever the intensity.
(iv) Cloud
(1) When the base of the lowest cloud of over 4 oktas (BKN or OVC) changes from one of the following ranges to another:
1500ft or more
1000ft to 1400ft
500ft to 900ft
300ft to 400ft
200ft
100ft
Less than 100ft*
(2) * this includes state of sky obscured.
(3) Additional change groups of 500 to 600 feet and 700 to 900 feet apply at aerodromes serving oil rig helicopter operations when a TREND service is provided: Aberdeen, Norwich and Scatsta.
(4) When the amount of the lowest cloud below 1500ft changes from half or less (nil, FEW, SCT) to more than half (BKN or OVC), and visa versa. A chnge to 'sky clear' should be shown as 'SKC' and a change to no cloud below 5000ft and no CB to 'NSC', unless in either case CAVOL applies.
(c) TAF Variants/Amendments
(i) Surface Wind
(1) A change in mean direction of 30° or more, the mean speed before or after the change being 20kt or more; a change in mean direction of 60° or more, the mean speed before or after the change being 10kt or more.
(2) A change in mean speed of 10kt or more.
(3) A change in gust speed of 10kt or more, the mean speed before or after the change being 15kt or more.
(ii) Surface Visibility
(1) A change in the prevailing visibility from one of the following ranges to another:
10km or more
5000m to 9km
1500m to 4900m
800m to 1400m
350m to 750m
300m or less
(2) Additional change groups of 1500m to 2900m and 3000m to 4900m as well as 5000m to 6km and 7 to 9km apply at aerodromes serving oil rig helicopter operations: Aberdeen, Benbecula, Blackpool, Humberside, Inverness, Kirkwall, Liverpool, Norwich, Scatsta, Sumburgh and Wick.

GEN 3.5.4.7 Criteria for Special Meteorological Reports and Forecasts

(iii) Weather
(1) Onset, cessation or change in intensity of:
– freezing precipitation;
– freezing fog;
– moderate or heavy: precipitation, including showers;
– low drifting: sand, dust or snow;
– blowing: sand, dust or snow;
– thunderstorm;
– squall, funnel cloud;
– other phenomena if associated with a significant change in visibility or cloud, whatever the intensity.
– CAVOK conditions.

(iv) Cloud
(1) When the base of the lowest cloud of over 4 oktas (BKN or OVC) changes from one of the following ranges to another:
5000ft or more
1500ft to 4900ft
1000ft to 1400ft
500ft to 900ft
200ft to 400ft
100ft or less*

(2) *This includes state of sky obscured.
(3) Additional change groups of 500ft to 600ft and 700ft to 900ft apply at aerodromes serving oil rig helicopter operations; Aberdeen, Benbecula, Blackpool, Humberside, Inverness, Kirkwall, Liverpool, Norwich, Scatsta, Sumburgh and Wick.
(4) When the amount of the lowest cloud below 1500ft changes from half or less (nil, FEW or SCT) to more than half (BKN or OVC) and visa versa. A change to 'sky clear' should be shown as 'SKC; and a change to no cloud below 500ft and no CB, to 'NSC', unless in either case CAVOK applies.

(d) Amended Route/Area Forecast (Advisory Criteria)
(i) Surface Wind. A change in direction of 30° or more, the speed before and/or after the change being at least 30kt. A change of speed of 20kt or more.
(ii) Temperature/Dew Point. 5°C or more.
(iii) Cloud Amount Changes in general forecast lowest cloud base below 1500ft from 4 oktas or less to more than 4 oktas, or more than 4 oktas to 4 oktas or less.

Element	Original Forecast	Revised Opinion
(iv) 2000ft wind	Less than 50kts	50kts or more
(v) Surface visibility (general visibility)	8km or more 5000m to 7000m 3700m to 4900m 1600m to 3600m 1500m or less	Less than 5000m Less than 3700m 8km or more or less than 1600m 5000m or more 3700m or more
(vi) Weather phenomena TS, SQ, GR, SA, RASN, SN, FZFG, FZRA, FZDZ	Not included Included	Now expected Not now expected
(vii) Cloud height (General forecast lowest cloud base)	2500ft or more 1500ft to 2400ft 700ft to 1400ft 300ft to 600ft 200ft or less	Less than 1500ft Less than 700ft More than 2500ft or less than 200ft More than 1500ft More than 700ft
(vii) Turbulence	Nil Light Moderate Severe	Moderate or severe Severe Nil Nil or light
(ix) Zero celsius isotherm	Below 500ft Above 500ft	Changes of 1000ft or more Changes of 25% or 2000ft whichever is smaller
(x) Airframe icing	Nil Light Moderate Severe	Moderate or severe severe Nil Nil or light
(xi) Area boundaries, significant fronts and tropical disturbances	Not included Included	Now expected Not now expected or Not now expected or 90nm different From forecast

GEN 3.5.5 NOTIFICATION REQUIRED FROM OPERATORS

1 Special Forecasts and Specialized Information

1.1 For departures where the standard pre-flight meteorological self-briefing material cannot be obtained or is inadequate for the intended flight, a Special Forecast may be issued on request to the appropriate Forecast Office for a specific period for a designated route, or an area which includes the route. Normally, a Special Flight Forecast will be supplied from the last UK departure point to the first transit aerodrome outside the coverage of standard documentation, at which point pilots should re-brief. However, by prior arrangement, a forecast may be prepared for other legs, provided initial ETD to final ETA does not exceed 6 hours and no stops longer than 60 minutes are planned.

1.2 The usual method of issuing Special Flight Forecasts is by AFS, Telex or Fax to the aerodrome of departure, but if the Flight Briefing Unit is not so equipped or will not be open, pilots may telephone the Forecasting Office for a dictation of the forecast. Similarly, Aerodrome Forecasts and reports for the destination and up to four alternates will be provided with the forecast, if not otherwise available.

2 Prior Notification for Special Forecasts

2.1 Forecast Offices normally require prior notification for Special Forecasts as follows:
(a) For flights up to 500nm, at least two hours before the time of collection;
(b) for flights of over 500nm, at least four hours before the time of collection.

2.1.1 Request for Special Forecasts must include details of the route, the period of the flight and where appropriate the ETD/ETA of each leg, the height to be flown and the time at which the forecast is required. Ideally a forecast should be collected no earlier than 90 minutes before departure.

2.2 It is in the interest of all concerned that the maximum possible period of notice is given. The Forecast Office will give priority to emergencies, in-flight forecast and to forecast requirements which have been properly notified. Other requests could be delayed at busy periods and might not comprise full forecasts. A Forecast collected a long time in advance of departure will be less specific and might be less accurate than one prepared nearer departure time.

2.3 Forecast Offices providing Special Forecasts are shown at Table 3.5.4.1. They are not provided for flights inbound to the UK.

2.4 Take-off Forecasts containing information on expected conditions over the runway complex in respect of surface wind, temperature and pressure can be made available from Forecast Offices. Prior notification is not normally required, and can be supplied up to three hours before the expected time of departure.

2.5 Meteorological information for specialized aviation use, as defined below, is not included in the AIRMET service or given as Special Forecasts but arrangements can be made for its provision on prior request:

(a) To enable glider, hang glider, microlight and balloon organizations to obtain surface wind and temperature, lee-wave, QNH and thermal activity forecasts;

(b) to provide meteorological information for special aviation events for which routine forecasts are not adequate;

(c) to provide helicopter operators in off-shore areas with forecast winds and temperatures at 1000ft amsl, information on airframe icing, and sea state and temperature.

2.6 Appropriate forecasts for (a) and (b) above will be made available up to twice in any 24 hour period. For (a), the initial request should be made to the nearest forecasting office designated as providing service 'C' at Table 3.5.4.1 at least 2 hours in advance of the forecast being required. For (b) and (c), application must be made to the Meteorological Authority (GEN 3.5.1 paragraph 2) for approval, giving at least 6 weeks notice of the requirement. The application must specify the nature of the aviation activity, the location(s) involved, the meteorological information required and the associated time periods. If appropriate an AFTN, Telex or Facsimile address should also be included. Applicants will be advised of the time at which the information will be available and of the means of collection/delivery.

3 Additional Meteorological Services

3.1 When specialist, non-standard, aviation meteorological services additional to those given above are required (eg forecaster briefings for aerial photography, test flying, crop spraying and for outlooks for over a day ahead), they may be obtained on a repayment basis by prior arrangement with the Met Office. Enquiries should be directed to the Met Office address at GEN 3.5.1 paragraph 2, or to one of the Forecast Offices listed at Table 3.5.4.1.

GEN 3.5.6 AIRCRAFT REPORTS

1 Routine Aircraft Observations

1.1 Routine Aircraft Observations are not required in the London or Scottish FIR/UIR or the Shanwick FIR, but in the Shanwick OCA, aircraft are to conform with the requirements for meteorological observations indicated in the ENR section or applicable NOTAM.

2 Special Aircraft Observations

2.1 Special Aircraft Observations are required in any UK FIR/UIR/OCA whenever:

(a) Severe turbulence or severe icing is encountered; or

(b) moderate turbulence, hail or cumulo-nimbus clouds are encountered during transonic or supersonic flight; or

(c) other meteorological conditions are encountered which, in the opinion of the pilot-in-command, might affect the safety or markedly affect the efficiency of other aircraft operations, for example, other en-route weather phenomena specified for SIGMET messages, or adverse conditions during the climb-out or approach not previously forecast or reported to the pilot-in-command. Observations are required if volcanic ash cloud is observed or encountered, or if pre-eruption volcanic activity or a volcanic eruption is observed to assist other users, ATS Providers and the Volcanic Ash Advisory Centre (VAAC); or

(d) exceptionally, they are requested by the meteorological office providing meteorological service for the flight; in which event the observation should be specifically addressed to that meteorological office; or

(e) exceptionally, there is an agreement to do so between the Meteorological Authority and the aircraft operator.

Table 3.5.6.1 – TURB and other Turbulence Criteria Table

Incidence:	Occasional – less than 1/3 of the time	Intermittent – 1/3 to 2/3	Continuous – more than 2/3
Intensity	**Aircraft Reaction (transport size aircraft)**		**Reaction inside aircraft**
Light	Turbulence that momentarily causes slight, erratic changes Altitude and/or attitude (pitch, roll, yaw) IAS fluctuates 5-15kt (<0.5g at the aircraft's centre of gravity) Report as **Light Turbulence** or Turbulence that causes slight, rapid and somewhat rhythmic bumpiness without appreciable changes in altitude or attitude. No IAS fluctuations. Report as **Light Chop.**		Occupants may feel a slight strain against seat belts or shoulder straps. Unsecured objects may be displaced slightly. Food service may be conducted and little or no difficulty is encountered in walking.
Moderate	Turbulence that is similar to Light Turbulence but of greater Intensity. Changes in altitude and/or attitude occur but the aircraft remains in positive control at all times IAS fluctuates 15-25kt. (0.5-1.0g at the aircraft's centre of gravity) Report as **Moderate Turbulence** or; Turbulence that is similar to Light Chop but of greater intensity. It causes rapid bumps or jolts without appreciable changes in altitude or attitude. IAS may fluctuate slightly. Report as **Moderate Chop.**		Occupants feel definite strains against seat belts or shoulder straps. Unsecured objects are dislodged. Food service and walking are difficult.
Severe	Turbulence that causes large, abrupt changes in altitude and/or attitude. Aircraft may be momentarily out of control. IAS fluctuates more than 25kt (>1.0g at the aircraft's centre of gravity). Report as **Severe Turbulence.**		Occupants are forced violently against seat belts or shoulder straps. Unsecured objects are tossed about. Food service and walking impossible.

Note 1 Pilots should report location(s), time(s), incidence, intensity, whether in or near clouds, altitude(s) and type of aircraft. All locations should be readily identifiable. Turbulence reports should be made on request, or in accordance with paragraph 2.

Example

Over Pole Hill 1230 intermittent Severe Turbulence in cloud FL310, B747

From 50 miles north of Glasgow to 30 miles west of Heathrow 1210 to 1250, occasional Moderate Chop TURB FL330, MD80.

Note 2 The UK does not use the term 'Extreme' in relation to turbulence.

3 Turbulence and Icing Reporting Criteria

3.1 Turbulence (TURB)

3.1.1 TURB remains an important operational factor at all levels but particularly above FL 150. The best information on TURB is obtained from pilots' Special Aircraft Observations; all pilots encountering TURB are requested to report time, location, level, intensity and aircraft type to the ATS Unit with whom they are in radio contact. High level turbulence (normally above FL 150 not associated with cumuliform cloud, including thunderstorms) should be reported as TURB, preceded by the appropriate intensity or preceded by Light or Moderate Chop.

3.2 Windshear Reporting Criteria

3.2.1 Pilots using navigation systems providing direct wind velocity readout should report the wind and altitude/height above and below the shear layer, and its location. Other pilots should report the loss or gain of airspeed and/or the presence of up-or-down draughts or a significant change in crosswind effect, the altitude/height and location, their phase of flight and aircraft type. Pilots not able to report windshear in these specific terms should do so in terms of its effect on the aircraft, the altitude/height and location and aircraft type, for example, 'Abrupt windshear at 500 feet QFE on finals, maximum thrust required, B747'. Pilots encountering windshear are requested to make a report even if windshear has previously been forecast or reported.

3.3 Airframe Icing

3.3.1 All pilots encountering unforecast icing are requested to report time, location, level, intensity, icing type* and aircraft type to the ATS Unit with whom they are in radio contact. It should be noted that the following icing intensity criteria are reporting definitions; they are not necessarily the same as forecasting definitions because reporting definitions are related to aircraft type and to the ice protection equipment installed, and do not involve cloud characteristics. For similar reasons, aircraft icing certification criteria might differ from reporting and/or forecasting criteria.

Table 3.5.6.2 – Airframe icing intensity criteria

Intensity	Ice Accumulation
Trace	Ice becomes perceptible. Rate of accumulation slightly greater than rate of sublimation. It is not hazardous even though De-icing/anti-icing equipment is not utilised, unless encountered for more than one hour.
Light	The rate of accumulation might create a problem if flight in this environment exceeds 1 hour. Occasional use of de-icing/anti-icing equipment removes/prevents accumulation. It does not present a problem if de-icing/anti-icing equipment is used.
Moderate	The rate of accumulation is such that even short encounters become potentially hazardous and use of de-icing/anti-icing equipment, or diversion, is necessary.
Severe	The rate of accumulation is such that de-icing/anti-icing equipment fails to reduce or control the hazard. Immediate diversion is necessary.

Rime Ice	Rough, milky, opaque ice formed by the instantaneous freezing of small super cooled water droplets.
Clear Ice:	A glossy, clear or translucent ice formed by the relatively slow freezing of large super cooled water droplets.

4 In-flight Procedures

4.1 Information to aircraft in flight is usually supplied in accordance with area Meteorological Watch procedures, supplemented when necessary by an En-route Forecast service. Information is also available from the appropriate ATS Unit at the commanders request, or from meteorological broadcasts.

4.2 An in-flight en-route service is available in exceptional circumstances by prior arrangement with the Meteorological Authority (GEN 3.5.1, paragraph 1). A meteorological office is designated to provide the aircraft in flight with the winds and temperatures for a specific route sector. Applications for this service should be made in advance, stating:

(a) The flight level(s) and the route sector required;
(b) the period of validity necessary;
(c) the approximate time and position in flight at which the request will be made;
(d) the ATS Unit with whom the aircraft is expected to be in contact.

4.3 Aircraft can obtain aerodrome weather information from any of the following:

(a) VOLMET broadcasts (see GEN 3.5.7);
(b) Automatic Terminal Information Service (ATIS) broadcasts;
(c) by request to an ATS Unit but whenever possible only if the information required is not available from a broadcast.

4.4 When an aircraft diverts, or proposes to divert, to an aerodrome along a route for which no forecast has been provided, the commander may request the relevant information from the ATS Unit serving the aircraft at the time, and the necessary forecasts will be provided by the associated Forecast Office.

Table 3.5.7.1 – Meteorological Radio Broadcasts (VOLMET)

Call Sign/ID	EM	Frequency MHZ	Operating Hours	Stations	Contents	Remarks
1	2	3	4	5	6	7
London Volmet (Main)	A3E	135.375	H24 continuous	Amsterdam Brussels Dublin Glasgow London Gatwick London Heathrow London Stansted Manchester Paris/Charles de Gaulle	(1) Half hourly reports (METAR) (2) The elements of each report broadcast in the following order: (a) Surface wind (b) Visibility (or CAVOK) (c) RVE if applicable (d) Weather (e) Cloud or (CAVOK) (f) Temperature (g) Dewpoint	The spoken word 'SNOCLO' will be to the end of the aerodrome report when that aerodrome is unusable for take-offs & landings due to heavy snow on runways or runway snow clearance
London Volmet (South)	A3E	128.600	H24 continuous	Birmingham Bournemouth Bristol Cardiff Jersey London Luton Norwich Southampton Southend	(h) QNH (I) Recent weather if applicable (j) Windshear if applicable (k) TREND if applicable (l) Runway contamination warning if applicable	
London Volmet (North) (Note 1)	A3E	126.600	H24 continuous	Durham Tees Valley Humberside Isle of Man Leeds Bradford Liverpool London Gatwick Manchester Newcastle Nottingham East Midlands	(3) Non-essential words such as surface wind, visibility etc are not spoken (4) Except for SNOCLO (see column 7) the Runway State Group is not broadcast	
Scottish Volmet	A3E	125.725	H24 Continuous	Aberdeen/Dyce Belfast Aldergrove Edinburgh Glasgow Inverness London Heathrow Prestwick Stornaway Sumburgh	(5) All broadcasts are in english	

Note 1 Broadcasting range extended to cover Southeast England and English Channel.
Note 2 An HF VOLMET broadcast for North Atlantic flights (Shannon VOLMET) is operated by the Republic of Ireland

GEN 3.5 GEN 3.5.9.1 Meteorological charts are available via facsimile from two automated services, Broadcast Fax and METFAX

GEN 3.5.8 SIGMET SERVICE

Table 3.5.8.1 – SIGMET and AIRMET Services

Name of MWO Location Indicators	Hours	FIR or CTA Served	Type of SIGMET validity	Specific SIGMET procedures	AIRMET procedures	ATS Unit served	Additional information
1	2	3	4	5	6	7	8
Met Office Exeter Exeter EGRR	H24	London FIR/UIR Scottish FIR/UIR Shanwick OCA Shanwick FIR	SIGMET 4 hours Volcanic Ash SIGMET 6 hours	Tropical cyclone SIGMET is not issued	ICAO Annex 3 AIRMET (low level en-route weather warning) is not issued in the UK	London ACC London TCC Manchester ACC Scottish ACC Shanwick OCA	Nil

1 MWOs are responsible for the preparation and dissemination of SIGMETs to appropriate ACC/FIC within their own and agreed adjacent FIRs. Aircraft in flight should be warned by the ACC/FIC of the occurrence or expected occurrence of one or more of the following SIGMET phenomena for the route ahead for up to 500nm or 2 hours flying time:

thunderstorm (See Note 2);

heavy hail (see Note 2);

tropical cyclone;

freezing rain;

severe turbulence (not associated with convective cloud);

severe icing (not associated with convective cloud);

severe mountain waves;

heavy sand/dust storm;

volcanic ash cloud.

Note 1: In general, SIGMET messages are identified by the letters WS at the beginning of the header line, but those referring to tropical cyclones and volcanic ash will be identified by WC and WV respectively. SIGMETs are usually valid for 4 hours (but exceptionally for up to 6 hours and for volcanic ash cloud and tropical cyclones, a further outlook for up to 12 hours may be included) and are re-issued if they are to remain valid after the original period expires. They can be cancelled or amended within the period of validity. SIGMETs are numbered sequentially from 0001 UTC each day.

Note 2: This refers only to thunderstorms (including if necessary, cumulo-nimbus cloud which is not accompanied by a thunderstorm) widespread within an area with little or no separation (FRQ), along a line with little of no separation (SQL), embedded in cloud layers (EMBD), or concealed in cloud layers or concealed by haze (OBSC), but does not refer to isolated or occasional thunderstorms not embedded in cloud layers or concealed by haze. Thunderstorms and tropical cyclones each imply moderate or severe turbulence, moderate or severe icing and hail. However, heavy hail (HVYGR) may be used as a further description of the thunderstorm as necessary.

Note 3: A volcanic ash cloud SIGMET will be issued based on advisory information provided by the relevant VAAC. The Met Watch Office listed in Table 3.5.8.1 will ensure that information included in SIGMET and NOTAM messages is consistent.

Note 4: With the exception of volcanic ash cloud SIGMET, SIGMET messages will be issued not more than 6 hours and usually not more than 4 hours before the expected time of occurrence of the phenomenon. Volcanic ash cloud SIGMET messages will be issued up to 12 hours before commencement of the period of validity, where practicable, and will be updates at least every 6 hours.

GEN 3.5.9 OTHER AUTOMATED METEOROLOGICAL SERVICES

1 Meteorological charts are available via facsimile from two automated services, Broadcast Fax and METFAX

1.1 Broadcast Fax is a routine broadcast service available to users requiring a minimum number of charts regularly each week.

(a) Charts routinely transmitted over the Broadcast Fax network cover:

(i) Low and medium level flights within the UK and Near Continent;

(ii) Medium and high level flights to Europe and the Mediterranean;

(iii) High level flights to North America;

(iv) High level flights to the Middle/Far East;

(v) High level flights to Africa.

(b) There are additional charts which are not routinely available by Broadcast Fax, for example EURSAM significant weather for high level flights to South America and upper winds/temperatures at other levels. These may be obtained on prior request from the Met Office's Operations Centre, Exeter. A charge to cover handling and transmission costs will be made for this facility, (and if Broadcast Fax is not used, an account must be set up in advance by application to the address at GEN 3.5.1 paragraph 2).

(c) Table 3.5.4.2 gives the geographical and vertical coverage, the times of issue and validity of charts which are routinely available by Broadcast Fax.

GEN 3.5

GEN 3.5.9.1 Meteorological charts are available via facsimile from two automated services, Broadcast Fax and METFAX

(d) It should be noted that forecasts may be amended at any time, which charts received via facsimile may not show. Therefore, it is advisable to check there are no changes to the forecast conditions prior to departure. Amendment criteria for forecasts are given in GEN 3.5.4 paragraph 7.

1.2 METFAX Services

1.2.1 METFAX is a dial up premium rate facsimile service designed primarily for the General Aviation sector, and enables low and medium/high level charts for the UK and Continental Europe to be obtained as required by the user.

1.2.2 The METAFX service also includes METAR and TAF bulletins, AIRMET area forecasts, planning forecasts and satellite images. The complete schedule can be found on the index page (dial 09060 700 501).

1.3 Broadcast Text Meteorological Information

1.3.1 This information is distributed in the by National Air Traffic Services Ltd (NATS) via the AFS. Data from these fixed groups of countries are available to broadcast text recipients; these are listed below. Alternatively, NATS offer tailored broadcasts to meet individual customer's needs. Information includes METAR, TAF and warnings of weather significant to flight safety (SIGMET) including volcanic activity reports.

(i) Contents of OPMET 1 teleprinter Broadcast – METARs, TAFs and SIGMETs for the following areas:

Belgium, Denmark, Faeros, Finland, France, Germany, Iceland, Ireland, Italy, Netherlands, Norway, Portugal, Spain/Canaries, Sweden, Switzerland, United Kingdom.

(ii) Contents of OPMET 2 teleprinter Broadcast – METARs, TAFs and SIGMETs for the following areas:

Algeria, Armenia, Austria, Azerbaijan, Bahrain, Belarus, Bulgaria, Croatia, Cyprus, Czech Republic, Eastern Europe, Egypt, Estonia, Georgia, Greece, Hungary, Iran, Iraq, Israel, Jordan, Kazakhstan, Kuwait, Kyrgyz, Latvia, Lebanon, Libya, Lithuania, Macedonia, Malta, Med – Eastern, Med – Central, Middle Europe, Moldova, Morocco, Near East, Poland, Portugal, Romania, Russian Fed East, Russian Fed West, Saudi Arabia, Serbia & Montenegro, Slovakia, Slovenia, Syria, Tajikistan, Tunisia, Turkey, Turkmenistan, Ukraine, Uzbekistan.

(iii) Contents of OPMET 3 teleprinter Broadcast – METARs, TAFs and SIGMETs for the following areas:

Austria, Belgium, Bulgaria, Denmark, Eastern Europe, Estonia, Faeros, Finland France, Germany, Greece, Hungary, Iceland, Ireland, Italy, Malta, Med – Central, Middle Europe, Netherlands, Norway, Portugal, Romania, Serbia & Montenegro, Sweden, Switzerland, Turkey, United Kingdom.

Note 1: METARs are broadcast as routine at half hourly (exceptionally hourly) intervals during aerodrome opening hours.

Note 2: TAFs valid for periods of less than 12 hours, usually 9 hours, (FC) are broadcast every three hours and TAFs valid for periods of 12 hours or more, usually 19 hours, (FT) are broadcast every six hours. Amendments are broadcast between routine times as required.

(c) Further details of the OPMET networks, including the reporting aerodromes in each area, are available from the Meteorological Authority.

Note 1: METARs are broadcast as routine at half-hourly (exceptionally hourly) intervals during aerodrome opening hours.

Note 2: TAFs valid for periods of less than 12 hours, usually for 9 hours, (FC) and QFA Alps are broadcast every three hours and TAFs valid for periods of 12 to 24 hours (FT) every six hours. Amendments are broadcast between routine times as required.

1.3.2 Additional Information:

(a) Local special meteorological reports (SPECIAL) are issued for operational use locally when conditions change through limits specified at GEN 3.5.4 paragraph 7.

(b) Special reports in the SPECI code form are defined as Special Reports disseminated beyond the aerodrome of origin (not applicable to UK civil aerodromes).

(c) In general TAFs are provided only for those aerodromes where official meteorological observations are available.

(d) Amended TAFs and AIRMET forecasts are issued when forecast conditions change significantly, see GEN 3.5.4 paragraph 7;

(e) The formats and codes used for METAR, SPECI, TREND, TAF and the METAR Runway State Group are described at GEN 3.5.10;

(f) The actual or forecast meteorological conditions for which a SIGMET warning is prepared are detailed at GEN 3.5.8.

1.3.3 Further actual or forecast meteorological conditions for which a SIGMET warning is prepared are detailed at GEN 3.5.8.

1.4 AIRMET Service

1.4.1 AIRMET is a general aviation weather briefing service. There are ten routine forecasts, in plain language, covering the UK and near continent. A map showing the coverage of the forecast areas is at GEN 3.5.26. Information is provided in text form via the AFS, facsimile and Internet. Additionally, the three main areas (Scottish, North and South) are available from the AIRMET forecast telephone service. These are plain language forecasts, provided in spoken form at dictation speed via the public telephone network..

1.4.2 The AIRMET Forecast telephone service is intended for use by pilots who do not have access to meteorological by other means. It is provided in a standard format to facilitate transcription on to a Pilots Proforma, examples of which are given at GEN 3.5.27 and 3.5.28. Spoken AIRMET forecasts are obtainable from the numbers given at GEN 3.5.29.

1.4.3 The forecasts will reflect the contents of SIGMETs which are current at the time of issue or amendments of the forecasts. Safety related amplification of an AIRMET forecast may be obtained from a forecaster by telephoning one of the forecast offices listed at GEN 3.5.4 Table 3.5.4.1 as providing service 'E'. Callers must be able to confirm that they have obtained a current AIRMET forecast on contacting the forecast office, otherwise no additional forecast information will be given.

GEN 3.5 GEN 3.5.9.1 Meteorological charts are available via facsimile from two automated services, Broadcast Fax and METFAX

1.4.4 Special Forecasts in accordance with GEN 3.5.5 paragraph 1 are not provided for flights within the coverage of AIRMET Forecasts. For flights which extend beyond the area of coverage, Special Forecasts will be available on request from selected forecast offices providing a service to civil aviation (paragraph 1 refers).

1.4.5 JERSEY AIRMET (Available from UK). The States of Jersey meteorological Department provides a 100nm radius Channel Islands Low Level telephone recorded forecast service called JERSEY AIRMET, with a format very similar to the UK AIRMET. Amplification or clarification of the current JERSEY AIRMET forecast may be obtained, following receipt of the recording, by consulting the Forecast Office at Jersey Airport. Telephone calls to JERSEY AIRMET and the Forecast Office are charged at a premium rate and are not available outside the Channel Islands and UK. The appropriate telephone numbers are shown at table 3.5.4.1. A fax copy of the Jersey AIRMET is available by dialling 01534 492256.

1.5 DIALMET

1.5.1 An automated METAR and TAF telephone service, giving weather reports and forecasts for the UK, near continent and Eire is available on 09063 800400. The aerodromes available and the three digits to be dialled for each individual aerodrome is published in 'HET MET', a pocket sized booklet that can be obtained by writing to the following address: Met Office, Fitzroy Road, Exeter, EX1 3PB. When using this service it should be remembered that the three selective digits should not be dialled before the call is answered; however, the <*> key followed by the three digits will provide quick access to the information once the recorded message begins.

1.6 Internet Services

1.6.1 A wide range of meteorological information is available on the world wide web for pre flight information. However, users should be aware of the risks of use of the public Internet in this regard. This includes, but not limited to, a browsers cache facility not providing the user with the very latest information; delays to, or irregular updates of sites; or the receipt of falsified data purporting to have come from a legitimate provider. Users should ensure, wherever possible, that the data is up to date and consistent with the general weather situation.

GEN 3.5 GEN 3.5.9.1.6

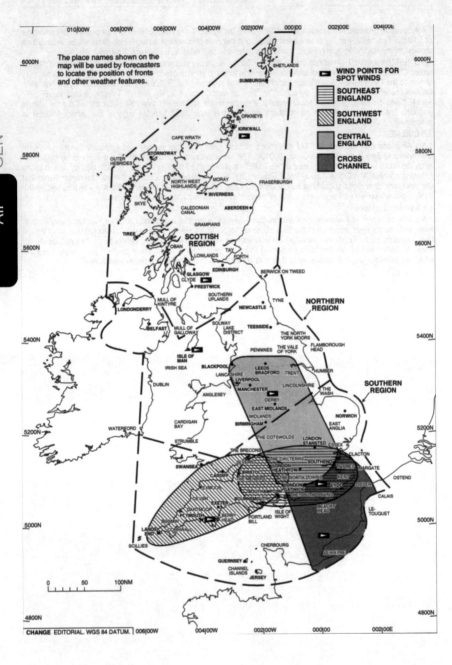

AIRMET COPY FORM
All temperatures degrees Celsius. All heights AMSL. All times UTC.

1   Forecast valid from:

            to:

2   Met Situation:

3   Winds

    (i)   Strong wind warnings (included if the surface wind anywhere in the region is forecast to exceed 20 knots).

    (ii) Winds at 1000 feet, 3000 feet and 6000 feet:

4   0°C level:

5   Weather conditions:

6   Remarks/Other warnings:

7   Outlook until:

GEN 3.5   GEN 3.5.9.1.6 Airmet Copy Form

Format of AIRMET Forecasts

1   Forecast valid from:

        to:

2   Met Situation:

3   Winds:

Location	Winds and Temperatures		
	1000 FT (UK Sig Wx) or 10000 FT (UK Upper Winds)	3000 FT (UK Sig Wx) or 18000 FT (UK Upper Winds)	6000 FT (UK Sig Wx) or 24000 FT (UK Upper Winds)
50N 00E			
GATWICK			
PLYMOUTH			
DERBY			
RONALDSWAY			
GLASGOW			
KIRKWALL			

Table 3.5.9.1 – Area Forecasts available on the AIRMET Telephone Menu Service
Dial: 09065 50 04 plus appropriate AIRMET code

AIRMET Code	Area Ident	Forecast	Period of Validity (UTC)	Available by (UTC)	Outlook Hours	To (UTC)
20	Southern Region	FAUK44	0500-1300	*	6	1900
		FAUK45	1100-1900	1030	6	0100 ND
21	Northern Region	FAUK54	0500-1300	*	6	1900
		FAUK55	1100-1900	1030	6	0100 ND
22	Scottish Region	FAUK70	0500-1300	*	6	1900
		FAUK71	1100-1900	1030	6	0100 ND

Note 1 *=will be recorded as soon as possible after 0500
Note 2 ND = Next day

GEN 3.5.10 METEOROLOGICAL CODES

1 Aerodrome Weather Report Codes (Actuals)

1.1 The content and format of an actual weather report is as shown in the following table:

Report Type	Location Identifier	Date/Time	Automatic Observation
METAR	EGSS	231020Z	AUTO
Wind	**Visibility**	**RVR**	**Present Weather**
31015G30KT	600	R24/P1500	SHRA
280V350	2500SW		
Cloud	**Temp/Dewpoint**	**QNH**	**Recent Weather**
FEW005	10/03	Q0995	RETS
SCT010CB			
BKN025			
Windshear	**Sea Surface Temperature And Sea State**	**Runway State**	**TREND**
WS RWY 23	W07/S4	88290592	NOSIG

2 Identifier

2.1 The identifier has three components as shown below:

(a) Report type

(i) METAR – Aviation routine weather report. These are compiled half-hourly or hourly at fixed times while the aeronautical station is open;

(ii) SPECI – Aviation selected special weather report. Special reports are prepared to supplement routine reports when improvements or deteriorations through certain criteria occur. However, by ICAO Regional Air Navigation agreement, they are not disseminated by either OPMET in UK or MOTNE in Europe.

(b) Location indicator
ICAO four-letter code letters (for UK aerodromes, see GEN 2.4).

(c) Date/Time
The date and time of observation in hours and minutes UTC, followed by the letter Z. Example: METAR EGSS 231020Z.

(d) AUTO
Where a report contains fully automated observations with no human intervention, it will be indicated by the code word 'AUTO', inserted immediately before the wind group. AUTO METARs shall only be disseminated when an aerodrome is closed or, at H24 aerodromes, when the accredited meteorological observer is on a duty break overnight. Users are reminded that in particular reports of visibility, present weather and cloud from automated systems should be treated with caution due to the limitations of the sensors themselves, the spatial area sampled by the sensors and the associated algorithms employed by the observing system. AUTO METAR shall indicate the limitations of the observing equipment through the use of additional codes, where applicable, given in paragraph 17.

3 Wind

3.1 Wind direction is given in degrees True (three digits) rounded to the nearest 10 degrees, followed by the windspeed (two digits, exceptionally three), both usually meaned over the ten minute period immediately preceding the time of observation. These are followed without a space by one of the abbreviations KT, KMH or MPS, to specify the unit used for reporting the windspeed. Example: 31015KT.

3.2 A further two or three digits preceded by a G gives the maximum gust speed in knots when it exceeds the mean speed by 10kt or more. Example: 31015G30KT.

3.3 Calm is indicated by '00000', followed by the units abbreviation, and variable wind direction by the abbreviation 'VRB' followed by the speed and unit.

3.4 If, during the 10 minute period preceding the time of the observation, the total variation in wind direction is 60° or more, the observed two extreme directions between which the wind has varied will be given in clockwise order, separated by the indicator letter V but only when the speed is greater than 3kt. Example: 31015G30KT 280V350.

4 Horizontal Visibility

4.1 In the METAR, the visibility reported is the prevailing visibility and, under certain conditions, the minimum visibility. Prevailing visibility is the visibility value that is reached or exceeded within at least half the horizon circle or within at least half of the surface of the aerodrome. These areas could comprise contiguous or non-contiguous sectors.

4.2 If the visibility in one direction which is not the prevailing visibility, is less than 1500 metres or less than 50% of the prevailing visibility, the lowest visibility observed should also be reported and its general direction in relation to the aerodrome indicated by reference to one of the eight points of the compass. If the lowest visibility is observed in more than one direction, then the most operationally significant direction should be reported. When the visibility is fluctuating rapidly and the prevailing visibility cannot be determined, only the lowest visibility should be reported, with no indication of direction.

Note: There is no requirement to report the lowest visibility if it is 10 km or more.

4.3 Visibility is recorded in metres (m) rounded down to:
(a) the nearest 50 m when the visibility is 800 m or less;
(b) the nearest 100 m when the visibility is greater than 800 m but less than or equal to 5000 m, and expressed in kilometres (km);
(c) The nearest 1 kilometre when the visibility is greater than 5000 m.

Note: The code 9999 indicates a visibility of 10 km or more; 0000 a visibility of less than 50 m.

4.4 Visibility reported in UK military aerodrome METARs will continue to be the minimum visibility.

5 RVR

5.1 An RVR group always includes the prefix R followed by the runway designator and a diagonal, in turn followed by the touch-down zone RVR in metres. If the RVR is assessed on two or more runways simultaneously, the RVR group will be repeated; parallel runways will be distinguished by appending, to the runway designator, L, C or R indicating the left, central or right parallel respectively. Examples: R24L/1100 R24R/0750.

5.2 When the RVR is greater than the maximum value which can be assessed the group will be preceded by the letter indicator P followed by the highest value which can be assessed.
Example: R24/P1500.

5.3 When the RVR is below the minimum value which can be assessed, the RVR will be reported as M followed by the appropriate minimum value assessed. Example: R24/M0050.

5.4 If it is possible to determine mean values of RVR, the mean value of RVR over the 10 minute period immediately preceding the observation will be reported; trends and significant variations may be reported as follows:

(a) Trends. If RVR values during the 10 minute period preceding the observation show a distinct increasing or decreasing tendency, such that the mean during the first five minutes varies by 100m or more from the mean during the second five minutes, this will be indicated by subscripts U or D for increasing or decreasing tendencies; otherwise, subscript N will indicate no distinct change during the period. Example: R24/1100D.

(b) Significant Variations. When the RVR at a runway varies significantly such that, during the 10 minute period preceding the observation, the 1 minute mean extreme values vary from the 10 minute mean value by either more than 50 metres or more than 20% of the 10 minute mean value (whichever is greater), the 1 minute mean minimum and maximum values will be given in that order, separated by V, instead of the 10 minute mean. Example: R24/0750V1100.

5.5 If the 10 minute period immediately preceding the observation includes a marked discontinuity in runway visual range values, only those values occurring after the discontinuity should be used to obtain mean values.

5.6 A complete RVR group may therefore be of the form. Example: R24L/0750V1100U.

Note: Until further notice, UK aerodromes will not be required to report RVR trends and significant variations. RVR is reported when the horizontal visibility or RVR is less than 1500m. For multi-site RVR/IRVR systems, the value quoted is that for the Touch Down Zone (TDZ). If the RVR is assessed for two or more runways simultaneously, the value for each runway is given.

6 Weather

6.1 Each weather group may consist of appropriate intensity indicators and letter abbreviations combined in groups of two to nine characters and drawn from the following table:

Table 3.5.10.1 – Significant Present and Forecast Weather Codes

Qualifier		Weather Phenomena		
Intensity or	Description	Precipitation	Obscuration	Other
Light	BC – Patches	DZ – Drizzle	BR – Mist	DS – Duststorm
Moderate	BL – Blowing	GR – Hail GS – Small hail (<5mm diameter and/or snow pellets)	DU – Widespread Dust	FC – Funnel Cloud (s) (tornado or water-spout)
+ Heavy (Well developed in case of FC and PO)	DR – Drifting FZ – Freezing (super cooled)		FG – Fog FU – Smoke	PO – Dust/Sand Whirls (Dust Devils)
VC In the vicinity (not at the aerodrome but not further away than approx 8km from the aerodrome perimeter)	MI – Shallow PR – Partial (covering part of aerodrome)	IC – Ice Crystals (Diamond Dust) PL – Ice Pellets RA – Rain SG – Snow Grains SN – Snow	HZ – Haze SA – Sand VA – Volcanic Ash	SQ – Squall SS – Sandstorm
	SH – Showers TS – Thunderstorms			

6.2 Mixture of precipitation types may be reported in combination as one group, but up to three separate groups may be inserted to indicate the presence of more than one independent weather type.
Examples: MIFG, VCSH, +SHRA, RASN, –DZ HZ.
Note 1: BR, HZ, FU, IC, DU and SA will not be reported when the visibility is greater than 5000m.
Note 2: Some codes are shown that will not be used in UK METARs and TAFs but may be seen in continental reports and when flying in Europe.

7 Cloud

7.1 A six character group will be given under normal circumstances. The first three to indicate cloud amount:
(a) **FEW** to indicate 1 to 2 oktas;
(b) **SCT** (scattered) to indicate 3 to 4 oktas;
(c) **BKN** (broken) to indicate 5 to 7 oktas;
(d) **OVC** (overcast) to indicate 8 oktas
and the last three characters indicate the height of the base of the cloud layer in hundreds of feet above aerodrome level. Example: FEW018.
7.2 Types of cloud other than significant convective clouds are not identified. Significant clouds are.
(a) **CB** Cumulonimbus;
(b) **TCU** Towering Cumulus.
Example: SCT018CB.
7.3 Reporting of layers or masses of cloud is made as follows:
(a) First Group: Lowest individual layer of any amount;
(b) Second Group: Next individual layer of more than 2 oktas;
(c) Third Group: Next higher layer of more than 4 oktas;
(d) Additional Group: Significant convective cloud if not already reported.
The cloud groups are given in ascending order of height. Example: FEW005 SCT010 SCT018CB BKN025.
7.4 When there is no cloud below 5000ft or below the highest minimum sector altitude (whichever is the greater) and there is no towering cumulus or cumulonimbus, 'NSC' (no significant cloud) is reported. However, the amount, height of cloud base and cloud type of towering cumulus or cumulonimbus shall be reported, irrespective of the cloud base height.
7.5 Sky obscured is coded by VV followed by the vertical visibility in hundreds of feet. When the vertical visibility cannot be assessed the group will read VV///: (See GEN 1.7) Example: VV003.

8 CAVOK

8.1 The visibility, RVR, weather and cloud groups are replaced by CAVOK when the following conditions exist:
(a) Visibility is 10km or more;
(b) No cloud below 5000ft or below the highest Minimum Sector Altitude, whichever is the greater, and no Cumulo-nimbus;
(c) No significant weather phenomena at or in the vicinity of aerodrome.

9 Air Temperature/Dewpoint

9.1 These are given in Degrees Celsius, M indicates a negative value. Examples: 10/03, 01/M01.
If the dew point is missing, the temperature would be reported as 10///.
9.2 Temperatures are reported to the nearest whole degree Celsius, with observed values involving 0.5°C rounded up to the next higher degree Celsius, for example +2.5°C is rounded off to +3°C, -2.5°C is rounded off to -2°C.

10 QNH

10.1 QNH is rounded down to the next whole millibar and reported as a four digit group preceded by the letter indicator Q. If the value of QNH is less than 1000 mbs the first digit will be 0. Example: Q0995.

10.2 Where reported in inches of mercury, the pressure is prefixed by 'A', and the pressure entered in hundredths of inches, viz with the decimal point omitted between the second and third figure. Example: A3027.

11 Supplementary Information

(a) Recent Weather. Recent Weather will be operationally significant weather observed since the previous observation (or in the last hour whichever period is the shorter), but not now. The appropriate present weather code will be used, preceded by the letter indicator RE; up to three groups may be inserted to indicate the former presence of more than one weather type. Example: RETS REGR.

(b) Windshear. The Windshear may be inserted if reported along the take-off or approach paths in the lowest 1600ft with reference to the runway. WS is used to begin the group. Examples: WS RWY20, WS ALL RWY.
Until further notice, UK aerodromes will not insert windshear groups.

(c) Sea surface temperature and state. The sea surface temperature is preceded by the letter indicator W and given in Degrees Celsius, with M indicating a negative value. The sea state is a single numerical value, 0-9, preceded by the letter indicator S and decodes to give the height in metres of well-developed wind waves over the open sea. Sea surface temperature and state are not used in the UK.

12 Runway State Group

12.1 An eight figure Runway State Group may be added to the end of the METAR (or SPECI) (following any TREND) when there is lying precipitation or other runway contamination. It is composed as follows:

(a) Runway Designator (First Two Digits)
27 = Runway 27 or 27L
77 = Runway27R (50 added to the designator for 'right' Runway)
88 = All runways
99 = A repetition of the last message received because no new information received.

(b) Runway Deposits (Third Digit)
0 = Clear and dry
1= Damp
2 =Wet or water patches
3 =Rime or frost covered (depth normally less than 1 mm)
4 = Dry Snow
5 = Wet Snow
6 = Slush
7 = Ice
8 = Compacted or rolled snow
9 = Frozen ruts or ridges
/ = Type of deposit not reported (eg due to runway clearance in progress).

(c) Extent of Runway Contamination (Fourth Digit)
1 = 10% or less
2 = 11%1o25%
5 = 26% to 50%
9 = 51 % to 100%
/ = Not reported (eg due to runway clearance in progress).

(d) Depth of Deposit (Fifth and Sixth Digits)
The quoted depth is the mean number of readings or, if operationally significant, the greatest depth measured.
00 = less than 1 mm
01 = 1 mm etc through to
90 = 90 mm
91 = not used
92 = 10 cm
93 = 15 cm
94 =20 cm
95 = 25 cm
96 = 30 cm
97 = 35 cm
98 =40 cm or more
99 = Runway(s) non-operational due to snow, slush, ice, large drifts or runway clearance, but depth not reported.
// = Depth of deposit operationally not significant or not measurable.

(e) Friction Coefficient or Braking Action (Seventh and Eighth Digits)
The mean value is transmitted or, if operationally significant, the lowest value. For example:
28 =Friction coefficient 0.28
35= Friction coefficient 0.35
91 = Braking action: Poor
92 = Braking action: Medium/Poor
93 = Braking action: Medium

94 = Braking action: Medium/Good
95 = Braking action: Good
99 = Figures unreliable (eg if equipment has been or used which does not measure satisfactorily in slush or loose snow)
// = Braking action not reported (eg runway not operational; closed; etc)

Note 1: CLRD. If contamination conditions on all runways cease to exist, a group consisting of the figures 88, the abbreviation CLRD, and the Braking Action, is sent.

Note 2: It should be noted that runways can only be inspected as frequently as conditions permit, so that a re-issue of a previous half hourly report does not necessarily mean that the runway has been inspected again during this period, but might mean that no significant change is apparent.

Note 3: It is emphasised that this reporting system is completely independent of the normal NOTAM system and these reports are not used by AIS for amending SNOWTAM received from originators.

12.2 If the aerodrome is closed due to contamination of runways, the abbreviation **SNOCLO** is used in place of a runway state group.

13 TREND

13.1 For selected aerodromes, this is a forecast of significant changes in conditions during the two hours after the observation time:

(a) Change Indicator:
BECMG (becoming) or TEMPO (temporary), which may be followed by a time group (hours and minutes UTC) preceded by one of the letter indicators FM (from), TL (until), AT (at);

(b) Weather:
Standard codes are used. NOSIG replaces the trend group when no significant changes are forecast to occur during the trend forecast period.

Examples: BECMG FM1100 25035G50KT; TEMPO FM0630 TL0830 3000 SHRA.

14 RMK

14.1 The indicator 'RMK' (remarks) denotes an optional section containing additional meteorological elements. It will be appended to METARs by national decision, and should not be disseminated internationally. In UK, the section will not be inserted without prior authority of the MET Authority.

15 Missing Information

15.1 Information that is missing in a METAR or SPECI may be replaced by diagonals.

16 Examples of METAR:

(a) METAR EGGX 301220Z 14005KT 1200 0600E R12/1000N DZ BCFG VV/// 08/07 Q1004 NOSIG=

(b) METAR EGLY 301220Z 24015KT 200V280 8000 -RA SCT010 BKN025 18/15 Q0983 TEMPO 3000 RA BKN008 OVC020=

(c) METAR EGPZ 301220Z 30025G37KT 270V360 1200 0800NE +SHSNRAGS FEW005 SCT010 BKN020CB 03/M01 Q0999 RETS BECMG AT1300 9999 NSW SCT015 BKN100=

The above METAR for 1220 UTC on the 30th of the month, in plain language:

EGGX: Surface wind: mean 140 Deg True, 5kt; Prevailing visibility 1200m, minimum visibility 600 metres (to east); mean RVR 1000 metres (at threshold Runway 12, no apparent tendency); moderate drizzle with fog patches; Sky obscured, vertical visibility not available; dry bulb temperature Plus 8 C, dew point Plus 7 C; Aerodrome QNH 1004m b; Trend: no significant change expected next two hours;

EGLY: Surface wind: mean 240 Deg True, 15kt; varying between 200 and 280 deg; Prevailing visibility 8km; Light rain; cloud 3-4 oktas base 1000ft, 5-7 oktas 2500ft, 8 oktas 8000ft; dry bulb: plus 18 C, dew point: plus 15 C; QNH 983m b; Trend: temporarily 3000m in moderate rain with 5-7 oktas 800ft, 8 oktas 2000ft;

EGPZ: Surface wind: mean 300 Deg True, 25kt; Gust 37kt, varying between 270 and 360 deg; Prevailing visibility 1200m minimum visibility 800m to northeast. Heavy shower of snow, rain and small hail; 1-2 oktas base 500ft, 3-4 oktas base 1000ft, 5-7 oktas CB base 2000ft; dry bulb: plus 3 C, dew point: minus 1 C; QNH 999m b; thunderstorm since previous report. Trend: improving at 1300 UTC to 1km or more, nil weather, 3-4 oktas 1500ft, 5-7oktas 10000ft.

17 AUTO METAR coding

17.1 Where the observation is generated by an automatic observing system without any human input or supervision, the code 'AUTO' shall be inserted between the date/time of the report group and the wind group.

17.2 Where the observation is generated by an automatic observing system without any human input or supervision, the letters 'NDV' (no directional variation reported) shall be appended to the prevailing visibility (without any space between the visibility value and the letters 'NDV').

17.3 Where the observation is generated by an automatic observing system without any human input or supervision and the present weather cannot be detected due to unserviceable or missing present weather sensors, the lack of present weather information should be indicated by two slashes (////). If the present weather sensor is unable to determine the state or form of the precipitation, 'UP' (un-identified precipitation) or 'FZUP' (freezing unidentified precipitation), together with any intensity qualifiers, should be reported as appropriate. If the present weather sensor is serviceable but not detecting any present weather, then no present weather group shall be reported in the METAR.

17.4 Where the observation is generated by an automatic observing system without any human input or supervision, the inability of the system to detect towering cumulus or cumulonimbus shall be indicated by three slashes (////) after each cloud group of cloud amount and height.

GEN 3.5 — GEN 3.5.10.17 AUTO METAR coding

17.5 Where the observation is generated by an automatic observing system without any human input or supervision, the letters 'NCD' (no cloud detected) shall be used to indicate that the observing system has determined that there are no clouds below 5000 feet present.

17.6 Recent unidentified precipitation ('REUP') shall be reported if moderate or heavy unidentified precipitation has ceased or decreased in intensity since being reported in the last routine report or within the last hour, whichever is the shorter.

17.7 A failure of the measuring equipment or processing system for visibility or cloud should be indicated by four slashes (////) for visibility and 9 slashes (/////////) for cloud.

17.8 Examples of AUTO METAR coding:
a) METAR EGZZ 292220Z AUTO 29010KT 600ONDVII FEW01O //i BKN025 /// 17/12 Q0996=
b) METAR EGZZ 300450Z AUTO VRB02KT 300ONDV BR NCD 10/09 Q1002=

18 Aerodrome Forecast (TAF) Codes

18.1 TAFs describe the forecast prevailing conditions at an aerodrome and usually cover a period of 9 to 24 hours. The validity periods of many of the latter do not start until 8 hours after the nominal time of origin and the forecast details only cover the last 18 hours. The 9 hour TAFs are updated and re-issued every 3 hours and those valid for 12 to 24 hours, every 6 hours. Amendments are issued as and when necessary. The forecast period of a TAF may be divided into two or more self contained parts by the use of the abbreviation FM followed by a time. TAFs are issued separately from the METAR or SPECI and do not refer to any specific report; however, many of the METAR groups are also used in TAFs and significant differences are detailed in paragraph 18.

18.2 The content and format of a TAF is as in the following table:

Report Type	Location Identifier	Date/Time of Origin	Validity Time	Wind	Visibility
TAF	EGZZ	130600Z	130716	31015KT	8000
Weather	Cloud	Variant	Validity Times		
-SHRA	FEW005 SCT018CB BKN025	TEMPO	1116		
Visibility	Weather	Cloud	Probability	Validity Time	Weather
4000	+SHRA	BKN010CB	PROB30	1416	TSRA

18.2.1 Example of TAF
(a) 9 hr TAF:
FCUK31 EGGY 300900
TAF EGGW 301019 2301OKT 9999 SCT010 BKNO18 BECMG1114 6000 - RA BKN012 TEMPO 1418 2000 RADZ OVCOO4FM1800 3002OG3OKT 9999 - SHRA BKN015CB=
(b) 18 hr TAF:
FTUK31 EGGY 302200
TAF EGLL 310624 13010KT 9000 BKNO10 BECMG 0608 BKN020 PROB30 TEMPO 0816 17025G4OKT 4000 TSRA BKNO15CB BECMG 1821 3000 BR SKC=

19 Differences from the METAR

(a) Identifier. In the validity period, the first two digits indicate the day on which the period begins, the next two digits indicate the time of commencement of the forecast in whole hours UTC and the last two digits are the time of ending of the forecast in whole hours.

(b) Wind. In the forecast of the surface wind, the expected prevailing direction will be given. When it is not possible to forecast a prevailing surface wind direction due to its expected variability, for example, during light wind conditions (3 kts or less) or thunderstorms, the forecast wind direction will be indicated by the use of the abbreviation 'VRB'.

(c) Horizontal Visibility. As with the METAR code, except that only one value (the prevailing visibility) will be forecast. Visibility is reported in steps detailed in para 4.3.

(d) Weather. If no significant weather is expected the group is omitted. However, after a change group, if the weather ceases to be significant, the abbreviation NSW is used for No Significant Weather.

(e) Cloud. When clear sky is forecast the cloud group is replaced by SKC (sky clear). When no cumulo-nimbus or clouds below 5000 ft or below the highest minimum sector altitude, whichever is the greater, are forecast and CAVOK or SKC are not appropriate, then NSC (No Significant Cloud) is used. Only CB cloud will be specified.

(f) Significant Changes. The abbreviation FM followed by the time to the nearest hour and minute UTC is used to indicate the beginning of a self contained part in a forecast. All conditions given before this group are superseded by the conditions indicated after the group. Example: FM1220 27017KT 4000 BKN010.
The change indicator BECMG followed by a four figure time group, indicates an expected permanent change in the forecast meteorological conditions, at either a regular or irregular rate, occurring at an unspecified time within the period. Example: BECMG 2124 1500 BR.
The change indicator TEMPO followed by a four figure time group indicates a period of temporary fluctuations to the forecast meteorological conditions which may occur at any time during the period given. The conditions following these groups are expected to last less than one hour in each instance and in aggregate less than half the period indicated. Example: TEMPO 1116 4000 +SHRA BKN010CB.

(g) **Probability.** The probability of occurrence happening will be given as a percentage, although only 30% and 40% will be used. The abbreviation PROB is used to introduce the group, followed by a time group, or an indicator and a time group.
Examples: (a) PROB30 0507 0800 FG BKN004: (b) PROB40 TEMPO 1416 TSRA.

(h) **Amendments.** When a TAF requires amendment, the amended forecast shall be indicated by inserting AMD after TAF in the identifier and this new forecast covers the remaining validity period of the original TAF.
Example: TAF AMD EGZZ 130820Z 130816 21007KT 9999 BKM020 BECMG 0912 4000 RADZ BKN008=
Any further amendments to a TAF that has already been amended will result in the same 'TAF AMD' coding being used; however the date and time of origin will be updated.

(i) **Corrections.** When a TAF requires correction, the corrected forecast shall be identified by inserting COR after TAF in the identifier.
Example: Original: TAF   EGZZ 140905Z 141019 27012KT 4000 RUDZ BKN012 TEMPO 1019 BKN008=
Update: TAF COR EGZZ 140918Z 141019 27012KT 4000 RADZ BKN012 TEMPO 1019 BKN008=
Note: A correction will be issued only to correct an obvious typographical errror. However, a TAF amendment shall be issued instead whenever such a change would result in the meteorological conditions forecast being better or worse than previously stated, for example, 3000 RADZ instead of 4000 RADZ.

(j) **Other Groups.** Three further TAF groups are not used for civil aerodromes in the UK but are shown here to assist in decoding overseas and UK military TAF.

(i) Forecast Temperature Txaa/ggZ TNbb/hhZ
aa   Maximum temperature preceded by the letter indicators TX and given in Degrees Celsius, with M indicating a negative value.
gg   Time of maximum temperature, in UTC
bb   Minimum temperature preceded by the letter indicators TN and given in Degrees Celsius, with M indicating a negative value.
hh   Time of minimum temperature, in UTC
Example: TX25/13Z TN09/05Z

(ii) Airframe Ice Accretion 6 Ic hhh tL
6   Group indicator
Ic   Type of airframe ice accretion:
0   none
1   light
2   light in cloud
3   light in precipitation
4   moderate
5   moderate in cloud
6   moderate in precipitation
7   severe
8   severe in cloud
9   severe in precipitation
hhh   Height above ground level of lowest icing level (hundreds of feet)
tL   Thickness of icing layer (for decode see turbulence layer below)
Example of an icing forecast: 650104
This decodes as moderate icing potential is expected in clouds (code 5) from 1000 feet (code 010 in hundreds of feet) to 5000 feet AGL (4000 feet thickness as denoted by the ending '4'). Icing groups can be repeated as often as necessary to indicate more than one layer or type of icing. If one layer of icing is expected to be greater than 9000 feet in thickness, two groups must be coded. Example: 630309 631203. This decodes as light icing is expected in precipitation (code 3 in both the first and second groups) and that the potential exists from 3000 feet to 15,000 feet. Notice that the ending '9 in the first group (indicating 9000 feet thickness) triggers the need for the second group.

(iii) Turbulence 5 B hhh tL
5   Group indicator
B   Turbulence:
0   none
1   light
2   moderate in clear air, infrequent
3   moderate in clear air, frequent
4   moderate in cloud, infrequent
5   moderate in cloud, frequent
6   severe in clear air, infrequent
7   severe in clear air, frequent
8   severe in cloud, infrequent
9   severe in cloud, frequent
hhh   Height above ground level of lowest level of turbulence (hundreds of feet)
tL   Thickness of turbulent layer:
0   up to top of clouds
1-9 thickness in thousands of feet

Example: 530804
This decodes as frequent moderate turbulence is expected in clear air (code 3 from the table) and that the turbulence will extend from 8000 feet to 12000 feet AGL. As with the icing forecast above, if a layer of turbulence is forecast to exceed 9000 feet in thickness, a second group would be required.

20 Reports in Abbreviated Plain Language
20.1 Some reports may be disseminated in abbreviated plain language. These will use:
(a) Standard ICAO abbreviations and
(b) numerical values of a self explanatory nature.
The abbreviations referred to under (a) are contained in the Procedures for Air Navigation Services – ICAO Abbreviations and Codes (ICAO Doc 8400)

GEN 3.6 SEARCH AND RESCUE
GEN 3.6.1 RESPONSIBLE AUTHORITIES

1 Responsibility for Search and Rescue (SAR) for civil aircraft within the UK Search and Rescue Region (SRR) rests jointly with the Department of Transport (DfT) and the Ministry of Defence (MOD)
1.1 The DfT is responsible for SAR policy for civil aviation.
Department of Transport
Civil Aviation Directorate
Great Minster House
76 Marsham Street
London
SW1P 4DR
Tel: 020 7944 6374
Fax: 020 7944 2193
1.2 The Civil Aviation Authority (CAA) acts as adviser on SAR to the DfT. Queries on SAR for civil aviation, including matters arising from this section of the AIP, should be addressed in the first instance to the following:
Civil Aviation Authority
Directorate of Airspace Policy
ORA K6 CAA House
45-59 Kingsway
London
WC2B 6TE
Tel: 020 7453 6541
Fax: 020 7453 6565
Telex: 883092 – EGGA
AFS: EGGAYFYG
1.4 The MOD is responsible for the implementation of SAR services for civil aviation throughout the UK SRR. This responsibility is discharged through a single Aeronautical Rescue Co-ordination Centre (ARCC) at Kinloss.
ARCC Kinloss
RAF Kinloss
Forres
Moray
IV36 0UH
Tel: 01343 836001
Tel: 01309 672161 Ex 6202
Fax: 01309 678308/9
Telex: 75193
AFS: EGQKYCYX

2 Applicable ICAO Documents
2.1 The Standards, Recommended Practices and, when applicable, the procedures contained in the following ICAO documents are applied:
Annex 2   Rules of the Air;
Annex 3   Meteorological Services for International Air Navigation;
Annex 6   Operation of Aircraft – Parts I, II & III;
Annex 10 Aeronautical Telecommunications – Volume I & II;
Annex 11 Air Traffic Services;
Annex 12 Search and Rescue;
Annex 13 Aircraft Accident Investigation;
Annex 15 Aeronautical Information Services;

GEN 3.6 GEN 3.6.3 TYPES OF SERVICE

Annex 17   Security;
Annex 18   The Safe Transport of Dangerous Goods by Air;
Doc 4444   Procedures for Air Navigation Services – Air Traffic Management;
ATM/501
Doc 7030   Regional Supplementary Procedures;
Doc 7754   Air Navigation Plan – European Region;
Doc 8755   Air Navigation Plan – North Atlantic;
Doc 9731   IAMSAR Manual.
2.2 Differences from ICAO Standards Recommended Practices and Procedures are given at GEN 1.7.

GEN 3.6.2 THE UK SEARCH AND RESCUE REGION (SRR)

1 In general, aeronautical SRR boundary coincides with FIR boundary. The UK SRR comprises the London, Scottish and Shanwick FIRs but to the North and East of the UK, the boundaries with Iceland, Norway, Denmark and Germany have been modified by a series of bilateral agreements. Under these agreements, the SRR boundary no longer follow the FIR boundaries but have been aligned with the median line. The UK SRR boundary co-ordinates are given below and a chart depicting the areas of responsibility is shown at GEN 3-6-10.

2 ARCC KINLOSS. The area of responsibility is enclosed by the lines joining the following points:
610000N 0300000W – 610000N 0040000W – 632833N 0004622W – 632833N 0000000E – 620000N 0000000E –
620000N 0012222E – 614310N 0013329E – 612122N 0014718E – 595346N 0020430E – 591722N 0014236E –
582546N 0012854E – 575416N 0015748E – 563540N 0023642E – 560510N 0031455E – 555458N 0032055E –
555004N 0032355E – 554552N 0032208E – 543715N 0025349E – 542245N 0024543E – 531803N 0030319E –
524657N 0031213E – 523715N 0031055E – 522457N 0030325E – 521721N 0025555E – 520557N 0024249E –
515857N 0023731E – 514815N 0022849E – 513000N 0020000E – 510700N 0020000E – 510000N 0012800E –
504000N 0012800E – 500000N 0001500W – 500000N 0020000W – 485000N 0080000W – 450000N 0080000W –
450000N 0300000W – 610000N 0300000W.

with the exception of the Shannon FIR which is bounded by lines joining the following points:
540000N 0150000W – 543400N 0100000W -544500N 0090000W – 552000N 0081500W – 552500N 0072000W –
552000N 0065500W – 542500N 0081000W – 535500N 0053000W – 522000N 0053000W – 510000N 0080000W
– 510000N 0150000W – 540000N 0150000W.

GEN 3.6.3 TYPES OF SERVICE

1 Locations and types of dedicated SAR facilities are listed below, and shown on the chart at GEN 3-6-10:
(a) RAF and RN fixed wing aircraft and helicopters ;
(b) DTLR helicopters;
(c) RAF Mountain Rescue Teams;
1.1 Specialized SAR landplanes carry droppable survival aids including multi-seat inflatable life rafts. Specialized SAR helicopters are equipped with winch gear.
Table 3.6.3.1 – State of readiness of individual SAR units

Unit	State of Readiness	Remarks
RAF Nimrod	60 min (H24)	
RAF Sea King	15 min (0800-2200 local time)	
	45 min (2200-0800 local time)	
RN Sea King whichever is earlier	15 min (Day)* 45 min (Night)	* ie until local time or end of civil twilight
HM Coastguard S61N (Portland)	15 min (0800-2000 UTC)	
HM Coastguard S61N (Others)	15 min (0730-2100 local time) 45 min (2100-0730 local time)	
RAF Mountain Rescue Teams (MRT)	60 min (H24)	

1.2 In addition, the ARCC can call upon the following resources for assistance:
(a) HM Coastguard (see GEN 3-6-11);
(b) Royal National Lifeboat Institution, through HM Coastguard;
(c) RN ships and helicopters not SAR-dedicated and RAF aircraft not SAR-dedicated;
(d) Civil Police, Fire and Ambulance services;
(e) Civil aircraft;
(f) Merchant vessels;
(g) Army, RN and RAF personnel;
(h) Civilian Mountain Rescue Teams;
(i) British Telecom International Coast Radio Stations;
(j) COSPAS/SARSAT satellite distress alerting system.

GEN 3.6

GEN 3.6.3 TYPES OF SERVICE

1.3 Distress Frequencies

1.3.1 SAR aircraft and other Military aircraft carry the distress frequencies shown in table 3.6.3.2:
Table 3.6.3.2 Distress frequencies carried by SAR aircraft and other military aircraft

Frequency	Speech Facility	Homing Facility
121.500 MHz	(a) RAF Long Range Maritime Patrol (LRMP) aircraft and helicopters (b) Certain other military aircraft (c) HMCG helicopters (d) RN Sea King helicopters	(a) RAF LRMP aircraft, RAF Sea King helicopters (b) RN Sea King helicopters (c) HMCG helicopters (d) Some other military aircraft
243.00 MHz	(a) RAF aircraft (b) RN aircraft (c) HMCG helicopters	(a) RAF LRMP aircraft, RAF Sea King helicopters (b) RN Sea King helicopters (c) HMCG helicopters (d) Some other military helicopters
500 kHz		RAF LRMP aircraft
2182 kHz	(a) RAF LRMP aircraft (b) Some helicopters	(a) RAF Sea King helicopters (b) HMCG helicopters

1.4 Scene of Search Frequencies

1.4.1 SAR aircraft may use any of the following frequencies as a Scene of Search frequency

123.100 MHz	Civil
282.800 MHz	NATO
* 5680 kHz	Civil/Military-Day (SAMAR 1)
* 3085 kHz	Military-Night (SAMAR 4)
156.300MHz	Channel 6 VHF/FM Marine
* 3023 kHz	Civil/Military-Night (SAMAR 1)
* 5695 kHz	Military-Day (SAMAR 4)
* 8364 kHz	International intercommunication
* Emission J3E or 6K00A3EJN	

Note: Other HF frequencies may be used as directed by the controlling ARCC.

GEN 3.6.4 SAR AGREEMENTS

1 As a Contracting State under the Convention on International Civil Aviation, the United Kingdom is committed to providing SAR services for international civil aviation throughout defined areas on a 24 hour basis. These areas consist of the UK overland area and adjacent sea areas to approximately midway to the European mainland to the east and to 030W over the North Atlantic, excluding the Shannon FIR (see chart at GEN 3-6-10).

1.1 The International Civil Aviation Organization's (ICAO) Regional Air Navigation Plans do not define the scale of effort which should be available but identify the required facilities by types of services which should be provided with due regard to the density of traffic and the size and passenger capacity of aircraft operating in the region.

1.2 As a member of NATO, a contracting State under the ICAO Convention and according to bilateral inter-RCC agreements, the United Kingdom can seek SAR assistance from the resources of other nations as necessary.

GEN 3.6.5 CONDITIONS OF AVAILABILITY

1 The availability of both British and NATO military SAR facilities is on the authority of the appropriate RCC which remains responsible for operational control of such facilities throughout the duration of the requirement (whether a Maritime RCC (MRCC) or an ARCC). DfT contracted SAR helicopters are under the operational control of HM Coastguard.

GEN 3.6.6 PROCEDURES AND SIGNALS USED

1 The Rescue Organization

1.1 When an Area Control Centre (ACC) has reason to believe that an aircraft is in a state of emergency, it will alert the ARCC and notify the local police, if appropriate. The ARCC will, in turn, alert SAR units and RAF Mountain Rescue Teams (MRT), and the police will notify civilian MRT, fire, ambulance and hospital services. At some places, arrangements are made for the ACC to notify the fire service directly. Should the first report of an accident be given to the police by a member of the public, the police will alert fire and other services. The police will also advise the ACC of the rescue action being taken and give full details.

1.2 The ARCC will alert the appropriate MRCCs should an Aviation SAR incident become a Maritime accident.

1.3 When the location of a civil aircraft which has crashed on land is known, and no air search is necessary, the civil ground organization (normally the police) takes responsibility for dealing with the incident. However, it is essential that both the ACC and the ARCC are informed to avoid duplication of effort and for expert consideration of any SAR back-up services which could be required.

1.4 In the vicinity of aerodromes it is not possible to define in specific terms where the responsibility of the SAR services begins and that of the aerodrome emergency services ends with respect to potential incidents, so the closest co-operation must be maintained between these two services.

1.5 Direct speech circuits exist between the UK ARCC and the RAF-manned Distress and Diversion (D & D) cells at the London and Scottish ACCs. Under normal circumstances, the quickest and most reliable means for an aerodrome to alert the ARCC is via the ACC and its D & D cell, but a direct call to the ARCC on any of its listed numbers may at times be more expedient.

1.6 DfT contracted SAR helicopters are located at four sites in the UK under the operational control of HM Coastguard. On occasions, aviation distress incidents are first reported to and initial action is taken by the Coastguard. However, responsibility for aeronautical SAR usually remains with the military ARCC so it is vital that it is notified irrespective of how or where an aviation incident is reported. Locations of HM Coastguard MRCCs are shown at GEN 3-6-11.

2 Alerting

2.1 The alerting service is available for all aircraft which are known by the air traffic services to be operating within the UK Flight Information Regions (FIRs). The responsibility for initiating action normally rests with the Air Traffic Service Unit (ATSU) which was last in communication with the aircraft in need of SAR assistance or which receives such information from an external source.

2.2 If a distress signal and/or message is intercepted by their aircraft, pilots are to inform the appropriate ATSU giving all available information. The UK ARCC does not maintain a listening watch on VHF.

2.3 Difficult Areas for SAR

2.3.1 Although the UK has not formally designated land or sea areas where SAR operations would be difficult, it is strongly recommended that General Aviation (GA) aircraft, when proposing to operate over mountainous or sparsely populated areas, should carry appropriate survival equipment, including an Emergency Locator Transmitter (ELT).

2.3.2 The following areas within the UK are considered to be difficult from a SAR aspect:

The Scottish Highlands
The Hebrides
Orkneys and Shetlands
The Pennine Range
The Lake District
The Yorkshire Moors
The Welsh Mountains
The Peak District of Derbyshire
Exmoor
Dartmoor

2.4 For flight over the North Atlantic (NAT), International GA (IGA) pilots should pay particular attention to the varying requirements of the NAT provider states. Canada, Iceland, Denmark and Greenland have requirements more stringent than those of the UK particularly regarding the carriage of ELTs, survival equipment, communications, aircraft inspections and navigation equipment, fuel reserves and instrument ratings. IGA pilots planning transoceanic flight are strongly recommended to read the North Atlantic International General Aviation Operations Manual to seek advice and guidance from the UK Aeronautical Information Service (see GEN 3.1).

3 Communications

3.1 Distress and urgency communications within the UK SRR are in accordance with standard international procedures.

3.2 Emergency Service

3.2.1 An emergency communications and aid service is continuously available on 121.500MHz. The service covers most of the UK above 3000ft, but in many areas reception is good below this altitude. In an emergency, pilots who have difficulty in establishing communication on the frequency in use should make use of the service on 121.500MHz.

3.2.2 South of N55, operational control is exercised from London ACC at West Drayton, call sign 'London Centre'. North of N55, operational control is exercised from Scottish ACC at Prestwick, call sign 'Scottish Centre'. Aerodromes capable of providing an emergency service on 121.500MHz are listed at paragraph 5.

3.2.3 It is not necessary for pilots in emergency to address the initial 'MAYDAY' or 'PAN PAN' call to a specific unit. The Emergency Controller at West Drayton or Prestwick will answer the call depending on the location of the transmission. If transponder fitted, in emergency select code 7700 (emergency) or 7600 (radio failure) to assist the Emergency Controller in determining your position and providing a timely response.

3.2.4 Pilots should be aware that the airspace west of Airway N864 has limited VHF Direction Finding (VDF) coverage and the Emergency Controller's ability to locate an aircraft using VDF is dependent on the aircraft's position and altitude. Furthermore, at the London ACC, due to the lack of primary radar in the Southwest, non-transponding aircraft will be difficult to locate and will not receive a radar service from the D & D Emergency Controller. Response time to any incident is governed by the amount of assistance received from other ATC units and the Coastguard authorities.

4 Emergency Satellite Voice Calls from Aircraft

4.1 For aircraft flying in the London, Scottish and Shanwick FIRs/UIRs, in the event that all other means of communication have failed, dedicated satellite voice telephone numbers for the London ATCC (Mil) and Scottish ACC D & D sections and for the Shanwick OAC have been programmed into the aeronautical Ground Earth Stations of the Inmarsat Signatories.

4.2 The allocated airborne numbers for use via the aircraft satellite voice equipment are as follows:

(a) Shanwick OAC 423201. To be used only in emergency situations, excluding communications failure.
(b) Shanwick Radio 425002. To be used for aircraft communications failure.
(c) London D & D 423202
(d) Scottish D & D 423203

4.3 It must be emphasised that these numbers are for emergency use only, when all other airborne means of communication with the appropriate ATSU have failed.

4.4 Almost instantaneous aircraft position fixing by auto-triangulation is available H24 on 121.500MHz over most of central and southeast England above 3000ft amsl but down to 2000ft amsl in the vicinity of the London airports. Outside the coverage of the auto-triangulation system, aircraft position can still be determined by DF bearing information on 121.500MHz but involves manual plotting and it may take several minutes before the aircraft location is determined.

4.5 Aircraft operating in the Shanwick OCA are required to maintain a continuous watch on 121.500MHz (see page ENR 2-2-4-8 paragraph 9). For operations over the sea, out of range of land based emergency communications, pilots, particularly IGA pilots at low altitude, should, if in difficulty, attempt to establish contact with aircraft at higher levels.

4.6 When an emergency is ended, it is important that the controlling authority is so informed and that the pilot states his intentions before leaving the frequency in use. This will ensure that SAR actions already underway are cancelled.

5 Units with Emergency Facilities on 121.500MHz

5.1 In addition to the 24 hour service provided by the London and Scottish ACCs, the following units provide a service on 121.500MHz. Belfast Aldergrove, Belfast City, Birmingham, Edinburgh, Jersey, London Heathrow, Manchester, Prestwick, Sumburgh. At Lyneham wich is a Military Emergency Diversion Aerodrome (MEDA), a 24 hour communications monitor is maintained on 121.500MHz. The civil aerodromes listed do not routinely monitor 121.500MHz, but this frequency can be activated as necessary to provide a clear channel for emergency traffic. This facility can be requested by direct call to the unit concerned on an in-use frequency, or, more usually, by request from an ACC.

5.2 Aircraft Not Equipped with Radio

5.2.1 A pilot of an aircraft not equipped with radio is advised to file a Flight Plan if he intends to fly more than 10nm from the coast or over sparsely populated or mountainous areas as this will assist rescue action should the aircraft be reported overdue. Pilots should particularly note that Flight Plans can only be delivered to destinations which are on the AFS or linked to the AFS by the parent station scheme and that search action can only be initiated if an aircraft is reported overdue. This action is performed by the destination aerodrome ATSU, when established, but pilots intending to fly to destinations which are not on the AFS or linked to them by a parent station should advise a responsible person at their destinations of the intended flight and arrange for that person to notify the ATS authorities in the event of non-arrival.

5.3 Procedure for a Pilot-in-Command Requiring SAR Escort Facilities

5.3.1 If the pilot-in-command of an aircraft, whilst flying over water or sparsely inhabited areas, believes the operating efficiency of his aircraft has become impaired he should notify the appropriate ACC to alert the SAR organization. If requesting escort facilities, the pilot should bear in mind that SAR aircraft are limited in numbers and that the facility should only be requested if absolutely necessary.

5.3.2 When provided, the SAR aircraft will be positioned as close as practicable to the aircraft being escorted and it is important that RTF contact be established between the two aircraft as early as possible and maintained during the entire operation.

5.3.3 This service is available, subject to the MOD operational requirements, throughout the SRR and no charge is normally raised against the individual for its legitimate use.

6 Flight in Areas in which Search and Rescue Operations are in Progress

6.1 To avoid interference with SAR operations and to avoid unnecessary collision hazard, pilots are strongly advised not to fly near an area where SAR operations are known to be in progress. Crews of aircraft involved in the SAR operation may be performing complex manoeuvres, often in poor weather conditions, and may not be able to maintain a good lookout for itinerant aircraft.

6.2 Pilots who consider it necessary to fly in a known area of SAR operations should:

(a) Contact the ARCC by telephone before departure;

(b) file a Flight Plan giving times of entering and leaving the area and the height to be flown, ensuring that the ARCC is included among the addressees;

(c) obtain the latest information about weather conditions en-route and in the search area;

(d) monitor the VHF International Distress (121.500MHz) and the Scene of Search (123.100MHz) frequencies when in the vicinity, but avoid transmitting on these frequencies.

6.3 Under certain circumstances, a Temporary Danger Area (TDA) may be established around the scene of an incident. This will normally be established by NOTAM. If such a measure fails to achieve its objective, Restriction of Flying (Emergency) Regulations may be invoked. These will make it an offence for an aircraft to be flown in the designated area. Such regulations will also be promulgated by NOTAM (see also ENR 1.1.5).

7 Action by Survivors

(See also paragraphs 8 and 10)

7.1 Basic procedures that can assist SAR operations are set out below.

7.2 Life Rafts

Survivors should use some or all of the following methods when search aircraft or surface craft are seen or heard:

(a) Fire distress flares or cartridges;

(b) use some object with a bright flat surface as a heliograph;

(c) flash torch;

(d) fly anything in the form of a flag and, if possible, make the international distress signal by flying a ball, or something resembling a ball, above or below it;

(e) use the fluorescent marker to leave a trail in the sea.

7.1.2 Crash Landing in Isolated Area

Survivors should use some or all of the following methods to attract attention when aircraft or surface craft are heard or seen:

(a) Visual and ground signals (see paragraphs 10 to 12 and the Ground to Air Signalling Code at GEN 3-6-9);

(b) make the aircraft as conspicuous as possible by spreading any parachutes or other material over the wings and fuselage;

(c) smoke or fire. A continuously burning fire is recommended, with material kept ready to hand to cause it to smoke at short notice. A quantity of green branches, leaves, oil or rubber from the aircraft should achieve the desired result. Three fires in the form of a triangle make a good signal especially at night.

7.1.3 Action at Night

The following technique is used when RAF aircraft are searching for survivors at night:

(a) The search aircraft will fire a single green pyrotechnic at intervals of 5 to 10 minutes;

(b) survivors should then allow 15 seconds after they see the signal (so that the search aircraft can pass out of the glare) and should then fire a red pyrotechnic followed after a short interval by a second. The object of the second signal is to enable the crew of the aircraft to check that they are heading towards the survivors;

(c) the survivors should fire additional pyrotechnics if the aircraft appears to be getting off-track and when it is approaching the overhead so that an accurate position can be obtained.

Note: Survivors should ensure that pyrotechnics are not aimed directly at search aircraft approaching the overhead; some rocket pyrotechnics have considerable energy and can hazard SAR aircraft.

8 Emergency Locator Transmitters (ELT)

8.1 ELT's operating on 121.500MHz and 243.00MHz transmit a distinctive warbling tone. This is a swept tone sweeping downwards over the range of 700Hz within the limits of 1600-300Hz and is repeated two to four times per second. 121.500MHz is not exclusively for aeronautical use and many Emergency Position Indicating Radio Beacons (EPIRBs) in use on both land and sea transmit on the frequency. Pilots of aircraft when listening out on the frequency of 121.500MHz and hearing a warbling note as transmitted by an ELT or EPIRB should report the incident immediately to the relevant ATCC. They should pass the following information to the ATCC:

(a) The time the transmission was first heard and last heard by the same aircraft;

(b) any bearing obtained;

(c) the callsign, position and altitude of the reporting aircraft.

8.2 The COSPAS/SARSAT (C/S) system uses 4 near-polar orbital satellites to detect and localize signals from ELTs. The UK has a Local User Terminal (LUT) at Combe Martin which processes the satellite downlink and passes ELT locations to the UK Mission Control Centre (UKMCC) which is co-located with ARCC Kinloss. C/S operates on 121.500, 243.00MHz and 406.00kHz. Location accuracy is normally better than 20km on 121.500 and 243.00MHz and better than 5km on 406.00kHz. The C/S system will detect transmissions on any of these 3 frequencies throughout the UK SRR. The maximum waiting time (ie the time between ELT activation and satellite detection) should not exceed 90 minutes and will normally be quicker.

8.3 Because the C/S system can detect ELT transmissions throughout the UK SRR and because much of the UK SRR covers an area of busy civil and military air traffic where the distress frequencies are routinely monitored, survivors should switch on an ELT without delay.

8.4 Valuable SAR assets can be expended searching for the source of inadvertent distress transmissions and thus delay the response to an actual emergency situation; great care should be taken to avoid inadvertent transmissions but should they occur, a report should be made to an ATCC as soon as possible to cancel an unnecessary SAR response. It is recommended that the emergency frequency (121.500 or 243.00MHz) is briefly monitored as part of aircraft shutdown checks to detect inadvertent transmissions.

8.5 If MF/HF survival radio equipment is carried, it should be set up as soon as possible and should be operated for a period of four minutes, at approximately ten minutes intervals. When an accurate watch is available, transmissions on 500kHz should be made during the international silence periods of three minutes starting at H + 15 and H + 45. Similarly, transmissions on 2182kHz should be made during the international silence periods of three minutes starting at H and H + 30.

9 SAR Callsigns

9.1 Within the UK SRR, forces engaged in SAR normally use callsigns assigned by the ARCC and prefixed by the root word 'RESCUE'. Fixed wing assets use a 2 figure number, for example 'RESCUE 41'. Military helicopters use the word 'HELICOPTER' plus a 3 figure number, for example 'RESCUE HELICOPTER 190' (which may be abbreviated to 'RESCUE 190' after initial contact). DFT contracted SAR helicopters use the prefix 'COASTGUARD RESCUE' plus their civil aircraft registration, for example 'COASTGUARD RESCUE G-BIMU (may be abbreviated to 'RESCUE MU' after initial contact).

10 Search and Rescue Signals

10.1 The SAR signals to be used are in accordance with international procedures. When signalling to surface craft, visual signals can be more effective than audio signals because of possibly high noise levels on board the surface craft.

10.2 Signals with Surface Craft

10.2.1 The following manoeuvres performed in sequence by an aircraft mean that the aircraft wishes to direct a surface craft towards an aircraft or a surface craft in distress.

(a) Circling the surface craft at least once.

(b) Crossing the projected course of the surface craft ahead at low altitudes and:

(i) rocking the wings; or

(ii) opening and closing the throttle; or

(ii) changing the propeller pitch.

10.2.2 The following manoeuvre by an aircraft means that the assistance of the surface craft to which the signal is directed is no longer required: Crossing the wake of the surface craft close astern at a low altitude by:

(a) rocking the wings; or

(b) opening and closing the throttle; or

GEN 3.6 — GEN 3.6.6.10 Search and Rescue Signals

(c) changing the propeller pitch.
10.3 The following replies may be made by surface craft to the signals in paragraph 10.2.1 (b).
10.3.1 To acknowledge receipt of signals:
(a) hoist the 'code pennant' (vertical red and white stripes) close up (meaning understood);
flash a succession of 'T's by signal lamp in the Morse Code;
(c) change heading to follow the aircraft.
10.3.2 To indicate inability to comply:
(a) hoist the international flag 'N' (a blue and white chequered square);
(b) flash a succession of 'N's in the Morse Code.

11 Ground-to-Air Emergency Signalling Code
11.1 The following guidelines should be employed when using the symbols set out below and detailed at GEN 3-6-9 Tables 1 and 2.
11.2 The symbols should be at least 2.5m (8ft) long and should be made as conspicuous as possible by attempting to provide the maximum colour contrast between the symbols and the background against which they are displayed
11.3 Care should be taken to lay out symbols exactly as depicted to avoid any possible confusion with other symbols.
11.4 Symbols may be formed by any means such as: strips of fabric, parachute material, pieces of wood, stones, by trampling, or staining the surface with oil etc. When the surface is snow covered, signals can be made by dragging, shovelling or trampling the snow. The symbols thus formed will appear to be black from the air.
11.5 In addition to using these symbols every effort should be made to attract attention to them by means of radio, flares, smoke, reflected light or by other available means.

12 Air-to-Ground Signals
12.1 The following signals by aircraft mean that the ground signals have been understood.
During the hours of daylight   rocking the aircraft's wings.
During the hours of darkness   flashing on and off twice the aircraft's landing lights or, if not so equipped, by switching its navigation lights on and off twice.
12.2 Lack of the above signals indicates that the ground signal is not understood.

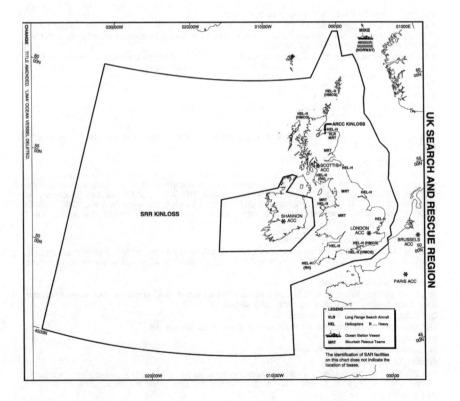

GROUND TO AIR EMERGENCY SIGNALLING CODE

Table 1:
GROUND-AIR VISUAL SIGNAL CODE FOR USE BY SURVIVORS

No.	Message	Code Symbol
1	Require assistance	V
2	Require medical assistance	X
3	No or negative	N
4	Yes or affirmative	Y
5	Proceeding in this direction	↑

Table 2:
GROUND-AIR VISUAL SIGNAL CODE FOR USE BY RESCUE UNITS

No.	Message	Code Symbol
1	Operation completed	LLL
2	We have found all personnel	LL
3	We have found only some personnel	++
4	We are not able to continue. Returning to base	XX
5	We have divided into two groups. Each proceeding in direction indicated	⤻
6	Information received that aircraft is in this direction	→→
7	Nothing found. Will continue to search	NN

GEN 3.6.6.11 HM Coastguard Maritime Rescue Centres

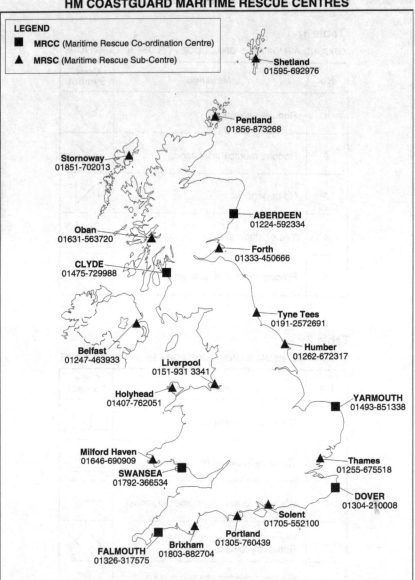

GEN 4 CHARGES FOR AERODROMES/HELIPORTS AND AIR NAVIGATION SERVICES

GEN 4.1 AERODROME/HELIPORT CHARGES

1 Landing charges for UK Aerodromes and Heliports are not established centrally and vary between Aerodrome Operators

2 Landing of Aircraft

2.1 Charges are generally calculated on the Aircraft's Maximum Total Weight Authorised (MTWA) in accordance with its Certificate of Airworthiness.

2.2 Additional Charges may be levied for:
(a) parking, hangarage and storage of aircraft;
(b) passenger load supplements;
(c) security;
(d) noise related items.

3 Conditions Applicable to the Landing, Parking or Storage of Aircraft

3.1 The conditions under which aircraft may land, be parked, housed or otherwise dealt with, are governed by the Acts of Parliament in respect of certain aerodromes and are established by the aerodrome authority in respect of others.

3.1.1 In general, it is a condition that the Aerodrome Authority shall have a lien on the aircraft, its parts and accessories, for fees and charges which become due and payable, and shall be at liberty, subject to certain provisions, to sell, remove, destroy or otherwise dispose of the aircraft and any of its parts and accessories in order to satisfy any such lien.

3.1.2 To obtain the precise legal terms and conditions of use for any specific aerodrome, reference should be made to the appropriate authorities at that aerodrome.

4 Aerodrome Operators

(a) British Airports Authority plc (BAA)
(i) BAA operates London Gatwick, Heathrow and Stansted, and Southampton Airports as subsidiary companies.
(ii) The Conditions of Use and Scale of Charges can be obtained from the Finance Departments at the addresses below:

BAA plc
130 Wilton Road
London
SW1V 1LQ
Tel: 020 7834 9449
Fax: 020 7932 6699
Telex: 919268 BAAPLC G

Gatwick Airport Ltd
PO Box 29
Horley
Surrey
RH6 0YP
Tel: 01293 503249
Fax: 01293 504700

Heathrow Airport Ltd
Hounslow
Middlesex
TW6 1JH
Tel: 020 8745 7970
Fax: 020 8745 7832

Stansted Airport Ltd
Stansted
Essex
CM24 9QW
Tel: 01279 662382
Fax: 01279 662974

Managing Director
Southampton International Airport Ltd
Southampton Airport
Southampton
Hants
SO9 1RH
Tel: 023 8062 9600
Fax: 023 8062 9300

(b) Scottish Airports Ltd
(i) Scottish Airports Ltd operate Aberdeen, Edinburgh and Glasgow Airports.
(ii) The Conditions of Use and Scale of Charges can be obtained from the Finance Departments at the addresses below:

Scottish Airports Ltd
St Andrew's Drive
Glasgow Airport
Paisley
Renfrewshire
PA3 2SW
Tel: 0141 887 1111
Fax: 0141 887 1669

Aberdeen Airport Ltd
Dyce
Aberdeen
AB2 0DU
Tel: 01224 722331
Fax: 01224 725724

Edinburgh Airport Ltd
Edinburgh
EH12 9DN
Tel: 0131 333 1000
Fax: 0131-335 3181

Glasgow Airport Ltd
Paisley
Renfrewshire
PA3 2ST
Tel: 0141 887 1111
Fax: 0141 848 4586

(c) Highlands and Islands Airports Ltd (HIAL)
(i) The airports managed by HIAL are:
Barra
Benbecula
Campbeltown
Inverness
Islay
Kirkwall
Stornoway
Sumburgh
Tiree
Wick

(ii) The Conditions of Use and Scale of Charges can be obtained from the Head Office
Highlands and Islands Airports Ltd
Head Office
Inverness Airport
Inverness
IV2 7JB
Tel: 01667 462445
Fax: 01667 464216
e-mail: hial@hial.co.uk

(d) Other Civil Aerodrome Operators
(i) Details of Conditions of Use and Scale of Charges at aerodromes not listed in this section may be obtained on application to the individual aerodrome operators, the addresses of which appear in the AD section.

(e) Ministry of Defence (Government Aerodromes)
(i) The Ministry of Defence is responsible for RAF Aerodromes, RN Stations and other Government Aerodromes
(ii) The Conditions of Use, and Scale of Charges can be obtained from the addressee below:

Ministry of Defence
CS (CTS) 1a
HQ Strike Command
RAF High Wycombe
Bucks
HP14 4UE
Tel: 01494 461461 Ex 7234
Fax: 01494 497277

GEN 4.2 AIR NAVIGATION SERVICES CHARGES
GEN 4.2.1 AERODROME AIR NAVIGATION SERVICE CHARGES

1 Charges at London Heathrow, London Gatwick, London Stansted, Aberdeen/Dyce, Edinburgh and Glasgow Aerodromes where National Air Traffic Services Limited (NATS) provides the Navigation Services

1.1 Consequent to the Civil Aviation Authority (Navigation Services Charges) Specification 2005 (as amended) the following are the charges payable for navigation services at the aerodromes to which the Specification applies from 1 April 2005.

1.2 The standard charge prescribed under the Specification for each complete metric tonne of the Maximum Take-off Weight Authorised of aircraft in respect of which the charge is made and for each fraction of a metric tonne is in accordance with the following table:

Aerodrome	Weight	Charge
London Gatwick	For each metric tonne and each fraction of a metric tonne	£1.42
London Heathrow	Up to 100 metric tonnes	
London Stansted	For each additional metric tonne and for each fraction Of a metric tonne over 100 metric tonnes	£0.57
Aberdeen/Dyce	For each metric tonne and each fraction of a metric tonne up to 20 metric tonnes	£5.63
	For each additional metric tonne and for each fraction of a metric tonne over 20 metric tonnes	£3.46
Edinburgh	All	£2.41
Glasgow		£1.91

Note 1: The above charges are exclusive of Value Added tax, See GEN 4.2.3 paragraph 3. Please note there is no distinction between domestic and international flights

Note 2: The minimum charge per landing is £10.00.

Note 3: A special rate of 50% of the standard navigation services charge may be applied for in the case of a flight carried out for the sole purpose of training or testing of flying personnel.

Note 4: Any reference to weight in these charges means the Maximum Take-off Weight (MTOW) as defined in the Air Navigation Order 2000.

Note 5: The charge at the London Airports (Gatwick, Heathrow and Stansted) include an element for London Approach which is:

(a) For each metric tonne and each fraction of a metric tonne up to 100 metric tonnes   £0.20
(b) For each additional metric tonne and for each fraction of a metric tonne over 100 metric tonnes   £0.08

(a) Upon each landing of the aircraft at that aerodrome within hours	the standard charge
(b) Upon each landing of the aircraft landing	the standard charge surcharged by 75% or by the specified amount, whichever is the greater at that aerodrome outside hours
(c) Upon each take-off of the aircraft at that	the specified amount or 75% of the standard charge, whichever is the greater aerodrome outside hours, being either
(i) a take-off which does not take place within 1 hour of landing or	
(ii) a take-off which takes place within 1 hour of a landing made within hours	

Note 1: 'The specified amount' means, in relation to a landing or take-off, the additional cost incurred by NATS in providing navigation services by reason of the landing or take-off, as the case may be, being made outside hours.

Note 2: 'Within hours' means within the notified hours of watch of the air traffic control unit at the aerodrome, and 'outside hours' shall be construed accordingly.

2 Rebates

2.1 The following rebates apply to navigation service charges at London Heathrow, London Gatwick, London Stansted, Aberdeen/Dyce, Edinburgh and Glasgow Aerodromes at which NATS provides the navigation services.

(a) The standard charge in respect of non-public transport flights on a journey not exceeding 185 kilometres is reduced by 50%.

(b) The standard charge at Aberdeen/Dyce, Edinburgh and Glasgow in respect of flights by fixed-wing aircraft on a scheduled journey which does not exceed 185 kilometres is reduced by 50%.

(c) a special rate of 50% of the standard navigation services charge may be applied for in the case of a flight carried out for the sole purpose of the training or testing of flying crew;

(d) reduced charges may be made, subject to prior application, in the case of light aircraft based at an aerodrome provided the airport authority itself makes comparable concessions in respect of landing fees.

Note: Prior written application for the grant of the special rates specified in (c) and (d) should be made to the Airport Director.

3 Exemptions

3.1 Exemptions from navigation services charges will not be granted when aircraft are diverted or obliged to land exceptionally (eg by reasons of bad weather, traffic congestion, etc). Exemptions from surcharges described in the Table of Charges at Section 1 paragraph 1.3 will however be granted provided the Airport Director is fully satisfied that a genuine emergency, arising from the development of engine trouble, illness, etc has necessitated an immediate landing at the first available aerodrome without prior arrangement; the exemption will only be allowed when the degree of urgency has precluded the aircraft operator from making prior arrangements for the landing. Aircraft engaged on such operations as emergency ambulance services are not therefore exempt from the surcharge.

4 Minimum charge

4.1 Navigation services payable in accordance with paragraphs 1.3 and 2 (a), (b) and (c) are subject to a minimum charge of £10.00 per landing.

5 Charges for Approach Services provided from an Aerodrome to aircraft which do not land at that Aerodrome

5.1 Subject to the provisions of the Civil Aviation Authority (Navigation Services Charges) Specification 2002 (as amended) the charge for every aircraft engaged on a flight which is not for the purpose of public transport for which navigation services are provided by NATS in connection with an approach to an aerodrome (which is not the aerodrome of intended landing of the aircraft) referred to in paragraph 1.2 (whether or not services are actually used or could be used with the equipment installed in the aircraft) is as follows:

for each approach to any aerodrome referred to in paragraph 1.2: 25% of the standard charge.

5.2 Minimum Charge

5.2.1 The minimum charge payable in accordance with paragraph 5.1 shall be £10.00.

GEN 4.2.2 EN-ROUTE AIR NAVIGATION SERVICE CHARGES

1 Charges for En-route Navigation Services made available by NATS in the Shanwick Oceanic Control Area

1.1 Subject to the provisions of the Civil Aviation Authority (Navigation Service Charges) Specification 2005 (amended by amendment 2005) the charge payable by the operator of every aircraft which flies within the Shanwick Oceanic Control Area and in respect of which a flight plan is communicated to the appropriate Air Traffic Control Unit is £56.44.

1.2 NATS currently grants dispensations from the Shanwick charges in the following cases:

(a) Flights made for the purposes of Search and Rescue operations;

(b) flights other than military flights made exclusively for the carriage on official business of a reigning Monarch or his immediate family, a Head of State, a Head of Government or a Government Minister;

(c) flights made for the purpose of checking or testing equipment used or intended to be used as aids to air navigation;

(d) test flights and flights made exclusively for the purposes of instruction or training of flight crew;

(e) flights made by aircraft of which the Maximum Total Weight Authorised is less than 5.7 metric tonnes.

2 Charges for the Navigation Services made available by NATS for flights made by Helicopter to a Vessel or Off-shore Installation in the North Sea

2.1 Subject to the provisions of the Civil Aviation Authority (Navigation Services Charges) Specification 2005 (amended by amendment 2005) the charge payable by the operator of every helicopter (whether or not registered in the United Kingdom), which flies from any point within the United Kingdom to a vessel or an off-shore installation within the area described below, is £228.00.

2.1.1 The area is bounded by straight lines joining successively the following points:

6300N 00500W – 632833N 000000EW – thence South along the UK Median Line to 5500N 00302E – 5500N 00100W –5600N 00230W – 5740N 00230W – 5740N 00400W – 5830N 00400W – 5830N 00500W – 6300N 00500W.

2.2 Subject to the provisions of the Civil Aviation Authority (Navigation Services Charges) Specification 2005 (amended by Second Amendment 2005) the charge payable by the operator of every helicopter (whether or not registered in the United Kingdom), which flies from any point within the United Kingdom to a vessel or an off-shore installation within the area described below, is £156.00.

2.2.1 The area is bounded by straight lines joining successively the following points:

5500N 00100W – 5500N 00300E – 5423N 00245E – 5256N 00309E – 5230N 00247E – 5226N 00137E – 5238N 00140E – 5251N 00124E – 5319N 00010E – 5500N 00100W.

2.3 NATS grants dispensation from the charge for flights made for the purpose of Search and Rescue operations.

3 Eurocontrol

3.1 Charges for En-route Navigation Services made available in the Airspace of Member and Contracting States of Eurocontrol

3.1.1 Under a Multilateral Agreement (Cmnd 4916) concluded between the Member States of the European Organisation for the Safety of Air Navigation 'EUROCONTROL' (Belgium, France, the Federal Republic of Germany, the Republic of Ireland, Luxembourg, the Netherlands and the United Kingdom), a system of charges for the use of route air navigation facilities and services in the airspace of the Member States of the Organisation was introduced on 1 November 1971. A Multilateral Agreement (Cmnd 8662) relating to route charges was concluded at Brussels on 12 February 1981 between the following Contracting States – Austria, Belgium, Federal Republic of Germany, France, the Republic of Ireland, Luxembourg, Netherlands, Portugal, Spain, Switzerland and the United Kingdom.

Since then the following States have also signed the Multilateral Agreement – Albania, Bulgaria, Croatia, Cyprus, Czech Republic, Denmark, Greece, Hungary, Italy, the Former Yugoslav Republic of Macedonia, Malta, Norway, Romania, Slovakia, Slovenia, Sweden and Turkey, Moldova and Monaco.

3.1.2 The Civil Aviation Authority (Eurocontrol Charges) Specification 2003, prescribes the charges payable to Eurocontrol by the operator of any aircraft (whether or not registered in the United Kingdom), for which navigation services (not being navigation services provided in connection with the use of an aerodrome) are made available in the airspace comprised in the following regions

London Flight Information Region;
London Upper Flight Information Region;
Scottish Flight Information Region;
Scottish Upper Flight Information Region.

3.1.3 The charges collected in respect of the defined airspace, less the costs of collection, are remitted to the United Kingdom Government by Eurocontrol in accordance with a bilateral agreement (Cmnd 9215).

3.1.4 The Specification prescribes that the following flights will not be subject to the United Kingdom charges:

(a) flights by military aircraft;
(b) flights made for the purposes of Search and Rescue operations;
(c) flights by aircraft of which the maximum total weight authorised is 5700kg or less made entirely in accordance with the Visual Flight Rules in the Rules of the Air Regulations 1996 (SI 1996/1393);
(d) flights terminating at the aerodrome from which the aircraft has taken off;
(e) flights other than the flights referred to in (a) of this paragraph made exclusively for the purpose of the carriage on official business of a reigning Monarch or his immediate family, a Head of State, a Head of Government or a Government Minister;
(f) flights made exclusively for the purpose of checking or testing equipment used or intended to be used as aids to air navigation;
(g) flights made exclusively for the purposes of the instruction or testing of flight crew within the specified airspace of the United Kingdom;
(h) flights made by aircraft of which the maximum total weight authorized is less than two metric tonnes;
(i) flights made by helicopters from any point in the United Kingdom to a vessel or an off-shore installation within the area bounded by straight lines joining successively the following points:
6300N 00500W – 632833N 000000EW – thence South along the UK Median Line to 5500N 00302E – 5500N 00100W – 5600N 00230W – 5740N 00230W – 5740N 00400W – 5830N 00400W – 5830N 00500W – 6300N 00500W.
(j) flights made by helicopters from any point in the United Kingdom to a vessel or an off-shore installation within the area bounded by straight lines joining successively the following points:
5500N 00100W – 5500N 00300E – 5423N 00245E – 5256N 00309E – 5230N 00247E – 5226N 00137E – 5238N 00140E – 5251N 00124E – 5319N 00010E – 5500N 00100W.

3.1.5 Further information concerning the Eurocontrol Route Charges System may be obtained from:
Eurocontrol
Central Route Charges Office
Rue de la Fusee 96
B-1130 – Brussels
BELGIUM
Tel: 00 32 2 729 9011
Fax: 00 32 2 729 9094

GEN 4.2.3 METHODS OF PAYMENT

1 Aerodrome, North Sea Helicopter and Shanwick Charges

1.1 Navigation Service charges are invoiced to the aircraft operators, and are payable to:
National Air Traffic Services Ltd
Osborne House
1-5 Osborne Terrace
Edinburgh
EH12 5HG

2 Eurocontrol Charges

2.1 The charges prescribed in the Specification are payable in euros to Eurocontrol at its principal office in Brussels. Arrangements have been made under which United Kingdom citizens and companies registered in the United Kingdom may pay either in sterling or euros into a Eurocontrol bank account in the United Kingdom. Other operators, who are nationals of a Member State of Eurocontrol or of a Contracting State, may pay the charges in euros or their national currencies in the State of which they are nationals. Arrangements have also been made that will enable operators, who are not nationals of a Member State, to pay the charges in euros into bank accounts designated by Eurocontrol in any of the Member States.

3 Value Added Tax

3.1 The charges quoted herein are exclusive of any Value Added Tax which may be chargeable in accordance with the provisions of the Finance Act 1972 or with any Orders or Regulations made thereunder or by virtue of any Act replacing or amending the same.

AIP
ENR section

ENR 1 GENERAL RULES AND PROCEDURES

ENR 1.1 GENERAL RULES
ENR 1.1.1 ATS ROUTES AND UPPER CONTROL AREAS (UTA)

1 ATS Routes Description

1.1 ATS Routes are predicated upon significant geographical points which may or may not coincide with the location of a radio navigation aid. These significant points are shown in column 1 of the table depicted in ENR 3. Any coincident radio navigation aid is depicted immediately underneath.

1.2 Except where stated otherwise the width of an Airway is 5nm either side of a straight line joining each two consecutive points shown in column 1 of the table. Upper ATS Routes and Advisory Routes have no declared width but for the purposes of ATS provision are deemed to be 5nm either side of a straight line joining each two consecutive points. The vertical extent is shown in column 3 of the table. Where lower limits of Airways and Advisory Routes are defined as Flight Levels an absolute minimum altitude of 3001ft applies unless otherwise stated in column 3 and the minimum cruising level shown in column 5 may not always be available.

1.3 Unless otherwise stated the ATS Routes catalogued in the table in ENR 3 are designed to contain aircraft navigating to RNP5 (ICAO DOC 9613 – AN/937 refers).

1.4 The ATS Route network, above FL 95, is hereby notified for the purposes of Articles 50 and 51 of the Air Navigation Order 2000.

1.5 When ATS Routes transit TMAs, the status of the TMA takes precedence in airspace classification and conditions of use. The carriage of RNAV is not expected to be mandated in TMAs before 2005. Unless otherwise notified, non-RNAV equipped aircraft may operate in TMAs at all levels up to the maximum published level of the TMA.

1.6 Conditional Routes (CDRs) are ATS Routes which are usable only under specified conditions. Three types of Conditional Routes are used as described below.

(a) Category One – A route which is permanently plannable during the times published in ENR 3.

(b) Category Two – A route which is only plannable in accordance with the conditions stated in the daily Conditional Route Availability Message (CRAM) issued by the Central Flow Management Unit (CFMU).

(c) Category Three – A route which is not plannable per se but may be used tactically at the discretion of ATC. A CDR may have more than one Category.

2 ATS Route Designators

2.1 In accordance with ICAO Annex 11, the following prefix designators are used to indicate European Regional RNAV Routes, L, M, N, P and for non Regional RNAV Routes Q, T, Y, Z. Routes designated with these prefixes are compulsory RNAV at all levels except when otherwise notified eg. sections of certain ADRs in the Scottish FIR. ADRs are designated with the Suffix D (not to be confused with Class D Airspace which is depicted as D on Aeronautical charts).

2.2 Because of the recent mandate for aircraft to carry B-RNAV equipment on all routes above FL 95 and the shortage of RNAV designators in the European Region, conventional designators will continue to be used on many RNAV routes for some time to come. These are A, B, G, R for Regional Routes and H, J, V, W for non-Regional Routes. Non-RNAV equipped aircraft may operate at and below FL 95 on these routes. Over a period of time, many such routes may be allocated a new RNAV designator when a change of alignment or international interface arrangements require such a change. Following such a change the availability for Non-RNAV equipped aircraft will be indicated in the notes.

3 Flight Planning Restrictions

3.1 Flight planning restrictions applicable to ATS routes within UK airspace are published monthly as an AIP Supplement titled **ROUTE AVAILABILITY DOCUMENT**. This data is used as the UK Annex (Annex EG) of the Eurocontrol CFMU Route Availability Document (RAD). Changes to the AIP Supplement required between AIRAC Cycle publication dates, due to urgent operational reasons, will be notified by NOTAM. The monthly AIP Supplement is published electronically on the UK AIS CD-ROM in Adobe pdf format. The data is also available through the CFMU RAD published monthly on the CFMU and Eurocontrol AIS AGORA websites. A paper copy of the AIP Supplement is available to subscribers of the UK Integrated Aeronautical Information Package on request from AIS:

Aeronautical Information Service (AIS)
National Air Traffic Services Limited
Room 160 Control Tower Building
London Heathrow Airport
Hounslow
Middlesex
TW6 1 JJ
Tel: + 44 (0)20 8745 3460
Fax: + 44 (0)20 8745 3453.

Note: For non-subscribers a request should be sent (with a stamped addressed envelope) to the above address.

3.2 For information on standard routeings within UK airspace, and alternative routeings during CDR closures, users should refer to the UK Standard Route Document as published via the UK AIS CD-ROM (bespoke database and Adobe PDF formats) or the Eurocontrol AIS AGORA website (Adobe pdf format only).

4 Rules and Procedures

4.1 Control Areas (Airways)

4.1.1 Radio Communications and Equipment

4.1.1.1 The requirements for radio communications and equipment are set out in GEN 1.5.

4.1.2 ATC Clearance

4.1.2.1 One of the following phrases may be included in the initial clearance when the air traffic situation necessitates the regulation of departing flights:

ENR 1.1   ENR 1.1.1.4 Rules and Procedures

(a) 'Clearance expires .... (time)' – this indicates that if the aircraft is not airborne by the time stated, a fresh clearance will need to be obtained;
(b) 'Take-off not before .... (time)' – this is given so that the pilot can calculate the best time to start engines;
(c) 'Unable to clear .... (level planned)' – when ATC is unable to clear the flight at the level planned an alternative will be offered whenever possible, the acceptance of which will avoid or reduce delay;
(d) 'Join Airways at .... (place and level) not before .... (time)' – may be used when an Airways clearance is given to an aircraft, the first part of whose flight from the origination aerodrome is in uncontrolled airspace.

4.1.3 Airborne Procedures (See also ENR 1-4-1, paragraph 2.1, Note 2)

4.1.3.1 When an aircraft is cleared to leave or join an Airway at a certain point, it should be flown so as to cross the actual boundary of the Airway as near to that point as is practicable.

4.1.3.2 All aircraft flying Airways are required to adhere to IFR procedures in all weather conditions. However, when radar cover is not available ATC may offer VMC climb or descent clearances in order to avoid excessive traffic delays. Such clearances will be offered subject to the following:

(a) By day only in Visual Meteorological Conditions;
(b) subject to the agreement of the pilot concerned;
(c) the pilot will be responsible for effecting his own separation;
(d) essential traffic information will be given;

4.1.3.3 Unless otherwise authorised by ATC, aircraft flying along Airways are required, in so far as practicable, to operate along the defined centre-line.

4.1.4 Flights Joining Airways

4.1.4.1 Pilots wishing to join an Airway are required to file a flight plan either before departure or when airborne, and to request joining clearance when at least 10 minutes flying time from the intended joining point. If the destination or any part of the route is subject to Air Traffic Flow Management, pilots must have received the required authorisation/approval from the appropriate Air Traffic Flow Management Unit (ENR 1.9).

4.1.4.2 Joining clearance should be obtained as follows: Initial call – '........ (identification) request joining clearance .... (Airway) at .... (position)'. When instructed by ATC the following flight details should be passed:

(a) Identification;
(b) Aircraft type;
(c) Position and heading;
(d) Level and flight conditions;
(e) Departure aerodrome;
(f) Estimated time at entry point;
(g) Route and point of first intended landing;
(h) True Airspeed;
(i) Desired level on Airway (if different from the above).

4.1.4.3 Requests for joining clearance of Airways for which the Controlling Authorities are London, Scottish or Manchester Control should be obtained as follows:

(a) From the ATSU with which the aircraft is already in communication; or
(b) from the appropriate FIR Controller (if different from (a)); or, if it is not possible to obtain any form of clearance using (a) or (b), then
(c) on the published frequency of the Airway Controlling Authority.

4.1.4.4 In order to prevent confliction with other Airways traffic, pilots should ensure that they are at the cleared flight level when they cross the Airway boundary, unless specific permission to do otherwise has been given by ATC.

4.1.5 Flights Crossing Airways in IFR

4.1.5.1 Aircraft may, without ATC clearance, fly at right angles across the base of an en-route section of an Airway where the lower limit is defined as a Flight Level.

4.1.5.2 Pilots wishing to cross an Airway are required to file a flight plan either before departure or when airborne, and to request crossing clearance when at least ten minutes flying time from the intended crossing point.

4.1.5.3 Crossing clearance should be obtained as follows: Initial call – '........ (identification) request crossing .... (Airway) at .... (position)'. When instructed by ATC the following flight details should be passed:

(a) Identification;
(b) Aircraft type;
(c) Position and heading;
(d) Level and flight conditions;
(e) Position of crossing;
(f) Requested crossing level;
(g) Estimated time of crossing.

4.1.5.4 Requests for joining clearance of Airways for which the Controlling Authorities are London, Scottish or Manchester Control should be obtained as follows:

(a) From the ATSU with which the aircraft is already in communication; or
(b) from the appropriate FIR Controller (if different from (a)); or, if it is not possible to obtain any form of clearance using (a) or (b), then
(c) on the published frequency of the Airway Controlling Authority.

4.1.5.5 Unless otherwise requested by ATC, aircraft crossing Airways will remain in communication with the FIR Controller and, after obtaining clearance, will report as follows when the aircraft is estimated to be at the boundary of the Airway: '.......... (identification) – Crossing (Airway) .... (position) .... (time) .... at .... (level)'.

4.1.5.6 Except where otherwise authorized by ATC, aircraft are required to cross the Airway by the shortest route (normally, at right angles) and to be in level flight at the cleared flight level on entering the Airway.

4.1.6 Airway Crossings or Penetrations in VMC – Civil Aircraft

4.1.6.1 Powered Aircraft – Airway Crossings (See also ENR 1.4.1, paragraph 2.1, Note 2)

4.1.6.1.1 Aircraft may, without ATC clearance, fly at right angles across the base of an en-route section of an Airway where the lower limit is defined as a Flight Level.

4.1.6.1.2 Powered aircraft may cross an Airway in VMC by day without compliance with the full IFR requirements in relation to the aircraft equipment provided that the pilot holds a valid Instrument Rating and that clearance is obtained from the appropriate ACC. This clearance must be obtained by RTF (normally on the FIR frequency); the request for clearance and a crossing report should be made as shown in paragraphs 4.1.5.3 and 4.1.5.5.

4.1.6.2 Powered Aircraft – Other penetrations of Airways (see also ENR 1.1.4.1, paragraph 1 and ENR 1.4.1, paragraph 2.1, Note 2).

4.1.6.2.1 Other flights in VMC, for example photographic survey flights, may also do so without compliance with full IFR requirements, provided that:

(a) Prior arrangements are made with the appropriate ACC;
(b) specific ATC clearance is obtained for individual flights;
(c) the aircraft can communicate by RTF on the appropriate Airways frequency.

4.1.7 Procedures for Military Aircraft

4.1.7.1 These procedures apply to military aircraft in all weather conditions.

4.1.7.1.1 Military aircraft flying along Airways will conform to the normal Airways procedures.

4.1.7.1.2 Military aircraft crossing Airways will do so either:

(a) Under the control of an approved Air Traffic Control Radar Unit; or
(b) under a positive Air Traffic Control Clearance.

4.1.7.1.3 In an emergency, where neither a radar nor a procedural crossing can be obtained, an Airway may be crossed at an intermediate 500ft level. The intermediate 500ft levels referred to are flight levels of whole thousands plus 500ft.

4.2 Air Traffic Service Advisory Routes

4.2.1 Introduction

4.2.1.1 The Air Traffic Service (ATS) Advisory Routes (ADRs) tabulated at ENR 3.1 and described in paragraph 1 are Class F Airspace.

4.2.1.2 An Air Traffic Advisory Service providing separation between participating IFR traffic is available and in the interests of safety, pilots are urged to make use of this service whenever flying on ADRs. Nevertheless, it is emphasised that, even if an aircraft is following these procedures, which are detailed below, **advice issued to participating aircraft relates only to known traffic.**

4.2.1.3 Certain ADRs pass through, originate from, or terminate in Controlled Airspace and pilots are reminded that the ATS acceptance of a flight plan to fly on such an ADR does not absolve the pilot from the responsibility for ensuring that the requisite qualifications and licences are held for flight in Controlled Airspace.

4.2.2 Procedure for Participating Aircraft

4.2.2.1 Flight Plans

4.2.2.1.1 A flight plan must be submitted if it is desired to participate in the Advisory Service on an ADR. ATC will acknowledge acceptance of the plan by issuing approval to depart and advising the levels to be flown. When the flight is planned to join an ATS Advisory Route from other than Controlled or Advisory Airspace, approval to join must be sought not less than 10 minutes before ETA at the entry point.

4.2.2.1.2 Formation flights along Advisory Routes are considered to be non-standard flights and prior notification is required before an ATC clearance will be issued. The general requirements for non-standard flights as detailed at ENR 1.1.4.1, paragraph 1 must be followed.

4.2.2.2 Carriage of Radio Communication and Navigation Equipment

4.2.2.2.1 When aircraft are operating on an ATS Advisory Route directly associated with CAS it is essential that they should also carry the radio communication and navigation equipment prescribed for the CAS concerned.

4.2.2.3 Listening Watch

4.2.2.3.1 Listening Watch must be maintained throughout flight within Advisory Airspace on the appropriate frequencies.

4.2.2.4 Priority on ATS Advisory Routes

4.2.2.4.1 Aircraft already flying on ATS Advisory Routes will have priority over other aircraft wishing to join at the same level.

4.2.2.5 Allocation of Levels

4.2.2.5.1 Level allocations will be in accordance with the Quadrantal Rule.

4.2.2.6 Separation Standards

4.2.2.6.1 Vertical separation is based on the Quadrantal Rule and on occasions 500ft vertical separation may be used between aircraft flying on the routes and aircraft cleared to cross them; when this occurs ATC will pass essential traffic information.

4.2.2.6.2 Minimum rates of climb or descent are to be in accordance with the instructions detailed at ENR 1.1.3.1, paragraph 2.

4.2.2.7 Minimum and Maximum Cruising Levels

4.2.2.7.1 On each advisory route segment, where a maximum cruising level is published in column 3 of the table, Advisory Service is not available above this level. A minimum cruising level is published in column 5 of the table after consideration of VHF RTF cover, terrain clearance and military air activity. However, the lowest level at which Advisory Service is available is the minimum usable cruising level based on the Regional QNH (but see paragraph 6.2).

4.2.2.8 Position Reports
4.2.2.8.1 Position Reports must be made on the appropriate RTF frequency at, or as soon as possible after passing:
(a) Each compulsory reporting point on the route;
(b) any other reporting point requested by ATS.
Note: Since separation is based on position reports, it is essential that these reports, particularly when positions and forward estimates are determined by dead reckoning, should be calculated with care and amended promptly if found to be inaccurate. (ENR 1.1.3.1, paragraph 1).
4.2.2.9 Leaving or Joining Advisory Routes
4.2.2.9.1 When an aircraft is cleared to leave or join an advisory route at a certain point, it should be flown in such a manner so as to cross the deemed boundary of the advisory route (see paragraph 1.2) as near to that point as is practicable.
4.2.3 Flights crossing ADRs or Published Holding Patterns under IFR
4.2.3.1 Aircraft participating in the Advisory Services may on occasions be flying at non-quadrantal levels, particularly near boundaries of TMAs or other Regulated Airspace. Pilots of aircraft wishing to cross an ATS Advisory Route or published Holding Pattern are strongly advised to adopt the following procedure:
(a) Contact the appropriate Controller at least 10 minutes flying time from the intended crossing point or holding pattern with the following details: ......(identification) crossing ...... Advisory Route ...... (designation) or approaching Holding Pattern ...... (Reporting Point) at ...... (position) ...... (time) ....(level);
(b) Request traffic information;
(c) Remain in contact, report crossing and clear.
4.2.3.2 Where possible, selected crossing points should be associated with a radio facility to ensure accurate navigation and Routes should be crossed at an angle of 90° to the direction of the Route or as close to this angle as possible.
4.2.3.3 Pilots unable to communicate in order to obtain traffic information should cross an ATS Advisory Route as near as possible at an angle of 90° to the direction of the Route and at the appropriate quadrantal level. Crossing published Holding Patterns under these circumstances should be avoided since holding aircraft will not necessarily be flying at the appropriate quadrantal flight level.
4.2.4 Other Traffic Using ADRs
4.2.4.1 Pilots are strongly advised not to fly along ATS Advisory Routes or in Advisory Areas unless participating in the Advisory Service but if they do so they should listen out on the appropriate frequency. Where this frequency is different from the FIR frequency pilots should monitor both.
4.2.5 Radar Advisory Service
4.2.5.1 Radar Advisory Service as described at ENR 1.6.1.1 is provided by Radar Units to aircraft on certain of the ATS Advisory Routes and Areas.
4.3 The Upper Airspace Control Area and the Hebrides UTA
4.3.1 Military Mandatory Radar Service Area (MRSA)
4.3.1.1 A large part of the Upper Airspace Control Area is covered by a Military Mandatory Radar Service Area (MRSA) within which military aircraft flying between FL 245 and FL 660 are required to operate under a radar control/procedural control service.
4.3.2 Upper Airspace Control Area
4.3.2.1 Rules. The following rules apply to aircraft flying in the Upper Airspace Control Area:
(a) A flight plan must be filed;
(b) ATC permission must be obtained before the Area is entered;
(c) a continuous RTF watch must be kept on the appropriate frequency;
(d) the flight must be conducted in accordance with ATC instructions.
4.3.2.2 Altimeter Setting Procedures
4.3.2.2.1 All aircraft flying in the Upper Airspace Control Area must use the standard altimeter setting of 1013.2mb.
4.3.2.3 Cruising Levels
4.3.2.3.1 Cruising levels will be allocated in accordance with the semi-circular rules depicted in the Table of Cruising Levels at ENR 1.7.4. ATC may allocate a level not appropriate to the aircraft track, eg to effect transition to and from Oceanic levels.
4.3.2.3.2 The providers of Air Traffic Services in the United Kingdom Upper Airspace may apply a reduced vertical separation minimum of 1000ft, between FL 290 and FL 410 inclusive, in the London and Scottish UIRs between aircraft that are RVSM approved. Aircraft that are not RVSM approved will be provided with a minimum of 2000ft separation.
4.3.2.4 Exemptions
4.3.2.4.1 By prior agreement, Research and Development flights may be exempted from some of the rules and procedures but ATC will co-ordinate such flights.
4.3.2.4.2 The above rules and procedures do not apply to gliders.
4.3.2.4.3 By prior agreement, civil aircraft operating on contract to the MoD, aircraft undergoing air tests, or aircraft calibrating navigation aids may be exempted from the RVSM requirements.
4.3.3 Procedure for Glider Operations within Temporary Reserved Areas inClass C Airspace above FL245
4.3.3.1 The gliding representative will initiate a request to the appropriate area control supervisor (see paragraph 4.3.3.2) two hours in advance of the intended flight advising the intention to use the designated area and confirm the following details:
(a) Temporary Reserved Area concerned (See charts ENR 6.3.0.3/4);
(b) Upper Limit, if known (Scottish Link has an upper limit of FL 270);
(c) expected launch time, time of entry into, and duration in Class C Airspace (negotiated if any other priority task);

(d) the number of gliders and associated callsign(s);
(e) name and telephone contact number.

4.3.3.2 For Scottish and Northumbria areas contact Scottish Area Control Centre Civil ATC Watch Supervisor (WS) on Tel: 01292 692842. For Yorkshire and Welsh areas contact Swanwick Military Supervisor (SMS) on Tel: 01489 612493

4.3.3.3 Following co-ordination, the Supervisor will contact the gliding representative to discuss the activity, issue the clearance and allocate the frequency to be employed, or, if the activity cannot be accommodated, advise the representative of the reason and negotiate a new period.

4.3.3.4 The glider pilot will establish 2-way RTF contact passing FL 200 in the climb, maintain a listening watch on the frequency, and report again when passing FL 245 in descent.

4.3.3.5 The military controller will initiate a radio check with the glider pilot on the hour and half hour whilst the aircraft is above FL 245 to confirm continuing RTF contact. In the event of not receiving a radio check call the glider pilot will immediately attempt to re-establish 2-way contact and if unsuccessful shall descend below FL 245 within 15 minutes.

Note: 15 minutes after the last unsuccessful "operations normal" radio check by the military controller the airspace above FL 245 will be deemed clear of gliders and GAT aircraft will be allowed access.

4.3.3.6 The glider pilot is responsible for remaining within the designated area. In addition, all gliders flying within Class C Airspace are to be fitted with appropriate radio and navigational equipment. In the event of either of these equipments becoming unserviceable, gliders are to leave Class C Airspace by descent.

4.3.3.7 Whilst operating within a designated area, glider pilots will be in receipt of a Flight Information Service with proximity warnings of co-ordinated traffic through the area. The glider pilots will be responsible for their own separation.

4.3.3.8 Whilst operating within a designated area all position reports are to be made in relation to Airway/Upper ATS Route Reporting Points.

4.3.4 Co-ordination of Civil and Military Aircraft

4.3.4.1 NATS radars cover most of the Upper Airspace. Within this cover, procedures exist for the co-ordination of civil and known military aircraft and they receive a radar control and/or a procedural ATC Service. Outside radar cover, a procedural ATC service is provided.

4.3.4.2 Military aircraft are normally under the control of NATS or autonomous radar Units but outside the Mandatory Radar Service Area, they are not obliged to receive an ATC Service. In these circumstances it is not always possible for ATC to offer avoiding action because the behaviour of such aircraft is unpredictable. However, whenever practicable, ATC will pass traffic information on them to aircraft under control.

4.3.4.3 Due to the routine operation of high speed military aircraft within the UIRs, civil aircraft operators should flight plan only on the published ATS Route Structure. When traffic conditions permit, ATC may authorise aircraft to fly more direct tracks.

4.3.4.3.1 For individual flights within the Scottish UIR, operators may file outside the published ATS Route Structure subject to authorisation by the Scottish ACC ATC Watch Manager (Tel: 01292 692763, Fax: 01292 692872). Authorisation for routine operations outside the published ATS Route Structure must be obtained from ATC Operational Support at Scottish ACC (Tel: 01292 692611, Fax: 01292 692610).

4.3.4.4 There is a military TACAN route system in the Upper Airspace. Some of the routes join the published Upper ATS Route Structure at certain reporting points as well as to a similar TACAN route network over the rest of Europe. See chart of the military TACAN routes at ENR 6.3.5.1.

4.3.5 Non-Standard Civil Flights and Unusual Aerial Activities in the UK Upper Airspace

4.3.5.1 Certain civil flying activities such as training and general test flying in the Upper Airspace (at and above FL 245) may require a specialised radar service that can best be provided by military ATS Units. However, it should be borne in mind that the aircraft handling capacity of military ATS Units may be committed to the Units primary tasks, and therefore, it is advisable that aircraft operators requiring a service should discuss their proposed task with the relevant ATS Unit prior to commencement of the flight.

4.3.5.2 Information concerning the military ATS Units may be obtained from the RAF en-route documents or from civil ATS Units.

4.3.5.3 The approval of an Unusual Aerial Activity (UAA) in the Upper Airspace can often only be given after extensive co-ordination and the request should be submitted at the earliest opportunity to the Airspace Utilisation Section (AUS) at Directorate of Airspace Policy, K7, CAA House, 45-59 Kingsway, London WC2B 6TE, as detailed at ENR 1.1.4.4, paragraph 3.

4.3.6 Flight Plans, ATC clearance and other procedures

The normal requirements for flight plans, ATC clearance, and the regulations, rules and procedures appropriate for flight in Control Areas apply.

4.3.6.1 Clearance to enter the Hebrides UTA

4.3.6.1.1 Directly from Shanwick OCA, Reykjavik OCA, Shannon UTA or from the Upper Airspace CTA: Aircraft will be cleared into the Hebrides UTA without specific entry clearance.

4.3.6.1.2 From outside Controlled Airspace:

Aircraft must obtain prior clearance from 'Scottish Control' in accordance with the procedures established for flight joining Airways.

4.3.6.2 Eastbound and Westbound Flights

4.3.6.2.1 Traffic transitting the Scottish UIR must Flight Plan along established Upper ATS Routes and exit via promulgated Reporting Points. Specified exemptions will be notified to the appropriate operating companies.

4.3.6.2.2 Traffic operating within the Scottish UIR will be cleared along selected tracks based upon the VOR and NDB facilities at Talla, Glasgow, Machrihanish, Belfast, Tiree, Benbecula and Stornoway. These tracks may be varied at the discretion of ATC depending upon the pattern of North Atlantic traffic. If at any time an aircraft within the UTA is found to be off its cleared track, the pilot shall at once inform ATC of his true position and take an immediate action to return to the cleared track as quickly as possible. Aircraft destined for the North Atlantic should, wherever possible, flight plan to use the routes contained within the Standard Routes Document published on the AIS CD-ROM.

4.3.6.3 Loss of Communication
4.3.6.3.1 In the event of radio communication failure, pilots will follow the procedures shown at ENR 1.1.3.2/7. Attempts should also be made to establish communication on other control channels available in the Scottish FIR/UIR or on NARTEL HF channels.

4.3.7 Military Training Area (MTA) and Military Temporary Reserved Airspace (MTRA)
4.3.7.1 MTAs and MTRA areas are established in the UK Upper Airspace for the operational freedom of military aircraft engaged in exercise or training, and the nature of this activity is incompatible with civil air traffic services procedures. Pilots should not therefore flight plan any route through active MTAs or MTRA areas nor make in flight requests for transit through these areas when active, as ATS cannot authorise such flights. Notwithstanding, some military air traffic control radar units are able to provide an Air Traffic Service within active MTAs or MTRA areas to the following categories of civil registered aircraft:
(a) Aircraft in emergency which may have to be routed through an active MTA or MTRA for flight safety reasons;
(b) aircraft sponsored by MOD(DPA);
(c) test flights by UK manufacturers of military and civil aircraft;
(d) aitests by civil or military aircraft departing from or arriving at UK aerodromes;
(e) special flights authorised by HQ 3 Group (ATC);
(f) air ambulance flights where the most expeditious routeing is justifiable on humanitarian grounds.
Note 1 Pilots of such aircraft requiring this service should make their request to the UK Civil ATS Unit with which they are in contact.
Note 2 Military controllers may provide aircraft in the above categories with RAS (and any associated Procedural Service), RIS or FIS.
4.3.7.2 If pilots inadvertently flight plan to transit MTAs or MTRA areas during notified periods of activity, ATS instructions will be issued re-routeing the aircraft to bypass those areas. To avoid such unexpected re-routeing, pilots are requested to ensure that account is taken of the published periods of activity of any MTA or MTRA area near the route (detailed in ENR 5.2), when planning flight in the Upper Airspace.
4.3.7.3 MTAs and MTRA areas are depicted on the charts at ENR 6.3.2.1 and ENR 6.5.1.1.

5 Reporting Points
5.1 Designated Reporting Points are marked with a. ▲ Reporting Points marked with a are 'on request' Reporting Points, at which a report will be made only when requested by the controlling authority.
5.1.1 RVSM Entry/Exit Points are marked with a
5.2 Definition of the Five Letter Reporting Points shown on ATS Routes can be found at ENR 4.3.
5.3 In addition to the designated Reporting Points in the Hebrides Upper Control Area, ATC may ask for a position report from aircraft when they cross specified VOR radials.

6 Terrain Clearance
6.1 Control Areas (Airways)
6.1.1 Where the lower limit of a section of an Airway is defined as a Flight Level and therefore varies in height, an absolute minimum altitude applies. This minimum altitude for the Airway base is at least 1000ft above any fixed obstacle within 15nm of the centre-line. The lowest usable level will always be at least 500ft above the Airway base, thus providing not less than 1500ft terrain clearance within 15nm of any position on the centre-line of the Airway.
6.1.2 On sections of Airways adjacent to Control Zones and Areas where the lower limit is established at not less than 700ft above terrain, ATC clearances are designed to enable aircraft to remain at least 500ft above the base of the Airway.
6.2 Air Traffic Service Advisory Routes
6.2.1 Those minimum flight levels/altitudes in column 5 of the table marked with a dagger † though largely over water, do not provide 1500ft terrain clearance within 15nm of track, and pilots are reminded that they are responsible for terrain clearance. However, controllers will set an altitude or flight level below which the advisory service will not be provided. For flights at and below 3000ft ALT along ATS Advisory Routes, altimeters should be set to the appropriate Regional QNH. Where Flight Levels are shown, the lowest may not be available unless the Regional QNH is of sufficiently high value.

7 En-Route Holding
7.1 Control Areas (Airways)
7.1.1 Except where otherwise instructed by ATC, holding en-route will be carried out on tracks parallel to the centre-line of the Airway, turning right at the Reporting Point. Exceptions are shown at ENR 3.6.
7.1.2 Whenever possible, pilots will be given a specific time at which to leave the Reporting Point and the holding pattern should be adjusted accordingly.
7.1.3 Pilots are required to report as follows:
(a) The time and level of reaching a specific holding point to which cleared;
(b) when leaving a holding point;
(c) when vacating a previously assigned level for a new assigned level.
7.2 Air Traffic Service Advisory Routes

7.2.1 Holding will normally be accomplished in accordance with Airways procedures. Pilots not participating in the Advisory Service and intending to cross these patterns are strongly advised to comply with the procedures detailed in paragraph 4.2.3. Holding patterns established at the following Reporting Points: BONBY; FIWUD; FOYLE; FYNER; GARVA; GAVEL; GUSSI; KISTA; Lands End VOR; MIKEL; Scotstown Head NDB; SMOKI; TROUT and VANIN and are shown at ENR 3.6.

7.3 En-Route High Level Holding

7.3.1 Within the Upper Airspace, en-route holding patterns have been established for Eastbound Atlantic traffic. The 1.5 minute holding patterns, based on ICAO recommended speeds, are as shown at ENR 3.6. Also, within the Upper Airspace, en-route holding patterns have been established at MALBY; Manchester (VOR MCT); MERLY; PLYMO and Southampton (VOR SAM) and are shown at ENR 3.6.

8 Hazards and Danger Areas
8.1 Air Traffic Service Advisory Routes

8.1.1 The hazards mentioned in column 6 of the tabulation are given as an approximate guide only. For the latest information, the Chart of United Kingdom Airspace Restrictions (as amended), and NOTAM and Supplement should always be consulted. Pilots should at all times keep a sharp lookout for aircraft not complying with the Advisory Procedure. Military activity may be intense in some localities. Reference should also be made to ENR 5.2 for Areas of Intense Air Activity.

8.2 The Upper Airspace Control Area and the Hebrides UTA

8.2.1 The Chart of the United Kingdom Airspace Restrictions at ENR 6.5.1.1 (as amended), NOTAM and AICs should always be consulted for information on activity in Danger Areas adjacent to the route being flown.

ENR 1.1 GENERAL RULES
ENR 1.1.2 CLASS G AIRSPACE

1 Airspace other than Controlled Airspace and Advisory Routes – Class G Airspace

1.1 The residual Airspace within the UK Flight Information Regions (FIRs) and Shanwick FIR which lies outside Controlled Airspace and Advisory Routes (Classes A to F) is designated Class G. Although ICAO Standards and Recommended Practices (SARPs) require only a Flight Information and Alerting Service to be provided, other services are available from Air Traffic Control Units in certain circumstances. These services are:

(a) An Air Traffic Service to participating flights arriving at, departing from and over flying aerodromes located within Class G Airspace. The service comprises Aerodrome Control, Approach Control and, where appropriate, Approach Radar Control Services. These, together with other Air Traffic Services, are described at GEN 3.3.

Note: Although these services are advisory in nature as the Airspace is not Controlled, participating flights are expected to comply with ATC instructions.

(b) Radar Advisory Service (RAS) and Radar Information Service (RIS), as described at ENR 1.6.1.1/2, are available from suitably equipped ATC Units subject to Controller workload and radar availability. In certain areas Radar Advisory Service Areas (RASAs), Lower Airspace Radar Services (LARS) and Military Middle Airspace Radar Services are established, as described at ENR 1.6, to provide RAS and RIS.

1.2 Flight Information Service as described in ICAO Annex 11 Chapter 4 is available to aircraft flying outside Controlled Airspace and Advisory Routes. It is provided by the appropriate ACC through Flight Information Service Officers (FISO) operating on specially allocated RTF channels. In addition to normal FIS, the FSO will:

(a) On receipt of a request for joining or crossing clearance of Controlled Airspace or Advisory Routes either:

(i) inform the pilot that he should change frequency in time to make the request direct to the appropriate ATC Unit at least ten minutes before ETA for the entry or crossing point; or

(ii) obtain the clearance from the appropriate ATC Unit himself/herself and pass it to the pilot on the FIR frequency.

(b) Pass ETA to destination aerodromes in special circumstances, such as diversions, or at particular locations when traffic conditions demand it. Normally, however, pilots who wish destination aerodromes outside Controlled Airspace to have prior warning of arrival should communicate direct with ATC at the aerodrome concerned, at least ten minutes before ETA.

(c) Accept airborne flight plans and pass the information to the appropriate authority.

(d) Operate a very limited warning system of proximity hazards. Whenever possible, the FISO will tell aircraft of known traffic in the vicinity and will also warn them when his/her information clearly suggests a possibility of dangerous proximity. However, he cannot assume responsibility for the accuracy or completeness of this information because:

(i) position reports passed to him/her may be unreliable in the absence of accurate navigational or position fixing aids;

(ii) many civil and military aircraft not communicating with ATC fly on a multiplicity of tracks and altitudes in the FIR.

1.3 It is emphasized that FIS is only informatory and the FISO will often stress this fundamental aspect of the service by prefacing his messages to the aircraft with the phrase 'You are informed that ....'; particularly when there is a possibility of conflicting air traffic. The FISO cannot:

(a) Exercise positive control over aircraft; or

(b) issue clearance to alter course, climb or descend; or

(c) give positive advice on the avoidance of collision.

1.4 Pilots should remain well clear of Controlled Airspace if they have no Air Traffic Control clearance to enter it. Acceptance of a request for a clearance does not imply that a clearance to enter Controlled Airspace has been given or granted. Pilots should avoid flying parallel to and near an Airway as this may lead to a risk of collision and can cause undesirable complications at international FIR Boundaries. In particular, flight at the base of Controlled and Upper Airspace Control Area Airspace should be avoided except that an aircraft may, without ATC clearance, fly at right angles across the base of an en-route section of an Airway where the lower limit is defined as a Flight Level.

2 Low Level Cross-Channel Operations – UK/France

2.1 Pilots undertaking Cross-Channel flights are reminded that a flight plan MUST be filed for all flights to or from the United Kingdom which will cross the United Kingdom/France FIR Boundary.

2.2 When filing the flight plan with the UK and French Authorities, pilots are to ensure that well defined significant points/features, at which the aircraft will cross the UK and French coast-lines, are included in Item 18 (Other Information) of the flight plan form (eg Beachy Head, Berck-sur-Mer, Lydd, Boulogne, Dover, Cap Gris Nez, etc). This is for Search and Rescue purposes but will also assist ATC.

2.3 Pilots should plan their flights, where possible, at such altitudes which would enable radio contact to be maintained with the appropriate ATC Unit whilst the aircraft is transiting the Channel. In addition, the French Authorities have requested that aircraft fly at altitudes which will keep them within Radar cover. The carriage of Secondary Surveillance Radar (SSR) equipment is recommended.

2.4 Position reports are required when crossing the coast outbound, inbound and when crossing the FIR Boundary.

2.5 Pilots undertaking Cross-Channel flights under IFR, are reminded that the normal IFR Rules will apply particularly regarding altitudes and flight levels. Pilots are also reminded that the IMC rating is not recognised by the French Authorities.

2.6 In UK Airspace a bi-directional Recommended VFR Route between the Solent CTA and the Channel Islands CTR routeing towards the Cherboug Peninsula is established (See AD 2 EGJJ.3.1). All traffic using the route above 3000ft amsl are advised to maintain the appropriate quadrantal flight level irrespective of the flight rules being observed. Pilots flying above 3000ft amsl are reminded of the requirement to maintain an appropriate semi-circular level whilst within the French FIR.

ENR 1.1 GENERAL RULES
ENR 1.1.3 GENERAL FLIGHT PROCEDURES

1 Position Reporting within the London and Scottish FIR/UIR

1.1 Pilots are to make a position report in the following circumstances:

(a) After transfer of communication;

(b) on reaching the limit of ATS clearance;

(c) when instructed by Air Traffic Control;

(d) when operating helicopters in the North Sea Low Level Radar Advisory and Flight Information areas of responsibility and on Helicopter Routes within the London Control Zone and London/City Control Zone (see ENR 1.15 and AD 2 EGLL 2.22, paragraph 17);

(e) when operating flights across the English Channel (see ENR 1.1.2.2, paragraph 2).

1.1.1 The initial call changing radio frequency shall contain only the aircraft identification and flight level. Any subsequent report shall contain aircraft identification, position and time except as provided for in respect of helicopter operations in the areas specified in paragraph 1.1 (d) above.

Note: When changing frequency between any of the London Control, Scottish Control or Manchester Control Centres, pilots are required to state their call sign and Flight Level/Altitudes only (plus any other details when specifically instructed by ATC). When the aircraft is in level flight but cleared to another FL/ALT, both FL/ALT should be passed. Similarly, when the aircraft is not in level flight, the pilot should state the FL/ALT through which the aircraft is passing and the FL/ALT to which it is cleared.

1.1.2 Certain Reporting Points on the UK/Amsterdam FIR Boundary are designated 'Compulsory' in the Netherlands AIP. Position Reports should therefore be made at these points when in communication with Amsterdam or Maastricht Control.

1.2 Omit Position Report Procedure

1.2.1 In order to reduce RTF communication a pilot may be instructed by Air Traffic Control to omit position reports provided that the aircraft is radar identified.

1.3 DME Distance Reports to ATC

1.3.1 Pilots, when requested by ATC to report their distance from a DME facility which they do not have displayed, should retune their equipment to that DME. If, for any reason, they are unable to report their distance from the requested DME, ATC is to be informed. Pilots should not calculate the distance based on the reading from another DME.

2 Climb and Descent

2.1 Vacating (Leaving) Levels

2.1.1 When pilots are instructed to report leaving a level, they should advise ATC that they have left an assigned level only when the aircraft's altimeter indicates that the aircraft has actually departed from that level and is maintaining a positive rate of climb or descent in accordance with published procedures.

2.2 Minimum Rates of Climb and Descent

2.2.1 In order to ensure that controllers can accurately predict flight profiles to maintain standard vertical separation, pilots of aircraft commencing a climb or descent in accordance with an ATC clearance should inform the controller if they anticipate that their rate of climb or descent during the level change will be less than 500ft per minute, or if at any time during such a climb or descent their vertical speed is, in fact, less than 500ft per minute.

2.2.2 This requirement applies to both the en-route phase of flight and to terminal holding above Transition Altitude.

Note: This is not a prohibition on the use of rates of climb or descent of less than 500ft per minute where necessary to comply with other operating requirements.

183

ENR 1.1   ENR 1.1.3.2 Climb and Descent

2.3 Noise Abatement Approach Techniques

2.3.1 The use of Continuous Descent Approach (CDA) and Low Power/Low Drag Approach (LP/LD) techniques (as defined at GEN 1.7.32) is required, subject to compliance with ATC requirements, at certain UK airports as detailed in the appropriate AD 2 sections. At other locations, although not required, these techniques are considered to be 'best practice' for the reduction of noise nuisance and emissions and should be adopted by pilots whenever operationally practicable, commensurate with the ATC clearance.

3 Radio Failure and Loss of Communication Procedures

3.1 General Procedures

3.1.1 The English Language is used for all communications between aircraft and ATC in the UK.

3.1.2 VHF/RTF is used for all air-ground communications throughout the airspace under UK jurisdiction except that HF is also used in the Shanwick Oceanic Control Area and that UHF is also available at the London ACC and at certain aerodromes (see ENR 1.15 and ENR 2.1 Sections for details).

3.1.3 So far as possible, pilots should make use of the ICAO standard RTF phraseology in ICAO DOC 4444 chapter 12 when communicating with ATC.

3.1.3.1 As a general principle all messages should be acknowledged by use of the aircraft call sign.

3.1.3.2 Messages containing any of the following items must be read back in full:

(a) Level instructions
(b) Heading instructions;
(c) Speed instructions;
(d) Airways or route clearances;
(e) Runway-in-use;
(f) Clearance to enter, land on, take-off, backtrack or cross and active runway;
(g) SSR operating instructions;
(h) Altimeter Settings;
(i) VDF information;
(j) Frequency changes.

3.2 Radio Failure Procedures For Pilots

3.2.1 Failure of Navigation Equipment

3.2.1.1 If part of an aircraft's radio navigation equipment fails but two-way communication can still be maintained with ATC, the pilot must inform ATC of the failure and report his altitude and approximate position. ATC may, at its discretion, authorise the pilot to continue his flight in or into Controlled Airspace. When radar is available it may, subject to workload, be used to provide navigational assistance to the pilot.

3.2.1.2 If no authorisation to proceed is given by ATC, the pilot should leave, or avoid Controlled Airspace and areas of dense traffic, and either:

(a) Go to an area in which he can continue his flight in VMC or (if this is not possible);

(b) select a suitable area in which to descend through cloud, fly visually to a suitable aerodrome and land as soon as practicable.

But before doing so, however, he should consult ATC who may be able to give him instructions or advice. He should also take into consideration the latest meteorological information and terrain clearance and should make full use of ground VHF D/F stations. He must at all times keep ATC informed of his intentions.

3.2.2 Failure of Two-way Radio Communications Equipment

3.2.2.1 As soon as ATC know that two-way communication has failed they will, as far as practical, maintain separation between the aircraft experiencing the communication failure and other aircraft, based on the assumption that the aircraft will operate in accordance with radio communication failure procedures described below.

3.2.2.2 It should be noted that for many aerodromes in the UK, the radio communications failure procedures published in the AD 2 section differ from, or amplify, the basic procedures published below.

3.2.2.3 For the purposes of these procedures, ATC will expect an IFR flight following the ATS route structure to adopt the IMC procedure in paragraph 3.2.4. If there is an overriding safety reason, the pilot may adopt the VMC procedure.

3.2.2.4 Flights operating outside controlled or advisory airspace, without reference to ATS, should only use these procedures when the pilot decides that there is a need to alert ATC that two-way radio communications failure has occurred.

3.2.2.5 It should be noted that the use of loss of two-way communications procedures may result in aircraft flying outside controlled airspace.

3.2.2.6 The procedures detailed in this section apply to two-way radio communications failure. In the event that an additional emergency situation develops, ATC will expect the pilot to select secondary radar transponder on Mode A, Code 7700.

3.2.2.7 The expression Expected Approach Time (EAT) will mean either an EAT given by the appropriate ATC Unit or, if the pilot has been given 'No delay expected', the ETA over the appropriate designated landing aid serving the destination aerodrome.

3.2.2.8 Pilots are given an EAT of 'Delay not determined' when the destination runways cannot be used for landing and it is not possible to accurately predict when they will become available. In some circumstances an EAT of 'Delay not determined' will also be given when a preceding flight has elected to remain over the holding facility pending an improvement in weather conditions at the destination. If 'Delay not determined' has been given, do not attempt to land at the destination aerodrome, divert to the alternate destination specified in the current flight plan or another suitable airfield.

ENR 1.1   ENR 1.1.3.3 Radio Failure and Loss of Communication Procedures

3.2.2.9 The 'current flight plan' is the flight plan, as filed and acknowledged with an ATC Unit, by the pilot or a designated representative.

3.2.2.10 The procedure that should be used by Special VFR Flights is detailed at ENR 1.2.2, paragraph 2.9.

3.2.2.11 Essential information may be relayed by ATC using the ACARS/Data Link. Pilots may endeavour to use alternative methods for communicating with ATC such as HF. The Distress and Diversion Cells (D&D) serving the London FIR/UIR and the Scottish FIR/UIR may be contacted by phone by aircraft that have approved installations that can access the UK telephone network. The telephone numbers are as follows:
London D&D Tel: 01895 426150
Scottish D&D Tel: 01292 692380

3.2.3 Visual Meteorological Conditions (VMC)

3.2.3.1 A VFR flight experiencing communication failure shall: When VMC can be maintained, the pilot should set transponder on Mode A, Code 7600 with Mode C and land at the nearest suitable aerodrome. Pilots should take account of visual landing aids and keep watch for instructions as may be issued by visual signals from the ground. The pilot should report arrival to the appropriate ATC unit as soon as possible. When VMC cannot be maintained, the pilot should adopt the procedures for IMC detailed below.

3.2.3.2 Subject to the provisions of paragraph 3.2.2.3, an IFR flight experiencing communication failure in VMC shall: When VMC can be maintained, the pilot should set transponder to Mode A, Code 7600 with Mode C and land at the nearest suitable aerodrome. Pilots should take account of visual landing aids and keep watch for instructions as may be issued by visual signals from the ground. The pilot should report arrival to the appropriate ATC unit as soon as possible. If it does not appear feasible to continue the flight in VMC, or if it would be inappropriate to follow this procedure, the pilot should adopt the procedures for flights in IMC detailed below.

Note: Pilots already in receipt of an ATC clearance may enter controlled airspace and follow the procedures referred to above. Those flights, that have not received an ATC clearance, should not enter controlled or advisory airspace unless an overriding safety reason compels entry.

3.2.4 Instrument Meteorological Conditions (IMC)

3.2.4.1 A flight experiencing communication failure in IMC shall:

(a) Operate secondary radar transponder on Mode A, Code 7600 with Mode C.

(b) (i) Maintain, for a period of 7 minutes, the current speed and last assigned level or minimum safe altitude, if this higher. The period of seven minutes begins when the transponder is set to 7600 and this should be done as soon as the pilot has detected communications failure.

(ii) If failure occurs when the aircraft is following a notified departure procedure such as a Standard Instrument Departure (SID) and clearance to climb, or re-routing instructions have not been given, the procedure should be flown in accordance with the published lateral track and vertical profile, including any stepped climbs, until the last position, fix, or waypoint, published for the procedure, has been reached. Then, for that part of the period of 7 minutes that may remain, maintain the current speed and last assigned level or minimum safe altitude, if this higher.

(iii) Thereafter, adjust the speed and level in accordance with the current flight plan and continue the flight to the appropriate designated landing aid serving the destination aerodrome. Attempt to transmit position reports and altitude/flight level on the appropriate frequency when over routine reporting points.

(c) (i) If being radar vectored, or proceeding offset according to RNAV, without a specified limit, continue in accordance with ATC instructions last acknowledged for 3 minutes only and then proceed in the most direct manner possible to rejoin the current flight planned route. Pilots should ensure that they remain at, or above, the minimum safe altitude.

(ii) If being radar vectored by an Approach Control Radar Unit (callsign DIRECTOR/RADAR/APPROACH), comply with the loss of communications procedures notified on the appropriate Radar Vectoring Chart as detailed in the AD 2 section of the UK AIP.

(d) (i) Arrange the flight to arrive over the appropriate designated landing aid serving the destination aerodrome as closely as possible to the ETA last acknowledged by ATC. If no such ETA has been acknowledged, the pilot should use an ETA derived from the last acknowledged position report and the flight-planned times for the subsequent sections of the flight.

(ii) Arrange the flight to arrive over the appropriate designated landing aid serving the destination aerodrome at the highest notified Minimum Sector Altitude taking account of en-route terrain clearance requirements.

(iii) If following a notified Standard Arrival Route (STAR), after the seven minute period detailed in paragraph (b) (i) has been completed, pilots should arrange descent as close as possible to the published descent planning profile. If no descent profile is published, pilots should arrange descent to be at the minimum published level at the appropriate designated Initial Approach fix.

(e) On reaching the appropriate designated landing aid serving the destination aerodrome, begin further descent at the last acknowledged EAT. If no EAT has been acknowledged, the descent should be started at the ETA calculated in (d) (i), above, or as close as possible to this time. If necessary, remain within the holding pattern until the minimum holding level, published for the facility, has been reached. The rate of descent in holding patterns should not be less than 500 ft per minute. If 'Delay not determined' has been given, do not attempt to land at the destination aerodrome, divert to the alternate destination specified in the current flight plan or another suitable airfield.

(f) Carry out the notified instrument approach procedure as specified for the designated navigational aid and, if possible, land within 30 minutes of the EAT or the calculated ETA. When practical, pilots should take account of visual landing aids and keep watch for instructions that may be issued by visual signals from the ground.

(g) If communications failure occurs during an approach directed by radar, continue visually, or by using an alternative aid. If this is not practical, carry out the missed approach procedure and continue to a holding facility appropriate to the airfield of intended landing for which an instrument approach is notified and then carry out that procedure.

3.3 Actions taken by ATC

(a) As far as is practical, ATC shall maintain separation between the aircraft experiencing the communication failure and other aircraft based on the assumption that the aircraft will operate in accordance with published radio communication failure procedures.

(b) ATC will assume that an aircraft's receiver may be functioning and will transmit instructions for routeing and other relevant information such as the EAT, weather information, altimeter settings and runway in use at destination (or alternate) aerodromes.
(c) ATC will use all means possible to monitor the flight's progress and inform other flights where necessary.
(d) ATC will attempt to re-establish communications with the pilot by monitoring standby frequencies (where available) and by contacting the aircraft operator, or handling agent or by use of ACARS/Data Link when available.
(e) ATC will co-ordinate the flight with other ATC agencies as required.
(f) If the flight re-establishes communications with an ATC unit during flight, or after the aircraft has landed, the ATC unit will relay the pilot's intentions, or that the aircraft has landed, to the ATC Unit that was providing an ATS when the communications failure occurred.
(g) If the aircraft's progress cannot be monitored by radar and there has been no other indication of the aircraft's progress, or landing, normal overdue action will commence 30 minutes after the ETA for the destination airfield.

4 Use of Airborne Collision Avoidance Systems (ACAS) in United Kingdom FIR and UIR

4.1 General
4.1.1 ACAS indications shall be used by pilots in the avoidance of potential collisions, enhancement of situational awareness, and the active search for, and visual acquisition of, conflicting traffic. The ability of ACAS to fulfil its role of assisting pilots in the avoidance of potential collisions is dependent on the correct and timely response by pilots to ACAS indications.
4.1.2 The Traffic Alert and Collision Avoidance System – TCAS II is accepted by the Civil Aviation Authority (CAA) as a suitable ACAS system provided its installation is certificated by the State of Registry, and that its operation by flight crew is in accordance with appropriate instructions

4.2 Procedures to be Established
4.2.1 An operator shall establish procedures to ensure that:
(a) When ACAS is installed and serviceable, it shall be used in flight in a mode that enables Resolution Advisories (RA) to be produced unless to do so would not be appropriate for conditions existing at the time, and
(b) when undue proximity to another aircraft is detected by ACAS, the commander or the pilot to whom conduct of the flight has been delegated shall ensure that corrective action is initiated immediately to establish safe separation.
(c) The circumstances when it is appropriate to operate ACAS in the Traffic Advisory (TA) only mode are specified in the Flight Operations Manual. This should be limited to particular in-flight failures (for example engine failure or emergency descent), during take-offs or landings in limiting performance conditions (for example at high altitude airports), and locations where States have approved specific procedures permitting aircraft to operate in close proximity only.

4.3 TCAS II Operating Characteristics
4.3.1 TCAS II will issue a TA only when another aircraft with a compatible operating transponder is close in both range and altitude. If the transponder in the potentially conflicting aircraft is providing altitude data, an RA may be issued.
4.3.2 TAs and RAs can be issued on the basis of 'time to closest point of approach (CPA)' or 'fixed distance' thresholds being penetrated. On most occasions, TAs and RAs will be issued on the 'time to CPA' basis, but in RVSM penetration of airspace fixed range and altitude thresholds are likely to be a more frequent reason.
Note: In cases where a vertical speed of closure causes RAs to be issued, TCAS in the climbing/descending aircraft may advise a reduction in the climb or descent rate, whilst TCAS II in the other aircraft may advise a 'Climb' or 'Descend' RA. If the climbing/descending aircraft in this pair is diverging in range at a slow rate, the 'Climb' or 'Descend' RA issued to the Flight Crew in the other aircraft may remain displayed for several minutes, even though the former has levelled off at its cleared flight level. Although this particular circumstance is likely to be rare, even when it does occur, excessive altitude excursions need not result.

4.4 Operation of Aircraft When ACAS II is Unserviceable
4.4.1 The current TCAS II Minimum Equipment List permits TCAS II equipped aircraft to operate for up to 10 days with the equipment out of service. This position will be kept under review.
4.4.2 Due to the safety benefits arising from TCAS operations and the collaborative way in which it arrives at collision avoidance solutions any aeroplane with an unserviceable transponder as well as an unserviceable TCAS will not be permitted in UK airspace for which mandatory carriage of a transponder is required.

4.5 Operation of TCAS II in RVSM Airspace
4.5.1 Above FL 290, TAs and RAs are most likely to occur in airspace where aircraft change altitude to reduce separation from 2000 ft to 1000 ft: this airspace is described as a 'Transition Area'. Specifically:
(a) TAs can be expected when aircraft vertically separated by 1000 ft pass each other. If the speed at which they pass is low, such as when one is overtaking the other, TAs may be intermittent or they may last for long periods.
(b) RAs can be expected when the vertical speed of closure, which may be the sum of the vertical speeds of both aircraft or the vertical speed of one of the aircraft, exceeds approximately 1500 ft/min. RAs might also be issued when either aircraft experiences turbulence sufficient to cause TCAS to project the vertical separation between both aircraft to be less than 800 ft at CPA, or when a 'soft altitude hold' function in either aircraft achieves the same result.

4.6 Guidance for Aircraft Operators and Flight Crews
4.6.1 Flight Crews can reduce the likelihood of TAs and RAs occurring above FL 290 where separation is less than 2000 ft vertically and 5 nm horizontally by confining vertical speeds to less than 1500 ft/min. Desirably, the vertical speed should be between 500 and 1000 ft/min.
4.6.2 The TCAS II function control selector should not be moved from the 'TA/RA' or 'Normal' position upon entering RVSM Airspace. Although it is implicit that such TAs and RAs as have been described could be termed 'unnecessary', this might not always be the case. For this reason, Flight Crews would be unwise either to disable an effective collision avoidance device without sound reason, or to assume that any TA or RA issued in this airspace is other than genuine.

ENR 1.1   ENR 1.1.3.4 Use of Airborne Collision Avoidance Systems (ACAS) in United Kingdom FIR and UIR

4.6.3 Flight Crews shall not manoeuvre an aircraft solely in response to a TA. TAs are intended to alert the pilot to the possibility of an RA, and to assist in visual acquisition of conflicting traffic. However, visually acquired traffic may not be the same traffic causing a TA, and visual perception of an encounter may be misleading, particularly at night.

4.6.4 In the event that an RA is issued, Flight Crews shall:

(a) Respond immediately and manoeuvre as indicated by the ACAS unless doing so would jeopardise the safety of the aircraft;

(b) follow the RA even if there is a conflict between that RA and an air traffic control (ATC) instruction to manoeuvre;

(c) not manoeuvre in the opposite sense or direction to that of the RA;

(d) limit RA manoeuvres to the minimum extent necessary to comply with the RA.

4.6.5 Flight Crews should note that:

(a) Other critical warnings such as Stall Warning, Windshear Warning and Ground Proximity Warning Systems have priority over ACAS.

(b) visually acquired traffic may not be that causing an RA, as the visual perception of an encounter may be misleading, particularly at night.

(c) ATC may not know when an ACAS system issues an RA. It is possible for ATC to issue instructions to an aircraft that are unknowingly contrary to RA instructions on that aircraft. Therefore, it is essential that ATC be notified when an ATC instruction is not being followed because it conflicts with an RA.

(d) a manoeuvre opposite to the sense of an RA may result in a reduction in vertical separation with the 'threat' aircraft and therefore must be avoided at all times; this is particularly true in the case of an ACAS-ACAS co-ordinated encounter, when the RAs complement each other in order to reduce the potential for collision. Manoeuvres, or lack of manoeuvres, that result in vertical rates opposite to the sense of an RA could result in a collision with the threat aircraft.

4.6.6 A pilot who has deviated from an air traffic control instruction or clearance in response to an RA shall:

(a) As soon as possible, as permitted by flight deck workload, notify the appropriate ATC unit of the RA, including the direction of any deviation from the current ATC instruction or clearance;

(b) when they are unable to comply with a clearance or instruction that conflicts with an RA, notify ATC as soon as possible consistent with flying the aircraft.

(c) promptly comply with any modified RAs.

(d) return to the terms of the ATC instruction or clearance when the conflict is resolved.

(e) after initiating a return to, or resuming the current clearance, notify ATC as soon as possible consistent with flying the aircraft.

4.6.6.1 Verbal reports should be made to Air Traffic Control at the first practicable moment and written reports submitted to the designated Authority as soon as possible after the flight has ended.

4.7 Guidance for Air Traffic Service Providers and for Air Traffic Controllers

4.7.1 The operation of TCAS II equipment will affect ATC operations to some extent, irrespective of the type of airspace. ATC will expect Flight Crew to react to RAs and to notify any manoeuvres initiated in response to RAs in accordance with guidance published in CAP 579. The Manual of Air Traffic Services Part 1 (Supplementary Instruction 3/2001) provides information on TCAS II to Air Traffic Controllers: it reiterates the phraseology that Flight Crews will use and the replies that Air Traffic Controllers should make.

4.7.2 It will be apparent from paragraph 4.5.1 that TAs will be more frequent in North Atlantic RVSM Airspace than elsewhere. Air Traffic Controllers should be aware of this and, where possible, be prepared to provide requested traffic information to Flight Crews.

4.7.3 As pilots are not required to take avoiding action on the basis of TA information alone, ATC does not expect requests for traffic information to be made unless the other aircraft cannot be seen and the pilots believe their aircraft is about to be endangered.

4.7.4 ATC expects pilots to respond immediately to an RA. Pilots are expected to restrict their RA manoeuvres to the minimum required to resolve the confliction, advise the Air Traffic Control Unit as soon as is practical thereafter and return to their original flight path as soon as it is safe to do so.

4.7.5 Pilots should be aware that any deviation from an ATC clearance has the potential to disrupt the controller's tactical plan and may result in a reduction of standard separation between aircraft other than those originally involved. It is vital that Flight Crew maintain a good look out and return to their original flight path as soon as it is safe and practical to do so.

5 Diversion

5.1 Diversion is the act of flying to an aerodrome other than the planned destination with the intention of landing there.

5.2 Normally diversion is made when one of the following circumstances occurs at the planned destination:

(a) The weather is reported to be below the operating company's minima;

(b) there are obstacles on the manoeuvring area constituting a hazard to landing aircraft which cannot be cleared within a reasonable time;

(c) there is a failure of an essential ground aid which is required for the landing;

(d) there is likely to be an unacceptable delay to landing.

5.3 Diversion may be originated by either the pilot or his operating company, or exceptionally by ATC.

5.3.1 When a pilot decides to divert he should inform ATC. ATC will, if possible, advise his operating company or a nominated addressee of his diversion when this is specifically requested by the pilot.

5.3.2 An operating company proposing to divert one of its aircraft should consult ATC before any decision on diversion is passed to the pilot. The message to the pilot will be in this form:

'Company advise divert to ÖÖ (aerodrome). Weather at …… (diversion aerodrome) …… Reason for diversion …… (clearance instructions). Acknowledge'. The pilot should either follow this advice or if he is unable to do so, give his reasons and state what he intends to do.

ENR 1.1 ENR 1.1.3.4 Use of Airborne Collision Avoidance Systems (ACAS) in United Kingdom FIR and UIR

5.3.3 In exceptional circumstances, it may be necessary for ATC to advise a pilot to divert before being able to consult his operating company. In such a case, the company will be told as soon as possible and the message to the pilot will be in the form: 'Request divert to .... (aerodrome). Weather at ...... (diversion aerodrome) ...... Reason for diversion ...... (clearance instructions). Acknowledge'. If the pilot is unable to comply with this request, he should give his reasons and state his intention.

ENR 1.1.4 ARRANGEMENTS FOR PARTICULAR TYPES OF FLIGHT (NON-STANDARD, NON-DEVIATING, UNUSUAL, ROYAL, OBSERVATION, SPECIAL VFR ACCESS TO CLASS C AIRSPACE ABOVE FL 246)

1 Non-Standard Flights (NSFs) in Controlled Airspace

1.1 Unless alternative arrangements have been made with the appropriate ATC Unit, an initial telephone call shall be made to the ATC Operations Department (as detailed below) to ascertain the notification required to operate flights in Controlled Airspace involving aerial tasks which do not follow published routes or notified procedures:

(a) for flights south of 5300N (except as at (b) and (c))
London Area Control Centre (LACC)
ATC Operations
Box 9
NATS
London Area Control Centre
Sopwith Way
Swanwick
Southampton
Hants
SO31 7AY
Tel: 01489 612340
Fax: 01489 612430
e-mail: nonstandard.flightapplications@nats.co.uk (see notes 1 and 3)

(b) for flights within the London Terminal Control Centre (LTCC) area of responsibility (except as at (c)
TC Operations
NATS Ltd
London Terminal Control Centre
Porters Way
West Drayton
Middlesex
UB7 9AX
Tel: 01895 426152
e-mail: ltccnsf.request:nats.co.uk (See notes 2 and 4)

(c) for flights within the London CTR and London City CTR
TC Operations
NATS Ltd
London Terminal Control Centre
Porters Way
West Drayton
Middlesex
UB7 9AX
Tel: 01895 426152
e-mail: ltccnsf.request:nats.co.uk (See notes 2 and 4)

(d) for flights between 5230N and 5415N
Manchester Area Control Centre (MACC)
ATC Operations
NATS Ltd
Room 406
Tower Block
Manchester Airport
Manchester
M90 2PL
Tel: 0161 499 5315
Fax: 0161 499 5312
e-mail: nonstandard.flightapplications@nats.co.uk (see notes 1, 3 and 4)

(e) for flights north of 5415N and over Northern Ireland
Scottish Area Control Centre (ScACC)
ATC Airspace Reservation Cell
NATS Ltd
Room 2.20
Scottish and Oceanic Area Control Centre
Sherwood Road
Prestwick
Ayrshire
KA9 2NR
Tel: 01292 692431
Fax: 01292 692610
e-mail: reservation.cell@nats.co.uk (see notes 3 and 4)

(f) for localised VFR flights above FL245 not requiring reserved airspace and outside of theATS route structure:
Military Airspace Manager (MAM)
Airspace Management Cell
London Area Control Centre
Sopwith Way
Swanwick
Southampton
Hants SQ31 7AY
Tel: 01489 422495
Fax: 01489 422497
E-mail: mabcc.mil@nats.co.uk
Hours: Mon-Fri (excluding PH) 0800-1700.
See also Airspace Division Chart at ENR 6.1.1.1

Note 1: For applications using the NSF e-mail address (Nonstandard.flightapplications@nats.co.uk) the ATC unit to which the application is being made should be placed in the 'subject' heading (ie. LACC). Applicants using the email facility should ensure that file sizes do not exceed 5MB. Zipped files are acceptable.

Note 2: For applications using NSF e-mail address (ltccnsf.request@nats.co.uk), the ATC unit to which the application is being made should be placed in the 'subject' heading (ie. LTCC or London CTR/London City CTR). Applications using e-mail facility should ensure file sizes do not exceed 5MB. Zipped files are acceptable. Applications are to be made using the LTCC Application Form available by e-mailed request from the aforementioned address.

Note 3: For flights north of 5415N, operations are strongly recommended to apply for NSF for operations within Class F airspace.

Note 4: For the purpose of this requirement, a formation flight of civil aircraft intending to operate in Controlled Airspace is considered to be a non-standard Flight.

Note 5: Applicants using the NSF E-mail address (maccnsf.request@nats.co.uk) should ensure that the file sizes do not exceed 5mb. Zipped files are acceptable.

ENR 1.1   ENR 1.1.4.1 Non-Standard Flights (NSFs) in Controlled Airspace

1.2 Applications should normally give a minimum of 21 days notice and include:
(a) Purpose of flight;
(b) the area of operation and proposed tracks to be flown – 1 copy of a suitable aeronautical chart with a list of National Ordnance Survey Grid and/or co-ordinates detailing the requested areas of operation in relation to Controlled Airspace;
(c) estimated duration of aerial task;
(d) operating heights;
(e) aircraft type, callsign and registration letters on any aircraft likely to be used;
(f) aerodrome of departure;
(g) planned date of operation and requested validity period.

Those applications which are agreed will be allocated a non-standard flight reference number. This is only an approval in principle and prior clearance must be obtained from the appropriate ATC Watch Supervisor on the day. This is normally obtained by telephone prior to departure. However, since many tasks are weather-dependent, some have to be abandoned after the aircraft is airborne. To overcome the particular difficulty of having to land and co-ordinate another detail by telephone, the following procedures may be adopted by pilots of those NSFs which have been previously allocated an NSF number by London ACC or London TCC, and who wish to abandon the original task co-ordinated prior to take-off and proceed to another location.

1.2.1 The aircraft commander will establish RTF contact on the London FIS frequency (callsign 'London Information') appropriate to the area of the country over which the new task is required to be flown, prefixing the message with the phrase 'Non-Standard Flight Request'. The following information will then be passed to the Flight Information Service Officer (FISO):
(a) The Non-Standard Flight number;
(b) the requested area of activity (this is essential as many NSF numbers refer to several sites);
(c) ETA at site;
(d) the requested Flight Level or Altitude for the task;
(e) the duration of the task;
(f) the aircraft callsign.

1.2.2 The FISO will relay these details to the appropriate ATC Unit and, in due course, will advise the pilot whether or not the NSF is approved, together with any special conditions and a contact frequency for the ATC Unit concerned. Pilots should not call for an approval directly on an operational ATC frequency. This is particularly important in the case of frequencies in use by London TCC or London ACC (London Control).

1.2.3 In the case of NSFs affecting Airspace for which London TCC is responsible, it may sometimes be necessary for the pilot to land at a convenient aerodrome and telephone Terminal Control Senior Watch Assistant to discuss the requirements of the task in detail.

1.2.4 Operators are to note that in no circumstances can any discussion be entered into on any frequency in the event that permission is refused.

1.3 ATC clearance does not imply exemption from the requirements of the current Air Navigation Order (ANO) or the Rules of the Air Regulations. Applications for flights which require exemption or written permission under the ANO are to be forwarded to:

The Civil Aviation Authority
Operating Standards Division
Aviation House
Gatwick Airport South
Gatwick
West Sussex
RH6 0YR

1.4 Because of the nature of ATC operations (and notwithstanding the requirements of GEN 1.5.3, paragraph 1.3 concerning the carriage of SSR transponders), the approval of an application for a Non-Standard Flight will depend on the carriage of SSR transponder equipment normally with Mode C.

1.5 Due to the inherent difficulties of handling a formation flight in a busy traffic situation, pilots should be aware that it may not always be possible to issue an ATC clearance at the time requested.

1.6 Enhanced Non-Standard Flights (ENSFs) – Entry into EG R157 (Hyde Park)/E1G R158 (City of London)/EG R169 (Isle of Dogs) Restricted Areas

1.6.1 For those aircraft not already exempted (see individual entry for Restricted Area at ENR 5.1), ENSFs are required for flight within EG R157, EG R158 and EG R159. Requests should be made using the 'Application for LTCC NSF & ENSF Approval' form available from TC Operations at para 1.1 (c) giving a minimum of 28 days notice. (See Note 2 to para 1.1).

1.6.2 ENSFs are subject to security considerations by the Metropolitan Police and may be refused on public interest grounds.

1.6.3 Once the security process is complete and LTCC provisional ATC approval in principle is granted, an 'ENSF Notification – Approval' form will be returned to the operator. Details of how to obtain a Metropolitan Police authorisation for an ENSF and the ATC tactical approval on the day of flight are detailed on the 'ENSF Notification – Approval' form.

1.7 Unusual Aerial Activities in Controlled Airspace

1.7.1 Normally, requests for the approval of Unusual Aerial Activities remaining within Controlled Airspace at all times are treated by the controlling authority as Non-Standard Flights.

189

ENR 1.1   ENR 1.1.4.1 Non-Standard Flights (NSFs) in Controlled Airspace

1.7.2 A request for the approval of an Unusual Aerial Activity (UAA) which crosses the boundary between Controlled Airspace and the Open FIR has to be treated differently from the Non-Standard Flight (NSF) type. The UAA which takes place both in and outside Controlled Airspace is more complex and time-consuming to resolve, because of the additional negotiation needed between ATC and airspace users, and should be notified to ATC at the earliest opportunity. A UAA of this nature is processed by the CAA Airspace Utilisation Section (AUS), which consults all agencies affected, arranges NOTAM action, and publishes an Airspace Co-ordination Notice (ACN). The ACN details the agreements reached about the route inside and outside Controlled Airspace and reflects the NSF approval issued by the relevant controlling authority.

1.7.3 Details of UAA / NSF which cross the boundaries of Controlled Airspace should be submitted as per paragraph 1.2 with a copy to:
Directorate of Airspace Policy
Airspace Utilisation Section (AUS)
K7 CAA House
45-59 Kingsway
London
WC2B 6TE
Tel: 020 7453 6599
Fax: 020 7453 6593

1.8 VFR Flight in Class C Airspace Above FL 245
1.8.1 VFR flight by civil aircraft above FL 245 shall not be permitted unless it has been accorded specific arrangements by the appropriate ATS authority. VIFIR flight shall only be authorised:
(a) In reserved airspace;
(b) Outside reserved airspace up to FL 285, and then only when authorised in accordance with the procedures detailed for Non-Standard Flights in Controlled Airspace.

1.8.2 If utilising permanently established reserved airspace, the established booking procedures for that airspace should be followed. If there is a need for the establishment of temporary reserved areas then procedures for conducting Unusual Aerial Activities in Controlled Airspace shall be followed as detailed in paragraph 1.7. Standing arrangements for temporary reserved areas for gliding in Class C airspace are shown at ENR 1.1.1.6 paragraph 4.3.3.

1.8.3 It is anticipated the demand for VFR access outside of an airspace reservation will be minimal. Such access will be accommodated within the context of safety, capacity and effect on the ATS network as a whole; consequently VFR access to the ATS route structure is only likely to be permitted in exceptional circumstances. In this case the appropriate civil ATC Unit will co-ordinate provision of ATS. Operators seeking to operate in such areas should contact the appropriate ACC Operations Department as detailed at paragraphs 1.1 to 1.5. Applications for WIR flight to avoid IFR ATS route flow restrictions will not be granted.

1.8.4 Operators seeking localised VFR flight above FL 245 not requiring reserved airspace and clear of the ATS route structure should contact the Military Airspace Manager (MAM) in the Airspace Management Cell located at LACC, who will co-ordinate access arrangements and military ATC provision within unit capacity. Contact details are shown at paragraph 1.1 (f). Such flights shall only be permitted where procedures are established with the controlling authority.

1.9 VFR Flight in Class C Areas of Delegated ATS
1.9.1 Charts depicting these areas are detailed at ENR 6.2 pages. These delegated areas of ATS are busy international interfaces. Consequently, approval for VFR flight will only be granted in exceptional circumstances and after co-ordination with and agreement of the respective ATS provider. Applications for VFR access to these areas should in the first instance be made to AUS as detailed in paragraph 1.7.

2 Non Deviating Status (NDS)

2.1 NDS may be agreed by prior arrangement with the appropriate controlling authorities for certain flights within Controlled Airspace excluding in the UIR active Danger and Military Training Areas.
2.2 The requirement for NDS may be expressed as all, or part, of a notified flight profile and not merely for a constant heading and, or, flight level. Application for NDS should only be made where an inability to maintain specific track(s) and or flight level(s) could render a task operationally ineffective. NDS would not be appropriate for, nor would it be granted to, GAT aircraft carrying freight or passengers between destinations or GA aircraft general handling etc or in transit.
2.3 NDS affords priority of passage over all other OAT and GAT except for: aircraft in emergency; Royal Flights; Air Defence Priority Flights; GAT with higher civil priority category or other higher priority Special flights.
2.4 The Airspace Utilisation Section (AUS), is the central authority for authorising NDS and is the focal point for NDS applications, inter unit negotiations and approvals.
2.5 AUS normally requires a minimum of 21 days notice of pre-flight requests for NDS in order to obtain agreement from the affected Air Traffic Service Units (ATSUs) and, or, Airborne Surveillance and Control (ASAC) Units. A request for NDS should include:
(a) Operating authority, including a point of contact;
(b) type of aircraft operation (e.g. flight trial, calibration etc);
(c) reference number or other discrete nomenclature;
(d) aircraft registration(s) and/or callsign(s);
(e) details of flight(s);
(i) departure aerodrome and destination;
(ii) route or area;
(iii) profiles (if appropriate);
(iv) times;
(v) altitudes or Flight Levels;
(vi) facilities to be used (if appropriate);
(vii) any non-ATC agencies involved;
(viii) any specific requirements eg frequencies to be used etc.

ENR 1.1 ENR 1.1.4.3 Unusual Aerial Activities (UAA) Outside Controlled Airspace

(f) details of any specific flexibility, limitation or critical aspect. Shorter notice applications may be considered on merit, but AUS may direct aircraft operating authorities to refer their requests for NDS direct to the appropriate ATSU(s) or ASAC unit(s) involved if the application cannot be processed in time by AUS.

2.6 Units will attempt to agree to short notice requests for NDS but, if given insufficient time, may decline or modify the request. Profile adjustments may, in any event, need to be negotiated if such changes would result in less disruption to other traffic.

2.7 Flights granted NDS will remain under radar control or procedural service. If in the interest of flight safety it should become necessary to give NDS flights avoiding action, such instructions from a controller are Mandatory.

3 Unusual Aerial Activities (UAA) Outside Controlled Airspace

3.1 A UAA may constitute a hazard if pilots of non-participating aircraft are not aware that it is taking place. The Civil Aviation Authority and in particular the Directorate of Airspace Policy, Airspace Utilisation Section (AUS) require appropriate prior notification of a UAA to enable either AUS to co-ordinate and notify the event, or for the Authority to issue a Permission or Exemption under the Air Navigation Order (ANO) and the Regulations. Event or display organisers are advised to utilise the CAA Publication CAP 403 'Flying Displays and Special Events: A Guide to Safety and Administrative Arrangements'. The document is available from The Stationery Office Tel: 0870 600 5522 and on the CAA web site: http//www.caa.co.uk/publications.

3.2 Individual participating pilots are advised to check that the event or display organiser has made proper application for any required Permission or Exemption.

3.3 While there are many types of UAA, most fall within one of the following categories:

(a) A concentration of aircraft significantly greater than normal, e.g. a Rally or Fly-in;

(b) Activities requiring the issue of a Permission or an Exemption from the ANO and the Regulations, eg low flying near assemblies of people, the dropping of articles or parachutists, or balloon or kite flying;

(c) Air Shows, Displays, Air Races and other aeronautical competitions, aerial surveys and avoidance of ground events and hazards;

(d) Activities requiring the establishment and approval of a temporary ATC Unit. (See CAP 670 'ATS Safety Requirements' and CAP 403 for requirements and recommendations). In the case of the provision of a Flight Information Service (FIS) at a temporary FIS Unit refer to CAP 410 'Manual of Flight Information Services' (Part A and B) and CAP 427 'Flight Information Service and the FISO Licence'. All are available on the CAA web site, as above.

3.4 All event or display organisers wishing to arrange a UAA are to use the standard notification forms SRG 1303 (flying Display Notification Form) or SRG 1304 (Special Events and Unusual Aerial Activity Application Form) as appropriate.

3.5 The length of notice required by the Authority and AUS is as follows:

(a) UAA at a licensed aerodrome or site where a temporary aerodrome licence is required – 60 days;

(b) UAA at an aerodrome or a site where an aerodrome licence is not necessary – 42 days;

(c) If an activity is intended to attract more than 100 aircraft it is essential that proposals are discussed with both the Aerodrome Standards Department (ASD) and the appropriate Regional Manager – Air Traffic Services (ATS) prior to any firm arrangements being made. These discussions should be initiated at least 90 days prior to the date of the Activity. If the organiser has any doubt on the level or type of Air Traffic Service that should be provided, he/she is strongly recommended to contact the relevant Regional Manager (ATS) for guidance;

(d) If it is intended to establish a Temporary Air Traffic Control Unit (ATCU) at an event, it is essential that organisers refer to the document CAP 670 'ATS Safety Requirements' which contains comprehensive information and requirements for the establishment of such a unit. This document is available on the CAA web site http//www.caa.co.uk/publications. A provider of Air Traffic Control must be nominated and he/she is required to apply to the appropriate CAA ATSSD Regional Office in advance of the event for unit approval. A copy of the proposed Manual of Air Traffic Services Part 2 (MATS Part 2) should be submitted as soon as possible but no later than 60 days before the event. The format of the MATS Part 2 is laid out in CAP 670 – Part B, Section 2, ATC 02, with further information in Part B, Section 1, APP 04 page 3. Established ATS Units intending to hold a Flying Display or Special Event are required to notify their ATSSD Regional office if the event requires changes to safety related procedures at that unit. Copies of Form SRG 1417, which also may be used for the application for a temporary VHF frequency, are available from the CAA ATSSD Regional Offices or on the CAA web site. A minimum of 90 days notice is required for a temporary assignment of VHF aeronautical channels.

Note: Air Traffic Services Standards Department (ATSSD) Regional Offices Regional Manager ATS

Southern Regional Office
Floor 2W
Aviation House
Gatwick Airport South
West Sussex
RH6 0YR
Tel: 01293 573426
Fax: 01293 573974
Regional Manager ATS

Central Regional Office
Manchester International Office Centre
Suite 5 Styal Road
Wythenshawe
Manchester
M22 5WB
Tel: 0161499 3055 Ex 242
Fax: 0161499 3048
Regional Manager ATS

Northern Regional Office
7 Melville Terrace
Stirling
FK8 2ND
Tel: 01786 431400
Fax: 01786 448030

ATS Licensing
ATS Standards Department
Aviation House
Gatwick Airport South
West Sussex
RH6 0YR
Tel: 01293 573329
Fax: 01293 573974

See chart at ENR 6.1.1.2 for Area of Responsibility of the ATSSD Regional Offices.

ENR 1.1   ENR 1.1.4.3 Unusual Aerial Activities (UAA) Outside Controlled Airspace

3.6 Display organisers and pilots are advised that, although every effort will be made to deal with late notification forms, no guarantee can be given that they will be processed in time for the event.

3.7 Forms SRG 1303 and SRG 1304 can be obtained from the CAA web site, and when completed should be returned to:

Civil Aviation Authority
General Aviation Department
1W Aviation House
Gatwick Airport South
West Sussex
RH6 0YR
Tel: 01293 573227/573517/573525
Fax: 01293 573973.

To arrive at least 28 days before the event, a copy of the form should be sent to:

Directorate of Airspace Policy
Airspace Utilisation Section (AUS)
K7 CAA House
45-59 Kingsway
London
WC2B 6TE
Tel: 020 7453 6599
Fax: 020 7453 6593

3.8 Event or display organisers should note that whenever military aircraft participate in a civil aviation event, the Ministry of Defence (MOD) requires the organiser to complete a special questionnaire which is separate from, and additional to, the notification required by the Authority. Event organisers will receive copies of the military questionnaire from the MOD when the military participation is confirmed. The completed questionnaires are to be sent to AUS.

4 Royal Flights

4.1 Introduction

4.1.1 A Royal Flight within UK airspace is defined as the movement of an aircraft specifically tasked to carry one or more of the following members of the Royal Family:

Her Majesty The Queen
His Royal Highness The Prince Philip, Duke of Edinburgh
His Royal Highness The Prince of Wales
His Royal Highness The Duke of York
His Royal Highness The Earl of Wessex
Her Royal Highness The Countess of Wessex
Her Royal Highness The Princess Royal

4.1.2 When so directed by the Directorate of Airspace Policy (DAP) Assistant Director Airspace Policy 1, certain flights within UK airspace by members of other Royal Families, other reigning Sovereigns, Prime Ministers and Heads of State of Commonwealth and foreign countries, may also be afforded Royal Flight Status.

4.2 Special ATC Arrangements for Royal Flights in Fixed-Wing Aircraft

4.2.1 Establishment of Temporary (Class A/C) Controlled Airspace (CAS-T)

4.2.1.1 Royal Flights in fixed-wing aircraft are, whenever possible, to take place within the national ATS route structure. Standard ATC procedures shall be applied to Royal Flights when operating in Class A/C airspace, with the exception that controllers may not authorise an aircraft to climb or descend in VMC in the vicinity of the Royal Flight aircraft. In all other instances, the airspace around the route will be designated CAS-T.

4.2.1.2 CAS-T of appropriate height/width bands and levels, will be established to encompass any portion of the track and flight level of the Royal aircraft which lies outside of permanent Class A/C airspace. Control Zones and Control Areas will be established around all airfields used for the departure or arrival of a Royal Flight.

4.2.1.3 Regardless of the prevailing meteorological conditions, aircraft may only fly within CAS-T when ATC clearance has been obtained from the controlling authorities specified in the following sub-paras:

(a) **Temporary Control Zones.** Temporary Control Zones will be established around airfields of departure and destination where no permanent control zone exist. Control Zones for Royal Flights will normally extend for 10 nm radius from the centre of the airfield from ground level to a flight level designated for each Royal Flight. The Control Zone will be established for a period (for outbound flights) of 15 minutes before, until 30 minutes after, the ETD of the Royal aircraft or (for inbound flights) for a period of 15 minutes before, until 30 minutes after, the ETA of the Royal aircraft at the airfield concerned, based on planned times. Overall control of these Control Zones is to be exercised, as appropriate, by the Commanding Officer of a military airfield or the ATS authority of a civil airfield.

(b) **Temporary Control Areas.** Temporary Control Areas will be established to meet the specific requirements of a Royal Flight. The lateral and vertical limits, the duration and the controlling authority of such areas will be promulgated via NOTAM. The controlling authority will be the appropriate civil ATCC.

(c) **Permanent Control Zones and Areas.** The controlling authority will be the designated controlling authority for the Permanent Zone or Area and the duration will be as laid down in sub paras 4.2.1.3 (a) and (b). Where an airfield has its own Control Zone, then the requirement to establish a Temporary Control Zone of the dimensions specified in para 4.2.1.3 (a) may be waived.

(d) **Temporary Controlled Airways.** Temporary Controlled Airways will be established to join temporary or permanent Control Zones or Control Areas, as appropriate, for 15 minutes before ETD at the departure airfield until 30 minutes after ETA at the destination. The lateral dimensions of such airways will be 5 nm each side of the intended track of the Royal Flight and vertical limits will be designated. The controlling authority will be the appropriate civil ATCC.

4.2.1.4 A Temporary Control Zone or Area may be cancelled at the discretion of the Military Commander or Civil ATC Supervisor, as appropriate, when the Royal aircraft has left the zone or area and is established en-route in a Temporary Controlled Airway, permanent Class A/C airspace, or has landed.

4.2.1.5 Training Flights, including parachute training flights, by any member of The Royal Family planned and carried out under VFR or IFR, and under the control of an ATCRU or aerodrome radar, will normally be classified as Royal Flights. CAS-T, if required, will be established as agreed by the aircraft operating organisation and the Directorate Airspace Policy, Airspace Utilisation Section.

4.2.2 Procedures Applicable to Royal Flight CAS-T

4.2.2.1 CAS-T will normally be notified as Class A airspace for the purpose of the Rules of the Air Regulations 1996.

4.2.2.2 CAS-T not already notified under Rule 21 of the Rules of the Air Regulations 1996, is hereby notified for the purpose of Rule 21 and IFR applies at all times.

4.2.2.3 CAS-T established outside of existing Class A/C airspace, is hereby notified respectively as either Control Zones or Control Areas (as appropriate) as defined in Article 155(1) of the Air Navigation Order 2005.

4.2.2.4 Clearances to climb or descend maintaining VMC will not be given to aircraft in CAS-T.

4.2.2.5 Gliders shall not fly in CAS-T.

4.2.3 Promulgation of Royal Flight information

4.2.3.1 Dissemination of information concerning a Royal Flight is made via a Notification Message on a Royal Flight Collective, giving full flight details. Information on the establishment of CAS-T, including vertical limits, is promulgated by NOTAM.

4.3 Royal Flights in Helicopters

4.3.1 CAS-T is not normally established for Royal Flights in helicopters.

4.3.2 For Royal helicopter flights a Royal Low Level Corridor (RLLC) marked by a series of check points will be promulgated by Notification Message. These check points will be approximately 20 minutes flying time apart and will coincide with turning points. The Notification Message will indicate the ETDs/ETAs for given check points. Within the RLLC, protected Zones applying to military aircraft only are established extending 5 nm either side of the helicopter's intended track and from ground level to 1000 ft above the maximum cruise altitude. Military flying within these zones is strictly controlled and, such aircraft, with the exception of military light aircraft and helicopters with an IAS of 140 kt or less, are to maintain a lateral separation of at least 5 nm from the Royal Helicopter. This may be reduced to 3 nm subject to the military ATC conditions for reduced radar separation being met. Military light aircraft or helicopters, with an IAS of 140 kt or less, and civilian pilots flying near the route should keep a good look out and maintain adequate separation from the Royal aircraft.

4.3.3 The Notification Message will include a list of call signs and frequencies of certain nominated aerodromes from which pilots may obtain information on the progress of the Royal helicopter.

4.4 Royal Flight Callsigns

4.4.1 The flight plan aircraft identification and the radiotelephony designators for flights flown in aircraft of No. 32 (The Royal) Squadron, the Queen's Helicopter Flight (TQHF) or in civilian chartered aircraft are as follows:

(a) Royal Flights. Royal flight callsigns are as follows:

(i) No. 32 (The Royal) Squadron (See note). The 3 letter operator designator KRF followed by an identification number and the letter R, eg KRF 1R, and the radiotelephony callsign KITTYHAWK followed by an identification number and the letter R.

(ii) TQHF The 3-letter designator TQF followed by an identification number and the letter R, eg TQF 1 R, and the radiotelephony callsign 'RAINBOW' followed by an identification number and the letter R.

(iii) Civilian Chartered Aircraft. The 3 letter designator KRH followed by an identification number and the letter R, eg KRH 1 R, and the radiotelephony callsign 'SPARROWHAWK' followed by an identification number and the letter R.

(b) Flights by Passengers entitled to CAA Priority. Callsigns for flights by aircraft carrying passengers entitled to CAA priority are as follows:

(i) No. 32 (The Royal) Squadron (See note). The 3 letter operator designator KRF and the radiotelephony callsign KITTYHAWK followed by an identification number.

(ii) TQHF The 3 letter operator designator TQF and the radiotelephony callsign 'RAINBOW' followed by an identification number and the letter S.

(iii) Civilian Chartered Fixed-wing Aircraft. The 3 letter operator designator KRH and the radiotelephony callsign SPARROWHAWK followed by an identification number.

(iv) Civilian Chartered Rotary-wing Aircraft. The 3 letter operator designator KRH and the radiotelephony callsign 'SPARROWHAWK' followed by an identification number and the letter S.

(c) Positioning Flights. Callsigns for positioning flights are as follows:

(i) No. 32 (The Royal) Squadron (See note). The 3 letter operator designator RRF and the radiotelephony callsign 'KITTY' followed by an identification number.

(ii) TQHF. The 3 letter operator designator will be TQF and the radiotelephony callsign 'RAINBOW' followed by an identification number.

(iii) Civilian Chartered Fixed-wing Aircraft. The normal aircraft callsign will be used.

(iv) Civilian Chartered Rotary-wing Aircraft. The 3 letter operator designator KRH and the radiotelephony callsign 'SPARROWHAWK' followed by an identification number.

(d) Other Flights by Aircraft of No. 32 (The Royal) Squadron (See note). All other flights carried out by No. 32 (The Royal) Squadron will use the 3 letter designator RRR and the radiotelephony callsign 'ASCOT' followed by the required identification number.

(e) Helicopters flown by HRH The Duke of York. For helicopters of TQHF flown by HRH The Duke of York, the 3 letter operator designator will be LPD and the radiotelephony callsign will be 'LEOPARD'.

Note: The rule also applies whenever No. 10 Squadron or No. 216 Squadron aircraft are being utilised for Royal/VIP flights.

5 Observation Flights Conducted Under the Treaty on Open Skies

5.1 Introduction

5.1.1 The Treaty on Open Skies was signed on 24 March 1992 by 25 Countries, including the UK, to promote greater transparency in military activities and thereby enhance international security. The Treaty has now been expanded to include 27 Countries. To fulfil its obligations under the terms of the Treaty, the UK is committed to accept Observation Flights by Observation Teams from any of the signatory Countries over any part of UK territory, including Controlled Airspace.

5.2 Observation Flights

5.2.1 Observation Flights may be conducted from either of the following two Open Skies Aerodromes:
(a) RAF Brize Norton;
(b) RAF Leuchars.

5.2.1.1 Occasionally, an Observation Flight may over fly several Western European Union (WEU) Countries in one mission – Combined Observation Flight. In this case, the Observation Flight could commence and/or end within the UK, or merely over fly the UK during the mission. Additionally, a refuelling stop within the UK may be required.

5.2.1.2 The aircraft used during the Observation Flight may be provided by either the UK or the visiting Country. In either case, a UK Flight Monitor will always be available on the flight deck to act as an interface with ATC agencies.

5.2.2 Although Her Majesty's Government will receive at least 3 days notice of the arrival of an Observation Team within the UK, the intended route and profile of the Observation Flight (Mission Plan) will not be known until approximately 24 hours prior to commencement. Upon receipt of the Mission Plan, the Airspace Utilisation Section (AUS) will initiate any Danger Area closure action and notify details of the route to the appropriate ATC agencies by means of an Airspace Co-ordination Notice (ACN). AUS will also take NOTAM action to notify other agencies and airspace users.

5.2.3 Under the terms of the Treaty, aircraft undertaking Observation Flights are to be afforded due priority over other aircraft. Observation Flights within UK airspace are therefore granted Category B Status (as detailed in MATS Part 1, Section 1, Chapter 4, para 9) when within Controlled Airspace and are to be afforded priority over all other aircraft (except those in emergency) when outside Controlled Airspace.

6 Special Flights

6.1 Introduction

6.1.1 Special Flight Notifications (SFNs) can be applied to a variety of special aerial tasks which may take place throughout an extended period of time. The most common are Police Authority Air Support Unit (ASU) and Air Operations Unit (AOU) flights, Helicopter Emergency Medical Service (HEMS) flights and HM Government sponsored flights, (including Ministry of Defence and other flights). The nature of SFN flights is such that they will often require to be afforded priority over most other flights.

6.1.2 The purpose of an SFN is to ensure that those ATC agencies likely to provide services to the subject aircraft are aware of any special handling requirements, and that aircraft operators are aware of the conditions under which priority over most other flights is afforded.

6.2 Content

6.2.1 SFNs will contain details of:
(a) The purpose of the subject flight.
(b) The priority(ies) of the subject flight, to be in accordance with flight categories defined at CAP 493 Manual of Air Traffic Services Pt 1 Section 1 Chapter 4 and as authorised by the CAA as follows:
(1) Police Authority ASU and AOU flights – Category A (Police Emergencies), Category B (the normal operational priority)
or Category Z (training, test and other flights involving Police Authority aircraft).
(2) HEMS flights – A, E or Z as described in AIC 99/2005 (Yellow 186).
(3) HM Government sponsored flights, subject to the nature of the activity but usually Category B for special surveys. Category E applies to time-critical test and training flights, Category Z to all routine training, test and other flights.
(c The period(s) during which flights may take place.
(d) The period of validity of the SFN.
(1) A Special Flight Notification should normally be issued for no longer than twelve months from the date of issue.
(2) Exceptionally, subject to the approval of the NATS SFN Co-ordinator and the CAA, this period may be extended to no longer than fifteen months from the date of issue when the period of operation of the flight is expected to be longer than twelve, but no longer than fifteen months from the date of issue and/or to assist in the timely administration of notices of renewal.
(e) The name of the operator and the type of aircraft.
(f) The callsign(s) to be used and the main operating base(s).
(g) The routine operating area of the activity.
(h) The operating level (or levels), where appropriate to include minimum and/or maximum levels and/or level bands.
(i) The minimum weather criteria required for the particular operation.
(j) Points of contact for the operator and ATC agencies responsible for the area in which the flight is to be undertaken.
(k) The rules under which the aircraft captain is to operate the aircraft.
(1) Available communications and (if applicable) discrete SSR codes to be used.
(m) Pre-flight notification and co-ordination requirements for the flight, to include the minimum pre-notification period where appropriate or possible.

(n) The action by the aircraft captain in the event of loss of communications (AIP ENR 1.1.3 refers to basic national procedures).
(o) Any special considerations.

6.3 Promulgation
6.3.1 Responsibility for the drafting, promulgation and distribution of SFNs is vested in the NATS SFN Co-ordinator. All queries concerning, and requests for, Special Flight Notification are to be submitted to the:
NATS SFN Co-ordinator
Room 3319/Box 9
National Air Traffic Services Ltd
London Area Control Centre
Sopwith Way,
Swanwick
Hants
S031 7AY
Tel: 01489 612030/612590
Fax: 01489 612430
e-mail: special.flights@nats.co.uk

6.4 Enquiries
6.4.1 All enquiries concerning SFN policy may be addressed to the CAA at:
Terminal Airspace
Directorate of Airspace Policy
K6 CAA House
45-59 Kingsway
London
WC213 6TE
Fax: 020 7453 6565
E-mail: terminal.airspace@dap.caa.co.uk

ENR 1.1.5 AIRSPACE RESTRICTIONS, DANGER AREAS AND HAZARDS TO FLIGHT

1 Airspace Restrictions

1.1 Restriction of Flying Regulations

1.1.1 The Secretary of State for the Department for Transport (DfT) is empowered under Article 96 of the Air Navigation Order (ANO) to make regulations prohibiting, restricting or imposing conditions on flight by civil aircraft in United Kingdom airspace and by any United Kingdom registered civil aircraft in any other airspace within which the United Kingdom, under international arrangements, has undertaken to provide navigational services to aircraft. Restriction of Flying Regulations are made only when the Secretary of State deems it necessary in the public interest.

1.1.2 Prohibited Area – An airspace of defined dimensions within which the flight of aircraft is prohibited.

1.1.3 Restricted Area – An airspace of defined dimensions within which the flight of aircraft is restricted in accordance with certain specified conditions.

1.1.4 Prohibited and Restricted Areas established under these Regulations may be temporary or permanent. When time permits, details of temporary Prohibited and Restricted Areas are promulgated by Supplements to the UK AIP or AIC but in the case of Emergency Restriction of Flying Regulations (see paragraph 1.2) the information will be promulgated by NOTAM. Permanent Prohibited and Restricted Areas are tabulated at ENR 5.1.1.1/2 and 5.1.2.1/8.

1.2 Emergency Restriction of Flying Regulations

1.2.1 An Emergency Controlling Authority (ECA) may seek to inhibit flight in the vicinity of an emergency incident on land or at sea within the United Kingdom Flight Information Regions if it considers it essential for the safety of life or property and particularly for the protection of those engaged in Search and Rescue action.

1.2.2 Depending upon the nature of the incident the initial action will normally be the establishment of a Temporary Danger Area (see paragraph 1.3.1) notified by NOTAM. However, if a Temporary Danger Area fails to meet the objective or is deemed to be inappropriate for a particular incident, Emergency Restriction of Flying Regulations may be introduced. The Regulations make it an offence to fly within the designated Temporary Restricted Area without the permission of the appropriate ECA. Notification of the coming into force of Emergency Restriction of Flying Regulations and details of the Temporary Restricted Area will be made by NOTAM and at the same time any previously established Temporary Danger Area will be withdrawn.

1.2.3 The ECA is the only authority which may grant permission for aircraft to be flown within the notified airspace. Subject to overriding considerations of safety, flights by aircraft directly associated with the emergency will invariably be given priority over those seeking to overfly for any other reason.

1.3 Danger Area – Airspace which has been notified as such within which activities dangerous to the flight of aircraft may take place or exist at such times as may be notified.

1.3.1 Areas within which activities dangerous to the flight of aircraft may take place or exist during the promulgated 'Hours of Activity (UTC)'. See ENR 5.1.3.1/22 column 3.

1.3.1.1 Danger Areas encompass for example; weapon ranges, including test and practice ranges for all types of weapons (guns, bombs, aircraft cannons and rockets etc) aerial combat training, parachutist training and demolition areas. It is emphasised that only the types of hazardous activities most likely to be encountered are listed. Areas will not be reserved for one type of activity only and various hazards may be encountered in one area simultaneously. Pilots are warned that, in addition to the hazards already mentioned, military aircraft may be towing targets with cable lengths which, although normally 6000ft, may extend to 24000ft. The target itself may be anything up to 2500ft below the towing aircraft and therefore the combination of towing aircraft, cable and target presents a considerable hazard. Pilots are reminded that aircraft in the towing configuration have right of way over other converging powered aircraft under the provisions of the Rules for Avoiding Aerial Collisions and pilots must realise that, although the cable and target may not be immediately apparent, this does not absolve them from giving way to the towing aircraft. The potential hazards of flying through active Danger Areas cannot be over stressed.

1.3.1.2 In the immediate vicinity of Danger Areas in which military aircraft operate many of those aircraft fly arrival, holding and departure patterns. Pilots of itinerant aircraft flying close to Danger Areas are advised to keep an especially sharp lookout for such aircraft and, by taking any necessary evasive action (unless the Rules for avoiding aircraft collisions require otherwise) in good time, permit them to continue their manoeuvres.

1.3.1.3 Byelaws. Unauthorised entry into many Danger Areas is prohibited within the Period of Activity of the Danger Area as listed at 5.1.3.1 to ENR 5.1.3.1/22 by reason of Bye-laws made under the Military Lands Act 1892 and associated legislation. For those Danger Areas where Bye-laws which prohibit entry apply, the Remarks column (3) of 5-1-3-1 to ENR 5.1.3.1/22 includes the year and number of the relevant Statutory Instruments (SI).

1.3.1.4 ENR 5.1.3.1/22 contains details only of those UK Danger Areas which have an upper limit in excess of 500ft above ground level. There are many ranges (rifle, small arms etc) with upper limits of 500ft or less above ground level, see paragraph 3.1 Small Arms Ranges and details as listed at ENR 5.3.1.1/4. Pilots should therefore satisfy themselves that they are clear of such Small Arms Ranges when flying at or below 500ft.

1.3.1.5 Temporary Danger Areas may be established at short notice around the scene of emergency incidents when it is considered that the activity associated with the incident could be hazardous to flight (see paragraph 1.2).

1.3.2 Danger Area Crossing Service

1.3.2.1 A Danger Area Crossing Service (DACS) is an inflight service available for over 24% of UK Danger Areas. Details of unit contact frequencies and availability are given for the applicable areas under the 'Remarks' Column 3 at ENR 5.1.3.1/22 and on the legend to chart ENR 6-5-1-1 (United Kingdom Airspace Restrictions and Hazardous Areas). The contact frequencies are also printed on the 1:500 000 UK ICAO Aeronautical Charts legends.

1.3.3 Danger Area Activity Information Service

1.3.3.1 A Danger Area Activity Information Service (DAAIS) is an inflight service available for over 68% of UK Danger Areas. For a few Danger Areas this includes periods of activity outside the hours of availability of a DACS.

1.3.3.2 The purpose of the DAAIS is to enable pilots to obtain, via a Nominated Service Unit (NSU), an airborne update of the activity status of a participating Danger Area whose position is relevant to the flight of the aircraft. Such an update will assist pilots in deciding whether it would be prudent, on flight safety grounds, to penetrate the area. It is strongly emphasised that information obtained from an NSU is only pertinent to the ACTIVITY STATUS of a Danger Area and is not a clearance to cross that Danger Area, whether or not it is active. The DAAIS does not absolve pilots from the responsibility of obtaining as much information as possible on a relevant Danger Area by existing methods of promulgation, as part of normal pre-flight briefing procedures. Details including frequencies of NSUs providing a DAAIS are tabulated in the 'Remarks' Column 3 of ENR 5.1.3.1/22 and on the legend to chart ENR 6-5-1-1 (United Kingdom Airspace Restrictions and Hazardous Areas). The contact frequencies are printed on the legend of the 1:500 000 UK ICAO Aeronautical Charts.

1.3.3.3 To obtain a DAAIS, pilots should call the appropriate NSU on the relevant frequency using the following phraseology: '(NSU callsign) this is (aircraft callsign) request activity status of (Danger Area)'. The reply from the NSU will depend upon:

(a) The promulgated activity status of the Danger Area;

(b) the actual state of activity at the time of call. Generally the reply will be: '(Aircraft callsign) this is (NSU callsign) (Danger Area) is active/not active'. The reply may be qualified by a statement indicating when or for what period of time the area will be active or when any temporary activity may restart.

1.3.3.4 If there is no reply from a NSU which is being called for information on activity, pilots should assume that the relevant Danger Area is active.

1.3.3.5 None of the provisions of the DAAIS apply to aircraft operating on:

(a) Airways and Upper Air Routes and;

(b) Advisory Routes when participating in the Advisory Service; where such Airways and Routes cross Danger Areas. For these situations procedures exist which are specifically detailed in relevant ATC Unit instructions.

1.4 Pilotless Target Aircraft

1.4.1 Pilotless target aircraft are operated and manoeuvred within certain Danger Areas as indicated in the list at ENR 5-1-3-1/22. Pilotless target aircraft, of the Meteor and Jindivik type require the use of a runway for take-off and landing, are painted orange and red and may be flown day and night in all weather conditions. At night, standard navigation lights are displayed.

1.4.2 Within the EG D201, EG D201A and EG D201B Aberporth Danger Areas, pilotless target aircraft are operated under the radar control of MOD Aberporth. Within the EG D202 Llanbedr and EG D701E Hebrides Danger Areas, pilotless target aircraft are operated under the control of MOD Llanbedr.

1.4.3 Pilots should note that in the landing configuration on recovery into Llanbedr or Benecula Aerodrome the manoeuvrability of a pilotless target aircraft is below that which can be achieved by a manned aircraft. Care should be taken to avoid Danger Area EG D202 when active unless a crossing clearance has been obtained from Llanbedr Radar on 122.500 MHz or 386.675 MHz. Similarly, EG D701E should be avoided if active, a Danger Area Activity Information Service is available from Scottish Information on 127.275 MHz.

2 Hazards to Flight

2.1 Military Training Area (MTA) or Military Temporary Reserved Airspace (MTRA) – An area of Upper Airspace of defined dimensions within which intense military flying training takes place.

2.1.1 In the Upper Airspace, intense military flying training normally takes place in delineated Military Training Areas or Military Temporary Reserved Airspace. Because of the random nature of the activity within these areas it is not possible to provide civil air traffic control service in an MTA during the published hours of activity or in an MTRA during a booked period of activity within the published hours. Details are at ENR 5.2 and further information is contained at ENR 1.1.1.7, paragraph 4.3.7.

2.2 Area of Intense Air Activity (AIAA) – Airspace within which the intensity of civil and/or military flying is exceptionally high or where aircraft, either singly or in combination with others, regularly participate in unusual manoeuvres.

2.2.1 Intense civil and/or military air activity takes place within the areas listed in ENR 5.2. Pilots of non-participating aircraft who are unable to avoid AIAAs are to keep a good lookout and are strongly advised to make use of a radar service if available; these areas are depicted at ENR 6.5.2.1.

2.3 Aerial Tactics Area – Airspace of defined dimensions designated for air combat training within which high energy manoeuvres are regularly practised by aircraft formations.

2.3.1 Air combat training by military aircraft practising high energy manoeuvres regularly takes place in the areas listed in ENR 5.2. Pilots unable to avoid these areas are strongly advised to make use of a radar service; these areas are depicted at ENR 6.5.2.1.

2.4 Air-to-Air Refuelling Area (AARA) – Airspace of defined dimensions within which air-to-air-refuelling takes place under radar service.

2.4.1 Areas in which air-to-air refuelling under radar service takes place are listed in ENR 5.2. Refuelling aircraft will not necessarily conform with the Quadrantal/Semi-circular Flight Rules and are unable to take rapid avoiding action.

2.5 Boscombe Down Advisory Radio Area (As depicted at ENR 6.5.2.1)

2.5.1 Test flight aircraft are routinely flown from MOD Boscombe Down in the Advisory Radio Area as shown at ENR 5.2.5. A test profile involves manoeuvres that are required to take place overland but which may place the aircraft at the limits of its flight envelope. Consequently, whilst the test pilot remains responsible for the safe conduct of the flight, there could be occasions when the pilot would be unable to manoeuvre the aircraft in compliance with the Rules of the Air.

2.5.2 Pilots of other aircraft flying in the area are strongly advised to call Boscombe Down (ENR 5.2.5), who will provide pilots with information on any relevant test flight activity and, if requested, advice on arranging a detour of the test area.

2.5.3 Participation in the Advisory Radio Area, which is highly recommended, is designed to enhance flight safety. It does not afford any form of increased separation or right of way for the test flights and is not intended to inhibit the passage of other aircraft in the area.

2.6 UK Military Low Flying System

2.6.1 Military low flying occurs in most parts of the United Kingdom at any height up to 2000ft above the surface. However, the greatest concentration is between 250ft and 500ft and civil pilots are advised to avoid flying in that height band whenever possible.

2.6.2 Military aircraft are considered to be low flying when:

(a) Fixed wing aircraft, except light propeller-driven aircraft, are flying below 2000ft above the surface;

(b) Light propeller-driven aircraft and helicopters are flying below 500ft above the surface.

2.6.3 Military helicopter operations in the Salisbury Plain Area

2.6.3.1 A considerable number of helicopters operate to and from the military establishments in, and around, the Salisbury Plain Area.

2.6.3.2 In addition to the intensive daytime activities, military helicopters may be encountered operating during the hours of darkness without, or with restricted, navigation lights within the area enclosed by the following co-ordinates: 513000N 0014200W – 513600N 0011336W thence anti-clockwise by an arc of a circle radius 5nm centred on 513654N 0010543W – 513324N 0010000W – 513000N 0010000W – 513000N 0010600W – 512400N 0010600W – 511821N 0010036W thence clockwise by an arc of a circle radius 5nm centred on 511403N 0005634W – 511114N 0005000W – 505336N 0005000W – 505654N 0011305W – 510115N 0011039W thence anti-clockwise by an arc of a circle radius 8nm centred on 505701N 0012124W (EGHI ATZ) – 510459N 0012017W – 510123N 0012722W – 505512N 0013047W – 505003N 0020205W – 505027N 0020549W – 505718N 0021200W – 511109N 0021749W – 512036N 0020922W – 512224N 0020257W – 512909N 0014402W – 513000N 0014200W.

2.6.4 Geographical details of military low flying activities within the United Kingdom are shown on the chart ENR 6.5.2.1 (which is updated annually by re-issue), copies of which may be obtained from:

CAA Chart Sales (AFE)
Unit 1a Ringway Trading Estate
Shadowmoss Road
Manchester
M22 5LH
Tel: 0161 499 0023
Fax: 0161 499 0298

2.7 Parachute Flares and Other Illuminants

2.7.1 For night training, ground illuminating parachute, rocket, mini and pistol flares may be launched during the hours of darkness from:

(a) Sandhurst – 512147N 0004635W;

(b) Castlelaw within the area bounded by straight lines joining the following co-ordinates: 555355N 0031651W – 555357N 0031301W – 555113N 0031646W – 552133N 0031256W – 555505N 0031651W;

(c) Various inland sites between Dover and Folkestone;

(d) Pippingford Park Training Area within a circle radius 1200m centred on 510346N 0000404E (excluding parachute flares).

2.7.2 Parachute flares reach a maximum height of 1000ft agl and burn with a bright white light for approximately 30 seconds. Rocket flares do not exceed 300ft agl and the mini and pistol flares reach a height of approximately 100ft agl. The non-parachute flares only burn for a few seconds.

3 Activities of a Dangerous Nature

3.1 Small Arms Ranges

3.1.1 Small arms ranges in the UK with a vertical hazard height of 500ft agl do not attract UK Danger Area Status. However, firing at some ranges can take place across open areas of ground over which an aircraft might legally be flown below 500ft agl.

3.1.2 Listed at ENR 5.3 are the details of the small arms ranges notified to the Authority which might pose a hazard to flight below 500ft agl. The small arms ranges may be in use at any time and pilots are strongly advised to avoid these areas. The list includes small arms ranges, located within the lateral boundaries of UK Danger Areas, which may be in use outside the activity hours of these Danger Areas.

3.2 High Intensity Radio Transmission Area (HIRTA) – Airspace of defined dimensions within which there is radio energy of an intensity which may cause interference with and on rare occasions damage to communications and navigation equipment.

3.2.1 Areas within which there is radio energy of an intensity which could cause interference with and on rare occasions, cause damage to, communications and navigation equipment such as Radio Altimeter, VOR, ILS and Doppler are listed at ENR 5.3. The intensity may be sufficient to detonate electrically initiated explosive devices carried or fitted in aircraft.

3.2.2 Only the most significant sources are listed and in some of these areas the intensity of the radio energy may be such that it would be injurious to remain for more than one minute in the immediate vicinity of the energy source. This is especially relevant to helicopter operations and the list contains appropriate warnings; however it would be prudent for helicopter pilots to avoid lingering closer than 100m to any radar aerial. Pilots approaching oil production platforms on which dish aerials can be observed should, wherever possible, approach from a direction out of the general line-of-shoot of such aerials.

3.2.3 Airborne Early Warning (AEW) aircraft operate within United Kingdom airspace and due to possible radiation hazards, all aircraft should maintain a minimum separation of 1000m lateral and 1000ft vertical from such aircraft. AEW aircraft can be identified as follows:

(a) RAF/NATO/USAF E-3 – a Boeing 707 with a large rotodome mounted on the upper fuselage (E-3 Orbit Areas are listed at ENR 5.3);

(b) USN E-2C – a medium size twin turboprop with a four-finned cantilever tail and a large rotodome mounted on the upper fuselage.

3.3 Gas Venting Operations

3.3.1 Severe turbulence and power fluctuations in turbine engines could be experienced over gas venting sites during venting of natural (methane) gas under high pressure. Locations of gas venting sites are listed at ENR 5.3.

3.4 Laser Sites

3.4.1 Laser sites, as listed at ENR 5.3, are locations where laser sources are located permanently and which have been notified to the Airspace Utilisation Section. Only those sites which radiate sufficient power to cause distraction or eye damage, and which intentionally emit laser beams into airspace or are likely to in the event of a malfunction, are included.

3.5 Radiosonde Balloon Ascents

3.5.1 The Met Office releases helium or hydrogen filled balloons from a number of locations throughout the United Kingdom which are listed at ENR 5.3. These balloons carry a small radio transmitter which sends back atmospheric information about temperature, pressure and humidity; by way of a tracking system the balloons also provide data on wind speed and direction at various levels. A typical installation consists of a balloon, diameter at launch approximately 1.5 metres, to which is attached a small parachute. The radiosonde is attached underneath the parachute on a suspension string of approximately 33 metres in length. The distance the balloons travel away from the launch site is dependant on the wind strength, but they can attain altitudes of up to 80000ft.

3.5.2 Balloon launches from all other sites by organisations and members of the public require written permission from the CAA in accordance with the Air Navigation Order before releasing meteorological balloons into notified airspace. Article 97 specifies the requirements for notification and permission for the launch of balloons; such permission may be conditional. Organisations and members of the public wishing to obtain permission for the above activity shall contact the Airspace Utilisation Section (AUS) at Directorate of Airspace Policy, K7, CAA House, 45-59 Kingsway, London WC2B 6TE, at least five working days in advance, to allow AUS to take appropriate notification action.

3.5.3 Radiosondes, minus the balloon, may also be air dropped; this activity will be promulgated by NOTAM.

4 Air Navigation Obstacle

4.1 Land-Based Air Navigation Obstacles

4.1.1 In the United Kingdom a land-based 'Air Navigation Obstacle' is defined as any building or work, including waste heaps, which attains or exceeds a height of 300ft agl. Details of those obstacles of which the Civil Aviation Authority has been informed are listed in ENR 5.4. In cases where a number of structures form the obstacle, the position of the highest is given. In the case of masts, the position of the centre of the mast is given (but it should be noted that the stays or guys may spread out for a considerable distance).

4.2 Aerodrome Obstacles

4.2.1 An Aerodrome obstacle is one that is located on an area intended for the surface movement of aircraft or that extends above a defined surface intended to protect aircraft in flight. Obstacle clearance surfaces can extend up to 15km from the runway thresholds. Such aerodrome obstacles are listed in the AD Section and are shown on Instrument Approach and Landing Charts where these have been published. The method of lighting Aerodrome Obstacles is detailed in Civil Aviation Publication CAP 168 and is briefly described in CAP 637 'Visual Aids Handbook' which is available from:

Documedia Solutions Ltd
37 Windsor Street
Cheltenham
Glos
GL52 2DG
Tel: 0870 8871410
Fax: 0870 8871411

4.3 En-Route Obstacles

4.3.1 En-Route Obstacles are those located outside or beyond the areas detailed in paragraph 4.2.1. It is recommended that they should be lit if:

(a) they are 150 metres (492ft) agl or more in height;

(b) they are less than 150 metres (492ft) agl in height, but are by virtue of their nature or location considered never-the-less to present a significant hazard to air navigation.

4.3.2 Advice on the scale of lighting to be displayed may be obtained from:

Directorate of Airspace Policy
ORA K6 CAA House
45-59 Kingsway
London
WC2B 6TE
Tel: 020 7453 6545.

4.3.3 Details of unserviceability and return to service of lights on such obstacles, when notified to the UK AIS, will be promulgated by NOTAM. Air navigation obstacles with a height of less than 150 metres are sometimes lit, but details of unserviceability of lights on these obstacles are not normally promulgated; those which, for operational reasons, will be promulgated are shown on the relevant ENR 5.4 pages. Obstacles listed at ENR 5.4 annotated 'FLR' in Column 2 are those which burn off high pressure gas; the flame, which may not be visible in bright sunlight, can extend for 600ft.

4.3.4 Details of all air navigation obstacles known at the date of the chart's preparation are shown on certain Aeronautical Charts published by the Civil Aviation Authority. These charts indicate whether or not the obstacle is normally lighted. Operators should be aware that obstruction lighting on lit en-route obstructions is not necessarily located at the structures highest point.

4.4 Off-shore Air Navigation Obstacles

4.4.1 Numerous fixed installations associated with off-shore exploration of oil and gas from the Continental Shelf sea bed, exist within the United Kingdom Off-shore Concession Area and Flight Information Regions. A part of the United Kingdom Concession Area lies within the Norwegian Flight Information Region and parts of some foreign Concession Areas lie within the United Kingdom Flight Information Regions (ENR 2.2 refers). These fixed installations vary in height up to 541ft amsl and display navigation warning lights. Details of those installations of which the Civil Aviation Authority has been informed which attains or exceeds an elevation of 300ft amsl within the United Kingdom Flight Information Regions and Concession Areas within the Norwegian Flight Information Region are listed in ENR 5.4.2. Most of the installations are equipped with a helideck which comes within the definition of an aerodrome. **Many installations burn off high pressure gas and the flame, which may not be visible in bright sunlight, can extend for 600ft. Pilots should be aware that even if no flame is visible there is still danger from the venting of high pressure gas.**

4.4.2 Pilots should also be aware of high intensity radio transmissions from some installations (see paragraph 3.2).

4.5 Area Codes

4.5.1 The lists of land based obstacles show in Column 1 a combined Area Code and Reference Number. The first three digits of the number refer to the area in which the obstacle is sited. These areas are shown on the chart at ENR 6.5.4.1. The final three digits of the number are a unique reference number for that obstacle. Obstacles are listed in area groups and in descending order of latitude within areas.

4.6 Mountains and Hills with Warning Lights

4.6.1 Those conspicuous mountains and hills listed below are regarded as hazardous to aviation and are marked by red obstacle lights:

(a) Bohill Mt, Co Antrim 543600N 0060554W 871ft amsl;

(b) Bredon Hill, Worcs 520328N 0020304W 1046ft amsl;

(c) Church Hill, Co Fermanagh 542745N 0075403W 1126ft amsl.

5 Aerial Sporting and Recreational Activities

5.1 Glider Launching Sites

5.1.1 Glider launching may take place from designated sites which are regarded as aerodromes. The sites are listed at ENR 5.5. Where launching takes place within the Aerodrome Traffic Zone of an aerodrome listed within the AD section, details are also shown at AD 2 and AD 3.

5.1.2 Gliders may be launched by towing aircraft, or by winch and cable or ground tow up to a height of 2000ft agl. At a few sites the height of 2000ft may be exceeded (see paragraph 5.3).

5.1.3 Sites are listed primarily to identify hazards to other airspace users and listing does not imply any right for a glider or powered aircraft to use the sites.

5.2 Hang Gliding, Paragliding and Parascending Sites

5.2.1 Hang Gliding and/or parascending may take place from sites which, because of the low speed characteristics of hang-gliders, paragliders and parascenders and the difficulty of seeing them in certain conditions, are listed as hazards to other airspace users.

5.2.2 The locations of cable-launched hang/paragliding sites are listed at ENR 5.5. Foot launched activity sites are severely affected by wind speed and direction existing at the time. Although activity is usually at a peak during weekends, hang-gliding and/or parascending may take place at any time, particularly in the summer months. Airspace users should be aware that single or groups of soaring and motorised hang/para-gliders can be found flying anywhere within the open FIR up to 15000ft, and are therefore not listed.

5.2.3 At certain sites hang gliders and/or parascenders may be launched by winch/auto-tow and cables may be carried up to 2000ft agl. At a few sites the height of 2000ft may be exceeded (see paragraph 5.3). The cable launching of the aircraft may be encountered within the airspace contained in a circle radius 1.5nm of the notified position of the site.

5.3 Cable Launching of Gliders, Hang Gliders and Parascending Parachutes

5.3.1 The launching of gliders, hang gliders and parascending parachutes by winch and cable or by ground tow to above 200ft (60 m) agl requires permission in writing under Article 86 of the Air Navigation Order from the Civil Aviation Authority.

5.3.2 At sites where cable launching is permitted, cables may be carried up to heights of 2000ft agl. At a few sites the heights of 2000ft may be exceeded. It is a condition of the permission that when cable launching is taking place, a white ground conspicuity signal as described in Rule 44(8) of the Rules of the Air Regulations 1996 shall be displayed.

5.3.3 Sites which have permission to cable launch above 200ft agl are listed at ENR 5.5.

5.4 Free-fall Parachuting Drop Zones

5.4.1 Intensive free-fall parachuting may be conducted up to FL 150 at any of the Drop Zones listed at ENR 5.5 and in several Danger Areas. Listing of a Drop Zone does not imply any right to a parachutist to use that Drop Zone. Some Government and licensed aerodromes where regular parachuting takes place are included in the list but parachuting may also take place during daylight hours at any Government or licensed aerodrome. Drop Zone activity information may be available from certain Air Traffic Service Units (ATSUs) but pilots are advised to assume a Drop Zone is active if no information can be obtained.

5.4.2 Parachuting also takes place at temporary sites, eg for display purposes, and will normally be notified by NOTAM as Temporary Navigation Warnings. Night parachuting may take place at any Drop Zone: Club Chief Instructors will notify in writing all forthcoming night parachuting, at least five working days in advance, to the Airspace Utilisation Section (AUS) at Directorate of Airspace Policy, K1, CAA House, 45-59 Kingsway, London WC2B 6TE, to allow AUS to take appropriate notification action.

5.4.3 Visual sighting of free-falling bodies is virtually impossible and the presence of an aircraft within the Drop Zone may be similarly difficult to detect from the parachutists' point of view. Parachute dropping aircraft and, on occasions, parachutists may be encountered outside the notified portion of airspace. Pilots are strongly advised to give a wide berth to all such Drop Zones where parachuting may be taking place.

5.4.4 Where permission is obtained for drops within Controlled Airspace, dropping aircraft are to have serviceable SSR with Mode C.

5.5 Microlight Flying sites

5.5.1 Those Microlight Flying Sites where flying is known to take place are listed at ENR 5.5 and are regarded as aerodromes. Sites are listed primarily as hazards to other airspace users and the listing does not imply any right for aircraft to use the sites. Microlight aircraft might be encountered at sites not included in the listing (See also AD Section).

5.6 Captive and Free Flight Manned Balloon Launch Sites

5.6.1 Frequent launchings of manned balloons take place at or near:

(a) Ashton Court, Bristol, Avon – 512639N 0023825W;

(b) Bath, Avon (Several sites in or around Bath).

(c) Marsh Benham, Newbury, Berks – 512310N 0013852W (Several sites in or around Newbury).

5.6.2 Flights by captive passenger carrying balloons will take place at:
Bournemouth Lower Gardens – 504308N 0015238W. Active daily 1000-2245 (Winter), one hour earlier during the Summer up to 488ft agl;

5.7 Kites

5.7.1 High flying kites may be hazardous to aircraft because of the possibility of collision with the towline. It is known that kites are flown higher than 200ft (60m) agl at: Graves Park, Sheffield, S Yorks – 531958N 0012811W. Up to 1200ft agl/1860ft amsl (Site elevation 660ft amsl).

6 Other Temporary Hazards

6.1 Hazards of a temporary nature will be notified, whenever time permits, by NOTAM as Temporary Navigation Warnings.

6.2 Activity of a hazardous nature may occur without notification within the Aerodrome Traffic Zones of active aerodromes not normally available to civil aircraft (see ENR 2.2).

7 Keevil Aerodrome

7.1 In addition to its use as a glider site and free-fall parachuting drop zone (ENR 5.5.3.2 refers), Keevil Aerodrome 511850N 0020643W, is used extensively as a military dropping zone by Hercules aircraft engaged in parachute heavy supply dropping. This activity may take place at any time, day or night and pilots are advised to avoid the aerodrome by 2nm laterally or 2000ft vertically whenever possible.

ENR 1.2 VISUAL FLIGHT RULES

1 VFR Flight

1.1 VFR flights shall be conducted so that the aircraft is flown in conditions of visibility and distance from clouds equal to or greater than those specified in Table 1.

Table 1

Airspace Class	B		C, D or E			F or G	
	FL100 or above	Below FL100	FL100 or above	Below FL100	FL100 or above	Below FL100	
Distance from Cloud	Clear of cloud	Clear of cloud	1500m Horizontally and 1000ft vertically	1500m Horizontally and 1000ft vertically (1)	1500m Horizontally and 1000ft vertically	1500m Horizontally and 1000ft vertically	
Flight visibility	8km	5km	8km	5km (2)	8km	5km (3)	

Notes:
(1) Or if at 3000ft or below and flying at 140kt or less: Clear of Cloud and in Sight of the Surface.
(2) Or if a Helicopter and flying at 3000ft or below: Clear of Cloud and in Sight of the Surface.
(3) Or if at 3000ft or below:
either: any aircraft flying at more than 140kt: Clear of Cloud and in Sight of the Surface in a Flight Visibility of 5km.
or: any aircraft flying at 140kt or less: Clear of Cloud and in Sight of the Surface in a Flight Visibility of 1500m.
or: helicopters flying at a reasonable speed for the actual visibility:
Clear of Cloud and in Sight of the Surface.

1.2 For the purposes of an aeroplane taking off from or approaching to land at an aerodrome within Class B, C or D Airspace, the visibility, if any, communicated to the commander of an aeroplane by the appropriate air traffic control unit shall be taken to be the flight visibility for the time being.

1.3 The minimum heights at which aircraft may be flown are detailed in Rule 5 of the Rules of the Air Regulations 1996, as amended.

1.4 Except where otherwise indicated in air traffic control clearances or specified by the appropriate ATS authority, it is not mandatory in the United Kingdom for VFR flights in level cruising flight when operated above 3000 ft (900 m) from the ground or water, or a higher datum as specified by the appropriate ATS authority, to adopt any particular cruising level system. Such flights are advised to adopt the table of cruising levels for IFR flights as given at ENR 1.7, paragraph 6.1 (b).

1.5 VFR flights shall comply with the provisions of ICAO Annex 2, paragraph 3.6, when operating in Classes B, C and D Airspace. The United Kingdom Regulations relating to VFR Flight Plan and Air Traffic Control Clearances are detailed in Rule 27 of the Rules of the Air Regulations 1996, as amended.

Note: A Special VFR clearance may be requested without the submission of a filed flight plan. Brief details of the proposed flight should be passed to the appropriate Air Traffic Control Unit.

1.6 ICAO Annex 2 precludes authorisation for VFR flights to operate above FL 290 where a vertical separation minimum of 300m (1000ft) is applied above FL 290. Therefore, for aircraft operating as General Air Traffic (GAT), VFR flights shall not be authorised within the London and Scottish UIRs above FL 290, as described in ENR 2.1

2 Special VFR Flight

2.1 Clearance for Special VFR flight in the UK is an authorization by ATC for a pilot to fly within a Control Zone although he is unable to comply with IFR. In exceptional circumstances, requests for Special VFR flight may be granted for aircraft with an all-up-weight exceeding 5700kg and capable of flight under IFR. Special VFR clearance is only granted when traffic conditions permit it to take place without hindrance to the normal IFR flights, but for aircraft using certain notified lanes, routes and local flying areas see paragraph 2.2. Without prejudice to existing weather limitations on Special VFR flights at specific aerodromes (as detailed within the AD 2 Section) ATC will not issue a Special VFR clearance to any fixed-wing aircraft intending to depart from an aerodrome within a Control Zone, when the official meteorological report indicates that the visibility is 1800m or less and/or the cloud ceiling is less than 600ft.

2.2 Aircraft using the access lanes and local flying areas notified for Denham, White Waltham and Fairoaks in the London CTR and any temporary Special Access Lanes which may be notified from time to time will be considered as Special VFR flights and compliance with the procedures published for the relevant airspace will be accepted as compliance with ATC clearance. Separate requests should not be made nor will separate clearances be given. Separation between aircraft which are using such airspace cannot be given, and pilots are responsible for providing their own separation from other aircraft in the relevant airspace.

2.3 When operating on a Special VFR clearance, the pilot must comply with ATC instructions and remain at all times in flight conditions which enable him to determine his flight path and to keep clear of obstacles. Therefore, it is implicit in all Special VFR clearances that the aircraft remains clear of cloud and in sight of the surface. It may be necessary for ATC purposes to impose a height limitation on a Special VFR clearance which will require the pilot to fly either at or not above a specific level

2.4 A full flight plan, Form CA48/RAF2919, is not required for Special VFR flight but ATC must be given brief details of the call sign, aircraft type and pilots intentions. These details may be passed either by RTF or, at busy aerodromes, through the Flight Clearance Office. A full flight plan must be filed if the pilot wishes the destination aerodrome to be notified of the flight.

2.5 Requests for Special VFR clearance to enter a Control Zone, or to transit a Control Zone, may be made to the ATC authority whilst airborne. Aircraft departing from aerodromes adjacent to a Control Zone boundary and wishing to enter may obtain Special VFR clearance either prior to take-off by telephone or by RTF when airborne. In any case, all such requests must specify the ETA for the selected entry point and must be made 5-10 minutes beforehand.

2.6 ATC will provide standard separation between all Special VFR flights and between such flights and other aircraft under IFR. However, pilots with a Special VFR clearance should note that they cannot be given separation from aircraft flying in the lanes, routes and local flying areas detailed in paragraph 2.2; nor from aircraft flying in any temporary Special Access Lanes which may be notified from time to time.

2.7 A Special VFR clearance within a Control Zone does not absolve the pilot from the responsibility for avoiding an Aerodrome Traffic Zone unless prior permission to penetrate the ATZ has been obtained from the relevant ATC Unit.

2.8 Because Special VFR flights are made at the lower levels, it is important for pilots to realise that a Special VFR clearance does not absolve them from the need to comply with the relevant low flying restrictions of Rule 5 of the Rules of the Air Regulations 1996 (other than the 1500ft rule where the clearance permits flight below that height). In particular, it does not absolve pilots from the requirement that an aircraft, other than a helicopter, flying over congested areas must fly at such a height as would enable it to clear the area and alight without danger to persons or property on the ground in the event of an engine failure and that a helicopter, whether flying over a congested area or not, must fly at such a height as would enable it to alight without danger to persons or property on the ground in the event of an engine failure. In addition there are special rules applicable to flight by helicopters over London (see AD 2- EGLL, AD 2.22, paragraph 15).

2.9 Radio Communication Failure Procedures

2.9.1 The procedures to be adopted by pilots experiencing two-way radio communication failure are:

(a) If the aircraft is suitably equipped, operate the Transponder on Mode A, Code 7600 and Mode C;

(b) if it is believed that the radio communication transmitter is functioning, transmit blind giving position reports and stating intentions;

(c) if, when radio communication failure occurs, the aircraft is not yet in the CTR, the pilot must in all cases remain clear even if Special VFR clearance has been obtained;

(d) if Special VFR clearance has been obtained and the aircraft is in the CTR when the radio communication failure occurs, proceed as follows:

(i) Aircraft inbound to an aerodrome in the CTR – proceed in accordance with Special VFR clearance to the aerodrome and land as soon as possible. When in aerodrome traffic circuit watch for visual signals;

(ii) Aircraft transiting a CTR – continue flight not above the cleared altitude to leave the CTR by the most direct route, taking into account weather limitations, obstacle clearance and areas of known dense traffic.

Note: In (i) and (ii), if flying on a heading advised by radar, when radio communication failure occurs, resume own navigation and carry out the appropriate procedure described. In all cases, notify the ATC Unit concerned as soon as possible after landing.

ENR 1.3 INSTRUMENT FLIGHT RULES

1 IFR Flight
ICAO Annex 2 Rules as applied within UK Airspace and incorporating UK Differences (GEN 1.7 refers).

1.1 Aircraft Equipment

1.1.1 Aircraft shall be equipped with suitable instruments and with navigation equipment appropriate to the route to be flown.

1.2 Minimum Levels

1.2.1 Except when necessary for take-off or landing, or except when specifically authorised by the appropriate authority, an IFR flight shall be flown at a level which is at least 1000ft (300m) above the highest obstacle located within 5nm (9.25km) of the estimated position of the aircraft; except that the United Kingdom regulations do not apply to an aircraft operating under IFR and flying at an altitude not exceeding 3000ft (900m) if that aircraft is clear of cloud and in sight of the surface.

Note 1: The estimated position of the aircraft will take account of the navigational accuracy which can be achieved on the relevant route segment, having regard to the navigational facilities available on the ground and in the aircraft.

Note 2: See also ICAO Annex 2, paragraph 3.1.2 and GEN 1.7 Differences.

Note 3: The United Kingdom has no statutory requirements relating specifically to minimum IFR altitude when operating over high terrain or mountainous territory.

1.3 Change from IFR flight to VFR flight

1.3.1 An aircraft electing to change the conduct of its flight from compliance with the instrument flight rules to compliance with the visual flight rules shall, if a flight plan was submitted, notify the appropriate air traffic services unit specifically that the IFR flight is cancelled and communicate thereto the changes to be made to its current flight plan.

1.3.2 When an aircraft operating under the instrument flight rules is flown in or encounters visual meteorological conditions it shall not cancel its IFR flight unless it is anticipated, and intended, that the flight will be continued for a reasonable period of time in uninterrupted visual meteorological conditions.

2 Rules applicable to IFR flights within Controlled Airspace

2.1 IFR flights shall comply with the provisions of ICAO Annex 2, paragraph 3.6 when operated in controlled airspace.

2.2 As specified in the ICAO EUR Regional Supplementary Procedures (DOC 7030/4-EUR), flights shall be conducted in accordance with Instrument Flight Rules when operated within or above the EUR RVSM airspace. Therefore, flights operating as General Air Traffic (GAT) within the London and Scottish UIRs at or above FL 290, as described in ENR 2.1, shall be conducted in accordance with the Instrument Flight Rules.

2.3 An IFR flight operating in cruising flight in controlled airspace shall be flown at a cruising level, or, if authorised to employ cruise climb techniques, between two levels or above a level, selected from the Tables of cruising levels at ENR 1.7 paragraph 6.1 (a), except that the correlation of levels to track prescribed therein shall not apply whenever otherwise indicated in air traffic control clearances or specified by the appropriate ATS authority in Aeronautical Information Publications.

3 Rules applicable to IFR flights outside Controlled Airspace

3.1 Cruising Levels
3.1.1 An IFR flight operating in level cruising flight outside of controlled airspace shall be flown at a cruising level appropriate to its track as specified in the Tables of cruising levels at ENR 1.7 paragraph 6.1 (b).
Note: This provision does not preclude the use of cruise climb techniques by aircraft in supersonic flight.

3.2 Communications
3.2.1 An IFR flight operating outside controlled airspace but within or into areas, or along routes, designated by the appropriate ATS authority in accordance with ICAO Annex 2, paragraph 3.3.1.2 (a) or (b), shall maintain a listening watch on the appropriate radio frequency and establish two-way communication, as necessary, with the air traffic services unit providing flight information service.
Note: See note following ICAO Annex 2, paragraph 3.6.5.1.

3.3 Position Reports
3.3.1 An IFR flight operating outside controlled airspace and required by the appropriate ATS authority to: submit a flight plan, maintain a listening watch on the appropriate radio frequency and establish two-way communication, as necessary, with the air traffic services unit providing flight information service, shall report position as specified in ICAO Annex 2, paragraph 3.6.3 for controlled flights.
Note: Aircraft electing to use the air traffic advisory service whilst operating IFR within specified advisory airspace are expected to comply with the provisions of ICAO Annex 2, paragraph 3.6, except that the flight plan and changes thereto are not subjected to clearances and that two-way communication will be maintained with the unit providing the air traffic advisory service.

ENR 1.4 ATS AIRSPACE CLASSIFICATION

1 Air Traffic Services Airspace Classification
1.1 Within the UK FIR and UIR, Airspace is classified as A, B, D, E, F and G in accordance with ICAO Standards, subject to the Differences notified at GEN 1.7. The Airspace Classifications are described in subsequent paragraphs.
1.2 All Class A, C, D and E Airspace is hereby notified for the purposes of Article 97 sub-paragraphs (4) (5) (6) and (11) (b)(i)of the Air Navigation Order 2005.

2 Airspace Classifications
2.1 Class A – Controlled Airspace

	IFR	VFR
Service	Air traffic Control Service	
Separation	Separation provided between all IFR flights by ATC	VFR FLIGHT
ATC Rules	Flight plan required (See Note 1)	NOT
	ATC clearance required	
	Radio communication required	PERMITTED
	ATC instructions are mandatory	
VMC Minima	Not applicable (See Note 2)	
Speed Limitations	As published in procedures or instructed by ATC	

Note 1: In certain circumstances, Flight Plan requirements may be satisfied by passing flight details on RTF (detailed at ENR 1.10).
Note 2: For the purposes of:
(a) Climbs and descents maintaining VMC;
(b) Powered aircraft – Airways crossings (ENR 1.1.1.3, paragraph 4.1.6.1); and
(c) Powered aircraft – Other penetrations of Airways (ENR 1.1.1.3, paragraph 4.1.6.2).
In Class A Airspace, the VMC minima are to be:

At or above FL 100:	8km flight visibility
	1500 m horizontal and 1000ft vertical distance from cloud
Below FL 100:	5km flight visibility
	1500 m horizontal and 1000ft vertical distance from cloud

2.1.1 Notifications
2.1.1.1 The following Airspace is notified Class A Airspace:
(i) All Control Areas (Airways) as notified within the UK FIR With the exceptions listed below:
(a) those parts which lie within the boundaries of the Belfast TMAs/CTRs and Scottish TMAs;
(b) part of Airway L1 0 (between Isle of Man VOR IOM and Belfast VOR BEL);
(c) part of Airway L1 8 (between BADSI and LIPGO);
(d) part of Airway N601 (GRICE to 552239N 0031545W and two parts of N601 (Area 1 bounded by 551735N 0025427W – 551724N 0024532W – 551241 N 0023052W – 545912N 0022555W – 54561 ON 0024159W – 551735N 0025427W and Area 2 bounded by 545912N 0022555W – 541 SOON 0021001 W – 544628N 0023625W – 54561 ON 0024159W – 545912N 0022555W;

(e) part of Airway N864 between the southern boundary of L9 and EXMOR below FL 105;
(f) part of Airway P6 (between abm Isle of Man VOR IOM and Belfast VOR BEL);
(g) Airway P18 (between TILNI and Newcastle VOR NEW below FL 125 and between NEW and Aberdeen VOR ADN at all levels);
(h) part of Airway P600 (between GELKI and 56140ON 0033819W\O;
(i) parts of the North Sea Control Area (CTA 2 GODOS and CTA 3 MOLIX) above FL 195; and
(j) the BANBA Control Area.
(ii) Channel Islands Control Zone and Control Area Except the Jersey Control Zone, the Guernsey Control Zone and the Alderney Control Zone; Outside the notified hours of watch of the Jersey Air Traffic Control Unit:
(a) Those parts of the Channel Islands CTR and CTA which lie within the Brest FIR are notified as Class E Airspace and are controlled by Brest ACC;
(b) Those parts of the Channel Islands CTR/CTA which lie within the London FIR are notified as Class G Airspace.
(iii) Clacton Control Area;
(iv) Cotswold Control Area;
(v) Daventry Control Area;
(vi) London Terminal Control Area;
(vii) London Control Zone; IFR procedures apply in all weather conditions in the London CTR except for flights made in accordance with certain special procedures detailed below:
(a) The access lanes/local flying areas for Denham, White Waltham and Fairoaks, are hereby notified for the purposes of Schedule 8, Private Pilots Licence (Aeroplanes), sub-para 2 (c) (ii) and Basic Commercial Pilots Licence (Aeroplanes) sub-para 3 (g) (ii) of the Air Navigation Order 2005 when there is a flight visibility of at least 3km;
(b) The Northolt Aerodrome Traffic Zone and the Radar Manoeuvring Area contained within the London Control Zone are hereby notified for the purposes of Schedule 8, Private Pilots Licence (Aeroplanes), sub-para 2 (c) (ii) and Basic Commercial Pilots Licence (Aeroplanes) sub-para 3 (g) (ii) , of the Air Navigation Order 2005 when there is a flight visibility of at least 4km.
(viii) Manchester Terminal Control Area;
(ix) North Sea Control Area; CTA1 (ROMPA) FL 215 TO FL 245; CTA2 (GODOS) and CTA3 (MOLIX) FL 175 to FL 195
(x) Shanwick Oceanic Control Area;
The Shanwick, Santa Maria, New York and Reykjavik Oceanic Control Areas are hereby notified pursuant to Article 155 (1) of the Air Navigation Order 2005 at and above FL 55 for the purpose of Rule 21 of the Rules of the Air Regulations 1996 (Flight in Class A Airspace).
(xi) Worthing Control Area.
2.2 Class B – Controlled Airspace

	IFR	VFR
Service	Air Traffic Control Service	
Separation	Separation provided between all flights by ATC	
ATC Rules	Flight plan required (See Note) ATC clearance required Radio Communications required ATC instructions are mandatory	
VMC Minima	Not applicable	**At or above FL100** 8km flight visibility Clear of cloud **Below FL100** 5km flight visibility Clear of cloud
Speed Limitations		As published in procedures or instructed by ATC

Note: In certain circumstances, Flight Plan requirements may be satisfied by passing flight details on RTF (detailed at ENR 1.10).

2.3 Class C – Controlled Airspace

	IFR	VFR
Service	Air Traffic Control Service	
Separation	Separation provided between all IFR flights	All VFR flights separated from all IFR flights by by ATC. Traffic information provided on other VFR flights to enable pilots to effect own traffic avoidance and integration
ATC Rules		Flight plan required (See Note) ATC clearance required Radio communication required ATC instructions are mandatory
VMC Minima	Not applicable	**At or above FL100** 8km flight visibility 1500m horizontal and 1000ft vertical distance from cloud **Below FL100** 5km flight visibility 1500m horizontal and 1000ft vertical distance from cloud **or** **At or below 3000ft**
(a) aircraft 140ft IAS or less (except helicopters) (b) helicopters		5km flight visibility and clear of cloud and in sight of the surface; clear of cloud and in sight of the surface
Speed Limitation	As published in procedures or instructed by ATC	**Below FL100** 250kt IAS **or** Lower when published in procedures or instructed by ATC

Note: In certain circumstances, Flight Plan requirements may be satisfied by passing flight details on RTF (detailed at ENR 1.10).

2.3.1 Notifications
2.3.1.1 The following airspace is notified as Class C Airspace:
(i) Upper Control Area – the London and Scottish UIRs between FL 245 and FL 660 (which includes the Hebrides Upper Control Area (UTA));
(ii) Airway L18 between BADSI and LIPGO;
(iii) BANBA Control Area;
(iv) Parts of the North Sea Control Area (CTA 2 GODOS and CTA 3 MOLIX) above FL 195.

2.3.2 VFR Flight in Class C Airspace
2.3.2.1 General arrangements for VFR flight in Class C airspace are specified at ENR 1.1.4.

2.3.2.2 Specific arrangements for glider operations within Temporary Reserved Areas in Class C airspace above FL 245 are detailed at ENR 1.1.1 paragraph 4.3.3 (Charts see ENR 6.3.0.3/4).

2.4 Class D – Controlled Airspace

	IFR	VFR
Service	Air Traffic Control Service	
Separation	Separation provided between all IFR flights by ATC Traffic information provided on conflicting ATC VFR flights	ATC separation not provided Traffic information provided on IFR and other VFR flights to enable pilots to effect own traffic avoidance and integration.
ATC Rules		Flight plan required (See Note 1) ATC clearance required Radio communication required ATC instructions are mandatory
VMC Minima	Not applicable	**At or above FL100** 8km flight visibility 1500m horizontal and 1000ft vertical distance from cloud **Below FL100** 5km flight visibility 1500m horizontal and 1000ft vertical distance from cloud **or** **At or below 3000ft** **(a)** aircraft (except helicopters) 140kt IAS or less 5km flight visibility and clear of cloud and in sight of the surface. **(b)** helicopters clear of cloud and in sight of the surface.
Speed Limitation	Below FL100 250kt IAS or Lower when published in procedures or instructed by ATC	

Note 1: In certain circumstances, Flight Plan requirements may be satisfied by passing flight details on RTF (detailed at ENR 1.10).

2.4.1 Notifications

2.4.1.1 The following airspace is notified as Class D Airspace during the notified hours of watch of the appropriate Air Traffic Control Unit. Some Class D Airspace is further notified (as annotated (†)) for the purposes of Rule 27 (4) (c) of the Schedule to the Rules of the Air Regulations 1996.

Aberdeen Control Zone/Control Area (†) (Note 3(a));
Alderney Control Zone;
Belfast Control Zone;
Belfast City Control Zone/Control Area;
Birmingham Control Zone/Control Area (Note 4);
Bournemouth Control Zone (†);
Bristol Control Zone/Control Area;
Brize Norton Control Zone;
Cardiff Control Zone/Control Area (Note 4);
Durham Tees Valley Control Zone/Control Area
Edinburgh Control Zone (†) (Notes 3(b) and 4);
Glasgow Control Zone (†) (Notes 3(b) and 4);
Guernsey Control Zone;
Isle of Man Control Zone/Control Area (†) (Note 2);
Jersey Control Zone;
Leeds Bradford Control Zone/Control Area (†);
Liverpool Control Zone (Note 4);

London City Control Zone (Note 3(d));
London Gatwick Control Zone/Control Area (Note 4);
London Luton Control Zone/Control Area (Note 4);
London Stansted Control Zone/Control Area (Note 4);
Lyneham Control Zone/Control Area;
Manchester Control Zone/Control Area (Note 4);
Newcastle Control Zone/Control Area (†) (Note 3(b));
Nottingham East Midlands Control Zone/Control Area (Note 4);
Prestwick Control Zone/Control Area (†) (Notes 3(b) and 4);
Scottish Terminal Control Area as shown at ENR 6.2.1.5;
Solent Control Area (†);
Southampton Control Zone (†);
Strangford Control Area;
Sumburgh Control Zone/Control Area (Note 3(c)).

Note 2: Notified as Class D Airspace for the purposes of the Rules of the Air Regulations 1996 during the notified hours of watch of the appropriate Air Traffic Control Unit and are further notified for the purposes of Rule 27(4)(c) of the Schedule to the Rules of the Air Regulations 1996 (As applied to the Isle of Man by the Civil Aviation (Subordinate Legislation) (Application) (No 2) Order 1996).

ENR 1.4 ENR 1.4.2 Airspace Classifications

Note 3: (a) The hours of watch of the Aberdeen ATC Unit are Mon-Sat 0615-2150 (Winter), 0515-2050 (Summer) and Sun 0635-2150 (Winter), 0535-2050 (Summer), and by arrangement. Pilots should note that movements may occur in the Aberdeen CTR/CTA outside these times, and that the Airspace is notified H24 as Class D.
(b) Except at night, these rules do not apply to a non-radio mechanically driven aircraft that has obtained the permission of the Controlling Authority for the flight, provided that it remains at least 1500m horizontally and 1000ft vertically from cloud, and in a flight visibility of at least 5km.
(c) Pilots are advised that, particularly during the winter months, frequent extensions to the published hours of watch of Sumburgh ATC Unit take place and will be notified by NOTAM. Pilots should therefore check current NOTAM when planning flights in the vicinity of Sumburgh CTR/CTA and, if in doubt, should call on the appropriate frequency to ascertain the current status of the Airspace.

Note 4: The following portions of Airspace are further notified:

(a) Birmingham Control Zone and Control Area.
The Birmingham Aerodrome Traffic Zone is hereby notified for the purposes of Schedule 8, Private Pilots Licence (Aeroplanes), sub-para 2 (c) (ii) and Basic Commercial Pilots Licence (Aeroplanes) sub-para 3 (g) (ii) of the Air Navigation Order 2005, when there is a flight visibility of at least 3km.

(b) Cardiff Control Zone and Control Area.
The Cardiff Aerodrome Traffic Zone is hereby notified for the purposes of Schedule 8, Private Pilots Licence (Aeroplanes), sub-para 2 (c) (ii) and Basic Commercial Pilots Licence (Aeroplanes) sub-para 3 (g) (ii), of the Air Navigation Order 2005, when there is a flight visibility of at least 3km.

(c) Edinburgh Control Zone.
The Aerodrome Traffic Zone and the entry/exit lanes at Edinburgh Airport, are hereby notified for the purposes of Schedule 8 of the Air Navigation Order 2005, Part A, Private Pilots Licence (Aeroplanes), sub-para 2 (c) (ii) and Basic Commercial Pilots Licence (Aeroplanes) sub-para 3 (g) (ii), when there is a flight visibility of at least 3km.

(d) Glasgow Control Zone.
The Aerodrome Traffic Zone and the entry/exit lanes at Glasgow Airport, are hereby notified for the purposes of Schedule 8 of the Air Navigation Order 2005, Part A, Private Pilots Licence (Aeroplanes), sub-para 2 (c) (ii) and Basic Commercial Pilots Licence (Aeroplanes) sub-para 3 (g) (ii) , when there is a flight visibility of at least 3km.

(e) Liverpool Control Zone and Control Area.
The access lanes and Aerodrome Traffic Zone/local flying area for Liverpool are hereby notified for the purposes of Schedule 8, Private Pilots Licence (Aeroplanes), sub-paragraph 2 (c) (ii) and Basic Commercial Pilots Licence (Aeroplanes) sub-paragraph 3 (g) (ii), of the Air Navigation Order 2005 when there is a flight visibility of at least 3km.

(f) London Gatwick Control Zone and Control Area.
That part of the Redhill local flying area that lies within the Gatwick CTR is hereby notified for the purposes of Schedule 8, Private Pilots Licence (Aeroplanes), sub-para 2 (c) (ii) and Basic Commercial Pilots Licence (Aeroplanes) sub-para 3 (g) (ii), of the Air Navigation Order 2005 when there is a flight visibility of at least 3km.

(g) London Luton Control Zone and Control Area.
The London Luton Aerodrome Traffic Zone and the entry/exit lanes in the London Luton Control Zone are hereby notified for the purposes of Schedule 8, Private Pilots Licence (Aeroplanes), sub-para 2 (c) (ii) and Basic Commercial Pilots Licence (Aeroplanes) sub-paragraph 3 (g) (ii), of the Air Navigation Order 2005 when there is a flight visibility of at least 3km.

(h) London Stansted Control Zone and Control Area.
(1) The London Stansted Aerodrome Traffic Zone is hereby notified for the purposes of Schedule 8, Private Pilots Licence (Aeroplanes), sub-para 2 (c) (ii) and Basic Commercial Pilots Licence (Aeroplanes) sub-para 3 (g) (ii), of the Air Navigation Order 2005 when there is a flight visibility of at least 3km.

(2) That part of the Andrewsfield Aerodrome Traffic Zone from ground level to 1500ft QNH which lies within the London Stansted Control Zone, is notified for the purposes of Schedule 8, Private Pilots Licence (Aeroplanes) sub-para 2 (c) (ii) and Basic Commercial Pilots Licence (Aeroplanes) sub-para 3 (g) (ii), of the Air Navigation Order 2005, when there is a flight visibility of at least 3km.

(3) The routes specified at EGSS AD 2.22, para 7 are hereby notified, pursuant to the provisions of Rule 5 (2) (a) (i) of the Rules of the Air Regulations 1996 for the purposes of Rule 5.

(i) Manchester Control Zone and Control Area.
(1) The access lanes and Aerodrome Traffic Zones/local flying areas for Manchester and Manchester Woodford are hereby notified for the purposes of Schedule 8, Private Pilots Licence (Aeroplanes), sub-paragraph 2 (c) (ii) and Basic Commercial Pilots Licence (Aeroplanes) sub-paragraph 3 (g) (ii), of the Air Navigation Order 2005 when there is a flight visibility of at least 3km.

(2) The Manchester Control Zone Special Low Level Route is hereby notified for the purposes of Schedule 8, Private Pilots Licence (Aeroplanes), sub-paragraph 2 (c) (ii) and Basic Commercial Pilots Licence (Aeroplanes) sub-paragraph 3 (g) (ii), of the Air Navigation Order 2005 when there is a flight visibility of at least 4km. The Low Level Route is illustrated at AD 2-EGCC-4-1.

(3) The Manchester Control Zone Special Low Level Route is hereby notified pursuant to the provisions of Rule 5 (2) (a) (i) of the Rules of the Air Regulations 1996 for the purposes of Rule 5.

(j) Nottingham East Midlands Control Zone and Control Area.
The Nottingham East Midlands Aerodrome Traffic Zone is hereby notified for the purposes of Schedule 8 Part A, Private Pilots Licence (Aeroplanes), sub-para 2 (c) (ii) and Basic Commercial Pilots Licence (Aeroplanes) sub-para 3 (g) (ii), of the Air Navigation Order 2005, when there is a flight visibility of at least 3km. The entry/exit lanes in the Nottingham East Midlands Control Zone are hereby notified for the purposes of Schedule 8 Part A, Private Pilots Licence (Aeroplanes), sub-para 2 (c) (ii) and Basic Commercial Pilots Licence (Aeroplanes) sub-para 3 (g) (ii), of the Air Navigation Order 2005, when there is a flight visibility of at least 4km.

ENR 1.4   ENR 1.4.2 Airspace Classifications

(k) Prestwick Control Zone.
The Aerodrome Traffic Zone and the entry/exit lanes at Prestwick Airport are hereby notified for the purposes of Schedule 8 of the Air Navigation Order 2005, Part A, Private Pilots Licence (Aeroplanes), sub-para 2 (c) (ii) and Basic Commercial Pilots Licence (Aeroplanes) sub-para 3 (g) (ii), when there is a flight visibility of at least 3km.

2.4.1.2 The following sections of Airways are notified as Class D Airspace during the notified hours of watch of the appropriate Air Traffic Control Unit with vertical and lateral limits as defined in ENR 3.1:

(a) B226   PIPAR to Talla VOR TLA;
(b) L10   Belfast VOR BEL to Isle of Man VOR IOM;
(c) L602   Talla VOR TLA to HAVEN;
(d) N57   Talla VOR TLA to 552112N 0032102W;
(e) N601   GRICE to 552239N 0031545W and two areas which are part of N601 (Area 1 bounded by 551735N 0025427W – 551724N 0024532W – 551241N 0023052W – 545912N 0022555W – 545610N 0024159W – 551735N 0025427W and Area 2 bounded by 545912N 0022555W – 541500N 0021001W – 544628N 0023625W – 545610N 0024159W – 545912N 0022555W) (See ENR 3.1.1.47).
(f) N615   Glasgow VOR GOW to 550826N 0040603W;
(g) An area of N864, below FL105, from a line joining 513946N 0032432W – 513838N 0031727W – 513743N 0030845W to EXMOR
(h) P6   541212N 0043605W (abm Isle of Man VOR IOM) to Belfast VOE BEL;
(i) P18   Aberdeen VOR AND to Newcastle VOR NEW (All levels) and Newcastle VOR NEW to TILNI (below FL 125);
(j) P600   561400N 0033819W to GELKI.

2.5 Class E – Controlled Airspace

Class E – Controlled Airspace

	IFR	VFR
Service	Air Traffic Control Service	Air Traffic Control Service to communicating flights
Separation	Separation provided between all IFR flights by ATC Traffic information provided on conflicting VFR flights	ATC separation not provided Traffic information provided on request, as far as practicable, on IFR and other known VFR flights to enable pilots to effect own traffic avoidance and integration
ATC Rules	Flight plan required (See Note) ATC clearance required Radio Communications required ATC instructions are mandatory	None However pilots are encouraged to contact ATC and comply with instructions
VMC Minima	Not applicable	**At or above FL100** 8km flight visibility 1500m horizontal and 1000ft vertical distance from cloud **Below FL100** 5km flight visibility 1500m horizontal and 1000ft vertical distance from cloud

Note: In certain circumstances, Flight Plan requirements may be satisfied by passing flight details on RTF (detailed at ENR 1.10).

2.5.1 Notifications

2.5.1.1 The following airspace is notified as Class E Airspace:
(i) Belfast Terminal Control Area;
(ii) Parts of the Scottish Terminal Control Area below 6000ft (See ENR 6.2.1.5);
(iii) A part of the Durham Tees Valley Control Zone.

ENR 1.4   ENR 1.4.2 Airspace Classifications

2.6 Class F – Advisory Airspace

	IFR	VFR
Service	Air Traffic Advisory Service to participating Flights	Air Traffic Services as appropriate
Separation	Separation provided between participating IFR flights by ATC	ATC separation not provided
ATC Rules	Participating flight Flight plan required (See Note) ATC clearance required Radio communication required ATC instructions are mandatory	None
VMC Minima	Not applicable	**At or above FL100** 8km flight visibility 1500m horizontal and 1000ft vertical distance from cloud **Below FL100** 5km flight visibility 1500m horizontal and 1000ft vertical distance from cloud or **At or below 3000ft** **(a)** aircraft (except helicopters) greater than 140kt IAS 5km flight visibility and clear of cloud and in sight of the surface **(b)** aircraft (except helicopters) 140kt IAS or less 1500m flight visibility clear of cloud and in sight of the surface **(c)** helicopters at a speed which, having regard to the visibility, is reasonable: clear of cloud and in sight of the surface
Speed Limitations	**Below FL100** 250kt IAS or lower when published in procedures or instructed by ATC	

Note: In certain circumstances, Flight Plan requirements may be satisfied by passing flight details on RTF (detailed at ENR 1.10).

2.6.1 Designation
All advisory routes within the UK FIR are Class F Airspace.

2.7 Class G Airspace

	IFR	VFR
Service		Air Traffic Services as appropriate
Separation		ATC separation not provided (See Note 1)
ATC Rules		None (See Note 2)
VMC Minima	Not applicable	**At or above FL100** 8km flight visibility 1500m horizontal and 1000ft vertical distance from cloud **Below FL100** 5km flight visibility 1500m horizontal and 1000ft vertical distance from cloud **or** **At or below 3000ft** **(a)** Aircraft (except helicopters) greater than 140kt IAS 5km flight visibility clear of cloud and in sight of the surface **(b)** Aircraft (except helicopters) 140kt IAS or less 1500m flight visibility clear of cloud and in sight of the surface **(c)** helicopters at a speed which, having regard to the visibility, is reasonable clear of cloud and in sight of the surface.
Speed Limitations	**Below FL100** 250kt IAS **or** lower when published in procedures or instructed by ATC	

Note 1: Where Air Traffic Control units provide ATS to traffic outside Controlled Airspace, separation may be provided between known flights.

Note 2: Aircraft receiving services from Air Traffic Control units are expected to comply with clearances and instructions unless the pilot advises otherwise.

2.7.1 Designation

2.7.1.1 All UK Airspace, including that above FL 660, not included in Classes A to F.

2.7.2 Aerodrome Traffic Zones

2.7.2.1 Aerodrome Traffic Zones (ATZs) are not included in the Airspace Classification System. An ATZ assumes the conditions associated with the Class of Airspace in which it is situated. As a minimum, when flying within an ATZ, the requirements of Rule 39 of the Rules of the Air Regulations 1996 must be complied with. Where the requirements of the Class of Airspace of which an ATZ forms a part are more stringent than Rule 39 then those must be complied with. Thus, in Class G Airspace Rule 39 will be the relevant requirement, but in Class A Airspace the more onerous requirements of Class A take precedence.

2.7.2.2 Aerodromes at which ATZs may be established are those which:

(a) Are government aerodromes; or
(b) have an Air Traffic Control Unit; or
(c) have an Aerodrome Flight Information Unit; or
(d) are licensed and have a means of two-way radio communication with aircraft; and whose hours of operation are notified for the purposes of Rule 39.

2.7.2.3 Pilots should be aware that in order to comply with the provisions of Rule 39 they must adopt the following procedures:

(a) Before taking off or landing at an aerodrome within an ATZ or transiting through the associated airspace, obtain the permission of the air traffic control unit, or where there is no air traffic control unit, obtain information from the flight information service unit or air/ground radio station to enable the flight to be conducted safely.
(b) Radio equipped aircraft must maintain a continuous watch on the appropriate radio frequency and advise the air traffic control unit, flight information unit or air/ground radio station of their position and height on entering the zone and immediately prior to leaving it.
(c) Non-radio aircraft operating within a notified ATZ must comply with any conditions prescribed by the air traffic control unit, flight information unit or air/ground radio station prior to the commencement of the flight with any instructions issued by visual means.

2.7.2.4 Failure to establish two-way radio communications with the air traffic control unit, flight information unit or air/ground radio station during their notified hours of operation must not be taken as an indication that the ATZ is inactive. In that event, except where the aircraft is in a state of emergency or is being operated in accordance with radio failure procedures, pilots should remain clear of the ATZ.

2.7.2.5 Rule 39 does not apply outside the notified hours of operation. **Permanent changes or temporary extensions to ATZ hours may be notified by United Kingdom NOTAM. Pilots should exercise caution, however, since some airfields may continue to operate outside of those notified hours.**

ENR 1.5 HOLDING, APPROACH AND DEPARTURE PROCEDURES

1 Holding and Approach to Land Procedures

1.1 General

1.1.1 UK Holding and Instrument Approach Procedures are designed using criteria contained in ICAO Document 8168-OPS/611 (PANS-OPS) VOL II. These criteria include:

(a) The use of Obstacle Clearance Height (OCH) as the basic obstacle clearance element in calculating minima;

(b) aeroplane categories related to speed, which can result in a reduction of Obstacle Clearance Heights for the more manoeuvrable aeroplanes;

(c) the definition of a Missed Approach Point for non-precision procedures;

(d) the use of the term 'Decision Height' in relation to precision procedures and 'Minimum Descent Height' in relation to non-precision and Visual (Circling) procedures.

1.1.2 The UK Holding and Instrument Approach Procedures appear at AD 2.24.

1.1.3 PANS-OPS stresses the need for flight crew and operational personnel to adhere strictly to the published procedures in order to achieve and maintain an acceptable level of safety in operations. In addition, within the UK, procedures contained within Controlled Airspace are subject to Rule 31(3) (a) of the Rules of the Air Regulations 1996 (IFR within Controlled Airspace).

2 Visual Manoeuvring (Circling) VM(C) in the Vicinity of the Aerodrome after Completing an Instrument Approach

2.1 Introduction

2.1.1 Visual Manoeuvring (Circling) VM(C) is the term used to describe the visual phase of flight, after completing an Instrument Approach, where an aircraft is manoeuvred into position for a landing on a runway which is not suitably located for a straight-in approach.

2.1.2 The VM(C) area is the area in which obstacle clearance has been considered for aircraft manoeuvring visually before landing. Aircraft performance has a direct effect on the airspace and visibility needed to perform the circling manoeuvre; since the most significant factor in performance is speed, the size of the VM(C) area varies with the category of the aircraft. The limits of the area applicable to each category of aircraft are defined by combining arcs centred on the threshold of each usable runway; the total area thus enclosed is the VM(C) area – see example at Diagram 1.

2.2 Obstacle Clearance

2.2.1 When the VM(C) area has been established, the Obstacle Clearance Height (OCH) is determined for each category of aircraft. The criteria used to determine the OCH is as follows:

Diagram 1: Construction of Visual Manoeuvring (Circling) Area for a Category E Aircraft

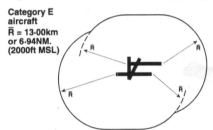

Category E aircraft
$\bar{R}$ = 13·00km or 6·94NM. (2000ft MSL)

2.2 Obstacle Clearance

2.2.1 When the VM(C) area has been established, the obstacle clearance Height (OCH) is determined for each category of aircraft. The criteria used to determine the OCH is as follows:

Table 1 – Visual Manoeuvring (Circling) – Criteria for Determining OCH.

Aircraft Category	Maximum Speeds for circling (kt)	Circling Area Maximum radii from RWY THR (nm)	Minimum Obstacle Clearance (ft)	Lowest Permissible OCH aal (ft)
A	100	1.68	300	400
B	135	2.66	300	500
C	180	4.20	400	600
D	205	5.28	400	700
E	240	6.94	500	800

2.3 Sectorization of Visual Manoeuvring (Circling) Area

2.3.1 It is permissible to eliminate from obstacle clearance consideration a particular sector, within the total VM(C) area, where the sector lies outside the final approach and missed approach areas; this 'eliminated' sector is determined by the dimensions of the ICAO Annex 14 instrument approach surfaces – see example at Diagram 2.

ENR 1.5 ENR 1.5.2 Visual Manoeuvring (Circling) VM(C) in the Vicinity of the Aerodrome after Completing an Instrument Approach

Diagram 2. Sectorized Visual Manoeuvring (Circling) Area

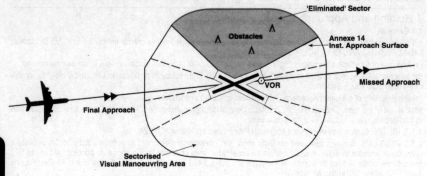

2.4 Obstacle Clearance Heights for Aerodromes with Published Instrument Approach Procedures

2.4.1 When the principle of sectorisation has been applied to a particular aerodrome, two OCHs for VM(C) will be promulgated at AD 2.24. The first OCH will allow a pilot to manoeuvre safely within the total VM(C) area and the second OCH is for the pilot who exercises the option to avoid manoeuvring within the eliminated sector and restricts his visual manoeuvring to the remaining sectorised area.

3 Summary of Holding and Approach to Land Procedures at Individual Aerodromes Introductory Notes

3.1 † For precision approach procedures (ILS) QFE values are above threshold elevation of the runway direction to which the procedure applies. For non-precision approach procedures QFE values are above aerodrome elevation except for those approaches indicated on the IAC by '†' which are above threshold elevation of the runway direction to which the procedure applies. QNH and amsl values are shown in bold and QFE values (other than OCHs) in parentheses ().

3.2 In the missed approach procedure the altitudes (heights) given are those to be attained on Go-Around in the absence of ATC instructions and should normally be flown using QNH altimeter setting for the aerodrome.

3.3 Where the term 'straight ahead' is used in missed approach procedures, pilots should maintain Final Approach Track (FAT) unless a different track is given.

3.4 There are two types of VDF procedure, QDM and QGH. In the QDM procedure the pilot calls for a series of QDM and uses them to follow the published approach pattern, making his own adjustment to heading and height. In the QGH procedure the controller obtains bearing from the aircrafts transmissions, interprets this information and passes to the pilot headings and heights to fly designed to keep the aircraft in the published pattern. Normally, at civil aerodromes, only QDM procedure is available; however, in some cases, for specific operational reasons, there will be provision for QGH procedure. Those aerodromes that have been approved to carry out both types of VDF procedure will have this provision shown against the procedure. Pilots are reminded that it is their responsibility to ensure with ATC that the correct procedure is being flown.

3.5 These procedures have been established in accordance with the ICAO PANS-OPS, except for those UK differences shown at GEN 1.7. While certain specified allowances for wind effect have been made in determining the areas which will contain the various procedures, it is emphasized that these Holding and Approach to Land procedures are based on still air conditions and in practice due allowances must be made for wind.

3.6 Shuttle Procedure: is a procedure designed to allow for descent and/or positioning after arrival and prior to the commencement of the instrument descent procedure.

3.7 Those procedures at aerodromes which lie within Controlled Airspace are notified for the purposes of Rule 31(3)(a)(ii) of the Rules of the Air Regulations 1996. Those procedures at aerodromes which do not lie within Controlled Airspace are notified for the purposes of Rule 40 of those Rules. These Rules require that, where an aerodrome is provided with one or more notified Instrument Approach Procedure, unless otherwise authorized by ATC, pilots requiring to use an Instrument Approach Procedure shall use only such notified procedures. This applies irrespective of whether the aerodrome is situated inside or outside Controlled Airspace.

3.8 Pilots should be aware that within Controlled Airspace ATC provide separation between aircraft carrying out Instrument Approach Procedures and other IFR or Special VFR traffic within that Airspace and will pass pertinent traffic information on known VFR flights.

However, in certain Controlled Airspaces there may be VFR traffic of which ATC has no knowledge.

At aerodromes situated outside Controlled Airspace separation can only be provided between known IFR flights and pertinent traffic information will be given on known VFR flights.

3.9 For availability of Instrument Approach Charts, see GEN 3.2.

3.10 Aircraft Categorization

Aircraft Category A	nominal V at less than 91kt IAS
Aircraft Category B	nominal V at 91kt to 120kt IAS
Aircraft Category C	nominal V at 121kt to 140kt IAS
Aircraft Category D	nominal V at 141kt to 165kt IAS
Aircraft Category E	nominal V at 166kt to 210kt IAS

Note: Nominal V at is defined as 1.3 x the stalling speed in the landing configuration at maximum certificated landing mass.

ENR 1.5   ENR 1.5.4 Radar Approach Procedures – Obstacle Clearance Heights and Missed Approach Procedures

3.11 Descent Gradient

3.11.1 To permit the approximate rates of descent required in the intermediate and final approach segments of non-precision Instrument Approach Procedures to be calculated, whenever an Intermediate Fix (IF) or Final Approach Fix (FAF) is included in a non-precision procedure, the descent path required in the corresponding approach segment will be specified, as a percentage gradient.

3.11.2 The approximate rate of descent required, in feet per minute, in a particular segment, is equal to the product of Groundspeed (kt) and Gradient (%).

3.11.3 A table of rounded rate of descent values for a selection of Groundspeed (G/S) and gradient values is given below:

Rate of Descent from Groundspeed and % Gradient, feet/minutes (Rounded values)

% Gradient	Groundspeed (kts)							
	60	80	100	120	140	160	180	200
0.1	5	10	10	10	15	15	20	20
0.2	10	15	20	25	30	30	35	40
0.3	20	25	30	35	40	50	55	60
0.4	25	30	40	50	55	65	70	80
0.5	30	40	50	60	70	80	90	100
0.6	35	50	60	70	85	95	110	120
0.7	45	55	70	85	100	115	130	145
0.8	50	65	80	100	115	130	145	165
0.9	55	75	90	110	130	145	165	185
1.0	60	80	100	120	140	160	180	200
2.0	120	160	200	240	280	320	370	410
3.0	180	240	300	360	420	480	550	610
4.0	240	320	400	490	570	650	730	810
5.0	300	400	510	610	710	810	910	1010
6.0	360	490	610	730	850	970	1090	1220

eg % gradient 5.2, G/S 120kt, Rate of Descent required 635ft/min

3.12 Visual Manoeuvring (Circling) OCHs

3.12.1 Visual Manoeuvring (Circling) (VM(C)) OCHs are listed for the total VM(C) area, and where operational advantage can be gained by sectorisation, for the relevant sectorised VM(C) area also.

3.12.2 Where the OCH, appropriate to Aircraft Category, of the Instrument Approach Procedure (IAP) which precedes the visual manoeuvre is higher than the relevant VM(C) OCH listed for that Aircraft Category, the OCH of the IAP shall be the lowest OCH for Visual Manoeuvring (Circling) following that approach.

3.13 Procedure Turns

3.13.1 Except where an 80°/260° procedure turn is specifically prescribed, procedure turns specified in instrument approach procedures may be flown as 45°/180° or 80°/260° type, at pilots discretion.

3.13.2 Since the 80°/260° procedure turn occupies less airspace along track than the 45°/180° type, aircraft completing an 80°/260° manoeuvre will normally return to the extended Final Approach Track approximately 1nm closer to the FAP or FAF (when provided) than if a 45°/180° manoeuvre had been used; pilots using an 80°/260° manoeuvre should allow for this factor when preparing for any subsequent inbound descent. Where an 80°/260° procedure turn is specifically prescribed, due allowance for this factor will already be provided in the procedure design.

3.14 Alternative Race-track Procedure

3.14.1 Extensions to holding patterns may be specified in some cases as alternative procedures for use after holding. Where such alternative procedures are joined without first entering the hold, they should be joined and flown as race-track procedures. However, because of the requirement to intercept the inbound track before returning to the facility following a Sector 1 parallel entry in such procedures, and its effect on the distance available for any subsequent descent, aircraft joining a race-track procedure from Sector 1 are recommended to join the hold prior to entering the race-track.

3.15 Established

3.15.1 Aircraft are considered to be 'established' when they are within half full scale deflection for the ILS and VOR, or within ± 5° of the required bearing for NDB(L).

3.16 Missed Approach Climb Gradient

3.16.1 Unless otherwise specified, the normal climb gradient upon which Missed Approach Procedures are based is 2.5% (1 in 40).

4 Radar Approach Procedures – Obstacle Clearance Heights and Missed Approach Procedures

4.1 Operating Information

4.1.1 During the Intermediate phase of the procedure the pilot will be asked to check both his Minima and the Missed Approach Point (MAPt); he will not be given the Obstacle Clearance Height (OCH) for his category of aircraft nor the location of the MAPt unless he specifically requests this information.

ENR 1.5   ENR 1.5.4 Radar Approach Procedures – Obstacle Clearance Heights and Missed Approach Procedures

4.1.2 SRA procedures designed to ICAO PANS-OPS criteria permit descent on final approach to the OCH (subject to any specified height/distance (Step Down Fix (SDF)) limitation), without regard to the 'advisory heights' given by the Controller. These 'advisory heights' are not essential for obstacle clearance and are only provided as a guide to pilots wishing to maintain a constant angle descent path.
(a) pilots flying an SRA requiring the alternative datum profile, will be passed advisory altitudes/heights based on the appropriate datum and rounded up to the nearest 10ft.
4.1.3 Pilots are recommended to fly to the radar advisory heights. However, where an SDF is specified as part of the procedure the SDF 'not-below' height is mandatory. Adherence to the nominal glidepath defined by advisory heights will ensure compliance with SDF minimum height requirements.
4.1.4 A MAPt is designated for each procedure; it is normally located at the point where the radar approach terminates (Radar Termination Range (RTR)). However, where operationally advantageous, the MAPt for the 2nm SRA may be designated as 1nm after RTR (ie 1nm before threshold).
Note: Where the MAPt is designated as 1nm after RTR, talkdown will still cease at 2nm (RTR), and it will be the pilots responsibility to determine when the MAPt has been reached.
4.1.5 QNH values in the Missed Approach Procedures are shown in bold and QFE values in (parentheses). QFE values are above aerodrome elevation except for those procedures indicated '†' which are above threshold elevation of the runway served by the procedure. Threshold elevations are shown against the relevant aerodrome in the aerodrome/airport schedules in the AD Section.
4.2 Advisory Heights for Surveillance Radar Approach Procedures
4.2.1 The nominal glidepaths upon which the advisory heights for particular procedures are based, are annotated in the procedure descriptions within the IACs.
4.2.2 The table below shows the advisory heights applicable to approximate 2.75°, 3°, 3.25° and 3.50° glidepath.

Range from touchdown (nm)	2.75° Glidepath	3° Glidepath	3.25° Glidepath	3.5° Glidepath
4.5	1350	1400	1575	1650
4	1200	1250	1400	1500
3.5	1050	1100	1225	1300
3	900	950	1050	1125
2.5	750	800	875	925
2	600	650	700	750
1.5	450	500	525	550
1	300	350	350	375

5 FM Broadcast Interference
5.1 Aircraft ILS/VOR equipment may experience interference from high powered FM broadcast stations in the radio frequency band 88-108 MHz (GEN 3.4 refers). Pilots are warned that instrument approach procedures based on ILS/VOR may not be available.

6 Protection of Instrument Approach Procedures at Aerodromes outside Controlled Airspace
6.1 Aerodromes located outside Controlled Airspace, for which Instrument Approach Procedures (IAP) are published, are identified by a 'cone' symbol on the Aeronautical Charts ICAO Scale 1: 500 000 United Kingdom and the Topographical Air Charts of the United Kingdom Scale 1: 250 000. Portrayal on these charts is provided in order to assist pilots of VFR flights to avoid confliction with IFR traffic at these aerodromes. Pilots are urged to take this information into account in their pre-flight planning.
6.2 A number of these aerodromes have notified Visual Reference Points (VRP) and Visual Routes (VR) which are geographically deconflicted from the instrument patterns and, notwithstanding that their use is voluntary, VFR pilots may be requested to route with reference to these. VRP and VR are described at AD 2 and AD 3.
6.3 Pilots should note for guidance at GEN 3.3 under Responsibilities of Approach Control.

7 ATC Assistance and Responsibilities
(Refer to GEN 3.3)

1.6 RADAR SERVICES AND PROCEDURES

1.6.1 USE OF RADAR IN AIR TRAFFIC SERVICES

1 General

1.1 The UK generally subscribes to the procedures for the use of radar in ATS which are given in ICAO Doc 4444 Part X with the important difference that the radar service provided outside Controlled Airspace will be either an advisory service or an information service as described below. In addition, in order to clarify the exact arrangements in use within each type of airspace in the UK FIR, the UK has found it necessary to amplify certain of the ICAO statements.

2 Types of Radar Service

2.1 ICAO References
Doc 4444, Part VI, paras 8 and 11.

2.2 The provision of Radar Control, Radar Advisory and Radar Information Services is dependent upon specific types of airspace. Details of the services provided are stated in the table below.

Table (see also maps at ENR 6-1-6-1/2 and ENR 6-3-2-1/2).

Type of Airspace	Type of Service	ATC action with regard to unknown aircraft
Class A Airspace Controlled Airspace subject to IFR at all times	Radar Control Service	Traffic information and avoiding action will not be given unless information has been received which indicates that a radar echo may be a particular aircraft which is lost or expecting radio failure
Class D Airspace Controlled airspace below FL245 in which all flights are subject to the authority of ATC		
Class E Airspace Controlled airspace in which VFR flight Without ATC clearance is permitted	Radar Control Service	Traffic information will be passed provided this does not compromise radar sequencing of traffic or separation of IFR flights. Avoiding action will be given at the request of pilots but to limits decided by the radar controller or if information has been received which indicates that a radar echo may be a particular aircraft which is lost or experiencing radio failure.
Class C Airspace Upper Airspace Control Area	Radar Control Service	(a) within the Military Mandatory Radar Service Areas (MRAS) Procedures exist to ensure separation between aircraft operating in the Upper Airspace but under the control of different ATS units. In general, all aircraft operating off the promulgated Upper ATS routes will be vectored clear of those operating on the routes. In order to eliminate the possibility of a radar induced confliction, neither traffic information nor avoiding action will be given unless information received indicates that an unknown aircraft is lost or has experienced radio failure. (b) Outside the Military Mandatory Radar Service Areas (MRSA) Whenever practicable traffic information will be given. Avoiding action will be given if controllers consider it necessary or if it is requested by pilots Note: Because of the sudden appearance and unpredictable movement of unknown aircraft it is not always possible to provide the requisite separation. See ENR 1.1.1.6 para 4.3.4
Class F Airspace, Advisory Routes	Radar Advisory Service or Radar Information Service	Traffic information will be passed followed by advice on avoiding action
Class G Airspace, All other Airspace		Traffic information will be passed but no avoiding action is to be given. The pilot is responsible for his own separation.

3 Radar Service Outside Controlled Airspace

3.1 Radar Advisory Service (RAS)

3.1.1 RAS is an air traffic radar service in which the controller will provide advice necessary to maintain prescribed separation between aircraft participating in the advisory service, and in which he will pass to the pilot the bearing, distance, and, if known, level of conflicting non-participating traffic, together with advice on action necessary to resolve the confliction. Where time does not permit this procedure to be adopted, the controller will pass advice on avoiding action followed by information on the conflicting traffic. Under a RAS, the following conditions apply:

(a) The service will only be provided to flights under IFR irrespective of meteorological conditions;

(b) controllers will expect the pilot to accept vectors or level allocations which may require flight in IMC. **Pilots not qualified to fly in IMC should accept a RAS only where compliance with ATC advice permits the flight to be continued in VMC;**

215

ENR 1.6

1.6.1.3 Radar Service Outside Controlled Airspace

(c) there is no legal requirement for a pilot flying outside Controlled Airspace to comply with instructions because of the advisory nature of the service. However, a pilot who chooses not to comply with advisory avoiding action must inform the controller. The pilot will then become responsible for initiating any avoiding action that may subsequently prove necessary;

(d) the pilot must advise the controller before changing heading or level;

(e) the avoiding action instructions which a controller may pass to resolve a confliction with non-participating traffic will, where possible, be aimed at achieving separation which is not less than 5nm or 3000ft, except when specified otherwise by the regulating authority. However, it is recognised that in the event of the sudden appearance of unknown traffic, and when unknown aircraft make unpredictable changes in flight path, it is not always possible to achieve these minima;

(f) information on conflicting traffic will be passed until the confliction is resolved;

(g) the pilot remains responsible for terrain clearance, although ATSUs providing a RAS will set a level or levels below which a RAS will be refused or terminated.

3.2 Radar Information Service (RIS)

3.2.1 RIS is an air traffic radar service in which the controller will inform the pilot of the bearing, distance, and, if known, the level of the conflicting traffic. No avoiding action will be offered. **The pilot is wholly responsible for maintaining separation from other aircraft whether or not the controller has passed traffic information.** Under a RIS, the following conditions apply:

(a) The service may be requested under any flight rules or meteorological conditions;

(b) the controller will only update details of conflicting traffic, after the initial warning, at the pilot's request or if the controller considers that the conflicting traffic continues to constitute a definite hazard;

(c) the controller may provide radar vectors for the purpose of tactical planning or at the request of the pilot. However, vectors will not be provided to maintain separation from other aircraft, which remains the responsibility of the pilot. There is no requirement for a pilot to accept vectors;

(d) the pilot must advise the controller before changing level, level band or route;

(e) RIS may be offered when the provision of RAS is impracticable;

(f) requests for a RIS to be changed to a RAS will be accepted subject to the controller's workload; prescribed separation will be applied as soon as practicable. If a RAS cannot be provided the controller will continue to offer a RIS;

(g) for manoeuvring flights which involve frequent changes of heading or flight level, RIS may be requested by the pilot or offered by the controller. Information on conflicting traffic will be passed with reference to cardinal points. The pilot must indicate the level band within which he wishes to operate and is responsible for selecting the manoeuvring area, but may request the controller's assistance in finding a suitable location. The controller may suggest re-positioning on his own initiative, but the pilot is not bound to comply;

(h) the pilot remains responsible for terrain clearance. ATSUs providing a RIS will set a level or levels below which vectors will not be provided, except when specified otherwise by the regulating authority.

3.3 Establishing a Service

3.3.1 In order to establish a radar service the pilot and controller must reach an 'accord'. When requesting a radar service the pilot must state the flight rules under which he is operating and whether he requires a RAS or RIS. If the controller is able to offer a service he will attempt to identify the aircraft. When he is satisfied that he has positively identified the aircraft, the controller will confirm the type of service he is about to provide, and the pilot must give a read-back of the service. **The identification procedure does not imply that a radar service is being provided and the pilot must not assume that he is in receipt of a RAS or a RIS until the controller makes a positive statement to that effect.** If a controller is unable to provide a service he will inform the pilot.

3.3.2 Should the pilot fail to specify the type of service required, the controller will ask the pilot which service he requires before endeavouring to provide any service.

3.3.3 London Control – Requests for RAS or RIS

3.3.3.1 In order to avoid excessive RTF conversations on the frequencies used by 'London Control', pilots who intend to request such a service from 'London Control' are to make their initial request on the FIS frequency ('London Information') appropriate to their geographical position. The FIS controller will co-ordinate with the appropriate Radar Sector and subsequently inform the pilot whether or not a RAS or RIS can be provided and, if so, on what frequency.

3.3.3.2 Pilots should note that no RAS or RIS will be available on any London Control Frequency below FL 70. In any case a serviceable transponder will be a pre-requisite for either service.

3.4 Limitations of Service

3.4.1 Outside Controlled Airspace any radar service may be limited. If a radar controller considers that he cannot maintain a full radar service he will warn the pilot of the nature of the limitations that may affect the service being provided. Thereafter, the pilot is expected to take the stated limitation into account in his general airmanship. In particular, warning of the limitation will be given to the pilot in the following circumstances:

(a) When the aircraft is close to lateral or vertical limits of solid radar cover;

(b) when the aircraft is close to areas of permanent echoes or weather returns;

(c) when the aircraft is operating in areas of high traffic density;

(d) when the controller considers that the performance of his radar is suspect;

(e) when the controller is providing service using SSR data only.

3.4.2 When a RAS or RIS is provided by an Approach Control Unit it is normally limited to a range of 40nm from the aerodrome.

4 Radar Vectoring Controlled Airspace

4.1 At certain aerodromes where the associated Controlled Airspace does not encompass the Radar Vectoring Area, aircraft may be vectored outside the notified airspace for approaches to certain runways. The aerodromes and runways to which this procedure may apply are listed below:

Aerodromes	Runways
Birmingham	24
Bournemouth	08
Bristol	27
Cardiff	12
Durham Tees Valley	05
Edinburgh	30
Leeds Bradford	14, 27
London City	28

Note: Whilst the aircraft is outside Controlled Airspace a Radar Advisory Service will be provided. To reduce RTF loading, pilots will not be advised of the changes of services given in these circumstances.

5 Radar Vectoring for ILS Approach

5.1 ICAO Reference
Doc 4444, Chapter 8, paragraph 8.9.4.

5.2 Aircraft being positioned for final approach will be given a heading to close with the localizer at a range of at least 5nm from the runway threshold and at a level below the glide path. The pilot will be told to complete the turn on and to report established on ILS, but at this point if he requests it, ATC will give another vector to bring the aircraft on to the localizer. If the pilot wishes to lock himself on to the localizer, he must ask permission from ATC when there is still time for the action to take place without crossing the localizer.

5.3 On occasions in order to maintain the correct spacing between aircraft, ATC will deliberately vector the aircraft through the localizer for approach from the other side. Pilots will be warned when this manoeuvre is being given.

6 Terrain Clearance

6.1 ICAO Reference
Doc 4444, Chapter 8, paragraphs 8.6.5.2 and 8.6.5.3.

6.2 Controllers will ensure that levels assigned to IFR flights when in receipt of a Radar Control Service and to flights in receipt of a RAS will provide at least the minimum terrain clearances given below:

6.2.1 Within 30nm of the radar antenna but excluding the Final and Intermediate Approach Area: 1000ft above any fixed obstacle which is closer than 5nm to the aircraft or which is situated within the area 15nm ahead of and 20 degrees either side of the aircraft's track. These distances may be reduced to 3nm and 10nm respectively where official CAA approval has been promulgated. Levels assigned to aircraft during initial approach will also provide this terrain clearance.

6.2.2 Outside 30nm from the Radar Antenna, for flights on Airways or Advisory Routes, 1000ft above any fixed obstacle within 15nm of the centre-line; otherwise 1000ft above any fixed obstacle within 30nm of the aircraft.

6.2.3 Radar Controllers have no responsibility for the terrain clearance of, and will not assign levels to, aircraft in receipt of a RIS or aircraft operating Special VFR or VFR which accept radar vectors.

Note 1: In sections of Airways where the base is defined as a Flight Level, the lowest usable level normally provides not less than 1500ft Terrain Clearance.

Note 2: ATSUs providing a RAS outside Controlled Airspace will set a level or levels below which the service will be refused or terminated.

6.2.4 Radar Vectoring Area Charts

6.2.4.1 A Radar Vectoring Area is a defined area in the vicinity of an aerodrome, in which the minimum safe levels allocated by a radar controller vectoring IFR flights have been predetermined.

6.2.4.2 Charts for individual aerodromes appear at the end of this Section. Each chart shows the following:
(a) Outline of the Radar Vectoring Area;
(b) significant obstructions and spot heights;
(c) minimum safe altitude within the Radar Vectoring Area;
(d) each Final Approach Track;
(e) aerodrome elevation;
(f) Transition Altitude;
(g) loss of communication procedures.

7 Navigational Assistance

7.1 ICAO Reference
Doc 4444, Chapter 8, paragraph 8.6.6.

7.2 Identified aircraft operating within Controlled Airspace are deemed to be separated from unknown aircraft flying in adjoining uncontrolled airspace. Whenever practicable, however, the radar controller will aim to keep aircraft under his control at least 2nm within the boundary of Controlled Airspace.

8 Weather Avoidance

8.1 ICAO Reference
Doc 4444, Chapter 8, paragraph 8.6.9

8.2 As far as possible aircraft which have planned to operate within Controlled Airspace will be vectored so that they remain at least 2nm inside the Controlled Airspace boundary. If on the evidence of clutter on the radar display ATC considers it expedient for the aircraft to leave Controlled Airspace in order to avoid weather, the pilot will be advised and will be responsible for accepting the detour into uncontrolled airspace. If a pilot using his aircraft radar intends to detour observed weather he should when operating within Controlled Airspace obtain clearance from the radar controller to do so, and if under these circumstances it is necessary to leave Controlled Airspace the pilot shall request permission to re-join.

1.6.2 SSR OPERATING PROCEDURES

1 General

1.1 In accordance with Article 20 and Schedule 5 of the Air Navigation Order 2005, the SSR transponder shall be operated within the airspace notified at paragraph GEN 1.5.3, paragraph 1.3.

1.2 In airspace where the operation of transponders is not mandatory pilots of suitably equipped aircraft should comply with paragraph 2.2 except when remaining within an aerodrome traffic pattern below 3000ft agl.

1.3 With the exceptions detailed in paragraph 2 pilots shall:

(a) If proceeding from an area where a specific Mode A code has been assigned to the aircraft by an ATS Unit, maintain that code setting unless otherwise instructed;

(b) select or reselect Mode A codes, or switch off the equipment when airborne only when instructed by an ATS Unit;

(c) acknowledge Mode A code setting instructions by reading back the code to be set;

(d) select Mode C simultaneously with Mode A unless otherwise instructed by an ATS Unit;

(e) when reporting levels under routine procedures or when requested by ATC, state the current altimeter reading to the nearest 100ft. This is to assist in the verification of Mode C data transmitted by the aircraft.

Note: If, on verification there is a difference of more than 200ft between the level readout and the reported level, the pilot will normally be instructed to switch off Mode C. If independent switching of Mode C is not possible the pilot will be instructed to select Mode A Code 0000 to indicate a transponder malfunction.

2 Special Purpose Mode A Codes

2.1 Some Mode A codes are reserved internationally for special purposes and should be selected as follows:

(a) Code 7700. To indicate an emergency condition, this code should be selected as soon as is practicable after declaring an emergency situation, and having due regard for the over-riding importance of controlling aircraft and containing the emergency. However, if the aircraft is already transmitting a discrete code and receiving an air traffic service, that code may be retained at the discretion of either the pilot or the controller;

(b) Code 7600. To indicate a radio failure;

(c) Code 7500. To indicate unlawful interference with the planned operation of a flight, unless circumstances warrant the use of Code 7700;

(d) Code 1000 to indicate an aircradft conducting IFR flight as GAT, where the down-linked aircraft identification is validated as matching the aircraft identification entered in the flight plan.

(e) Code 7007. This code is allocated to aircraft engaged on airborne observation flights under the terms of the Treaty on Open Skies See ENR 1-6-2-1. Flight Priority Category B status has been granted for such flights and details will be published by NOTAM. Mode C should be operated with all of the above codes.

2.2 Mode A Conspicuity Code

2.2.1 When operating at and above FL 100 pilots shall select Mode A code 7000 and Mode C except:

(a) When receiving a service from an ATS Unit or Air Surveillance and Control System Unit which requires a different setting;

(b) when circumstances require the use of one of the Special Purpose Mode A Codes or one of the other specific Mode A conspicuity codes assigned in accordance with the UK SSR Code Assignment Plan as detailed in the table at ENR 1.6.2.4 to ENR 1.6.2.8

2.2.2 When operating below FL 100 pilots shall select Mode A Code 7000 and Mode C except as above.

2.2.3 Pilots are warned of the need for caution when selecting Mode A code 7000 due to the proximity of the Special Purposes Mode A Codes.

2.3 Parachute Dropping

2.3.1 Unless a discrete Mode A code has already been assigned, pilots of transponder equipped aircraft should select Mode A Code 0033, together with Mode C, five minutes before the drop commences until the parachutists are estimated to be on the ground.

2.4 Aerobatic Manoeuvres

2.4.1 The use of Special Purpose Code 7004 shall be for solo or formation aerobatics, whilst displaying, practising or training for a display or for aerobatics training or general aerobatic practice. Any civil or military pilot may use this code whilst conducting aerobatic manoeuvres.

2.4.2 Unless a discrete code has already been assigned, pilots of transponder equipped aircraft should select Code 7004 together with Mode C, five minutes before commencement of their aerobatic manoeuvres until they cease and resume normal operations.

2.4.3 Pilots are encouraged to contact ATCUs and advice them of the vertical, lateral and temporal limits within which they will be operating and using the SSR Code 7004.

2.4.4 Controllers are reminded that SSR Code 7004 must be considered as unvalidated and the associated Mode C unverified. Traffic information will be passed to aircraft receiving a service as follows:
'Unknown aerobatic traffic, (number) O'clock (distance) miles opposite direction/crossing left/right indicating (altitude) unverified (if Mode C displayed)'.

3 Mode S Aircraft Identification

3.1 To comply with ICAO airborne equipment requirements, all Mode S transponder equipped aircraft engaged in international civil aviation must incorporate an Aircraft Identification Feature (sometimes referred to as Flight Identity or Flight ID). Correct setting of Aircraft Identification is essential for the correlation of radar tracks with flight plan data in Air Traffic Management (ATM) and Airport Operator ground systems. Data analysed by the Euro-control AMP has shown that many Mode S compliant aircraft are transmitting an incorrect Aircraft Identification, for example, incorrectly setting ABC_123 instead of ABC123.

3.2 Incorrect Aircraft Identification settings compromise the safety and ATM benefits of Mode S and will prohibit automatic flight plan correlation, which could affect subsequent ATC clearances and sequencing.

3.3 In accordance with ICAO Doc 8168 (PANS-OPS) Vol I, Part VIII, para 1.3, flight crew of aircraft equipped with Mode S having an Aircraft Identification Feature shall set the Aircraft Identification in the transponder. This setting shall correspond to the Aircraft Identification in the transponder. This setting shall correspond to the Aircraft Identification specified in item 7 of the ICAO flight plan, or if no flight plan has been filed, the aircraft registration.

3.4 Aircraft Identification, not exceeding 7 alphanumeric characters, is to be entered in item 7 of the flight plan and set in the aircraft as follows:

3.4.1 Either

(a) the ICAO three letter designator (not the IATA two letter designator) for the aircraft operating agency followed by the flight identification, for example, BAW213 or JTR25 when:

(i) In radiotelephony, the call sign used consists of the ICAO telephony designator for the operating agency followed by the flight identification, for example, SPEEDBIRD 213 or HERBIE 25.

Or

(b) The registration marking of the aircraft, for example, G-INFO or EIAKO, when:

(i) In radiotelephony, the call sign used comprises the registration marking alone, for example, G-INFO, or is preceded by the ICAO telephony designator for the operating agency, for example, SVENAIR EIAKO;

(ii) the aircraft is not equipped with radio.

Note 1: When the Aircraft Identification consists of less than 7 characters, no zeros, dashes or spaces are to be added either before or between the characters. Only alphanumeric characters are to be used. Nothing should be added after the final character of the registration. For example, an aircraft registered G-INFO would be input as GINFO.

Note 2: Appendix 2 to ICAO Doc 4444 (PANS-ATM) refers. ICAO designators and telephony designators for aircraft operating agencies are contained in ICAO Doc 8585.

4 Transponder Failure

4.1 Failure before intended departure.

4.1.1 If the transponder fails before intended departure and cannot be repaired pilots shall:

(a) Plan to proceed as directly as possible to the nearest suitable aerodrome where repair can be made;

(b) inform ATS as soon as possible preferably before the submission of a flight plan. When granting clearance to such aircraft, ATC will take into account the existing and anticipated traffic situation and may have to modify the time of departure, flight level or route of the intended flight;

(c) insert in item 10 of the ICAO flight plan under SSR the letter N for complete unserviceability of the transponder or in the case of partial failure, the letter corresponding to the remaining transponder capability as specified in ICAO Doc 4444, Appendix 2.

4.2 Failure after departure.

4.2.1 If the transponder fails after departure or en-route, ATS Units will endeavour to provide for continuation of the flight in accordance with the original flight plan. In certain traffic situations this may not be possible particularly when the failure is detected shortly after take-off. The aircraft may then be required to return to the departure aerodrome or to land at another aerodrome acceptable to the operator and to ATC. After landing, pilots shall make every effort to have the transponder restored to normal operation. If the transponder cannot be repaired then the provisions in paragraph 4.1.1 apply.

4.3 At present the temporary failure of SSR Code C alone would not restrict the normal operation of the flight.

5 Radio Telephony Phraseology For Use With SSR

This is in accordance with ICAO Doc 4444, Chapter 12, para 12.4.3.

6 UK SSR Code Assignment Plan

(as detailed in the table at ENR 1.6.2.4 to ENR 1.6.2.8 (See also notes below)).

Note 1: Pilots are not to pre-select Mode A code settings for discrete codes until instructed to do so by the appropriate controlling agency.

Note 2: The codes or series annotated with *are used for conspicuity, co-ordination or special purposes and, unless procedures have been agreed with the UK CAA Directorate of Airspace Policy, the Mode A and associated Mode C data must be considered unvalidated and unverified.

Note 3: These codes are assigned by Aberdeen for helicopters operating Northern North Sea Off-shore area and out to 5° West.

Note 4: Non-discrete assignment to military aircraft operating within the Vale of York AIAA. Leeming and Linton-on-Ouse are authorised by UK Directorate of Airspace Policy, to apply standard vertical separation between military aircraft under service and military aircraft assigned this code.

1.6.2.5 UK SSR Code Assignment Plan

Note 5: Ground based transponder equipment.
Note 6: For use within a 25nm radius of Colerne in the FIR, up to FL 100, outside of Controlled Airspace. Although both codes are for conspicuity purposes, the Mode A code 4577 will be validated and verifies.
Note 7: For use by military fixed-wing aircraft on passing 2000ft MSD in the descent to the UK Low Flying System (LFS) and retained whilst operating in the LFS. When a radar service is required on climb-out from the LFS, the code will be retained by the aircraft until alternative instructions are passed by an ATC unit.
Note 8: These codes will be assigned to aircraft on air policing missions within the UK FIR/UIR operating under the Air Defence Priority Flight (ADPF) status. NATO Air Surveillance and Control System (ASACS) units, NATO Airborne Early Warning aircraft or military ATCCs will normally control the aircraft. The actual controlling unit for a particular flight will be notified by the relevant NATA ASACS unit to the military ATC supervisor whose area of responsibility contains the intended route of the ADPF aircraft. The military ATC supervisor will then inform the relevant civil ATCC. Where any doubt about the controlling agency exists, the relevant military ATC supervisor with the responsibility for the area within which the ADPF aircraft is operating should be contacted in the first instance.
Note 9: This Code is assigned by the FISO to aircraft requesting entry into Controlled Airspace for the purpose of joining or crossing the ATS Route network. Pilots shall only select the code when they are within two-way communication with the London FIS. When available, Mode C should also be selected. If communication is lost or the aircraft leaves the FIS frequency, pilots shall deselect this special purpose code. The assignment of this code does not imply the provision of a radar service and the code and any associated Mode C must be considered to be unvalidated and unverified.
Note 10: the first code in each octal block (ie. 4610, 4630, 4640, 4650 and 4660) will be assigned to aircraft under control of the Scottish Military Allocator.

Codes/Series	Controlling Authority/Funtcion
*0000	SSR data unreliable
*0001	Height Monitoring Unit (See note 5)
*0002	Ground Transponder Testing (See note 5)
0003–0013	Not allocated
*0014–0015	Thames Radar (Flying Eye)
*0016–0017	Thames Valley Air Ambulance
*0020	Air Ambulance Helicopter Emergency Medivac
*0021	Fixed-wing aircraft (Receiving service from a ship)
*0022	Helicopter(s) (Receiving service from a ship)
*0023	Aircraft engaged in actual SAR Operations
*0024	Radar Flight Evaluation/Calibration
0025	Not allocated
*0026	Special Tasks (Mil) – activated under Special Flight Notification (SFN)
*0027	London ACC FIS Ops (See note 9)
*0030	FIR Lost
*0031	An aircraft receiving a radar service from D & D centre
*0032	Aircraft engaged in police air support operations
*0033	Aircraft Paradropping
*0034	Antenna trailing/target towing
*0035	Selected Flights – Helicopters
*0036	Helicopter Pipeline/Powerline Inspection Flights
*0037	Royal Flights – Helicopters
*0040	Civil Helicopters North Sea
*0041-0042	Greater Manchester Police ASU
*0043-0044	Metropolitan Police ASU
*0045	Sussex police ASU
*0046	Essex Police ASU
*0047	Surrey Police ASU
*0050	Chiltern Police ASU (Western Base)
*0051	Chiltern Police ASU (Eastern Base)
*0052	Norfolk Police ASU
	West Yorkshire Police ASU
*0053	Suffolk Police ASU
	South Yorkshire Police ASU

Codes/Series	Controlling Authority/Funtcion
*0054	Cambridgeshire police ASU
	Merseyside Police ASU
	South & East Wales police ASU
*0055	Northumbria Police ASU
	Cheshire Police ASU
*0057	Strathclyde Police ASU
	Humberside Police ASU
	East Midlands Police ASU
*0060	West Midlands Police ASU
*0061	Lancashire Police ASU
	Western Counties Police ASU
0062-0077	Not allocated
0100	NATO – No 1 Air Control Centre
0101-0117	Transit (ORCAM) Brussels
0120-0137	Transit (ORCAM) Germany
0140-0177	Transit (ORCAM) Amsterdam
0200	NATO – no 1 Air Control Centre
0201-0223	Stansted Approach (TC)/Essex Radar
0201-0217	RAF Leuchars
0201-0257	Ireland Domestic
	RNAS Yeovilton
*0220	RAF Leuchars Conspicuity
0220-0237	RAF Shawbury
0221-0247	RAF Leuchars
0224-0243	Anglia Radar
*0240	RAF Shawbury Conspicuity
0241-0246	RAF Shawbury
*0224	North Denes Conspicuity
*0245-0267	Anglia Radar
*0247	Cranfield Airport – IFR Conspicuity Purposes
*0252	Sheffield City Airport Conspicuity
*0260	Liverpool Airport Conspicuity
0260-0261	Oil Survey Helicopters – Faeroes/Iceland Gap
*0260-0267	Westland Helicopters Yeovil
*0260-0267	Coventry Approach
0261-0267	Liverpool Airport
0270-0277	Superdomestic – Italy to UK
*0300	NATO – No 1 Air Control Centre
0301-0377	Transit (ORCAM) UK

ENR 1.6 1.6.2.5 UK SSR Code Assignment Plan

Codes/Series	Controlling Authority/Funtcion
*0400	NATO – No 1 Air Control Centre
0401-0416	RAF Leeming
0401-0420	Birmingham Approach
0401-0430	Exeter Approach
0401-0437	Ireland Domestic
0401-0467	RAF Lakenheath
*0417	RAF Leeming Conspicuity
0420-0427	RAF Leeming
0421-0446	Farnborough Radar/LARS
0430-0443	Edinburgh Approach
*0440-0467	ACMI (NSAR – Waddington)
*0447	Farnborough LARS
	Blackbushe Departures
0450-0456	Blackpool Approach
	Farnborough Radar/LARS
*0457	Blackpool Approach (Liverpool Bay and Morecambe Bay Helicopters)
	Farnborough LARS – Fairoaks Departures
0460-0466	Blackpool Approach
0460-0467	Farnborough Radar/LARS
*0467	Blackpool Approach (Liverpool Bay and Morecambe Bay Helicopters)
0470-0477	Not allocated
0500	NATO – No 1 Air Control Centre
0501-0577	Transit (ORCAM) UK
0600	NATO – No 1 Air Control Centre
0601-0637	Transit (ORCAM) Germany
0640-0677	Transit (ORCAM) Paris
0700	NATO – No 1 Air Control Centre
0701-0777	Transit (ORCAM) Maastricht
1000	IFR GAT flights operating in designated Mode S Airspace
1001-1077	Transit (ORCAM) Spain
1100	NATO – No 1 Air Control Centre
1101-1137	Transit (ORCAM) Rhein
1140-1177	Transit (ORCAM) UK
1200	NATO – No 1 Air Control Centre
1201-1277	Channel Islands Domestic
1300	NATO – No 1 Air Control Centre
1301-1327	NATO – Air Policing (Air Defence Priority Flights) (See note 8)
1330-1357	Transit (ORCAM) Bremen
1360-1377	Transit (ORCAM) Munich
1400	NATO – No 1 Air Control Centre
1401-1407	UK Domestic
1410-1437	Superdomestic – Shannon to UK
1470-1477	Superdomestic – Dublin to UK
1500	NATO – No 1 Air Control Centre
*1501-1537	NATO – CRC Scampton
*1540-1577	NATO – CAOC 9 Exercises (activated by NOTAM)
1600	NATO – No 1 Air Control Centre
*1601-1677	NATO – CAOC 9 Exercises (activated by NOTAM)
1700	NATO – No 1 Air Control Centre

Codes/Series	Controlling Authority/Funtcion
1701-1727	NATO – CAOC 9 Exercises (activated by NOTAM)
1730-1747	Glasgow Approach
1730-1757	RAF Coningsby
1730-1750	RAF St Mawgan
*1751	RAF St Mawgan Conspicuity
1760-1777	RNAS Yeovilton Fighter Control
*2000	Aircraft from non SSR environment
2001-2077	Transit (ORCAM) Shannon
*2100	NATO – No 1 Air Control Centre
2101-2177	Transit (ORCAM) Amsterdam
*2200	NATO – CAOC 9 Exercises (activated by NOTAM)
2201-2277	Superdomestic UK to LF, LE, LP, GC and FA
*2300	NATO – CAOC 9 Exercises (activated by NOTAM)
2301-2337	Transit (ORCAM) Bordeaux
2340-2377	Transit (ORCAM) Brest
*2400	NATO – CAOC 9 Exercises (activated by NOTAM)
2401-2477	NATO – CRC Boulmer
*2500	NATO – CAOC 9 Exercises (activated by NOTAM)
2501-2577	Transit (ORCAM) Karlsruhe
*2600	NATO – CAOC 9 Exercises (activated by NOTAM)
2601-2616	Aberdeen Approach
2601-2637	RAF Cranwell
*2601-2640	RAF Spadeadam
2601-2645	MOD Boscombe Down
2601-2657	Irish Domestic Westbound departures and Eastbound arrivals
*2617	Aberdeen Airport Conspicuity
2620	Aberdeen Approach
2621-2630	Aberdeen (Sumburgh Approach)
2631-2637	Aberdeen (Northern North Sea Off-shore) (See note 3)
2640-2657	Aberdeen (Northern North Sea Off-shore – Sumburgh Sector) (See note 3)
2641-2642	RAF Cranwell – Lincolnshire AIAA
2646-2647	MOD Boscombe Down – High Risks Trial
*2650	MOD Boscombe Down Conspicuity
2650-2653	Leeds Bradford Approach
2651-2657	MOD Boscombe Down
*2654	Leeds Bradford Conspicuity
2655-2677	Leeds Bradford Approach
2660-2675	Middle Wallop
2677-2677	Aberdeen (Northern North Sea Off-shore) (See note 3)
*2676-2677	Middle Wallop Conspicuity
*2700	NATO – CAOC 9 Exercises (activated by NOTAM)
2701-2737	Transit (ORCAM) Shannon
2740-2777	Transit (ORCAM) Zurich
3000	NATO – Aircraft receiving a service from AEW aircraft
3001-3077	Transit (ORCAM) Zurich

ENR 1.6 1.6.2.5 UK SSR Code Assignment Plan

Codes/Series	Controlling Authority/Funtcion
*3100	NATO – Aircraft receiving a service from AEW aircraft
3101-3127	Transit (ORCAM) Germany
3130-3177	Transit (ORCAM) Amsterdam
3200	NATO – Aircraft receiving a service from AEW aircraft
3201-3216	UK Domestic (LACC Special Sector Codes)
3217-3220	UK Domestic
3221-3257	Superdomestic – UK to Oceanic via Shannon/Dublin
3260-3277	UK Domestic
3300	NATO – Aircraft receiving a service from AEW aircraft
3301-3304	Swanwick (Military) – Special Tasks
3305-3307	London D & D Cell
3310-3377	Swanwick (Military)
3400	NATO – Aircraft receiving a service from AEW aircraft
3401-3457	Superdomestic – UK to Germany, Netherlands and Benelux
3460-3477	Not allocated
3500	NATO – Aircraft receiving a service from AEW aircraft
3501-3507	Transit (ORCAM) Luxembourg
3510-3537	Transit (ORCAM) Maastricht
3540-3577	Transit (ORCAM) Berlin
3600	NATO – Aircraft receiving a service from AEW aircraft
3601-3623	RAF Benson
3601-3632	Scottish ATSOCA Purposes
3601-3634	RAF Waddington
3601-3647	Jersey Approach
3601-3657	Cardiff Approach
*3624	RAF Benson Conspicuity
3653-3653	RAF Odiham
3640-3665	RAF Marham
3640-3677	Aberdeen (Northern North Sea Off-shore) (See note 3)
3641-3677	BAe Warton
*3645	Cardiff Approach – St Athan Conspicuity
3646-3657	Cardiff Approach
3660-3677	Southampton Approach
*3666	RAF Marham – Visual Recovery
*3667	RAF Marhsm – FIS Conspicuity
*3700	NATO – Aircraft receiving a service from AEW aircraft
*3701-3710	BAe Woodford
3701-3710	Norwich Approach
3701-3736	RAF Brize Norton
3701-3747	Guernsey Approach
	RAF Lossiemouth
*3711	Norwich Approach Conspicuity
	Woodford Entry/Exit Lane (Woodford Inbounds and Outbounds)
*3712	Woodford Entry/Exit Lane (Manchester Inbounds)
*3713	Manchester VFR/SVFR (Outbounds)
3720-3727	RAF Valley
3720-3754	RAF Cottesmore

Codes/Series	Controlling Authority/Funtcion
3730-3766	Newcastle Approach
3730-3736	RAF Valley
*3737	RAF Valley – Visual Recovery
	RAF Brize Norton Approach Conspicuity
3740-3745	RAF Brize Norton
3740-3747	RAF Valley
*3750	RAF Valley – VFR VATA East
3750-3763	Gatwick Approach (TC)
*3751	RAF Valley – VFR VATA West
3752	RAF Valley – RIFA
*3753	RAF Valley – Low Level Helicopters
*3754	RAF Valley – Special Tasks
3755	RAF Valley
3755-3762	RAF Wittering Approach
3756-3765	RAF Valley – VATA IFR Traffic
3764-3767	Gatwick Tower
*3767	Newcastle Approach Conspicuity
3770-3777	RAF Northolt
4000	NATO – Aircraft receiving a service from AEW aircraft
4001-4077	Transit (ORCAM) Aix-en-Provence
4100	NATO – Aircraft receiving a service from AEW aircraft
4101-4127	Transit (ORCAM) Frankfurt
4130-4177	Transit (ORCAM) Dusseldorf
4200	NATO – Aircraft receiving a service from AEW aircraft
4201-4214	Heathrow Domestic
4215-4247	Superdomestic – Shannon inbound UK
*4250	Manston Conspicuity
	Humberside Approach
*4251	Manston Special/Test Flight Conspicuity
4252-4267	Manston Approach
4300	NATO – Aircraft receiving a service from AEW aircraft
4301-4307	UK Domestic
4310-4323	UK Domestic (Gatwick Special Sector Codes)
4324-4337	UK Domestic (Manchester Special Sector Codes)
4340-4353	UK Domestic (SCoACC Special Sector Codes)
4354-377	UK Domestic
4400	NATO – Aircraft receiving a service from AEW aircraft
4401-4427	Superdomestic – Brussels FIR to UK FIR
4430-4447	Superdomestic – UK to Eire and Oceanic
4500	NATO – Aircraft receiving a service from AEW aircraft
4501-4515	RAF Lyneham
4501-4547	RAF Linton-on-Ouse
	Wattisham
*4510	Prestwick Approach Conspicuity
4510-4520	Prestwick Approach
*4516-4517	RAF Lyneham FIS Conspicuity
4520-4524	RAF Lyneham
4530-4542	MOD Aberporth
4530-4567	Plymouth (Military) Radar

ENR 1.6

1.6.2.5 UK SSR Code Assignment Plan

Codes/Series	Controlling Authority/Funtcion
4550-4567	Isle of Man
4550-4573	East Midlands Approach
4574	Not allocated
*4575	RAF Leeming/RAF Linton-on-Ouse (See note 4)
	Southend Airport Conspicuity
*4576-4577	RAF Colerne Conspicuity (See note 6)
4576-4577	Vale of York AIAA Conspicuity
4600	NATO – Aircraft receiving a service from AEW aircraft
4601	Hawarden Conspicuity
4602	Hawarden IFR Procedural
4603-4607	Not allocated
4601-4667	Scottish (Military) Radar (See note 10)
4670-4676	TC Luton
4670-4677	MOD Llanbedr
*4677	Carlisle Airport Conspicuity
	Luton Airport Tower Conspicuity
4700	NATO – Aircraft receiving a service from AEW aircraft
*4701-4777	Special Events (activated by NOTAM)
5000	NATO – Aircraft receiving a service from AEW aircraft
5001-5012	LTCC Special Tasks
5013-5077	UK Domestic
5100	NATO – Aircraft receiving a service from AEW aircraft
5101-5177	UK Domestic
5200	NATO – Aircraft receiving a service from AEW aircraft
5201-5260	Transit (ORCAM) UK
5261-5270	Transit (ORCAM) Dublin to Europe
5271-5277	Transit (ORCAM) Channel Islands
5300	NATO – Aircraft receiving a service from AEW aircraft
5301-5377	Transit (ORCAM) Barcelona
5400	NATO – Aircraft receiving a service from AEW aircraft
5401-5477	UK Domestic
5500	NATO – Aircraft receiving a service from AEW aircraft
5501-5577	Transit (ORCAM) Barcelona
5600	NATO – Aircraft receiving a service from AEW aircraft
5601-5647	Transit (ORCAM) Paris
5650-5657	Transit (ORCAM) Luxembourg
5660-5677	Transit (ORCAM) Reims
5700	NATO – Aircraft receiving a service from AEW aircraft
5701-5777	Transit (ORCAM) Geneva
*6000	NATO – CAOC 9 Exercises (activated by NOTAM)
6001-6006	Not allocated
6007-6077	London (Military) Radar
*6100	NATO – CAOC 9 Exercises (activated by NOTAM)
6101-6157	London (Military) Radar
*6160	Doncaster Sheffield Conspicuity
6160-6177	Plymouth (Military) Radar

Codes/Series	Controlling Authority/Funtcion
6161-6177	Doncaster Sheffield Approach
*6200	NATO – CAOC 9 Exercises (activated by NOTAM)
6201-6227	Superdomestic – Dublin inbound UK
6230-6247	Superdomestic – UK to Scandinavia and Russia
6250-6257	Superdomestic – UK to Amsterdam
6260-6277	Superdomestic – Amsterdam to UK, Eire and Iceland
*6300	NATO – CAOC 9 Exercises (activated by NOTAM)
6301-6377	Superdomestic – UK to France
*6400	NATO – CAOC 9 Exercises (activated by NOTAM)
6401-6457	Swanwick (Military) Radar
6460-6476	Cambridge Approach
*6477	Cambridge Conspicuity
*6500	NATO – CAOC 9 Exercises (activated by NOTAM)
6501-6507	Superdomestic Ireland to UK
6510-6517	Superdomestic – UK to USA Canada & Carribbean
6520-6537	UK domestic
6540-6577	CRC Scampton
*6600	NATO – CAOC 9 Exercises (activated by NOTAM)
6601-6677	Transit (ORCAM) Germany
*6700	NATO – CAOC 9 Exercises (activated by NOTAM)
6701-6747	Transit (ORCAM) Reims
6750-6777	Transit (ORCAM) Aix-en-Provence
*7000	Conspicuity code
*7001	Military Fixed Wing Low Level Conspicuity/Climbout (See note 7)
*7002	Danger Areas General
*7003	Red Arrows Transit/Display
*7004	Conspicuity Aerobatics and Display
*7005	Autonomous military operations within active TRAs
7006	Not allocated
7007	Open Skies Observation Aircraft
7010-7027	Not allocated
7030-7045	RNSA Culdrose
7030-7047	Thames Radar – Special VFR Aldergrove Approach
7077-7077	Durham Tees Valley Airport
7077-7077	Aberdeen (Northern North Sea Off-shore) (See note 3)
*7046-7047	RNAS Culdrose Conspicuity
7050-7057	Thames Radar – London City Approach
7050-7077	RNAS Culdrose
*7067	Durham Tees Valley Airport Conspicuity
7070-7073	Thames Radar – Heathrow
7074-7076	TC Heathrow Approach
7077	Thames Radar – Battersea Helicopters
*7100	LTCC and LACC Saturation Code
7101-7177	Transit (ORCAM) Brussels
7200	RN Ships
7201-7247	Transit (ORCAM) Vienna

223

ENR 1.6

ENR 1.6.3.1 Availability of Service

Codes/Series	Controlling Authority/Funtcion
7250-7257	UK Superdomestic for destinations in France and Barcelona FIR
7260-7267	Superdomestic – Shannon/Dublin to France and Spain
7270-7277	Plymouth Radar Superdomestic for destinations in UK and France
7300	Not allocated
7301-7307	Superdomestic – Shannon Eastbound landing UK
7310-7327	Superdomestic – UK to Netherlands
7330-7347	Superdomestic – Netherlands to UK
7350-7357	RAF Coltishall
7350-7361	MOD Ops in EG D701 (Hebrides)
7350-7367	RNAS Culdrose
7350-7373	Manchester Approach
7350-7377	Bournemouth Approach/LARS
*7360	RAF Coltishall Conspicuity
7361-7377	RAF Coltishall
*7362	MOD Ops in EG D702 (Fort George)
*7363	MOD Ops in EG D703 (Tain)
*7374	Dundee Airport Conspicuity
*7375	Manchester TMA and Woodvale Local Area (Woodvale UAS Conspicuity)
*7400	MPA/DEFRA/Fishery Protection Conspicuity

Codes/Series	Controlling Authority/Funtcion
7401-7437	UK Domestic
7440-7477	Superdomestic Spain and France to UK, Ireland, Iceland and North America
*7500	Special Purpose Code – Hi-Jacking
7501-7537	Transit (ORCAM) Geneva
7540-7547	Transit (ORCAM) Bremen
7550-7577	Transit (ORCAM) Paris
*7600	Special Purpose Code – Radio Failure
7601-7607	Superdomestic – Shannon/Dublin to Nordic States
7610-7617	Superdomestic – Ireland to UK
7610-7657	Superdomestic – UK to USA, Canada & Carribean
7620-7657	Superdomestic – UK to USA, Canada, Canaries & Carribean
*7700	Special Purpose Code – Emergency
7701-7717	Superdomestic – UK to France and Spain
7720-7727	Transit (ORCAM) Munich
7730-7757	Superdomestic – Shannon Eastbound landing UK
7760-7775	Superdomestic – UK to Channel Islands
7776-7777	SSR Monitors (See note 5)

ENR 1.6.3 LOWER AIRSPACE RADAR SERVICE

1 Availability of Service

1.1 The service is available to all aircraft flying outside Controlled Airspace up to and including FL 95, within the limits of radar/radio cover. The service will be provided within approximately 30nm of each participating ATS Unit. Unless a participating ATS Unit is H24, the service will normally be available between Winter 0800 and 1700, Summer 0700 and 1600, Mondays to Fridays. However, as some participating Units may remain open to serve evening, night or weekend flying, pilots are recommended to call for the service irrespective of the published hours of ATS. If no reply is received after three consecutive calls, it should be assumed that the service is not available. Information on the operation of aerodromes outside their published hours may be obtained by telephone from the appropriate Military Air Traffic Control Centre:

(a) North of 5430N – ScATCC (Mil) Prestwick 01292 479800 Ex 6704;

(b) South of 5430N – LATCC (Mil) West Drayton 01895 426150.

1.2 LARS will not normally be available from non-H24 Units at weekends and during public holidays.

1.3 Pilots intending to operate above FL 95 may be advised to contact an appropriate ATCRU and request a RAS or RIS. However, as VHF frequencies at Military ATCRUs are not continuously monitored, unless in use, civil pilots may ask controllers to arrange a frequency on which to call the appropriate Unit.

2 Description of Service

2.1 The service provided will be a Radar Advisory Service (RAS) or Radar Information Service (RIS) as detailed at ENR 1.6.1.1/2.

2.2 Request for a RIS to be upgraded to a RAS will be accepted subject to the controller's workload and standard separation will be applied as soon as practicable. If a RAS cannot be provided the controller will continue to offer a RIS. If for any reason a RAS cannot be provided on initial pilots request the controller may offer a RIS.

2.3 Outside regulated airspace any radar service may be limited. If a radar controller considers that he cannot maintain a full radar service he will warn the pilot of the nature of the limitations which may affect the service being provided. Thereafter the pilot is expected to take the stated limitations into account in his general airmanship. In particular, warning of the limitations will be given to the pilot in the following circumstances:

(a) When the aircraft is close to the lateral or vertical limits of solid radar cover;

(b) when the aircraft is close to areas of permanent echoes or weather returns;

(c) when the aircraft is operating in areas of high traffic density;

(d) when the controller considers the performance of his radar suspect;

(e) when the controller is using SSR only.

2.4 In areas of high traffic density, controllers may have to limit RAS to the extent that standard separation from all traffic cannot be maintained and advisory avoiding action cannot be given. In these circumstances, pilots will be so advised. However standard separation will be applied between participating traffic.

2.5 Emergency Service. In emergency, pilots will be given all possible assistance.

3 Procedures

3.1 Pilots intending to use the Lower Airspace Radar Service should note the participating ATS Units close to their intended track and comply with the following procedures:

(a) When within approximately 40nm of a participating ATS Unit, establish two-way RTF communication on the appropriate frequency using the phraseology:

(b) '......(Participating ATS Unit), this is ...... (Aircraft callsign), request Lower Airspace Radar Service';

(b) The controller may be engaged on another frequency; pilots may, therefore, be asked to 'stand-by for controller'. When asked to go ahead, pilots should pass the following information:
(i) Callsign and type of aircraft;
(ii) estimated position;
(iii) heading;
(iv) Flight Level or Altitude;
(v) intention (next reporting/turning point, destination etc);
(vi) request for Radar Service (RAS or RIS);

(c) maintain a listening watch on the allocated RTF frequency;

(d) follow advice issued by controllers, or if unable to do so, advise controller of non-compliance;

(e) advise the controller when service is no longer required. Reporting of flight conditions is not required unless requested by controllers.

3.2 Aircraft will be identified and pilots so informed before radar service is given.

3.3 Under a RAS or RIS, participating LARS aircraft will be given the service in accordance with ENR 1.6.1 (Use of Radar in Air Traffic Services). Aircraft receiving RAS under LARS must be flown in accordance with the Quadrantal Rule, except during short term manoeuvres, as advised by the controller for separation against known participating traffic, or avoiding action against non participating traffic.

3.4 Whenever possible, aircraft will be handed over from controller to controller in an area of over-lapping radar cover and pilots told to 'Contact' the next Unit. When this cannot be effected, pilots will be informed of their position and advised which Unit to call for further service.

3.5 If a pilot wishes to enter regulated airspace, even though he may be in receipt of a LARS beforehand, he remains responsible for obtaining the required clearances before entry. LARS Controllers may assist in obtaining clearance, if workload permits, but pilots must be prepared to carry out this task independently.

4 Terrain Clearance

4.1 Terrain clearance will be the responsibility of pilots. However, LARS Units will set a level or levels below which a RAS is to be refused or terminated.

5 Advice to Pilots

5.1 The provision of LARS is at the discretion of the controllers concerned because they may be fully engaged in their primary tasks. Therefore, occasionally, the service may not be available.

5.2 While every effort will be made to ensure safe separation for pilots complying with RAS procedures, since compliance is not compulsory, some aircraft may not be known to controllers. Pilots should therefore keep a careful look-out at all times.

5.2.1 Farnborough and Boscombe Down Service Limitations

5.2.1.1 Due to periodic traffic congestion and high ATC workload, only a limited radar service (ENR 1.6.1.1, paragraph 2) may be available from the following ATS Units:

(a) Farnborough – Limited RIS – At all altitudes/Flight Levels. Aircraft inbound to Farnborough Approach on 134.350 MHz. On weekdays (excluding PHs) LARS/MATZ service is not normally available on 125.250 MHz after 2000hrs (one hour earlier in summer). Traffic inbound to Odiham should contact Odiham Approach on 131.300 MHz;

(b) Boscombe Down – Limited RIS – At and below FL 40. Subject to ATC workload, pilots, will be informed of any limitations to RAS and standard separation will be provided whenever possible.

6 ATS Units Participating in the Lower Airspace Radar Service
See next page

ENR 1.6

ENR 1.6.3.6 ATS Units Participating in the Lower Airspace Radar Service

Unit	Position	Frequency to be used (MHz)	Service Radius (nm)	Availability/ Remarks (P) – Denotes hours of watch as published
Boscombe Down	510912N 0014504W	126.70	30	P Not available weekends & PH
Bournemouth	504648N 0015033W	119.475	30	0730-2100 winter (summer 1hr earlier)
Bristol	512258N 0024309W	125.650	40	H24
Bristol Filton	513110N 0023527W	122.725	30	Mon-Fri 0800-1800 winter (summer 1hr earlier). Other times Bristol. (See para 7)
Brize Norton	514500N 0013500W	124.275	60	H24
Cardiff	512348N 0032036W	126.625	40	0545-2300 winter (summer 1hr earlier)
Coltishall	524518N 0012127E	125.900	40	P
Coningsby	530535N 0000958W	120.800	30	P
Cottesmore	524409N 0003856W	130.200	30	P
Culdrose	500508N 0051515W	134.050	30	P
Durham Tees Valley	543033N 0012546W	118.850	40	H24
Exeter	504404N 0032450W	128.975	30	Mon 0001-0100 & 0700-2359 Tue-Fri 0001-0200 & 0700-2359 Sat 0001-0200 & 0800-1700 Sun 0830-2359 (winter)
				Mon 0600-2359 Tue-Fri 0001-0100 & 0600-2359 Sat 0001-0100 & 0530-2000 Sun 0700-2359 (summer)
Farnborough	511633N 0004635W	125.250	30	0800-2000 winter (summer 1hr earlier)
Humberside	533428N 0002103W	119.125	30	Sun-Fri 0630-2015 Sat 0630-2000 winter (summer 1hr earlier)
Leeming	541733N 0013207W	127.750	30	P
Leuchars	562229N 0025140W	126.500	40	H24
Linton on Ouse	540258N 0011513W	118.550	30	P
Lossiemouth	574225N 0032015W	118.90	40	Mon-Fri, during Inverness Airport notified hours. Lossiemouth also provide a radar service to aircraft operating to/from Inverness Airport. Pilots should contact Lossiemouth Radar for a service. Outwith these hours a service may be available to Inverness traffic. Such occasions will be promulgated by NOTAM.
Manston	512032N 0012046E	126.350	25	0900-1700 winter (summer 1hr earlier)
Marham	523854N 0003302E	124.150	30	P
Newcastle	550217N 0014123W	124.375	40	H24

ENR 1.6 ENR 1.6.3.6 ATS Units Participating in the Lower Airspace Radar Service

Unit	Position	Frequency to be used (MHz)	Service Radius (nm)	Availability/ Remarks (P) – Denotes hours of watch as published
Plymouth Military	501900N 0040700W	121.250	40	P Plymouth Military Radar utilises radar and radios located at Portland to provide LARS in the Portland area. Pilots operating east of the western edge of AWYs N864/N862 should call Plymouth Military on frequency 124.50MHz.
Plymouth Military	503405N 0022659W	124.150	40	Pilots operating west of the western edge of AWYs N864/N862 should call Plymouth Military on frequency 121.250MHz
Shawbury	524737N 0024004W	120.775	40	P
Southend	513417N 0004144E	130.775	25	0900-1800 winter (summer 1hr earlier)
St Mawgan	502626N 0045943W	128.725	30	P
Valley	531450N 0043201W	125.225	40	P
Waddington	530958N 0003126W	127.350	30	H24
Warton	534442N 0025302W	129.525	40	Mon-Thu 0730-2000 Fri 0730-1700 winter (summer 1hr earlier)
Yeovilton	510029N 0023845W	127.350	30	P

7 Bristol Filton and Bristol LARS

7.1 Due to the impracticalities caused by the position of Bristol Controlled Airspace with regard to the LARS area served by Bristol Filton, it has been agreed that during the hours of service published for Bristol Filton LARS the following procedures will apply:

(a) Aircraft requiring a LARS north of a line between Avonmouth and the M4 junction 18, ie north of a line 513017N 0024298W and 513013N 0022093W, will receive the service from Bristol Filton.

(b) Aircraft requiring a LARS south of a line between Avonmouth and the M4 junction 18 will receive the service from Bristol.

(c) Aircraft contacting either Unit, but within the other's agreed area of responsibility, will be instructed to contact the appropriate Unit.

ENR 1.6.4 MILITARY MIDDLE AIRSPACE RADAR SERVICE

1 Availability of Service

1.1 This service is available to all aircraft flying outside Controlled airspace in the UK FIR except for flight along advisory routes and for flight within the Sumburgh FISA. It is available from FL 100 to FL 240. This service is subject to Unit capacity.

1.2 The military Units providing this service together with their boundaries are depicted on the chart at ENR 6-1-6-4. The table below shows their hours of operation, the RTF operating frequency on which this service is normally provided and a telephone number for pre-flight contact.

Unit & Callsign	Operating Hours	Initial frequency (ICF)	Contact telephone number
London Military	H24	135.275 MHz	01895 426464
Swanwick Military	H24	135.150 MHz 128.700 MHz 127.450 MHz	01489 612417
Scottish Military	H24	134.300 MHz	01292 479800 Ex 6020 or 6002

1.3 Participating aircraft must be equipped with a serviceable transponder.

2 Type of Service

2.1 The service provided will be a Radar Advisory Service or Radar Information Service (See ENR 1.6.1.1/2).

3 Procedures

3.1 In order to comply with the requirements of the FPPS at London/Swanwick Military captains of aircraft requiring a radar service in the Upper, Middle or Lower Airspace within the London/Swanwick Military area of responsibility are to pre-notify their intended flight details to London/Swanwick Military by one of the following methods:

(a) Pre-flight Notification – Flight Plans. As the preferred method of notification flight plans (F2919/CA48) should be submitted as far in advance of ETD as possible and in any case not less than 30 minutes before service is required. The London/Swanwick Military signals address – EGWDZQZX – must be included on the flight plan. When appropriate these additions to the standard flight plan format must also be included:

(i) Item 18. The point and the time at which a radar service is required to commence;

(ii) Item 15. The point of entry into the area and the point of exit.

Note: Item 15. If a flight is planned to enter any Controlled Airspace (CAS) within the London/Swanwick Military area of responsibility and a service is required before joining or after leaving CAS, both parts of the route may be entered in Item 15 of the same flight plan. In this case both IFPS – EGZYIFPS – and London/Swanwick Military EGWDZQZX must appear as addressees.

(b) Pre-Flight Notification – Military Prenote. When it has not been possible to file a flight plan, as sub-paragraph (a), relevant details of the intended flight should be telephoned by the pilot or by his aerodrome operations or ATC to London/Swanwick Military, Main Flight Plan Reception Section, (ATOTN Telephone Ext 6710) at least 15 minutes before service is required. Flight details should be passed in this order:

(i) Callsign;

(ii) number of aircraft (if more than 1) and aircraft type(s);

(iii) position and time at which service is required to commence;

(iv) speed and flight level at commencement of service;

(v) route (including any required speed or level changes);

(vi) position of leaving the delineated area (if applicable); and

(vii) destination (ICAO Location Indicator).

(c) In-Flight Notification (Air Filing). Exceptionally, when neither form of pre-flight notification has been made the flight details listed in sub-paragraph (b) above, may be notified in flight (Air Filed) by radio to:

(i) ATCRU. Airfile to the ATCRU, currently providing a service for onward transmission by them to London/Swanwick Military at least 15 minutes in advance of service being required;

(ii) London/Swanwick Military. Request radar service by calling London Radar on the appropriate (ICF), at least 5 minutes before service is required passing the details listed in sub-paragraph (b) above.

3.2 Changes to Flight Details
(a) Pre-Flight Notification. Changes to pre-flight notifications are to be passed to LATCC (Mil) as soon as possible by:
(i) Amended flight plan if time permits (as in paragraph 3.1 (a) (ii)); otherwise
(ii) by telephone (as in paragraph 3.1 (b)).
(b) In-Flight Notification (Air Filing). By RT as soon as possible (as in paragraph 3.1 (c))

ENR 1.6.5 WARTON RADAR ADVISORY SERVICE AREA (RASA)

1 Services
1.1 The service is provided by Warton Radar, who will, during the notified hours, offer aircraft a Radar Advisory Service (RAS) or a Radar Information Service (RIS) within the designated area shown in paragraph 2 below. A flight Information Service (FIS) only may be offered because of workload and/or traffic intensity.

2 Service Area

Name	Lateral Limits	Vertical Limits	Controlling Authority
Warton Radar Advisory Service Area (RASA)	544230N 0030808W – 544230N 0023409W – 541201N 0021700W – 541113N 0021858W – 535557N 0020827W – 535539N 0020919W – 534529N 0023244W – 534127N 0023240W – 533223N 0025907W – 533213N 0031406W – 535127N 0040000W – 541902N 0040000W – 544230N 0030808W	FL 245 FL 55	Warton Radar 01722-852392

The area is depicted at ENR 6.1.6.5.

3 Callsigns, Frequencies and Availability
3.1 The service is available Monday to Friday between the hours of 0730 to 2000 Winter, 0630 to 1900 Summer and Friday between the hours of 0730 to 1700 Winter, 0630 to 1600 Summer, excluding Public Holidays, on frequency 129.525MHz, callsign 'Warton Radar'.

4 Procedures
4.1 Pilots of aircraft intending to fly within the area involved during the promulgated hours:
(a) If in communication with an ATC Unit, should contact 'Warton Radar' when advised;
(b) if flying in the FIR and not in communication with an ATS Unit, are advised to contact Warton Radar prior to entering the Airspace detailed in paragraph 2.
4.2 Manchester remains the controlling authority for Advisory Route W2D.
4.3 Pilots wishing to join or cross Controlled Airspace should provide Warton Radar with an estimate for the joining or crossing points and their requested route. Warton will endeavour to obtain the appropriate clearance from 'Manchester Control'.
4.4 Traffic routeing from/to Warton/Blackpool via WAL should expect to be advised to fly via Reporting Point ESTRY, defined as 533952N 0031542W (WAL VOR/DME fix 350°/17nm). This route is designed so as to alleviate the risk of potential conflictial with intensive flying training activity associated with RAF Woodvale.
4.5 Pilots should note that a RAS or RIS will be provided below FL 55 within the limits of the Warton LARS

5 Radio Communication
5.1 Pilots receiving a service who wish to leave the frequency temporarily (for example to listen to VOLMET), and as a result will be unable to maintain two-way communications, should advise 'Warton Radar' of their intention to leave the frequency and also of their return to it.
5.2 If radio communication with 'Warton Radar' is lost, attempts should be made to establish contact in the first instance with 'Warton Radar' on frequency 129.525 MHz. If this fails then attempts should be made to establish contact with 'London Information' on 125.475 MHz.
5.3 If complete radio failure occurs, pilots should follow the standard radio failure procedure detailed at ENR 1.1.3.2/4.

6 Radar Failure
6.1 In the event of temporary radar failure, a Flight Information Service will be available on the notified frequency until the radar is restored.
6.2 In the event of protracted radar failure, 'Warton Radar' will advise pilots to contact the appropriate ATS Unit.

7 Meteorological Information
7.1 Aerodrome weather reports will not normally be given on the radar service frequency; however, weather broadcasts are available from VOLMET.

ENR 1.6   ENR 1.6.6.1 Area of Responsibility

ENR 1.6.6 – (Former) PENNINE RADAR AREA OF RESPONSIBILITY

1 Area of Responsibility

1.1 Provision of Air Traffic Services Outside Controlled Airspace (ATSOCAS) within the area defined below is a NATS licensed task which is provided, subject to unit workload, by the LATCC (Mil).
(a) 550000N 0015420W – 550000N 0004444W – 534153N 0002245E – 534134N 0010443W – 534007N 0011937W – 535348N 0013100W – 535955N 0014027W – 540236N 0014900W – 535557N 0020827W – 541113N 0021858W – 541201N 0021700W – 545127N 0023916W – 550000N 0015420W.

(b) But excluding:
(i) Aerodrome Traffic Zones (ATZ);
(ii) Danger Areas;
(iii) Managed Danger Areas (MDA) (when promulgated as active);
(iv) Newcastle CTR/CTA;
(v) Durham Tees Valley CTR/CTA;
(vi) That portion of the area derogated to Warton as part of the Warton RASA (ENR 1.6.5.1);
(vii) Services to aircraft operating to/from Newcastle and Durham Tees Valley Airports via OTBED.

1.2 Vertical limits between FL 55 and FL 245. RAS shall not be provided below 4000 ft Regional Pressure Setting (RPS). Exceptionally, a RIS may be provided below 4000 ft RPS, however, radar vectors will not be provided. (If there are high traffic levels in the Vale of York AIAA this may preclude this may preclude standard separation from being maintained and a RIS or a re-route may be offered).

1.3 Core operating hours are Mon-Fri 0700-2030 (Summer 1hr earlier), however, outside of these hours a service may still be provided H24 subject to unit capacity.

1.4 Other ATSUs providing a radar service, within the Area of Responsibility, are:
(a) Lower Airspace Radar Service (LARS) areas of Warton, Durham Tees Valley, Newcastle, Leeming, Linton-on-Ouse and Humberside;
(b) Anglia Radar Area of Responsibility (ENR 6.1.15.3).

2 Radio Communication

2.1 Pilots receiving a service who wish to leave the frequency temporarily (for example to listen to VOLMET), and as a result will be unable to maintain two-way communication, must inform the Controlling Authority of their intention to leave the frequency and also of their return to it.

2.2 The initial Contact Frequency for LATCC (Mil) is 135.275 MHz.
West of Greenwich Meridian – 5000 ft amsl and above;
East of Greenwich Meridian to Longitude 001 E – 10000 ft amsl and above;
East of Longitude 00 1 E – 15000 ft amsl and above.
The precise radio coverage obtainable at any particular time will vary with weather conditions and other factors including the airborne equipment.

2.3 If radio communication is lost, attempts should be made to establish contact with either Newcastle, Durham Tees Valley, London, Manchester or Scottish Control as appropriate to the planned route.

2.4 If complete radio failure occurs, pilots should follow the standard radio failure procedure detailed at ENR 1.1.3.2/4.

ENR 1.7 ALTIMETER SETTING PROCEDURES

1 Notification

1.1 The Selected Transition Altitudes listed in paragraph 4 are notified for the purposes of Rule 30 of the Rules of the Air Regulations 1996.

2 Introduction

2.1 The Altimeter Setting Procedures in use in the UK generally conform to those contained in ICAO Doc 8168-OPS/611. Differences are in bold.

2.2 The purpose of these procedures is to provide pilots with suitable pressure information which will assist them in maintaining adequate terrain clearance and also to ensure a safe standard of flight separation by the general use of altimeters set at 1013.2 mb.

3 General Procedures

3.1 The Transition Altitude within the UK is 3000ft except in, or beneath, that Airspace specified at paragraph 4.1.

3.2 Transition Altitudes are shown in the aerodrome directory in AD 2.17 as well as on aerodrome Approach Charts.

3.3 QNH and temperature reports for certain aerodromes are given in MET broadcasts and can also be obtained from ATS Units. These QNH values are rounded down to the nearest whole millibar but are available at certain aerodromes in tenths of millibars for landing aircraft on request.

3.4 Vertical positioning of aircraft when at, or below, any Transition Altitude will normally be expressed in terms of Altitude. Vertical positioning at, or above, any Transition Level will normally be expressed in terms of Flight Level. When descending through the Transition Layer vertical position will be expressed in terms of Altitude, and when climbing in terms of Flight Level. It should not be assumed that separation exists between the Transition Altitude and the Transition Level.

3.5 Flight Levels are measured with reference to the Standard Pressure datum of 1013.2 millibars. In the UK consecutive Flight Levels above the Transition Level are separated by pressure intervals corresponding to 500ft in the ISA and above FL 250 by pressure levels corresponding to 1000ft. FL 250 is not normally used outside the ATS route structure.

230

3.6 Altimeter setting procedures at military aerodromes may vary from those detailed in this section.

3.7 Altimeter Setting Regions (ASR). To make up for any lack of stations reporting actual QNH, the UK has been divided into a number of ASRs for each of which the National Meteorological Office calculates the lowest forecast QNH (Regional Pressure Setting) for each hour. These values are available hourly for the period H+1 to H+2 and may be obtained from all aerodromes having an Air Traffic Service, from the London, Manchester and Scottish ACCs, or by telephone.

3.8 The ASRs are listed below together with the MET Office Codes. The area covered by the Regions are shown on the combined Flight Information Region (FIR) and ASR Chart at ENR 6.1.7.1.

Skerry (01)	Holyhead (07)	Chatham (12)	Orkney (17)
Portree (02)	Barnsley (08)	Portland (13)	Marlin (18)
Rattray (03)	Humber (09)	Yarmouth (14)	Petrel (19)
Tyne (04)	Scillies (10)	Cotswold (15)	Skua (20)
Belfast (05)	Wessex (11)	Shetland (16)	Puffin (21)

3.9 Airspace within all Control Zones (CTRs), and within and below all Terminal Control Areas (TMAs), Control Areas (CTAs) except Airways and the Daventry and Worthing Control Areas, during their notified hours of operation, do not form part of the ASR Regional Pressure Setting system.

3.10 When flying in Airspace below TMAs and CTAs detailed above, pilots should use the QNH of an adjacent aerodrome when flying below the Transition Altitude. It may be assumed that for aerodromes located beneath such Areas, the differences in the QNH values are insignificant. When flying beneath Airways whose base levels are expressed as Altitudes pilots are recommended to use the QNH of an adjacent aerodrome in order to avoid penetrating the base of Controlled Airspace.

3.11 Within the Channel Islands Control Zone, the lowest forecast QNH value is available for terrain clearance purposes.

3.12 Pilots operating north of 6130N within the Airspace detailed at ENR 2.2.1.1 when not receiving a service from Sumburgh Radar are advised to set the Puffin Regional Pressure Setting as the pressure datum whilst flying at or below 3000ft.

3.13 The QNH settings to be used within the Northern North Sea Radar Service Areas are shown at ENR 6.1.15.1.

4 Selected Transition Altitudes

4.1 The following Transition Altitudes apply to flights within or beneath the following Airspace:

Aberdeen CTR/CTA	6000ft
Belfast CTR/TMA	6000ft
Birmingham CTR/CTA	4000ft
Cardiff CTR/CTA	4000ft
Durham Tees Valley CTR/CTA	6000ft †
Edinburgh CTR/CTA	6000ft
Glasgow CTR	6000ft
Leeds Bradford CTR/CTA	5000ft †
London TMA	6000ft
Manchester TMA	5000ft
Newcastle CTR/CTA	6000ft
Nottingham East Midlands CTR/CTA	4000ft
Scottish TMA	6000ft
Solent CTA	4000ft †
Sumburgh CTR/CTA	6000ft †

Note: † Outside the notified hours of operation the Transition Altitude is 3000ft.

5 Detailed Procedures

5.1 Take-off and climb

5.1.1 A QNH altimeter setting is given with the taxiing clearance prior to take-off.

5.1.2 At UK aerodromes the designated location for pre-flight altimeter checks is the apron.

5.1.3 For all major UK aerodromes, the apron elevation (or the elevation of various parts of an apron where there is significant variation between them) has been determined and the value is displayed in the flight clearance office at the aerodrome concerned. It is also given at AD 2.8.

5.1.4 Within Controlled Airspace a pilot should set one altimeter to the latest Aerodrome QNH prior to take-off. While flying at, or below, the Transition Altitude vertical position will be expressed in terms of altitude based upon the Aerodrome QNH. When cleared for climb to a Flight Level, vertical position will be expressed in terms of Flight Level, unless intermediate altitude reports have been specifically requested by Air Traffic Control.

5.1.5 Outside Controlled Airspace, a pilot may use any desired setting for take-off and climb. However, when under IFR, vertical position must be expressed in terms of Flight Level on climbing through the Transition Altitude.

5.1.6 Pilots taking-off at aerodromes beneath Terminal Control Areas and Control Areas should use aerodrome QNH when flying below the Transition Altitude and beneath these Areas, except that the aerodrome QFE may be used when flying within the circuit. It may be assumed that for aerodromes beneath the same TMA or CTA the differences in their QNH values are insignificant

5.2 En-route

5.2.1 Within Controlled Airspace

5.2.1.1 At and above the transition level and during en-route flight the aircraft should be flown at Flight Levels. The latest and most appropriate Regional Pressure Setting value is to be used for checking terrain clearance in flight. Aircraft flying in a Control Zone or TMA at an Altitude at or below the Transition Altitude will be given the appropriate QNH setting in their clearance to enter the Zone/TMA.

5.2.2 Outside Controlled Airspace

5.2.2.1 In flight at or below 3000ft amsl, pilots may use any desired setting. However, pilots flying beneath a TMA or CTA should use the QNH of an aerodrome situated beneath that area when flying below the Transition Altitude. It may be assumed that for aerodromes beneath the same TMA or CTA, the differences in the QNH values are insignificant. References to vertical position in flight plans and communications with ATC are to be expressed in terms of altitude. Pilots in flight at or below 3000ft amsl on an Advisory Route should set altimeters to the appropriate Regional Pressure Setting.

5.2.2.2 When flying under IFR above the Transition Altitude pilots must have 1013.2 mb set on an altimeter and conform to the Quadrantal Rule in accordance with ENR 1.7.4, paragraph 6.1 when flying at/below FL 245, and the Semi-circular Rule when above FL 250. When flying under VFR pilots are recommended to conform to the Quadrantal Rule and Semi-circular Rule as appropriate. The latest and most appropriate lowest forecast Regional Pressure Setting value should be used for checking terrain clearance.

5.3 Approach and Landing

5.3.1 When an aircraft is descended from a Flight Level to an Altitude preparatory to commencing approach for landing, ATC will pass the appropriate aerodrome QNH. On vacating the Flight Level, the pilot will change to the aerodrome QNH unless further Flight Level vacating reports have been requested by ATC, in which case, the aerodrome QNH will be set following the final Flight Level vacating report. Thereafter, the pilot will continue to fly on the aerodrome QNH until established on final approach when QFE or any other desired setting may be used. However, ATC (except at certain military aerodromes (See Note)) will assume that an aircraft is using QFE on final approach when carrying out a radar approach and any heights passed by the radar controller will be related to QFE datum. A reminder of the assumed setting will be included in the RTF phraseology. To ensure the greatest possible degree of safety and uniformity, it is recommended that all pilots use QFE but, if the pilot advises that he is using QNH, heights will be amended as necessary and 'Altitude' will be substituted for 'height' in the RTF phraseology. It should be noted that the Obstacle Clearance Height is always given with reference to the aerodrome or threshold elevation.

5.3.2 If it is known that a particular company uses, or a pilot has clearly indicated that he will use QNH during the final approach, the controller may omit QFE and substitute QNH and the relevant elevation in the appropriate messages.

5.3.3 Vertical positioning of aircraft during approach will, below transition level, be controlled by reference to Altitudes and then to heights. The Transition Altitude is not normally given in the approach and landing clearance.

5.3.4 Pilots landing at aerodromes beneath Terminal Control Areas and Control Areas should use aerodrome QNH when flying below the Transition Altitude and beneath these Areas, except that the aerodrome QFE may be used when flying within the circuit. It may be assumed that for aerodromes within the boundary of the same TMA or CTA, differences in their QNH values are insignificant.

5.3.5 The threshold elevation of each instrument runway that is 7ft or more below the aerodrome elevation is given at AD 2.12. The barometric pressure setting to be used for landing on such a runway will be passed by ATC as QNH …… threshold elevation …… , or QFE …… threshold ……

Note: At USAF operated aerodromes QFE is not used. All procedures below the Transition Altitude will be based on aerodrome QNH, and all vertical displacements given as altitudes. Aerodrome QFE will be available on request

5.4 Missed Approach

5.4.1 In the event of a missed approach, pilots may continue to use the altimeter setting selected for final approach, but reference to the vertical position of the aircraft exchanged in communication with ATC should be expressed in terms of altitude on aerodrome QNH, unless otherwise instructed by ATC.

5.5 Flight Planning

5.5.1 The levels at which a flight is to be conducted are to be specified in the flight plan:

(a) In terms of flight level numbers for that part of the flight to be conducted at or above the transition level; and

(b) in terms of altitudes (ft) for that part of the flight at or below the Transition Altitude.

5.5.2 The flight level or levels selected for a flight should ensure adequate terrain clearance at all points along the route to be flown, should meet Air Traffic Service requirements and should comply with the Quadrantal or Semi-Circular Rule, where applicable.

5.5.3 The information required to determine the lowest flight level to ensure adequate terrain clearance may be obtained from the appropriate Air Traffic Service Unit or Meteorological Office. A chart for converting QNH values to flight levels for this purpose is shown at ENR 1.7.5.

5.5.4 The Transition Altitude applicable to the aerodromes of departure and destination and for alternate aerodromes should be noted.

5.5.5 Flight levels are to be specified in the flight plan by number and not in terms of ft as is the case with altitudes.

5.5.6 When flight plans are filed for IFR flights above 3000ft amsl outside Controlled Airspace in the UK, the cruising level must be selected from the appropriate table at paragraph 6.1.

6 Tables of Cruising Levels

6.1 The cruising levels to be observed when so required are as follows:

(a) Within Controlled Airspace in areas where, on the basis of regional air navigation agreement and in accordance with conditions specified therein, a vertical separation minimum (VSM) of 1000ft (300 m) is applied between FL 290 and FL 410 inclusive (†);

IFR Flights at levels below 24500ft		IFR Flights at levels above 24500ft	
Magnetic track °	Cruising level	Magnetic track °	Cruising level
Less than 180	1000ft and above at intervals of 2000ft	Less than 180	25000ft and above at intervals of 2000ft 41000ft and above at intervals of 4000ft
180 but less than 360	2000ft and above at intervals of 2000ft	180 but less than 360	26000ft and above at intervals of 2000ft to 40000ft 43000ft and above at intervals of 4000ft

(†) Except when, on the basis of regional air navigation agreements, a modified table of cruising levels based on a nominal vertical separation minimum of 300 m (1000ft) is prescribed for use, under specified conditions, by aircraft operating above FL 410 within designated portions of the airspace.

(b) IFR Flights Outside Controlled Airspace above 3000ft amsl;

IFR Flights at levels below 24500ft	
Magnetic track °	Cruising level
Less than 090°	Odd thousands of ft
090° but less than 180°	Odd thousands plus 500ft
180° but less than 270°	Even thousands of ft
270° but less than 360°	Even thousands plus 500ft

ENR 1.7 ENR 1.7.6 Tables of Cruising Levels

FLIGHT LEVEL GRAPH

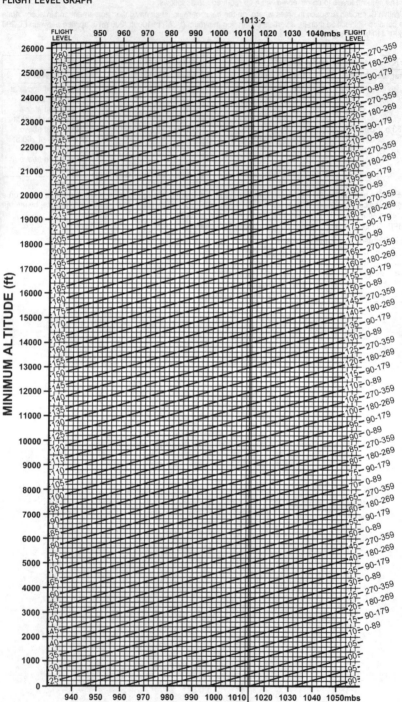

ENR 1.8 REGIONAL SUPPLEMENTARY PROCEDURES (DOC 7030) REGIONAL SUPPLEMENTARY PROCEDURES ARE APPLIED IN ACCORDANCE WITH ICAO DOC 7030/4 REGIONAL SUPPLEMENTARY PROCEDURES

1 RSVM
1.1 The airspace within the London and Scottish UIRs between FL 290 and FL 410 inclusive is an element of EUR RVSM airspace. Furthermore, the airspace within the London and Scottish UIRs between FL 290 and FL 410 inclusive is designated for the purpose of transitioning non-RVSM approved aircraft operating to/from the North Atlantic Region.
1.2 Within this airspace, the vertical separation minimum shall be:
(a) 300 metres (1000ft) between RVSM approved aircraft;
(b) 600 metres (2000ft) between:
(i) non-RVSM approved aircraft and any other aircraft operating within the EUR RVSM airspace;
(ii) formation flights of State aircraft and any other aircraft operating within the EUR RVSM airspace; and
(iii) an aircraft experiencing a communications failure in flight and any other aircraft, when both aircraft are operating within the EUR RVSM airspace.

ENR 1.10 FLIGHT PLANNING

1 General Procedures
1.1 Reference Documents
1.1.1 – ICAO Annex 2, Chapter 3.
– ICAO Doc 4444 Air Traffic Management – PANS ATM/501/14, Chapter 4,
– Chapter 10 and Appendices 2 and 3.
– ICAO Doc 7030/4 Regional Supplementary Procedures, Part EUR.
– Central Flow Management Unit (CFMU) Handbook.
– Integrated Initial Flight Plan Processing System (IFPS) Users Manual.
– CAP 694 The UK Flight Plan Guide

1.2 Types and Categories of Flight Plan
1.2.1 There are two types of flight plan:
(a) Visual Flight Rules (VFR) flight plan;
(b) Instrument Flight Rules (IFR) flight plan.
1.2.2 Flight plans fall into three categories:
(a) Full flight plans: the information filed on Form CA48/RAF 2919;
(b) Repetitive Flight Plans (see paragraph 5);
(c) Abbreviated Flight Plans: the limited information required to obtain a clearance for a portion of flight (eg: flying in a Control Zone, crossing an Airway) filed either by telephone prior to take-off or by RTF when airborne. The destination aerodrome will be advised of the flight only if the flight plan information covers the whole route of the flight.
1.2.3 Full and Abbreviated flight plans may be filed by RTF with the appropriate controlling Air Traffic Service Unit (ATSU).
1.3 A guide to filing a flight plan is shown at page ENR 1.10.3.

1.4 When to file a Flight Plan
1.4.1 A flight plan may be filed for any flight.
1.4.2 A flight plan must be filed for the following:
(a) for all flights within Class A Airspace;
(b) for all flights within any Controlled Airspace in IMC or at night, except for those operating under SVFR;
(c) for all flights within any Controlled Airspace in VMC if the flight is to be conducted in accordance with IFR;
(d) for all flights within Class B, C & D Controlled Airspace irrespective of weather conditions;
(e) for any flight from an aerodrome in the United Kingdom, being a flight whose destination is more than 40km from the aerodrome of departure and the aircraft Maximum Total Weight Authorised exceeds 5700kg;
(f) for all flights to or from the United Kingdom which will cross the United Kingdom FIR Boundary;
(g) for any flight in Class F Airspace wishing to participate in the Air Traffic Advisory Service.
1.4.3 The occasions on which a VFR flight plan must be filed are specified at paragraph 1.4.2, sub-paras (d), (e), (f) and (g) (further details on VFR flight plans are at paragraph 3).
1.4.4 It is advisable to file a flight plan if the flight involves flying over the sea, more than 10nm from the UK coastline, or over sparsely populated areas where Search and Rescue operations would be difficult.

1.5 Booking Out
1.5.1 Rule 20 of the Rules of the Air Regulations 1996 requires that a pilot intending to make a flight shall inform the Air Traffic Service Unit (ATSU) at the aerodrome of departure; the filing of a flight plan constitutes compliance with this Rule. In the absence of an ATSU at the departure aerodrome, the pilot may submit his flight plan through the Parent Unit (see paragraph 2). However, the requirements of Rule 20 must be complied with irrespective of whether or not a flight plan has been filed. Therefore, on those occasions when there is no necessity to submit a flight plan, the pilot remains responsible for notifying the ATSU at the departure aerodrome of his intention to fly. This action is known as 'Booking Out' but unlike the normal flight plan procedure, the information will not be transmitted to any other ATSU.

1.6 Submission Time Parameters

1.6.1 Normally, flight plans should be filed on the ground at least 60 minutes before clearance to start up or taxi is requested; however, for North Atlantic and flights subject to Air Traffic Flow Management (ATFM) measures a minimum of 3 hours is required. (When completing the flight plan the departure time entered in Field 13 must be the Estimated Off Block Time (EOBT) not the planned airborne time). Exceptionally, in cases where it is impossible to meet this requirement, operators should give as much notice as possible and never less than thirty minutes. Otherwise, if this is not possible, a flight plan can be filed when airborne with any ATSU, but normally with the FIR Controller responsible for the area in which the aircraft is flying. If the airborne flight plan contains an intention to enter Controlled Airspace or certain Control Zones/Control Areas, at least 10 minutes prior warning of entry must be given. In all cases, the message should start with the words 'I wish to file an airborne flight plan'. It should be noted that passing an airborne flight plan over the RT may, due to the controller's workload, result in a delay in the message being filed.

1.7 Submitting a Flight Plan Through the Departure Aerodrome ATSU

1.7.1 A written flight plan, which is filed through the ATSU at the departure aerodrome, must be submitted on Form CA 48/RAF 2919. The local ATSU may assist in compiling the flight plan details and checking them; however, the ultimate responsibility for filing an accurate flight plan rests with the pilot or the operator. If the departure aerodrome is not connected to the AFTN, the pilot is responsible for arranging for the details of the flight plan to be passed to the appropriate Parent Unit.

1.8 Persons On Board

1.8.1 The number of persons on board a flight for which a plan has been filed must be available to ATSUs for SAR purposes for the period up to the ETA at the destination plus one hour. If this information has been sent to the Operators handling agency at destination, no further action is required. Otherwise, this information is to be made available as follows:

(a) Where the operator or departure handling agency closes down before the ETA of a flight at destination plus one hour, the operator or departure handling agency will lodge the number of persons on board with the ATSU serving the aerodrome of departure;

(b) where the departure ATSU closes down before the ETA plus one hour, the ATSU will lodge the number of persons on board directly with the appropriate ACC;

(c) at aerodromes without an ATSU, where the aerodrome closes down before ETA destination plus one hour, the aerodrome operator or departure handling agency will lodge the name and address of officials who have access to flight departure records with the appropriate ACC, so that they can be contacted as necessary, either direct, or through the local police.

1.9 Action in the Event of Diversion

1.9.1 If a pilot lands at an aerodrome other than the destination specified in the flight plan, he must ensure that the ATSU at the original destination is informed within 30 minutes of his flight planned ETA, to avoid unnecessary action being taken by the Alerting Services.

1.10 Cancelling an IFR Flight Plan in Flight

1.10.1 If a pilot has begun a flight in Controlled Airspace under an IFR flight plan, he may decide, on entering VMC, that he will cancel his IFR flight plan and VFR (Rule 31(3) of the Rules of the Air Regulations 1996). However, it must be stressed that a pilot cannot exercise this choice in Controlled Airspace which is notified as Class A Airspace and, therefore, in which all flights in all weather conditions are subject to IFR procedures. In Controlled Airspace where the exercise of the pilots choice is possible, pilots may request the cancellation of IFR flight plans by notifying the ACC, provided that they are operating in VMC. An IFR flight plan may be cancelled by transmitting the following message: '.......(identification) – Cancel IFR flight plan'. ATC cannot approve or disapprove cancellation of an IFR flight plan but, when in possession of information that IMC is likely to be encountered along the intended route of flight, will advise the pilot accordingly as follows:

'IMC reported (or forecast) in the vicinity of ....'

The fact that a pilot reports that he is flying in VMC does not in itself constitute cancellation of an IFR flight plan. Unless cancellation action is taken, the flight will continue to be regulated in relation to other IFR traffic.

ENR 1.10.2 UK Parent Unit System

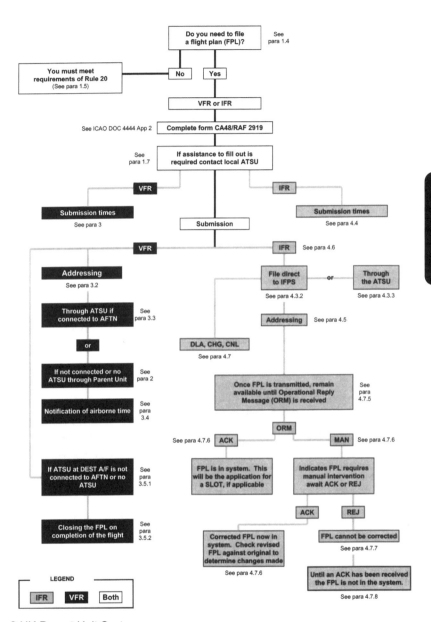

2 UK Parent Unit System

2.1 Facilities exist within the UK for the interchange of messages for aerodromes not connected to the AFTN, and also for aerodromes without an ATSU, through the use of nominated ATSUs which have the capabilities to act as Parent ATSUs (Parent Units).

2.2 Areas of Responsibility

2.2.1 The map at ENR 6.1.10.1 shows the associated area of responsibility for each Parent Unit which provides the services specified in the above paragraphs. Any operator, at an aerodrome which does not have an ATSU, or is not on the AFTN, wishing to file a flight plan should pass details of the flight plan to the Parent Unit within whose area of responsibility the aerodrome lies. The staff at the Parent Unit will assist in the completion of the flight plan and will address it appropriately for processing through the AFTN. When specific addresses are required by the pilot or the operator, in addition to those normally inserted by the ATSU for the flight being planned, it should be ensured that such requirements are notified at the time of filing the flight plan. Operators and pilots are reminded that paragraph 1.6 (time requirement for filing a flight plan) is most important when filing with the Parent Unit.

2.3 Departure Time

2.3.1 The FIR Controller will accept departure times from pilots who have departed from aerodromes where there is no ATSU, or it is outside the ATSUs hours of operation. The pilot is to advise the FIR Controller to pass the departure time to the ATSU to which the flight plan was submitted.

2.3.2 When it is known by a pilot that the ATSU at the departure aerodrome is going to be closed at the time of departure, the flight plan is to be filed with the Parent Unit and the airborne time passed as described in paragraph 2.3.1.

2.4 Changes, Delays or Cancellation of a Flight Plan

2.4.1 It is essential that ATC is advised of cancellations, delays over 30 minutes and changes to flight plan details. A second flight plan cannot be used to amend the first. The original flight plan must first be cancelled and then a revised flight plan filed.

2.5 Contact Numbers for Parent Units

Parent Unit	Telephone Number	Fax Number
London Heathrow	020 8745 3111/3163	020 8745 3491/3492
Manchester	0161 499 5502/5500	0161 499 5504
Scottish ACC	01292 692679/692663	01292 671048

3 VFR Flight Plans

3.1 When to File a VFR Flight Plan

3.1.1 A VFR flight plan may be filed for any flight.

3.1.2 A VFR flight plan must be filed for the following:

(a) For all flights to or from the United Kingdom which will cross the United Kingdom FIR Boundary;

(b) for all flights within Class B – D Controlled Airspace (this requirement may be satisfied by passing flight details on RTF);

(c) for any flight in Class F Airspace wishing to participate in the Air Traffic Advisory Service;

(d) for any flight from an aerodrome in the United Kingdom, being a flight whose destination is more than 40km from the aerodrome of departure and the aircraft Maximum Total Weight Authorised exceeds 5700kg.

3.1.3 It is advisable to file a VFR flight plan if the flight involves flying over the sea, more than 10nm from the UK coastline, or over sparsely populated areas where Search and Rescue operations would be difficult.

3.2 Addressing of VFR Flight Plans

3.2.1 In addition to addressing a VFR flight plan to the Destination Aerodrome, and when applicable the appropriate adjacent foreign FIR(s), it must also be addressed to the appropriate UK FIR(s) as listed below:

    EGZYVFRP   Scottish and Oceanic FIRs;
    EGZYVFRT   London FIR.

3.2.2 VFR Flight Plans with portion(s) of flight operated as IFR

3.2.2.1 IFPS is the only source for the distribution of IFR/General Air Traffic (GAT) flight plans and associated messages to ATSUs within the participating European States – the IFPS Zone. Although IFPS handles IFR flight plans, it will not process the VFR portions of any mixed VFR/IFR flight plan. Therefore, in order to ensure that all relevant ATSUs are included in the flight plan message distribution, pilots or Aircraft Operators should make certain that whenever a flight plan contains portions of the flight operated under VFR, in addition to IFR, the FPL must be addressed to:

IFPS (EGZY1FS)
Aerodrome of departure
Aerodrome of destination
All FIRs that the flight will route through as VFR (in UK address to EG2YVFRP for Scottish/Oceanic FIRs and/or EG2YVFRT for London FIR)
Any additional addressees specifically required by State or Aerodrome Authorities

3.3 Submission Time Parameters

3.3.1 VFR flight plans should be submitted to the ATSU at the departure aerodrome on Form CA 48/RAF 2919 at least 60 minutes before clearance to start up or taxi is requested. The local ATSU, if required, will assist in compiling the flight plan. If the departure aerodrome is not connected to the AFTN, the pilot is responsible for arranging for the ATSU to despatch the completed flight plan via the Parent Unit (see paragraph 2). If the departure aerodrome has no ATSU, the pilot will arrange for the flight plan to be passed to the aerodromes Parent Unit for onward transmission.

3.4 Airborne Time

3.4.1 The pilot is responsible for ensuring that the airborne time of the flight is passed to the ATSU with whom the flight plan has been filed. The ATSU will ensure that the departure message, if required, is sent to the appropriate addressees. The pilot should try to arrange for a 'responsible person' on the ground to telephone the airborne time to the ATSU, as passing it over the RTF may, due to controller workload, lead to a delay in sending a departure message. Failure to pass the airborne time will result in the flight plan remaining inactive; consequently, this could result in the destination aerodrome not being aware that alerting action should be taken.

3.5 Action When the Destination Aerodrome has no ATSU or AFTN Link

3.5.1 If a pilot has filed a VFR flight plan to a destination which does not have an active ATSU and is not connected to the AFTN, he is required to pass the ETA, prior to departure, to a 'responsible person' at the destination aerodrome. In the event of the aircraft failing to arrive at the destination aerodrome within 30 minutes of the notified ETA, the 'responsible person' must immediately advise the Parent Unit. This action is the trigger by which the Parent Unit will commence alerting action.

3.5.2 Exceptionally, where a pilot is unable to find someone to act as a 'responsible person' at the destination aerodrome, he must contact the appropriate Parent Unit prior to departure and request that it acts in this capacity. Should a pilot need to take this course of action, he will be required to contact the Parent Unit within 30 minutes of landing at the destination to confirm his arrival. Failure to complete this action will automatically result in the Parent Unit initiating alerting action.

4 IFR Flight Plans
4.1 Introduction
4.1.1 The UK is a participating State in the Integrated Initial Flight Plan Processing System (IFPS), which is an integral part of the Eurocontrol centralised Air Traffic Flow Management initiative. The IFPS is the sole source for distribution of IFR/General Air Traffic (GAT) flight plan information to ATSUs within the participating European States, which collectively comprise the IFPS Zone (see map at ENR 6.1.10.2). Additionally, IFPS provides accurate flight data to the Air Traffic Flow Management (ATFM) elements of the Central Flow Management Unit (CFMU), located at Haren, Brussels.

4.1.2 IFPS will not handle VFR flight plans or Operational Air Traffic (OAT) flights; however, it will process the GAT portion(s) of a mixed OAT/GAT flight plan and, similarly, the IFR portion(s) of a VFR/IFR flight plan.

4.1.3 Full details of the procedures relating to IFPS and ATFM are contained within the relevant sections of 'IFPS Users Manual' and 'The CFMU Handbook' which are available, free of charge, from:

Eurocontrol Library
Rue de la Fusee
96 B-1130 Brussels
Belgium
Tel No: +32 2 729 36 39
Fax No: +32 2 729 91 09

4.2 General Description of IFPS
4.2.1 IFPS comprises 2 IFPS Units (IFPU) sited within the Eurocontrol facilities at Haren, Brussels and at Bretigny, Paris. The IFPS Zone is divided into 2 separate geographical areas, each IFPU having a primary responsibility for one area and a secondary role, for contingency purposes, for the other. All IFR/GAT flight plans and associated messages must be addressed to both IFPUs. The primary IFPU will process the flight plan, or associated message, whilst the other will hold both the raw and processed data, to be used in the event of failure of the primary Unit. Following successful processing the flight plan will be delivered, at the appropriate time, to all the ATSU addressees on the flight profiled route within the IFPS Zone.

4.3 Filing of Flight Plans and Associated Messages
4.3.1 Aircraft Operators (AO) are ultimately responsible for the complete filing of their IFR/GAT flight plans and associated messages. This encompasses compilation (including addressing), accuracy and submission of flight plans and also for the reception of an Acknowledgement (ACK) message from IFPS (see paragraph 4.7.5).

4.3.2 AOs who have the facilities and are prepared to file their own flight plans and associated messages directly with IFPS and any other non – IFPS States affected by the flight (see paragraph 4.5.4) may do so. This is the standard IFPS IFR/GAT flight plan filing procedure and is termed 'direct filing'.

4.3.3 AOs who, for whatever reason, are unable to conform to the direct filing procedure should make local arrangements to file their IFR/GAT flight plans using one of the following methods:
(a) Through the ATSU at the aerodrome of departure; or
(b) for operators at aerodromes where the ATSU is not connected to the AFTN, or alternatively where there is no ATSU, through the designated Parent Unit.

4.3.4 The occasions on which an Arrival (ARR) message must be issued are minimal (ie, when an aircraft has diverted or when a controlled flight has experienced loss of radio communication). In each instance the responsibility for issuing an ARR message will rest with the ATSU at the landing aerodrome.

4.3.5 Within the UK ATSUs at the aerodrome of departure will continue, when appropriate, to assist in the compilation of flight plans. However, the responsibility, as specified at paragraph 4.3.1, continues to rest with the AO.

4.4 Submission of Flight Plans
4.4.1 Flight Plans should be filed a minimum of 3 hours before Estimated Off Block Time (EOBT) for North Atlantic flights and those flights subject to ATFM measures, and a minimum of 60 minutes before EOBT for other flights. (The CFMU has made it known that there is limited flexibility within the system to allow for the handling of special or late notice flights).

4.5 Addressing of IFR Flight Plans
4.5.1 Flights Wholly Within the IFPS Zone. For IFR/GAT flight plans and associated messages, for flights conducted wholly within the IFPS Zone, it will be necessary to address these messages only to the two IFPUs. To further simplify AFTN addressing a single collective address, EGZYIFPS, which covers both IFPUs, has been established. The individual IFPU addresses are:

	AFTN	SITA
Haren	EBBDZMFP	BRUEP7X
Bretigny	LFPYZMFP	PAREP7X

All flight plans and associated messages must be addressed to both IFPUs; this can be achieved by using either the AFTN collective or individual addresses or, alternatively, by using the individual SITA addresses.

4.5.2 Flights Entering or Over flying the IFPS Zone. For that portion of the flight within the IFPS Zone, only the two IFPUs need to be addressed as in paragraph 4.5.1.

4.5.3 Flights Departing from an Aerodrome Within, and then Exiting, the IFPS Zone. For that portion of the flight within the IFPS Zone, only the two IFPUs need be addressed as indicated in paragraph 4.5.1. For any portion(s) of the flight outside the IFPS Zone, the AO is responsible for ensuring that the flight plan or associated message is addressed to all appropriate ATSUs, in accordance with ICAO procedures. The procedure described at paragraph 4.5.4 is the preferred method of addressing.

4.5.4 The Re-Addressing Function. The purpose of the re-addressing function is to ensure consistency between messages distributed both within and outside the IFPS Zone. This consistency is achieved by ensuring that data is not distributed to external addresses until it is successfully processed by IFPS. Any additional addresses to be included should be inserted after the Originator Information line and immediately before the open bracket which indicates the beginning of the message text (see Notes). An example of an AFTN message with such additional addresses included is shown below:

ZCZC BOC548 250925 MB
FF EBBDZMFP LFPYZMFP
250920 EGLLZPZX
AD ADDRESS1 ADDRESS2 ADDRESS3 ...... ADDRESS7
AD ADDRESS8
(FPL-BAW83-IS-B767/H-SXR/C-EGLL1430...... RMK/TCAS EQUIPPED)
NNNN

Note 1: The extra address lines must begin with the keyword **AD** to distinguish them from other comment lines which may be present.
Note 2: The extra address lines must be consecutive (no other comment lines between them), and they must be immediately before the line containing the open bracket.
Note 3: There must be no more than 7 addresses per line and each must be of 8 characters.
Note 4: Specific addresses for any VFR or OAT portion(s) of the flight also have to be added, preferably by using the re-addressing function.
Note 5: Changes or amendments made to a flight plan by an IFPS Operator, following co-ordination with the originator of the message, will be present in the flight plan message which is distributed by IFPS to the appropriate addresses within the IFPS Zone and to those outside, if the re-addressing function is used. Additional addresses which appear in the normal address line of the flight plan will not receive such amendments from IFPS.

4.5.5 IFR Flight Plans with portion(s) of flight operated as VFR
4.5.5.1 IFPS is the only source for the distribution of IFR/General Air Traffic (GAT) flight plans and associated messages to ATSUs within the participating European States – the IFPS Zone. Although IFPS handles IFR flight plans, it **will not** process the VFR portions of any mixed VFR/IFR flight plan. Therefore, in order to ensure that all relevant ATSUs are included in the flight plan message distribution, pilots or Aircraft Operators should make certain that whenever a flight plan contains portions of the flight operated under VFR, in addition to IFR, the FPL must be addressed to:

IFPS (EGZYIFPS)
Aerodrome of departure
Aerodrome of destination
All FIRs that the flight will route through as VFR (in UK address to EGZYVFR for Scottish/Oceanic FIRs and/or EGZYVFRT for London FIR)
Any additional addressees specifically required by State or Aerodrome Authorities

4.6 Compilation of Flight Plans
4.6.1 The compilation of filed flight plans must be in accordance with the procedures specified in ICAO Doc 4444 and the IFPS Users Manual.
4.6.2 IFPS will not accept flight plans submitted more than 144 hours in advance of the flight taking place. If a flight plan is submitted for a flight which will not commence within the next 24 hours, the Estimated Off Block Date (EOBD) must be indicated in Field 18 of the flight plan using the ICAO convention, i.e. DOF/020905 (Year; Month; Day).
4.6.3 To indicate the necessity for 'special handling', the appropriate Status Indicator (STS) should be inserted in Field 18 of the flight plan.
4.6.3.1 The following standardised abbreviations should be used:

STS/EMER	for flights in a state of emergency;
STS/HOSP	for medical flights specifically declared by the medical authorities;
STS/SAR	for flights engaged in Search and Rescue missions;
STS/HUM	for flights operating for humanitarian reasons;
STS/HEAD	for flights with 'Head of State' status;
STS/STATE	for flights other than 'Head of State' specifically required by State authorities;
STS/PROTECTED	for use in flight plans which should only be available to those who 'need to know'. Normally flights that are security sensitive.
STS/ATFMEXEMPTAPPROVED	for use only when approval has been obtained from the appropriate State authority for exemption from flow regulation.

If more than one designator is to be used, each should be inserted as a separate STS/entry within field 18 of the Flight Plan form.
4.6.3.2 The following STS/indicators will be recognised by the CFMU and will be provided with automatic exemption from flow regulation:
STS/EMER; STS/HEAD; STS/SAR and STS/ATFMEXEMPTAPPROVED.
4.6.3.3 The following STS/indicators require approval for exemption from flow regulation from the appropriate State authorities, in accordance with the requirements detailed in the ATFM Users Handbook and in the ENR 1.9 section:
STS/HUM; STS/HOSP and STS/STATE.
4.6.3.4 In addition to military operations, operators of customs or police aircraft shall insert the letter M in Item 8 of the ICAO Flight Plan Form.

4.6.3.5 For formation flights that intend to operate – for any part – as GAT, it is essential for en-route ATC Providers to have as much notification as possible in order for planning to take place. Although use of FPL Item 9 in the current ICAO standard Flight Plan Proforma provides for indication of the number of aircraft (if more than one operating under the same callsign), the ability of some ATC Flight Data Processing Systems to detect and highlight this to control staff may not be robust; this is especially the case where Air-to-air refuelling tanker aircraft file as singleton, only to include an FPL Item 18 remark that it will be joined by other aircraft which have filed separate flight plans. To this end, commanders of all planned GAT formation flights are requested to enter RMK/Formation flight in FPL Item 18 of their flight plan to ensure that ATC Flight Data Processing Systems can detect and promulgate such information correctly to control staff. Any queries should be directed to Head of ATC Operational Support, London ACC +44 (0) 1489 612590.

4.6.4 Replacement Flight Plan Procedure. If, within 4 hours of the EOBT, an alternative routeing is selected between the same points of departure and destination, the procedure shall be as follows:

(a) The original Flight Plan **must be cancelled** by submitting a CNL message using the DD priority indicator;

(b) the replacement Flight Plan shall be filed not less than 5 minutes after the CNL message (It is recommended that the replacement Flight Plan is not submitted until the ACK for the CNL message has been received);

(c) the replacement Flight Plan shall contain in Field 18 the indication RFP/Qn where:

(1) **RFP/Q** refers to the replacement Flight Plan; and

(2) **n** corresponds to the sequence number relating to the replacement Flight Plan.

Example: First replacement Flight Plan – ICAO Field 18 – **RFP/Q1**;
Second replacement Flight Plan – ICAO Field 18 – **RFP/Q2**.

4.7 Compilation of Associated Messages

4.7.1 The compilation of DEP, ARR, CHG, DLA and CNL messages is detailed in ICAO Doc 4444, Appendix 3. However, as IFPS identifies messages by 'Key Fields', the following cannot be changed by use of a CHG message;
AIRCRAFT CALLSIGN
DEPARTURE AERODROME
DESTINATION AERODROME
EOBT
EOBD.

4.7.2 To change one of the fields listed at paragraph 4.7.1 (except by means of a DLA to EOBT) it will be necessary to cancel the original flight plan and, after a time lapse of **at least** 5 minutes, refile a flight plan containing the corrected data.

4.7.3 It is a European ATFM requirement that all controlled flights that are departing, arriving or over-flying Europe that have a change (+ or –) in an EOBT of more than 15 minutes shall be notified to the CFMU through IFPS. Modification procedures are, therefore, necessary to enable Aircraft Operators (AOs) to meet this requirement whenever they know that a flight will not meet its EOBT.

Note 1: AOs should not modify the EOBT simply as a result of an ATFM delay. The EOBT is to be only if the original EOBT established by the AOs cannot be met. It is not possible to amend the EOBT to an earlier time than the EOBT given in the flight plan. The procedure to be followed to modify the EOBT of a flight is shown in ENR 1.9 and full details are contained in the IFPS Users Manual, pages 4.1 and 4.2. Some states outside the CFMU area of responsibility still require AOs to update the EOBT, regardless of why the flight's original EOBT may have changed. AOs should bear in mind the formula (as shown in ENR 1.9) for calculating the new EOBT when doing this. Where it is known that ATC send departure messages (DEP) for all flights, then this DEP message will suffice.

Note 2: Extreme care should be exercised when compiling a DLA message; the time specified in the message must be the EOBT, not the planned airborne time and or the Calculated Take Off Time (CTOT).

4.7.4 IFPS processes Repetitive Flight Plans (RPL) internally as daily flight plans; therefore short term (one calendar day) amendments to RPLs should be processed through the submission of appropriate associated messages.

4.7.5 There are three additional messages which are exclusive to IFPS operations. These are the Operational Reply Messages (ORM) and consist of the ACK (Acceptance Acknowledgement Message); MAN (Referred for Manual Repair) and REJ (Message Rejected) messages.

4.7.6 Message originators will receive an ACK message from IFPS if the filed flight plan or associated message is automatically accepted into the system. Alternatively, receipt of a MAN message will indicate that the flight plan or associated message has not been accepted and is awaiting manual intervention by an IFPS Operator. Dependent upon the success, or otherwise, of the operator to repair the message an ACK or a REJ will be received. In this instance the ACK message will include the repaired message and will thus enable the originator to check where changes have been made. (**Note:** It is essential that the flight crew are provided with details of the accepted flight plan route, where this has been repaired by IFPS). A REJ message will advise the originator that the message filed has NOT been accepted into the IFPS. The REJ message will indicate the errors in the message which need to be resolved and will also include a copy of the message received by IFPS; this will enable the originator to confirm that the message has/has not been corrupted during transmission.

4.7.7 If a flight plan, or associated message, is rejected by IFPS (i.e. REJ message has been received) the corrected message, must be resubmitted without delay.

4.7.8 Until an ACK message has been received by the message originator, the requirement to submit a valid flight plan for an IFR/GAT flight intending to operate within the IFPS Zone will not have been satisfied. The flight details will NOT have been processed by IFPS and consequently the flight data will NOT have been distributed to the relevant ATSUs within the IFPS Zone. Similarly, processed data will NOT have been sent to the Tactical database (TACT) of the CFMU to be considered for ATFM purposes. Therefore, errors in the flight plan, or associated message(s), may result in the flight concerned being delayed. **4.7.9** Any queries of an urgent operational nature may be directed to the IFPU Supervisor at:

Haren	IFPU 1	Tel No: +32 2 745 19 50
Bretigny	IFPU 2	Tel No: +33 16 988 17 50

4.8 Supplementary Flight Plan Information

4.8.1 As an alternative to ICAO procedure that Supplementary Information should not be transmitted in a flight plan message (ICAO Doc 4444: Appendices 2 and 3) it should be noted that IFPS is able to process and store Field 19 – Supplementary Flight Plan Information. Where such information is supplied as part of a flight plan submission to IFPS it will be extracted and stored for later retrieval, if required, in the event of an emergency situation arising. Supplementary flight plan information will not be included in the normal flight plan distribution by IFPS.

4.8.2 Whilst the ICAO procedure should normally be followed by flight plan originators in the UK, they may avail themselves of the IFPS facility if they so wish.

4.8.3 ATS Authorities, or other relevant bodies, requiring Supplementary flight plan information on a particular flight and for urgent operational reasons may contact the Supervisor at the appropriate IFPU; assistance will be provided by either:

(a) giving information on Field 19 where such information has been submitted to and stored by IFPS;

(b) giving advice on a contact name/Tel No. of the AO and/or originator of the flight plan, which may be stored in the CFMU database;

(c) giving any additional information which may be contained in Field 18.

4.9 Contingency

4.9.1 The IFPS structure, incorporating two individual IFPUs, has been designed, and has the capacity, to permit one unit to act as a full back-up for the other in the event of outage of a single unit.

4.9.2 As all IFR/GAT flight plans within the IFPS Zone are addressed to both IFPUs, the effect of one unit being out of action will be transparent to flight plan originators.

4.9.3 The likelihood of a simultaneous outage of both IFPUs is considered to be extremely low. In such an event, flight plan originators will be alerted, by NOTAM, to reinstate the filing of messages, for flight plan and RPL operations, to all appropriate addresses, both within and outside the IFPS Zone.

5 Submission of Repetitive Flight Plan (RPL) Data to Eurocontrol CFMU – Brussels

5.1 Introduction

5.1.1 As part of the continuing development of the Central Flow Management Unit (CFMU), Eurocontrol will assume full responsibility for the reception, processing and distribution of Repetitive Flight Plan (RPL) data within the IFPS Zone (See the chart at ENR 6.1.10.2).

5.1.2 Flights within the IFPS Zone shall be filed solely with Eurocontrol at the CFMU, Brussels, in accordance with the requirements and procedures detailed below.

5.1.3 RPLs for flights affecting the IFPS Zone, but which have a route portion outside the IFPS Zone, shall continue to be filed to the National Authorities of those external States in accordance with existing procedures (see paragraph 5.5.2). It should be noted in particular that all affected National Administrations outside the Zone which are on the route of the flights must have agreed to the use of RPLs.

5.1.4 Attention is drawn to the fact that the Shanwick (EGGX) and Santa Maria (LPPO) OACCs are NOT within the IFPS Zone.

5.2 Types of Submission

5.2.1 RPL data submission may be in the form of a New List or a Revised List.

5.2.2 A **New List (NLST)** is a submission that contains only new information (typically the start of a new Winter or Summer period).

5.2.3 A **Revised List (RLST)** is a submission that contains revised information to a previously submitted list. This revised or amended information could be a combination of any of the following:

(a) changes;

(b) cancellations; or

(c) additional new flights.

5.3 RPL Submission Criteria

5.3.1 A NLST must be received by Eurocontrol with a minimum of 14 days before the intended first flight.

5.3.2 A RLST must be received by Eurocontrol such that:

(a) there is a minimum of 7 working (see paragraph 5.6.2) days between reception of the file by Eurocontrol and the activation of the first flight affected by the amendment; and

(b) there must be two Mondays between reception of the file and activation of the first flight affected by the amendment.

5.4 RPL Submission Procedure

5.4.1 RPLs may be submitted in any of the following formats:

IFPS RPL format (former DBO/DBE format)   via diskette, SITATEX or electronic file transfer (see Note)

ICAO format (hard copy) on paper (ICAO Doc 4444)

Note: The method of electronic file transfer shall be agreed between Aircraft Operators (AOs) and the CFMU.

5.4.2 Details of IFPS RPL format may be found in the IFPS User Manual section of the CFMU Handbook. Copies can be obtained from the Eurocontrol Library at the address given in paragraph 5.6.3.

5.4.3 On receipt of an RPL file, Eurocontrol will send an acknowledgement of receipt (as shown below) SITA or Fax as appropriate.

Example of Acknowledge of Reception Sent to RPL Originators (SITA or FAX)

ZCZC 001 251220
QN MADWEZZ
BRUER7X ddhhmm

ENR 1.10

ENR 1.10.5 Submission of Repetitive Flight Plan (RPL) Data to Eurocontrol CFMU – Brussels

FROM: EUROCONTROL/CFMU
TO: ZZZ
ATTN: Mrs. Brown
SUBJ: ACK OF YR RPL SUBMISSION 96-01
Nr.RPL: 12
– INITIAL CHECK OF FORMAT OK.
– FURTHER PROCESSING IN PROGRESS. WE WILL CONTACT YOU IF NECESSARY
BRGDS
D.TAYLOR/RPL TEAM

5.4.4 If no acknowledgement is received from Eurocontrol within 2 working days of dispatch, the originator must contact the RPL Team to confirm that the file has been received.

5.4.5 Following the acknowledgement the RPL Team will process the file and will contact the originator again only if there are any problems, such as the route or validity periods. It follows, therefore, that if no subsequent query is initiated by Eurocontrol, the originator can assume that the file has been successfully processed into the RPL database.

5.4.6 Any change to the address or contact number of the AO (for example, a change of contact number/address for obtaining supplementary information) must be advised to the RPL Team immediately.

5.4.7 Eurocontrol is able to accept RPL data which covers more than one Winter/Summer period but Originators must ensure that any such data is amended to reflect any changes of the clock (ie to reflect Summer/Winter time).

5.5 Specific Eurocontrol requirements for RPL operations

5.5.1 The basic principles for the submission of Repetitive Flight Plans are contained in ICAO Docs 4444 and 7030. The following paragraphs detail the differences between the ICAO Standard and the Eurocontrol requirement, which permits a more flexible approach within the basic rules. Full details are contained in the IFPS User Manual section of the CFMU Handbook.

5.5.2 RPLs shall cover the entire flight from the departure aerodrome to the destination aerodrome. Therefore, an RPL shall be submitted by the flight plan originator for its entire route. A mixture of both RPL and FPL message shall not be permitted. RPL procedures shall be applied only when ALL ATS authorities concerned with the flights have agreed to accept RPLs. In this respect, all States of the IFPS Zone accept RPLs. It is the responsibility of the AO to ensure that RPLs for flights which are partly outside the Zone are properly co-ordinated and addressed to the relevant external ATS authorities.

5.5.3 For Eurocontrol purposes an RLST may be submitted which contains only changes, cancellations and additions (ie '-' and '+'). Details of unchanged flights (ie 'blanks') are not required.

5.5.4 The '-' must come before the '+'.

5.5.5 For a cancellation or change, the '-' must be an exact duplicate of the original '+' that it is to cancel, in order for it to be accepted by the RPL processing system.

5.5.6 The NLSTs and RLSTs are to be numbered sequentially as this enables Eurocontrol to ensure that the lists are entered into the RPL database in the correct order. It also provides a double check for possible missing submissions. The first NLST of the season should be numbered 001 and each following list, regardless of whether it is a NLST or RLST, is to be numbered sequentially.

5.5.7 The numbering of the RPL submissions is done on line '0' (sender record) starting at character 37 of the diskette file and in field 'E' of an ICAO hard copy file (on paper).

5.5.8 To suspend an RPL the originator should send the information in the format shown in paragraph 5.5.8.1. However, originators should note that flights cannot be suspended for less than 3 days. If the suspension is for less than 3 days, individual daily cancellation messages must be sent by the originator to the IFPS in order not to waste ATC capacity by leaving 'ghost' flights in the CFMU and ATC databases.

5.5.8.1 To suspend an RPL(s) the RPL Originator must send by SITA, FAX a letter to the Eurocontrol RPL office with an instruction which contains the following information:
Please suspend the following flights with effect from ddmm until ddmm:
AIRCRAFT-ID VAL-FROM VAL-UNTIL DAYS-OF-OPERATION ADEP EOBT ADES
Note:
(a) Flights cannot be suspended for periods of less than 3 days.
(b) a suspension message shall be received not less than 48 hours before the EOBT of the earliest affected flight(s). When sufficient notice cannot be given, individual CNL messages must be filed.
(c) if the 'UNTIL' is not filled in, then a Recovery message will have to be sent in order to re-instate the flights.

5.5.8.2 The RSUS message is an ADEXP message which has not been implemented in the RPL system. This message shall not be used. Originators should use the media and layout described in paragraph 5.5.8.1.

5.5.9 To cancel an RPL for a specific day, the originator need only send a normal ICAO CNL message to both of the IFPS units (EBBDZMFP and LFPYZMFP or RUEP7X and PAREP7X) but not earlier than 20 hours before the EOBT of the flight. The same rule applies for a change (CHG) or delay (DLA) message since at 20 hours before EOBT the RPL is transferred to the IFPS and the RPL effectively becomes an FPL.

5.5.10 To recover any RPL which has been suspended for an undefined period, the originator must send the instruction in the format shown in paragraph 5.5.10.1.

5.5.10.1 To recover an RPL(s), the RPL Originator must send by SITA, FAX a letter to the Eurocontrol RPL Office with an instruction which contains the following information:
Please recover the following flights with effect from ddmm:
AIRCRAFT-ID VAL-FROM VAL-UNTIL DAYS-OF-OPERATION ADEP EOBT ADES
Note: A recovery message shall be received not loss than 48 hours before the EOBT of the earliest affected flight(s). When sufficient notice cannot be given, individual FPL messages must be filed.

ENR 1.10 ENR 1.10.5 Submission of Repetitive Flight Plan (RPL)
Data to Eurocontrol CFMU – Brussels

5.5.10.2 The RREC message is an ADEXP message which has not been implemented in the RPL system. This message shall not be used. Originators should use the media and layout described in paragraph 5.5.10.1.

5.5.11 For route portions outside the IFPS Zone, normal ICAO procedures apply.

5.6 General Information

5.6.1 RPL data at Eurocontrol is handled by a dedicated section known as the RPL Team.

5.6.2 The RPL Team working day is from 0800 to 1715 (Central European Time) Monday to Friday, including Public Holidays but excluding 25 December. Originators of RPL data should take these operating hours into account when submitting RPL data to Eurocontrol.

5.6.3 RPL data files may be sent to Eurocontrol by any of the following means of communication:
Eurocontrol CFMU
FDO/RPL Team
Rue de la Fusee 96
B-1130 Brussels
Belgium
SITATEX: BRUER7X
FAX: 32 2 729 9042

5.6.4 The RPL Team may be contacted by telephone on 32 2 729 9847/61/66.

5.6.5 The use of hard copy via post is discouraged. Submission via diskette, SITATEX or electronic file transfer removes the chance of an RPL operator making any typographical errors when copying the data from the hard copy into the IFPS RPL system.

5.6.6 In order to assist AOs in submitting RPL programs via diskette, Eurocontrol has developed a PC based program which runs under Windows 3.1/Windows 95 and which has been designed to enable RPL files to be output in ICAO format or IFPS RPL format. The PC based program is available free of charge from Eurocontrol CFMU, User Requirement Section at the address given in paragraph 5.6.3.

6 Low Level Civil Aircraft Notification Procedures (CANP)

6.1 Introduction

6.1.1 Many military and civil aircraft operate in Class G Airspace below 2000ft agl, where ground radio and radar coverage is not always available to assist pilots in avoiding collisions. Collision avoidance must necessarily, therefore, be based on the 'see and avoid' principle, assisted as far as possible by information on known activity. Whereas a variety of civil aviation activities take place within this airspace, military activity consists mainly of low flying training.

6.1.2 It is not practicable to obtain and disseminate traffic information on all civil flights below 2000ft agl, nor is it possible to disseminate details of military low level flights within the UK Low Flying System (UKLFS) to civil operators. Nevertheless, the greatest conflict of interests occurs at or below 1000ft agl where the majority of military low level operations take place and where civil aircraft may be engaged upon activities, as defined at paragraph 6.2.1, which might inhibit pilot look-out or reduce aircraft manoeuvrability. In addition, certain recreational and other civil flying activity, away from licensed aerodromes, needs to be considered.

6.1.3 A system exists to collect information on civil aerial activities for distribution to military operators to assist in flight planning. This system is known as the Low Level Civil Aircraft Notification Procedures (CANP).

6.1.4 Before commencing any low flying sortie, military pilots receive a comprehensive brief on all factors likely to affect their flight, including relevant CANP details. Hence, maximum participation in CANP by those planning to conduct the qualifying activities is essential if full benefit is to be obtained from the procedure.

6.1.5 Pilots/operators, or their representatives, intending to embark upon aerial activities described below, should notify details of the flights to the Low Flying Booking Cell (LFBC) at the London Air Traffic Control Centre (Military), (LATCC(Mil)). For the purposes of CANP, direct-dial, Freephone and E-mail facilities are available as follows:

  Monday to Thursday: 0700 – 2300 (Local);
  Friday to Sunday: 0700 – 1700 (Local).

6.1.6 E-mail or Fax notification is preferred for CANP requests as this allows the LFBC to E-mail, 'faxback' or telephone confirmation of fax receipt and issue a reference number to the aircraft operating authority. Contact numbers are as follows:
Fax: 0500 300120;
Tel: 0800 515544;
E-mail: if.bookings@nats.co.uk

6.2 Commercial Aerial Activity

6.2.1 The following civil aerial activities at and below 1000ft agl, with an expected duration in excess of 20 minutes at a specific location, should be notified to the LFBC:
(a) Aerial Crop Spraying (this includes all agricultural tasks carried out by an aircraft);
(b) underslung aerial load lifting;
(c) aerial photography;
(d) aerial survey/air surveillance.

6.2.1.1 Pipeline/powerline inspection activity is the subject of an AIC. However, aircraft carrying out powerline inspections and which are able to operate within a limited geographical area may apply for warning status under CANP. Any requests for such protection should be made as far in advance as is possible through the E-mail address, Freefax or Freephone numbers shown at paragraph 6.1.6. The manager of the UKLFS at HQ Strike Command Detachment, LATCC (Mil) will consider requests of this nature on a case-by-case basis.

6.2.2 Procedure

6.2.2.1 CANP fax and telephone messages should provide details of the intended activity in the following format:

ENR 1.10   ENR 1.10.6 Low Level Civil Aircraft Notification Procedures (CANP)

(a) Type of activity;
(b) Location(s): Preferably as a 2-letter, 6-figure grid reference taken from an OS 1:50,000 map, although latitude and longitude will be accepted. The name of a nearby village or town is also required;
(c) Area of Operation(s): (See paragraph 6.2.4.1);
(d) Date and Time of intended operation(s): Start/finish in local time;
(e) Maximum Operating Height(s) agl;
(f) Number and Type(s) of aircraft;
(g) Contact fax and/or telephone numbers;
(h) Operating Company and fax/telephone number(s) (if applicable).
Example: CANP NOTIFICATION
A – UNDERSLUNG LOADS
B – SU 561310 – OVINGTON
C – 2 NM RADIUS
D – 12 SEPTEMBER – 1000 to 1300
E – 1000 FEET AGL
F – SINGLE MB105 HELICOPTER
G – Contact fax and telephone number of the site
H – ROTARY HELICOPTERS LTD – fax and telephone number of operator

6.2.2.2 Once a notification has been accepted, the LFBC will allocate a reference number which pilots/operators should retain. Operators are advised that, in the interests of safety and accuracy, all telephone calls to the LFBC are recorded.

6.2.2.3 Operators should, where possible, use the E-mail, Freefax facility as the primary method of filing a notification. Requests should be submitted using Form CA 2366 as shown at ENR 1.10.16. Customised variations of this form are acceptable if they contain all the information required at paragraph 6.2.2.1. A contact E-mail, fax and telephone number must be provided in order that notification can be confirmed and a reference number issued. (Additional copies of Form CA 2366 can be obtained from the Directorate of Airspace Policy (DAP) at the address shown at paragraph 6.4.1). Users will receive a CANP reference number from the LFBC by 'faxback' or return telephone call. This reference number should be retained until the termination of the activity with which it is associated.

6.2.3 Pre-notification Required

6.2.3.1 Pre-notification of intended operations should be communicated by fax if possible, to the LFBC not less than 4 hours before commencement of the activity. Fax requests will receive an E-mail, 'faxback' or telephone call from the LFBC with time authentication and reference number. Notifications by telephone will receive a time authentication followed by a return call from the LFBC with a reference number. Successful transmission of the CA 2366, or a time authenticator for notification by telephone, not less than 4 hours from the start of the CANP activity can be considered as confirmation that a CANP avoid for the period requested will be issued.

6.2.3.2 Whenever possible, pre-notification of operations due to take place up to 1300 hours (local time) should be made the previous day and those due to take place after 1300 hours (local time) should be pre-notified on the morning of the same day. It is accepted there will be occasions when the minimum pre-notification cannot be met. Nevertheless, late notifications should still be made and every effort will be made to distribute the information as widely as possible. However, reports received less than 4 hours before operations are due to commence are, progressively as the time diminishes, less likely to reach all military pilots before they depart on their low level sorties and will, therefore, only be issued as a warning to military aircrew.

6.2.3.3 CANP operators who are aware of commercial activities well in advance are encouraged to contact the manager of the UKLFS, as far in advance as possible (Fax: 01895 426381, Tel: 01895 426686), with as many details of the activity as are available at the time.

6.2.4 Operating Area Boundaries

6.2.4.1 The airspace notified under CANP should not exceed an area bounded by a 2nm radius circle. If more than one area is to be notified, these areas are not to be activated concurrently. In the case of underslung aerial load lifting operations the area should be defined as a corridor extending 2nm either side of intended track from ground level to a maximum of 1000ft agl. When the route of an underslung load exceeds 20nm it should, wherever possible, be divided into sections not exceeding 20nm in length; an overlap of 20 minutes is acceptable in such circumstances.

6.2.4.2 Pilots of military fixed-wing aircraft flying at an IAS greater than 140kt will avoid areas reported under CANP either laterally or vertically. CANP users should note that military pilots may overfly the reported area by a minimum of 500ft. Thus, for example, if the height of the CANP area is 1000ft agl, military aircraft may overfly the area at a minimum height of 1500ft agl. Therefore, the lateral and vertical boundaries which define the area of activity should equate only to the parameters within which the activity is planned to take place and should not build in an allowance as a safety factor.

6.2.4.3 Pilots/operators should note that, other than in exceptional circumstances, the dimensions of a CANP 'avoidance' as defined at paragraph 6.2.4.1 are generally not negotiable. Any request for a CANP of non-standard dimensions should be made, as far in advance as possible, to the manager of the UKLFS at the contact numbers shown at paragraph 6.2.3.3.

6.2.5 Cancellation and Re-submissions

6.2.5.1 Activities reported under CANP may considerably restrict the airspace available for military low flying training. Thus, in order to maintain the integrity of the CANP system, every reasonable attempt should be made to inform the LFBC as soon as it becomes obvious that an activity previously notified will no longer take place or that activity has been completed. Notification of a completed activity should be made irrespective of the time remaining on the CANP

6.2.5.2 To eliminate the possibility of error, an application must be made in accordance with paragraph 6.2.2.1 on each occasion. Re-submission by reference to a previously issued CANP Reference Number will not be accepted by the LFBC.

ENR 1.10.6 Low Level Civil Aircraft Notification Procedures (CANP)

6.2.6 Infringements of CANP Airspace
6.2.6.1 Infringements of CANP airspace will be fully investigated. If it is considered that CANP airspace has been infringed by military aircraft, and more than 4 hours pre-notification has been given in accordance with paragraph 6.2.3.1, then pilots/operators should contact the LFBC as soon as possible with the following information:
(a) Reference Number (paragraph 6.2.2.2 refers);
(b) date/time of incident;
(c) number and type of aircraft involved;
(d) position and estimated profile (heading/height) of aircraft involved.
6.2.6.2 Pilots/operators should note that military light aircraft flying at an IAS of 140kt or less, helicopters and any aircraft flying within a MATZ, need not avoid CANP airspace. However, pilots of such military aircraft will be aware of the notified activity, subject to the minimum notifying period indicated at paragraph 6.2.3.2.

6.3 Recreational and Other Aerial Activities
6.3.1 Recreational Aerial Activities
6.3.1.1 The LFBC invite notifications concerning certain recreational aerial activities planned to occur at and below 1000ft agl. Such notifications will be granted warning status under CANP and will be promulgated to military aircrew. Notifications are only required, however, when 5 or more Gliders, Hang and Paragliders, Free-flight Balloons, Microlight aircraft or model aircraft will be operating:
(a) From a site not listed in the UK AIP for such activity; or
(b) from a site listed in the UK AIP but outside the published operating hours of the site, where these are detailed.
6.3.1.2 Notwithstanding the provisions of paragraph 6.3.1.1(b), operators will be aware that Permissions for cable launched gliding, hang-gliding and paragliding activities, to a height of more than 60 metres agl, are issued by DAP. Individual Permissions will stipulate that, if activity is during a weekday, it is conditional on compliance with the CANP system.

6.3.2 Other Aerial Activities
6.3.2.1 The LFBC also invites notification of the following activities:
(a) Tethered and Captive Balloons (to a height greater than 60 metres agl);
(b) kite flying, involving 5 or more kites from a specified site, (to a height greater than 60 metres);
(c) operations of aircraft from water;
(d) any other aerial activity likely to create an exceptional concentration of aircraft at a site not listed in the UK AIP.

6.3.3 Procedure
6.3.3.1 E-mail, Fax or telephone notification should provide details of the intended activity as at paragraph 6.2.2.1.
Example: RECREATIONAL ACTIVITY
A – HANG GLIDING
B – ST 187101 – UPOTTERY AERODROME, DEVON
C – 2NM RADIUS
D – 19 NOVEMBER – 0900 to 1500 (local time)
E – N/A
F – EXPECTED NUMBER OF HANG GLIDERS – 6
G – Telephone number of the site
H – DISCOVER AIR HANG GLIDING GROUP (Telephone number if different to that at G)
6.3.3.2 Once a notification has been accepted, the LFBC will allocate a reference number which pilots/operators should retain.
6.3.3.3 The E-mail or Freefax facility detailed in paragraph 6.1.6 should be used where possible for the notification of recreational activities.

6.3.4 Pre-notification is required as in paragraph 6.2.3.1.
6.3.5 Operating Area Boundaries.
6.3.5.1 The airspace notified should not exceed an area bounded by a 2nm radius circle, from ground level to 1000ft agl.
6.3.5.2 Recreational and other aerial activities will not normally attract CANP avoidance areas; however, warnings of such activities will be promulgated to military aircrew.

6.3.6 Cancellation
6.3.6.1 Every reasonable attempt should be made to inform the LFBC as soon as it becomes obvious that an activity previously notified will no longer take place or that the activity has been completed. Notification of a completed activity should be made irrespective of the time remaining on the CANP.

6.4 Comments/Recommendations
6.4.1 Users are invited to forward comments on CANP, or recommend improvements to the procedure, to DAP at the following address:
Directorate of Airspace Policy (DAP)
Off-route Airspace
K6 Gate 3 CAA House
45-59 Kingsway
London
WC2B 6TE
Tel: 020 7453 6543
Fax: 020 7453 6565

ENR 1.10 ENR 1.10.6 Low Level Civil Aircraft Notification Procedures (CANP)

DIRECTORATE OF AIRSPACE POLICY

CIVIL LOW FLYING/AERIAL ACTIVITY

COMMERCIAL* / RECREATIONAL* / OTHER*

Description of Activity..

| Timing: | **START** | Date .... | Time .... | (Local) |
| | **FINISH** | Date .... | Time .... | (Local) |

Location: **1:50,000 OS Grid Ref**...

or

Lat and Long N .... *E/W ....

Radius ....NM

Route or Multiple Positions

Position	Radius of Operation	Times (Local)

Height: Upper Height.. FT AGL

Type of Aircraft .... No of Aircraft ....

Contact No: Fax.. Tel ....

Name .... Company Tel No ....

Company*/Club*/Individual*

Official Use Only –

Time Authentication:	Reference Number:	LFBC Ops Initials:

CONFIRMATION MUST BE OBTAINED FROM LFBC BEFORE FLIGHT

*Delete as required

CA 2366

ENR 1.11 ADDRESSING OF FLIGHT PLAN MESSAGES

1 The United Kingdom operates a collective addressing system which is detailed in Civil Aviation Publication CAP 550 'Random Flight Plan AFTN Address Book' which is available from:

Documedia Solutions Limited
37 Windsor Street
Cheltenham
Glos
GL52 2DG
Telephone: 0870 8871410.

Category of Flight (IFR, VFR or both)	Route (Into or via FIR and/or TMA)	Message Address
IFR	Flight wholly within the IFPS Zone (ENR 1.10 refers)	EGZYIFPS (IFPS collective addresses)
VFR	Entering or remaining within Scottish FIR	EGZYVFRP
	Entering or remaining within London FIR	EGZYVFRT
IFR/VFR	Within the Shanwick OCA and FIR	EGGXZOZX
	Flights which will transit the SOTA	EBBDZMFP and LFPYZMFP
	Flights which will transit the BOTA	LFRRZQZX
	Inbound to Guernsey and Alderney and over flying the Channel Islands up to FL 195	EGJJZRZX (Channel Island Control Zone)

1.1 In addition a VFR Flight Plan is to be addressed to the destination aerodrome (EG-ZTZX).

1.2 For VFR flights departing UK FIRs, use the appropriate destination collective, not VFRP/VFRT.

1.3 For flights departing UK FIRs that contain VFR and IFR elements. Although IFPS handles IFR flight plans it will not process the VFR portions of any mixed VFR/IFR flight plan, therefore the FPL must be addressed to:

(a) IFPS (EGZYIFPS), AD of departure and AD of destination;
(b) all FIRs that the flight will route through as VFR;
(c) any additional addresses required by state or aerodrome authorities.

2 The following table, which is not exhaustive, lists the principle exceptions and additions which, unless otherwise specified, apply to VFR flights.

Aerodrome	Requirement
EGAC (Belfast City)	Add EGAAZPZX (Aldergrove ARO)
EGBE (Coventry)	Add EGBBZTZX (Birmingham Approach)
EGCB (Manchester Barton)	Add EGCCZQZX (Manchester ACC)
EGCC (Manchester)	Add EGCCZQZX ONLY
EGCD (Manchester Woodford)	Add EGCCZQZX
EGDG (St Mawgan)	Add EGDGYXYW
EGGP (Liverpool)	Add EGCCZQZX
EGHH (Bournemouth)	Add EGHIZTZX (Southampton Zone)
EGKK (London Gatwick)	Add EGKKZPZX (not ZTZX)
EGLC (London City)	Add EGLCZGZX and EGLCZPZX ONLY (ZTZX not required)
EGLF (Farnborough)	Add ZTZX for all flights (VFR/IFR)
EGLK (Blackbushe)	Add EGLFZTZX (Farnborough ATC)
EGMH (Manston)	Add EGMHZPZO
EGNR (Hawarden)	Add EGCCZQZX
EGPB (Sumburgh)	Add EGPDZHZR
EGPH (Edinburgh)	Add EGPHZGZX and EGQTYXYX (Military flights only) EGPHZXZX
EGTF (Fairoaks)	Add EGLFZTZX
Northern Ireland Aerodromes	Add EGPXZQZX (Scottish ACC)

ENR 1.11 ENR 1.12 INTERCEPTION OF CIVIL AIRCRAFT 1 Notification

3 Collectives are available for the following VFR flights departing UK Airspace.

Destination Country/FIR	Collective Address	Other Requirements
Austria – Vienna (LOVV)	EGZYVFRA	VFR Flight Plans must not be sent more than 10 hrs before EOBT
Belgium – Burssels inc Lux (EBBU)	EGZYVFRB	
Czech Republic – Prague (LKAA)	EGZTVFRL	
Denmark – Copenhagen (EKDK)	EGZYVFRK	If flight enters BELGIAN airspace add EBBUZFZX If flight enters FRENCH airspace add LFFFZFZX and LFQQZPZX
France – Bordeaux (LFBB)	EGZYVFRO	Use appropriate FIR collective Address for destination AD
France – Reims (LFEE)	EGZYVFRE	Add destination ZPZX and ZTZX
France – Paris (LFFF)	EGZYVFRF	Add ZFZX of each FIR entered
France – Marseille (LFMM)	EGZYVFRM	
France – Brest inc Channel Is (LFRR)	EGZYVFRR	
Germany – Berlin (EGBB)	EGZYVFRG	Add ZFZX address for FIR/FIC/FIS of destination AD
Germany – Frankfurt (EDFF)		Add ZQZX of each FIR entered
Germany – Dusseldorf (EDLL)		If flight enters FRENCH Airspace add LFFFZFZX and LFQQZPZX
Germany – Munich (EDMM)		If flight enters MAASTRICHT TMA add EHBKZTZX
Germany – Bremen (EDWW)		
Ireland – Shannon (EISN)	EGZYVFRI	Add ZQZX of CTRs entered or transitted
Italy – Brindisi (LIBB)	EGZYVFRD	Add ZQZX of CTRs entered or transitted
Italy – Milan (LIMM)		
Italy – Rome (LIRR)		
Netherlands – Amsterdam (EHAA)	EGZYVFRH	If flight enters BELGIAN airspace add EBBUZFZX If flight enters FRENCH airspace add LFFFZFZX and LFQQZPZX
Norway – Norway (ENOR)	EGZYVFRW	If flight enters DANISH airspace add EKDKZFZX
Spain – Barcelona (LECB)	EGZYVFRC	If flight enters LFRR FIR add EGJJZFZX and LFRRZFZX
Spain – Madrid (LECM)		
Sweden – Sweden (ESAA)	EGZYVFRV	If flight enters DUTCH airspace add EHAAZFZX and EHMCZFZX Add ESSAZPZX if AD does not have a location identifier
Switzerland – Geneva (LSAG)	EGZYVFRS	If flight enters LFRR FIR add LFRRZFZX and EGJJZFZX
Switzerland – Zurich (LSAZ)		

For all VFR Flights: Add destination ZTZX
The following collective addresses are available to aid VFR flights departing UK Airspace, however, it is recommended that aircraft operators consult the AIPs of the respective states whose airspace they plan to use.

ENR 1.12 INTERCEPTION OF CIVIL AIRCRAFT

1 Notification

1.1 The procedure to be followed by the pilot-in-command of an intercepted aircraft and visual signals for use by intercepting and intercepted aircraft listed below are hereby notified for the purposes of Schedule 11 of the Air Navigation Order 2000.

1.2 Under Article 9 of the Convention on International Civil Aviation, each contracting State reserves the right for reasons of military necessity or public safety, to restrict or prohibit the aircraft of other States from flying over certain areas of its territory.

1.3 The Regulations of a State may prescribe the need to investigate the identity of aircraft. Accordingly, it may be necessary to lead an aircraft of another nation, which has been intercepted, away from a particular area (such as a prohibited area) or, an intercepted aircraft may be required to land for security reasons at a designated aerodrome.

1.4 To avoid the interception of civil aircraft, adherence to flight plans and ATC procedures and the maintenance of a listening watch on the appropriate ATC frequency, make the possibility of interception highly improbable. If the identity of an aircraft is in doubt, all possible efforts will be made to secure identification through the appropriate Air Traffic Service Units.

1.5 As interception of civil aircraft are, in all cases, potentially hazardous, the interception procedures will only be used as a last resort.

1.6 The word 'interception' does not include the intercept and escort service provided on request to an aircraft in distress in accordance with Search and Rescue procedures.

1.7 An aircraft which is intercepted by another aircraft shall immediately:

(a) Follow the instructions given by the intercepting aircraft, interpreting and responding to visual signals in accordance with the tables at ENR 1.12.2;

(b) notify, if possible, the appropriate Air Traffic Services Unit;

(c) attempt to establish radio communication with the intercepting aircraft or with the appropriate intercept control unit, by making a general call on the emergency frequency 121.500 MHz, giving the identity of the intercepted aircraft and the nature of the flight; and if no contact has been established and if practicable, repeating this call on the emergency frequency 243 MHz;

(d) if equipped with SSR transponder, select Mode A, Code 7700 and Mode C, unless otherwise instructed by the appropriate Air Traffic Services Unit.

1.8 If radio contact with the intercepting aircraft is established but communication in a common language is not possible, attempts shall be made to convey essential information and acknowledgement of instructions by using the following phrases and pronunciations:

Phrase	Pronunciation	Meaning
Call Sign	Kol sa-In	My call sign is (call sign)
Wilco	Vill-Ko	Understood will comply
Can Not	Kann Nott	Unable to comply
Repeat	Ree-Peet	Repeat your instruction
Am Lost	Am Losst	Position unknown
Mayday	Mayday	I am in distress
Hijack	Hi-Jack	I have been hijacked
Land (place name)	Laand (place name)	I request to land at (place name)
Descend	Dee-Send	I require descent

Note 1: The call sign required to be given is that used in radiotelephony communications with Air Traffic Services Units and corresponding to the aircraft identification in the flight plan.

Note 2: Circumstances may not always permit, nor make desirable, the use of the phrase 'HIJACK'.

1.9 The following phrases are expected to be used by the intercepting aircraft in the circumstances described above: (ICAO Annex 2, Appendix 2 – paragraph 3, Table 2.1 and Attachment A, Table A1 refer).

Phrase	Pronunciation	Meaning
Call Sign	Kol sa – In	What is you call sign?
Follow	Fol-Lo	Follow me
Descend	Dee-Send	Descend for landing
You Land	You-Laand	Land at this aerodrome
Proceed	Pro-Seed	You may proceed

1.10 If any instructions received by radio from any sources conflict with those given by the intercepting aircraft by signals, the intercepted aircraft shall request immediate clarification while continuing to comply with the visual instructions given by the intercepting aircraft.

1.11 If any instructions received by radio from any sources conflict with those given by the intercepting aircraft by radio, the intercepted aircraft shall request immediate clarification while continuing to comply with the radio instructions given by the intercepting aircraft.

ENR 1.12 INTERCEPTION OF CIVIL AIRCRAFT

Signals for use in the event of interception

Signals initiated by intercepting aircraft and responses by intercepted aircraft

Series	INTERCEPTING Aircraft Signals	Meaning	INTERCEPTED Aircraft Responds	Meaning
1	DAY-Rocking wings from a position slightly above and ahead of, and normally to the left, of the intercepted aircraft and, after acknowledgement, a slow level turn, normally to the left, on to the desired heading.	You have been intercepted Follow me	**AEROPLANES:** DAY-Rocking wings and following	Understood will comply
	NIGHT-Same and, in addition, flashing navigational lights at irregular intervals		NIGHT-Same and, in addition, flashing navigational lights at irregular intervals	
	Note 1: Meteorological conditions or terrain may require the intercepting aircraft to take up a position slightly above and ahead of, and to the right of the intercepted aircraft and to make the subsequent turn to the right.		**HELICOPTERS:** Day or NIGHT-Rocking aircraft, flashing navigational lights at irregular intervals and following: **Note:** Additional action required to be taken by intercepted Aircraft is prescribed in paragraphs 1.7, 1.8, 1.9, 1.10 and 1.11 and ENR 1.12.1.	
	Note 2: If the intercepted aircraft is not able to keep pace with the intercepting aircraft, the latter is expected to fly a series of race-track patterns and to rock its wings each time it passes the intercepted aircraft.			
2	DAY or NIGHT – An abrupt breakaway manoeuvre from the intercepted aircraft consisting of a climbing turn of 90 degrees or more without crossing the line of flight of the intercepted aircraft	You may proceed	**AEROPLANES:** DAY or NIGHT – Rocking wings **HELICOPTERS:** DAY or NIGHT – Rocking wings	Understood will comply

ENR 1.12 INTERCEPTION OF CIVIL AIRCRAFT

DAY – Circling aerodrome, lowering landing gear and over flying runway in the directions of landing or, if the intercepted aircraft is a helicopter, over flying the helicopter landing area.
NIGHT – Same and, in addition, showing steady Landing lights.

Series	INTERCEPTING Aircraft Signals	Meaning	INTERCEPTED Aircraft Responds	Meaning
	AEROPLANES: DAY-Raising landing gear while passing over landing runway at a height exceeding 300m (1000ft) but not exceeding 600m (2000ft) above the aerodrome level, continuing to circle the aerodrome	Aerodrome you have designated is inadequate	DAY or NIGHT – It is desired that the intercepted aircraft follow the intercepting aircraft to an alternate aerodrome, the intercepting aircraft raises its landing gear and uses the Series 1 signals prescribed for intercepting aircraft.	Understood follow me
	NIGHT-Flashing landing lights while passing over the anding runway at lat height exceeding 300m (1000ft) but not exceeding 600m (2000ft) above the aerodrome level and continuing to circle the aerodrome. If unable to flash landing lights, flash any other lights available		If it is decided to release the intercepted aircraft, the intercepting aircraft uses the Series 2 signals prescribed for intercepting aircraft.	Understood you may proceed
	AEROPLANES: DAY or NIGHT – Regular switching on and off all available lights but in such a manner as to be distinct from flashing lights.	Cannot comply	DAY or NIGHT – Use series 2 signals prescribed for intercepting aircraft.	Understood
	AEROPLANES: DAY or NIGHT – Irregular flashing of all available Lights	In distress	DAY or NIGHT – Use Series 2 signals prescribed for intercepting aircraft.	Understood
	HELICOPTERS: DAY or NIGHT – Irregular flashing of all available lights			

ENR 1.14 AIR TRAFFIC INCIDENTS

1 AIRPROX Reporting – General

1.1.1 An AIRPROX Report should be made whenever a pilot or a controller considers that the distance between aircraft as well as their relative positions and speed have been such that the safety of the aircraft involved was or may have been compromised.

2 AIRPROX in UK Airspace

2.1 AIRPROX reports may be initiated by pilots or controllers and will be co-ordinated subsequently by the UK AIRPROX Board (UKAB). Where the event occurs in controlled airspace or otherwise meets the MOR criteria, an MOR investigation will be initiated by the Safety Data Department (SDD) of the Safety Regulation Group (SRG).

3 AIRPROX Reporting Procedures

3.1 Investigations are sometimes made difficult because the correct reporting procedure has not been followed or those involved were not made aware at the time that an AIRPROX report was being filed. In some cases it has not been possible to trace the other aircraft involved, owing to the time taken for the initial details of the occurrence to reach the UKAB. Pilots and controllers are therefore reminded that the appropriate procedure for reporting an AIRPROX occurrence is as follows.

3.2 Initial Report – Pilots

3.2.1 An initial report of an AIRPROX by a pilot should be made immediately by radio to the ATS Unit with which the pilot is in communication, prefixing the message with the word AIRPROX.

3.2.2 The essential information required should follow the sequence of items shaded on the AIRPROX Report Form CA 1094/RAF 765A. (Forms available from UKAB, paragraph 10.1 refers).

3.2.3 If the AIRPROX cannot be reported on the radio at the time, an initial report should be made by the pilot immediately after landing by telephone or other means to any UK ATS Unit but preferably to an ACC.

3.2.4 The AFTN may be used to make the initial report from places abroad when the AIRPROX could not be reported on radio at the time. This AFTN message should be sent either to the UK ACC in whose airspace the incident took place or to the UK office of the pilot's operating company, which should without delay, telephone or use the AFTN to make the initial report to the appropriate ATS Unit.

3.3 Initial Report – Controllers

3.3.1 Whenever a Controller operating within the UK FIRs or Shanwick OCA initiates an AIRPROX report or receives an AIRPROX from a pilot, it should be sent either by signal/telex/fax using the Initial Report Form CA 1094A in accordance with MATS Part 1, Section 6, Chapter 2. Pilots of aircraft involved in an AIRPROX initiated by a controller should be informed by controller or his/her Unit management as soon as possible that an AIRPROX is being reported by ATC.

3.4 Confirmation Report

3.4.1 Initial reports must be confirmed within seven days, by completing the full AIRPROX reporting procedure – pilots (Form CA 1094/RAF 765A.) and controllers (Form CA 1261). AIRPROX which occur inside controlled airspace or which are otherwise reportable under the auspices of the MOR Scheme will be subject to a MOR investigation, initiated and handled by the Safety Data Department (SDD) of the CAA. Reports should be sent to:

Director UK AIRPROX Board
Hillingdon House
RAF Uxbridge
Middlesex
UB10 0RU
Tel: 01895 815121/2/5/8
Fax: 01895 815124
AFTN: EGGFYTYA
Telex: 934725
E-mail: ops@airproxboard.org.uk

3.4.2 A pilot leaving the UK for a period exceeding seven days must use the AFTN to transmit his confirmatory AIRPROX report to his company. The message will be accepted as Class A traffic. The operating company should complete the Form CA 1094/RAF 765A. and send it to the UKAB without delay.

3.4.3 Providing that the initial reporting and confirmatory procedure has been followed, all AIRPROX reports will receive immediate and thorough investigation. The originator of the report and all aircraft operators and ATS agencies involved will be advised that the report has been received and that the investigation is under way. A request will be made by appropriate handling agency for all those involved to submit reports on their version of the AIRPROX.

4 Investigation of AIRPROX

4.1 The primary reason for investigation (by the appropriate authority) is to determine the cause of an AIRPROX, thereby leading to action to reduce the risk of collisions. Within the UK, those AIRPROX which occur in Controlled Airspace (CAS), or otherwise involve British operated Commercial Air Transport aircraft, or turbine powered aircraft, or any other AIRPROX where a civil pilot voluntarily reports the incident under the MOR Scheme, will be investigated by the SRG as an MOR. After investigation, such AIRPROX will be assessed by the UKAB, in common with all other AIRPROX reports. Details of AIRPROX investigated as MORs will appear in the relevant SDD Safety Occurrence 'Listing' document which is published monthly. Any enquiries regarding MOR investigation of AIRPROX are to be directed to the CAA at the following address:

Safety Data Department (SDD)
Civil Aviation Authority
Aviation House
Gatwick Airport South
West Sussex
RH6 0YR
Tel: 01293 573221
Fax: 01293 573972
AFTN: EGGRYAYD
Telex: 878753

5 AIRPROX Assessment

5.1 Following appropriate investigation, the UKAB (a panel of civil and military pilots, controllers and operators from diverse aviation backgrounds) will assess each AIRPROX report submitted to determine cause, degree of collision risk and make any safety recommendations as appropriate.

5.2 Once each AIRPROX case is finalised (i.e. the investigation and assessment stage has been completed), the pilots, controllers and their respective operating bodies involved in the AIRPROX will be advised of the findings. Additionally, all AIRPROX reports and assessments are published at regular intervals.

Note: The conclusions reached by the UKAB have no legal significance, and the anonymity of individuals and companies involved in an AIRPROX is preserved throughout the assessment and the subsequent publication process.

5.3 Operators, pilots and controllers who seek further information on any AIRPROX are requested only to contact the UKAB or SDD and no other organisation involved.

6 AIRPROX in Foreign Airspace

6.1 Whilst the CAA has no authority to investigate any AIRPROX in foreign airspace, it is concerned about them particularly when UK public transport aircraft are involved. Accordingly, for these aircraft, copies of confirmatory reports made to foreign authorities and details of any response received from them are to be sent to SDD. This satisfies the MOR requirements. The CAA expects commanders/operators of the aircraft to initiate, confirm and follow through AIRPROX reports directly with the foreign authorities themselves in accordance with the appropriate national procedures. The Authority will, however, assist reportees where they have difficulty in following national procedures or in obtaining a response. Assistance may be obtained from:

International Services Department
Civil Aviation Authority
1E Aviation House
Gatwick Airport South
West Sussex
RH6 0YR
Tel: 01293 573386
Fax: 01293 573992

The CAA may also take action of its own accord with a foreign authority on receipt of a report or follow-up where, for instance, from its knowledge of previous occurrences, it considers this necessary.

6.2 Initial Report

6.2.1 The procedure for reporting AIRPROX, as described in the appropriate State's AIP, should be followed.

6.3 Confirmation Report

6.3.1 The State's confirmation procedure should be adhered to. In addition a copy of the AIRPROX report on the State's form, an approved Company form (ASR) or the CAA AIRPROX Report Form CA 1094/RAF 765A. (amended as necessary to take account of its use for foreign airspace) should be sent to SDD at the address shown at paragraph 4.1.

6.3.2 If there is likely to be an appreciable delay in the transmission of a report, or the AIRPROX is considered particularly serious, the report to SDD should be sent via fax, telex or the AFTN. The message should be prefixed AIRPROX, sent via the AFTN and allocated a priority of at least 'GG' to enable the Authority to take action promptly when it considers it necessary.

6.3.3 It is essential that where the form used may apply to an AIRPROX or ATC incident, the originator clearly annotates it as an AIRPROX.

6.3.4 AIRPROX and ATC Incident occurrences involving UK public transport aircraft in foreign airspace will be published in the SDD monthly listings for the information and action, where appropriate, of other UK operators likely to use the same airspace.

6.4 Investigation of the Occurrence

6.4.1 Providing that the initial and confirmatory reporting procedure has been followed, an investigation should be carried out by the appropriate foreign authority and the commander/operator advised of the findings and any preventive action taken. When a response is received, the commander/operator should pass the details to SDD indicating whether or not they are satisfied with the outcome.

6.4.2 If the commander or operator is not satisfied and/or the CAA does not consider the outcome adequate, then further action may be taken with the authority concerned.

ENR 1.15 ENR 1.15.1 Southern North Sea Low Level Radar Advisory, Flight Information Service and Helicopter Operating Procedures

7 ATC Incidents in Foreign Airspace
7.1 The procedures in the State's AIP should be followed.
7.2 If a State or approved Company form for incidents is not used, CAA Form 1673 (Occurrence Report) must be used instead. CAA Form CA 1094/RAF 765A. (AIRPROX Report) must not be employed for incidents. Form should be clearly annotated 'ATC Incident'.
7.3 The amount of attention accorded to incidents varies from State to State and it may be advisable on occasions to seek the assistance of the CAA in follow-up.

8 ATC incidents in UK Airspace – Foreign Pilots/Operators
8.1 Foreign operators/pilots may submit a report, in regard to an ATC incident (which is not an AIRPROX) occurring in UK Airspace, to the Safety Data Department (SDD) of the CAA (address at paragraph 4.1). Such reports will be handled in accordance with the provisions of the Mandatory Occurrence Reporting Scheme of the CAA.

9 Military Personnel – Reporting of AIRPROX
9.1 Military pilots and controllers should refer to the appropriate regulations within JSP 318A for AIRPROX reporting.

10 AIRPROX Report Forms CA1094/RAF765A, CA1094A and CA1261
10.1 CA1094/RAF765A forms are available from the UKAB website www.airporxboard.org.uk, SDD and all CAA ATS Units.
10.2 The CA1261 form is available from SDD, all CAA ATS Units and Alpha Office Solutions.
Alpha Office Solutions
The Birches Estate
Imberhorne Lane
East Grinstead
West Sussex
RH19 1XL
Tel: 01342 310523
Fax: 01342 311866

ENR 1.15 OFF-SHORE OPERATIONS

1 Southern North Sea Low Level Radar Advisory, Flight Information Service and Helicopter Operating Procedures
1.1 Introduction
1.1.1 To enhance flight safety and expedite Search and Rescue in the Southern North Sea Airspace, a Radar Advisory, Radar Information, Flight Information and Alerting Service is available from Air Traffic Service Units (ATSUs) at Aberdeen Airport (Anglia Radar) and RAF Coltishall. These services are available to helicopters operating in support to the off-shore oil and gas industry and to civil and military aircraft transitting the area at and below FL 65.
1.2 Description

ENR 1.15   ENR 1.15.1 Southern North Sea Low Level Radar Advisory, Flight Information Service and Helicopter Operating Procedures

Lateral Limits	Vertical Limits	Controlling Authorities Callsign and frequencies Hours
Southern North Sea Airspace 550000N 0010000W – 550000N 0030301E – 543715N 0025349E – 542245N 0024543E – 535745N 0025155E – 534003N 0025719E – 533503N 0025913E – 532809N 0030055E – 531803N 0030319E – 52551N 0030936E – 523606N 0025307E – 523606N 0014323E – then clockwise by an arc radius 2nm centred on Great Yarmouth/North Denes Aerodrome (523806N 001034E) to 523702N 0014036E – 525055N 0012611E to north of Blakeney Point (530000N 0010000E) to Strubby (531836N 0001034E) to Easlington (533919N 0000706E) – 550000N 0010000W	FL 65 SFC	Anglia Radar (Aberdeen ATSU) (Excluding that area delegated to RAF Coltishall during its hours of operation as shown below) 'Anglia Radar' Primary: 125.275 MHz Secondary: 128.925 MHz **Note**: When Anglia Radar is not providing a RAS, RIS, FIS and alerting service on their secondary frequency, helicopter pilots operating off-shore, in support of oil and gas exploration, will use the secondary frequency for traffic information and broadcasting advisory messages Hours: 0630-2030 daily (Winter) (Summer 1hr earlier)
RAF Coltishall Area of Responsibility 524506N 0014254E – 530203N 0014952E then along the north eastern boundary of the Hewett HPZ to 530631N 0014808E – 531042N 0014453E to north of Blakeney Point (530000N 0010000E) to 525055N 0012611E – 524244N 0013443E then clockwise by an arc radius 7nm centred on Great Yarmouth/North Denes Aerodrome (523806N 0014323E) to 524506N 0014254E	FL 65 SFC	RAF Coltishall. 'Coltishall Radar' 125.900 or 297.150 MHz **Note**: Outside the published operating hours of RAF Coltishall, an ATS will be provided by Anglia Radar. Hours: 0830 to 1730 Mon-Thu and 0830 to 1700 Fri Winter (Summer 1hr earlier). The hours of operation may be varied at short notice due to operational requirements.

1.3 Each ATSU will provide, within its specified area of responsibility, RAS/RIS within the limits of radar cover. Outside of radar cover or in the event of radar failure, a Flight Information and Alerting Service will be provided within the limits of VHF cover. These services will be provided to helicopter pilots routeing:

(a) To off-shore installations, until the time that the pilot is in contact with the destination rig/platform;

(b) from off-shore installations, from the time two-way communications is established with the appropriate ATSU, until the time that the pilot is in contact with the destination landing pad or other agency.

Note: Under the terms of a Memorandum of Understanding between National Air Traffic Services Limited and the helicopter companies Bristows and CHC-Scotia, aircraft operated by these companies will be provided with a modified RAS.

ENR 1.15 ENR 1.15.1 Southern North Sea Low Level Radar Advisory, Flight Information Service and Helicopter Operating Procedures

1.4 Airspace Structure

1.4.1 Helicopter Main Routes (HMR)

1.4.1.1 A Helicopter Main Route is a route indicating where helicopters are operating on a regular and frequent basis, and where an Alerting Service, Flight Information Service or other Advisory Services may be provided. HMR has no lateral dimensions but in the Southern North Sea Airspace the vertical operational limits are from 1500ft amsl up to and including FL 60. However, should helicopter icing conditions or other flight safety considerations dictate, helicopters may be forced to operate below 1500ft amsl. In these circumstances, where possible, pilots will endeavour to follow HMRs and advise the appropriate ATSU of the new altitude giving the reason for operating below 1500ft amsl.

1.4.1.2 Helicopter pilots operating along HMRs normally navigate by use of area navigation equipment. In the general interest of flight safety, civil helicopter pilots are strongly recommended to use the HMR track structure whenever possible.

1.4.1.3 Other traffic operating in proximity of these routes are advised to maintain and alert look out.

1.4.2 The Southern North Sea HMR Track Structure is as follows:

HMR 1 Norwich (524033N 0011658E) to Indefatigable 49/24N (*531718N 0024317E);
HMR 2 North Denes (523806N 0014323E) to Valiant North/Loggs (*532320N 0020016E) to Murdoch (*541605N 0021920E) (Note);
HMR 3 Norwich (524033N 0011658E) to Clipper (*532731N 0014352E);
HMR 4 North Denes (523806N 0014323E) to Pickerill 'B' (533127N 0010933E);
HMR 5 Amethyst 'A1D' (*533639N 0004322E) to Rough 'A' (*534928N 000281 IE) (Note);
HMR 6 LAGER (533640N 0000849W to Clipper (*532731N 0014352E) to Viking 'BD' (*532648N 0021955E);
HMR 7 Humberside (533429N 0002103W to MILDE (533531N 0001516W) to LAGER (533640N 0000849W) to West Sole 'A' (*534210N 0010858E) (Note);
HMR 8 LAGER (533640N 0000849W) to Rough 'A' (*534928N 0002811E) to Cleeton PQ (*540159N 0004336E);
HMR 9 Rough 'A' (*534928N 0002811E) to Ravenspurn (*540150N 0010603E);
HMR 10 West Sole 'A' (*534210N 0010858E) to Cleeton 'PQ' (*540159N 0004336E) (Note).

Note: The maximum cruising level on all HMRs, beneath EG D323B and EG D323C, is restructed to 4000ftALT unless cleared by Anglia Radar.

1.4.3 Oversea Corridor. The oversea corridor consists of the Airspace from 750ft to 2500ft ALT within the area bounded by straight lines joining in succession:

533544N 0015732E – 533328N 0021621E – 532200N 0023900E – 531143N 0025505E – 523632N 0014526E then clockwise by an arc radius 2nm centred on Great Yarmouth/North Denes Aerodrome (523806N 0014323E) to 523702N 0014036E – 525055N 0012611E then clockwise by an arc radius 1nm centred on Bacton (525127N 0012734E) to 525159N 0012610E – 530354N 0013836E – 531507N – 0015421E – 532132N 0013545E – 532838N 0014150E – 533544N 0015732E.

1.4.4 Helicopter Protected Zone (HPZ). HPZs are established to safeguard helicopters approaching and departing platforms and for helicopters engaged on extensive unco-ordinated inter-platform flying. Inter-platform flying by civil helicopters within HPZs contained within the Oversea Corridor will be conducted on the company or field discrete frequency. HPZs consist of the Airspace from sea level to 2000ft ALT contained within tangential lines, not exceeding 5nm in length, joining the neighbouring circumferences of circles 1.5nm radius around each individual platform helideck.

1.4.4.1 The position of individual platforms, together with their maximum height amsl and helideck height amsl, within their parent field complex are detailed at paragraph 5.2.

1.4.4.2 Special ATC co-ordination procedures apply to helicopters operating in the Hewett HPZ as shown at paragraph 1.5.1.3.

1.5 Operating Procedures

1.5.1 General

1.5.1.1 Helicopter Procedures. Helicopter pilots wishing to use this service must establish two way R/T communication with the appropriate ATSU. For flights within the same or adjacent field complex, helicopter pilots should remain on the field frequency. Pilots should advise ATC before changing frequency and/or altitude.

1.5.1.2 Fixed-Wing Procedures. Pilots of civil and military fixed-wing aircraft intending to fly within the areas of responsibility of RAF Coltishall and Anglia Radar are strongly advised to make use of the services provided. Pilots are also advised that the helicopters on inter-platform flights in the same field complex normally operate at 500ft amsl and frequently carry underslung loads which limit the pilots ability to take sudden avoiding action.

1.5.1.3 Hewett HPZ. Helicopters operating within or climbing from Hewett HPZ are not to climb above a maximum altitude of 1000ft RPS without clearance from ATC Coltishall during its hours of operation. All helicopters operating within the Hewett HPZ are to set their transponders to their discrete Anglia Radar assigned SSR code.

1.5.2 Helicopter Main Routes (HMR)

1.5.2.1 Helicopter pilots operating along HMRs are to make position reports at 20nm intervals.

1.5.2.2 Position reports on initial contact will include the following information:

(a) Callsign'
(b) Type;
(c) Point of departure;
(d) Point of next landing;
(e) Altitude/requested Altitude;
(f) Total number of people on board.

ENR 1.15   ENR 1.15.1 Southern North Sea Low Level Radar Advisory, Flight Information Service and Helicopter Operating Procedures

1.5.2.3 Subsequent position reports will include the following abbreviated information:
(a) Callsign;
(b) Position;
(c) Altitude.

1.5.2.4 Helicopter pilots will, prior to leaving a frequency, inform the ATSU of the next off-shore sector to be flown.

1.5.3 Helicopter Corridor

1.5.3.1 Pilots of helicopters entering the Corridor from Great Yarmouth/North Denes should contact Anglia Radar before departing the ATZ. Helicopters operating within the Corridor should not normally be flown below 750ft amsl unless forced to fly beneath by weather or for essential operating reasons.

1.5.3.2 Civil Fixed-Wing Procedures. Pilots of fixed-wing aircraft are recommended to avoid the Corridor Airspace, however, if penetration or underflight is essential, contact should be made with Anglia Radar no later than 10 nm before entering the area giving their position, altitude, squawk, heading and intentions.

1.5.4 Cruising Altitude

1.5.4.1 Helicopters will normally plan to fly at the following en-route altitudes:
(a) Outbound (land to sea) 2000ft and 3000ft amsl;
(b) Inbound (sea to land) 1500ft and 2500ft amsl.
(c) Inter-field:
(i) Northbound (270° to 089° MAG track) 1000ft, 2000ft and 3000ft amsl;
(ii) Southbound (090° to 269° MAG track) 500ft, 1500ft and 2500ft amsl. Above the Transition Altitude (3000ft amsl) all aircraft should conform to the Quadrantal Rule.

1.5.5 Altimeter Setting Procedures

1.5.5.1 En-route altitudes of 3000ft amsl and below will be flown with reference to the appropriate Regional Pressure Setting (RPS). Anglia Radar will give the appropriate pressure setting on first contact. Helicopters operating along HMRs crossing from one Altimeter Setting Region (ASR) to another, and when in contact with Anglia Radar will not change the RPS datum until instructed to do so. This procedure is to enable the Controller to plan and control vertical separation in the vicinity of an ASR boundary.

Note: When instructed by Anglia Radar to squawk the allocated code, helicopter pilots are requested to state their level to the nearest hundred feet in order that the Mode C transponder information can be verified.

1.5.6 Use of GPS for North Sea Operations

1.5.6.1 UK AOC holders intending to use GPS for en-route navigation for North Sea flight operations are to use GPS equipment that meets or exceeds CAA Specification 22. AOC holders requiring further information should contact their assigned flight operations inspector. Non UK AOC holders are recommended to operate to at least the Specification 22 standard.

1.6. Out of Hours Helicopter Operations

1.6.1 Helicopter off-shore support activity is not confined to published ATS hours and helicopters may be operating in VMC or IMC at all levels and times.

2 Northern North Sea and Atlantic Rim Low Level Radar Advisory, Flight Information Service and Helicopter Operating Procedures

2.1 To enhance flight safety and expedite Search and Rescue in the Northern North Sea Airspace, including Atlantic Rim Airspace and the East Shetland Basin, a Radar Advisory and Flight information and Alerting Service is available from the Air Traffic Service Unit (ATSU) at Aberdeen Airport. These services are available to helicopters operating in support to the off-shore oil and gas industry and to civil and military aircraft transitting the area at and below FL 85.

2.2 Description. Within the areas of responsibility specified below, a Radar Advisory Service within the limits of radar cover and, outside radar cover or in the event of radar failure, a Flight Information and Alerting Service within the limits of VHF cover will be provided. Outside the hours of service notified above, a Flight Information and Alerting Service is available within the limits of VHF RTF cover from the FIR Sector at Scottish ACC, callsign Scottish information on 129.225 MHz. The above services will be provided to helicopter pilots routeing:

(a) To off-shore installations until two-way communication is established with their destination; and
(b) from off-shore installations from the time two-way communication is established with the appropriate Off-shore Sector.

ENR 1.15 ENR 1.15.2 Northern North Sea and Atlantic Rim Low Level Radar Advisory, Flight Information Service and Helicopter Operating Procedures

Lateral Limits (A chart depicting the sectors is at ENR 6.1.15.1)	Vertical Limits	Controlling Authorities Callsign and frequencies Hours
Sumburgh Radar Sector – Cormorant QNH Area 602856N 0004429W – 610000N 0002000W – 610000N 000000E 605109N 000000E thence clockwise by the arc of a circle radius 70nm centred on 595244N 0011712W (SUM VOR) to – 600000N 0010112E – 600000N 000059E thence anti-clockwise by the arc of a circle radius 40nm centred on 595244N 0011712W (SUM VOR) to 602856N 0004429W	FL 85 SL	Sumbrugh Radar (Aberdeen ATSU) 'Sumburgh Radar' – 131.300 MHz Due to restricted coverage in the vicinity of the Beryl Field (SUM 102 R95 DME), two way communication with Sumburgh Radar on 131.300MHz may not be established below 2000ft ALT
Sumburgh Radar Sector – Sumburgh QNH Area This area of responsibility, excluding that portion of W5D contained within the area, is enclosed by straight lines joining in succession The following points: 600000N 0000059E thence anti-clockwise by the arc of a circle radius 40nm centred on 595244N 0011712W (SUM VOR) to 602856N 0004429W – 601515N 0005500W – 595244N 0011712W – 593000N 0013800W – 590000N 0021602W – 590000N 0013754E – 591722N 0014236E – 595346N 0020430E – 600000N 0020320E – 600000N 00000059E	FL 85 SL	Hours: (Both Sumburgh Areas) 0700-2130 Winter (Summer 1hr earlier) Services occasionally extend beyond published hours of operation

259

ENR 1.15 ENR 1.15.2 Northern North Sea and Atlantic Rim Low Level Radar Advisory, Flight Information Service and Helicopter Operating Procedures

Lateral Limits (A chart depicting the sectors is at ENR 6.1.15.1)	Vertical Limits	Controlling Authorities Callsign and frequencies Hours
Aberdeen Sectors The area of responsibility, excluding that portion of WSD and the Aberdeen CTR/CTA contained within the area, is enclosed by lines Joining the following points: 590000N 0021602W – 571838N 0021602E (ADN VOR) 560000N 0013000W – 560000N 0000000E – 560510N 0031455E – 563540N 0023642E – 573628N 0020654E – 575416N 0015748E – 582546N 0012854E – 590000N 0013754E – 590000N 0021602W	FL85 SL	Aberdeen ATC 'Aberdeen Radar' – 134.100 MHz within the defined sector, out to AND 90 DME from the north western boundary to the 044 HMR, and then out to AND 80 DME clockwise from the 044 HMR. Hours: 0700-2200 Winter (Summer 1hr earlier) Sumburgh Radar (Aberdeen ATSU) 'Sumburgh Radar - 131.300MHz within that part of the Aberdeen sector which lies beyond AND 090 DME and anti-clockwise from the 044 HMR. (This area is delegated to Sumburgh Radar to optimise available R/T coverage) Hours: 0700-2130 Winter (Summer 1hr earlier) Aberdeen ATC 'Aberdeen Information' – 135.175 MHz within the Aberdeen sector outwith the areas described above. Hours: 0700-2200 Winter (summer 1hr earlier) Outside these hours a service may be available from Aberdeen Approach/Radar
Brent Radar Sector – Cormorant QNH Area Area enclosed by arcs of circles joining in turn the following points: 605109N 0000000E – 620000N 0000000E – 620000N 0012222E – 612122N 0014718E – 600000N 0020320E – 600000N 0010112E thence anti-clockwise by the arc of a circle radius 70nm centred on 595244N 0011712W (SUM VOR) to 605109N 0000000E	FL85 SL	Aberdeen ATSU 'Brent Radar' – 122.250 MHz is the primary ATC frequency 129.950 MHz is the secondary frequency Hours: 0700 to 2100 Winter (Summer 1hr earlier) Services occasionally extended beyond published hours of operation
East Shetland Basin (ESB) To co-ordinate the inter-platform and transit helicopter traffic within the ESB an ATS is established in the area which will provide a radar service. Pilots intending to over fly the ESB below FL 85 are strongly recommended to Contact Brent Radar before penetrating the airspace. Area enclosed by arcs of circles joining in succession the Following points: 614410N 0003900E – 612412N 0003900E (Gate Golf) – 610807N 0004319E (Gate Hotel) – 605828N 0004555E (Gate Juliet) – 604846N 0010430E (Gate Lima) – 604116N 0011454E (Gate Mike) – 603808N 0012042E (Gate November) – 603300N 0014257E (Gate Oscar) – 603300N 0015659E – 612122N 0014718E – 614410N 0013329E – 614410N 0003900E	FL 85 SL	Aberdeen ATSU 'Brent Radar' – 122.250 MHz is the primary ATC frequency. 129.950MHz is the secondary frequency Hours: 0700 to 2100 Winter (Summer 1hr earlier) Services occasionally extended beyond published hours of operation

ENR 1.15 ENR 1.15.2 Northern North Sea and Atlantic Rim Low Level Radar Advisory, Flight Information Service and Helicopter Operating Procedures

2.3 Helicopter Main Routes – En-Route Structure

2.3.1 A Helicopter Main Route (HMR) is a route where helicopters operate on a regular and frequent basis, and where Alerting Service, Flight Information Service or Advisory Service may be provided. HMRs have no lateral dimensions but over the Northern North Sea (55°N to 62°N) the vertical operational limits are from 1500ft amsl to FL 85. However, should helicopter icing conditions or other flight safety considerations dictate, helicopters may be requested to operate below 1500ft amsl and where possible pilots shall endeavour to follow the HMR and advise the appropriate ATS Unit of the new altitude giving reasons for operating below 1500ft amsl. Military operations near HMRs are normally conducted at or below 1000ft amsl or above FL 85 and with due regard for civil helicopter operations when crossing HMRs. Helicopter pilots operating along HMRs normally maintain track by use of Area navigation equipment and in the general interest of flight safety, civil helicopter pilots are strongly recommended to use the HMR track structure whenever possible.

2.3.2 The HMR track structure between Sumburgh and the East Shetland Basin is as follows:
(a) HMR 'Golf' (eastbound) from 40 DME SUM on HMR 'Hotel' direct to Gate 'Golf';
(b) HMR 'Hotel' (eastbound) from Gate 'Hotel' parallel to HMR 'Juliet' to 40 DME SUM thence to Sumburgh VOR;
(c) HMR 'Juliet' (westbound) from Gate 'Juliet' direct track to Sumburgh VOR;
(d) HMR 'Lima' (eastbound) from Sumburgh VOR direct to Gate 'Lima';
(e) HMR 'Mike' (westbound) from Gate 'Mike' parallel to HMR 'Lima' to 40 DME SUM thence to Sumburgh VOR.

2.3.3 Aberdeen based pilots use HMRs based on a radial track system centred on 571838N 0021602W (ADN VOR). The system is keyed to the outbound master HMR 029 commencing at 575506N 0014423W and terminating at Gate 'Lima' (604846N 0010430E) on the ESB boundary. The other HMRs, spaced at 3 degree intervals, are designated alternately 'inbound' or 'outbound' and are identified by three figures. They terminate at either the appropriate Gate on the ESB boundary or the Median Line. The table below shows co-ordinates for the Gates and the points at which the HMRs intercept the Median Line. When operating to or from permanent off-shore installations or mobile vessels, pilots normally use the nearest HMR. In the Aberdeen area, HMRs 023 to 086 (inclusive) are joined to two parallel bi-directional 'feeder/funnel' HMRs, HMRs Whiskey and Echo. The direction of use of HMRs Whiskey and Echo is dependent upon the runway in use at Aberdeen. BALIS (571403N 0020354) and NOBAL (571518N 0020336W) are used as reporting points within the Aberdeen CTR.

HMR	Coord	HMR	Coord
HMR 023	612412N 0003900E (G Gate)	HMR 071	580923N 0014403E
HMR 026	605828N 0004555E (J Gate)	HMR 074	580248N 0015004E
HMR 029	604846N 0010430E (L Gate)	HMR 077	575606N 0015608E
HMR 032	603808N 0012042E (N Gate)	HMR 080	574904N 0020028E
HMR 035	603300N 0014257E (O Gate)	HMR 083	574153N 0020408E
HMR 038	602326N 0015851E	HMR 086	573435N 0020749E
HMR 041	600658N 0020200E	HMR 089	572709N 0021133E
HMR 044	595102N 0020249E	HMR 092	571931N 0021521E
HMR 047	593014N 0015015E	HMR 095	571139N 0021915E
HMR 050	591341N 0014136E	HMR 098	570328N 0022316E
HMR 053	590103N 0013811E	HMR 101	565455N 0022726E
HMR 056	584942N 0013510E	HMR 104	564556N 0023147E
HMR 059	583924N 0013227E	HMR 107	563633N 0023535E
HMR 062	582958N 0012959E	HMR 110	562416N 0025108E
HMR 065	582221N 0013204E	HMR 113	560957N 0030900E
HMR 068	581553N 0013804E	HMR 116	555535N 0032034E

2.3.3.1 HMR Whiskey is aligned along the axis MOCHA (593256N 0012159W) to Scotstown Head NDB (573333N 0014902W) thence to Hackley Head (571949N 0015717W). When Runway 16 is in use at Aberdeen. HMR Whiskey is the designated inbound track and when Runway 34 is in use HMR Whiskey is the designated outbound track for traffic planning to operate at 2000ft and above.

2.3.3.2 HMR Echo is aligned along the axis GORSE (571037N 0015351W) to SPIKE (573226N 0013953W) to TYSTI (592223N 0011440W) to TIRIK (593219N 0011214W). When Runway 34 is in use at Aberdeen, HMR Echo is the designated inbound track but will also be used by outbound traffic planning to operate below 2000ft. When Runway 16 is in use at Aberdeen, HMR Echo is the designated outbound track for traffic at all levels.

2.3.3.3 Helicopter traffic inbound to Aberdeen following the HMRs 026 to 086 (inclusive) will be required to maintain the inbound HMR until intercepting either HMR Whiskey (Runway 16 in use at Aberdeen) or HMR Echo (Runway 34 in use at Aberdeen) and then follow the the designated inbound track as directed by Aberdeen ATC. Inbound HMRs south of the HMR 086 terminate at 30 DME AND, from whence helicopter traffic will be directed by Aberdeen ATC.

2.3.3.4 Outbound helicopter traffic utilising the HMRs 023 to 113 (inclusive) will be directed by Aberdeen to join the desired outbound HME via either HMR Whiskey/Echo or as directed to a specific position on the required HMR, dependent upon the runway in use at Aberdeen. (Outbound HMRs south of HMR 086 commence at 40 DME ADN).

2.3.3.5 Within the Aberdeen Sectors Aberdeen ATC will allocate on a systematic basis the following ALTs/FLs:
(a) Outbound – 3000ft, Quadrantal FLs to provide a minimum of 1000ft vertical separation. (1000ft is below the base of the HMRs and is therefore not allocated routinely, but is used when conditions dictate – eg icing. Planned departures below 2000ft will not be allocated via HMR Whiskey – Runway 34 in use – but will be routed via HMR Echo. 2000ft may be allocated to outbound traffic in the early morning period when there is no inbound traffic).
(b) Inbound – 2000ft, Quadrantal FLs to provide a minimum of 1000ft vertical separation.

ENR 1.15   ENR 1.15.2 Northern North Sea and Atlantic Rim Low Level Radar Advisory, Flight Information Service and Helicopter Operating Procedures

(c) Inter-field – Pilots on inter-rigging flights over a significant distance are advised to cruise at appropriate quadrantal levels or altitudes above 3000ft to avoid aircraft following the HMR structure. Pilots on short flights should abide by the 500ft and 1000ft en-route altitude convention detailed at ENR 1.15.3, para 1.5.4.1 (c).

Note: Under certain operating and meteorological conditions civil helicopters may operate at lower altitudes.

2.3.4 Aberdeen – Atlantic Rim (West of Shetland Operations).

2.3.4.1 The HMR tracks between Aberdeen and the Atlantic Rim (as depicted at ENR 6.1.15.6) are as follows:

(a) HMR X-Ray (Outbound) – ADN VOR (571838N 0021602W) to SMOKI (574637N 0023556W) to WIK VOR (582732N 0030601W) to SODKI (584751N 0033753W) to VAMLA (600000N 0040000W);

(b) HMR Yankee (Inbound) – NESTA (600000N 0041000W) to MADOX (584343N 0034639W) to WIK VOR (582732N 0030601W).

Note: HMR X-Ray is bi-directional between Aberdeen and Wick.

2.3.4.2 Altimeter Setting:

(a) Within 30 DME ADN – Aberdeen QNH, or as directed by ATC;

(b) Outside 30 DME ADN – The appropriate Regional Pressure Setting (RPS), or as directed by ATC.

Note: To enable controllers to plan and control vertical separation near the boundaries of Altimeter Setting Regions (ASR), helicopters approaching an ASR boundary, and in radio contact with ATC, are not to change RPS until instructed to do so.

2.3.4.3 Cruising altitudes;
Northbound – 3000ft;
Southbound – 2000ft to SMOKI and then as instructed by ATC.

Note: Under certain operating and meteorological conditions civil helicopters may operate at lower altitudes.

2.3.5 Charts depicting the Aberdeen Area HMR track structure are at AD 2-EGPD-3-1/2. The chart at ENR 6.1.15.5 depicts the en-route HMR track structure and shows the bi-directional HMR established between the Ekofisk Hotel platform and platform P37/4, a route normally used by Norwegian helicopters.

2.4 Operating Procedures

2.4.1 Helicopter Procedures. Helicopter pilots wishing to use the service specified in paragraph 2.3 must file a flight plan. Pilots who have established two-way communication with the appropriate Sector and subsequently do not receive acknowledgement of a scheduled position report, should make every effort to relay the report via another aircraft or agency. For flights within the same or adjacent field complexes, helicopter pilots should maintain RTF contact on the field, company frequency or Traffic Area frequency. Position reports by civil helicopter pilots operating on HMRs are to be based on distance from either Aberdeen or Sumburgh VORs, according to the departure or destination aerodrome. On outbound HMRs an initial report is to be made at 40nm and then at 20nm intervals, subject to the limitations of VHF cover. For inbound flights subsequent to the initial call, reports are to be made at the same 20nm intervals according to the destination aerodrome. If the elapsed time between two reporting points exceeds 15 minutes, an additional report is to be made after 15 minutes elapsed time since the last report. Position reports are to include the following information:

(a) Callsign;

(b) Position (HMR and range);

(c) ALT or FL;

(d) Persons on board (POB) (Inbound flights only).

2.4.2 Fixed-Wing Operating Procedures. Pilots of civil and military fixed-wing aircraft intending to fly within the areas of responsibility of the above Sectors are strongly advised to make use of the services provided. Whenever possible civil aircraft should be flown above the Transition Altitude at the appropriate quadrant level. Pilots are advised that helicopters engaged on inter-platform flights within the same filed complex normally operate at about 500ft amsl and frequently carry under slung loads which limit the pilot's ability to take sudden avoidance action.

2.4.3 General Procedures. Aircraft operating within the Sumburgh Radar Sector and the Brent Radar Sector will be allocated cruising levels as follows:

(a) Aircraft operating below the Transition level:

(i) Eastbound – 1000ft and 3000ft;

(ii) Westbound – 2000ft.

(b) Aircraft operating at or above the Transition level will be allocated cruising levels according to the quadrantal rule. Transition level will be flown with reference to the QNH as directed by ATC.

2.4.4 Off-shore Operations. Helicopter off-shore support activity is not confined to published ATS hours and helicopters may be operating in VMC or IMC at all levels and times.

2.4.4.1 The positions of individual platforms within their parent field complex are detailed at paragraph 5.2.

2.4.5 Within the North Sea Radar Service Areas as described in paragraph 2.2, during the published hours of operation, the following QNH values will be used:

(a) Brent Radar Sector (122.250MHz) and northern area of the Sumburgh Radar Sector (131.300MHz) – Cormorant QNH. If the actual Cormorant QNH data is lost, the Basin QNH will be referred to. (This being the lowest of the Puffin or Marlin as appropriate).

(b) Southern area of the Sumburgh Radar Sector (131.300MHz) – Sumburgh QNH;

(c) Aberdeen Sectors (except the southeast part) – Aberdeen QNH;

(d) Southeast part of Aberdeen Information Sector (south of HMR 098 and at ranges greater than 100 DME ADN) – Fulmar QNH. If the actual Fulmar QNH data is lost, the McCabe QNH will be referred to. (This being the lowest of the Rattray and Skua).

Note: See chart at ENR 6.1.15.1 for areas.

2.4.5.1 During their hours of operation, Aberdeen ATC will instruct outbound aircraft operating south of HMR 098 to set the Fulmar Forecast QNH at 100 DME ADN. Inbound aircraft will be instructed to set the Aberdeen QNH at 100 DME ADN.

ENR 1.15 ENR 1.15.3 Morecambe Bay and Liverpool Bay Gas Fields
– Helicopter Support Flights

2.4.5.2 Transition level within the above areas will be determined with reference to the actual pressure settings.

2.4.6 Use of GPS for North Sea Operations

2.4.6.1 UK AOC Holders intending to use GPS for enroute navigation for North Sea flight operations are to use GPS equipment that meets or exceeds CAA Specification 22. AOC holders requiring further information should contact their assigned flight operations Inspector. Non UK AOC holders are recommended to operate to at least the Specification 22 standard.

2.5 Sumburgh CTR Helicopter Procedures

2.5.1 Standard Arrival and Departure Routes

2.5.1.1 The Standard Arrival and Departure Routes are established for use in conjunction with the HMRs and are shown at AD 2-EGPB-3-1/2. The route orientation is dependant on the runway-in-use as follows:

(a) Westerly Operation (Runways 27 and 33 in use):

Departures for the East Shetland Basin are to route via BODAM (595506N 0011606W) and those for HMRs Whiskey/Echo are to route via SILOK (594612N 0012900W);

Arrivals from the East Shetland Basin are to route via IZACK (595327N 0010113W) and those from HMRs Whiskey/Echo are to route via BENTY (594615N 0010810W).

(b) Easterly Operation (Runways 09 and 15 in use):

Departures for the East Shetland Basin are to route via IZACK and those for HMRs Whiskey/Echo are to route via BENTY; Arrivals from the East Shetland Basin are to route via BODAM and those from HMRs Whiskey/Echo are to route via SILOK.

2.5.1.2 Additionally, to assist in the integration of flights operating under VFR or, at night, in accordance with a Special VFR clearance, VFR/SVFR flights may be instructed to route via Mousa Visual Reference Point (600000N 0010936W) (See AD 2-EGPB-2.22 paragraph 2).

2.5.2 Helicopter Holding Patterns

2.5.2.1 IFR Holding Patterns. IFR holding patterns are established at Juliet 20, Mike 20, MOCHA and TIRIK. The holds are 1-minute left hand patterns aligned on the HMR inbound track.

2.5.2.2 Visual Holding Patterns. Visual holding patterns are established at BODAM, Mousa VRP, IZACK, BENTY and SILOK. Holding is to be conducted clear of cloud and in sight of the surface and helicopters will adopt a 2 minutes left hand orbit, except at BENTY which is right hand. Maximum holding altitude 1000ft amsl.

3 Morecambe Bay and Liverpool Bay Gas Fields – Helicopter Support Flights

3.1 Permanent platforms positioned on the Morecambe Bay and Liverpool Bay Gas Fields are shown at ENR 6.1.15.7.

3.2 Helicopter Protected Zone (HPZ)

3.2.1 A Helicopter Protected Zone (HPZ), to safeguard helicopters approaching or departing platforms or when engaged on uncoordinated inter-platform flying, is established around the Morecambe Bay and Liverpool Bay Gas Fields. A HPZ consists of the Airspace from sea level to 2000ft amsl contained within the tangential lines, not exceeding 5nm in length, joining the neighbouring circumferences of circles 1.5nm radius around each individual platform helideck.

3.3 Airspace Structure – Morecambe Bay

3.3.1 The helicopter support land base is Blackpool Airport. Low level flights, normal operating height 1000ft amsl or above on the Blackpool QNH, operate daily between Blackpool Airport and the helidecks. Flights normally route direct, however, when two or more helicopters are operating, an anti-clockwise route structure is used to provide lateral separation.

3.3.2 The route structure is:

(a) Outbound – Blackpool to Point 'N' (534922N 0030858W) to DP4;

(b) Inbound – DP3 to Point 'S' (534648N 0031040W) to Blackpool;

Note: Points 'N' and 'S' may be used as holding points.

3.3.3 Helicopters frequently operate between:

(a) Morecambe Bay Gas Field and Heysham Helipad (540223N 0025423W);

(b) Blackpool and the Heysham Helipad.

Note: Under certain operating and meteorological conditions civil helicopters may operate below 1000ft amsl on the Blackpool QNH.

3.3.4 Helicopter traffic information is available from Blackpool Approach during published hours of operation.

3.3.5 Pilots are warned that gas release and burn-off operations may take place at any time without prior notification from offshore gas installations.

3.4 Airspace Structure- Liverpool Bay

3.4.1 The helicopter support land base is Blackpool Airport. Low level flights, normal operating height 1000ft amsl on the Blackpool QNH, operate daily between Blackpool Airport and the helidecks. Transit height to/from the Lennox platform is 500ft amsl. Flights between helidecks are normally conducted between 500ft and 1000ft.

3.4.2 The route structure is:

(a) Blackpool to Gate G (534449N 0030441W) to Douglas (533210N 0033449W);

(b) Blackpool to Gate G to Off-Shore Storage Installation (534102N 0033248W);

(c) Blackpool to Gate G to Lennox (533719N 0031037W).

Note: All routes are bi-directional.

3.4.3 Helicopter traffic information is available from Warton Approach during the Warton ATC published hours of operation. Outside these hours, information is available from Blackpool Approach.

ENR 1.15 ENR 1.15.3 Morecambe Bay and Liverpool Bay Gas Fields
– Helicopter Support Flights

3.4.4 Gas release and burn-off operations may take place at any time without prior notification from off-shore gas installations.

3.5 Altimeter Setting

3.5.1 The Blackpool QNH will be used for flights between Blackpool/Heysham Helipad and the Morecambe Bay HPZ and for flights between Blackpool Airport and the Liverpool Bay HPZ.

4 Flight Plan Procedures for Helicopter Operations over Sea Areas around the United Kingdom

4.1 Pilots are warned of the need to consider application of the procedures detailed at ENR 1.10.5 paragraph 3.5, when operating under both VFR and IFR in support of off-shore facilities (particularly over the Southern and Northern North Sea, Atlantic Rim and Morecambe and Liverpool Bays). When flying to a location without an ATSU or AFTN link, nomination of a responsible person is vital to guarantee the initiation of alerting action in the event of non-arrival.

5 RTF and NDB Frequencies Used on Off-shore Installations

5.1 General

5.1.1 All Operators wishing to establish an aeronautical radio station within the UK off-shore areas under concession are required to obtain regulatory approval from the Civil Aviation Authority before establishing that radio station.

5.1.1.1 Application for approval to establish and operate an aeronautical radio station on fixed or mobile installations must be made in writing to:

Civil Aviation Authority
ATS Standards Dept
Safety Regulation Group
2W Aviation House
London Gatwick Airport South
West Sussex
RH6 OYR
Tel: 01293 573692
Fax: 01293 573974

Applicants should note:

(a) Co-ordination with other European states may be necessary before a frequency assignment can be made; applications for approval should be made as early as possible and can be made up to six months before operationally required.

(b) Operators of radio stations are also required to obtain a radio licence from the Radio-communications Agency, approval by the Civil Aviation Authority is a pre-requisite to obtaining a radio licence (Information on aeronautical radio stations is shared between the Aeronautical Licensing, CAA House, 45-59 Kingsway, London, WC2B 6TE).

5.1.2 The frequencies listed for fixed installations and the frequency plan for mobile installations apply to the UK Areas under Concession.

(a) Fixed installations are individually listed in paragraph 5.2;

(b) Mobiles must operate in accordance with the frequency plan shown in paragraph 5.3;

(c) Aeronautical RTF operations for all installations north of 56°N and east of 5°W and within the UK Off-shore Areas Under Concession, with the exception of the East Shetland Basin that is Offshore RTF Areas A to I (Note 1) have each been provided with two RTF frequencies, one for Traffic (Note 2) RTF calls and the other for Logistics (Note 3) RTF calls, as listed in paragraph 5.3.1. No suffix will be added to the Civil Aviation Authority approved callsign when traffic information is to be passed by the operator of the aeronautical radio station. When logistics information is to be passed the suffix 'LOG' shall be added to the approved callsign. Simultaneous operation of the traffic and the LOG frequencies may be required, installations are therefore required to carry radio facilities to support this. The six British Petroleum Forties Field installations have a system of operation that has been demonstrated as being suitable to support their operation using only one frequency, these installations will continue to operate using the Traffic frequency for both Traffic and LOG RTF calls (The Traffic Frequency DOC applies). (See note 2)

Note 1: Operators wishing to operate installations in Area X on the RTF frequency allocations chart should contact the Authority at least 90 days before operation is required in order to obtain a temporary RTF frequency assignment.

Note 2: The Traffic frequency is to be used for giving information on aircraft positions, obtaining deck clearance, and for lifting calls.

Note 3: The logistics (LOG) frequency should be used for all other calls permitted within the terms of an OPC service such as, departure messages, in-field routeings, payload information, and ordering fuel and refreshments.

RTF Operation – Areas A to P.

Frequency assignments stated in paragraphs 5.2 and 5.3 may only be used for communication between installations and aircraft that are both within the same offshore RTF frequency assignment area. The Traffic frequencies may be used for communication with aircraft that are below 2000 feet and the LOG frequencies may be used for communication with aircraft that are below 7000 feet. Published Air Traffic Service procedures should be followed where available. A listening watch should be maintained with the Air Traffic Service Unit (ATSU) where possible. Prior to leaving the ATSU's operating area, pilots should advise the ATSU of the aircraft's movement intentions. Where maintaining a listening watch with the ATSU is not possible and at all times when the aircraft is below 1500 feet a listening watch on the RTF Area Traffic Frequency shall be maintained.

Outbound – To the off-shore installation.

Pilots who are in communication with the ATSU should, once established in the descent and still above 1500 feet, establish contact with their destination on the Area Traffic Frequency, and hand over the flight watch. They should then advise the ATSU on passing 1500 feet. Landing clearances should be obtained from the HLO who will be operating on the Traffic frequency.

264

ENR 2.2 ENR 2.2.2 Procedures for Penetration of a MATZ by Civil Aircraft

Inbound – From the off-shore installation.
Lifting calls should be made on the Traffic frequency. Once airborne establish communication with the appropriate ATSU whilst below 1000 feet or as soon as practical. Published Air Traffic Service procedures should be followed where available. The appropriate LOG frequency should be used as required to exchange information with the destination. When operating below 1500 feet or outside the ATSU's area of operation pilots should retune the ATSU radio to the LOG frequency for passing Logistics information. When not being used for LOG messages the radio should be retuned to monitor the appropriate ATSU's frequency.

5.1.3 Offshore installation's VHF RTF facilities and aeronautical Non Directional Beacons operate by arrangement only; the un-serviceability of these facilities will NOT be promulgated by NOTAM.

5.1.4 NDBs on both fixed and mobile installations generally operate on shared frequencies. NDBs that share frequencies should only be switched on if requested by the helicopter pilot and then only after the frequency has been monitored by the pilot and found to be vacant immediately prior to switching on. The pilot should advise the NDB operators as soon as they no longer require the use of the NDB. When no longer required the NDB should be switched off. Additionally, in order to assist helicopters transiting fields a small number of NDBs on fixed installations have been assigned frequencies that enable them to operate as close to H24 as practical. Pilots may find these NDBs already on and that they remain on after they have advised the installation that they no longer require their use.

5.1.5 An Air Traffic Control Service is provided in the East Shetland Basin by NATS Limited from Aberdeen.

5.1.6 A Flight Information Service is provided for parts of the Off-shore Areas Under Concession by NATS Limited from Aberdeen.

ENR 2.2 OTHER REGULATED AIRSPACE MILITARY AERODROME TRAFFIC ZONES

1 Description

1.1 Military Aerodrome Traffic Zones (MATZ) are established at the locations listed at paragraph 4 and shown on the chart at ENR 6.2.2.4.1. The purpose of the MATZ is to provide a volume of airspace within which increased protection may be given to aircraft in the critical stages of circuit, approach and climb-out. Normally, these zones comprise:

(a) The airspace within 5nm radius of the mid-point of the longest Runway from the surface to 3000ft aal.

(b) The airspace within a 'stub' (or at some aerodromes 2 stubs) projected from the above airspace having a length of 5nm along its centre-line, aligned with a selected final approach path, and a width of 4nm (2nm either side of the centre-line), from 1000ft aal to 3000ft aal.

1.1.1 The dimensions of the zone and associated stub(s) may vary. See paragraph 4 and ENR 6.2.2.3.1 (chart) for details.

1.2 Often, two or more neighbouring MATZ are amalgamated, with one of the aerodromes being designated as the controlling authority of the combined MATZ (CMATZ). In these instances, the upper limit is measured from the higher or highest aerodrome forming part of the combined CMATZ.

1.3 An ATZ (Aerodrome Traffic Zone as defined in the Air Navigation Order currently in force) exists within a MATZ and is based upon the same reference point as defined in paragraph 1.1 (a). Although civil recognition of a MATZ is not mandatory, pilots are to comply with the provisions of the current Rules of the Air Regulations in respect of the ATZ. The notified hours of operation of an ATZ may vary from the notified hours of watch of a MATZ.

2 Procedures for Penetration of a MATZ by Civil Aircraft

2.1 A MATZ Penetration Service is available from the controlling aerodromes listed at paragraph 4 for the provision of increased protection to VHF RTF equipped civil aircraft. Pilots wishing to penetrate a MATZ are requested to observe the following procedures:

(a) When 15nm or 5 minutes flying time from the zone boundary, whichever is the greater, establish two-way RTF communication with the controlling aerodrome on the appropriate frequency using the phraseology: '..... (controlling aerodrome), this is ..... (aircraft callsign), request MATZ penetration.'

(b) when the call is acknowledged and when asked to 'pass your message', the pilot should pass the following information:

(i) Callsign;

(ii) Type of aircraft;

(iii) Position;

(iv) Heading;

(v) Altitude;

(vi) Intentions (eg destination);

(c) comply with any instructions issued by the controller;

(d) maintain a listening watch on the allocated RTF frequency until the aircraft is clear of the MATZ;

(e) advise the controller when the aircraft is clear of the MATZ. Flight conditions are not required unless requested by the controller.

2.2 The ATS Unit providing the MATZ Penetration Service will give traffic information and any instructions necessary to achieve safe separation from known or observed traffic in the zone. The service will, whenever possible, be based on radar observations and either a Radar Advisory or Radar Information Service will be given. When radar separation cannot be applied, vertical separation of at least 500ft between known traffic will be applied. When safe lateral or vertical separation cannot be achieved, pilots will be advised to avoid the MATZ.

2.3 If appropriate, controllers will endeavour to co-ordinate flights with the controlling authority of an adjacent zone, but pilots should not assume clearance to penetrate another MATZ until it is explicitly given.

ENR 2.2 — ENR 2.2.2 Procedures for Penetration of a MATZ by Civil Aircraft

2.4 To ensure safe vertical separation is applied, all aircraft will be given an altimeter setting to use within the zone. Normally this will be the aerodrome QFE. Exceptionally, within the Odiham MATZ the transit pressure setting will be the Farnborough QNH, and within the Warton MATZ the setting will be the Warton QNH and within the Lakenheath/Mildenhall MATZ the setting will be the Lakenheath QNH.

2.5 In the case of overlapping MATZs, the altimeter setting to be used will, except as detailed below, be the QFE of the higher or highest aerodrome of the combined zone. This will be passed as the 'Clutch QFE'. At Royal Naval Air Stations Yeovilton and Culdrose and their satellite aerodromes of Merryfield and Predannack, the altimeter settings will be the QFE values for Yeovilton and Culdrose respectively.

2.6 Whilst every effort will be made to ensure the safe separation of aircraft complying with these procedures. Since compliance is not compulsory, some civil aircraft within the MATZ may not be known to the controller. Pilots should therefore keep a good lookout at all times.

2.7 Terrain clearance will be the responsibility of pilots.

3 Availability of the MATZ Penetration Service

3.1 A MATZ Penetration Service will be available during the published hours of watch of the respective ATS Units. However, as many units are often open for flying outside normal operating hours, pilots should call for the penetration service irrespective of the hours of watch published. If, outside normal operating hours, no reply is received after two consecutive calls, pilots are advised to proceed with caution. Information on the operation of aerodromes outside their normal operating hours may be obtained by telephone from the appropriate Military Air Traffic Control Centre:

(a) North of 5430N – Telephone: Scottish ACC (Mil) 01292 479800 Ex 6703/4
(b) South of 5430N – Telephone: London ACC (Mil) 01895 426150.

4 MATZ Participating Aerodromes

4.1 Details of participating aerodromes are given on ENR 2.2.3.3.

Note 1: These aerodromes are opened on very limited occasions when advised by NOTAM or Supplement.
Note 2: Non-standard north easterly stub SFC to 3000ft.
Note 3: Helicopters tasked to operate in EG D208 are required to call Lakenheath ATC to notify intended entry to EG D208 prior to penetrating the CMATZ. No restrictions will be imposed by Lakenheath on helicopters which operate within that portion of their north-easterly stub which is also within the lateral limits of EG D208, provided that the aircraft remain at or below 800ft amsl.
Note 4: Non-standard extension to both stubs – 5nm south of extended centre-lines.
Note 5: Non-standard demarcation of the 5nm circles which are joined by a straight line at their most easterly points.
Note 6: Non-standard reference point aligned with common radar touchdown point.
Note 7: Non-standard MATZ with the following dimensions:
Lateral A rectangle of airspace, 20nm x 6nm. The major axis is centred on the Aerodrome Reference Point (ARP), aligned with the major runway headings 071° (T)/251° (T) and off-set 1nm to the south.
Vertical The portion of the rectangle contained within the part circle radius 5nm centred on the ARP extends from the surface to 3000ft AAL. The remainder extends from 1000ft AAL to 3000ft AAL.
Warning The northern sector of the ATZ is not wholly contained within the MATZ.
Note 8: Warning 5nm radius portion of MATZ co-incident with EG R313.
Note 9: If Boscombe Down is closed but Middle Wallop remains open, a CMATZ penetration service will be provided by Wallop Approach on 126.700 MHz.

ENR 2.2   ENR 2.2.4 MATZ Participating Aerodromes

MATZ	Mid-point Of the longest runway	AD Elevation (ft)	Stub Heading(s) T° to AD	Controlling Aerodrome	Frequency to be used (MHz)	Remarks
1	2	3	4	5	6	7
Barkston Heath	525748N 0003337W	367	058 (2nm stub)	Cranwell	119.375	MATZ 3nm radius Stub extends from SFC to 3000ft aal
Benson	513659N 0010545W	203	188	Benson	120.900	
Boscombe Down	510912N 0014504W	407	230/050	Boscombe Down	126.700	Note 9
Church Fenton	535004N 0011143W	29	235	Church Fenton	126.500	
Coltishall	524518N 0012127E	66	216	Colitshall	125.900	
Coningsby	530535N 0000958W	25	252	Coningsby	120.800	
Cottesmore	524409N 0003656W	461	221/041	Cottesmore	130.200	
Cranwell	530148N 0002934W	218	263	Cranwell	119.375	
Culdrose	500508N 0051515W	267	293	Culdrose	134.050	
Dishforth	540814N 0012513W	117	–	Leeming	127.750	MATZ 3nm radius
Fairford	514101N 0001472W	286	268	Brize Norton	119.000	Note 1
Honington	522033N 0004623E	174	261	Lakenheath	128.900	Note 1
Kinloss	573858N 0033338W	22	–	Lossiemouth	119.350	
Lakenheath	522433N 0003340E	32	056/236	Lakenheath	128.900	Notes 2 & 3
Leeming	541733N 0013207W	132	156	Leeming	127.750	
Leuchars	562229N 0025140W	38	262/082	Leuchars	126.500	
Linton on Ouse	540258N 0011513W	53	211	Linton on Ouse	118.550	
Lossiemouth	574225N 0032015W	42	224	Lossiemouth	119.350	
Marham	523854N 00033302E	75	235/055	Marham	124.150	
Merryfield	505747N 0025613W	146	–	Yeovilton	127.350	MATZ 3nm radius
Middle Wallop	510822N 0013407W	297	256 (3nm stub)	Boscombe Down	126.700	Note 6
Mildenhall	522142N 0002911E	33	103/283	Lakenheath	128.900	Notes 4 & 5
Mona	531533N 0042227W	202	–	Valley	125.225	
Odiham	511403N 0005634W	405	093	Farnborough	125.250	
Predannack	500006N 0051355W	295	–	Culdrose	134.050	
Scampton	531828N 0003303W	202	041	Waddington	127.350	Note 8
Sculthorpe	525045N 0004557E	214	234	Marham	124.150	Note 1
Shawbury	524737N 0024004W	249	180/360	Shawbury	120.775	
St Mawgan	502626N 0045944W	390	301	St Mawgan	128.725	
Ternhill	525225N 0023155W	272	225 (2nm stub)	Shawbury	120.775	
Topcliffe	541220N 0012255W	92	201	Leeming	127.750	
Valley	531450N 0043201W	37	130	Valley	125.225	
Waddington	530958N 0003126W	231	202	Waddington	127.350	
Warton	534442N 0025302W	54	–	Warton	129.525	Note 7
Wattisham	520738N 0005721E	284	228/048	Wattisham	125.800	
Wittering	523645N 0002836W	275	253/073	Cottesmore	130.200	
Yeovilton	510029N 0023845W	75	263/083	Yeovilton	127.350	

ENR 5.5 AERIAL SPORTING AND RECREATIONAL ACTIVITIES
ENR 5.5.1 GLIDER LAUNCHING SITES

Designation	Launch Type	Elevation ft amsl	Vertical limits ft agl	Operator Tel No.	Hours
Glider Launching Sites					
Abingdon, Oxon 514115N 0011858W	T	261			HJ Sat, Sun & PH
Aboyne, Glamorgan 570430N 0025005W	T	460		Deeside Gliding Club 01339 885339	HJ
Andreas, Isle of Man 542210N 0042524W	W	110	2000		HJ
Arbroath, Angus, Tayside 563455N 0023716W	W	160	2000		HJ
Aston Down, Stroud 514228N 0020750W	W & T	600	3000	Cotswold Gliding Club 01285 760415	HJ
Aylesbury/Thame, Bucks 514633N 0005625W	W	289	2000	Upward Bound Trust Gliding Club	HJ
Barrow/Walney Island, Cumbria 540752N 0031549W	W & T	47	2000	Lakes Gliding Club 01229 471458	HJ
Bellarena, Co Londonderry 550836N 0065757W	T	15	2000	Ulster Gliding Club 028 7776 3321 028 7775 0301 (weekends only)	HJ
Bembridge, Isle of Wight 504041N 0010633W	T	55		Vectis Gliding Club	HJ
Benone Strand, Co Londonderry 551000N 0065133W	W	SL	2000	Ulster Gliding Club 028 7775 50301	HJ
Bicester, Oxon 515458N 0010756W	W & T	267	3000	Windrushes Gliding Club	HJ
Bidford, Warwicks 520803N 0015103W	T	135		Bidford Gliding Centre 01789 772606	HJ
Bleese Hall, Kendall 541707N 0024105W	W	330	2000	Lakes Gliding Club 01229 471458	HJ Sat, Sun & PH
Brent Tor, Tavistock, Devon 503517N 0040850W	W	820	2000	Dartmoor Gliding Club 01822 810712	HJ
Burn, Selby, Humberside 534445N 0010504W	W & T	20	2000	Burn Gliding Club 01757 270296	HJ

ENR 5.5.1 GLIDER LAUNCHING SITES

Designation	Launch Type	Elevation ft amsl	Vertical limits ft agl	Operator Tel No.	Hours
Camphill, Buxton, Derby 531818N 0014353W	W	1350	2000	Derby & Lancs Gliding Club 01298 871270	HJ
Carlton Moor, Cleveland 542429N 0011206W	W	1200	2000	Carlton Moor Gliding Club 01624 778234 (weekends only)	HJ
Challock, Ashford, Kent 511230N 0004945E	W & T	600	2000	Kent Gliding Club 01233 740274	HJ
Chipping, Preston, Lancs 535301N 0023714W	W	600	3000	Bowland Forest Gliding Club 01995 61267	HJ
Chivenor, Devon 510514N 0040901W	T	27			HJ
Cosford, West Midlands 523824N 0021820W	W & T	271	3000		HJ Sat, Sun PH & 1700-SS Mon-Fri Winter (Summer 1Hr earlier)
Cranwell (North), Lincs 530231N 0002936W	W & T	220	3000		HJ Sat, Sun, PH & Wed 1 Oct-30Apr
Cross Hayes, Staffs 524740N 0014914W	W & T	320	2000	Needwood Forest Gliding Club	HJ 1 May-30 Sep
Crowland, Lincs 524233N 0000834W	T	10		Peterborough & Spalding Gliding Club 01733 210463	HJ
Culdrose, Cornwall 500510N 0051521W	W & T	267	2000		HJ
Currock Hill, Northumberland 545602N 0015043W	W & T	800	2000	Northumbria Gliding Club 01207 561286	HJ
Dishforth, Yorks 540826N 0012506W	W & T	117	3000		HJ
Drumshade, Kirriemuir, Angus 563838N 0030126W	W	230	2000	Angus Gliding Club	HJ
Dunstable Downs, Beds 515200N 0003254W	W & T	500	2000	London Gliding Club 01582 663419	HJ
Eaglescott, Devon 505542N 0035922W	W & T	655	2000	North Devon Gliding Club 01769 520404	HJ
Easterton, Elgin, Grampian 573508N 0031841W	W & T	361	2000	Highland Gliding Club 01343 860272	HJ

ENR 5.5

ENR 5.5.1 GLIDER LAUNCHING SITES

Designation	Launch Type	Elevation ft amsl	Vertical limits ft agl	Operator Tel No.	Hours
Edge Hill/Shenington, Oxon 520507N 0012828W	W & T	642	2500	Shennington Gliding Club 01295 688121	HJ
Eyres Field, Gallows Hill, Dorset 504233N 0021310W	W & T	205	2000	Dorset Gliding Club 01929 405599	HJ Sat, Sun PH, Wed & 1200-SS Fri Winter (Summer 1Hr earlier)
Falgunzeon, Dalbeattie, Dumfries 545638N 0034424W	W	600	2000	Dumfries & District Gliding Club 01387 760601	HJ
Feshiebridge, Highlands 570613N 0035330W	W & T	860	2000	Cairngorm Gliding Club 01540 651317	HJ
Gransden Lodge, Cambridge 521041N 0000653W	W & T	254	3000	Cambridge Gliding Club 01767 677077	HJ
Grayrigg, Kendall, Cumbria 542246N 0023820W	W	820	1000	Eden Soaring Society	Sat, Sun, Mon & PH & as notified by NOTAM
Halesland, Avon 511544N 0024356W	W & T	870	2000	Mendip Gliding Club 01749 870312	HJ
Halton, Bucks 514733N 0004416W	W & T	370	2000		HJ
Henlow, Beds 520110N 0001806W	T	170			HJ
Hinton-in-the-Hedges, Oxon 520145N 0011229W	W & T	500	2000	Aquila Gliding Club 01295 811056	HJ
Hullavington, Wilts 513147N 0020814W	W	343	2000		HJ Sat, Sun & PH & as notified by NOTAM
Husbands Bosworth, Leics 522626N 0010238W	W & T	505	3000	Coventry Gliding Club 01858 880521	HJ
Jurby, Isle of Man 542114N 0043108W	T	89		Manx Gliding Club	HJ
Keevil, Wilts 511850N 0020643W	W & T	200	3000		HJ
Kenley, Surrey 511820N 0000537W	W	566	1700	Surrey Hills Gliding Club 020 8763 0091	HJ
Kinloss, Grampian 573858N 0033338W	W & T	22	3000		HJ
Kirknewton, Lothian 555227N 0032405W	W	652	2000		HJ Sat, Sun, PH & 1700-SS Fri Winter (Summer 1hr earlier)& as notified by NOTAM

ENR 5.5.1 GLIDER LAUNCHING SITES

Designation	Launch Type	Elevation ft amsl	Vertical limits ft agl	Operator Tel No.	Hours
Kirton-in-Lindsey, Lincs 532745N 0003436W	W & T	203	2000	Trent Valley Gliding Club 01652 648777	HJ
Lasham, Alton, Hants 511112N 0010155W	W & T	618	3000	Lasham Gliding Society 01256 381322	HJ
Lee-on-Solent, Hants 504855N 0011225W	W & T	32	2000		HJ
Linton-on-Ouse, Yorks 540256N 0011510W	T	53			HJ
Little Rissington, Glos 515203N 0014143W	T	730			HJ
Lleweni Parc, Denbigh, Clwyd 531239N 0032312W	W & T	200	3000	Glyndwr Soaring Club 01745 813774	HJ
Llantisilio, Llandegla, Wrexham 530239N 0031315W	W	1120	2000	Vale of Clwd Gliding Club	HJ
Long Mynd, Salop 523108N 0025233W	W & T	1411	3000	Midland Gliding Club 01588 650206	HJ
Lyveden, Northants 522758N 0003430W	W	279	2000	Welland Gliding Club 01832 205237	HJ
Manby 532130N 0000459E	W	60	2000	Spilsby Soaring Trust 01754 830221	HJ
Marham, Norfolk 523854N 0003302E	W & T	75	3000		HJ
Merryfield 505729N 0025614W	W	146	2000		HJ
Milfield, Northumberland 553514N 0020510W	T	150	2000	Borders Gliding Club 01688 216284	HJ Sat, Sun, PH & as notified by NOTAM
North Hill, Devon 505107N 0031639W	W & T	921	2000	Devon & Somerset Gliding club 01404 841386	HJ
North Weald, Essex 514318N 0000915E	W & T	321	2000	Essex Gliding Club 01992 522222	HJ
Nympsfield, Glos 514251N 0021701W	W & T	700	3000	Bristol & Gloucester Gliding Club 01453 860342	HJ

ENR 5.5.1 GLIDER LAUNCHING SITES

Designation	Launch Type	Elevation ft amsl	Vertical limits ft agl	Operator Tel No.	Hours
Oban/North Connel, Strathclyde 562740N 0052410W	W & T	20	2000	Connel Gliding Club 01631 710428	HJ
Odiham, Hants 511403N 0005634W	W & T	405	2500	DRA Farnborough Gliding Club 01256 703157	HJ
Parham, Sussex 505532N 0002828W	W & T	110	2000	Southdown Gliding Club 01903 746706	HJ
Perranporth, Cornwall 501947N 0051039W	W & T	330	2000	Cornish Gliding & Flying Club 01872 572124	HJ
Pocklington, Humberside 535541N 0004751W	W & T	87	2000	Wolds Gliding Club 01759 303579	HJ
Portmoak, Tayside 561121N 0031945W	W & T	360	2000	Scottish Gliding Union 01592 840243	HJ
Predannack, Cornwall 500006N 0051355W	W	295	2000		HJ Sat, Sun & PH as notified by NOTAM
Rattlesden, Suffolk 521001N 0005216E	W & T	305	2000	Rattlesden Gliding Club 01449 737789	HJ
Retford/Gamston, Notts 531650N 0005705W	W	87	2000	Dukeries Gliding Club	HJ
Rhigos, Powys 514434N 0033505W	W & T	780	2000	Vale of Neath Gliding Club	HJ
Ridgewell, Essex 520253N 0033303E	W & T	273	2000	Essex Gliding Club 01440 785103	HJ
Ringmer, Kitsons Field, Sussex 505423N 0000618E	W & T	72	2500	East Sussex Gliding Club 01825 840347	HJ
Rivar Hill, Wilts 512038N 0013235W	W	730	3000	Shalbourne Soaring Society 01264 731204	HJ
Sackville Lodge, Risely, Beds 521551N 0002905W	W	250	2000	Sackville Gliding Club 01234 708877	HJ
Saltby, Leics 524947N 0004245W	W & T	480	2000	Buckminster Gliding Club 01476 860385	HJ
Salmesbury, Lancs 534626N 0023359W	T	269			HJ

ENR 5.5   ENR 5.5.1 GLIDER LAUNCHING SITES

Designation	Launch Type	Elevation ft amsl	Vertical limits ft agl	Operator Tel No.	Hours
Sandhill Farm, Wilts 513614N 0014030W	W & T	350	2000	Vale of White Horse Gliding Club 01793 783685	HJ
Sealand, Cheshire 531309N 0030055W	W	15	2000		HJ Sat, Sun, PH & 1700-SS Mon-Fri Winter (Summer 1Hr earlier) & as notified by NOTAM
Seighford, Staffs 524940N 0021212W	W & T	321	2000	Staffordshire Gliding Club 01785 282575	HJ
Shobdon, Hereford 521429N 0025253W	T	328		Herefordshire Gliding Club 01568 708908	HJ
Sleap, Salop 525002N 0024618W	T	275			HJ
Snitterfield, Warwicks 521406N 0014310W	W	375	2000	Stratford on Avon Gliding Club 01789 731095	Hj
Spilsted Farm, Stream Lane, Sussex 505613N 0003109E	W	160	2000		HJ Sat, Sun & Thu
St Athan, Glams 512417N 0032609W	W	163	2000		HJ Sat, Sun, PH & 1700-SS Mon-Fri Winter (Summer 1hr earlier) & as notified by NOTAM
Strathaven, South Lanarkshire 554048N 0040620W	W & T	847	2000	Strathclyde Gliding Club 01357 520235	HJ
Strubby, Lincs 531836N 0001034E	W & T	47	3000	Lincolnshire Gliding Club 01507 450698	HJ
Sutton Bank, Yorks 541338N 0011249W	W & T	920	2000	Yorkshire Gliding Club 01845 597237	HJ
Swansea 513619N 004040W	W	299	1500	636 VGS	HJ
Syerston, Notts 530121N 0005447W	W & T	224	3000		HJ
Talgarth, Powys 515848N 0031215W	T	970		Black Mountain Gliding Club 01874 711463	HJ
Ternhill, Salop 525216N 0023200W	T	272			HJ
The Park, Kingston Deverill 510742N 0021445W	W	697	3000	Bath, Wilts & North Dorset Gliding Club 01985 844095	HJ
Tibenham, Norfolk 522724N 0010915E	W & T	186	3000	Norfolk Gliding Club 01379 677207	HJ

ENR 5.5 — ENR 5.5.1 GLIDER LAUNCHING SITES

Designation	Launch Type	Elevation ft amsl	Vertical limits ft agl	Operator Tel No.	Hours
Topcliffe, Yorks 541220N 0012255W	T	92			HJ Sat, Sun, PH & 1700-SS Fri Winter (Summer 1hr earlier) & as notified by NOTAM
Upavon, Wilts 511712N 0014700W	W & T	575	2000		HJ Sat, Sun, PH & 1700-SS Mon-Fri Winter (Summer 1hr earlier)
Upwood, Cambs 522612N 0000836W	W	75	2000	Nene Valley Gliding Club 0860 693479	HJ
Usk, Gwent 514306N 0025101W	W & T	80	2000	South Wales Gliding Club 01291 690536	HJ
Waldershare Park, Kent 511020N 0011636E	W	375	2000	Channel Gliding Club 01304 824888	HJ
Wattisham, Suffolk 520739N 0005722E	W & T	284	3000		HJ
Watton, Norfolk 523344N 0005145E	W	207	3000		HJ Sat, Sun & other times by NOTAM
Weston-on-the-Green, Oxon 5152497N 0011311N	W	282	3000	Oxford Gliding Club 01869 343265	HJ
Wethersfield, Essex 515827N 0003014E	W	321	2000		HJ Sat, Sun, PH & 1700-SS Mon-Fri Winter (Summer 1hr earlier) & as notified by NOTAM
Winthorpe, Notts 530544N 0004616W	W & T	60	2000	Newark & Notts Gliding Club 01636 707151	HJ
Wittering, Cambs	W & T	273	3000	Four Counties Gliding Club	HJ Sat, Sun & PH & after 1700 on Fri (Summer 1 Hr earlier)
Wormingford, Essex 515630N 0004723E	W & T	236	3000	Essex & Suffolk Gliding Club 01206 242596	HJ
Wycombe Air Park, Bucks 513642N 0004830W	T	520		Booker Gliding Club 01494 442501	HJ
Yeovilton, Somerset 510034N 0023820W	W & T	75	2000	Heron Gliding Club	HJ
York/Rufforth, Yorks 535651N 0011116W	W & T	65	2000	York Gliding Centre 01904 738694	HJ

ENR 5.5.2 Hang Gliding and Parascending

Designation	Launch Type	Elevation ft amsl	Vertical limits ft agl	Operator Tel No.	Hours
Hang Gliding and Parascending					
Berriewood Farm, Condover, Salop 523800N 0024337W		400	2000		HJ Sat, Sun & PH
Bloreheath Farm, Almington, Salop 525435N 0022624W		300	800		HJ, Sat, Sun & PH
Bradwell Moor, Derbyshire 531913N 0014707W		1500	2000		HJ
Bromwich Park Farm, Oswestry, Salop 524925N 0030011W		280	500		HJ Sat, Sun, PH & before 0900 & after 1700 Mon-Fri Winter (Summer 1hr earlier)
Brown Wardle, Lancs 533952N 0020921W		1312	2000		HJ
Brunton Aerodrome, Northumberland 553129N 0014035W		100	2000		HJ
Buckley Farm, Pentre, Salop 524452N 0025648W		180	500		HJ Sat, Sun, PH & before 0900 & after 1700 Mon-Fri Winter (Summer 1hr earlier)
Chale Bay, Isle of Wight 503632N 0012024W		130	2000		HJ
Chetwynd Aerodrome, Shropshire 524846N 0022434W		280	2000		HJ Sat, Sun & PH
Chirk, Shropshire 525650N 0030252W		430	1200		HJ Sat, Sun, PH & Before 0900 & after 1700 Mon-Fri Winter (Summer 1 hr earlier)
Cockle Park, Northumberland 551214N 0014018W		248	2000		HJ
Cravens Gorse, Charlton Abbots, Glos 515452N 0015528W		886	1000		HJ
Culdrose, Cornwall 500502N 0051505W		270	2000		HJ Sat, Sun & PH
Darley Moor, Derbyshire 525841N 0014435W		600	2000		HJ
Devils Dyke, Saddlescombe Farm, Sussex 505233N 0001303W		666	2000		HJ
East Kirkby, Lincs 530811N 0000020E		50	2000		HJ Sat, Sun, PH & 1700-SS Mon-Fri Winter (Summer 1hr earlier)

ENR 5.5.2 Hang Gliding and Parascending

Designation	Launch Type	Elevation ft amsl	Vertical limits ft agl	Operator Tel No.	Hours
Great Fransham, Norfolk 524131N 0004857E		210	2000		HJ
Haimwood Farm, Melverley, Salop 524437N 0030021W		180	1200		HJ Sat, Sun & PH Occasioanl activation by NOTAM
Horspath, Oxford 514433N 0011059W		320	1000		HJ
Kennel Farm, Warlingham, Surrey 511849N 0000229W		590	500		HJ
Manchester Barton, Lancs 532817N 0022321W		73	1000		HJ
Manor Farm, Drayton St Leonard, Oxon 514020N 0010816W		180	1500		HJ Sat, Sun, PH & 1700-SS Mon-Fri Winter (Summer 1hr earlier)
Mendip Forest, Somerset 511718N 0024327W		800	1500		HJ
Mendlesham, Suffolk 521342N 0010729E		221	2000		HJ Sat, Sun, PH & between May-Seo 1600-SS Mon-Fri
Metfield, Suffolk 522151N 0012330E		182	2000		HJ Sat, Sun & PH
Middle Wallop, Hants 510858N 0013413W		297	2000		HJ
Monks Field, Shadoxhurst, Kent 510634N 0005004E		140	2000		HJ
North Luffenham, Leics 523754N 0003627W		352	1500		HJ Sat, Sun & PH
Parham, Framlingham, Suffolk 521146N 0012418E		120	2000		HJ Sat, Sun & PH
Park Farm Down, Berkshire		540	1500		HJ Sat, Sun & PH & before 0900-1700 Mon-Fri (Summer 1Hr earlier)
513138N 0013429W					
Sculthorpe, Norfolk 525049N 0004548E		214	3000		HJ Sat, Sun & PH
Shrewesbury School 524211N 0024600W		250	500		HJ Sat, Sun, PH & 0900-17 Mon-Fri Winter (Summer 1hr earlier)
South Ambersham, Sussex 505835N 0004102W		80	1000		HJ Sat, Sun & PH

ENR 5.5.3 FREE-FALL DROP ZONES

Designation	Launch Type	Elevation ft amsl	Vertical limits ft agl	Operator Tel No.	Hours
South Cerney, Glos 514115N 0015515W		360	2000		HJ
Spitalgate, Lincs 525359N 0003547W		420	2000		HJ
Thorney Island, Sussex 504858N 0005511W		18	2000		HJ Sat, Sun & PH
Trehayne Vean, Cornwall 501824N 0045959W		300	2000		HJ
Upfield Farm, Gwent 513308N 0025344W		20	2000		HJ
Upottery, Devon 505305N 0030921W		835	2000		HJ
FREE-FALL DROP ZONES					
Abingdon, Oxon Circle 1.5nm radius of 514115N 0011858W			FL 85 †	Benson ATC 01491 837766 Ex 7017 London TCC 01895 426422	Activity notified on the day to Benson ATC London TCC outside hrs of Benson † Exceptionally, drops may be made from up to FL 150 with London TCC permission Hours: As notified
Ballykelly, Co. Londonderry Circle 1.5nm radius of 550340N 0070048W			FL 150	Londonderry ATC 028 7772 1472 Scottish ACC 01292 692763	Activity subject to permission from Londonderry ATC or notified on the day to Scottish ACC outside hrs of Londonderry Alternative contact: 129.950 MHz Hours: Normally during daylight hrs
Ballyrogan, Co. Londonderry Circle 1.5nm radius of 545929N 0064526W			FL 150	Londonderry ATC 028 7181 1099 Scottish ACC 01292 692763	Activity notified on the day to Londonderry ATC or Scottish ACC outside Normal hrs of Londonderry Hours: Normally during daylight hrs
Boscombe Down, Wilts Circle 1.5nm radius of 510900N 0014500W			FL 150	Boscombe Down ATC 01980 663051 01980 663052	Activity subject to permission from Boscombe Down ATC
Bridlington, Yorks Circle 1.5nm radius of 540713N 0001411W			FL 150	London ACC 01262 677367	Activity notified on the day to London ACC Alternative contact: 129.900 MHz Hours: Normally during daylight hrs

ENR 5.5.3 FREE-FALL DROP ZONES

Designation	Launch Type	Elevation ft amsl	Vertical limits ft agl	Operator Tel No.	Hours
Brize Norton, Oxon Circle 1.5nm radius of 514516N 0013421W			FL 150	Brize Norton ATC 01993 897878	Activity subject to permission from Brize Norton ATC Hours: Normally during daylight hrs
Brunton, Northumberland Circle 1.5nm radius of 553128N 0014027W			FL 150	Newcastle ATC 0191 286 0966	Activity notified on the day to Newcastle ATC Alternative contact: 129.900 MHz Hours: normally during daylight hrs
Cark, Cumbria Circle 1.5nm radius of 540946N 0025737W			FL 145	London ACC 01489 612420	Activity notified on the day to London ACC Alternative contact: 129.900 MHz Hours: Normally during daylight hrs Sat, Sun & PH
Carlisle, Cumbria Circle 1.5nm radius of 545615N 0024833W			FL 95	Carlisle ATC 01228 573629 Scottish ACC 01292 692763	Activity subject to permission from Carlisle ATC or notified on the day to Scottish ACC outside hrs of Carlisle Alternative contact: 123.600 MHz Hours: Normally during hrs Sat, Sun & PH Also normally during daylight hrs Mon-Fri evenings
Chalgrove, Oxon Circle 1.5nm radius of 514032N 0010459W			ALT 5500	Benson ATC 01491 837766 Ex 7017 London TCC 01895 426422	Activity notified on the day to Benson ATC or London TCC outside hrs of Benson
Chatteris, Cumbria Circle 1.5nm radius of 522919N 0000512E			FL 150	London ACC 01489 612420	Activity notified on the day to London ACC Alternative contact: 129.900 MHz Hours: Normally during daylight hrs Tue-Sun & PH
Cockerham, Lancs Circle 1.5nm radius of 535744N 0025007W			FL 95	London ACC 01489 612420	Activity notified on the day to London ACC Alternative contact: 129.900 MHz Hours: Normally during daylight hours
Dunkeswell, Devon Circle 1.5nm radius of 505138N 0031406W			FL 150	Exeter ATC 01392 367433 Ex 215 London ACC 01489 612420	Activity notified on the day to Exeter ATC or London ACC outside hrs of Exeter Alternative contact: 129.900 MHz or 123.475 MHz Hours: Normally during daylight hrs
Errol, Tayside Circle 1.5nm radius of 562418N 0031055W			ALT 5500 †	Scottish ACC 01292 692763	Activity notified on the day to Scottish ACC † Drops may be made from above ALT 5500ft with Scottish ACC permission Alternative contact: 129.900 MHz Hours: Normally during daylight hrs Wed-Sun & PH

ENR 5.5.3 FREE-FALL DROP ZONES

Designation	Launch Type	Elevation ft amsl	Vertical limits ft agl	Operator Tel No.	Hours
Henlow, Beds Circle 1.5nm radius of 520113N 0001812W			ALT 3500 †	Luton ATC 01582 395455	Activity notified on the day to Luton ATC † Drops may be made from above ALT 3500ft with Luton ATC permission
Hibaldstow, Humberside Circle 1.5nm radius of 532956N 0003048W			FL 120 †	Humberside ATC 01652 688456	Activity subject to permission from Humberside ATC † Drops may be made from up to FL 150 with Manchester ACC permission Alternative contact: 129.925 MHz Hours: Normally during daylight hrs
Hinton-in-the-Hedges, Banbury, Oxon Circle 1.5nm radius of 520136N 0011216W			FL 65 †	Brize Norton ATC 01993 897878	Activity notified on the day to Brize Norton ATC † Drops may be made from up to FL 150 with London TCC permission Alternative contact: 119.450 MHz (up to 3000ft) then 129.900 MHz Hours: Normally during daylight hrs Tue-Sun & PH
Keevil, Wilts Circle 2nm radius of 511851N 00260637W			FL 150	Lyneham ATC 01249 890381 Ex 6522	Activity notified on the day to Lyneham ATC. See also ENR 1.1.5.7 for details on Heavy supply dropping by Hercules aircraft Hours: As notified
Killykergan, Co Londonderry Circle 1.5nm radius of 550103N 0063951W			FL 150	Londonderry ATC 028 2955 8609 Scottish ACC 01292 592763	Activity notified on the day to Londonderry ATC or Scottish ACC outside Hrs of Londonderry. Area overlaps Movenis drop zone Alternative contact: 129.900 MHz Hours: Normally during daylight hrs
Kingsmur, Fife Circle 1.5nm radius of 561604N 0024503W			FL 150	Leuchars ATC 01334 839471 Ex 7284	Activity must be notified to, and co-ordinated with Leuchars ATC Alternative contact: 129.900 MHz Hours: Normally during daylight hrs Fri, Sat, Sun & PH and other times as notified to RAF Leuchars

ENR 5.5.3 FREE-FALL DROP ZONES

Designation	Launch Type	Elevation ft amsl	Vertical limits ft agl	Operator Tel No.	Hours
Langar, Notts Circle 1.5nm radius of 525338N 0005416W			FL 150	Cottesmore ATC 01572 812241 Ex 7330	Activity notified on the day to Cottesmore ATC or London TCC outside hrs of Cottesmore Alternative Hrs: 129.900 MHz Hours: Normally during daylight hrs
Lashenden/Headcorn, Kent Circle 1.5nm radius of 510925N 0003902E			ALT 3500 †	Manston ATC 01843 823351 Ex 6302 London TCC 01895 426422	Activity notified on the day to Manston ATC or London TCC outside hrs of Manston † Drops may be made from up to FL 150 with London TCC permission Alternative contact: 122.00 MHz Hours: Normally during daylight hrs
Lewknor, Oxon Circle 1.5nm radius of 514015N 0005849W			ALT 5000 †	Benson ATC 01491 837766 Ex 7554 London TCC 01895 426422	Activity notified on the day to Benson ATC or London TCC outside hrs of Benson † Drops may be made from up to FL 150 with London TCC permission Alternative contact: 129.900 MHz Hours: Normally during daylight hrs Sat, Sun & PH
Middle Wallop, Hants Circle 1.5nm radius of 510902N 0013405W			FL 150	Boscombe Down ATC 01980 663051 01980 663052	Activity notified on the day to Boscombe Down ATC Hours: As notified
Movenis, Co. Londonderry Circle 1.5nm radius of 545915N 0063853W			FL 150	Londonderry ATC 028 7181 1099 Scottish ACC 01292 692763	Activity notified on the day to Londonderry ATC or Scottish ACC outside hrs of Londonderry Alternative contact: 129.900 MHz Hours: Normally during daylight hrs
Netheravon, Wilts Circle 1.5nm radius of 511423N 0014615W			FL 150	Salisbury Ops 01980 674710	Activity notified on the day to Salisbury Ops Alternative contact: 128.300 MHz Hours: Normally during daylight hrs
Old Buckenham, Norfolk Circle 1.5nm radius of 522950N 0010303E			FL 150	Marham ATC 01760 338643 London ACC 01489 612420	Activity notified on the day to Marham ATC or London ACC outside hrs of Marham Alternative contact: 124.400 MHz Hours: Normally during daylight hrs

ENR 5.5.3 FREE-FALL DROP ZONES

Designation	Launch Type	Elevation ft amsl	Vertical limits ft agl	Operator Tel No.	Hours
Perranporth, Cornwall			FL 150	07885 628772 or 01872 552266 St Mawgan ATC 01637 857234	Activity notified on the day to St Mawgan ATC 128.725 and Culdrose ATC
Peterborough Sibson, Cambs Circle 1.5nm radius of 523335N 0002346W			FL 150	Cottesmore ATC 01572 812241 Ex 7330 London TCC 01895 426422	Activity notified on the day to Cottesmore ATC or London TCC outside hrs of Cottesmore Alternative contact: 129.900 MHz or 122.300 MHz Hours: Normally during daylight hrs
Peterlee, Co Durham			FL 150	Newcastle ATC 0191 2860966	Activity notified on the day to Newcastle ATC Circle 1.5nm radius of 544606N 0012300W Alternative contact: 129.900 MHz Hours: Normally during daylight hrs
Redlands, Wiltshire Circle 1.5nm radius of 513352N 0014205W			FL 65	Lyneham ATC 01249 890381	Activity must be notified on the day to Lyneham ATC Alternative contact: 129.900 MHz or 129.825 MHz Hours: Normally during daylight hrs
South Cerney, Glos Circle 1.5nm radius of 514114N 0015519W			FL 150	Brize Norton ATC 01993 897878	Activity notified on the day to Brize Norton ATC (All drops subject to Permission from Brize Norton NATC prior to take off) Alternative contact: 129.900 MHz Hours: Normally during daylight hrs
Strathallan, Tayside Circle 1.5nm radius of 561930N 0034455W			ALT 5000 †	Scottish ACC 01292 692763	Activity notified on the day to Scottish ACC † Drops may be made from above ALT 5000ft with Scottish ACC permission Alternative contact: 129.900 MHz Hours: Normally during daylight hrs Sat, Sun & PH
Tilstock, Shropshire Circle 1.5nm radius of 525551N 0023905W			FL 85 †	Shawbury ATC 01939 250351 Ex 7232 London ACC 01489 612420	Activity notified on the day to Shawbury ATC or London ACC outside hrs of Shawbury Alternative contact: 118.100 MHz † Drops may be made from up to FL 150 with Manchester ACC permission Hours: Normally during daylight hrs Fri from 1400 & Sat & PH Winter (Summer 1hr earlier); and other times as notified

ENR 5.5.4 MICROLIGHT SITES

Designation	Launch Type	Elevation ft amsl	Vertical limits ft agl	Operator Tel No.	Hours
Weston-on-the-Green, Oxford Circle 2nm radius of 515246N 0011320W			FL 85 †	Brize Norton ATC 01993 897878	Activity notified on the day to Brize Norton ATC † Drops may be made from up to FL 150 with London TCC permission Alternative contact: 133.650 MHz Hours: Normally during daylight hrs. Night parachuting frequent
MICROLIGHT SITES					
Andreas, Isle of Man 542210N 0042524W		110		Manx Eagle Club 01624 861300	
Arclid, Sandbach 530828N 0021900W		262		Cheshire Microlight Centre 01270 764713	
Auchinleck 552827N 0042212W		380			
Belle Vue 505829N 0040524W		675		01805 623113	
Bracklesham Bay, Chichester 504514N 0004950W		5			
Brooklands Farm 522123N 0001432W		50			
Bucknall 531159N 0001516W		50		01526 388249	
Caltonmoor, Derbys 530201N 0014959W		990			
Catton, Derbys 524340N 0014006W		230			
Caunton 530713N 0005318W		160		01623 883802	
Chatteris, Mount Pleasant 522901N 0000624E		5		Microlight Sport Aviation 01354 742340	
Chiltern Park 513302N 0010605W		180		Thames Valley Microlight Club 01491 872163 01691 772659	
Chirk 525649N 0030243W		448		01691 774137	

ENR 5.5   ENR 5.5.4 MICROLIGHT SITES

Designation	Launch Type	Elevation ft amsl	Vertical limits ft agl	Operator Tel No.	Hours
Clench Common 512314N 0014411W		623		G S Aviation 01672 515535	
Colemore Common, Petersfield, Hants 510338N 0010035W		610			
Cromer, Northrepps 525410N 0011943E		165		01263 513015	
Darley Moor, Derbys 525841N 0014435W		600		Airways Airsports Ltd 01335 344308	
Davidstow Moor 503815N 0043708W		969		Moorland Flying Club 01840 261517	
East Fortune 560003N 0024404W		120	2000	East of Scotland Microlights 01620 880332	HJ (PPR)
Finmere 515907N 0010323W		395		Microflights 01296 712705	
Glassonby (Penrith) 544418N 0023905W		600		01768 898382	
Glidden Farm, Portsmouth, Hants 505603N 0010318W		450			
Graveley 515628N 0001212W		395		01438 317112	
Great Orton, Cumbria 545233N 0030438W		234			
Halwell, South Devon 502155N 0034235W		623		South Hams Flying Club 01803 839046	
Hoghton (Higher Barn Farm) 534425N 0023505W		329			
Hougham (Glebe Farm) 530022N 0004114W		80			
Hewish		20			
512248N 0025208W					
Hougham (Glebe Farm) 530022N 0004114W		120		01400 250293	
Hunsdon 514811N 0000339E		254		Jay Airsports 07956 434958	

ENR 5.5.4 MICROLIGHT SITES

Designation	Launch Type	Elevation ft amsl	Vertical limits ft agl	Operator Tel No.	Hours
Husthwaite, Baxby 540925N 0011354W		132		Baxby Airsports 01347 868443	
Ince 533158N 0030139W		10		West Lancashire Microlight School 0151 9293319	
Jurby, Isle of Man 542114N 0043108W		89			
Kirkbride 545256N 0031220W		38			
Linton 511217N 0003042E		70			
Long Marston 520825N 0014513W		154		Aerolight Flight Training 01789 299229	
Manton 512543N 0014627W		610			
Meikle Endovie, Aberdeen 571349N 0024003W		475			
Middleton Sands 535952N 0025346W		17			
Milson (Cleobury Mortimer) 522141N 0023245W		500		01584 890486	
Misk Hill 530311N 0011455W		581			
Monmouth (Yew Tree Farm) 515401N 0024602W		650			
Montrose 564351N 0022704W		25			
Movenis (McMasters Farm) 545915N 0063853W		180		02829 558609	
Newton Peveril 504738N 0020603W		9		01258 857205	
Oakley 514705N 0010428W		249			
Otherton 524231N 0020541W		340		Staffordshire Microlight Centre 01543 673075	

ENR 5.5.4 MICROLIGHT SITES

Designation	Launch Type	Elevation ft amsl	Vertical limits ft agl	Operator Tel No.	Hours
Oxton 530241N 0010006W		273			
Packington 524258N 0012817W		320			
Pilling Sands 535631N 0025608W		17			
Plaistows 514341N 0002247W		395		Jay Airsports 01727 866066	
Pound Green, Worcs 522414N 0022115W		360		01299 401447	
Redlands, Swindon 513320N 0014125W		320		Redlands Microlight Flying School 01793 791014	
Roddige 524241N 0014435W		171		The Microlight School Lichfield 01283 792193	
Rogart (Rovie Farm) 575926N 0041019W		33			
Rollesby 524134N 0013708E		35			
Rossall Field (Cockerham) 535603N 0025038W		15		Ribble Valley Microlight Club 01282 436280	
Saddington 523039N 0010102W		460			
Sandy, Beds 520744N 0001835W		80		01767 691616	
St Michael's 535106N 0024735W		30		Northern Microlight School 01995 641058	
Stair 552750N 0042557W		380			
Stoke, Isle of Grain, Kent 512702N 0003814E		10			
Sutton Meadows 522306N 0000336E		3		Pegasus Flight Training 01487 842360	
Swinford 522536N 0010937W		492		01455 558791	

Designation	Launch Type	Elevation ft amsl	Vertical limits ft agl	Operator Tel No.	Hours
Tarn Farm, Cockerham 535603N 0025038W		15		Ribble Valley Microlight Club	
Thorness Bay 504415N 0012149W		10			
Thornhill 560856N 0041109W		45			
Water Eaton 513826N 0014705W		270			
Waverton 531004N 0024718W		115			
Weston Zoyland 510618N 0025411W		33		Weston Zoyland Microlight Training School 07970 710453	
Wombleton, Pickering 541401N 0005808W		120		01751 432356	
Wyton 522125N 0000628W		135		01480 52451 Ex 6412	
Yatesbury 512602N 0015405W		525			

ENR 5.6 BIRD MIGRATION AND AREAS WITH SENSITIVE FAUNA

1 Bird Migration

1.1 There are no well defined heavily used bird migratory routes within United Kingdom airspace at any time of the year. Bird migration to and from the United Kingdom occurs largely in the autumn and winter on broad fronts in streams running E-W and N-S. The density of bird movements is greatest in SE England where these two fronts cross. Most migrating birds fly below 5000ft.

2 Areas with Sensitive Fauna

2.1 As elsewhere in the world, offshore islands, headlands, cliffs, inland waters and shallow estuaries attract flocks of birds for breeding, roosting and feeding at various times of the year. Within 20nm or so of such locations concentrations of birds flying mostly below 1500ft may be encountered (see ENR 6.5.6.1 and ENR 6.5.6.2).

2.2 In order to lessen the risk of bird strikes, pilots of low flying aircraft should, whenever possible, avoid flying at less than 1500 ft above surface level over areas where birds are likely to concentrate. Where it is necessary to fly lower than this, pilots should bear in mind that the risk of a bird strike increases with speed (it is a fact that birds rarely hit an object moving slower than 80 knots). Apart from endangering aircraft by flying close to bird colonies, the breeding of the birds may be upset and the practice should be avoided on conservation grounds. It should also be appreciated that, especially in the case of sea bird colonies, concentrations of birds may be soaring on lee waves downwind of the areas where they breed.

3 Bird Sanctuaries

3.1 A Bird Sanctuary is Airspace of defined dimensions within which large colonies of birds are known to breed.

3.2 Pilots are specifically requested to avoid the Bird Sanctuaries listed below, especially during any stated breeding season.

Identification and Name Lateral Limits	Upper Limit (ft) Lower Limit (ft)	Remarks
1	2	3
Isle of May A circle 1nm radius centred on 561109N 0023327W	2000ft ALT SFC	Pilots are requested to avoid the area throughout the year. Peak activity April to September.
Martin Mere A circle 1nm radius centred on 533726N 0025242W	2000ft ALT SFC	Pilots are requested to avoid the area throughout the year. Peak activity September to March
Gibraltar Point A circle 1.5nm radius centred on 530616N 0001949E	2000ft ALT SFC	Pilots are requested to avoid the area throughout the year.
Minsmere A circle 1nm radius centred on 521445N 0013701E	2000ft ALT SFC	Pilots are requested to avoid the area throughout the year
Havergate A circle 1nm radius centred on 520423N 0013113E	2000ft ALT SFC	Pilots are requested to avoid the area throughout the year
Otmoor A circle 1nm radius centred on 514918N 0011038W	2000ft ALT SFC	Pilots are requested to avoid the area throughout the year
Severn A circle 3nm radius centred on 514429N 0022405W	4000ft ALT SFC	Pilots are requested to avoid the area from mid September to early April
Fleet A circle 3nm radius centred on 503914N 0023610W	4000ft ALT SFC	Pilots are requested to avoid the area throughout the year

4 Release of Racing Pigeons

4.1 In agreement with the Royal Pigeon Racing Association concerning the release of large numbers of racing pigeons:

(a) There will be no release of racing pigeons within a radius of 7nm from the following airports:

Aberdeen Dyce	Humberside	Manchester †
Belfast Aldergrove	Inverness	Newcastle
Birmingham	Kirkwall	Norwich
Bournemouth	Leeds Bradford	Nottingham East Midlands
Bristol	Liverpool	Prestwick
Cardiff	London Gatwick	Southampton
Durham Tees Valley	London Heathrow	Southend
Edinburgh	London Luton	Sumburgh
Glasgow	London Stansted	

† See AD Section for further information

(b) For other aerodromes in the United Kingdom all liberations shall be notified to the SATCO in writing fourteen days prior to the date of release and additionally by telephone 30 minutes before release time. On receipt of the 30 minute warning the SATCO or senior controller on duty may delay the liberation by up to 30 minutes for air traffic control purposes. Exceptionally, the liberation time may be delayed for a longer period.

ENR 6.1   ENR 6.1.6.3 Lower Airspace Radar Service

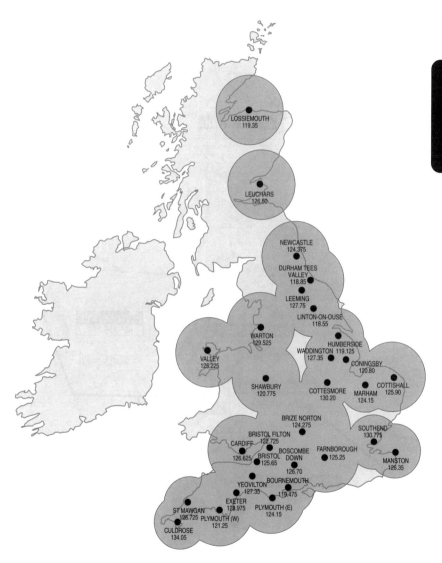

ENR 6.1   ENR 6.1.6.4 Military Middle Airspace Radar Service

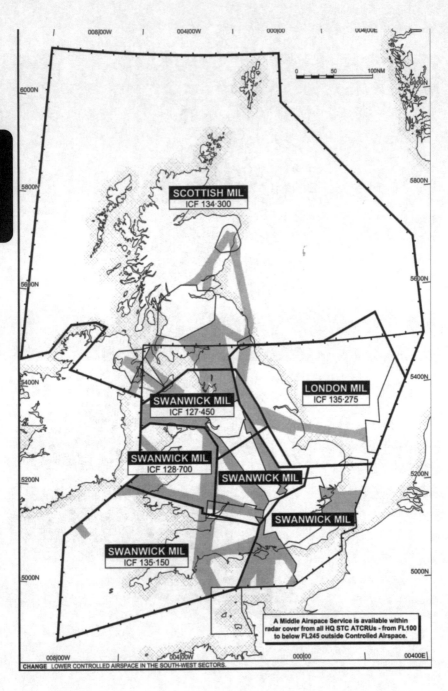

ENR 6.1

ENR 6.1.6.5 Warton Radar Advisory Service Area (RASA)

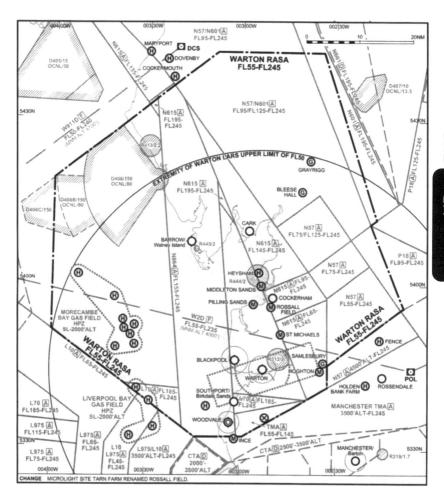

ENR 6.1 ENR 6.1.7.1 A United Kingdom Altimeter Setting and Flight Information Regions

ENR 6.1 ENR 6.1.7.1 B United Kingdom Altimeter Setting and Flight Information Regions

ENR 6.1

ENR 6.2.0.1 London ACC – FIS Sectors

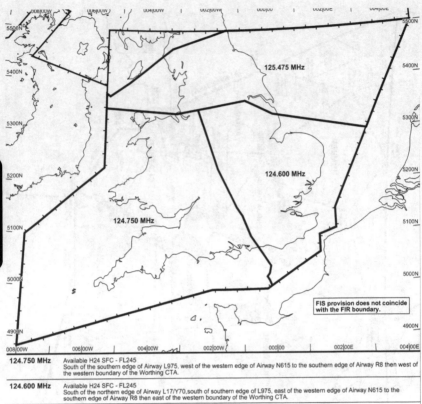

124.750 MHz	Available H24 SFC - FL245 South of the southern edge of Airway L975, west of the western edge of Airway N615 to the southern edge of Airway R8 then west of the western boundary of the Worthing CTA.
124.600 MHz	Available H24 SFC - FL245 South of the northern edge of Airway L17/Y70, south of southern edge of L975, east of the western edge of Airway N615 to the southern edge of Airway R8 then east of the western boundary of the Worthing CTA.
125.475 MHz	Available H24 SFC - FL245 North of the northern edge of L17/Y70, north of southern edge of L975 - 534600N 0053000W - 534358N 0052505W - 540007N 0044030W - 543852N 0031617W - 550000N 0012555W - 550000N 0050000E.
CHANGE	AIRWAY ROUTE DESIGNATOR N615.

ENR 6.1

ENR 6.2.0.2 Scottish ACC – FIS Sectors

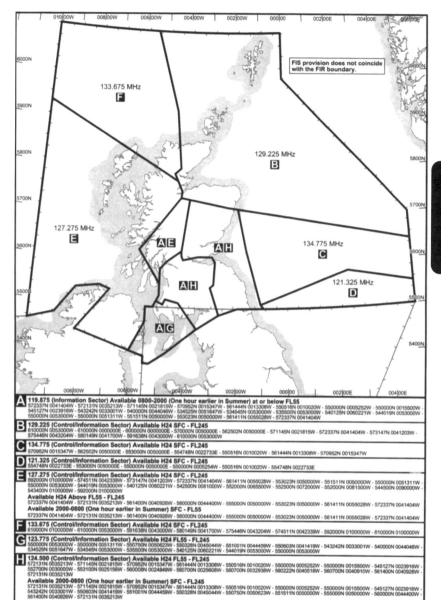

ENR 6.1

ENR 6.2.2.3.1 Military Aerodrome Traffic Zones

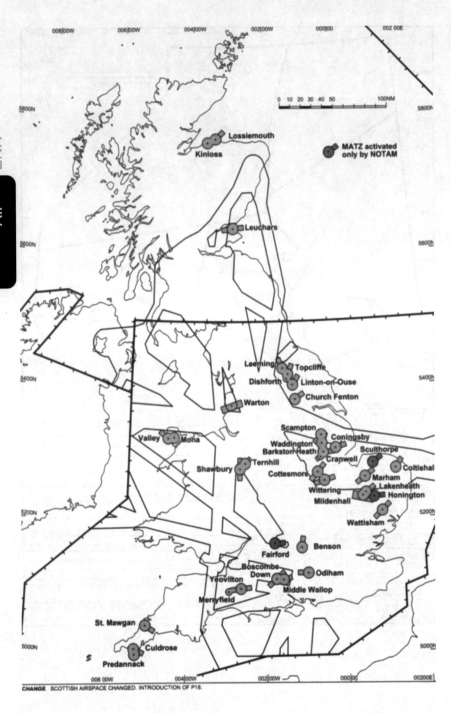

ENR 6.1 ENR 6.3.0.3 Class B Gliding Areas South of 5520N

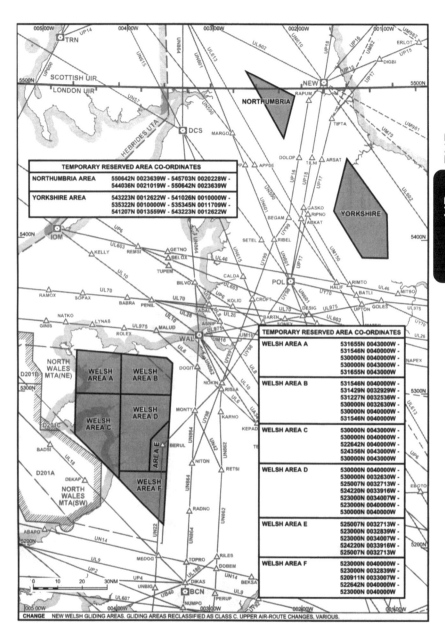

ENR 6.1

ENR 6.3.0.4 Class B Gliding Areas North of 5520 N

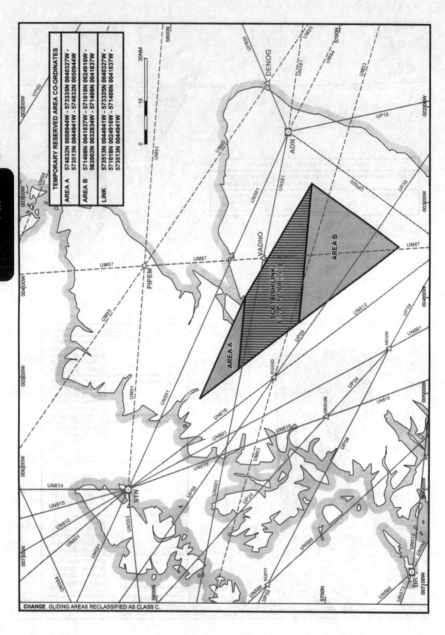

ENR 6.1   ENR 6.5.6.1 Bird Concentration Areas March – July

CHANGE REVISED BIRD CONCENTRATION AREAS.

ENR 6.1 ENR 6.5.6.2 Bird Concentration Areas October – March

CHANGE REVISED BIRD CONCENTRATION AREAS.

AIP
AD section

AD 1 – AERODROME/HELIPORTS INTRODUCTION

AD 1.1 – AERODROME/HELIPORT AVAILABILITY
AD 1.1.1 – General Conditions

1 General Conditions under which Aerodromes/Heliports and Associated Facilities are Available for Use

1.1 Civil aircraft are not permitted to land at aerodromes in the United Kingdom other than those mentioned in this publication, except in cases of genuine emergency in flight or where special permission has been obtained from the aerodrome operator.

2 Applicable ICAO Documents

2.1 The Standard Recommended Practices and Procedures contained in the following ICAO documents are applied with the Differences noted hereunder:

Annex 14 Chapter 3	Aerodromes	
DOC 9137 AN/898	Airport Services Manual	
DOC 7030	Regional Supplementary Procedures	
DOC 7754	ANP – EUMED Region	

3 Prior Permission Requirements

3.1 There are two classes of civil aerodromes at which landings require the prior permission of the owner or the authority concerned. The persons in charge of most of these aerodromes require application for permission to land to be made by telephone prior to take-off; in cases where permission by RTF is acceptable or where application in writing is required this is indicated in AD 2 or AD 3 as appropriate. The two classes of aerodrome are as follows:

(a) Licensed Aerodrome (Ordinary Licence). Prior arrangements with the person in charge are necessary as these aerodromes are licensed only for the use of the licensee or by persons specifically authorised by him.

(b) Unlicensed Aerodromes

(i) Prior arrangements with the owner or person in charge are necessary.

(ii) Unlicensed aerodromes are not inspected by the Civil Aviation Authority.

3.2 Use of Government aerodromes is also subject to prior permission. (See paragraph 12)

4 Availability of Ground Services

4.1 Details of the various ground services available at an aerodrome (ie fuel, hangarage, repair facilities, etc) may be found at AD 2 and AD 3.

4.2 It should be noted, however, that as some of the services are not the direct responsibility of the aerodrome management, the hours during which such services are available do not necessarily coincide with the hours during which the aerodrome is open to visiting aircraft.

4.3 It is the responsibility of the aircraft operator to ensure, before departure for an aerodrome, that such services are available.

5 Noise Abatement Requirements

5.1 Section 78(1) of the Civil Aviation Act 1982 enables the Secretary of State to publish a notice imposing a duty on operators of aircraft at designated aerodromes to comply, after take-off or before landing, with specified requirements for the purpose of limiting or mitigating the effect of noise and vibration connected with the taking-off or landing of aircraft. Notices under Section 78 are set out in AD 2.21 for London Gatwick, London Heathrow and London Stansted.

6 Operational Hours

6.1 The hours indicated against AD 2.3/3.3 item 1 (AD/Heliport) are to be considered as the aerodrome/heliport operating hours or hours of availability unless otherwise stated. Note that revised operating hours may be notified by NOTAM.

6.2 All times are in UTC unless otherwise stated. Where 'SS' or 'SS+' is shown as an alternative closing time, the aerodrome service closes at whichever time is earlier, unless otherwise stated.

6.3 The commencement and cessation of the Summer Period (UTC + 1 hour) and the Winter Period (UTC) is shown at GEN 2.1.

7 Closure of Aerodromes

7.1 Pilots will not be refused permission to land or take-off at public-licensed aerodromes (AD 1.4 refers) solely because of bad weather conditions. Pilots of public and non-public transport aircraft should bear in mind, however, that Articles 38, 39 and 40 of the Air Navigation Order 2000, require that they do not infringe the applicable aerodrome operating minima.

7.2 The only circumstances in which a public-licensed aerodrome will be closed to normal air traffic during its published hours of availability are:

(a) when the surface of the landing area is unfit;

(b) at times and in conditions specified in NOTAM or AIP Supplements;

(c) if essential aerodrome facilities are unserviceable.

7.3 In an emergency, pilots will be allowed to land regardless of the conditions of the aerodrome and aerodrome facilities.

8 Operations Outside Published Operating Hours

8.1 Aerodromes/Heliports may not be used outside the published hours of availability (see AD 2.3/3.3) without the prior permission of the appropriate authority. This applies not only to aerodromes of origin and destination but also to alternates.

8.2 Any application for operations outside published operating hours should be made to the authority controlling the aerodrome/heliport.

9 Declared Distances

The distance shown for Take-off Run Available, Emergency Distance, Take-off Distance Available and Landing Distance Available in Items AD 2.13 and AD 3.13, are notified for the purposes of Regulation 5(4)(c) of the Air Navigation (General) Regulations 1993.

9.1 'Take-off Run Available' (TORA) is defined as the length of runway which is available and suitable for the ground run of an aeroplane taking-off.

9.2 'Accelerate-Stop Distance Available' (ASDA) is defined as the length of the declared take-off run plus the length of stop way available. Known in the United Kingdom as 'Emergency Distance' (ED).

9.3 'Take-off Distance Available' (TODA) is defined as the length of the declared take-off run plus the length of clear way available.

9.4 'Landing Distance Available' (LDA) is defined as the length of runway (or surface, when this is unpaved) which is available and suitable for the ground landing run of the aeroplane commencing at the landing threshold or displaced landing threshold.

10 Obstacles

10.1 For the purposes of ICAO Type C Charts, operators should assume that, in addition to the obstacles listed in the AD Section of the UK AIP and those which are shown on any ICAO Type A Charts, obstacles of up to 300ft above ground level may exist:

(a) outside the coverage of any ICAO Type A Chart; and

(b) beyond 4nm from the Aerodrome Reference Point.

10.2 Information for licensed aerodromes is limited to significant obstacles and the lists are not comprehensive, they do not necessarily include all shadowed obstacles.

10.3 To determine the existence of obstacles 400ft or more above aerodrome level, operators should thus combine the above advice on the possible existence of obstacles with the detailed contour information which is given on the Ordnance Survey 1:50 000 Maps; relating the outcome to the published aerodrome elevation.

11 Visual Ground Aids

11.1 Details of visual ground aids, lighting, paved runways markings and self manoeuvring stand markings are given in Civil Aviation Publication CAP 637.

12 Use of Government Aerodromes

12.1 The term Government Aerodrome is applied to those aerodromes in the United Kingdom which are in the occupation of any government department or visiting force. Therefore all UK military aerodromes are so designated.

12.2 It is the Ministry of Defence policy to encourage the use of active Government aerodromes by United Kingdom civil aircraft on inland flights, provided this is consistent with defence requirements and local interests. Those aerodromes available to civil aircraft are listed in AD 1.3, but permission to use any government aerodrome must be obtained from the operating authority at the aerodrome concerned before take-off (in this connection filing of a flight plan does not constitute prior permission). Unless there are special circumstances, permission will not normally be given for flights outside normal working hours of the aerodrome concerned.

12.3 Where permission is sought to use an inactive Government aerodrome, or for overseas flights, at least three weeks notice should be given, except where the flight is of a humanitarian nature or is for agricultural aviation.

12.4 Foreign aircraft are not normally permitted to use UK Government aerodromes. However, those aerodromes identified as Military Emergency Diversion Aerodromes in GEN 3.6 may be used by foreign aircraft when the destination aerodrome, as indicated in the flight plan, is unusable because of adverse conditions.

12.5 The pilot of a civil aircraft wishing to use a military aerodrome will be required to comply with the following instructions:

(a) Prior to flight departure, to obtain permission to land there from the authority at the aerodrome concerned; in this connection the filing of a flight plan does not constitute obtaining prior permission; also to provide details of the aircraft, crew and passengers. Permission to use an aerodrome may be withheld for operational or administrative reasons; rejection of a request for such reasons must be accepted as final;

(b) all aircraft must comply with the ATC procedures in force at the aerodrome. Aeronautical information which is not published in the UK AIP may be obtained from: No 1 AIDU, RAF Northolt, West End Road, Ruislip, Middlesex, HA4 6NG. Tel: 020 8845 2300 Ex 7209, Fax: 020 8841 1078.

(c) if making an instrument approach, pilots of civil aircraft should calculate their Aerodrome Operating Minima in accordance with AD 1.1.2 paragraph 9.

(d) after landing, to report personally to Air Traffic Control and give details of the aircraft, crew and passengers. Note that control of entry precautions normal to a particular military aerodrome will, in the interests of security, be applied to all persons arriving or departing by civil aircraft.

(e) before taking-off, to report personally to Air Traffic Control and give particulars of flight crew and passengers, and obtain taxiing instructions. The pilot is responsible for deciding whether or not to proceed after receipt of clearance to taxi and take-off.

12.6 Hangarage for civil aircraft may be provided subject to agreement and, if provided, will be entirely at the owners risk.

12.7 Fuel, oil, or similar products, may be provided at the Commanding Officer's discretion but servicing or loading of civil aircraft cannot be undertaken except in cases of distress or exceptional circumstances. Pilots may make their own arrangements with civil petrol agents to refuel their civil aircraft on the aerodrome, provided that they furnish adequate cover against damage or loss arising from the presence of the agent's equipment and that prior permission for such arrangements is obtained from the Commanding Officer of the station. No permanent positioning of civil refuelling equipment or parties will be permitted, unless the approval of the Ministry of Defence has been obtained.

12.8 Liability will not be accepted by the controlling Department, its servants or agents, or by any servant or agent of the Crown for the loss or damage, by accident, fire, flood, tempest, explosion or any other cause to aircraft, or for loss or damage, from whatever cause, arising to goods, mail or other articles belonging to any person, even if such loss or damage is caused by or arises from negligence on the part of the Department's servants or agents or any servant or agent of the Crown. The use of military aerodromes will be permitted only upon the understanding that the controlling Department and the Crown will be held indemnified against all claims whatsoever made by third parties in respect of personal injuries (whether fatal or otherwise), damage to or loss of property howsoever caused, which may arise as a result of the facilities granted.

12.9 The use of any apparatus such as tractors, cranes, chocks, starter trolleys, etc, belonging to or under the charge of the controlling Department, by the personnel of aircraft or other persons making use of the aerodrome, will be entirely at the risk of the person using such apparatus, and no liability will be accepted for any loss, damage or injury caused by or arising from the use of any such apparatus (whether under the control or management of any servant or agent of the controlling Department or of the Crown or otherwise) other than liability for personal injuries or death caused by or arising from negligence on the part of any servant or agent of the controlling Department or the Crown. The use of such apparatus will be permitted only upon the understanding that the controlling Department and the Crown will be held indemnified against all claims whatsoever which may result from such use. It must, further, be clearly understood that the controlling Department does not in any way guarantee the safety or fitness of any such apparatus or of any equipment, petrol or oil, or similar products, supplied.

12.10 The civil use of Government aerodromes is, in addition to the limitations referred to above, subject to the appropriate charges being paid at the time. Use is also subject to the availability of appropriate Air Traffic Control and crash/rescue services as laid down by Ministry of Defence regulations.

12.11 The Government aerodromes listed at AD 1.3 are hereby notified for the purposes of Articles 101 (1)(b) and 102 of the Air Navigation Order 2000, subject to the conditions so specified as available for the take-off and landing of aircraft engaged on flights for the purpose of the public transport of passengers or for the purpose of instruction in flying.

13 CAT II/III Operations at Aerodromes

13.1 Promulgation of an aerodrome/runway as available for Category II or Category III operations means that it is suitably equipped and that procedures appropriate to such operations have been determined and are applied when relevant.

13.2 Promulgation implies that at least the following facilities are available:

ILS – certificated to relevant performance category.

Lighting – suitable for Category promulgated.

RVR System – may be automatic or manned system for Category II; will be automatic system for Category III.

13.3 Special procedures and safeguards will be applied during Category II and III operations. In general, these are intended to provide protection for aircraft operating in low visibilities and to avoid disturbance to the ILS signals. The details of any special taxi routes, runway turn-off points and runway holding points are shown in AD 2.

13.4 Protection of ILS signals during Category II or III operations may dictate that pre-take-off holding positions are more distant from the runway than the holding positions used in good weather. Taxiways lying within the ILS Sensitive Area are marked by colour coded taxiway centre-line (alternate yellow/green lights). Pilots should avoid stopping their aircraft within the ILS Sensitive Area and should make their 'Runway Vacated' call only after the aircraft is clear of the Sensitive Area.

13.5 In actual Category II or III weather conditions pilots will be informed by ATC of any unserviceabilities in the promulgated facilities so that they can amend their minima, if necessary, according to their operations manual. Pilots who wish to carry out a practice Category II or Category III approach are to request Practice Category II (or Category III) Approach on initial contact with Approach Control. For practice approaches there is no guarantee that the full safeguarding procedures will be applied and pilots should anticipate the possibility of resultant ILS signal disturbance.

13.6 Details of aircraft operator requirements for low minima operations, Category II and III, are contained in document JAR-OPS 1, sub-part E, which is obtainable on pre-payment from: Documedia Solutions Ltd, 37 Windsor Street, Cheltenham, Glos, GL52 2DG.

14 Runway utilisation procedures are detailed at GEN 3.3.

15 Runway Surface Condition Reporting

15.1 The following paragraphs describe the method by which the presence, or otherwise, of water and other contaminants on a runway is reported at UK aerodromes. Additional information relating to runways affected by slush, snow and ice can be found at AD 1.2.2 Snow Plan and guidance on the risks and factors associated with aircraft operations on runways contaminated with snow, slush and water is published in an Aeronautical Information Circular.

15.2 The prescence or otherwise of surface water on a runway will be reported on RTF using the following descriptions:

Reporting Term – Surface Conditions

DRY The surface is not affected by water, slush, snow or ice.

Note: Reports that the runway is dry are not normally passed to pilots. If no runway surface report is passed, pilots will assume the surface to be dry.

DAMP The surface shows a change of colour due to moisture.
Note: If there is sufficient moisture to produce a surface film or the surface appears reflective, the runway will be reported as WET.
WET The surface is soaked but no significant patches of standing water are visible.
Note: Standing water is considered to exist when water on the runway surface is deeper than 3mm. Patches of standing water covering more than 25% of the assessed area will be reported as WATER PATCHES.
For JAR-OPS performance purposes, runways reported as DRY, DAMP or WET should be considered as NOT CONTAMINATED.
WATER PATCHES Significant patches of standing water are visible.
Note: Water patches will be reported when more than 25% of the assessed area is covered by water more than 3mm deep.
FLOODED Extensive patches of standing water are visible.
Note: Flooded will be reported when more than 50% of the assessed area is covered by water more than 3mm deep.
For JAR-OPS performance purposes, runways reported as WATER PATCHES or FLOODED should be considered as CONTAMINATED.

15.3 When reported, the presence or otherwise of surface water on a runway will be assessed over the most significant portion of the runway. Details of the assessed area should be available from the aerodrome authority.

15.4 Runway surface condition reports will be given sequentially for each third of the runway to be used, for example, 'Runway surface is WET, WATER PATCHES, WET' or 'Runway surface is WET. WET, WET'.

15.5 A brief description of any water patches greater than 9mm in depth, which may affect engine performance, will be appended to a runway surface condition report. In such conditions, further information on the location, extent and depth of the water patches should be available from the aerodrome authority.

15.6 A brief description of any notable quantity of water outside the assessed area (eg: water collected on the runway edge) will be appended to a runway surface condition report.

15.7 When the runway surface is affected by dry or compacted snow or ice a braking action report (see AD 1.2.2 SNOW PLAN paragraph 5.4.1) will normally be available.

15.8 When a runway is contaminated by water (ie more than 3mm), wet snow or slush, a braking action report will not be available due to limitations of existing friction measuring equipment (see AS 1.2.2 SNOWPLAN, paragraph 5.4.2), however, a runway surface condition report will normally be available (see paragraph 15.2 for runways contaminated with water and AIP AD 1.2.2 paragraph 5 for other contaminants) stating the type of contaminant and its respective depth (see AD 1.2.2 SNOWPLAN, paragraph 5.2 and 6).

15.9 At Government aerodromes, runway surface conditions will be described in plain language, and where a braking action measuring device has been used, braking action will be described as GOOD, MEDIUM or POOR. The report may also include the type of measuring devise used and the Mu value.

16 Runway Friction Assessment

16.1 Full details of this procedure can be found in CAP 683. The assessment of runway friction.

16.2 Aerodrome authorities are required to conduct periodic surveys of the friction characteristics of their runway surfaces. The purpose of these surveys is to predict the need for maintenance of the runway surface to prevent an unacceptable deterioration of friction as detailed in Table 1. The recognised Continuous Friction Measuring Equipment (CFME) devices in the UK are the Mu Meter and Grip tester.

Mu-Meter and Grip Tester Friction Levels

Continuous Friction Measuring Equipment (CFME)	Design Objective Level (DOL)	Maintenance Planning Level (MPL)	Minimum Friction Level (MFL)
Mu-Meter	0.72 or above	0.57	0.50
Grip Tester	0.80 or above	0.63	0.55

Note: Friction Levels correspond to the CAP 683 procedure.

16.3 If a survey indicates that the runway surface friction characteristics have deteriorated below the specified Minimum Friction Level (MFL), then that runway will be notified by NOTAM as a runway that 'may be slippery when wet'.

16.4 When a runway is notified as 'may be slippery when wet', aircraft operators may request additional information relating to that notification from the aerodrome operator. However, any performance calculations or adjustment made as a result of this information is the responsibility of the aircraft operator.

AD 1.1.2 – AERODROME OPERATING MINIMA

1 Introduction

1.1 This section of AIP specifies the notified method of calculating Aerodrome Operating Minima (AOM) for the purpose of Article 39 (Public Transport Aircraft not registered in the United Kingdom) and Article 40 (non-public Transport Aircraft). The method of calculating minima is compliant with JAR-OPS 1 Commercial Air Transportation-Aeroplanes or JAR-OPS 3 Commercial Air Transportation-Helicopters and is therefore the same method used to calculate minima published on commercially available flight guides. For public transport aircraft registered in the UK the declared minima in relation to Article 38(3) shall be those in JAR-OPS Subpart E unless more restrictive minima are notified in respect of a particular aerodrome. For Category 2 and 3, operations advice should be obtained from the Civil Aviation Authority, Flight Operations Department, Gatwick.

1.2 It should be noted that, the privileges of pilot licences, Rules of the Air and limitations in the aircraft Flight Manual can be more restrictive than the aerodrome operating minima contained in this section. In establishing the aerodrome operating minima that will apply, full account must be taken of:

(a) the type and handling characteristics of the aircraft.
(b) the composition of the flight crew and their competence and experience.
(c) the dimensions and characteristics of the runway which may be selected for use.
(d) the adequacy and performance of the available visual and non-visual ground aids.
(e) the equipment available on the aircraft for the purpose of navigation and/or control of the flight path, as appropriate during the take-off, the approach, the flare, the landing, roll-out and missed approach.
(f) the obstacles in the approach, missed approach and climb out areas, required for the execution of contingency procedures and necessary clearance.
(g) the obstacle clearance altitude/height for the instrument approach procedures; and
(h) the means to determine and report meteorological conditions.

1.3 Under the provisions of the Air Navigation Order 2000,public transport aircraft shall observe AOM, non-public transport aircraft shall observe AOM when conducting an approach to a runway with a notified Instrument Approach Procedure (IAP).

1.4 For take-off, the minima published in this section are mandatory for public transport flights. Operators and commanders of non-public transport flights should ensure that appropriate take-off minima are applied, and it is strongly recommended that the take-off minima used should be no lower than the minima specified for public transport flights.

1.5 In this section paragraphs 2,3,6,7,8,9,10,11 apply to all aircraft, paragraph 4 applies only to aeroplanes and paragraph 5 only to helicopters.

2 Glossary of Terms

2.1 These explanations of terms are offered as a practical guide. For formal definitions refer to Article 129 in the Air Navigation Order 2000 (ANO).

2.2 Aerodrome Operating Minima (AOM). The minimum conditions of cloud ceiling and Runway Visual Range for take-off, and of Decision Height/Minimum Descent Height, Runway Visual Range and visual reference for landing.

2.3 Aeroplane Category. The criterion for the classification of aeroplanes by categories is the indicated airspeed at threshold (VAT) which is equal to the stalling speed (VSO) multiplied by 1.3 or VS1g multiplied by 1.23 in the landing configuration at the maximum certificated landing mass. If both VSO and VS1g are available, the higher resulting VAT shall be used. The aeroplane categories corresponding to VAT values are in the Table below:

Aeroplane Category	VAT
A	Less than 91kt
B	From 91 to 120kt
C	From 121 to 140kt
D	From 141 to 165kt
E	From 166 to 210kt

2.3.1 The landing configuration which is to be taken into consideration shall be defined by the operator or by the aeroplane manufacturer.

Note: Helicopters use aeroplane category A procedures.

2.4 Cloud Ceiling in relation to an aerodrome means the vertical distance from the elevation of the aerodrome to the lowest part of any cloud visible from the aerodrome that is sufficient to obscure more than one-half of the sky so visible.

2.5 Commercial Air Transportation. The transportation by air of passengers, cargo or mail for remuneration or hire (Commercial Air Transportation is not intended to cover Aerial Work or Corporate Aviation). In the UK the term Public Transport is used, as defined in Article 130 of the Air Navigation Order.

2.6 Decision Height (DH). The height in a precision approach at which a missed approach must be initiated if the required visual reference to continue the approach has not been established.

2.7 Final Approach and Take-off area (FATO). A defined area over which the final phase of the approach manoeuvre to hover or land by a helicopter is completed and from which the take-off manoeuvre is commenced and, where the FATO is to be used by helicopters operating to Performance Class 1, includes the rejected take-off area available.

2.8 Helicopter Performance. Helicopter operations can be within 3 performance classes:

(a) Performance Class 1 Operations
Helicopters with performance such that, in case of critical power-unit failure, it is able to land on the rejected take-off area or safely continue the flight to an appropriate landing area, depending on when the failure occurs.

(b) Performance Class 2 Operations
Helicopters with performance such that, in case of critical power-unit failure, it is able to safely continue the flight, except when the failure occurs prior to a defined point after take-off or after a defined point before landing, in which cases a forced landing may be required.

(c) Performance Class 3 Operations
Helicopters with performance such that, in case of power-unit failure at any point in the flight profile, a forced landing must be performed.

2.9 Heliport. An aerodrome or a defined area of land, water or a structure used or intended to be used wholly or in part for the arrival, departure and surface movement of helicopters.

2.10 Instrument Approaches. Instrument approaches are divided into non-precision approaches and precision approaches:

(a) Non-precision approach
An instrument approach using non-visual aids for guidance in azimuth or elevation but which is not a precision approach.

(b) Precision approach
An instrument approach to landing using ILS, MLS or PAR for guidance in both azimuth and elevation.
Categories of precision operation
Category 1 operation
A precision instrument approach and landing with a decision height not lower than 200ft (60m) and with a runway visual range not less than 550m (500m for helicopters).
Category 2 operation
A precision instrument approach and landing with a decision height below 200ft (60m) but not lower than 100ft (30m) and a runway visual range of not less than 300m.
Category 3 operations are sub-divided as follows:
(i) **Category 3A** A precision instrument approach and landing with:
(1) a decision height lower than 100ft (30m); and
(2) a runway visual range not less than 200m.
(ii) **Category 3B** A precision instrument approach and landing with:
(1) a decision height lower than 50ft (15m), or no decision height; and
(2) a runway visual range lower than 200m but not less than 75m.
2.11 Instrument Approach Procedure (IAP). A series of pre-determined manoeuvres by reference to flight instruments with specified protection from obstacles from the Initial Approach Fix or, where applicable, from the beginning of a defined arrival route, to a point from which a landing can be completed and thereafter if a landing is not completed, to a position at which holding or en-route obstacle clearance criteria apply.
2.12 Low Visibility Procedure (LVP). Procedures applied at an aerodrome for the purpose of ensuring safe operation during Category 2 and 3 approaches and Low Visibility Take-off. (LVPs are initiated when the cloud base lowers to 300ft and is expected to lower further or the RVR falls to 1200m and is expected to deteriorate further. They should be fully in place by the time the cloud base reaches 200ft or the RVR falls to 600m.)
2.13 Low Visibility Take-off. A take-off where the runway visual range is less than 400m.
2.14 Minimum Descent Height (MDH). The height in a non-precision approach below which descent may not be made without the required visual reference.
2.15 Missed Approach Point (MAPt). That point in a non-precision approach procedure at or before which the prescribed missed approach procedure must be initiated in order to ensure that the minimum obstacle clearance is not infringed.
2.16 Notified. Information set forth in a document published by or with the authority of the CAA and entitled Supplement (NOTAM) or AIP and for the time being in force.
2.17 Runway Visual Range (RVR). The distance in the direction of take-off or landing over which the runway lights or surface markings can be seen, calculated either by human observation or instruments.
2.18 Visual Approach. An approach when either part or all of an instrument approach procedure is not completed and the approach is executed with visual reference to the runway.
2.19 Visual Reference. A view of the section of the runway and/or the approach area and/or their visual aids, which the pilot must see in sufficient time to assess whether or not a safe landing can be made from the type of approach being conducted.

3 General
3.1 Altimeter Error
3.1.1 When calculating DH, account must be taken of the errors of indicated height which occur when the aircraft is in the approach configuration. Details of the Pressure Error Correction (PEC) should be available from the aircraft flight manual or handbook. In the absence of this information a PEC of +50ft has been found to be suitable for a wide range of light aircraft and should be used. This addition of 50ft need only be applied to DH. The required RVR should be calculated prior to applying the PEC.
3.1.2 The use of a radio altimeter is only applicable to approved Category 2 and Category 3 operations. For an aircraft flying a Category 1 or non-precision IAP, DH/MDH is indicated on the pressure altimeter. At DH/MDH any readings from a radio altimeter may be unreliable because of the large area of terrain providing return signals to the instrument.
3.1.3 Temperature Error. Pressure altimeters are calibrated to indicate true altitude under International Standard Atmosphere (ISA) conditions. Any deviation from ISA will therefore result in an erroneous reading on the altimeter. The altimeter will over read for temperatures below ISA and the following table details the values to correct this error.

Temperature Error

Aerodrome	Altitude Above Altimeter Source Elevation (ft) (normally destination elevation)												
Temp °C	200	300	400	500	600	700	800	900	1000	1500	2000	3000	4000	5000
0	20	20	30	30	40	40	50	50	60	90	120	170	230	280
-10	20	30	40	50	60	70	80	90	100	150	200	290	390	490
-20	30	50	60	70	90	100	120	130	140	210	280	420	570	710
-30	40	60	80	100	120	140	150	170	190	280	380	570	760	950
-40	50	80	100	120	150	170	190	220	240	360	480	720	970	1210
-50	60	90	120	150	180	210	240	270	300	450	590	890	1190	1500

Values to be added to published altitudes (ft)

3.2 Obstacle Clearance Height (OCH)

3.2.1 The DH/MDH for an approach is determined in part by considering obstacles that could affect the approach of a particular aircraft. In broad terms, the more accurate the aid supporting the approach to be flown and the slower the approach speed of the aircraft the smaller the area in which obstacles need to be considered. The height which is calculated to clear all obstacles by a defined margin within a particular area is called Obstacle Clearance Height and is the lowest height above the elevation of the relevant runway threshold or above the aerodrome elevation used in establishing compliance with the appropriate obstacle clearance criteria.

3.2.2 The OCH is listed for individual aerodrome approaches on the relevant instrument Approach Chart in section AD 2. For the purposes of calculating OCH helicopters are categorised as Aeroplane Category A.

3.3 Determination of DH/MDH

3.3.1 Instrument Rating Holders in Current Practice

3.3.1.1 A pilot with a valid instrument rating and who is in current practice may use the minima calculated in accordance with this section.

3.3.2 IMC Rating Holder in Current Practice

3.3.2.1 Pilots with a valid Instrument Meteorological Conditions (IMC) Rating are recommended to add 200ft to the minimum applicable DH/MDH, but with absolute minima of 500ft for a precision approach and 600ft for a non-precision approach. The UK IMC Rating may not be valid outside UK territorial airspace, therefore IMC Rated pilots should check the validity of their rating for the State in which they intend to fly. If the rating is not valid pilots must comply with the basic licence privileges, subject to the regulations of that State.

3.3.3 Pilots not in Current Practice

3.3.3.1 If a pilot is not in current practice he should try to avoid having to make an instrument approach in bad weather. If he has to make such an approach, even if he is fully confident of his abilities, he is advised to add 100ft to his calculated DH/MDH. Further increments should be added depending on when the pilot was last in full practice, and his or her familiarity with the aircraft, the procedure and the aerodrome environment.

3.4 Determination of Take-off Minima

3.4.1 The take-off minima must be selected to ensure sufficient guidance to control the aircraft in the event of both a discontinued take-off in adverse circumstances and a continued take-off after failure of the critical power unit.

3.4.2 For an aircraft flown for the purpose of Public Transport, it must be ensured that take-off minima calculations are:
(a) In accordance with the procedures and requirements of the State of Registry of the aircraft; and
(b) not less than those required in the following paragraphs.

3.4.3 For an aircraft being flown for a purpose other than public transport, it must be ensured that, the take-off minima selected have been calculated in accordance with the procedures and requirements of the State of Registry of the aircraft. In addition, it is strongly recommended that the take-off minima selected should not be less than those given in paragraphs 4.1 or 5.1. It should be noted that the weather minima limitations in pilot licence privileges may override some of the calculated minima.

3.4.4 For night operations by helicopters conducting public transport, some ground lighting must be available to illuminate the Final Approach and Take-off area (FATO) and any obstacles unless otherwise agreed by the CAA.

3.5 Non-Precision approach

3.5.1 System minima

System Minima for Non-Precision Approach Aids

System Minima	
Facility	Lowest MDH (ft)
ILS/MLS (no glidepath – LLZ)	250
SRA (terminating at 0.5nm)	250
SRA (terminating at 1nm)	300
SRA (Terminating at 2nm)	350
VOR	300
VOR/DME	250
NDB	300
VDF (QDM and QGH)	300

3.5.2 Minimum Descent Height
3.5.2.1 The minimum descent height for a non-precision approach must not be lower than:
(a) The OCH for the category of aircraft; and
(b) the system minimum.

3.5.3 Visual reference
3.5.3.1 The approach must not be continued below MDA/MDH unless at least one of the following visual references for the intended runway is distinctly visible and identifiable to the pilot:
(a) elements of the approach light system
(b) the threshold
(c) the threshold markings
(d) the threshold lights
(e) the threshold identification lights
(f) the visual glide slope indicator
(g) the touchdown zone or touchdown zone markings
(h) the touchdown zone lights
(i) FATO/runway edge lights; or
(j) other visual references accepted by the CAA.

3.6 Precision approach – Category I Operations
3.6.1 System Minima
3.6.1.1 The system minima for the Category I precision approach aids ILS, MLS and PAR is 200ft.

3.6.2 Decision Height
3.6.2.1 The decision height to be used for a Category I precision approach must not be lower than:
(a) The minimum decision height specified in the Aeroplane Flight Manual (AFM) if stated;
(b) the minimum height to which the precision approach aid can be used without the required visual reference;
(c) the OCH for the category of aeroplane; or
(d) 200ft.

3.6.3 Visual Reference
3.6.3.1 A pilot may not continue an approach below the Category I decision height, determined in accordance with sub-paragraph 3.6.2, unless at least one of the following visual references for the intended runway is distinctly visible and identifiable to the pilot:
(a) elements of the approach light system
(b) the threshold
(c) the threshold markings
(d) the threshold lights
(e) the threshold identification lights
(f) the visual glide slope indicator
(g) the touchdown zone or touchdown zone markings
(h) the touchdown zone lights; or
(i) FATO/runway edge lights

4 Aerodrome Operating Minima – Aeroplanes
4.1 Take-off
4.1.1 Single engine aeroplanes
4.1.1.1 The minima for take-off by single-engine aeroplanes when flying for public transport are 1000ft cloud ceiling and 1800m RVR.
4.1.1.2 The minima selected for all flights by single engine aeroplanes should be adequate to ensure a high probability of a successful forced landing being made should a failure of the engine occur after take-off.

4.1.2 Multi-engine aeroplanes operating in accordance with Performance Class A
4.1.2.1 For multi-engined aeroplanes, whose performance is such that, in the event of a critical power unit failure at any point during take-off, the aeroplane can either stop or continue the take-off to a height of 1500ft above the aerodrome while clearing obstacles by the required margins, the take-off minima must not be less than the RVR/Visibility values given in Table 2:

Table 2 – RVR/Visibility for Take-Off

Facilities	RVR/Visibility (Note 3)
Nil (Day only)	500m
Runway edge lighting and/or centre line marking	250/300m (Notes 1 & 2)
Runway edge and centre line lighting	200/250m (Note 1)
Runway edge and centre line lighting and multiple RVR information	150/200m (Notes 1 & 4)

AD 1.1.2.4 Aerodrome Operating Minima – Aeroplanes

Note 1: The higher values apply to Category D aeroplanes.
Note 2: For night operations at least runway edge and runway end lights are required.
Note 3: The reported RVR/Visibility value representative of the initial part of the take-off run can be replaced by pilot assessment.
Note 4: The required RVR value must be achieved for all of the relevant RVR reporting points with the exception given in Note 3.

4.1.3 Multi-engine aeroplanes operating in accordance with Performance Class B or C

4.1.3.1 For multi-engined aeroplanes with performance such that they cannot comply with the performance conditions in sub-paragraph 4.1.2 in the event of a critical power unit failure, there may be a need to re-land immediately and to see and avoid obstacles in the take-off area. Such aeroplanes may be operated to the following take-off minima provided they are able to comply with the applicable obstacle clearance criteria, assuming engine failure at the height specified. The take-off minima must be established, based upon the height from which the one engine inoperative net take-off flight path can be constructed. The RVR minima used may not be lower than either of the values given in Table 2 above or Table 3.

Table 3 – Assumed Engine Failure Height Above the Runway Versus RVR/Visibility

Assumed engine failure height above the take-off runway	RVR/Visibility (Note 2)
50ft or less	200m
51-100ft	300m
101-150ft	400m
151-200ft	500m
201-300ft	1000m
>300ft	1500m (Note 1)

Note 1: 1500m is also applicable if no positive take-off flight path can be constructed.
Note 2: The reported RVR/Visibility value representative of the initial part of the take-off run can be replaced by pilot assessment.

4.1.4 Multi-engine aeroplanes which do not meet the applicable one engine inoperative climb requirements must be treated as single engine aeroplanes.

4.1.5 When reported RVR is not available, the commander shall not commence take-off unless he can determine that the actual conditions satisfy the applicable take-off minima.

4.2 Non-precision approach

4.2.1 The lowest minima for conducting non-precision approaches are:

Table 4a – RVR for Non-Precision Approach – Full Facilities

MDH (ft)	Non-precision Approach Minima – Full Facilities (Notes 1, 5,6 & 7)			
	RVR/Aeroplane Category (m)			
	A	B	C	D
250-299	800	800	800	1200
300-449	900	1000	1000	1400
450-649	1000	1200	1200	1600
650 and above	1200	1400	1400	1800

Table 4b – RVR for Non-Precision Approach – Intermediate Facilities

MDH (ft)	Non-precision Approach Minima – Intermediate Facilities (Notes 2, 5, 6 & 7)			
	RVR/Aeroplane Category (m)			
	A	B	C	D
250-299	1000	1100	1200	1400
300-449	1200	1300	1400	1600
450-649	1400	1500	1600	1800
650 and above	1500	1500	1800	2000

AD 1.1.2.4 Aerodrome Operating Minima – Aeroplanes

Table 4c – RVR for Non-Precision Approach – Basic Facilities

MDH (ft)	Non-precision Approach Minima – Basic Facilities (Notes 3, 5 ,6 & 7)			
	RVR/Aeroplane Category (m)			
	A	B	C	D
250-299	1200	1300	1400	1600
300-449	1300	1400	1600	1800
450-649	1500	1500	1800	2000
650 and above	1500	1500	2000	2000

Table 4d – RVR for Non-Precision Approach – Nil Approach Light Facilities

MDH (ft)	Non-precision Approach Minima – Nil Approach light facilities (Notes 4, 5, 6 & 7)			
	RVR/Aeroplane Category (m)			
	A	B	C	D
250-299	1500	1500	1600	1800
300-449	1500	1500	1800	2000
450-649	1500	1500	2000	2000
650 and above	1500	1500	2000	2000

Note 1: Full facilities comprise runway markings, 720m or more of HI/MI approach lights, runway edge lights, threshold lights and runway end lights. Lights must be on.
Note 2: Intermediate facilities comprise runway markings, 420-719m of HI/MI approach lights, runway edge lights, threshold lights and runway end lights. Lights must be on.
Note 3: Basic facilities comprise runway markings, <420m of HI/MI approach lights, any length of LI approach lights, runway edge lights, threshold lights and runway end lights. Lights must be on.
Note 4: Nil approach light facilities comprise runway markings, runway edge lights, threshold lights, runway end lights or no lights at all.
Note 5: The tables are only applicable to conventional approaches with a nominal descent slope of not greater than 4°. Greater descent slopes will usually require that visual glide slope guidance (eg PAPI) is also visible at the Minimum Descent Height.
Note 6: The above figures are either reported RVR or meteorological visibility converted to RVR as in Table 10 at paragraph 7.
Note 7: The MDH mentioned in Table 4a, 4b, 4c and 4d refers to the initial calculation of MDH. When selecting the associated RVR, there is no need to take account of a rounding up to the nearest 10ft, which may be done for operational purposes, eg conversion to MDA.

4.2.2 Night operations. For night operations by aeroplanes at least runway edge, threshold and runway end lights must be on.

4.3 Precision approach

4.3.1 The lowest minima to be used by an operator or pilot for Category I operations are:

Table 5 RVR for Cat I Approach

Decision Height (ft)	Category I Minima			
	Facilities/RVR (m) (Note 5)			
	Full (Notes 1 & 6)	Intermediate (Notes 2 & 6)	Basic (Notes 3 & 6)	Nil (Notes 4 & 6)
200	550	700	800	1000
201-250	600	700	800	1000
251-300	650	800	900	1200
301 and above	800	900	1000	1200

Note 1: Full facilities comprise runway markings, 720m or more of HI/MI approach lights, runway edge lights, threshold lights and runway end lights. Lights must be on.
Note 2: Intermediate facilities comprise runway markings, 420-719m of HI/MI approach lights, runway edge lights, threshold lights and runway end lights. Lights must be on.
Note 3: Basic facilities comprise runway markings, <420m of HI/MI approach lights, any length of LI approach lights, runway edge lights, threshold lights and runway end lights. Lights must be on.
Note 4: Nil approach light facilities comprise runway markings, runway edge lights, threshold lights, runway end lights or no lights at all.

Note 5: The above figures are either the reported RVR or meteorological visibility converted to RVR in accordance with Table 10 below.

Note 6: The Table is applicable to conventional approaches with a glide slope angle up to and including 4°

Note 7: The DH mentioned in Table 5 refers to the initial calculation of DH. When selecting the associated RVR, there is no need to take account of a rounding up to the nearest ten feet, which may be done for operational purposes, (eg conversion to DA).

4.3.2 Night operations. For night operations by aeroplanes at least runway edge, threshold and runway end lights must be on.

4.4 Single pilot operation

4.4.1 For single pilot operations in an aeroplane, the minimum RVR for all approaches in accordance with the above, except that an RVR of less than 800m is not permitted unless using a suitable autopilot coupled to an ILS or MLS, in which case normal minima apply. The Decision Height applied must not be less than 1.25 x the minimum use height for the autopilot.

4.5 Circling

4.5.1 Circling. This is the visual phase of an instrument approach to bring an aircraft into position for landing on a runway which is not suitably located for a straight-in approach.

4.5.2 The minima to be used for aeroplanes circling (including prescribed tracks) will never be less than the values tabulated below.

Table 6 – Visibility and MDH for Circling vs Aeroplane Category

	Aeroplane Category			
	A	B	C	D
MDH not less than	400	500	600	700
Minimum meteorological visibility	1500	1600	2400	3600

5 Aerodrome Operating Minima – Helicopters

5.1 Take-off

5.1.1 Performance Class 1 operations

For Performance Class 1 operations, the minimum RVR/Visibility (RVR/VIS) in the following table, must be observed:

Table 7 – Performance Class 1 Operations

Onshore Heliports with	RVR/Visibility IFR Departure Procedures
No lighting and no markings (Day)	250m or the rejected take-off distance whichever is greater
No markings (Night)	800m
Runway edge/FATO lighting and centre-line marking	200m
Runway edge/FATO lighting and centre-line marking and RVR information	150m
Runway edge/FATO lighting and centre-line lighting and RVR information	150m
Offshore Helideck	
Two pilot operations	250m (See Note 1)
Single pilot operations	250m (See Note 1)

Note 1: The commander must establish that the take-off flight path is free of obstacles.

5.1.2 Performance Class 2 operations

5.1.2.1 For Performance Class 2 operations, the minimum take-off minima of 800m RVR/VIS must be applied and the helicopter must remain clear of cloud during the take-off manoeuvre, or until reaching Performance Class 1 capabilities.

5.1.3 Performance Class 3 operations

5.1.3.1 For Performance Class 3 operations the minimum take-off minima of 600ft cloud ceiling and 800m RVR/VIS must be applied.

5.2 Onshore non-precision approaches

5.2.1 For onshore non-precision approaches by performance Class 1 and 2 helicopters, the minima given in the following Table shall apply:

Table 8 – Onshore Non-Precision Approach Minima

MDH (ft)	Onshore Non-Precision Approach Minima (m) (Notes 5, 6, & 7)			
	Facilities			
	Full (Note 1)	Intermediate (Note 2)	Basic (Note 3)	Nil (Note 4)
250-299	600	800	1000	1000
300-449	800	1000	1000	1000
450 and above	1000	1000	1000	1000

Note 1: Full facilities comprise FATO/runway markings, 720m or more of HI/MI approach lights, FATO/runway edge lights, threshold lights and runway end lights. Lights must be on.
Note 2: Intermediate facilities comprise FATO/runway markings, 420-719m of HI/MI approach lights, FATO/runway edge lights, threshold lights and runway end lights. Lights must be on.
Note 3: Basic facilities comprise FATO/runway markings, <420m of HI/MI approach lights, FATO/runway edge lights, threshold lights and runway end lights. Lights must be on.
Note 4: Nil approach light facilities comprise FATO/runway markings, FATO/runway edge lights, threshold lights, FATO/runway end lights or no lights at all.
Note 5: The table is only applicable to conventional approaches with a nominal descent slope of not greater than 4°. Greater descent slopes will usually require that visual glide slope guidance (eg PAPI) is also visible at the Minimum Descent Height.
Note 6: The above figures are either reported RVR or meteorological visibility converted to RVR as in Table 10 at paragraph 7.
Note 7: The MDH mentioned in Tables 4a, 4b, 4c and 4d refers to the initial calculation of MDH. When selecting the associated RVR, there is no need to take account of a rounding up to the nearest ten feet, which may be done for operational purposes, eg, conversion to MDA.

5.2.2 For night public transport operations, some ground lighting must be available to illuminate the FATO and any obstacles unless otherwise agreed by the CAA, however, the notified landing minima are mandatory to all aircraft where there is a published IAP.

5.3 Airborne Radar Approach (ARA) for Over Water Operations

5.3.1 General

(a) A helicopter shall not conduct ARAs unless authorised by the CAA.
(b) ARAs are not permitted for single pilot operations to rigs or vessels under way.
(c) A helicopter shall not undertake an ARA unless the radar can provide course guidance to ensure obstacle clearance.
(d) Before commencing the final approach the commander shall ensure that a clear path exists on the radar screen for the final and missed approach segments. If lateral clearance from any obstacle will be less than 1.0nm, the commander shall:
(i) Approach to a nearby target structure and thereafter proceed visually to the destination structure; or
(ii) Make the approach from another direction leading to a circling manoeuvre.
(e) The Commander shall ensure that the cloud ceiling is sufficiently clear above the helideck to permit a safe landing.

5.3.2 Minimum Descent Height

Not withstanding the minima at sub-paragraphs (a) and (b) below, the MDH shall not be less than 50ft above the elevation of the heli deck.
(a) The MDH is determined from a radio altimeter. The MDH for an ARA shall not be lower than:
(i) 200ft by day
(ii) 300ft by night
(b) The MDH for an approach leading to a circling manoeuvre shall not be lower than:
(i) 300ft by day
(ii) 500ft by night

5.3.3 Minimum Descent Altitude (MDA). An MDA may only be used if the radio altimeter is unserviceable. The MDA shall be a minimum of MDH + 200ft and shall be based on a calibrated barometer at the destination or on the lowest forecast QNH for the region.

5.3.4 Decision range. The Decision Range shall not be less than 0.75nm unless an operator has demonstrated to the Authority that a lesser Decision Range can be used at an acceptable level of safety.

5.3.5 Visual reference. A helicopter shall not continue an approach beyond Decision Range or below MDH/MDA unless he has visual contact with the destination.

5.3.6 Single pilot operations. The MDH/MDA for a single pilot ARA shall be 100ft higher than that calculated using sub-paragraphs 5.3.2 and 5.3.3 above. The Decision Range shall not be less than 1.0nm.

5.4 Precision approach

5.4.1 The lowest minima to be used for Category I operations by helicopters operating to Performance Class 1 and 2 are:

Table 9 – Category 1 Onshore Approach Minima

Category 1 Onshore Approach Minima (m) (Notes 5, 6 & 7)				
Facilities				
DH (ft)	Full	Intermediate	Basic	Nil
	(Note 1)	(Note 2)	(Note 3)	(Note 4)
200	500	600	700	1000
201-250	550	650	750	1000
251-300	600	700	800	1000
301 and above	750	800	900	1000

Note 1: Full facilities comprise FATO/runway markings, 720m or more of HI/MI approach lights, FATO/runway edge lights, threshold lights and FATO/runway end lights. Lights must be on.
Note 2: Intermediate facilities comprise FATO/runway markings, 420-719m of HI/MI approach lights, FATO/runway edge lights, threshold lights and FATO/runway end lights. Lights must be on.
Note 3: basic facilities comprise FATO/runway markings, <420m of HI/MI approach lights, any length of LI approach lights, FATO/runway edge lights, threshold lights and FATO/runway end lights. Lights must be on.
Note 4: Nil approach light facilities comprise FATO/runway markings, FATO/runway edge lights, threshold lights, FATO/runway end lights or no lights at all.
Note 5: The above figures are either the reported RVR or meteorological visibility converted to RVR in accordance with Note 10.
Note 6: The table is applicable to conventional approaches with a glide slope angle up to and including 4°.
Note 7: The DH mentioned in Table 5 refers to the initial calculation of DH. When selecting the associated RVR, there is no need to take account of a rounding up to the nearest 10ft, which may be done for operational purposes, (eg conversion to DA).

5.4.2 For night public transport operations some ground lighting must be available to illuminate the FATO and any obstacles unless otherwise agreed by the CAA, however, the notified landing minima are mandatory to all aircraft where there is a published IAP.

5.5 Single pilot operation

5.5.1 For single pilot operations in a helicopter, the minimum RVR for all approaches must be in accordance with the above. An RVR of less than 800m is not permitted except when using a suitable autopilot coupled to an ILS or MLS, in which case normal minima apply. The Decision Height applied must not be less than 1.25 x the minimum use height for the autopilot.

5.6 Circling

5.6.1 Circling. The visual phase of an instrument approach, to bring a helicopter into position for landing on a heliport/runway that is not suitably located for a straight in approach.

5.6.2 For circling the specified MDH shall not be less than 250ft, and the meteorological visibility shall not be less than 800m.

6 Visual approach – All aircraft (see definition in paragraph 2)

6.1 An aircraft shall not conduct a visual approach with an RVR of less than 800m.
Note: VFR flights must be conducted in accordance with the Visual Flight Rules applicable to the airspace, and taking into account the limitations of the pilot's licence privileges.

7 Conversion of Reported Meteorological Visibility to RVR – All aircraft

7.1 Meteorological visibility to RVR conversion must not be used for calculating take-off minima, Category II or III minima or when a reported RVR is available.
7.2 When converting meteorological visibility to RVR in all circumstances other than those in sub-paragraph 7.1 above, the following Table is used:

Table 10 – Conversion of Visibility to RVR

Lighting Elements in Operation	RVR = Reported Met visibility multiplied by	
	Day	Night
Hi approach and runway lighting	1.5	2.0
Any type of lighting installation other than above	1.0	1.5
No lighting	1.0	Not applicable

8 Approach Ban – All aircraft

8.1 The requirements for the commencement and continuation of an approach are defined in Articles 38, 39 and 40 of the Air Navigation Order 2000.
8.2 An aircraft may commence an instrument approach regardless of the reported RVR/Visibility but the approach shall not be continued below 1000ft above the aerodrome if the relevant RVR/Visibility for that runway is at the time less than the specified minimum for landing.

8.3 Where RVR is not available, RVR values may be derived by converting the reported visibility in accordance with paragraph 7.

8.4 If, after passing 1000ft in accordance with paragraph 8.2, the reported RVR/Visibility falls below the applicable minimum, the approach may be continued to DA/H or MDA/H.

8.5 The approach may be continued below DA/H or MDA/H and the landing may be completed provided that the required visual reference is established at the DA/H or MDA/H and is maintained.

9 Determination of minima – Additional Cases

9.1 Military aerodromes in UK Territorial Airspace

9.1.1 Military aerodromes do not use PANS-OPS design criteria but do publish a Procedure Minimum for each IAP, shown on the Royal Air Force Approach Chart in a Table of Aircraft Categories (CAT); the words 'Procedure Minimum' are not shown. The Procedure Minimum shown in bold print is a minimum height (minimum with QFE set on the altimeter) with the minimum altitude shown in light print beside to the left. The Procedure Minimum (minimum height) will also be passed by ATC who will request the pilots DH/MDH and intentions. The Procedure Minimum can be converted to an equivalent to OCH by following the procedures in the following paragraphs. The equivalent OCH can then be used to calculate the MDH and RVR in the normal way in accordance with the procedures above.

9.1.2 Precision Approaches, ILS and PAR, for which the absolute minimum is 200ft above touchdown elevation, are normally based on a 3° glide path. The glide path angle, also shown on the chart, may be as low as 2.5°. The following increments should be made to the given Procedure Minimum to obtain the equivalent of OCH. There is no provision for the use of radio altimeters.

Nominal Glidepath Angle °	Aeroplane Categories (ft)			
	A	B	C	D
2.5	Nil	10	20	30
2.6	10	20	30	40
2.7	10	20	30	40
2.8	20	30	40	50
2.9	20	30	40	50
3.0	30	40	50	60

9.1.3 For non-precision approaches the Procedure Minimum may be taken to be the OCH.

9.1.4 DH/MDH should then be determined as previously described. The related RVR continues to be obtained from the Tables in paragraphs 4 and 5.

9.1.5 For Circling, the OCH should be determined by adjusting the published Royal Air Force values, shown on the Approach Charts, as follows:

Aeroplane Categories	Increment (ft)
A and B	Zero
C and D	+ 100ft

The minimum visibility should be determined as described in paragraph 4.5.

9.1.6 OCH information will not be used at British Military Aerodromes.

9.1.7 Instrument Approaches at British Military Aerodromes (except Royal Navy Aerodromes) are flown on aerodrome QNH until approaching the Final Approach Fix or Point, when QFE is set for the remainder of the approach. All flying in the visual circuit is on QFE. At Royal Navy Aerodromes instrument and visual approaches are flown on QFE. Aircraft which are unable to comply with the above procedure should inform ATC and will be accommodated wherever possible.

10 Overseas Aerodromes

10.1 AOM at overseas aerodromes should not be lower than those published as State minima or permitted by the privileges of the pilot's licence by the UK or overseas State.

11 Aerodromes without published Instrument Approach Procedures

For an aircraft landing at an aerodrome without an instrument approach procedure either:

(a) A descent should be made in VMC until in visual contact with the ground, then fly to the destination.

(b) An IAP at a nearby aerodrome should be flown and proceed as in (a); or

(c) if neither (a) nor (b) is possible, first obtain an accurate fix and then descend not lower than 1000ft above the highest obstacle within 5nm (8km) of the aircraft. If visual contact (as at (a) above) has not been established at this height, the aircraft should divert to a suitable alternate with a published IAP.

AD 1.2 – RESCUE AND FIRE FIGHTING SERVICES AND SNOW PLAN
AD 1.2.1 – RESCUE AND FIRE FIGHTING SERVICES

1 The categories of fire and rescue equipment given at AD 2.6, their relevance to individual aircraft, and the minimum scales of equipment required to meet the respective categories are contained in Civil Aviation Publication – 'Licensing of Aerodromes' – CAP 168. They approximate closely to those contained in the relevant ICAO publications.

2 For the convenience of aircraft operators the relationship of the fire and rescue equipment categories to individual aircraft is summarised as follows:

Aerodrome Fire and Rescue Category	Aircraft Overall Length (m)	Maximum Fuselage Width (m)
‡Special	0 up to but not including 9	2
1	0 up to but not including 9	2
2	9 up to but not including 12	2
3	12 up to but not including 18	3
4	18 up to but not including 24	4
5	24 up to but not including 28	4
6	28 up to but not including 39	5
7	39 up to but not including 49	5
8	49 up to but not including 61	7
9	61 up to but not including 76	7

‡ Aerodromes which are generally licensed solely in order that flying instruction may take place.

3 Category of land based helicopter landing sites for Rescue and Fire Fighting purposes is as follows:

Site Category	†Aircraft Overall Length (m)
‡Special	0 up to but not including 15
H1	0 up to but not including 15
H2	15 up to but not including 24
H3	24 up to but not including 35

† Overall length includes rotors and tail boom
‡ Aerodromes which are licensed solely in order that flying instruction may take place.

4 It should be noted that aircraft requiring ONE Category of fire and rescue equipment higher than that given at AD 2.6 are not automatically precluded from using the aerodrome. Operators of such aircraft should first consult with the aerodrome management.

5 Two tables have been produced to assist with determination of adequacy when comparing military and civil RFF categories. Each table uses different criteria in forming a comparison and commanders should only use the table appropriate to their flight details

6 The following table compares ICAO minimum standards with those likely to be available at Government aerodromes. It is to be used by civil pilots wishing to use Government facilities.

CAA Category	RAF Equivalent
1	2
2	4
3	4
4	4
5	6
6	7
7	No equivalent
8	No equivalent
9	No equivalent

6.1 The above table compares minimum standards recommended by ICAO with those valid at Government aerodromes. Whilst compliance with these minimum standards is desirable, the Authority accepts that the slight shortfall in the facilities provided at RAF Category Two and Three aerodromes has been compensated for by significant over provision of facilities in some other areas. Therefore, operators may undertake, as a minimum, operation of Civil Category Two and Three aircraft from RAF Category Two and Three aerodromes respectively.

7 The following table compares crash/fire requirements for Government aircraft with those facilities likely to be available at Civil aerodromes. It is to be used by pilots of Government aircraft wishing to use civil facilities.

AD 1   AD 1.2.1 – RESCUE AND FIRE FIGHTING SERVICES

RAF Category	Civil/ICAO Equivalent
Nil	1
1	2
2	3
3	5
4	5
5	6
6	7
7	8
–	8
9	

7.1 Although no direct comparison can be made between RAF and the other fire categories, the above table is an approximation of the relationship between the categories.

AD 1.2.2 – SNOW PLAN

1 Introduction

1.1 Agreement has been reached through the International Civil Aviation Organisation to apply standard procedures at international aerodromes within the European/Mediterranean Region, for the clearance of winter contaminants (snow, ice, slush and associated water) from aerodrome surfaces and for the measuring and reporting of aerodrome surface conditions during the winter period.

1.2 The following paragraphs describe the procedures, equipment and techniques that are employed at specified aerodromes in the United Kingdom for the clearance of winter contaminants from pavements. The methods used for the measurement and reporting of aerodrome surface conditions in accordance with internationally agreed procedures are included. Other United Kingdom aerodromes complying with these procedures, in whole or in part, are also included in the Snow Plan.

2 Responsibility for Planning and Implementation

2.1 The clearance of winter contaminants and the measurement and reporting of surface conditions is the responsibility of the aerodrome authority, assisted as necessary by other agencies.

2.2 Prior to the onset of winter conditions, aerodrome authorities prepare a plan to effect efficient clearance and measurement procedures intended to ensure maximum availability of the aerodrome. The plan is formulated in co-operation with ATS and the aerodrome users. Arrangements are made to ensure that the plan can be implemented as soon as meteorological forecasts indicate the likelihood of surface contamination. The first priority is to clear operational runways and other essential parts of the movement area. Provision for measurement and reporting procedures are made. Subsequently, the surfaces cleared are maintained free of contaminant as far as is reasonably practicable.

3 Clearance Techniques

3.1 Whenever possible, the full length and width of runways is cleared completely. Various methods are employed and brief details of those available at individual aerodromes are given in the aerodrome entry of the AIP at AD 2.7.

3.2 Mechanical snow clearing equipment; blowers, sweepers, ploughs and rotary brushes form the main part of the contaminant clearance equipment used at most large aerodromes. As far as practicable, clearance techniques employed prevent the build-up of snow banks. Where this is unavoidable, every effort is made to restrict snow banks to such a height and distance apart as to ensure safe manoeuvring of the most critical aircraft, in this context, normally using the particular aerodrome.

3.3 Slush and associated standing water is cleared whilst it is forming. Clearance may have to be repeated at intervals and some interruption of operations may be inevitable.

3.4 Salt is only used if it is found to be essential and is restricted to areas around edge drains to prevent slush build-up and to ensure continuous drainage. Liquid or other chemicals used for clearing ice are non-toxic and should have no detrimental effects on aircraft, aerodrome surfaces or the friction value of aerodrome pavements.

4 Operational Priorities for the Clearance of Movement Areas

4.1 The order in which the various parts of the movement area of an aerodrome are cleared will depend on many factors and is the subject of local consultation between the aerodrome authority and users. However, as a general policy, clearance is carried out in accordance with the standard order of priority given below:
(a) One main runway, including rapid exits, appropriate to the weather conditions prevailing at the time
(b) appropriate run-up areas where needed
(c) aprons
(d) essential associated taxiways, priority being given to those presenting gradient difficulties
(e) airport roads.

4.2 Other runways and taxiways are cleared as conditions allow.

5 Assessment and Notification of Runway Surface and Allied Conditions

5.1 General

5.1.1 At participating United Kingdom aerodromes, ATS or the aerodrome authority according to local organisation, assess and report runway surface conditions. Information on runway conditions will be notified by SNOWTAM, OPMET RUNWAY STATE MESSAGE (where applicable) or by RTF on request.

5.1.2 Until a satisfactory method has been found to determine accurately and quickly the density of a contaminant on a runway, the nature of the surface covering is described using the following categories based on subjective assessment by the personnel making the inspection:

Ice – water in its solid state, it takes many forms including sheet ice, hoar frost and rime (assumed specific gravity 0.92)

Dry snow – a condition where snow can be blown loose, or if compacted by hand, will fall apart again upon release (assumed specific gravity less than 0.35).

Compacted snow – snow which has been compressed into a solid mass that resists further compression and will hold together or break up into chunks if picked up. (assumed specific gravity 0.35 to 0.50).

Wet snow – a composition which, if compacted by hand, will stick together and tend to, or does form a snowball (assumed specific gravity greater than 0.5).

Slush – a water saturated snow which, with a heel and toe slap down action with the foot against the ground, will be displaced with a splatter (assumed specific gravity 0.50 to 0.80).

Associated standing water – standing water produces as a result of melting contaminant in which there are no visible traces of slush or ice crystal (assumed specific gravity 1.0).

5.2 Depth of Snow or Slush

5.2.1 A standard depth gauge is used to measure the depth of snow or slush on runways. Readings are taken at approximately 300 metre intervals between 5 and 10 metres on each side of the centre-line and clear of the effects of rutting. By international agreement depth information is given in millimetres representing the mean of readings obtained for each third of the total runway length.

5.3 Snow Banks

5.3.1 The height and distance apart of snow banks is reported as soon as these are likely to affect safe manoeuvring by the most critical aircraft, in this context, normally using the aerodrome.

5.4 Runways Affected by Snow and Ice

5.4.1 On runways affected by compacted snow or ice the braking action assessment is made by use of either of the following methods:

(a) Continuous Recording Friction Measuring Trailer

This method employs a trailer towed by a vehicle at 40mph. The equipment provides a continuous register of the mean of friction values either on paper trace or by means of a digital read-out that is used in conjunction with a hand computer. The principle employed is one of the following methods:

(i) The measurement of the side-force generated between the surface and a pair of pneumatic tyres set at a fixed toe-out angle.

(ii) The measurement of the load and drag on a single wheel chain driven from the axle of a double wheel and made to slip at approximately 14.5% of the forward speed.

(b) Brake Testing Decelerometer

An assessment is made of the coefficient of friction using a brake testing decelerometer fitted in a car, van or light truck, the brakes being applied at 25-30mph ensuring that the vehicle wheels are momentarily locked. A standard procedure is followed to ensure uniformity in technique. The principle employed is the assessment of the friction between skidding pneumatic tyres and selected points on the surface being tested.

5.4.2 The methods described in paragraph 5.4.1 are limited to use on ice (gritted or un-gritted) and dry or compacted snow. They are likely to produce misleading high readings in slush or un-compacted, wet snow, or water and it will not detect, for example, that the possibility of 'slush planning' exists.

5.4.3 Braking action tests, where appropriate, are made over the usable length of the runway at approximately 3 metres each side of the centre-line and in such a manner as to produce mean values for each third of the length available. Assessment of Stopway braking action where applicable should also be available on request.

5.4.4 The results of braking action testing on compacted snow or ice are interpreted by reference to the Snow and Ice Table below.

Friction Number	Estimated Braking Action	OPMET Snowtam Code
.40 and above	Good	95
.39 – .36	Medium/Good	94
.35 – .30	Medium	93
.29 – .26	Medium/Poor	92
.25 and below	Poor	91
If for any reason, the reading Is considered unreliable		99

AD 1   AD 1.2.1.5 Assessment and Notification of Runway Surface and Allied Conditions

5.4.5 It is important to remember that the braking action assessment obtained form the Snow and Ice Table is only a rough indication of the relative slipperiness of a contaminated runway. The description 'Good' is used in comparative sense – good for an icy surface –and is intended to indicate that aircraft generally, but nor specifically, should not be subject to undue directional control or braking difficulties, but clearly a surface affected by ice and/or snow is not as good as a clean dry or even a wet runway. The description 'Good' should not be used for braking action on untreated ice but may be used, where appropriate, when ice has been gritted. 'Poor' will almost invariably mean that conditions are extremely slippery, and probably acceptable only, if at all, to aircraft needing little or no braking or steering. Where 'Poor' braking assessment exists, landings should only be attempted if the Landing Distance Available exceeds the Landing Distance Required on a 'very slippery' or icy runway as given in the aircraft Flight Manual. The intermediate values of 'Medium/Good', and 'Medium/Poor' have been included only to amplify the description when conditions are found to be Medium. The procedure is insufficiently refined to be able to discriminate accurately in the narrow numerical bands as set out in the table.

5.4.6 In exceptional circumstances, grit may have to be used to increase the friction value of manoeuvring areas affected by ice or snow but it will be left on the surface only for so long as the ice or snow persists. The specification of the grit used has been agreed internationally and is selected as providing the best compromise between improving the coefficient of friction and presenting the least hazard to aircraft. However, the risk of ingestion into jet engines or of damage to the control surfaces of propeller driven aircraft, particularly where reverse thrust is used, cannot be entirely discounted. Caution in using reverse thrust is therefore advised, particularly when a sudden thaw has resulted in the grit lying on an otherwise bare surface.

5.5 Runways Affected by Slush

5.5.1 Aircraft operations on runways affected by slush can be particularly hazardous and every effort is made to clear the surface, as far as is reasonably practicable, of all slush contaminant prior to aircraft movement. However, the practical difficulties of ensuring that a runway is totally slush free are significant and success depends heavily on the prevailing meteorological conditions, the resources and time available. In such conditions, up to date runway condition reports are provided. However, because of the effects of drag, friction measuring machines can produce misleading readings when operated in slush. In addition, because of the infinitely variable characteristics of the contaminant, no satisfactory method of assessing braking action in slush exists. For these reasons reports containing estimates of braking action derived from readings in these conditions do not include plain language and pilots will be informed on the RTF of the extent and depth of the contamination only:

5.5.2. Aqua planing conditions should be assumed to exist whenever depths of water or slush exceeding approximately 3mm effect a significant portion of the available runway.

6 Availability of Information

6.1 Information on the current state of progress of snow clearance and on the conditions of the movement areas is available from a designated authority at the aerodrome concerned. Information on pavement conditions is also available by RTF from the aerodrome concerned.

6.2 Information on current surface conditions at United Kingdom and other European aerodromes generally is also be available from the following sources:

(a) Flight Briefing Units at aerodromes
(b) SNOWTAM
(c) Locations served by the OPMET system. (see GEN 3.5.9)

6.3 The SNOWTAM provides a standard report which includes an assessment of each third of the runway. Whilst appropriate conditions prevail, participating authorities issue SNOWTAMs as follows:

(a) A new (not revised) SNOWTAM whenever there is a significant change in conditions
(b) The maximum validity of a SNOWTAM will in no case exceed 24 hours.

6.4 Risks and factors associated with operations on runways affected by snow, slush or water are detailed within an Aeronautical Information Circular.

6.5 Runway surface conditions are reported every half hour for as long as conditions warrant by those aerodromes contributing to the OPMET Broadcast system of disseminating information in the following format:

runway designator; type, extent and depth of deposit; and, where appropriate, braking action.

The report is included as an eight digit code at the end of routine aerodrome meteorological reports (METARs). See GEN 3.5.

6.6 RTF reports to pilots provide an assessment in plain language of the available runway length, including a description of the prevailing conditions ie ice, snow or slush, and where appropriate braking action, together with the time of the measurement.

6.7 Aerodromes currently participating in the Snowplan are listed below – details of clearance methods employed, priorities and braking action assessment are given in the individual aerodrome entry at paragraph 2.7.

Aberdeen Dyce	Exeter	London Luton
Belfast Aldergrove	Glasgow	London Stansted
Belfast City	Guernsey	Lydd
Biggin Hill	Hawarden	Manchester
Birmingham	Humberside	Newcastle
Blackpool	Inverness	Norwich
Bournemouth	Isle of Man	Nottingham East Midlands
Bristol	Jersey	Prestwick
Cambridge	Kirkwall	Southampton
Cardiff	Leeds Bradford	Southend
Carlisle	Liverpool	Stornoway
Coventry	London City	Sumburgh
Durham Tees Valley	London Gatwick	Wick
Edinburgh	London Heathrow	

6.8 Information on the distribution of SNOWTAM and the division of the aerodromes into distribution lists is published annually as an Aeronautical Information Circular before the onset of winter.

AD 1.4 GROUPING OF AERODROMES/HELIPORTS

1 Aerodrome Categorisation

1.1 Responsible Authority

1.1.1 The Civil Aviation Authority, through its Directorate of Aerodrome Standards, is responsible for the general surveillance of all licensed civil aerodromes in the United Kingdom. Enquiries relating to aerodromes and licensing should be made to Aerodrome Standards Department. Tel: 01293 573283, Fax: 01293 573971. See GEN 1.1 for address.

1.2 Licence Types

1.2.1 In the United Kingdom there are two types of civil aerodrome licence, namely Public Use Licence and Ordinary Licence.

1.2.2 Aerodromes or Heliports operated in accordance with a Public Use Licence must have their hours of availability notified in the UK AIP and the aerodrome/heliport must be available to all operators on certain equal terms and conditions. However, this does not necessarily mean that the aerodrome is available to all flights without limitation. Aircraft operators must check and comply with the requirements and conditions of use indicated at AD 2 or AD 3.

1.2.3 Aerodromes or Heliports operated in accordance with an Ordinary Licence may accept flights operated by the holder of the licence or by those specifically authorised by that licence holder. This normally means that prior permission is required for most flights but it does not exclude the possibility of scheduled or non-scheduled public transport flights being arranged after the formal agreement of the licence holder

1.2.4 The annotations within column 4 of the table at AD 1.3 reflect the aerodrome/heliport licence type and must be interpreted in conjunction with the above descriptions and with the conditions of use indicated at AD 2 and AD 3.

1.3 Government Aerodromes

1.3.1 Detailed descriptions of Government aerodromes are not listed in the UK AIP. The one exception to this exclusion is Northolt which is operated by the Royal Air Force and regularly used by civilian business aviation.

1.3.2 Several Government aerodromes accept civil traffic subject to prior approval. These aerodromes are listed at AD 1.3 and detailed information about them is published in the UK Military documentation which is available to civil pilots on application to the RAF AIDU at Northolt (GEN 3.2.3 for address and contact details).

1.3.3 Refer to AD 1.1.1 paragraph 12 for general conditions of use.

1.4 Unlicensed Aerodromes

1.4.1 Unlicensed aerodromes are not detailed in the UK AIP.

The Air Navigation Order

UK AIM
The UK Aeronautical Information Manual

AIR NAVIGATION ORDER

ARTICLES TO THE AIR NAVIGATION ORDER
CITATION, COMMENCEMENT AND REVOCATION

Article 1 Citation and commencement
Article 2 Revocation

PART 1 REGISTRATION AND MARKING OF AIRCRAFT
Article 3 Aircraft to be registered
Article 4 Registration of aircraft in the UK
Article 5 Nationality and registration marks

PART 2 AIR OPERATORS'CERTIFICATES
Article 6 Grant of air operators' certificates
Article 7 Grant of police air operators' certificates

PART 3 AIRWORTHINESS AND EQUIPMENT OF AIRCRAFT
Article 8 Certificate of airworthiness to be in force
Article 9 Issue, renewal, etc., of certificates of airworthiness
Article 10 Validity of certificate of airworthiness
Article 11 Issue, renewal etc; of permits to fly
Article 12 Issue of EASA permits fly
Article 13 Issue etc; certificates of validation of permits to fly or equivalent document
Article 14 Certificate of maintenance review
Article 15 Technical log
Article 16 Requirement for a certificate of release to service
Article 17 Requirement for a certificate of release to service under Part 145
Article 18 Licensing of maintenance engineers
Article 19 Equipment of aircraft
Article 20 Radio equipment of aircraft
Article 21 Minimum equipment requirements
Article 22 Aircraft, engine and propeller log books
Article 23 Aircraft weight schedule
Article 24 Access and inspection for airworthiness purposes

PART 4 AIRCRAFT CREW AND LICENSING
Article 25 Composition of crew of aircraft
Article 26 Members of flight crew – requirement for licence
Article 27 Grant, renewal and effect of flight crew licences
Article 28 Maintenance of privileges of aircraft ratings in United Kingdom licences
Article 29 Maintenance of privileges of aircraft ratings in JAR-FCL licences, United Kingdom licences for which there are JAR-FCL equivalents, United Kingdom Basic Commercial Pilot's licences and United Kingdom Flight Engineer's Licences
Article 30 Maintenance of privileges of aircraft ratings in National Private Pilot's Licences
Article 31 Maintenance of privileges of other ratings
Article 32 Medical requirements
Article 33 Miscellaneous licensing provisions
Article 34 Validation of licences
Article 35 Personal flying logbook
Article 36 Instruction in flying
Article 37 Glider pilot – minimum age

PART 5 OPERATION OF AIRCRAFT
Article 38 Operations Manual
Article 39 Police operations manual
Article 40 Training Manual
Article 41 Flight data monitoring, accident prevention and flight safety programme
Article 42 Public transport – operator's responsibilities
Article 43 Loading – public transport aircraft and suspended loads
Article 44 Public transport – aeroplanes -operating conditions and performance requirements
Article 45 Public transport – helicopters - operating conditions and performance requirements
Article 46 Public transport operations at night or in instrument meteorological conditions aeroplanes with one power unit which are registered elsewhere than in the United Kingdom
Article 47 Public transport aircraft registered in the United Kingdom – aerodrome operating minima
Article 48 Public transport aircraft registered elsewhere than in the United Kingdom – aerodrome operating minima
Article 49 Non-public transport aircraft – aerodrome operating minima

Article 50 Pilots to remain at controls
Article 51 Wearing of survival suits by crew
Article 52 Pre-flight action by commander of aircraft
Article 53 Passenger briefing by commander
Article 54 Public transport of passengers – additional duties of commander
Article 55 Operation of radio in aircraft
Article 56 Minimum navigation performance
Article 57 Height keeping performance – aircraft registered in the United Kingdom
Article 58 Height keeping performance – aircraft registered elsewhere than in the United Kingdom
Article 59 Area navigation and required navigation performance capabilities – aircraft registered in the United Kingdom
Article 60 Area navigation and required navigation performance capabilities – aircraft registered elsewhere than in the United Kingdom
Article 61 Use of airborne collision avoidance system
Article 62 Use of flight recording systems and preservation of records
Article 63 Towing of gliders
Article 64 Operation of self-sustaining gliders
Article 65 Towing, picking up and raising of persons and articles
Article 66 Dropping of articles and animals
Article 67 Dropping of persons and grant of parachuting permissions
Article 68 Grant of aerial application certificates
Article 69 Carriage of weapons and of munitions of war
Article 70 Carriage of dangerous goods
Article 71 Method of carriage of persons
Article 72 Exits and break-in markings
Article 73 Endangering safety of an aircraft
Article 74 Endangering safety of any person or property
Article 75 Drunkenness in aircraft
Article 76 Smoking in aircraft
Article 77 Authority of commander of an aircraft
Article 78 Acting in a disruptive manner
Article 79 Stowaways
Article 80 Flying displays

PART 6 FATIGUE OF CREW AND PROTECTION OF CREW FROM COSMIC RADIATION 353

Article 81 Application and interpretation of Part 6
Article 82 Fatigue of crew – operator's responsibilities
Article 83 Fatigue of crew – responsibilities of crew
Article 84 Flight times - responsibilities of flight crew
Article 85 Protection of aircrew from cosmic radiation

PART 7 DOCUMENTS AND RECORDS

Article 86 Documents to be carried
Article 87 Keeping and production of records of exposure to cosmic radiation
Article 88 Production of documents and records
Article 89 Production of air traffic service equipment documents and records
Article 90 Power to inspect and copy documents and records
Article 91 Preservation of documents, etc.
Article 92 Revocation, suspension and variation of certificates, licences and other documents
Article 93 Revocation, suspension and variation of permissions, etc. granted under article 138 or article 140
Article 94 Offences in relation to documents and records

PART 8 MOVEMENT OF AIRCRAFT

Article 95 Rules of the air
Article 96 Power to prohibit or restrict flying
Article 97 Balloons, kites, airships, gliders and parascending parachutes
Article 98 Regulation of small aircraft
Article 99 Regulation of rockets

PART 9 AIR TRAFFIC SERVICES

Article 100 Requirement for an air traffic control approval
Article 101 Duty of person in charge to satisfy himself as to competence of controllers
Article 102 Manual of air traffic services
Article 103 Provision of air traffic services

Article 104 Making of an air traffic direction in the interests of safety
Article 105 Making of a direction for airspace policy purposes
Article 106 Use of radio call signs at aerodromes

PART 10 LICENSING OF AIR TRAFFIC CONTROLLERS
Article 107 Prohibition of unlicensed air traffic controllers and student air traffic controllers
Article 108 Grant and renewal of air traffic controller's and student air traffic controller's licences
Article 109 Privileges of an air traffic controller licence or a student air traffic controller licence
Article 110 Maintenance of validity of ratings and endorsements
Article 111 Obligation to notify rating ceasing to be valid and change of unit
Article 112 Requirement for medical certificate
Article 113 Appropriate licence
Article 114 Incapacity of air traffic controllers
Article 115 Fatigue of air traffic controllers – air traffic controller's responsibilities
Article 116 Prohibition of drunkenness etc. of controllers
Article 117 Failing exams
Article 118 Use of simulators
Article 119 Approval of courses and persons
Article 120 Acting as an air traffic controller and a student air traffic controller

PART 11 FLIGHT INFORMATION SERVICES AND LICENSING OF FLIGHT INFORMATION SERVICE OFFICERS
Article 121 Prohibition of unlicensed flight information service officers
Article 122 Licensing of flight information service officers
Article 123 Flight information service manual

PART 12 AIR TRAFFIC SERVICE EQUIPMENT
Article 124 Air traffic service equipment
Article 125 Air traffic service equipment records

PART 13 AERODROMES, AERONAUTICAL LIGHTS AND DANGEROUS LIGHTS
Article 126 Aerodromes – public transport of passengers and instruction in flying
Article 127 Use of Government aerodromes
Article 128 Licensing of aerodromes
Article 129 Charges at aerodromes licensed for public use
Article 130 Use of aerodromes by aircraft of Contracting States and of the Commonwealth
Article 131 Noise and vibration caused by aircraft on aerodromes
Article 132 Aeronautical lights
Article 133 Lighting of en-route obstacles
Article 134 Lighting if wind turbine generators in United Kingdom territorial waters
Article 135 Dangerous lights
Article 136 Customs and Excise aerodromes
Article 137 Aviation fuel at aerodromes

PART 14 GENERAL
Article 138 Restriction with respect to carriage for valuable consideration in aircraft registered elsewhere than the UK
Article 139 Filing and approval of tariffs
Article 140 Restriction on aerial photography, aerial survey and aerial work in aircraft registered elsewhere than in the UK
Article 141 Flights over any foreign country
Article 142 Mandatory reporting of occurrences
Article 143 Mandatory reporting of birdstrikes
Article 144 Power to prevent aircraft flying
Article 145 Right of access to aerodromes and other places
Article 146 Obstruction of persons
Article 147 Directions
Article 148 Penalties
Article 149 Extra-territorial effect of the Order
Article 150 Aircraft in transit over certain United Kingdom territorial waters
Article 151 Application of Order to British -controlled aircraft registered elsewhere than in the United Kingdom
Article 152 Application of Order to the Crown and visiting forces, etc.
Article 153 Exemption from Order
Article 154 Appeal to County Court or Sheriff Court
Article 155 Interpretation

Article 156 Meaning of aerodrome traffic zone
Article 157 Public transport and aerial work – general rules
Article 158 Public transport and aerial work – exceptions – flying displays etc.
Article 159 Public transport and aerial work – exceptions – charity flights
Article 160 Public transport and aerial work – exceptions – cost sharing
Article 161 Public transport and aerial work – exceptions – recovery of direct costs
Article 162 Public transport and aerial work – exceptions – jointly owned aircraft
Article 163 Public transport and aerial work – exceptions – parachuting
Article 164 Exceptions from application of provisions of the order for certain classes of aircraft
Article 165 Approval of persons to furnish reports
Article 166 Certificates, authorisations, approvals and permissions
Article 167 Competent authority
Article 168 Saving

PART I REGISTRATION AND MARKING OF AIRCRAFT

CITATION, COMMENCEMENT AND REVOCATION

Citation and Commencement
1 This Order may be cited as the Air Navigation Order 2005 and shall come into force on 20 August 2005.

Revocation
2 The Orders specified in Schedule 1 are hereby revoked.

Article 3: Aircraft to be registered
(1) Subject to paragraph (2), (3) and (4) an aircraft shall not fly in or over the United Kingdom unless it is registered in:

(a) some part of the Commonwealth;

(b) a Contracting State; or

(c) some other country in relation to which there is in force an agreement between Her Majesty's Government in the United Kingdom and the Government of that country which makes provision for the flight over the United Kingdom of aircraft registered in that country.

(2) A non-EASA glider may fly unregistered, and shall be deemed to be registered in the United Kingdom for the purposes of articles 19, 20, 26 and 52 of this Order, on any flight which:

(a) begins and ends in the United Kingdom without passing over any other country, and

(b) is not for the purpose of public transport or aerial work other than aerial work which consists of the giving of instruction in flying or the conducting of flying tests in a glider owned or operated by a flying club of which the person giving the instruction or conducting the test and the person receiving the instruction or undergoing the test are both members.

(3) Any non-EASA aircraft may fly unregistered on any flight which:

(a) begins and ends in the United Kingdom without passing over any other country, and

(b) is in accordance with the 'B Conditions'.

(4) Paragraph (1) shall not apply to any non-EASA kite or non-EASA captive balloon.

(5) If an aircraft flies over the United Kingdom in contravention of paragraph (1) in such manner or circumstances that if the aircraft had been registered in the United Kingdom an offence against this Order or any regulations made thereunder would have been committed, the like offence shall be deemed to have been committed in respect of that aircraft.

Article 4: Registration of aircraft in the United Kingdom
(1) The CAA shall be the authority for the registration of aircraft in the United Kingdom and shall be responsible for maintaining the register and may record therein the particulars specified in paragraph (7) in a legible or a non-legible form so long as the recording is capable of being reproduced in a legible form.

(2) Subject to the provisions of this article, an aircraft shall not be registered or continue to be registered in the United Kingdom if it appears to the CAA that:

(a) the aircraft is registered outside the United Kingdom and that such registration does not cease by operation of law upon the aircraft being registered in the United Kingdom;

(b) an unqualified person holds any legal or beneficial interest by way of ownership in the aircraft or any share therein;

(c) the aircraft could more suitably be registered in some other part of the Commonwealth; or

(d) it would be inexpedient in the public interest for the aircraft to be registered in the United Kingdom.

(3) The following persons and no others shall be qualified to hold a legal or beneficial interest by way of ownership in an aircraft registered in the United Kingdom or a share therein:

(a) the Crown in right of Her Majesty's Government in the United Kingdom;

(b) Commonwealth citizens;

(c) nationals of any EEA State;

(d) British protected persons;

(e) bodies incorporated in some part of the Commonwealth and having their principal place of business in any part of the Commonwealth;

(f) undertakings formed in accordance with the law of an EEA State and having their registered office, central administration or principal place of business within the European Economic Area; or

(g) firms carrying on business in Scotland and in this sub-paragraph 'firm' has the same meaning as in the Partnership Act 1890(a).

(4) If any unqualified person;

(a) residing or having a place of business in the United Kingdom holds a legal or beneficial interest by way of ownership in an aircraft, or a share therein, the CAA, upon being satisfied that the aircraft may otherwise be properly so registered, may register the aircraft in the United Kingdom.

(b) has registered an aircraft in pursuance of this paragraph he shall not cause or permit the aircraft, while it is registered to be used for the purpose of public transport or aerial work.

(5) If an aircraft is chartered by demise to a person qualified as aforesaid the CAA may, whether or not an unqualified person is entitled as owner to a legal or beneficial interest therein, register the aircraft in the United Kingdom in the name of the charterer by demise upon being satisfied that the aircraft may otherwise be properly so registered, and subject to the provisions of this article the aircraft may remain so registered during the continuation of the charter.

(6) Application for the registration of an aircraft in the United Kingdom shall be made in writing to the CAA, and shall
(a) include or be accompanied by such particulars and evidence relating to the aircraft and the ownership and chartering thereof as it may require to enable it to determine whether the aircraft may properly be registered in the United Kingdom and to issue the certificate referred to in paragraph (8) and
(b) in particular, include the proper description of the aircraft according to column 4 of the 'Classification of aircraft' in Part A of Schedule 2.

(7) Upon receiving an application for the registration of an aircraft in the United Kingdom and being satisfied that the aircraft may properly be so registered, the CAA shall register the aircraft, wherever it may be, and shall include in the register the following particulars:
(a) the number of the certificate;
(b) the nationality mark of the aircraft, and the registration mark assigned to it by the CAA;
(c) the name of the constructor of the aircraft and its designation;
(d) the serial number of the aircraft;
(e) the name and address of every person who is entitled as owner to a legal interest in the aircraft or a share therein, or, in the case of an aircraft which is the subject of a charter by demise, the name and address of the charterer by demise; and
(f) in the case of an aircraft registered in pursuance of paragraphs (4) or (5), an indication that it is so registered.

(8) The CAA:
(a) shall subject to sub-paragraph (b) furnish to the person in whose name the aircraft is registered (here in after in this article referred to as 'the registered owner') a certificate of registration, which shall include the foregoing particulars and the date on which the certificate was issued.
(b) shall not be required to furnish a certificate of registration if the registered owner is the holder of an aircraft dealer's certificate granted under this Order who has made to the CAA and has not withdrawn a statement of his intention that the aircraft is to fly only in accordance with the conditions set forth in Part C of Schedule 2 to this Order, and in that case the aircraft shall fly only in accordance with those conditions.

(9) The CAA may grant to any person qualified as aforesaid an aircraft dealer's certificate if it is satisfied that he has a place of business in the United Kingdom for buying and selling aircraft.

(10) Subject to paragraphs (4), (5) and (17), if at any time after an aircraft has been registered in the United Kingdom an unqualified person becomes entitled to a legal or beneficial interest by way of ownership in the aircraft or a share therein, the registration of the aircraft shall thereupon become void and the certificate of registration shall forthwith be returned by the registered owner to the CAA.

(11) Any person who is the registered owner of an aircraft registered in the United Kingdom shall forthwith inform the CAA in writing of:
(a) any change in the particulars which were furnished to the CAA upon application being made for the registration of the aircraft;
(b) the destruction of the aircraft, or its permanent withdrawal from use; or
(c) in the case of an aircraft registered in pursuance of paragraph (5), the termination of the demise charter.

(12) Any person who becomes the owner of an aircraft registered in the United Kingdom shall within 28 days inform the CAA in writing to that effect.

(13) The CAA may, whenever it appears to it necessary or appropriate to do so for giving effect to this Part of this Order or for bringing up to date or otherwise correcting the particulars entered on the register, amend the register or, if it thinks fit, may cancel the registration of the aircraft, and shall cancel that registration within 2 months of being satisfied that there has been a change in the ownership of the aircraft.

(14) The Secretary of State may, by regulations, adapt or modify the foregoing provisions of this article as he deems necessary or expedient for the purpose of providing for the temporary transfer of aircraft to or from the United Kingdom register, either generally or in relation to a particular case or class of cases.

(15) In this article references to an interest in an aircraft do not include references to an interest in an aircraft to which a person is entitled only by virtue of his membership of a flying club and the reference in paragraph (11) to the registered owner of an aircraft includes, in the case of a deceased person, his legal personal representative, and in the case of a body corporate which has been dissolved, its successor.

(16) Nothing in this article shall require the CAA to cancel the registration of an aircraft if in its opinion it would be inexpedient in the public interest to do so.

(17) The registration of an aircraft which is the subject of an undischarged mortgage entered in the Register of Aircraft Mortgages kept by the CAA under an Order in Council made under Section 86 of the Civil Aviation Act 1982(a) shall not become void by virtue of paragraph (10), nor shall the CAA cancel the registration of such an aircraft pursuant to this article, unless all persons shown in the Register of Aircraft Mortgages as mortgagees of that aircraft have consented to the cancellation.

Article 5: Nationality and registration marks

(1) An aircraft (other than an aircraft permitted by or under this Order to fly without being registered) shall not fly unless it bears painted thereon or affixed thereto, in the manner required by the law of the country in which it is registered, the nationality and registration marks required by that law.

(2) The marks to be borne by aircraft registered in the United Kingdom shall comply with Part B of Schedule 2.

(3) Subject to paragraph (4), an aircraft shall not bear any marks which purport to indicate:
(a) that the aircraft is registered in a country in which it is not in fact registered; or
(b) that the aircraft is a State aircraft of a particular country if it is not in fact such an aircraft, unless the appropriate authority of that country has sanctioned the bearing of such marks.

(4) Marks approved by the CAA for the purposes of flight in accordance with the 'B Conditions' shall be deemed not to purport to indicate that the aircraft is registered in a country in which it is not in fact registered.

PART 2 AIR OPERATORS' CERTIFICATES

Article 6: Issue of air operators' certificates

(1) Subject to article 7, an aircraft registered in the United Kingdom shall not fly on any flight for the purpose of public transport, otherwise than under and in accordance with the terms of an air operator's certificate granted to the operator of the aircraft under paragraph (2), certifying that the holder of the certificate is competent to secure that aircraft operated by him on such flights as that in question are operated safely.

(2) The CAA shall grant an air operator's certificate if it is satisfied that the applicant is competent, having regard in particular to:

(a) his previous conduct and experience; and

(b) his equipment, organisation, staffing, maintenance and other arrangements;

to secure the safe operation of aircraft of the types specified in the certificate on flights of the description and for the purposes so specified.

Article 7: Grant of police air operators' certificates

(1) A flight by an aircraft registered in the United Kingdom in the service of a police authority shall, for the purposes of this order, de deemed to be a flight for the purpose of public transport.

(2) If any passenger is carried on such a flight it shall be deemed to be for the purpose of public transport of passengers, and save as otherwise expressly provided, the provisions of this order and of any regulations made thereunder shall be compiled with in relation to a flight in the service of a police authority as if that flight was for the purpose of public transport or public transport of passengers as the case may be.

(3) An aircraft registered in the United Kingdom shall not fly on any flight in the service of a police authority otherwise than under and in accordance with either the terms of an air operator's certificate granted to the operator of the aircraft under article 6(2) or the terms of a police air operator's certificate granted to the operator of the aircraft under paragraph (4).

(4) The CAA shall grant a police air operator's certificate if it is satisfied that the applicant is competent having regard in particular to:

(a) his previous conduct and experience; and

(b) his equipment, organisation, staffing, maintenance and other arrangements;

to secure that the operation of aircraft of the types specified in the certificate shall be as safe as is appropriate when flying on flights of the description and for the purposes so specified.

PART 3 AIRWORTHINESS AND EQUIPMENT OF AIRCRAFT

Article 8: Certificate of airworthiness to be in force

(1) Subject to paragraph (2) an aircraft shall not fly unless there is in force in respect thereof a certificate of airworthiness duly issued or rendered valid under the law of the country in which the aircraft is registered or the State of the operator, and any conditions subject to which the certificate was issued or rendered valid are complied with.

(2) The foregoing prohibition shall not apply to flights, beginning and ending in the United Kingdom without passing over any other country, of:

(a) a non-ESA glider, if it is not being used for the public transport of passengers or aerial work other than aerial work which consists of the giving of instruction in flying or the conducting of flying tests in a glider owned or operated by a flying club of which the person giving the instruction or conducting the test and the person receiving the instruction or undergoing the test are both members;

(b) a non-ESA balloon flying on a private flight;

(c) a non-ESA kite;

(d) an non-ESA aircraft flying in accordance with the A Conditions or the B Conditions or

(e) an aircraft flying in accordance with of a national permit to fly, an EASA permit to fly issued by the CAA or a certificate of validation issued by the CAA under article 13.

(3) In the case of:

(a) a non-EASA aircraft registered in the United Kingdom the certificate of airworthiness referred to in paragraph (1) shall be a national certificate of airworthiness.

(b) an EASE aircraft registered in the United Kingdom the certificate of airworthiness referred to in paragraph (1) shall be an EASA certificate of airworthiness issue by the CAA.

(4) For the purpose of paragraph (1) a certificate of airworthiness:

(a) shall include an EASE restricted certificate of airworthiness issued by the CAA and

(b) shall include an EASA restricted certificate of airworthiness issued by the competent authority of a State other than the United Kingdom which does not contain a condition restricting the aircraft to flight within the airspace of the issuing state; but

(c) shall not include an EASA restricted certificate of airworthiness issued by the competent authority of a state other than the United Kingdom which contains a condition restricting the aircraft to flight within the airspace of the issuing state.

(5) An aircraft registered in the United Kingdom with an EASA certificate of airworthiness shall not fly otherwise than in accordance with any condition or limitations contained in its flight manual unless otherwise permitted by the CAA.

Article 9: Issue, renewal, etc., of certificates of airworthiness

(1) Subject to paragraph (2). the CAA shall issue in respect of any non-EASA aircraft a national certificate of airworthiness if it is satisfied that the aircraft is fit to fly having regard to:

(a) the design, construction, workmanship and materials of the aircraft (including in particular any engines fitted therein), and of any equipment carried in the aircraft which it considers necessary for the airworthiness of the aircraft; and
(b) the results of flying trials, and such other tests of the aircraft as it may require.
(2) If the CAA has issued a certificate of airworthiness in respect of an aircraft which in its opinion, is a prototype aircraft or a modification of a prototype aircraft, it may dispense with flying trials in the case of any other aircraft if it is satisfied that it conforms to such prototype or modification.
(3) Every national certificate of airworthiness shall specify such category which is, in the opinion of the CAA, appropriate to the aircraft in accordance with Part B of Schedule 3 and the certificate shall be issued subject to the condition that the aircraft shall be flown only for the purposes indicated in that part in relation to that category.
(4) Any Certificate of airworthiness issued by the CAA prior to the date on which this Order comes into force which is specified as being in the Transport Category (Passenger), Transport Category (Cargo), Aerial work or Private Category shall be deemed to be:
(a) in the case of a non-EASA aircraft a national certificate of airworthiness in the standard category referred to in Part B of Schedule 3; and
(b) in the case of an EASA aircraft an EASA certificate of airworthiness.
(5) The CAA may issue a national certificate of airworthiness subject to such other conditions relating to the airworthiness of the aircraft as it thinks fit.
(6) The CAA may issue a certificate of validation rendering valid for the purpose of this Order a certificate of airworthiness issued in respect of any aircraft registered elsewhere than in the United Kingdom under the law of any country other than the United Kingdom.
(7) Nothing in this Order shall oblige the CAA to accept an application for the issue of a certificate of airworthiness or certificate of validation or for the variation or renewal of any such certificate when the application is not supported by such reports from such approved persons under article 165 as the CAA may specify (either generally or in a particular case or class of cases).

10 Validity of certificate of airworthiness

A certificate of airworthiness or a certificate of validation issued in respect of an aircraft registered in the United Kingdom shall cease to be in force:
(a) if the aircraft, or such of its equipment as is necessary for the airworthiness of thr aircraft, is overhauled, repaired or modified, or if any part of the aircraft or of such equipment is removed or is replaced, otherwise than in a manner and with material of a type approved by EASA in the case of an EASA aircraft or the CAA in the case of a non-EASA aircraft either generally or in relation to a class of aircraft or to the particular aircraft;
(b) until the satisfactory completion of any inspection made for the purpose of ascertaining whether the aircraft remains airworthy or maintenance of the aircraft or any equipment described in sub-paragraph (a) which inspection or maintenance has;
(i) been made mandatory by EASA or the CAA; or
(ii) become required by a maintenance schedule approved by the CAA in relation to that aircraft; or
(c) until the completion to the satisfaction of EASA or the CAA as the case may be of any modification of the aircraft or of any equipment necessary for the airworthiness of the aircraft, being a modification required by EASA or the CAA for the purpose of ensuring that the aircraft remains airworthy.

Article 11: Issue, validity, etc., of national permits to fly

(1) The CAA shall
(a) Subject to sub-paragraph (b) issue in respect of any non-EASA aircraft registered in the United Kingdom a national permit to fly if it is satisfied that the aircraft is fit to fly having regard to the airworthiness of the aircraft and the conditions to be attached to the permit.
(b) refuse to issue a national permit to fly in respect of a non-EASA aircraft registered in the United Kingdom if it appears to the CAA that the aircraft is eligible for and ought to fly under and in accordance with a certificate of airworthiness.
(2)(a) Subject to paragraph (4), an aircraft flying in accordance with a national permit to fly shall not fly for the purpose of public transport or aerial work other than aerial work which consists of flights for the purpose of flying displays, associated practice, test and positioning flights or the exhibition or demonstration of the aircraft.
(3) No person shall be carried during flights for the purpose of flying displays or demonstration flying, except the minimum flight crew unless the prior permission of the CAA has been obtained.
(4) With the permission of the CAA, an aircraft flying in accordance with a national permit to fly may fly for the purpose of aerial work which consists of the giving of instruction in flying or the conduct of flying tests, subject to the aircraft being owned or operated under arrangements entered into by a flying club of which the person giving the instruction or conducting the test and the person receiving the instruction or undergoing the test are both members.
(5) The CAA may issue a national permit to fly subject to such conditions relating to the airworthiness, operation or maintenance of the aircraft as it thinks fit.
(6) A national permit to fly issued in respect of an aircraft shall cease to be in force:
(a) until the satisfactory completion of any inspection made for the purpose of ascertaining whether the aircraft remains airworthy, modification or maintenance of the aircraft or any of its equipment, which inspection, modification or maintenance has:
(i) been made mandatory by the CAA; or
(ii) become required as a condition of the permit to fly;
(b) if any other conditions of the permit are not complied with;
(c) if the aircraft, engines or propellers, or such of its equipment as is necessary for the airworthiness of the aircraft, are modified or repaired; unless the repair, or modification has been approved by the CAA or by a person approved by the CAA for the purpose;

(d) unless the permit includes a current certificate of validity issued by the CAA or by a person approved by the CAA for the purpose.

(7) A placard shall be affixed to any aircraft flying in accordance with a permit to fly in full view of the occupants which shall be worded as follows:

"Occupant Warning This aircraft has not been certificated to an International Requirement"

(8) An aircraft flying in accordance with a permit to fly shall only be flown by day and in accordance with the Visual Flight Rules unless the prior permission of the CAA has been obtained.

(9) Nothing in this Order shall oblige the CAA to accept an application for the issue, variation or renewal of a permit to fly when the application is not supported by such reports from such approved persons as the CAA may specify (either generally or in a particular case or class of cases).

Article 12: Issue of EASA permits to fly

Where the CAA is authorised so to do under Commission Regulation No 1702/2003 (a) it shall in respect of an EASA aircraft registered in the United Kingdom issue an EASA permit to fly in rhe same circumstances as it would issue a national permit to fly to a non-EASA aircraft.

Article 13: Issue etc. of certificates of validation of permits to fly or equivalent documents

(1) The CAA shall issue in respect of any aircraft registered elsewhere than the United Kingdom a certificate of validation if it is satisfied that there is in respect of the aircraft a permit to fly or equivalent document issued or validated by the competent authority of the country in which the aircraft is registered which applies standards which are substantially equivalent to those required for the issue of a permit to fly by the CAA.

(2) An aircraft flying in accordance with a certificate of validation shall not fly for the purpose of public transport or aerial work other than aerial work which consists of flights for the purpose of flying displays, associated practice, test and positioning flights or the exhibition or demonstration of the aircraft.

(3) The CAA may issue a certificate of validation subject to such other conditions relating to the airworthiness, operation or maintenance of the aircraft as it thinks fit.

Article 14: Certificate of maintenance review

(1) An aircraft registered in the United Kingdom:

(a) in respect of which a certificate of airworthiness is in force shall not fly unless the aircraft (including in particular its engines), together with its equipment and radio station, is maintained in accordance with a maintenance schedule approved by the CAA in relation to that aircraft.

(b) which is public transport or an aerial work aircraft shall not fly unless there is in force a certificate (in this order referred to as a "certificate of maintenance review") issued in respect of the aircraft in accordance with the provisions of this article and the certification certifies the date on which the maintenance review was carried out and the date when the next review is due.

(2) A maintenance schedule approved under paragraph (1)(a) in relation to a public transport or aerial work aircraft shall specifiy the occasions on which a review must be carried out for the purpose of issuing a certificate of maintenance review.

(3) A certificate of maintenance review may be issued for the purposes of this article only by:

(a) the holder of an aircraft maintenance engineer's licence:

(i) granted under this Order, being a licence which entitles him to issue that certificate;

(ii) granted under the law of a country other than the United Kingdom and rendered valid under this Order in accordance with the privileges endorsed on the licence; or

(iii) granted under the law of any such country as may be prescribed in accordance with the privileges endorsed on the licence and subject to any conditions which may be prescribed;

(b) a person whom the CAA has authorised to issue a certificate of maintenance review in a particular case, and in accordance with that authority;

(c) a person approved by the CAA as being competent to issue such a certificate, and in accordance with that approval; or

(d) the holder of an aircraft maintenance licence granted by the CAA under Part 66 in accordance with the privileges endorsed on the licence.

(4) In approving a maintenance schedule, the CAA may direct that certificates of maintenance review relating to that schedule, or to any part thereof specified in its direction, may be issued only by the holder of such a licence as is so specified.

(5) A person referred to in paragraph (3) shall not issue a certificate of maintenance review unless he has first verified that:

(a) maintenance has been carried out on the aircraft in accordance with the maintenance schedule approved for that aircraft;

(b) inspections and modifications required by the CAA as provided in article 10 have been completed as certified in the relevant certificate of release to service issued under this Order or under Part 145;

(c) defects entered in the technical log or approved record of the aircraft in accordance with article 15 have been rectified or the rectification thereof has been deferred in accordance with procedures approved by the CAA; and

(d) certificates of release to service have been issued:

(i) under this Order or in accordance with paragraph 21A.163 (d) of Part 21 in respect of an aircraft falling within article 16 (1); or

(ii) under Part 145 in respect of an aircraft required to be maintained in accordance with Part 145;

and for this purpose the operator of the aircraft shall make available to that person such information as is necessary.

(6) A certificate of maintenance review shall be issued in duplicate.

(7) One copy of the most recently issued certificate shall be carried in the aircraft when article 86 so requires, and the other shall be kept by the operator elsewhere than in the aircraft.
(8) Subject to article 91, each certificate of maintenance review shall be preserved by the operator of the aircraft for a period of 2 years after it has been issued.

Article 15: Technical Log

(1) This article applies to public transport and aerial work aircraft registered in the United Kingdom.
(2) Subject to paragraph (3), a technical log shall be kept in respect of every aircraft to which this article applies.
(3) In the case of an aircraft of which the maximum total weight authorised is 2730kg or less and which is not operated by the holder of an air operator's certificate granted by the CAA under article 6 (2) a record approved by the CAA (in this article, article 14 (5) (c) and in Schedule 6 called "an approved record") may be kept instead of a technical log.
(4) Subject to paragraph (5), at the end of every flight by an aircraft to which this article applies the commander shall enter in the technical log or the approved record as the case may be:
(a) the times when the aircraft took off and landed;
(b) particulars of any defect which is known to him and which affects the airworthiness or safe operation of the aircraft, or if no such defect is known to him, an entry to that effect; and
(c) such other particulars in respect of the airworthiness or operation of the aircraft as the CAA may require, and he shall sign and date the entries.
(5) In the case of two or more consecutive flights each of which begins and ends:
(a) within the same period of 24 hours;
(b) at the same aerodrome, except where each such flight is for the purpose of dropping or projecting any material for agricultural, public health or similar purposes; and
(c) with the same person as commander of the aircraft; the commander of an aircraft;
the commander may except where he becomes aware of a defect during an earlier flight, make the entries specified in paragraph (4) at the end of the last of such consecutive flights.
(6) Upon the rectification of any defect which has been entered in a technical log or approved record in accordance with paragraphs (4) and (5) a person issuing a certificate of release to service issued under this Order or under Part 145 in respect of that defect shall enter the certificate in the technical log or approved log in such a position as to be readily identifiable with the defect to which it relates.
(7) Subject to paragraph (8) the technical log or approved record shall be carried in the aircraft when article 86 so requires and copies of the entries referred to in this article shall be kept on the ground.
(8) In the case of an aeroplane of which the maximum total weight authorised is 2730kg or less, or a helicopter, if it is not reasonably practicable for the copy of the technical log or approved record to be kept on the ground it may be carried in the aeroplane or helicopter, as the case may be, in a container approved by the CAA for that purpose.
(9) Subject to article 91, a technical log or approved record required by this article shall be preserved by the operator of the aircraft to which it relates for a period of at least 2 years after the aircraft has been destroyed or has been permanently withdrawn from use, or for such shorter period as the CAA may permit in a particular case.

Article 16: Requirement for a certificate of release to service

(1) This article shall apply to any aircraft registered in the United Kingdom in respect of which a certificate of airworthiness is in force except any such aircraft required to be maintained in accordance with Part 145.
(2) Except as provided in paragraphs (3) (5) and (8) an aircraft to which this article applies shall not fly unless there is in force a certificate of release to service issued under this Order if the aircraft or any part of the aircraft or such of its equipment as is necessary for the airworthiness of the aircraft has been overhauled, repaired, replaced, modified, maintained, or has been inspected as provided in article 10 (b).
(3) If a repair or replacement of a part of a non-EASA aircraft or its equipment is carried out when the aircraft is at such a place that it is not reasonably practicable:
(a) for the repair or replacement to be carried out in such a manner that a certificate of release to service under this Order can be issued in respect thereof; or
(b) for such a certificate to be issued while the aircraft is at that place;
it may fly to a place which satisfies the criteria in paragraph (4) and in such case the commander of the aircraft shall cause written particulars of the flight, and the reasons for making it, to be given to the CAA within 10 days thereafter
(4) A place satsfies the criteria in this paragraph of it is:
(a) the nearest place at which a certificate of release to service under this order can be issued;
(b) a place to which the aircraft can, in the reasonable opinion of the commander, safely fly by a route for which it is properly equipped; and
(c) a place to which it is reasonable to fly having regard to any hazards to the liberty or health of any person on board.
(5) A certificate of release to service shall not be required to be in force in respect of an aircraft to which this article applies of which the maximum total weight authorised does not exceed 2730kg and in respect of which a certificate of airworthiness in the special category referred to in Part B of Schedule 3 is in force, unless the CAA gives a direction to the contrary in a particular case.
(6) A certificate of release to service shall not be required to be in force in respect of an aircraft to which this applies of which the maximum total weight authorised does not exceed 2730kg and which is a private aircraft if it flies in the circumstances specified in paragraph (7).
(7) The circumstances referred to in paragraph (6) are:
(a) the only repairs or replacements in respect of which a certificate of release to service is not in force are of such a description as may be prescribed.

(b) Such repairs or replacements have been carried out personally by the holder of a pilot's licence granted or rendered valid under this Order who is the owner or operator of the aircraft.

(c) the person carrying out the repairs or replacements shall keep in the aircraft log book kept in respect of the aircraft pursuant to article 22 a record which identifies the repairs or replacement and shall sign and date the entries; and

(d) any equipment or parts used in carrying out such repairs or replacements shall be of a type approved by EASA or the CAA either generally or in relation to a class of aircraft or one particular aircraft.

(8) A certificate of release to service issued under this order shall not be required to be in force in respect of an aircraft to which this article applies if there a certificate of release to service issued in accordance with paragraph 21A.163 (d) of Part 21.

(9) Neither:

(a) equipment provided in compliance with Schedule 4 (except equipment specified in paragraph 4 of the Schedule); nor

(b) radio communication and radio navigation equipment provided for use in an aircraft or in any survival craft carried in an aircraft, whether or not such apparatus is provided in compliance with this Order or any regulations made thereunder;

shall be installed or placed on board for use in an aircraft to which this article applies after being overhauled, repaired, modified or inspected, unless there is in force in respect thereof at the time when it is installed or placed on board a certificate of release to service issued under this Order.

(10) A certificate of release to service issued under this Order shall:

(a) certify that the aircraft or any part thereof or its equipment has been overhauled, repaired, replaced, modified or maintained, as the case may be, in a manner and with material of a type approved by EASA or the CAA either generally or in relation to a class of aircraft or the particular aircraft and shall identify the overhaul, repair, replacement, modification or maintenance to which the certificate relates and shall include particulars of the work done; or

(b) certify in relation to any inspection required by the CAA that the aircraft or the part thereof or its equipment, as the case may be, has been inspected in accordance with the requirements of the CAA and that any consequential repair, replacement or modification has been carried out.

(11) A certificate of release to service issued under this Order may be issued only by:

(a) the holder of an aircraft maintenance engineer's licence:

(i) granted under this Order, being a licence which entitles him to issue that certificate;

(ii) granted under the law of a country other than the United Kingdom and rendered valid under this Order, in accordance with the privileges endorsed on the licence:

(b) the holder of an aircraft maintenance engineer's licence or authorisation as such an engineer granted or issued by or under the law of any Contracting State other than the United Kingdom in which the overhaul, repair, replacement, modification or inspection has been carried out, but only in respect of aircraft to which this article applies of which the maximum total weight authorised does not exceed 2730kg and in accordance with the privileges endorsed on the licence;

(c) a person approved by the CAA as being competent to issue such certification, and in accordance with that approval;

(d) a person whom the CAA has authorised to issue the certificate in a particular case, and in accordance with that authority;

(e) in relation only to the adjustment and compensation of direct reading magnetic compasses, the holder of a United Kingdom Airline Transport Pilot's Licence (Aeroplanes) or a JAR-FCL Airline Transport Pilot License (Aeroplane) or a Flight Navigator's Licence granted or rendered valid under this Order;

(f) a person approved in accordance with Part 145, and in accordance with that approval.

(g) the holder of an aircraft maintenance licence granted by the CAA under Part 66 in accordance with the privileges endorsed on the licence.

(12) In this article, the expression 'repair' includes in relation to a compass the adjustment and compensation thereof and the expression 'repaired' shall be construed accordingly.

Article 17: Requirement for a certificate of release to service under Part 145

An EASA aircraft to which Part 145 applies shall not fly when a certificate of release to service is required by or under Part 145 unless such a certificate is in force.

Article 18: Licensing of maintenance engineers

(1) The CAA shall grant an aircraft maintenance engineers' licences, subject to such conditions as it thinks fit, upon being satisfied that the applicant is a fit person to hold the licence and is qualified by reason of his knowledge, experience, competence and skill in aeronautical engineering, and for that purpose the applicant shall furnish such evidence and undergo such examinations and tests as the CAA may require of him.

(2) An aircraft maintenance engineer's licence shall authorise the holder, subject to such conditions as may be specified in the licence, to issue:

(a) certificates of maintenance review in respect of such aircraft as may be so specified;

(b) certificates of release to service under this Order in respect of such overhauls, repairs, replacements, modifications, maintenance and inspections of such aircraft and such equipment as may be so specified; or

(c) certificates of fitness for flight under 'Paragraph 1 (4) of the A Conditions in respect of such aircraft as may be so specified.

(3) A licence shall, subject to article 92, remain in force for the period specified therein, not exceeding 5 years, but may be renewed by the CAA from time to time upon being satisfied that the applicant is a fit person and is qualified as aforesaid.

(4) The CAA may issue a certificate rendering valid for the purposes of this Order any licence as an aircraft maintenance engineer granted under the law of any country other than the United Kingdom.

(5) An aircraft maintenance engineer's licence granted under this article shall not be valid unless it bears the ordinary signature of the holder in ink or in indelible pencil; provided that if the licence is annexed to an aircraft maintenance licence issued under Part 66 it shall be sufficient if that Part 66 licence bears such a signature.

(6) Without prejudice to any other provision of this Order the CAA may, for the purposes of this article:

(a) approve any course of training or instruction;

(b) authorise a person to conduct such examinations or tests as it may specify; and

(c) approve a person to provide or conduct any course of training or instruction.

(7) The holder of an aircraft maintenance engineer's licence granted under paragraph (1) or of an aircraft maintenance licence granted under Part 66 shall not exercise the privileges of such a licence if he knows or suspects that his physical or mental condition renders him unfit to exercise such privileges.

(8) The holder of an aircraft maintenance engineer's licence sgranted under paragraph (1) or of an aircraft maintenance licence granted under Part 66 shall not, when exercising the privileges of such a licence, be under the influence of drink or a drug to such an extent as to impair his capacity to exercise such privileges.

Article 19: Equipment of aircraft

(1) An aircraft shall not fly unless it is so equipped as to comply with the law of the country in which it is registered, and to enable lights and markings to be displayed, and signals to be made, in accordance with this Order and any regulations made thereunder.

(2) In the case of any aircraft registered in the United Kingdom the equipment required to be provided (in addition to any other equipment required by or under this Order) shall:

(a) be that specified in such parts of Schedule 4 to this Order as are applicable in the circumstances;

(b) comply with the provisions of that Schedule;

(c) except that specified in paragraph 4 of the said Schedule, be of a type approved by EASA or the CAA either generally or in relation to a class of aircraft or in relation to that aircraft; and

(d) be installed in a manner approved by EASA in the case of an EASA-aircraft and the CAA in the case of a non-EASA aircraft.

(3) In any particular case the CAA may direct that an aircraft registered in the United Kingdom shall carry such additional or special equipment or supplies as it may specify for the purpose of facilitating the navigation of the aircraft, the carrying out of search and rescue operations, or the survival of the persons carried in the aircraft.

(4) The equipment carried in compliance with this article shall be so installed or stowed and kept stowed, and so maintained and adjusted, as to be readily accessible and capable of being used by the person for whose use it is intended.

(5) The position of equipment provided for emergency use shall be indicated by clear markings in or on the aircraft.

(6) In every public transport aircraft registered in the United Kingdom there shall be provided individually for each passenger or, if the CAA so permits in writing, exhibited in a prominent position in every passenger compartment, a notice which complies with paragraph (7)

(7) A notice complies with this paragraph if it:

(a) is relevant to the aircraft in question.

(b) contains pictorial instructions on the brace position to be adopted in the event of an emergency landing.

(c) contains pictorial instructions on the method of use of the safety belts and safety harnesses as appropriate.

(d) contains pictorial information as to where emergency exits are to be found and instructions as to how they are to be used.

(e) contains pictorial information as to where the lifejackets, escape slides, liferafts and oxygen masks, if required to be provided by paragraph (2), are to be found and instructions as to how they are to be used.

(8) All equipment installed or carried in an aircraft, whether or not in compliance with this article, shall be so installed or stowed and so maintained and adjusted as not to be a source of danger in itself or to impair the airworthiness of the aircraft or the proper functioning of any equipment or services necessary for the safety of the aircraft.

(9) Without prejudice to paragraph (2), all navigational equipment capable of establishing the aircraft's position in relation to its position at some earlier time by computing and applying the resultant of the acceleration and gravitational forces acting upon it when carried in an aircraft registered in the United Kingdom (whether or not in compliance with this order or any regulations made thereunder) shall be of a type approved by EASA or the CAA either generally or in relation to a class of aircraft or in relation to that aircraft and shall be installed in a manner so approved.

(10) This article shall not apply in relation to radio communication and radio navigation equipment except that specified in Schedule 2.

Article 20: Radio equipment of aircraft

(1) An aircraft shall not fly unless it is so equipped with radio communication and radio navigation equipment as to comply with the law of the country in which the aircraft is registered or the State of the operator and to enable communications to be made and the aircraft to be navigated, in accordance with the provisions of this Order and any regulations made thereunder.

(2) Without prejudice to paragraph (1), the aircraft shall be equipped with radio communication and radio navigation equipment in accordance with Schedule 5.

(3) In any particular case the CAA may direct that an aircraft registered in the United Kingdom shall carry such additional or special radio communication or radio navigation equipment as it may specify for the purpose of facilitating the navigation of the aircraft, the carrying out of search and rescue operations or the survival of the persons carried in the aircraft.

(4) Subject to such exceptions as may be prescribed the radio communication and radio navigation equipment provided in compliance with this article in an aircraft registered in the United Kingdom shall always be maintained in serviceable condition.

(5) All radio communication and radio navigation equipment installed in an aircraft registered in the United Kingdom or carried on such an aircraft for use in connection with the aircraft (whether or not in compliance with this Order or any regulations made thereunder) shall;
(a) be of a type approved by EASA or the CAA in relation to the purpose for which it is to be used, and
(b) except in the case of a non-EASA glider which is permitted by article 3(2) to fly unregistered, be installed in a manner approved by EASA in the case of an EASA aircraft and the CAA in the case of a non-EASA aircraft.
(6) Neither the equipment referred to in Paragraph (5) nor the manner in which it is installed shall be modified except with the approval of EASA in the case of an EASA aircraft or the CAA in the case of a non-EASA aircraft..

Article 21: Minimum equipment requirements

(1) The CAA may grant in respect of any aircraft or class of aircraft registered in the United Kingdom a permission permitting such aircraft to commence a flight in specified circumstances notwithstanding that any specified item of equipment required by or under this Order to be carried in the circumstances of the intended flight is not carried or is not in a fit condition for use.

(2) An aircraft registered in the United Kingdom shall not commence a flight if any of the equipment required by or under this Order to be carried in the circumstances of the intended flight is not carried or is not in a fit condition for use unless:

(a) the aircraft does so under and in accordance with the terms of a permission under this article which has been granted to the operator; and

(b) in the case of an aircraft to which article 38 or 39 applies, the operations manual or police operations manual repsectively contains the particulars of that permission.

Article 22: Aircraft, engine and propeller log books

(1) In addition to any other log books required by or under this Order, the following log books shall be kept in respect of aircraft registered in the United Kingdom:
(a) an aircraft log book;
(b) a separate log book in respect of each engine fitted in the aircraft; and
(c) a separate log book in respect of each variable pitch propeller fitted to the aircraft.

(2) The log books shall include the particulars respectively specified in Schedule 6 and in the case of an aircraft having a maximum total weight authorised not exceeding 2730kg shall be of a type approved by the CAA.

(3) Each entry in the log book:
(a) other than such an entry as is referred to in sub-paragraphs 2(4)(b) or 3(4)(b) of Schedule 6, shall be made as soon as practicable after the occurrence to which it relates, but in no event more than 7 days after the expiration of the certificate of maintenance review (if any) in force in respect of the aircraft at the time of the occurrence.

(b) being such an entry as is referred to in sub-paragraphs 2(4)(d) or 3(4)(d) of Schedule 6 shall be made upon each occasion that any maintenance, overhaul, repair, replacement, modification or inspection is undertaken on the engine or propeller, as the case may be.

(4) Any document which is incorporated by reference in a log book shall be deemed, for the purposes of this Order, to be part of the log book.

(5) It shall be the duty of the operator of every aircraft in respect of which log books are required to be kept to keep them or cause them to be kept in accordance with the foregoing provisions of this article.

(6) Subject to article 91 every log book shall be preserved by the operator of the aircraft until a date 2 years after the aircraft, the engine or the variable pitch propeller, as the case may be, has been destroyed or has been permanently withdrawn from use.

Article 23: Aircraft weight schedule

(1) Every flying machine and glider in respect of which a certificate of airworthiness issued or rendered valid under this Order is in force shall be weighed, and the position of its centre of gravity determined, at such times and in such manner as the CAA may require or approve in the case of that aircraft.

(2) Upon the aircraft being weighed the operator of the aircraft shall prepare a weight schedule showing:

(a) either the basic weight of the aircraft, that is to say, the empty weight of the aircraft established in accordance with the type certification basis of the aircraft or such other weights as may be approved by the CAA or EASA in the case of that aircraft; and

(b) either the position of the centre of gravity of the aircraft at its basic weight or such other position of the centre of gravity as may be approved by the CAA or EASA in the case of that aircraft.

(3) Subject to article 91 the weight schedule shall be preserved by the operator of the aircraft until the expiration of a period of six months following the next occasion on which the aircraft is weighed for the purposes of this article.

Article 24: Access and inspection for airworthiness purposes

(1) The CAA may cause such inspections, investigations, tests, experiments and flight trials to be made as it deems necessary for the purposes of this Part of this Order or for the purposes of Part 21, Part 145 or Part M and any person authorised to do so in writing by the CAA may at any reasonable time inspect any part of, or material intended to be incorporated in or used in the manufacture of any part of, an aircraft or its equipment or any document relating thereto and may for that purpose go upon any aerodrome or enter any aircraft factory.

PART 4 AIRCRAFT CREW AND LICENSING

Article 25: Composition of crew of aircraft

(1) An aircraft shall not fly unless it carries a flight crew of the number and description required by the law of the country in which it is registered.

(2) An aircraft registered in the United Kingdom:

(a) shall carry a flight crew adequate in number and description to ensure the safety of the aircraft;

(b) which has a flight manual, shall carry a flight crew of at least the number and description specified in the flight manual;

(c) which does not now have a flight manual but has done in the past, shall carry a flight crew of at least the number and description specified in that flight manual.

(3) A flying machine registered in the United Kingdom and flying for the purpose of public transport having a maximum total weight authorised exceeding 5700kg shall carry not less than two pilots as members of the flight crew thereof.

(4) Subject to paragraph (6), an aeroplane registered in the United Kingdom shall carry at least two pilots as members of its flight crew if it:

(a) has a maximum total weight authorised of 5700 kg or less;

(b) is flying for the purpose of public transport;

(c) is flying in circumstances where the commander is required to comply with the Instrument Flight Rules; and

(d) comes within paragraph (5).

(5) For the purpose of paragraph (4)(d) an aeroplane comes with this paragraph if it has:

(a) one or more turbine jets;

(b) one or more turbine propeller engines and provided with a means of pressurising the personnel compartments;

(c) two or more turbine propeller engines and a maximum approved passenger seating configuration of more than nine;

(d) two or more turbine propeller engines and a maximum approved passenger seating configuration of fewer than 10 and not provided with the means of pressurising the personnel compartments, unless it is equipped with an autopilot which has been approved by the CAA for the purposes of this article and which is serviceable on take-off; or

(e) two or more piston engines, unless it is equipped with an autopilot which has been approved by the CAA for the purposes of this article and which is serviceable on take-off.

(6) An aeroplane

(a) described in paragraphs (5)(d) or (5)(e) which is equipped with an approved autopilot shall not be required to carry two pilots notwithstanding that before take-off the approved autopilot is found to be unserviceable, if the aeroplane flies in accordance with arrangements approved by the CAA.

(b) described in paragraphs (5)(c), (d) or (e) which is flying under and in accordance with the terms of a police air operator's certificate shall not be required to carry two pilots.

(7) Subject to paragraph (8), a helicopter registered in the United Kingdom shall carry at least two pilots as members of its flight crew if it:

(a) has a maximum total weight authorised of 5700kg or less;

(b) has a maximum approved passenger seating configuration of 9 or less;

(c) is flying for the purpose of public transport; and

(d) is flying in circumstances where the commander is required to comply with the Instrument Flight Rules or which is flying by night with visual ground reference.

(8) A helicopter described in paragraph (7) shall not be required to carry two pilots if it –

(a) is equipped with an autopilot with, at least, altitude hold and heading mode which is serviceable on take-off.

(b) is equipped with such an autopilot notwithstanding that before take-off the approved autopilot is found to be unserviceable, if the helicopter flies in accordance with arrangements approved by the CAA; orr

(c) is flying under and in accordance with the terms of a police air operators certificate.

(9) An aircraft registered in the United Kingdom engaged on a flight for the purpose of public transport shall carry:

(a) a flight navigator as a member of the flight crew; or

(b) navigational equipment suitable for the route to be flown;

if on the route or any diversion therefrom, being a route or diversion planned before take-off, the aircraft is intended to be more than 500 nautical miles from the point of take-off measured along the route to be flown, and to pass over part of an area specified in Schedule 7.

(10) A flight navigator carried in compliance with paragraph 9 shall be carried in addition to any person who is carried in accordance with this article to perform other duties.

(11) An aircraft registered in the United Kingdom which is required by article 20 to be equipped with radio communications apparatus shall carry a flight radiotelephony operator as a member of the flight crew.

(12) Paragraphs (13) and (14) apply to any flight for the purpose of public transport by an aircraft registered in the United Kingdom which has a maximum approved passenger seating configuration of more than 19 and on which at least one passenger is carried.

(13) The crew of an aircraft on a flight to which this paragraph applies shall include cabin crew carried for the purposes of performing in the interests of the safety of passengers, duties to be assigned by the operator or the commander of the aircraft but who shall not act as members of the flight crew.

(14) On a flight to which this paragraph applies -

(a) there shall, subject to sub-paragraph (b), be carried not less than one member of the cabin crew for every 50 or fraction of 50 passenger seats installed in the aircraft.

(b) the number of members of the cabin crew calculated in accordance with sub-paragraph (a) need not be carried if the CAA has granted written permission to the operator to carry a lesser number on that flight and the operator carries the number specified in that permission and complies with any other terms and conditions subject to which such permission is granted.

(15) The CAA may in the interests of safety direct the operator of any aircraft registered in the United Kingdom that all or any aircraft operated by him when flying in circumstances specified in the direction shall carry, in addition to the cabin crew required to be carried therein by the foregoing provisions of this article, such additional persons as members of the cabin crew as it may specify in the direction.

Article 26: Members of flight crew – requirement for licence

(1) Subject to the provisions of this article, a person shall not act as a member of the flight crew of an aircraft registered in the United Kingdom unless he is the holder of an appropriate licence granted or rendered valid under this Order.

(2) A person may within the United Kingdom, the Channel Islands, and the Isle of Man without being the holder of such a licence:

(a) act as a flight radiotelephony operator if

(i) he does so as the pilot of a glider on a private flight and he does not communicate by radiotelephony with any air traffic control unit; or

(ii) he does so as a person being trained in an aircraft registered in the United Kingdom to perform duties as a member of the flight crew of an aircraft and;

(aa) he is authorised to operate the radiotelephony station by the holder of the licence granted in respect of that station under any enactment;

(bb) messages are transmitted only for the purposes of instruction, or of the safety or navigation of the aircraft;

(cc) messages are transmitted only on a frequency exceeding 60 MHz assigned by the CAA for the purposes of this sub-paragraph;

(dd) the operation of the transmitter requires the use only of external switches; and

(ee) the stability of the frequency radiated is maintained automatically by the transmitter;

(b) act as pilot in command of an aircraft for the purpose of becoming qualified for the grant or renewal of a pilot's licence or the inclusion or variation of any rating in a pilot's licence if:

(i) he is at least 16 years of age;

(ii) he is the holder of a valid medical certificate to the effect that he is fit so to act issued by a person approved by the CAA;

(iii) he complies with any conditions subject to which that medical certificate was issued;

(iv) no other person is carried in the aircraft;

(v) the aircraft is not flying for the purpose of public transport or aerial work other than aerial work which consists of the giving of instruction in flying or the conducting of flying tests; and

(vi) he so acts in accordance with instructions given by a person holding a pilot's licence granted under this Order or a JAA licence, being a licence which includes a flight instructor rating, a flying instructor's rating or an assistant flying instructor's rating entitling him to give instruction in flying the type of aircraft being flown;

(c) act as pilot of an aircraft in respect of which the flight crew required to be carried by or under this Order does not exceed one pilot for the purpose of becoming qualified for the grant or renewal of a pilot's licence or the inclusion or variation of any rating in a pilot's licence if:

(i) the aircraft is not flying for the purpose of public transport or aerial work other than aerial work which consists of the giving of instruction in flying or the conducting of flying tests;

(ii) he so acts in accordance with instructions given by a person holding a pilot's licence granted under this Order or a JAA licence, being a licence which includes a flight instructor rating, or an assistant flying instructor's rating entitling him to give instruction in flying the type of aircraft being flown; and

(iii) the aircraft is fitted with dual controls and he is accompanied in the aircraft by the said instructor who is seated at the other set of controls or the aircraft is fitted with controls designed for and capable of use by two persons and he is accompanied in the aircraft by the said instructor who is seated so as to be able to use the controls;

(d) act as pilot in command of a helicopter or gyroplane at night if:

(i) he is the holder of an appropriate licence granted or rendered valid under this Order in all respects save that the licence does not include an instrument rating and he has not within the immediately preceding 13 months carried out as pilot in command not less than 5 take-offs and 5 landings at a time when the depression of the centre of the sun was not less than 12° below the horizon;

(ii) he so acts in accordance with instructions given by a person holding a pilot's licence granted under this Order or a JAA licence, being a licence which includes a flight instructor rating, or a assistant flying instructor's rating entitling him to give instruction in flying the type of helicopter or gyroplane being flown by night;

(iii) no person other than that specified in sub-paragraph (ii) above is carried; and

(iv) the helicopter or gyroplane is not flying for the purpose of public transport or aerial work other than aerial work which consists of the giving of instruction in flying or the conducting of flying tests;

(e) act as pilot in command of a balloon if:

(i) he is the holder of an appropriate licence granted or rendered valid under this Order in all respects save that he has not within the immediately preceding 13 months carried out as pilot in command 5 flights each of not less than 5 minutes duration;

(ii) he so acts in accordance with instructions given by a person authorised by the CAA to supervise flying in the type of balloon being flown;

(iii) no person other than that specified in sub-paragraph (ii) above is carried; and

(iv) the balloon is not flying for the purpose of public transport or aerial work other than aerial work which consists of the giving of instruction in flying or the conducting of flying tests.

(3) Subject as aforesaid, a person shall not act as a member of the flight crew required by or under this Order to be carried in an aircraft registered in a country other than the United Kingdom unless:

(a) in the case of an aircraft flying for the purpose of public transport or aerial work, he is the holder of an appropriate licence granted or rendered valid under the law of the country in which the aircraft is registered or the State of the operator; or

ANO — Article 26: Members of flight crew – requirement for licence

(b) in the case of any other aircraft, he is the holder of an appropriate licence granted or rendered valid under the law of the country in which the aircraft is registered or under this Order, and the CAA does not in the particular case give a direction to the contrary.

(4) For the purposes of this Part of this Order

(a) subject to sub-paragraph (b), a licence granted either under the law of a Contracting State other than the United Kingdom but which is not a JAA licence or a licence granted under the law of a relevant overseas territory, purporting in either case to authorise the holder to act as a member of the flight crew of an aircraft, not being a licence purporting to authorise him to act as a student pilot only, shall, unless the CAA in the particular case gives a direction to the contrary, be deemed to be a licence rendered valid under this Order but does not entitle the holder:

(i) to act as a member of the flight crew of any aircraft flying for the purpose of public transport or aerial work or on any flight in respect of which he receives remuneration for his services as a member of the flight crew; or

(ii) in the case of a pilot's licence, to act as pilot of any aircraft flying in controlled airspace in circumstances requiring compliance with the Instrument Flight Rules or to give any instruction in flying.

(b) a JAA licence shall, unless the CAA in the particular case gives a direction to the contrary, be deemed to be a licence rendered valid under this Order.

(5) Notwithstanding paragraph (1), a person may, unless the certificate of airworthiness in force in respect of the aircraft otherwise requires, act as pilot of an aircraft registered in the United Kingdom for the purpose of undergoing training or tests for the grant or renewal of a pilot's licence or for the inclusion, renewal or extension of a rating therein without being the holder of an appropriate licence, if the conditions specified in paragraph (6) are complied with.

(6) The conditions referred to in paragraph (5) are -

(a) No other person shall be carried in the aircraft or in an aircraft being towed thereby except:

(i) a person carried as a member of the flight crew in compliance with this Order;

(ii) a person authorised by the CAA to witness the training or tests or to conduct the said tests; or

(iii) if the pilot in command of the aircraft is the holder of an appropriate licence, a person carried for the purpose of being trained or tested as a member of the flight crew of an aircraft; and

(b) the person acting as the pilot of the aircraft without being the holder of an appropriate licence either:

(i) within the period of six months immediately preceding was serving as a qualified pilot of an aircraft in any of Her Majesty's naval, military or air forces, and his physical condition has not, so far as he is aware, so deteriorated during that period as to render him unfit for the licence for which he intends to qualify; or

(ii) holds a pilot's, a flight navigator's or a flight engineer's licence granted under article 27 and the purpose of the training or test is to enable him to qualify under this Order for the grant of a pilot's licence or for the inclusion of an additional type in the aircraft rating in his licence and he acts under the supervision of a person who is the holder of an appropriate licence.

(7) Notwithstanding paragraph (1), a person may act as a member of the flight crew (otherwise than a pilot) of an aircraft registered in the United Kingdom for the purposes of undergoing training or tests for the grant or renewal of a flight navigator's or a flight engineer's licence or for the inclusion, renewal or extension of a rating therein, without being the holder of an appropriate licence if he acts under the supervision and in the presence of another person who is the holder of the type of licence or rating for which the person undergoing the training or tests is being trained or tested.

(8) Notwithstanding paragraph (1), a person may act as a member of the flight crew of an aircraft registered in the United Kingdom without being the holder of an appropriate licence if, in so doing, he is acting in the course of his duty as a member of any of Her Majesty's naval, military or air forces.

(9) An appropriate licence for the purposes of this article means a licence which entitles the holder to perform the functions which he undertakes in relation to the aircraft concerned and the flight on which it is engaged.

(10) This article shall not require a licence to be held by a person by reason of his acting as a member of the flight crew of a glider unless:

(a) he acts as a flight radiotelephony operator otherwise than in accordance with paragraph (2)(a)(i); or

(b) the flight is for the purpose of public transport or aerial work, other than aerial work which consists of the giving of instruction in flying or the conducting of flying tests in a glider owned or operated by a flying club of which the person giving the instruction or conducting the test and the person receiving the instruction or undergoing the test are both members.

(11) Notwithstanding anything in this article:

(a) the holder of a licence granted or rendered valid under this Order, being a licence endorsed to the effect that the holder does not satisfy in full the relevant minimum standards established under the Chicago Convention, shall not act as a member of the flight crew of an aircraft registered in the United Kingdom in or over the territory of a Contracting State other than the United Kingdom except in accordance with permission granted by the competent authorities of that State;

(b) the holder of a licence granted or rendered valid under the law of a Contracting State other than the United Kingdom, being a licence endorsed as aforesaid, shall not act as a member of the flight crew of any aircraft in or over the United Kingdom except in accordance with permission granted by the CAA, whether or not the licence is or is deemed to be rendered valid under this Order.

Article 27: Grant, renewal and effect of flight crew licences

(1) Subject to paragraph (2), the CAA shall grant licences, subject to such conditions as it thinks fit, of any of the classes specified in Part A of Schedule 8 to this Order authorising the holder to act as a member of the flight crew of an aircraft registered in the United Kingdom, upon being satisfied that the applicant is:

(a) a fit person to hold the licence, and

(b) is qualified by reason of his knowledge, experience, competence, skill and physical and mental fitness to act in the capacity to which the licence relates,

and for that purpose the applicant shall furnish such evidence and undergo such examinations and tests (including in particular medical examinations) and undertake such courses of training as the CAA may require of him.

(2) The CAA shall not grant:

(a) a United Kingdom Private Pilot's Licence (Aeroplanes) to any person who was not on 30th June 2000 the holder of such a licence;

(b) a United Kingdom Basic Commercial Pilot's Licence (Aeroplanes) to any person who was not on 30th June 2000 the holder of such a licence;

(c) a United Kingdom Private Pilots Licence (Helicopters) to any person who was not on 31st December 2000 the holder of such a licence

(d) a United Kingdom Commercial Pilot's Licence (Aeroplanes) or a United Kingdom Airline Pilot's Transport Pilot's Licence (Aeroplanes) to any person who was not on 30th June 2002 respectively the holder of such a licence;

(e) a United Kingdom Commercial Pilot's Licence (Helicopters) or a United Kingdom Airline Transport Pilot's Licence (Helicopters) to any person who was not on 31st December 2002 respectively the holder of such a licence.

(3) A licence granted under this article

(a) shall not be valid unless it bears thereon the ordinary signature of the holder in ink or indelible ink.

(b) Subject to article 92 shall:

(i) remain in force for the period indicated in the licence, not exceeding the period specified in respect of a licence of that class Part A of Schedule 8, and may be renewed by the CAA from time to time upon being satisfied that the applicant is a fit person and qualified as aforesaid;

(ii) If no period is indicated in the licence it shall remain in force, subject as aforesaid for the lifetime of the holder.

(c) shall not be granted to any person who is under the minimum age specified for that class of licence in Part A Schedule 8.

(5) Subject to paragraph (5), the CAA shall not on or after 1st July 2000 grant a United Kingdom Private Pilot's Licence (Aeroplanes) to any person who was not on 30th June 2000 the holder of such a licence.

(ii) Subject to sub-paragraph (iii), the CAA may include in a licence a rating or qualification, subject to such conditions as it thinks fit, of any of the classes specified in Part B of Schedule 8, upon being satisfied that the applicant is qualified as aforesaid to act as the capacity to which the rating or qualification relates, and such rating or qualification shall be deemed to form part of the licence.

(5) The CAA shall not:

(a) grant a flying instructor's rating (aeroplanes), an assistant flying instructor's rating (aeroplanes), an assistant flying instructor's rating (aeroplanes), a flying instructor's rating (helicopters) or an assistant flying instructor's rating (helicopters).

(b) include in a United Kingdom Private Pilot's Licences (Aeroplanes) containing only a microlight class rating (in this Part of this Order and in Schedule 8 called "a Microlight Licence") or only an SLMG class rating (in this Part and in Schedule 8 called "an SLMG Licence") granted on or after 1st July 2000 any additional class or type rating;

(c) include in a National Private Pilots Licence (Aeroplanes) any rating or qualification other than an aircraft rating which included only one or more of a simple single engine aeroplane (NPPL) class rating, a Microlight class rating or an SLMG class rating.

(d) include a simple single engine aeroplane (NPPL) class rating, a Microlight class rating or an SLMG class rating in an aircraft rating included in any United Kingdom licence.

(6) Noting in this Order shall oblige the CAA to accept an application for the issue of a National Private Pilots Licence (Aeroplanes) when the application is not supported by such reports from such persons as the CAA may approve (either generally or in a particular case or class of cases).

(7) Subject to any conditions of the licence including those specified in Part A of Schedule 8 and to any other provisions of this Order, a licence of any class shall entitle the holder to perform the functions specified in respect of that licence in Section 1 of Part A of the said Schedule under the heading 'Privileges' or Section 2 or Section 3 off Part A of the said Schedule under the heading 'Privileges and conditions', and a rating or qualification of any class shall entitle the holder of the licence in which such rating or qualification is included to perform the functions specified in respect of that rating or qualification in Part B of the said Schedule.

Article 28: Maintenance of privileges of aircraft ratings in United Kingdom licences for which there are no JAR-FCL equivalents, except for Basic Commercial Pilot's Licences and United Kingdom Flight Engineer's Licences

(1) This article applies to any United Kingdom aeroplane licence for which there is no JAR-FCL equivalent other than a United Kingdom Basic Commercial Pilot's Licence and a United Kingdom Flight Engineer's Licence.

(2) Subject to paragraphs (3) and (4), the holder of a pilot's licence to which this article applies shall not be entitled to exercise the privileges of an aircraft rating contained in the licence on a flight unless the licence bears a valid certificate or a valid certificate of experience to the functions he is to perform on that flight in accordance with Section 1 of Part C of Schedule 8 and shall otherwise comply with that section.

(3) The holder of a Private Pilot's Licence (Balloons and Airships) to which this article applies shall be entitled to exercise the privileges of an aircraft rating contained in the licence on a flight when the licence does not bear a certificate referred to in paragraph (2)

(4) The holder of a Microlight Licence, an SLMG Licence or a United Kingdom Private Pilot's Licence (Gyroplanes) shall not be entitled to exercise the privileges of an aircraft rating contained in the licence on a flight unless the certificate of test or certificate of experience required by paragraph (2) is included in the personal flying log book required to be kept by him under article 25.

(5) The holder of a flight navigator's licence to which this article applies shall not be entitled to perform functions on a flight to which article 25(9) applies unless the licence bears a valid certificate of experience which certificate shall be appropriate to the functions he is to perform on that flight in accordance with Section2 of Part C of the said Schedule and shall otherwise comply with that Part.

Article 29 Maintenance of privileges of aircraft ratings in JAR-FCL licences, United Kingdom licences for which there are JAR-FCL equivalents, United Kingdom Basic Commercial Pilot's Licences and United Kingdom Flight Engineer's Licences

(1) This article applies to:
(a) JAR-FCL licences;
(b) United Kingdom aeroplane licences for which there are JAR-FCL equivalents;
(c) United Kingdom Basic Commercial Pilot's Licences;
(d) United Kingdom Flight Engineer's Licences; and
(e) United Kingdom helicopter licences to which there are JAR-FCL equivalents.

(2) The holder of a pilot's licence to which this article applies shall not be entitled to exercise the privileges of an aircraft rating contained in the licence on a flight unless –
(a) the licence bears a valid certificate of revalidation in respect of the rating; and
(b) the holder has undertaken differences training in accordance with paragraph 1.235 of Section 1 of JAR-FCL 1 in the case of an aeroplane and paragraph 2.235 of Section 1 of JAR-FCL 2 in the case of a helicopter and has had particulars thereof entered in his personal flying log book in accordance with the relevant paragraph.

(3) The holder of a United Kingdom Flight Engineer's Licence shall not be entitled to exercise the privileges of an aircraft rating contained in the licence on a flight unless the licence bears a valid certificate in respect of the rating.

Article 30: Maintenance of privileges of aircraft ratings in National Private Pilot's Licences

(1) The holder of a National Private Pilot's Licence (Aeroplanes) shall not be entitled to exercise the privileges of a simple single engine aeroplane (NPPL) class rating contained in the licence on a flight unless the rating is valid in accordance with Section 3 of Part C of Schedule 8.

(2) The holder of a National Private Pilot's Licence (Aeroplanes) shall not be entitled to exercise the privileges of an SLMG class rating or a Microlight class rating contained in the licence on a flight unless the licence includes a valid certificate of test or a valid certificate of experience in respect of the rating, which certificate shall in either case be appropriate to the functions he is to perform on that flight in accordance with Section 1 of Part C of Schedule 8 and shall otherwise comply with that Section.

Article 31: Maintenance of privileges of other ratings

(1) A person shall not be entitled to perform the functions to which a flying instructor's rating (gyroplanes), an assistant flying instructor's rating (gyroplanes) or an instrument meteorological conditions rating (aeroplanes) relates unless his licence bears a valid certificate of test, which certificate shall be appropriate to the functions to which the rating relates in accordance with Section 1 of Part C of Schedule 8 to this Order and shall otherwise comply with that Part.

(2) A person shall not be entitled to perform the functions to which an instrument rating or an instructor's rating (other than a flying instructor's rating (gyroplanes) or an assistant flying instructor's rating (gyroplanes)) relates unless his licence bears a valid certificate of revalidation in respect of the rating.

Article 32: Medical requirements

(1) The holder of a licence granted under article 27, other than a Flight Radiotelephony Operator's Licence, shall not be entitled to perform any of the functions to which his licence relates unless it includes an appropriate valid medical certificate issued under paragraph (3).

(2) Every applicant for or holder of a licence granted under article 27 shall upon such occasions as the CAA may require himself to medical examination by a person approved by the CAA, either generally or in a particular case or class of cases, who shall make a report to the CAA in such form as the CAA may require.

(3) On the basis of such medical examination, the CAA or any person approved by it as competent to do so may issue a medical certificate subject to such conditions as it or he thinks fit to the effect that it or he is assessed the holder of the licence as meeting the requirements specified in respect of the certificate and the certificate shall, without prejudice to paragraph (6), be valid for such period as is therein specified and shall be deemed to form part of the licence.

(4) A person shall not be entitled to act as a member of the flight crew of an aircraft registered in the United Kingdom if he knows or suspects that his physical or mental condition renders him temporarily or permanently unfit to perform such functions or to act in such capacity.

(5) Every holder of a medical certificate issued under this article who:
(a) suffers any personal injury involving incapacity to undertake his functions as a member of the flight crew;
(b) suffers any illness involving incapacity to undertake those functions throughout a period of 21 days or more; or
(c) in the case of a woman, has reason to believe that she is pregnant;
shall inform the CAA in writing of such injury, illness or pregnancy, as soon as possible in the case of injury or pregnancy, and as soon as the period of 21 days has expired in the case of illness.

(6) The medical certificate shall be deemed to be suspended upon the occurrence of such injury or the expiry of such period of illness or the confirmation of the pregnancy and –
(a) in the case of injury or illness the suspension shall cease upon the holder being medically examined under arrangements made by the CAA and pronounced fit to resume his functions as a member of the flight crew or upon the CAA exempting, subject to such conditions as it thinks fit, the holder from the requirement of a medical examination; and
(b) in the case of pregnancy, the suspension may be lifted by the CAA for such period and subject to such conditions as it thinks fit and shall cease upon the holder being medically examined under arrangements made by the CAA after the pregnancy has ended and pronounced fit to resume her functions as a member of the flight crew.

Article 33: Miscellaneous licensing provisions

(1) A person who, on the last occasion when he took a test for the purposes of articles 28 29, 30 or 31, failed that test shall not be entitled to fly in the capacity for which that test would have qualified him had he passed it.

(2) Nothing in this Order shall prohibit the holder of a pilot's licence from acting as pilot of an aircraft certificated for single pilot operation when, with the permission of the CAA, he is testing any person for the purposes of articles 27(1), 27(4), 28(2), 29(2) or 31, not with standing that:

(a) the type of aircraft in which the test is conducted is not specified in an aircraft rating included in his licence; or

(b) that the licence or personal flying log book, as the case may be, does not include a valid certificate of test, experience or revalidation in respect of the type of aircraft.

(3) Without prejudice to any other provision of this Order the CAA may, for the purpose of this Part of this Order:

(a) approve any course of training or instruction;

(b) authorise a person to conduct such examinations or tests as it may specify; and

(c) approve a person to provide any course of training or instruction.

Article 34: Validation of Licences

(1) Subject to paragraphs (2) and (6), the CAA may issue a certificate of validation rendering valid for the purposes of this Order any flight crew licence granted under the law of any country other than the United Kingdom.

(2) Pursuant to Council Directive 91/670 EEC on mutual acceptance of personnel licences for the exercise of functions in civil aviation as it has effect in accordance with the EEA Agreement as amended by the Decision of the EEA Joint Committee No. 7/94 of 21st March 1994 the CAA shall, subject to paragraphs (4) and (5), issue a certificate of validation rendering valid a relevant licence granted under the law of an EEA State.

(3) For the purposes of this article, a relevant licence is one based on requirements equivalent to those for the equivalent licence granted by the CAA under article 27.

(4) The CAA

(a) may ask the Commission for an opinion on the equivalence of a licence submitted for validation pursuant to paragraph (2) of this article within three weeks of receipt by the CAA of all necessary information in respect of an application for validation.

(b) shall, if it does not ask the Commission for such an opinion, within three months of receipt of all necessary information in respect of the application either issue the certificate of validation or inform the applicant of any additional requirements or tests which are necessary to enable the CAA to grant the certificate of validation.

(5) If after the examination of a licence the CAA has reasonable doubts as to the equivalence of that licence;

(a) the CAA may stipulate additional requirements or tests (or both) as necessary to enable the certificate of validation to be issued.

(b) Any such additional requirements or tests (or both) shall as soon as reasonably practicable to the licence holder, the authority which issued the licence and to the Commission.

(6) Pursuant to the said Council Directive, the CAA shall issue a certificate of validation rendering valid any licence issued in accordance with the requirements of Annex 1 to the Chicago Convention if the bearer satisfies the special validation requirements laid down in the annex to the said Council Directive.

Article 35: Personal flying logbook

(1) Every member of the flight crew of an aircraft registered in the United Kingdom and every person who engages in flying for the purpose of qualifying for the grant or renewal of a licence under this Order shall keep a personal flying log book in which the following particulars shall be recorded:

(a) the name and address of the holder of the log book;

(b) particulars of the holder's licence (if any) to act as a member of the flight crew of an aircraft; and

(c) the name and address of his employer (if any).

(2) Particulars of each flight during which the holder of the log book acted either as a member of the flight crew of an aircraft or for the purpose of qualifying for the grant or renewal of a licence under this Order, as the case may be, shall be recorded in the log book at the end of each flight or as soon thereafter as is reasonably practicable, including:

(a) the date, the places at which the holder embarked on and disembarked from the aircraft and the time spent during the course of a flight when he was acting in either capacity;

(b) the type and registration marks of the aircraft;

(c) the capacity in which the holder acted in flight;

(d) particulars of any special conditions under which the flight was conducted, including night flying and instrument flying; and

(e) particulars of any test or examination undertaken whilst in flight.

(3) For the purposes of this article, a helicopter shall be deemed to be in flight from the moment the helicopter first moves under its own power for the purpose of taking off until the rotors are next stopped.

(4) Particulars of any test or examination undertaken whilst in a flight simulator shall be recorded in the log book, including:

(a) the date of the test or examination;

(b) the type of simulator;

(c) the capacity in which the holder acted; and

(d) the nature of the test or examination.

Article 36: Instruction in flying

(1) A person shall not give any instruction in flying to which this article applies unless:

(a) he holds a licence, granted or rendered valid under this Order or a JAA licence, entitling him to act as pilot in command of the aircraft for the purpose and in the circumstances under which the instruction is to be given; and

(b) his licence includes an instructor's rating entitling the holder to give the instruction.
(2) This article applies to instruction in flying given to any person flying or about to fly a flying machine or glider for the purpose of becoming qualified for:
(a) the grant of a pilot's licence; and
(b) the inclusion or variation of any rating in his licence.

Article 37: Glider pilot – minimum age
A person under the age of 16 years shall not act as pilot in command of a glider.

PART 5 OPERATION OF AIRCRAFT

Article 38: Operations manual
(1) This article:
(a) Shall, subject to sub-paragraph (b), this article shall apply to public transport aircraft registered in the United Kingdom except aircraft used for the time being solely for flights not intended to exceed 60 minutes in duration, which are either:
(i) flights solely for training persons to perform duties in an aircraft; or
(ii) flights intended to begin and end at the same aerodrome.
(b) shall not apply to an aircraft flying, or intended by the operator of the aircraft to fly solely under and in accordance with the terms of a police air operator's certificate.
(2) The operator of every aircraft to which this article applies shall:
(a) make available to each member of his operating staff an operations manual;
(b) ensure that each copy of the operations manual is kept up to date; and
(c) ensure that on each flight every member of the crew has access to a copy of every part of the operations manual which is relevant to his duties on the flight.
(3) An operations manual:
(a) shall, subject to sub-paragraph (b), contain all such information and instructions as may be necessary to enable the operating staff to perform their duties as such including in particular information and instructions relating to the matters specified in Part A of Schedule 9; but
(b) shall not be required to contain any information or instructions available in a flight manual accessible to the persons by whom the information or instructions may be required.
(4) An aircraft to which this article applies shall not fly unless, at least 30 days prior to such flight, the operator of the aircraft has furnished to the CAA a copy of the whole of the operations manual for the time being in effect in respect of the aircraft.
(5) Subject to paragraph (6), any amendments or additions to the operations manual shall be furnished to the CAA by the operator before or immediately after they come into effect.
(6) Where an amendment or addition relates to the operation of an aircraft to which the operations manual did not previously relate, that aircraft shall not fly for the purpose of public transport until the amendment or addition has been furnished to the CAA.
(7) Without prejudice to paragraphs (4) and (5), the operator shall make such amendments or additions to the operations manual as the CAA may require for the purpose of ensuring the safety of the aircraft or of persons or property carried therein or the safety, efficiency or regularity of air navigation.
(8) If in the course of a flight on which the equipment specified in Scale 0 in paragraph 6 of Schedule 4 hereto is required to be provided the said equipment becomes unserviceable, the aircraft shall be operated on the remainder of that flight in accordance with any relevant instructions in the operations manual.

Article 39: Police operations manual
(1) This article shall apply to aircraft flying, or intended by the operator of the aircraft to fly, solely under and in accordance with the terms of a police air operator's certificate.
(2) An aircraft to which this article applies shall not fly except under and in accordance with the terms of Part I and Part II of a police operations manual, Part I of which shall have been approved in respect of the aircraft by the CAA.
(3) The operator of every aircraft to which this article applies shall:
(a) make available to each member of its operating staff a police operations manual;
(b) ensure that each copy of the operations manual is kept up to date; and
(c) ensure that on each flight every member of the crew has access to a copy of every part of the operations manual which is relevant to his duties on the flight.
(4) Each police operations manual shall contain all such information and instructions as may be necessary to enable the operating staff to perform their duties as such.
(5) An aircraft to which this article applies shall not fly unless, not less than 30 days prior to such flight, the operator of the aircraft has furnished to the CAA a copy of Part II of the police operations manual for the time being in effect in respect of the aircraft.
(6) Subject to paragraph (7), any amendments or additions to Part II of the police operations manual shall be furnished to the CAA by the operator before or immediately after they come into effect.
(7) Where an amendment or addition relates to the operation of an aircraft to which the police operations manual did not previously relate, that aircraft shall not fly in the service of a police authority under and in accordance with the terms of a police operator's certificate until the amendment or addition has been furnished to the CAA.
(8) Without prejudice to paragraph (5), the operator shall make such amendments or additions to the police operations manual as the CAA may require for the purpose of ensuring the safety of the aircraft, or of persons or property carried therein, or the safety, efficiency or regularity of air navigation.

Article 40: Training manual

(1) Subject to paragraph (2), the operator of every aircraft registered in the United Kingdom and flying for the purpose of public transport shall:

(a) make a training manual available to every person appointed by the operator to give or to supervise the training, experience, practice or periodical tests required under article 42(3) of this Order; and

(b) ensure that each copy of that training manual is kept up to date.

(2) This article shall not apply to aircraft flying, or intended by the operator of the aircraft to fly, solely under and in accordance with the terms of a police air operator's certificate.

(3) Each training manual shall contain all such information and instructions as may be necessary to enable a person appointed by the operator to give or to supervise the training, experience, practice and periodical tests required under article 42(3) to perform his duties as such including in particular information and instructions relating to the matters specified in Part B of Schedule 9.

(4) An aircraft to which this article applies shall not fly unless not less than 30 days prior to such flight the operator of the aircraft has furnished to the CAA a copy of the whole of his training manual relating to the crew of that aircraft.

(5) Subject to paragraph (6), any amendments or additions to the training manual shall be furnished to the CAA by the operator before or immediately after they come into effect.

(7) Where an amendment or addition relates to training, experience, practice or periodical tests on an aircraft to which the training manual did not previously relate, that aircraft shall not fly for the purpose of public transport until the amendment or addition has been furnished to the CAA.

(8) Without prejudice to paragraphs (4) and (5), the operator shall make such amendments or additions to the training manual as the CAA may require for the purpose of ensuring the safety of the aircraft, or of persons or property carried therein, or the safety, efficiency or regularity of air navigation.

Article 41: Flight data monitoring, accident prevention and flight safety programme

(1) The operator of an aircraft registered in the United Kingdom flying for the purpose of public transport shall establish and maintain an accident prevention and flight safety programme.

(2) The operator of an aeroplane registered in the United Kingdom with a maximum total weight authorised of more than 27,000kg flying for the purpose of public transport shall include a flight data monitoring programme as part of its accident prevention and flight safety programme.

(3) The sole objective of an accident prevention and flight safety programme shall be the prevention of accidents and incidents and each programme shall be designed and managed to meet that objective.

(4) It shall not be the purpose of an accident prevention and flight safety programme to apportion blame or liability.

Article 42: Public transport – operator's responsibilities

(1) The operator of an aircraft registered in the United Kingdom shall not permit the aircraft to fly for the purpose of public transport without first:

(a) designating from among the flight crew a pilot to be the commander of the aircraft for the flight;

(b) satisfying himself by every reasonable means that the aeronautical radio stations and navigational aids serving the intended route or any planned diversion therefrom are adequate for the safe navigation of the aircraft; and

(c) subject to paragraph (2), satisfying himself by every reasonable means that

(i) every place (whether or not an aerodrome) at which it is intended to take-off or land and any alternate place (whether or not an aerodrome) at which a landing may be made are suitable for the purpose; and

(ii) in particular that they will be adequately manned and equipped at the time at which it is reasonably estimated such a take-off or landing will be made (including such manning and equipment as may be prescribed) to ensure so far as practicable the safety of the aircraft and its passengers.

(2) Without prejudice to any conditions imposed pursuant to article 6 of this Order, the operator of an aircraft shall not be required for the purposes of this article to satisfy himself as to the adequacy of fire-fighting, search, rescue or other services which are required only after the occurrence of an accident.

(3) The operator of an aircraft registered in the United Kingdom shall not permit any person to be a member of the crew during any flight for the purpose of public transport (except a flight for the sole purpose of training persons to perform duties in aircraft) unless:

(a) such person has had the training, experience, practice and periodical tests specified in Part B of Schedule 9 in respect of the duties which he is to perform; and

(b) the operator has satisfied himself that such person is competent to perform his duties, and in particular to use the equipment provided in the aircraft for that purpose.

(4) The operator shall maintain, preserve, produce and furnish information respecting records relating to the matters specified in paragraph (3) in accordance with Part C of the said Schedule 9.

(5) The operator of an aircraft registered in the United Kingdom shall not permit any member of the flight crew thereof, during any flight for the purpose of the public transport of passengers, to simulate emergency manoeuvres and procedures which the operator has reason to believe will adversely affect the flight characteristics of the aircraft.

Article 43: Loading – public transport aircraft and suspended loads

(1) The operator of an aircraft registered in the United Kingdom shall not cause or permit it to be loaded for a flight for the purpose of public transport, or any load to be suspended therefrom, except under the supervision of a person whom he has caused to be furnished with written instructions as to the distribution and securing of the load so as to ensure that:

(a) the load may safely be carried on the flight; and

(b) any conditions subject to which the certificate of airworthiness in force in respect of the aircraft was issued or rendered valid, being conditions relating to the loading of the aircraft, are complied with.

(2) Subject to paragraph (3), the instructions shall indicate the weight of the aircraft prepared for service, that is to say the aggregate of the weight of the aircraft (shown in the weight schedule referred to in article 23) and the weight of such additional items in or on the aircraft as the operator thinks fit to include; and the instructions shall indicate the additional items included in the weight of the aircraft prepared for service, and show the position of the centre of gravity of the aircraft at that weight.

(3) Paragraph (2) shall not apply in relation to a flight if:

(a) the aircraft's maximum total weight authorised does not exceed 1150kg;

(b) the aircraft's maximum total weight authorised does not exceed 2730kg and the flight is intended not to exceed 60 minutes in duration and is either:

(i) a flight solely for training persons to perform duties in an aircraft; or

(ii) a flight intended to begin and end at the same aerodrome; or

(iii) the aircraft is a helicopter the maximum total weight authorised of which does not exceed 3000kg, and the total seating capacity of which does not exceed 5 persons.

(4) The operator of an aircraft shall not cause or permit it to be loaded for a flight for the purpose of public transport in contravention of the instructions referred to in paragraph (1).

(5) Subject to paragraphs (6) and (7), the person supervising the loading of the aircraft shall, before the commencement of any such flight, prepare and sign a load sheet in duplicate conforming to the prescribed requirements, and shall (unless he is himself the commander of the aircraft) submit the load sheet for examination by the commander of the aircraft who shall sign his name thereon.

(6) The requirements of paragraph (5) shall not apply if:

(a) the load and the distributing and securing thereof upon the next intended flight are to be unchanged from the previous flight; and

(b) the commander of the aircraft makes and signs an endorsement to that effect upon the load sheet for the previous flight, indicating:

(i) the date of the endorsement;

(ii) the place of departure upon the next intended flight; and

(iii) the next intended place of destination; or

(7) The requirements of paragraph (5) shall not apply if paragraph (2) does not apply in relation to the flight.

(8) Subject to paragraph (9), one copy of the load sheet shall be carried in the aircraft when article 86 so requires until the flights to which it relates have been completed and one copy of that load sheet and of the instructions referred to in this article shall be preserved by the operator until the expiration of a period of six months thereafter and shall not be carried in the aircraft.

(9) In the case of an aeroplane of which the maximum total weight authorised does not exceed 2730kg, or a helicopter, if it is not reasonably practicable for the copy of the load sheet to be kept on the ground it may be carried in the aeroplane or helicopter, as the case may be, in a container approved by the CAA for that purpose.

(10) The operator of an aircraft registered in the United Kingdom and flying for the purpose of the public transport of passengers shall not cause or permit baggage to be carried in the passenger compartment of the aircraft unless:

(a) such baggage can be properly secured; and

(b) in the case of an aircraft capable of seating more than 30 passengers, such baggage (other than baggage carried in accordance with a permission issued pursuant to article 54(6)(b)) shall not exceed the capacity of the spaces in the passenger compartment approved by the CAA for the purpose of stowing baggage.

Article 44: Public transport – aeroplanes - operating conditions and performance requirements

(1) Subject to paragraph (4) an aeroplane registered in the United Kingdom and flying for the purpose of public transport shall comply with subpart F of Section 1 of JAR-OPS1.

(2) The assessment of the ability of an aeroplane to comply with paragraph (1) shall be based on the information as to its performance approved by the state of design and contained in the flight manual for the aeroplane.

(3) In the event of the approved information in the flight manual being insufficient for that purpose such assessment shall be based on additional data acceptable to the CAA.

(4) An aeroplane need not comply with paragraph (1) if its is flying under and in accordance with a permission granted to the operator by the CAA under paragraph (5)

(5) The CAA may grant in respect of any aeroplane a permission authorising it to comply with the applicable provisions of Schedule 2 to the Air Navigation (General) Regulation 2005.

(6) The applicable provisions for an aeroplane in respect of which such a permission has been granted shall be those provisions of the said Schedule applicable to an aeroplane of the performance group specified in the permission.

(7) A aeroplane registered in the United Kingdom when flying under and in accordance with a permission granted by the CAA under paragraph (5) when flying over water for the purpose of public transport shall fly, except as may be necessary for the purpose of take-off or landing, at such an altitude as would enable the aeroplane:

(a) if it has one engine only, in the event of the failure of that engine; or

(b) if it has more than one engine, in the event of the failure of one of those engines and with the remaining engine or engines operating within the maximum continuous power conditions specified in the certificate of airworthiness relating to the aircraft; to reach a place at which it can safely land at a height sufficient to enable it to do so.

(8) Without prejudice to paragraph (7), an aeroplane flying under and in accordance with a permission granted by the CAA under paragraph (5) in respect of which either that permission or the certificate of airworthiness of the aeroplane designates the aeroplane as being of performance group X shall not fly over water for the purpose of public transport so as to be more than 60 minutes flying time from the nearest shore, unless the aeroplane has more than 2 power units.

(9) For the purposes of paragraph (8), flying time shall be calculated at normal cruising speed with one power unit inoperative.

Article 45: Public transport – helicopters – operating conditions and performance requirements

(1) A helicopter registered in the United Kingdom shall not fly for the purpose of public transport, except for the sole purpose of training persons to perform duties in a helicopter unless such requirements as may be prescribed in respect of its weight and related performance and flight in specified meteorological conditions or at night are complied with.

(2) The assessment of the ability of a helicopter to comply with paragraph (1) shall be based on the information as to its performance approved by the state of design and contained in the flight manual for the helicopter.

(3) In the event of the approved information in the flight manual being insufficient for that purpose such assessment shall be based on additional data acceptable to the CAA.

(4) A helicopter registered in the United Kingdom when flying over water for the purpose of public transport shall fly, except as amay be necessary for the purpose of take-off or landing, at such an altitude as would enable the helicopter:

(a) if it has one engine only, in the event of the failure of that engine; or

(b) if it has more than one engine, in the event of the failure of one of those engines and with the remaining engine or engines operating within the maximum continuous power conditions specified in the certificate of airworthiness or flight manual for the helicopter; to reach a place at which it can safely land at a height sufficient to enable it to do so.

(5) Without prejudice to paragraph (4), a helicopter carrying out Performance Class 3 operations:

(a) shall not fly over water for the purpose of public transport in the specified circumstances unless it is equipped with the required apparatus;

(b) which is equipped with the required apparatus and which is flying under and in accordance with the terms of an air operator's certificate granted by the CAA under article 6(2), shall not fly in the specified circumstances on any flight for more than three minutes except with the permission in writing of the CAA;

(c) which is equipped with the required apparatus and which is flying under and in accordance with the terms of a police air operator's certificate:

(i) on which is carried any passenger who is not permitted passengers, shall not fly in the specified circumstances on any flight for more than 20 minutes;

(ii) on which no passenger is carried other than a permitted passenger, shall not fly over water on any flight for more than 10 minutes so as to be more than 5 minutes from a point it can make an autorotative descent to land suitable for an emergency landing;

(d) shall not fly for the purpose of public transport over that part of the bed of the River Thames which lies between the following points:

    (i) Hammersmith Bridge   (51° 29'18"N) (00° 13'51"W)

    (ii) Greenwich Reach   (51° 29'06"N) (00° 00'43"W)

between the ordinary high water marks on each of its banks unless it is equipped with the required apparatus.

(6) For the purposes of paragraph (5) flying time shall be calculated on the assumption that a helicopter is flying in still air at the speed specified in the flight manual for the helicopter as the speed for compliance with regulations governing flights over water.

(7) Without prejudice to paragraph (4), a helicopter carrying out Performance Class 1 or Performance Class 2 operations:

(a) which is flying under and in accordance with the terms of an air operator's certificate granted by the CAA under article 6(2), shall not fly over water for the purpose of public transport for more than 15 minutes during any flight unless it is equipped with the required apparatus;

(b) which is not equipped with the required apparatus and which is flying under and in accordance with the terms of a police air operator's certificate on which any passenger is carried who is not a permitted passenger, shall not fly over any water on any flight for more than 15 minutes.

(8) Notwithstanding paragraph (1) , a helicopter specified in its flight manual as being in either Group A or Category A may fly for the purpose of public transport in accordance with the weight and related performance requirements prescribed for helicopters carrying out:

(a) Performance Class 2 operations if:

(i) the maximum total weight authorised of the helicopter is less than 5700kg; and

(ii) the total number of passengers carried on the helicopter does not exceed 15; or

(b) Performance Class 3 operations if:

(i) the maximum total weight authorised of the helicopter is less than 3175kg; and

(ii) the total number of passengers carried does not exceed 9.

(9) For the purposes of this article

(a) 'permitted passenger' means:

(i) a police officer;

(ii) an employee of a police authority in the course of his duty;

(iii) a medical attendant;

(iv) the holder of a valid pilot's licence who intends to act as a member of the flight crew of an aircraft flying under and in accordance with the terms of a police air operator's certificate and who is being carried for the purpose of training or familiarisation; or

(v) a CAA Flight Operations Inspector;

(vi) a Home Office police aviation adviser;

(vii) an employee of the fire and rescue authority under the Fire and Rescue Services Act 2004;

(viii) an officer of revenue and customs;

(ix) an employee of the Ministry of Defence in the course of his duty; or
(x) such other person being carried for purposes connected with police operations as may be permitted in writing by the CAA.
(b) "required apparatus" means apparatus approved by the CAA enabling the helicopter to which it is fitted to land safely on water; and
(c) "specified circumstances" means circumstances in which a helicopter is more than 20 seconds flying time from a point from which it can make an autorotative descent to land suitable for an emergency landing.

Article 46: Public transport operations at night or in Instrument Meteorological Conditions by aeroplanes with one power unit which are registered elsewhere than in the United Kingdom

An aeroplane which is registered elsewhere than in the United Kingdom and is powered by one power unit only shall not fly for the purpose of public transport at night or when the cloud ceiling or visibility prevailing at the aerodrome of departure or forecast for the estimated time of landing at the aerodrome at which it is intended to land or at any alternate aerodrome are less than 1,000 feet and 1 nautical mile respectively.

Article 47: Public transport aircraft registered in the United Kingdom – aerodrome operating minima

(1) This article shall apply to public transport aircraft registered in the United Kingdom.

(2) Subject to paragraph (3), the operator of every aircraft to which this article applies shall establish and include in the operations manual or the police operations manual relating to the aircraft the particulars (in this sub-article called 'the said particulars') of the aerodrome operating minima appropriate to every aerodrome of intended departure or landing and every alternate aerodrome.

(3) In relation to any flight wherein:

(a) neither an operations manual nor a police operations manual is required or

(b) it is not practicable to include the said particulars in the operations manual or the police operations manual;

the operator of the said aircraft shall, prior to the commencement of the flight, cause to be furnished in writing to the commander of the aircraft the said particulars calculated in accordance with the required data and instructions provided in accordance with paragraph (4) or (5) and the operator shall cause a copy of the said particulars to be retained outside the aircraft for a minimum period of three months.

(4) The operator of every aircraft to which this article applies for which an operations manual or a police operations manual is required by this Order, shall include in that operations manual such data and instructions (in this article called 'the required data and instructions') as will enable the commander of the aircraft to calculate the aerodrome operating minima appropriate to aerodromes the use of which cannot reasonably have been foreseen by the operator prior to the commencement of the flight.

(5) The operator of every aircraft to which this article applies for which neither an operations manual nor a police operations manual is required by this Order shall, prior to the commencement of the flight, cause to be furnished in writing to the commander of the aircraft the required data and instructions; and the operator shall cause a copy of the required data and instructions to be retained outside the aircraft for a minimum period of three months after the flight.

(6) The specified aerodrome operating minima shall not permit a landing or take-off in circumstances where the relevant aerodrome operating minima declared by the competent authority would prohibit it, unless that authority otherwise permits in writing.

(7) In establishing aerodrome operating minima for the purposes of this article the operator of the aircraft shall take into account the following matters:

(a) the type and performance and handling characteristics of the aircraft and any relevant conditions in its certificate of airworthiness;

(b) the composition of its crew;

(c) the physical characteristics of the relevant aerodrome and its surroundings;

(d) the dimensions of the runways which may be selected for use; and

(e) whether or not there are in use at the relevant aerodrome any aids, visual or otherwise, to assist aircraft in approach, landing or take-off, being aids which the crew of the aircraft are trained and equipped to use, the nature of any such aids that are in use, and the procedures for approach, landing and take-off which may be adopted according to the existence or absence of such aids;

and shall establish in relation to each runway which may be selected for use such aerodrome operating minima as are appropriate to each set of circumstances which can reasonably be expected.

(8) An aircraft to which this article applies shall not commence a flight at a time when:

(a) the cloud ceiling or the runway visual range at the aerodrome of departure is less than the relevant minimum specified for take-off; or

(b) according to the information available to the commander of the aircraft it would not be able without contravening paragraphs (9) or (10), to land at the aerodrome of intended destination at the estimated time of arrival there and at any alternate aerodrome at any time at which according to a reasonable estimate the aircraft would arrive there.

(9) An aircraft to which article 38 applies, when making a descent to an aerodrome, shall not descend from a height of 1000ft or more above the aerodrome to a height less than 1000ft above the aerodrome if the relevant runway visual range at the aerodrome is at the time less than the specified minimum for landing.

(10) An aircraft to which this article applies, when making a descent to an aerodrome, shall not:

(a) continue an approach to landing at any aerodrome by flying below the relevant specified decision height; or

(b) descend below the relevant specified minimum descent height;

unless in either case from such height the specified visual reference for landing is established and is maintained.

(11) If, according to the information available, an aircraft would as regards any flight be required by the Rules of the Air Regulations 1996 to be flown in accordance with the Instrument Flight Rules at the aerodrome of intended landing, the commander of the aircraft shall select prior to take-off an alternate aerodrome unless no aerodrome suitable for that purpose is available.

(12) In this article 'specified' in relation to aerodrome operating minima means such particulars of aerodrome operating minima as have been specified by the operator in, or are ascertainable by reference to, the operations manual relating to that aircraft, or furnished in writing to the commander of the aircraft by the operator in accordance with paragraph (3).

Article 48: Public transport aircraft registered elsewhere in the United Kingdom – aerodrome operating minima

(1) This article shall apply to public transport aircraft registered elsewhere than in the United Kingdom.

(2) An aircraft to which this article applies shall not fly in or over the United Kingdom unless the operator thereof has made available to the flight crew aerodrome operating minima which comply with paragraph (3) in respect of every aerodrome at which it is intended to land or take-off and every alternate aerodrome.

(3) The aerodrome operating minima provided in accordance with paragraph (2) shall be no less restrictive than either:
(a) minima calculated in accordance with the notified method for calculating aerodrome operating minima; or
(b) minima which comply with the law of the country in which the aircraft is registered;
whichever are the more restrictive.

(4)(a) An aircraft to which this article applies shall not:
(i) conduct a Category II, Category IIIA or Category IIIB approach and landing; or
(ii) take-off when the relevant runway visual range is less than 150 metres;
otherwise than under and in accordance with the terms of an approval to do so granted in accordance with he law of the country in which it is registered.

(5) An aircraft to which this article applies shall not take-off from or land at an aerodrome in the United Kingdom in contravention of the specified aerodrome operating minima.

(6) Without prejudice to paragraphs (4) and (5), an aircraft to which this article applies, when making a descent to an aerodrome, shall not descend from a height of 1000 feet or more above the aerodrome to a height of less than 1000 feet above the aerodrome if the relevant runway visual range at the aerodrome is at the time less than the specified minimum for landing.

(7) Without prejudice to paragraphs (4) and (5), an aircraft to which this article applies, when making a descent to an aerodrome, shall not:
(a) continue an approach to landing at any aerodrome by flying below the relevant specified decision height; or
(b) descend below the relevant specified minimum descent height,
unless in either case from such height the specified visual reference for landing is established and is maintained.

(8) In this article:
(a) 'specified' means specified by the operator in the aerodrome operating minima made available to the flight crew under paragraph (2);
(b) 'a Category II approach and landing' means a landing following a precision approach using an Instrument Landing System or Microwave Landing System with:
(i) a decision height below 200 feet but not less than 100 feet; and
(ii) a runway visual range of not less than 300 metres;
(c) 'a Category IIIA approach and landing' means a landing following a precision approach using an Instrument Landing System or Microwave Landing System with:
(i) a decision height lower than 100 feet; and
(ii) a runway visual range of not less than 200 metres; and
(d) 'a Category IIIB approach and landing' means a landing following a precision approach using an Instrument Landing System or Microwave Landing System with:
(i) a decision height lower than 50 feet or no decision height; and
(ii) a runway visual range less than 200 metres but not less than 75 metres.

Article 49: Non-public transport aircraft – aerodrome operating minima

(1) This article shall apply to any aircraft which is not a public transport aircraft.

(2) An aircraft to which this article applies shall not:
(a) conduct a Category II, Category IIIA or Category IIIB approach and landing; or
(b) take-off when the relevant runway visual range is less than 150 metres;
otherwise than under and in accordance with the terms of an approval so to do granted in accordance with the law of the country in which it is registered.

(3) In the case of an aircraft registered in the United Kingdom, the approval referred to in paragraph (2) shall be issued by the CAA;

(4) Without prejudice to paragraph (2) an aircraft to which this article applies when making a descent at an aerodrome to a runway in respect of which there is a notified instrument approach procedure shall not descend from a height of 1000ft or more above the aerodrome to a height less than 1000ft above the aerodrome if the relevant runway visual range for that runway is at the time less than the specified minimum for landing.

(5) Without prejudice to paragraph (2) an aircraft to which this article applies when making a descent to a runway in respect of which there is a notified instrument approach procedure shall not:
(a) continue an approach to landing on such a runway by flying below the relevant specified decision height; or

(b) descend below the relevant specified minimum descent height;
unless in either case from such height the specified visual reference for landing is established and is maintained.
(6) If, according to the information available, an aircraft would as regards any flight be required by the Rules of the Air Regulations 1996 to be flown in accordance with the Instrument Flight Rules at the aerodrome of intended landing, the commander of the aircraft shall select prior to take-off an alternate aerodrome unless no aerodrome suitable for that purpose is available.
(7) In this article 'specified' in relation to aerodrome operating minima means such particulars of aerodrome operating minima as have been notified in respect of the aerodrome or if the relevant minima have not been notified such minima as are ascertainable by reference to the notified method for calculating aerodrome operating minima.
(8) In this article Category II, Category IIIA and Category IIIB approach and landing have the same meaning as in article 48(8).

Article 50: Pilots to remain at controls

(1) The commander of a flying machine or glider registered in the United Kingdom shall cause one pilot to remain at the controls at all times while it is in flight.
(2) If the flying machine or glider is required by or under this Order to carry two pilots, the commander shall cause both pilots to remain at the controls during take-off and landing.
(3) If the flying machine or glider carries two or more pilots (whether or not it is required to do so) and is engaged on a flight for the purpose of the public transport of passengers, the commander shall remain at the controls during take-off and landing.
(4) Each pilot at the controls shall be secured in his seat by either a safety belt with or without one diagonal shoulder strap, or a safety harness except that during take-off and landing a safety harness shall be worn if it is required by article 19 and Schedule 4 to be provided.

Article 51: Wearing of survival suits by crew

(1) Subject to paragraph (2), each member of the crew of an aircraft registered in the United Kingdom shall wear a survival suit if such a suit is required by article 19 and Schedule 4 to be carried.
(2) This article shall not apply to any member of the crew of such an aircraft flying under and in accordance with the terms of a police air operator's certificate.

Article 52: Pre-flight action by commander of aircraft

The commander of an aircraft registered in the United Kingdom shall reasonably satisfy himself before the aircraft takes off:
(a) that the flight can safely be made, taking into account the latest information available as to the route and aerodrome to be used, the weather reports and forecasts available and any alternative course of action which can be adopted in case the flight cannot be completed as planned;
(b) either that:
(i) the equipment required by or under this Order to be carried in the circumstances of the intended flight is carried and is in a fit condition for use; or
(ii) that the flight may commence under and in accordance with the terms of a permission granted to the operator pursuant to article 21
(c) that the aircraft is in every way fit for the intended flight, and that where a certificate of maintenance review is required by article 14(1) to be in force, it is in force and will not cease to be in force during the intended flight;
(d) that the load carried by the aircraft is of such weight, and is so distributed and secured, that it may safely be carried on the intended flight;
(e) in the case of a flying machine or airship, that sufficient fuel, oil and engine coolant (if required) are carried for the intended flight, and that a safe margin has been allowed for contingencies, and, in the case of a flight for the purpose of public transport, that the instructions in the operations manual relating to fuel, oil and engine coolant have been complied with;
(f) in the case of an airship or balloon, that sufficient ballast is carried for the intended flight;
(g) in the case of a flying machine, that having regard to the performance of the flying machine in the conditions to be expected on the intended flight, and to any obstructions at the places of departure and intended destination and on the intended route, it is capable of safely taking off, reaching and maintaining a safe height thereafter and making a safe landing at the place of intended destination; and
(h) that any pre-flight check system established by the operator and set forth in the operations manual or elsewhere has been complied with by each member of the crew of the aircraft; and
(i) in the case of a balloon, that the balloon will be able to land clear of any congested area.

Article 53: Passenger briefing by commander

(1) Subject to paragraph (2), the commander of an aircraft registered in the United Kingdom shall take all reasonable steps to ensure:
(a) before the aircraft takes off on any flight, that all passengers are made familiar with the position and method of use of emergency exits, safety belts (with diagonal shoulder strap where required to be carried), safety harnesses and (where required to be carried) oxygen equipment, lifejackets and the floor path lighting system and all other devices required by or under this Order and intended for use by passengers individually in the case of an emergency occurring to the aircraft; and
(b) that in an emergency during a flight, all passengers are instructed in the emergency action which they should take.
(2) This article shall not apply to the commander of an aircraft registered in the United Kingdom in relation to a flight under and in accordance with the terms of a police air operator's certificate.

Article 54: Public transport of passengers – additional duties of commander

(1) This article applies to flights for the purpose of the public transport of passengers by aircraft registered in the United Kingdom other than flights under and in accordance with the terms of a police air operator's certificate.

(2) In the case of an aircraft which is not a seaplane, on a flight to which this article applies on which it is intended to reach a point more than 30 minutes flying time (while flying in still air at the speed specified in the relevant certificate of airworthiness as the speed for compliance with regulations governing flights over water) from the nearest land, the commander shall subject to paragraph (9), take all reasonable steps to ensure that before take-off all passengers are given a demonstration of the method of use of the lifejackets required by or under this Order for the use of passengers;

(3) In the case of an aircraft which is not a seaplane but is required by article 25(13) to carry cabin crew, the commander shall, subject to paragraph (9), take all reasonable steps to ensure that, before the aircraft takes off on a flight to which this article applies on which:

(a) it is intended to proceed beyond gliding distance from land; or

(b) in the event of any emergency occurring during take-off or during the landing at the intended destination or any likely alternate destination it is reasonably possible that the aircraft would be forced to land onto water;

all passengers are given a demonstration of the method of use of the lifejackets required by or under this order for the use of passengers.

(4) In the case of an aircraft which is a seaplane, the commander shall take all reasonable steps to ensure that before the aircraft takes off on a flight to which this article applies all passengers are given a demonstration of the method of use of the lifejackets required by or under this order for the use of passengers.

(5) Before the aircraft takes off on a flight to which this article applies, and before it lands, the commander shall take all reasonable steps to ensure that the crew of the aircraft are properly secured in their seats and that any persons carried in compliance with article 25(3) and (14) are properly secured in seats which shall be in a passenger compartment and which shall be so situated that those persons can readily assist passengers;

(6) From the moment when, after the embarkation of its passengers for the purpose of taking off on a flight to which this article applies, it first moves until after it has taken off, and before it lands until it comes to rest for the purpose of the disembarkation of its passengers, and whenever by reason of turbulent air or any emergency occurring during the flight he considers the precaution necessary the commander shall take all reasonable steps to ensure that:

(a) all passengers of 2 years of age or more are properly secured in their seats by safety belts (with diagonal shoulder strap, where required to be carried) or safety harnesses and that all passengers under the age of 2 years are properly secured by means of a child restraint device; and

(b) those items of baggage in the passenger compartment which he reasonably considers ought by virtue of their size, weight or nature to be properly secured are properly secured and, in the case of an aircraft capable of seating more than 30 passengers, that such baggage is either stowed in the passenger compartment stowage spaces approved by the CAA for the purpose or carried in accordance with the terms of a permission granted by the CAA.

(7) in the case of aircraft in respect of which a certificate of airworthiness was first issued (whether in the United Kingdom or elsewhere) on or after 1st January 1989 except in a case where a pressure greater than 700 hectopascals is maintained in all passenger and crew compartments throughout the flight, the commander shall take all reasonable steps to ensure that on a flight to which this article applies:

(a) before the aircraft reaches flight level 100 the method of use of the oxygen provided in the aircraft in compliance with the requirements of article 19 and Schedule 4 is demonstrated to all passengers;

(b) when flying above flight level 120 all passengers and cabin crew are recommended to use oxygen; and

(c) during any period when the aircraft is flying above flight level 100 oxygen is used by all the flight crew of the aircraft;

(8) in the case of aircraft in respect of which a certificate of airworthiness was first issued (whether in the United Kingdom or elsewhere) prior to 1st January 1989, except in the case where a pressure greater than 700 hectopascals is maintained in all passenger and crew compartments throughout the flight, the commander shall take all reasonable steps to ensure that on a flight to which this article applies:

(a) before the aircraft reaches flight level 130 the method of use of the oxygen provided in the aircraft in compliance with the requirements of article 19 of Schedule 4 is demonstrated to all passengers;

(b) when flying above flight level 130 all passengers and cabin crew are recommended to use oxygen; and

(c) during any period when the aircraft is flying above flight level 100 oxygen is used by all the flight crew of the aircraft;

provided that he comply instead with paragraph (7)

(9) Where the only requirement to give a demonstration required by paragraph (2) or (3) arises because it is reasonably possible that the aircraft would be forced to land onto water at one or more of the likely alternate destinations the demonstration need not be given until after the decision has been taken to divert to such a destination.

Article 55: Operation of radio in aircraft

(1) The radio station in an aircraft shall not be operated, whether or not the aircraft is in flight, except in accordance with the conditions of the licence issued in respect of that station under the law of the country in which the aircraft is registered or the State of the operator and by a person duly licensed or otherwise permitted to operate the radio station under that law.

(2) Subject to paragraph (3), whenever an aircraft is in flight in such circumstances that it is required by or under this Order to be equipped with radio communications apparatus, a continuous radio watch shall be maintained by a member of the flight crew listening to the signals transmitted upon the frequency notified, or designated by a message received from an appropriate aeronautical radio station, for use by that aircraft.

(3) The radio watch:

(a) may be discontinued or continued on another frequency to the extent that a message as aforesaid so permits.

(b) may be kept by a device installed in the aircraft if:

(i) the appropriate aeronautical radio station has been informed to that effect and has raised no objection; and

(ii) that station is notified, or in the case of a station situated in a country other than the United Kingdom, otherwise designated as transmitting a signal suitable for that purpose.

(4) Whenever an aircraft is in flight in such circumstances that it is required by or under this Order to be equipped with radio or radio navigation equipment a member of the flight crew shall operate that equipment in such a manner as he may be instructed by the appropriate air traffic control unit or as may be notified in relation to any notified airspace in which the aircraft is flying.

(5) The radio station in an aircraft shall not be operated so as to cause interference which impairs the efficiency of aeronautical telecommunications or navigational services, and in particular emissions shall not be made except as follows:

(a) emissions of the class and frequency for the time being in use, in accordance with general international aeronautical practice, in the airspace in which the aircraft is flying;

(b) distress, urgency and safety messages and signals, in accordance with general international aeronautical practice;

(c) messages and signals relating to the flight of the aircraft, in accordance with general international aeronautical practice; and

(d) such public correspondence messages as may be permitted by or under the aircraft radio station licence referred to in paragraph (1).

(6) In any flying machine registered in the United Kingdom which is engaged on a flight for the purpose of public transport the pilot and the flight engineer (if any) shall not make use of a hand-held microphone (whether for the purpose of radio communication or of intercommunication within the aircraft) whilst the aircraft is flying in controlled airspace below flight level 150 or is taking off or landing.

Article 56: Minimum navigation performance

(1) An aircraft registered in the United Kingdom shall not fly in North Atlantic Minimum Navigation Performance Specification airspace unless it is equipped with navigation systems which enable the aircraft to maintain the prescribed navigation performance capability.

(2) The equipment required by paragraph (1) shall

(a) be approved by EASA or the CAA;

(b) be installed in a manner approved by EASAvin the case of an EASA aircraft and the CAA in the case of a non-EASA aircraft;

(c) be maintained in a manner approved by the CAA; and

(d) while the aircraft is flying in the said airspace, be operated in accordance with procedures approved by the CAA.

Article 57: Height keeping performance – aircraft registered in the United Kingdom

(1) Unless otherwise authorised by the appropriate air traffic control unit, an aircraft registered in the United Kingdom shall not fly in reduced vertical separation minimum airspace notified for the purpose of this article, unless it is equipped with height keeping systems which enable the aircraft to maintain the prescribed height keeping performance capability.

(2) The equipment required by paragraph (1) shall:

(a) be approved by EASA or the CAA;

(b) be installed in a manner approved by EASA in the case of an EASA aircraft and the CAA in the case of a non-EASA aircraft.

(c) be maintained in a manner approved by the CAA; and

(d) while the aircraft is flying in the said airspace, be operated in accordance with procedures approved by the CAA.

Article 57: Height keeping performance – aircraft registered in the United Kingdom

(1) Unless otherwise authorised by the appropriate air traffic control unit an aircraft registered in the United Kingdom shall not fly in reduced vertical separation minimum airspace notified for the purpose of this article, unless it is equipped with height keeping systems which enable the aircraft to maintain the prescribed height keeping performance capability.

(2) The equipment required by paragraph (1) shall:

(a) be approved by EASA or the CAA;

(b) be installed in a manner approved by EASA in the case of an EASA aircraft and the CAA in the case of a non-EASA aircraft.

(c) be maintained in a manner approved by the CAA; and

(d) while the aircraft is flying in the said airspace, be operated in accordance with procedures approved by the CAA.

Article 58: Height keeping performance – aircraft registered elsewhere than in the United Kingdom

Unless other wise authorised by the appropriate air traffic control unit an aircraft registered elsewhere than in the United Kingdom shall not fly in United Kingdom reduced vertical separation minimum airspace unless:

(a) it is so equipped with height keeping systems as to comply with the law of the country in which the aircraft is registered in so far as that law requires it to be so equipped when flying in any specified areas; and

(b) the said equipment is capable of being operated so to enable the aircraft to maintain the height keeping performance prescribed in respect of the airspace in which the aircraft is flying, and it is so operated.

Article 59: Area navigation and required navigation performance capabilities – aircraft registered in the United Kingdom

(1) Subject to paragraph (3) an aircraft registered in the United Kingdom shall not fly in designated required navigation performance airspace unless it is equipped with area navigation equipment which enables the aircraft to maintain the navigation performance capability specified in respect of that airspace.

(2) The equipment required by paragraph (1) shall:

(a) be approved by EASA or the CAA;

(b) be installed in a manner approved by EASA in the case of an EASA aircraft and the CAA in the case of a non-EASA aircraft.
(c) be maintained in a manner approved by the CAA; and
(d) while the aircraft is flying in the said airspace, be operated in accordance with procedures approved by the CAA.
(3) An aircraft need not comply with the requirements of paragraph (1) and (2) where the flight has been authorised by the appropriate air traffic control unit not withstanding the lack of compliance and provided that the aircraft complies with any instructions the air traffic control unit may give in the particular case.

Article 60: Area navigation and required navigation performance capabilities – aircraft registered elsewhere than in the United Kingdom

(1) An aircraft registered elsewhere than in the United Kingdom shall not fly in designated required navigation performance airspace in the United Kingdom unless it is equipped with area navigation equipment so as to comply with the law of the country in which the aircraft is registered in so far as that law requires it to be so equipped when flying within designated required navigation performance airspace.

(2) Subject to paragraph (3) the said navigation equipment shall be capable of being operated so as to enable the aircraft to maintain the navigation performance capability notified in respect of the airspace in which the aircraft is flying, and shall be so operated.

(3) An aircraft need not comply with the requirements of paragraph (2) where the flight has been authorised by the appropriate United Kingdom air traffic control unit notwithstanding the lack of compliance and provided that the aircraft complies with any instructions the air traffic control unit may give in the particular case.

Article 61: Use of airborne collision avoidance system

On any flight on which an airborne collision avoidance system is required by article 20 and Schedule 5 to be carried in an aeroplane, the system shall be operated:
(a) in the case of an aircraft to which article 38 applies, in accordance with procedures contained in the Operations Manual for the aircraft;
(b) in the case of an aircraft registered in the United Kingdom to which article 38 does not apply, in accordance with procedures which are suitable having regard to the purposes of the equipment; or
(c) in the case of an aircraft which is registered elsewhere than in the United Kingdom, in accordance with any procedures with which it is required to comply under the law of the country in which the aircraft is registered.

Article 62: Use of flight recording systems and preservation of records

(1) On any flight on which a flight data recorder, a cockpit voice recorder or a combined cockpit voice recorder/flight data recorder is required by paragraph 5(4), (5), (6) or (7) of Schedule 4 to be carried in an aeroplane, it shall always be in use from the beginning of the take-off run to the end of the landing run.

(2) The operator of the aeroplane shall at all times, subject to article 91, preserve:
(a) the last 25 hours of recording made by any flight data recorder required by or under this Order to be carried in an aeroplane; and
(b) a record of not less than one representative flight, that is to say, a recording of a flight made within the last 12 months which includes a take-off, climb, cruise, descent, approach to landing and landing, together with a means of identifying the record with the flight to which it relates;
and shall preserve such records for such period as the CAA may in a particular case direct.

(3) On any flight on which a cockpit voice recorder, a flight data recorder or a combined cockpit voice recorder/flight data recorder is required by paragraph 5(16) of Schedule 4 to be carried in a helicopter, it shall always be in use from the time the rotors first turn for the purpose of taking off until the rotors are next stopped.

(4) The operator of the helicopter shall at all times, subject to article 91, preserve:
(a) the last 8 hours of recording made by any flight data recorder specified at paragraph (1) or (2) of Scale SS of paragraph 6 of Schedule 4 and required by or under this Order to be carried in the helicopter;
(b) in the case of a combined cockpit voice recorder/flight data recorder specified at paragraph (3) of the said Scale SS and required by or under this Order to be carried in a helicopter either:
(i) the last 8 hours of recording; or
(ii) the last 5 hours of recording or the duration of the last flight, whichever is the greater, together with an additional period of recording for either:
(aa) the period immediately preceding the last five hours of recording or the duration of the last flight, whichever is the greater; or
(bb) such period or periods as the CAA may permit in any particular case or class of cases or generally.

(5) The additional recording retained pursuant to sub-paragraphs (b)(ii)(aa) and (bb) of paragraph (4) shall, together with the recording required to be retained under sub-paragraph (b)(ii) of paragraph (4), total a period of 8 hours and shall be retained in accordance with arrangements approved by the CAA.

Article 63: Towing of gliders

(1) An aircraft in flight shall not tow a glider unless the flight manual for the towing aircraft includes an express provision that it may be used for that purpose.
(2) The length of the combination of towing aircraft, tow rope and glider in flight shall not exceed 150 metres.
(3) The commander of an aircraft which is about to tow a glider shall satisfy himself, before the towing aircraft takes off:
(a) that the tow rope is in good condition and is of adequate strength for the purpose, and that the combination of towing aircraft and glider, having regard to its performance in the conditions to be expected on the intended flight and to any obstructions at the place of departure and on the intended route, is capable of safely taking off, reaching and maintaining a safe height at which to separate the combination and that thereafter the towing aircraft can make a safe landing at the place of intended destination;

(b) that signals have been agreed and communication established with persons suitably stationed so as to enable the glider to take-off safely; and
(c) that emergency signals have been agreed between the commander of the towing aircraft and the commander of the glider, to be used, respectively, by the commander of the towing aircraft to indicate that the tow should immediately be released by the glider, and by the commander of the glider to indicate that the tow cannot be released.
(4) The glider shall be attached to the towing aircraft by means of the tow rope before the aircraft takes off.

Article 64: Operation of self-sustaining gliders
A self-sustaining glider shall not take-off under its own power.

Article 65: Towing, picking up and raising of persons and articles
(1) Subject to the provisions of this article, an aircraft in flight shall not, by means external to the aircraft, tow any article, other than a glider, or pick up or raise any person, animal or article, unless the certificate of airworthiness issued or rendered valid in respect of that aircraft under the law of the country in which the aircraft is registered includes an express provision that it may be used for that purpose.
(2) An aircraft shall not launch or pick up tow ropes, banners or similar articles other than at an aerodrome.
(3) An aircraft in flight shall not tow any article, other than a glider, at night or when flight visibility is less than one nautical mile.
(4) The length of the combination of towing aircraft, tow rope, and article in tow, shall not exceed 150 metres.
(5) A helicopter shall not fly at any height over a congested area of a city, town or settlement at any time when any article, person or animal is suspended from the helicopter.
(6) A passenger shall not be carried in a helicopter at any time when an article, person or animal is suspended therefrom, other than a passenger who has duties to perform in connection with the article, person or animal or a passenger who has been picked up or raised by means external to the helicopter or a passenger who it is intended shall be lowered to the surface by such means.
(7) Nothing in this article shall:
(a) prohibit the towing in a reasonable manner by an aircraft in flight of any radio aerial, any instrument which is being used for experimental purposes, or any signal, apparatus or article required or permitted by or under this Order to be towed or displayed by an aircraft in flight;
(b) prohibit the picking up or raising of any person, animal or article in an emergency or for the purpose of saving life;
(c) apply to any aircraft while it is flying in accordance with the B Conditions; or
(d) be taken to permit the towing or picking up of a glider otherwise than in accordance with article 63.

Article 66: Dropping of articles and animals
(1) Articles and animals (whether or not attached to a parachute) shall not be dropped, or permitted to drop, from an aircraft in flight so as to endanger persons or property.
(2) Subject to paragraph (3), except under and in accordance with the terms of an aerial application certificate granted under article 68 of this Order, articles and animals (whether or not attached to a parachute) shall not be dropped, or permitted to drop, to the surface from an aircraft flying over the United Kingdom.
(3) Paragraph (2) shall not apply to the dropping of articles by, or with the authority of, the commander of the aircraft in any of the following circumstances:
(a) the dropping of articles for the purpose of saving life;
(b) the jettisoning, in case of emergency, of fuel or other articles in the aircraft;
(c) the dropping of ballast in the form of fine sand or water;
(d) the dropping of articles solely for the purpose of navigating the aircraft in accordance with ordinary practice or with the provisions of this Order;
(e) the dropping at an aerodrome of tow ropes, banners, or similar articles towed by aircraft;
(f) the dropping of articles for the purposes of public health or as a measure against weather conditions, surface icing or oil pollution, or for training for the dropping of articles for any such purposes, if the articles are dropped with the permission of the CAA and in accordance with any conditions subject to which that permission may have been given; or
(g) the dropping of wind drift indicators for the purpose of enabling parachute descents to be made if the wind drift indicators are dropped with the permission of the CAA and in accordance with any conditions subject to which that permission may have been given.
(4) For the purposes of this article 'dropping' includes projecting and lowering.
(5) Nothing in this article shall prohibit the lowering of any article or animal from a helicopter to the surface, if the certificate of airworthiness issued or rendered valid in respect of the helicopter under the law of the country in which it is registered includes an express provision that it may be used for that purpose.

Article 67: Dropping of persons and grant of parachuting permissions
(1) A person shall not drop, be dropped or permitted to drop to the surface or jump from an aircraft flying over the United Kingdom except under and in accordance with the terms of either a police air operator's certificate or a written permission granted by the CAA under this article.
(2) For the purposes of this article 'dropping' includes projecting and lowering.
(3) Notwithstanding the grant of a police air operator's certificate or a parachuting permission, a person shall not drop, be dropped or be permitted to drop from an aircraft in flight so as to endanger persons or property.
(4) An aircraft shall not be used for the purpose of dropping persons unless:
(a) the certificate of airworthiness issued or rendered valid in respect of that aircraft under the law of the country in which the aircraft is registered and that certificate or the flight manual for the aircraft includes an express provision that it may be used for that purpose and the aircraft is operated in accordance with a written permission granted by the CAA under this article; or

(b) the aircraft is operated under and in accordance with the terms of a police air operator's certificate.

(5) Every applicant for and holder of a parachuting permission shall make available to the CAA if requested to do so a parachuting manual and shall make such amendments or additions to such manual as the CAA may require.

(6) The holder of a parachuting permission shall make the manual available to every employee or person who is or may engage in parachuting activities conducted by him.

(7) The manual shall contain all such information and instructions as may be necessary to enable such employees or persons to perform their duties.

(8) Nothing in this article shall apply to the descent of persons by parachute from an aircraft in an emergency.

(9) Nothing in this article shall prohibit the lowering of any person in an emergency or for the purpose of saving life.

(10) Nothing in this article shall prohibit the lowering of any person from a helicopter to the surface if the certificate of airworthiness issued or rendered valid in respect of the helicopter under the law of the country in which it is registered and that certificate or the flight manual for the helicopter includes an express provision that it may be used for that purpose.

Article 68: Grant of aerial application certificates

(1) An aircraft shall not be used for the dropping of articles for the purposes of agriculture, horticulture or forestry or for training for the dropping of articles for any of such purposes, otherwise than under and in accordance with the terms of an aerial application certificate granted to the operator of the aircraft under paragraph (2).

(2) The CAA:

(a) shall grant an aerial application certificate if it is satisfied that the applicant is a fit person to hold the certificate and is competent, having regard in particular to his previous conduct and experience, his equipment, organisation, staffing and other arrangements, to secure the safe operation of the aircraft specified in the certificate on flights for the purposes specified in paragraph (1).

(b) may grant such a certificate subject to such conditions as the CAA thinks fit including, without prejudice to the generality of the foregoing, conditions for ensuring that the aircraft and any article dropped from it do not endanger persons or property in the aircraft or elsewhere.

(3) Every applicant for and holder of an aerial application certificate shall make available to the CAA upon application and to every member of his operating staff upon the certificate being granted, an aerial application manual.

(4) The manual shall contain all such information and instructions as may be necessary to enable the operating staff to perform their duties as such.

(5) The holder of an aerial certificate shall make such amendments of or additions to the manual as the CAA may require.

Article 69: Carriage of weapons and of munitions of war

(1) Subject to paragraph (6) an aircraft shall not carry any munition of war unless:

(a) such munition of war is carried with the written permission of the CAA and;

(b) subject to paragraph (2), the commander of the aircraft is informed in writing by the operator before the flight commences of the type, weight or quantity and location of any such munition of war on board or suspended beneath the aircraft and any conditions of the permission of the CAA; and

(2) in the case of an aircraft which is flying under and in accordance with the terms of a police air operator's certificate the commander of the aircraft is informed of the matters referred to in sub-paragraph (1)(b) but he need not be so informed in writing.

(3) Subject to paragraph (5), it shall be unlawful for an aircraft to carry any sporting weapon or munition of war in any compartment or apparatus to which passengers have access.

(4) Subject to paragraph (5), it shall be unlawful for a person to carry or have in his possession or take or cause to be taken on board an aircraft, to suspend or cause to be suspended beneath an aircraft or to deliver or cause to be delivered for carriage thereon any sporting weapon or munition or war unless:

(a) the sporting weapon or munition of war:

(i) is either part of the baggage of a passenger on the aircraft or consigned as cargo to be carried thereby;

(ii) is carried in a part of the aircraft, or in any apparatus attached to the aircraft inaccessible to passengers; and

(iii) in the case of a firearm, is unloaded;

(b) particulars of the sporting weapon or munition of war have been furnished by that passenger or by the consignor to the operator before the flight commences; and

(c) without prejudice to paragraph (1) the operator consents to the carriage of such sporting weapon or munition of war by the aircraft.

(5) Paragraphs (3) and (4) shall not apply to or in relation to an aircraft which is flying under and in accordance with the terms of a police air operator's certificate.

(6) Nothing in this article shall apply to any sporting weapon or munition of war taken or carried on board an aircraft registered in a country other than the United Kingdom if the sporting weapon or munition of war, as the case may be, may under the law of the country in which the aircraft is registered be lawfully taken or carried on board for the purpose of ensuring the safety of the aircraft or of persons on board.

(7) For the purposes of this article:

(a) 'munition of war' means

(i) any weapon or ammunition;

(ii) any article containing an explosive, noxious liquid or gas; or

(iii) any other thing which is designed or made for use in warfare or against persons, including parts, whether components or accessories, for such weapon, ammunition or article.

(b) 'sporting weapon' means

(i) any weapon or ammunition;

(ii) any article containing an explosive, noxious liquid or gas; or
(iii) any other thing, including parts, whether components or accessories, for such weapon, ammunition or article; which is not a munition of war.

Article 70: Carriage of dangerous goods

(1) Without prejudice to any other provisions of this Order, the Secretary of State may make regulations prescribing:
(a) the classification of certain articles and substances as dangerous goods;
(b) the categories of dangerous goods which an aircraft may not carry;
(c) the conditions which apply to the loading on, suspension beneath and carriage by an aircraft of dangerous goods;
(d) the manner in which dangerous goods must be packed, marked, labelled and consigned before being loaded on, suspended beneath or carried by an aircraft;
(e) any other provisions for securing the safety of aircraft and any apparatus attached thereto, and the safety of persons and property on the surface in relation to the loading on, suspension beneath or carriage by an aircraft of dangerous goods;
(f) the persons to whom information about the carriage of dangerous goods must be provided;
(g) the documents which must be produced to the CAA or an authorised person on request; and
(h) the powers to be conferred on an authorised person relating to the enforcement of the regulations made hereunder.
(2) It shall be an offence to contravene or permit the contravention of or fail to comply with any regulations made hereunder.
(3) The provisions of this article and of any regulations made thereunder shall be additional to and not in derogation from article 69

Article 71: Method of carriage of persons

(1) A person shall not:
(a) subject to paragraphs (2) and (3) be in or on any part of an aircraft in flight which is not a part designed for the accommodation of persons and in particular a person shall not be on the wings or undercarriage of an aircraft.
(b) be in or on any object, other than a glider or flying machine, towed by or attached to an aircraft in flight.
(2) A person may have temporary access to:
(a) any part of an aircraft for the purpose of taking action necessary for the safety of the aircraft or of any person, animal or goods therein; and
(b) any part of an aircraft in which cargo or stores are carried, being a part which is designed to enable a person to have access thereto while the aircraft is in flight.
(3) This article shall not apply to a passenger in a helicopter flying under and in accordance with a police air operator's certificate who is disembarking in accordance with a procedure contained in the police operations manual for the helicopter.

Article 72: Exits and break-in markings

(1) This article shall apply to every public transport aeroplane or helicopter registered in the United Kingdom.
(2) Whenever an aeroplane or helicopter to which this article applies is carrying passengers, every exit therefrom and every internal door in the aeroplane or helicopter shall be in working order, and, subject to paragraph (3), during take-off and landing and during any emergency, every such exit and door shall be kept free from obstruction and shall not be fastened by locking or otherwise so as to prevent, hinder or delay its use by passengers.
(3) In the case of:
(a) an exit which, in accordance with arrangements approved by the CAA either generally or in relation to a class of aeroplane or helicopter or a particular aeroplane or helicopter, is not required for use by passengers, the exit may be obstructed by cargo;
(b) a door between the flight crew compartment and any adjacent compartment to which passengers have access, the door may be locked or bolted if the commander of the aeroplane or helicopter so determines, for the purpose of preventing access by passengers to the flight crew compartment;
(c) any internal door which is so placed that it cannot prevent, hinder or delay the exit of passengers from the aeroplane or helicopter if an emergency if it is not in working order, paragraph (2) shall not apply.
(4) Every exit from the aeroplane or helicopter shall be marked with the words 'exit' or 'emergency exit' in capital letters, which shall be red in colour and if necessary shall be outlined in white to contrast with the background.
(5) Every exit from the aeroplane or helicopter shall be marked with instructions in English and with diagrams to indicate the correct method of opening the exit, which shall be red in colour and located on a background which provides adequate contrast.
(6) The markings required by paragraph (5) shall be placed on or near the inside surface of the door or other closure of the exit and, if it is openable from the outside of the aeroplane or helicopter, on or near the exterior surface.
(7) An operator of an aeroplane or helicopter shall ensure that if areas of the fuselage suitable for break-in by rescue crews in emergency are marked on aeroplanes or helicopters, such areas shall be marked upon the exterior surface of the fuselage with markings to show the areas (in this paragraph referred to as 'break-in areas') which can, for purposes of rescue in an emergency, be most readily and effectively broken into by persons outside the aeroplane or helicopter.
(8) The markings required by paragraph (7) shall:
(a) be red or yellow, and if necessary shall be outlined in white to contrast with the background
(b) if the corner markings are more than 2 metres apart, have intermediate lines 9 cenitmetres x 3 centimetres inserted so that there is no more than 2 metres between adjacent marks.
(9) The markings required by this article shall:

(a) be painted, or affixed by other equally permanent means; and

(b) be kept at all times clean and unobscured

(10) Subject to compliance with paragraph (11), if one, but not more than one, exit from an aeroplane or helicopter becomes inoperative at a place where it is not reasonably practicable for it to be repaired or replaced, nothing in this article shall prevent that aeroplane or helicopter from carrying passengers until it next lands at a place where the exit can be repaired or replaced.

(11) On any flight on which this paragraph must be complied with:

(a) the number of passengers carried and the position of the seats which they occupy shall be in accordance with arrangements approved by the CAA either in relation to the particular aeroplane or helicopter or to a class of aeroplane or helicopter; and

(b) in accordance with arrangements so approved, the exit shall be fastened by locking or otherwise, the words 'Exit' or 'Emergency Exit' shall be covered, and the exit shall be marked by a red disc at least 23 centimetres in diameter with a horizontal white bar across it bearing the words 'No Exit' in red letters.

Article 73: Endangering safety of an aircraft

A person shall not recklessly or negligently act in a manner likely to endanger an aircraft, or any person therein.

Article 74: Endangering safety of any person or property

A person shall not recklessly or negligently cause or permit an aircraft to endanger any person or property.

Article 75: Drunkenness in aircraft

(1) A person shall not enter any aircraft when drunk, or be drunk in any aircraft.

(2) A person shall not, when acting as a member of the crew of any aircraft or being carried in any aircraft for the purpose of so acting, be under the influence of drink or a drug to such an extent as to impair his capacity so to act.

Article 76: Smoking in aircraft

(1) Notices indicating when smoking is prohibited shall be exhibited in every aircraft registered in the United Kingdom so as to be visible from each passenger seat therein.

(2) A person shall not smoke in any compartment of an aircraft registered in the United Kingdom at a time when smoking is prohibited in that compartment by a notice to that effect exhibited by or on behalf of the commander of the aircraft.

Article 77: Authority of commander of an aircraft

Every person in an aircraft shall obey all lawful commands which the commander of that aircraft may give for the purpose of securing the safety of the aircraft and of persons or property carried therein, or the safety, efficiency or regularity of air navigation.

Article 78: Acting in a disruptive manner

No person shall while in an aircraft:

(a) use any threatening, abusive or insulting words towards a member of the crew of the aircraft;

(b) behave in a threatening, abusive, insulting or disorderly manner towards a member of the crew of the aircraft; or

(c) intentionally interfere with the performance by a member of the crew of the aircraft of his duties.

Article 79: Stowaways

A person shall not secrete himself for the purpose of being carried in an aircraft without the consent of either the operator or the commander thereof or of any other person entitled to give consent to his being carried in the aircraft.

Article 80: Flying Displays

(1) No person shall act as the organiser of a flying display (in this article referred to as 'the flying display director') unless he has obtained the permission in writing of the CAA under paragraph (5) for that flying display.

(2) The commander of an aircraft who is:

(a) intending to participate in a flying display shall take all reasonable steps to satisfy himself before he participates that:

(i) the flying display director has been granted an appropriate permission under paragraph (5);

(ii) the flight can comply with any relevant conditions subject to which that permission may have been granted; and

(iii) the pilot has been granted an appropriate pilot display authorisation; or

(b) participating in a flying display for which a permission has been granted shall comply with any conditions subject to which that permission may have been granted.

(3) No person shall act as pilot of an aircraft participating in a flying display unless he holds an appropriate pilot display authorisation and he complies with any conditions subject to which the authorisation may have been given.

(4) The flying display director shall not permit any person to act as pilot of an aircraft which participates in a flying display unless such person holds an appropriate pilot display authorisation.

(5) The CAA:

(a) shall grant a permission required by virtue of paragraph (1) if it is satisfied that the applicant is a fit and competent person, having regard in particular to his previous conduct and experience, his organisation, staffing and other arrangements, to safely organise the proposed flying display.

(b) may grant such a permission subject to such conditions, which may include conditions in respect of military aircraft, as the CAA thinks fit.

(6) The CAA shall, for the purposes of this article:

(a) grant a pilot display authorisation authorising the holder to act as pilot of an aircraft taking part in a flying display upon it being satisfied that the applicant is a fit person to hold the authorisation and is qualified by reason of his knowledge, experience, competence, skill, physical and mental fitness to fly in accordance therewith and for that purpose the applicant shall furnish such evidence and undergo such examinations and tests as the CAA may require; and

(b) authorise a person to conduct such examinations or tests as it may specify.

(7) A pilot display authorisation granted in accordance with this article shall, subject to article 92, remain in force for the period indicated in the pilot display authorisation.

(8) Subject to paragraph (9), for the purposes of this article, an appropriate pilot display authorisation shall mean such an authorisation which is valid and appropriate to the intended flight and which has been either:

(a) granted by the CAA pursuant to paragraph (6)(a); or

(b) granted by the competent authority of a JAA Full Member State.

(9) A pilot display authorisation granted by the competent authority of a JAA Full Member State shall not be an appropriate pilot display authorisation for the purposes of this article if the CAA has given a direction to that effect.

(10) A direction may be issued under paragraph (9) either in respect of a particular authorisation, a specified category of authorisations or generally.

(11) Paragraph (1) shall not apply to either:

(a) a flying display which takes place at an aerodrome in the occupation of the Ministry of Defence or of any visiting force or any other premises in the occupation or under the control of the Ministry of Defence; or

(b) a flying display at which the only participating aircraft are military aircraft.

(12) The flying display director shall not permit any military aircraft to participate in a flying display unless he complies with any conditions specified in respect of military aircraft subject to which permission for the flying display may have been granted.

(13) Nothing in this article shall apply to an aircraft race or contest or to an aircraft taking part in such a race or contest or to the commander or pilot thereof whether or not such race or contest is held in association with a flying display.

PART 6 FATIGUE OF CREW AND PROTECTION OF CREW FROM COSMIC RADIATION

Article 81: Application and interpretation of Part 6

(1) Subject to paragraph (2), articles 82 and 83 shall apply to any aircraft registered in the United Kingdom which is either:

(a) engaged on a flight for the purpose of public transport; or

(b) operated by an air transport undertaking.

(2) Articles 82 and 83 shall not apply in relation to a flight made only for the purpose of instruction in flying given by or on behalf of a flying club or flying school, or a person who is not an air transport undertaking.

(3) For the purposes of this Part:

(a) 'flight time', in relation to any person, means all time spent by that person in:

(i) a civil aircraft whether or not registered in the United Kingdom (other than such an aircraft of which the maximum total weight authorised does not exceed 1600kg and which is not flying for the purpose of public transport or aerial work); or

(ii) a military aircraft (other than such an aircraft of which the maximum total weight authorised does not exceed 1600kg and which is flying on a military air experience flight);

while it is in flight and he is carried therein as a member of the crew thereof;

(b) 'day' means a continuous period of 24 hours beginning at midnight Co-ordinated Universal Time;

(c) a helicopter shall be deemed to be in flight from the moment the helicopter first moves under its own power for the purpose of taking off until the rotors are next stopped; and

(d) a military air experience flight is a flight by a military aircraft operated under the auspices of the Royal Air Force Air Cadet Organisation for the purpose of providing air experience to its cadets.

Article 82: Fatigue of crew-operator's responsibilities

(1) The operator of an aircraft to which this article applies shall not cause or permit that aircraft to make a flight unless:

(a) he has established a scheme for the regulation of flight times for every person flying in that aircraft as a member of its crew;

(b) the scheme is approved by the CAA;

(c) either:

(i) the scheme is incorporated in the operations manual required by article 38; or

(ii) in any case where an operations manual is not required by that article, the scheme is incorporated in a document, a copy of which has been made available to every person flying in that aircraft as a member of its crew; and

(d) he has taken all such steps as are reasonably practicable to ensure that the provisions of the scheme will be complied with in relation to every person flying in that aircraft as a member of its crew.

(2) The operator of an aircraft to which this article applies shall not cause or permit any person to fly therein as a member of its crew if he knows or has reason to believe that the person is suffering from, or, having regard to the circumstances of the flight to be undertaken, is likely to suffer from, such fatigue while he is so flying as may endanger the safety of the aircraft or of its occupants.

(3) The operator of an aircraft to which this article applies shall not cause or permit any person to fly therein as a member of its flight crew unless the operator has in his possession an accurate and up-to-date record in respect of that person and in respect of the 28 days immediately preceding the flight showing:

(a) all his flight times; and
(b) brief particulars of the nature of the functions performed by him in the course of his flight times.

(4) The record referred to in paragraph (3) shall, subject to article 91, be preserved by the operator of the aircraft until a date 12 months after the flight referred to in that paragraph.

Article 83: Fatigue of crew – responsibilities of crew

(1) A person shall not act as a member of the crew of an aircraft to which this article applies if he knows or suspects that he is suffering from, or, having regard to the circumstances of the flight to be undertaken, is likely to suffer from, such fatigue as may endanger the safety of the aircraft or of its occupants.

(2) A person shall not act as a member of the flight crew of an aircraft to which this article applies unless he has ensured that the operator of the aircraft is aware of his flight times during the period of 28 days preceding the flight.

Article 84: Flight times – responsibilities of flight crew

(1) Subject to paragraph (2), a person shall not act as a member of the flight crew of an aircraft registered in the United Kingdom if at the beginning of the flight the aggregate of all his previous flight times:

(a) during the period of 28 consecutive days expiring at the end of the day on which the flight begins exceeds 100 hours; or
(b) during the period of twelve months expiring at the end of the previous month exceeds 900 hours.

(2) This article shall not apply to a flight which is:
(a) A private flight in an aircraft of which the maximum total weight does not exceed 1600kg; or
(b) a flight which is not for the purpose of public transport and is not operated by an air transport undertaking where, at the time when the flight begins, the aggregate of all the flight times of the member of the flight crew concerned since he was last medically examined and found fit by a person approved by the CAA for the purpose of article 32(2) does not exceed 25 hours.

Article 85: Protection of aircrew from cosmic radiation

(1) A relevant undertaking shall take appropriate measures to:
(a) assess the exposure to cosmic radiation when in flight of those air crew who are liable to be subject to cosmic radiation in excess of 1 milliSievert per year;
(b) take into account the assessed exposure when organising work schedules with a view to reducing the doses of highly exposed air crew; and
(c) inform the workers concerned of the health risks their work involves.

(2) A relevant undertaking shall ensure that in relation to a pregnant air crew member, the conditions of exposure to cosmic radiation when she is in flight are such that the equivalent dose to the foetus will be as low as reasonably achievable and is unlikely to exceed 1 milliSievert during the remainder of the pregnancy.

(3) Nothing in paragraph (2) shall require the undertaking concerned to take any action in relation to an air crew member until she has notified the undertaking in writing that she is pregnant.

(4) The definition in article 155 of 'crew' shall not apply for the purposes of this article.

(5) In this article and in article 87
(a) 'air-crew' has the same meaning as in article 42 of Council Directive 96/29/Euratom of 13th May 1996; and
(b) 'undertaking' includes a natural or legal person and 'relevant undertaking' means an undertaking established in the United Kingdom which operates aircraft.

(6) In this article:
(a) 'highly exposed air crew' and 'milliSievert' have the same respective meanings as in article 42 of Council Directive 96/29/Euratom of 13th May 1996; and
(b) 'year' means any period of twelve months

PART 7 DOCUMENTS AND RECORDS

Article 86: Documents to be carried

(1) An aircraft shall not fly unless it carries the documents which it is required to carry under the law of the country in which it is registered.

(2) Subject to paragraph (3), an aircraft registered in the United Kingdom shall, when in flight, carry documents in accordance with Schedule 10.

(3) If the flight is intended to begin and end at the same aerodrome and does not include passage over the territory of any country other than the United Kingdom, the documents may be kept at that aerodrome instead of being carried in the aircraft.

Article 87: Keeping and production of records of exposure to cosmic radiation

(1) A relevant undertaking shall keep a record for the period and in the manner prescribed of the exposure to cosmic radiation of air crew assessed under article 85 and the names of the air crew concerned.

(2) A relevant undertaking shall, within a reasonable period after being requested to do so by an authorised person, cause to be produced to that person the record required to be kept under paragraph (1).

(3) A relevant undertaking shall, within a reasonable period after being requested to do so by a person in respect of whom a record is required to be kept under paragraph (1), supply a copy of that record to that person.

Article 88: Production of documents and records

(1) The commander of an aircraft shall, within a reasonable time after being requested to do so by an authorised person, cause to be produced to that person:
(a) the certificates of registration and airworthiness in force in respect of the aircraft;
(b) the licences of its flight crew; and
(c) such other documents as the aircraft is required by article 86 to carry when in flight.

(2) The operator of an aircraft registered in the United Kingdom shall, within a reasonable time after being requested to do so by an authorised person, cause to be produced to that person such of the following documents or records as have been requested by that person being documents or records which are required, by or under this Order, to be in force or to be carried, preserved or made available:
(a) the documents referred to in Schedule 10 to this Order as Documents A, B and G;
(b) the aircraft log book, engine log books and variable pitch propeller log books required under this Order to be kept;
(c) the weight schedule, if any, required to be preserved under article 23(3);
(d) in the case of a public transport aircraft or aerial work aircraft, the documents referred to in Schedule 10 as Documents D, E, F and H;
(e) any records of flight times, duty periods and rest periods which he is required by article 82(4) to preserve, and such other documents and information in the possession or control of the operator, as the authorised person may require for the purpose of determining whether those records are complete and accurate;
(f) any such operations manuals as are required to be made available under article 38(2)(a);
(g) the record made by any flight data recorder required to be carried by or under this Order.

(3) The holder of a licence granted or rendered valid under this Order or of a medical certificate required under article 26(2)(b)(ii) shall, within a reasonable time after being requested to do so by an authorised person, cause to be produced to that person his licence, including any certificate of validation.

(4) Every person required by article 35 to keep a personal flying log book shall cause it to be produced within a reasonable time to an authorised person after being requested to do so by him within 2 years after the date of the last entry.

Article 89: Production of air traffic service equipment documents and records

The holder of an approval under articles 124 or 125 shall within a reasonable time after being requested to do so by an authorised person, cause to be produced to that person any documents and records relating to any air traffic service equipment used or intended to be used in connection with the provision of a service to an aircraft.

Article 90: Power to inspect and copy documents and records

An authorised person shall have the power to inspect and copy any certificate, licence, log book, document or record which he has the power pursuant to this Order and any regulations made thereunder to require to be produced to him.

Article 91: Preservation of documents, etc.

(1) Subject to paragraphs (2), (3), (4) and (5) a person required by this Order to preserve any document or record by reason of his being the operator of an aircraft shall, if he ceases to be the operator of the aircraft, continue to preserve the document or record as if he had not ceased to be the operator, and in the event of his death the duty to preserve the document or record shall fall upon his personal representative.

(2) If another person becomes the operator of the aircraft, the first-mentioned operator or his personal representative shall deliver to that person upon demand the certificates of maintenance review and release to service, the log books and the weight schedule and any record made by a flight data recorder and preserved in accordance with article 62(2) and (4) which are in force or required to be preserved in respect of that aircraft.

(3) If an engine or variable pitch propeller is removed from the aircraft and installed in another aircraft operated by another person the first-mentioned operator or his personal representative shall deliver to that person upon demand the log book relating to that engine or propeller.

(4) If any person in respect of whom a record has been kept by the first mentioned operator in accordance with article 82(4) becomes a member of the flight crew of a public transport aircraft registered in the United Kingdom and operated by another person the first-mentioned operator or his personal representative shall deliver those records to that other person upon demand.

(5) It shall be the duty of the other person referred to in paragraphs (2), (3) and (4) to deal with the document or record delivered to him as if he were the first-mentioned operator.

Article 92: Revocation, suspension and variation of certificates, licences and other documents

(1) Subject to paragraphs (5) and (6), the CAA may, if it thinks fit, provisionally suspend or vary any certificate, licence, approval, permission, exemption, authorisation or other document issued, granted or having effect under this Order, pending inquiry into or consideration of the case.

(2) The CAA may, on sufficient ground being shown to its satisfaction after due inquiry, revoke, suspend or vary any such certificate, licence, approval, permission, exemption, authorisation or other document.

(3) The holder or any person having the possession or custody of any certificate, licence, approval, permission, exemption or other document which has been revoked, suspended or varied under this Order shall surrender it to the CAA within a reasonable time after being required to do so by the CAA.

(4) The breach of any condition subject to which any certificate, licence, approval, permission, exemption or other document, other than a licence issued in respect of an aerodrome, has been granted or issued or which has effect under this Order shall, in the absence of provision to the contrary in the document, render the document invalid during the continuance of the breach.

(5) The provisions of article 93 shall have effect, in place of the provisions of this article, in relation to permits to which that article applies.

(6) Notwithstanding paragraph (1), a flight manual, performance schedule or other document incorporated by reference in the certificate of airworthiness may be varied on sufficient ground being shown to the satisfaction of the CAA, whether or not after due inquiry.

Article 93: Revocation, suspension and variation of permissions, etc. granted under article 138 or article 140

(1) Subject to the provisions of this article, the Secretary of State may revoke, suspend or vary any permit to which this article applies.

(2) Save as provided by paragraph (3), the Secretary of State may exercise his powers under paragraph (1) only after notifying the permit-holder of his intention to do so and after due consideration of the case.

(3) If, by reason of the urgency of the matter, it appears to the Secretary of State to be necessary for him to do so, he may provisionally suspend or vary a permit to which this article applies without complying with the requirements of paragraph (2); but he shall in any such case comply with those requirements as soon thereafter as is reasonably practicable and shall then, in the light of his due consideration of the case, either:

(a) revoke the provisional suspension or variation of the permit; or

(b) substitute therefore a definitive revocation, suspension or variation, which, if a definitive suspension, may be for the same or a different period as the provisional suspension (if any) or, if a definitive variation, may be in the same or different terms as the provisional variation (if any).

(4) The powers vested in the Secretary of State by paragraph (1) or paragraph (3) may be exercised by him whenever, in his judgement and whether or not by reason of anything done or omitted to be done by the permit-holder or otherwise connected with the permit-holder, it is necessary or expedient that the permit-holder should not enjoy, or should no longer enjoy, the rights conferred on him by a permit to which this article applies or should enjoy them subject to such limitations or qualifications as the Secretary of State may determine.

(5) In particular, and without prejudice to the generality of the foregoing, the Secretary of State may exercise his said powers if it appears to him that:

(a) the person to whom the permit was granted has committed a breach of any condition to which it is subject;

(b) any agreement between Her Majesty's Government in the United Kingdom and the Government of any other country in pursuance of which or in reliance on which the permit was granted is no longer in force or that that other Government has committed a breach thereof;

(c) it appears to him that the person to whom the permit was granted, or a Government or another country which is a party to such an agreement referred to in sub-paragraph (b), or the aeronautical authorities of the country concerned have:

(i) acted in a manner which is inconsistent with or prejudicial to the operation in good faith, and according to its object and purpose, of any such agreement as aforesaid, or

(ii) have engaged in unfair, discriminatory or restrictive practices to the prejudice of the holder of an Air Transport Licence granted under section 65 of the Civil Aviation Act 1982 or the holder of a route licence granted under that section as applied by section 69A of that Act in his operation of air services to or from points in the country concerned; and

(d) the person to whom the permit was granted, having been granted it as a person designated by the Government of a country other than the United Kingdom for the purposes of any such agreement referred to in sub-paragraph (b), is no longer so designated or that that person has so conducted himself, or that such circumstances have arisen in relation to him, as to make it necessary or expedient to disregard or qualify the consequences of his being so designated.

(6) The permit-holder or any person having the possession or custody of any permit which has been revoked, suspended or varied under this article shall surrender it to the Secretary of State within a reasonable time of being required by him to do so.

(7) The breach of any condition subject to which any permit to which this article applies has been granted shall render the permit invalid during the continuance of the breach.

(8) The permits to which this article applies are permissions granted by the Secretary of State under article 138 or article 140 and any approvals or authorisations of, or consents to, any matter which the Secretary of State has granted, or is deemed to have granted, in pursuance of a permission which he has so granted.

(9) References in this article to the 'permit-holder' are references to the person to whom any permit to which this article applies has been granted or is deemed to have been granted.

Article 94: Offences in relation to documents and records

(1) A person shall not with intent to deceive:

(a) use any certificate, licence, approval, permission, exemption or other document issued or required by or under this Order or by or under Part 21, 66, 145, 147 or M which has been forged, altered, revoked or suspended, or to which he is not entitled

(b) lend any certificate, licence, approval, permission, exemption or any other document issued or having effect or required by or under this Order or by or under Part 21, 66, 145, 147 or M to, or allow it to be used by, any other person; or

(c) make any false representation for the purpose of procuring for himself or any other person the grant, issue, renewal or variation of any such certificate, licence, approval, permission or exemption or other document;

and in this paragraph a reference to a certificate, licence, approval, permission, exemption or other document includes a copy or purported copy thereof.

(2) A person shall not intentionally damage, alter or render illegible any log book or other record required by or under this Order or by or under Part 21, 66, 145, 147 or M to be maintained or any entry made therein, or knowingly make, or procure or assist in the making of, any false entry in or material omission from any such log book or record or destroy any such log book or record during the period for which it is required under this Order to be preserved.

(3) All entries made in writing in any log book or record referred to in paragraph (2) shall be made in ink or indelible pencil.

(4) A person shall not knowingly make in a load sheet any entry which is incorrect in any material particular, or any material omission from such a load sheet.

(5) A person shall not purport to issue any certificate for the purposes of this Order, of any regulations made thereunder or of Part 21, 66, 146, 147 or M unless he is authorised to do so under this Order or Part 21, 66, 145, 147 or M as the case may be

(6) A person shall not issue any such certificate as aforesaid unless he has satisfied himself that all statements in the certificate are correct.

PART 8 MOVEMENT OF AIRCRAFT

Article 95: Rules of the Air

(1) Without prejudice to any other provision of this Order, the Secretary of State may make regulations (in this article called the 'Rules of the Air') prescribing:

(a) the manner in which aircraft may move or fly including in particular provision for requiring aircraft to give way to military aircraft;

(b) the lights and other signals to be shown or made by aircraft or persons;

(c) the lighting and marking of aerodromes; and

(d) any other provisions for securing the safety of aircraft in flight and in movement and the safety of persons and property on the surface.

(2) Subject to the provisions of paragraph (3), it shall be an offence to contravene, to permit the contravention of, or to fail to comply with, the Rules of the Air.

(3) It shall be lawful for the Rules of the Air to be departed from to the extent necessary:

(a) for avoiding immediate danger;

(b) for complying with the law of any country other than the United Kingdom within which the aircraft then is; or

(c) for complying with Military Flying Regulations (Joint Service Publication 550) or Flying Orders to Contractors (Aviation Publication 67) issued by the Secretary of State in relation to an aircraft of which the commander is acting as such in the course of his duty as a member of any of Her Majesty's naval, military or air forces.

(4) If any departure from the Rules of the Air is made for the purpose of avoiding immediate danger, the commander of the aircraft shall cause written particulars of the departure, and of the circumstances giving rise to it, to be given within 10 days thereafter to the competent authority of the country in whose territory the departure was made or if the departure was made over the high seas, to the CAA.

(5) Nothing in the Rules of the Air shall exonerate any person from the consequences of any neglect in the use of lights or signals or of the neglect of any precautions required by ordinary aviation practice or by the special circumstances of the case.

Article 96: Power to prohibit or restrict flying

(1) Where the Secretary of State deems it necessary in the public interest to restrict or prohibit flying by reason of:

(a) the intended gathering or movement of a large number of persons;

(b) the intended holding of an aircraft race or contest or of a flying display; or

(c) national defence or any other reason affecting the public interest;

the Secretary of State may make regulations prohibiting, restricting or imposing conditions on flights by aircraft specified in paragraph (2) flying in the circumstances specified in paragraph (2).

(2) The aircraft and circumstances referred to in paragraph (1) are:

(a) aircraft, whether or not registered in the United Kingdom, in any airspace over the United Kingdom or in the neighbourhood of an offshore installation; and

(b) aircraft registered in the United Kingdom, in any other airspace, being airspace in respect of which Her Majesty's Government in the United Kingdom has in pursuance of international arrangements undertaken to provide navigation services for aircraft.

(3) Regulations made under this article may apply either generally or in relation to any class of aircraft.

(4) It shall be an offence to contravene or permit the contravention of or fail to comply with any regulations made hereunder.

(5) If the commander of an aircraft becomes aware that the aircraft is flying in contravention of any regulations which have been made for any of the reasons referred to in paragraph (1)(c) he shall, unless otherwise instructed pursuant to paragraph (6), cause the aircraft to leave the area to which the regulations relate by flying to the least possible extent over such area and the aircraft shall not begin to descend while over such an area.

(6) The commander of an aircraft flying either within an area for which regulations have been made for any of the reasons referred to in paragraph (1)(c) or within airspace notified as a Danger Area shall forthwith comply with instructions given by radio by the appropriate air traffic control unit or by, or on behalf of, the person responsible for safety within the relevant airspace.

Article 97: Balloons, kites, airships, gliders and parascending parachutes

(1) The provisions of this article shall apply only to or in relation to aircraft within the United Kingdom.

(2) A balloon in captive or tethered flight shall not be flown within 60 metres of any vessel, vehicle or structure except with the permission of the person in charge of any such vessel, vehicle or structure.

(3) Without the permission of the CAA:

(a) a glider or parascending parachute shall not be launched by winch and cable or by ground tow to a height of more than 60 metres above ground level;

(b) a balloon in captive flight shall not be flown within the aerodrome traffic zone of a notified aerodrome during the notified operating hours of that aerodrome;

(c) a balloon in captive or tethered flight shall not be flown at a height measured to the top of the balloon of more than 60 metres above ground level;

(d) a kite shall not be flown at a height of more than 30 metres above ground level within the aerodrome traffic zone of a notified aerodrome during the notified operating hours of that aerodrome;

(e) a kite shall not be flown at a height of more than 60 metres above ground level; and

(f) a parascending parachute shall not be launched by winch and cable or by ground tow within the aerodrome traffic zone of a notified aerodrome during the notified operating hours of that aerodrome.

(4) An uncontrollable balloon in captive or released flight shall not be flown in airspace notified for the purposes of this paragraph without the permission in writing of the CAA.

(5) A controllable balloon shall not be flown in free controlled flight:
(a) within airspace notified for the purposes of this paragraph; or
(b) within the aerodrome traffic zone of a notified aerodrome during the notified operating hours of that aerodrome; except during the day and in visual meteorological conditions.

(6) A controllable balloon shall not be flown in tethered flight:
(a) within airspace notified for the purposes of this paragraph; or
(b) within the aerodrome traffic zone of a notified aerodrome;
except with the permission of the appropriate air traffic control.

(7) A balloon when in captive flight shall be securely moored and shall not be left unattended unless it is fitted with a device which ensures its automatic deflation if it breaks free of its moorings.

(8) An airship with a capacity exceeding 3000 cubic metres shall not be moored other than at a notified aerodrome except with the permission in writing of the CAA.

(9) An airship with a capacity not exceeding 3000 cubic metres, unless it is moored on a notified aerodrome, shall not be moored:
(a) within 2km of a congested area; or
(b) within the aerodrome traffic zone of a notified aerodrome;
except with the permission in writing of the CAA..

(10) An airship when moored in the open shall be securely moored and shall not be left unattended.

(11) A person shall not cause or permit:
(a) a group of small balloons exceeding 1000 in number to be simultaneously released at a single site wholly or partly within the aerodrome traffic zone of a notified aerodrome during the notified operating hours of that aerodrome unless that person has given to the CAA not less than 28 days previous notice in writing of the release.
(b) A person shall not cause or permit a group of small balloons exceeding 2000 but not exceeding 10000 in number to be simultaneously released at a single site:
(i) within airspace notified for the purposes of this sub-paragraph;or
(ii) within the aerodrome traffic zone of a notified aerodrome during the notified operating hours of that aerodrome without the permission in writing of the CAA.
(c) a group of small balloons greater than 10000 in number to be simultaneously released at a single site except with the permission in writing of the CAA.

(12) For the purposes of this article:
(a) in paragraph (5) "day" means the time from half an hour before sunrise until half an hour after sunset (both times exclusive), sunset and sun rise being determined at surface level;
(b) the "notified operating hours" means the times notified in respect of an aerodrome during which rule 39 of the Rules of the Air Regulations 1996 apples;
(c) simultaneously released at a single site' means the release of a specified number of balloons during a period not exceeding 15 minutes from within an area not exceeding 1km square.

Article 98: Regulation of small aircraft

(1) A person shall not cause or permit any article or animal (whether or not attached to a parachute) to be dropped from a small aircraft so as to endanger persons or property.

(2) The person in charge of a small aircraft which weighs more than 7kg without its fuel but including any articles or equipment installed in or attached to the aircraft at the commencement of its flight shall not fly such an aircraft:
(a) unless the person in charge of the aircraft has reasonably satisfied himself that the flight can safely be made;
(b) in Class A, C, D or E airspace unless the permission of the appropriate air traffic control unit has been obtained;
(c) within an aerodrome traffic zone during the notified hours of watch of the air traffic control unit (if any) at that aerodrome unless the permission of any such air traffic control unit has been obtained;
(d) at a height exceeding 400ft above the surface unless it is flying in airspace described in sub-paragraphs (b) or (c) and in accordance with the requirements thereof; or
(e) for aerial work purposes other than in accordance with a permission issued by the CAA.

Article 99: Regulation of rockets

(1) Subject to paragraph (2), this article applies to:
(a) small rockets of which the total impulse of the motor or combination of motors exceeds 160 Newton-seconds: and
(b) large rockets.

(2) This article shall not apply to
(a) an activity to which the Outer Space Act 1986 applies: or
(b) a military rocket.

(3) No person shall launch a small rocket to which this article applies unless the condition in paragraph (4), and any of the conditions in paragraph (5) which are applicable, are satisfied.

(4) The condition first mentioned in paragraph (3) is that he has reasonably satisfied himself that:
(a) the flight can be safely made: and
(b) the airspace within which the flight will take place is, and will throughout the flight, remain clear of any obstructions including any aircraft in flight.

(5) The conditions mentioned secondly in paragraph (3) are that:

(a) for a flight within controlled airspace, he has obtained the permission of the appropriate air traffic control unit for aircraft flying in that airspace;
(b) for a flight within an aerodrome traffic zone at any of the times specified in Column 2 of the Table in rule 39(1) of the Rules of the Air Regulations 1996.
(i) he has obtained the permission of the air traffic control unit at the aerodrome; or
(ii) where there is no air traffic control unit, he has obtained from the aerodrome flight information service unit at that aerodrome information to enable the flight within the zone to be conducted safely; or
(iii) where there is no air traffic control unit nor aerodrome flight information service unit, he has obtained information from the air/ground radio station at that aerodrome to enable the flight to be conducted safely
(c) a flight for aerial work purposes is carried out under and in accordance with a permission granted by the CAA.
(6) No person shall launch a large rocket unless he does so under and in accordance with a permission granted by the CAA.

PART 9 AIR TRAFFIC SERVICES

Article 100: Requirement for air traffic control approval

(1) No person in charge of the provision of an air traffic control service shall provide such a service in respect of United Kingdom airspace or airspace outside the United Kingdom for which the United Kingdom has, in pursuance of international arrangements, undertaken to provide air navigation services otherwise than under and in accordance with the terms of an air traffic control approval granted to him by the CAA.

(2) The CAA shall grant an air traffic control approval if it is satisfied that the applicant is competent, having regard to his organisation, staffing, equipment, maintenance and other arrangements, to provide a service which is safe for use by aircraft.

Article 101: Duty of person in charge to satisfy himself as to competence of controllers

The holder of an approval under article 100 shall not permit any person to act as an air traffic controller or to act as a student air traffic controller in the provision of the service under the approval unless:
(a) such person holds an appropriate licence; and
(b) the holder has satisfied himself that such person is competent to perform his duties.

Article 102: Manual of Air Traffic Services

A person shall not provide an air traffic control service at any place unless:
(a) the service is provided in accordance with the standards and procedures specified in a manual of air traffic services in respect of that place;
(b) the manual is produced to the CAA within a reasonable time after a request for its production is made by the CAA; and
(c) such amendments or additions have been made to the manual as the CAA may from time to time require.

Article 103: Provision of air traffic services

(1) In the case of an aerodrome (other than a Government aerodrome) in respect of which there is equipment for providing holding aid, let-down aid or approach to landing by radio or radar, the person in charge of the aerodrome shall:
(a) inform the CAA in advance of the periods during and times at which any such equipment is to be in operation for the purpose of providing such aid as is specified by the said person; and
(b) during any period and at such times as are notified, cause an approach control service to be provided.

Article 104: Making of an air traffic direction in the interests of safety

(1) The CAA may, in the interests of safety, direct the person in charge of an aerodrome that there shall be provided in respect of any aerodrome (other than a Government aerodrome) such an air traffic control service, a flight information service or a means of two way radio communication as the CAA considers appropriate.

(2) The CAA may, in the interests of safety, direct the holder of a licence to provide air traffic services granted under Part I of the Transport Act 2000 that there shall be provided, in respect of United Kingdom airspace or airspace outside the United Kingdom for which the United Kingdom has in pursuance of international arrangements undertaken to provide air navigation services, otherwise than in respect of an aerodrome, such an air traffic control service, a flight information service or a means of two way radio communication as the CAA considers appropriate.

(3) The CAA may specify in a direction made under this article the periods during which, the times at which, the manner in which and the airspace within which such service or such means shall be provided.

(4) The person who has been directed shall cause such a service or means to be provided in accordance with any such direction.

(5) A provisional air traffic direction:
(a) may, if it thinks fit, make a provisional air traffic direction in accordance with paragraphs (1) or (2) pending inquiry into or consideration of the case.
(b) shall have effect as though it was an air traffic direction made in accordance with paragraph (1) or (2) as the case may be.

Article 105: Making of a direction for airspace policy purposes

(1) After consultation with the Secretary of State the CAA may direct in accordance with paragraphs (2) and (3) any person in charge of the provision of air traffic services to provide air traffic services in respect of United Kingdom airspace or airspace outside the United Kingdom for which the United Kingdom has undertaken in pursuance of international arrangements to provide air traffic services.

(2) A direction under paragraph (1) may be made:

(a) in the interests of ensuring the efficient use of airspace; or

(b) to require that air traffic services are provided to a standard considered appropriate by the CAA for the airspace classification.

(3) The CAA may specify in a direction under paragraph (1) the air traffic services and the standard to which they are to be provided and the periods during which, the times at which, the manner in which, and the airspace within which such services shall be provided.

(4) The person to whom a direction is given shall cause such a service to be provided in accordance with the direction.

Article 106: Use of radio call signs at aerodromes

The person in charge of an aerodrome provided with means of two-way radio communication shall not cause or permit any call sign to be used for a purpose other than a purpose for which that call sign has been notified.

PART 10 LICENSING OF AIR TRAFFIC CONTROLLERS

Article 107: Prohibition of unlicensed air traffic controllers and student air traffic controllers

(1) Subject to paragraphs (3) and (4), a person shall not act as an air traffic controller or hold himself out, whether by use of a radio call sign or in any other way, as a person who may so act unless he is the holder of, and complies with the privileges and conditions of:

(a) a valid student air traffic controller's licence granted under this order;

(b) an appropriate air traffic controller's license granted under this order; or

(c) a valid air traffic controller's licence so granted which is not appropriate but he is supervised as though he were the holder of a student air traffic controller's license.

(2) A person shall not act as an air traffic controller unless he has identified himself in such a manner as may be notified.

(3) A licence shall not be required by any person who, acting in the course of his employment, passes on such instructions or advice as he has been instructed so to do by the holder of an air traffic controller's licence which entitles that holder to give such instructions or advice.

(4) A licence shall not be required by any person who acts in the course of his duty as a member of any of Her Majesty's naval, military or air forces or a visiting force.

Article 108: Grant and renewal of air traffic controllers and student air traffic controller's licenses

(1) Subject to the provisions of this article the CAA shall grant licences, subject to any conditions it thinks fit, of either of the classes specified in Part A of Schedule 11, authorising an applicant to act as an air traffic controller or a student air traffic controller in the United Kingdom.

(2) Before granting a licence the CAA must be satisfied that the applicant is:

(a) a fit person to act asin the capacity to which the licence relates, and

(b) is qualified by reason of his knowledge, experience, skill and physical and mental fitness to act in the capacity to which the licence relates, for which purpose he shall furnish such evidence and undergo such examinations, assessments and tests (including in particular medical examinations) and undertake such courses of training as the CAA may require of him.

(3) Such a licence:

(a) shall, subject to article 92, remain in force for the period indicated therein, not exceeding any period specified in Schedule 11for that licence.

(b) shall, if no period is indicated in the licence it shall remain in force, subject to article 92, for the lifetime of the holder.

(c) may be renewed by the CAA from time to time upon its being satisfied that the applicant is a fit person and qualified as specified in paragraph (2);.

(d) shall not be granted to any person who is under the minimum age specified for that licence in Part A of Schedule 11.

(e) shall not be valid unless the holder has signed it in ink with his ordinary signature.

(4) The CAA may include in an air traffic controller's licence, subject to such conditions as it thinks fit, any of the ratings and endorsements specified in Part B of Schedule 11, upon its being satisfied that the applicant is qualified as specified in paragraph (2) (b) to act in the capacity to which the rating or endorsement relates, and such rating or endorsement shall be deemed to form part of the licence.

(5) The holder of an air traffic controller's or a student air traffic controller's licence shall, upon such occasions as the CAA may require, submit himself for such examinations, assessments and tests (Including in particular medical examinations) and furnish such evidence as to his knowledge, experience, competence and skill and undergo such courses of training, as the CAA may require.

Article 109: Privileges of an air traffic controller's licence and a student air traffic controller's licence

(1) Subject to article 110, and to any conditions of the licence granted under article 108, an air traffic controller's licence shall entitle the holder to:
(a) exercise the privileges specified in paragraph 1(3) of Part A of Schedule 11; and
(b) exercise the privileges of any rating or endorsement included in the licence as specified in Part B of the said Schedule.
(2) Subject to article 110 and to any conditions of the licence granted under article 1-8 a student air traffic controller's licence shall entitle the holder to exercise the privileges specified in paragraph 2(3) of Part A of Schedule 11.

Article 110: Maintenance of validity of ratings and endorsements

(1) The holder of an air traffic controller's licence shall not be entitled to exercise the privileges of a rating or endorsement contained in such a licence unless the licence includes a curent unit licence endorsement specifying that the rating or endorsement is valid for:
(a) the aerodrome or place at which he so act;
(b) the sector on which or the operational position at which he so acts;
(c) the surveillance equipment (if any) with which, he so acts.
(2) A unit licence endorsement may be entered in a licence either by the CAA or by the holder of an air traffic controller licence which includes an examiner licence endorsement.

Article 111: Obligation to notify rating ceasing to be valid and change of unit

(1) Subject to paragraph (2), when a rating ceases to be valid for a sector or operational position the holder of the licence shall forthwith inform the person who is approved pursuant to article 100 to provide an air traffic control service for that sector or operational position to that effect.
(2) When a rating ceases to be valid for a sector or operational position and is not valid for any other sector or operational position the holder of the licence shall notify the CAA and forward the licence to the CAA, for the purpose, who shall endorse the licence accordingly and return it to the holder.
(3) Whenever a person ceases to act as an air traffic controller at a particular unit he shall notify the CAA and forward the licence to the CAA, or a person approved by the CAA for the purpose, who shall endorse the licence accordingly and return it to the holder.

Article 112: Requirement for medical certificate

(1) On the basis of the medical examination referred to in article 108(2) (b) and (5), the CAA, or any person approved by it as competent to do so, may issue a medical certificate subject to such conditions as it or he thinks fit to the effect that the holder of a licence has been assessed as fit to perform the functions to which the licence relates.
(2) The certificate shall be deemed to form part of the licence.
(3) The holder of an air traffic controller's licence shall not act as an air traffic controller unless his licence includes a medical certificate issued and in force under paragraph (1).
(4) The holder of a student air traffic controller's licence shall not act as a student air traffic controller unless his licence includes a medical certificate issued and in force under paragraph (1).

Article 113: Appropriate licence

An air traffic controller's licence shall not be an appropriate licence for the purposes of this Part of this Order unless it includes valid ratings, endorsements and certificates which authorise the holder to provide, at the aerodrome or place, the type of air traffic control service for the sector on which or the operational position for which it is being provided and with the type of surveillance equipment being used (if any).

Article 114: Incapacity of air traffic controllers

(1) Every holder of an air traffic controller's licence granted under article 108 who:
(a) suffers any personal injury or illness involving incapacity to undertake the functions to which his licence relates throughout a period of 20 consecutive days; or
(b) in the case of a woman, has reason to believe that she is pregnant;
shall inform the CAA in writing of such injury, illness or pregnancy as soon as possible.
(2) An air traffic controller's medical certificate shall cease to be in force on the expiry of the period of injury or illness referred to in paragraph (1)(a) and shall come into force again (provided it has not expires):
(a) upon the holder being medically examined under arrangements made by the CAA and pronounced fit to resume his functions under the licence; or
(b) upon the CAA exempting the holder from the requirement of a medical examination subject to such conditions as the CAA may think fit.

Article 115: Fatigue of air traffic controllers – air traffic controllers' responsibilities

A person shall not act as an air traffic controller if he knows or suspects that he is suffering from or, having regard to the circumstances of the period of duty to be undertaken, is likely to suffer from, such fatigue as may endanger the safety of any aircraft to which an air traffic control service may be provided.

Article 116: Prohibition of acting under the influence of drink or a drug

A person shall not act as an air traffic controller or a student air traffic controller whilst under the influence of drink or a drug to such an extent as to impair his capacity to act as such.

Article 117: Failing exams

A person who, on the last occasion when he was examined, assessed or tested for the purposes of this Part, failed the examination, assessment or test shall not be entitled to act in the capacity for which that examination, assessment or test would have qualified him had he passed it.

Article 118: Use of simulators
No part of any examination, assessment or test undertaken for the purposes of this Part or Schedule 11 or any training which has been approved under article 119 shall be undertaken in a simulator unless that simulator has been approved by the CAA.

Article 119: Approval of courses, persons and simulators
Without prejudice to any other provision of this order the CAA may for the purposes of this Part:
(a) approve any course of training or instruction;
(b) authorise a person to conduct such examinations, assessments or tests as it may specify;
(c) approve a person to provide any course of training or instruction; and
(d) approve a simulator.

Article 120: Acting as an air traffic controller etc
(1) For the purposes of this Part and Schedule 11.
(a) " to act as an air traffic controller" shall mean either:
(i) giving an air traffic control service; or
(ii) the supervision of a student ar traffic controller;
or both; and
(b) "acting as a student air traffic controller" shall mean acting as an air traffic controller under the supervision of an air traffic controller.

PART 11 FLIGHT INFORMATION SERVICES AND LICENSING OF FLIGHT INFORMATION SERVICE OFFICERS

Article 121 Prohibition of unlicensed flight information service officers
(1) A person shall not act as a flight information service officer at any aerodrome or area control centre or hold himself out, whether by use of a radio call sign or in any other way, as a person who may so act unless he if the holder of and complies with the terms of a flight information service officer's license granted under this Ordr authorising him to act as such at that aerodrome or area control centre.

(2) A person shall not act as a flight information service officer unless he has identified himself in such a manner as may be notified.

(3) For the purposes of this Part and Schedule 11 "acting as a flight information service officer" shall mean giving a flight information service.

Article 122 Licensing of flight information officers
(1) The CAA shall grant a license subject to such conditions as it thinks fit to any person aged 18 years or more to act as a flight information service officer upon its being satisfied that the applicant is a fit person to hold the license and is qualified by reason of his knowledge, experience, competence, skill and physical and mental fitness so to act, and for that purpose the applicant shall furnish such evidence and undergo such examinations and tests and undertake such courses of training as the CAA may require him.

(2) A license to act as a flight information officer:
(a) may be renewed by the CAA from time to time, upon being satisfied that the applicant is a fit person and is qualified as aforesaid;
(b) shall remain in force, subject to article 92, for the period indicated in the license or if no period is indicarted, for the lifetime of the holder.

(3) A flight information service officer's license shall not authorise the giving of a flight information service at an aerodrome or area control centre unless that aerodrome or area control centre has been specified in the license by a person authorised by the CAA for the purpose and the licence has been validated in respect of that aerodrome or area control centre by a person authorised for the purpose by the CAA.

(4) If, throughout any period of 90 days the holder of the license has not at any time given such a service at a particular aerodrome or area control centre, the license shall cease to be valid for that aerodrome or area control centre at the end of that period until the licence has been revalidated in respect of that aerodrome or area control centre by a person authorised by the CAA for the purpose.

(5) A license to act as a flight information service officer shall not be valid unless the holder of the license has signed his name thereon in ink or indelible pencil with his ordinary signature.

(6) Every holder of a flight information service officer's license shall upon such occasions as the CAA may require, submit himself to such examinations and tests and furnish such evidence as to his knowledge, experience, competence and skill and undergo such courses of training as the CAA may require;

Article 123 Flight information service manual
A person shall not provide a flight information service at any aerodrome or area control centre unless:
(a) the service pis provided in accordance with the standards and procedures specified in a flight information service manual in respect of that aerodrome or area control centre;
(b) the manual is produced to the CAA within a reasonable time after a request for its production is made by the CAA; and
(c) such amendments or additions have been made to the manual as the CAA may from time to time require.

PART 12 AIR TRAFFIC SERVICE EQUIPMENT

Article 124: Air traffic service equipment

(1) A person shall not cause or permit any air traffic service equipment to be established or used in the United Kingdom otherwise than under and in accordance with an approval granted by the CAA to the person in charge of the equipment.

(2) An approval shall be granted pursuant to paragraph (1) upon the CAA being satisfied:
(a) as to the intended purpose of the equipment;
(b) that the equipment is fit for its intended purpose; and
(c) that the person is competent to operate the equipment.

(3) The person in charge of an aeronautical radio station at an aerodrome for which a licence for public use has been granted shall cause to be notified in relation to that aeronautical radio station the type and availability of operation of any service which is available for use by any aircraft.

(4) An approval granted pursuant to paragraph (1) may include a condition requiring a person in charge of an aeronautical radio station at any other aerodrome or place to cause the information specified in paragraph (3) to be notified.

(5) An approval granted under paragraph (1) may in addition to any other conditions which may be imposed include a condition requiring the person in charge of the equipment to use a person approved by the CAA under paragraph (6) for the provision of particular services in connection with the equipment and in particular but without limitation may include a condition requiring that the equipment be flight checked by such an approved person.

(6) The CAA may approve a person to provide particular services in connection with approved equipment.

(7) For the purpose of paragraphs (1) and (6) an approval may be granted in respect of one or more than one person or generally.

(8) The provisions of this article shall not apply in respect of any air traffic service equipment of which the person solely in charge is the Secretary of State.

Article 125: Air traffic service equipment records

(1) The person in charge of any air traffic service equipment and any associated apparatus required under paragraph (2) or (3) shall keep in respect of such equipment or apparatus records in accordance with Part A of Schedule 12, and shall preserve such records for a period of one year or such longer period as the CAA may in a particular case direct.

(2) The person in charge of an aeronautical radio station which is used for the provision of an air traffic control service by an air traffic control unit shall provide recording apparatus in accordance with paragraph (4)

(3) The CAA may direct the person in charge of any other air traffic service equipment to provide recording apparatus in accordance with paragraph (4).

(4) The person in charge of the air traffic service equipment in respect of which recording apparatus is required to be provided under paragraph (2) or (3) shall, subject to paragraph (7):
(a) ensure that when operated the apparatus is capable of recording and replaying the terms or content of any message or signal transmitted or received by or through that equipment or in the case of an aeronautical radio station is capable of recording and replaying the terms or content of any voice radio message or signal transmitted to an aircraft either alone or in common with other aircraft or received from an aircraft by the air traffic control unit;
(b) ensure that the apparatus is in operation at all times when the equipment is being used in connection with the provision of a service provided for the purpose of facilitating the navigation of the aircraft;
(c) ensure that each record made by the apparatus complies with Part B of Schedule 12;
(d) not cause or permit that apparatus to be used unless it is approved by the CAA; and
(e) comply with the terms of such an approval.

(5) The CAA may in considering whether or not to grant an approval, without limitation, have regard to the matters specified in Part C of Schedule 12.

(6) An approval may be granted:
(a) in addition to any other conditions which are imposed, subject to conditions relating to the matters to which the CAA may have had regard under paragraph (5);
(b) in respect of one or more than one person or generally.

(5) If any apparatus provided in compliance with paragraph (2) ceases to be capable of recording the matters required by this article to be included in the records, the person required to provide that apparatus shall ensure that, so far as practicable, a record is kept which complies with Part B of Schedule 15 and on which the particulars specified therein are recorded together with, in the case of apparatus provided in compliance with paragraph (2)(a), a summary of voice communications exchanged between the aeronautical radio station and any aircraft.

(7) If any apparatus provided in compliance with paragraph (2) or (3) ceases to be capable of recording the matters required by this article to be included in the records, the person required to provide that apparatus shall ensure that, so far as practicable, a record is kept which complies with Part B of Schedule 12 and on which the particulars specified therein are recorded together with, in the case of apparatus provided in compliance with paragraph (2), a summary of voice communications exchanged between the aeronautical radio station and any aircraft.

(8) If any apparatus provided in compliance with paragraph (2) or (3) becomes unserviceable, the person in charge of the air traffic service equipment shall ensure that the apparatus is rendered serviceable again as soon as reasonably practicable.

(9) The person in charge of any air traffic service equipment shall preserve any record made in compliance with paragraphs (4) or (7) for a period of 30 days from the date on which the terms or content of the message or signal were recorded or for such longer period as the CAA may in a particular case direct.

(10) Subject to paragraph (11), a person required by this article to preserve any record by reason of his being the person in charge of the air traffic service equipment shall, if he ceases to be such a person, continue to preserve the record as if he had not ceased to be such a person, and in the event of his death the duty to preserve the record shall fall upon his personal representative.

(11) If another person becomes the person in charge of the air traffic service equipment the previous person in charge or his personal representative shall deliver the record to that other person on demand, and it shall be the duty of that other person to deal with the record delivered to him as if he was that previous person in charge.

(12) The person in charge of any air traffic service equipment shall within a reasonable time after being requested to do so by an authorised person, produce any record required to be preserved under this article to that authorised person.

(13) The provisions of this article shall not apply in respect of any air traffic service equipment of which the person solely in charge is the Secretary of State.

PART 13 AERODROMES, AERONAUTICAL LIGHTS AND DANGEROUS LIGHTS

Article 126: Aerodromes – public transport of passengers and instruction in flying

(1) An aircraft to which this paragraph applies shall not take-off or land at a place in the United Kingdom other than:

(a) an aerodrome licensed under this Order for the take-off and landing of such aircraft; or

(b) a Government aerodrome, notified as available for the take-off and landing of such aircraft, or in respect of which the person in charge of the aerodrome has given his permission for the particular aircraft to take-off or land as the case may be;

and in accordance with any conditions subject to which the aerodrome may have been licensed or notified, or subject to which such permission may have been given.

(2) Subject to paragraph (4), paragraph (1) applies to:

(a) any aeroplane of which the maximum total weight authorised exceeds 2730kg flying on a flight::

(i) for the purpose of the public transport of passengers;

(ii) for the purpose of instruction in flying given to any person for the purpose of becoming qualified for the grant of a pilot's licence or the inclusion of an aircraft rating, a night rating or a night qualification in a licence; or

(iii) for the purpose of carrying out flying tests in respect of the grant of a pilot's licence or the inclusion of an aircraft rating or a night rating in a licence;

(b) any aeroplane of which the maximum total weight authorised does not exceed 2730kg flying on a flight:

(i) which is a scheduled journey for the purpose of the public transport of passengers;

(ii) for the purpose of the public transport of passengers beginning and ending at the same aerodrome;

(iii) for the purpose of:

(aa) instruction in flying given to any person for the purpose of becoming qualified for the grant of a pilot's licence or the inclusion of an aircraft rating, a night rating or a night qualification in a licence; or

(bb) a flying test in respect of the grant of a pilot's licence or the inclusion of an aircraft rating, a night rating or a night qualification in a licence; or

(iv) flights for the purpose of the public transport of passengers at night;

(c) any helicopter or gyroplane engaged flying on a flight specified in sub-paragraphs (b)(i) and (iii); and

(d) any glider (other than a glider being flown under arrangements made by a flying club and carrying no person other than a member of the club) flying on a flight for the purpose of the public transport of passengers or for the purpose of instruction in flying.

(3) Subject to paragraph (4):

(a) the person in charge of any area in the United Kingdom intended to be used for the take off or landing of helicopters at night other than such a place as is specified in paragraph (1) shall cause to be in operation, whenever a helicopter flying for the purpose of the public transport of passengers is taking off or landing at that area by night, such lighting as will enable the pilot of the helicopter:

(i) in the case of landing, to identify the landing area in flight, to determine the landing direction and to make a safe approach and landing; and

(ii) in the case of taking off, to make a safe take-off.

(b) a helicopter flying for the purpose of the public transport of passengers at night shall not take-off or land at a place to which sub-paragraph (a) applies unless there is in operation such lighting.

(4) Paragraph (1) shall not apply to or in relation to an aircraft flying under and in accordance with the terms of a police air operator's certificate.

Article 127: Use of Government aerodromes

With the concurrence of the Secretary of State, the CAA may cause to be notified subject to such conditions as it thinks fit, any Government aerodrome; as an aerodrome available for the take-off and landing of aircraft engaged on flights for the purpose of the public transport of passengers or for the purpose of instruction in flying or of any classes of such aircraft.

Article 128: Licensing of aerodromes

(1) The CAA shall grant a licence in respect of any aerodrome in the United Kingdom if it is satisfied that:

(a) the applicant is competent, having regard to his previous conduct and experience, his equipment, organisation, staffing, maintenance and other arrangements, to secure that the aerodrome and the airspace within which its visual traffic pattern is normally contained are safe for use by aircraft;

(b) the aerodrome is safe for use by aircraft, having regard in particular to the physical characteristics of the aerodrome and of its surroundings; and

(c) the aerodrome manual submitted pursuant to paragraph (6) is adequate.

(2) If the applicant so requests or if the CAA considers that an aerodrome should be available for the take-off or landing of aircraft to all persons on equal terms and conditions, it may grant a licence (in this Order referred to as 'a licence for public use') which in addition to any other conditions which it may impose shall be subject to the condition that the aerodrome shall at all times when it is available for the take-off or landing of aircraft be so available to all persons on equal terms and conditions.

(3) The holder of an aerodrome licence granted under this Order (in this article called 'an aerodrome licence holder') shall:
(a) furnish to any person on request information concerning the terms of the licence; and
(b) in the case of a licence for public use, cause to be notified the times during which the aerodrome will be available for the take-off or landing of aircraft engaged on flights for the purpose of the public transport of passengers or instruction in flying.

(4) An aerodrome licence holder shall not contravene or cause or permit to be contravened any condition of the aerodrome licence at any time in relation to an aircraft flying on a specified flight in article 126(2), but the licence shall not cease to be valid by reason only of such a contravention.

(5) An aerodrome licence holder shall take all reasonable steps to secure that the aerodrome and the airspace within which its visual traffic pattern is normally contained are safe at all times for use by aircraft.

(6) Upon making an application for an aerodrome licence the applicant shall submit to the CAA an aerodrome manual for that aerodrome.

(7) An aerodrome manual required pursuant to this article shall contain all such information and instructions as may be necessary to enable the aerodrome operating staff to perform their duties as such including, in particular, information and instructions relating to the matters specified in Schedule 13.

(8) Every aerodrome licence holder shall:
(a) furnish to the CAA any amendments or additions to the aerodrome manual before or immediately after they come into effect;
(b) without prejudice to sub-paragraph (a), make such amendments or additions to the aerodrome manual as the CAA may require for the purpose of ensuring the safe operation of aircraft at the aerodrome or the safety of air navigation; and
(c) maintain the aerodrome manual and make such amendments as may be necessary for the purposes of keeping its contents up to date.

(9) Every aerodrome licence holder shall make available to each member of the aerodrome operating staff a copy of the aerodrome manual, or a copy of every part of the aerodrome manual which is relevant to his duties and shall ensure that each such copy is kept up to date.

(10) Every aerodrome licence holder shall take all reasonable steps to secure that each member of the aerodrome operating staff:
(a) is aware of the contents of every part of the aerodrome manual which is relevant to his duties as such; and
(b) undertakes his duties as such in conformity with the relevant provisions of the manual.

(11) For the purposes of this article:
(a) 'aerodrome operating staff' means all persons, whether or not the aerodrome licence holder and whether or not employed by the aerodrome licence holder, whose duties are concerned either with ensuring that the aerodrome and airspace within which its visual traffic pattern is normally contained are safe for use by aircraft, or whose duties require them to have access to the aerodrome manoeuvring area or apron;
(b) 'visual traffic pattern' means the aerodrome traffic zone of the aerodrome, or, in the case of an aerodrome which is not notified for the purposes of rule 39 of the Rules of the Air Regulations 1996, the airspace which would comprise the aerodrome traffic zone of the aerodrome if it was so notified.

Article 129: Charges at aerodromes licensed for public use

The licensee of any aerodrome in respect of which a licence for public use has been granted shall, when required by the Secretary of State, furnish to the Secretary of State such particulars as he may require of the charges established by the licensee for the use of the aerodrome or of any facilities provided at the aerodrome for the safety, efficiency or regularity of air navigation.

Article 130: Use of aerodromes by aircraft of Contracting States and of the Commonwealth

The person in charge of any aerodrome in the United Kingdom which is open to public use by aircraft registered in the United Kingdom (whether or not the aerodrome is a licensed aerodrome) shall cause the aerodrome, and all air navigation facilities, to be available for use by aircraft registered in other Contracting States or in any part of the Commonwealth on the same terms and conditions as for use by aircraft registered in the United Kingdom.

Article 131: Noise and vibration caused by aircraft on aerodromes

(1) The Secretary of State may prescribe the conditions under which noise and vibration may be caused by aircraft (including military aircraft) on Government aerodromes, licensed aerodromes or on aerodromes at which the manufacture, repair or maintenance of aircraft is carried out by persons carrying on business as manufacturers or repairers of aircraft.

(2) Section 77(2) of the Civil Aviation Act 1982 shall apply to any aerodrome in relation to which the Secretary of State has prescribed conditions in accordance with paragraph (1).

Article 132: Aeronautical lights

(1) Except with the permission of the CAA and in accordance with any conditions subject to which the permission may be granted, a person shall not establish, maintain or alter the character of:
(a) an aeronautical beacon within the United Kingdom; or
(b) any aeronautical ground light (other than an aeronautical beacon) at an aerodrome licensed under this Order, or which forms part of the lighting system for use by aircraft taking off from or landing at such an aerodrome.

(2) In the case of an aeronautical beacon which is or may be visible from the waters within an area of a general lighthouse authority, the CAA shall not give its permission for the purpose of this article except with the consent of that authority; or

(3) A person shall not intentionally or negligently damage or interfere with any aeronautical ground light established by or with the permission of the CAA.

Article 133: Lighting of en-route obstacles

(1) For the purposes of this article, an "en-route obstacle" means any building. structure or erection which is 150 metres or more above ground level but it does not include a building, structure or erection:
(a) which is in the vicinity of a licensed aerodrome; and
(b) to which section 47 of the Civil Aviation Act 1982 applies.

(2) The person in charge of an en-route obstacle shall ensure that it is fitted with medium intensity steady red lights positioned as close as possible to the top of the obstacle and at intermediate levels spaced so far as practicable equally between the top lights and ground level with an interval not exceeding 52 metres.

(3) Subject to paragraph (4), the person in charge of an en-route obstacle shall ensure that, by night, the lights required to be fitted by this article shall be displayed.

(4) In the event of the failure of any light which is required by this article to be displayed by night the person in charge shall repair or replace the light as soon as is reasonably practicable.

(5) At each level on the obstacle where lights are required to be fitted, sufficient lights shall be fitted and arranged so as to show when displayed in all directions.

(6) In any particular case the CAA may direct that an en-route obstacle shall be fitted with and shall display such additional lights in such positions and at such times as it may specify.

(7) This article shall not apply to any en-route obstacle in respect of which the CAA has granted permission for the purposes of this article to the person in charge.

(8) A permission may be granted for the purposes of this article in respect of a particular case or class of cases or generally.

Article 134 Lighting of Wind Turbine Generators in United Kingdom Territorial Waters

(1) This article shall apply to any wind turbine generator which is situated in waters within or adjacent to the United Kingdom up to the seaward limits of the territorial sea and the height of which is 60 metres or more above the level of the sea at the highest astronomical tide.

(2) Subject to paragraph (3) the person in charge of a wind turbine generator to which this article applies shall ensure that it is fitted with at least one medium intensity steady red light positioned as close as reasonably practicable to the top of the fixed structure.

(3) Where four or more wind turbine generators to which this article applies are located together in the same group, with the permission of the CAA only those on the periphery of the group need be fitted with a light in accordance with paragraph (2).

(4) The light or lights required by paragraph (2) shall, subject to paragraph (5), be so fitted as to show when displayed in all directions without interruption.

(5) When displayed:
(a) the angle of the plane of the beam of peak intensity emitted by the light shall be elevated to between 3 and 4 degrees above the horizontal plane;
(b) not more than 45% or less than 20% of the minimum peak intensity specified for a light of this type shall be visible at the horizontal plane;
(c) not more than 10% of the minimum peak intensity specified for a light of this type shall be visible at a depression of 1.5 degrees or more below the horizontal plane.

(6) The person in charge of a wind turbine generator to which this article applies shall:
(a) subject to sub-paragraph (b) ensure that by night, any light required to be fitted by this article shall be displayed;
(b) in the event of the failure of the light which is required by this article to be displayed by night, repair or replace the light as soon as is reasonably practicable.

(7) When visibility in all directions from every wind turbine generator to which this article applies in a group is more than 5 km the light intensity for any light required by this article to be fitted to any generator in the group and displayed may be reduced to not less than 10% of the minimum peak intensity specified for a light of this type.

(8) In any particular case the CAA may direct that a wind turbine generator to which this article applies shall be fitted with and shall display such additional lights in such positions and at such times as it may specify.

(9) This article shall not apply to any wind turbine generator in respect of which the CAA has granted a permission for the purposes of this article to the person in charge.

(10) A permission may be granted for the purposes of this article in respect of a particular case or class of cases or generally.

(11) In this article:
(a) "wind turbine generator" is a generating station which is wholly or mainly driven by wind;
(b) the height of a wind turbine generator is the height of the fixed structure or if greater the maximum vertical extent of any blade attached to that structure; and
(c) a wind turbine generator is in the same group as another wind turbine generator if the same person is in charge of both and:
(i) it is within 2 km of that other wind turbine generator; or
(ii) it is within 2 km of a wind turbine generator which is in the same group as that other wind turbine generator.

Article 135: Dangerous lights

(1) A person shall not exhibit in the United Kingdom any light which:
(a) by reason of its glare is liable to endanger aircraft taking off from or landing at an aerodrome; or
(b) by reason of its liability to be mistaken for an aeronautical ground light is liable to endanger aircraft.
(2) If any light which appears to the CAA to be such a light as aforesaid is exhibited the CAA may cause a notice to be served upon the person who is the occupier of the place where the light is exhibited or has charge of the light, directing that person, within a reasonable time to be specified in the notice, to take such steps as may be specified in the notice for extinguishing or screening the light and for preventing for the future the exhibition of any other light which may similarly endanger aircraft.
(3) The notice may be served either personally or by post, or by affixing it in some conspicuous place near to the light to which it relates.
(4) In the case of a light which is or may be visible from any waters within the area of a general lighthouse authority, the power of the CAA under this article shall not be exercised except with the consent of that authority.

Article 136: Customs and Excise airports

(1) The Secretary of State may, with the concurrence of the Commissioners of Revenue and Customs and subject to such conditions as they may think fit, by order designate any aerodrome to be a place for the landing or departure of aircraft for the purpose of the enactments for the time being in force relating to customs and excise.
(2) The Secretary of State may, with the concurrence of the Commissioners of Revenue and Customs by order revoke any designation so made.

Article 137: Aviation fuel at aerodromes

(1) Subject to paragraph (2), a person who has the management of any aviation fuel installation on an aerodrome in the United Kingdom shall not cause or permit any fuel to be delivered to that installation or from it to an aircraft unless:
(a) when the aviation fuel is delivered into the installation he is satisfied that:
(i) the installation is capable of storing and dispensing the fuel so as not to render it unfit for use in aircraft;
(ii) the installation is marked in a manner appropriate to the grade of fuel stored or if different grades are stored in different parts each part is so marked; and
(iii) in the case of delivery into the installation or part thereof from a vehicle or vessel, the fuel has been sampled and is of a grade appropriate to that installation or that part of the installation as the case may be and is fit for use in aircraft;
(b) when any aviation fuel is dispensed from the installation he is satisfied as the result of sampling that the fuel is fit for use in aircraft.
(2) Paragraph (1) shall not apply in respect of fuel which has been removed from an aircraft and is intended for use in another aircraft operated by the same operator as the aircraft from which it has been removed.
(3) A person to whom paragraph (1) applies shall keep a written record in respect of each installation of which he has the management, which record shall include:
(a) particulars of the grade and quantity of aviation fuel delivered and the date of delivery;
(b) particulars of all samples taken of the aviation fuel and of the results of tests of those samples; and
(c) particulars of the maintenance and cleaning of the installation;
and he shall preserve the written record for a period of 12 months or such longer period as the CAA may in a particular case direct and shall, within a reasonable time after being requested to do so by an authorised person, produce such record to that person.
(4) A person shall not cause or permit any aviation fuel to be dispensed for use in an aircraft if he knows or has reason to believe that the aviation fuel is not fit for use in aircraft.
(5) If it appears to the CAA or an authorised person that any aviation fuel is intended or likely to be delivered in contravention of any provision of this article, the CAA or that authorised person may direct the person having the management of the installation not to permit aviation fuel to be dispensed from that installation until the direction has been revoked by the CAA or by an authorised person.
(6) For the purpose of this article:
(a) 'aviation fuel' means fuel intended for use in aircraft; and
(b) 'aviation fuel installation' means any apparatus or container, including a vehicle, designed, manufactured or adapted for the storage of aviation fuel or for the delivery of such fuel to an aircraft.

PART 14 GENERAL

Article 138: Restriction on carriage for valuable consideration in aircraft registered elsewhere than in the United Kingdom

(1) An aircraft registered in a Contracting State other than the United Kingdom, or in a foreign country, shall not take on board or discharge any passengers or cargo in the United Kingdom where valuable consideration is given or promised in respect of the carriage of such persons or cargo unless:
(a) it does so with the permission of the Secretary of State granted under this article to the operator or the charterer of the aircraft or to the Government of the country in which the aircraft is registered, and in accordance with any conditions to which such permission may be subject; or
(b) it is exercising traffic rights permitted by virtue of Council Regulation 2408/92 on access for Community air carriers to intra-community air routes (as that Regulation has effect in accordance with the EEA Agreement as amended by the Decision of the EEA Joint Committee No. 7/94 of 21st March 1994).
(2) Without prejudice to article 93 or of paragraph (1), any breach by a person to whom a permission has been granted under this article of any condition to which that permission was subject shall constitute a contravention of this article.

Article 139: Filing and approval of tariffs

(1) Where a permission granted under article 138(1) contains a tariff provision, the operator or charterer of the aircraft concerned shall file with the CAA the tariff which it proposes to apply on flights to which the said permission relates and the CAA shall consider the proposed tariff and may, if it thinks fit, approve or disapprove it.

(2) For the purposes of this article, 'tariff provision' means a condition as to any of the following matters:

(a) the price to be charged for the carriage of passengers, baggage or cargo on flights to which a permission granted under article 138(1) relates;

(b) any additional goods, services or other benefits to be provided in connection with such carriage;

(c) the prices, if any, to be charged for any such additional goods, services or benefits; and

(d) the commission, or rates of commission, to be paid in relation to the carriage of passengers, baggage or cargo; and includes any condition as to the applicability of any such price, the provision of any such goods, services or benefits or the payment of any such commission or of commission at any such rate.

(3) The CAA shall act on behalf of the Crown in performing the functions conferred on it by this article.

Article 140: Restriction on aerial photography, aerial survey and aerial work in aircraft registered elsewhere than in the United Kingdom

(1) An aircraft registered in a Contracting State other than the United Kingdom, or in a foreign country, shall not fly over the United Kingdom for the purpose of aerial photography or aerial survey (whether or not valuable consideration is given or promised in respect of the flight or the purpose of the flight) or for the purpose of any other form of aerial work except with the permission of the Secretary of State granted under this article to the operator or the charterer of the aircraft and in accordance with any conditions to which such permission may be subject.

(2) Without prejudice to article 93 or to paragraph (1), any breach by a person to whom a permission has been granted under this article of any condition to which that permission was subject shall constitute a contravention of this article.

Article 141: Flights over any foreign country

(1) The operator and the commander of an aircraft registered in the United Kingdom (or, if the operator's principal place of business or permanent residence is in the United Kingdom, any other aircraft) which is being flown over any foreign country shall not allow that aircraft to be used for a purpose which is prejudicial to the security, public order or public health of, or to the safety of air navigation in relation to, that country.

(2) A person does not contravene paragraph (1) if he neither knew nor suspected that the aircraft was being or was to be used for a purpose referred to in paragraph (1).

(3) The operator or commander of an aircraft registered in the United Kingdom (or, if the operator's principal place of business or permanent residence is in the United Kingdom, any other aircraft) which is being flown over any foreign country shall comply with any directions given by the appropriate aeronautical authorities of that country whenever:

(a) the flight has not been duly authorised; or

(b) there are reasonable grounds for the appropriate aeronautical authorities to believe that the aircraft is being or will be used for a purpose which is prejudicial to the security, public order or public health of, or to the safety of air navigation in relation to, that country;

unless the lives of persons on board or the safety of the aircraft would thereby be endangered.

(4) A person does not contravene paragraph (3) if he neither knew nor suspected that directions were being given by the appropriate aeronautical authorities.

(5) The requirement in paragraph (3) is without prejudice to any other requirement to comply with directions of an aeronautical authority.

(6) In this article 'appropriate aeronautical authorities' includes any person, whether a member of a country's military or civil authorities, authorised under the law of the foreign country to issue directions to aircraft flying over that country.

Article 142: Mandatory reporting of occurrences

(1) The objective of this article is to contribute to the improvement of air safety by ensuring that relevant information on safety is reported, collected, stored, protected and disseminated.

(2) The sole objective of occurrence reporting is the prevention of accidents and incidents and not to attribute blame or liability.

(3) This article shall apply to occurrences which endanger or which, if not corrected, would endanger an aircraft, its occupants or any other person.

(4) Without prejudice to the generality of paragraph (3), a list of examples of these occurrences is set out in Annexes 1 and 11 (and their Appendices) of Directive 2003/42 of the European Parliament and of the Council of 13th June 2003 on occurrence reporting in civil aviation

(5) Every person listed below shall report to the CAA any event which constitutes an occurrence for the purposes of paragraph (3) and which comes to his attention in the exercise of his functions:

(a) the operator and the commander of a turbine-powered aircraft which has a certificate of airworthiness issued by the CAA;

(b) the operator and the commander of an aircraft operated under an air operator's certificate granted by the CAA;

(c) a person who carries on the business of manufacturing a turbine-powered or a public transport aircraft, or any equipment or part thereof, in the United Kingdom;

(d) a person who carries on the business of maintaining or modifying a turbine powered aircraft, which has a certificate of airworthiness issued by the CAA, and a person who carries on the business of maintaining or modifying any equipment or part of such an aircraft;

(e) a person who carries on the business of maintaining or modifying an aircraft, operated under an air operator's certificate granted by the CAA, and a person who carries on the business of maintaining or modifying any equipment or part of such an aircraft;

(f) a person who signs an airworthiness review certificate, or a certificate of release to service in respect of a turbine powered aircraft, which has a certificate of airworthiness issued by the CAA, and a person who signs an airworthiness review certificate or a certificate of release to service in respect of any equipment or part of such an aircraft;

(g) a person who signs an airworthiness review certificate, or a certificate of release to service in respect of an aircraft, operated under an air operator's certificate granted by the CAA, and a person who signs an airworthiness review certificate or a certificate of release to service in respect of any equipment or part of such an aircraft;

(h) a person who performs a function which requires him to be authorised by the CAA as an air traffic controller or as a flight information service officer;

(i) a licensee and a manager of a licensed aerodrome or a manager of an airport to which Council Regulation (EEC) No. 2408/92 of 23rd July 1992 on access for Community air carriers to intra-Community air routes applies;

(j) a person who performs a function in respect of the installation, modification, maintenance, repair, overhaul, flight checking or inspection of air navigation facilities which are utilised by a person who provides an air traffic control service under an approval issued by the CAA;

(k) a person who performs a function in respect of the ground-handling of aircraft, including fuelling, servicing, loadsheet preparation, loading, de-icing and towing at an airport to which Council Regulation (EEC) No. 2408/92 of 23rd July 1992 on access for Community air carriers to intra-Community air routes applies.

(6) Reports of occurrences shall be made within such time, by such means and containing such information as may be prescribed and shall be presented in such form as the CAA may in any particular case approve.

(7) A person listed in paragraph (5) shall make a report to the CAA within such time, by such means, and containing such information as the CAA may specify in a notice in writing served upon him, being information which is in his possession or control and which relates to an occurrence which has been reported by him or another person to the CAA in accordance with this article.

(8) A person shall not make any report under this article if he knows or has reason to believe that the report is false in any particular.

(9) The CAA shall put in place a mechanism to collect, evaluate, process and store occurrences reported in accordance with paragraphs (5) to (7).

(10) The CAA shall store in its databases the reports which it has collected of occurrences, accidents and serious incidents.

(11) The CAA shall make all relevant safety-related information stored in the databases mentioned in paragraph (10) available to the competent authorities of the other Member States and the Commission.

(12) The CAA shall ensure that the databases referred to in paragraph (10) are compatible with the software developed by the European Commission for the purpose of implementing Directive 2003/42 of the European Parliament and of the Council of 13th June 2003 on occurrence reporting in civil aviation.

(13) The CAA, having received an occurrence report, shall enter it into its databases and notify, whenever necessary: the competent authority of the Member State where the occurrence took place; where the aircraft is registered; where the aircraft was manufactured, and where the operator's air operator's certificate was granted.

(14) The CAA shall provide any entity entrusted with regulating civil aviation safety or with investigating civil aviation accidents and incidents within the Community with access to information on occurrences collected and exchanged in accordance with paragraphs (9) to (13) to enable it to draw the safety lessons from the reported occurrences.

(15) The CAA and the Chief Inspector of Air Accidents shall use any information received in accordance with the terms of this article solely for the purposes set out in this article.

(16) The names or addresses of individual persons shall not be recorded on the databases referred to in paragraph (10).

(17) Without prejudice to the rules of criminal law, no proceedings shall be instituted in respect of unpremeditated or inadvertent infringements of the law which come to the attention of the relevant authorities only because they have been reported under this article as required by Article 4 of Directive 2003/42 of the European Parliament and of the Council of 13th June 2003 on occurrence reporting in civil aviation, except in cases of gross negligence.

(18) The provisions in paragraphs (15) to (17) shall apply without prejudice to the right of access to information by judicial authorities.

(19) The CAA shall put in place a system of voluntary reporting to collect and analyse information on observed deficiencies in aviation which are not required to be reported under the system of mandatory reporting, but which are perceived by the reporter as an actual or potential hazard.

(20) Voluntary reports presented to the CAA under paragraph (19) shall be subjected to a process of disidentification by it where the person making the report requests that his identity is not recorded on the databases.

(21) The CAA shall ensure that relevant safety information deriving from the analysis of reports, which have been subjected to disidentification, are stored and made available to all parties so that they can be used for improving safety in aviation.

Article 143: Mandatory reporting of birdstrikes

(1) Subject to the provisions of this article, the commander of an aircraft shall make a report to the CAA of any birdstrike occurrence which occurs whilst the aircraft is in flight within the United Kingdom.

(2) The report shall be made within such time, by such means and shall contain such information as may be prescribed and it shall be presented in such form as the CAA may in a particular case approve.

(3) Nothing in this article shall require a person to report any occurrence which he has reported under article 142 or which he has reason to believe has been or will be reported by another person to the CAA in accordance with that article.

(4) A person shall not make any report under this article if he knows or has reason to believe that the report is false in any particular.

(5) In this article "birdstrike occurrence" means an incident in flight in which the commander of an aircraft has reason to believe that the aircraft has been in collision with one or more than one bird.

Article 144: Power to prevent aircraft flying

(1) If it appears to the CAA or an authorised person that any aircraft is intended or likely to be flown:

(a) in such circumstances that any provision of article 3, 5, 6, 8, 25, 26,43, 62. 69, 70 or 75 (2) would be contravened in relation to the flight;

(b) in such circumstances that the flight would be in contravention of any other provision of this Order, of any regulations made thereunder or of Part 21, 145 or M and be a cause of danger to any person or property whether or not in the aircraft; or

(c) while in a condition unfit for the flight, whether or not the flight would otherwise be in contravention of any provision of this Order, of any regulations made thereunder or of Part 21, 145 or M;

the CAA or that authorised person may direct the operator or the commander of the aircraft that he is not to permit the aircraft to make the particular flight or any other flight of such description as may be specified in the direction, until the direction has been revoked by the CAA or by an authorised person, and the CAA or that authorised person may take such steps as are necessary to detain the aircraft.

(2) For the purposes of paragraph (1) the CAA or any authorised person may enter upon and inspect any aircraft.

(3) If it appears to the Secretary of State or an authorised person that any aircraft is intended or likely to be flown in such circumstances that any provision of article 138, 140 or 141 would be contravened in relation to the flight, the Secretary of State or that authorised person may direct the operator or the commander of the aircraft that he is not to permit the aircraft to make a particular flight or any other flight of such description as may be specified in the direction until the direction has been revoked by the Secretary of State or by an authorised person, and the Secretary of State or any authorised person may take such steps as are necessary to detain the aircraft.

(4) For the purposes of paragraph (3) the Secretary of State or any authorised person may enter upon any aerodrome and may enter upon and inspect any aircraft.

Article 145: Right of access to aerodromes and other places

(1) Subject to paragraph (2), the CAA and any authorised person shall have the right of access at all reasonable times:

(a) to any aerodrome, for the purpose of inspecting the aerodrome;

(b) to any aerodrome for the purpose of inspecting any aircraft on the aerodrome or any document which it or he has power to demand under this Order, or for the purpose of detaining any aircraft under the provisions of this Order;

(c) to any place where an aircraft has landed, for the purpose of inspecting the aircraft or any document which it or he has power to demand under this Order and for the purpose of detaining the aircraft under the provisions of this Order; and

(d) to any building or place from which an air traffic control service is being provided or where any air traffic service equipment requiring approval under article 124 is situated for the purpose of inspecting:

(i) any equipment used or intended to be used in connection with the provision of a service to an aircraft in flight or on the ground, or

(ii) any document or record which it or he has power to demand under this Order.

(2) Access to a Government aerodrome shall be obtained with the permission of the person in charge of the aerodrome.

Article 146: Obstruction of persons

A person shall not intentionally obstruct or impede any person acting in the exercise of his powers or the performance of his duties under this Order.

Article 147: Directions

(1) Where any provision of this order or any regulations made thereunder gives to a person the power to direct, the person to whom such a power is given shall also have the power to revoke or vary any such direction.

(2) Any person who without reasonable excuse fails to comply with any direction given to him under any provision of this Order or any regulations made thereunder shall be deemed for the purposes of this Order to have contravened that provision.

Article 148: Penalties

(1) If any provision of this Order, of any regulations made thereunder or of Part 21, 145 or M is contravened in relation to an aircraft, the operator of that aircraft and the commander and, in the case of a contravention of article 138, the charterer of that aircraft, shall (without prejudice to the liability of any other person for that contravention) be deemed for the purposes of the following provisions of this article to have contravened that provision unless he proves that the contravention occurred without his consent or connivance and that he exercised all due diligence to prevent the contravention.

(2) If it is proved that an act or omission of any person which would otherwise have been a contravention by that person of a provision of this Order, of any regulations made thereunder or of Part 21, 66 145, 147 or M was due to any cause not avoidable by the exercise of reasonable care by that person the act or omission shall be deemed not to be a contravention by that person of that provision.

(3) Where a person is charged with contravening a provision of this Order or of any regulations made thereunder by reason of his having been a member of the flight crew of an aircraft on a flight for the purpose of public transport or aerial work the flight shall be treated (without prejudice to the liability of any other person under this Order) as not having been for that purpose if he proves that he neither knew nor suspected that the flight was for that purpose.

(4) If any person contravenes any provision of this Order, of any regulations made thereunder or of Part 21, 66, 145, 147 or M not being a provision referred to in paragraphs (5) (6) or (7), he shall be guilty of an offence and liable on summary conviction to a fine not exceeding Level 3 on the standard scale.

(5) If any person contravenes any provision specified in Part A of Schedule 14 he shall be guilty of an offence and liable on summary conviction to a fine not exceeding Level 4 on the standard scale.

(6) If any person contravenes any provision specified in Part B of the said Schedule he shall be guilty of an offence and liable on summary conviction to a fine not exceeding the statutory maximum and on conviction on indictment to a fine or imprisonment for a term not exceeding two years or both.

(7) If any person contravenes any provision specified in Part C of the said Schedule he shall be guilty of an offence and liable on summary conviction to a fine not exceeding the statutory maximum and on conviction on indictment to a fine or imprisonment for a term not exceeding five years or both.

Article 149: Extra-territorial effect of the Order
(1) Except where the context otherwise requires, the provisions of this Order:
(a) in so far as they apply (whether by express reference or otherwise) to aircraft registered in the United Kingdom, shall apply to such aircraft wherever they may be;
(b) in so far as they apply as aforesaid to other aircraft shall apply to such other aircraft when they are within the United Kingdom or on or in the neighbourhood of an offshore installation;
(c) in so far as they prohibit, require or regulate (whether by express reference or otherwise) the doing of anything by persons in, or by any of the crew of, any aircraft registered in the United Kingdom, shall apply to such persons and crew, wherever they may be;
(d) in so far as they prohibit, require or regulate as aforesaid the doing of anything in relation to any aircraft registered in the United Kingdom by other persons shall, where such persons are Commonwealth citizens, British protected persons or citizens of the Republic of Ireland, apply to them wherever they may be; and
(e) in so far as they prohibit, require or regulate as aforesaid the doing of anything in relation to any aircraft on or in the neighbourhood of an offshore installation, shall apply to every person irrespective of his nationality or, in the case of a body corporate, of the law under which it was incorporated and wherever that person or body may be.
(2) Nothing in this article shall be construed as extending to make any person guilty of an offence in any case in which it is provided by section 3(1) of the British Nationality Act 1948 that that person shall not be guilty of an offence.

Article 150: Aircraft in transit over certain United Kingdom territorial waters
(1) Where an aircraft, not being an aircraft registered in the United Kingdom, is flying over the territorial waters adjacent to the United Kingdom within part of a strait referred to in paragraph (4) solely for the purpose of continuous and expeditious transit of the strait, only the following articles and Schedules shall apply to that aircraft: article 20 and Schedule 5, to the extent necessary for the monitoring of the appropriate distress radio frequency, article 95(2) (3) and (4), together with the regulations made thereunder, article 148, article 153 and Part A of Schedule 14.
(2) The powers conferred by the provisions referred to in paragraph (1) shall not be exercised in a way which would hamper the transit of the strait by an aircraft not registered in the United Kingdom, but without prejudice to action needed to secure the safety of aircraft.
(3) In this article 'transit of the strait' means over flight of the strait from an area of high seas at one end of the strait to an area of high seas at the other end, or flight to or from an area of high seas over some part of the strait for the purpose of entering, leaving or returning from a State bordering the strait and 'an area of high seas' means any area outside the territorial waters of any State.
(4) The parts of the straits to which this article applies are specified in Schedule 15.

Article 151: Application of Order to British controlled aircraft registered elsewhere than in the United Kingdom
The CAA may direct that such of the provisions of this Order and of any regulations made or having effect thereunder as may be specified in the direction shall have effect as if reference in those provisions to aircraft registered in the United Kingdom included references to the aircraft specified in the direction, being an aircraft registered elsewhere than in the United Kingdom but for the time being under the management of a person who, or of persons each of whom is qualified to hold a legal or beneficial interest by way of ownership in an aircraft registered in the United Kingdom.

Article 152: Application of Order to the Crown and visiting forces etc.
(1) Subject to the provisions of this article, the provisions of this Order shall apply to or in relation to aircraft belonging to or exclusively employed in the service of Her Majesty as they apply to or in relation to other aircraft.
(2) For the purposes of such application, the Department or other authority for the time being responsible on behalf of Her Majesty for the management of the aircraft shall be deemed to be the operator of the aircraft and, in the case of an aircraft belonging to Her Majesty, to be the owner of the interest of Her Majesty in the aircraft.
(3) Nothing in this article shall render liable to any penalty any Department or other authority responsible on behalf of Her Majesty for the management of any aircraft.
(4) Save as otherwise expressly provided the naval, military and air force authorities and members of any visiting force and any international headquarters and the members and property held or used for the purpose of such a force or headquarters shall be exempt from the provisions of this Order and of any regulations made thereunder to the same extent as if that force or headquarters formed part of the forces of Her Majesty raised in the United Kingdom and for the time being serving there.
(5) Save as otherwise provided by paragraph (6), article 80(5) and (12), article 81(3), article 95(1)(a) and article 131, nothing in this Order shall apply to or in relation to any military aircraft.
(6) Where a military aircraft is flown by a civilian pilot and is not commanded by a person who is acting in the course of his duty as a member of any of Her Majesty's naval, military or air forces or as a member of a visiting force or international headquarters, the following provisions of this Order shall apply on the occasion of that flight, that is to say, articles 73, 74, 75 and 96 and in addition article 95 (so far as applicable) shall apply unless the aircraft is flown in compliance with Military Flying Regulations (Joint Service Publication 550) or Flying Orders to Contractors (Aviation Publication 67) issued by the Secretary of State.

Article 153: Exemption from Order
The CAA may exempt from any of the provisions of this Order (other than articles 85, 87, 93, 138, 139, 140, 141 or 154) or any regulations made thereunder, any aircraft or persons or classes of aircraft or persons, either absolutely or subject to such conditions as it thinks fit.

Article 154: Appeal to County Court or Sheriff Court

(1) Subject to paragraphs (2), (3) and (4), an appeal shall lie to a county court from any decision of the CAA that a person is not a fit person to hold a licence to act as:

(a) an aircraft maintenance engineer,

(b) a member of the flight crew of an aircraft,

(c) an air traffic controller;

(d) a student air traffic controller; or

(e) a flight information service officer, and

if the court is satisfied that on the evidence submitted to the CAA it was wrong in so deciding, the court may reverse the CAA's decision and the CAA shall give effect to the court's determination.

(2) An appeal shall not lie from a decision of the CAA that a person is not qualified to hold the licence by reason of a deficiency in his knowledge, experience, competence, skill, physical or mental fitness.

(3) If the appellant resides or has his registered or principal office in Scotland the appeal shall lie to the sheriff within whose jurisdiction he resides and the appeal shall be brought within 21 days from the date of the CAA's decision or within such further period as the sheriff may in his discretion allow.

(4) Notwithstanding any provision to the contrary in rules governing appeals to a county court in Northern Ireland, if the appellant resides or has his registered or principal office in Northern Ireland the appeal shall lie to the county court held under the County Courts (Northern Ireland) Order 1980.

(5) The CAA shall be a respondent to any appeal under this article.

(6) For the purposes of any provision relating to the time within which an appeal may be brought, the CAA's decision shall be deemed to have been taken on the date on which the CAA furnished a statement of its reasons for the decision to the applicant for the licence, or as the case may be, the holder or former holder of it.

(7) In the case of an appeal to the sheriff:

(a) the sheriff may, if he thinks fit, and shall on the application of any party, appoint one or more persons of skill and experience in the matter to which the proceedings relate to act as assessor, but where it is proposed to appoint any person as an assessor objection to him either personally or in respect of his qualification may be stated by any party to the appeal and shall be considered and disposed of by the sheriff;

(b) the assessors for each sheriffdom shall be appointed from a list of persons approved for the purposes by the sheriff principal and such a list:

(i) shall be published in such manner as the sheriff principal shall direct; and

(ii) shall be in force for 3 years only, but persons entered in any such list may be again approved in any subsequent list; it shall be lawful for the sheriff principal to defer the preparation of such a list until application has been made to appoint an assessor in an appeal in one of the courts in his sheriffdom;

(c) the sheriff before whom an appeal is heard with the assistance of an assessor shall make a note of any question submitted by him to such assessor and of the answer thereto;

(d) an appeal shall lie on a point of law from any decision of a sheriff under this article to the Court of Session.

Article 155: Interpretation

(1) In this Order:

'**A Conditions**' means the conditions so entitled set out in paragraph 1 of Part A of Schedule 3;

'**Accident prevention and flight safety programme**' means a programme designed to detect and eliminate or avoid hazards in order to improve the safety of flight operations;

'**Aerial work**' has the meaning assigned to it by article 157;

'**Aerial work aircraft**' means an aircraft (other than a public transport aircraft) flying, or intended by the operator to fly, for the purpose of aerial work;

'**Aerial work undertaking**' means an undertaking whose business includes the performance of aerial work;

'**Aerobatic manoeuvres**' includes loops, spins, rolls, bunts, stall turns, inverted flying and any other similar manoeuvre;

'**Aerodrome**' means any area of land or water designed, equipped, set apart or commonly used for affording facilities for the landing and departure of aircraft and includes any area or space, whether on the ground, on the roof of a building or elsewhere, which is designed, equipped or set apart for affording facilities for the landing and departure of aircraft capable of descending or climbing vertically, but shall not include any area the use of which for affording facilities for the landing and departure of aircraft has been abandoned and has not been resumed;

'**Aerodrome control service**' means an air traffic control service for any aircraft on the manoeuvring area or apron of the aerodrome in respect of which the service is being provided or which is flying in, or in the vicinity of, the aerodrome traffic zone of that aerodrome by visual reference to the surface or any aircraft transferred from approach control in accordance with procedures approved by the CAA.

'**Aerodrome operating minima**' in relation to the operation of an aircraft at an aerodrome means the cloud ceiling and runway visual range for take-off, and the decision height or minimum descent height, runway visual range and visual reference for landing, which are the minimum for the operation of that aircraft at that aerodrome;

'**Aerodrome traffic zone**' has the meaning assigned to it by article 156;

'**Aeronautical beacon**' means an aeronautical ground light which is visible either continuously or intermittently to designate a particular point on the surface of the earth;

'**Aeronautical ground light**' means any light specifically provided as an aid to air navigation, other than a light displayed on an aircraft;

'**Aeronautical radio station**' means a radio station on the surface, which transmits or receives signals for the purpose of assisting aircraft;

'**Air Control**' means an aerodrome control service excluding that part of the aerodrome control service provided by ground movement control;

'**Air/ground communications service**' means a service provided from an aerodrome to give information to pilots of aircraft flying in the vicinity of the aerodrome by means of radio signals and 'air ground communications service unit' shall be construed accordingly;

'**Air traffic control service**' means the giving of instructions, advice or information by means of radio signals to aircraft in the interests of safety;

'**Air traffic control unit**' means a person appointed by a person maintaining an aerodrome or place to provide an air traffic control service;

'**Air traffic service equipment**' means ground based equipment, including an aeronautical radio station, used or intended to be used in connection with the provision of a service to an aircraft in flight or on the ground which equipment is not otherwise approved by or under this order but excluding:

(a) any public electronic communications network; and

(b) any equipment in respect of which the CAA has made a direction that it shall be deemed not to be air traffic service equipment for the purposes of articles 124 and 125;

'**Air transport undertaking**' means an undertaking whose business includes the undertaking of flights for the purposes of public transport of passengers or cargo;

'**Alternate aerodrome**' means an aerodrome to which an aircraft may proceed when it becomes either impossible or inadvisable to proceed to or to land at the aerodrome of intended landing;

'**Altitude hold and heading mode**' means aircraft autopilot functions which enable the aircraft to maintain an accurate height and an accurate heading;

'**Annual costs**' in relation to the operation of an aircraft means the best estimate reasonably practicable at the time of a particular flight in respect of the year commencing on the first day of January preceding the date of the flight, of the costs of keeping and maintaining and the indirect costs of operating the aircraft, such costs in either case excluding direct costs and being those actually and necessarily incurred without a view to profit;

'**Annual flying hours**' means the best estimate reasonably practicable at the time of a particular flight by an aircraft of the hours flown or to be flown by the aircraft in respect of the year commencing on the first day of January preceding the date of the flight;

'**Approach control service**' means an air traffic control service for any aircraft which is not receiving an aerodrome control service, which is flying in, or in the vicinity of the aerodrome traffic zone of the aerodrome in respect of which the service is being provided, whether or not the aircraft is flying by visual reference to the surface;

'**Approach to landing**' means that portion of the flight of the aircraft, when approaching to land, in which it is descending below a height of 1000ft above the relevant specified decision height or minimum descent height;

'**Appropriate aeronautical radio station**' means in relation to an aircraft an aeronautical radio station serving the area in which the aircraft is for the time being;

'**Appropriate air traffic control unit**' means in relation to an aircraft either the air traffic control unit notified as serving the area in which the aircraft is for the time being, or the air traffic control unit notified as serving the area which the aircraft intends to enter and with which unit the aircraft is required to communicate prior to entering that area, as the context requires;

'**Apron**' means the part of an aerodrome provided for the stationing of aircraft for the embarkation and disembarkation of passengers, for loading and unloading of cargo and for parking;

'**Area control centre**' means an air traffic control unit established to provide an area control service to aircraft flying within a notified flight information region which are not receiving an aerodrome control service or an approach control service;

'**Area control service**' means an air traffic control service for any aircraft which is flying neither in nor in the vicinity of an aerodrome traffic;

'**Area navigation equipment**' means equipment carried on board an aircraft which enables the aircraft to navigate on any desired flight path within the coverage of appropriate ground based navigation aids or within the limits of that on-board equipment or a combination of the two;

'**Authorised person**' means:

(a) any constable;

(b) in article 144(3) and (4) any person authorised by the Secretary of State (whether by name, or by class or description) either generally or in relation to a particular case or class of cases; and

(c) in article 144(1) and (2) and in any article other than article 144, any person authorised by the CAA (whether by name or by class or description) either generally or in relation to a particular case or class of cases;

'**B Conditions**' means the conditions so entitled set out in paragraph 2 of Part A of Schedule 3;

'**Basic EASA Regulation**' means Regulation (EC) No 1592/2002 of the European Parliament and of the Council of 15 July 2002 on common rules in the field of civil aviation and establishing a European Aviation Safety Agency;

'**Beneficial interest**' includes interests arising under contract and other equitable interests;

'**British protected person**' has the same meaning as in Section 50 of the British Nationality Act 1981;

'**Cabin crew**' in relation to an aircraft means those persons on a flight for the purpose of public transport carried for the purpose of performing in the interests of the safety of passengers duties to be assigned by the operator or the commander of the aircraft but who shall not act as a member of the flight crew;

'**Captive balloon**' means a balloon which when in flight is attached by a restraining device to the surface;

'**Captive flight**' means flight by an uncontrollable balloon during which it is attached to the surface by a restraining device;

'**Cargo**' includes mail and (for the avoidance of doubt) animals;

'**Certificate of airworthiness**' includes in the case of a national certificate of airworthiness any flight manual, performance schedule or other document, whatever its title, incorporated by reference in that certificate relating to the certificate of airworthiness;;

'**Certificate of maintenance review**' has the meaning assigned to it by article 14(1)(b);

'**Certificate of release to service issued under Part 145**' means a certificate of release to service issued in accordance with Part 145;

'**Certificate of release to service issued under the Order**' means a certificate issued by a person specified in article 16(11) which conforms with article 16(10);

'**Certificate of revalidation**' means a certificate issued in accordance with Section 2 of Part C of Schedule 8 for the purpose of maintaining the privileges of a flight crew licence;

'**Certificate of validation**' means a certificate issued by the CAA rendering valid for the purposes of this order a certificate of airworthiness or a permit to fly issued in respect of an aircraft registered elsewhere than in the United Kingdom or a flight crew licence granted under the law of a country other than the United Kingdom;

'**Certificate of validity**' means a certificate issued under article 11(6)(d) for the purpose of maintaining the validity of a permit to fly issued by the CAA;

'**Certificated for single pilot operation**' means an aircraft which is not required to carry more than one pilot by virtue of any one or more of the following:

(a) the certificate of airworthiness duly issued or rendered valid under the law of the country in which the aircraft is registered;

(b) if no certificate of airworthiness is required to be in force, the certificate of airworthiness, if any, last in force in respect of the aircraft or the related flight manual;

(c) if no certificate of airworthiness is or has previously been in force but the aircraft is identical in design with an aircraft in respect of which such a certificate is or has been in force, the certificate of airworthiness which is or has been in force in respect of such an identical aircraft; or the related flight manual; or

(d) in the case of an aircraft flying in accordance with the conditions of a permit to fly issued by the CAA, that permit to fly;

'**Class A airspace**', '**Class B airspace**', '**Class C airspace**', '**Class D airspace**' and '**Class E airspace**' mean airspace respectively notified as such;

'**Class rating**' in respect of aeroplanes has the meaning specified in paragraph 1.220 of JAR-FCL 1;

'**Cloud ceiling**' in relation to an aerodrome means the vertical distance from the elevation of the aerodrome to the lowest part of any cloud visible from the aerodrome which is sufficient to obscure more than one-half of the sky so visible;

'**Commander**' in relation to an aircraft means the member of the flight crew designated as commander of that aircraft by the operator thereof, or, failing such a person, the person who is for the time being the pilot in command of the aircraft;

'**the Commonwealth**' means the United Kingdom, the Channel Islands, the Isle of Man, the countries mentioned in Schedule 3 to the British Nationality Act 1981 and all other territories forming part of Her Majesty's dominions or in which Her Majesty has jurisdiction and 'Commonwealth citizen' shall be construed accordingly;

'**Competent authority**' means subject to article 167, in relation to the United Kingdom, the CAA, and in relation to any other country the authority responsible under the law of that country for promoting the safety of civil aviation;

'**Conditional sale agreement**' has the same meaning as in section 189 of the Consumer Credit Act 1974;

'**Congested area**' in relation to a city, town or settlement, means any area which is substantially used for residential, industrial, commercial or recreational purposes;

'**Contracting State**' means any State (including the United Kingdom) which is party to the Chicago Convention;

'**Controllable balloon**' means a balloon, not being a small balloon, which is capable of free controlled flight;

'**Controlled airspace**' means airspace which has been notified as Class A, Class B, Class C, Class D or Class E airspace;

'**Control area**' means controlled airspace which has been further notified as a control area and which extends upwards from a notified altitude or flight level;

'**Control zone**' means controlled airspace which has been further notified as a control zone and which extends upwards from the surface;

'**Co-pilot**' in relation to an aircraft means a pilot who in performing his duties as such is subject to the direction of another pilot carried in the aircraft;

'**Country**' includes a territory;

'**Crew**' means a member of the flight crew, a person carried on the flight deck who is appointed by the operator of the aircraft to give or to supervise the training, experience, practice and periodical tests required in respect of the flight crew under article 42(3) or a member of the cabin crew;

'**Critical power unit**' means the power unit whose failure would most adversely affect the performance or handling qualities of an aircraft;

'**Danger area**' means airspace which has been notified as such within which activities dangerous to the flight of aircraft may take place or exist at such times as may be notified;

'**Decision height**' in relation to the operation of an aircraft at an aerodrome means the height in a precision approach at which a missed approach must be initiated if the required visual reference to continue that approach has not been established;

'**Declared distances**' has the meaning which has been notified;

'**Designated required navigation performance airspace**' means airspace which has been notified, prescribed or otherwise designated by the competent authority for the airspace as requiring specified navigation performance capabilities to ne met by aircraft flying within it;

'**Direct costs**' means, in respect of a flight, the costs actually and necessarily incurred in connection with that flight without a view to profit but excluding any remuneration payable to the pilot for his services as such;

'**Director**' has the same meaning as in section 53(1) of the Companies Act 1989;

'**Disidentification**' means removing from reports submitted all personal details pertaining to the reporter and technical details which might lead to the identity of the reporter, or of third parties, being inferred from the information;

'**EASA**' means the European Aviation Safety Agency;

'**EASA aircraft**' means an aircraft which is required by virtue of the Basic EASA Regulation and any implementing rules adopted by the Commission in accordance with the Regulation to hold an EASA certificate of airworthiness, an EASA restricted certificate of airworthiness or an EASA permit to fly;
'**EASA certificate of airworthiness**' means a certificate of airworthiness issued in respect of an EASA aircraft under and in accordance with subpart H of Part 21;
'**EASA permit to fly**' means a permit to fly issued in respect of an EASA aircraft under and in accordance with subpart H of Part 21;
'**EASA restricted certificate of airworthiness**' means a restricted certificate of airworthiness issued in respect of an EASA aircraft under and in accordance with subpart H of Part 21;
'**European Aviation Safety Agency**' means the Agency established under the Basic EASA Regulation;
'**Flight**' and '**to fly**' have the meanings respectively assigned to them by paragraph (2);
'**Flight check**' means a check carried out by an aircraft in flight of the accuracy and reliability of signals transmitted by an aeronautical radio station;
'**Flight crew**' in relation to an aircraft means those members of the crew of the aircraft who respectively undertake to act as pilot, flight navigator, flight engineer and flight radiotelephony operator of the aircraft;
'**Flight data monitoring programme**' means a programme of analysing recorded flight data in order to improve the safety of flight operations;
'**Flight information service unit**' means
(a) in the case of an aerodrome:
(i) the giving of information by means of radio signals to aircraft flying in or intending to fly within the aerodrome traffic zone of that aerodrome; and
(ii) the grant or refuse permission, pursuant to Rule 35 or 36(2) of the Rules of the Air;
(b) in the case of an area control centre, to give information by means of radio signals to aircraft;
and 'aerodrome flight information service' shall be construed accordingly;
'**Flight information service unit**' means a person appointed by the CAA, or by any other person maintaining an aerodrome or area control centre and 'aerodrome flight information service unit' shall be construed accordingly;
'**Flight level**' means one of a series of levels of equal atmospheric pressure, separated by notified intervals and each expressed as the number of hundreds of feet which would be indicated at that level on a pressure altimeter calibrated in accordance with the International Standard Atmosphere and set to 1013.2 hectopascals;
'**Flight manual**' means a document provided for an aircraft stating the approved limitations within which the aircraft is considered airworthy as defined by the appropriate airworthiness requirements, and additional instructions and information necessary for the safe operation of the aircraft;
'**Flight recording system**' means a system comprising either a flight data recorder or a cockpit voice recorder or both;
'**Flight simulator**' means apparatus by means of which flight conditions in an aircraft are simulated on the ground;
'**Flight visibility**' means the visibility forward from the flight deck of an aircraft in flight;
'**Flying display**' means any flying activity deliberately performed for the purpose of providing an exhibition or entertainment at an advertised event open to the public;
'**Flying machine**' means an aeroplane, a powered lift tilt rotor aircraft, a self-launching motor glider, a helicopter or a gyroplane;
'**Free balloon**' means a balloon which when in flight is not attached by any form of restraining device to the surface;
'**Free controlled flight**' means flight during which a balloon is not attached to the surface by any form of restraining device (other than a tether not exceeding 5 metres in length which may be used as part of the take-off procedure) and during which the height of the balloon is controllable by means of a device attached to the balloon and operated by the commander of the balloon or by remote control;
'**General lighthouse authority**' has the same meaning as in section 193 of the Merchant Shipping Act 1995;
'**Glider**' means:
(a) a non-power driven heavier than air aircraft, deriving its lift in flight chiefly from aerodynamic reactions on surfaces which remain fixed under given conditions of flight; and
(b) a self-sustaining glider; and
(c) a self-propelled hang-glider;
and a reference in this Order to a glider shall include a reference to a self-sustaining glider and a self-propelled hang-glider;
'**Government aerodrome**' means any aerodrome in the United Kingdom which is in the occupation of any Government Department or visiting force;
'**Ground Movement Control**' means that part of an aerodrome control service provided to an aircraft while it is on the manoeuvring area or apron of an aerodrome;
'**Hire-purchase agreement**' has the same meaning as in section 189 of the Consumer Credit Act 1974;
'**Holding**' means, in respect of an aircraft approaching an aerodrome to land, a manoeuvre in the air which keeps that aircraft within a specified volume of airspace;
'**Instructor's rating**' means a flying instructor's rating, an assistant flying instructor's rating, a flight instructor rating (aeroplane), a flight instructor rating (helicopter), a type rating instructor rating (multi-pilot aeroplane), a type rating instructor rating (helicopter), a class rating instructor rating (single pilot aeroplane), an instrument rating instructor rating (aeroplane) or an instrument rating instructor rating (helicopter);
'**Instrument Flight Rules**' means Instrument Flight Rules prescribed by Section VI of the Rules of the Air Regulations 1996;

'**Instrument Landing System**' means a ground-based radio system designed to transmit radio signals at very high frequency that allow the pilot of an aircraft to accurately determine the aircraft's position relative to a defined approach path whilst carrying out an approach to land;
'**Instrument Meteorological Conditions**' means weather precluding flight in compliance with the Visual Flight Rules;
'**International headquarters**' means an international headquarters designated by Order in Council under section 1 of the International Headquarters and Defence Organisations Act 1964;
'**JAA**' means the Joint Aviation Authorities, an associated body of the European Civil Aviation Conference;
'**JAA Full Member State**' means a State which is full member of the JAA;
'**JAA licence**' means a flight crew licence granted under JAR-FCL 1 or 2 by the competent authority of a JAA full member state in accordance with a procedure which has been assessed as satisfactory following an inspection by a licensing and a medical standardisation team of the JAA;
'**JAR-FCL 1**' means the Joint Aviation Requirement of the JAA bearing that title including Amendment 3 adopted by the JAA on 1st July 2003;
'**JAR-FCL 2**' means the Joint Aviation Requirement of the JAA bearing that title including Amendment 3 adopted by the JAA on 1st September 2003;
'**JAR-FCL** licence' means a licence included in Section 2 of Part A of Schedule 8;
'**JAR-OPS 1**' means the Joint Aviation Requirement of the JAA bearing that title including Amendment 7 adopted by the JAA on 1st September 2004;
'**JAR-OPS 3**' means the Joint Aviation Requirement of the JAA bearing that title including Amendment 4 adopted by the JAA on 1st April 2004;
'**Kg**' means kilogramme or kilogrammes as the context requires;
'**km**' means kilometre or kilometres as the context requires;
'**To land**' in relation to aircraft includes alighting on the water;
'**Landing Decision Point**' means the latest point in the course of a landing at which following recognition of a power unit failure, the helicopter will be able to safely abort the landing and perform a go-around;
'**Large rocket**' means a rocket of which the total impulse of the motor or combination of motors is more than 10,240 Newton-seconds;
'**Legal personal representative**' means the person so constituted executor, administrator, or other representative, of a deceased person;
'**Let down**' means, in respect of an aircraft approaching an aerodrome to land a defined procedure designed to enable an aircraft safely to descend to a point at which it can continue the approach visually;
'**Licence**' in relation to a flight crew licence includes any certificate of competency or certificate of validity issued with the licence or required to be held in connection with the licence by the law of the country in which the licence is granted;
'**Licence for public use**' has the meaning assigned to it by article 128(2);
'**Licensed aerodrome**' means an aerodrome licensed under this Order;
'**Lifejacket**' includes any device designed to support a person individually in or on the water;
'**Log book**' in the case of an aircraft log book, engine log book or variable pitch propeller log book, or personal flying log book, includes a record kept either in a book, or by any other means approved by the CAA in the particular case;
'**Maintenance**' means in relation to an aircraft any one or combination of overhaul, repair, inspection, replacement, modification or defect rectification of an aircraft or component, with the exception of pre-flight inspection;
'**Manoeuvring area**' means the part of an aerodrome provided for the take-off and landing of aircraft and for the movement of aircraft on the surface, excluding the apron and any part of the aerodrome provided for the maintenance of aircraft;
'**Maximum approved passenger seating configuration**' means:
(a) in the case of an aircraft to which article 38 applies the maximum approved passenger seating configuration specified in the operations manual of the aircraft; and
(b) in any other case, the maximum number of passengers which may be carried in the aircraft under and in accordance with its certificate of airworthiness, its flight manual and this order;
'**Maximum total weight authorised**' in relation to an aircraft means the maximum total weight of the aircraft and its contents at which the aircraft may take-off anywhere in the world, in the most favourable circumstances in accordance with the certificate of airworthiness in force in respect of the aircraft;
'**Medical attendant**' means a person carried on a flight for the purpose of attending to any person in the aircraft in need of medical attention, or to be available to attend to such a person;
'**Medium intensity steady red light**' means a red light which complies with the characteristics described for a medium intensity Type C light as specified in Volume 1 (Aerodrome Design and Operations) of Annex 14 (Fourth Edition July 2004) to the Chicago Convention;
'**Microlight aeroplane**' means an aeroplane designed to carry not more than two persons which has:
(a) a maximum total weight authorised not exceeding:
(i) 300kg for a single seat landplane, (or 390kg for a single seat landplane in respect of which a United Kingdom permit to fly or certificate of airworthiness was in force prior to 1st January 2003),
(ii) 450kg for a two seat landplane,
(iii) 330kg for a single seat amphibian or floatplane, or
(iv) 495kg for a two seat amphibian or floatplane; and
(b) a stalling speed at the maximum total weight authorised not exceeding 35 knots calibrated airspeed;
'**Microwave Landing System**' means a ground based radio system designed to transmit radio signals at super high frequency that allow the pilot of an aircraft to accurately determine the aircraft's position within a defined volume of airspace whilst carrying out an approach to land;

'**Military aircraft**' means the naval, military or air force aircraft of any country and:

(a) any aircraft being constructed for the naval, military or air force of any country under a contract entered into by the Secretary of State; and

(b) any aircraft in respect of which there is in force a certificate issued by the Secretary of State that the aircraft is to be treated for the purposes of this Order as a military aircraft;

'**Military rocket**' means:

(a) any rocket being constructed for the naval, military or air force of any country under a contract entered into by the Secretary of State; and

(b) any rocket in respect of which there is in force a certificate issued by the Secretary of State that the rocket is to be treated for the purposes of the Order as a military rocket;

'**Minimum descent height**' in relation to the operation of an aircraft at an aerodrome means the height in a non-precision approach below which descent may not be made without the required visual reference;

'**Multi-crew co-operation**' means the functioning of the flight crew as a team of co-operating members led by the pilot in command;

'**National certificate of airworthiness**' means a certificate of airworthiness issued under and in accordance with Part 3 of this order and which is not an EASA certificate of airworthiness;

'**National permit to fly**' means a permit to fly issued under and in accordance with Part 3 of this Order and which is not an EASA permit to fly;

'**Nautical mile**' means the International Nautical Mile, that is to say, a distance of 1852 metres;

'**Night**' means the time from half an hour after sunset until half an hour before sunrise (both times inclusive), sunset and sunrise being determined at surface level;

'**Non-EASA aircraft**' means an aircraft which is not required by virtue of the Basic EASA Regulation and any implementing rules adopted by the Commission in accordance with that Regulation to hold an EASA certificate or airworthiness, an EASA restricted certificate of airworthiness or an EASA permit to fly; and a non-EASA balloon, a non-EASA glider and non-EASA kite shall be construed accordingly;

'**Non-precision approach**' means an instrument approach using non-visual aids for guidance in azimuth or elevation but which is not a precision approach;

'**Non-revenue flight**' means:

(a) in the case of a flight by an aeroplane, any flight which the holder of a United Kingdom Private Pilot's Licence (Aeroplanes) may undertake pursuant to paragraph (2)(a) and (b) of the privileges of that licence set out in Section 1 of Part A of Schedule 8;

(b) in the case of a flight by a helicopter, any flight which the holder of a United Kingdom Private Pilot's Licence (Helicopters) may undertake pursuant to paragraph (2)(a) and (b) of the privileges of that licence set out in Section 1 of Part A of Schedule 8; and

(c) in the case of a flight by a gyroplane, any flight which the holder of a United Kingdom Private Pilot's Licence (Gyroplanes) may undertake pursuant to paragraph (2)(a) and (b) of the privileges of that licence set out in Section 1 of Part A of Schedule 8;

'**North Atlantic Minimum Navigation Performance Specification airspace**' means the airspace prescribed as such;

'**Notified**' means set out with the authority of the CAA in a document published by or under an arrangement entered into with the CAA and entitled 'United Kingdom Notam' or 'Air Pilot' and for the time being in force;

'**Notified aerodrome**' means an aerodrome which is notified for the purpose of rule 39 of the Rules of the Air Regulations 1996;

'**Obstacle limitation surfaces**' has the same meaning as in 'CAP 168 Licensing of aerodromes' published by the CAA in May 2004;

'**Occurrence**' means an operational interruption, defect, fault or other irregular circumstance that has or may have influenced flight safety and that has not resulted in an accident or serious incident as those terms are defined in regulation 2 of the Civil Aviation (Investigation of Air Accidents and Incidents) Regulations 1996.

'**Offshore service**' means an air traffic control service for any aircraft flying to or from offshore oil and gas installations and for the other aircraft operating in the vicinity of these aircraft in airspace specified for this purpose in the manual of air traffic services;

'**Operating staff**' means the servants and agents employed by an operator of an aircraft, whether or not as members of the crew, to ensure that flights of the aircraft are conducted in a safe manner, and includes an operator who himself performs those functions;

'**Operational position**' means a position provided and equipped for the purpose of providing a particular type of air traffic control service;

'**Operator**' has the meaning assigned to it by paragraph (3);

'**Parascending parachute**' means a parachute which is towed by cable in such a manner as to cause it to ascend;

'**Part 21**' means the annex so entitled to Commission Regulation (EC) No. 1702/2003;

'**Part 66**' means annex III so entitled to Commission Regulation (EC) No; 2042/2003

'**Part 147**' means annex II so entitled to Commission Regulation (EC) No; 2042/2003

'**Part M**' means annex I so entitled to Commission Regulation (EC) No. 2042/2003;

'**Passenger**' means a person other than a member of the crew;

'**Performance Class 1 Operations**' means flights where, in the event of the failure of a power unit, the helicopter will be able to safely continue the flight and land at an appropriate landing area unless the power unit failure recognition occurs during take-off at or prior to reaching the take-off decision point in which case the helicopter will be able to safely land back within the area from which it has taken off;

'**Performance Class 2 Operations**' means flights where, in the event of the failure of a power unit, the helicopter will be able to safely continue the flight to an appropriate landing area or, where the failure occurs at a point during the take-off manoeuvre or the landing manoeuvre when it cannot do so, the helicopter will be able to carry out a forced landing;

'**Performance Class 3 Operations**' means flights where, in the event of the failure of a power unit at any time during the flight, the helicopter will be required to carry out a forced landing;

'**Period of duty**' means the period between the commencement and end of a shift during which an air traffic controller performs, or could be called upon to perform, any of the functions specified in respect of a rating included in his licence;

'**Pilot in command**' in relation to an aircraft means a person who for the time being is in charge of the piloting of the aircraft without being under the direction of any other pilot in the aircraft;

'**Police air operator's certificate**' means a certificate granted by the CAA under article 7(4);

'**Police authority**' means a Chief Officer of police for any area of England or Wales, a Chief Constable for any area or Scotland and the Chief Constable of the Northern Ireland Police Service;

'**Police officer**' means any person who is a member of a police force or of the Northern Ireland Police Service (including for the avoidance of doubt, the Northern Ireland Police Service Reserve) and any special constable;

'**Pre-flight inspection**' means the inspection carried out before flight to ensure that the aircraft is fit for the intended flight;

'**Precision approach**' means an instrument approach using Instrument Landing System, Microwave Landing System or Precision Approach Radar for guidance in both azimuth and elevation;

'**Precision approach radar**' means radar equipment designed to enable an air traffic controller to determine accurately an airaft's position whilst it is carrying out an approach to land so that the air traffic controller can provide instructions and guidance to the pilot to enable him to manoeuvre the aircraft relative to a defined approach path;

'**Pressurised aircraft**' means an aircraft provided with means of maintaining in any compartment a pressure greater than that of the surrounding atmosphere;

'**Private aircraft**' means an aircraft which is neither an aerial work nor a public transport aircraft;

'**Private flight**' means a flight which is neither for the purpose of aerial work nor public transport;

'**Proficiency check**' has the meaning specified in paragraph 1.001 of Section 1 of JAR-FCL 1 in respect of aeroplanes and paragraph 2.001 in Section 1 of JAR-FCL 2 in respect of helicopters;

'**Public electronic communications network**' has the same meaning as in section 151 of the communications Act 2003;

'**Public transport**' has the meaning assigned to it by article 157;

'**Public transport aircraft**' means an aircraft flying, or intended by the operator of the aircraft to fly, for the purpose of public transport;

'**Record**' has the same meaning as in section 81(6) of the Transport Act 2000;

'**Reduced vertical separation minimum airspace**' means any airspace between flight level 290 and flight level 410 inclusive designated by the relevant competent authority as being airspace within which a vertical separation minimum of 1000 feet or 300 metres shall be applied.

'**Released flight**' means flight by an uncontrollable balloon during which it is not attached to the surface by any form of restraining device;

'**Relevant overseas territory**' means any colony and any country or place outside Her Majesty's dominions in which for the time being Her Majesty has jurisdiction;

'**Replacement**' in relation to any part of an aircraft or its equipment includes the removal and replacement of that part whether or not by the same part, and whether or not any work is done on it, but does not include the removal and replacement of a part which is designed to be removable solely for the purpose of enabling another part to be inspected, repaired, removed or replaced or cargo to be loaded;

'**Rocket**' means a device which is propelled by ejecting expanding gasses generated in its motor from self contained propellant and which is not dependent on the intake of outside substances. It includes any part of the device intended to become separated during operation;

'**Runway visual range**' in relation to a runway means the distance in the direction of take-off or landing over which the runway lights or surface markings may be seen from the touchdown zone as calculated by either human observation or instruments in:

(a) the vicinity of the touchdown zone; or

(b) where this is not reasonably practicable in the vicinity of the mid-point of the runway;

and the distance, if any, communicated to the commander of an aircraft by or on behalf of the person in charge of the aerodrome as being the runway visual range shall be taken to be the runway visual range for the time being;

'**Scheduled journey**' means one of a series of journeys which are undertaken between the same two places and which together amount to a systematic service;

'**Seaplane**' has the same meaning as for the purpose of section 97 of the Civil Aviation Act 1982;

'**Sector**' means part of the airspace controlled from an area control centre or other place;

'**Self launching motor glider**' means an aircraft with the characteristics of a non-power driven glider and which is fitted with one or more power units which is designed or intended to take-off under its own power;

'**Self-propelled hang-glider**' means an aircraft comprising an aerofoil wing and a mechanical propulsion device which:

(a) is foot launched;

(b) has a stall speed or minimum steady flight speed in the landing configuration not exceeding 35 knots calibrated airspeed;

(c) carries a maximum of two persons;

(d) has a maximum fuel capacity of 10 litres; and

(e) has a maximum unladen weight, including full fuel, of 60kg for single place aircraft and 70kg for two place aircraft;

'**Self-sustaining glider**' means an aircraft with the characteristics of a non-power driven glider which is fitted with one or more power units capable of sustaining the aircraft in flight but which is not designed or intended to take-off under its own power;

'**Simple single engine aeroplane**' means for the purposes of the National Private Pilot's Licence a single engine piston aeroplane with a maximum take-off weight authorised not exceeding 2000kg and which is not a microlight aeroplane or a self launching motor glider;

'**Skill test**' has the meaning specified in paragraph 1.001 of Section 1 of JAR-FCL 1 in respect of aeroplanes and paragraph 2.001 of Section 1 of JAR-FCL 2 in respect of helicopters;

'**SLMG**' means a self launching motor glider;

'**Small aircraft**' means any unmanned aircraft, other than a balloon or a kite, weighing not more than 20kg without its fuel nor including any articles or equipment installed in or attached to the aircraft at the commencement of its flight;

'**Small balloon**' means a balloon not exceeding 2 metres in any linear dimension at any stage of its flight, including any basket or other equipment attached to the balloon;

'**Small rocket**' means a rocket of which the total impulse of the motor or combination of motors does not exceed 10,240 Newton-seconds;

'**Special tasks service**' means an air traffic control service:

(a) for any aircraft flying for the purposes of research and development of aircraft, aircraft equipment or aircraft systems and which is not flying in accordance with normal aviation practice; and

(b) for other aircraft in the vicinity of these aircraft;

'**Special VFR flight**' means a flight which is a special VFR flight for the purposes of the Rules of the Air Regulations 1996

'**State aircraft**' means an aircraft engaged in military, customs, police or similar services;

'**State of design**' means the state having jurisdiction over the organisation responsible for the type design of an aircraft;

'**State of the operator**' means the State in which the operator of an aircraft has his principal place of business or, if he has no such place of business, his permanent residence, in circumstances where:

(a) that aircraft is registered in another Contracting State;

(b) the operator is operating that aircraft pursuant to an agreement for its lease, charter or interchange or any similar arrangement;

(c) the State in which the aircraft is registered has, by agreement with the State in which the operator of the aircraft has his principal place of business or, if he has no such place of business, his permanent residence, agreed to transfer to it its functions and duties as State of registry in respect of that aircraft in relation to, in the case of article 8(1), airworthiness, in the case of article 20(1), aircraft radio equipment, in the case of article 26(3), flight crew licensing or in the case of article 55(1), radio licensing; and

(d) the agreement has been registered with the Council of the International Civil Aviation Organisation or the existence and scope of the agreement have been directly communicated to the CAA.

'**Take-off decision point**' means the latest point in the take-off at which, following recognition of a power unit failure, the helicopter will be able to carry out a rejected take-off;

'**Technical log**' means a record containing the information specified in paragraph 1.915 of Section 2 of JAR-OPS 1.

'**Terminal control service**' means an air traffic control service for any aircraft flying in, departing, or intending to fly within a terminal control area while it is in the terminal control area or any sector adjacent thereto and is specified for this purpose in the manual of air traffic services;

'**Tethered flight**' means flight by a controllable balloon throughout which it is flown within limits imposed by a restraining device which attaches the balloon to the surface;

'**Touring motor glider**' has the meaning specified in paragraph 1.001 of Section 1 of JAR-FCL 1;

'**Type rating**' in respect of aeroplanes has the meaning specified in paragraph 1.215 of Section 1 of JAR-FCL 1;

'**Type rating**' in respect of helicopters has the meaning specified in paragraph 2.215 of Section 1 of JAR-FCL 2;

'**Uncontrollable balloon**' means a balloon, not being a small balloon, which is not capable of free controlled flight;

'**United Kingdom licence**' means a licence included in Section 1 of Part A of Schedule 8;

'**United Kingdom licence for which there is a JAR-FCL equivalent**' means the following licences included in Section 1 of Part A of Schedule 8;

Private Pilot's Licence (Aeroplanes);

Commercial Pilot's Licence (Aeroplanes);

Airline Transport Pilot's Licence (Aeroplanes);

Private Pilot's Licence (Helicopters);

Commercial Pilot's Licence (Helicopters and Gyroplanes);

Airline Transport Pilot's Licence (Helicopters and Gyroplanes);

'**United Kingdom licence for which there is no JAR-FCL equivalent**' means any licence included in Section 1 of Part A of Schedule 8 other than any such licence which is a United Kingdom licence in respect of which there is a JAR-FCL equivalent;

'**United Kingdom reduced vertical separation minimum airspace**' means United Kingdom airspace which has been notified as reduced vertical separation minimum airspace for the purposes of article 58;

'**Valuable consideration**' means any right, interest, profit or benefit, forbearance, detriment, loss or responsibility accruing, given, suffered or undertaken pursuant to an agreement, which is of more than a nominal nature;

'**Visiting force**' means any such body, contingent or detachment of the forces of any country as is a visiting force for the purpose of the provisions of the Visiting Forces Act 1952:

(a) which apply to that country by virtue of paragraph (a) of section 1(1) of that Act; or

(b) which from time to time apply to that country by virtue of paragraph (b) of the said section 1(1) and of any Order in Council made or hereafter to be made under the said section 1 designating that country for the purposes of all the provisions of that Act following section 1(2) of that Act;

'**Visual Flight Rules**' means Visual Flight Rules prescribed by Section V of the Rules of the Air Regulations 1996
'**Visual Meteorological Conditions**' means weather permitting flight in accordance with the Visual Flight Rules.

(2) An aircraft shall be deemed to be in flight:

(a) in the case of a piloted flying machine, from the moment when, after the embarkation of its crew for the purpose of taking off, it first moves under its own power until the moment when it next comes to rest after landing;

(b) in the case of a pilotless flying machine, or a glider, from the moment when it first moves for the purpose of taking off until the moment when it next comes to rest after landing;

(c) in the case of an airship, from the moment when it first becomes detached from the surface until the moment when it next becomes attached thereto or comes to rest thereon;

(d) in the case of a free balloon, from the moment when the balloon, including the canopy and basket, becomes separated from the surface until the moment it next comes to rest thereon; and

(e) in the case of a captive balloon, from the moment when the balloon, including the canopy and basket, becomes separated from the surface, apart from a restraining device attaching it to the surface, until the moment when it next comes to rest thereon;

and the expressions 'a flight' and 'to fly' shall be construed accordingly.

(3) Subject to paragraph (4), references in this Order to the operator of an aircraft are, for the purposes of the application of any provision of this Order in relation to any particular aircraft, references to the person who at the relevant time has the management of that aircraft, and cognate expressions shall be construed accordingly.

(4) For the purposes of the application of any provision in Part III of this Order, when by virtue of any charter or other agreement for the hire or loan of an aircraft a person other than an air transport undertaking or an aerial work undertaking has the management of that aircraft for a period not exceeding 14 days, paragraph (3) shall have effect as if that agreement had not been entered into.

(5) References in the Order to:

(a) a certificate of airworthiness include both a national certificate or airworthiness and an EASA certificate or airworthiness unless otherwise stated;

(b) an aircraft, aeroplane, powered lift tilt rotor aircraft, self-launching motor glider, helicopter, gyroplane, airship, balloon or kite include both EASA and non-EASA examples unless otherwise stated;

(6) The expressions appearing in the 'General classification of aircraft' in Part A of Schedule 2 shall have the meanings thereby assigned to them.

Article 156: Meaning of aerodrome traffic zone

(1) The aerodrome traffic zone of a notified aerodrome which is not on an offshore installation and at which the length of the longest runway is notified as 1850 metres or less shall be, subject to paragraphs(2) and (5), the airspace extending from the surface to a height of 2000 feet above the level of the aerodrome within the area bounded by a circle centred on the notified mod-point of the longest runway and having a radius of 2 nautical miles.

(2) Where the aerodrome traffic zone specified in paragraph (1) would extend less than 11/2 nautical miles beyond the end of any runway at the aerodrome and this paragraph is notified as being applicable, the aerodrome traffic zone shall be that specified in paragraph (3) as though the length of the longest runway at the aerodrome were notified as greater than 1850 metres.

(3) The aerodrome traffic zone of a notified aerodrome which is not on an offshore installation and at which the length of the longest runway is notified as greater than 1850 metres shall be, subject to paragraph (5), the airspace extending from the surface to a height of 2000 feet above the level of the aerodrome within the area bounded by a circle centred on the notified mid-point of the longest runway and having a radius of 21/2 nautical miles.

(4) The aerodrome traffic zone of a notified aerodrome which is on an offshore installation shall be, subject to paragraph (5), the airspace extending from mean sea level to 2000 feet above mean sea level and within 11/2 nautical miles of the offshore installation.

(5) The aerodrome traffic zone of a notified aerodrome shall exclude any airspace which is within the aerodrome traffic zone of another aerodrome which is notified for the purposes of this article as being the controlling aerodrome.

Article 157: Public transport and aerial work – general rules

(1) Subject to the provisions of this article and articles 158 to 163, aerial work means any purpose (other than public transport) for which an aircraft is flown if valuable consideration is given or promised in respect of the flight or the purpose of the flight.

(2) If the only such valuable consideration consists of remuneration for the services of the pilot the flight shall be deemed to be a private flight for the purposes of Part 3 of this Order.

(3) Subject to the provisions of this article and articles 158 to 163, an aircraft in flight shall for the purposes of this Order be deemed to fly for the purposes of public transport:

(a) if valuable consideration is given or promised for the carriage of passengers or cargo in the aircraft on that flight;

(b) if any passengers or cargo are carried gratuitously in the aircraft on that flight by an air transport undertaking, not being persons in the employment of the undertaking (including, in the case of a body corporate, its directors and, in the case of the CAA, the members of the CAA), persons with the authority of the CAA either making any inspection or witnessing any training, practice or test for the purposes of this Order, or cargo intended to be used by any such passengers as aforesaid, or by the undertaking; or

(c) for the purposes of Part 3 of this Order (other than articles 19(2) and 20(2)), if valuable consideration is given or promised for the primary purpose of conferring on a particular person the right to fly the aircraft on that flight (not being a single-seat aircraft of which the maximum total weight authorised does not exceed 910kg) otherwise than under a hire-purchase or conditional sale agreement.

(4) Notwithstanding that an aircraft may be flying for the purpose of public transport by reason of paragraph (3)(c), it shall not be deemed to be flying for the purpose of the public transport of passengers unless valuable consideration is given for the carriage of those passengers.

(5) A glider shall not be deemed to fly for the purpose of public transport for the purposes of Part 3 of this Order by virtue of paragraph (3)(c) if the valuable consideration given or promised for the primary purpose of conferring on a particular person the right to fly the glider on that flight is given or promised by a member of a flying club and the glider is owned or operated by that flying club.

(6) Notwithstanding the giving or promising of valuable consideration specified in paragraph (3)(c) in respect of the flight or the purpose of the flight it shall:

(a) subject to sub-paragraph (b), for all purposes other than Part 3 of this Order; and

(b) for the purposes of articles 19(2) and 12(2);

be deemed to be a private flight.

(7) Where under a transaction effected by or on behalf of a member of an association of persons on the one hand and the association of persons or any member thereof on the other hand, a person is carried in, or is given the right to fly, an aircraft in such circumstances that valuable consideration would be given or promised if the transaction were effected otherwise than aforesaid, valuable consideration shall, for the purposes of this Order, be deemed to have been given or promised, notwithstanding any rule of law as to such transactions.

(8) For the purposes of

(a) paragraph (3)(a), there shall be disregarded any valuable consideration given or promised in respect of a flight or the purpose of a flight by one company to another company which is:

(i) its holding company;

(ii) its subsidiary; or

(iii) another subsidiary of the same holding company.

(b) For the purposes of this article 'holding company' and 'subsidiary' have the meanings respectively specified in Section 736 of the Companies Act 1985.

Article 158: Public transport and aerial work – exceptions – flying displays etc.

(1) A flight shall, for the purposes of Part 4 of this Order, be deemed to be a private flight if:

(a) the flight is:

(i) wholly or principally for the purpose of taking part in an aircraft race, contest or flying display;

(ii) for the purpose of positioning the aircraft for such a flight as is specified in sub-paragraph (i) and is made with the intention of carrying out such a flight; or

(iii) for the purpose of returning after such a flight as is specified in sub-paragraph (i) to a place at which the aircraft is usually based;

(b) the only valuable consideration in respect of the flight or the purpose of the flight other than:

(i) valuable consideration specified in article 157(3)(c); or

(ii) in the case of an aircraft owned in accordance with article 162(2), valuable consideration which falls within article 162(3): falls within paragraph (2)(a) or (2)(b) or both.

(2) Valuable consideration falls within this paragraph if it either is:

(a) that given or promised to the owner or operator of an aircraft taking part in such a race, contest or flying display and such valuable consideration does not exceed the direct costs of the flight and a contribution to the annual costs of the aircraft which contribution shall bear no greater proportion to the total annual costs of the aircraft than the duration of the flight bears to the annual flying hours of the aircraft; or

(b) one or more prizes awarded to the pilot in command of an aircraft taking part in an aircraft race or contest to a value which shall not exceed £500 in respect of any one race or contest except with the permission in writing of the CAA granted to the organiser of the race or contest;

or falls within both sub-paragraphs (a) and (b).

(3) Any prize falling within paragraph (2)(b) shall be deemed for the purposes of this Order not to constitute remuneration for services as a pilot.

Article 159: Public transport and aerial work – exceptions – charity flights

(1) Subject to paragraph (2), a flight shall be deemed to be a private flight if the only valuable consideration given or promised in respect of the flight or the purpose of the flight other than:

(a) valuable consideration specified in article 157(3)(c); or

(b) in the case of an aircraft owned in accordance with article 162(2), valuable consideration which falls within article 162(3); or

is given or promised to a registered charity which is not the operator of the aircraft and the flight is made with the permission in writing of the CAA and in accordance with any conditions therein specified.

(2) If valuable consideration specified in article 157(3)(c) is given or promised the flight shall for the purposes of Part 3 of this Order (other than articles 19(2) and 20(2)) be deemed to be for the purposes of public transport.

Article 160: Public transport and aerial work – exceptions – cost sharing

(1) Subject to paragraph (4), a flight shall be deemed to be a private flight if the only valuable consideration given or promised in respect of the flight or the purpose of the flight falls within paragraph (2) and the criteria in paragraph (3) are satisfied.

(2) valuable consideration falls within this paragraph if it is:

(a) valuable consideration specified in article 157(3)(c);

(b) in the case of an aircraft owned in accordance with article 162(2), valuable consideration which falls within article 162(3); or

(c) is a contribution to the direct costs of the flight otherwise payable by the pilot in command.;

or falls within any two or all three sub-paragraphs.

(3) The criteria in this paragraph are satisfied if:
(a) no more than 4 persons (including the pilot) are carried;
(b) the proportion which the contribution referred to in paragraph 2(c) bears to the direct costs shall not exceed the proportion which the number of persons carried on the flight (excluding the pilot) bears to the number of persons carried (including the pilot);
(c) no information shall have been published or advertised prior to the commencement of the flight other than, in the case of an aircraft operated by a flying club, advertising wholly within the premises of such a flying club in which case all the persons carried on such a flight who are aged 18 years or over shall be members of that flying club; and
(d) no person acting as a pilot shall be employed as a pilot by, or be a party to a contract for the provision of services as a pilot with the operator of the aircraft being flown.
(4) If valuable consideration specified in article 157(3)(c) is given or promised, the flight shall for the purposes of Part 3 of this Order (other than articles 19(2) and 20(2)) be deemed to be for the purpose of public transport.

Article 161: Public transport and aerial work – exceptions – recovery of direct costs

(1) Subject to paragraph (2), a flight shall be deemed to be a private flight if the only valuable consideration given or promised in respect of the flight or the purpose of the flight other than:
(a) valuable consideration specified in article 157(3)(c); or
(b) in the case of an aircraft owned in accordance with article 162(2), valuable consideration which falls within article 162(3); is the payment of the whole or part of the direct costs otherwise payable by the pilot in command by or on behalf of the employer of the pilot in command, or by or on behalf of a body corporate of which the pilot in command is a director, provided that neither the pilot in command nor any other person who is carried is legally obliged, whether under a contract or otherwise, to be carried.
(2) If valuable consideration specified in article 157(3)(c) is given or promised the flight shall for the purposes of Part 3 of this Order (other than articles 19(2) and 20(2)) be deemed to be for the purposes of public transport.

Article 162: Public transport and aerial work – exceptions – jointly owned aircraft

(1) A flight shall be deemed to be a private flight if the aircraft falls within paragraph (2) and the only valuable consideration given or promised in respect of the flight or the purpose of the flight fall within paragraph (3).
(2) An aircraft falls within this paragraph if it is owned:
(a) jointly by persons (each of whom is a natural person) who each hold not less than a 5% beneficial share and:
(i) the aircraft is registered in the names of all the joint owners; or
(ii) the aircraft is registered in the name or names of one or more of the joint owners as trustee or trustees for all the joint owners and written notice has been given to the CAA of the names of all the persons beneficially entitled to a share in the aircraft; or
(b) by a company in the name of which the aircraft is registered and the registered shareholders of which (each of whom is a natural person) each hold not less than 5% of the shares in that company; and
(3) Valuable consideration falls within this paragraph if it is either:
(a) in respect of and is no greater than the direct costs of the flight and is given or promised by one or more of the joint owners of the aircraft or registered shareholders of the company which owns the aircraft; or
(b) in respect of the annual costs and given by one or more of such joint owners or shareholders (as aforesaid); or falls within both sub-paragraphs (a) and (b).

Article 163: Public transport and aerial work – exceptions - parachuting

A flight shall be deemed to be for the purposes of aerial work if it is a flight in respect of which valuable consideration has been given or promised for the carriage of passengers and which is for the purpose of:
(a) the dropping of persons by parachute and which is made under and in accordance with the terms of a parachuting permission granted by the CAA pursuant to article 67;
(b) positioning the aircraft for such a flight as is specified in sub-paragraph (a) and which is made with the intention of carrying out such a flight and on which no person is carried who it is not intended shall be carried on such a flight and who may be carried on such a flight in accordance with the terms of a parachuting permission granted by the CAA under article 67; or
(c) returning after such a flight as is specified in sub-paragraph (a) to the place at which the persons carried on such a flight are usually based and on which flight no persons are carried other than persons carried on the flight specified in sub-paragraph (a);

Article 164: Exceptions from application of provisions of the Order for certain classes of aircraft

The provisions of this Order other than articles 68, 74, 96(1), 97, 98, 144(1)(b) and (c), 155(1) and (2) shall not apply to or in relation to:
(a) any small balloon;
(b) any kite weighing not more than 2kg;
(c) any small aircraft; or
(d) any parachute, including a parascending parachute.

Article 165: Approval of persons to furnish reports

In relation to any of its functions under any of the provisions of this Order the CAA may approve a person as qualified to furnish reports to it and may accept such reports

Article 166: Certificates, authorisations, approvals and permissions

Wherever in this Order there is provision for the issue or grant of a certificate, authorisation, approval or permission by the CAA, unless otherwise provided, such a certificate, authorisation, approval or permission:

(a) shall be in writing;

(b) may be issued or granted subject to such conditions as the CAA thinks fit;

(c) may be issued or granted, subject to article 92, for such periods as the CAA thinks fit.

Article 167: Competent authority

(1) The CAA shall be:

(a) the national aviation authority of the United Kingdom for the purposes of Regulation (EC) No. 1592/2002 of the European Parliament and of the Council of 15th July 2002 on common rules in the field of civil aviation and establishing a European Aviation Safety Agency, and

(b) the competent authority of the United Kingdom for the purposes of:

(i) Commission Regulation (EC) No. 1702/2003 of 24th September 2003 laying down implementing rules for the airworthiness and environmental certification of aircraft and related products, parts and appliances, as well as for the certification of design and production organisations; and

(ii) Commission Regulation (EC) No. 2042/2003 of 20th November 2003 on the continuing airworthiness of aircraft and aeronautical products, parts and appliances, and on the approval of organisations and personnel involved in these tasks.

(c) The Secretary of State shall be the competent authority under article 15 of the Council Directive 96/29/Euratom of 13th May 1996 for the purposes of article 42 of the Directive.

Article 168: Saving

(1) Subject to articles 128 and 130, nothing in this Order or any regulations made thereunder shall confer any right to land in any place as against the owner of the land or other persons interested therein.

(2) Nothing in this Order shall oblige the CAA to accept an application form the holder of any current certificate, licence, approval, permission or other document, being an application for the renewal of that document, or for the granting of another document in continuation of or in substitution for the current document, if the application is made more than 60 days before the current document is due to expire.

SCHEDULE 1 ORDERS REVOKED

Article 2

References

	The Air Navigation Order 200	S.I. 2000/1562
	The Air Navigation (Amendment) Order 2001	S.I. 2001/397
	The Air Navigation (Amendment) Order 2002	S.I. 2002/264
	The Air Navigation (Amendment) (No. 2) Order 2002	S.I. 2002/1628
	The Air Navigation (Amendment) Order 2003	S.I. 2003/777
	The Air Navigation (Amendment) (No. 2) Order 2003	S.I. 2003/2905
	The Air Navigation (Amendment) Order 2004	S.I. 2004/705

PART A
Articles 4(6) and 155(6)

Table of general classification of aircraft

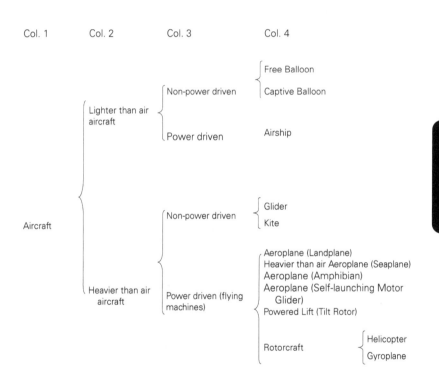

PART B Article 5(2)
Nationality and registration marks of aircraft registered in the United Kingdom

1 General

(1) The nationality mark of the aircraft shall be the capital letter 'G' in Roman character and the registration mark shall be a group of four capital letters in Roman character assigned by the CAA on the registration of the aircraft. The letters shall be without ornamentation and a hyphen shall be placed between the nationality mark and the registration mark.

(2) The nationality and registration marks shall be displayed to the best advantage, taking into consideration the constructional features of the aircraft and shall always be kept clean and visible.

ANO — Schedule 1 Article 5(2) Nationality and registration marks of aircraft registered in the United Kingdom

(3) The letters constituting each group of marks shall be of equal height and they, and the hyphen, shall all be of the same single colour which shall clearly contrast with the background on which they appear.
(4) The nationality and registration marks shall also be inscribed on a fireproof metal plate affixed in a prominent position:
(a) in the case of a microlight aeroplane, either in accordance with sub-paragraph (c) or on the wing;
(b) in the case of a balloon, on the basket or envelope; or
(c) in the case of any other aircraft on the fuselage or car as the case may be.
(5) The nationality and registration marks shall be painted on the aircraft or shall be affixed thereto by any other means ensuring a similar degree of permanence in the manner specified in paragraphs 2 and 3 of this Part.

2 Position and size of marks

(1) The position and size of marks on heavier than air aircraft (excluding kites) shall be as follows:
(a) on the horizontal surfaces of wings:
(i) on aircraft having a fixed wing surface, the marks shall appear on the lower surface of the wing structure and shall be on the port wing unless they extend across the whole surface of both wings. So far as is possible the marks shall be located equidistant from the leading and trailing edges of the wings. The tops of the letters shall be towards the leading edge of the wing;
(ii) the height of the letters shall be:
(aa) subject to sub-paragraph (bb), at least 50 centimetres;
(bb) if the wings are not large enough for the marks to be 50 centimetres in height, marks of the greatest height practicable in the circumstances;
(b) on the fuselage (or equivalent structure) and vertical tail surfaces:
(i) the marks shall also appear either:
(aa) on each side of the fuselage (or equivalent structure), and shall, in the case of fixed wing aircraft be located between the wings and the horizontal tail surface; or
(bb) on the vertical tail surfaces;
(ii) when located on a single vertical tail surface, the marks shall appear on both sides. When located on multi-vertical tail surfaces, the marks shall appear on the outboard sides of the outer-surfaces. Subject to sub-paragraphs (iv) and (v), the height of the letters constituting each group of marks shall be at least 30 centimetres;
(iii) if one of the surfaces authorised for displaying the required marks is large enough for those marks to be 30 centimetres in height (whilst complying with sub-paragraph (v)) and the other is not, marks of 30 centimetres in height shall be placed on the largest authorised surface;
(iv) if neither authorised surface is large enough for marks of 30 centimetres in height (whilst complying with sub-paragraph (v)), marks of the greatest height practicable in the circumstances shall be displayed on the larger of the two authorised surfaces;
(v) the marks on the vertical tail surfaces shall be such as to leave a margin of at least 5 centimetres along each side of the vertical tail surface;
(vi) on rotary wing aircraft where owing to the structure of the aircraft the greatest height practicable for the marks on the side of the fuselage (or equivalent structure) is less than 30 centimetres, the marks shall also appear on the lower surface of the fuselage as close to the line of symmetry as is practicable and shall be placed with the tops of the letters towards the nose. The height of the letters constituting each group of marks shall be:
(aa) subject to sub-paragraph (bb), at least 50 centimetres; or
(bb) if the lower surface of the fuselage is not large enough for the marks to be of 50 centimetres in height, marks of the greatest height practicable in the circumstances;
(c) wherever in this paragraph marks of the greatest height practicable in the circumstances are required, that height shall be such as is consistent with compliance with paragraph 3 of this Part.
(2) The position and size of marks on airships and free balloons shall be as follows:
(a) in the case of airships the marks shall be placed on each side of the airship. They shall be placed horizontally either on the hull near the maximum cross-section of the airship or on the lower vertical stabiliser;
(b) in the case of free balloons, the marks shall be in two places on diametrically opposite sides of the balloon;
(c) in the case of both airships and free balloons the side marks shall be so placed as to be visible from the sides and from the ground. The height of the letters shall be at least 50 centimetres.

3 Width, spacing and thickness of marks

(1) For the purposes of this paragraph:
(a) 'standard letter' shall mean any letter other than the letters I, M and W;
(b) the width of each standard letter and the length of the hyphen between the nationality mark and the registration mark shall be two thirds of the height of a letter;
(c) the width of the letters M and W shall be neither less than two thirds of their height nor more than their height; and
(d) the width of the letter I shall be one sixth of the height of the letter.
(2) The thickness of the lines comprising each letter and hyphen shall be one sixth of the height of the letters forming the marks.
(3) Each letter and hyphen shall be separated from the letter or hyphen which it immediately precedes or follows by a space equal to either one quarter or one half of the width of a standard letter. Each such space shall be equal to every other such space within the marks.

PART C

4 Conditions in aircraft dealer's certificate

The conditions in an aircraft dealer's certificate shall be as follows:

(1) The operator of the aircraft shall be the registered owner of the aircraft, who shall be the holder of an aircraft dealer's certificate granted under this Order.
(2) The aircraft shall fly only for the purpose of:
(a) testing the aircraft;
(b) demonstrating the aircraft with a view to the sale of that aircraft or of other similar aircraft;
(c) proceeding to or from a place at which the aircraft is to be tested or demonstrated as aforesaid, or overhauled, repaired or modified;
(d) delivering the aircraft to a person who has agreed to buy, lease or sell it; or
(e) proceeding to or from a place for the purpose of storage.
(3) Without prejudice to article 52 the operator of the aircraft shall satisfy himself before the aircraft takes off that the aircraft is in every way fit for the intended flight.
(4) The aircraft shall fly only within the United Kingdom.

SCHEDULE 3

A and B Conditions and categories of certificate of airworthiness
Articles 3(3), 8(2), and 65(7)
PART A

A and B Conditions

A Conditions

(1) An non-EASA aircraft registered in the United Kingdom may fly for a purpose set out in paragraph (2) subject to the conditions contained in paragraphs (3) to (8) when either:
(a) it does not have a certificate of airworthiness duly issued or rendered valid under the law of the United Kingdom; or
(b) the certificate of airworthiness or certificate of validation issued in respect of the aircraft has ceased to be in force by virtue of any of the matters specified in Article 10.
(2) The purposes referred to in paragraph (1) are:
(a) In the case of an aircraft falling within paragraph (1)(a) the aircraft shall fly only for the purpose of enabling it to:
(i) qualify for the issue, renewal or validation of a certificate of airworthiness after an application has been made for such issue, renewal or validation as the case may be, or carry out a functional check of a previously approved modification of the aircraft (and for the purpose of this Schedule 'a previously approved modification' shall mean a modification which has previously been approved by the CAA or by an organisation approved for that purpose by the CAA in respect of that aircraft or another aircraft of the same type);
(ii) proceed to or from a place at which any inspection, repair, modification, maintenance, approval, test or weighing of, or the installation of equipment in, the aircraft is to take place or has taken place for a purpose referred to in sub-paragraph (i), after any relevant application has been made, or at which the installation of furnishings in, or the painting of, the aircraft is to be undertaken; or
(iii) proceed to or from a place at which the aircraft is to be or has been stored.
(b) In the case of an aircraft falling within paragraph (1)(b), the aircraft shall fly only so as to enable it to:
(i) proceed to a place at which any inspection or maintenance required by virtue of Article 10(b)(ii) is to take place; or
(ii) proceed to a place at which any inspection, maintenance or modification required by virtue of Article 10(b)(i) or (c) is to take place and in respect of which flight the CAA has given permission in writing; or
(iii) carry out a functional check, test or in-flight adjustment in connection with the carrying out in a manner approved by the CAA of any overhaul, repair, previously approved modification, inspection or maintenance required by virtue of Article 10.
(3) The aircraft, including any modifications, shall be of a design which previously has been approved by the CAA, or by an organisation approved for that purpose by the CAA, as being compliant with a standard accepted by the CAA as appropriate for the issue of a United Kingdom certificate of airworthiness.
(4) The aircraft and its engines shall be certified as fit for flight by the holder of an aircraft maintenance engineer's licence granted under this Order, being a licence which entitles him to issue that certificate or by a person approved by the CAA for the purpose of issuing certificates under this condition, and in accordance with that approval.
(5) The aircraft shall carry the minimum flight crew specified in any certificate of airworthiness or validation which has previously been in force under the Order in respect of the aircraft, or is or has previously been in force in respect of any other aircraft of identical design.
(6) The aircraft shall not carry any persons or cargo except persons performing duties in the aircraft in connection with the flight or persons who are carried in the aircraft to perform duties in connection with a purpose referred to in paragraph (2).
(7) The aircraft shall not fly over any congested area of a city, town or settlement except to the extent that it is necessary to do so in order to take-off or land.

ANO — SCHEDULE 3 Articles 3(2), 8(2), and 55(7)
PART B

B Conditions

(1) A non-EASA aircraft may fly for a purpose set out in paragraph (2) subject to the conditions set out in paragraphs (3) to (8) whether or not it is registered in accordance with article 3(1) and when there is not in force:

(a) in the case of an aircraft which is so registered, a certificate of airworthiness duly issued or rendered valid under the law of the country in which the aircraft is registered or,

(b) in the case of an aircraft which is not so registered, either a certificate of airworthiness duly issued or rendered valid under the law of the United Kingdom or a permit to fly issued by the CAA in respect of that aircraft.

(2) The purposes referred to in paragraph (1) are:

(a) experimenting with or testing the aircraft (including any engines installed thereon) or any equipment installed or carried in the aircraft;

(b) enabling it to qualify for the issue of a certificate of airworthiness or the validation thereof or the approval of a modification of the aircraft or the issue of a permit to fly;

(c) demonstrating and displaying the aircraft, any engines installed thereon or any equipment installed or carried in the aircraft with a view to the sale thereof or of other similar aircraft, engines or equipment;

(d) demonstrating and displaying the aircraft to employees of the operator;

(e) the giving of flying training to or the testing of flight crew employed by the operator or the training or testing of other persons employed by the operator and who are carried or are intended to be carried pursuant to paragraph (7)(a); or

(f) proceeding to or from a place at which any experiment, inspection, repair, modification, maintenance, approval, test or weighing of the aircraft, the installation of equipment in the aircraft, demonstration, display or training is to take place for a purpose referred to in sub-paragraphs (a), (b), (c), (d) or (e) or at which installation of furnishings in, or the painting of, the aircraft is to be undertaken.

(3) The flight shall be operated by a person approved by the CAA for the purposes of these Conditions and subject to any additional conditions which may be specified in such an approval.

(4) If not registered in the United Kingdom the aircraft shall be marked in a manner approved by the CAA for the purposes of these Conditions, and articles 20, 22, 52, 55, 86 and 88 shall be complied with in relation to the aircraft as if it was registered in the United Kingdom.

(5) If not registered in the United Kingdom, the aircraft shall carry such flight crew as may be necessary to ensure the safety of the aircraft.

(6) No person shall act as pilot in command of the aircraft except a person approved for the purpose by the CAA.

(7) The aircraft shall not carry any cargo, or any persons other than the flight crew except the following:

(a) persons employed by the operator who during the flight carry out duties or are tested or receive training in connection with a purpose specified in paragraph (2);

(b) persons acting on behalf of the manufacturers of component parts of the aircraft (including its engines) or of equipment installed in or carried in the aircraft for carrying out during the flight duties in connection with a purpose so specified;

(c) persons approved by the CAA under article 165r as qualified to furnish reports for the purposes of article 9;

(d) persons other than those carried under the preceding provisions of this paragraph who are carried in the aircraft in order to carry out a technical evaluation of the aircraft or its operation;

(e) cargo which comprises equipment carried in connection with a purpose specified in paragraph (2)(f); or

(f) persons employed by the operator or persons acting on behalf of the manufacturers of component parts of the aircraft (including its engines) or of equipment installed in or carried in the aircraft in connection with a purpose specified in paragraph (2)(f) which persons have duties in connection with that purpose.

(8) The aircraft shall not fly, except in accordance with procedures which have been approved by the CAA in relation to that flight, over any congested area of a city, town or settlement.

PART B
Article 9

Categories of certificate or airworthiness and purposes for which aircraft may fly

Categories of certificate of airworthiness	Purpose for which the aircraft may fly
Standard	Any purpose
Special Category	Any purpose, other than public transport, specified in the certificate of airworthiness but not including the carriage of passengers unless expressly permitted

SCHEDULE 4

Articles 16(9) and 19(2)
Aircraft equipment

1 Every aircraft of a description specified in the first column of the Table set forth in paragraph 5 of this Schedule and which is registered in the United Kingdom shall be provided, when flying in the circumstances specified in the second column of the said Table, with adequate equipment, and for the purpose of this paragraph the expression 'adequate equipment' shall mean, subject to paragraph (2), the scales of equipment respectively indicated in the third column of that Table.

2 (1) If the aircraft is flying in a combination of such circumstances, the scales of equipment shall not on that account be required to be duplicated.

(2) The equipment carried in an aircraft as being necessary for the airworthiness of the aircraft shall be taken into account in determining whether this Schedule is complied with in respect of that aircraft.

(3) (1) For the purposes of the Table in paragraph 5, flying time in relation to a helicopter or gyroplane shall be calculated on the assumption that it is flying in still air at the speed specified in the relevant flight manual as the speed for compliance with regulations governing flights over water.

(2) In this Schedule "day" means the time from half an hour before sunrise until half an hour after sunset (both times exclusive), sunset and sunrise being determined at surface level.

4 The following items of equipment shall not be required to be of a type approved by EASA or the CAA:

(a) The equipment referred to in Scale A (2).
(b) First aid equipment and handbook, referred to in Scale A.
(c) Time-pieces, referred to in Scale F.
(d) Torches, referred to in Scales G, H, K and Z.
(e) Whistles, referred to in Scale;
(f) Sea anchors, referred to in Scales J and K.
(g) Rocket signals, referred to in Scale J.
(h) Equipment for mooring, anchoring or manoeuvring aircraft on the water, referred to in Scale J.
(i) Paddles, referred to in Scale K.
(j) Food and water, referred to in Scales K, U and V.
(k) First aid equipment, referred to in Scales K, U and V.
(1) Stoves, cooking utensils, snow shovels, ice saws, sleeping bags and Arctic suits, referred to in Scale V.
(m) Megaphones, referred to in Scale Y.

5 Table

Description of aircraft	Circumstances of flight	Scale of equipment required
(1) Gliders	(a) flying for purposes other than public transport or aerial work; and when flying by night	A(2)
	(b) flying for the purpose of public transport or aerial work; and	A, B(1), (2), (3), (4), (5), (6) and (7)
	(i) when flying by night	D and F(1)
	(ii) when carrying out aerobatic manoeuvres	C and G, B(8) and (9)
(2) Aeroplanes	(a) flying for purposes other than public transport; and	A(1) and (2) and B(1), (2), (3), (4), (5) and (6)
	(i) when flying by night	C and D
	(ii) when flying under Instrument Flight Rules:	
	(aa) outside controlled airspace	D
	(bb) within Class A, B or C airspace	E with E (4) duplicated and F
	(cc) within Class D and E airspace	E and F
	(iii) when carrying out aerobatic manoeuvres	B(8) and (9)
	(b) flying for the purpose of public transport; and	A, B(1), (2), (3), (4), (5), (6) and (7), D and F(1)
	(i) when flying under Instrument Flight Rules except flights outside controlled airspace in the case of aeroplanes having a maximum total weight authorised not exceeding 1150kg	E with E(4) duplicated and F
	(ii) when flying by night; and in the case of aeroplanes of which the maximum total weight authorised exceeds 1150kg	C and G, E with E(4) duplicated and F
	(iii) when flying over water beyond gliding distance from land	H
	(iv) on all flights on which in the event of any emergency occurring during the take-off or during the landing at the intended destination or any likely alternate destination it is reasonably possible that the aeroplane would be forced to land onto water	H

ANO

SCHEDULE 4 Articles 12(6) and 14(2)
Aircraft equipment

Description of aircraft	Circumstances of flight	Scale of equipment required
	(v) when flying over water: (aa) in the case of an aeroplanes capable of continuing the flight to an aerodrome with the critical power unit becoming inoperative, at a greater distance from land suitable for making an emergency landing than that corresponding to 120 minutes at cruising speed or 400 nautical miles whichever is the lesser; or	H and K
	(bb) in the case of all other aeroplanes, at a greater distance from and suitable for making an emergency landing than that corresponding to 30 minutes at cruising speed or 100 nautical miles, whichever is the lesser	H and K
	(vi) having a certificate of airworthiness first issued (whether in the United Kingdom or elsewhere) before 1st January 2002	KK(1) or (2)
	(vii) having a certificate of airworthiness first issued (whether in the United Kingdom or elsewhere) on or after 1st January 1989	KK(2)
	(viii) on all flights which involve manoeuvres on water	H, J and K
	(ix) when flying at a height of 10,000ft or more above mean sea level;	
	(aa) having a certificate of airworthiness first issued (whether in the United Kingdom or elsewhere) before 1st January 1989	L1 or L2
	(bb) having a certificate of airworthiness first issued (whether in the United Kingdom or elsewhere) on or after 1st January 1989	L2
	(x) on flights when the weather reports or forecasts available at the aerodrome at the time of departure indicate that conditions favouring ice formation are likely to be met	M
	(xi) when carrying out aerobatic manoeuvres	B(8) and (9)
	(xii) on all flights on which the aircraft carries a flight crew of more than one person	N
	(xiii) on all flights for the purpose of the public transport of passengers	Q and Y(1), (2) and (3)
	(xiv) on all flights by a pressurised aircraft	R
	(xv) when flying over substantially uninhabited land areas where, in the event of an emergency landing, tropical conditions are likely to be met	U
	(xvi) when flying over substantially uninhabited land or other areas where, in the event of an emergency landing, polar conditions are likely to be met	V
	(xvii) when flying at an altitude of more than 49,000ft	W
(3) Turbine-jet aeroplanes having a maximum total weight authorised exceeding 5700kg or pressurised aircraft having a maximum total weight authorised exceeding 11 400kg	when flying for the purpose of public transport	O
(4) Turbine-engined aeroplanes having a maximum total weight authorised exceeding 5700kg and piston-engined aeroplanes having a maximum total weight authorised exceeding 27,000kg except for such aeroplanes falling within paragraphs (5) or (6):		
(a) which are operated by an air transport undertaking; or	When flying on any flight	P
(b) which are public transport aeroplanes in respect of which application has been made and not withdrawn or refused for a certificate of airworthiness, and which fly under an EASA permit to fly, "A Conditions" or under a certificate of airworthiness in the Special Category described in Part B of Schedule 3.	When flying on any flight	P

ANO

SCHEDULE 4 Articles 12(6) and 14(2)
Aircraft equipment

Description of aircraft	Circumstances of flight	Scale of equipment required
(5) Public transport aeroplanes in respect of which there is in force a certificate of airworthiness and public transport aeroplanes in respect of which an application has been made, and not withdrawn or refused, for such a certificate of airworthiness and which fly under an EASA permit to fly 'A Conditions' or under a certificate of airworthiness in the Special Category described in Part B of Schedule 3 except for such aeroplanes falling within paragraph (6):		
(a) which conform to a type first issued with a type certificate (whether in the United Kingdom or elsewhere) on or after 1st April 1971 and which have a maximum total weight authorised exceeding 5700kg but not exceeding 11 400kg; or	When flying on any flight	S(1)
(b) which conform to a type first issued with a type certificate (whether in the United Kingdom or elsewhere) on or after 1st April 1971 and which have a maximum total weight authorised exceeding 11 400kg but not exceeding 27,000kg; or	When flying on any flight	S(2)
(c) which conform to a type first issued with a type certificate (whether in the United Kingdom or elsewhere) on or after 1st April 1971 and which have a maximum total weight authorised exceeding 27,000kg but not exceeding 230,000kg; or	When flying on any flight	S(3)
(d) which conform to a type first issued with a type certificate in the United Kingdom on or after 1st January 1970 and which have a maximum total weight authorised exceeding 230,000kg;	When flying on any flight	S(3)
(6) Public transport aeroplanes in respect of which there is in force a certificate of airworthiness and public transport aeroplanes in respect of which application has been made, and not withdrawn or refused, for a certificate of airworthiness and which fly under an EASA permit to fly, 'A Conditions' or under a certificate of airworthiness in the Special Category described in Part B of Schedule 3;		
(a) for which an individual certificate of airworthiness was first issued (whether in the United Kingdom or elsewhere) on or after 1st June 1990 and which have a maximum total weight authorised not exceeding 5700kg are powered by 2 or more turbine engines and are certified to carry more than 9 passengers; or		S(4)
(b) for which an individual certificate of airworthiness was first issued (whether in the United Kingdom or elsewhere) on or after 1st June 1990 and which have a maximum total weight authorised exceeding 5700kg but not exceeding 27,000kg; or		S(5)

SCHEDULE 4 Articles 12(6) and 14(2)
Aircraft equipment

Description of aircraft	Circumstances of flight	Scale of equipment required
(c) for which an individual certificate of airworthiness was first issued (whether in the United Kingdom or elsewhere) on or after 1st June 1990 and which have a maximum total weight authorised exceeding 27,000kg.	When flying on any flight	S(6)
(7) Aerial Work and Private aeroplanes for which an individual certificate of airworthiness was first issued (whether in the United Kingdom or elsewhere) on or after 1st June 1990 and which have a maximum total weight authorised exceeding 27,000kg.	When flying on any flight	S(6)
(8) Public transport eroplanes:		
(a) which conform to a type first issued with a type certificate (whether in the United Kingdom or elsewhere) on or after 1st April 1971 and having a maximum total weight authorised exceeding 27,000kg; or	When flying on any flight	T
(b) which conform to a type first issued with a type certificate in the United Kingdom on or after 1st January 1970 and which have a maximum total weight authorised exceeding 230,000kg and in respect of which there is in force such a certificate of airworthiness; or	When flying on any flight	T
(c) having a maximum total weight authorised exceeding 27,000kg which conform to a type first issued with a type certificate on or after 1st April 1971 (or 1st January 1970 in the case of an aeroplane having a maximum total weight authorised exceeding 230,000kg) in respect of which an application has been made, and not withdrawn or refused for a certificate of airworthiness and which fly under an EASA permit to fly, 'A Conditions or under a certificate of airworthiness in the Special Category described in Part B of Schedule 3	When flying on any flight	T
(9) Aeroplanes powered by one or more turbine jets or one or more turbine propeller engines and which have a maximum total weight authorised exceeding 15,000kg or which a maximum approved passenger seating configuration of more than 30 passengers:	When flying for the purpose of public transport	X(1)
(10) Aeroplanes which are powered by one or more turbine jets or one or more turbine propeller engines and which have a maximum total weight authorised exceeding 5 700kg but not exceeding 15,000kg or with a maximum approved passenger configuration or more than 9 but not exceeding 30:	When flying for the purpose of public transport except when flying under and in accordance with the terms of a police air operator's certificate	X(1)

ANO

SCHEDULE 4 Articles 12(6) and 14(2)
Aircraft equipment

Description of aircraft	Circumstances of flight	Scale of equipment required
(11) Aeroplanes which are powered by one or more turbine jets or one or more turbine propeller engines and which have a maximum total weight authorised exceeding 5700kg or with a maximum approved passenger seating configuration or more than 9:		
(a) in respect of which there is in force a certificate of airworthiness except any such aeroplanes as come within sub-paragraph (b)	When flying for purposes other than public transport	X(1) or X(2)
(b) in respect of which there is in force a certificate of airworthiness and which have equipment capable of giving warning to the pilot of the potentially hazardous proximity of ground or water installed before 1st April 2000	When flying for purposes other than public transport on or after 1st January 2007	X(1) or X(2)
(12) Aeroplanes:		
(a) powered by one or more turbo-jets and which have a maximum total weight authorised exceeding 22 700kg; or	When flying by night for the purpose of the public transport of passengers	Z(1) and (2)
(b) having a maximum total weight authorised exceeding 5700kg and which conform to a type for which a certificate of airworthiness was first applied for (whether in the United Kingdom or elsewhere) after 30th April 1972 but not including any aeroplane which in the opinion of the CAA is identical in all matters affecting the provision of emergency evacuation facilities to an aeroplane for which a certificate of airworthiness was first applied for before that date; or	When flying by night for the purposes of the public transport of passengers	Z(1) and (2)
(c) with a maximum approved passenger seating configuration of more than 19 or	When flying by night for the purpose of the public transport of passengers	Z(1)
(d) having a maximum total weight authorised exceeding 5700kg and which conform to a type for which a certificate of airworthiness was first applied for (whether in the United Kingdom or elsewhere) after 30th April 1972 but not including any aeroplane which in the opinion of the CAA is identical in all matters affecting the provision of emergency evacuation facilities to an aeroplane for which a certificate of airworthiness was first applied for before that date; or	When flying for the purpose of the public transport of passengers	Z(3)
(e) powered by one or more turbo-jets and which have a maximum total weight authorised exceeding 22 700kg; or	When flying for the purposes of the public transport of passengers	Z(3)
(f) first issued with a type certificate (whether in the United Kingdom or elsewhere) on or after 1st January 1958 and with a maximum approved passenger seating configuration of more than 19	When flying for the purposes of the public transport of passengers	Z(3)
(13) Aeroplanes:		

SCHEDULE 4 Articles 12(6) and 14(2)
Aircraft equipment

Description of aircraft	Circumstances of flight	Scale of equipment required
(a) powered by one or more turbine jets	When flying on any flight	AA
(b) powered by one or more turbine propeller engines and having a maximum total weight authorised exceeding 5700kg and first issued with a certificate of airworthiness in the United Kingdom on or after 1st April 1989	When flying on any flight	AA
(14) Public transport aeroplanes	When flying for the purpose of public transport of passengers	Y(4)
(15) Helicopters and Gyroplanes	(a) flying for purposes other than public transport; and	A (1)and (2) B (1), (2), (3), (4), (5) and (6)
	(i) when flying by day under Visual Flight Rules with visual ground reference	D
	(ii) when flying by day under Instrument Flight Rules or without visual ground reference	
	(aa) outside controlled airspace	E with E (2) duplicated
	(bb) within controlled airspace	E with both E (2) and E(4) duplicated and F with F (4) for all weights
	(iii) when flying at night	
	(aa) with visual ground reference	C, E, G (iii) and G (v)
	(bb) without visual ground reference outside controlled airspace	C, E with E (2) duplicated G (3) and G (5) and G(6)
	(cc) without visual ground reference within controlled airspace	C, E with both E (2) and E (4) duplicated, F with F (4) for all weights, G (3) and G (5) and (6)
	(b) flying for the purpose of public transport; and	A, B(1), (2), (3), (4), (5), (6) and (7) F (1) and F (4) for all weights
	(i) when flying by day under Visual Flight Rules with visual ground reference	D
	(ii) when flying by day under Instrument Flight Rules or without	E with both E(2) and E(4) duplicated
	visual ground reference	F(2), F(3) and F(5)
	(iii) when flying by night with visual ground reference	
	(aa) when flying with one pilot	C, E with E(2) duplicated and either E(4) duplicated or a radio altimeter, F(2), F(3), F(5) and G
	(bb) when flying in circumstances where two pilots are required	C, E, F(2), F(3), F(5) and G for each pilots station
	(iv) when flying by night without visual ground reference	C, E with both E(2) and E(4) duplicated, F(2), F(3), F(5) and G
	(v) when flying over water	
	(aa) in the case of a helicopter carrying out Performance Class 2 or 3 operations or a gyroplane classified in its certificate of airworthiness as being of Performance group A2 or B when beyond auto-rotational gliding distance from land suitable for an emergency landing	E and H

Description of aircraft	Circumstances of flight	Scale of equipment required
	(bb) on all flights on which in the event of any emergency occurring during the take-off or during the landing at the intended destination or any likely alternate destination it is reasonably possible that the helicopter or gyroplane would be forced to land onto water	H
	(cc) in the case of a helicopter carrying out Performance Class 1 or 2 operations or a gyroplane classified in its certificate or airworthiness as being of performance group A2 when beyond 10 minutes flying from land	E, H, K and T
	(dd) for more than a total of 3 minutes in any flight	EE
	(ee) in the case of a helicopter carrying out Performance Class 1 or 2 operations or a gyroplane classified classified in its certificate of airworthiness as being of performance group A2 which is intended to fly beyond 10 minutes flying time from land or which actually flies beyond 10 minutes flying time from land, on a flight which is either in support of or in connection with the offshore exploitation, or exploration of mineral resources (including gas) or is on a flight under and in accordance with the terms of a police air operator's certificate, when in either case the weather reports or forecasts available to the commander of the aircraft indicate that the sea temperature will be less than plus 10°C during the flight or when any part of the flight is at night	I
	(vi) when flying on Performance Class 1 or 2 operations over water beyond 10 minutes flying time from land and not required to comply with sub-paragraph (ix)	KK(2)
	(vii) when flying on Performance Class 3 operations beyond auto-rotational or safe forced landing distance from land	KK(2)
	(viii) when flying over land areas which have been designated by the State concerned as areas in which search and rescue would be especially difficult	KK(2)
	(ix) when flying on performance Class 1 or 2 operations over water in a hostile environment at a distance from land corresponding to more than ten minutes flying time at normal cruising speed in support of or in connection with the offshore exploitation or exploration of mineral resources (including gas)	KK(3)
	(x) on all flights which involve manoeuvres on water	H, J and K
	(xi) when flying at a height of 10,000ft or more above mean sea level:	
	(aa) having a certificate of airworthiness first issued (whether in the United Kingdom or elsewhere) before 1st January 1989	L1 or L2
	(bb) having a certificate of airworthiness first issued (whether in the United Kingdom or elsewhere) on or after 1st January 1989	L2
	(xii) on flights when the weather reports or forecasts available at the aerodrome at the time of departure indicate that conditions favouring ice formation are likely to be met	M
	(xiii) on all flights on which the aircraft carries a flight crew of more than one person	N
	(xiv) on all flights for the purpose of the public transport of passengers	Y(I), (ii) and (iii)
	(xv) when flying over substantially uninhabited land areas where in the event of an emergency landing, tropical conditions are likely to be met	U
	(xvi) when flying over substantially uninhabited land or other areas where, in the event of an emergency landing, polar conditions are likely to be met	V
(16) Helicopters and Gyroplanes:	When flying by night for the purpose of the public transport of	Z(1) and (2)

SCHEDULE 4 Articles 12(6) and 14(2)
Aircraft equipment

Description of aircraft	Circumstances of flight	Scale of equipment required
(a) having a maximum total weight authorised exceeding 5700kg and which conform to a type for which a certificate of airworthiness was first applied for (whether in the United Kingdom or elsewhere) after 30th April 1972 but not including any helicopter or gyroplane which in the opinion of the CAA is identical in all matters affecting the provision of emergency evacuation facilities to a helicopter or gyroplane for which a certificate of airworthiness was first applied for before that date; or	passengers	
(b) with a maximum approved passenger seating configuration or more than 19; or	When flying by night for the purposes of the public transport of passengers	Z(1)
(c) which are public transport helicopters or gyroplanes in respect of which there is in force a certificate of airworthiness and public transport helicopters or gyroplanes in respect of which application has been made and not withdrawn or refused for a certificate of airworthiness, and which fly under an EASA permit to fly, "A conditions" or under a certificate of airworthiness in the Special Category described in Part B of Schedule 3; and		
(i) which have a maximum total weight authorised exceeding 2730kg but not exceeding 7000kg or with a maximum approved passenger seating configuration of more than 9 or both	When flying on any flight	SS(1) or (3)
(ii) which have a maximum total weight authorised exceeding 7000kg	When flying on any flight	SS(2) or (3)

6 The scales of equipment indicated in the foregoing Table shall be as follows:

Scale A

(1) Spare fuses for all electrical circuits the fuses of which can be replaced in flight, consisting of 10 per cent of the number of each rating or three of each rating, whichever is the greater.

(2) Maps, charts, codes and other documents and navigational equipment necessary, in addition to any other equipment required under this Order, for the intended flight of the aircraft including any diversion which may reasonably be expected.

(3) First aid equipment of good quality, sufficient in quantity, having regard to the number of persons on board the aircraft, and including the following:

(a) roller bandages,
(b) triangular bandages;
(c) adhesive plaster;
(d) absorbent gauze and wound dressings;
(e) burn dressings;
(f) safety pins;
(g) haemostatic bandages or tourniquets;
(i) scissors;
(j) antiseptic;
(k) analgesic and stimulant drugs;
(l) splints, in the case of aeroplanes the maximum total weight authorised of which exceeds 5700kg;
(m) a handbook on first aid.

(4) In the case of a flying machine used for the public transport of passengers in which, while the flying machine is at rest on the ground, the sill of any external door intended for the disembarkation of passengers, whether normally or in an emergency:

(a) is more than 1.82 metres from the ground when the undercarriage of the machine is in the normal position for taxiing; or

(b) would be more than 1.82 metres from the ground if the whole or any part or the undercarriage should collapse, break or fail to function;
apparatus readily available for use at each such door consisting of a device or devices which will enable passengers to reach the ground safely in an emergency while the flying machine is on the ground, and can be readily fixed in position for use.

Scale AA

(1) Subject to sub-paragraph (2), an altitude alerting system capable of alerting the pilot upon approaching a pre-selected altitude in either ascent or descent, by a sequence of visual and aural signals in sufficient time to establish level flight at that pre-selected altitude and when deviating above or below that pre-selected altitude, by a visual and an aural signal.

(2) If the system becomes unserviceable, the aircraft may fly or continue to fly, until it first lands at a place at which it is reasonably practicable for the system to be repaired or replaced.

Scale B

(1) If the maximum total weight authorised of the aircraft is 2730kg or less, for every pilot's seat and for any seat situated alongside a pilot's seat, either a safety belt with one diagonal shoulder strap or a safety harness, or with the permission of the CAA, a safety belt without a diagonal shoulder strap which permission may be granted if the CAA is satisfied that it is not reasonably practicable to fit a safety belt with one diagonal shoulder strap or a safety harness.

(2) If the maximum total weight authorised of the aircraft exceeds 2730kg, either a safety harness for every pilot's seat and for any seat situated alongside a pilot's seat, or with the permission of the CAA, a safety belt with one diagonal shoulder strap which permission may be granted if the CAA is satisfied that it is not reasonably practicable to fit a safety harness.

(3) For every seat in use (not being a seat referred to in paragraphs (1), (2), (5) and (6)) a safety belt with or without one diagonal shoulder strap or a safety harness.

(4) In addition and to be attached to or secured by the equipment required in paragraph (3) above, a child restraint device for every child under the age of two years on board.

(5) On all flights for the public transport of passengers by aircraft, for each seat for use by cabin crew who are required to be carried under this Order, a safety harness.

(6) On all flights in aeroplanes in respect of which a certificate of airworthiness was first issued (whether in the United Kingdom or elsewhere) on or after 1st February 1989 the maximum total weight authorised of which does not exceed 5700kg with a maximum approved passenger seating configuration of more than 9 (otherwise than in seats referred to under paragraph (1) or (2), a safety belt with one diagonal shoulder strap or a safety harness for each seat intended for use by a passenger.

(7) If the commander cannot, from his own seat, see all the passengers' seats in the aircraft, a means of indicating to the passengers that seat belts should be fastened.

(8) Subject to paragraph (9), a safety harness for every seat in use.

(9) In the case of an aircraft carrying out aerobatic manoeuvres consisting only of erect spinning, the CAA may permit a safety belt with one diagonal shoulder strap to be fitted if it is satisfied that such restraint is sufficient for the carrying out of erect spinning in that aircraft and that it is not reasonably practicable to fit a safety harness in that aircraft.

Scale C

(1) Equipment for displaying the lights required by the Rules of the Air Regulations 1996.

(2) Electrical equipment, supplied from the main source of supply in the aircraft, to provide sufficient illumination to enable the flight crew properly to carry out their duties during flight.

(3) Unless the aircraft is equipped with radio, devices for making the visual signal specified in the Rules of the Air Regulations 1996 as indicating a request for permission to land.

Scale D

(1) In the case of a helicopter or gyroplane, a slip indicator.

(2) In the case of any other flying machine either:

(a) a turn indicator and a slip indicator; or

(b) a gyroscopic bank and pitch indicator and a gyroscopic direction indicator.

(c) A sensitive pressure altimeter adjustable for any sea level barometric pressure which the weather report or forecasts available to the commander of the aircraft indicate is likely to be encountered during the intended flight.

Scale E

(1) In the case of:

(a) a helicopter or gyroplane, a slip indicator.

(b) any other flying machine, a slip indicator and either a turn indicator or, at the option of the operator, an additional gyroscopic bank and pitch indicator.

(2) A gyroscopic bank and pitch indicator.

(3) A gyroscopic direction indicator.

(4) A sensitive pressure altimeter adjustable for any sea level barometric pressure which the weather report or forecasts available to the commander of the aircraft indicate is likely to be encountered during the intended flight.

Scale EE

(1) Subject to paragraph (2), a radio altimeter with an audio voice warning operating below a preset height and a visual warning capable of operating at a height selectable by the pilot.

ANO — SCHEDULE 4 Articles 12(6) and 14(2)
Aircraft equipment Scale EE

(2) A helicopter flying under and in accordance with the terms of a police air operator's certificate may instead be equipped with a radio altimeter with an audio warning and a visual warning each capable of operating at a height selectable by the pilot.

Scale F

(1) A timepiece indicating the time in hours, minutes and seconds.
(2) A means of indicating whether the power supply to the gyroscopic instrument is adequate.
(3) A rate of climb and descent indicator.
(4) A means of indicating in the flight crew compartment the outside air temperature calibrated in degrees celsius.
(5) If the maximum total weight authorised of the aircraft exceeds 5700kg two air speed indicators.

Scale G

(1) In the case of an aircraft other than a helicopter or gyroplane landing lights consisting of 2 single filament lamps, or one dual filament lamp with separately energised filaments.
(2) An electrical lighting system to provide illumination in every passenger compartment.
(3) Either:
(a) One electric torch for each member of the crew of the aircraft; or
(b) one electric torch
(i) for each member of the flight crew of the aircraft; and
(ii) affixed adjacent to each floor level exit intended for the disembarkation of passengers whether normally or in an emergency, provided that such torches shall:
(aa) be readily accessible for use by the crew of the aircraft at all times; and
(bb) number in total not less than the minimum number of members of the cabin crew required to be carried with a full passenger complement.
(4) In the case of an aircraft other than a helicopter or gyroplane of which the maximum total weight authorised exceeds 5700kg, means of observing the existence and build up of ice on the aircraft.
(5) In the case of a helicopter carrying out performance Class 1 or 2 operations or a gyroplane in respect of which there is in force a certificate of airworthiness designating the gyroplane as being of performance group A, either:
(a) 2 landing lights both of which are adjustable so as to illuminate the ground in front of and below the helicopter or gyroplane and one of which is adjustable so as to illuminate the ground on either side of the helicopter or gyroplane; or
(b) one landing light or, if the maximum total weight authorised of the helicopter or gyroplane exceeds 5700kg, one dual filament landing light with separately energised filaments, or 2 single filament lights, each of which is adjustable so as to illuminate the ground in front of and below the helicopter or gyroplane, and 2 parachute flares.
(6) In the case of a helicopter carrying out Performance Class 3 operations or gyroplane in respect of which there is in force a certificate of airworthiness designating the helicopter or gyroplane as being of performance group B, either:
(a) one landing light and 2 parachute flares; or
(b) if the maximum total weight authorised of the helicopter or gyroplane exceeds 5700kg, either one dual filament landing light with separately energised filaments or 2 single filament landing lights, and 2 parachute flares.

Scale H

(1) Subject to paragraph (i2, for each person on board, a lifejacket equipped with a whistle and waterproof torch.
(2) Lifejackets constructed and carried solely for use by children under three years of age need not be equipped with a whistle.

Scale I

A survival suit for each member of the crew.

Scale J

(1) Additional flotation equipment, capable of supporting one-fifth of the number of persons on board, and provided in a place of stowage accessible from outside the flying machine.
(2) Parachute distress rocket signals capable of making, from the surface of the water, the pyrotechnical signal of distress specified in the Rules of the Air Regulations 1996 and complying with Part III of Schedule 15 to the Merchant Shipping (Life-Saving Appliances) Regulations 1980.
(3) A sea anchor and other equipment necessary to facilitate mooring, anchoring or manoeuvring the flying machine on water, appropriate to its size, weight and handling characteristics.

Scale K

(1) In the case of:
(a) a flying machine, other than a helicopter or gyroplane carrying 20 or more persons, liferafts sufficient to accommodate all persons on board.
(b) In the case of a helicopter or gyroplane carrying 20 or more persons, a minimum of 2 liferafts sufficient together to accommodate all persons on board.
(2) Each liferaft shall contain the following equipment:
(a) means for maintaining buoyancy;
(b) a sea anchor;
(c) life-lines, and means of attaching one liferaft to another;
(d) paddles or other means of propulsion;
(e) means of protecting the occupants from the elements;

(f) a waterproof torch;
(g) marine type pyrotechnical distress signals;
(h) means of making sea water drinkable, unless the full quantity of fresh water is carried as specified in sub-paragraph (i);
(i) for each 4 or proportion of 4 persons the liferaft is designed to carry:
(i) 100 grammes of glucose toffee tablets; and
(ii) 1/2 litre of fresh water in durable containers or in any case in which it is not reasonably practicable to carry the quantity of water above specified, as large a quantity of fresh water as is reasonably practicable in the circumstances. In no case however shall the quantity of water carried be less than is sufficient, when added to the amount of fresh water capable of being produced by means of the equipment specified in sub-paragraph (h) to provide 1/2 litre of water for each 4 or proportion of 4 persons the liferaft is designed to carry; and
(j) first aid equipment.
(3) Items (2)(f) to (j) inclusive shall be contained in a pack.
(4) The number of survival beacon radio apparatus carried when the aircraft is carrying the number of liferafts specified in column 1 of the following Table shall be not less than the number specified in, or calculated in accordance with, column 2.

Column 1	Column 2
Not more than 8 liferafts	2 survival beacon radio apparatus
For every additional 4 or proportion of 4 liferafts	1 additional survival beacon radio apparatus

Scale KK

(1) A survival emergency locator transmitter capable of operating in accordance with the relevant provisions of Annex 10 to the Chicago Convention, Volume III (Fifth Edition July 1995) and of transmitting on 121.5 MHz and 406 MHz.
(2) An automatic emergency locator transmitter capable of operating in accordance with the relevant provisions of Annex 10 to the Chicago Convention, Volume III (Fifth Edition July 1995) and transmitting on 121.5 MHz and 406 MHz.
(3) An automatically deployable emergency locator transmitter capable of operating in accordance with the relevant provisions of Annex 10 to the Chicago Convention, Volume III (Fifth Edition July 1995) and transmitting on 121.5 MHz and 406 MHz.

Scale L1

Part I

(1) In every flying machine which is provided with means for maintaining a pressure greater than 700 hectopascals throughout the flight in the flight crew compartment and in the compartments in which the passengers are carried:
(a) a supply of oxygen sufficient, in the event of failure to maintain such pressure, occurring in the circumstances specified in columns 1 and 2 of the Table set out in Part II, for continuous use, during the periods specified in column 3 of the said Table, by the persons for whom oxygen is to be provided in accordance with column 4 of that Table; and
(b) in addition, in every case where the flying machine flies above flight level 350, a supply of oxygen in a portable container sufficient for the simultaneous first aid treatment of 2 passengers;
together with suitable and sufficient apparatus to enable such persons to use the oxygen.
(2) In any other flying machine:
(a) a supply of oxygen sufficient for continuous use by all the crew other than the flight crew, and if passengers are carried, by 10% of the number of passengers, for any period exceeding 30 minutes during which the flying machine flies above flight level 100 but not above flight level 130 and the flight crew shall be supplied with oxygen sufficient for continuous use for any period during which the flying machine flies above flight level 100; and
(b) a supply of oxygen sufficient for continuous use by all persons on board for the whole time during which the flying machine flies above flight level 130;
together with suitable and sufficient apparatus to enable such persons to use the oxygen.
(3) The quantity of oxygen required for the purpose of complying with paragraphs (1) and (2) of this Part shall be computed in accordance with the information and instructions relating thereto specified in the operations manual relating to the aircraft under paragraph 1(f) of Part A of Schedule 9.

Part II SCHEDULE 4 Articles 12(6) and 14(2)
Aircraft equipment Scale L1

Column 1 Vertical displacement of the flying machine in relation to flight levels	Column 2 Capability of flying machine to descend (where relevant)	Column 3 Period of supply of oxygen	Column 4 Persons for whom oxygen is to be provided
Above flight level 100	–	30 minutes or the period specified at A hereunder whichever is the greater	In addition to any passengers for whom oxygen is provided as specified below, all the crew.
Above flight level 100 but not above flight level 300	Flying machine is either flying at or below flight level 150 or is capable of descending & continuing to destination as specified at X hereunder	30 minutes or the period specified at A hereunder whichever is the greater	10% of number of passengers
	Flying machine is flying above flight level 150 and is not so capable	10 minutes or the period specified at B hereunder whichever is the greater and in addition 30 minutes or the period specified at C hereunder whichever is greater	All passengers 10% of number of passengers
Above flight level 300 but not above flight level 350	Flying machine is capable of descending and continuing to destination as specified at Y hereunder	30 minutes or the period specified at A hereunder whichever is the greater	15% of number of passengers
	Flying machine is not so capable	10 minutes or the period specified at B hereunder whichever is the greater and in addition 30 minutes or the period specified at C hereunder whichever is the greater	All passengers 15% of number of passengers
Above flight level 350	–	10 minutes or the period specified at B hereunder whichever is the greater and in addition 30 minutes or the period specified at C hereunder whichever is the greater	All passengers 15% of number of passengers

A The whole period during which, after a failure to maintain a pressure greater than 700 hectopascals in the control compartment and in the compartments in which passengers are carried has occurred, the flying machine flies above flight level 100.

B The whole period during which, after a failure to maintain such pressure has occurred, the flying machine flies above flight level 150.

C The whole period during which, after a failure to maintain such pressure has occurred, the flying machine flies above flight level 100, but not above flight level 150.

X The flying machine is capable, at the time when a failure to maintain such pressure occurs, of descending in accordance with the emergency descent procedure specified in the relevant flight manual and without flying below the minimum altitudes for safe flight specified in the operations manual relating to the aircraft, to flight level 150 within 6 minutes, and of continuing at or below that flight level to its place of intended destination or any other place at which a safe landing can be made.

Y The flying machine is capable, at the time when a failure to maintain such pressure occurs, of descending in accordance with the emergency descent procedure specified in the relevant flight manual and without flying below the minimum altitudes for safe flight specified in the operations manual relating to the aircraft, to flight level 150 within 4 minutes, and of continuing at or below that flight level to its place of intended destination or any other place at which a safe landing can be made.

Scale L2

(1) A supply of oxygen and the associated equipment to meet the requirements set out in Part I in the case of unpressurised aircraft and Part II in the case of pressurised aircraft.

(2) The duration for the purposes of this Scale shall be whichever is the greater of:

(a) that calculated in accordance with the operations manual prior to the commencement of the flight, being the period or periods which it is reasonably anticipated that the aircraft will be flown in the circumstances of the intended flight at a height where the said requirements apply and in calculating the said duration account shall be taken of:

(i) in the case of pressurised aircraft, the possibility of depressurisation when flying above flight level 100;
(ii) the possibility of failure of one or more of the aircraft engines;
(iii) restrictions due to required minimum safe altitude;
(iv) fuel requirement; and
(v) the performance of the aircraft; or

(b) the period or periods during which the aircraft is actually flown in the circumstances specified in the said Parts;

Part I

Unpressurised aircraft

(1) When flying at or below flight level 100: Nil.

(2) When flying above flight level 100 but not exceeding flight level 120:

Supply for	Duration
(a) Members of the flight crew	Any period during which the aircraft flies above flight level 100
(b) Members of the cabin crew and 10% of passengers	For all continuous period exceeding 30 minutes during which the aircraft flies above flight level 100 but not exceeding flight level 120, the duration shall be the period by which 30 minutes is exceeded.

(3) When flying above flight level 120:

Supply for	Duration
(a) Members of the flight crew	Any period during which the aircraft flies above flight level 120
(b) Members of the cabin crew and all of passengers	Any period during which the aircraft flies above flight level 120.

ANO   SCHEDULE 4 Articles 12(6) and 14(2)
Aircraft equipment Scale L2

Part II
Pressurised aircraft
(1) When flying at or below flight level 100: Nil.
(2) When flying above flight level 100 but not exceeding flight level 250:

Supply for	Duration
(a) Members of the flight crew	30 minutes or whenever the cabin pressure altitude exceeds 10,000ft whichever is the greater
(b) Members of the cabin crew and 10% of passengers	(i) When the aircraft is capable of descending and continuing to its destination as specified at a hereunder, 30 minutes or whenever the cabin pressure altitude exceeds 10,000ft, whichever is the greater.
	(ii) When the aircraft is not so capable, whenever the cabin pressure altitude is greater than 10,000ft but does not exceed 12,000ft
(c) Members of the cabin crew and passengers	(i) When the aircraft is capable of descending and continuing to its destination as specified at A hereunder, no requirement other than that at (2)(b)(i) of this part of this scale.
	(ii) When the aircraft is not so capable and the cabin pressure altitude exceeds 12,000ft, the duration shall be the period when the cabin pressure altitude exceeds 12,000ft or 10 minutes, whichever is the greater.

(3) When flying above flight level 250:

Supply for	Duration
(a) Members of the flight crew	2 hours or whenever the cabin pressure altitude exceeds 10,000ft, whichever is the greater
(b) Members of the cabin crew	Whenever the cabin pressure altitude exceeds 10,000ft and a portable supply for 15 minutes
(c) 10% of passengers	Whenever the cabin pressure altitude exceeds 10,000ft but does not exceed 12,000ft
(d) 30% of passengers	Whenever the cabin pressure altitude exceeds 12,000ft but does not exceed 15,000ft
(e) All passengers	If the cabin pressure altitude exceeds 15,000ft, the duration shall be the period when the cabin pressure altitude exceeds 15,000ft or 10 minutes whichever is greater
(f) 2% of passengers or 2 passengers, whichever is the greater, being a supply of first aid oxygen which must be available for simultaneous first aid treatment of 2% or 2 passengers wherever they are seated in the aircraft	Whenever, after decompression, the cabin pressure altitude exceeds 8000ft

A The flying machine is capable, at the time when a failure to maintain cabin pressurisation occurs, of descending in accordance with the emergency descent procedure specified in the relevant flight manual and without flying below the minimum altitudes for safe flight specified in the operations manual relating to the aircraft, to flight level 120 within 5 minutes and of continuing at or below that flight level to its place of intended destination or any other place at which a safe landing can be made.

Scale M
Equipment to prevent the impairment through ice formation of the functioning of the controls, means of propulsion, lifting surfaces, windows or equipment of the aircraft so as to endanger the safety of the aircraft.

Scale N
An intercommunication system for use by all members of the flight crew and including microphones, not of a hand-held type, for use by the pilot and flight engineer (if any).

Scale O
(1) Subject to paragraph (2), a radar set capable of giving warning to the pilot in command of the aircraft and to the co-pilot of the presence of cumulo-nimbus clouds and other potentially hazardous weather conditions.
(2) A flight may commence if the set is unserviceable or continue if the set becomes unserviceable thereafter:
(a) so as to give the warning only to one pilot, so long as the aircraft is flying only to the place at which it first becomes reasonably practicable for the set to be repaired; or
(b) when the weather report or forecasts available to the commander of the aircraft indicate that cumulo-nimbus clouds or other potentially hazardous weather conditions, which can be detected by the set when in working order, are unlikely

ANO SCHEDULE 4 Articles 12(6) and 14(2)
Aircraft equipment Scale S

to be encountered on the intended route or any planned diversion therefrom or the commander has satisfied himself that any such weather conditions will be encountered in daylight and can be seen and avoided, and the aircraft is in either case operated throughout the flight in accordance with any relevant instructions given in the operations manual.

Scale P

(1) Subject to paragraphs (2) and (5), a flight data recorder which is capable of recording, by reference to a time-scale, the following data:
(a) indicated airspeed;
(b) indicated altitude;
(c) vertical acceleration;
(d) magnetic heading;
(e) pitch attitude, if the equipment provided in the aeroplane is of such a nature as to enable this item to be recorded;
(f) engine power, if the equipment provided in the aeroplane is of such a nature as to enable this item to be recorded;
(g) flap position;
(h) roll attitude, if the equipment provided in the aeroplane is of such a nature as to enable this item to be recorded.
(2) Subject to paragraph (5), any aeroplane having a maximum total weight authorised not exceeding 11400kg may be provided with:
(a) a flight data recorder capable of recording the data described in paragraph (1)(a) to (1)(h); or
(b) a 4 channel cockpit voice recorder.
(3) Subject to paragraph (5), in addition, on all flights by turbine-powered aeroplanes having a maximum total weight authorised exceeding 11400kg, a 4 channel cockpit voice recorder.
(4) The flight data recorder and cockpit voice recorder referred to above shall be so constructed that the record would be likely to be preserved in the event of an accident to the aeroplane.
(5) An aeroplane shall not be required to carry the said equipment, if before take-off the equipment is found to be unserviceable and the aircraft flies in accordance with arrangements approved by the CAA.

Scale Q

If the maximum total weight authorised of the aeroplane exceeds 5700kg and it was first registered, whether in the United Kingdom or elsewhere, on or after 1st June 1965, a door between the flight crew compartment and any adjacent compartment to which passengers have access, which door shall be fitted with a lock or bolt capable of being worked from the flight crew compartment.

Scale R

(1) In respect of:
(a) aeroplanes having a maximum total weight authorised exceeding 5700kg, equipment sufficient to protect the eyes, nose and mouth of all members of the flight crew required to be carried by virtue of article 25 for a period of not less than 15 minutes and, in addition, were the minimum flight crew required as aforesaid is more than one and a member of the cabin crew is not required to be carried by virtue of article 25, portable equipment sufficient to protect the eyes, nose and mouth of one member of the flight crew for a period of not less than 15 minutes.
(b) In respect of aeroplanes having a maximum total weight authorised not exceeding 5700kg, either the equipment specified in paragraph (1)(a) or, in the case of such aeroplanes restricted by virtue of the operator's operations manual to flight at or below flight level 250 and capable of descending as specified at paragraph (4) such equipment sufficient to protect the eyes only.
(2) In respect of:
(a) aeroplanes having a maximum total weight authorised exceeding 5700kg, portable equipment to protect the eyes, nose and mouth of all members of the cabin crew required to be carried by virtue of article 25 for a period of not less than 15 minutes.
(b) In respect of aeroplanes having a maximum total weight authorised not exceeding 5700kg, subject to paragraph (3), the equipment specified in paragraph (2)(a).
(3) Sub-paragraph (2)(b) shall not apply to such aeroplanes restricted by virtue of the operator's operations manual to flight at or below flight level 250 and capable of descending as specified at paragraph (4).
(4) The aeroplane is capable of descending in accordance with the emergency descent procedure specified in the relevant flight manual and without flying below the minimum altitudes for safe flight specified in the operations manual relating to the aeroplane, to flight level 100 within 4 minutes and of continuing at or below that flight level to its place of intended destination or any other place at which a safe landing can be made.

Scale S

(1) Subject to paragraphs (7) and (8), either a 4 channel cockpit voice recorder or a flight data recorder capable of recording by reference to a time scale the data required to determine the following matters accurately in respect of the aeroplane: the flight path, attitude and the basic lift, thrust and drag forces acting upon it;
(2) Subject to paragraph (7) and (8) a 4 channel cockpit voice recorder and a flight data recorder capable of recording by reference to a time scale the data required to determine the following matters accurately in respect of the aeroplane: the information specified in paragraph (1) together with use of VHF transmitters;
(3) Subject to paragraphs (7) and (8) a 4 channel cockpit voice recorder and a flight data recorder capable of recording by reference to a time scale the data required to determine the following matters accurately in respect of the aeroplane: the flight path, attitude, the basic lift, thrust and drag forces acting upon it, the selection of high lift devices (if any) and airbrakes (if any), the position of primary flying control and pitch trim surfaces, outside air temperature, instrument landing deviations, use of automatic flight control systems, use of VHF transmitters, radio altitude (if any), the level or availability of essential AC electricity supply and cockpit warnings relating to engine fire and engine shut-down, cabin pressurisation, presence of smoke and hydraulic/pneumatic power supply;

ANO   SCHEDULE 4 Articles 12(6) and 14(2)
Aircraft equipment Scale S

(4) Subject to paragraph (7) and (8) either a cockpit voice recorder and a flight data recorder or a combined cockpit voice recorder/flight data recorder capable in either case of recording by reference to a time scale the data required to determine the following matters accurately in respect of the aeroplane:
(a) the flight path
(b) speed
(c) attitude
(d) engine power
(e) outside air temperature
(f) configuration of lift and drag devices
(g) use of VHF transmitters and
(h) use of automatic flight control systems;

(5) Subject to paragraph (7) and (8) a cockpit voice recorder and a flight data recorder capable of recording by reference to a time scale the data required to determine the following matters accurately in respect of the aeroplane:
(a) the flight path
(b) speed
(c) attitude
(d) engine power
(e) outside air temperature
(f) configuration of lift and drag devices
(g) use of VHF transmitters, and
(h) use of automatic flight control systems;

(6) Subject to paragraph (7) and (8) a cockpit voice recorder and a flight data recorder capable of recording by reference to a time scale the data required to determine the following matters accurately in respect of the aeroplane:
(a) the flight path
(b) speed
(c) attitude
(d) engine power
(e) outside air temperature
(f) instrument landing system deviations
(g) marker beacon passage
(h) radio altitude
(i) configuration of the landing gear and lift and drag devices
(j) position of primary flying controls
(k) pitch trim position
(l) use of automatic flight control systems
(m) use of VHF transmitters
(n) ground speed/drift angle or latitude/longitude if the navigational equipment provided in the aeroplane is of such a nature as to enable this information to be recorded with reasonable practicability,
(o) cockpit warnings relating to ground proximity, and
(p) the master warning system;

(7) An aircraft shall not be required to carry the said equipment specified in paragraphs (1) to (6) if, before take-off the equipment is found to be unserviceable and the aircraft flies in accordance with arrangements approved by the CAA.

(8) The cockpit voice recorder or flight data recorder or combined cockpit voice recorder/flight data recorder, as the case may be, shall be so constructed that the record would be likely to be preserved in the event of an accident.

Scale SS

(1) Subject to paragraph (4) and (5), a 4 channel cockpit voice recorder capable of recording and retaining the data recorded during at least the last 30 minutes of its operation and a flight data recorder capable of recording and retaining the data recorded during at least the last 8 hours of its operation being the data required to determine by reference to a time scale the following matters accurately in respect of the helicopter or gyroplane:
(a) flight path
(b) speed
(c) attitude
(d) engine power
(e) main rotor speed
(f) outside air temperature
(g) position of pilot's primary flight controls
(h) use of VHF transmitter
(j) use of automatic flight controls (if any)
(k) use of stability augmentation system (if any)
(l) cockpit warnings relating to the master warning system; and
(m) selection of hydraulic system and cockpit warnings of failure of essential hydraulic systems

SCHEDULE 4 Articles 12(6) and 14(2)
Aircraft equipment Scale V

(2) Subject to paragraph (4) and (5), a 4 channel cockpit voice recorder capable of recording and retaining the data recorded during at least the last 30 minutes of its operation and a flight data recorder capable of recording and retaining the data recorded during at least the last 8 hours of its operation being the data required to determine by reference to a time scale the information specified in paragraph (i) together with the following matters accurately in respect of the helicopter or gyroplane:

(a) landing gear configuration

(b) indicated sling load force if an indicator is provided in the helicopter or gyroplane of such a nature as to enable this information to be recorded with reasonable practicability

(c) radio altitude

(d) instrument landing system deviations

(e) marker beacon passage

(f) ground speed/drift angle or latitude/longitude if the navigational equipment provided in the helicopter or gyroplane is of such a nature as to enable this information to be recorded with reasonable practicability; and

(g) main gear box oil temperature and pressure

(3) Subject to paragraph (4) and (5):

(a) A combined cockpit voice recorder/flight data recorder which meets the following requirements:

(i) in the case of a helicopter or gyroplane which is otherwise required to carry a flight data recorder specified at paragraph (1) the flight data recorder shall be capable of recording the data specified therein and retaining it for the duration therein specified;

(ii) in the case of a helicopter or gyroplane which is otherwise required to carry a flight data recorder specified at paragraph (2), the flight data recorder shall be capable of recording the data specified therein and retaining it for the duration therein specified;

(iii) the cockpit voice recorder shall be capable of recording and retaining at least the last hour of cockpit voice recording information on not less than three separate channels.

(b) In any case when a combined cockpit voice recorder/flight data recorder specified at paragraph (3i) (a) is required to be carried by or under this Order, the flight data recorder shall be capable of retaining:

(i) as protected data the data recorded during at least the last 5 hours of its operation or the maximum duration of the flight, whichever is the greater; and

(ii) additional data as unprotected data for a period which together with the period for which protected data is required to be retained amounts to a total of 8 hours.

Provided that the flight data recorder need not be capable of retaining the said additional data if additional data is retained which relates to the period immediately preceding the period to which the required protected data relates or for such other period or periods as the CAA may permit pursuant to article 62 and the additional data is retained in accordance with arrangements approved by the CAA.

(4) A helicopter or gyroplane shall not be required to carry the said equipment specified in paragraphs (1) to (3) if, before take-off, the equipment is found to be unserviceable and the aircraft flies in accordance with arrangements approved by the CAA.

(5) With the exception of flight data which it is expressly stated above may be unprotected, the cockpit voice recorder, flight data recorder or combined cockpit voice recorder and flight data recorder, as the case may be, shall be so constructed and installed that the record (herein referred to as 'protected data') would be likely to be preserved in the event of an accident and each cockpit voice recorder, flight data recorder or combined cockpit voice recorder/flight data recorder required to be carried on the helicopter or gyroplane shall have attached an automatically activated underwater sonar location device or an emergency locator radio transmitter.

Scale T

An underwater sonar location device except in respect of those helicopters or gyroplanes which are required to carry equipment in accordance with Scale SS.

Scale U

(1) 1 survival beacon radio apparatus;

(2) Marine type pyrotechnical distress signals;

(3) For each 4 or proportion of 4 persons on board, 100 grammes of glucose toffee tablets;

(4) For each 4 or proportion of 4 persons on board, 1/2 litre of fresh water in durable containers;

(5) First aid equipment.

Scale V

(1) 1 survival beacon radio apparatus

(2) Marine type pyrotechnical distress signals

(3) For each 4 or proportion of 4 persons on board, 100 grammes of glucose toffee tablets

(4) For each 4 or proportion of 4 persons on board, 1/2 litre of fresh water in durable containers

(5) First aid equipment

(6) For every 75 or proportion of 75 persons on board, 1 stove suitable for use with aircraft fuel

(7) 1 cooking utensil, in which snow or ice can be melted

(8) 2 snow shovels

(9) 2 ice saws

(10) single or multiple sleeping-bags, sufficient for the use of one-third of all persons on board

(11) 1 Arctic suit for each member of the crew of the aircraft.

ANO   SCHEDULE 4 Articles 12(6) and 14(2)
                                                              Aircraft equipment Scale W

Scale W

(1) Subject to paragraph (2), cosmic radiation detection equipment calibrated in millirems per hour and capable of indicating the action and alert levels of radiation dose rate.

(2) An aircraft shall not be required to carry the said equipment if before take-off the equipment is found to be unserviceable and it is not reasonably practicable to repair or replace it at the aerodrome of departure and the radiation forecast available to the commander of the aircraft indicates that hazardous radiation conditions are unlikely to be encountered by the aircraft on its intended route or any planned diversion therefrom.

Scale X

(1) Subject to paragraph (3), a Terrain Awareness Warning System known as Class A, being equipment capable of giving warning to the pilot of the potentially hazardous proximity of ground or water, including excessive closure rate to terrain, flight into terrain when not in landing configuration, excessive downward deviation from an instrument landing system glideslope, ad predictive terrain hazard warning function and a visual display.

(2) Subject to paragraph (3), a Terrain Awareness and Warning System known as Class B, being equipment capable of giving warning to the pilot of the potentially hazardous proximity of ground or water, including a predictive terrain hazard warning function.

(3) If the equipment becomes unserviceable, the aircraft may fly or continue to fly until it first lands at a place at which it is reasonably practicable for the equipment to be repaired or replaced.

Scale Y

(1) If the aircraft may in accordance with its certificate of airworthiness carry more than 19 and less than 100 passengers, one portable battery-powered megaphone capable of conveying instructions to all persons in the passenger compartment and readily available for use by a member of the crew.

(2) If the aircraft may in accordance with its certificate of airworthiness carry more than 99 and less than 200 passengers, 2 portable battery-powered megaphones together capable of conveying instructions to all persons in the passenger compartment and each readily available for use by a member of the crew.

(3) If the aircraft may in accordance with its certificate of airworthiness carry more than 199 passengers, 3 portable battery-powered megaphones together capable of conveying instructions to all persons in the passenger compartment and each readily available for use by a member of the crew.

(4) If the aircraft may in accordance with its certificate of airworthiness carry more than 19 passengers:

(a) a public address system; and

(b) an interphone system of communication between members of the flight crew and the cabin crew.

Scale Z

(1) An emergency lighting system to provide illumination in the passenger compartment sufficient to facilitate the evacuation of the aircraft notwithstanding the failure of the lighting systems specified in paragraph (2) of Scale G.

(2) An emergency lighting system to provide illumination outside the aircraft sufficient to facilitate the evacuation of the aircraft.

(3) An emergency floor path lighting system in the passenger compartment sufficient to facilitate the evacuation of the aircraft notwithstanding the failure of the lighting systems specified in paragraph (2) of Scale G; provided that if the equipment becomes unserviceable the aircraft may fly or continue to fly in accordance with arrangements approved by the CAA.

SCHEDULE 5

Article 20(2)

Radio communication and radio navigation equipment to be carried in aircraft

1 Subject to paragraph 3, every aircraft shall be provided, when flying in the circumstances specified in the first column of the Table in paragraph 2 of this Schedule, with the scales of equipment respectively indicated in that column of the Table: Provided that, if the aircraft is flying in a combination of such circumstances the scales of equipment shall not on that account be required to be duplicated.

2 TABLE

ANO — SCHEDULE 5 Article 20(2)
Radio communication and radio navigation equipment to be carried in aircraft

Aircraft and circumstances of flight	Scale of equipment required								
	A	B	C	D	E	F	G	H	J
(1) All aircraft (other than gliders) within the United Kingdom									
(a) When flying under Instrument Flight Rules within controlled airspace	A*				E*	F*#			
(b) When flying within controlled airspace	A*						G*		
(c) When making an approach to landing at an aerodrome notified for the purpose of this sub-paragraph									
(d) When flying for the purpose of public transport					E*				
(2) All aircraft within the United Kingdom									
(a) When flying at or above flight level 245	A*								
(b) When flying within airspace notified for the purpose of this sub-paragraph	B*								
(3) All aircraft (other than gliders) within the United Kingdom									
(a) When flying at or above flight level 245					E*	F*			
(b) When flying within airspace notified for the purpose of this sub-paragraph					E*				
(c) When flying at or above flight level 100					E*				
(4) When flying under Instrument Flight Rules within airspace notified for the purposes of this paragraph, on or after 31 March 2005:									
(a) all aeroplanes having a maximum take-off weight authorised not exceeding 5700kg and a maximum cruising true airspeed capability not exceeding 250 knots					E2				
(b) all rotorcraft					E2				
(c) all aeroplanes having either a maximum take-off weight authorised of more than 5700kg or a maximum cruising true airspeed capability of more than 250 knots					E3				
(d) all aircraft required to carry scale E2 or E3					EE				
(5) All aircraft registered in the United Kingdom, wherever they may be									
(a) When flying for the purpose of public transport under Instrument Flight Rules									
(i) While making an approach to landing	A		C					H	
(ii) On all other occasions	A		C					H	
(b) when flying for the purpose of public transport				D					
(c) multi-engined aircraft when flying for the purpose of public transport under Visual Flight Rules	A				E*				
(d) single engined aircraft when flying for the purpose of public transport under Visual Flight Rules	A								
(i) Over a route on which navigation is effected solely by visual reference to landmarks	A	B							
(ii) On all other occasions	A*								
(e) When flying under Instrument Flight Rules within controlled airspace and not required to comply with paragraph (5)(a) above									

ANO Schedule 5 — Article 20(2)

Radio communication and radio navigation equipment to be carried in aircraft

Aircraft and circumstances of flight	A	B	C	D	E	F	G	H	J
(6) All aeroplanes registered in the United Kingdom, wherever they may be and all aeroplanes wherever registered when flying in the United Kingdom, powered by one or more turbine jets or turbine propeller engines and either having a maximum take-off weight exceeding 15,000kg or which in accordance with the certificate of airworthiness in force in respect thereof may carry more than 30 passengers									J
(7) All aeroplanes powered by one or more turbine jet or turbine propeller engines and either having a maximum take-off weight exceeding 5700kg or a maximum approved passenger seating configuration of more than 19, and									J
(a) registered in the United Kingdom and flying for the purposes of public transport;									J
(b) registered in the United Kingdom and flying within the airspace of member states of the European Civil Aviation Conference; or									J
(c) flying in the United Kingdom									
3 (1) In the case of sub-paragraphs (1), (2), (3), (4)(a), (4)(c) and (5)(e) of paragraph 2, the specified equipment need not be carried if the appropriate air traffic control unit otherwise permits in relation to the particular flight and the aircraft complies with any instructions which the air traffic control unit may give in the particular case.									J
(2) An aircraft which is not a public transport aircraft and which is flying in Class D or Class E airspace shall not be required to be provided with distance measuring equipment in accordance with paragraph (b) of Scale F when flying in the circumstances specified in sub-paragraph (1)(a) of paragraph (2).									

4 The scales of radio and radio navigation equipment indicated in the foregoing Table shall be as follows:

Scale A
Radio equipment capable of maintaining direct two-way communication with the appropriate aeronautical radio stations.

Scale B
Radio navigation equipment capable of enabling the aircraft to be navigated on the intended route including such equipment as may be prescribed.

Scale C
Radio equipment capable of receiving from the appropriate aeronautical radio stations meteorological broadcasts relevant to the intended flight.

Scale D
Radio navigation equipment capable of receiving signals from one or more aeronautical radio stations on the surface to enable the aircraft to be guided to a point from which a visual landing can be made at the aerodrome at which the aircraft is to land.

Scale E1
Secondary surveillance radar equipment which includes a pressure altitude reporting transponder capable of operating in Mode A and Mode C and of being operated in accordance with such instructions as may be given to the aircraft by the air traffic control unit.

Scale E2
Secondary surveillance radar equipment which includes a pressure altitude reporting transponder capable of operating in Mode A and Mode C and has the capability and functionality prescribed for Mode S Elementary Surveillance and is capable of being operated in accordance with such instructions as may be given to the aircraft by the air traffic control unit.

Scale E3
Secondary surveillance radar equipment which includes a pressure altitude reporting transponder capable of operating in Mode A and Mode C and has the capability and functionality prescribed for Mode S Elementary Surveillance and is capable of being operated in accordance with such instructions as may be given to the aircraft by the air traffic control unit.

Scale EE
The aircraft shall, in the circumstances specified in paragraph 2.1.5.3 of Column IV (Third Edition July 2002) of Annex 10 to the Chicago Convention, comply with the requirements for antenna diversity set out in that paragraph.

Scale F
Radio and radio navigation equipment capable of enabling the aircraft to be navigated along the intended route including:
(i) automatic direction finding equipment;
(ii) distance measuring equipment; and
(iii) VHF omni-range equipment.

Scale G
Radio navigation equipment capable of enabling the aircraft to make an approach to landing using the Instrument Landing System.

Scale H
(1) Subject to paragraphs (2) and (3), radio navigation equipment capable of enabling the aircraft to be navigated on the intended route including:
(a) automatic direction finding equipment;
(b) distance measuring equipment;
(c) duplicated VHF omni-range equipment; and
(d) a 75 MHz marker beacon receiver.
(2) An aircraft may fly notwithstanding that it does not carry the equipment specified in this Scale if it carries alternative radio navigation equipment or navigational equipment approved in accordance with article 19(9).
(3) Where not more than one item of equipment specified in this Scale is unserviceable when the aircraft is about to begin a flight, the aircraft may nevertheless take-off on that flight if:
(a) it is not reasonably practicable for the repair or replacement of that item to be carried out before the beginning of the flight;
(b) the aircraft has not made more than one flight since the item was last serviceable; and
(c) the commander of the aircraft has satisfied himself that, taking into account the latest information available as to the route and aerodrome to be used (including any planned diversion) and the weather conditions likely to be encountered, the flight can be made safely and in accordance with any relevant requirements of the appropriate air traffic control unit.

SCHEDULE 5 Article 20(2) 4 The scales of radio and radio navigation equipment indicated in the foregoing Table shall be as follows:Scale J

Scale J
An airborne collision avoidance system.

5 In this Schedule:

(1) "Airborne collision avoidance system" means an aeroplane system which conforms to requirements prescribed for the purpose; is based on secondary surveillance radar transponder signals; operates independently of ground based equipment and which is designed to provide advice and appropriate avoidance manoeuvres to the pilot in relation to other aeroplanes which are equipped with secondary surveillance radar and are in undue proximity;

(2) 'Automatic direction finding equipment' means radio navigation equipment which automatically indicates the bearing of any radio station transmitting the signals received by such equipment;

(3) 'Distance measuring equipment' means radio equipment capable of providing a continuous indication of the aircraft's distance from the appropriate aeronautical radio stations;

(4) 'Mode A' means replying to an interrogation from secondary surveillance radar units on the surface to elicit transponder replies for identity and surveillance with identity provided in the form of a 4 digit identity code;

(5) 'Mode C' means replying to an interrogation from secondary surveillance radar units on the surface to elicit transponder replies for automatic pressure-altitude transmission and surveillance.

(6) 'Secondary surveillance radar equipment' means such type of radio equipment as may be notified as being capable of

(a) replying to an interrogation from secondary surveillance radar units on the surface and

(b) being operated in accordance with such instructions as may be given to the aircraft by the appropriate air traffic control unit;

(7) 'VHF omni-range equipment' means radio navigation equipment capable of giving visual indications of bearings of the aircraft by means of signals received from very high frequency omni-directional radio ranges;

SCHEDULE 6

Article 22
Aircraft, engine and propeller log books

1 Aircraft log book
The following entries shall be included in the aircraft log book:

(1) the name of the constructor, the type of the aircraft, the number assigned to it by the constructor and the date of the construction of the aircraft;

(2) the nationality and registration marks of the aircraft;

(3) the name and address of the operator of the aircraft;

(4) the date of each flight and the duration of the period between take-off and landing, or, if more than one flight was made on that day, the number of flights and the total duration of the periods between take-offs and landings on that day;

(5) subject to paragraph (8), particulars of all maintenance work carried out on the aircraft or its equipment;

(6) subject to paragraph (8), particulars of any defects occurring in the aircraft or in any equipment required to be carried therein by or under this Order, and of the action taken to rectify such defects including a reference to the relevant entries in the technical log required by article 15(2) and (3).

(7) subject to paragraph (8), particulars of any overhauls, repairs, replacements and modifications relating to the aircraft or any such equipment as aforesaid;

(8) entries shall not be required to be made under paragraphs (5), (6) and (7) in respect of any engine or variable pitch propeller.

2 Engine log book
The following entries shall be included in the engine log book:

(1) the name of the constructor, the type of engine, the number assigned to it by the constructor and the date of the construction of the engine;

(2) the nationality and registration marks of each aircraft in which the engine is fitted;

(3) the name and address of the operator of each such aircraft;

(4) either:

(a) the date of each flight and the duration of the period between take-off and landing or, if more than one flight was made on that day, the number of flights and the total duration of the periods between take-offs and landings on that day; or

(b) the aggregate duration of periods between take-off and landing for all flights made by that aircraft since the immediately preceding occasion that any maintenance, overhaul, repair, replacement, modification or inspection was undertaken on the engine;

(5) particulars of all maintenance work done on the engine;

(6) particulars of any defects occurring in the engine, and of the rectification of such defects, including a reference to the relevant entries in the technical log required by article 15(2) and (3);

(7) particulars of all overhauls, repairs, replacements and modifications relating to the engine or any of its accessories.

ANO SCHEDULE 7 Article 25(9) Areas specified in connection with the carriage of flight navigators as members of the flight crews or suitable navigational equipment on public transport aircraft

3 Variable pitch propeller log book

The following entries shall be included in the variable pitch propeller log book:

(1) the name of the constructor, the type of propeller, the number assigned to it by the constructor and the date of the construction of the propeller;

(2) the nationality and registration marks of each aircraft, and the type and number of each engine, to which the propeller is fitted;

(3) the name and address of the operator of each such aircraft;

(4) either:

(a) the date of each flight and the duration of the period between take-off and landing or, if more than one flight was made on that day, the number of flights and the total duration of the periods between take-offs and landings on that day; or

(b) the aggregate duration of periods between take-off and landing for all flights made by that aircraft since the immediately preceding occasion that any maintenance, overhaul, repair, replacement, modification or inspection was undertaken on the propeller;

(5) particulars of all maintenance work done on the propeller;

(6) particulars of any defects occurring in the propeller, and of the rectification of such defects, including a reference to the relevant entries in the technical log required by article 15(2) and (3).

(7) particulars of any overhauls, repairs, replacements and modifications relating to the propeller.

SCHEDULE 7

Article 25(9)
Areas specified in connection with the carriage of flight navigators as members of the flight crews or suitable navigational equipment on public transport aircraft

The following areas are hereby specified for the purposes of article 25(9):

Area A – Arctic
All that area north of latitude 68° north, but excluding any part thereof within the area enclosed by rhumb lines joining successively the following points:
68° north latitude 00° east/west longitude
73° north latitude 15° east longitude
73° north latitude 30° east longitude
68° north latitude 45° east longitude
68° north latitude 00° east/west longitude

Area B – Antarctic
All that area south of latitude 55° south.

Area C – Sahara
All that area enclosed by rhumb lines joining successively the following points:
30° north latitude 05° west longitude
24° north latitude 11° west longitude
14° north latitude 11° west longitude
14° north latitude 28° east longitude
24° north latitude 28° east longitude
28° north latitude 23° east longitude
30° north latitude 15° east longitude
30° north latitude 05° west longitude

Area D – South America
All that area enclosed by rhumb lines joining successively the following points:
04° north latitude 72° west longitude
04° north latitude 60° west longitude
08° south latitude 42° west longitude
18° south latitude 54° west longitude
18° south latitude 60° west longitude
14° south latitude 72° west longitude
05° south latitude 76° west longitude
04° north latitude 72° west longitude

Area E – Pacific Ocean
All that area enclosed by rhumb lines joining successively the following points:
60° north latitude 180° east/west longitude
20° north latitude 128° east longitude
04° north latitude 128° east longitude
04° north latitude 180° east/west longitude
55° south latitude 180° east/west longitude
55° south latitude 82° west longitude
25° south latitude 82° west longitude
60° north latitude 155° west longitude
60° north latitude 180° east/west longitude

Area F – Australia
All that area enclosed by rhumb lines joining successively the following points:
18° south latitude 123° east longitude
30° south latitude 118° east longitude
30° south latitude 135° east longitude
18° south latitude 123° east longitude

Area G – Indian Ocean
All that area enclosed by rhumb lines joining successively the following points:
35° south latitude 110° east longitude
55° south latitude 180° east/west longitude
55° south latitude 10° east longitude
40° south latitude 10° east longitude
25° south latitude 60° east longitude
20° south latitude 60° east longitude
05° south latitude 43° east longitude
10° north latitude 55° east longitude
10° north latitude 73° east longitude
04° north latitude 77° east longitude
04° north latitude 92° east longitude

ANO   SCHEDULE 7 Article 25(9) Areas specified in connection with the carriage of flight navigators as members of the flight crews or suitable navigational equipment on public transport aircraft

10° south latitude 100° east longitude
10° south latitude 110° east longitude
35° south latitude 110° east longitude

Area H – North Atlantic Ocean
All that area enclosed by rhumb lines joining successively the following points:
55° north latitude 15° west longitude
68° north latitude 28° west longitude
68° north latitude 60° west longitude
45° north latitude 45° west longitude
40° north latitude 60° west longitude
40° north latitude 19° west longitude
55° north latitude 15° west longitude

Area I – South Atlantic Ocean
All that area enclosed by rhumb lines joining successively the following points:
40° north latitude 60° west longitude
18° north latitude 60° west longitude
05° south latitude 30° west longitude
55° south latitude 55° west longitude
55° south latitude 10° east longitude
40° south latitude 10° east longitude
02° north latitude 05° east longitude
02° north latitude 10° east longitude
15° north latitude 25° west longitude
40° north latitude 19° west longitude
40° north latitude 60° west longitude

Area J – Northern Canada
All that area enclosed by rhumb lines joining successively the following points:
68° north latitude 130° west longitude
55° north latitude 115° west longitude
55° north latitude 70° west longitude
68° north latitude 60° west longitude
68° north latitude 130° west longitude

Area K – Northern Asia
All that area enclosed by rhumb lines joining successively the following points:
68° north latitude 56° east longitude
68° north latitude 160° east longitude
50° north latitude 125° east longitude
50° north latitude 56° east longitude
68° north latitude 56° east longitude

Area L – Southern Asia
All that area enclosed by rhumb lines joining successively the following points:
50° north latitude 56° east longitude
50° north latitude 125° east longitude
40° north latitude 110° east longitude
30° north latitude 110° east longitude
30° north latitude 80° east longitude
35° north latitude 80° east longitude
35° north latitude 56° east longitude
50° north latitude 56° east longitude

SCHEDULE 8

Flight Crew of Aircraft – Licences, Ratings, Qualifications and Maintenance of Licence Privileges
Articles 27, 28, 29, 30 and 31
PART A – FLIGHT CREW LICENCES

Section 1 – United Kingdom Licences
Minimum age, period of validity, privileges

Sub-Section 1 – AEROPLANE PILOTS

Private Pilot's Licence (Aeroplanes)
Minimum age – 17 years
No maximum period of validity

Privileges:
(1) Subject to paragraph (2), the holder of a Private Pilot's Licence (Aeroplanes) shall be entitled to fly as pilot in command or co-pilot of an aeroplane of any of the types or classes specified or otherwise falling within an aircraft rating included in the licence.
(2) He shall not:
(a) fly such an aeroplane for the purpose of public transport or aerial work save as hereinafter provided:
(i) he may fly such an aeroplane for the purpose of aerial work which consists of
(aa) the giving of instruction in flying, if his licence includes a flying instructor's rating, class rating instructor rating, flight instructor rating or an assistant flying instructor's rating; or
(bb) the conducting of flying tests for the purposes of this Order;
in either case in an aeroplane owned, or operated under arrangements entered into, by a flying club of which the person giving the instruction or conducting the test and the person receiving the instruction or undergoing the test are both members;
(ii) he may fly such an aeroplane for the purpose of aerial work which consists of:
(aa) towing a glider in flight; or

ANO   SCHEDULE 8 Flight Crew of Aircraft – Licences, Ratings, Qualifications and Maintenance of Licence Privileges Articles 27, 28, 29, 30 and 31
PART A – FLIGHT CREW LICENCES – Basic Commercial Pilot's Licence (Aeroplanes)

(bb) a flight for the purpose of dropping of persons by parachute;
in either case in an aeroplane owned, or operated under arrangements entered into, by a club of which the holder of the licence and any person carried in the aircraft or in any glider towed by the aircraft are members.

(b) receive any remuneration for his services as a pilot on a flight save that if his licence includes a flying instructor's rating, a flight instructor rating or an assistant flying instructor's rating by virtue of which he is entitled to give instruction in flying microlight aircraft or self-launching motor gliders he may receive remuneration for the giving of such instruction or the conducting of such flying tests as are specified in sub-paragraph (a)(i) in a microlight aircraft or a self-launching motor glider.

(c) unless his licence includes an instrument rating (aeroplane) or an instrument meteorological conditions rating (aeroplanes), fly as pilot in command of such an aeroplane:
(i) on a flight outside controlled airspace when the flight visibility is less than 3km;
(ii) on a special VFR flight in a control zone in a flight visibility of less than 10km except on a route or in an aerodrome traffic zone notified for the purpose of this sub-paragraph; or
(iii) out of sight of the surface.

(d) fly as pilot in command of such an aeroplane at night unless his licence includes a night rating (aeroplanes) or a night qualification (aeroplanes).

(e) unless his licence includes an instrument rating (aeroplane), fly as pilot in command or co-pilot of such an aeroplane flying in Class A, B or C airspace in circumstances which require compliance with the Instrument Flight Rules.

(f) unless his licence includes an instrument rating (aeroplane) or an instrument meteorological conditions rating (aeroplanes), fly as pilot in command or co-pilot of such an aeroplane flying in Class D or E airspace in circumstances which require compliance with the Instrument Flight Rules.

(g) fly as pilot in command of such an aeroplane carrying passengers unless within the preceding 90 days he has made three take-offs and three landings as the sole manipulator of the controls of an aeroplane of the same type or class and if such a flight is to be carried out at night and his licence does not include an instrument rating (aeroplane) at least one of those take-offs and landings shall have been at night.

Basic Commercial Pilot's Licence (Aeroplanes)
Minimum age – 18 years
Maximum period of validity – 10 years
Privileges:
(1) The holder of a Basic Commercial Pilot's Licence (Aeroplanes) shall be entitled to exercise the privileges of a United Kingdom Private Pilot's Licence (Aeroplanes).

(2) Subject to paragraph (3) and (7), he shall be entitled to fly as pilot in command of an aeroplane of a type or class on which he is so qualified and which is specified in an aircraft rating included in the licence when the aeroplane is engaged on a flight for any purpose whatsoever.

(3) He shall not:
(a) fly such an aeroplane on a flight for the purpose of public transport if he has less than 400 hours of flying experience as pilot in command of aeroplanes other than self-launching motor gliders or microlight aircraft.

(b) fly such an aeroplane on a flight for the purpose of public transport if its maximum total weight authorised exceeds 2300kg.

(C) fly such an aeroplane on any scheduled journey.

(d) such an aeroplane on a flight for the purpose of public transport except a flight beginning and ending at the same aerodrome and not extending beyond 25 nautical miles from that aerodrome.

(e) fly such an aeroplane on a flight for the purpose of public transport after he attains the age of 60 years unless the aeroplane is fitted with dual controls and carries a second pilot who has not attained the age of 60 years and who holds an appropriate licence under this Order entitling him to act as pilot in command or co-pilot of that aeroplane.

(f) fly such an aeroplane at night, unless his licence includes a night rating (aeroplanes) or a night qualification (aeroplane).

(g) unless his licence includes an instrument rating (aeroplane) or an instrument meteorological conditions rating (aeroplanes), fly as pilot in command of such an aeroplane:
(i) on a flight outside controlled airspace when the flight visibility is less than 3km;
(ii) on a special VFR flight in a control zone in a flight visibility of less than 10km except on a route or in an aerodrome traffic zone notified for the purposes of this sub-paragraph; or
(cc) out of sight of the surface.

(h) unless his licence includes an instrument rating (aeroplane), fly as pilot in command or co-pilot of such an aeroplane flying in Class A, B or C airspace in circumstances which require compliance with the Instrument Flight Rules

(i) unless his licence includes an instrument rating (aeroplane) or an instrument meteorological conditions rating (aeroplanes), fly as pilot in command or co-pilot of such an aeroplane flying in Class D or E airspace in circumstances which require compliance with the Instrument Flight Rules.

(j) fly as pilot in command of such an aeroplane carrying passengers unless within the preceding 90 days he has made three take-offs and three landings as the sole manipulator of the controls of an aeroplane of the same type or class and if the flight is to be undertaken at night and his licence does not include an instrument rating (aeroplane) at least one of those take-offs and landings shall have been at night.

(4) Subject to paragraph (5), he shall be entitled to fly as pilot in command of an aeroplane of a type or class specified in an instructor's rating included in the licence on a flight for the purpose of aerial work which consists of:
(a) the giving of instruction in flying; or

417

ANO SCHEDULE 8 Flight Crew of Aircraft – Licences, Ratings, Qualifications and Maintenance of Licence Privileges Articles 27, 28, 29, 30 and 31
PART A – FLIGHT CREW LICENCES – Basic Commercial Pilot's Licence (Aeroplanes)

(b) the conducting of flying tests for the purposes of this Order;
in either case in an aeroplane owned, or operated under arrangements entered into, by a flying club of which the person giving the instruction or conducting the test and the person receiving the instruction or undergoing the test are both members.
(5) He shall not be entitled to exercise the privileges contained in paragraph (4) other than in an aeroplane which he is entitled to fly as pilot in command on a private flight, an aerial work flight or a public transport flight pursuant to the privileges set out in paragraph (1) or (2) of these privileges.
(6) Subject to paragraph (7) he shall be entitled to fly as co-pilot of any aeroplane of a type specified in an aircraft rating included in the licence when the aeroplane is engaged on a flight for any purpose whatsoever provided that he shall not be entitled to fly as co-pilot of an aeroplane which is engaged on a flight for the purpose of public transport unless he has more than 400 hours of flying experience as pilot in command of aeroplanes other than self-launching motor gliders and microlight aircraft and the aeroplane is certificated for single pilot operation.
(7) He shall not at any time after he attains the age of 65 years act as pilot in command or co-pilot of any aeroplane on a flight for the purpose of public transport.

Commercial Pilot's Licence (Aeroplanes)
Minimum age – 18 years
Maximum period of validity – 10 years
Privileges:
(1) The holder of a Commercial Pilot's Licence (Aeroplanes) shall be entitled to exercise the privileges of a United Kingdom Private Pilot's Licence (Aeroplanes) which includes an instrument meteorological conditions rating (aeroplanes) and a night rating (aeroplanes) or night qualification (aeroplane), and shall be entitled to fly as pilot in command of an aeroplane:
(a) on a special VFR flight notwithstanding that the flight visibility is less than 3km;
(b) when the aeroplane is taking off or landing at any place notwithstanding that the flight visibility below cloud is less than 1800 metres.
(2) Subject to paragraphs (3) and (7), he shall be entitled to fly as pilot in command of an aeroplane of a type or class on which he is so qualified and which is specified in an aircraft rating included in the licence when the aeroplane is engaged on a flight for any purpose whatsoever.
(3) He shall not:
(a) unless his licence includes an instrument rating (aeroplane), fly such an aeroplane on any scheduled journey.
(b) fly as pilot in command of an aeroplane carrying passengers unless he has carried out at least three take-offs and three landings as pilot flying in an aeroplane of the same type or class or in a flight simulator, approved for the purpose, of the aeroplane type or class to be used, in the preceding 90 days;
(c) as co-pilot serve at the flying controls in an aeroplane carrying passengers during take-off and landing unless he has served as a pilot at the controls during take-off and landing in an aeroplane of the same type or in a flight simulator, approved for the purpose, of the aeroplane type to be used, in the preceding 90 days; or
(d) as the holder of a licence which does not include a valid instrument rating (aeroplane) act as pilot in command of an aeroplane carrying passengers at night unless during the previous 90 days at least one of the take-offs and landings required in sub-paragraph (b) above has been carried out at night.
(e) unless his licence includes an instrument rating (aeroplane), fly any such aeroplane of which the maximum total weight authorised exceeds 2300kg on any flight for the purpose of public transport, except a flight beginning and ending at the same aerodrome and not extending beyond 25 nautical miles from that aerodrome.
(f) fly such an aeroplane on a flight for the purpose of public transport unless it is certificated for single pilot operation.
(g) fly such an aeroplane on any flight for the purpose of public transport after he attains the age of 60 years unless the aeroplane is fitted with dual controls and carries a second pilot who has not attained the age of 60 years and who holds an appropriate licence under this Order entitling him to act as pilot in command or co-pilot of that aeroplane.
(h) unless his licence includes an instrument rating (aeroplane), fly as pilot in command or co-pilot of such an aeroplane flying in Class A, B or C airspace in circumstances which require compliance with the Instrument Flight Rules.
(4) Subject to paragraph (5), he shall be entitled to fly as pilot in command of an aeroplane of a type or class specified in an instructor's rating included in the licence on a flight for the purpose of aerial work which consists of:
(a) the giving of instruction in flying; or
(b) the conducting of flying tests for the purposes of this Order;
in either case in an aeroplane owned, or operated under arrangements entered into, by a flying club of which the person giving the instruction or conducting the test and the person receiving the instruction or undergoing the test are both members.
(5) He shall not be entitled to exercise privileges contained in paragraph (4) other than in an aeroplane which he is entitled to fly as pilot in command on a private flight, an aerial work flight or a public transport flight pursuant to the privileges set out in paragraph (1) or (2) of these privileges.
(6) Subject to paragraph (7) he shall be entitled to fly as co-pilot of any aeroplane of a type specified in an aircraft rating included in the licence when the aeroplane is engaged on a flight for any purpose whatsoever.
(7) He shall not at any time after he attains the age of 65 years act as pilot in command or co-pilot of any aeroplane on a flight for the purpose of public transport.

SCHEDULE 8 Flight Crew of Aircraft – Licences, Ratings, Qualifications and Maintenance of Licence Privileges Articles 27, 28, 29, 30 and 31
PART A – FLIGHT CREW LICENCES – Commercial Pilot's Licence (Helicopters and Gyroplanes)

Airline Transport Pilot's Licence (Aeroplanes)
Minimum age – 21 years
Maximum period of validity – 10 years
Privileges:
The holder of an Airline Transport Pilot's Licence (Aeroplanes) shall be entitled to exercise the privileges of a United Kingdom Commercial Pilot's Licence (Aeroplanes) except that sub-paragraph (3) (f) of those privileges shall not apply.

Sub-Section 2 – HELICOPTER AND GYROPLANE PILOTS
Private Pilot's Licence (Helicopters)
Minimum age – 17 years
No maximum period of validity
Privileges:
(1) Subject to paragraph (2), the holder of the Private Pilot's Licence (Helicopter) shall be entitled to fly as pilot in command or co-pilot of any helicopter of a type specified in an aircraft rating included in the licence.
(2) He shall not
(a) fly such a helicopter for the purpose of public transport or aerial work other than aerial work which consists of:
(i) the giving of instruction in flying if his licence includes a flying instructor's rating, flight instructor rating or an assistant flying instructor's rating; or
(ii) the conducting of flying tests for the purposes of this Order;
in either case in a helicopter owned, or operated under arrangements entered into, by a flying club of which the person giving the instruction or conducting the test and the person receiving the instruction or undergoing the test are both members.
(b) receive any remuneration for his services as a pilot on a flight other than remuneration for the giving of such instruction or the conducting of such flying tests as are specified in sub-paragraph (a).
(c) fly as pilot in command of such a helicopter at night unless his licence includes a night rating (helicopters) or a night qualification (helicopter).
(d) unless his licence includes an instrument rating (helicopter) fly as pilot in command or co-pilot of such a helicopter flying in Class A, B or C airspace in circumstances which require compliance with the Instrument flight Rules.
(e) fly as pilot in command of such a helicopter carrying passengers unless:
(i) within the preceding 90 days he has made three circuits, each to include take-offs and landings as the sole manipulator of the controls of a helicopter of the same type; or
(ii) if the privileges are to be exercised by night and his licence does not include an instrument rating, within the preceding 90 days he has made three circuits, each to include take-offs and landings by night as the sole manipulator of the controls of a helicopter of the same type.

Private Pilot's Licence (Gyroplanes)
Minimum age – 17 years
No maximum period of validity
Privileges:
(1) Subject to paragraph (2), the holder of a Private Pilot's Licence (Gyroplanes) shall be entitled to fly as pilot in command or co-pilot of any gyroplane of a type specified in the aircraft rating included in the licence.
(2) He shall not
(a) fly such a gyroplane for the purpose of public transport or aerial work other than aerial work which consists of:
(i) the giving of instruction in flying if his licence includes a flying instructor's rating, flight instructor rating or an assistant flying instructor's rating; or
(ii) the conducting of flying tests for the purposes of this Order;
in either case in a gyroplane owned, or operated under arrangements entered into, by a flying club of which the person giving the instruction or conducting the test and the person receiving the instruction or undergoing the test are both members.
(b) receive any remuneration for his services as a pilot on a flight other than remuneration for the giving of such instruction or the conducting of such flying tests as are specified in sub-paragraph (a).
(c) fly as pilot in command of such a gyroplane at night unless his licence includes a night rating (gyroplanes) and he has within the immediately preceding 13 months carried out as pilot in command not less than 5 take-offs and five landings at a time when the depression of the centre of the sun was not less than 12° below the horizon.

Commercial Pilot's Licence (Helicopters and Gyroplanes)
Minimum age – 18 years
Maximum period of validity – 10 years
Privileges:
(1) Subject to paragraphs (2) and (5), the holder of a Commercial Pilot's Licence (Helicopters and Gyroplanes) shall be entitled to
(a) exercise the privileges of a United Kingdom Private Pilot's Licence (Helicopters) or a United Kingdom Private Pilot's Licence (gyroplanes) which includes respectively either a night rating (helicopters) or night qualification (helicopter) or a night rating (gyroplanes); and
(2) to fly as pilot in command of any helicopter or gyroplane on which he is so qualified and which is of a type specified in an aircraft rating included in the licence when the helicopter or gyroplane is engaged on a flight for any purpose whatsoever.
(2) He shall not:
(a) unless his licence includes an instrument rating (helicopter) fly such a helicopter on any scheduled journey or on any flight for the purpose of public transport other than in visual meteorological conditions.

419

ANO SCHEDULE 8 Flight Crew of Aircraft – Licences, Ratings, Qualifications and Maintenance of Licence Privileges Articles 27, 28, 29, 30 and 31
PART A – FLIGHT CREW LICENCES – Commercial Pilot's Licence (Helicopters and Gyroplanes)

(b) fly such a helicopter on a flight for the purpose of public transport unless it is certificated for single pilot operation.
(c) fly such a helicopter on any flight for the purpose of public transport after he attains the age of 60 years unless the helicopter is fitted with dual controls and carries a second pilot who has not attained the age of 60 years and who holds an appropriate licence under this Order entitling him to act as pilot in command or co-pilot of that helicopter.
(d) unless his licence includes an instrument rating (helicopter) fly as pilot in command of such a helicopter flying in Class A, B or C airspace in circumstances which require compliance with the Instrument Flight Rules.
(e) fly as pilot in command of a helicopter carrying passengers unless he has carried out at least three circuits, each to include take-offs and landings, as pilot flying in a helicopter of the same type or a flight simulator of the helicopter type to be used, in the preceding 90 days; or
(f) as the holder of a helicopter licence which does not include a valid instrument rating (helicopter) act as pilot in command of a helicopter carrying passengers at night unless during the previous 90 days at least one of the take-offs and landings required in sub-paragraph (e) above has been carried out at night.
(g) fly such a gyroplane on a flight for the purpose of public transport unless it is certificated for single pilot operation.
(h) fly such a gyroplane at night unless he has within the immediately preceding 13 months carried out as pilot in command not less than 5 take-offs and 5 landings at a time when the depression of the centre of the sun was not less than 12° below the horizon; or
(i) fly such a gyroplane on any flight for the purpose of public transport after he attains the age of 60 years unless the gyroplane is fitted with dual controls and carries a second pilot who has not attained the age of 60 years and who holds an appropriate licence under this Order entitling him to act as pilot in command or co-pilot of that gyroplane.
(3) Subject to paragraphs (4) and (5) he shall be entitled to fly as co-pilot of any helicopter or gyroplane of a type specified in an aircraft rating included in the licence when the helicopter or gyroplane is engaged on a flight for any purpose whatsoever.
(4) He shall not:
(a) unless his licence includes an instrument rating (helicopter) fly as co-pilot of a helicopter flying in Class A, B or C airspace in circumstances which require compliance with the Instrument flight Rules.
(b) as co-pilot serve at the flying controls in a helicopter carrying passengers during take-off and landing unless he has served as a pilot at the controls during take-off and landing in a helicopter of the same type or in a flight simulator of the helicopter type to be used, in the preceding 90 days.
(c) unless his licence includes an instrument rating (helicopter) fly as co-pilot of a helicopter on any scheduled journey or on a flight for the purpose of public transport other than in visual meteorological conditions.
(5) He shall not at any time after he attains the age of 65 years act as pilot in command or co-pilot of any helicopter or gyroplane on a flight for the purpose of public transport.

Airline Transport Pilot's Licence (Helicopters and Gyroplanes)
Minimum age – 21 years
Maximum period of validity – 10 years
Privileges:
The holder of an Airline Transport Pilot's Licence (Helicopters and Gyroplanes) shall be entitled to exercise the privileges of a United Kingdom Commercial Pilot's Licence (Helicopters and Gyroplanes) except that sub-paragraphs (2)(b) and (2)(g) of those privileges shall not apply.

Sub-Section 3 – BALLOON AND AIRSHIP PILOTS

Private Pilot's Licence (Balloons and Airships)
Minimum age – 17 years
No maximum period of validity
Privileges:
(1) Subject to paragraph (2), the holder of Private Pilot's Licence (Balloons and Airships) shall be entitled to fly as pilot in command of any type of balloon or airship on which he is so qualified and which is specified in an aircraft rating in the licence and co-pilot of any type of balloon or airship specified in such a rating.
(2) He shall not:
(a) fly such a balloon or airship for the purpose of public transport or aerial work, other than aerial work which consists of the giving of instruction in flying or the conducting of flying tests in either case in a balloon or airship owned, or operated under arrangements entered into, by a flying club of which the person giving the instruction or conducting the test and the person receiving the instruction or undergoing the test are both members.
(b) receive any remuneration for his services as a pilot on a flight other than remuneration for the giving of such instruction or the conducting of such flying tests as are specified in sub-paragraph (a): or
(c) fly such a balloon unless he has within the immediately preceding 13 months carried out as pilot in command in a free balloon 5 flights each of not less than 5 minutes duration.

Commercial Pilot's Licence (Balloons)
Minimum age – 18 years
Maximum period of validity – 10 years*
Privileges:
(1) The holder of a Commercial Pilot's Licence (Balloons) shall be entitled to exercise the privileges of a United Kingdom Private Pilot's Licence (Balloons and Airships).
(2) Subject to paragraph (3), he shall be entitled to fly, when the balloon is flying for any purpose whatsoever, as pilot in command or co-pilot of any type of balloon specified in the aircraft rating included in the licence.
(3) He shall not act as pilot in command on a flight for the purpose of the public transport of passengers unless he has within the immediately preceding 90 days carried out as pilot in command in a free balloon 3 flights each of not less than 5 minutes duration.

Commercial Pilot's Licence (Airships)
Minimum age – 18 years
Maximum period of validity – 10 years
Privileges:
(1) The holder of a Commercial Pilot's Licence (Airships) shall be entitled to exercise the privileges of a United Kingdom Private Pilot's Licence (Balloons and Airships).
(2) He shall be entitled to fly, when the airship is flying for any purpose whatsoever, as pilot in command of any type of airship on which he is so qualified and which is specified in an aircraft rating included in the licence and as co-pilot of any type of airship specified in such a rating.

Sub-Section 4 – GLIDER PILOTS
Commercial Pilot's Licence (Gliders)
Minimum age – 18 years
Maximum period of validity – 10 years
Privileges:
The holder of a Commercial Pilot's Licence (Gliders) shall be entitled to fly for any purpose as pilot in command or co-pilot of:
(a) any glider of which the maximum total weight authorised does not exceed 680kg;
(b) any glider of which the maximum total weight authorised exceeds 680kg and which is of a type specified in the rating included in the licence.

Sub-Section 5 – OTHER FLIGHT CREW
Flight Navigator's Licence
Minimum age – 21 years
Maximum period of validity – 10 years
Privileges:
The holder of a Flight Navigator's Licence shall be entitled to act as flight navigator in any aircraft.

Flight Engineer's Licence
Minimum age – 21 years
Maximum period of validity – 10 years
Privileges:
The holder of a Flight Engineer's Licence shall be entitled to act as flight engineer in any type of aircraft specified in an aircraft rating included in the licence.

Flight Radiotelephony Operator's Licence
Minimum age – 16 years
Maximum period of validity – 10 years
Privileges:
The holder of a Flight Radiotelephony Operator's Licence shall be entitled to operate radiotelephony apparatus in any aircraft if the stability of the frequency radiated by the transmitter is maintained automatically but shall not be entitled to operate the transmitter, or to adjust its frequency, except by the use of external switching devices.

Section 2 – JAR-FCL Licences
AEROPLANE PILOTS
Private Pilot Licence (Aeroplane)
Minimum age – 17 years
Maximum period of validity – 5 years
Privileges and conditions:
(1) Subject to any conditions specified in respect of the licence, the privileges of the holder of a Private Pilot Licence (Aeroplane) are to act, but not for remuneration, as pilot in command or co-pilot of any aeroplane specified in a class or type rating included in Part XII of the licence engaged in non-revenue flights.
(2) The licence is subject to the conditions and restrictions specified in paragraph 1.175 of JAR-FCL 1.
(3) The holder shall not:
(a) unless his licence includes an instrument rating (aeroplane) or an instrument meteorological conditions rating (aeroplanes), fly as pilot in command of such an aeroplane:
(i) on a flight outside controlled airspace when the flight visibility is less than 3km;
(ii) on a special VFR flight in a control zone in a flight visibility of less than 10km except on a route or in an aerodrome traffic zone notified for the purpose of this sub-paragraph; or
(iii) out of sight of the surface.
(b) unless his licence includes an instrument meteorological conditions rating (aeroplanes), fly as pilot in command or co-pilot of such an aeroplane flying in Class D or E airspace in circumstances which require compliance with the Instrument Flight Rules.
(c) fly as pilot in command of such an aeroplane at night unless his licence includes a night rating (aeroplanes) or a night qualification (aeroplane).
(d) fly as pilot in command of such an aeroplane carrying passengers unless within the preceding 90 days he has made three take-offs and three landings as the sole manipulator of the controls of an aeroplane of the same type or class and if such a flight is to be carried out at night and his licence does not include an instrument rating (aeroplanes) at least one of those take-offs and landing shall have been at night.

ANO SCHEDULE 8 Flight Crew of Aircraft – Licences, Ratings, Qualifications and Maintenance of Licence Privileges Articles 27, 28, 29, 30 and 31
Section 2 – JAR-FCL Licences AEROPLANE PILOTS – Commercial Pilot Licence (Aeroplane)

Commercial Pilot Licence (Aeroplane)

Minimum age -18 years
Maximum period of validity – 5 years

Privileges and conditions:

(1) Subject to any conditions specified in respect of the licence, the privileges of the holder of a Commercial Pilot Licence (Aeroplane) are to:

(a) exercise all the privileges of the holder of a JAR-FCL Private Pilot Licence (Aeroplane) which includes a night qualification;

(b) act as pilot in command or co-pilot of any aeroplane specified in a type or class rating included in Part XII of the licence on a flight other than a public transport flight;

(c) act as pilot in command on a public transport flight of any aeroplane included in Part XII of the licence certificated for single pilot operation.

(d) act as co-pilot on a public transport flight of any aeroplane included in Part XII of the licence.

(2) The licence is subject to the conditions and restrictions specified in paragraph 1.175 of JAR-FCL 1.

(3) The holder shall not:

(a) fly as pilot in command on a flight for the purpose of public transport unless he complies with the requirements of paragraph 1.960(a)(1) and (2) of Section 1 of JAR-OPS 1;

(b) unless his licence includes an instrument rating (aeroplane), fly such an aeroplane on any scheduled journey.

(c) fly as pilot in command of an aeroplane carrying passengers unless he has carried out at least three take-offs and three landings as pilot flying in an aeroplane of the same type or class or in a flight simulator, approved for the purpose, of the aeroplane type or class to be used, in the preceding 90 days;

(d) as co-pilot serve at the flying controls in an aeroplane carrying passengers during take-off and landing unless he has served as a pilot at the controls during take-off and landing in an aeroplane of the same type or in a flight simulator, approved for the purpose, of the aeroplane type to be used, in the preceding 90 days; or

(e) as the holder of a licence which does not include a valid instrument rating (aeroplane) act as pilot in command of an aeroplane carrying passengers at night unless during the previous 90 days at least one of the take-offs and landings required in sub-paragraph (c) has been carried out at night.

(f) unless his licence includes an instrument rating (aeroplane), fly any such aeroplane of which the maximum total weight authorised exceeds 2300kg on any flight for the purpose of public transport, except a flight beginning and ending at the same aerodrome and not extending beyond 25 nautical miles from that aerodrome.

(4) The holder shall be entitled subject to paragraph (5), to fly as pilot in command of an aeroplane of a type or class specified in any flying instructor's rating, class rating instructor rating, flight instructor rating or assistant flying instructor's rating included in the licence on a flight for the purpose of aerial work which consists of:

(a) the giving of instruction in flying; or

(b) the conducting of flying tests for the purposes of this Order;

in either case in an aeroplane owned, or operated under arrangements entered into, by a flying club of which the person giving the instruction or conducting the test and the person receiving the instruction or undergoing the test are both members.

(5) The holder shall not be entitled to exercise privileges contained in paragraph (4) other than in an aeroplane which he is entitled to fly as pilot in command on a private flight, an aerial work flight or a public transport flight pursuant to the privileges set out in paragraph (1) or (2) of these privileges.

Curtailment of privileges of licence holders aged 60 years or more

(7) The holder of a licence who has attained the age of 60 years but not attained the age of 65 years shall not act as a pilot of an aeroplane on a public transport flight except where the holder is:

(a) as a member of a multi-pilot crew; and

(b the only pilot in the flight crew who has attained age 60.

(8) The holder of a licence who has attained the age of 65 years shall not act as a pilot of an aeroplane on a public transport flight.

Airline Transport Pilot Licence (Aeroplane)

Minimum age – 21 years
Maximum period of validity – 5 years

Privileges and conditions:

(1) Subject to any conditions specified in respect of the licence, the privileges of the holder of an Airline Transport Pilot Licence (Aeroplane) are to:

(a) exercise all the privileges of the holder of a JAR-FCL Private Pilot Licence (Aeroplane), a JAR-FCL Commercial Pilot Licence (Aeroplane) and an instrument rating (aeroplane); and

(b) act as pilot in command or co-pilot of any aeroplane specified in a type rating included in Part XII of the licence on a public transport flight.

(2) The licence is subject to the conditions and restrictions specified in paragraph 1.175 of JAR-FCL 1.

Curtailment of privileges of licence holders aged 60 years or more

(1) Age 60-64. The holder of a licence who has attained the age of 60 years shall not act as a pilot of an aeroplane on a public transport flight except:

(a) as a member of a multi-pilot crew and provided that;

(b) such holder is the only pilot in the flight crew who has attained age 60.

(2) Age 65. The holder of a licence who has attained the age of 65 years shall not act as a pilot of an aeroplane on a public transport flight.

ANO   SCHEDULE 8 Flight Crew of Aircraft – Licences, Ratings, Qualifications and Maintenance of Licence Privileges Articles 27, 28, 29, 30 and 31
Section 2 – JAR-FCL Licences AEROPLANE PILOTS – Commercial Pilot Licence (Helicopter)

Airline Transport Pilot Licence (Aeroplane)
Minimum age – 21 years
Maximum period of validity – 5 years
Privileges and conditions:
(1) Subject to any conditions specified in respect of the licence, the privileges of the holder of an Airline Transport Pilot Licence (Aeroplane) are to:
(a) exercise all the privileges of the holder of a JAR-FCL Private Pilot Licence (Aeroplane), a JAR-FCL Commercial Pilot Licence (Aeroplane) and an instrument rating (aeroplane); and
(b) act as pilot in command or co-pilot of any aeroplane specified in a type rating included in Part XII of the licence on a public transport flight.
(2) The licence is subject to the conditions and restrictions specified in paragraph 1. 175 of Section 1 of JAR-FCL 1.
Curtailment of privileges of licence holders aged 60 years or more
(3) The holder of a licence who has attained the age of 60 years but not attained the age of 65 years shall not act as a pilot of an aeroplane on a public transport flight except where the holder is:
(a) a member of a multi-pilot crew; and
(b) the only pilot in the flight crew who has attained the age of 60.
(4) The holder of a licence who has attained the age of 65 years shall not act as a pilot of an aeroplane on a public transport flight.

Sub-Section 2 HELICOPTER PILOTS
Private Pilot Licence (Helicopter)
Minimum age – 17 years
Maximum period of validity – 5 years
Privileges and conditions:
(1) Subject to any conditions specified in respect of the licence, the privileges of the holder of a Private Pilot Licence (Helicopter) are to act, but not for remuneration, as pilot in command or co-pilot of any helicopter included in a type rating in Part XII of the licence engaged in non-revenue flights.
(2) The licence is subject to the conditions and restrictions specified in paragraph 2.175 of Section 1 of JAR-FCL 2.
(3) The holder shall not:
(a) fly as pilot in command of such a helicopter at night unless his licence includes a night rating (helicopters) or a night qualification (helicopter).
(b) fly as pilot in command of such a helicopter carrying passengers unless:
(i) within the preceding 90 days he has made three solo circuits, each to include take-offs and landings as the sole manipulator of the controls of a helicopter of the same type; or
(ii) if the privileges are to be exercised by night and his licence does not include an instrument rating, within the preceding 90 days he has made three circuits, each to include take-offs and landings by night as the sole manipulator of the controls of a helicopter of the same type.

Commercial Pilot Licence (Helicopter)
Minimum age – 18 years
Maximum period of validity – 5 years
Privileges and conditions:
(1) Subject to any conditions specified in respect of the licence, the privileges of the holder of a Commercial Pilot Licence (Helicopter) are to:
(a) exercise all the privileges of the holder of a JAR-FCL Private Pilot Licence (Helicopter);
(b) act as pilot in command or co-pilot of any helicopter included in a type rating in Part XII of the licence on a flight other than a public transport flight;
(c) act as pilot in command on a public transport flight of any helicopter certificated for single-pilot operation included in Part XII of the licence;
(d) act as co-pilot on a public transport flight in any helicopter included in Part XII of the licence required to be operated with a co-pilot.
(2) The licence is subject to the conditions and restrictions specified in paragraph 2.175 of Section 1 of JAR-FCL 2.
(3) The holder shall not fly as pilot in command on a flight for the purpose of public transport unless he complies with the requirements of paragraph 3.960(a)(2) of Section 1 of JAR-OPS 3.
(4) The holder shall not:
(a) unless his licence includes an instrument rating (helicopter), fly such a helicopter on any scheduled journey or on any flight for the purpose of public transport other than in visual meteorological conditions.
(b) fly as pilot in command of a helicopter carrying passengers unless he has carried out at least three circuits, each to include take-offs and landings, as pilot flying in a helicopter of the same type or a flight simulator of the helicopter type to be used, in the preceding 90 days; or
(c) as the holder of a helicopter licence which does not include a valid instrument rating (helicopter) act as pilot in command of a helicopter carrying passengers at night unless during the previous 90 days at least one of the take-offs and landings required in sub-paragraph (b) above has been carried out at night.
Curtailment of privileges of licence holders aged 60 years or more
(5) The holder of a licence who has attained the age of 60 years but not attained the age of 65 years shall not act as a pilot of a helicopter on a public transport flight except where the holder is:
(a) a member of a multi-pilot crew: and,

423

ANO SCHEDULE 8 Flight Crew of Aircraft – Licences, Ratings, Qualifications and Maintenance of Licence Privileges Articles 27, 28, 29, 30 and 31
Section 2 – JAR-FCL Licences AEROPLANE PILOTS – Commercial Pilot Licence (Aeroplane)

(b) the only pilot in the flight crew who has attained age 60.
(2) The holder of a licence who has attained the age of 65 years shall not act as a pilot of a helicopter on a public transport flight.

Airline Transport Pilot Licence (Helicopter)
Minimum age – 21 years
Maximum period of validity – 5 years
Privileges and conditions:
(1) Subject to any conditions specified in respect of the licence, the privileges of the holder of an Airline Transport Pilot Licence (Helicopter) are to:
(a) exercise all the privileges of the holder of a JAR-FCL Private Pilot Licence (Helicopter) and a JAR-FCL Commercial Pilot Licence (Helicopter);and
(b) subject to paragraph (2), act as pilot in command or co-pilot in any helicopter included in a type rating in Part XII of the licence on a public transport flight.
(2) The holder shall not fly as pilot in command on a flight for the purpose of public transport unless he complies with the requirements of paragraph 3.960(a)(2) of Section 1 of JAR-OPS 3.
Curtailment of privileges of licence holders aged 60 years or more
(3) The holder of a licence who has attained the age of 60 years but not attained the age of 65 years shall not act as a pilot of a helicopter on a public transport flight except where the holder is:
(a) a member of a multi-pilot crew; and
(b) the only pilot in the flight crew who has attained age 60.
(4) The holder of a licence who has attained the age of 65 years shall not act as a pilot of a helicopter on a public transport flight.

Section 3 – National Private Pilot's Licence (Aeroplanes)

National Private Pilot's Licence (Aeroplanes)
Minimum age – 17 years
No maximum period of validity

Privileges and conditions:
(1) Subject to paragraphs (2), (3), (4), (5), (6) and (7) the holder of the licence shall be entitled to fly as pilot in command of any simple single engine aeroplane, microlight aeroplane or SLMG specified or otherwise failing within an aircraft rating included in the licence.
Flight outside the United Kingdom
(2) He shall not fly:
(a) such a simple single engine aeroplane or a microlight aeroplane outside the United Kingdom except with the permission of the competent authority for the airspace in which he flies;
(b) such a SLMG in or over the territory of a Contracting State other than the United Kingdom except in accordance with permission granted by the competent authorities of that State provided that he may fly a SLMG outside the United Kingdom if his licence includes a SLMG rating and a medical certificate appropriate for such a flight.
Flight for purpose of public transport and aerial work
(2) He shall not fly any such an aeroplane for the purpose of public transport or aerial work except in the circumstances specified in paragraph (4).
(3) The circumstances referred to in paragraph (3) are that he flies such an aeroplane for the purpose of aerial work which consists of towing another aeroplane or glider in flight:
(a) in an aeroplane owned, or operated under arrangements entered into, by a flying club of which the holder of the licence and any person carried in the towing aeroplane or in any aeroplane or glider being towed are members; or
(b) in an aeroplane owned, or operated under arrangements entered into, by an organisation approved by the CAA for the purpose of this provision when:
(i) the holder of the licence is a member of an organisation approved by the CAA for the purpose of this provision; and
(ii) any person carried in the towing aeroplane or in any aeroplane or glider being towed is a member of an organisation approved by the CAA for the purpose of this provision.
Prohibitions on flight in specified conditions
(5) He shall not fly:
(a) as pilot in command of such a simple single engine aeroplane on a flight outside controlled airspace when the flight visibility is less than 5 km.
(b) as pilot in command of such a SLMG or microlight aeroplane on a flight outside controlled airspace when the flight visibility is less than 3km.
(c) as pilot in command of such an aeroplane:
(i) on a special VFR flight in a control zone in a flight visibility of less than 10 km; or
(ii) out of sight of the surface.
(d) as pilot in command of such a simple single engine aeroplane in circumstances which require compliance with the Instrument Flight Rules.
Carriage of persons
(6) He shall not fly as pilot in command of such an aeroplane:
(a) when the total number of persons carried (including the pilot) exceeds four; or
(b) when carrying passengers unless within the preceding 90 days he has made at least three take-offs and three landings as the sole manipulator of the controls of an aeroplane of the same class as that being flown.
Differences training

(7) He shall not fly:
(a) as pilot in command of such a simple single engine aeroplane where:
(i) the aeroplane is fitted with a tricycle undercarriage;
(ii) the aeroplane is fitted with a tailwheel;
(iii) the engine is fitted with either a supercharger or turbo-charger;
(iv) the engine is fitted with a variable pitch propeller;
(v) the landing gear is retractable;
(vi) a cabin pressurisation system is fitted; or
(vii) the aeroplane has a maximum continuous cruising speed in excess of 140 knots indicated airspeed;
unless appropriate differences training has been completed and recorded in his personal flying log book.
(b) as pilot in command of such a microlight aeroplane where:
(i) the aeroplane has 3 axis controls and his previous training and experience has only been in an aeroplane with flexwing controls; or
(ii) the aeroplane has flexwing controls and his previous training and experience has only been in an aeroplane with 3 axis controls; unless appropriate differences training has been completed and recorded in his personal flying logbook.

PART B – RATINGS AND QUALIFICATIONS

1 The following ratings may be included in a pilot's licence granted under Part 4 and, subject to the provisions of this Order and of the licence, the inclusion of a rating in a licence shall have the consequences respectively specified as follows:

Aircraft rating:
The rating shall entitle the holder of the licence to act as pilot of aircraft of the types and classes specified in an aircraft rating included in the licence and different types and classes of aircraft may be specified in respect of different privileges of a licence.

Instrument meteorological conditions rating (aeroplanes):
(1) Subject to paragraph (2), the raring shall within the United Kingdom
(a) entitle the holder of a United Kingdom Private Pilot's Licence (Aeroplanes) or a United Kingdom Basic Commercial Pilot's Licence (Aeroplanes) to fly as pilot in command of an aeroplane without being subject to the restrictions contained respectively in paragraphs (2)(c) or (f) of the privileges of the United Kingdom Private Pilot's Licence (Aeroplanes) or (3)(g) or (i) of the privileges of the United Kingdom Basic Commercial Pilot's Licence (Aeroplanes).
(b) entitle the holder of a JAR-FCL Private Pilot Licence (Aeroplane) to fly as pilot in command of an aeroplane in Class D or E airspace in circumstances which require compliance with the Instrument Flight Rules.
(2) The rating shall not entitle the holder of the licence to fly:
(a) on a special VFR flight in a control zone in a flight visibility of less than 3km; or
(b) when the aeroplane is taking off or landing at any place if the flight visibility below cloud is less than 1800 metres.

Instrument rating (aeroplane) shall entitle the holder of the licence to act as pilot in command or co-pilot of an aeroplane flying in controlled airspace in circumstances which require compliance with the Instrument Flight Rules.

Instrument rating (helicopter) shall entitle the holder of the licence to act as pilot in command or co-pilot of a helicopter flying in controlled airspace in circumstances which require compliance with the Instrument Flight Rules.

Microlight class rating shall, when included in the aircraft rating of a National Private Pilot's Licence (Aeroplanes) or a United Kingdom Private Pilot's Licence (Aeroplanes) and subject to the conditions of the licence in which it is included, entitle the holder to act as pilot in command of any microlight aeroplane.

Night rating (aeroplanes) shall entitle the holder of a United Kingdom Private Pilot's Licence (Aeroplanes) or a United Kingdom Basic Commercial Pilot's Licence (Aeroplanes) to act as pilot in command of an aeroplane at night.

Night qualification (aeroplane) shall entitle the holder of a United Kingdom Private Pilot's Licence (Aeroplanes), a JAR-FCL Private Pilot Licence (Aeroplane) or a United Kingdom Basic Commercial Pilot's Licence (Aeroplanes) to act as pilot in command of an aeroplane at night.

Night rating (helicopters) shall entitle the holder of a United Kingdom Private Pilot's Licence (Helicopters) to act as pilot in command of a helicopter at night.

Night qualification (helicopter) shall entitle the holder of either a United Kingdom Private Pilot's Licence (Helicopters) or a JAR-FCL Private Pilot Licence (Helicopter) to act as pilot in command of a helicopter at night.

Night rating (gyroplanes) shall entitle the holder of a United Kingdom Private Pilot's Licence (Gyroplanes) to act as pilot in command of a gyroplane at night.

Simple single engine aeroplane (NPPL) class rating shall, when included in the aircraft rating of a National Private Pilot's Licence (Aeroplanes) and subject to the conditions of that licence, entitle the holder to act as pilot in command of any simple single engine aeroplane with a maximum take-off weight authorised not exceeding 2000kgs excluding any such aeroplane which is a self launching motor glider or a microlight aeroplane.

SLMG class rating shall, when included in the aircraft rating of a National Private Pilot's Licence (Aeroplanes) or a United Kingdom Private Pilot's Licence (Aeroplanes) and subject to the conditions of the licence in which it is included, entitle the holder to act as pilot in command of any SLMG.

Towing rating (flying machines) shall entitle the holder of the licence to act as pilot of a flying machine while towing a glider in flight for the purposes of public transport or aerial work.

Flying instructor's rating shall entitle the holder of the licence to give instruction in flying aircraft of such types and classes as may be specified in the rating for that purpose.

Assistant flying instructor's rating shall entitle the holder of the licence to give instruction in flying aircraft of such types and classes as may be specified in the rating for that purpose provided that:
(a) such instruction shall only be given under the supervision of a person present during the take-off and landing at the aerodrome at which the instruction is to begin and end and holding a pilot's licence endorsed with a flying instructor's rating;

ANO   SCHEDULE 8 PART B – RATINGS AND QUALIFICATIONS
Instrument meteorological conditions rating (aeroplanes):

(b) such a rating shall not entitle the holder of the licence to give directions to the person undergoing instruction in respect of the performance by that person of:
(i) his first solo flight;
(ii) his first solo flight by night;
(iii) his first solo cross-country flight otherwise than by night; or
(iv) his first solo cross-country flight by night.

Flight instructor rating (aeroplane) shall entitle the holder of the licence to give instruction in flying aircraft of such types and classes as may be specified in the rating for that purpose subject to the restrictions specified below.

Flight instructor rating (aeroplane) – Restrictions
Restricted period
(1) Until the holder of a flight instructor (aeroplane) rating has completed at least 100 hours flight instruction and, in addition, has supervised at least 25 solo flights by students, the privileges of the rating shall be restricted.
(2) The restrictions shall be removed from the rating when the above requirements have been met and on the recommendation of the supervising flight instructor (aeroplane).

Restricted Privileges
(3) The privileges shall be restricted to carrying out under the supervision of the holder of a flight instructor (aeroplane) rating approved for this purpose:
(a) flight instruction for the issue of the Private Pilot Licence (Aeroplane) or those parts of integrated courses at Private Pilot Licence (Aeroplane) level and class and type ratings for single-engine aeroplanes, excluding approval of first solo flights by day or by night and first solo cross country flights by day or by night; and
(b) night flying instruction.

Flight instructor rating (helicopter) shall entitle the holder of the licence to give instruction in flying helicopters of such types as may be specified in the rating for that purpose subject to the restrictions specified below.

Flight instructor rating (helicopter) Restrictions
Restricted period.
(1) Until the holder of a flight instructor (helicopter) rating has completed at least 100 hours flight instruction and, in addition, has supervised at least 25 solo flights by students, the privileges of the rating shall be restricted.
(2) The restrictions shall be removed from the rating when the above requirements have been met and on the recommendation of the supervising flight instructor (helicopter).

Restricted Privileges.
(3) The privileges shall be restricted to carrying out under the supervision of the holder of a flight instructor (helicopter) rating approved for this purpose:
(a) flight instruction for the issue of the Private Pilot Licence (Helicopter) or those parts of integrated courses at Private Pilot Licence (Helicopter) level and class and type ratings for single-engine helicopters, excluding approval of first solo flights by day or by night and first solo cross country flights by day or by night; and
(b) night flying instruction.

Type rating instructor rating (multi-pilot aeroplane) shall entitle the holder to instruct licence holders for the issue of a multi-pilot aeroplane type rating, including the instruction required for multi-crew co-operation.

Type rating instructor rating (helicopter) shall entitle the holder to instruct licence holders for the issue of a type rating, including the instruction required for multi-crew co-operation as applicable.

Class rating instructor rating (single-pilot aeroplane) shall entitle the holder to instruct licence holders for the issue of a type or class rating for single-pilot aeroplanes.

Instrument rating instructor rating (aeroplane) shall entitle the holder to conduct flight instruction for the issue of an instrument rating (aeroplane) or an instrument meteorological conditions rating (aeroplanes).

Instrument rating instructor rating (helicopter) shall entitle the holder to conduct flight instruction for the issue of an instrument rating (helicopter).

2 An aircraft rating included in a flight engineer's licence shall entitle the holder of the licence to act as flight engineer only of aircraft of a type specified in the aircraft rating.

3 For the purposes of this Schedule:
'Solo flight' means a flight on which the pilot of the aircraft is not accompanied by a person holding a pilot's licence granted or rendered valid under this Order.
'Cross-country flight' means any flight during the course of which the aircraft is more than 3 nautical miles from the aerodrome of departure.

SCHEDULE 8
PART C – SECTION 1 – REQUIREMENT FOR CERTIFICATE OF TEST OR EXPERIENCE

1 Appropriateness of certificate
(a) A certificate of test or a certificate of experience required by article 28, 30(20 or 31(1) shall not be appropriate to the functions to be performed on a flight unless it is a certificate appropriate to the description of the flight according to the following table:

ANO SCHEDULE 8 PART C – SECTION 1 – REQUIREMENT FOR CERTIFICATE OF TEST OR EXPERIENCE 1 Appropriateness of certificate

Case	Class of national Licence	Description of flight	Certificate required
A	Microlight Licence Private Pilots Licence (Gyroplanes)	Any flight within the privileges of SLMG Licence	Certificate of test or certificate of the licence experience
B	Commercial Pilots Licence (Helicopters and Gyroplanes) Commercial Pilots Licence (Balloons) Commercial Pilots Licence (Gliders) Commercial Pilots Licence (Airships) Airline Transport pilots Licence (Helicopters and Gyroplanes)	Carriage of passengers on a flight in respect of which the holder of the licence receives renumeration	Certificate of test
C	Commercial Pilots Licence (Helicopters and Gyroplanes) Commercial Pilots Licence (Balloons) Commercial Pilots Licence (Gliders) Commercial Pilots Licence (Airships) Airline Transport pilots Licence (Helicopters and Gyroplanes)	For public transport	Certificate of test
D	Commercial Pilots Licence (Helicopters and Gyroplanes) Commercial Pilots Licence (Balloons) Commercial Pilots Licence (Gliders) Commercial Pilots Licence (Airships) Airline Transport pilots Licence (Helicopters and Gyroplanes)	For aerial work	Certificate of test or certificate of experience
E	Commercial Pilots Licence (Helicopters and Gyroplanes) Commercial Pilots Licence (Balloons) Commercial Pilots Licence (Gliders) Commercial Pilots Licence (Airships) Airline Transport pilots Licence (Helicopters and Gyroplanes)	Any flight within the privileges of a	Certificate of test or certificate Private Pilots licence of experience
F	Flight Navigators Licence	Flights to which article 25(9) applies	Certificate of experience

ANO   SCHEDULE 8 PART C – SECTION 1 – REQUIREMENT FOR CERTIFICATE OF TEST OR EXPERIENCE 1 Appropriateness of certificate

(b) For the purposes of this Part of this Schedule, references to Cases are references to the Cases indicated in the first Column of the Table in paragraph 1(a) of this Part of this Schedule.

2 Certificate of test

A certificate of test required by article 28, 30(2) or 31(1) shall be signed by a person authorised by the CAA to sign certificates of this kind and shall certify the following particulars:

(a) the functions to which the certificate relates;

(b) that the person signing the certificate is satisfied that on a date specified in the certificate the holder of the licence or personal flying logbook of which the certificate forms a part' as the case may be, passed an appropriate test of his ability to perform the functions to which the certificate relates;

(c) the type of aircraft or flight simulator in or by means of which the test was conducted; and

(d) the date on which it was signed.

3 Nature of test

The appropriate test referred to in paragraph 2 above shall be:

(a) in the case of a test which entitles the holder of the licence of which the certificate forms part to act as pilot in command and/or co-pilot of aircraft of the type, types or class specified in the certificate, a test of the pilot's competence to fly the aircraft as pilot in command and/or co-pilot and shall, where the CAA so specifies in respect of the whole or part of a test, be conducted in an aircraft in flight or by means of a flight simulator approved by the CAA;

(b) in the case of a test which entitles the holder of the licence of which the certificate forms part to perform the functions to which a flying instructor's rating (gyroplanes), an assistant flying instructor's rating (gyroplanes) or an instrument meteorological conditions rating (aeroplanes) relates, a test of his ability to perform the functions to which the rating relates and shall, where the CAA so specifies in respect of the whole or part of the test, be conducted in an aircraft in flight.

4 Period of validity of certificate of test

A certificate of test:

(a) required by article 28 in respect of a Commercial Pilot's Licence (Balloons) shall not be valid in relation to a flight made more than 13 months after the date of the test which it certifies and, required by article 28 or 30(2) in respect of any other licence, shall not be valid in relation to a flight made more than 13 months in Cases A, B and E or more than 6 months in Cases C and D after the date of the test which it certifies; provided that in the case of Cases C and D, 2 certificates of test shall together be deemed to constitute a valid certificate of test if they certify flying tests conducted on 2 occasions within the period of 13 months preceding the flight on which the functions are to be performed, such occasions being separated by an interval of not less than 4 months, and if both certificates are appropriate to those functions.

(b) required by article 31(1) in respect of an instrument meteorological conditions rating (aeroplanes) shall not be valid in relation to a flight made more than 25 months after the date of the test which it certifies.

(c) required by article 31(1) in respect of an assistant flying instructor's rating (gyroplanes) and a flying instructor's rating (gyroplanes) shall not be valid in relation to a flight made more than 3 years after the date of the test which it certifies.

5 Certificate of experience

A certificate of experience required by article 28 or 30(2) shall be signed by a person authorised by the CAA to sign such a certificate and shall certify the following particulars:

(a) the functions to which the certificate relates;

(b) in the case of a pilot, that on the date on which the certificate was signed the holder of the licence or personal flying log book of which it forms part, as the case may be, produced his personal flying log book to the person signing the certificate and satisfied him that he had appropriate experience in the capacity to which his licence relates within the appropriate period specified in paragraph 6 of this Part of this Schedule;

(c) in the case of a flight navigator, that on the date on which the certificate was signed the holder of the licence of which it forms part produced his navigation logs, charts and workings of astronomical observations to the person signing the certificate and satisfied him that he had appropriate experience in the capacity to which the licence relates within the appropriate period specified in paragraph 6 of this Part of this Schedule;

(d) in the case of a pilot or flight engineer, the type or types of aircraft in which the experience was gained;

(e) the date on which it was signed.

6 Period of experience

A certificate of experience shall not be valid unless the experience was gained within the period of 13 months preceding the signing of the certificate in the case of Cases A, E and F, or 6 months preceding the signing of the certificate in the case of Case D.

7 Period of validity of certificate of experience

A certificate of experience in respect of a Commercial Pilot's Licence (Balloons) shall not be valid for more than 13 months after it was signed and in respect of any other licence shall not be valid for more than 6 months after it was signed for Case D nor for more than 13 months after it was signed for any other case.

SCHEDULE 8
SECTION 2 – REQUIREMENT FOR CERTIFICATE OF REVALIDATION

1 Appropriate certificate of revalidation
A certificate of revalidation required by article 29 or 31(2) shall not be appropriate to the exercise of the privileges of a flight crew licence unless it is a certificate which accords with this Section.

2 Type and class ratings
(1) Aeroplane type and class ratings
(a) Type ratings and multi-engine class ratings, aeroplane
(i) Validity: Type ratings and multi-engine class ratings for aeroplanes are valid for one year beginning with the date of issue, or the date of expiry if revalidated within the period of three months preceding the date of expiry.
(ii) Revalidation: For revalidation of type ratings and multi-engine class ratings, aeroplane, the applicant shall satisfy the requirements specified in paragraph 1.245(a) and (b) of JAR-FCL 1.
(b) Single-pilot single-engine class ratings
(i) Validity: Single-pilot single-engine class ratings are valid for two years from the date of issue, or the date of expiry if revalidated within the period of three months preceding the date of expiry.
(ii) Revalidation of all single-engine piston aeroplane class ratings (land) and all touring motor glider ratings: For revalidation of single-pilot single-engine piston aeroplane class ratings (land) class ratings or touring motor glider class ratings (or both) the applicant shall on single-engine piston aeroplanes (land) or touring motor gliders (as the case may be) satisfy the requirements specified in paragraph 1.245(c)(1) of JAR-FCL 1.
(iii) Revalidation of single-engine turbo-prop aeroplanes (land) single-pilot: For revalidation of single-engine turbo-prop (land) class ratings the applicant shall within the three months preceding the expiry date of the rating, pass a proficiency check with an authorised examiner on an aeroplane in the relevant class.
(iv) Revalidation of single-engine piston aeroplanes (sea): For revalidation of single pilot single engine piston aeroplane (sea) class ratings the applicant shall:
(aa) within the three months preceding the expiry date of the rating, pass a proficiency check with an authorised examiner on a single-engine piston aeroplane (sea); or
(bb) within 12 months preceding the expiry of the rating complete 12 hours of flight time including 6 hours of pilot in command time on either a single engine piston aeroplane (sea) or a single engine piston aeroplane (land) and 12 water take-offs and 12 alightings on water; and either complete a training flight of at least 1 hours duration with a flight instructor or pass a proficiency check or skill test for any other class or type rating.
(c) Expired ratings
(i) If a type rating or multi-engine class rating has expired, the applicant shall meet the requirements in paragraph (b) above and meet any refresher training requirements as determined by the CAA. The rating will be valid from the date of completion of the renewal requirements.
(ii) If a single-pilot single-engine class rating has expired, the applicant shall complete the skill test in accordance with the requirements specified at Appendix 3 to paragraph 1.240 of JAR-FCL 1.
(2) Helicopter type ratings
(a) Type ratings, helicopter – validity
Type ratings for helicopters are valid for one year beginning with the date of issue, or the date of expiry if revalidated within the period of three months preceding the date of expiry.
(b) Type ratings, helicopter – revalidation
For revalidation of type ratings, helicopter, the applicant shall complete the requirements specified in paragraph 2.245(b) of JAR-FCL 2.
(c) Expired ratings. If a type rating has expired, the applicant shall meet the requirements in sub-paragraph (b) above and meet any refresher training requirements as determined by the CAA. The rating will be valid from the date of completion of the renewal requirements.
(3) Flight engineer type ratings
(a) Type ratings – validity
Flight engineer type ratings are valid for one year beginning with the date of issue, or the date of expiry if revalidated within the period of three months preceding the date of expiry.
(b) Type ratings – revalidation
For revalidation of flight engineer type ratings the applicant shall, within the three months preceding the expiry date of the rating, pass a proficiency check with an authorised examiner on the relevant type of aircraft.

3 Forms of certificate of revalidation
(1) A certificate of revalidation required by article 29 or 31(2) shall be signed by a person authorised by the CAA to sign certificates of this kind and shall certify:
(a) the functions to which the certificate relates;
(b) that the person signing the certificate is satisfied that on a date specified in the certificate, the holder of the licence of which the certificate forms a part met the appropriate requirements for revalidation specified in respect of the rating, in the case of an aircraft rating in paragraph 2 and in the case of any other rating specified in the Table at sub-paragraph (2) below, to exercise the privileges of the licence or rating to which the certificate relates;
(c) the type of aircraft or flight simulator in or by means of which the test was conducted; and
(d) the date on which it was signed.

ANO   SCHEDULE 8 SECTION 2 – REQUIREMENT FOR CERTIFICATE OF REVALIDATION – 2 Type and class ratings

(2) The requirements for revalidation of a rating listed in Column1 are those set out in Column 2 of the following Table

Rating	Paragraph in Section 1 of JAR-FCL 1 or 2
Instrument rating (aeroplane)	1.185
Instrument rating (helicopter)	2.185
Flight instructor (aeroplane)	1.355
Flying instructor's rating (aeroplanes)	
Assistant flying instructor's rating (aeroplanes)	
Flight instructor (helicopter)	2.355
Flying instructor's rating (helicopters)	
Assistant flying instructor's rating (helicopters)	
Type rating instructor rating (multi-pilot aeroplane)	1.370
Type rating instructor rating (helicopter)	2.370
Class rating instructor rating (single pilot aeroplane)	1.385
Instrument rating instructor rating (aeroplane)	1.400
Instrument rating instructor rating (helicopter)	2.400

Section 3 – Maintenance of validity of National Private Pilot's Licence (Aeroplanes)

1 A simple single engine aeroplane (NPPL) class rating included in a National Private Pilot's Licence (Aeroplanes) shall not be valid for the purposes of article 30(1) unless the provisions of this Section have been complied with.

2 A simple single engine aeroplane (NPPL) class rating shall be valid if either:

(a) the holder has within the 12 months preceding the flight flown not less than six hours in an aeroplane falling within the simple single engine aeroplane (NPPL) class rating, four hours of which shall have been as pilot in command and he has carried out a training flight of at least 1 hour duration with a flying instructor within the previous 24 months; or

(b) he has within the three months preceding the expiry of the rating undertaken a simple single engine aeroplane (NPPL) General Skills Test.

SCHEDULE 9

Articles 38(3), 40(3) and 42(3)
Public transport – operational requirements
PART A – OPERATIONS MANUAL

(1) Information and instructions relating to the following matters shall be included in the operations manual referred to in article 38(2):

(a) the number of the crew to be carried in the aircraft, on each stage of any route to be flown, and the respective capacities in which they are to act, and instructions as to the order and circumstances in which command is to be assumed by members of the crew;

(b) the respective duties of each member of the crew and the other members of the operating staff;

(c) the scheme referred to in article 82(1)(c)(i);

(d) such technical particulars concerning the aircraft, its engines and equipment and concerning the performance of the aircraft as may be necessary to enable the flight crew of the aircraft to perform their respective duties;

(e) the manner in which the quantities of fuel and oil to be carried by the aircraft are to be computed and records of fuel and oil carried and consumed on each stage of the route to be flown are to be maintained; the instructions shall take account of all circumstances likely to be encountered on the flight including the possibility of failure of one or more of the aircraft engines;

(f) the manner in which the quantity, if any, of oxygen and oxygen equipment to be carried in the aircraft for the purpose of complying with Scale L1 or L2 in Schedule 4 is to be computed;

(g) the check system to be followed by the crew of the aircraft prior to and on take-off, on landing and in an emergency, so as to ensure that the operating procedures contained in the operations manual and in the flight manual or performance schedule forming part of the relevant certificate of airworthiness are complied with;

(h) the circumstances in which a radio watch is to be maintained;

(i) the circumstances in which oxygen is to be used by the crew of the aircraft, and by passengers;

(j) subject to paragraph 2, communication, navigational aids, aerodromes, local regulations, in-flight procedures, approach and landing procedures and such other information as the operator may deem necessary for the proper conduct of flight operations; the information referred to in this paragraph shall be contained in a route guide, which may be in the form of a separate volume;

(k) the reporting in flight to the notified authorities of meteorological observations;

(l) subject to paragraph 2, the minimum altitudes for safe flight on each stage of the route to be flown and any planned diversion therefrom, such minimum altitudes being not lower than any which may be applicable under the law of the United Kingdom or of the countries whose territory is to be flown over;

(m) the particulars referred to in article 47(2);

(n) emergency flight procedures, including procedures for the instruction of passengers in the position and use of emergency equipment and procedures to be adopted when the commander of the aircraft becomes aware that another aircraft or a vessel is in distress and needs assistance;

(o) in the case of aircraft intended to fly at an altitude of more than 49,000ft the procedures for the use of cosmic radiation detection equipment;

(p) the labelling and marking of dangerous goods, the manner in which they must be loaded on or suspended beneath an aircraft, the responsibilities of members of the crew in respect of the carriage of dangerous goods and the action to be taken in the event of emergencies arising involving dangerous goods;

(q) such particulars of any permission granted to the operator pursuant to article 21 as may be necessary to enable the commander of the aircraft to determine whether he can comply with article 52(b)(ii);

(r) procedures for the operation of any airborne collision avoidance system carried on the aircraft.

(s) the establishment and maintenance of an accident prevention and flight safety programme.

(2) In relation to any flight which is not one of a series of flights between the same two places it shall be sufficient if, to the extent that it is not practicable to comply with sub-paragraphs (j) and (l), the manual contains such information and instructions as will enable the equivalent data to be ascertained before take-off.

SCHEDULE 10

Public transport – operational requirements
PART B – TRAINING MANUAL

The following information and instructions in relation to the training, experience, practice and periodical tests required under article 42(3) shall be included in the training manual referred to in article 40(3):

(a) the manner in which the training, practice and periodical tests required under article 42(3) and specified in Part C of this Schedule are to be carried out;

(b) the minimum qualifications and experience which the operator requires of persons appointed by him to give or to supervise the said training, practice and periodical tests;

(c) the type of training, practice and periodical tests which each such person is appointed to give or to supervise;

(d) the type of aircraft in respect of which each such person is appointed to give or to supervise the said training, practice and periodical tests;

(e) the minimum qualifications and experience required for each member of the crew undergoing the said training, practice and periodical tests;

(f) the current syllabus for, and specimen forms for recording, the said training, practice and periodical tests;

(g) the manner in which instrument flight conditions and engine failure are to be simulated in the aircraft in flight;

ANO — SCHEDULE 10 – Public transport – operational requirements
PART B – TRAINING MANUAL

(h) the extent to which the said training and testing is permitted in the course of flights for the purpose of public transport;
(i) the use to be made in the said training and testing of apparatus approved for the purpose by the CAA.

SCHEDULE 10
PART C – CREW TRAINING AND TESTS

1 The training, experience, practice and periodical tests required under article 42(3) for members of the crew of an aircraft engaged on a flight for the purpose of public transport shall be as specified in paragraph 2.

2 Crew

(1) Every member of the crew shall:
(a) have been tested within the relevant period by or on behalf of the operator as to his knowledge of the use of the emergency and life saving equipment required to be carried in the aircraft on the flight; and
(b) have practised within the relevant period, under the supervision of the operator or of a person appointed by him for the purpose, the carrying out of the duties required of him in case of an emergency occurring to the aircraft, either in an aircraft of the type to be used on the flight or in apparatus approved by the CAA for the purpose and controlled by persons so approved.

2 Pilots

(a) Every pilot included in the flight crew who is intended by the operator to fly as pilot in circumstances requiring compliance with the Instrument Flight Rules shall within the relevant period have been tested by or on behalf of the operator:
(i) as to his competence to perform his duties while executing normal manoeuvres and procedures in flight, in an aircraft of the type to be used on the flight, including the use of the instruments and equipment provided in the aircraft; and
(ii) as to his competence to perform his duties in instrument flight conditions while executing emergency manoeuvres and procedures in flight, in an aircraft of the type to be used on the flight, including the use of the instruments and equipment provided in the aircraft.
(b) A pilot's ability to carry out normal manoeuvres and procedures shall be tested in the aircraft in flight.
(c) The other tests required by sub-paragraph (a) may be conducted either in the aircraft in flight, or under the supervision of a person approved by the CAA for the purpose by means of a flight simulator approved by the CAA.
(d) The tests specified in sub- paragraph (a)(ii) when conducted in the aircraft in flight shall be carried out either in actual instrument flight conditions or in instrument flight conditions simulated by means approved by the CAA.
(e) Every pilot included in the flight crew whose licence does not include an instrument rating or who, notwithstanding the inclusion of such a rating in his licence, is not intended by the operator to fly in circumstances requiring compliance with the Instrument Flight Rules, shall within the relevant period have been tested, by or on behalf of the operator in flight in an aircraft of the type to be used on the flight:
(i) as to his competence to act as pilot of the aircraft, while executing normal manoeuvres and procedures; and
(ii) as to his competence to act as pilot of the aircraft, while executing emergency manoeuvres and procedures.
(f) Every pilot included in the flight crew who is seated at the flying controls during the take-off or landing and who is intended by the operator to fly as pilot in circumstances requiring compliance with the Instrument Flight Rules shall within the relevant period have been tested as to his proficiency in using instrument approach-to-land systems of the type in use at the aerodrome of intended landing and any alternate aerodromes, such test being carried out either in flight in instrument flight conditions or in instrument flight conditions simulated by means approved by the CAA or under the supervision of a person approved by the CAA for the purpose by means of a flight simulator approved by the CAA.
(g) In the case of a helicopter, every pilot included in the flight crew whose licence does not include an instrument rating but who is intended to fly at night under visual flight conditions, shall within the relevant period have been tested, by or on behalf of the operator, in a helicopter of the type to be used on the flight:
(i) as to his competence to act as pilot of that helicopter, while executing normal manoeuvres and procedures; and
(ii) as to his competence to act as pilot of that helicopter, while executing specified manoeuvres and procedures in flight in instrument flight conditions by means approved by the CAA.
(h) Every pilot included in the flight crew and who is seated at the flying controls during take-off or landing shall within the relevant period have carried out, when seated at the flying controls not less than three take-offs and three landings in aircraft of the type to be used on the flight.

3 Flight engineers

(a) Every flight engineer included in the flight crew shall within the relevant period have been tested by or on behalf of the operator:
(i) as to his competence to perform his duties while executing normal procedures in flight, in an aircraft of the type to be used on the flight;
(ii) as to his competence to perform his duties while executing emergency procedures in flight, in an aircraft of the type to be used on the flight.
(b) A flight engineer's ability to carry out normal procedures shall be tested in an aircraft in flight and the other tests required by this sub-paragraph may be conducted either in the aircraft in flight, or under the supervision of a person approved by the CAA for the purpose by means of a flight simulator approved by the CAA.

4 Flight navigators and flight radiotelephony operators

Every flight navigator and flight radiotelephony operator whose inclusion in the flight crew is required under article 25(9) and (11) respectively shall within the relevant period have been tested by or on behalf of the operator as to his competence to perform his duties in conditions corresponding to those likely to be encountered on the flight:

(a) in the case of a flight navigator, using equipment of the type to be used in the aircraft on the flight for purposes of navigation;

(b) in the case of a flight radiotelephony operator using radio equipment of the type installed in the aircraft to be used on the flight, and including a test of his ability to carry out emergency procedures.

5 Aircraft commanders

(a) The pilot designated as commander of the aircraft for the flight shall within the relevant period have demonstrated to the satisfaction of the operator that he has adequate knowledge of the route to be taken, the aerodromes of take-off and landing, and any alternate aerodromes, including in particular his knowledge of:

(i) the terrain;

(ii) the seasonal meteorological conditions;

(iii) the meteorological, communications and air traffic facilities, services and procedures;

(iv) the search and rescue procedures; and

(v) the navigational facilities;

relevant to the route.

(b) In determining whether a pilot's knowledge of the matters referred to in sub-paragraph (a) is sufficient to render him competent to perform the duties of aircraft commander on the flight, the operator shall take into account the pilot's flying experience in conjunction with the following:

(i) the experience of other members of the intended flight crew;

(ii) the influence of terrain and obstructions on departure and approach procedures at the aerodromes of take-off and intended landing and at alternate aerodromes;

(iii) the similarity of the instrument approach procedures and let-down aids to those with which the pilot is familiar;

(iv) the dimensions of runways which may be used in the course of the flight in relation to the performance limits of aircraft of the type to be used on the flight;

(v) the reliability of meteorological forecasts and the probability of difficult meteorological conditions in the areas to be traversed;

(vi) the adequacy of the information available regarding the aerodrome of intended landing and any alternate aerodromes;

(vii) the nature of air traffic control procedures and the familiarity of the pilot with such procedures;

(viii) the influence of terrain on route conditions and the extent of the assistance obtainable en route from navigational aids and air-to-ground communication facilities; and

(ix) the extent to which it is possible for the pilot to become familiar with unusual aerodrome procedures and features of the route by means of ground instruction and training devices.

6 Definitions and validity periods

For the purposes of this part:

(a) 'visual flight conditions' means weather conditions such that the pilot is able to fly by visual reference to objects outside the aircraft;

(b) 'instrument flight conditions' means weather conditions such that the pilot is unable to fly by visual reference to objects outside the aircraft;

(c) 'relevant period' means a period which immediately precedes the commencement of the flight, being, subject to sub-paragraph (d), a period:

(i) in the case of sub-paragraph (2)(h), of 3 months;

(ii) in the case of sub-paragraphs (2)(a)(ii), (2)(e)(ii), (2)(f), (2)(g)(ii) and (3)(a)(ii), of 6 months;

(iii) in the case of sub-paragraphs (1), (2)(a)(i), (2)(e)(i), (2)(g)(i), (3)(a)(i), (4) and (5)(a), of 13 months.

(d) Any pilot of the aircraft to whom the provisions of sub-paragraphs (2)(a)(ii), (2)(e)(ii) or (2)(f) and any flight engineer of the aircraft to whom the provisions of sub-paragraph (3)(a)(ii) apply shall for the purposes of the flight be deemed to have complied with such requirements respectively within the relevant period if he has qualified to perform his duties in accordance therewith on 2 occasions within the period of 13 months immediately preceding the flight, such occasions being separated by an interval of not less than 4 months.

(e) The requirements of sub-paragraph (5)(a) shall be deemed to have been complied with within the relevant period by a pilot designated as commander of the aircraft for the flight if, having become qualified so as to act on flights between the same places over the same route more than 13 months before commencement of the flight, he has within the period of 13 months immediately preceding the flight flown as pilot of an aircraft between those places over that route.

(3) The records required to be maintained by an operator under article 42(4) of this Order shall be accurate and up-to-date records so kept as to show, on any date, in relation to each person who has during the period of 2 years immediately preceding that date flown as a member of the crew of any public transport aircraft operated by that operator:

(a) the date and particulars of each test required by this Part undergone by that person during the said period including the name and qualifications of the examiner;

(b) the date upon which that person last practised the carrying out of duties referred to in paragraph 2(1)(b) of this Part;

(c) the operator's conclusions based on each such test and practice as to that person's competence to perform his duties; and

(d) the date and particulars of any decision taken by the operator during the said period in pursuance of paragraph 2(5)(a) of this Part including particulars of the evidence upon which that decision was based.

(4) The operator shall whenever called upon to do so by any authorised person produce for the inspection of any person so authorised all records referred to in paragraph 3 and furnish to any such person all such information as he may require in connection with any such records and produce for his inspection all log books, certificates, papers and other documents, whatsoever which he may reasonably require to see for the purpose of determining whether such records are complete or of verifying the accuracy of their contents.

(5) The operator shall at the request of any person in respect of whom he is required to keep records as aforesaid furnish to that person, or to any operator of aircraft for the purpose of public transport by whom that person may subsequently be employed, particulars of any qualifications in accordance with this Schedule obtained by such person whilst in his service.

SCHEDULE 10
Articles 86 and 88
Documents to be carried

1 Circumstances in which documents are to be carries

(1) On a flight for the purpose of public transport Documents A, B, C, D, E, F, H and, if the flight is international air navigation, Documents G and I shall be carried

(2) On a flight for the purpose of aerial work Documents A, B, C, E, F and, if the flight is international air navigation, Documents G and I shall be carried.

(3) On a private flight, being international air navigation Documents A, B, C, G and I shall be carried.

(4) On a flight made in accordance with the terms of a permission granted to the operator under article 21, Document J shall be carried.

2 Description of documents

For the purposes of this Schedule:

(1) 'Document A' means the licence in force under the Wireless Telegraphy Act 1949 in respect of the aircraft radio station installed in the aircraft;

(2) 'Document B' means the certificate of airworthiness in force in respect of the aircraft; provided that, where the certificate of airworthiness includes the flight manual for the aircraft, with the permission of the CAA, an aircraft to which article 38 applies need not carry the flight manual as part of this document;

(3) 'Document C' means the licences of the members of the flight crew of the aircraft;

(4) 'Document D' means one copy of the load sheet, if any, required by article 43 in respect of the flight;

(5) 'Document E' means one copy of each certificate of maintenance review, if any, in force in respect of the aircraft;

(6) 'Document F' means the technical log, if any, in which entries are required to be made under article 15;

(7) 'Document G' means the certificate of registration in force in respect of the aircraft;

(8) 'Document H' means those parts of the operations manual, if any, required by article 38(2)(c) to be carried on the flight;

(9) 'Document I' means a copy of the notified procedures to be followed by the pilot in command of an intercepted aircraft, and the notified visual signals for use by intercepting and intercepted aircraft;

(10) 'Document J' means the permission, if any, granted in respect of the aircraft under article 21, provided that, with the permission of the CAA, an aircraft to which article 38 applies need not carry such a permission if it carries an operations manual which includes the particulars specified at sub-paragraph (1)(q) of Part A of Schedule 9.

3 Definitions

For the purposes of this Schedule:

'International air navigation' means any flight which includes passage over the territory of any country other than the United Kingdom, except any of the Channel Islands, the Isle of Man, any country to which there is power to extend the Civil Aviation Act 1982 under section 108(1) of that act.

SCHEDULE 11
Article 108
Air Traffic Controllers – Licences, Ratings, Endorsements and Maintenance of Licence Privileges
PART A AIR TRAFFIC CONTROLLER LICENCES

1 Air Traffic Controller's Licence

(1) The minimum age at which a person may be granted an Air Traffic Controller's Licence shall be 20 years.

(2) There shall be no maximum period of validity for an Air Traffic Controller's Licence.

(3) The privileges of an Air Traffic Controller's Licence are to:

(a) act as an air traffic controller for any sector or operational position for which a valid rating and endorsement and current unit licence endorsement are included in the licence; and

(b) exercise the privileges of a Student Air Traffic Controller Licence.

2 Student Air Traffic Controller's Licence

(1) The minimum age at which a person may be granted a Student Air Traffic Controller's Licence shall be 18 years.

(2) The maximum period of validity for a Student Air Traffic Controller's Licence shall be two years.

SCHEDULE 11 Article 108 Air Traffic Controllers – Licences, Ratings, Endorsements and Maintenance of Licence Privileges – PART B
3 Ratings and rating endorsements

(3) The privileges of a Student Air Traffic Controller's Licence are to act as an air traffic controller under the supervision of another person who is present at the time and who:

(a) is the holder of an air traffic controller's licence entitling him to provide unsupervised the type of air traffic control service which is being provided by the student air traffic controller; and

(b) holds an On the Job Training Instructor Licence Endorsement.

PART B RATINGS, RATING ENDORSEMENTS AND LICENCE ENDORSEMENTS

1 Inclusion of ratings, rating endorsements and licence endorsements

Ratings, rating endorsements and licence endorsements of the classes contained in paragraphs 3 and 4 below may be included in an air traffic controller's licence granted under article 108 and, subject to the provisions of this Order and of the licence, the inclusion of a rating, rating endorsement or licence endorsement shall have the consequences respectively specified.

2 Exercise of more than one function

(1) Subject to sub-paragraphs (2) and (3), the holder of a licence which includes ratings of two or more of the classes specified in paragraph 3 shall not at any one time perform the functions specified in respect of more than one of those ratings.

(2) The functions of the following ratings may be excercised at the same time:

(a) an Aerodrome Control Instrument Rating and an Approach Control Procedural Rating; and

(b) an Aerodrome Control Instrument Rating and an Approach Control Surveillance Rating, provided that the holder shall not exercise the functions of any Radar Rating Endorsement, Surveillance Radar Approach Rating Endorsement or Precision Approach Radar Rating Endorsement included in the Approach Control Surveillance Rating.

(3) When a surveillance radar approach terminating at a point less than 2 nautical miles from the point of intersection of the glide path with the runway is being provided under an Approach Control Surveillance Rating, no other function under the Approach Control Surveillance Rating shall be exercised at the same time.

3 Ratings and rating endorsements

(1) There shall be the following classes of aerodrome control ratings and endorsements:

(a) An **Aerodrome Control Visual Rating** shall entitle the holder to act as an air traffic controller in the course of the provision of an aerodrome control service at an aerodrome with no instrument approach or departure procedures.

(b) An **Aerodrome Control Instrument Rating** shall entitle the holder to act as an air traffic controller in the course of the provision of an aerodrome control service in accordance with the provisions of one or more of the following Rating Endorsements:

(i) A **Tower Control Rating Endorsement** shall entitle the holder to provide an aerodrome control service at an aerodrome where the aerodrome control service is not divided into air control and ground movement control;

(ii) A **Ground Movement Control Rating Endorsement** shall entitle the holder to provide a ground movement control service at an aerodrome where the aerodrome control service is divided into ground movement control and air control;

(iii) A **Ground Movement Surveillance Control Rating Endorsement** shall entitle the holder of a Tower Control Rating Endorsement or a Ground Movement Control Rating Endorsement to use aerodrome surface movement and guidance systems in the provision of an aerodrome control service;

(iv) An **Air Control Rating Endorsement** shall entitle the holder to provide an air control service at an aerodrome where the aerodrome control service is divided into ground movement control and air control; provided that nothing in this Order shall prevent the holder of an Air Control Rating Endorsement from using aerodrome surface movement and guidance systems in the provision of an air control service;

(v) An **Aerodrome Radar Control Rating Endorsement** shall entitle the holder of an Air Control Rating Endorsement or a Tower Control Rating Endorsement to use radar in the provision of an aerodrome control service to aircraft flying in the vicinity of the aerodrome; provided that nothing in this Order shall prevent the holder of an Air Control Rating Endorsement or a Tower Control Rating Endorsement from using an aerodrome traffic monitor in the provision of an aerodrome control service.

(2) There shall be the following classes of approach control ratings and endorsements:

(a) an **Approach Control Procedural Rating** shall entitle the holder to act as an air traffic controller in the course of the provision of an approach control service, without the use of any surveillance equipment.

(b) An **Approach Control Surveillance Rating** shall entitle the holder to act as an air traffic controller in the course of the provision of an approach control service with the use of surveillance equipment in accordance with the provisions of one or more of the following Rating Endorsements:

(i) a **Radar Rating Endorsement** shall entitle the holder to use radar in the provision of an approach control service except for anything authorised by a specific rating endorsement below;

(ii) a **Surveillance Radar Approach Rating Endorsement** shall entitle the holder of a Radar Rating Endorsement to provide ground controlled non-precision radar approaches with the use of surveillance radar equipment;

(iii) a **Precision Approach Radar Rating Endorsement** shall entitle the holder of a Radar Rating Endorsement to provide ground controlled precision approaches using precision approach radar equipment;

(iv) a **Terminal Control Rating Endorsement** shall entitle the holder of a Radar Rating Endorsement to provide a terminal control service;

(v) an **Offshore Rating Endorsement** shall entitle the holder of a Radar Rating Endorsement to provide an offshore service;

SCHEDULE 11 Article 108 Air Traffic Controllers – Licences, Ratings, Endorsements and Maintenance of Licence Privileges – PART B
3 Ratings and rating endorsements

(vi) a **Special Tasks Rating Endorsement** shall entitle the holder of a Radar Rating Endorsement to provide a special tasks service.

(3) There shall be the following classes of area control ratings and endorsements:

(a) an **Area Control Procedural Rating** shall entitle the holder to act as an air traffic controller in the course of the provision of an area control service without the use of any surveillance equipment except for anything authorised by the rating endorsement below:

(i) an **Oceanic Control Rating Endorsement** shall entitle the holder to provide an area control service in the Shanwick Oceanic Control Area;

(b) an **Area Control Surveillance Rating** shall entitle the holder to act as an air traffic controller in the course of the provision of an area control service with the use of surveillance equipment in accordance with the provisions of one or more of the following Rating Endorsements:

(i) a **Radar Rating Endorsement** shall entitle the holder to use radar in the provision of an area control service;

(ii) a **Terminal Control Rating Endorsement** shall entitle the holder of a Radar Rating Endorsement to provide a terminal control service;

(iii) an **Offshore Rating Endorsement** shall entitle the holder of a Radar Rating Endorsement to provide an offshore service;

(v) a **Special Tasks Rating Endorsement** shall entitle the holder of a Radar Rating Endorsement to provide a special tasks service.

4 Licence Endorsements

(1) **An Examiner Licence Endorsement** shall entitle the holder to sign a unit licence endorsement in respect of:

(a) the air traffic control services that his air traffic controller licence entitles him to provide; or

(b) such other air traffic control services as the CAA may authorise for that holder.

(2) An **On the Job Training Instructor Licence Endorsement** shall entitle the holder to supervise and give operational air traffic control instruction to the holder of a Student Air Traffic Controller's Licence or Air Traffic Controller's Licence in relation to an air traffic control service which his Air Traffic Controller's Licence entitles him to provide.

(3) **A Unit Licence Endorsement:**

(a) specifies the aerodrome or place at which the holder is entitled to exercise the privileges of his licence and the validity of any ratings, rating endorsements or licence endorsements included in the licence; and

(b) is valid for the period of 12 months beginning with the date of issue or date of renewal.

SCHEDULE 12

Article 125
Air traffic service equipment – records required and matters to which the CAA may have regard
PART A – Records to be kept in accordance with article 125(1)

(1) A record of any functional tests, flight checks and particulars of any maintenance, repair, overhaul, replacement or modification.

(2) Subject to paragraph (3), the record shall be kept in a legible or a non-legible form so long as the record is capable of being reproduced by the person required to keep the record in a legible form and it shall be so reproduced by that person if requested by an authorised person.

(3) In any particular case the CAA may direct that the record is kept or be capable of being reproduced in such a form as it may specify.

PART B – Records required in accordance with article 125(4)(c)

Each record made by the apparatus provided in compliance with article 125(2) or (3) shall be adequately identified and in particular shall include:

(a) the identification of the aeronautical radio station;

(b) the date or dates on which the record was made;

(c) a means of determining the time at which each message or signal was transmitted or received;

(d) the identity of the aircraft to or from which and the radio frequency on which the message or signal was transmitted or received; and

(e) the time at which the record started and finished.

PART C – Matters to which the CAA may have regard in granting an approval of apparatus in accordance with article 125(5)

(1) The purpose for which the apparatus is to be used.

(2) The manner in which the apparatus has been specified and produced *in relation to the purpose for which it is to be used.

(3) The adequacy, in relation to the purpose for which the apparatus is to be used, of the operating parameters of the apparatus (if any).

(4) The manner in which the apparatus has been or will be operated, installed, modified, maintained, repaired and overhauled.

(5) The manner in which the apparatus has been or will be inspected.

SCHEDULE 13

AERODROME MANUAL
Article 128(7)

Information and instructions relating to the following matters shall be included in the aerodrome manual referred to in article 128:

(a) the name and status of the official in charge of day to day operation of the aerodrome together with the names and status of other senior aerodrome operating staff and instructions as to the order and circumstances in which they may be required to act as the official in charge;

(b) the system of aeronautical information service available;

(c) procedures for promulgating information concerning the aerodrome's state;

(d) procedures for the control of access, vehicles and work in relation to the aerodrome manoeuvring area and apron;

(e) procedures for complying with article 142 and for the removal of disabled aircraft;

(f) in the case of an aerodrome which has facilities for fuel storage, procedures for complying with article 137;

(g) plans to a scale of 1:2500 depicting the layout of runways, taxiways and aprons, aerodrome markings, aerodrome lighting if such lighting is provided, and the siting of any navigational aids within the runway strip; provided that in the case of copies or extracts of the manual provided or made available to a member of the aerodrome operating staff, the plans shall be of a scale reasonably appropriate for the purposes of article 128(9);

(h) in respect of an aerodrome in relation to which there is a notified instrument approach procedure, survey information sufficient to provide data for the production of aeronautical charts relating to that aerodrome;

(i) description, height and location of obstacles which infringe standard obstacle limitation surfaces, and whether they are lit;

(j) data for and method of calculation of declared distances and elevations at the beginning and end of each declared distance;

(k) method of calculating reduced declared distances and the procedure for their promulgation;

(l) details of surfaces and bearing strengths of runways, taxiways and aprons;

(m) the system of the management of air traffic in the airspace associated with the aerodrome, including procedures for the co-ordination of traffic with adjacent aerodromes, except any such information or procedures already published in any manual of air traffic services;

(n) operational procedures for the routine and special inspection of the aerodrome manoeuvring area and aprons;

(o) if operations are permitted during periods of low visibility, procedures for the protection of the runways during such periods;

(p) procedures for the safe integration of all aviation activities undertaken at the aerodrome;

(q) procedures for the control of bird hazards;

(r) procedures for the use and inspection of the aerodrome lighting system, if such a system is provided; and

(s) the scale of rescue, first aid and fire service facilities, the aerodrome emergency procedures and procedures to be adopted in the event of temporary depletion of the rescue and fire service facilities.

SCHEDULE 14

PENALTIES
PART A – PROVISIONS REFERRED TO IN ARTICLE 148

Article Of order	Subject matter
3	Aircraft flying unregistered
5	Aircraft flying with false or incorrect markings
14(1)(a)	Flight without appropriate maintenance
14(1)(b)	Flight without a certificate of maintenance review
15	Failure to keep a technical log
16	Flight without a certificate of release to service issued under the Order or under paragraph 21A.163(d) of Part 21
17	Flight without a certificate of release to service issued under Part 145
18(7)&(8)	Exercise of privileges of aircraft maintenance engineer's licence whilst unfit or drunk

SCHEDULE 14 PENALTIES
PART A – PROVISIONS REFERRED TO IN ARTICLE 148

19	Flight without required equipment
20	Flight without required radio communication or radio navigation equipment
21	Minimum equipment requirements
22	Failure to keep log books
23	Requirement to weigh aircraft and keep weight schedule
25	Crew requirement
28, 30(2) & 31(1)	Requirement for appropriate certificate of test or experience
29&31(2)	Requirement for appropriate certificate of revalidation
30(1)	Requirement for valid rating
321)	Flight without valid medical certificate
32(4)	Flight in unfit condition
33(1)	Prohibition of flight after failure of test
36	Instruction in flying without appropriate licence and rating
38	Operations manual requirement
39	Police operations manual requirement
40	Training manual requirement
42	Operator's responsibilities in connection with crew
43	Requirements for loading aircraft
44 & 45	Operational restrictions on aircraft and helicopters
46	Prohibition on public transport flights at night or in Instrument Meteorological Conditions by non-UK registered single engined aeroplanes
47	Aerodrome operating minima – UK registered public transport aircraft
48	Aerodrome operating minima – public transport aircraft registered elsewhere than in the United Kingdom
49	Aerodrome operating minima – non-public transport aircraft
50	Requirement for pilot to remain at controls
52	Pre-flight action by commander of aircraft
53	Requirement for passenger briefing
54	Additional duties of commander on flight for public transport of passengers
55	Requirements for radio station in aircraft to be licensed and for operation of same
56	Requirement for minimum navigation performance equipment
57	Requirement for height keeping performance equipment-aircraft registered in the United Kingdom
58	Requirement for height keeping performance equipment-aircraft registered elsewhere than in the United Kingdom
59	Requirement for area navigation equipment and required navigation performance -aircraft registered in the United Kingdom
60	Requirement for area navigation equipment and required navigation performance -aircraft registered elsewhere than in the United Kingdom
61	Requirement for an airborne collision avoidance system
62	Use of flight recording systems and preservation of records
63	Towing of gliders
65	Towing, picking up and raising of persons and articles by aircraft
66	Dropping of articles and animals from aircraft
67	Dropping of persons
68	Requirement for aerial application certificate
71	Carriage of persons in or on any part of an aircraft not designed for that purpose
72	Requirement for exits and break-in markings
76	Prohibition of smoking in aircraft
77	Requirement to obey lawful commands of aircraft commander
78(a)&(b)	Acting in a disruptive manner
79	Prohibition of stowaways
80	Flying displays
82(3)	Operator's obligation to obtain flight time records of flight crew
83(2)	Flight crew member's obligation to inform operator of flight times
84	Flight time limitations for flight crew
95	Breach of the Rules of the Air

SCHEDULE 14 PENALTIES
PART A – PROVISIONS REFERRED TO IN ARTICLE 148

96	Flight in contravention of restriction of flying regulations
97	Flight by balloons, kites, airships, gliders and parascending parachutes
98	Flight by small aircraft
101	Requirement for an approved provider of air traffic services to be satisfied as to competence of air traffic controllers
103	Provision of air traffic services
104	Requirement to comply with an air traffic direction
105	Requirement to comply with an airspace policy direction
106	Use of radio call signs at aerodromes
107	Requirement for licensing of air traffic controllers
121	Requirement for licensing of flight information service officers
123	Requirement for aerodrome information service manual
124	Requirement for licensed aerodrome
125	Requirements to keep air traffic service equipment records
126	Requirement for use of licensed aerodromes
128(4)	Contravention of conditions of aerodrome licence
132	Use of aeronautical lights
133	Requirement to light en-route obstacles
134	Requirement to light offshore wind turbine generators
135(1)	Prohibition of dangerous lights
135(2)	Failure to extinguish or screen dangerous lights
137(1)&(3)	Management of aviation fuel at aerodromes
146	Obstruction of persons performing duties under the Order

SCHEDULE 14

PENALTIES
PART B – PROVISIONS REFERRED TO IN ARTICLE 148(6)

Article Of order	Subject matter
6	Flight for the purpose of public transport without an air operator's certificate
7	Flight in the service of a chief officer of police without a police air operator's certificate
8	Flight without a certificate of airworthiness
26	Requirement to hold an appropriate flight crew licence
69	Prohibition of carriage of weapons and munitions of war
70(2)	Prohibition of carriage of dangerous goods
74	Endangering safety of persons or property
75	Prohibition of drunkenness in aircraft
78(c)	Intentional interference
82(1)	Operator's obligation to regulate flight times of flight crew
82(2)	Operator's obligation not to allow flight by crew in dangerous state of fatigue
83(1)	Crew's obligation not to fly in dangerous state of fatigue
85	Protection of air crew from cosmic radiation
87	Keeping and production of records of exposure to cosmic radiation
94 (except (3))	Use of false or unauthorised documents and records
100	Provision of an air traffic control service without an approval
115	Controller's obligation not to act in a dangerous state of fatigue
116	Prohibition of acting under the influence of drink or a drug
137(4)	Use of aviation fuel which is unfit for use in aircraft
138	Restriction of carriage for valuable consideration by aircraft registered elsewhere than in the United Kingdom
140	Restriction of flights for aerial photography, aerial survey and aerial work by aircraft registered elsewhere than in the –United Kingdom
141	Operators' or commanders' obligations in respect of flights over any foreign country
142(5), (6) & (7)	Failure to report an occurrence
142(8)	Making false occurrence reports
144	Flight in contravention of directions not to fly

PART C – PROVISIONS REFERED TO IN ARTICLE 148(7)

Article Of order	Subject matter
73	Endangering safety of aircraft

SCHEDULE 15

PARTS OF STRAITS SPECIFIED IN CONNECTION WITH THE FLIGHT OF AIRCRAFT IN TRANSIT OVER UNITED KINGDOM TERRITORIAL WATERS
Article 150

(1) The following parts of the straits named hereafter are hereby specified for the purposes of Article 150(4):
(a) In the Straits of Dover, the territorial waters adjacent to the United Kingdom which are:
(i) to the south of a rhumb line joining
position 51°08'23" north latitude: 1°23'00" east longitude and
position 51°22'41" north latitude: 1°50'06" east longitude; and
(ii) to the east of a rhumb line joining
position 50°54'33" north latitude: 0°58'05" east longitude and
position 50°43'15" north latitude: 0°51'39" east longitude;
(2) In the North Channel, the territorial waters adjacent to the United Kingdom which are:
(a) to the north of a rhumb line joining
position 54°13'30" north latitude: 5°39'28" west longitude and
position 54°09'02" north latitude: 5°18'07" west longitude;
(b) to the west of a rhumb line joining
position 54°26'02" north latitude: 4°51'37" west longitude and
position 54°38'01" north latitude: 4°51'16" west longitude; and
(c) to the east of a rhumb line joining
(i) position 55°40'24" north latitude: 6°30'59" west longitude and
position 55°29'24" north latitude: 6°40'3 1" west longitude;
(ii) position 55°24'54" north latitude: 6°44'33" west longitude and
position 55°10' 15" north latitude: 6°44'33" west longitude;
(3) In the Fair Isle Channel, the territorial waters adjacent to the United Kingdom which are:
(a) to the north of a rhumb line joining
position 59°10'54" north latitude: 2°01'32" west longitude and
position 59°33'27" north latitude: 2°38'35" west longitude; and
(b) to the south of a rhumb line joining
position 59°51'06" north latitude: 0°52'10" west longitude and
position 59°51'06" north latitude: 1°46'36" west longitude.
(2) The parts of each of the Straits specified in paragraph (1) are shown hatched on Charts A, B and C respectively.

RULES OF THE AIR

Section I – Interpretation
1 Interpretation

Section II – General
2 Application of Rules to aircraft
3 Misuse of signals and markings
4 Reporting hazardous conditions
5 Low flying
6 Simulated instrument flight
7 Practice instrument approaches

Section III – Lights and signals to be shown or made by aircraft
8 General
9 Display of lights by aircraft
10 Failure of navigation and anti-collision lights
11 Flying machines
12 Gliders
13 Free balloons
14 Captive balloons and kites
15 Airships

Section IV – General flight rules
16 Weather reports and forecasts
17 Rules for avoiding aerial collisions
18 Aerobatic manoeuvres
19 Right-hand traffic rule
20 Notification of arrival and departure
21 Flight in Class A airspace
22 Choice of VFR or IFR
23 Speed limitation

Section V – Visual Flight Rules
24 Visual flight and reported visibility
25 Flight within controlled airspace
26 Flight outside controlled airspace
27 VFR flight plan and air traffic control clearance

Section VI – Instrument flight rules
28 Instrument Flight Rules
29 Minimum height
30 Quadrantal rule and semi-circular rule
31 Flight plan and air traffic control clearance
32 Position reports

Section VII – Aerodrome traffic rules
33 Application of aerodrome traffic rules
34 Visual signals
35 Movement of aircraft on aerodrome
36 Access to and movement of persons and vehicles on the aerodrome
37 Right of way on the ground
38 Launching, picking up and dropping of tow ropes, etc.
39 Flight within aerodrome traffic zones

Section VIII – Special Rules
40 Use of radio navigation aids

Section IX – Aerodrome signals and markings – visual and aural signals
41 General
42 Signals in the signals area
43 Markings for paved runways and taxiways
44 Markings on unpaved manoeuvring areas
45 Signals visible from the ground
46 Lights and pyrotechnic signals for control of aerodrome traffic
47 Marshalling signals (from marshaller to an aircraft)
48 Marshalling signals (from a pilot of an aircraft to a marshaller)
49 Distress, urgency and safety signals

SECTION I INTERPRETATION

Rule 1: Interpretation
(1) In these Rules, unless the context otherwise requires:
"air traffic control clearance" means authorisation by an air traffic control unit for an aircraft to proceed under conditions specified by that unit;
"anti-collision light" means:
(a) in relation to rotorcraft a flashing red light;
(b) in relation to any other aircraft a flashing red or flashing white light;
and in either case showing in all directions for the purpose of enabling the aircraft to be more readily detected by the pilots of distant aircraft;
"ground visibility" means the horizontal visibility at ground level;
"IFR flight" means a flight conducted in accordance with the Instrument Flight Rules in Section VI of these Rules;
"the Order" means the Air Navigation (No 2) Order 1995;
"runway" means an area, whether or not paved, which is provided for the take-off or landing run of aircraft;
"special VFR flight" means a flight made at any time in a control zone which is Class A airspace, or in any other control zone in Instrument Meteorological Conditions or at night, in respect of which the appropriate air traffic control unit has given permission for the flight to be made in accordance with special instructions given by that unit instead of in accordance with the Instrument Flight Rules and in the course of which flight the aircraft complies with any instructions given by that unit and remains clear of cloud and in sight of the surface;
"VFR flight" means a flight conducted in accordance with the Visual Flight Rules in Section V of these Rules.
(2) In these Rules, unless the context otherwise requires, any reference to:
(a) a numbered rule is a reference to the rule in these Rules so numbered;
(b) a numbered paragraph or sub-paragraph is a reference to the paragraph or sub-paragraph so numbered in the rule or paragraph, as the case may be, in which that reference appears.
(3) Subject to the provisions of paragraph (1) expressions used in these Rules shall, unless the context otherwise requires, have the same respective meanings as in the Order.

SECTION II GENERAL

Rule 2: Application of rules to aircraft
These Rules, in so far as they are applicable in relation to aircraft, shall, subject to the provisions of rule 33, apply in relation to:
(a) all aircraft within the United Kingdom and, for the purposes of rule 5, in the neighbourhood of an offshore installation; and
(b) all aircraft registered in the United Kingdom, wherever they may be.

Rule 3: Misuse of signals and markings
(1) A signal or marking to which a meaning is given by these Rules, or which is required by these Rules to be used in circumstances, or for a purpose therein specified, shall not be used except with that meaning, or for that purpose.
(2) A person in an aircraft or on an aerodrome or at any place at which an aircraft is taking off or landing shall not make any signal which may be confused with a signal specified in these Rules, and, except with lawful authority, shall not make any signal which he knows or ought reasonably to know to be a signal in use for signalling to or from any of Her Majesty's naval, military or air force aircraft.

Rule 4: Reporting hazardous conditions
The commander of an aircraft shall, on meeting with hazardous conditions in the course of a flight, or as soon as possible thereafter, send to the appropriate air traffic control unit by the quickest means available information containing such particulars of the hazardous conditions as may be pertinent to the safety of other aircraft.

Rule 5: Low flying
(1) Prohibitions to be observed are:
(a) an aircraft shall comply with the low flying prohibitions set out in paragraph (2) subject to the low flying exemptions set out in paragraph (3).
(b) where an aircraft is flying in circumstances such that more than one of the low flying prohibitions apply it must fly at the greatest height required by any of the applicable prohibitions.
(2) The low flying prohibitions
(a) failure of power unit
An aircraft shall not be flown below such height as would enable it, in the event of a power unit failure, to make an emergency landing without causing danger to persons or property on the surface.
(b) The 500 feet rule
Except with the permission in writing of the CAA, an aircraft shall not be flown closer than 500 feet to any person, vessel, vehicle or structure.
(c) The 1000 feet rule
Except with the permission in writing of the CAA, an aircraft flying over a congested area of a city town or settlement shall not fly below a height of 1,000 feet above the highest fixed obstacle within a horizontal radius of 600 metres of the aircraft.

(d) The land clear rule
An aircraft flying over a congested area of a city town or settlement shall not fly below such height as will permit, in the event of a power unit failure, the aircraft to land clear of the congested area.

(e) Flying over open air assemblies
Except with the permission in writing of the CAA, an aircraft shall not fly over an organised open-air assembly of more than 1,000 persons below:

(i) a height of 1,000 feet, or

(ii) such height as will permit, in the event of a power unit failure, the aircraft to alight clear of the assembly, whichever is the higher.

(f) Landing and taking off near open air assemblies
An aircraft shall not land or take-off within 1,000 metres of an organised open-air assembly of more than 1,000 persons, except

(i) at an aerodrome, in accordance with procedures notified by the CAA, or

(ii) at a landing site other than an aerodrome, in accordance with procedures notified by the CAA and with the written permission of the organiser of the assembly.

(3) Exemptions from the low flying prohibitions

(a) Landing and taking off

(i) Any aircraft shall be exempt from any low flying prohibition in so far as it is flying in accordance with normal aviation practice for the purpose of taking off from, landing at or practising approaches to landing at or checking navigational aids or procedures at a Government or licensed aerodrome.

(ii) Any aircraft shall be exempt from the 500 feet rule when landing and taking-off in accordance with normal aviation practice.

(b) Captive balloons and kites
None of the low flying prohibitions shall apply to any captive balloon or kite.

(c) Special VFR clearance and notified routes
Any aircraft shall be exempt from the 1000 feet rule when flying on a special VFR flight, or when operating in accordance with the procedures notified for the route being flown; provided that when flying in accordance with this exemption landings may not be made other than at a licensed or Government aerodrome, unless the permission of the CAA has been pbtained.

(d) Balloons and helicopters over congested areas

(i) A balloon shall be exempt from the 1000 feet rule when landing because it is becalmed.

(ii) Any helicopter flying over a congested area shall be exempt from the land clear rule.

(e) Police air operator's certificate
Any aircraft flying in accordance with the terms of a police air operator's certificate shall be exempt from the 500 feet rule, the 1000 feet rule, the prohibition on flying over open air assemblies and the prohibition on landing and taking off near open air assemblies.

(f) Flying displays etc
An aircraft taking part in a flying display, air race or contest shall be exempt from the 500 feet rule when within a horizontal distance of 1,000 metres of the gathering of persons assembled to witness the event.

(g) Glider hill soaring
A glider when hill-soaring shall be exempt from the 500 feet rule.

(h) Picking up and dropping at an aerodrome
Any aircraft picking up or dropping tow ropes, banners or similar articles at an aerodrome shall be exempt from the 500 feet rule.

(i) Manoeuvring helicopters
A helicopter shall be exempt from the 500 feet rule when conducting manoeuvres in accordance with normal aviation practice, within the boundaries of a licensed or Government aerodrome, or at other sites with the permission of the CAA; provided that when flying in accordance with this exemption the helicopter must not be operated closer than 60 metres to persons, vessels vehicles or structures located outside the aerodrome or site.

(j) Dropping articles with CAA permission

(i) Any aircraft shall be exempt from the 500 feet rule when flying in accordance with article 56(3)(f) of the Order, and

(ii) Any aircraft shall be exempt from the 500 feet rule when flying in accordance with an aerial application certificate issued by the CAA under article 58 of the Order.

Rule 6: Simulated instrument flight

(1) An aircraft shall not be flown in simulated instrument flight conditions unless:

(a) the aircraft is fitted with dual controls which are functioning properly;

(b) an additional pilot (in this rule called a "safety pilot") is carried in a second control seat of the aircraft for the purpose of rendering such assistance as may be necessary to the pilot flying the aircraft; and

(c) if the safety pilot's field of vision is not adequate both forward and to each side of the aircraft, a third person, being a competent observer, occupies a position in the aircraft which from his field of vision makes good the deficiencies in that of the safety pilot, and from which he can readily communicate with the safety pilot.

(2) For the purposes of this rule the expression "simulated instrument flight" means a flight during which mechanical or optical devices are used in order to reduce the field of vision or the range of visibility from the cockpit of the aircraft.

Rule 7: Practice instrument approaches

(1) Within the United Kingdom an aircraft shall not carry out instrument approach practice when flying in Visual Meteorological Conditions unless:

(a) the appropriate air traffic control unit has previously been informed that the flight is to be made for the purpose of instrument approach practice; and

(b) if the flight is not being carried out in simulated instrument flight conditions, a competent observer is carried in such a position in the aircraft that he has an adequate field of vision and can readily communicate with the pilot flying the aircraft.

(2) For the purposes of this rule the expression "simulated instrument flight" shall have the same meaning as in rule 6.

SECTION III LIGHTS AND OTHER SIGNALS TO BE SHOWN OR MADE BY AIRCRAFT

Rule 8: General

(1) For the purposes of this section of these Rules the horizontal plane of a light shown in an aircraft means the plane which would be the horizontal plane passing through the source of that light, if the aircraft were in level flight.

(2) Where by reason of the physical construction of an aircraft it is necessary to fit more than one lamp in order to show a light required by this section of these Rules, the lamps shall be so fitted and constructed that, so far as is reasonably practicable, not more than one such lamp is visible from any one point outside the aircraft.

(3) Where in these Rules a light is required to show through specified angles in the horizontal plane, the lamps giving such light shall be so constructed and fitted that the light is visible from any point in any vertical plane within those angles throughout angles of 90° above and below the horizontal plane, but, so far as is reasonably practicable, through no greater angle, either in the horizontal plane or the vertical plane.

(4) Where in these Rules a light is required to show in all directions, the lamps giving such light shall be so constructed and fitted that, so far as is reasonably practicable, the light is visible from any point in the horizontal plane and on any vertical plane passing through the source of that light.

Rule 9: Display of lights by aircraft

(1) (a) By night an aircraft shall display such of the lights specified in these Rules as may be appropriate to the circumstances of the case, and shall not display any other lights which might obscure or otherwise impair the visibility of, or be mistaken for, such lights.

(b) By day an aircraft fitted with an anti-collision light shall display such a light in flight.

(2) A flying machine on a United Kingdom aerodrome shall:

(a) display by night either the lights which it would be required to display when flying or the lights specified in rule 11(2)(c) unless it is stationary on the apron or part of the aerodrome provided for the maintenance of aircraft;

(b) subject to paragraph (3), display when stationary on the apron by day or night with engines running a red anti-collision light, if fitted.

(3) A helicopter to which article 27 of the Order applies may, when stationary on an offshore installation, switch off the anti-collision light required to be shown by paragraph (2) provided that is done in accordance with a procedure contained in the operations manual of the helicopter as a signal to ground personnel that it is safe to approach the helicopter for the purpose of embarkation or disembarkation of passengers or the loading or unloading of cargo.

(4) Notwithstanding the provisions of this section of these Rules the commander of an aircraft may switch off or reduce the intensity of any flashing light fitted to the aircraft if such a light does or is likely to:

(a) adversely affect the performance of the duties of any member of the flight crew; or

(b) subject an outside observer to unreasonable dazzle.

Rule 10: Failure of navigation and anti-collision lights

(1) In the United Kingdom, in the event of the failure of any light which is required by these Rules to be displayed at night, if the light cannot be immediately repaired or replaced the aircraft shall not depart from the aerodrome and, if in flight, shall land as soon as in the opinion of the commander of the aircraft it can safely do so, unless authorised by the appropriate air traffic control unit to continue its flight.

(2) In the United Kingdom, in the event of a failure of an anti-collision light when flying by day, an aircraft may continue to fly by day provided that the light is repaired at the earliest practicable opportunity.

Rule 11: Flying machines

(1) A flying machine when flying at night shall display lights as follows:

(a) in the case of a flying machine registered in the United Kingdom having a maximum total weight authorised of more than 5700kg or any other flying machine registered in the United Kingdom which conforms to a type first issued with a type certificate on or after 1st April 1988 the system of lights in paragraph (2)(b);

(b) in the case of a flying machine registered in the United Kingdom which conforms to a type first issued with a type certificate before 1st April 1988 having a maximum total weight authorised of 5700kg or less, any one of the following systems of lights:

(i) that specified in paragraph (2)(a), or that specified in paragraph (2)(b); or

(ii) that specified in paragraph (2)(d), excluding sub-paragraph (ii);

(c) in the case of any other flying machine one of the systems of lights specified in paragraph (2).

(2) The systems of lights referred to in paragraph (1) are as follows:

(a) (i) a steady green light of at least five candela showing to the starboard side through an angle of 110° from dead ahead in the horizontal plane;

(ii) a steady red light of at least five candela showing to the port side through an angle of 110° from dead ahead in the horizontal plane; and

(iii) a steady white light of at least three candela showing through angles of 70° from dead astern to each side in the horizontal plane;

(b) (i) the lights specified in sub-paragraph (a); and

(ii) an anti-collision light;

(c) the lights specified in sub-paragraph (a), but all being flashing lights flashing together;

(d) the lights specified in sub-paragraph (a), but all being flashing lights flashing together in alternation with one or both of the following:

(i) a flashing white light of at least twenty candela showing in all directions;

(ii) a flashing red light of at least twenty candela showing through angles of 70° from dead astern to each side in the horizontal plane.

(3) If the lamp showing either the red or the green light specified in paragraph (2)(a) is fitted more than 2 metres from the wing tip, a lamp may, notwithstanding the provisions of rule 9(1), be fitted at the wing tip to indicate its position showing a steady light of the same colour through the same angle.

Rule 12: Gliders

A glider while flying at night shall display either a steady red light of at least five candela, showing in all directions, or lights in accordance with rule 11(2) and (3).

Rule 13: Free balloons

A free balloon while flying at night shall display a steady red light of at least five candela showing in all directions, suspended not less than 5 metres and not more than 10 metres below the basket, or if there is no basket, below the lowest part of the balloon.

Rule 14: Captive balloons and kites

(1) A captive balloon or kite while flying at night at a height exceeding 60 metres above the surface shall display lights as follows:

(a) a group of two steady lights consisting of a white light placed 4 metres above a red light, both being of at least five candela and showing in all directions, the white light being placed not less than 5 metres or more than 10 metres below the basket, or if there is no basket, below the lowest part of the balloon or kite;

(b) on the mooring cable, at intervals of not more than 300 metres measured from the group of lights referred to in sub-paragraph (a), groups of two lights of the colour and power and in the relative positions specified in that sub-paragraph, and, if the lowest group of lights is obscured by cloud, an additional group below the cloud base; and

(c) on the surface, a group of three flashing lights arranged in a horizontal plane at the apexes of a triangle, approximately equilateral, each side of which measures at least 25 metres; one side of the triangle shall be approximately at right angles to the horizontal projection of the cable and shall be delimited by two red lights; the third light shall be a green light so placed that the triangle encloses the object on the surface to which the balloon or kite is moored.

(2) A captive balloon while flying by day at a height exceeding 60 metres above the surface shall have attached to its mooring cable at intervals of not more than 200 metres measured from the basket, or, if there is no basket, from the lowest part of the balloon, tubular streamers not less than 40 centimetres in diameter and 2 metres in length, and marked with alternate bands of red and white 50 centimetres wide.

(3) A kite flown in the circumstances referred to in paragraph (2) shall have attached to its mooring cable either:

(a) tubular streamers as specified in paragraph (2), or

(b) at intervals of not more than 100 metres measured from the lowest part of the kite, streamers not less than 80 centimetres long and 30 centimetres wide at their widest point and marked with alternate bands of red and white 10 centimetres wide.

Rule 15: Airships

(1) Except as provided in paragraph (2), an airship while flying at night shall display the following lights:
(a) a steady white light of at least five candela showing through angles of 110° from dead ahead to each side in the horizontal plane;
(b) a steady green light of at least five candela showing to the starboard side through an angle of 110° from dead ahead in the horizontal plane;
(c) a steady red light of at least five candela showing to the port side through an angle of 110° from dead ahead in the horizontal plane;
(d) a steady white light of at least five candela showing through angles of 70° from dead astern to each side in the horizontal plane; and
(e) an anti-collision light.
(2) (a) Subject to sub-paragraph (b), an airship while flying at night shall display, if it is not under command, or has voluntarily stopped its engines, or is being towed, the following steady lights:
(i) the white lights referred to in paragraph (1)(a) and (d);
(ii) two red lights, each of at least five candela and showing in all directions suspended below the control car so that one is at least 4 metres above the other and at least 8 metres below the control car; and
(iii) if the airship is making way but not otherwise, the green and red lights referred to in paragraph (1)(b) and (c).
(b) An airship while picking up its moorings, notwithstanding that it is not under command, shall display only the lights specified in paragraph (1).
(3) An airship, while moored within the United Kingdom by night, shall display the following steady lights:
(a) when moored to a mooring mast, at or near the rear a white light of at least five candela showing in all directions;
(b) when moored otherwise than to a mooring mast:
(i) a white light of at least five candela showing through angles of 110° from dead ahead to each side in the horizontal plane; and
(ii) a white light of at least five candela showing through angles of 70° from dead astern to each side in the horizontal plane.
(4) An airship while flying by day, if it is not under command, or has voluntarily stopped its engines, or is being towed, shall display two black balls suspended below the control car so that one is at least 4 metres above the other and at least 8 metres below the control car.
(5) For the purposes of this rule:
(a) an airship shall be deemed not to be under command when it is unable to execute a manoeuvre which it may be required to execute by or under these Rules;
(b) an airship shall be deemed to be making way when it is not moored and is in motion relative to the air.

SECTION IV GENERAL FLIGHT RULES

Rule 16: Weather reports and forecasts

(1) Immediately before an aircraft flies the commander of the aircraft shall examine the current reports and forecasts of the weather conditions on the proposed flight path, being reports and forecasts which it is reasonably practicable for him to obtain, in order to determine whether Instrument Meteorological Conditions prevail or are likely to prevail during any part of the flight.
(2) An aircraft which is unable to communicate by radio with an air traffic control unit at the aerodrome of destination shall not begin a flight to an aerodrome within a control zone if the information which it is reasonably practicable for the commander of the aircraft to obtain indicates that it will arrive at that aerodrome when the ground visibility is less than 10km or the cloud ceiling is less than 1500 feet, unless the commander of the aircraft has obtained from an air traffic control unit at that aerodrome permission to enter the aerodrome traffic zone.

Rule 17: Rules for avoiding aerial collisions

(1) General
(a) Notwithstanding that the flight is being made with air traffic control clearance it shall remain the duty of the commander of an aircraft to take all possible measures to ensure that his aircraft does not collide with any other aircraft.
(b) An aircraft shall not be flown in such proximity to other aircraft as to create a danger of collision.
(c) Subject to sub-paragraph (g), aircraft shall not fly in formation unless the commanders of the aircraft have agreed to do so.
(d) An aircraft which is obliged by these Rules to give way to another aircraft shall avoid passing over or under the other aircraft, or crossing ahead of it, unless passing well clear of it.
(e) Subject to sub-paragraph (g), an aircraft which has the right-of-way under this rule shall maintain its course and speed.
(f) For the purposes of this rule a glider and a flying machine which is towing it shall be considered to be a single aircraft under the command of the commander of the towing flying machine.
(g) Sub-paragraphs (c) and (e) shall not apply to an aircraft flying under and in accordance with the terms of a police air operator's certificate.
(2) Converging
(a) Subject to the provisions of paragraphs (3) and (4), an aircraft in the air shall give way to other converging aircraft as follows:
(i) flying machines shall give way to airships, gliders and balloons;
(ii) airships shall give way to gliders and balloons;
(iii) gliders shall give way to balloons.
(b) (i) Subject to the provisions of sub-paragraphs (a) and (b)(ii), when two aircraft are converging in the air at approximately the same altitude, the aircraft which has the other on its right shall give way.
(ii) Mechanically driven aircraft shall give way to aircraft which are towing other aircraft or objects.

(3) Approaching head-on
When two aircraft are approaching head-on or approximately so in the air and there is danger of collision, each shall alter its course to the right.

(4) Overtaking
(a) Subject to sub-paragraph (b), an aircraft which is being overtaken in the air shall have the right-of-way and the overtaking aircraft, whether climbing, descending or in horizontal flight, shall keep out of the way of the other aircraft by altering its course to the right, and shall not cease to keep out of the way of the other aircraft until that other aircraft has been passed and is clear, notwithstanding any change in the relative positions of the two aircraft.
(b) A glider overtaking another glider in the United Kingdom may alter its course to the right or to the left.

(5) Flight in the vicinity of an aerodrome
Without prejudice to the provisions of rule 39, a flying machine, glider or airship while flying in the vicinity of what the commander of the aircraft knows or ought reasonably to know to be an aerodrome, or moving on an aerodrome, shall, unless in the case of an aerodrome having an air traffic control unit that unit otherwise authorises:
(a) conform to the pattern of traffic formed by other aircraft intending to land at that aerodrome, or keep clear of the airspace in which the pattern is formed; and
(b) make all turns to the left unless ground signals otherwise indicate.

(6) Order of landing
(a) An aircraft while landing or on final approach to land shall have the right-of-way over other aircraft in flight or on the ground or water.
(b) (i) Subject to sub-paragraph (ii), in the case of two or more flying machines, gliders or airships approaching any place for the purpose of landing, the aircraft at the lower altitude shall have the right-of-way, but it shall not cut in front of another aircraft which is on final approach to land or overtake that aircraft.
(ii) (aa) When an air traffic control unit has communicated to any aircraft an order of priority for landing, the aircraft shall approach to land in that order.
(bb) When the commander of an aircraft is aware that another aircraft is making an emergency landing, he shall give way to that aircraft, and at night, notwithstanding that he may have received permission to land, shall not attempt to land until he has received further permission so to do.

(7) Landing and take-off
(a) A flying machine, glider or airship shall take-off and land in the direction indicated by the ground signals or, if no such signals are displayed, into the wind, unless good aviation practice demands otherwise.
(b) A flying machine or glider shall not land on a runway at an aerodrome if the runway is not clear of other aircraft unless, in the case of an aerodrome having an air traffic control unit, that unit otherwise authorises.
(c) Where take-offs and landings are not confined to a runway:
(i) a flying machine or glider when landing shall leave clear on its left any aircraft which has landed or is already landing or about to take-off; if such a flying machine or glider is about to turn it shall turn to the left after the commander of the aircraft has satisfied himself that such action will not interfere with other traffic movements; and
(ii) a flying machine about to take-off shall take up position and manoeuvre in such a way as to leave clear on its left any aircraft which has already taken off or is about to take-off.
(d) A flying machine after landing shall move clear of the landing area as soon as it is possible to do so unless, in the case of an aerodrome having an air traffic control unit, that unit otherwise authorises.

Rule 18: Aerobatic manoeuvres
An aircraft shall not carry out any aerobatic manoeuvre:
(a) over the congested area of any city, town or settlement; or
(b) within controlled airspace except with the consent of the appropriate air traffic control unit.

Rule 19: Right-hand traffic rule
(1) Subject to paragraph (2), an aircraft which is flying within the United Kingdom in sight of the ground and following a road, railway, canal or coastline, or any other line of landmarks, shall keep such line of landmarks on its left.
(2) Paragraph (1) shall not apply to an aircraft flying within controlled airspace in accordance with instructions given by the appropriate air traffic control unit.

Rule 20: Notification of arrival and departure
(1) The commander of an aircraft who has caused notice of its intended arrival at any aerodrome to be given to the air traffic control unit or other authority at that aerodrome shall ensure that the air traffic control unit or other authority at that aerodrome is informed as quickly as possible of any change of intended destination and any estimated delay in arrival of 45 minutes or more.
(2) The commander of an aircraft arriving at or departing from an aerodrome in the United Kingdom shall take all reasonable steps to ensure upon landing or prior to departure, as the case may be, that notice of that event is given to the person in charge of the aerodrome, or to the air traffic control unit or aerodrome flight information service unit at the aerodrome.
(3) Without prejudice to the provisions of rules 27 and 31, before taking off on any flight from an aerodrome in the United Kingdom, being a flight whose intended destination is more than 40km from the aerodrome of departure, the commander of an aircraft of which the maximum total weight authorised exceeds 5700kg shall cause a flight plan containing such particulars of the intended flight as may be necessary for search and rescue purposes to be communicated to the air traffic control unit notified for the purpose of this rule.
(4) Without prejudice to the provisions of rules 20(3), 27 and 31, the commander of an aircraft who intends to fly or who flies across any boundary of airspace notified as either the London or Scottish Flight Information Region other than the boundary common to each, shall before so flying, cause a flight plan, containing such particulars of the intended flight as may be necessary for search and rescue purposes, to be communicated to the appropriate air traffic control unit within the London or Scottish Flight Information Region.

Rule 21: Flight in Class A airspace

(1) Subject to paragraph (2), in relation to flights in Visual Meteorological Conditions in Class A airspace, the commander of an aircraft shall comply with rules 31 and 32 as if the flights were IFR flights but shall not elect to continue the flight in compliance with the Visual Flight Rules for the purposes of rule 31(3).

(2) Paragraph (1) shall not apply to the commander of a glider which is flying in Class A airspace which is notified for the purpose of this paragraph if the glider is flown in accordance with conditions such as may also be notified for the purpose of this paragraph in respect of that airspace.

Rule 22: Choice of VFR or IFR

(1) Subject to paragraph (2) and to the provisions of rule 21 an aircraft shall always be flown in accordance with the Visual Flight Rules or the Instrument Flight Rules.

(2) In the United Kingdom an aircraft flying at night:

(a) outside a control zone shall be flown in accordance with the Instrument Flight Rules;

(b) in a control zone shall be flown in accordance with the Instrument Flight Rules unless it is flying on a special VFR flight.

Rule 23: Speed Limitation

(1) Subject to paragraph (3), an aircraft shall not fly below flight level 100 at a speed which according to its air speed indicator is more than 250 knots unless it is flying in accordance with the terms of a written permission of the Authority.

(2) The Authority may grant a permission for the purpose of this rule subject to such conditions as it thinks fit and either generally or in respect of any aircraft or class of aircraft.

(3) Paragraph (1) shall not apply to:

(a) flight in Class A airspace;

(b) VFR flight or IFR flight in Class B airspace;

(c) IFR flight in Class C airspace;

(d) VFR flight in Class C airspace or VFR flight or IFR flight in Class D airspace when authorised by the appropriate air traffic control unit;

(e) the flight of an aircraft taking part in an exhibition of flying for which a permission under article 61 of the Order is required, if the flight is made in accordance with the terms of a permission granted to the organiser of the exhibition of flying under article 61 of the Order, and in accordance with the conditions of a display authorisation granted to the pilot under article 61 of the Order; or

(f) the flight of an aircraft flying in accordance with the "A Conditions" or the "B Conditions" set forth in Schedule 2 to the Order.

SECTION V VISUAL FLIGHT RULES

Rule 24: Visual flight and reported visibility

(1) In relation to flights within controlled airspace rules 25 and 27 shall be the Visual Flight Rules.

(2) In relation to flights outside controlled airspace rule 26 shall be the Visual Flight Rules.

(3) For the purposes of an aeroplane taking off from or approaching to land at an aerodrome within Class B, C, or D airspace, the visibility, if any, communicated to the commander of an aeroplane by the appropriate air traffic control unit shall be taken to be the flight visibility for the time being.

Rule 25: Flight within controlled airspace

(1) Within Class B airspace:

(a) an aircraft flying within Class B airspace at or above flight level 100 shall remain clear of cloud and in a flight visibility of at least 8km;

(b) an aircraft flying within Class B airspace below flight level 100 shall remain clear of cloud and in a flight visibility of at least 5km.

(2) Within Class C, Class D or Class E airspace:

(a) an aircraft flying within Class C, Class D or Class E airspace at or above flight level 100 shall remain at least 1500 metres horizontally and 1000 feet vertically away from cloud and in a flight visibility of at least 8km;

(b) subject to sub-paragraph (c), an aircraft flying within Class C, Class D or Class E airspace below flight level 100 shall remain at least 1500 metres horizontally and 1000 feet vertically away from cloud and in a flight visibility of at least 5km;

(c) sub-paragraph (b) shall be deemed to be complied with if:

(i) the aircraft is not a helicopter and is flying at or below 3000 feet above mean sea level at a speed which, according to its airspeed indicator, is 140 knots or less and it remains clear of cloud, in sight of the surface and in a flight visibility of at least 5km; or

(ii) the aircraft is a helicopter flying at or below 3000 feet above mean sea level and it remains clear of cloud and in sight of the surface.

Rule 26: Flight outside controlled airspace

(1) An aircraft flying outside controlled airspace at or above flight level 100 shall remain at least 1500 metres horizontally and 1000 feet vertically away from cloud and in a flight visibility of at least 8km.

(2) (a) Subject to sub-paragraph (b), an aircraft flying outside controlled airspace below flight level 100 shall remain at least 1500 metres horizontally and 1000 feet vertically away from cloud and in a flight visibility of at least 5km.

(b) Sub-paragraph (a) shall be deemed to be complied with if:

(i) the aircraft is flying at or below 3000 feet above mean sea level and remains clear of cloud and in sight of the surface and in a flight visibility of at least 5km;

(ii) the aircraft, other than a helicopter, is flying at or below 3000 feet above mean sea level at a speed which according to its air speed indicator is 140 knots or less and remains clear of cloud and in sight of the surface and in a flight visibility of at least 1500 metres; or

(iii) in the case of a helicopter the helicopter is flying at or below 3000 feet above mean sea level flying at a speed, which having regard to the visibility is reasonable, and remains clear of cloud and in sight of the surface.

Rule 27: VFR flight plan and air traffic control clearance

(1) Unless otherwise authorised by the appropriate air traffic control unit before an aircraft flies within Class B, Class C or Class D airspace during the notified hours of watch of the appropriate air traffic control unit, the commander of the aircraft shall cause a flight plan to be communicated to the appropriate air traffic control unit and shall obtain an air traffic control clearance to fly within the said airspace.

(2) The flight plan shall contain such particulars of the flight as may be necessary to enable the air traffic control unit to issue a clearance and for search and rescue purposes.

(2A) Any flight plan for a flight within United Kingdom reduced vertical separation minimum airspace shall also state whether or not the aircraft is equipped with height keeping systems as required by articles 48 or 49 of the Air Navigation Order 2000.

(3) Whilst flying within the said airspace during the notified hours of watch of the appropriate air traffic control unit the commander of the aircraft shall:

(a) cause a continuous watch to be maintained on the notified radio frequency appropriate to the circumstances; and

(b) comply with any instructions which the appropriate air traffic control unit may give in a particular case.

(4) Paragraphs (1), (2) and (3) shall not apply in respect of:

(a) any glider flying or intending to fly in Class B airspace notified for the purpose of this sub-paragraph;

(b) any glider flying during the day in controlled airspace notified for the purpose of this sub-paragraph which remains at least 1500 metres horizontally and 1000 feet vertically away from cloud and in a flight visibility of at least 8km; or

(c) any mechanically driven aircraft without radio equipment flying during the day in controlled airspace notified for the purpose of this sub-paragraph which remains at least 1500 metres horizontally and 1000 feet vertically away from cloud and in a flight visibility of at least 5km the commander of which has previously obtained the permission of the appropriate air traffic control unit to fly within the said airspace.

SECTION VI INSTRUMENT FLIGHT RULES

Rule 28: Instrument Flight Rules
(1) In relation to flights within controlled airspace rules 29, 31 and 32 shall be the Instrument Flight Rules.
(2) In relation to flights outside controlled airspace rules 29 and 30 shall be the Instrument Flight Rules.

Rule 29: Minimum height
Without prejudice to the provisions of rule 5, in order to comply with the Instrument Flight Rules an aircraft shall not fly at a height of less than 1000 feet above the highest obstacle within a distance of 5 nautical miles of the aircraft unless:
(a) it is necessary for the aircraft to do so in order to take-off or land;
(b) the aircraft is flying on a route notified for the purposes of this rule;
(c) the aircraft has been otherwise authorised by the competent authority; or
(d) the aircraft is flying at an altitude not exceeding 3000 feet above mean sea level and remains clear of cloud and in sight of the surface.

Rule 30: Quadrantal rule and semi-circular rule
(1) Subject to paragraph (2), in order to comply with the Instrument Flight Rules, an aircraft when in level flight above 3000 feet above mean sea level or above the appropriate transition altitude, whichever is the higher, shall be flown at a level appropriate to its magnetic track, in accordance with the appropriate Table set forth in this rule. The level of flight shall be measured by an altimeter set:
(a) in the case of a flight over the United Kingdom, to a pressure setting of 1013.2 hectopascals; or
(b) in the case of any other flight, according to the system published by the competent authority in relation to the area over which the aircraft is flying.
(2) An aircraft may be flown at a level other than the level required by paragraph (1) if it is flying in conformity with instructions given by an air traffic control unit or in accordance with notified en-route holding patterns or in accordance with holding procedures notified in relation to an aerodrome.
(3) For the purposes of this rule "transition altitude" means the altitude so notified in relation to flight over such area or areas as may be notified.

TABLE I – Flights at levels below 24,500ft

Magnetic Track	Cruising Level
Less than 90°	Odd thousands of feet
90° but less than 180°	Odd thousands of feet + 500ft
180° but less than 270°	Even thousands of feet
270° but less than 360°	even thousands of feet + 500ft

TABLE II – Flights at levels above 24,500ft

Magnetic Track	Cruising Level
Less than 180°	25,000ft 27,000ft 29,000ft 31,000ft 33,000ft 35,000ft 37,000ft 39,000ft 41,000ft or higher levels at intervals of 4000ft
180° but less than 360°	26,000ft 28,000ft 30,000ft 32,000ft 34,000ft 36,000ft 38,000ft 40,000ft 43,000ft or higher levels at intervals of 4000ft

Rule 31: Flight plan and air traffic control clearance

(1) In order to comply with the Instrument Flight Rules, before an aircraft either takes off from a point within any controlled airspace or otherwise flies within any controlled airspace the commander of the aircraft shall cause a flight plan to be communicated to the appropriate air traffic control unit and shall obtain an air traffic control clearance based on such flight plan.

(2) The flight plan shall contain such particulars of the intended flight as may be necessary to enable the air traffic control unit to issue an air traffic control clearance, and for search and rescue purposes.

(a) Any flight plan for a flight within United Kingdom reduced vertical separation minimum airspace shall also state whether or not the aircraft is equipped with height keeping systems as required by articles 48 or 49 of the Air Navigation Order 2000.

(3) (a) Subject to sub-paragraph (b), the commander of the aircraft shall fly in conformity with:

(i) the air traffic control clearance issued for the flight, as amended by any further instructions given by an air traffic control unit; and

(ii) (aa) the instrument departure procedures notified in relation to the aerodrome of departure, unless he is otherwise authorised by the appropriate air traffic control unit; and

(bb) the holding and instrument approach procedures notified in relation to the aerodrome of destination, unless he is otherwise authorised by the air traffic control unit.

(b) The commander of the aircraft shall not be required to comply with sub-paragraph (a) if:

(i) he is able to fly in uninterrupted Visual Meteorological Conditions for so long as he remains in controlled airspace; and

(ii) he has informed the appropriate air traffic control unit of his intention to continue the flight in compliance with Visual Flight Rules and has requested that unit to cancel his flight plan.

(4) If for the purpose of avoiding immediate danger any departure is made from the provisions of paragraph (3) (as is permitted by article 74(3) of the Order) the commander of the aircraft shall, in addition to causing particulars to be given in accordance with article 74(4) of the Order, as soon as possible inform the appropriate air traffic control unit of the deviation.

(5) The commander of the aircraft after it has flown in controlled airspace shall, unless he has requested the appropriate air traffic control unit to cancel his flight plan, forthwith inform that unit when the aircraft lands within or leaves the controlled airspace.

Rule 32: Position reports

In order to comply with the Instrument Flight Rules the commander of an aircraft in IFR flight who flies in or is intending to enter controlled airspace shall report to the appropriate air traffic control unit the time, and the position and level of the aircraft at such reporting points or at such intervals of time as may be notified for this purpose or as may be directed by the air traffic control unit.

SECTION VII AERODROME TRAFFIC RULES

Rule 33: Application of aerodrome traffic rules

The rules in this section of these Rules which are expressed to apply to flying machines shall also be observed, so far as is practicable, in relation to all other aircraft.

Rule 34: Visual signals

(1) Subject to paragraph (2), the commander of a flying machine on, or in the pattern of traffic at, an aerodrome shall observe such visual signals as may be displayed at, or directed to him from the aerodrome by the authority of the person in charge of the aerodrome and shall obey any instructions which may be given to him by means of such signals.

(2) The commander of such a flying machine shall not be required to obey the signals referred to in rule 47 (Marshalling Signals) if in his opinion it is inadvisable to do so in the interests of safety.

Rule 35: Movement of aircraft on aerodromes

An aircraft shall not taxi on the apron or the manoeuvring area of an aerodrome without the permission of the person in charge of the aerodrome or, where the aerodrome has an air traffic control unit or an aerodrome flight information service unit for the time being notified as being on watch, without the permission of that unit.

Rule 36: Access to and movement of persons and vehicles on the aerodrome

(1) A person or vehicle shall not go onto any part of an aerodrome (not being a part of the aerodrome which is a public right of way) without the permission of the person in charge of that part of the aerodrome, and except in accordance with any conditions subject to which that permission may have been granted.

(2) A vehicle or person shall not go or move on the manoeuvring area of an aerodrome having an air traffic control unit without the permission of that unit, and except in accordance with any conditions subject to which that permission may have been granted.

(3) Any permission granted for the purposes of this rule may be granted whether in respect of persons or vehicles generally, or in respect of any particular person or vehicle or any class of person or vehicle.

Rule 37: Right of way on the ground

(1) This rule shall apply to flying machines and vehicles on any part of a land aerodrome provided for the use of aircraft and under the control of the person in charge of the aerodrome.

(2) Notwithstanding any air traffic control clearance it shall remain the duty of the commander of an aircraft to take all possible measures to ensure that his aircraft does not collide with any other aircraft or with any vehicle.

(3) (a) Flying machines and vehicles shall give way to aircraft which are taking off or landing.

(b) Vehicles, and flying machines which are not taking off or landing, shall give way to vehicles towing aircraft.

(c) Vehicles which are not towing aircraft shall give way to aircraft.
(4) Subject to the provisions of paragraph (3) and of rule 17(7)(c), in case of danger of collision between two flying machines:
(a) when the two flying machines are approaching head-on or approximately so, each shall alter its course to the right;
(b) when the two flying machines are on converging courses, the one which has the other on its right shall give way to the other and shall avoid crossing ahead of the other unless passing well clear of it;
(c) a flying machine which is being overtaken shall have the right-of-way, and the overtaking flying machine shall keep out of the way of the other flying machine by altering its course to the left until that other flying machine has been passed and is clear, notwithstanding any change in the relative positions of the two flying machines.
(5) Subject to the provisions of paragraph (3)(b) a vehicle shall:
(a) overtake another vehicle so that the other vehicle is on the left of the overtaking vehicle;
(b) keep to the left when passing another vehicle which is approaching head-on or approximately so.

Rule 38: Launching, picking up and dropping of tow ropes, etc.

(1) Tow ropes, banners or similar articles towed by aircraft shall not be launched at an aerodrome except in accordance with arrangements made with the air traffic control unit at the aerodrome or, if there is no such unit, with the person in charge of the aerodrome.
(2) Tow ropes, banners or similar articles towed by aircraft shall not be picked up by or dropped from aircraft at an aerodrome except:
(a) in accordance with arrangements with the air traffic control unit at the aerodrome or, if there is no such unit, with the person in charge of the aerodrome; or
(b) in the area designated by the marking described in rule 44(7), and the ropes, banners or similar articles shall be picked up and dropped when the aircraft is flying in the direction appropriate for landing.

Rule 39: Flight within aerodrome traffic zones

(1) Paragraphs (2) and (3) shall apply only in relation to such of the aerodromes described in Column 1 of the following Table as are notified for the purposes of this rule and at such times as are specified in Column 2 thereof.

Column 1	Column 2
(a) Government aerodrome	at such times as are notified
(b) An aerodrome having an air traffic control unit or an aerodrome flight information service unit	during the notified hours of watch of the air traffic control unit or the aerodrome flight information service unit
(c) A licenced aerodrome having a means of two-way radio communication with aircraft	during the notified hours of watch of the air/ground radio station

(2) An aircraft shall not fly, take-off or land within the aerodrome traffic zone of an aerodrome to which this paragraph applies unless the commander of the aircraft has obtained the permission of the air traffic control unit at the aerodrome or, where there is no air traffic control unit, has obtained from the aerodrome flight information unit at that aerodrome information to enable the flight within the zone to be conducted with safety or, where there is no air traffic control unit nor aerodrome flight information service unit, has obtained information from the air/ground radio station at that aerodrome to enable the flight to be conducted with safety.
(3) The commander of an aircraft flying within the aerodrome traffic zone of an aerodrome to which this paragraph applies shall:
(a) cause a continuous watch to be maintained on the appropriate radio frequency notified for communications at the aerodrome or, if this is not possible, cause a watch to be kept for such instructions as may be issued by visual means;
(b) where the aircraft is fitted with means of communication by radio with the ground, communicate his position and height to the air traffic control unit, the aerodrome flight information service unit or the air/ground radio station at the aerodrome (as the case may be), on entering the zone and immediately prior to leaving it.

SECTION VIII SPECIAL RULES

Rule 40: Use of radio navigation aids

(1) Subject to paragraph (2), the commander of an aircraft shall not make use of any radio navigation aid without complying with such restrictions and appropriate procedures as may be notified in relation to that aid unless authorised by an air traffic control unit.
(2) The commander of an aircraft shall not be required to comply with this rule if he is required to comply with rule 31.

SECTION IX AERODROME SIGNALS AND MARKINGS – VISUAL AND AURAL SIGNALS

Rule 41: General

(1) Whenever any signal specified in this section of these Rules is given or displayed, or whenever any marking so specified is displayed, by any person in an aircraft, or at an aerodrome, or at any other place which is being used by aircraft for landing or take-off, it shall, when given or displayed in the United Kingdom, have the meaning assigned to it in this section.

(2) All dimensions other than those in rule 45(6), of signals or markings specified in this section of these Rules (but not distances at which markings must be placed) shall be subject to a tolerance of 10 per cent, plus or minus.

Rule 42: Signals in the signals area

(1) When any signal specified in the following paragraphs of this rule is displayed it shall be placed in a signals area, which shall be a square visible from all directions bordered by a white strip 30 centimetres wide the internal sides measuring 12 metres.

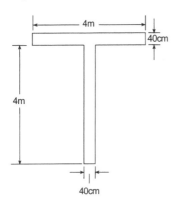

(2) A white landing T, as illustrated in this paragraph, signifies that aeroplanes and gliders taking off or landing shall do so in a direction parallel with the shaft of the T and towards the cross arm, unless otherwise authorised by the appropriate air traffic control unit.

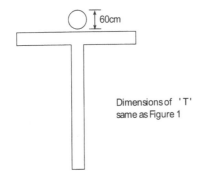

Dimensions of 'T' same as Figure 1

(3) A white disc 60 centimetres in diameter displayed alongside the cross arm of the T and in line with the shaft of the T, as illustrated in this paragraph, signifies that the direction of landing and take-off do not necessarily coincide.

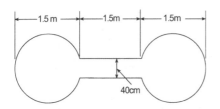

(4) A white dumb-bell, as illustrated in this paragraph, signifies that movements of aeroplanes and gliders on the ground shall be confined to paved, metalled or similar hard surfaces.

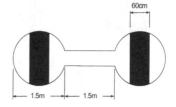

(5) A white dumb-bell as described in (4) above but with a black strip 60 centimetres wide across each disc at right angles to the shaft of the dumb-bell, as illustrated in this paragraph, signifies that aeroplanes and gliders taking off or landing shall do so on a runway but that movement on the ground is not confined to paved, metalled or similar hard surfaces.

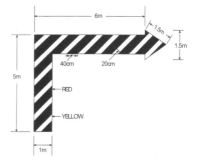

(6) A red and yellow striped arrow, as illustrated in this paragraph, the shaft of which is one metre wide placed along the whole or a total of 11 metres of two adjacent sides of the signals area and pointing in a clockwise direction signifies that a right-hand circuit is in force.

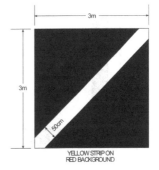

(7) A red panel 3 metres square with a yellow strip along one diagonal 50 centimetres wide, as illustrated in this paragraph, signifies that the state of the manoeuvring area is poor and pilots must exercise special care when landing.

ANO Rule 43: Markings for paved runways and taxiways

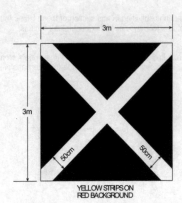

YELLOW STRIPS ON RED BACKGROUND

(8) A red panel 3 metres square with a yellow strip, 50 centimetres diagonal, as illustrated in this paragraph, signifies that the aerodrome is unsafe for the movement of aircraft and that landing on the aerodrome is prohibited.

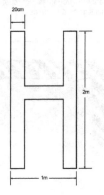

(9) A white letter H, as illustrated in this paragraph, signifies that helicopters shall take-off and land only within the area designated by the marking specified in rule 44(5).

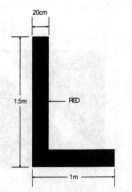

(10) A red letter L displayed on the dumb-bell specified in paragraphs (4) and (5), as illustrated in this paragraph, signifies that light aircraft are permitted to take-off and land either on a runway or on the area designated by the marking specified in rule 44(6).

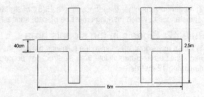

(11) A white double cross, as illustrated in this paragraph, signifies that glider flying is in progress.

Rule 43: Markings for paved runways and taxiways

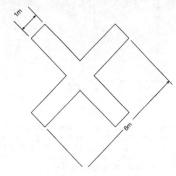

(1) Two or more white crosses, as illustrated in this paragraph, displayed on a runway or taxiway, with the arms of the crosses at an angle of 45° to the centre line of the runway, at intervals of not more than 300 metres signify that the section of the runway or taxiway marked by them is unfit for the movement of aircraft.

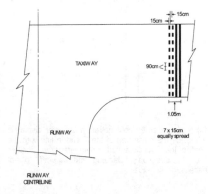

(2)(a) Two yellow broken lines and two continuous lines, as illustrated in this paragraph, signify the holding position closest to the runway beyond which no part of a flying machine or vehicle shall project in the direction of the runway without permission from the air traffic control unit at the aerodrome during the notified hours of watch of that unit. Outside the notified hours of watch of that unit or where there is no air traffic control unit at the aerodrome the markings signify the position closest to the runway beyond which no part of a flying machine or vehicle shall project in the direction of the runway when the flying machine or vehicle is required by virtue of rule 37(3)(a) of these Rules to give way to aircraft which are taking off from or landing on that runway.

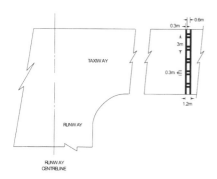

(b) A yellow marking, as illustrated in this paragraph, signifies a holding position other than that closest to the runway beyond which no part of a flying machine or vehicle shall project in the direction of the runway without permission from the air traffic control unit at the aerodrome during the notified hours of watch of that unit. Outside the notified hours of watch of that unit or where there is no air traffic control unit at the aerodrome the marking may be disregarded.

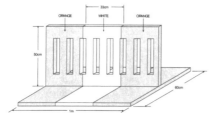

(3) Orange and white markers, as illustrated in this paragraph, spaced no more than 15 metres apart, signify the boundary of that part of a paved runway, taxiway or apron which is unfit for the movement of aircraft.

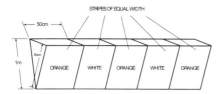

Rule 44: Markings on unpaved manoeuvring areas

(1) Markers with orange and white stripes of an equal width of 50 centimetres, with an orange stripe at each end, as illustrated in this paragraph, alternating with flags 60 centimetres square showing equal orange and white triangular areas, indicate the boundary of an area unfit for the movement of aircraft and one or more white crosses as specified in rule 43(1) indicate the said area. The distance between any two successive orange and white flags shall not exceed 90 metres.

(2) Striped markers, as specified in paragraph (1), spaced not more than 45 metres apart, indicate the boundary of an aerodrome.

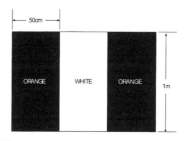

(3) On structures, markers with orange and white vertical stripes, of an equal width of 50 centimetres, with an orange stripe at each end, as illustrated in this paragraph, spaced not more than 45 metres apart, indicate the boundary of an aerodrome. The pattern of the marker shall be visible from inside and outside the aerodrome and the marker shall be affixed not more than 15 centimetres from the top of the structure.

(4) White flat rectangular markers 3 metres long and 1 metre wide at intervals not exceeding 90 metres, flush with the surface of the unpaved runway or stopway, as the case may be, indicate the boundary of an unpaved runway or of a stopway.

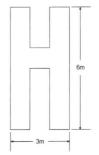

(5) A white letter H, as illustrated in this paragraph, indicates an area which shall be used only for the taking off and landing of helicopters.

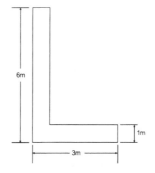

(6) A white letter L, an illustrated in this paragraph, indicates a part of the manoeuvring area which shall be used only for the taking off and landing of light aircraft.

(7) A yellow cross with two arms each 6 metres long by 1 metre wide at right angles, indicates that tow ropes, banners and similar articles towed by aircraft shall only be picked up and dropped in the area in which the cross is placed.

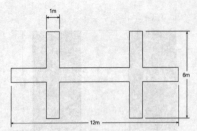

(8) A white double cross, as illustrated in this paragraph, indicates an area which shall be used only for the taking off and landing of gliders.

(9) A white landing T as specified in rule 42(2) placed at the left-hand side of the runway when viewed from the direction of landing indicates the runway to be used, and at an aerodrome with no runway it indicates the direction for take-off and landing.

Rule 45: Signals visible from the ground

(1) A black ball 60 centimetres in diameter suspended from a mast signifies that the directions of take-off and landing are not necessarily the same.

(2) A checkered flag or board, 1.2 metres by 90 centimetres containing twelve equal squares, 4 horizontally and 3 vertically, coloured red and yellow alternately, signifies that aircraft may move on the manoeuvring area and apron only in accordance with the permission of the air traffic control unit at the aerodrome.

(3) Two red balls 60 centimetres in diameter, disposed vertically one above the other, 60 centimetres apart and suspended from a mast, signify that glider flying is in progress at the aerodrome.

(4) Black arabic numerals in two-figure groups and, where parallel runways are provided the letter or letters L (left), LC (left centre), C (centre), RC (right centre) and R (right), placed against a yellow background, indicate the direction for take-off or the runway in use.

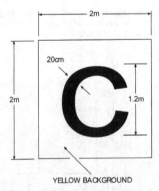

YELLOW BACKGROUND

(5) A black letter C against a yellow background, as illustrated in this paragraph, indicates the position at which a pilot can report to the air traffic control unit or to the person in charge of the aerodrome.

(6) A rectangular green flag of not less than 60 centimetres square and not more than 66 centimetres square flown from a mast indicates that a right-hand circuit is in force.

Rule 46: Lights and pyrotechnic signals for control of aerodrome traffic

Each signal described in the first column of Table A, when directed from an aerodrome to an aircraft or to a vehicle, or from an aircraft, shall have the meanings respectively appearing in the second, third and fourth columns of that Table opposite the description of the signal.

Table A – Meaning of lights and pyrotechnic signals

Characteristic and colour of light Beam or pyrotechnic	To an aircraft in flight	From an aerodrome To an aircraft or vehicle on the aerodrome	From an aircraft in flight to an aerodrome
(a) Continuous red light	Give way to other aircraft and continue circling	Stop	–
(b) Red pyrotechnic light, or red flare	Do not land; wait for permission	–	Immediate assistance is requested
(c) Red flashes	Do not land; aerodrome not available for landing	Move clear of landing area	–
(d) Green flashes	Return to aerodrome; wait for permission to land	To an aircraft; you may move on the manoeuvring area and apron. To a vehicle; you may move on the Manoeuvring area	–
(e) Continuous green light	You may land	You may take-off (not applicable to a vehicle)	–
(f) Continuous green light, or green Flashes, or green pyrotechnic light	–	–	By night; may I land? By day; may I land in direction different from that indicated by landing T? I am compelled to land
(g) White flashes	Land at this aerodrome after receiving continuous green light, and then after receiving green flashes, proceed to the apron	Return to starting point on the aerodrome	I am compelled to land
(h) White pyrotechnic lights switching on and off the navigation lights. Switching on and off the landing lights	–	–	–

Rule 47: Marshalling signals (from a marshaller to an aircraft)

Each of the signals for the guidance of aircraft manoeuvring on or off the ground, described in the first column of Table B, paragraphs (a) to (x) shall, in the United Kingdom, have the meanings set forth in the second column of that Table opposite the description of the signal. By day any such signals shall be given by hand or by circular bats and by night by torches or illuminated wands (save that the signals described at paragraphs (v) and (w) shall not be given at night).

Description of signal	Meaning of signal	In Daylight	By Night
(a) Right or left arm down, the other arm moved across body and extended to indicate position of the other marshaller	Proceed under guidance of another marshaller		
(b) Arms repeatedly moved upward and backward, beckoning onwards	Move ahead		
(c) Right arm down, left arm repeatedly moved upward and backward. The speed of arm movement indicates the rate of turn	Open up starboard engine or turn to port		
(d) Left arm down, the right arm repeatedly moved upward and backward. The speed of arm movement indicates the Rate of turn	Open up port engine or turn to starboard		
(e) Arms repeatedly crossed above the head. The speed of the arm movement indicates the urgency of the stop	Stop		
(f) A circular motion of the right hand at head level, with the left arm pointing to the appropriate engine	Start engines		
(g) Arms extended, the palms facing inwards, then swung from the extended position inwards	Chocks inserted		
(h) Arms down, the palms facing outwards, then swung outwards	Chocks away		
(j) Either arm and hand placed level with the chest, then moved laterally with the palm downwards	Cut engines		
(k) Arms placed down, with the palms towards the ground, then moved up and down several times	Slow down		
(l) Arms placed down, with the palms towards the ground, then either the right or left arm moved up and down indicating that the motors on the left or right side, as the case may be should be slowed down	Slow down engines on indicated side		
(m) Arms placed above the head in a vertical position	This bay		

Rule 47: Marshalling signals (from a marshaller to an aircraft)

Description of signal	Meaning of signal
(n) The right arm raised at the elbow, with the arm facing forward	All clear: marshalling finished
(o) Arms placed horizontally sideways	Hover
(p) Arms placed down and crossed in front of the body	Land
(q) Arms placed horizontally sideways with the palms up beckoning upwards. The speed of the arm movement indicates the rate of ascent	Move upwards
(r) Arms place horizontally sideways with the palms towards the ground beckoning downwards. The speed of arm movement indicates the rate of descent	Move downwards
(s) Either arm placed horizontally sideways, then the other arm moved in front of the body to the side, in the direction of the movement, indicating that the helicopter should move horizontally to the left or right side, as the case may be; repeated several times	Move horizontally
(t) Arms placed down, the palms facing forward, then repeatedly swept up and down to shoulder level	Move back
(u) Left arm extended horizontally forward, then right arm making a horizontal slicing movement below left arm	Release load
(v) Raise arm, with fist clenched horizontally in front of body, then extended fingers	Release brakes
Raise arm and hand, with fingers Extended, horizontally in front of Body, then clench fist	Engage brakes
(w) Left hand overhead with the number of fingers extended, to indicate the number of the engine to be started, and circular motion of right hand at head level	Start engines(s)
(x) Point left arm down, move right arm down from overhead, vertical position to horizontal forward position, repeating right arm movement	Back aircraft's tail to starboard
Point right arm down, move left arm down from overhead, vertical position to horizontal forward position, repeating left arm movement	Back aircraft's tail to port

Rule 48: Marshalling signals (from a pilot of an aircraft to a marshaller)

The following signals made by a pilot in an aircraft to a marshaller on the ground shall respectively have the following meanings:

Description of Signal	Meaning of signal
(a) Raise arm and hand with fingers extended horizontally in front of face, then clench fist	Brakes engaged
(b) Raise arm with fist clenched horizontally in front of face, then extend fingers	Brakes released
(c) Arms extended palms facing outwards, move hands inwards to cross in front of face.	Insert chocks
(d) Hands crossed in front of face, palms facing outwards, move arms outwards.	Remove chocks
(e) Raise the number of fingers on one hand indicating the number of the engine to be started. For this purpose the aircraft engines shall be numbered in relation to the marshaller facing the aircraft, from his right to his left, for example, No. 1 engine shall be the port outer engine, No. 2 engine shall be the port inner engine, No. 3 engine shall be the starboard inner engine No. 4 engine shall be the starboard outer engine.	Ready to start engines

Rule 49: Distress, urgency and safety signals

(1) The following signals, given either together or separately before the sending of a message, signify that an aircraft is threatened by grave and imminent danger and requests immediate assistance:
(a) by radiotelephony: the spoken word "MAYDAY";
(b) visual signalling:
(i) the signal SOS (• • • – – – • • •);
(ii) a succession of pyrotechnic lights fired at short intervals each showing a single red light;
(iii) a parachute flare showing a red light;
(c) by sound signalling other than radiotelephony:
(i) the signal SOS (• • • – – – • • •);
(ii) a continuous sounding with any sound apparatus.

(2) The following signals, given either together or separately, before the sending of a message, signify that the commander of the aircraft wishes to give notice of difficulties which compel it to land but that he does not require immediate assistance:
(a) a succession of white pyrotechnic lights;
(b) the repeated switching on and off of the aircraft landing lights;
(c) the repeated switching on and off of its navigation lights, in such a manner as to be clearly distinguishable from the flashing navigation lights described in rule 11.

(3) The following signals, given either together or separately, indicate that the commander of the aircraft has an urgent message to transmit concerning the safety of a ship, aircraft, vehicle or other property or of a person on board or within sight of the aircraft from which the signal is given:
(a) by radiotelephony: the repeated spoken word, "PAN PAN";
(b) by visual signalling: the signal XXX (– • • – – • • – – • • –);
(c) by sound signalling other than radiotelephony: the signal XXX – – • • • •

THE AIR NAVIGATION GENERAL REGULATIONS 2005

SECTION 3 THE AIR NAVIGATION (GENERAL) REGULATIONS 2005
Coming into force 20th August 2005
ARRANGEMENT OF REGULATIONS

PART 1 GENERAL
1 Citation and commencement
2 Revocation
3 Interpretation

PART 2 LOAD SHEETS
4 Particulars and weighing requirements

PART 3 AIRCRAFT PERFORMANCE
5 Aeroplanes to which article 44(5) applies
6 Helicopters to which article 45(1) applies
7 Weight and performance: general provisions

PART 4 NOISE AND VIBRATION, MAINTENANCE AND AERODROME FACILITIES
8 Noise and vibration caused by aircraft on aerodromes
9 Pilots maintenance prescribed repairs or replacements
10 Aeroplanes flying for the purpose of public transport of passengers-aerodrome facilities for approach to landing and landing

PART 5 MANDATORY REPORTING
11 Reportable occurrences, time and manner of reporting and information to be reported
12 Mandatory reporting of birdstrikes-time and manner of reporting and information to be reported

PART 6 NAVIGATION PERFORMANCE AND EQUIPMENT
13 Minimum navigation performance and height keeping specifications
14 Airborne Collision Avoidance System
15 Mode S Transponder

SCHEDULE 1
REGULATIONS REVOKED

SCHEDULE 2
AEROPLANE PERFORMANCE
1 Weight and performance of public transport aeroplanes designated as aeroplanes of performance group A or performance group B
2 Weight and performance of public transport aeroplanes designated as aeroplanes of performance group C
3 Weight and performance of public transport aeroplanes designated as aeroplanes of performance group D
4 Weight and performance of public transport aeroplanes designated as aeroplanes of performance group E
5 Weight and performance of public transport aeroplanes designated as aeroplanes of performance group F
6 Weight and performance of public transport aeroplanes designated as aeroplanes of performance group X
7 Weight and performance of public transport aeroplanes designated as aeroplanes of performance group Z

SCHEDULE 3
HELICOPTER PERFORMANCE
1 Weight and performance of public transport helicopters carrying out Performance Class 1 operations
2 Weight and performance of public transport helicopters carrying out Performance Class 2 operations
3 Weight and performance of public transport helicopters carrying out Performance Class 3 operations

The Secretary of State for Transport, in exercise of his powers under articles 16(7)(a), 42(1)(c)(ii), 43(5), 44(5), 45(1), 56(1), 57(1), 58(1)(b), 131, 142(6), 143(2), 155(1) and paragraphs 3 and 4(5) of Schedule 5 to the Air Navigation Order 2005(a), hereby makes the following Regulations:

1 Citation and commencement
These Regulations may be cited as the Air Navigation (General) Regulations 2005, and shall come into force on 20 August 2005.

2 Revocation
The following Regulations specified in Schedule 1 are revoked.

3 Interpretation
In these Regulations:
(a) "the Order" means the Air Navigation Order 2005;

(b) references to an "article" means (unless inconsistent with the context) an article of the Order; or;
(c) other expressions used in these Regulations shall have the same respective meanings as in the Order.

PART 2 LOAD SHEETS

4 Particulars and weighing requirements

(1) Every load sheet required by article 43(5) shall contain the following particulars
(a) the nationality mark of the aircraft to which the load sheet relates, and the registration mark assigned to that aircraft by the CAA;
(b) particulars of the flight to which the load sheet relates;
(c) the total weight of the aircraft as loaded for that flight;
(d) the weights of the several items from which the total weight of the aircraft, as so loaded, has been calculated including in particular the weight of the aircraft prepared for service and the respective total weights of the crew (unless included in the weight of the aircraft prepared for service), passengers, baggage and cargo intended to be carried on the flight;
(e) the manner in which the load is distributed;
(f) the position of the centre of gravity of the aircraft resulting from the particulars mentioned in sub-paragrpahs (c), (d) and (e) which may be given approximately if and to the extent that the relevant certificate of airworthiness so permits; and
(g) a certificate at the foot or end of the load sheet, signed by the person referred to in article 43(1) as responsible for the supervision of the loading of the aircraft, that the aircraft has been loaded in accordance with the written instructions furnished to him by the operator of the aircraft pursuant to article 43(1).

(2) Subject to paragraph (3) for the purpose of calculating the total weight of the aircraft the respective total weights of the passengers and crew together with their hand baggage entered in the load sheet shall be computed from the actual weight of each person and for that purpose each person and their hand baggage shall be separately weighed;
(3) The total weights of the passengers and crew together with their hand baggage may, in accordance with paragraphs (4) and (9) and subject to the provisions of paragraphs (13), (14) and (15), be calculated at not less than the appropriate weights shown in Tables 1 and 2 and the load sheet shall bear a notation to that effect.

	Passenger seats available		
	20 or more Male	30 or more Female	All adult
Passengers on all flights except holiday charters	88kg	70kg	84kg
Passengers on holiday charters	83kg	69kg	76kg
Children (between 2-12yrs) or infants under 2yrs of age if occupying a separate seat	35kg	35kg	
Infants under 2yrs of age if sharing a seat with an adult	0kg	0kg	
Flight crew	85kg	85kg	
Cabin Crew	75kg	75kg	

(4) Subject to paragraph (5) where the total number of passenger seats available on an aircraft is 20 or more, the weights for males and females in Table 1 are applicable.
(5) Where the total number of passenger seats available is 30 or more, the all adult weights in Table 1 may be used for passengers over the age of 12 years.
(6) For the purpose of Table 1, 'holiday charter' means a flight by an aircraft for the carriage of passengers each of whom carried pursuant to an agreement which provides for carriage by air to a place outside the United Kingdom and back from that place, or form another place to the United Kingdom (whether or not on the same aircraft) and for accommodation at a place outside the United Kingdom.
(7) Where the total number of passenger seats available on an aircraft is 19 or less the weights in Table 2 are applicable.

	Passenger seats available		
	1-5	6-9	10-19
Male passengers	104kg	96kg	92kg
Female passengers	86kg	78kg	74kg
Children (between 2-12yrs) or infants under 2yrs of age if occupying a separate seat	35kg	35kg	35kg
Infants under 2yrs of age if sharing a seat with an adult	0kg	0kg	0kg
Flight crew	85kg	85kg	85kg
Cabin crew	75kg	75kg	75kg

(8) On flights where no hand baggage is carried or where such hand baggage is accounted for separately, 6kg may be deducted from the weight of passengers over 12 years of age when using Table 2.
(9) Where an immersion suit is worn or carried by a passenger or crew member, 3kg shall be added to the appropriate weight shown in Table 1 or 2 in each such case.
(10) Subject to paragraph (11) for the purposes of calculating the total weight of the aircraft the respective total weights of the hold baggage and cargo entered in the load sheet shall be computed from the actual weight of each piece of baggage, cargo or cargo container and for that purpose each piece or container shall be separately weighed.
(11) In the case of an aircraft where the total number of passenger seats available is 20 or more, the total weights of the hold baggage may, subject to the provisions of paragraphs (13), (14) and (15), be calculated at not less than the weights shown in Table 3 and the load sheet shall bear a notation to that effect.

Journey made by aircraft	Hold baggage per piece
Domestic	11kg
European	13kg
Intercontinental	15kg

(12) For the purposes of Table 3
(a) a journey made by an aircraft shall be treated as domestic if it is confined within an area enclosed by rhumb lines joining successively the following points:
N6100.00 W01100.00
N6100.00 E00200.00
N5105.00 E00200.00
N4930.00 W00400.00
N4930.00 W01100.00
N6100.00 W01100.00
but excluding any journey to or from Shannon.
(b) a journey made by an aircraft, not being a domestic journey, shall be treated as European if it is confined within an area enclosed by rhumb lines joining successively the following points:
N7200.00 E04500.00
N4000.00 E04500.00
N3500.00 E03700.00
N3000.00 E03700.00
N3000.00 W00600.00
N2700.00 W00900.00
N2700.00 W03000.00
N6700.00 W03000.00
N7200.00 W01000.00
N7200.00 E04500.00
(c) a journey made by an aircraft shall be treated as intercontinental if it is neither domestic nor European.
(13) If it appears to the person supervising the loading of the aircraft that any of the circumstances described in paragraph (14) arise he shall, if he considers it necessary in the interests of the safety of the aircraft, or if the CAA has so directed in the particular case, require any such person and his hand baggage, passenger or hold baggage, as the case may be, to be weighed for the purpose of the entry to be made in the load sheet.
(14) The circumstances referred to in paragraph (13) are:
(a) that any person and his hand baggage to be carried exceeds the weights set out in Table 1 or 2; or
(b) where paragraph (8) applies, any passenger to be carried exceeds the weights set out in Table 2 as adjusted in accordance with that paragraph; or
(15) If any person and his hand baggage, passenger of any hold baggage has been weighed pursuant to paragraph (13), the weights entered in the load sheet shall take account of the actual weight of that person and his hand baggage, that passenger or that hold baggage, as the case may be, or of the weight determined in accordance with paragraphs (3) or (11), whichever weight shall be the greater.

PART 3 AIRCRAFT PERFORMANCE

5 Aeroplanes to which article 44(5) applies
(1) Aeroplanes to which this regulation applies shall comply with Schedule 2.
(2) This regulation applies to aeroplanes to which the CAA has granted a permission under article 44(5) except any aeroplane flying solely for the purpose of training persons to perform duties in aeroplanes

6 Helicopters to which article 45(1) applies
(1) Helicopters to which this regulation applies shall comply with Schedule 3.
(2) This regulation applies to helicopters to which article 45(1) applies except any helicopter flying solely for the purpose of training persons to perform duties in helicopters.

7 Weight and performance: general provisions

(1) The assessment of the ability of an aeroplane to comply with the requirements of Schedule 2 and of a helicopter to comply with the requirements of Schedule 3 (relating in either case to weight, performance and flights in specified meteorological conditions or at night) shall be based on the specified information as to its performance:

(2) In assessing the ability of an aeroplane to comply with sub-paragraphs (8), (12), (16), and (18) or paragraph 1 of Schedule 2, with sub-paragraphs (8) and (11) of paragraph 6 of Schedule 2 and with sub-paragraph (19) of paragraph 7 of Schedule 2, account may be taken of any reduction of the weight of the aeroplane which may be achieved after the failure of a power unit by such jettisoning of fuel as is feasible and prudent in the circumstances of the flight and in accordance with the flight manual included in the certificate of airworthiness relating to the aircraft.

(3) In this part and in Schedules 2 and 3:

"specified" in relation to an aircraft means specified in, or ascertainable by reference to

(a) the certificate of airworthiness in force under the Order in respect of that aircraft; or

(b) the flight manual or performance schedule included in that certificate, or other document, whatever its title, incorporated by reference in that certificate; or

(c) if there is no flight manual or performance schedule or other such document so incorporated in that certificate, the flight manual for that aircraft;

"the accelerate-stop distance" means the distance from the point on the surface of the aerodrome at which the aeroplane can commence its take-off run to the nearest point in the direction of take-off at which the aeroplane cannot roll over the surface of the aerodrome and be brought to rest in an emergency without the risk of accident;

"the landing distance available" means the distance from the point on the surface of the aerodrome above which the aeroplane can commence its landing, having regard to the obstructions in its approach path, to the nearest point in the direction of landing at which the surface of the aerodrome is incapable of bearing the weight of the aeroplane under normal operating conditions or at which there is an obstacle capable of affecting the safety of the aeroplane;

"the take-off distance available" means either the distance from the point on the surface of the aerodrome at which the aeroplane can commence its take-off run to the nearest obstacle in the direction of take-off projecting above the surface of the aerodrome and capable of affecting the safety of the aeroplane, or one and one half times the take-off run available, whichever is the less;

"the take-off run available" means the distance from the point on the surface of the aerodrome at which the aeroplane can commence its take-off run to the nearest point in the direction of take-off at which the surface of the aerodrome is incapable of bearing the weight of the aeroplane under normal operating conditions.

(4) For the purposes of Schedules 2 and 3:

(a) the weight of the aircraft at the commencement of the take-off run or of the take-off shall be taken to be its gross weight including everything and everyone carried in or on it at the commencement of the take-off run or of the take-off;

(b) the landing weight of the aircraft shall be taken to be the weight of the aircraft at the estimated time of landing allowing for the weight of the fuel and oil expected to be used on the flight to the aerodrome at which it is intended to land or alternate aerodrome, as the case may be;

(c) where any distance referred to in paragraph (3) has been declared in respect of any aerodrome by the authority responsible for regulating air navigation over the territory of the Contracting State in which the aerodrome is situated, and in the case of an aerodrome in the United Kingdom, notified, that distance shall be deemed to be the relevant distance.

PART 4 NOISE AND VIBRATION, MAINTENANCE AND AERODROME FACILITIES

8 Noise and vibration caused by aircraft on aerodromes

For the purposes of article 131, the conditions under which noise and vibration may be caused by aircraft (including military aircraft) on Government aerodromes, licensed aerodromes or on aerodromes at which the manufacture, repair or maintenance of aircraft is carried out by persons carrying on business as manufacturers or repairers of aircraft, shall be as follows;

(a) the aircraft is taking off or landing; or

(b) the aircraft is moving on the ground or water; or

(c) the engines are being operated in the aircraft;

(i) for the purpose of ensuring their satisfactory performance;

(ii) for the purpose of bringing them to a proper temperature in preparation for, or at the end of, a flight; or

(iii) for the purpose of ensuring that the instruments, accessories or other components of the aircraft are in a satisfactory condition.

9 Pilots maintenance – prescribed repairs or replacements

For the purposes of article 16(7)(a), the following repairs or replacements are prescribed;

(a) replacement of landing gear tyres, landing skids or skid shoes;

(b) replacement of elastic shock absorber cord units on landing gear where special tools are not required;

(c) replacement of defective safety wiring or split pins excluding those in engine, transmission, flight control and rotor systems;

(d) patch-repairs to fabric not requiring rib stitching or the removal of structural parts or control surfaces, if the repairs do not cover up structural damage and do not include repairs to rotor blades;

(e) repairs to upholstery and decorative furnishing of the cabin or cockpit interior when repair does not require dismantling of any structure or operating system or interfere with an operating system or affect the structure of the aircraft;

(f) repairs, not requiring welding, to fairings, non-structural cover plates and cowlings;
(g) replacement of side windows where that work does not interfere with the structure or with any operating system.
(h) replacement of safety belts or safety harness;
(i) replacement of seats or seat parts not involving dismantling of any structure or of any operating system;
(j) replacement of bulbs, reflectors, glasses, lenses or lights;
(k) replacement of any cowling not requiring removal of the propeller, rotors or disconnection of engine or flight controls;
(l) replacement of unserviceable sparking plugs;
(m) replacement of batteries;
(n) replacement of wings and tail surfaces and controls, the attachments of which are designed to provide for assembly immediately before each flight and dismantling after each flight;
(o) replacement of main rotor blades that are designed for removal where special tools are not required;
(p) replacement of generator and fan belts designed for removal where special tools are not required;
(q) replacement of VHF communication equipment, being equipment which is not combined with navigation equipment.

10 Aeroplanes flying for the purpose of public transport of passengers – aerodrome facilities for approach to landing and landing

(1) This regulation shall apply to every aeroplane registered in the United Kingdom engaged on a flight for the purpose of public transport of passengers on a scheduled journey and to every aeroplane so registered whose maximum total weight authorised exceeds 5,700kg engaged on such a flight otherwise than on a scheduled journey.

(2) For the purposes of article 42(1)(c)(ii), the following manning and equipment are prescribed in relation to aerodromes intended to be used for landing or as an alternate aerodrome by aircraft to which this regulation applies;

(a) air traffic control service or aerodrome flight information service, including the reporting to aircraft of the current meteorological conditions at the aerodrome;

(b) very high frequency radiotelephony;

(c) at least one of the following radio navigation aids, either at the aerodrome or elsewhere, and in either case for the purpose of assisting the pilot in locating the aerodrome and in making an approach to landing there:

(i) radio direction finding equipment utilising emissions in the very high frequency bands;
(ii) a non-directional radio beacon transmitting signals in the low or medium frequency bands;
(iii) very high frequency omni-directional radio range;
(iv) radar equipment.

(3) Subject to paragraph (4) an aircraft to which this regulation applies shall not land or make an approach to landing at any aerodrome unless:

(a) the services and equipment according with paragraph (2) are provided and are in operation at that aerodrome; and

(b) such services and equipment can be made use of by that aircraftl; and

(c) in the case of the navigation aids specified in paragraph (2)(c), instructions and procedures for the use of the aid are included in the operations manual relating to the aircraft.

(4) A person shall be deemed not to have contravened the provisions of paragraph (3) if he proves that;

(a) for the time being use could not be made of the radio navigation aids provided under sub-paragraph (2)(c) whether by reason of those aids not being in operation or of the unserviceability of equipment in the aircraft itself; and

(b) the approach to landing was made in accordance with instructions and procedures appropriate to that circumstance and included in the operations manual.

PART 5 MANDATORY REPORTING

11 Reportable occurrences – time and manner of reporting and information to be reported

(1) For the purposes of article 142(6) but subject to paragraph (2) it is prescribed that s report containing the information referred to in paragraph (5) shall be made to the CAA by post, telex, electronic, facsimile transmission or other similar means which produces a document containing a text of the communication (written in English) within 96 hours of the reportable occurrence coming to the knowledge of the person making the report.

(2) If at the time referred to in paragraph (1) any of the information referred to in that paragraph is not in the possession of the person making the report, he shall despatch the information to the CAA by post, telex, electronic, facsimile transmission or other similar means which produces a document containing text of the communication (written in English) within 96 hours of the information coming into his possession.

(3) For the purposes of article 142(6), a report shall, as far as possible, contain the following information:

(a) the type, series and registration marks of the aircraft concerned;
(b) the name of the operator of the aircraft;
(c) the date of the reportable occurrence;
(d) if the person making the report has instituted an investigation into the reportable occurrence, whether or not this has been completed;
(e) a description of the reportable occurrence, including its effects and any other relevant information;
(f) in the case of a reportable occurrence which occurs during flight;

ANO Regulations – 11 Reportable occurrences – time and manner of reporting and information to be reported

(i) the Co-ordinated Universal Time of the occurrence;
(ii) the last point of departure and the next point of intended landing of the aircraft at that time; and
(iii) the geographical position of the aircraft at that time;
(g) in the case of a defect in or malfunctioning of an aircraft or any part or equipment of an aircraft, the name of the manufacturer of the aircraft, part or equipment, as the case may be, and, where appropriate, the part number and modification standard of the part or equipment and its location on the aircraft;
(h) the signature and name in block capitals of the person making the report, the name of his employer and the capacity in which he acts for that employer; and
(i) in the case of a report made by the commander of an aircraft or a person referred to in sub-paragraph (f), (g) or (h) of article 142(5), the address or telephone number at which communications should be made to him, if different from that of his place of employment.

12 Mandatory reporting of bird strikes – time and manner of reporting and information to be reported

(1) Without prejudice to the CAA's power in a particular case to approve the form of presentation and subject to paragraph (2), for the purposes of article 143(2) a report containing the information referred to in paragraph (3) shall be made to the CAA by post, telex, electronic, facsimile transmission or other similar means which produce a document containing the text of the communication, (written in English) within 96 hours of the bird strike occurrence coming to the knowledge of the person making the report.

(2) If at that time any of the said information is not in the possession of that person a report containing that information shall be made to the CAA by post, telex, electronic, facsimile transmission or other similar means which produce a document containing the text of the communication (written in English), within 96 hours of the information coming into his possession.

(3) For the purposes of article 143(2) of the Order, a report shall, as far as possible, contain the following information:
(a) the type, series and registration marks of the aircraft concerned;
(b) the name of the operator of the aircraft;
(c) the date and the co-ordinated universal time of the bird strike occurrence;
(d) the last point of departure and the next point of intended landing of the aircraft at that time;
(e) a description of the bird strike occurrence, including the part(s) of the aircraft affected, the effect on flight and any other relevant information;
(f) the bird species/description;
(g) the weather at the time of the occurrence;
(h) the runway in use (where relevant);
(i) the height and speed of the aircraft;
(j) the phase of flight;
(k) the position (if en route) of the aircraft at the time of the bird strike;
(l) any other reporting action taken;
(m) the signature and name in block capitals of the person making the report;
(n) the name of his employer and the capacity in which he acts for that employer; and
(o) the address or telephone number at which communications should be made to him.

PART 6 NAVIGATION PERFORMANCE AND EQUIPMENT

13 Minimum navigation performance and height specifications

(1) For the purposes of article 56(1), the following navigation performance capability is hereby prescribed, that is to say, a capability to ensure that:
(a) the standard deviation of lateral errors in the track of the aircraft is not more than 6.3 nautical miles; and
(b) the proportion of the flight time of the aircraft during which the actual track of the aircraft is 30 nautical miles or more off the track along which it has been given an air traffic control clearance to fly is less than 5.3×10^{-4}; and
(c) the proportion of the flight time of the aircraft during which the actual track of the aircraft is between 50 and 70 nautical miles off the track along which it has been given an air traffic control clearance to fly is less than 13×10^{-5}.

(2) For the purposes of articles 56(1) and 155(1), the following airspace is hereby prescribed as North Atlantic Minimum Navigation Performance Specified airspace, that is to say, the airspace from flight level 285 to flight level 420 within the area defined by rhumb lines joining successively the following points;

N3410.00 W01748.00	N6100.00 W00000.00	N5700.00 W05900.00	N2700.00 W06000.00
N3630.00 W01500.00	N8200.00 00000.00	N5300.00 W05400.00	N2700.00 W04000.00
N4200.00 W01500.00	N8200.00 E03000.00	N4900.00 W05100.00	N2218.00 W04000.00
N4300.00 W01300.00	North Pole	N4500.00 W05100.00	N1700.00 W03730.00
N4500.00 W01300.00	N8200.00 W06000.00	N4500.00 W05300.00	N2400.00 W02500.00
N4500.00 W00800.00	N7800.00 W07500.00	N4336.00 W06000.00	N3000.00 W02500.00
N5100.00 W00800.00	N7600.00 W07600.00	N4152.00 W06700.00	N3000.00 W02000.00
N5100.00 W01500.00	N6500.00 W05745.00	N3900.00 W06700.00	N3139.00 W01725.00
N5400.00 W01500.00	N6500.00 W06000.00	N3835.00 W06853.00	
N5434.00 W01000.00	N6400.00 W06300.00	N3830.00 W06915.00	
N6100.00 W01000.00	N6100.00 W06300.00	N3830.00 W06000.00	

And from there by that part of the arc of a circle radius 100 nautical miles centered on N3304.00 W01621.00 to N3410.00 W01748.00.

(3) For the purposes of article 57(1), the following height keeping performance capability is prescribed, that is to say, a capability to ensure that:

(a) altimetry system error shall be in compliance with paragraph 2.1.1(2) of Document 7030/4-NAT Part 1 Rules of the Air, Air Traffic Services and Search and Rescue (ICAO Regional Procedures Fourth Edition – 1987).

(b) in respect of aircraft first registered in a Contracting State on or after 1st January 1997 altitude can be automatically controlled with a tolerance band of +/- 65 feet; and

(c) in respect of aircraft registered in a Contracting State before 1st January 1997 altitude can be automatically controlled within a tolerance band of +/- 130 feet.

14 Airborne collision avoidance system

For the purposes of paragraphs 3 and 4 of Schedule 5 to the Order, the prescribed requirements for an airborne collision avoidance system shall be the requirements for ACAS II equipment set out in Volume IV (Third Edition – July 2004) Chapter 4 of Annex 10 to the Chicago Convention.

15 Mode S Transponder

(1) For the purposes of paragraph 3 of Schedule 5 to the Order, the capability and functionality prescribed for Mode S Elementary Surveillance shall be that specified for a level 2 transponder in Volumes III (First Edition-July 1995) and IV (Third Edition July 2002) of Annex 10 (Third Edition) to the Chicago Convention together with the additional functionality specified in paragraph (3).

(2) For the purposes of paragraph 3 of Schedule 5 to the Order, the capability and functionality prescribed for Mode S Enhanced Surveillance shall be that specified for a level 2 transponder in Volumes III (First Edition-July 1995) and IV (Third Edition July 2002) of Annex 10 (Third Edition) to the Chicago Convention together with the additional functionality specified 'in paragraph (3) and the additional down linked parameters specified in paragraph (4).

(3) This is the additional functionality referred to in paragraphs (1) and (2):

(a) capability to support the Interrogator Identifier (11) Code and Surveillance Identifier (S0 Code functionality.

(b) extended Squitter functionality which, for this purpose, means functionality that supports Mode S Elementary Surveillance and Mode S Enhanced Surveillance to provide Automic Dependant Surveillance-Broadcast, using unsolicited Transponder broadcasts; and

(c) ACAS Active Resolution Advisory.

(4) These are the additional down linked parameters referred to in paragraph (2):

(a) reporting of the Magnetic Heading Down linked Aircraft Parameter;

(b) reporting of the indicated Airspeed Down linked Aircraft Parameter;

(c) reporting of the Mach Number Down linked Aircraft Parameter;

(d) reporting of the Vertical Rate Down linked Aircraft Parameter;

(e) reporting of the Roll Angle Down linked Aircraft Parameter;

(f) reporting of the Track Angle Rate Down linked Aircraft Parameter;

(g) reporting of the True Track Angle Down linked Aircraft Parameter;

(h) reporting of the Ground Speed Down linked Aircraft Parameter;

(i) reporting of the Selected Vertical intent Down linked Aircraft Parameter (including Barometric Pressure Setting).

SCHEDULE 2
REGULATION 5
AEROPLANE PERFORMANCE

1 Weight and performance of public transport aeroplanes designated as aeroplanes of performance group A or performance group B

For the purposes of article 44(6), an aeroplane registered in the United Kingdom in respect of which there is in force under article 44(5) a permission in which the aeroplane is designated as being of performance group A or performance group B shall not fly for the purpose of public transport unless the weight of the aeroplane at the commencement of the take-off run is such that the following conditions are satisfied

(1) That weight does not exceed the maximum take-off weight specified for the altitude and the air temperature at the aerodrome at which the take-off is to be made.

(2) Subject to sub-paragraph (3) the take-off run, take-off distance and the accelerate-stop distance required for take-off, specified as being appropriate to

(a) the weight of the aeroplane at the commencement of the take-off run;

(b) the altitude at the aerodrome;

(c) the air temperature at the aerodrome;

(d) the condition of the surface of the runway from which the take-off will be made;

(e) the slope of the surface of the aerodrome in the direction of take-off over the take-off run available, the take-off distance available and the accelerate-stop distance available, respectively; and

(f) not more than 50 per cent, of the reported wind component opposite to the direction of take-off or not less than 150 per cent of the reported wind component in the direction of take-off;

Shall not exceed the take-off run, the take-off distance and the accelerate-stop distance available, respectively, at the aerodrome at which the take-off is to be made.

ANO

SCHEDULE 2 REGULATION 5 AEROPLANE PERFORMANCE
1 Weight and performance of public transport aeroplanes designated as aeroplanes of performance group A or performance group B

(3) In ascertaining the accelerate-stop distance available required pursuant to sub-paragraph (2), the point at which the pilot is assumed to decide to discontinue the take-off shall not be nearer to the start of the take-off run than the point at which, in ascertaining the take-off run required and the take-off distance required, he is assumed to decide to continue the take-off, in the event of power unit failure.

(4) The net take-off flight path with one power unit inoperative, specified as being appropriate to;
(a) the weight of the aeroplane at the commencement of the take-off run;
(b) the altitude at the aerodrome;
(c) the air temperature at the aerodrome; and
(d) not more than 50 per cent of the reported wind component opposite to the direction of take-off or not less than 150 per cent of the reported wind component in the direction of take-off; and
which is plotted from a point 35 feet or 50 feet, as appropriate, above the end of the take-off distance required at the aerodrome at which the take-off is to be made to a height of 1,500 feet above the aerodrome, shall show that the aeroplane will meet the requirement of sub-paragraph (5).

(5) Subject to sub-paragraphs (6) and (7) the requirement referred to in sub-paragraph (4) is that the aeroplane will clear any obstacle in its path by a vertical interval of at least 35 feet; and if it is intended that the aeroplane shall change its direction of flight by more than 15° before reaching 1,500 feet the vertical interval shall not be less than 50 feet during the change of direction.

(6) For the purpose of sub-paragraph (5) an obstacle shall be deemed to be in the path of the aeroplane if the distance from the obstacle to the nearest point on the ground below the intended line of flight of the aeroplane does not exceed
(a) a distance of 60 metres plus half the wing span of the aeroplane plus one eighth of the distance from such point to the end of the take-off distance available measured along the intended line of flight of the aeroplane; or
(b) 900 metres,
whichever is the less.

(7) In assessing the ability of the aeroplane to satisfy sub-paragraph (5), it shall not be assumed to make a change of direction of a radius less than the specified radius of steady turn.

(8) The aeroplane shall meet the requirements referred to in sub-paragraph (9) in the meteorological conditions expected for the flight, in the event of any one power unit becoming inoperative at any point on its route or on any planned diversion therefrom and with the other power unit or units operating within the maximum continuous power conditions specified.

(9) Subject to sub-paragraph (10) and (11) the requirements referred to in sub-paragraph (8) are that the aeroplane shall:
(a) be capable of continuing the flight, clearing by a vertical interval of at least 2,000 feet obstacles within 10 nautical miles either side of the intended track, to an aerodrome at which it can comply with sub-paragraphs (20) or (21), as appropriate, relating to an alternate aerodrome; and
(b) on arrival over such aerodrome the gradient of the specified net flight path with one power unit inoperative shall not be less than zero at 1,500 feet above the aerodrome.

(10) In assessing the ability of the aeroplane to satisfy sub-paragraph (9) it shall not be assumed to be capable of flying at an altitude exceeding the specified maximum permissible altitude for power unit restarting:

(11) If the operator of the aeroplane is satisfied, taking into account the navigation aids which can be made use of by the aeroplane on the route, that the commander of the aeroplane will be able to maintain his intended track on that route within a margin of 5 nautical miles, sub-paragraph (9) shall have effect as if 5 nautical miles were substituted for 10 nautical miles.

(12) If the aeroplane has three or more power units, it shall meet the requirements referred to in sub-paragraph 13;
(a) in the meteorological conditions expected for the flight; and
(b) in the event of any two power units becoming inoperative at any point:
(i) along the route; or
(ii) on any planned diversion therefrom more than 90 minutes flying time in still air at the all power units operating economical cruise speed from the nearest aerodrome at which it can comply with sub-paragraphs (20) or (21), as appropriate, relating to an alternate aerodrome,.

(13) Subject to sub-paragraphs (14) and (15), the requirements referred to in sub-paragraph (12) are that the aeroplane shall:
(a) be capable of continuing the flight with all other power units operating within the specified maximum continuous power conditions, clearing by a vertical interval of at least 2,000 feet obstacles within 10 nautical miles either side of the intended track to such an aerodrome; and
(b) on arrival over such an aerodrome the gradient of the specified net flight path with two power units inoperative shall not be less than zero at 1,500 feet above the aerodrome.

(14) In assessing the ability of the aeroplane to satisfy sub-paragraph (13) it shall not be assumed to be capable of flying at an altitude exceeding the specified maximum permissible altitude for power unit restarting:

(15) If the operator of the aeroplane is satisfied, taking into account the navigation aids which can be made use of by the aeroplane on the route, that the commander of the aeroplane will be able to maintain his intended track on that route within a margin of 5 nautical miles, sub-paragraph (13) shall have effect as if 5 nautical miles were substituted for 10 nautical miles.

(16) Unless it is flying under, and in accordance with the terms of any written permission granted by the CAA to the operator under this paragraph, an aeroplane having:
(a) two power units and a maximum total weight authorised which exceeds 5,700kg; and
(b) which is not limited by its certificate of airworthiness to the carriage of less than 20 passengers,
shall meet the requirements of sub-paragraph (17).

2 Weight and performance of public transport aeroplanes designated as aeroplanes of performance group C

(17) The requirements referred to in sub-paragraph (16) are that the aeroplane shall, in the meteorological conditions expected for the flight:

(a) be more than 60 minutes flying time at the normal one engine inoperative cruise speed in still air from the nearest aerodrome at which it can comply with sub-paragraphs (20) or (21), relating to an alternate aerodrome, at any point along:

(b) the route; or

(c) any planned diversion from it.

(c) in the case of an aeroplane having two power units and a maximum total weight authorised of 5,700kg or less or in the case of **(18)** An aeroplane having either:

(a) two power units and a maximum total weight authorised of more than 5,700kg or less; or

(b) two power units and a maximum total weight authorised of more than 5,700kg but which is limited by its certificate of airworthiness to the carriage of less than 20 passengers,

shall in the meteorological conditions expected for the flight, not be more than 90 minutes flying time in still air at the all power units operating economical cruise speed from the nearest aerodrome at which it can comply with sub-paragraphs (20) or (21), as appropriate, relating to an alternate aerodrome.

(19) The landing weight of the aeroplane will not exceed the maximum landing weight specified for the altitude and the expected air temperature for the estimated time of landing at the aerodrome at which it is intended to land and at any alternate aerodrome.

(20) Subject to sub-paragraphs (22) and (23), in the case of a turbine-jet powered aeroplane, the landing distance required shall not exceed at the aerodrome at which it is intended to land or at any alternate aerodrome, as the case may be, the landing distance available on

(a) the most suitable runway for a landing in still air conditions; and

(b) the runway that may be required for landing because of the forecast wind conditions.

(21) Subject to sub-paragraphs (22) and (23), in the case of an aeroplane powered by turbine propeller or piston engines, the landing distances required, respectively specified as being appropriate to the aerodrome at which it is intended to land and any such alternate aerodromes, at which it is intended to land or at any alternate aerodrome, as the case may be, the landing distance available on

(a) the most suitable runway for a landing in still air conditions; and

(b) the runway that may be required for landing because of the forecast wind conditions:

(22) If an alternate aerodrome is designated in the flight plan, the specified landing distance required may be that appropriate to an alternate aerodrome when assessing the ability of the aeroplane to satisfy sub-paragraphs (20) and (21) at the aerodrome of destination.

(23) For the purposes of sub-paragraphs (20), (21) and (22) the landing distance required shall be that specified as being appropriate to

(a) the landing weight;

(b) the altitude of the aerodrome;

(c) the temperature in the specified international standard atmosphere appropriate to the altitude at the aerodrome;

(d) a level surface in the case of runways usable in both directions; or the average slope of the runway in the case of runways usable in only one direction; and

(e) still air conditions in the case of the most suitable runway for a landing in still air conditions and not more than 50 per cent of the forecast wind component opposite to the direction of landing or not less than 150 per cent of the forecast wind component in the direction of landing in the case of the runway that may be required for landing because of the forecast wind conditions.

2 Weight and performance of public transport aeroplanes designated as aeroplanes of performance group C

For the purposes of article 44(6), an aeroplane registered in the United Kingdom in respect of which there is in force under article 44(5) a permission in which the aeroplane is designated as being of performance group C shall not fly for the purpose of public transport unless the weight of the aeroplane at the commencement of the take-off run is such that the following conditions are satisfied;

(1) That weight does not exceed the maximum take-off weight specified for the altitude and the air temperature at the aerodrome at which the take-off is to be made.

(2) The take-off run required and the take-off distance required, specified as being appropriate to

(a) the weight of the aeroplane at the commencement of the take-off run;

(b) the altitude at the aerodrome;

(c) the air temperature at the aerodrome;

(d) the average slope of the surface of the aerodrome in the direction of take-off over the accelerate-stop distance available; and

(e) not more than 50 per cent of the reported wind component opposite to the direction of take-off or not less than 150 per cent of the reported wind component in the direction of take-off,

shall not exceed the take-off run available and the accelerate-stop distance available, respectively, at the aerodrome at which the take-off is to be made.

(3) Subject to sub-paragraph (7), the net take-off flight path with all power units operating specified as being appropriate to

(a) the weight of the aeroplane at the commencement of the take-off run;

(b) the altitude of the aerodrome;

(ci) the air temperature at the aerodrome; and

ANO   SCHEDULE 2 REGULATION 5 AEROPLANE PERFORMANCE
2 Weight and performance of public transport aeroplanes designated as aeroplanes of performance group C

(d) not more than 50 per cent of the reported wind component opposite to the direction of the take-off or not less than 150 per cent of the reported wind component in the direction of take-off,

and plotted from a point 50 feet above the end of the take-off distance required at the aerodrome at which the take-off is to be made to a height of 1,500 feet above the aerodrome, shall show that the aeroplane will meet the requirement of sub-paragraph (4).

(4) Subject to sub-paragraphs (5) and (6) the requirement referred to in sub-paragraph (3) is that the aeroplane shall clear any obstacle in its path by a vertical interval of not less than 35 feet; but if it is intended that the aeroplane shall change its direction of flight by more than 15° before reaching 1,500 feet the vertical interval shall be not less than 50 feet during the change of direction.

(5) For the purpose of sub-paragraph (4) an obstacle shall be deemed to be in the path of the aeroplane if the distance from the obstacle to the nearest point on the ground below the intended line of flight of the aeroplane does not exceed 75 metres.

(6) In assessing the ability of the aeroplane to satisfy sub-paragraph (4), it shall not be assumed to make a change of direction of a radius less than the specified radius of steady turn.

(7) In the case of an aeroplane which is intended to be flown for any period before reaching a height of 1,500 feet above the aerodrome from which the take-off is to be made in conditions which will not ensure that any obstacles can be located by means of visual observation, the net take-off flight path with one power unit inoperative, which is:

(a) specified as being appropriate to the factors contained in paragraphs (a) to (d) of sub-paragraph (3); and

(b) plotted from the point of the net take-off flight path with all power units operating specified as being appropriate to those factors at which in the meteorological conditions expected for the flight, the loss of visual reference would occur, shall show that the aeroplane will meet the requirement of sub-paragraph (8).

(8) Subject to sub-paragraphs (9) and (10) the requirement referred to in sub-paragraph (7) is that the aeroplane will clear by a vertical interval of not less than 35 feet any obstacle in its path; but if it is intended that the aeroplane shall change its direction of flight by more than 15° before reaching 1,500 feet the vertical interval shall not be less than 50 feet during the change of direction.

(9) For the purpose of sub-paragraph (8) an obstacle shall be deemed to be in the path of the aeroplane if the distance from the obstacle to the nearest point on the ground below the intended line of flight of the aeroplane does not exceed

(a) 75 metres plus one-eighth of the distance from such point to the end of the emergency distance available measured along the intended line of flight of the aeroplane; or

(b) 900 metres, whichever is less.

(10) In assessing the ability of the aeroplane to satisfy sub-paragraph (8) it shall not be assumed the aeroplane will make a change of direction of a radius of less than the specified radius of steady turn.

(11) In the meteorological conditions expected for the flight the aeroplane shall:

(a) at any time after it reaches a height of 1,500 feet above the aerodrome from which the take-off is made will;

(b) in the event of any one power unit becoming inoperative at any point on its route or on any planned diversion from it;

(c) with the other power unit or power units operating within the specified maximum continuous power conditions, meet the requirements of sub-paragraph (12).

(12) Subject to sub-paragraph (13) the requirements referred to in sub-paragraph (11) are:

(a) the aeroplane shall be capable of continuing the flight at altitudes not less than the relevant minimum altitude for safe flight stated in, or calculated from the information contained in, the operations manual relating to the aeroplane to a point 1,500 feet above an aerodrome at which a safe landing can be made; and

(b) after arrival at that point be capable of maintaining that height:

(13) In assessing the ability the ability of the aeroplane to satisfy sub-paragraph (12) it shall not be assumed to be capable of flying at any point on its route at an altitude exceeding the performance ceiling, with all power units operating, specified as being appropriate to its estimated weight at that point.

(14) The landing weight of the aeroplane shall not exceed the maximum landing weight specified for the altitude and the expected air temperature for the estimated time of landing at the aerodrome at which it is intended to land and at any alternate aerodrome.

(15) Subject to sub-paragraph (17), the distance required by the aeroplane to land at the aerodrome at which it is intended and any alternative aerodrome from a height of 50 feet, otherwise than in accordance with specified data for short field landing shall not, exceed 70 per cent of the landing distance available on:

(a) the most suitable runway for a landing in still air conditions; or

(b) on the runway that may be required for landing because of the forecast wind conditions.

(16) For the purposes of sub-paragraph (15) the distance required to land from a height of 50 feet shall be taken to be that specified as being appropriate to

(a) the landing weight;

(b) the altitude at the aerodrome;

(c) the temperature in the specified international standard atmosphere appropriate to the altitude at the aerodrome;

(d) the level surface in the case of runways usable in both directions or the average slope of the runway in the case of runways usable in only one direction;

(e) still air conditions in the case of the most suitable runway for landing in still air conditions; or

(f) not more than 50 per cent of the forecast wind component opposite to the direction of landing or not less than 150 per cent of the forecast wind component in the direction of landing in the case of the runway that may be required for landing because of the forecast wind conditions;

(17) As an alternative to sub-paragraph (15) but subject to sub-paragraph (18), (19) and (20), the distance required by the aeroplane to land in accordance with specified data for short field landing, with all power units operating or with one power unit inoperative shall exceed the landing distance available:

(a) on the most suitable runway for a landing in still air conditions; and

(b) on the runway that may be required for landing because of the forecast wind conditions,

at both the aerodrome of intended destination and at any alternate aerodrome.

(18) For the purposes of sub-paragraph (17) the distance required to land from the appropriate height shall be taken to be that specified as being appropriate to the factors set forth in sub-paragraphs (a) to (f) of sub-paragraph (16) and, subject to sub-paragraph (19), the appropriate height shall be

(a) for a landing with all power units operating – any height between 30 and 50 feet in the United Kingdom, and 50 feet elsewhere; and

(b) for a landing with one power unit inoperative – 50 feet in the United Kingdom and elsewhere:

(19) If the specified distance required to land with one power unit inoperative from a height of 50 feet at the aerodrome of intended destination exceeds the landing distance available, it shall be sufficient compliance with paragraph (b) of sub-paragraph (18) if an alternate aerodrome, which has available the specified landing distance required to land with one power unit inoperative from such a height, is designated in the flight plan.

(20) The distance required by the aeroplane to land shall be determined in accordance with sub-paragraph (16) and not in accordance with sub-paragraph (17), if either:

(a) if it is intended to land at night; or

(b) it is intended to land when the cloud ceiling or ground visibility forecast for the estimated time of landing at the aerodrome of intended destination and at any alternate aerodrome at which it is intended to land in accordance with specified data for short field landing with all power units operating, are less than 500 feet and one nautical mile respectively.

3 Weight and performance of public transport aeroplanes designated as aeroplanes of performance group D

(1) For the purposes of article 44(6), an aeroplane registered in the United Kingdom in respect of which there is in force under article 44(5) a permission in which the aeroplane is designated as being of performance group D shall not fly for the purpose of public transport:

(a) at night; or

(b) when the cloud ceiling or visibility prevailing at the aerodrome of departure or forecast for the estimated time of landing at the aerodrome at which it is intended to land or at any alternate aerodrome are less than 1,000 feet and one nautical mile respectively.

(2) Such an aeroplane shall not fly for the purpose of public transport at any other time unless the weight of the aeroplane at the commencement of the take-off run is such that the following conditions are satisfied

(a) That weight does not exceed the maximum take-off weight specified for the altitude and air temperature at the aerodrome at which the take-off is to be made.

(b) The take-off run required and the take-off distance required specified as being appropriate to

(i) the weight of the aeroplane at the commencement of the take-off run;

(ii) the altitude of the aerodrome;

(iii) the air temperature at the aerodrome;

(iv) the average slope of the surface of the aerodrome in the direction of take-off over the emergency distance available; and

(v) not more than 50 per cent, of the reported wind component opposite to the direction of take-off or not less than 150 per cent, of the reported wind component in the direction of take-off,

shall not exceed the take-off run available and the accelerate-stop distance available, respectively, at the aerodrome at which the take-off is to be made.

(3) Subject to sub-paragraphs (4), (5) and (6) the net take-off flight path with all power units operating, specified as being appropriate to

(a) the weight of the aeroplane at the commencement of the take-off run;

(b) the altitude at the aerodrome;

(c) the air temperature at the aerodrome; and

(d) not more than 50 per cent. of the reported wind component opposite to the direction of take-off or not less than 150 per cent of the reported wind component in the direction of take-off,

and plotted from a point of 50 feet above the end of the take-off distance required at the aerodrome at which the take-off is to be made to the point at which the aeroplane reaches a height of 1,000 feet above the aerodrome shall show that the aeroplane will clear any obstacle in its path by a vertical interval of not less than 35 feet.

(5) For the purpose of sub-paragraph (3) an obstacle shall be deemed to be in the path of the aeroplane if the distance from the obstacle to the nearest point on the ground below the intended line of flight of the aeroplane does not exceed 75 metres.

(6) In assessing the ability of the aeroplane to satisfy sub-paragraph (3) it shall not be assumed to make a change of direction of a radius less than the specified radius of steady turn.

(7) Subject to sub-paragraph (8) the aeroplane shall:

(a) at any time after it reaches a height of 1,000 feet above the aerodrome from which the take-off is to be made;

(b) in the meteorological conditions expected for the flight;

(c) in the event of any one power unit becoming inoperative at any point on its route or on any planned diversion from it; and

ANO SCHEDULE 2 REGULATION 5 AEROPLANE PERFORMANCE
3 Weight and performance of public transport aeroplanes designated as aeroplanes of performance group D

(d) with the other power unit or power units, if any, operating within the maximum specified continuous power conditions, be capable of continuing the flight at altitudes not less than the relevant minimum altitudes for safe flight stated in, or calculated from the information contained in, the operations manual relating to the aeroplane to a point 1,000 feet above a place at which a safe landing can be made:

(8) In assessing the ability of the aeroplane to satisfy sub-paragraph (7) it shall not be assumed to be capable of flying at any point on its route at an altitude exceeding the performance ceiling with all power units operating specified as being appropriate to its estimated weight at that point.

(9) The landing weight of the aeroplane will not exceed the maximum landing weight specified for the altitude and the expected air temperature for the estimated time of landing at the aerodrome at which it is intended to land and at any alternate aerodrome.

(10) Subject to sub-paragraph (11) the distance required by the aeroplane to land from a height of 50 feet does not, at the aerodrome at which it is intended to land and at any alternate aerodrome, exceed 70 per cent of the landing distance available on the most suitable runway for a landing in still air conditions, and on the runway that may be required for landing because of the forecast wind conditions,

(11) For the purposes of sub-paragraph (10) the distance required to land from a height of 50 feet shall be taken to be that specified as being appropriate to

(a) the landing weight;
(b) the altitude at the aerodrome;
(c) the temperature in the specified international standard atmosphere appropriate to the altitude at the aerodrome;
(d) a level surface in the case of runways usable in both directions or the average slope of the runway in the case of runways usable in only one direction; and
(e) still air conditions in the case of the most suitable runway for a landing in still air conditions and not more than 50 per cent. of the forecast wind component opposite to the direction of landing or not less than 150 per cent of the forecast wind component in the direction of landing in the case of the runway that may be required for the landing because of the forecast wind conditions.

4 Weight and performance of public transport aeroplanes designated as aeroplanes of performance group E

(1) For the purposes of article 44(6) and subject to sub-paragraph (2), an aeroplane registered in the United Kingdom in respect of which there is in force under article 44(5) a permission in which the aeroplane is designated as being of performance group E shall not fly for the purpose of public transport unless the weight of the aeroplane at the commencement of the take-off run is such that the following conditions are satisfied;

(a) the weight for the altitude and the air temperature at the aerodrome at which the take-off is to be made does not exceed the maximum take-off weight specified as being appropriate to:

(i) the weight at which the aeroplane is capable, of climb of 700 feet per minute if it has retractable landing gear and of 500 feet per minute if it has fixed landing gear; in the en route configuration and with all power units operating within the specified maximum continuous power conditions, and

(ii) the weight at which the aeroplane is capable, with one power unit inoperative, of a rate of climb of 150 feet per minute in the enroute configuration and if necessary for it to be flown solely be reference to instruments for any period before reaching the minimum altitude for safe flight on the first stage of the route to be flown, as stated in, or calculated from, the information contained in the operations manual relating to the aeroplane..

(b) Subject to paragraph (c), with all power units operating within the maximum take-off power conditions specified, when multiplied by a factor of 1:33 the distance required by the aeroplane to attain a height of 50 feet shall not exceed the accelerate-stop distance available at the aerodrome at which the take-off is to be made.

(c) for the purposes of paragraph (b) the distance required by the aeroplane to attain a height of 50 feet shall be that appropriate to:

(i) the weight of the aeroplane at the commencement of the take-off run;
(ii) the altitude at the aerodrome,
(iii) the air temperature at the aerodrome; and
(iv) not more than 50 per cent, of the reported wind component opposite to the direction of take-off or not less than 150 per cent of the reported wind component in the direction of take-off.

(d) Subject to paragraphs (f) and (g) and in the circumstances and conditions referred to in paragraph (e) the aeroplane, shall be capable of continuing the flight at altitudes not less than the relevant minimum altitude for safe flight stated in, or calculated from, the information contained in, the operations manual relating to the aerodrome to a point 1000 feet above a place at which a safe landing can be made.

(e) these are the circumstances and conditions referred to in paragraph (d):

(i) at any time after the aeroplane reaches a height of 1000 feet above the aerodrome from which take-off is to be made;
(ii) in the meteorological conditions expected for the flight; and
(iii) in the event of any power unit becoming inoperative at any point on its route or on any planned diversion from such route, and with the other power unit or units if any, operating within the specified maximum continuous power conditions.

(f) Subject to paragraph (g) in assessing the ability of the aeroplane to satisfy paragraph (d) it shall not be assumed to be capable of flying at any point on its route or on any planned diversion from such route, at an altitude exceeding that at which it is capable of a rate of climb with all power units operating within the maximum continuous power conditions specified of 150 feet per minute.

(g) if it is necessary for the aircraft to be flown solely by reference to instruments, it shall be assumed to be capable, with one power unit inoperative, of a rate of climb of 100 feet per minute.

ANO   SCHEDULE 2 REGULATION 5 AEROPLANE PERFORMANCE
5 Weight and performance of public transport aeroplanes designated as aeroplanes of performance group F

(h) The landing weight of the aeroplane for the altitude and the expected air temperature for the estimated time of landing at the aerodrome at which it is intended to land and at any alternate aerodrome shall not exceed the maximum landing weight specified in paragraph (i).
(i) The maximum landing weights referred to in paragraph (h) are:
(i) those specified at which the aeroplane is capable, in the en route configuration and with all power units operating within the specified maximum continuous power conditions, of a rate of climb of 700 feet per minute if it has retractable landing gear and of 500 feet per minute if it has fixed landing gear; and
(ii) those specified at which the aeroplane is capable:
(aa) in the en route configuration and if it is necessary for it to be flown solely by reference to instruments for any period after leaving the minimum altitude for safe flight on the last stage of the route to be flown, stated in, or calculated from the information contained in, the operations manual relating to the aeroplane; and
(bb) with one power unit inoperative, of a rate of climb of 150 feet per minute.
(j) Subject to paragraph (k) the landing distance required shall not exceed 70 per cent, of the landing distance available on the most suitable runway for a landing in still air conditions at the aerodrome at which it is intended to land and at any alternate aerodrome.
(k) For the purposes of this paragraph (j) the distance required to land from a height of 50 feet shall be taken to be that specified as being appropriate to;
(i) the landing weight;
(ii) the altitude at the aerodrome; and
(iii) the temperature in the specified international standard atmosphere appropriate to the altitude at the aerodrome.
(2) An aeroplane designated by a permission granted under article 44(5) as an aeroplane of performance group E shall not fly for the purpose of public transport:
(a) at night; or
(b) when the cloud ceiling or visibility prevailing at the aerodrome of departure or forecast for the estimated time of landing at the aerodrome at which it is intended to land or at any alternate aerodrome are less than 1,000 feet and one nautical mile respectively: unless the aeroplane is capable, in the en route configuration and with one power unit inoperative, of a rate of climb of 150 feet per minute.

5 Weight and performance of public transport aeroplanes designated as aeroplanes of performance group F

For the purposes of article 44(6) and subject to sub-paragraph (2), an aeroplane registered in the United Kingdom in respect of which there is in force under article 44(5) in which the aeroplane as being of performance group F shall not fly for the purpose of public transport unless the weight of the aeroplane at the commencement of the take-off run is such that the following conditions are satisfied;
(1) That weight does not exceed the maximum take-off weight specified for the altitude and the air temperature at the aerodrome at which the take-off is to be made.
(2) The take-off distance required specified as being appropriate to;
(a) the weight of the aeroplane at the commencement of the take-off run;
(b) the altitude at the aerodrome;
(c) the air temperature at the aerodrome; and
(d) the average slope of the surface of the aerodrome in the direction of take-off over the take-off run available; and not more than 50 per cent of the reported wind component opposite to the direction of take-off or not less than 150 per cent of the reported wind component in the direction of take-off,
shall not exceed the take-off run available at the aerodrome at which the take-off is to be made.
(3) Subject to sub-paragraph (5), in the circumstances and conditions referred to in paragraph (4) the aeroplane shall be capable of continuing the flight at altitudes not less than the relevant minimum altitude for safe flight stated in, or calculated from the information contained in, the operations manual relating to the aeroplane to a point 1000 feet above:
(a) in the case of an aeroplane having one power unit, a place at which a safe landing can be made; and
(b) in the case of an aeroplane having two or more power units, an aerodrome at which it can comply with sub-paragraph (7).
(4) The circumstances and conditions referred to in sub-paragraph (3) are:
(a) at any time after the aeroplane reaches a height of 1000 feet above the aerodrome from which take-off is to be made;
(b) in the meteorological conditions expected for the flight; and
(c) in the event of any one power unit becoming inoperative at any point on its route or on any planned diversion from it, and with the other power unit or power units, if any, operating within the specified maximum continuous power conditions.
(5) In assessing the ability of the aeroplane to satisfy sub-paragraph (3):
(a) the aeroplane shall not be assumed to be capable of flying, at any point on its route or on any planned diversion from it, at an altitude exceeding that at which it is capable of a gradient of climb of 2 per cent, with all power units operating within maximum continuous power conditions specified, and
(b) the aeroplane shall be required to be capable of a gradient of climb of 1 per cent at the relevant minimum safe altitude with one power unit inoperative and with the other power unit or power units operating within the specified maximum continuous power conditions, over those parts of the route or any planned diversion, where in the meteorological conditions expected it is expected that the aeroplane will be out of sight of the surface due to cloud cover at or below the relevant minimum safe altitude.

SCHEDULE 2 REGULATION 5 AEROPLANE PERFORMANCE

5 Weight and performance of public transport aeroplanes designated as aeroplanes of performance group F

(6) The landing weight of the aeroplane will not exceed the maximum landing weight specified for the altitude and the expected air temperature for the estimated time of landing at the aerodrome at which it is intended to land and at any alternate aerodrome.

(7) Subject to sub-paragraph (8) the landing distance required shall not exceed the landing distance available on the most suitable runway for a landing in still air conditions at the aerodrome at which it is untended to land or at any alternative aerodrome, as the case may be.

(8) For the purposes of sub-paragraph (7) the landing distance required shall be that specified as being appropriate to;
(a) the landing weight;
(b) the altitude at the aerodrome;
(c) the temperature in the specified international standard atmosphere appropriate to the altitude at the aerodrome;
(d) a runway with a level surface; and
(e) still air conditions.

(9) An aeroplane with one power unit designated by a permission granted under article 44(5) as an aeroplane of performance group F shall not fly for the purpose of public transport:
(a) at night: or
(b) when the cloud ceiling or visibility prevailing at the aerodrome of departure or forecast for the estimated time of landing at the aerodrome at which it is intended to land or at any alternate aerodrome are less than 1000 feet and one nautical mile respectively..

6 Weight and performance of public transport aeroplanes designated as aeroplanes of performance group X

For the purposes of article 44(6), an aeroplane in respect of which there is in force under article 44(5) a permission designating the aeroplane as being of performance group X shall not fly for the purpose of public transport unless the weight of the aeroplane at the commencement of the take-off run is such that the following conditions are satisfied;

(1) The weight does not exceed the maximum take-off weight specified for the altitude at the aerodrome at which the take-off is to be made, or for the altitude and the air temperature at such aerodrome, as the case may be.

(2) The minimum effective take-off runway length required, specified as being appropriate to;
(a) the weight of the aeroplane at the commencement of the take-off run;
(b) the altitude at the aerodrome;
(c) the air temperature at the time of take-off;
(d) the condition of the surface of the runway from which the take-off will be made;
(e) the overall slope of the take-off run available; and
(f) not more than 50 per cent of the reported wind component opposite to the direction of take-off or not less than 150 per cent of the reported wind component in the direction of take-off,
shall not exceed the take-off run available at the aerodrome at which the take-off is to be made.

(3) Subject to sub-paragraphs (4) and (5) the take-off flight path with one power unit inoperative, specified as being appropriate to;
(a) the weight of the aeroplane at the commencement of the take-off run;
(b) the altitude at the aerodrome; and
(c) not more than 50 per cent of the reported wind component opposite to the direction of take-off or not less than 150 per cent of the reported wind component in the direction of take-off, and plotted from a point 50 feet above the end of the minimum effective take-off runway length required at the aerodrome at which the take-off is to be made,
shall show that the aeroplane will clear any obstacle in its path by a vertical interval of not less than the greater of 50 feet or 35 feet plus one-hundredth of the distance from the point on the ground below the intended line of flight of the aeroplane nearest to the obstacle to the end of the take-off distance available, measured along the intended line of flight of the aeroplane.

(4) For the purpose of sub-paragraph (3) an obstacle shall be deemed to be in the path of the aeroplane if the distance from the obstacle to the nearest point on the ground below the intended line of flight does not exceed;
(a) a distance of 60 metres plus half the wing span of the aeroplane plus one-eighth of the distance from such point to the end of the take-off distance available measured along the intended line of flight; or
(b) 900 metres,
whichever is the less.

(5) In assessing the ability of the aeroplane to satisfy sub-paragraph (3), in so far as it relates to flight path, it shall not be assumed to make a change of direction of a radius less than the radius of steady turn corresponding to an angle of bank of 15°.

(6) Subject to sub-paragraph (8) and in circumstances mentioned in sub-paragraph (7), the weight of the aeroplane, shall be such that the aeroplane, will be capable of a rate of climb of at least $K(Vso/100)^2$ feet per minute at an altitude not less than the minimum altitude for safe flight stated in or calculated from the information contained in the operations manual relating to the aeroplane, where Vso is in knots and K has the value of 797-1060/N, N being the number of power units installed.

(7) The circumstances mentioned in sub-paragraph (6) are:
(a) at any point on the route or any planned diversion from the route, having regard to the fuel and oil expected to be consumed up to that point; and
(b) with one power unit operative and the other power unit or units operating within the maximum continuous power conditions specified.

ANO SCHEDULE 2 REGULATION 5 AEROPLANE PERFORMANCE
7 Weight and performance of public transport aeroplanes designated as aeroplanes of performance group Z

(8) As a alternative to sub-paragraph (6), but subject to sub-paragraph (9), the aeroplane may be flown at an altitude from which, in the event of failure of one power unit, it is capable of reaching an aerodrome where a landing can be made in accordance with sub-paragraph (14), relating to an alternative aerodrome. In that case the weight of the aeroplane shall be such that, with the remaining power unit or units operating within the maximum continuous power conditions specified, it is capable of maintaining a minimum altitude on the route to such aerodrome of 2000 feet above all obstacles within 10 nautical miles on either side of the intended track.

(9) If the operator of the aeroplane is satisfied, taking into account the navigation aids which can be made use of by the aeroplane on the route, that the commander of the aeroplane will be able to maintain his intended track on that route within a margin of 5 nautical miles, sub-paragraph (8) shall have effect as if 5 nautical miles were substituted therein for 10 nautical miles and sub-paragraph (10) shall apply.

(10) In the circumstances referred to in sub-paragraph (9);

(a) the rate of climb, specified for the appropriate weight and altitude, used in calculating the flight path shall be reduced by an amount equal to K(Vso/100)2 feet per minute;

(b) the aeroplane shall comply with the climb requirements of subparagraph (6) at 1,000 feet above the chosen aerodrome;

(c) account shall be taken of the effect of wind and temperature on the flight path; and

(d) the weight of the aeroplane may be assumed to be progressively reduced by normal consumption of fuel and oil.

(11) Subject to sub-paragraph (12), if:

(a) any two power units of an aeroplane having four power units shall become inoperative at any point along the route or along any planned diversion from the route; and

(b) that point more than 90 minutes flying time (assuming all power units were to be operating) from the nearest aerodrome at which a landing can be made in compliance with sub-paragraph (14) relating to an alternate aerodrome, the aeroplane shall be capable of continuing the flight at an altitude of not less than 1000 feet above ground level to a point above that aerodrome.

(12) In assessing the ability of the aeroplane to satisfy sub-paragraph (11) it shall be assumed that the remaining, operative power units will operate within the specified maximum continuous power conditions, and account shall be taken of the temperature and wind conditions expected for the flight.

(13) The landing weight of the aeroplane will not exceed the maximum landing weight specified for the altitude at the aerodrome at which it is intended to land and at any alternate aerodrome.

(14) Subject to sub-paragraph (15) the required landing runway lengths respectively specified as being appropriate to the aerodrome of intended destination and the alternate aerodromes shall not exceed the landing distance available on;

(a) the most suitable runway for landing in still air conditions; and

(b) the runway that may be required for landing because of the forecast wind conditions,

at the aerodrome at which it is intended to land or at any alternate aerodrome, as the case may be.

(15) For the purpose of sub-paragraph (14) the required landing runway lengths shall be taken to be those specified as being appropriate to:

(a) the landing weight;

(b) the altitude at the aerodrome;

(c) still air conditions in the case of the most suitable runway for a landing in still air conditions; and

(d) not more than 50 per cent of the forecast wind component opposite to the direction of landing or not less than 150 per cent of the forecast wind component in the direction of landing in the case of the runway that may be required for landing because of the forecast wind conditions.

7 Weight and performance of public transport aeroplanes designated as aeroplanes of performance group Z

(1) For the purposes of article 44(6), an aeroplane registered in the United Kingdom, in respect of which there is in force under article 44(5) a permission designating the aeroplane as being of performance group Z, shall not fly for the purpose of public transport unless the weight of the aeroplane at the commencement of the take-off run is such that the following sub-paragraphs as apply to that aeroplane are satisfied.

(2) Sub-paragraphs (5) and (6) apply to all aeroplanes to which this paragraph applies.

(3) Sub-paragraphs (7) to (24) inclusive apply to all aeroplanes to which this paragraph applies:

(a) of which the specified maximum total weight authorised exceeds 5 700 kg; or

(b) of which the specified maximum total weight authorised does not exceed 5 700 kg, and which comply with neither paragraphs (a) nor (b) of sub-paragraph (5).

(4) Sub-paragraphs (25) to (36) apply to all aeroplanes to which this paragraph applies of which the specified maximum total weight authorised does not exceed 5 700 kg, and which comply with either or both of paragraphs (a) and (b) of sub-paragraph (5).

(5) For the purposes of sub-paragraphs (1) and (2) either:

(a) the wing loading of the aeroplane shall not exceed 20lb per square foot; or

(b) the stalling speed of the aeroplane in the landing configuration does not exceed 60 knots; or

(c) the aeroplane shall be capable of a gradient of climb of at least 1 in 200 at an altitude of 5 000 feet in the specified international standard atmosphere, with any one of its power units inoperative and the remaining power unit or units operating within the maximum continuous power conditions specified.

(6) The weight of the aeroplane at the commencement of the take-off run shall not exceed the maximum take-off weight, if any, specified for the altitude and the air temperature at the aerodrome at which the take-off is to be made.

(7) Subject to sub-paragraph (9) the distance required by the aeroplane to attain a height of 50 feet, with all power units operating within the maximum take-off power conditions specified shall not exceed the take-off run available at the aerodrome at which the take-off is to be made.

477

ANO SCHEDULE 2 REGULATION 5 AEROPLANE PERFORMANCE
7 Weight and performance of public transport aeroplanes designated as aeroplanes of performance group Z

(8) Subject to sub-paragraph (9) the distance required by the aeroplane to attain a height of 50 feet with all power units operating within the maximum take-off power conditions specified, when multiplied by a factor of either 1.33 for aeroplanes having two power units or by a factor of 1.18 for aeroplanes having four power units, shall not exceed the accelerate-stop distance available at the aerodrome at which the takeoff is to be made.

(9) For the purposes of sub-paragraphs (7) and (8) the distance required by the aeroplane to attain a height of 50 feet shall be that appropriate to:

(a) the weight of the aeroplane at the commencement of the take-off run;
(b) the altitude at the aerodrome;
(c) the air temperature at the aerodrome;
(d) the condition of the surface of the runway from which the take-off will be made;
(e) the slope of the surface of the aerodrome in the direction of take-off over the take-off run available and the accelerate-stop distance available, respectively; and
(f) not more than 50 per cent. of the reported wind component opposite to the direction of take-off or not less than 150 per cent. of the reported wind component in the direction of take-off.

(10) Subject to sub-paragraphs (12), (13) and (14), in the conditions mentioned in subparagraph (11), the take-off flight path of the aeroplane shall show that the aeroplane will clear any obstacle in its path by a vertical interval of at least 35 feet except that if it is intended that an aeroplane shall change its direction by more than 15 the vertical interval shall be not less than 50 feet during the change of direction.

(11) The conditions mentioned in sub-paragraph (10) are:

(a) that one power unit is inoperative and the remaining power unit or units are operating within the maximum take-off power conditions specified in subparagraph (12); and
(b) the take-off path is plotted from a point 50 feet above the end of the appropriate factored distance required for take-off under sub-paragraph (8) at the aerodrome at which the take-off is to be made.

(12) The maximum take-off power conditions specified in sub-paragraph (11) are those appropriate to:

(a) the weight of the aeroplane at the commencement of the take-off run;
(b) the altitude at the aerodrome;
(c) the air temperature at the aerodrome; and
(d) not more than 50 per cent. of the reported wind component opposite to the direction of take-off or not less than 150 per cent. of the reported wind component in the direction of take-off.

(13) For the purpose of sub-paragraph (10) an obstacle shall be deemed to be in the path of the aeroplane if the distance from the obstacle to the nearest point on the ground below the intended line of flight does not exceed:

(a) a distance of 60 metres plus half the wingspan of the aeroplane, plus one-eighth of the distance from such point to the end of the take-off distance available, measured along the intended line of flight; or
(b) 900 metres,

whichever is the less.

(14) In assessing the ability of the aeroplane to satisfy sub-paragraph (10), it shall not be assumed the aeroplane will make a change of direction of a radius less than a radius of steady turn corresponding to an angle of bank of 15'.

(15) Subject to sub-paragraph (17), in the circumstances and conditions referred to in subparagraph (16) the aeroplane shall be capable of continuing the flight, clearing obstacles within 10 nautical miles either side of the intended track, by a vertical interval of at least:

(a) 1 000 feet when the gradient of the flight path is not less than zero; or
(b) 2 000 feet when the gradient of the flight path is less than zero,

to an aerodrome at which it can comply with sub-paragraph (10), and on arrival over such aerodrome the flight path shall have a gradient of not less than zero at 1,500 feet above the aerodrome.

(16) The following are the circumstances and conditions referred to in sub-paragraph (15):

(a) the meteorological conditions expected for the flight;
(b) in the event of any one power unit becoming inoperative at any point on its route or on any planned diversion from it; and
(c) with the other power unit or units, if any, operating within the maximum continuous power conditions specified.

(17) For the purpose of sub-paragraph (15) the gradient of climb of the aeroplane shall be taken to be one per cent. less than that specified.

(18) In the meteorological conditions expected for the flight and at any point on its route or on any planned diversion from it the aeroplane shall be capable of climbing at a gradient of at least 1 in 50, with all power units operating within the maximum continuous power conditions specified at the following altitudes:

(a) the minimum altitudes for safe flight on each stage of the route to be flown or of any planned diversion from it specified in, or calculated from the information contained in, the operations manual relating to the aeroplane; and
(b) the minimum altitudes necessary for compliance with sub-paragraphs (15) and (19), as appropriate.

(19) If, on the route to be flown or on any planned diversion, the aeroplane will:

(a) be engaged in a flight over water; and
(b) at any point during such flight it maybe more than 90 minutes' flying time in still air from the nearest shore;

it shall be capable of complying with the requirements if sub-paragraph (20) in the event of two power units becoming inoperative during such time and with the other power units, if any, operating within the maximum' continuous power conditions specified.

(20) The requirements referred to in sub-paragraph (19) are:

(a) that the aeroplane is capable of continuing the flight, having regard to the meteorological conditions expected for the flight;

ANO SCHEDULE 2 REGULATION 5 AEROPLANE PERFORMANCE
7 Weight and performance of public transport aeroplanes designated as aeroplanes of performance group Z

(b) clearing all obstacles within 10 nautical miles either side of the intended track by a vertical interval of at least 1,000 feet; and

(c) to an aerodrome at which a safe landing can be made.

(21) The landing weight of the aeroplane shall not exceed the maximum landing weight, if any, specified for the altitude and the expected air temperature for the estimated time of landing at the aerodrome at which it is intended to land and at any alternate aerodrome.

(22) Subject to sub-paragraph (23) the distance required by the aeroplane to land at the aerodrome at which it is intended to land from a height of 50 feet shall not exceed 60 per cent of the landing distance available on:

(a) the most suitable runway for a landing in still air conditions; and

(b) the runway that may be required for landing because of the forecast wind conditions; provided that if an alternate aerodrome is designated in the flight plan, the landing distance required at the aerodrome at which it is intended to land shall not exceed 70 per cent. of that available on the runway.

(23) For the purpose of sub-paragraph (22) the distance required to land from a height of 50 feet shall be taken to be that appropriate to:

(a) the landing weight;

(b) the altitude at the aerodrome;

(c) the temperature in the specified international standard atmosphere appropriate to the altitude at the aerodrome;

(d) a level surface in the case of runways usable in both directions;

(e) the average slope of the runway in the case of runways usable in only one direction;

(f) still air conditions in the case of the most suitable runway for a landing in still air conditions; and

(g) not more than 50 per cent. of the forecast wind component opposite to the direction of landing or not less than 150 per cent. of the forecast wind component in the direction of landing in the case of the runway that may be required for landing because of the forecast wind conditions.

(24) Subject to paragraph (25), the distance required by the aeroplane to land from a height of 50 feet, at any alternate aerodrome shall not exceed 70 per cent of the landing distance available on:

(a) the most suitable runway for a landing in still air conditions; and

(b) the runway that may be required for landing because of the forecast wind conditions.

(25) For the purpose of sub-paragraph (24) the distance required to land from a height of 50 feet shall be determined in the manner provided in sub-paragraph (23).

(26) If the aeroplane is engaged:

(a) on a flight at night; or

(b) when the cloud ceiling or visibility prevailing at the aerodrome of departure or forecast for the estimated time of landing at the aerodrome of destination at which it is intended to land or at any alternate aerodrome are less than 1 000 feet and one nautical mile respectively,

it shall be capable of climbing at a gradient of at least 1 In 200 at an altitude of 2 500 feet in the specified international standard atmosphere, with any one of its power units inoperative and the remaining power unit or units, if any, operating within the maximum continuous power conditions specified.

(27) Subject to sub-paragraph (29), the distance required by the aeroplane to attain a height of 50 feet with ail power units operating within the maximum take-off power conditions specified, shall not exceed the take-off run available at the aerodrome at which the take-off is to be made.

(28) Subject to sub-paragraph (29), the distance required by the aeroplane to attain a height of 50 feet, with all power units operating within the maximum take-off power conditions specified, when multiplied by a factor of 1.33 shall not exceed the accelerate-stop distance available at the aerodrome at which the take-off is to be made.

(29) For the purposes of sub-paragraphs (27) and (28) the distance required by the aeroplane to attain a height of 50 feet shall be that appropriate to:

(a) the weight of the aeroplane at the commencement of the take-off run;

(b) the altitude at the aerodrome;

(c) the temperature in the specified international standard atmosphere appropriate to the altitude at the aerodrome or, if greater, the air temperature at the aerodrome less 15' centigrade;

(d) the slope of the surface of the aerodrome in the direction of take-off over the take-off run available and the accelerate-stop distance available respectively; and

(e) not more than 50 per cent. of the reported wind component opposite to the direction of take-off or not less than 150 per cent. of the reported wind component in the direction of take-off.

(30) The take-off flight path of the aeroplane, with all power units operating within the maximum take-off power conditions specified, appropriate to:

(a) the weight of the aeroplane at the commencement of the take-off run;

(b) the altitude at the aerodrome;

(c) the temperature in the specified international standard atmosphere appropriate to the altitude at the aerodrome, or, if greater, the air temperature at the aerodrome less 15' centigrade;

(d) not more than 50 per cent. of the reported wind component opposite to the direction of take-off or not less than 150 per cent. of the reported wind component in the direction of take-off; and

(e) plotted from a point 50 feet above the end of the factored distance required for take-off under sub-paragraph (28), at the aerodrome at which the take-off is to be made,

shall show that the aeroplane will meet the requirements of sub-paragraph (31).

ANO — SCHEDULE 2 REGULATION 5 AEROPLANE PERFORMANCE
7 Weight and performance of public transport aeroplanes designated as aeroplanes of performance group Z

(31) The requirements referred to in sub-paragraph (30) are that the aeroplane shall clear any obstacle lying within 60 metres plus half the wing span of the aeroplane on either side of its path by a vertical interval of at least 35 feet. In assessing the ability of the aeroplane to satisfy this sub-paragraph it shall not be assumed to make a change of direction of a radius less than a radius of steady turn corresponding to an angle of bank of 1 W.

(32) The aeroplane shall be capable of continuing the flight so as to reach a point above a place at which a safe landing can be made at a suitable height for such landing:
(a) in the meteorological conditions expected for the flight;
(b) in the event of any power unit becoming inoperative at any point on its route or on any planned diversion from it; and
(c) with the other power unit or units, if any, operating within the maximum continuous power conditions specified.

(33) The aeroplane shall be capable of climbing at a gradient of at least 1 in 50, with all power units operating within the maximum continuous power conditions specified at the altitudes referred to in paragraph (34) in the meterological conditions expected for the flight, and at any point on its route or any planned diversion.

(34) The altitudes referred to in paragraph (33) are:
(a) the minimum altitudes for safe flight on each stage of the route to be flown or on any planned diversion from it specified in, or calculated from, the information contained in the operations manual relating to the aeroplane; and
(b) the minimum altitudes necessary for compliance with paragraph (32).

(35) If on the route to be flown or any planned diversion from it the aeroplane will:
(a) be engaged on a flight over water;
during which at any point it may be more than 30 minutes' flying time in still air from the nearest shore; and
(c) in the event of one power unit becoming inoperative during such time and with the other power unit or units, if any, operating within the maximum continuous power conditions specified,
the aeroplane shall be capable of climbing at a gradient of at least 1 in 200 at an altitude of 5 000 feet in the specified international standard atmosphere.

(36) The landing weight of the aeroplane shall not exceed the maximum landing weight, if any, specified for the altitude and the expected air temperature for the estimated time of landing at the aerodrome at which it is intended to land and at any alternate aerodrome.

(37) Subject to sub-paragraph (38) the distance required by the aeroplane to land at the aerodrome at which it is intended to land and at any alternate aerodrome from a height of 50 feet shall not exceed 70 per cent, or, if a visual approach and landing will be possible in the meteorological conditions forecast for the estimated time of landing, 80 per cent of the landing distance available on:
(a) the most suitable runway for a landing in still air conditions; and
(b) the runway that may be required for landing because of the forecast wind conditions.

(38) For the purposes of sub-paragraph (37) the distance required to land from a height of 50 feet shall be taken to be that appropriate to: the landing weight;
(b) the altitude at the aerodrome;
(c) the temperature in the specified international standard atmosphere appropriate to the altitude at the aerodrome;
(d) a level surface in the case of runways usable in both directions or the average slope of the runway in the case of runways usable in only one direction; and either
still air conditions in the case of the most suitable runway for a landing in still air conditions; or
not more than 50 per cent of the forecast wind component opposite to the direction of landing or not less than 150 per cent. of the forecast wind component in the direction of landing in the case of the runway that may be required for landing because of the forecast wind conditions.

SCHEDULE 3
REGULATION 6
HELICOPTER PERFORMANCE

1 Weight and performance of public transport helicopters carrying out Performance Class 1 operations

For the purposes of article 45(1), a helicopter registered in the United Kingdom when carrying out Performance Class 1 operations shall not fly for the purpose of public transport unless the weight of the helicopter at the commencement of take-off is such that the following requirements are satisfied;
(a) the weight does not exceed the maximum take-off weight specified for the altitude and the air temperature at the site from which the take-off is to be made; and
(b) the landing weight of the helicopter will not exceed the maximum landing weight specified for the altitude and the expected air temperature for the estimated time of landing at the site at which it is intended to land and at any alternate site.

2 Weight and performance of public transport helicopters carrying out Performance Class 2 operations

(1) For the purposes of article 45(1) but subject to sub-paragraph (3), a helicopter registered in the United Kingdom when carrying out Performance Class 2 operations shall not fly for the purpose of public transport when the cloud ceiling or visibility prevailing at the departure site and forecast for the estimated time of landing at the site at which it is intended to land and at any alternate site are less than 500 feet and 1000 metres respectively and shall not fly for the purpose of public transport at any other time unless all of the requirements specified in sub-paragraph (2) are satisfied.

SCHEDULE 3 REGULATION 6 HELICOPTER PERFORMANCE
3 Weight and performance of public transport helicopters carrying out Performance Class 3 operations

(2) The requirements referred to in sub-paragraph (10 are that the weight of the helicopters at the commencement of take-off shall be such that:

(a) the weight does not exceed the maximum take-off weight specified for the altitude and the air temperature at the site from which the take-off is to be made; and

(b) the landing weight of the helicopter will not exceed the maximum landing weight specified for the altitude and the expected air temperature for the estimated time of landing at the site at which it is intended to land and at any alternate site.

(3) This paragraph shall not apply to a helicopter flying under and in accordance with the terms of a police air operator's certificate.

3 Weight and performance of public transport helicopters carrying out Performance Class 3 operations

(1) For the purposes of article 45(1) but subject to sub-paragraph (3), a helicopter registered in the United Kingdom when carrying out Performance Class 3 operations shall not fly for the purpose of public transport at night or out of sight of the surface or when the cloud ceiling or visibility prevailing at the departure site and forecast for the estimated time of landing at the site at which it is intended to land are less than 600 feet and 1000 metres respectively and shall not fly for the purpose of public transport at any other time unless all of the requirements specified in sub-paragraph (2) are complied with.

(2) The requirements referred to in sub-paragraph (1) are that the weight of the helicopter at the commencement to take-off shall be such that:

(a) the weight does not exceed the maximum take-off weight specified for the altitude and the air temperature at the site from which the take-off is to be made.

(b) the landing weight of the helicopter will not exceed the maximum landing weight specified for the altitude and the expected air temperature for the estimated time of landing at the site at which it is intended to land and at any alternate site.

(3) This paragraph shall not apply to a helicopter flying under and in accordance with the terms of a police air operator's certificate.

THE AIR NAVIGATION (RESTRICTION OF FLYING) (SCOTTISH HIGHLANDS) REGULATIONS 1981

Whereas the Secretary of State, for a reason affecting the public interest, deems it necessary in the public interest to restrict flying in the Scottish Highlands and certain approaches thereto by reason of intended military training flights:

Now, therefore, the Secretary of State, in exercise of the powers conferred on him by Article 66 of the Air Navigation Order 1980 and of all other powers enabling him in that behalf, hereby makes the following Regulations

1 These Regulations may be cited as the Air Navigation (Restriction of Flying) (Scottish Highlands) Regulations 1981 and shall come into operation on 1st October 1981.

2 (1) In Regulation 3 of these Regulations all times referred to are Local Time.

(2) The notified authority referred to in these Regulations is the London Air Traffic Control Centre (Military) Tactical Booking Cell.

3 (1) Between 1500 and 2300 hours each day on or between Monday and Thursday, an aircraft shall not fly at or below a height of 5000 feet above mean sea level within the area bounded by straight lines joining successively the following points:

N5830.00 W00327.00
N5825.24 W00325.09
N5803.45 W00412.34
N5803.00 W00430.00
N5800.00 W00437.00
N5747.00 W00425.00
N5739.00 W00430.00
N5738.00 W00445.00
N5730.00 W00438.00
N5718.00 W00452.00
N5711.00 W00453.00
N5709.00 W00500.00
N5700.00 W00502.00
N5654.00 W00505.00
N5656.00 W00547.00
N5713.00 W00535.00
N5750.00 W00543.00
N5800.00 W00515.00
N5830.00 W00449.00
N5830.00 W00430.00
N5825.00 W00430.00
N5830.00 W00420.00
N5830.00 W00327.00

Unless the aircraft is flying in accordance with an authorisation given by the notified authority.

(2) Between 1500 and 2300 hours each day on or between Monday and Thursday, an aircraft shall not fly at or between 750 feet and 5,000 feet above mean sea level within the area bounded by straight lines joining successively the following points:

N5750.00 W00543.00
N5740.05 W00540.50
N5738.41 W00557.35
N5700.00 W00556.40
N5700.00 W00615.00
N5747.16 W00616.33
N5750.00 W00543.00

Unless the aircraft is flying in accordance with an authorisation given by the notified authority.

(3) Between 1500 and 2300 hours each day on or between Monday and Thursday, an aircraft shall not fly below 2,000 feet above mean sea level within

(a) the area bounded by straight lines joining successively the following points:
N5822.18 W00332.00
N5814.35 W00319.26
N5811.22 W00326.50
N5819.00 W00339.18
N5822.18 W00332.00W

or

(b) the area bounded by straight lines joining successively the following points:
N5749.00 W00406.06
N5745.00 W00402.54
N5742.35 W00410.51
N5739.00 W00430.00
N5747.00 W00425.00
N5749.00 W00406.06

Unless the aircraft is flying in accordance with either:
(i) an authorisation given by the notified authority; or
(ii) an authorisation to cross either area referred to in sub-paragraph (a) or (b) of this paragraph given by the person in charge of Tain Range Danger Area on the notified frequency.

4 Nothing in these Regulations shall apply to any aircraft flying in accordance with an authorisation given by the notified authority to the commander of the aircraft before the flight commences when the purpose of the flight is to save life or property.

THE AIR NAVIGATION (RESTRICTION OF FLYING) (NUCLEAR INSTALLATIONS) REGULATIONS 2002

Whereas the Secretary of State for Transport, deems it necessary in the public interest to restrict flying in the vicinity of certain nuclear installations specified in the Second Schedule hereto:

Now therefore the Secretary of State for Transport in the exercise of the powers conferred by article 85 of the Air Navigation Order 2000, hereby makes the following Regulations:

1 These Regulations may be cited as the Air Navigation (Restriction of Flying) (Nuclear Installations) Regulations 2002 and shall come into force on 5th September 2002.

2 The Regulations specified in the First Schedule hereto are hereby revoked.

3 (1) This regulation applies to each of the nuclear installations specified in column 1 of the Second Schedule hereto each of which for the purpose of this regulation shall be taken to comprise an area bounded by a circle of the radius specified in Column 2 of the said Schedule opposite its name centred on the position so specified in Column 3 of the said Schedule.

(2) Subject to paragraph (3), an aircraft shall not fly over a nuclear installation to which this regulation applies below such height above mean sea level as is specified in Column 4 of the said Schedule opposite its name.

(3) Nothing in this regulation shall prohibit:

(a) in relation to the nuclear installations at Aldermaston, Barrow in Furness, Berkley, Bradwell, Burghfield, Capenhurst, Chapelcross, Dungeness, Hartlepool, Harwell, Heysham, Hunterston, Hinkley Point, Oldbury, Sellafield , Sizewell, Springfields, Torness, Trawsfynydd and Wylfa, flight for the purpose of landing at or taking off from the helicopter landing area at the installation with the permission of the person in charge of the installation and in accordance with any conditions to which that permission is subject;

(b) in relation to the nuclear installation at Devonport:

(i) flight by a helicopter for the purpose of landing at or taking off from HMS Drake Helicopter Landing Site with the permission of the Commodore of HMS Drake and in accordance with any conditions to which that permission is subject; or

(ii) flight by a helicopter for the purpose of landing at or taking off from any ship within the Devonport Dockyard with the permission of the Captain of the Port and Queen's Harbour Master for Devonport and in accordance with any conditions to which that permission is subject;

(c) in relation to the nuclear installation at Dungeness, flight by an aircraft which has taken off from or intends to land at London Ashford (Lydd) Airport flying in accordance with normal aviation practice which remains at least 1.5 nautical miles from the position specified in column 3 of the Schedule;

(d) in relation to the nuclear installation at Hinkley Point, flight by a helicopter flying within the Bridgewater Bay Danger Area with the permission of the person in charge of that Area and in accordance with any conditions to which that permission is subject which remains at least 1 nautical mile from the position specified in Column 3 of the Schedule.

(e) in relation to the nuclear installation at Rosyth, flight within the route notified as the Kelty Lane for the purpose of making an approach to land at, or a departure from Edinburgh Airport;

(f) in relation to the nuclear installation at Springfields;

(i) flight at a height of not less than 1,670 feet above mean sea level for the purpose of landing at Blackpool Airport; or

(ii) flight in airspace lying south of a straight line drawn from 534644N 0024454W to 534513N 0025044W for the purpose of landing at or taking off from Warton Aerodrome; and

(g) in relation to the nuclear installation at Wylfa, flight at a height of not less than 2000 feet above ground level whilst operating under and in accordance with a clearance from the air traffic control unit at RAF Valley.

(h) in relation to the nuclear installation at Bradwell, flight at a height of not less than 1500 feet above mean sea level whilst conducting an instrument approach procedure at London Southend Airport; and

(i) in relation to the nuclear installation at Hartlepool, flight at a height of not less than 1800 feet above mean sea level whilst conducting an instrument approach procedure at Teeside International Airport.

SCHEDULE 1: REVOCATIONS

Regulation 2

Regulations Revoked	References
The Air Navigation (Restriction of Flying) (Nuclear Installations) Regulations 2001	S.I. 2001/1607
The Air Navigation (Restriction of Flying) (Nuclear Installations) (No. 2) Regulations 2001	S.I. 2001/3600
The Air Navigation (Restriction of Flying) (Nuclear Installations) (No. 2) (Amendment) Regulations 2001	S.I. 2001/3768
The Air Navigation (Restrictions of Flying) (Nuclear Installations) (No. 2) (second Amendment) Regulations 2002	S.I. 2002/1382

SCHEDULE 2

Regulation 3

Column 1 Name of Nuclear Installation	Column 2 Radius in Nautical Miles	Column 3 Position	Column 4 Height in feet
Aldermaston	1.5	512203N 0010847W	2400
Barrow in Furness	0.5	540635N 0031410W	2000
Berkeley	2	514134N 0022936W	2000
Bradwell	2	514432N 0005352E	2000
Burghfield	1	512424N 0010125W	2400
Capenhurst	2	531550N 0025708W	2200
Chapelcross	2	550059N 0031334W	2400
Coulport/Faslane	2	560331N 0045159W	2200
Devonport	1	502326N 0041119W	2000
Dounreay	2	583435N 0034434W	2100
Dungeness	2	505449N 0005717E	2000
Hartlepool	2	543807N 0011049W	2000
Harwell	2	513430N 0011905W	2500
Heysham	2	540147N 0025452W	2000
Hinkley Point	2	511233N 0030749W	2000
Hunterston	2	554317N 0045338W	2000
Oldbury	2	513852N 0023415W	2000
Rosyth	0.5	560121N 0032709W	2000
Sellafield	2	542505N 0032944W	2200
Sizewell	2	521250N 0013707E	2000
Springfields	2	534634N 0024815W	2100
Torness	2	555806N 0022431W	2100
Trawsfynydd	2	525529N 0035655W	2700
Winfrith	1	504052N 0021535W	1000
Wylfa	2	532458N 0042852W	2100

THE AIR NAVIGATION (RESTRICTION OF FLYING) (PRISONS) REGULATIONS 2001

Whereas the Secretary of State, for a reason affecting the public interest, deems it necessary in the public interest to restrict flying in the vicinity of certain prisons specified in the Second Schedule hereto:

Now therefore the Secretary of State, in exercise of the powers conferred on him by article 85 of the Air Navigation Order 2000 and of all other powers enabling him in that behalf, hereby makes the following Regulations:

1 These Regulations may be cited as the Air Navigation (Restriction of Flying) (Prisons) Regulations 2001 and shall come into force on 11th May 2001.

2 The Regulations specified in the First Schedule hereto are hereby revoked.

3 (1) This regulation applies to each of the prisons specified in column 1 of the Second Schedule hereto, each of which for the purpose of this regulation shall be taken to comprise either an area bounded by a circle of the radius specified in column 2 of the said Schedule opposite its name centred on the position so specified in column 3 of the said Schedule or in the case of Belmarsh and Frankland the area described in column 2 of the said Schedule.

(2) Subject to paragraph (3), no helicopter shall fly over a prison to which this regulation applies below such height above mean sea level as is specified in column 4 of the said Schedule opposite its name.

(3) Nothing in this regulation shall prohibit:

(a) (deleted);

(b) flight by any helicopter operated by or on behalf of a police force for any area of the United Kingdom; or

(c) in relation to the high security prison at Belmarsh; flight by any helicopter flying in accordance with the Instrument Flight Rules and in accordance with instructions given by the air traffic control unit at London City airport.

ANO Restrictions – THE AIR NAVIGATION (RESTRICTION OF FLYING) (PRISONS) REGULATIONS 2001

THE FIRST SCHEDULE: REVOCATIONS

Regulation 2

Regulations Revoked	References
The Air Navigation (Restriction of Flying) (High Security Prisons) Regulations 1989	S.I. 1989/2118
The Air Navigation (Restriction of Flying) (High Security Prisons) (Amendment) Regulations 1991	S.I. 1991/1679
The Air Navigation (Restriction of Flying) (High Security Prisons) Amendment No. 2) Regulations 1992	S.I. 1992/1876
The Air Navigation (Restriction of Flying) (High Security Prisons) (Amendment No. 3) Regulations 1993	S.I. 1993/2123

THE SECOND SCHEDULE

Regulation 3

Column 1 Name of High Security Prison	Column 2 Radius in Nautical Miles or Description of Area	Column 3 Position	Column 4 Height in feet
Belmarsh	513020N 0000529E, thence by a straight line to 512943N 0000454E; thence in anti clockwise direction by an arc of a circle having a radius of 0.5 nautical miles and centred on 512951N 0000541E to 513020N 0000529E	–	2,000
Full Sutton	2	535837N 0005224W	2,000
Frankland	544859N 0013617W; thence clockwise by an arc of a circle having a radius of 2 nautical miles whose centre is at 544820N 0013301W to 544741N 0012945W; thence by a straight line to 544544N 0013054W; thence clockwise by an arc of a circle having a radius of 2 nautical miles whose centre is at 544623N 0013410W; to 544702N 0013726W; thence by a straight line to 54459N 0013617W.	–	2,200
Long Lartin	2	520627N 0015119W	2,200
Manchester	1	532934N 0021450W	1,700
Shotts	2	554950N 0034935W	2,800
Wakefield	1.3	534057N 0013034W	1,600
Whitemoor	2	523430N 0000446E	2,000
Woodhill	1.5	520049N 0004813W	2,400

The Air Navigation (Restriction of Flying) (Highgrove House) Regulations 1991

Whereas the Secretary of State for Transport, for a reason affecting the public interest deems it necessary in the public interest to restrict flying in the vicinity of Highgrove House:
Now therefore the Secretary of State for Transport, in exercise of the powers conferred on him by article 74 of the Air Navigation Order 1989 and of all other powers enabling him in that behalf, hereby makes the following regulations:
1 These Regulations may be cited as the Air Navigation (Restriction of Flying) (Highgrove House) Regulations 1991 and shall come into force on 7th March 1991.
2 Helicopters and microlight aeroplanes shall not fly below 2000 feet above mean sea level within the area bounded by a circle having a radius of 1.5 nautical miles centred on N5137.30 W00210.75 (Highgrove House):
Provided that nothing in this regulation shall apply to a flight by any helicopter:
(a) in the service of the Gloucestershire Constabulary;
(b) belonging to or exclusively employed in the service of Her Majesty; or
(c) visiting Highgrove House at the invitation of the person in charge of the household at Highgrove House, where the Gloucestershire Constabulary have been informed in advance of such visit.

The Air Navigation (Restriction of Flying) (Specified Area) Regulations 2005

Whereas the Secretary of State for Transport deems it necessary in the public interest to restrict flying in the area specified in the Schedule hereto:
Now, therefore, the Secretary of State for Transport, in exercise of the powers conferred upon him by article 85 of the Air Navigation Order(a) hereby makes the following Regulations:

Citation and commencement
1 These Regulations may be cited as the Air Navigation (Restriction of Flying) (Specified Area) Regulations 2005 and shall come into force on 1 April 2005.

Flight within specified area
2 Except with the permission in writing of the Civil Aviation Authority and in accordance with any conditions therein specified a helicopter shall not fly over the area specified in the Schedule below such height as would enable it to alight clear of the area in the event of failure of a power unit.

SCHEDULE
Regulation 2

The Specified Area
The area bounded by straight lines joining successively the following points: Kew Bridge (N5129.18 W0001 7.17); The Eastern extremity of Brent Reservoir (N5134.30 W00014.02); Gospel Oak Station (N5133.27 W00008.97); The South East corner of Springfield Park (N5134.12 W00003.20); Bromley-by-Bow Station (N5131.47 WOOOOO.65); The South West corner of Hither Green (N5126.72 WOOOOO.63); Herne Hill Station (N5127.18 W00006.07); Wimbledon Station (N5125.23 W00012.27); The North West corner of Casteinau Reservoir (N5128.87 W00014.03); Kew Bridge (N5129.18 W0001 7.17);
excluding so much of the bed of the River Thames as lies within that area between the ordinary high water marks on each of its banks.

The Air Navigation (Restriction of Flying) (Hyde Park) Regulations 2004

Whereas the Secretary of State for Transport deems it necessary in the public interest to restrict flying in airspace in the vicinity of Hyde Park, London;

Now therefore the Secretary of State for Transport, in exercise of the powers conferred by article 85 of the Air Navigation Order 2000, hereby makes the following Regulations:

1 These Regulations may be cited as the Air Navigation (Restriction of Flying) (Hyde Park) Regulations 2004 and shall come into force on 28th October 2004.

2 (1) Subject to paragraph (2), no aircraft shall fly below 1400 feet above mean sea level within the area bounded by:

(a) straight lines joining the following points:
513212N 0000911 W; and
51302ON 0000648M

(b) the clockwise arc of a circle having a radius of 0.55 nautical miles centred at 51300ON 0000730W between 51302ON 0000648W and 513001 N 0000637W;

(c) straight lines joining the following points:
513001 N 0000637W; and
512917N 0000634W;

(d) the clockwise arc of a circle having a radius 0.55 nautical miles centred at 512915N 0000726W between 512917N 0000634W and 512917N 0000819W;

(e) straight lines joining the following points:
512917N 0000819W; and
512939N 0001 132W;

(f) the clockwise arc of a circle having a radius 0.55 nautical miles centred at 513011 N 0001 123W between 512939N 0001 132W and 513028N 0001 209W;

(g) straight lines joining the following points:
513028N 0001209W; and
513208N 0001 038W;

(h) the clockwise are of a circle having a radius 0.55 nautical miles centred at 513151 N 0000952W between 513208N 0001 038W and 513212N 0000911 W;

(2) Paragraph (1) shall not apply to:

(a) any aircraft in the service of the Chief Officer of Police for the Metropolitan Police District;

(b) any aircraft flying in accordance with a Special Flight Notification issued by the appropriate air traffic control unit;

(c) any helicopter flying on Helicopter Route 4; and

(d) any aircraft flying in accordance with an Enhanced Non-Standard Flight clearance issued by the appropriate air traffic control unit.

The Air Navigation (Restriction of Flying) (Isle of Dogs) Regulations 2004

Whereas the Secretary of State for Transport deems it necessary in the public interest to restrict flying in airspace in the vicinity of Isle of Dogs, London:

Now therefore the Secretary of State for Transport, in exercise of the powers conferred by article 85 of the Air Navigation Order 2000, hereby makes the following Regulations:

1 These Regulations may be cited as the Air Navigation (Restriction of Flying) (Isle of Dogs) Regulations 2004 and shall come into force on 28th October 2004.

2 (1) Subject to paragraph (2), no aircrat shall fly below 1400 feet above mean sea level within the area bounded by:

(a) straight lines joining the following points:
513035N 0000025W;
512954N 0000033M and
512938N 0000022M

(b) the clockwise arc of a circle having a radius of 0.3 nautical miles centred at 512931 N 0000049W between 512938N 0000022W and 512921 N 00001 13W;

(c) straight lines joining the following points:
512921 N 00001 13W; and
513000N 00001 54W;

(d) the clockwise arc of a circle having a radius of 0.55 nautical miles centred at 513018N 000011 0W between 513000N 00001 54W and 513035N 0000025W;

(2) Paragraph (1) shall not apply to:

(a) any aircraft in the service of the Chief Officer of Police for the Metropolitan Police District;

(b) any aircraft flying in accordance with a Special Flight Notification issued by the appropriate air traffic control unit;

(c) any helicopter flying on Helicopter Route 4;

(d) any aircraft flying in accordance with an Enhanced Non-Standard Flight clearance issued by the appropriate air traffic control unit; and

(e) any aircraft approaching to, or departing from, London (City) Airport.

The Air Navigation (Restriction of Flying) (City of London) Regulations 2004

Whereas the Secretary of State for Transport deems it necessary in the public interest to restrict flying in airspace in the vicinity of the City of London:

Now therefore the Secretary of State for Transport, in exercise of the powers conferred by article 85 of the Air Navigation Order 2000), hereby makes the following Regulations:

1 These Regulations may be cited as the Air Navigation (Restriction of Flying) (City of London) Regulations 2004 and shall come into force on 28th October 2004.

2 (1) Subject to paragraph (2), no aircraft shall fly below 1400 feet above mean sea level within the area bounded by straight lines joining the following points: 513125N 0000547M 513118N 0000439M and 513043N 0000418W; 513016N 0000433W; 513037N 0000704W; 513108N 0000653W; and 513125N 0000547W

(2) Paragraph (1) shall not apply to:

(a) any aircraft in the service of the Chief Officer of Police for the Metropolitan Police District;

(b) any aircraft flying in accordance with a Special Flight Notification issued by the appropriate air traffic control unit;

(c) any helicopter flying on Helicopter Route 4; and

(d) any aircraft flying in accordance with an Enhanced Non-Standard Flight clearance issued by the appropriate air traffic control unit.

THE AIR NAVIGATION (DANGEROUS GOODS) REGULATIONS 2002

Made 11th November 2002, Coming into force 2nd Dec 2002

ARRANGEMENT OF REGULATIONS

PART I – PRELIMINARY
1 Citation and commencement
2 Revocation
3 Interpretation

PART II – REQUIREMENTS FOR CARRIAGE OF DANGEROUS GOODS
4 Requirement for approval of operator
5 Prohibition of carriage of dangerous goods

PART III – OPERATOR'S OBLIGATIONS
6 Provision of information by the operator to crew etc
7 Acceptance of dangerous goods by the operator
8 Method of loading by the operator
9 Inspections by the operator for damage, leakage or contamination
10 Removal of contamination by the operator

PART IV – SHIPPER'S RESPONSIBILITIES
11 Shipper's responsibilities

PART V – COMMANDER'S OBLIGATIONS
12 Commander's duty to inform air traffic services

PART VI – TRAINING
13 Provision of training

PART VII – PROVISION OF INFORMATION TO PASSENGERS AND IN RESPECT OF CARGO
14 Provision of information to passengers
15 Provision of information in respect of cargo

PART VIII – DOCUMENTS AND RECORDS, ENFORCEMENT POWERS AND GENERAL
16 Keeping of documents and records
17 Production of documents and records
18 Powers in relation to enforcement of the Regulations
19 Occurrence reporting
20 Dropping articles for agricultural, horticultural, forestry or pollution control purposes
21 Police aircraft

SCHEDULE
Schedule Regulations Revoked

The Secretary of State for Transport, in exercise of the powers conferred by articles 60(1) and 129(5) of the Air Navigation Order 2000 and of all other powers enabling him in that behalf, hereby makes the following Regulations:

PART 1 – PRELIMINARY

1 Citation and commencement
These Regulations may be cited as the Air Navigation (Dangerous Goods) Regulations 2002 and shall come into force on 2nd December 2002.

2 Revocation
The Regulations specified in the Schedule hereto are hereby revoked.

3 Interpretation
(1) In these Regulations:

'**acceptance check list**' means a document used to assist in carrying out a check on the external appearance of packages of dangerous goods and their associated documents to determine that all appropriate requirements have been met;

'**cargo aircraft**' means any aircraft which is carrying goods or property but not passengers and for the purposes of these Regulations the following are not considered to be passengers:

(a) a crew member;

(b) an operator's employee permitted to be carried by, and carried in accordance with, the instructions contained in the Operations Manual;

(c) an authorised representative of a competent national aviation authority;

(d) a person with duties in respect of a particular shipment on board;

'**dangerous goods**' means any article or substance which is identified as such in the Technical Instructions;

489

'**dangerous goods accident**' means an occurrence associated with and related to the carriage of dangerous goods by air which results in fatal or serious injury to a person or major property damage;

'**dangerous goods incident**' means an occurrence, other than a dangerous goods accident, which:

(a) is associated with and related to the carriage of dangerous goods by air, not necessarily occurring on board an aircraft, which results in injury to a person, property damage, fire, breakage, spillage, leakage of fluid or radiation or other evidence that the integrity of the packaging has not been maintained; or

(b) relates to the carriage of dangerous goods and which seriously jeopardises the aircraft or its occupants;

'**dangerous goods transport document**' means a document which is specified by the Technical Instructions and contains information about those dangerous goods;

'**freight container**' means an article of transport equipment for radioactive materials, designed to facilitate the carriage of such materials, either packaged or unpackaged, by one or more modes of transport, but does not include a unit load device;

'**handling agent**' means an agent who performs on behalf of the operator some or all of the functions of the latter including receiving, loading, unloading, transferring or other processing of passengers or cargo;

'**ID number**' means an identification number specified in the Technical Instructions for an item of dangerous goods which has not been assigned a UN number;

'**non-United Kingdom operator**' means an aircraft operator who holds an air operator's certificate issued otherwise than by the CAA;

'**overpack**' means an enclosure used by a single shipper to contain one or more packages and to form one handling unit for convenience of handling and stowage, but does not include a unit load device;

'**package**' means the complete product of the packing operation consisting of the packaging and its contents prepared for carriage;

'**packaging**' means the receptacles and any other components or materials necessary for the receptacle to perform its containment function;

'**proper shipping name**' means the name to be used to describe a particular article or substance in all shipping documents and notifications and, where appropriate, on packagings;

'**serious injury**' means an injury which is sustained by a person in an accident and which:

(a) requires hospitalisation for more than 48 hours, commencing within seven days from the date the injury was received; or

(b) results in a fracture of any bone (except simple fractures of fingers, toes or nose); or

(c) involves lacerations which cause severe haemorrhage, nerve, muscle or tendon damage; or

(d) involves injury to any internal organ; or

(e) involves second or third degree burns, or any burns affecting more than 5 per cent of the body surface; or

(f) involves verified exposure to infectious substances or injurious radiation;

'**Technical Instructions**' means the 2005-2006 English language edition of the Technical Instructions for the Safe Transport of Dangerous Goods by Air, approved and published .by decision of the Council of the International Civil Aviation Organisation;

'**UN number**' means the four-digit number assigned by the United Nations Committee of Experts on the Transport of Dangerous Goods to identify a substance or a particular group of substances;

'**unit load device**' means any type of container or pallet designed for loading onto an aircraft but does not include a freight container for radioactive materials or an overpack; and

'**United Kingdom operator**' means an aircraft operator who holds an air operator's certificate issued by the CAA.

(2) (a) Other expressions used in these Regulations shall have the same respective meanings as in the Air Navigation Order 2000.

(3) For the avoidance of doubt any reference in the Technical Instructions or these Regulations to the taking on board, loading onto or carriage of dangerous goods in or on an aircraft shall for the purpose of these Regulations be interpreted as applying also to the placing, suspending or carriage of such goods beneath an aircraft unless the context makes it otherwise apparent.

PART II – REQUIREMENTS FOR CARRIAGE OF DANGEROUS GOODS

4 Requirement for approval of operator

(1) An aircraft shall not carry or have loaded onto it any dangerous goods unless;

(a) the operator is approved under this regulation; and

(b) such goods are carried or loaded in accordance with:

(i) any conditions to which such approval may be subject; and

(ii) in accordance with the Technical Instructions.

(2) An approval under this regulation:

(a) shall be granted by the CAA if it is satisfied the operator 'is competent to carry dangerous goods safely;

(b) shall be in writing; and

(c) may be subject to such conditions as the CAA thinks fit.

5 Prohibition of carriage of dangerous goods

(1) Subject to paragraphs (2) and (3) a person shall not:

(a) deliver or cause to be delivered for carriage in, or
(b) take or cause to be taken on board;
an aircraft any dangerous goods, which he knows or ought to know or suspect to be goods capable of posing a risk to health, safety, property or the environment when carried by air, unless the Technical Instructions have been complied with and the package of those goods is in a fit condition for carriage by air.

(2) Subject to paragraph (3), these Regulations shall not apply to those dangerous goods specified in the Technical Instructions as being:
(a) for the proper navigation or safety of flight;
(b) to provide, during flight, medical aid to a patient;
(c) to provide, during flight, veterinary aid or a humane killer for an animal;
(d) to provide, during flight, aid in connection with search and rescue operations;
(e) permitted for carriage by passengers or crew members; or
(f) intended for use or sale during the flight in question.

(3) (a) The goods specified in paragraph (2) shall only be carried provided they comply with the following sub-paragraphs and Part 8 and the applicable provisions in paragraphs 1.1.3 and 2.2 of Part 1 of the Technical Instructions.
(b) The goods specified in sub-paragraph (2)(a) shall only be carried if:
(i) they are required to be carried on an aircraft by or under the Air Navigation Order 2000 or are otherwise intended for use on an aircraft for the purpose of the good order of the flight in accordance with the normal practice whether or not, in either case, such goods are required to be carried or intended to be used on that particular flight;
(ii) when they are intended as replacements or have been removed for replacement, they comply with paragraph 2.2.2 of Part 1 of the Technical Instructions;
(c) The goods specified in sub-paragraph (2)(b) and (2)(c) shall only be carried if:
(i) they are or may be required for use during the flight;
(ii) they are or may be required for use during a subsequent flight by the same aircraft and it will not be practicable to load the goods onto the aircraft in the intervening period before the commencement of that subsequent flight; or
(iii) they were used or might have been required for use during a previous flight by the same aircraft and it has not been practicable to unload them from the aircraft since that flight;
(d) The goods specified in sub-paragraph 2(e) shall only be carried by passengers or crew members if they comply with the provisions in Part 8 of the Technical Instructions;
(e) The goods specified in sub-paragraph (2)(f) shall only be carried if the Technical Instructions identify them as being items which can be carried on an aircraft for sale or use during a flight or, when they are intended as replacements for such items or have been removed for replacement, they are carried in accordance with paragraph 2.2.3 of Part 1 of the Technical Instructions.

PART III – OPERATOR'S OBLIGATIONS

6 Provision of information by the operator to crew etc.

(1) (a) The operator of an aircraft flying for the purposes of public transport shall ensure that all appropriate manuals, including the Operations Manual, contain information about dangerous goods so that ground staff and crew members can carry out their responsibilities in regard to the carriage of dangerous goods, including the actions to be taken in the event of emergencies involving dangerous goods.
(b) Where applicable, the operator shall ensure such information is also provided to his handling agent.

(2) The operator of an aircraft in which dangerous goods are to be carried as cargo shall ensure that, before the flight begins, the commander of the aircraft is provided with:
(a) written information about the dangerous goods as specified in paragraph 4.1 of Part 7 of the Technical Instructions; and
(b) information for use in responding to an in-flight emergency as specified in paragraph 4.8 of Part 7 of the Technical Instructions.

(3) The operator of an aircraft which is involved in an aircraft accident or an aircraft incident in the United Kingdom shall notify the CAA without delay of any dangerous goods carried as cargo on the aircraft.

7 Acceptance of dangerous goods by the operator

(1) The operator of an aircraft in which dangerous goods are to be carried shall ensure that no package, overpack or freight container which contains dangerous goods is accepted for carriage in an aircraft until such package, overpack or freight container has been inspected to determine that:
(a) insofar as it is reasonable to ascertain, the goods are not forbidden for carriage by air in any circumstances by the provisions of the Technical Instructions;
(b) insofar as it is reasonable to ascertain, the goods are classified as required by the Technical Instructions;
(c) insofar as it is reasonable to ascertain, the goods are packed as required by the Technical Instructions;
(d) the package, overpack or freight container is marked and labelled in accordance with the provisions of Chapters 2 and 3 of Part 5 of the Technical Instructions;
(e) the package, overpack or freight container is not leaking or damaged so that the contents may escape.

(2) The operator of an aircraft in which dangerous goods are to be carried shall ensure that no package, overpack or freight container which contains dangerous goods is accepted for carriage in that aircraft unless it is accompanied by a dangerous goods transport document, except where the Technical Instructions indicate that such a document is not required, and shall inspect such a document to determine that it complies with the provisions of the Technical Instructions.

(3) (a) For the purpose of each of the inspections required by paragraphs (1) and (2) an acceptance check list shall be used and the results of the inspection shall be recorded thereon.
(b) The acceptance check list shall be in such form and shall provide for the entry of such details as will enable the relevant inspection to be fully and accurately made by reference to the completion of that list.

8 Method of loading by the operator

(1) The operator of an aircraft in which dangerous goods are to be carried as cargo shall ensure that dangerous goods are not carried in any compartment occupied by passengers or on the flight deck, except in circumstances permitted by the provisions in paragraph 2.1 of Part 7 of the Technical Instructions.
(2) The operator of an aircraft in which dangerous goods are to be carried shall ensure that any package, overpack or freight container which contains dangerous goods is loaded, segregated, stowed and secured on an aircraft in accordance with the provisions in Chapter 2 of Part 7 of the Technical Instructions.
(3) The operator of an aircraft in which dangerous goods are to be carried shall ensure that packages, overpacks or freight containers bearing an indication that they can only be carried on a cargo aircraft are loaded and stowed in accordance with the provisions in paragraph 2.4.1 of Part 7 of the Technical Instructions and are not loaded on an aircraft carrying passengers.

9 Inspections by the operator for damage, leakage or contamination

(1) The operator of an aircraft in which dangerous goods are to be carried shall ensure packages, overpacks or freight containers which contain dangerous goods are inspected for evidence of damage or leakage before being loaded on an aircraft or placed in a unit load device.
(2) The operator of an aircraft in which dangerous goods are to be carried shall ensure a unit load device containing dangerous goods is not loaded unless it has been inspected and found free from any evidence of leakage from or damage to the packages, overpacks or freight containers contained in it.
(3) The operator of an aircraft in which dangerous goods are to be carried shall ensure that any package, overpack or freight container which contains dangerous goods which appears to be leaking or damaged is not loaded on an aircraft.
(4) The operator of an aircraft in which dangerous goods are to be carried shall ensure that any package, overpack or freight container which contains dangerous goods which is found to be leaking or damaged on an aircraft is removed and that other cargo or baggage loaded on that aircraft is in a fit state for carriage by air and has not been contaminated.
(5) The operator of an aircraft in which dangerous goods have been carried shall ensure after unloading that all packages, overpacks or freight containers which contain dangerous goods are inspected for signs of damage or leakage and if there is such evidence shall ensure that any part of the aircraft where the package, overpack or freight container was stowed, or any sling or other apparatus which has been used to suspend goods beneath the aircraft is inspected for damage or contamination.

10 Removal of contamination by the operator

(1) The operator of an aircraft in which dangerous goods are to be carried shall ensure that any contamination found as a result of leaking or damaged packages, overpacks or freight containers is removed without delay.
(2) The operator of an aircraft shall ensure that an aircraft is not permitted to fly for the purpose of carrying passengers or cargo if it is known or suspected that radioactive materials have leaked in or contaminated the aircraft, unless the radiation level resulting from the fixed contamination at any accessible surface and the non-fixed contamination are not more than the values specified in paragraph 3.2 of Part 7 of the Technical Instructions.

PART IV – SHIPPER'S RESPONSIBILITIES

11 Shipper's responsibilities

Before consigning any dangerous goods for carriage by air the shipper shall ensure that:

(a) the goods are not forbidden for carriage by air in any circumstances under the provisions in paragraph 2.1 of Part 1 of the Technical Instructions;

(b) if the goods are forbidden for carriage by air without approval, all such approvals have been obtained where the Technical Instructions indicate it is the responsibility of the shipper to so obtain them;

(c) the goods are classified according to the classification criteria contained in Part 2 of the Technical Instructions;

(d) the goods are packed according to paragraphs 2.2 and 2.4 of Part 1, Part 2, chapters 2 and 3 and paragraphs 4.2, 4.3 and 4.4 of Part 3 and Part 4 of the Technical Instructions and the packagings used are in accordance with such provisions of those paragraphs, chapters and Parts and Part 6 of the Technical Instructions as apply to those goods;

(e) the package is marked and labelled in English in addition to any other language required by the State of Origin as specified for those goods in paragraph 2.4 of Part 1, chapters 2 and 3 and paragraph 4.5 of Part 3, chapters 2 and 3 of Part 5 and chapter 2 of Part 6 of the Technical Instructions;

(f) the package is in a fit condition for carriage by air;

(g) when one or more packages are placed in an overpack, the overpack only contains packages of goods permitted to be carried by paragraph 1.1 of Part 5 of the Technical Instructions and the overpack is marked and labelled as required by paragraphs 2.4 and 3.2 of Part 5 of the Technical Instructions;

(h) a dangerous goods transport document:

(i) has been completed in English in addition to any other language required by the State of Origin as required by paragraph 4.1 of Part 5 of the Technical Instructions; and

(ii) contains a declaration signed by or on behalf of the shipper stating that the Technical Instructions have been complied with in that the dangerous goods:

(aa) are fully and accurately described;

(bb) are correctly classified, packed, marked and labelled; and (cc) are in a proper condition for carriage by air;

(i) the operator of the aircraft has been furnished with the dangerous goods transport document required by paragraph (h) and such other documents in respect of dangerous goods as are required by Part 4 and paragraphs 4.3 and 4.4 of Part 5 of the Technical Instructions.

PART V – COMMANDER'S OBLIGATIONS

12 Commander's duty to inform air traffic services

The commander of an aircraft carrying dangerous goods as cargo shall, in the event of an inflight emergency and as soon as the situation permits, inform the appropriate air traffic services unit of those dangerous goods in detail or as a summary or by reference to the location from where the detailed information can be obtained immediately.

PART VI – TRAINING

13 Provision of training

(1) The shipper, and any agent thereof, shall ensure that before a consignment of dangerous goods is offered by him for carriage by air all persons involved in its preparation have received training as specified in Chapter 4 of Part 1 and paragraph 1.5 of Part 5 of the Technical Instructions, to enable them to carry out their responsibilities with regard to the carriage of dangerous goods by air.

(2) (a) A United Kingdom operator, and any agent thereof, shall ensure that all relevant staff involved with the carriage of passengers or cargo by air have received training which complies with sub-paragraphs (c) and (d).

(b) The operator of an aircraft shall ensure that the staff of his handling agent have received that training.

(c) The training shall be as specified in Chapter 4 of Part 1 and paragraph 4.9 of Part 7 of the Technical Instructions.

(d) The training has been granted a training approval under paragraph (8).

(3) (a) The agent for a non-United Kingdom operator shall ensure that all relevant staff involved with the carriage of passengers or cargo by air have received training which complies with sub-paragraphs (b) and (c).

(b) The training shall be as specified in Chapter 4 of Part 1 and paragraph 4.9 of Part 7 of the Technical Instructions.

(c) The training has been granted a training approval under paragraph (8).

(4) The content of training programmes shall be as specified in Chapter 4 of Part 1 of the Technical Instructions.

(5) Initial and recurrent training programmes shall be established and maintained by or on behalf of the shipper and any agent thereof, the operator of an aircraft if it is a United Kingdom operator and the agent of the operator whether the operator is a United Kingdom operator or a non-United Kingdom operator in accordance with paragraph 4.1 of Part 1 of the Technical Instructions and recurrent training shall take place not less than every two years.

(6) Records of training shall be maintained as specified in paragraph 4.2 of Part 1 of the Technical Instructions.

(7) No person shall offer or provide training required by this part of these Regulations unless:

(a) (i) he has been granted a training instructor approval under paragraph (9); and

(ii) the training has been approved under paragraph (8); or

(b) he is employed by a shipper or an agent thereof and the training is offered or provided to other employees of that shipper or agent; or

(c) (i) he is employed by a United Kingdom operator or by an agent of a United Kingdom operator or of a non-United Kingdom operator;

(ii) the training has been approved under paragraph (8); and

(iii) the training is offered or provided to other employees of that United Kingdom aircraft operator or agent.

(8) A training approval under this regulation shall be:

(a) granted by the CAA If it is satisfied that the form and content of the training is adequate for its purpose;

(b) in writing; and

(c) subject to such conditions as the CAA thinks fit, and may in particular include a condition requiring that the training be provided by a person who holds a training instructor approval granted by the CAA under paragraph (9).

(9) A training instructor approval under this regulation shall be:

(a) granted by the CAA if it is satisfied that the person is competent to carry out the training specified in the approval granted under paragraph (8);

(b) in writing; and

(c) subject to such conditions as the CAA thinks fit.

PART VII – PROVISION OF INFORMATION TO PASSENGERS AND IN RESPECT OF CARGO

14 Provision of information to passengers

(1) An airport operator and the operator of an aircraft flying for the purpose of public transport of passengers or his agent shall ensure that persons who are or may become passengers on an aircraft flying for the purposes of public transport are warned as to the types of dangerous goods which they are forbidden from carrying on an aircraft as checked baggage or with them by displaying notices sufficient in number and prominence for this purpose:

(a) at each of the places at an airport where tickets are issued;

(b) at each of the areas at an airport maintained to assemble passengers to board an aircraft; and

(c) at any location where a passenger may be checked in.

(2) The operator of an aircraft flying for the purpose of the public transport of passengers or his agent shall ensure that passengers are warned as to the type of dangerous goods which they are forbidden from carrying on an aircraft as checked baggage or with them either by providing information with each passenger ticket, sufficient in prominence for this purpose, or by some other appropriate means such that passengers receive a warning in addition to that required by paragraph (1).

(3) Any person who, in the United Kingdom, makes available flight accommodation shall ensure that persons who are or may become passengers on an aircraft flying for the purposes of public transport are warned as to the types of dangerous goods which they are forbidden from carrying on an aircraft as checked baggage or with them by displaying notices sufficient in number and prominence for this purpose at any place where flight accommodation is offered for sale.

15 Provision of information in respect of cargo

The operator of an aircraft in which cargo is to be carried and any agent thereof shall ensure that notices giving information about the carriage of dangerous goods are displayed in sufficient number and prominence for this purpose at those places where cargo is accepted for carriage.

PART VIII – DOCUMENTS AND RECORDS, ENFORCEMENT POWERS AND GENERAL

16 Keeping of documents and records

(1) The operator of an aircraft carrying dangerous goods as cargo shall ensure that a copy of the dangerous goods transport document required by regulation 7(2) and the written information to the commander required by regulation 6(2)(a) are retained at a readily accessible location until after the full period of the flight on which the goods were carried.

(2) The operator of an aircraft in which dangerous goods are carried shall preserve for not less than three months:

(a) any dangerous goods transport document or other document in respect of dangerous goods which has been furnished to him by the shipper in accordance with regulation 7(2);

(b) the record of any acceptance check list completed in accordance with regulation 7(3); and

(c) the written information to the commander as required by regulation 6(2)(a).

(3) The record referred to in paragraph (2)(b) may be in a legible or a non-legible form so long as the recording is capable of being reproduced in a legible form.

17 Production of documents and records

(1) The operator of an aircraft on which dangerous goods are to be or have been carried and any agent thereof shall, within a reasonable time after being requested so to do by an authorised person, cause to be produced to that person such of the following documents as may have been requested by that person:

(a) the written approval referred to in regulation 4(1);

(b) the dangerous goods transport document or other document in respect of any dangerous goods, referred to in regulation 7(2);

(c) the completed acceptance checklist in a legible form in respect of any dangerous goods, referred to in regulation 7(3); and

(d) a copy of the written information provided to the commander of the aircraft in respect of any dangerous goods, referred to in regulation 6(2)(a).

(2) The aircraft operator, shipper and any agent of either of them shall, within a reasonable time after being requested so to do by an authorised person, cause to be produced to that person any document which relates to goods which the authorised person has reasonable grounds to suspect may be dangerous goods in respect of which the provisions of these Regulations have not been complied with.

18 Powers in relation to enforcement of the Regulations

(1) An authorised person may examine, take samples of and seize any goods which the authorised person has reasonable grounds to suspect may be dangerous goods in respect of which the provisions of these Regulations have not been complied with.

(2) An authorised person may open or require to be opened any baggage or package which the authorised person has reasonable grounds to suspect may contain dangerous goods in respect of which the provisions of these Regulations have not been complied with.

(3) (a) Subject to paragraph (5), any sample taken or goods seized by an authorised person under this regulation shall be retained or detained respectively for so long as the CAA considers necessary in all the circumstances and shall be disposed of in such manner as the CAA considers appropriate in all the circumstances.

(b) Without prejudice to the generality of sub-paragraph (a) any sample taken or goods seized under this regulation may be retained or detained respectively:

(i) for use as evidence at a trial for an offence; or

(ii) for forensic examination or for investigation in connection with an offence.

(4) (a) The person from whom any goods have been seized by an authorised person under this regulation may apply to the CAA for the item to be released to him.

(b) An application under this paragraph shall be made in writing and shall be accompanied by evidence of ownership by the applicant.

(c) The function of deciding a case where such an application as is referred to in sub-paragraph (a) has been made is hereby prescribed for the purposes of section 7(1) of the Civil Aviation Act 1982: and for the purpose of making any decision in such a case a quorum of the CAA shall be one member.

(d) Where the CAA is satisfied that the applicant is the owner of the item concerned and that further retention of the item is not necessary for the purposes of any criminal proceedings it shall arrange for the goods concerned to be returned to the applicant.

(5) Where further retention of goods is, in the opinion of the CAA no longer necessary and no application has been made under paragraph (4) or any such application has been unsuccessful the goods shall be destroyed or otherwise disposed of in accordance with the directions of the CAA.

19 Occurrence reporting

(1) A United Kingdom operator shall ensure that any dangerous goods accident, dangerous goods incident or the finding of undeclared or misdeclared dangerous goods in cargo or passenger's baggage, wherever it occurs, is reported to the CAA.

(2) A non-United Kingdom operator shall ensure that any dangerous goods accident, dangerous goods incident or the finding of undeclared or misdeclared dangerous goods in cargo or passenger's baggage which occurred in the United Kingdom is reported to the CAA.

(3) A report required under paragraph (1) or (2) shall contain such of the following information as is appropriate to the occurrence:
(a) date of the occurrence;
(b) location of the occurrence, flight number and flight date;
(c) description of the goods and the reference number of the air waybill, pouch, baggage tag and ticket;
(d) proper shipping name (including the technical name, if applicable);
(e) UN/ID number;
(f) class or division in accordance with the Technical Instructions and any subsidiary risk(s);
(g) type of packaging and the packaging specification marking;
(h) quantity of dangerous goods;
(i) name and address of the shipper or passenger;
(j) suspected cause of the occurrence;
(k) action taken;
(l) any other reporting action taken;
(m) name, title, address and contact number of the reporter;
(n) any other relevant details.

(4) (a) Subject to sub-paragraph (b) a report containing as much of the information referred to above as is in his possession shall be despatched in writing, or in such other form as the CAA may approve, and by the quickest available means to the CAA within 72 hours of the occurrence coming to the knowledge of the person making the report.

(b) If at that time any of the said information is not in the possession of that person, he shall despatch the information to the CAA in writing, or in such other form as the CAA may approve, and by the quickest available means within 72 hours of the information coming into his possession.

(5) Nothing in this regulation shall require a person to report any occurrence which he has reported under article 117 of the Air Navigation Order 2000 or which he has reason to believe has been or will be reported by another person to the CAA in accordance with that article.

20 Dropping articles for agricultural, horticultural, forestry or pollution control purposes

Subject to the provisions of regulation 4(1)(a) nothing in these Regulations shall apply to any aircraft flying solely for the purpose of dropping articles for the purpose of agriculture, horticulture, forestry or pollution control.

21 Police aircraft

Nothing in these Regulations other than regulation 4(1)(a) shall apply to the carriage of dangerous goods by an aircraft flying under and in accordance with the terms of a police air operator's certificate.

SCHEDULE

REGULATIONS REVOKED – REGULATION 2

	References
The Air Navigation (Dangerous Goods) Regulations 1994	S.I. 1994/3187
The Air Navigation (Dangerous Goods) (Amendment) Regulations 1996	S.I. 1996/3100
The Air Navigation (Dangerous Goods) (Second Amendment) Regulations 1997	S.I. 1997/2666
The Air Navigation (Dangerous Goods) (Third Amendment) Regulations 1998	S.I. 1998/2536
The Air Navigation (Dangerous Goods) (Fourth Amendment) Regulations 2001	S.I. 2001/918

CIVIL AVIATION AUTHORITY REGULATIONS 1991

ARRANGEMENT OF REGULATIONS

PART I – GENERAL
1 Citation and commencement
2 Revocation
3 Interpretation
4 Service of documents
5 Publication by the Authority

PART II – FUNCTIONS CONFERRED ON THE AUTHORITY BY OR UNDER AIR NAVIGATION ORDERS
6 Regulation of the conduct of the Authority
7 Reasons for decision
8 Inspection of aircraft register
9 Dissemination of reports of reportable occurrences
10–14 Substitution of a public use aerodrome licence for an ordinary aerodrome licence or of an ordinary aerodrome licence for a public use aerodrome licence

PART III – AIR TRANSPORT LICENSING
15 Regulation of the conduct of the Authority
16 Application for the grant, revocation, suspension or variation of licences
17 Revocation, suspension or variation of licences without application being made
18 Variation of schedules of terms
19 Environmental cases
20 Objections and representations
21 Consultation by the Authority
22 Furnishing of information by the Authority
23 Preliminary meetings
24 Preliminary hearings of allegations of behaviour damaging to a competitor
25 Hearings in connection with licences
26 Procedure at hearings
27 Appeals to the Secretary of State
28 Appeals from decisions after preliminary hearings of allegations of behaviour damaging to a competitor
29 Decisions on appeals
30 Transfer of licences
31 Surrender of licences

PART IIIA – REFERENCES IN RESPECT OF AN AIR TRAFFIC SERVICES LICENCE
31A Determination by the Authority
31B Representations
31C Hearings in connection with licence
31D Procedure at hearings
31E Determination by Authority and appeal to the Secretary of State
31F Decision by Secretary of State on appeal

PART IV – OTHER FUNCTIONS OF THE AUTHORITY
32 Participation in civil proceedings

The Secretary of State for Transport, in exercise of his powers under sections 2(3), 7(1) and (2), 11(2), 64(3), 65(1) and (6), 66(1) and (4), 67(1), (2) and (5), 84(1), 85(1) and 102(1) and (2) of and paragraph 15 of Schedule 1 to the Civil Aviation Act 1982 and of all other powers enabling him in that behalf, and after consultation with the Council on Tribunals under section 10 of the Tribunals and Inquiries Act 1971, hereby makes the following Regulations:

PART I – GENERAL

1 Citation and commencement
These Regulations may be cited as the Civil Aviation Authority Regulations 1991 and shall come into force on 1st September 1991.

2 The Regulations specified in the Schedule hereto are hereby revoked.
The Regulations specified in the Schedule hereto are hereby revoked.

3 Interpretation
(1) In these Regulations unless the context otherwise requires;
"the Act" means the Civil Aviation Act 1982;
"the Authority" means the Civil Aviation Authority;
"decision date" and **"transcript date"** have the meanings respectively assigned to them in regulation 26(8);

"**environmental application**" and "**environmental proposal**" have the meanings respectively assigned to them in regulation 19(1);

"**hearing**" or "**preliminary hearing**" means a hearing or preliminary hearing at which oral evidence or argument may be heard and "to hear" shall be construed accordingly;

"**operating licence**" means an operating licence granted by the CAA in accordance with Council Regulation 2407/92 on licensing of air carriers (as the Regulation has effect in accordance with the EEA Agreement as amended by the Decision of the EEA Joint Committee No 7/94 of 21 March 1994);

"**ordinary aerodrome licence**" means an aerodrome licence granted under an Air Navigation Order which does not include a public use condition;

"**party**" in relation to a case before the Authority means for the purposes of Part II of these Regulations, a person having the right to be heard pursuant to regulation 13(1) and, for the purposes of Part III of these Regulations, a person having the right to be heard pursuant to regulation 25(1);

"**party**" in relation to an appeal to the Secretary of State means any of the persons specified in regulation 27(3)(c) and (d);

"**the person concerned**" means, in relation to the registration of aircraft, the applicant for registration or the person in whose name the aircraft is registered, as the case may be, and in relation to a certificate, licence, approval, authorisation, validation or rating, the holder or former holder of or applicant for the certificate, licence, approval, authorisation, validation or rating, as the case may be; and in relation to making air traffic directions or airspace policy directions, the person who has been directed;

"**personnel licence**" means a licence, authorising a person to act as a member of a flight crew, an aircraft maintenance engineer, an air traffic controller, a student air traffic controller or an aerodrome flight information service officer;

"**public use aerodrome licence**" means an aerodrome licence granted under an Air Navigation Order which includes a public use condition;

"**public use condition**" means a condition included in an aerodrome licence granted under an Air Navigation Order that the aerodrome shall at all times when it is available for the take off or landing of aircraft be so available to all persons on equal terms and conditions;

"**rating**" means a rating on a personnel licence;

"**reportable occurrence**" has the same meaning as in article 94(2) of the Air Navigation Order 1989;

"**statement of policies**" means the publication of the Authority referred to in section 69 of the Act;

"**statutory duties**" means the duties of the Authority set out in sections 4 and 68 of the Act;

"**to substitute an ordinary aerodrome licence for a public use aerodrome licence**" means to vary a public use aerodrome licence by removing the public use condition and "**to substitute a public use aerodrome licence for an ordinary aerodrome licence**" means to vary an ordinary aerodrome licence by adding a public use condition.

(2) Any reference in these Regulations to a numbered regulation shall be construed as a reference to the regulation bearing that number in these Regulations.

(3) Any period of time specified in these Regulations by reference to days, working days or months;

(a) where such period is expressed to begin after a particular date, shall begin on the first day after that date, and shall be inclusive of the last day unless that day falls on a Saturday, Sunday, Christmas Day, Good Friday or any other day appointed by law to be a bank holiday in any part of the United Kingdom, in which case the period shall run to the immediately following working day; and

(b) where such period is expressed to run to or expire before a particular date or event, the period shall be calculated to expire on the last working day before the particular date or the date of that event.

(4) In computing any period of time specified in these Regulations by reference to hours or working days the whole of any Saturday, Sunday, Christmas Day, Good Friday or bank holiday shall be disregarded, and for that purpose any day which is appointed by law to be a bank holiday in any part of the United Kingdom shall be treated as a bank holiday.

(5) For the purposes of Part III of these Regulations, a need to allocate scarce bilateral capacity arises when the Authority has been notified by the Secretary of State that in his opinion by virtue of any provision made by or under the terms of an air services agreement or other international agreement or arrangement, the United Kingdom's share of the capacity on air transport services between the United Kingdom and another State which may be provided by British airlines within the meaning given by section 4(1) of the Act, (whether capacity is expressed in terms of the number of passenger seats or the amount of cargo carrying space which may be offered for sale by such operators, or otherwise) will, within 6 months of the date of notification, be insufficient to enable all persons holding air transport licences authorising them to operate such air transport services to make available all the capacity which they plan to provide.

4 Service of documents

(1) Subject to regulation 31A(5) (d) anything required to be served on any person under these Regulations or under section 66(4) or 84(1) of the Act shall be set out in a notice in writing which may be served either:

(a) by delivering it to that person;

(b) by leaving it at his proper address;

(c) by sending it by post to that address; or

(d) by sending it to him at that address by telex or other similar means which produce a document containing a text of the communication, in which event the document shall be regarded as served when it is received;

and where the person is a body corporate the document may be served upon the secretary of that body.

(2) For the purposes of this regulation the proper address of any person shall, in the case of a body corporate, be the registered or principal office of that body and in any other case be the last known address of the person to be served.

5 Publication by the Authority

Subject to regulation 31A(5) (b) any notice or other matter (not being a schedule of terms referred to in regulation 18 required by these Regulations, or by section 11(2), 64(3), 65(1) or (6) or 85(1) of the Act, to be published shall be published by the Authority in its Official Record.

PART II – FUNCTIONS CONFERRED ON THE AUTHORITY BY OR UNDER AIR NAVIGATION ORDERS

6 Regulation of the conduct of the Authority

(1) The functions conferred on the Authority by or under Air Navigation Orders with respect to:
(a) registration of aircraft;
(b) certification of operators of aircraft;
(c) certification of airworthiness of aircraft;
(d) noise certification;
(e) certification of compliance with the requirements for the emission by aircraft engines of unburned hydrocarbons;
(f) personnel licensing;
(g) licensing of aerodromes;
(h) validation of any certificate or licence;
(i) approval of equipment and approval or authorisation of persons;
(j) approval of schemes for the regulation of the flight times of aircraft crew;
(k) receiving reports of reportable occurrences;
(l) making air traffic directions;
(m) making airspace policy directions;
are hereby prescribed for the purposes of section 7(2) of the Act.

(2) Subject to paragraphs (8) and (9) of this regulation, a decision with respect to any of the matters referred to in paragraph (1) of this regulation, being a decision to register, refuse to register, cancel or amend the registration of an aircraft or to grant, refuse to grant, validate, refuse to validate, revoke, suspend, vary or refuse to vary a certificate, licence, approval, authorisation or rating, or make an air traffic direction or an airspace policy direction may be made on behalf of the Authority only by a member or employee of the Authority.

(3) Subject to paragraphs (8) (9) and (10) of this regulation, where;
(a) it is decided that it would be inexpedient in the public interest for an aircraft to be registered in the United Kingdom; or
(b) an application for the grant, validation or variation of a certificate, licence approval, authorisation or rating has been refused or granted in terms other than those requested by the applicant;
the Authority shall serve on the applicant a notice stating the reasons for the decision, and the applicant may within 14 days after the date of service of that notice request that the case be reviewed by the Authority.

(4) Subject to paragraphs (8), (9) and (10) of this regulation, where it is proposed to;
(a) cancel the registration of an aircraft on the grounds that it would be inexpedient in the public interest for it to continue to be registered in the United Kingdom; or
(b) revoke, suspend or vary a certificate, licence, approval, authorisation, validation or rating or make an air traffic direction or an airspace policy direction under an Air Navigation Order otherwise than on the application of the holder;
the Authority shall serve on the person concerned notice of the proposal together with the reasons for it, and the person concerned may within 14 days after the date of service of that notice, serve on the Authority a request that the case be decided by the Authority and not by any other person on its behalf.

(5) Any person who has failed any test or examination which he is required to pass before he is granted or may exercise the privileges of a personnel licence may within 14 days after being notified of his failure request that the Authority determine whether the test or examination was properly conducted.

(6)(a) The function of deciding a case where such a request as is referred to in paragraph (3), (4) or (5) of this regulation has been duly served on the Authority is hereby prescribed for the purposes of section 7(1) of the Act: and for the purpose of making any decision in such a case a quorum of the Authority shall be one member.
(b) The Authority shall sit with such technical assessors to advise it as the Authority may appoint, but the Authority shall not appoint as an assessor any person who participated in the decision or proposal or in giving or assessing the test or examination which is to be the subject of the Authority's decision.

(7) Where a request under paragraph (3), (4) or (5) has been duly served, the Authority shall, before making a decision:
(a) consider any representations which may have been served on it by the person concerned within 21 days after the date of service of the notice under that paragraph given by the Authority; and
(b) where the person concerned has requested the opportunity to make oral representations in his representations under sub-paragraph (a) above, afford him an opportunity to make such representations and consider them.

(7A)(a) Where an oral hearing is held it shall be held in public except where the Authority is satisfied that, in the interests of morals, public order, national security, juveniles or the protection of the private lives of the parties a private hearing is required, or where it considers that publicity would prejudice the interests of justice.
(b) The following persons shall be entitled to attend the hearing of an appeal, whether or not it is in private:
(i) a member of the Council on Tribunals or of the Scottish Committee of that Council; and
(ii) any other person which the Authority, with the consent of the parties, permits to attend the hearing.

(8) Nothing in this regulation shall:

(a) prevent the Authority or any person authorised so to act on behalf of the Authority from provisionally cancelling the registration of an aircraft or provisionally suspending or varying any certificate, licence, approval, authorisation, validation or rating granted or having effect under an Air Navigation Order or making a provisional air traffic direction pending inquiry into or consideration of the case.
(b) apply to the variation of a flight manual, performance schedule or other document incorporated by reference in a certificate of airworthiness.
(c) apply where the Authority refuses to register or cancels or amends the registration of an aircraft or refuses to grant or validate, grants or validates in terms other than those requested by the applicant, revokes, suspends or varies a certificate, licence, approval, authorisation or rating pursuant to a direction given by the Secretary of State.
(9) Nothing in paragraphs (2), (3) or (4) of this regulation shall apply:
(a) in respect of a medical certificate or certificate of test of experience relating to a personal licence;
(b) where pursuant to its duty under section 5 of the Act, the Authority refuses an application for the grant of an aerodrome licence or grants such an application in terms other than those requested by the applicant or proposes to revoke, suspend or vary an aerodrome licence otherwise than on the application of the holder.
(10) Nothing in paragraphs (3) or (4) of this regulation shall apply where the Authority:
(a) refuses an application by the holder of an aerodrome licence for the substitution of an ordinary aerodrome licence for a public use aerodrome licence; or
(b) proposes otherwise than on the application of the licence holder, to substitute a public use aerodrome licence for an ordinary aerodrome licence.

7 Reasons for decisions
Where the Authority makes a decision pursuant to regulation 6(6) it shall be the duty of the Authority to serve a statement of its reasons for the decision on the person concerned.

8 Inspection of aircraft register
The Authority shall, at all reasonable times and upon payment to it of any applicable charge under section 11 of the Act for inspecting the register, make the register of aircraft available for inspection by any person.

9 Dissemination of reports of reportable occurrences
The Authority shall make available, upon payment to it of any applicable charge under section 11 of the Act, reports of reportable occurrences or a summary of such reports, to any person who is:
(a) the operator or member of the flight crew of any aircraft;
(b) engaged in the design, manufacture, repair, manufacture, repair, maintenance or overhaul of aircraft, or of parts or equipment therefor;
(bb) the provider of an air traffic control service;
(c) the aeronautical authority of a country other than the United Kingdom, or the representative in the United Kingdom of such an authority;
(d) engaged in writing about civil aviation for publication in any newspaper, periodical, book or pamphlet;
(e) engaged in preparing a programme about civil aviation for television or radio;
(f) engaged in the study of civil aviation for any academic purpose; or
(g) any other person whose functions include the furthering of the safety of civil aviation:
Provided that the Authority shall not be required to make available any report or summary thereof to any person if it is satisfied that to do so will not further the safety of civil aviation.

10 Substitution of a public use aerodrome licence for an ordinary aerodrome licence or of an ordinary aerodrome licence for a public use aerodrome licence
(1) The Authority shall refuse to consider an application for the substitution of an ordinary aerodrome licence for a public use aerodrome licence unless:
(a) the application is made by the holder of the licence;
(b) it contains a statement of the grounds on which the application is made; and
(c) the application is accompanied by any applicable charge under section 11 of the Act.
(2) The Authority shall refuse to consider an application for the substitution of a public use aerodrome licence for an ordinary aerodrome licence unless
(a) it is made by;
(i) the holder of the licence;
(ii) any other holder of an aerodrome licence granted under an Air Navigation Order;
(iii) the holder of any air operator's certificate granted under an Air Navigation Order;
(iv) the holder of any air transport licence granted under the Act or of any operating licence; or
(v) the operator of any aircraft who satisfies the Authority that an aircraft operated by him has, during the 12 months immediately preceding the date on which the application is made, been granted or refused permission to land at or take off from the aerodrome to which the licence relates;
(b) it contains a statement of the grounds on which the application is made;
(c) the application is accompanied by any applicable charge under section 11 of the Act; and
(d) if made by someone other than the holder of the ordinary aerodrome licence, a copy of the application has been served on the holder within 24 hours after it has been served on the Authority.
(3) The Authority shall as soon as may be after an application has been served upon it in accordance with this regulation publish such particulars of the application as it thinks necessary for indicating the substance of the application and shall

ANO Civil Aviation Authority Regulations 1991 – Arrangement of Regulations –
10 Substitution of a public use aerodrome licence for an ordinary aerodrome licence

make a copy of the application available at its principal office for inspection by any person at any reasonable time: Provided that nothing herein shall require the Authority to publish an application for the substitution of a public use aerodrome licence for an ordinary aerodrome licence which is made by the holder of the licence.

(4) If the Authority proposes to substitute a public use aerodrome licence for an ordinary aerodrome licence it shall serve on the holder of that licence particulars of the proposal and of the reasons for it and shall publish those particulars and reasons.

11

Any person may serve on the Authority an objection to or representation about an application or proposal published pursuant to regulation 10 if;
(a) he does so within 21 days after the date of publication;
(b) he serves a copy of his objection or representation on the applicant and on the holder of the aerodrome licence to which the application or proposal relates within 24 hours after it has been served on the Authority; and
(c) he states the grounds of his objection or representation.

12

Before the date fixed for the hearing of a case pursuant to regulation 13, the Authority shall serve on any person who has the right to be heard in connection with the case or whom the Authority proposes to hear a copy of, or a summary of, any information in the possession of the Authority which has been provided in connection with the case or which the Authority has reason to believe will be referred to at the hearing of the case:

Provided that before serving such information which has been provided by any other person (not being a person who has provided information in connection with the case but does not wish to be heard), the Authority shall consult that person and shall not serve any information which in its opinion relates to the commercial or financial affairs of the person who has provided it and cannot be disclosed to the prospective recipient without disadvantage to the person who has provided it which, by comparison with the advantage to the public and the prospective recipient of disclosure to him, is unwarranted.

13

(1) Before any decision is made on an application or proposal published pursuant to regulation 10 the following persons shall have a right to be heard:
(a) the applicant;
(b) the holder of any air transport licence or operating licence;
(c) the holder of any air operator's certificate granted under an Air Navigation Order;
(d) the holder of any aerodrome licence granted under an Air Navigation Order;
(e) such persons (being persons who wish to be heard and who have served objections or representations pursuant to regulation 11 expressing the views of operators of aircraft described in regulation 10(2)(a)(v) as appear to the Authority to be representative of those who have served such objections or representations:

Provided that no person (other than the applicant and the holder of the licence to which the decision will relate) shall have a right to be heard unless he has served an objection or representation pursuant to regulation 11 and in so doing has stated that he wishes to be heard.

(2) Notwithstanding that a person does not have a right to be heard, the Authority may, if it thinks fit, hear him:

Provided that no person shall be heard pursuant to this paragraph unless he has served an objection or representation pursuant to regulation 11.

(3) No hearing shall be held pursuant to this regulation unless the Authority has served on all persons having a right to be heard and whom it proposes to hear in connection with the case not less than 14 days' notice of the date, time and place of the hearing, and the notice shall clearly identify the application or proposal to which it relates: a similar notice shall be published not less than 7 days before the date of the hearing, and shall be exhibited in a public place in the Authority's principal office during the 7 days immediately preceding the date of the hearing.

14

(1) The function of making a decision on an application or proposal published pursuant to regulation 10 is hereby prescribed for the purposes of section 7(1) of the Act, and for the purposes of making such a decision and of conducting a hearing pursuant to regulation 13 a quorum of the Authority shall be two members:

Provided that the quorum shall be one member, if the persons having the right to be heard in connection with the case have so consented.

(2) Hearings shall be conducted by the Authority, sitting with such employees of the Authority acting as advisers as it thinks fit.

(3) At a hearing every party to the case may appear in person or be represented by any other person whom he may have authorised to represent him and may produce oral and written evidence and may examine any other party to the case, any person whom the Authority hears pursuant to regulation 13(2) and any witnesses produced by any such party or person: the Authority may, to such extent as it thinks fit, permit any person heard by it pursuant to regulation 13(2) to exercise at the hearing the rights set out in this paragraph of a party to the case.

(4) Any person who has served an objection or representation pursuant to regulation 11 but who does not wish to be heard, may make a written submission which he shall serve on the Authority not less than 3 working days before the date fixed for the hearing of the case.

(5) Every hearing shall be held in public unless the Authority shall otherwise decide in relation to the whole or part of a particular case, but nothing in this regulation shall prevent a member of the Council on Tribunals or of its Scottish Committee from attending a hearing in his capacity as such.

(6) The failure of the Authority or of any person to give notice or publish any particulars in the time or manner provided for in these Regulations or any other procedural irregularity shall not invalidate the action taken by the

Authority; and the Authority may, and shall if it considers that any person may have been prejudiced, take such steps as it thinks fit before reaching its decision to cure the irregularity, whether by the giving of notice or otherwise.

(7) All the proceedings at a hearing of the Authority in connection with the case shall be recorded by a shorthand writer or by some other means, and if any person requests a record of the proceedings the Authority shall cause a mechanical recording or transcript of the shorthand or other record to be made available for purchase by that person at a reasonable price:

Provided that;

(a) the Authority shall not be required to make available a mechanical recording or transcript of the record of the proceedings at any time after the expiry of one year from the day of publication of its decision of the case; and

(b) a mechanical recording or transcript of the record of proceedings conducted otherwise than in public shall only be required to be made available for purchase by any party to the case or by any other person heard by the Authority at those proceedings.

(8) The Authority shall furnish a statement of its reasons for the decision to the parties to the case and to any person whom it has heard in connection with the case:

Provided that no statement of reasons need be furnished when an application is granted on the application of the holder of the licence to which the decision relates and no objections or representations have been served pursuant to regulation 11.

(9) The Authority may exclude from its statement of reasons furnished to any person (hereinafter referred to as "the relevant person") any matter if it considers it necessary to do so for the purpose of withholding from the relevant person information which in the opinion of the Authority relates to the commercial or financial affairs of another person and cannot be disclosed to the relevant person without disadvantage to the other person which, by comparison with the advantage to the public and the relevant person of disclosure to him, is unwarranted.

(10) The Authority may publish in such manner as it thinks fit particulars of, and its reasons for, any decision taken by it with respect to an application published pursuant to regulation 10.

PART III – AIR TRANSPORT LICENSING

15 Regulation of the conduct of the Authority

(1) The function of making a decision to;

(a) grant, revoke, suspend or vary other than provisionally an air transport licence in a case where an objection has been served pursuant to regulation 20;

(b) grant or vary other than provisionally a licence in terms other than those requested by the applicant;

(c) refuse to grant a licence;

(d) hold or refuse to hold a preliminary hearing;

(e) provisionally vary a licence where representations have been served pursuant to regulation 24(2);

(f) revoke or suspend an operating licence otherwise that at the request of the holder; or

(g) refuse to grant an operating licence;

is hereby prescribed for the purposes of section 7(1) of the Act.

(2) For the purposes of making any such decision as is referred to in paragraph (1)(a) to (c), (f) and (g) of this regulation and of conducting a hearing pursuant to regulation 26(1) a quorum of the Authority shall be two members unless;

(a) the Authority has dispensed with publication of the application or proposal in accordance with the proviso to regulation 16(3) or to regulation 17(2);

(b) the decision is to suspend the licence in accordance with regulation 17(3);

(c) the Authority has, in accordance with regulation 20(1), specified less than 21 days for serving an objection to or representation about the application or proposal;

(d) the Authority has, in accordance with the proviso to regulation 25(3) given the persons having a right to be heard and whom it proposes to hear in connection with the case less than 14 days' notice of the date of the hearing; or

(e) the persons having the right to be heard in connection with the case have consented to a quorum of the Authority being one member; in which case the quorum shall be one member.

(3) For the purposes of making such a decision as is referred to in paragraph (1)(d) or (e) of this regulation and of conducting a preliminary hearing pursuant to regulation 24 a quorum of the Authority shall be one member.

(4) Any other decision to grant, revoke, suspend or vary an air transport licence or a route licence and any other decision to grant, revoke or suspend an operating licence may be made on behalf of the Authority only by a member or employee of the Authority.

(5) The functions conferred upon the Authority by sections 64 to 67 of the Act are hereby prescribed for the purposes of section 7(2) thereof.

16 Application for the grant, revocation, suspension or variation of licences

(1) The Authority may refuse to consider an application for the grant, revocation, suspension or variation of an air transport licence or a route licence unless;

(a) subject to regulation 24, in the case of an application for the grant of a licence it has been served on the Authority not less than 6 months before the beginning of the period for which the licence is proposed to be in effect, and in any other case it has been served on the Authority not less than 6 months before the date on which it is proposed that the revocation, suspension or variation shall take effect;

(b) in the case of an application for the grant of a licence, the application contains all the particulars specified by the Authority in accordance with section 65(1) of the Act; and

(c) the application is accompanied by any applicable charge under section 11 of the Act.

(2) The Authority shall refuse to consider an application for the revocation, suspension or variation of an air transport licence pr a route licence made by a person other than the holder of the licence unless a copy of the application has

been served on the holder within 24 hours after it has been served on the Authority.

(3) The Authority shall as soon as may be after an application for the grant, revocation, suspension or variation of an air transport licence or a route licence has been served upon it in accordance with this regulation publish such particulars of the application as it thinks necessary for indicating the substance of the application, and shall make a copy of the application available at its principal office for inspection by any person at any reasonable time:

Provided that, except in the case of an environmental application

(i) the Authority may dispense with publication in any case where it is satisfied that for reasons of urgency it is desirable to do so and it is of the opinion that to do so is unlikely to prejudice the interests of any persons of a description specified in regulation 25(1);

(ii) the Authority may dispense with publication in the case of an application:

(a) for the grant of a licence for not more than four flights in any one direction between the same two places;

(b) made by its holder for the revocation or suspension of a licence;

(c) for the variation of a licence if in its opinion to do so is unlikely to prejudice the interests of any persons of a description specified in regulation 25(1).

(4) If within 12 months after the date on which objections to and representations about an application for the grant, variation, suspension or revocation of an air transport licence or a route licence must have been served on the Authority pursuant to regulation 20, the Authority has neither made a decision on the application nor given notice pursuant to regulation 25(3) of the date, time and place of the hearing of such application, it shall as soon as may be republish such particulars of the application as it thinks necessary for indicating the substance of the application and shall republish such particulars at 12 monthly intervals thereafter until such time as a decision has been made on the application or notice has been given as aforesaid pursuant to regulation 25(3).

(5) The Authority may direct that an application shall be treated as being such number of separate applications as it may specify in the direction, and the application shall be treated accordingly.

(6) Any person of a description specified in regulation 25(1)(b) to (d) may apply to the Authority for the variation, suspension or revocation of an air transport licence or a route licence but, except as provided in regulation 18(3), no person may apply for the variation of such a schedule of terms as is mentioned in regulation 18.

17 Revocation, suspension or variation of licences without application being made

(1) Subject to paragraph (3) of this regulation, if the Authority proposes to revoke, suspend, or vary an air transport licence (other than in pursuance of an application made to it in that regard) on the ground that it is not or is no longer satisfied as to the matters specified in paragraph (a) or (b) of section 66(3) of the Act, it shall

(a) serve on the holder of the licence not less than 21 days' notice of its intention to publish particulars of the proposal together with the reasons for its proposal;

(b) consider any representations which may be made to it by the holder of the licence before the expiration of the said notice; and

(c) as soon as may be after the expiration of the said notice or at such earlier time as the Authority and the holder of the licence may agree, publish particulars of the proposal unless it has abandoned the proposal:

Provided that the Authority may:

(i) with the consent of the holder of the licence dispense with publication of its proposal to revoke or suspend the licence;

(ii) dispense with publication of its proposal to vary the licence if it is satisfied that the variation is unlikely to prejudice the interests of any person of a description specified in regulation 25(1).

(2) Subject to paragraph (3) of this regulation, if the Authority proposes to revoke, suspend or vary an air transport licence on grounds other than those referred to in paragraph (1) of this regulation and otherwise than in pursuance of an application made to it in that regard, or proposes to revoke, suspend or vary a route licence otherwise than in pursuance of an application made to it in that regard, it shall publish particulars of the proposal and of the reasons for it, unless:

(a) the Secretary of State has directed the Authority to revoke, suspend or vary the licence as proposed or the proposal is made pursuant to a direction made by the Secretary of State under regulation 29(1) to re-hear the case;

(b) the Authority's duty under section 31(2) of the Airports Act 1986 (being a duty so to perform its air transport licensing functions as to secure that any traffic distribution rules in force under section 31 of the said Act are complied with) requires it to revoke, suspend or vary the licence as proposed;

(c) except in the case of an environmental proposal, the Authority is satisfied that to dispense with publication is unlikely to prejudice the interests of any person of a description specified in regulation 25(1) and the holder of the licence consents to the proposal not being published.

(3) The Authority may suspend an air transport licence notwithstanding that it has not complied with the requirements of paragraph (1) or (2) of this regulation if it has served on the holder of the licence not less than 6 working days' notice of its proposal to suspend the licence, together with its reasons for the proposal, and if, after considering any representations which may be made to it by the holder of the licence before the expiration of such notice it is not, or is no longer, satisfied as mentioned in section 66(3)(a) or (b) of the Act.

(4) Before reaching a decision that it has reason to believe that the holder of an air transport licence or a route licence is neither a United Kingdom national nor such a body as is mentioned in section 65(3)(b) of the Act, the Authority shall:

(a) serve on the holder of the licence not less than 21 days' notice of its intention to consider the matter; and

(b) consider any representations which may be made to it by the holder of the licence before the expiration of the said period.

18 Variation of schedules of terms

(1) If the Authority establishes any schedule of terms and includes in any air transport licence or a route licence a term that the holder of the licence shall comply with terms set out in that schedule as varied from time to time by the

Authority, the Authority may at any time propose to vary that schedule or any part thereof, and any such proposal shall for the purposes of these Regulations be treated as a proposal for the variation of every air transport licence or a route licence which contains such a term as aforesaid relating to that schedule or that part of that schedule, as the case may be.

(2) When any air transport licence or a route licence contains such a term as aforesaid relating to a schedule, the Authority shall publish that schedule and any variation to it in its Official Record or otherwise.

(3) An application for the variation of the schedule of terms set out in a document published by the Authority and entitled the United Kingdom Cabotage Air Passenger Tariff may be made by the holder of any air transport licence or a route licence which includes a term requiring the holder of the licence to comply with that schedule of terms as varied from time to time by the Authority, and any such application shall for the purposes of these Regulations be treated as an application for the variation of every air transport licence or a route licence which contains such a term as aforesaid.

(4) The Authority shall maintain a list of the names and addresses of all persons who hold a licence which includes such a term as is referred to in paragraph (3) of this regulation and shall serve copies of that list on any person who so demands.

19 Environmental cases

(1) For the purposes of this part of these Regulations, "environmental application" and "'environmental proposal" mean respectively an application and a proposal for the grant or variation of an air transport licence, being a licence which authorises, or which if granted would authorise, the holder to operate;

(a) a helicopter at a height of less that 3000 feet above the surface for the greater part of the distance which it flies over land; or

(b) any aircraft in circumstances which, in the opinion of the Authority, will or may cause an exceptional amount of noise, vibration, pollution or other disturbance"

but do not include;

(i) any such application or proposal which is not, in the opinion of the Authority, an application or proposal relating to a licence to operate a regular and frequent service; or

(ii) any such proposal made by the Authority to vary a licence on the ground that it is not or is no longer satisfied as to the matters specified in paragraph (a) or (b) of section 66(3) of the Act.

(2) If the Authority receives an environmental application or makes an environmental proposal it shall designate the case as an environmental case and publish notice of the designation.

20 Objections and representations

(1) Any person may serve on the Authority an objection to or representation about an application or proposal for the grant, revocation, suspension or variation (other than the provisional variation) of an air transport licence or a route licence if he does so:

(a) where an application or proposal is published within such period (being, subject to paragraph (2) of this regulation, not more than 21 days nor less than 7 days) as the Authority may specify when publishing the application or proposal;

(b) where the application or proposal is not published, but he has been notified by the Authority that the application or proposal has been made and will not be published, within 3 working days after being so notified;

Provided that nothing herein shall;

(i) permit the Authority to specify a period of less than 21 days for the service of objections or representations unless it is satisfied that for reasons of urgency it is desirable to do so;

(ii) permit the Authority to specify a period of less than 21 days for the service of objections or representations in a case where it has made a proposal pursuant to regulation 17, unless it has proposed to;

(a) revoke, suspend or vary an air transport licence or a route licence in accordance with a direction given by the Secretary of State;

(b) vary an air transport licence or a route licence for the sole reason that there is a need to allocate scarce bilateral capacity.

(2) If the Authority receives an environmental application or makes an environmental proposal it shall specify 42 days from the date of publication of the notice of designation pursuant to regulation 19 as the period for service of objections or representations on grounds of noise, vibration, pollution or other disturbance.

(3) Where the person making the objection or representation is the holder of an air transport licence or a route licence he shall, within 24 hours after it has been served on the Authority, serve a copy of it on:

(a) the applicant;

(b) any other person who is the holder of the licence to which the application or proposal relates; and

(c) any person whom the Authority is obliged by regulation 21 to consult in respect of the application or proposal;

and where the person making the objection or representation is not the holder of an air transport licence or route licence, the Authority shall within 7 days after the day on which the objection or representation has been served on the Authority serve a copy of it on the said persons, indicating whether the person making the objection or representation wishes to be heard pursuant to regulation 25.

(4) Upon being served as aforesaid, the applicant shall, if so required in writing by the person making the objection or representation, serve him with a copy of the application within 3 working days after being required so to do.

(5) References in this regulation to publication include references to republication pursuant to regulation 16(4), but when an application is republished, nothing in this regulation shall require a person who has served an objection to or representation about the application when it was previously published to re-serve that objection or representation.

21 Consultation by the Authority

The Authority shall not grant, refuse to grant, revoke, suspend or vary any air transport licence or route licence authorising flights to, from or within:

(b) the Isle of Man, without consulting the Isle of Man Department of Highways. Ports and Properties; or
(c) Gibraltar, without consulting the Secretary of State;
and subject to regulation 26(5) such consultations shall be completed before the date fixed for the hearing of the case pursuant to regulation 25:
Provided that consultation as aforesaid shall not be required in a case where;
(i) the application or licence in question is for not more than four flights in any one direction between the same two places;
(ii) the Authority is acting in pursuance of its duty under section 65(2) or (3) or 66(3) of the Act;
(iii) the Authority's duty under section 31(2) of the Airports Act 1986 requires it to refuse to grant or to revoke, suspend or vary the licence; or
(iv) the Authority is acting in pursuance of its duty under section 69A(4) of the Act.

22 Furnishing of information by the Authority

Before the date fixed for the hearing of a case pursuant to regulation 25, the Authority shall serve on any person who has the right to be heard in connection with the case or whom the Authority proposes to hear or is required to consult pursuant to regulation 21 a copy of, or a summary of, any information in the possession of the Authority which has been provided in connection with the case or which the Authority has reason to believe will be referred to at the hearing of the case:
Provided that;
(i) the Authority shall not serve any such information which has been provided by the Secretary of State if the Secretary of State has certified to the Authority that it would not be in the public interest for it to be disclosed;
(ii) before serving such information which has been provided by any other person (not being a person who has provided information in connection with the case but does not wish to be heard) the Authority shall consult that person and shall not serve any information which in its opinion relates to the commercial or financial affairs of the person who has provided it and cannot be disclosed to the prospective recipient without disadvantage to the person who has provided it which, by comparison with the advantage to the public and the prospective recipient of its disclosure to him, is unwarranted.

23 Preliminary Meetings

(1) Before the date fixed for the hearing of a case pursuant to regulation 25, the Authority may hold a preliminary meeting to discuss the conduct of the case.
(2) The Authority shall give to every party to the case and to every person whom the Authority proposes to hear in connection with the case and to any person consulted by the Authority pursuant to regulation 21 who has responded in writing notice of the date, time and place of the preliminary meeting and any such person may attend in person or be represented by any person whom he may have authorised to represent him.
(3) Preliminary meetings shall be conducted on behalf of the Authority only by a member or employee of the Authority.

24 Preliminary hearings of allegations of behaviour damaging to a competitor

(1) This regulation applies where the holder of any air transport licence or a route licence (hereinafter in this regulation referred to as "the applicant")
(a) has applied to the Authority for the variation of an air transport licence or a route licence held by another person (hereinafter in this regulation referred to as "the respondent") for the purpose of restraining the respondent from engaging in behaviour damaging to the applicant's business;
(b) has included in his application a statement giving particulars of the behaviour complained of and of the extent to which the applicant's business is being or is likely to be damaged thereby;
(c) has asked for a preliminary hearing of the application with a view to the respondent's air transport licence or route licence being provisionally varied pending a hearing pursuant to regulation 25; and
(d) has served a copy of his application on the respondent on the same day as he has served it on the Authority.
(2) The respondent shall, within 5 working days after the date of service of the application, serve on the Authority and on the applicant any representations he may wish the Authority to take into account in determining whether to hold a preliminary hearing.
(3) The Authority shall within 10 working days after the date of service of the application notify the applicant, the respondent and any person it is obliged by regulation 21 to consult in respect of the application of the date (which shall be within 20 working days after the date of service of the application), time and place of the preliminary hearing or of the fact that it has decided not to hold a preliminary hearing.
(4) Notice of the date, time and place of a preliminary hearing shall be of such length as is reasonably practicable and shall be given by such means (whether oral or written) as the Authority thinks fit.
(5) The Authority shall hold a preliminary hearing only if, having considered the terms of the application and of any representations served on it pursuant to paragraph (2) of this regulation, it is of the opinion that:
(a) there is prima facie evidence that the behaviour complained of by the applicant is being engaged in by the respondent and that behaviour has or is likely to have the effect of seriously damaging the business of the applicant; and
(b) having regard to its statutory duties, its statement of policies and to the urgency of the matter such a hearing is warranted.
(6) The applicant and the respondent shall have a right to be heard at a preliminary hearing and the Authority may hear such other persons as it thinks fit.
(7) Regulations 21, 22 and 26(1), (4), (6) and (7) shall apply in relation to a preliminary hearing as they apply in relation to a hearing pursuant to regulation 25.
(8) At a preliminary hearing the applicant and the respondent shall have the same rights as a party to a case in a hearing pursuant to regulation 25 and the Authority may, to such extent as it thinks fit, permit any other person whom it decides to hear to exercise the same rights.

(9) Where any person whom the Authority is obliged by regulation 21 to consult in respect of the application attends the preliminary hearing the Authority shall give him opportunity at the preliminary hearing to make observations on the evidence and arguments advanced by the applicant and the respondent and by any other person whom the Authority has decided to hear: and where any such observations are made the Authority shall give the applicant, the respondent and any other person it has decided to hear opportunity at the preliminary hearing to respond to them.

(10) Within 5 working days after the end of the preliminary hearing the Authority shall notify the applicant and the respondent and any person it is obliged by regulation 21 to consult in respect of the application;
(a) whether or not it has decided provisionally to vary the respondent's licence;
(b) if so, the terms of the provisional variation; and
(c) the date, time and place of the hearing to be held pursuant to regulation 25;
and shall furnish its reasons for the decision, as required by section 67(2) of the Act within 10 working days after the end of the preliminary hearing.

(11) The only decision which may be taken by the Authority after a preliminary hearing is a decision provisionally to vary or to refuse provisionally to vary the respondent's air transport licence or a route licence: and if the Authority provisionally varies the respondent's licence it shall in so doing provide that the provisional variation will cease to have effect when the decision reached by the Authority following a hearing pursuant to regulation 25 takes effect.

25 Hearings in connection with licences

(1) Before any decision to grant, refuse to grant, revoke, suspend or vary (other than provisionally) an air transport licence pr a route licence is made, the following persons shall have a right to be heard:
(a) the applicant;
(b) the holder of an operating licence; any air transport licence or a route licence;
(c) the holder of any air operator's certificate granted under an Air Navigation Order;
(d) the holder of any aerodrome licence granted under an Air Navigation Order;
(e) such persons (being persons who wish to be heard and who have served objections or representations pursuant to regulation 20 expressing the views of passengers or shippers of cargo) as appear to the Authority to be representative of those who have served such objections or representations:
(f) where the Authority has designated the case as an environmental case pursuant to regulation 19, such persons (being persons who wish to be heard and who have served objections or representations on grounds of noise, vibration, pollution or other disturbance pursuant to regulation 20(2)) as appear to it to be representative of those who have served such objections or representations:
Provided that;
(i) no person (other than the applicant and the holder of the licence to which the decision will relate) shall have a right to be heard unless he has served an objection or representation pursuant to regulation 20 and (unless he is a person of a description specified in subparagraph (e) or (f) of this regulation) in so doing has stated that he wishes to be heard;
(ii) no person shall be heard before a decision is made by the Authority in a case where the Secretary of State has directed that the licence be granted, refused, revoked, suspended or varied or where the Authority's duty under section 31(2) of the Airports Act 1986 requires that the licence be refused, revoked, suspended or varied.

(2) Notwithstanding that a person does not have the right to be heard, the Authority may, if it thinks fit, hear him:
Provided that no person shall be heard pursuant to this paragraph;
(i) unless he has served an objection or a representation pursuant to regulation 20;
(ii) in a case where such a direction as is referred to in proviso (ii) to paragraph (1) of this regulation has been given.

(3) No hearing shall be held pursuant to this regulation unless the Authority has served on all persons having a right to be heard and whom it proposes to hear in connection with the case and on any person who has been consulted by the Authority pursuant to regulation 21 and who has responded in writing not less than 14 days' notice of the date, time and place of the hearing, and the notice shall clearly identify the application or proposal to which it relates: a similar notice shall be published not less than 7 days before the date of the hearing, and shall be exhibited in a public place in the Authority's principal office during the 7 days immediately preceding the date of the hearing:
Provided that in cases where the Authority is satisfied that for reasons of urgency it is desirable to do so, a hearing may be held without such notice having been served, published and exhibited as aforesaid if the Authority has given notice of the date, time and place of the hearing, being notice of such length and by such means (whether oral or written) as it thinks fit, to the applicant and any person of a description specified in paragraph (1) of this regulation whose interests are in the opinion of the Authority likely to be prejudiced by the granting of the application and to any person consulted by the Authority pursuant to regulation 21.

(4) Two or more cases may be heard together, if the Authority thinks fit, but a party to one case shall not on that account be deemed to be a party to any other case.

26 Procedure at hearings

(1) Hearings shall be conducted by the Authority, sitting with such employees of the Authority acting as advisers as it thinks fit.

(2) At a hearing every party to a case may appear in person or be represented by any other person whom he may have authorised to represent him, and may produce oral and written evidence and may examine any other party to that case, any person whom the Authority hears pursuant to regulation 25(2) and any witnesses produced by any such party or person: the Authority may, to such extent as it thinks fit, permit any person heard by it pursuant to regulation 25(2) to exercise at the hearing the rights set out in this paragraph of a party to the case.

(3) Any person who has served an objection or representation pursuant to regulation 20 but who does not wish to be heard, may make a written submission which he shall serve on the Authority not less than 3 working days before the date fixed for the hearing of the case.

(4) Every hearing shall be held in public unless the Authority shall otherwise decide in relation to the whole or part of a particular case, but nothing in this regulation shall prevent a member of the Council on Tribunals or of its Scottish Committee from attending a hearing in his capacity as such.

(5) Where any person consulted by the Authority pursuant to regulation 21 has responded in writing and that person or a person acting on behalf of that person attends the hearing the Authority shall give him opportunity at the hearing to make observations on the evidence and arguments advanced by the parties to the case and by any persons heard by the Authority pursuant to regulation 25(2), and where any such observations are made the Authority shall give the parties to the case, and any person heard pursuant to regulation 25(2), opportunity at the hearing to respond to them.

(6) The failure of the Authority or of any person to give notice or publish any particulars in the time or manner provided for in the Act or in these Regulations or any other procedural irregularity shall not invalidate the action taken by the Authority; and the Authority may, and shall if it considers that any person may have been prejudiced, take such steps as it thinks fit before reaching its decision to cure the irregularity, whether by the giving of notice or otherwise.

(7) All the proceedings at a hearing of the Authority in connection with a case shall be recorded by a shorthand writer or by some other means, and if any person requests a record of the proceedings the Authority shall cause a mechanical recording or transcript of the shorthand or other record to be made available for purchase by that person at a reasonable price:

Provided that;

(a) the Authority shall not be required to make available a mechanical recording or transcript of the record of the proceedings at any time after the expiry of one year from the day of publication of its decision of the case; and

(b) a mechanical recording or transcript of the record of proceedings conducted otherwise than in public shall only be required to be made available for purchase by any party to the case or by any other person heard by the Authority at those proceedings.

(8) When the Authority provides to a person having a right of appeal pursuant to regulation 27(1);

(a) notification in writing of its decision of the case, the notification shall specify a date, being not less than 3 working days after the date on which a copy of the notification was available for collection by or despatch to that person (which date is hereinafter referred to as "the decision date");

(b) a mechanical recording or transcript of the record of proceedings in the case pursuant to a request made by that person within 7 days after the decision date, the recording or transcript shall be accompanied by a statement specifying a date, being not less than 3 working days after the date on which the recording or transcript was available for collection by or despatch to that person (which date is hereinafter referred to as "the transcript date");

and the Authority shall as soon as may be thereafter publish the decision date and the transcript date.

27 Appeals to the Secretary of State

(1) Every party to a case before the Authority (not being a person having a right to be heard by virtue only of regulation 25(1)(e) or (f)) shall have a right of appeal to the Secretary of State in accordance with the provisions of this regulation from the Authority's decision with respect to an air transport licence or an application for a licence.

(2) An appeal to the Secretary of State shall be made by a notice signed by or on behalf of the appellant and clearly identifying the case to which it relates and stating the grounds on which the appeal is based and the arguments on which the appellant relies.

(3) The appellant shall serve the notice of appeal on:

(a) the Secretary of State;

(b) the Authority;

(c) each of the parties to the case before the Authority;

(d) each person whom, pursuant to regulation 25(2), the Authority had decided to exercise its discretion to hear in connection with the case, whether that person was heard or not; and

(e) any person consulted by the Authority, pursuant to regulation 21, in connection with the case.

(4) Subject to paragraph (9) of this regulation, the notice of appeal shall be served within 21 days after the decision date or, if the appellant has made such a request as is referred to in regulation 26(8) and has within 24 hours after making his request to the Authority served notice on each of the persons referred to in paragraph (3)(a), (c), (d) and (e) of this regulation that he has done so, not later than 21 days from the transcript date.

(5) Any person having the right to appeal against a decision of the Authority may require it to furnish him with the names and addresses of the persons of the description specified in paragraph (3)(c), (d) or (e) of this regulation.

(6) Subject to paragraph (9) of this regulation, any party to the appeal (other than the appellant) and any person who has been served with notice of the appeal pursuant to paragraph 3(e) of this regulation may within 14 days after service thereof serve on the Secretary of State a submission giving reasons why the Authority's decision should or should not be upheld and shall within such period serve copies of any such submission on the Authority, the appellant and the persons who have been served with notice of the appeal pursuant to paragraph (3)(c), (d) and (e) of this regulation.

(7) Subject to paragraph (9) of this regulation, within 28 days after receiving notice of an appeal, the Authority shall serve on the Secretary of State any submission it may wish to make in connection with the appeal, including, if it thinks fit, an amplification and explanation of the reasons for its decision, and shall, within such period, serve copies of any such submission on the appellant and on the persons who have been served with notice of the appeal pursuant to paragraph (3)(c), (d) and (e) of this regulation.

(8) Subject to paragraph (9) of this regulation, within 14 days after the expiry of the period of 28 days referred to in the preceding paragraph of this regulation, the appellant may serve on the Secretary of State a reply to any submission made pursuant to paragraph (6) or (7) of this regulation and shall within such period serve copies of any such reply on the Authority and on the persons who have been served with notice of the appeal pursuant to paragraph (3)(c), (d) and (e) of this regulation.

(9) Where a case has come before the Authority solely because of a need to allocate scarce bilateral capacity, the references in paragraph (4) to 21 days shall be taken as references to 5 working days, the reference in paragraph

(6) to 14 days shall be taken as a reference to 5 working days, the reference in paragraph (7) to 28 days shall be taken as a reference to 8 working days and in paragraph (8) for the words "within 14 days after the expiry of the period of 28 days" there shall be substituted "within 4 working days after the expiry of the period of 8 working days".

(10) Before deciding an appeal the Secretary of State may;

(a) ask the appellant, any other person who has made a submission pursuant to the preceding paragraphs of this regulation, or the Authority, to amplify or explain any point made by them or to answer any other question, the answer to which appears to the Secretary of State necessary to enable him to determine the appeal, and the Secretary of State shall as the case may be give the appellant, the other parties to the appeal and the Authority an opportunity of replying to such amplification, explanation or answer;

(b) obtain from the Authority any information which is in the possession of the Authority but which, pursuant to paragraph (ii) of the proviso to regulation 22 the Authority did not furnish to any person having the right to be heard by the Authority in connection with the case: the Secretary of State shall give the Authority and the person who provided the information to the Authority an opportunity of making written submissions in connection with any information so obtained: a copy of any submission of the Authority made pursuant to this subparagraph shall be served only on the person who provided the information to the Authority and a copy of any submission of that person or body made pursuant to this subparagraph shall be served only on the Authority.

(11) In the appeal proceedings no person may submit to the Secretary of State evidence which was not before the Authority when it decided the case.

28 Appeal from decisions after preliminary hearings of allegations of behaviour damaging to a competitor

(1) Regulation 27 shall apply in relation to appeals from decisions of the Authority after preliminary hearings of allegations of behaviour damaging to a competitor as it applies in relation to any other case but with the modifications herein set out.

(2) Those modifications are;

(a) in paragraph (1) the reference to every party shall be taken as a reference to the applicant and the respondent;

(b) in paragraph (3) the reference in subparagraph (c) to each of the parties shall be taken as a reference to the applicant or respondent, as the case may be, and the reference in subparagraph (d) to regulation 25(2) shall be taken as a reference to regulation 24(6);

(c) in paragraph (4) the first reference to 21 days shall be taken as a reference to 5 working days and the reference to the decision date shall be taken as a reference to the date upon which the Authority furnished reasons for its decision; and all the subsequent words in that paragraph (which relate to a request for a transcript and a time from the transcript date) shall be deleted;

(d) in paragraph (6) the reference to 14 days shall be taken as a reference to 5 working days;

(e) in paragraph (7) the reference to 28 days shall be taken as a reference to 8 working days;

(f) in paragraph (8) for "within 14 days after the expiry of the period of 28 days" there shall be substituted "within 4 working days after the expiry of the period of 8 working days".

29 Decisions on appeals

(1) The Secretary of State may if he thinks fit uphold the decision of the Authority or direct it to re-hear the case which is the subject of the appeal or to reverse or vary its decision.

(2) The Secretary of State shall notify the Authority, the appellant and the persons who have been served with the notice of appeal pursuant to regulation 27(3) of his decision and of the reasons for it and the Authority shall publish the Secretary of State's notification.

(3) Where the Secretary of State directs the Authority to re-hear a case he shall at the same time notify the Authority and persons referred to in paragraph (2) of this regulation whether the Authority's decision is to have effect pending the further decision of the Authority.

(4)(a) Subject to paragraph (5) of this regulation, in determining an appeal the Secretary of State may, if he thinks fit, order the appellant to pay to any other party thereto either a specified sum in respect of the costs incurred by him in connection with the appeal, or the taxed amount of those costs or any part thereof;

(b) any costs required by an order under the foregoing subparagraph to be taxed may be taxed in the county court on such scale as may be directed by the order;

(c) any sum payable by virtue of an order under subparagraph (a) of this paragraph shall, if the county court so orders, be recoverable by execution issued from the county court or otherwise as if payable under an order of that court;

(d) the powers of the county court under the foregoing provisions of this paragraph may be exercised by the District Judge, or in Northern Ireland by the clerk of the Crown and Peace.

(5)(a) In determining an appeal where the appellant resides or has his registered or principal office in Scotland the Secretary of State may, if he thinks fit, order the appellant to pay to any other party thereto either a specified sum in respect of the expenses incurred by him in connection with the appeal, or the taxed amount of those expenses or any part thereof;

(b) any expenses required by an order under the foregoing subparagraph to be taxed may be taxed by the Auditor of the Court of Session on such a scale as may be directed by the order;

(c) any award of expenses by the Secretary of State under the foregoing provisions of this paragraph may be enforced in like manner as a recorded decree arbitral.

(6) An appeal to the Secretary of State shall not preclude him from consulting the competent authorities of any country or territory outside the United Kingdom for the purposes of section 6(2)(a) to (d) of the Act (which relates to national security, relations with other countries and territories and similar matters) notwithstanding that the consultation may relate to matters affecting the appeal.

(7) The failure of any person (other than the appellant in serving notice of appeal on the Secretary of State within the time prescribed in regulation 27(4)) to serve any notice, submission or reply, or copies thereof or to furnish any particulars in the time or manner provided for in the Act or in these Regulations or any other procedural irregularity shall not invalidate the decision of the Secretary of State; and the Secretary of State may, and shall if he considers that any person may have been prejudiced, take such steps as he thinks fit before deciding the appeal to cure the irregularity.

30 Transfer of licences

(1) Subject to the provisions of this regulation:

(a) if the sole holder of an air transport licence or a route licence (being an individual) shall die, the licence shall be treated from the time of his death as if it had then been granted to his legal personal representative;

(b) if in connection with the reconstruction of any body corporate or the amalgamation of any bodies corporate the whole of the business of the holder of a licence (being a body corporate), or such part thereof as includes the provision of carriage by air for reward of passengers or cargo, is transferred or sold to another body corporate, the licence shall be treated, from the date of the transfer or sale of the whole or the relevant part of the business, as if it had been granted to that other body corporate.

(2) The person required by paragraph (1) of this regulation to be treated as the holder of the licence may apply to the Authority

(a) if he is the legal personal representative of an individual licence holder who has died, for the transfer of the licence to any person entitled to a beneficial interest in the deceased's estate (including himself in his personal capacity if he is in that capacity entitled to such an interest); and

(b) in any other case, for the substitution of his own name in the licence for the name of the person by whom the licence was held.

(3) The application shall state the grounds on which it is based and shall be served on the Authority within 21 days after the date on which the applicant first became entitled to make it; and if no application as aforesaid is made within that period the licence shall cease at the expiration of that period to be treated as if granted to a person other than the person to whom it was granted.

(4) The application shall, for the purposes of these Regulations be treated as if it were an application for the variation of the licence, and the provisions of regulations 27 and 29 as to appeals shall apply accordingly.

(5) The Authority shall not grant an application for the transfer of a licence to, or the substitution of the name of, any person if;

(a) in the case of an air transport licence it would be bound under section 65(2) of the Act, and

(b) in the case of an air transport licence or a route licence it would be bound under section 65(3) or 69A(4) of the Act. To refuse that application if it were an application for the grant of a licence to that person.

(6) For the purposes of this regulation "legal personal representative" means a person constituted executor, administrator or other representative of a deceased person by probate, administration or other instrument.

31 Surrender of licences

If revocation or variation of an air transport licence or a route licence has taken effect, the Authority may require any person who has the licence in his possession or control to surrender it for cancellation or variation, as the case may be and any person who fails, without reasonable cause, to comply with any such requirement, shall be guilty of an offence and shall be liable on summary conviction to a fine not exceeding level 2 on the standard scale.

PART IIIA – REFERENCES IN RESPECT OF AN AIR TRAFFIC SERVICES LICENCE

31A Determination by the authority

(1) Where, pursuant to a condition of an air traffic services licence, any matter or question may be referred for determination by one or more Members of the Authority pursuant to these Regulations and such a matter or question is referred, it shall be determined in accordance with the provisions of this Part of these Regulations.

(2) The function of the Authority under section 7(5) of the Transport Act 2000 with respect to the modification of a licence, to the extent it gives rise to a matter or question referred for determination by one or more members of the Authority, is hereby prescribed for the purpose of section 7(2) of the Act.

(3) For the purpose of making any determination in such a case a quorum of the Authority shall be two Members.

(4) The Authority shall sit with such technical assessors to advise it as the Authority may appoint, but the Authority shall not appoint as an assessor any person who participated in the development of any notice or counter-notice in relation to the matter or question to be determined.

(5) For the purposes of this Part of these Regulations:

(a) "an air traffic services licence" means a licence granted pursuant to Section 6 of the Transport Act 2000;

(b) where the Authority is required to publish any information it may do so electronically or otherwise;

(c) where the Authority is required to make any information available at its principal office for inspection it may do so electronically or otherwise;

(d) anything which is required to be served on the Authority may be served electronically by sending it to an e-mail address which it has published for the purpose or in accordance with regulation 4.

31B Representations

(1) The Authority shall within one calendar month after a reference has been made serve notice of the reference on the licence holder and publish such particulars of the reference as it thinks necessary for indicating the substance of the reference, and shall make a copy of the reference available at its principal office for inspection by any person at any reasonable time.

(2) The licence holder or any other person may serve on the Authority a representation about a reference if he does so within 21 days of, in the case of the licence holder the date of service of notice or, in the case of any other person, publication of the reference by the Authority.

(3) The Authority shall make a copy of any representation which has been served on the Authority available at its principal office for inspection by any person at any reasonable time and shall serve a copy of any representations received from persons other than the licence holder on the licence holder.

31C Hearings in connection with licences

(1) Where a matter or question referred to in regulation 31A(1) has been referred the Authority shall, before making a determination:

(a) consider any representations which may have been served on it by the licence holder or any other person within the time permitted by regulation 31B(2);

(b) consider any written submissions served pursuant to regulation 31D(3); and

(c) conduct a hearing in accordance with regulation 31D and consider any representations made and evidence submitted at such a hearing.

(2) The following persons shall have a right to be heard at the hearing held pursuant to paragraph (1)(c):

(a) the licence holder; and

(b) such persons (being persons who wish to be heard and who have served representations within the time permitted by regulation 31B(2)) as appear to the Authority to be users of services provided by the licence holder or to be representative of such persons.

(3) Notwithstanding that a person does not have the right to be heard, the Authority may, if it thinks fit, hear him.

(4) No hearing shall be held pursuant to this regulation unless the Authority has served on the licence holder and any other person whom it proposes to hear in connection with the case not less than 14 days' notice of the date, time and place of the hearing, and the notice shall clearly identify the reference to which it relates: a similar notice shall be published not less than 7 days before the date of the hearing; and shall be exhibited in a public place in the Authority's principal office during the 7 days immediately preceding the date of the hearing.

31D Procedure at hearings

(1) Hearings shall be conducted by the Authority.

(2) At a hearing any person entitled to be heard may appear in person or be represented by any other person whom he may have authorised to represent him, and may produce oral and written evidence and may examine any other person whom the Authority hears pursuant to regulation 31C and any witnesses produced by any such person.

(3) Any person who has served a representation within the time permitted by regulation 31B(2) but who does not wish to be heard, may make a written submission which he shall serve on the Authority not less than 3 working days before the date fixed for the hearing of the case.

(4)(a) Where an oral hearing is held it shall be held in public unless, having regard to the subject matter of the hearing and any representations from the licence holder the Authority directs that the hearing or any part of the hearing shall take place in private.

(b) Nothing in this regulation shall prevent a member of the Council on Tribunals or of its Scottish Committee from attending a hearing in his capacity as such.

(5) The failure of the Authority or of any person to give notice or publish any particulars in the time or manner provided for in the Act or in these Regulations or any other procedural irregularity shall not invalidate the action taken by the Authority; and the Authority may, and shall if it considers that any person may have been prejudiced, take such steps as it thinks fit before reaching its determination to cure the irregularity, whether by the giving of notice or otherwise.

(6)(a) Subject to sub-paragraph (b) all the proceedings at a hearing of the Authority in connection with a case shall be recorded by a shorthand writer or by some other means, and if any person requests a record of the proceedings the Authority shall cause a mechanical recording or transcript of the shorthand or other record to be made available for purchase by that person at a reasonable price.

(b)(i) The Authority shall not be required to make available a mechanical recording or transcript of the record of the proceedings at any time after the expiry of one year from the day of publication of its determination of the case.

(ii) A mechanical recording or transcript of the record of proceedings conducted otherwise than in public shall only be required to be made available for purchase by any person heard by the Authority at those proceedings.

31E Determination by Authority and Appeal to the Secretary of State

(1) Where the Authority makes a determination pursuant to this Part of these Regulations the Authority shall serve a notice of its determination and a statement of its reasons for the determination on the licence holder.

(2)(a) Where the relevant condition of an air traffic services licence provides in respect of a matter or question that the decision of the Authority shall be definitive there shall be no appeal to the Secretary of State.

(b) Where the relevant condition of an air traffic services licence provides in respect of a matter or question that the licence holder may appeal to the Secretary of State, he may do so in accordance with this regulation.

(3)(a) An appeal to the Secretary of State shall be made by a notice signed by or on behalf of the licence holder and clearly identifying the matter or question to which it relates and stating the grounds on which the appeal is based and the arguments on which the licence holder relies.

(b) The licence holder shall serve the notice on the Secretary of State and a copy on the Authority within 14 days of receipt by the licence holder of the notice of determination and statement of reasons pursuant to paragraph (1).

(4) The Authority shall within one calendar month after receiving notice of appeal publish such particulars of the appeal as it thinks necessary to indicate the substance of the appeal.

(5) Within 14 days after receiving notice of an appeal, the Authority shall serve on the Secretary of State any submission it may wish to make in connection with the appeal including, if it thinks fit, an amplification and explanation of the

reasons for its determination, and shall, within such period, serve a copy of any such submission on the licence holder.
(6) Within 14 days after publication of the notice of the appeal by the Authority any person who appeared at the hearing before the Authority may serve on the Secretary of State a submission giving reasons why the Authority's determination should or should not be upheld and shall within such period serve copies of any such submission on the Authority and the licence holder.
(7) Within 14 days after receipt of any submission made pursuant to the preceding two paragraphs the licence holder may serve on the Secretary of State a reply and shall within such period serve a copy of any such reply on the Authority
(8) Before deciding an appeal the Secretary of State may ask the licence holder, the Authority or any other person who appeared at the hearing held by the Authority to amplify or explain any point made by them or to answer any other question, the answer to which appears to the Secretary of State necessary to enable him to determine the appeal, and the Secretary of State shall as the case may be give the licence holder, the Authority and any other person who appeared at the hearing held by the Authority an opportunity of replying to such amplification, explanation or answer.
(9)(a) Where any person is obliged to serve on the Authority any notice, representation, submission or other material pursuant to this regulation the Authority shall as soon as may be after receipt thereof make a copy available at its principal office for inspection by any person at any reasonable time.
(b) The Authority shall also make a copy of any representation, submission or other material which it is obliged to serve on the licence holder or the Secretary of State pursuant to this regulation available at its principal office for inspection by any person at any reasonable time.
(10) In the appeal proceedings none of the Authority, the licence holder or any other person may submit to the Secretary of State evidence which was not before the Authority when it decided the case.

31F Decision by Secretary of State on appeal
(1) The Secretary of State may if he thinks fit uphold the determination of the Authority in whole or in part or reverse or vary the whole or any part of its determination.
(2) The Secretary of State shall notify the Authority and the licence holder of his decision and of the reasons for it and the Authority shall publish the Secretary of State's notification.
(3) The failure of any person (other than the licence holder in serving notice of appeal on the Secretary of State within the time prescribed in regulation 31E(3)(b)) to serve any notice, representation, submission or reply, or copies thereof or to furnish any particulars in the time or manner provided for in the Act or in these Regulations or any other procedural irregularity shall not invalidate the decision of the Secretary of State; and the Secretary of State may, and shall if he considers that any person may have been prejudiced, take such steps as he thinks fit before deciding the appeal to cure the irregularity.

PART IV – OTHER FUNCTIONS OF THE AUTHORITY

32 Participation in civil proceedings
(1) The function of the Authority of being a party to civil proceedings is hereby prescribed for the purposes of section 7(2) of the Act.
(2) In any civil proceedings to which the Authority is or becomes a party, the Authority shall disclose to the Court and any other party to the proceedings information in its possession which, apart from section 23(1) of the Act, it would have been under a duty to disclose for the purpose of those proceedings.

SCHEDULE: REVOCATION
Regulation 2

Regulations Revoked	References
The Civil Aviation Authority Regulations 1983	1983/550
The Civil Aviation Authority (Amendment) Regulations 1987	1987/379
The Civil Aviation Authority (Amendment) Regulations 1989	1989/1826
The Civil Aviation Authority (Amendment) Regulations 1990	1990/9

Table of comparison

The following Table shows, in relation to each regulation of the Civil Aviation Authority Regulations 1983, as amended, the regulations of the 1991 Regulations in which it is reproduced.

1983 Regulations as amended	1991 Regulations
1	1
2	2
3	3
4	4
5	5
6	6
7	7
8	8
9	9
9A	10
9B	11
9C	12
9D	13
9E	14
10	15
11	16
12	17
13	18
14	19
15	20
16	21
17	22
18	23
18A	24
19	25
20	26
21	27
21A	28
22	29
23	30
24	31
25	32

General
UK AIM
The UK Aeronautical Information Manual

Aviation Organisations

AAACF
Airline, Aviation & Aerospace Christian Fellowships
Unit 5 CBC
Frimley Road
Camberley
Surrey
GU15 3EN
Tel & Fax: 01276 709474
e-mail: aaacf@lineone.net
Contact: Capt Christopher Cowell

AATA
Animal Air Transportation Association
Harris Associates Ltd
PO Box 251
Redhill
Surrey
RH1 5FU
Tel: 01737 822954
Fax: 01737 822249
e-mail: 100257.1720@compuserve.com
website: www.tim-harris.co.uk
Contact: Tim Harris

AAU
Association of Aerospace Universities
c/o Coventry University
School of Engineering (SE-Q)
Priory Street
Coventry
CV1 5FB
Tel: 024 7683 8655
Fax: 024 7683 8949
e-mail: mike.west@cov.ac.uk
Contact: Dr Mike West

ABTA
Association of British Travel Agents
68-71 Newman Street
London
W1T 3AH
Tel: 0207 637 2444
Fax: 0207 637 0713
e-mail: information@abta.co.uk
website: www.abta.com
Contact: Keith F Betton

ACAT
Association of Colleges of Aerospace Technology
c/o Brooklands College
Aero Avionics Department
Heath Road
Weybridge
Surrey
KT13 8TT
Tel: 01932 797772
Fax: 01932 797804
e-mail: acrzzz@aol.com
Contact: A C Russell

Acorne Sports Ltd
Bank House
Finings Road
Lane End
Bucks
HP14 3ES
Tel: 01494 880000
Fax: 01494 883377
website: www.acorne.co.uk
Contact: Fiona Falconer

Action for Airfields
PO Box 1015
Ipswich
IP1 1UJ
Tel: 0870 321 2015
Fax: 0870 321 2013
e-mail: info@airfields.org.uk
website: www.airfields.org.uk

Advantage Travel Centres
National Association of Independent Travel Agents
Kenilworth House
79-80 Margaret Street
London
W1N 7HB
Tel: 0207 323 3408
Fax: 0207 323 5189
e-mail: normang@advantagebusinesstravel.com
website: www.advantagebusinesstravel.com
Contact: Debbie Weston

AGIFORS
Airline Group, International Federation of Operational Research Societies
Mr G K Rand
c/o Department of Management Science
The Management School
Lancaster University
Lancaster
LA1 4YX
Tel: 01524 593849
Fax: 01524 844885
e-mail: g.rand@lancs.ac.uk
website: www.agifors.org
Contact: Graham K Rand

AIOA
Aviation Insurance Offices' Association
49 Leadenhall Street
London
EC3A 2BE
Tel: 0207 488 2424
Fax: 0207 702 3010
e-mail: dave.matcham@iua.co.uk
Contact: Dave Matcham

Air-Britain (Historians) Ltd
2 The Green
Edlesborough
Dunstable
Beds
LU6 2JF
Tel: 01525 220901
Fax: 01732 838969
e-mail: ron_webb@lineone.net
website: www.air-britain.com
Contact: A T Jones

The Aircrew Association
37 Picardy Road
Belvedere
Kent
DA17 5QH
Tel: 01322 437426
Fax: 01322 439238
e-mail: secretary@aircrew.org.uk
website: www.aircrew.org.uk

Air Display Association Europe
The Grange
Dunston
Lincs
LN4 2ET
Tel & Fax: 01526 320726
e-mail: adae@globalnet.co.uk
Contact: Dick Roberts

General — Aviation Organisations The Air League

The Air League
Broadway House
Tothill Street
London
SW1H 9NS
Tel: 0207 222 8463
Fax: 0207 222 8462
e-mail: exec@airleague.co.uk
website: www.airleague.co.uk
Contact: Edward Cox

Airship Association
6 Kings Road
Cheriton
Folkestone
Kent
CT20 3LG
Tel & Fax: 01303 277560
e-mail: secretary@airship-association.org
website: www.airship-association.org
Contact: Arnold W L Nayler

ALAE
Association of Licensed Aircraft Engineers (1981)
Bourn House
8 Park Street
Bagshot
Surrey
GU19 5AQ
Tel: 01276 474888
Fax: 01276 452767
e-mail: alae@bagshot.sagehost.co.uk
website: www.lae.mcmail.com
Contact: John Sawyer

AOA
Airport Operators Association
3 Birdcage Walk
London
SW1H 9JJ
Tel: 0207 222 2249
Fax: 0207 7976 7405
website: www.aoa.org.uk
Contact: Keith Jowett

AOPA
Aircraft Owners and Pilots Association
50a Cambridge Street
London
SW1V 4QQ
Tel: 0207 834 5631
Fax: 0207 834 8623
e-mail: aopa@easynet.co.uk
website: www.aopa.co.uk
Contact: Martin Robinson

AOPA
Channel Islands Region
The Cottage
Anne Port
St Martin
Jersey
JE3 6DT
Channel Islands
Tel: 01534 851681
Fax: 01534 854559
Contact: Charles Strasser

AOPA
Ireland
Loughlinstown Road
Celbridge
Co Kildare
Ireland
Tel: 00 353 1 700 1826

APRA
Air Public Relations Association
Kings Grant
2 Tattam Close
Woolstone
Milton Keynes
MK15 0NB
Tel: 01908 670770
Fax: 01908 235649
e-mail: nield@dial.pipex.com
Contact: Ron Neild

APRO
Airline Public Relations Organisation
33 Gossops Green Lane
Crawley
West Sussex
RH11 8BJ
Tel: 01293 520650
Fax: 01293 571003
e-mail: pkpr@cwcom.net
Contact: Pauline Kirkman

ASAAS
Association of Suppliers to Airlines, Airports & Shipping
2 Cunningham Avenue
Guildford
Surrey
GU1 2PE
Tel: 01483 563125
Fax: 01503 272670
e-mail: asaas.dutyfree@ntlworld.com
Contact: Roger Morris

ASG
Air Safety Group
Birchwood House
Letchbridge Park
Bishops Lydeard
Taunton
Somerset
TA4 3QU
Tel: 01823 430161
e-mail: russwms@talk21.com
Contact: Capt Russell Williams

ATA
Air Transport Auxiliary Association
40 Goldcrest Road
Chipping Sodbury
South Glos
BS17 6XG
Tel & Fax: 01454 319175
Contact: Diana Barnato Walker

ATA
Aviation Training Association
Dralda House
Crendon Street
High Wycombe
Bucks
HP13 6LS
Tel: 01494 445262
Fax: 01494 439984
e-mail: mail@aviation-training.org
Contact: Tony Hines

AUC
Air Transport Users Council
CAA House
45-59 Kingsway
London
WC2B 6TE
Tel: 0207 240 6061
Fax: 0207 240 7071
website: www.auc.org.uk
Contact: Dr Simon Evans

The Aviation Club of the UK
19 Priory Road
Hampton
Middx
TW12 2NR
Tel: 0208 941 8199
Fax: 0208 979 9428
e-mail: cknaomi@aol.com
Contact: Naomi King

Aviation in the Community
Green Hawk Trust
55 Plantation Road
Harrogate
HG2 0DB
Tel & Fax: 01423 509777
e-mail: 100630.457@compuserve.com
Contact: Dr Tony Denson

Balloon & Airship Society of Ireland
5 Cill Cais
Old Bawn
Tallaght
Dublin 24
Ireland

BAAC
British Association of Aviation Consultants
Carlyle House
235-237 Vauxhall Bridge Road
London
SW1V 1EJ
Tel: 0207 630 5358
Fax: 0207 828 0667
website: www.baac.org.uk
Contact: R J Collis

BAAC
British Aviation Archaeological Council
8 Holly Road
Oulton Broad
Lowestoft
Suffolk
NR32 3NH
Tel: 01502 585421
Contact: M Evans

BABO
The British Association of Balloon Operators
Cross Lanes Farm
Walcote
Alcestor
Warwicks
B49 6NA
Tel: 01672 563379
Fax: 01672 564148
e-mail: secretary@babo.org.uk
website: www.babo.org.uk
Contact: Sandra Hossack

BACA
Baltic Air Charter Association
c/o The Baltic Exchange
St Mary Axe
London
EC3A 8BH
Tel: 0207 623 5501
Fax: 0207 369 1622/3
e-mail: enquiries@balticexchange.co.uk
website: www.baca.org.uk
Contact: W van der Pol

BACD
British Association of Conference Destination
6th floor
Charles House
148-149 Great Charles Street
Birmingham
West Midlands
B3 3HT
Tel: 0121 212 1400
Fax: 0121 212 3131
website: www.bacd.org.uk
Contact: Sarah Smart

BAEA
British Aerobatic Association
White Waltham Airfield
Maidenhead
Berks
SL6 3NJ
Tel: 01234 713245
e-mail: info@aerobatics.org.uk
website: www.aerobatics.org.uk
Contact: Gareth Roberts

BAG
British Airports Group
Duxbury House
60 Petty France
London
SW1 9EU
Tel: 0207 227 1102
Fax: 0207 227 1105
e-mail: bag@sbac.co.uk
website: www.britishairportsgroup.co.uk
Contact: Richard Loveday

BALPA
British Air Line Pilots Association
81 New Road
Harlington
Hayes
Middx
UB3 5BG
Tel: 0208 476 4000
Fax: 0208 476 4077
e-mail: balpa@balpa.org.uk
website: www.balpa.org.uk
Contact: Chris Drake

BAPC
British Aviation Preservation Council
c/o Museum of Science & Industry
Liverpool Road
Castlefield
Manchester
M3 4FP
Tel: 0161 606 0121
Fax: 0161 606 0186
e-mail: n.forder@msim.org.uk
Contact: Nick Forder

BAR UK LTD
Board of Airline Representatives in the UK
200 Buckingham Palace Road
London
SW1W 9TA
Tel: 0207 707 4147
Fax: 0207 707 4182
e-mail: office@bar-uk.org
website: www.bar-uk.org
Contact: Peter N North

BASEA
British Airport Services & Equipment Association
Homelife House
26-32 Oxford Road
Bournemouth
Dorset
BH8 8EZ
Tel: 01202 299088
Fax: 01202 508234
e-mail: basea@dial.pipex.com
website: www.basea.org.uk
Contact: Janet Rose

BATA
British Air Transport Association
200 Buckingham Palace Road
London
SW1W 9TA
Tel: 0207 730 6931
Fax: 0207 823 6726
e-mail: batauk@talk21.com
website: www.bata.uk.com
Contact: Roger Wiltshire

BAUA
Business Aircraft Users Association
Crossmount House
Kinloch Rannoch
Perths
PH16 5QF
Scotland
Tel: 01882 632252
Fax: 01882 632454
Contact: Derek C Leggett

BBAC
British Balloon & Airship Club
Wellington House
Lower Icknield Way
Longwick
Princes Risborough
Bucks
HP27 9RZ
Tel: 01604 870025
e-mail: info@bbac.org
website: www.bbac.org
Contact: Patricia Pedler

BFTS ASSOCIATION
British Flying Training School Association
5 Thornton Crescent
Old Coulsdon
Surrey
CR3 1LJ
Tel: 01737 554496
Contact: A J Allam

BGA
British Gliding Association
Kimberley House
Vaughan Way
Leicester
LE1 4SE
Tel: 0116 253 1051
Fax: 0116 251 5939
e-mail: bga@gliding.co.uk
website: www.gliding.co.uk
Contact: Merri Head

BHAB
British Helicopter Advisory Board
Graham Suite
West Entrance
Fairoaks Airport
Chobham
Surrey
GU24 8HX
Tel: 01276 856100
Fax: 01276 856126
e-mail: info@bhab.demon.co.uk
website: www.bhab.demon.co.uk
Contact: Collette Warrington

BHPA
British Hang Gliding & Paragliding Association
The Old School Room
Loughborough Road
Leicester
LE4 5PJ
Tel: 0116 261 1322
Fax: 0116 261 1323
e-mail: office@bhpa.co.uk
website: www.bhpa.co.uk
Contact: Dave Wootton

BIS
British Interplanetary Society
27-29 South Lambeth Road
London
SW8 1SZ
Tel: 0207 735 3160
Fax: 0207 820 1504
e-mail: bis.bis@virgin.net
website: www.bis-spaceflight.com
Contact: Miss S A Jones

BMAA
British Microlight Aircraft Association
Bullring
Deddington
Banbury
Oxon
OX15 0TT
Tel: 01869 338888
Fax: 01869 337116
e-mail: general@bmaa.org
website: www.bmaa.org
Contact: Christopher Finnigan

BMFA
British Model Flying Association
Chacksfield House
31 St Andrews Road
Leicester
LE2 8RE
Tel: 0116 244 0028
Fax: 0116 244 0645
e-mail: admin@bmfa.org
website: www.bmfa.org
Contact: John Henderson

BNSC
British National Space Centre
151 Buckingham Palace Road
London
SW1W 9SS
Tel: 0207 215 0807
Fax: 0207 821 0936
e-mail: maria.bazell@bnsc.gsi.gov.uk
website: www.bnsc.gov.uk
Contact: Maria Bazell

BPA
British Parachute Association
5 Wharf Way
Glen Parva
Leicester
LE2 9TF
Tel: 0116 278 5271
Fax: 0116 247 7662
e-mail: skydive@bpa.org.uk
website: www.bpa.org.uk
Contact: Kieran Brady

BPC
Beagle Pup Club
20 Primrose Close
Flitwick
Bedford
MK45 1PJ
Contact: Phil Abbott

BPC
British Paramotoring Club
PO Box 275
Winchester
Hants
SO21 3WR
Tel & Fax: 01962 774443
e-mail: office@paramotoring.co.uk
website: www.paramotoring.org

BPPA
British Precision Pilots Association
Hill Farm
Yoxford
Saxmundham
Suffolk
IP1 HJ
Tel: 01728 668354
Fax: 01728 668402
e-mail: acdriver@email.msn.com
Contact: Rodney Blois

The British Cargo Airline Alliance
100 Rochester Row
London
SW1P 1JP
Tel: 0115 958 5131
Fax: 0115 958 7814
Contact: Steve Guynan

British Institute of Non-destructive Testing Certification Services Division
1 Spencer Parade
Northampton
NN1 5AA
Tel: 01604 259056
Fax: 01604 231489
e-mail: info@bindt.org
website: www.bindt.org
Contact: Mrs M Wood

BSI
British Standards Institution
389 Chiswick High Road
London
W4 4AL
Tel: 0208 996 9000
Fax: 0208 996 7400
website: www.bsi-global.com
Contact: Vivian E Thomas

BVQI
Bureau Veritas Quality International
224-226 Tower Bridge Court
Tower Bridge Road
London
SE1 2TX
Tel: 0207 661 0700
Fax: 0207 661 0790
e-mail: info@bvqi.com
website: www.bvqi.com
Contact: Olivier Guize

BWPA
British Women Pilots' Association
Brooklands Museum
Brooklands Road
Weybridge
Surrey
KT13 0QN
Tel: 01342 892739
e-mail: enquiries@bwpa.demon.co.uk
website: www.bwpa.com
Contact: Sue Chase

CC89 – AEEU Officers
Cabin Crew 89
28 Pegasus Court
Herschel Street
Slough
Berks
SL1 1PA
Tel: 01753 578439
Fax: 01753 693906
e-mail: cabincrew89@compuserve.com
Contact: Jim Welsh

CCCA
Conference of City Centre Airports
Indosuez House
122 Leadenhall Street
London
EC3V 4QH
Tel: 0207 971 4048
Fax: 0207 577 3212
Contact: William T Charnock

Chipmunk Club
DHC-1 Chipmunk Club
4 Coleman Court
Rosedale Close
Stanmore
Middx
HA7 3QF
Tel & Fax: 0208 954 5080
Contact: Ralph Steiner

CIMTIG
Chartered Institute of Marketing Travel Industry Group
Home Cottage
Old Lane
Tatsfield
Westerham
Kent
TN16 2LN
Tel & Fax: 01959 577469
e-mail: ugo@cimtig.org
website: www.cimtig.org
Contact: Jill Ugo

CSFA
Chartered Surveyors' Flying Association
11 Richford Street
London
W6 7HJ
Tel: 0207 629 8171

Aviation Organisations CTT

CTT
Council for Travel & Tourism
Vigilant House
120 Wilton Road
London
SW1V 1JZ
Tel: 0207 630 6686
Fax: 0207 630 6656
Contact: Mrs M Goddard

DEHMC
de Havilland Moth Club Ltd
Staggers
23 Hall Park Hill
Berkhamsted
Herts
HP4 2NH
e-mail: dhmoth@dhmothclub.co.uk
website: www.dhmothclub.co.uk
Contact: Stuart McKay

DMA
The Defence Manufacturers Association
Marlborough House
Headley Road
Grayshott
Surrey
GU26 6LG
Tel: 01428 607788
Fax: 01428 604567
e-mail: enquiries@the-dma.org.uk
website: www.the-dma.org.uk
Contact: Brinley M Salzmann

EAC
European Airshow Council
PO Box 1942
Fairford
Glos
GL7 4LW
Tel: 01285 713300 Ex 3490
e-mail: bob@european-airshow.com
website: www.european-airshow.com
Contact: Bob Dixon

EASA
European Aviation Suppliers Association
Huntingdon House
278-280 Huntingdon Street
Nottingham
Notts
NG1 3LY
Tel: 0115 952 4333
Fax: 0115 955 0516
e-mail: quality@easa.org.uk
website: www.easa.org.uk
Contact: Paul D Evans

EEF
Broadway House
Tothill Street
London
SW1H 9NQ
Tel: 0207 222 7777
Fax: 0207 222 2782
e-mail: enquiries@eef-fed.org.uk
website: www.eef.org.uk
Contact: Martin Temple

EEF South
Station Road
Hook
Hants
RG27 9TL
Tel: 01256 763969
Fax: 01256 768530
e-mail: enquiries@eef-south.org.uk
website: www.wwf.org.uk/south
Contact: Martin Temple

EMTA
Engineering & Marine Training Authority
EMTA House
14 Upton House
Watford
Herts
WD18 0JT
Tel: 01923 238441
Fax: 01923 256086
Contact: Michael D Sanderson

EPFA
The European Property Flying Association
Bryn Heulwen
Llangattock
Monmouth
NP5 4NG
Wales
Tel: 01600 715441
Contact: Nelson E R Whaley

ERA
European Regions Airline Association
The Baker Suite
Fairoaks Airport
Chobham
Woking
Surrey
GU24 8HX
Tel: 01276 856495
Fax: 01276 857038
e-mail: info@eraa.org
website: www.eraa.org
Contact: Ms Lesley Shepherd

ETOA
European Tour Operators Association
6 Weighhouse Street
London
W1Y 1YL
Tel: 0207 499 4412
Fax: 0207 499 4413
e-mail: info@etoa.org
website: www.etoa.org
Contact: Tom Jenkins

FAA – UK
Foreign Airlines Association (UK)
'Kingswood'
1 Vineyard Drive
Bourne End
Bucks
SL8 5PD
Tel & Fax: 01628 521355
Contact: Thomas F Kingston

FAC
Farnborough Aerospace Consortium
Council Offices
Farnborough Road
Farnborough
Hants
GU14 7JU
Tel: 01962 732266
Fax: 01962 736066
e-mail: fac@dial.pipex.com
website: www.fac.org.uk
Contact: Martin Best

FAGSA
Federation of Airline General Sales Agents
Arbor House
Broadway
North Walsall
WS1 2AN
Tel: 0121 782 1133
Fax: 0121 782 1386
e-mail: secretary@fagsa.org
website: www.fagsa.org
Contact: Malcolm J Warner

FAI
Fédération Aeronautique Internationale
24 Avenue Mon Repos
CH-1005 Lausanne
Switzerland
Tel: +41 21 345 1070
Fax: +41 21 345 1077
e-mail: info@fai.org
website: www.fai.org
Contact: Max Bishop

FARA
Formula Air Racing Association
49 Berkeley Square
London
W1X 5DB
Tel: 0207 495 6587
Fax: 0207 499 2782
Contact: Andrew Chadwick

FAST Association
Farnborough Air Sciences Trust Association
16 Laurel Road
Locksheath
Southampton
SO31 6QG
e-mail: anyname@fasta.co.uk
website: www.fasta.co.uk
Contact: Richard Gardner

FFA
Flying Farmers Association
Moor Farm
West Heslerton
Malton
North Yorks
YO17 8RU
Tel: 01944 738281
Fax: 01944 738240
e-mail:gbsdg@farmline.com
website: www.members.farmline.com/flyfarm
Contact: Paul A Stephens

GAAC
General Aviation Awareness Council
The British Light Aviation Centre
50a Cambridge Street
London
SW1V 4QQ
Tel: 0207 834 5631
Fax: 0207 834 8623
website: www.gaac.co.uk
Contact: David F Ogilvy

GAMTA
General Aviation Manufacturers Association
19 Church Street
Brill
Aylesbury
Bucks
HP18 9RT
Tel: 01844 238020
Fax: 01844 238087
e-mail: ga@gamta.org
website: www.gamta.org
Contact: Ms Anne Seckington

GAPAN
Guild of Air Pilots & Air Navigators
Cobham House
9 Warwick Court
Gray's Inn
London
WC1R 5DJ
Tel: 0207 404 4032
Fax: 0207 404 4035
e-mail: gapan@gapan.org
website: www.gapan.org
Contact: John Stoy

GASCO
General Aviation Safety Council
Rochester Airport
Maidstone Road
Chatham
Kent
ME5 9SD
e-mail: john.campbell@gen-av-safety.demon.co.uk
Contact: John Campbell

GATCO
UK Guild of Air Traffic Control Officers
24 The Greenwood
Guildford
Surrey
GU1 2ND
Tel & Fax: 01483 578347
e-mail: gatcocaf@msn.com
website: www.gatco.org
Contact: Kathleen Nuttall

GAVA
Guild of Aviation Artists
Unit 4.18
Bondway Business Centre
71 Bondway
Vauxhall Cross
London
SW8 1SQ
Tel & Fax: 0207 735 0634
e-mail: admin@gava.org.uk
Contact: Richard Gardner

GBTA
Guild of Business Travel Agents
Artillery House
Artillery Row
London
SW1P 1RT
Tel: 0207 222 2744
Fax: 0207 7976 7094
e-mail: info@gbta-guild.com
website: www.gbta-guild.com
Contact: Philip H Carlisle

Girls Venture Corps Air Cadets
Redhill Aerodrome
Kings Mill Lane
South Nutfield
Redhill
Surrey
RH1 5JY
Tel & Fax: 01737 823345
website: www.gvcac.org.uk
Contact: Mrs G Bennett

HAA
The Historic Aircraft Association
17 Ravensgate Avenue
Leamington Spa
Warwicks
CV32 6NQ
Tel: 01926 831324
Fax: 0247 6305748
Contact: Darrol Stinton

HCGB
Helicopter Club of Great Britain
Ryelands House
Aynho
Banbury
Oxon
OX17 3AT
Tel: 01869 810646
Fax: 01869 810755
Contact: Jeremy James

Aviation Organisations HPA

HPA
Handley Page Association
16 Guernsey Drive
Fleet
Hants
GU13 8TG
Tel: 01252 626996
Contact: B Bowen

HUMAN POWERED AIRCRAFT GROUP
4 Hamilton Place
London
W1V 0BQ
Tel: 0207 499 3515
Fax: 0207 499 6230
e-mail: conference@raes.org.uk
website: www.raes.org.uk/human-pow.htm
Contact: J F Low

THE HUNTER FLYING CLUB
Hangar 43
Exeter Airport
Exeter
Devon
EX5 2BA
Tel: 01392 445649
Fax: 01179 673858
e-mail: john.rodd@virgin.net
website: www.classicjets.co.uk
Contact: John Rodd

IAP
The Institution of Analysts & Programmers
Charles House
36 Culmington Road
London
W13 9NH
Tel: 0208 567 2118
Fax: 0208 567 4379
e-mail: dg@iap.org.uk
website: www.iap.org.uk

IAPA
International Airline Passenger Association
PO Box 380
Croydon
London
CR9 2ZQ
Tel: 0208 681 6555
Fax: 0208 681 0234
e-mail: info@iapa.co.uk
website: www.iapa.com

IARO
International Air Rail Organisation
Room B217
MacMillan House
Paddington Station
London
W2 1FT
Tel: 0208 750 6232
Fax: 0208 750 6647
e-mail: intl_airrail@baa.co.uk
website: www.iaro.com
Contact: Andrew Sharp

IATA – HQ
International Air Transport Association
Head Office
800 Place Victoria
PO Box 113
Montreal
Quebec
H4A 1MI
Canada
Tel: +1 514 874 0202
Fax: +1 514 874 9652
Website: www.iata.org

IATA – European
International Air Transport Association
IATA Centre
Route de L'Aeroport 33
PO Box 146
CH-1215 Geneva 15 Airport
Switzerland
Tel: +41 22 799 2525
Fax: +41 22 798 3553

IATA – UK
International Air Transport Association
UK Office
Lampton Block
Central House
Lampton Road
Hounslow
Middx
TW3 1HY
Tel: 0208 607 6200
Fax: 0208 607 6409
website: www.iata.org
Contact: Sam Goonetillake

IEE AEROSPACE GROUP
Institution of Electrical Engineers
Savoy Place
London
WC2R 0BL
Tel: 0207 344 5420
Fax: 0207 497 3633
e-mail: industry@iee.org.uk
website: www.iee.org.uk/industry/Aerospce
Contact: Heather Wade

IFA
International Federation of Airworthiness
Suite 1a
Dralda House
Crendon Street
High Wycombe
Bucks
HP13 6LS
Tel: 01444 530404
Fax: 01494 439984
e-mail: mail@aviation-training.org
website: www.ifairworthy.org
Contact: Graham Harris

IFALPA
International Federation of Air Line Pilots' Associations
Interpilot House
Gogmore Lane
Chertsey
Surrey
KT16 9AP
Tel: 01932 571711
Fax: 01932 570920
e-mail: davidpilcher@ifalpa.org
website: www.ifalpa.org
Contact: David Pilcher

IFCA
International Flight Catering Association
Surrey Place
Mill Lane
Godalming
Surrey
GU7 1EY
Tel: 01483 419449
Fax: 01483 419780
e-mail: associationservices@ifcanet.com
Contact: W M Seeman
website: www.ifcanet.com

IGA
Irish Gliding Association
1 Oakdown Road
Dublin 14
Ireland

IHGPA
Irish Hang Gliding & Paragliding Association
House of Sport
Longmile Road
Dublin 12
Ireland

ILT
The Institute of Logistics & Transport
Logistics & Transport Centre
PO Box 5787
Corby
Northants
NN17 4XQ
Tel: 01536 740100
Fax: 01536 740101
e-mail: enquiry@iolt.org.uk
website: www.iolt.org.uk
Contact: Graham A Ewer

IMECHE
Institution of Mechanical Engineers
1 Birdcage Walk
Westminster
London
SW1H 9JJ
Tel: 0207 222 7899
Fax: 0207 222 4557
e-mail: enquiries@imeche.org.uk
website: www.imeche.org.uk
Contact: David Atton

IPA
Independent Pilots Association
The Old Refectory
The Priory
Haywards Heath
West Sussex
RH16 3LB
Tel: 01444 441149
Fax: 01444 441192
e-mail: office@ipapilot.demon.co.uk
website: www.ipapilot.demon.co.uk
Contact: Capt Trevor Newton

ISAAC
Irish Society of Amateur Aircraft Constructors
15 Herbert Park
Bray
Co. Wicklow
Ireland

ITF
International Transport Workers' Federation
49-60 Borough High Street
London
SE1 1DS
Tel: 0207 403 2733
Fax: 0207 357 7871
e-mail: mail@itf.org.uk
website: www.itf.org.uk
Contact: Sam Dawson

ITM
Institute of Travel Management in Industry & Commerce
Easton House
Easton-on-the-Hill
Stamford
Lincs
PE9 3NZ
Tel: 01780
e-mail: secretariat@itm.org.uk
website: www.itm.org.uk
Contact: Loraine Holdcroft

IUAI
International Union of Aviation Insurers
6 Lovat Lane
London
EC3R 8DT
Tel: 0207 626 5314
Fax: 0207 929 3534
e-mail: dg.iuai@virgin.net
website: www.iuai.org
Contact: D R Gasson

JAA
Joint Aviation Authorities
8-10 Saturnustraat
PO Box 3000
NL-2130 K A Hoofddorp
Netherlands
Tel: +31 23 5679700
Fax: +31 23 5621714
website: www.jaa.nl
Contact: Klaus Koplin

Jodel Club
Brooklands
Church Lane
Brafield-on-the-Green
Northampton
NN7 1BA
Tel: 01604 890512
website: www.decollage.org/jodel

LAUA
Lloyd's Aviation Underwriters' Association
Lloyd's
Room 662
Lime Street
London
EC3M 7DQ
Tel: 0207 327 4045
Fax: 0207 327 4711
Contact: I S Macfarlane

LFA
Lawyers Flying Association
Dibb Lupton Alsop
2 Minster Court
Mincing Lane
London
EC3R 7XW
Tel: 0207 796 6516
Fax: 0207 796 6783
e-mail: tony@stapley.co.uk
website: www.stapley.co.uk/lfa.htm
Contact: Tim Scorer

LOCAL GOVERNMENT ASSOCIATION
26 Chapter Street
London
SW1P 4ND
Tel: 0207 664 3212
Fax: 0207 664 3232
website: www.lga.gov.uk
Contact: Ben Clayden

LRQA
Lloyd's Register Quality Assurance
LRQA Centre
Hirarnford
Middlemarch Office Village
Siskin Drive
Coventry
CV3 4FJ
Tel: 024 7688 2373
Fax: 024 7630 6055
e-mail: enquiries@lrqa.com
website: www.lrqa.com

523

Aviation Organisations MAF

MAF
Mission Aviation Fellowship (Europe)
The MAF Europe Operation Centre
Henwood
Kent
TN24 8BH
Tel: 01233 895503
Fax: 01233 895570
e-mail: info@maf-europe.org
website: www.maf-europe.org
Contact: Miss M Evans

MTTA
The Machine Tool Technologies Association
62 Bayswater Road
London
W2 3PS
Tel: 0207 298 6400
Fax: 0207 298 6430
e-mail: mtta@mtta.co.uk
website: www.mtta.co.uk
Contact: Clare Kelly

NAAAS
National Association of Air Ambulance Services
PO Box 6339
Basingstoke
Hants
RG21 4XP
Tel: 01256 493646
Fax: 01256 492780
website: www.naaas.co.uk
Contact: R D Emmans

NAAC
National Association of Agricultural Contractors
Samuelson House
Paxton Road
Orton Centre
Peterborough
Cambs
PE2 5LT
Tel: 01733 362920
Fax: 01733 362921
e-mail: naac@aea.uk.com
Contact: Jill Hewitt

NIAC
Northern Ireland Aerospace Consortium
Vinegar Court
14-16 Gordon Street
Belfast
Co Antrim
BT1 2LG
Northern Ireland
Tel: 028 9033 3734
Fax: 028 9033 3736
e-mail: sharon.devlin@niac.org.uk
www.niac.org.uk
Contact: Sharon Devlin

NQA
National Quality Assurance Ltd
Houghton Hall Park
Houghton Regis
Dunstable
Beds
LU5 5ZX
Tel & Fax: 01582 539000
e-mail: enquiries@nqa.com
website: www.nqa.com
Contact: Jim Speirs

NWAA
North West Aerospace Alliance
Pendle Business Centre
Commercial Road
Nelson
Lancs
BB9 9BT
Tel: 01282 604444
Fax: 01282 604000
e-mail: info@aerospace.co.uk
website: www.aerospace.co.uk
Contact: John Whalley

OPSA
Oxford Air Training School Past Students Association
Oxford Air Training School
Oxford Airport
Kidlington
Oxon
OX5 1RA
Tel: 01865 841234
Fax: 01865 841165
e-mail: jnicholson@cse-aviation.com
Contact: Anne North

PFA
The Popular Flying Association
Turweston Aerodrome
Nr Brackley
Northants
NN13 5YD
Tel: 01280 846786
Fax: 01280 846780
e-mail: office@pfa.org.uk
website: www.pfa.org.uk
Contact: Penny Sharpe

PFIA
Professional Flight Instructors Association
c/o Oxford Aviation training
Oxford Airport
Kidlington
Oxford
OX5 1RA
Tel: 01865 844299
Fax: 01865 844237
e-mail: phickley@oxfordaviation.net
Contact: Paul Hickley

PPR
Professional Pilots Register
28 Salehurst Road
Worthing
West Sussex
RH10 7GL
Tel: 01243 882390

The Queens Flight Association
11 Mill Close
Charlton-on-Otmoor
Oxon
OX5 2UE
Tel: 01865 331373
e-mail: jacksaf@uk.packardbell.org
website: www.tqft-kittyhawk.net
Contact: S A Jack Frost

RAEC
The Royal Aero Club of the United Kingdom
Kimberley House
Vaughan Way
Leicester
LE1 4SG
Tel: 0116 253 1051
Fax: 0116 251 5939
e-mail: bga@gliding.co.uk
website: www.royalaeroclub.org
Contact: Barry N Rolfe

RAEC
Records Racing & Rally Association
20 Woodlands Drive
Colsterworth
Grantham
Lincs
NG33 5NH
Tel & Fax: 01476 860606
e-mail: millermfs@aol.com
Contact: Judy Hanson

RAES
The Royal Aeronautical Society
4 Hamilton Place
London
W1J 7BQ
Tel: 0207 670 4300
Fax: 0207 499 6230
e-mail: info@raes.org.uk
website: www.aerosociety.com
Contact: Antonia Price

RAFA
The Royal Air Forces Association
43 Grove Park Road
Chiswick
London
W4 3RX
Tel: 0208 994 8504
Fax: 0208 742 1927
e-mail: rafa_chq@compuserve.com
website: www.rafa.org.uk
Contact: Ms Claire Connaughton

RAFBF
The Royal Air Force Benevolent Fund
67 Portland Place
London
W1N 4AR
Tel: 0207 580 8343
Fax: 0207 636 7005
e-mail: mail@raf-benfund.org.uk
website: www.raf-benfund.org.uk
Contact: Flt Lt A Dewar

RAFBFE/RIAT
The RAF Benevolent Fund Enterprises
Royal International Air Tattoo
Douglas Bader House
RAF Fairford
Glos
GL7 4DL
Tel: 01285 713300 ext 3000
Fax: 01285 713268
e-mail: postmaster@rafbfe.co.uk
website: www.rafbfe.co.uk
Contact: Ms Patti Heady

RAF CLUB
128 Piccadilly
London
W1V 0PY
Tel: 0207 399 1000
Fax: 0207 355 1516
e-mail: admin@rafclub.org.uk
website: www.rafclub.org.uk
Contact: Peter N Owen

RIN
The Royal Institute of Navigation
1 Kensington Gore
London
SW7 2AT
Tel: 0207 591 3130
Fax: 0207 591 3131
e-mail: rindir@atlas.co.uk
website: www.rin.org.uk
Contact: Gp Capt David W Broughton

Rolls Royce Heritage Trust
PO Box 31
Derby
DE24 8BJ
Tel: 01332 249118
Fax: 01332 249727
e-mail: richard.r.haigh.@rolls-royce.btx400.co.uk
website: www.rolls-royce.co.uk
Contact: Dave Piggott

RRMPSA
Rolls-Royce Management and Professional Staff Association
c/o Hayes Court
West Common Road
Bromley
Kent
BR2 7AU
Tel: 0208 462 7755
Fax: 0208 315 8234
Contact: John Kearney

SACC
Society of All Cargo Correspondents
2a West Street
Ewell
Surrey
KT17 1UU
Tel: 0208 393 2833
Fax: 0208 393 9046
Contact: Clive Woodbridge

SBAC
The Society of British Aerospace Companies Ltd
Duxbury House
60 Petty France
Victoria
London
SW1H 9EU
Tel: 0207 227 1000
Fax 0207 227 1067
e-mail: post@sbac.co.uk
website: www.sbac.co.uk
Contact: Lindsey Hart

SET
The Space Education Trust
c/o The Royal Aeronautical Society
4 Hamilton Place
London
W1V 0BQ
Tel: 01795 521784
Fax: 01795 520880
website: www.gbnet.net/orgs/set
Contact: Mrs Sue Bayford

SITA
Airline Telecommunications and Information Services
Capital Place
120 Bath Road
Hayes
Middx
UB3 5AN
Tel: 0208 230 3117
e-mail: info.eumea@sita.int
website: www.sita.int
Contact: Karl Moore

UKAPE
The United Kingdom Association of Professional Engineers
Hayes Court
West Common Road
Hayes
Bromley
Kent
BR2 7AU
Tel: 0208 462 7755
Fax: 0208 315 8234
Contact: John Kearney

UKFSC
UK Flight Safety Committee
Graham Suite
Fairoaks Airport
Chobham
Woking
Surrey
GU24 8HX
Tel: 01276 855193
Fax: 01276 855195
e-mail: uksfc@freezone.co.uk
website: www.ukfsc.co.uk
Contact: Ed Paintin

UKISC
United Kingdom Industrial Space Committee
PO Box 14
Wisbech
Cambs
PE13 1JZ
Tel: 01945 464975
Fax: 01945 461988
e-mail: hicks.ukisc@btinternet.com
website: www.ukspace.com/trade/ukisc.htm
Contact: Alan Hicks

UKSEDS
UK Students for the Exploration & Development of Space
c/o Royal Aeronautical Society
4 Hamilton Place
London
W1V 0BQ
Tel: 01795 521784
Fax: 01795 520880
e-mail: info@uk.seds.org
website: www.uk.seds.org
Contact: Louise Thorn

WAEO
World Aerospace Education Organization
118 Lutterworth Road
Aylestone
Leicester
LE2 8PG
Tel: 0116 291 0281
Contact: Dr J John Parker

WARBIRDS WORLDWIDE
PO Box 99
Mansfield
Notts
NG18 2AH
Tel: 01623 624288
Fax: 01623 622659
e-mail: enq@warbirdsworldwide.com
website: www.warbirdsworldwide.com
Contact: Paul A Coggan

WEAF
West of England Aerospace Forum
16 Clifton Park
Clifton
Bristol
BS8 3BY
Tel: 0117 915 2321
Fax: 0117 915 2321
e-mail: weaf@aeroforum.co.uk
website: www.aeroforum.co.uk
Contact: Julian Leonard

WORLDCHOICE ARTAC
The Alliance of Independent Travel Agents
Malborne House
Herlington
Orton Malborne
Peterborough
Cambs
PE2 5PR
Tel: 01733 390900
Fax: 01733 390997
e-mail: directors@worldchoice.co.uk
website: www.worldchoice.co.uk
Contact: Chris Fife

CAA contact details

CAA
Civil Aviation Authority
CAA House
45-59 Kingsway
London WC2B 6TE
Tel: 0207 379 7311 (Switchboard)
Tel: 0207 453 plus ext (Direct Dial)
Fax: 0207 453 4784

SRG
Safety Regulation Group
CAA
Aviation House
South Area
London Gatwick Airport
West Sussex
RH6 0YR
Tel: 01293 567171 (Switchboard)
Tel: 01293 57 plus ext (Direct Dial)
Fax: 01293 573999

NATS
National Air traffic Services
Fifth Floor
Brettenham House South
Lancaster Place
London
WC2E 7EN
Tel: 0207 309 8666
Website: www.nats.co.uk

A & C Section

Application for issue & renewal of certificates of airworthiness
Tel: 01293 768374
Fax: 01293 573860

Approval of maintenance organisations
Tel: 01293 573152 or 01293 573860

Approval of Maintenance schedules/programmes
Tel: 01293 573152 or 01293 573860

CAA additional directive, mandatory modifications and inspections
Tel: 01293 573168/61
Fax: 01293 573860

A/C Systems

Administration & equipment approval records
Tel: 01293 573132
Fax: 01293 573975

Avionics systems & equipment including instruments
Tel: 01293 573117
Fax: 01293 573975

Electrical generation & utilisation systems & equipment
Tel: 01293 573123
Fax: 01293 573975

Environmental & cabin safety systems & equipment
Tel: 01293 573138
Fax: 01293 573975

Mechanical & hydraulic systems & equipment
Tel: 01293 573126
Fax: 01293 573975

Powerplant installations & equipment
Tel: 01293 573131
Fax: 01293 573975

Requirements & continued airworthiness co-ordination
Tel: 01293 573142
Fax: 01293 573975

Aerodrome Standards

Aerodrome inspections & fire services
Tel: 01293 573279
Fax: 01293 573971

Aerodrome policy, standardisation & development
Tel: 01293 573575
Fax: 01293 573971

Aerodrome safeguarding & development advice & aspects
Tel: 01293 573276
Fax: 01293 573971

Aircraft Projects

Co-ordination of aircraft certification investigation
Tel: 01293 573284
Fax: 01293 573976

Company approval (design) investigation
Tel: 01293 573120
Fax: 01293 573976

Flight manuals
Tel: 01293 573189
Fax: 01293 573976

Helicopter certification
Tel: 01293 573284
Fax: 01293 573976

Master minimum equipment list
Tel: 01293 573189
Fax: 01293 573976

Aircraft Registrations

Additions, deletions & changes of registered ownership
Tel: 0207 453 6666
Fax: 0207 453 6670

Aircraft mortgage registration
Tel: 0207 453 6666
Fax: 0207 453 6670

Births, deaths & missing persons on UK registered aircraft
Tel: 0207 453 6666
Fax: 0207 453 6670

Exemptions from article 5 (display of nationality & registration marks)
Tel: 0207 453 6666
Fax: 0207 453 6670

Information from the UK register database
Tel: 0207 453 6660
Fax: 0207 453 6670

Registered owner address changes
Tel: 0207 453 6666
Fax: 0207 453 6670

ATS Standards

ATC/Engineering approvals
Tel: 01293 573426
Fax: 01293 573974

ATC/Engineering inspections
Tel: 01293 573426
Fax: 01293 573974

ATC/Licensing Policy
Tel: 01293 573436
Fax: 01293 573974

En route regulations
Tel: 01293 573259
Fax: 01293 573974

ATS Standards continued

Engineering requirement/systems
Tel: 01293 573137
Fax: 01293 573974

Quality
Tel: 01293 573732
Fax: 01293 573974

Regulations of all aspects of civil air traffic services
Tel: 01293 573424
Fax: 01293 573974

Engineer Standards

Aircraft maintenance engineer licences
Tel: 01293 573700
Fax: 01293 573779

Aircraft maintenance engineer training courses
Tel: 01293 573700
Fax: 01293 573779

Flight Crew Standards

Appointment of PPL examiners
Tel: 01293 573700
Fax: 01293 573996

Approval of training courses (professional & private)
Tel: 01293 573700
Fax: 01293 573996

Arrangements
Tel: 01293 573700
Fax: 01293 573996

Bookings
Tel: 01293 573700
Fax: 01293 573996

European pilot licensing enquiries
Tel: 01293 573700
Fax: 01293 573996

Fees & charges
Tel: 01293 573700
Fax: 01293 573996

Flying test bookings & arrangements
Tel: 01293 573700
Fax: 01293 573996

Ground examinations
Tel: 01293 573700
Fax: 01293 573996

Professional pilots, flight engineers & flight navigators
Tel: 01293 573700
Fax: 01293 573996

Flight Department

All weather certification
Tel: 01293 573097
Fax: 01293 573977

Fixed wing aircraft flight testing & simulators
Tel: 01293 573089
Fax: 01293 573977

Flight requirements
Tel: 01293 573098
Fax: 01293 573977

Helicopter flight testing & simulators
Tel: 01293 573107
Fax: 01293 573977

Noise certification
Tel: 01293 573095
Fax: 01293 573977

Flight Operations Department

Aircraft equipment
Tel: 01293 573508
Fax: 01293 573991

All weather operations/CAT II & III
Tel: 01293 573414
Fax: 01293 573991

Cabin safety
Tel: 01293 573341
Fax: 01293 573991

Carriage by air of dangerous goods, munitions of war & animals
Tel: 01293 573800
Fax: 01293 573991

Development of public transport legislation
Tel: 01293 573412
Fax: 01293 573991

Inspections & approval of flight simulators for use in airline training
Tel: 01293 573517
Fax: 01293 573991

Inspection & supervision of airline pilots training & testing procedure
Tel: 01293 573498
Fax: 01293 573991

Inspection of AOC maintenance
Tel: 01293 573711
Fax: 01293 573991

Interchange of aircraft/leasing
Tel: 01293 573422
Fax: 01293 573991

Policy on public transport operations
Tel: 01293 573412
Fax: 01293 573991

General Aviation

Administration & finance
Tel: 01293 573506
Fax: 01293 573973

Aerial work
Tel: 01293 573528
Fax: 01293 573973

Agricultural aviation
Tel: 01293 573528
Fax: 01293 573973

Air displays, display authorisations & exemptions for the dropping of articles & low flying
Tel: 01293 573510/40
Fax: 01293 572973

AOC ballooning
Tel: 01293 573717
Fax: 01293 573973

Balloon congested area take-off permissions & private & aerial work ballooning
Tel: 01293 573526/40
Fax: 01293 573973

Charity flight permissions
Tel: 01293 573526/40
Fax: 01293 573973

GASIL & safety evening
Tel: 01293 573225
Fax: 01293 573973

Gliding, hand-gliding, parascending, microlight aeroplanes, airships, gyroplanes, foot launched power flying machines & permissions for high speed flights
Tel: 01293 573526/40
Fax: 01293 573973

General — CAA contact details

Model aircraft
Tel: 01293 573526/40
Fax: 01293 573973

Operation of ex-military aircraft
Tel: 01293 573651/40
Fax: 01293 573973

Parachuting
Tel: 01293 573528/29
Fax: 01293 573973

Policy for general aviation
Tel: 01293 573512/24
Fax: 01293 573973

Private & corporate helicopters
Tel: 01293 573528/12
Fax: 01293 573973

Information Management

CAA publications
Tel: 01293 573373
Fax: 01293 573820

Medical Division

Medical examination & assessment of flight crew for licences
Tel: 01293 573700
Fax: 01293 573677

Medical research & human factors
Tel: 01293 573663
Fax: 01293 573870

Occupational health advice
Tel: 01293 573670
Fax: 01293 573567

Training in aviation medicine
Tel: 01293 573798
Fax: 01293 573567

Licensing Department

Issues & renewals of licenses, rating, conversions & validations
Tel: 01293 573700
Fax: 01293 573996

Propulsion engines, auxiliary power units, propellers & rotorcraft transmissions
Tel: 01293 573193
Fax: 01293 573979

CAA Regional Offices

SRFAC
Safety Regulation Finance Advisory Committee
Aviation House
Gatwick Airport South
West Sussex
RH6 0YR
Tel: 01293 573662

OAC
Operations Advisory Committee
Aviation House
South Area
Gatwick Airport
West Sussex
RH6 0YR
Tel: 01293 567171

SACP
Standing Advisory Committee on Pilot Licencing
Aviation House
South Area
Gatwick Airport
West Sussex
RH6 0YR
Tel: 01293 58530

ARB
Airworthiness Requirements Board
Aviation House
South Area
Gatwick Airport
West Sussex
RH6 0YR
Tel: 01293 573062
Fax: 01293 573838

Scotland
Aberdeen
Hangar 1
Aberdeen Airport
Dyce
Aberdeenshire
AB21 7DU
Scotland
Tel: 01224 793530

UK
East Midlands
Building 65
Ambassador Road
Castle Donington
Leics
DE74 2SA
Tel: 01332 811245

Gatwick
Ground Floor
Consort House
Consort Way
Horley
Surrey
RH6 7AF
Tel: 01293 828220

Gatwick – Head Office
Aviation House
South Area
Gatwick
W. Sussex
RH6 0YR
Tel: 01293 567171

CAA Regional Offices (continued)

Heathrow
Sipson House
595 Sipson Road
West Drayton
Middlesex
UB7 0JD
Tel: 0208 759 0205

Irvine
Galt House
Bank Street
Irvine
Ayrshire
KA12 0LL
Scotland
Tel: 01294 312192

Luton
1st Floor
Barratt House
668 Hitchin Road
Stopsley
Luton
Beds
LU2 7XH
Tel: 01582 410304

Manchester
Suite 5
Manchester International Office Centre
Styal Road
Wythenshawe
Manchester
M22 5WB
Tel: 0161 499 3055

Stansted
Walden Court
Parsonage Lane
Bishop's Stortford
Hers
CM23 5DB
Tel: 01279 466747

Stirling
7 Melville Terrace
Stirling
FK8 2ND
Tel: 01786 431400

Weston-super-Mare
Unit 101
Parkway
Worle
Weston-super-Mare
North Somerset
BS22 0WB
Tel: 01934 522260

AM RADIO STATIONS BROADCASTING IN THE UK

Frequency	Name	Frequency	Name
198	BBC Radio Four	1161	Radio Tay AM
252	Atlantic 252	1161	Brunel Classic Gold
603	Capital Gold	1170	Eleven-Seventy
630	BBC Three Counties Radio	1170	Signal Two
648	BBC World Service	1170	Magic AM
657	BBC Radio Cornwall	1170	Swansea Sound
666	Westward Radio	1170	Amber Classic Gold
666	BBC Radio York	1170	Capital Gold
693	BBC Radio Five Live	1197	Virgin Radio
720	BBC Radio Four	1215	Virgin Radio
729	BBC Essex	1233	Virgin Radio
756	Radio Maldwyn	1242	Virgin Radio
765	BBC Essex	1242	Capital Gold
774	BBC Radio Leeds	1251	Amber Classic Gold
774	BBC Radio Kent	1260	BBC Radio York
792	Classic Gold	1260	Virgin Radio
810	BBC Radio Scotland	1278	W Yorkshire's Classic Gold
828	Magic	1296	Radio XL
828	Classic Gold	1305	Premier Radio
828	BBC Asian Network	1305	Touch Radio
828	Gold Beat	1305	Magic AM
837	BBC Asian Network	1323	BBC Somerset Sound
855	Sunshine	1323	Capital Gold
873	BBC Radio Norfolk	1332	Classic Gold
882	BBC Radio Wales	1332	BBC Wiltshire Sound
909	BBC Radio Five Live	1341	BBC Radio Ulster
936	Brunel Classic Gold	1359	Touch Radio
945	Classic Gold Gem	1359	BBC Radio Solent for Dorset
945	Capital Gold	1359	The Breeze
954	Westward Radio	1359	Classic Gold
954	Classic Gold	1368	BBC Southern Counties
963	Asian Sound Radio	1368	BBC Wiltshire Sound
963	Liberty Radio	1377	Asian Sound Radio
990	Classic Gold WABC	1413	BBC Radio Gloucestershire
999	Magic	1413	Premier Radio
999	Classic Gold Gem	1413	Yorkshire Dales Radio
999	Valleys Radio	1431	The Breeze
1017	Classic Gold WABC	1431	Classic Gold
1026	BBC Radio Cambridgeshire	1449	BBC Radio Cambridgeshire
1026	Downtown Radio	1458	BBC Radio Cumbria
1035	West Sound AM	1458	Sunrise Radio
1035	Northsound Two	1476	County Sound Radio
1053	Talk Radio UK	1485	BBC Radio Merseyside
1089	Talk Radio UK	1530	W Yorkshire's Classic Gold
1107	Talk Radio UK	1530	BBC Essex
1107	Moray Firth Radio	1530	Classic Gold
1116	BBC Radio Guernsey	1548	Magic AM
1116	BBC Radio Derby	1548	Capital Gold
1116	Valleys Radio	1557	Capital Gold
1152	Plymouth Sound AM	1557	Classic Gold
1152	Magic AM	1584	BBC Radio Shropshire
1152	LBC	1584	London Turkish Radio
1152	Amber Classic Gold	1584	Radio Tay AM
1161	BBC Three Counties Radio	1584	BBC Radio Kent
1161	Magic AM Frequency Name		

UK CAA Medical Examiners

Medical Examiners with numbers in the series 10000 can examine Class 1 and Class 2 standard
Medical Examiners with numbers in the series 20000 can examine Class 2 standard only

Aberdeenshire

10058
Dr F P Howarth
The Rectory
38 Oldmeldrum Road
Bucksburn
AB21 9DU
Tel: 01224 715146
e-mail: peter@fphowarth.com

10201
Dr P M Rhodes
Scotstown Medical Centre
Cairnfold Road
Bridge of Don
AB22 8LD
Tel: 01224 702149
Fax: 01224 706688
e-mail: paul.rhodes@scotstown.grampian.scot.nhs.uk

20089
Dr D A Caughey
Capita Health Solutions
Forest Grove House
Foresterhill Road
Aberdeen
AB25 2ZP
Tel: 01224 669000
Fax: 01224 669030
e-mail: david.caughey@capita.co.uk

Avon

10089
Dr D J Short
12 Whitesfield Road
Nailsea
Bristol
BS48 2DT
Tel: 01275 855294
e-mail: ds007h2776@blueyonder.co.uk

10209
Dr A Coulson
Whiteladies Medical Group
Whiteladies Health Centre
Whatley Road
Clifton
Avon
BS8 2PU
Tel: 0117 973 1201
Fax: 0117 946 6850

Ayrshire

10021
Dr D J Cattanach
The Surgery
2 Station Road
Prestwick
KA9 1AQ
Tel: 01292 671444
Fax: 01292 678023
ATCOs only

10097
Dr M J Timmons
Dunbeith
44 London Road
Kilmarnock
KA3 7AE
Tel: 01563 540510 (Eves)
Tel: 01563 523593 (Surgery)
Mobile: 07766 883528
e-mail: mjt@mjtimmons.freeserve.co.uk

Bedfordshire

10091
Dr R F Sowerby
2 Troon Close
Bedford
MK41 8AY
Tel & Fax: 01234 214045

10105
Dr P J Ward
Liverpool Road Health Centre
9 Mersey Place
Luton
LU1 1HH
Tel: 01582 722525
Fax: 01582 421602
e-mail: peter.ward@gp-e81032.nhs.uk

10105
Dr P J Ward
Luton Aviation Medical Centre
131 Bushmead Road
Luton
LU2 7YT
Tel: 01582 484883
Fax: 01582 485128
e-mail: peter.ward@gp-e81032.nhs.uk

20037
Dr A M D Mitchell
179 Old Bedford Road
Luton
LU1 7EH
Tel: 01582 731754
Fax: 01582 434237
e-mail: drmitch@nildram.co.uk

20037
Dr A M D Mitchell
SKF (UK) Ltd
Sundon Park Road
Luton
LU3 3BL
Tel: 01582 731754
Fax: 01582 434237
e-mail: drmitch@nildram.co.uk

Berkshire

20014
Dr R W C Collett
The Surgery
6 London Hall Road
Reading
RG10 9JA
Tel: 0118 934 6680/1
Fax: 0118 934 6690
e-mail: bobcollett@aol.com

20063
Dr R C F Symons
The Symons Medical Centre
25 All Saints Avenue
Maidenhead
SL6 6EL
Tel & Fax: 01628 471711

Buckinghamshire

10211
Dr V P Robinson
Iver House
78 High Street
Iver
SL0 9NG
Tel: 01895 274353

10254
Dr K J Maxwell
British Airways Flying Club
Wycombe Air Park
Marlow
SL7 3DP
Tel: 01494 529262
Fax: 01494 461237
Mobile: 07890 285011
e-mail: kenmaxwell@eurobell.co.uk

Cambridgeshire

10103
Dr B A Wallace
24a Orchard Road
Melbourn
SG8 6HH
Tel: 01763 262034
e-mail: drbrianwallace@aol.com

Channel Islands

10015
Dr P Biggins
Cobo & St Martins Medical Practice
Castel
Guernsey
GY5 7HA
Tel: 01481 56404
e-mail: docbstresa@aol.com

10044
Dr I B Gee
Cobo Health Centre
Clos De Carteret
Cobo
Castel
Guernsey
GY5 7HA
Tel: 01481 256404
Fax: 01481 251121
e-mail: antibody2@aol.com

10087
Dr M J Rosser
Lister House
35 The Parade
Jersey
JE2 3QQ
Tel: 01534 736236
Mobile: 07797 723515
e-mail: rosdoc@jerseymail.co.uk

10087
Dr M J Rosser
Harewood
La Ruelle du Clos du Parcq
Jersey
JE3 8AQ
Tel: 01534 863916
Mobile: 07797 723515
e-mail: rosdoc@jerseymail.co.uk

Cheshire

10032
Dr I G Donnan
Owl House
97 Buxton Old Road
Disley
SK12 2BN
Tel: 01663 766946
Fax: 01663 766981
e-mail: iandonnan@aol.com

10059
Dr M F Hudson
Grasmere Medical Services
Goostrey
CW4 8NU
Tel: 01477 532527
Fax: 01477 544059
e-mail: martin-hudson@lineone.net

10107
Dr J E Wray
Aviation Medical Facilities (Manchester)
Business Aviation Centre
Hangar 7
Western Maintenance Area
Manchester Airport West
M90 5NE
Tel & Fax: 0161 436 0129
e-mail: avmed2000@btopenworld.com

10212
Dr R A Yuill
Countess of Chester Hospital
NHS Trust
Chester
CH2 1UL
Tel: 01244 355901
Tel: 0161 4277409 (Home)
Mobile: 07947 003934
e-mail: ra.yuill@ntlworld.com

10241
Mr A W M C Owen
General Consultant Surgeon
Upton Village Surgery
Upton-by-Chester
Chester
CH2 1HD
Tel: 01244 382238
Fax: 01244 381576
e-mail: c.tarrant@virgin.net

10305
Dr R Reisler
Aviation Medical Facilities (Manchester)
Business Aviation Centre
Hangar 7
Western Maintenance Area
Manchester Airport West
M90 5NE
Tel & Fax: 0161 436 0129
e-mail: avmed2000@northernexec.com

Cheshire (continued)

20050
Dr I Reid-Entwistle
Knollwood
42 Well Lane
Gayton
Wirral
CH60 8NG
Mobile: 07050 261980
Fax: 0151 342 5135
e-mail: ianreidentwistle@tiscali.co.uk

Cleveland

20001
Dr D B Acquilla
South Grange Medical Centre
Trunk Road
Eston
Middlesborough
TS6 9QG
Tel: 01642 467001
Fax: 01642 43334

20088
Dr J A N Slade
Woodlands Family Medical Centre
106 Yarm Lane
Stockton-on-Tees
TS18 1YE
Tel: 01642 607398
Fax: 01642 677846/604603

Clwyd

20094
Dr P D Saul
Health Centre
Beech Avenue
Rhos
Wrexham
LL14 1AA
Tel: 01978 845955
Fax: 01978 846757
e-mail: peter.saul@dr.com

Cornwall

10084
Dr A A G Quinton
Penzance Heliport (Brintel)
Eastern Green
Penzance
TR18 3AP
Tel: 01736 64296
Fax: 01736 64293
Brintel employees only

10084
Dr A A G Quinton
Isles of Scilly Sky Bus Ltd
(Westward)
Lands End
TR19 7RL
Tel: 01736 788771
Fax: 01736 787274
Westward employees only

10192
Dr R O Sills
Buttercup Wells
Lawhitton
Launceston
PL15 9PE
Tel: 01566 773915
e-mail: rosills@medicalhistory.com

10266
Dr A G Hillary
Helston Medical Centre
Trelawney Road
Helston
TR13 8AU
Tel: 01326 569700
Fax: 01326 565525
e-mail: hmcenq@hmc.cornwall.nhs.uk

Derbyshire

10070
Dr M F McGhee
53 Borough Street
Castle Donington
DE74 2LB
Tel: 01332 810241
Fax: 01332 811748
e-mail: mcghee_mf@gp-c82007.nhs.uk

10094
Dr K R Sumner
53 Borough Street
Castle Donington
DE74 2LB
Tel: 01332 810241
Fax: 01332 811748
e-mail: keith.sumner@ntlworld.co.uk

10292
Dr H Tailor
DHL Aviation
Medical Centre
Building 121
Cargo West
East Midlands Airport
Castle Donington
DE74 2TR
Tel: 01332 857318
DHL pilots only

Devon

10084
Dr A A G Quinton
Suite D Metropolitan House
The Millfields
Plymouth
PL1 3JB
Tel: 01752 229116
Fax: 01752 269098
e-mail: draaquinton@eclipse.co.uk

10192
Dr R O Sills
Cardio Analytics
Tamar Science Park
Derriford
Plymouth
PL6 8BU
Tel: 01752 201144
Fax: 01752 201145
e-mail: rosills@medicalhistory.com

10198
Dr D H McFadyen
Mount Pleasant Health Centre
Mount Pleasant Road
Exeter
EX4 7BW
Tel & Fax: 01392 255722

10200
Dr A C Renouf
Chapel Platt Surgery
1901 Fore Street
Exeter
EX3 0HE
Tel: 01392 875777
Fax: 01392 875770
Mobile: 07768 261559
e-mail: adrian.renouf@gp-l83661.nhs.uk

10200
Dr A C Renouf
Exeter Medical Services
Airways Flight Training
Building 12
Exeter Airport
Exeter
EX5 2DB
Tel: 01392 364216
Fax: 01392 368255
Mobile: 07768 261559
e-mail: adrian.renouf@gp-l83661.nhs.uk

10013
Dr W A Beck
Airways Flight Training (Exeter) Ltd
Building 12
Exeter Airport
EX5 2BD
Tel: 01392 364216
Fax: 01392 368255

Dorset

10043
Dr N Gamper
Burton Medical Centre
(Stable Block)
Christchurch
BH23 7JN
Tel: 01202 487887
Fax: 01202 484412
Mobile: 07771 562537

10078
Dr R M Odbert
Highcliffe Medical Centre
Helia House
248 Lymington Road
Highcliffe
Christchurch
BH23 5ET
Tel: 01425 272203

20022
Dr J M Evans
The Surgery
White-Cliff-Mill Street
Blandford Forum
DT11 7BH
Tel: 01258 452501
e-mail: jonathan.evans@gp-J811019.nhs.uk

10166
Dr M R Groom
The Melbury Clinic
Higher Barn
Holt Mill
DT2 0LX
Tel: 01935 873951
Mobile: 07968 160581
e-mail: mrg@woolcotts.demon.co.uk

10311
Dr G W Hickish
New Medical Centre
Ringwood Road
Bransgore
Christchurch
BH23 8AD
Tel: 01202 474311
Fax: 01202 484412
e-mail: postmater@gp-j815050.nhs.uk

Dumfries & Galloway

20029
Dr P G Hutchison
Greyfriars Medical Centre
33-37 Castle Street
Dumfries
DG1 1DL
Tel: 01387 257752
Fax: 01387 257020

Durham

10077
Dr M J Neville
Gainford Surgery
Main Road
Gainford
Darlington
DL2 3BE
Tel & Fax: 01325 730204
e-mail: michael.neville@gp-a83061.nhs.uk

10291
Dr D L Hamilton
Woodlands Private Hospital
Morton Park Way
Darlington
DL1 4PL
Mobile: 07900 430557
E-mail: dlhamiltonfrcsed@hotmail.com

East Dunbartonshire

10036
Dr D Doyle
35 Thorn Road
Bearsden
G61 4BS
Tel: 0141 942 7438
Mobile: 07766 332404
e-mail: daviddoyle@thebrain.fsnet.co.uk

East Sussex

10290
Dr E J D Henderson
The Surgery
14 Burwash Road
Hove
BN3 8GQ
Tel: 01273 739271
Fax: 01273 727786
e-mail: drhenderson@doctors.org.uk

20064
Dr A Tabor
5 Woodland Drive
Hove
BN3 6DH
Tel: 01273 557779

535

East Sussex (continued)

10264
Dr T G Nash
Consultant OB/GYN
Westlands
36 Collington Avenue
Bexhill on Sea
TN39 3NE
Tel: 01424 221886
Fax: 01424 213249

10167
Dr M F P Marshall
The Apollo Centre for Health
Wartling Road
Eastbourne
BN22 7PF
Tel: 01323 731160
e-mail: fmarshal@btinternet.com

20075
Dr P G Williams
1 Arlington Road
Eastbourne
BN21 1DH
Tel: 01323 727511
Fax: 01323 417985
e-mail: peter.williams12@virgin.net

East Yorkshire

10052
Dr W A Hart
67 Ferry Road
South Cave
Brough
HU15 2JG
Tel: 01430 424764
Fax: 01430 421553
e-mail: hart@enterprise.net

Edinburgh

10095
Dr A Thores
Murrayfield Hospital
122 Corstophine Road
Edinburgh
EH12 6UD
Tel: 0131 334 4806
Fax: 01383 872846
e-mail: esculapion@aol.com

Essex

10016
Dr D A Bishop
Occ Health Department
Room 6/1305
Stamping Operations
Ford Motor Co Ltd
Dagenham
RM9 6SA
Tel: 0208 526 1347
Fax: 0208 526 5773
e-mail: dbishop1@ford.com
Ford employees only

10208
Dr P K Orton
Matching Parsonage Farm
Newmans End
Matching
CM17 0QX
Tel: 01279 731536
Fax: 01279 661564
Mobile: 07747 023460
e-mail: matching@aviation-medica.com

10208
Dr P K Orton
Stapleford Flying Club
Stapleford Aerodrome
Stapleford Tawney
RM4 1SJ
Tel: 01708 688380
Fax: 01708 688421
Mobile: 07747 023460
e-mail: stapleford@aviation-medic4.com

10208
Dr P K Orton
Andrewsfield Aviation
Saling Airfield
Great Dunmow
CM6 3TH
Tel: 01371 856744
Fax: 01371 856500
Mobile: 07747 023460
e-mail: andrewsfield@aviation-medica.com

10208
Dr P K Orton
Aviation Medica
Inflite
Hangar One
First Avenue
London Stansted Airport
CM24 1RY
Tel: 01279 661580
Fax: 01279 661564
Mobile: 07747 023460
e-mail: stansted@aviation-medica.com

10253
Dr J H J Cockcroft
Stapleford Aerodrome
Stapleford Tawney
RM4 1SJ
Tel: 01708 688380
Fax: 01708 688421
Mobile: 07710 371472
e-mail: jhjcockcroft@skyflame.co.uk

10253
Dr J H J Cockcroft
The Health Centre
Stock Road
Billericay
CM12 0BJ
Tel: 01277 658071
Mobile: 07710 371472

10253
Dr J H J Cockcroft
Integrated Medical Care Limit
94 Station Lane
Hornchurch
RM12 6LX
Tel: 01708 445156

10276
Dr P S Cornes
Riverside Medical Centre
175 Ferry Road
Hullbridge
SS5 6JH
Tel: 01702 230555
Fax: 01702 231207
e-mail: user240132@aol.com

20044
Dr P J Nightingale
Terling
Oakley Road
Little Oakley
Harwich
CO12 5DR
Tel: 01255 508397

20076
Dr M J Murphy
18 North Hill
Colchester
CO1 1DZ
Fax: 01206 769880
Mobile: 07798 925728

20084
Dr S M E Solomon
White Cottage
Black Horse Lane
North Weald
Epping
CM16 6EP
Tel: 01992 522951
Fax: 01992 524486

10040
Dr G Fearnley
195 Thorpe Hall Avenue
Thorpe Bay
SS1 3AP
Tel: 01702 586028
e-mail: geoffrey.fearnley@btinternet.com

Fife

10095
Dr A Thores
5 Haven's Edge
Limekilns
Dunfermline
KY11 3LJ
Tel & Fax: 01383 872846
e-mail: esculapion@aol.com

10123
Dr M F Fawzi
North Glen Medical Practice
Pitcoudie Surgery
1 Huntsman Court
Glenrothes
KY7 6SX
Tel: 01592 620062

Glamorgan

10055
Dr J A Hill
Pontardawe Primary Care Centre
Industrial Estate
Pontardawe
Swansea
SA8 4JU
Tel: 01792 865271
Fax: 01792 865400

10086
Dr T A Reilly
Loxleigh
Southerndown
Bridgend
CF32 0RW
Tel: 01656 881081
Fax: 01656 881963
e-mail: terry@fit2fly.net
e-mail: loxleigh@tiscali.co.uk

20038
Dr R N W Morgan
The Terminal Building
Pembrey Airfield
Pembrey
SA16 0HZ
Tel & Fax: 01792 206732
e-mail: arcs.morgan@ntlworld.com

20038
Dr R N W Morgan
The Medical Examination Room
The Terminal Buildings
Swansea Airport
Fairwood Common
Swansea
SA2 7JU
Tel & Fax: 01792 206732
e-mail: arcs.morgan@ntlworld.com

Glasgow

10036
Dr D Doyle
35 Thorn Road
Bearsden
Glasgow
G61 4BS
Tel: 0141 942 7438
e-mail: daviddoyle@thebrain.fsnet.co.uk

Gloucestershire

10014
Dr A S Benzie
Beaumoor Barn House
East End
Fairford
GL7 4AP
Tel: 01285 713395
e-mail: stewart.benzie@gp-L84053.nhs.uk
e-mail: fairfordsurgery@gp-L84053.nhs.uk

10085
Dr I D Ramsay
Portland Practice
St Paul's Medical Centre
Cheltenham
GL50 4DP
Tel: 01242 215437
Mobile: 08444 771869
e-mail: ianramsay23@gawab.com

Greater Manchester

10107
Dr J E Wray
Aviation Medical Facilities
(Manchester)
Business Aviation Centre
Hangar 7
West Maintenance Area
Manchester Airport West
M90 5NE
Tel & Fax: 0161 436 0129
e-mail: avmed2000@btopenworld.com

537

Greater Manchester (continued)

20073
Dr T Sinclair
67 Chew Valley Road
Greenfield
Oldham
OL3 7JG
Tel & Fax: 01457 873100

10305
Dr R Reisler
Aviation Medical Facilities
(Manchester)
Business Aviation Centre
Hangar 7
West Maintenance Area
Manchester Airport West
M90 5NE
Tel & Fax: 0161 436 0129
e-mail: avmed2000@northernexec.com

Gwent

20069
Dr J Watkins
National Public Health Service for Wales
Mamhalid House
Mamhalid Park Estate
Pontypool
NP9 0YP
Tel: 01495 332316
Fax: 01495 753403
e-mail: watkinsj8@cf.ac.uk

Gwynedd

10054
Dr M Hickey
Minfor
Barmouth
LL42 1PL
Tel: 01341 280521
e-mail: malcolmhickey@compuserve.com

Hampshire

10080
Dr I C Perry
The Old Farm House
Grateley
Andover
SP11 8JR
Tel: 01264 889659
Fax: 01264 889639
e-mail: ian@ianperry.com

10112
Dr D E Cook
The Hampshire Clinic
Basing Road
Basingstoke
RG24 7AL
Tel: 0118 9810699
Fax: 0709 2006345
Mobile: 07702 151654
e-mail: donaldcook@beeb.net

10112
Dr D E Cook
The Wessex Nuffield Hospital
Winchester Road
Eastleigh
SO53 2DW
Tel: 0118 9810699
Fax: 0709 2006345
Mobile: 07702 151654
e-mail: donaldcook@beeb.net

10113
Dr J A East
Concept 2000
250 Farnborough Road
Farnborough
GU14 7LU
Tel: 01628 666236
e-mail: john_east@talk21.com

10114
Dr D A Evans
Gosport Health Centre
Bury Road
Gosport
PO12 3PN
Tel: 02392 583302

10165
Sqn Cdr P J Waugh
Bristow Helicopters
Search & Rescue Flight
Lee-on-Solent Airfield
Argus Gate
Broom Way
Lee-on-Solent
PO13 9YA
Tel: 01705 550142
Fax: 01705 550390
Bristow Helicopters personnel only

10277
Dr A W Cairns
Swan Surgery
Petersfield
GU32 3AB
Tel: 01730 264011
Fax: 01730 264546
e-mail: andrew.cairns@gp-j82098.nhs.uk

20004
Dr N P Arney
Forest Gate Surgery
Hazel Farm Road
Totton
SO40 8WU
Tel: 01703 663839
Fax: 01703 667090

20031
Dr A T Lloyd-Davies
St Mary's Surgery
Church Close
Andover
SP10 1DP
Tel: 01264 352983
e-mail: llodav@btopenworld.com

10002
Dr S Stork
National Air Traffic Services
London Area Control Centre
Sopwith Way
Swannick
SO31 7AY
Tel: 0208 7453331
Fax: 0208 7453179
e-mail: sheila.stork@nats.co.uk
NATS ATCO's only

10206
Dr P M Norris
Group Medical Adviser
British Aerospace Plc
Farnborough Aerospace Centre
Farnborough
GY14 6YU
British Aerospace personnel only

10012
Dr M Bagshaw
National Air Traffic Services
London Area Control Centre
Sopwith Way
Swanwick
SO31 7AY
Tel: 0208 74533331
Fax: 0208 7453179
e-mail: mikebagshaw@doctors.org.uk
NATS ATCO's only

Herefordshire

10115
Dr R N Ovenden
The Dairy
Church Walk
Eardisland
HR6 9BP
Tel: 01544 388503
e-mail: rnovenden@amserve.net

20030
Dr R B Laird
Moorfield Surgery
Edgar Street
Hereford
HR4 9JP
Tel: 01432 272175

Hertfordshire

10026
Dr D Cranston
363 Luton Road
Harpenden
AL5 3LZ
Tel: 01582 764673
Fax: 01582 765409
e-mail: davidcranston@btconnect.com

10098
Dr F D Trevarthen
10 Hertford Road
Digswell
Welwyn
AL6 0EB
Tel: 01438 714056
Fax: 01438 840617
e-mail: dtrevar934@aol.com

10103
Dr B A Wallace
Keffords
Barley
Royston
SG8 8LB
Tel: 01763 848287
Fax: 01763 849616
e-mail: drbrianwallace@aol.com

10110
Dr A Yardley-Jones
Corner Field
North Church Common
Berkhamsted
HP4 1LR
Tel & Fax: 01442 878322
e-mail: caa@yardley-jones.com

20017
Dr G H Cooray
8 Letchmore Road
Radlett
WD7 8HT
Tel & Fax: 01923 856017
e-mail: cooray@medimail.net

10302
Dr J T Wallace
40 Hadham Road
Bishops Stortford
CM23 2QT
Tel: 01279 654053
Fax: 01279 658459
e-mail: james.wallace6@ntlworld.com

20109
Dr M G Sadler
Dr Stranders & Partners
Davenport House
Bowers Way
Harpenden
AL5 4HX
Tel: 01582 767821
Fax: 01582 769285
e-mail: mark.sandler@gp-E82077.nhs.uk

Humberside

10197
Dr J H Loose
Sutton Manor Surgery
St Ives Close
Wawne Road
Hull
HU7 4PT
Tel: 01482 826457
Tel: 01482 879485
Fax: 01482 824182
Mobile: 07770 740768
Pager: 01523 143844
e-mail: jim@jloose.karoo.co.uk

20015
Dr S R Ell
Academic ENT Department
Ward 11
Hull Royal Infirmary
Anlaby Road
Hull
HU3 2JZ
Tel: 01482 675971

Invernesshire

10024
Dr C M Cook
142 Manse Road
Ardersier
IV1 2SR
Tel: 01667 462240
e-mail: cmalcolmcook@btinternet.com

Isle of Lewis

10235
Michie A B
The Group Practice
Health Centre
Stornoway
HS1 2PS
Tel: 01851 703775
Fax: 01851 706138
e-mail: brianmichie@doctors.org.uk

539

Isle of Man

10082
Dr A C Pilling
Kensington Group Practice
Kensington Road
Douglas
Tel: 01624 676774
Fax: 01624 614668

Isle of Wight

10116
Dr M K De Belder
The Medical Centre
Upper Green Road
St Helens
Tel: 01983 872772
Fax: 01983 874800

Kent

10045
Dr M D Glanfield
71 Aylesford Avenue
Beckenham
BR3 3RZ
Tel: 0208 402 8565
Fax: 0208 663 6446
Fax: 0208 777 2152

10199
Dr P V Player
Headcorn Aerodrome
Shenley Farm
Headcorn
Ashford
TN27 9HX
Tel: 01622 890226
Fax: 01580 754452
e-mail: ringden@aol.com

10199
Dr P V Player
North Ridge
Rye Road
Hawkhurst
TN18 4EX
Tel: 01580 753935
Fax: 01580 754452
e-mail: ringden@aol.com

20024
Dr R W Fry
The Surgery
42 The High Street
Chislehurst
BR7 5AX
Tel: 0208 467 5551/2
e-mail: rolandfry@aol.com

20082
Dr L Smith
Milestone Lodge
Canterbury Road
Elham
CT4 6UE
Tel: 01303 840149
Fax: 01303 840150

20082
Dr L Smith
Premier Occupational Health
Unit 3 Shearway Business Park
Folkestone
CT19 4RH
Tel: 0870 444 1399
Fax: 0870 444 2908

10069
Dr B J McAvoy
Waldhurst
14 Waldron Road
Broadstairs
CT10 1TB
Tel & Fax: 01843 862378
e-mail: brian_j.mcavoy@tiscali.co.uk

Lancashire

10079
Dr I T Owen
Chief Medical Officer
BAE Systems
Warton Site
Preston
PR4 1AX
Tel: 01772 852798
Fax: 01772 855282
BAE staff only

10212
Dr R A Yuill
14 St John Street
Off Deansgate
Manchester
M3 4EA
Tel: 01244 355901
Tel: 0161 427 7409
Mobile: 07947 003934
e-mail: ra.yuill@ntlworld.com

10107
Dr J R Wray
AVMED2000
Aviation & General Medical Services
9 Flaxfield Road
Formby
L37 8BH
Tel & Fax: 01704 878060
Mobile: 07770 798 353
e-mail: avmed2000@btopenworld.com

10079
Dr I T Owen
Royal Preston Hospital
Occupational Health Dept
Sharoe Green Lane
Fulwood
Preston
PR2 9HT
Tel: 01772 522276

Leicestershire

20056
Dr N R Seymour
Empingham Medical Centre
37 Main Street
Oakham
Rutland
LE15 8PR
Tel: 01780 460202
E-mail: noel_seymour@uwclub.net

10070
Dr M F McGhee
53 Borough Street
Castle Donington
DE74 2LB
Tel: 01332 810241
Fax: 01332 811748
e-mail: mcghee_mf@gp-C82007.nhs.uk

10094
Dr K R Sumner
53 Borough Street
Castle Donington
DE74 2LB
Tel: 01332 810241
Fax: 01332 811748
e-mail: keith.sumner@ntlworld.co.uk

Lincolnshire

10067
Dr J M Lunn
The Surgery
Silver Street
Coningsby
LN4 4SG
Tel & Fax: 01526 354088
e-mail: jlun@doctors.org.uk

10068
Dr P J Mansfield
Good Health keeping
Garrod House
Manby Park
Louth
LN11 8UT
Tel: 01507 329100
Fax: 01507 329111
Mobile: 07957 861775
e-mail: doctorpetermansfield@yahoo.co.uk

10265
Dr C M Evans
The Swineshead Medical Group
The Surgery
Church Lane
Swineshead
Boston
PE20 3JA
Tel: 01205 723354
Fax: 01205 724429
e-mail: carole.evans9@btinternet.com

20007
Dr S J Bell
Landor
7 Watery Lane
Dunholme
LN2 3QW
Tel: 01673 860103
e-mail: aeromedsjbell@aol.com

London

10007
Dr K Edgington
The Rood Lane Medical Group
164 Bishopsgate
London
EC2M 4LX
Tel: 0207 377 4646
Fax: 0207 377 4247
e-mail: enquiries@roodlane.co.uk

10080
Dr I C Perry
19 Cliveden Place
London
SW1W 8HD
Tel: 0207 730 8045/9328
Fax: 0207 730 1985
e-mail: ian@ianperry.com

10231
Dr A C E Stacey
Finsbury Healthcare
5 London Wall Buildings
Finsbury Circus
London
EC3N 5NS
Tel: 0207 64488480
Fax: 0207 63748793
e-mail: tony.stacey@genmed.org.uk

20101
Mr J F A Pitts
148 Harley Street
London
W1G 7LG
Tel: 01708 445156

20101
Mr J F A Pitts
515 Liverpool Road
London
N7 8NS
Tel: 01708 445156

Merseyside

10107
Dr J E Wray
AVMED2000
Aviation & General Medical Services
9 Flaxfield Road
Formby
L37 8BH
Tel & Fax: 01704 878060
Mobile: 07770 798 353
e-mail: avmed2000@btopenworld.com

10111
Dr A Zsigmond
43 Rodney Street
Liverpool
L1 9EW
Tel: 0151 709 7441
Fax: 0151 708 0526
e-mail: doctor@zsigmond.co.uk

10278
Dr E J H Byrne
The Village Medical Centre
20 Quarry Street
Woolton Village
Liverpool
L25 6HE
Tel: 0151 428 4282
Fax: 0151 421 0884
Mobile: 07802 876 752
e-mail: edward.byrne@livgp.nhs.uk

Middlesex

10002
Dr S Stork
National Air Traffic Services
Medical Unit
Control Tower Building
London Heathrow Airport
Hounslow
TW6 1JJ
Tel: 0208 745 3351
Fax: 0208 745 3179

541

Middlesex (continued)

10007
Dr K Edgington
Arora International
The Grove
Bath Road
Heathrow
UB7 0DG
Tel: 01293 775336
Fax: 01293 775344
e-mail: reception@amsgatwick.com

10012
Dr M Bagshaw
National Air Traffic Services Medical Unit
Control Tower Building
London Heathrow Airport
Hounslow
TW6 1JJ
Tel: 0208 745 3351
Fax: 0208 745 3179
e-mail: mikebagshaw@doctors.org.uk

10019
Dr N Byrne
British Airways Health Service
Waterside (HMAG)
Harmondsworth
UB7 0GB
Tel: 0208 738 7747
Fax: 0208 738 9754
BA STAFF ONLY

10034
Dr N P Dowdall
British Airways Health Service
Waterside (HMAG)
PO Box 365
Harmondsworth
UB7 0GB
Tel: 0208 738 7747
Fax: 0208 738 9754
BA STAFF ONLY

10048
Dr M I H Gray
National Air Traffic Services
London ATC Centre (LATCC)
Porters Way
West Drayton
UB7 9AX
Tel: 0208 745 3351
Fax: 0208 745 3179
Tuesdays only

10117
Dr G J Cresswell
BMI Aeromedical Service
Ground Floor
The Queens Building
London Heathrow Airport
Hounslow
TW6 1DY
Tel: 0208 990 5237
Mobile: 07785 316050
e-mail: graham.cresswell@flybmi.com

10203
Dr M Popplestone
British Airways Health Service
Waterside (HMAG)
PO Box 365
Harmondsworth
UB7 0GB
Tel: 0208 738 7747
Fax: 0208 738 9754
BA STAFF ONLY

10211
Dr V P Robinson
The Consulting Rooms
Bishopswood Private Hospital
Rickmansworth Road
Northwood
HA6 2JW
Tel: 01923 835814

10260
Dr A J Roberts
National Air Traffic Services
Medical Department
Control Tower Building
London Heathrow Airport
Hounslow
TW6 1JJ
Tel: 0208 745 3321
Fax: 0208 745 3352/3330

10261
Dr D E L McIvor
National Air Traffic Services
Medical Department
Control Tower Building
London Heathrow Airport
Hounslow
TW6 1JJ
Tel: 0208 745 3321
Fax: 0208 745 3352/3330

10093
Dr J R R Stott
British Airways Health Services
Waterside (HMAG)
PO Box 365
Harmondsworth
UB7 0GB
Tel: 0208 738 7747
Fax: 0208 738 9754
BA STAFF ONLY

10100
Dr A J Wagner
British Airways Health Services
Waterside (HMAG)
PO Box 365
Harmondsworth
UB7 0GB
Tel: 0208 738 7747
Fax: 0208 738 9754
BA STAFF ONLY

10010
Dr N C Lee
Aviation Medical Services
Weekly House Medical Centre
Padbury Oaks
Old Bath Road
Longford
UB7 0EH
Tel: 01753 681978
Fax: 01753 685658
e-mail: medicals@heathrow.avmed.org.uk

Norfolk

10033
Dr R F Dorling
Plowright Medical Centre
1 Jack Boddy Way
Swaffham
PE37 7HJ
Tel: 01760 722797
Fax: 01760 720025
e-mail: bob_dorling@msn.com

10033
Dr R F Dorling
1 North Pickenham Road
Necton
Swaffham
PE37 8EF
Tel: 01760 441344
Fax: 01760 441511
e-mail: bob_dorling@msn.com

10064
Dr P W Lawrence
Department Diagnostic Imaging
James Paget Hospital
Gorleston
Great Yarmouth
NR31 6LA
Tel: 01493 452395

10109
Dr R C J Wyndham
Bramfield House
Barford Road
Marlingford
Norwich
NR9 5HU
Tel: 01603 880517
Fax: 01603 880510

20049
Dr A P J Preece
The Old Buck
Church Lane
Sedgeford
Hunstanton
PE36 5NA
Tel: 01485 570905
Fax: 01485 570981
e-mail: janie.insideout@btopenworld.com

Northamptonshire

20066
Dr S Thompson
Abington Medical Centre
Beech Avenue
Northampton
NN3 2JG
Tel: 01604 791999

10187
Dr K C Herbert
The Tollgate
Staverton Road
Daventry
NN11 4NN
Tel: 01327 708108
Fax: 01327 877201
e-mail:kevinherbert@davmed.com

Northern Ireland

10061
Dr B J Ireland
Brookhill
Upper Ballinderry
Co Antrim
BT28 2NS
Tel: 0289 262 2289

20040
Dr P Munro
Talisker Lodge
54b Templepatrick Road
Ballyclare
Co Antrim
BT39 9TX
Tel & Fax: 02893 323421
e-mail: talisker.lodge@btopenworld.com

10284
Dr S J Houston
Belfast City Airport
Belfast
BT3 9JH
Mobile: 0772 907 8000
e-mail: stephen@bma-houston.fsnet.co.uk

10285
Dr J Flatt
Apartment 8
Rockfield Court
Magheralin
Craigavon
Co Armagh
BT67 0RU
Tel: 07703 561766
e-mail: justinflatt@doctors.org.uk

Northumberland

10018
Dr C R Brown
Bondgate Practice
Infirmary Close
Alnwick
NE66 2NL
Tel: 01665 510888
Fax: 01665 510581
e-mail: colin2@alncom.net

10018
Dr C R Brown
The Medical Centre
Newcastle Int Airport
Woolsington
Newcastle on Tyne
NE13 8BZ
Tel: 01665 5108888
Fax: 01665 510581
e-mail: colin@aldersyde.freeserve.co.uk

10202
Dr T A White
Falcon House Surgery
17 Heaton Road
Newcastle on Tyne
NE6 1SA
Tel: 0191 265 3361
Fax: 0191 224 3209

20019
Dr M J Dodd
The Consulting Rooms
Infirmary Drive
Alnwick
NE66 2NR
Tel: 01665 602388

Nottinghamshire

10222
Dr W Muir
Lowdham Medical Centre
Francklin Road
Nottingham
NG14 7BG
Tel: 0115 9663 633
e-mail: william.muir@gp-C84613.nhs.uk

20100
Mr K P Gibbin
11 Regent Street
Nottingham
NG1 5BS
Tel: 0115 947 5475
Fax: 0115 924 1606

543

Nottinghamshire (continued)

10070
Dr M F McGhee
53 Borough Street
Castle Donington
DE74 2LB
Tel: 01332 810241
Fax: 01332 811748
e-mail: mcghee_mf@go-C82007.nhs.uk

10094
Dr K R Sumner
53 Borough Street
Castle Donington
DE74 2LB
Tel: 01332 810241
Fax: 01332 811748
e-mail: keith.sumner@ntlworld.co.uk

Oxfordshire

10100
Dr A J Wagner
Woodlands Medical Centre
Woodlands Road
Didcot
OX11 0BB
Tel & Fax: 01235 835425

10118
Dr J R Jones
Langford Medical Practice
9 Nightingale Place
Bicester
OX6 0XX
Tel: 01869 840377
e-mail: richard.jones@gp-K84613.nhs.uk

10118
Dr J R Jones
Langford Hall
Oxford Airport
Kidlington
OX5 1RA
Tel: 01865 840377
e-mail: richard.jones@gp-K84613.nhs.uk

20021
Dr M J Elliott
The Surgery
Hambleden
Henley-on-Thames
RG9 6RT
Tel: 01491 571305
Fax: 01491 411089

Renfrewshire

10035
Dr A R Downie
The Consulting Rooms
21 Neilston Road
Paisley
PA2 6LW
Tel: 0141 889 5277

Shetland

10060
Dr M D Hunter
Gord
Levenwick
ZE2 9HX
Tel: 01950 422210
e-mail: levenwick.surgery@shetland-hb.scot.nhs.uk

Shropshire

10115
Dr R N Ovenden
31 Broad Street
Ludlow
SY8 1NJ
Mobile: 07779 024924
e-mail: movenden@amsrve.net

10063
Dr D Latto
The Surgery
Poynton Road
Shawbury
SY4 4JS
Tel: 01939 250237
Fax: 01939 250093
Mobile: 07979 770030
e-mail: david.latto@nhs.net

Somerset

10236
Dr F J Knight
The Park Medical Practice
Cannards Grave Road
Shepton Mallet
BA4 5RT
Tel: 01749 334383
Fax: 01749 334393
e-mail: fiona.knight@parkmedicalpractice.nhs.uk

10259
Dr P E Gibbins
Westland Helicopters Ltd
Lysander Road
Yeovil
BA20 2YB
Tel: 01935 703221
Fax: 01935 703223
e-mail: ochealth@wh1.co.uk
GKN Westland staff only

10287
Dr P L d'Ambrumenil
Managing Director & Senior Flight Surgeon
Aeromedical Ltd
17 Manor Road
Bridgewater
TA7 9JE
Tel: 01278 723330
Fax: 01278 723188

Staffordshire

10196
Dr J D Hill
Peel Croft Surgery
Lichfield Street
Burton on Trent
DE14 3RH
Tel: 01283 511546
Fax: 01283 515761

20065
Dr M T Tanner
The Nook
3a High Street
Kinver
Stourbridge
DY7 6HG
Tel: 01384 872200
e-mail: drmttanner@aol.com

Stirlingshire

10088
Dr G R Sharp
Saltire Room
Express Holiday Inn
Stirling
FK7 7XH
Tel & Fax: 01381 622451
Mobile: 07931 915125

Strathclyde

20033
Dr G Martin
Maryhill Health Centre
41 Shawpark Street
Maryhill
Glasgow
G20 9DR
Tel: 0141 531 8830
Fax: 0141 531 8863
e-mail: administrator@gp43311.glasgow-hb.scot.nhs.uk

Suffolk

10064
Dr P W Lawrence
7 The Woodlands
Lowestoft
NR32 5EZ
Tel: 01502 731640

10119
Dr R A Davenport
The Central Surgery
201 Hamilton Road
Felixstowe
IP11 7DT
Tel: 01394 283197
Fax: 01394 270304
e-mail: robert.davenport@gp-D83348.nhs.uk

10195
Dr S R J Feltwell
Central Surgery
201 Hamilton Road
Felixstowe
IP11 7DT
Tel: 01394 283197
Fax: 01394 270304
e-mail: stephen.feltwell@gp-D83348.nhs.uk

Surrey

10002
Dr S Stork
Guildford Nuffield Hospital
Stirling Road
Guildford
GU2 5RF
Tel: 01428 684284

10004
Dr D N Tallent
Brookdale Medical Centre
79 Povey Cross Road
Horley
RH6 0AE
Tel: 01293 776996
Fax: 01293 823649
e-mail: medicals@gatwick.avmed.org.uk

10007
Dr K Edgington
Airport Medical Services
The Holiday Inn Gatwick
Gatwick Airport
Horley
RH6 0BA
Tel: 01293 775336
Fax: 01293 775344
e-mail: reception@amsgatwick.com

10046
Dr S A Goodwin
Airport Medical Services
The Holiday Inn Gatwick
Gatwick Airport
Horley
RH6 0BA
Tel: 01293 775336
Fax: 01293 775344
e-mail: reception@amsgatwick.com

10053
Dr G B Hey
The Surgery
143 Park Road
Camberley
GU15 2NN
Tel: 01276 26171
Fax: 01276 685383

10075
Dr D R Morgan
Little Tithe
Seale
Guildford
GU10 1JA
Tel: 01428 682413
e-mail: dewimorgan@ukgateway.net

10080
Dr I C Perry
Clerklands Surgery
Horley
RH6 8AR
Tel: 0207 730 8045/9328
Fax: 0207 730 1985
e-mail: ian@ianperry.com
www.ianperry.com

10092
Dr J P G Spencer
North Downs Hospital
46 Tupwood Lane
Caterham
CR3 6DP
Tel: 01883 348981

10120
Dr J S Hamill
Vine Medical Centre
69 Pemberton Road
East Molesey
KT8 9LG
Tel: 0208 979 7766
Fax: 0208 941 9827
e-mail: jsh@aviationmedicals.co.uk

10206
Dr P M Norris
Group Medical Adviser
British Aerospace Plc
212 Richmond Road
Kingston upon Thames
KT2 5HF
Tel: 0208 546 0400
Mobile: 0468 617 344
BAE employees only

545

Surrey (continued)

10046
Dr S A Goodwin
Flat 2 Little Grange
15 Portley Wood Road
Whyteleafe
CR3 0BQ
Tel: 01293 775336
Fax: 01293 775344
e-mail: reception@amsgatwick.com

10301
Dr P J C Chapman
Brookdale Medical Centre
79 Povery Cross Road
Horley
RH6 0AE
Tel: 01293 776996
Fax: 01293 823649
e-mail: medicals@gatwick.avmed.org.uk

20090
Dr M M Stack
Farnborough Int Travel & Aviation Clinic
FITAC House
100 Mychett Road
Mychett
GU16 6ET
Tel: 01252 373755
Fax: 01252 373799
e-mail: fitac2000@aol.com

11012
Dr M Bagshaw
Brookdale Medical Centre
79 Povey Cross Road
Horley
RH6 0AE
Tel: 01293 776996
Fax: 01293 823649
e-mail: medicals@gatwick.avmed.org.uk

Sussex

10007
Dr K Edgington
Beech Lodge
Old Hollow
Worth
RH10 4TA
Tel: 01293 775336
Fax: 01293 775344
e-mail: reception@amsgatwick.com

10099
Dr R S Waddy
AXA-PPP OHS Ltd
23 St Leonard's Road
Eastbourne
BN21 3PX
Tel: 01323 724889
Fax: 01323 721161
e-mail: rosemary.waddy@axa-pppohs.co.uk

10210
Dr C J King
Barnhouse Surgery
Barnhouse Close
Pulborough
RH20 2HQ
Tel: 01798 872815/873709
Fax: 01798 872123

10251
Dr A J Tobias
Shoreham Airport
Shoreham by Sea
BN43 5FF
Tel & Fax: 01903 761088
Mobile: 07941 579068
e-mail: andy@andytobias.freeserve.co.uk

10251
Dr A J Tobias
11 Browning Road
Lancing
BN15 0PY
Tel & Fax: 01903 761088
e-mail: andy@andytobias.freeserve.co.uk

10264
Dr T G Nash
Consultant OBS/GYN
Westlands
36 Collington Avenue
Bexhill on Sea
TN39 3NE
Tel: 01424 221886
Fax: 01424 213249

10290
Dr E J D Henderson
14 Burwash Road
Hove
BN3 8GQ
Tel: 01273 739271
Fax: 01273 727786
e-mail: drhenderson@doctors.org.uk

20064
Dr A Tabor
5 Woodland Drive
Hove
BN3 6DH
Tel: 01273 557779

20068
Dr F C E Walker
Langley House
West Street
Chichester
PO19 1RW
Tel: 01243 782266
Fax: 01243 779188
e-mail: charles.walker@gp-H82013.nhs.uk

20075
Dr P G Williams
1 Arlington Road
Eastbourne
BN21 1DH
Tel: 01323 727531
Fax: 01323 417985
e-mail: peter.williams12@virgin.net

20093
Dr D W Holwell
Park Surgery
Denne Road
Horsham
RH12 1JF
Tel: 01403 217100
Fax: 01403 214639
e-mail: admin@parksurgery.com

CAA
Civil Aviation Authority
Aviation House
Gatwick Airport South
Gatwick
RH6 0YR
Tel: 01293 573700
Fax: 01293 573995
e-mail: caa.aeromedsect@srg.caa.co.uk

Tayside

10009
Dr N W McAdam
Fernbrae Private Hospital
329 Perth Road
Dundee
DD2 1LJ
Tel & Fax: 01575 540250

Warwickshire

10020
Dr J W Busby
Melgreen
Wasperton
Warwick
CV35 8EB
Tel: 01926 620131/624469
Fax: 01926 620131
e-mail: johnbusby@wasperton.plus.com

10126
Dr R T Courtenay
Sherbourne Medical Centre
40 Oxford Street
Leamington Spa
CV32 4RA
Tel: 01926 333500
Fax: 01926 470884
e-mail: richard.courtenay@sherbourne.nhs.uk

West Midlands

10030
Dr K G Dawson
Aeomedical Services Ltd
BUPA Parkway Hospital
Damson Parkway
Solihull
B91 2PP
Tel: 01675 442069
Fax: 01675 442064
Mobile: 07768 955037
e-mail: admin@aeromedicalservices.co.uk

10066
Dr S J Lim
Wordsley Green Health Centre
Wordsley Green
Stourbridge
DY8 5PD
Tel: 01384 277591
Fax: 01384 401156
e-mail: stephen.lim@dudley.nhs.uk

10066
Dr S J Lim
ASM Consulting
Room 22 Forward House
Birmingham Airport Cargo Centre
Coventry Road
Elmdon
B26 3QT
Tel: 0121 780 4778

10121
Dr P Blackford
2 Edgehill Road
Four Oaks
Sutton Coldfield
B74 4NU
Tel: 0121 353 1854
Fax: 0121 353 1167
Mobile: 07956 141692
e-mail: patrickblackford@yahoo.co.uk

10125
Dr R A Pearson
122 Wychwood Avenue
Knowle
Solihull
B93 9DH
Tel: 01564 776541

10125
Dr R A Pearson
ASM Consulting
Room 22 Forward House
Birmingham Airport Cargo Centre
Coventry Road
Elmdon
B26 3QT
Tel: 0121 780 4778

20006
Dr J Batten
109 Moseley Avenue
Coventry
CV6 1HS
Tel: 02476 592201
Fax: 02476 601226
e-mail: john.batten@nhs.net

20041
Dr G W Murray
80 Tettenhall Road
Wolverhampton
WV1 4TF
Tel: 01902 422677/421005
e-mail: murray@gp-M92042.ms.nhs.uk

20048
Dr R T Pomeroy
The Medical Centre
Craig Croft
Chelmsley Wood
Birmingham
B37 7TR
Tel: 0121 770 5656
e-mail: richard.pomeroy@gp-M69012.nhs.uk

West Sussex

10007
Dr K Edgington
Beech Lodge
Old Hollow
Worth
RH10 4TA
Tel: 01293 775336
Fax: 01293 775344
e-mail: reception@amsgatwick.com

10050
Dr P C Harborow
Cygnus Mentoring & Professional Development
79a High Street
East Grinstead
RH19 3DD
Tel: 01342 321172
Fax: 01825 321199
Mobile: 07702 368055
e-mail: peter.harborow@cygnusmentoring.co.uk

10210
Dr C J King
Barnhouse Surgery
Barnhouse Close
Pulborough
RH20 2HQ
Tel: 01798 872815/873709
Fax: 01798 872123

10251
Dr A J Tobias
11 Browning Road
Lancing
BN15 0PY
Tel & Fax: 01903 761088
e-mail: andy@andytobias.freeserve.co.uk

547

West Sussex (continued)

10251
Dr A J Tobias
Shoreham Airport
Shoreham by Sea
BN43 5FF
Tel: 01903 761088
Mobile: 0794 1579068
e-mail: andy@andytobias.freeserve.co.uk

20068
Dr F C E Walker
Langley House
West Street
Chichester
PO19 1RW
Tel: 01243 782266
Fax: 01243 779188
e-mail: charles.walker@gp-H82013.nhs.uk

20093
Dr D W Holwell
Park Surgery
Denne Road
Horsham
RH12 1JF
Tel: 01403 217100
Fax: 01403 214639
e-mail: admin@parksurgery.com

Wiltshire

10023
Dr P J E Chinneck
Meadows
Whittonditch Road
Ramsbury
Marlborough
SN8 2PX
Tel: 01672 520474

10048
Dr M I H Gray
Nut Tree Cottage
Lower Chicksgrove
Tisbury
SP3 6NB
Tel: 01722 714382
Mobile: 07980 430623
e-mail: mihgray@v21.me.uk

10075
Dr D R Morgan
Silver Birches
West Grimstead
Salisbury
SP5 3RE
Tel: 01722 710225
Fax: 01722 710559
e-mail: dewimorgan@ukgateway.net

20071
Dr N O Yerbury
New Court Surgery
Borough Fields
Wootton Bassett
SN4 7AX
Tel: 01793 852302

Yorkshire

10207
Dr J G Benjamin
Sheffield City Airport
Sheffield Business Park
Europa Link
Sheffield
S9 1XZ
Mobile: 07968 543217
Fax: 0114 229 2201
Voice Mail: 0114 292 2202
e-mail: mail@aviationmedicalexaminer.co.uk

10207
Dr J G Benjamin
Mulitflight Ltd
South Side Aviation
Leeds Bradford Int Airport
Leeds
LS19 7UG
Mobile: 07968 543217
e-mail: mail@aviationmedicalexaminer.co.uk

10226
Dr K A Welch
Moor Grange Surgery
60 Moor Grange View
Leeds
LS16 5BJ
Tel: 0113 295 4880
Fax: 0113 295 4881
e-mail: katewelch@nhs.net

20057
Dr M E Sheikh
Bentley Health Centre
Askern Road
Doncaster
DN5 0JX
Tel: 01302 820494 ext 6317
Fax: 01302 820496

10296
Dr P M Brown
Folkton Manor
Folkton
Scarborough
YO11 3UQ
Tel: 01723 890383
Fax: 01723 890701
e-mail: drpmbrown@aol.com

20095
Dr A M Jepson
Outwood Park Medical Centre
Potovens Lane
Outwood
Wakefield
WF1 2PE
Tel: 01924 822626
Fax: 01924 786248
e-mail: dr.jepson@btinternet.com

CAA Medical Department

The CAA Medical Department is designed as a contact point where all medical information can be obtained relevant to aviation. It is easy to contact them – Tel: 0207 3797311 Ext 3674. All enquiries are confidential and are dealt with in a helpful and professional way. If you ever need to phone the CAA Medical Department, always provide as much information as possible, this helps to prevent delays in dealing with your case. When extra medical evidence is required, always send original documents of what ever is requested, eg, X-rays. Failure to send the requested documents or copies can provide invalid results. All extra medical evidence supplied is reviewed by a Consultant Medical Examiner at Gatwick. When the investigations are complete all medical evidence supplied will be returned to the person in question.

JAA Medical Certification Summary & minimum periodic requirements

LICENCE	CLASS 1 ATPL & CPL	CLASS 2 PPL
Initial examination	AMC	AMC or AME
Issue of medical certificate	Initial: AMS renewal; AMC or AME With AMS validation	AMC or AME with AMS validation
Validity of certificate for routine	<40yrs – 12 months	<30yrs – 60 months
Medical examination	>40yrs – 6 months	20-49yrs – 24 months
		>50yrs – 12 months
ECG	At initial examination	At initial examination
	<30yrs – 60 months	40-49yrs – 24 months
	30-39yrs – 24 months	>50yrs – 12 months
	40-49yrs – 12 months	
	>50yrs – 6 months	
Audiogram	At initial examination	Initial instrument rating
	<40yrs – 60 months	<40yrs – 60 months
	>40yrs – 24 months	>40yrs – 24 months
Lipid Profile	At initial examination	If 2 or more risk factors at initial
	Then >40yrs	examination then >40yrs
Peak flow	At initial examination	At initial examination
	Then 30, 35, & 40yrs	>40yrs – 48 months
	>40yrs – 48 months	

Recommended CAA Medical prices

Microlight Medical Examination Can be conducted by a General Practitioner or AME. It consists of a basic physical examination, eye and colour blindness tests, blood and urine tests.

JAA Medical charges Medication

AME Basic PPL	£75-130
Initial PPL	£150-200
CPL Basic Medical	£90
Blood Tests	£10
ECG	£60

General Practitioner

Basic medical £90-100 (BMMA recommended rate)
All additional tests required cost extra to the basic price.

	JAR CL 1 Initial	JAR CL 2 Initial	UK CL 1 Initial	UK CL 2 Initial	UK CL 3 Initial
Medical Examination	65.00	65.00	65.00	65.00	65.00
Administration Fee	36.00	18.00	18.00	18.00	18.00
ECG Recording	32.00	32.00	32.00	32.00	
ECG Reading	27.00	27.00	27.00	27.00	
Audiogram & Report	26.00		26.00	26.00	
Chest X-ray	36.00		36.00	36.00	
Chest X-ray Reporting	8.00		8.00	8.00	
Heamoglobin Estimation & Report	14.00	14.00			
Lipid Estimation & Report	16.00				
Spirometry & Report	27.00		27.00		
Extended Ophthalmology	50.00				
TOTAL	337.00	156.00	239.00	212.00	83.00

Additional Fees for CAA Medicals

	£
JAA Class 1 Extended Renewal Examination	250.00
Colour Vision Test	28.00
Electroencephalogram & Report	85.00
Exercise ECG Test & Report	70.00
24hr ECG	70.00
24hr Blood Pressure Monitoring	50.00
Initial Consultant Review - Cardiology	150.00
Follow up Consultant Review - Cardiology	120.00
Initial Consultant Review – Medical	80.00
Follow up Consultant Review – Medical	50.00
Ophthalmic Consultation for Refraction	50.00
Respiratory Peak Flow Assessment & Report	5.00
Non UK License Holder Seeing CAA Consultant	140.00
FAA Medical Examination with ECG	160.00
FAA Medical Examination without ECG	101.00
South African Medical Examination with ECG	337.00
South African Medical Examination only	101.00
New Zealand Medical Examination only – Professional/Initial	253.00
New Zealand Medical Examination only – Private/Initial	115.00
New Zealand Medical Examination only – Renewal	83.00

What can I take?

A list of drugs is available from your local AME, which are permitted for use whilst flying. In each case, it is necessary to consider an individual's susceptibility to the medication and all other aspects of the clinical condition before a drug is deemed safe to take whilst flying.

ICAO Aircraft designators

ICAO designator	ACFT type
A6	Grumman Intruder
A7	LTV Corsair
A9	Aero Commander Quail/Sparrow
A109	Agusta A109
A119	Agusta A119 Koala
A124	Antonov An-124
A139	Augusta AB-139
A140	Antonov An-140
A225	Antonov An-225
A306	Airbus A300 Freighter
A310	Airbus A310
A319	Airbus A319
A320	Airbus A320
A321	Airbus A321
A332	Airbus A330-200
A333	Airbus A330-300
A342	Airbus A340-200
A343	Airbus A340-300
A345	Airbus A340-500
A346	Airbus A340-600
A388	Airbus A380-800
A748	Hawker Siddeley HS-748
AA1	Grumman American Yankee
AA5	Grumman American Cheetah, Tiger, Traveler
AC6L	North American Rockwell Commander 680/685
AC11	North American Rockwell Commander 112/114/115
AC50	North American Rockwell Commander 500
AC52	North American Rockwell Commander 520
AC56	North American Rockwell Commander 560
AC68	North American Rockwell Commander 680
AC72	North American Rockwell Commander 720
AC80	North American Rockwell Commander 680 Turbo
AC90	North American Rockwell Commander 690 Jetprop
AC95	North American Rockwell Commander 695/980/1000
ACAR	Auster J-5
AN2	Antonov 2
AN3	Antonov 3
AN8	Antonov 8
AN12	Antonov 12
AN22	Antonov 22
AN24	Antonov 24
AN26	Antonov 26
AN28	Antonov 28
AN30	Antonov 30
AN38	Antonov 38
AN70	Antonov 70
AN72	Antonov 72
ANSN	Avro Anson
AR15	Aeronca Sedan
AS32	Aeropsatiale Super Puma/Cougar/Tiger
AS50	Aeropsatiale Ecureuil/Fennec/Super Star/Squirrel/Esquilo
AS55	Aeropsatiale Twin Star/Ecureuil/Twin Squirrel
AS65	Aeropsatiale Dauphin/Dauphin 2/Dolphin/Panther/Atalef
AUJ4	Auster J-4
AUS5	Auster 5

General — ICAO ACFT designators B772

ICAO designator	ACFT type
AUS6	Auster 6
AUS7	Auster 7
AUS9	Auster B5
AVID	Avid
B0105	Eurocopter Bo1O5
B06	Bell 206 Jet Ranger/Long Ranger/Creek/Sea Ranger
B06T	Bell 206LT Twin Ranger
B17	Boeing Flying Fortress
B18T	Beech 18
B25	North American Mitchell
B29	Boeing Super Fortress
B36T	Beech Bonanza
B47G	Bell 47G/D/Trooper
B47J	Bell Ranger
B74D	Boeing 747-400
B74R	Boeing 747SR
B74S	Boeing 747SP
B105	Bolkow BO-105
B209	Bolkow BO-209
B212	Bell 212
B222	Bell 222
B230	Bell 230
B350	Beech Super King Air 350
B407	Bell 407
B412	Bell 412
B427	Bell 427
B430	Bell 430
BAC1-11	British Aerospace BAC1-11
B461	British Aerospace 146-100
B462	British Aerospace 146-200
B463	British Aerospace 146-300
B609	Bell 609
B701	Boeing 707-100
B712	Boeing 717-200
B720	Boeing 720
B721	Boeing C-22/727-100
B722	Boeing 727-200
B731	Boeing 737-100
B732	Boeing 737-200
B733	Boeing 737-300
B734	Boeing 737-400
B735	Boeing 737-500
B736	Boeing 737-600
B737	Boeing 737-700
B738	Boeing 737-800
B739	Boeing 737-900
B741	Boeing 747-100
B742	Boeing 747-200
B743	Boeing 747-300
B744	Boeing 747-400
B752	Boeing 757-200
B753	Boeing 757-300
B762	Boeing 767-200
B763	Boeing 767-300
B764	Boeing 767-400
B772	Boeing 777-200

ICAO designator	ACFT type
B773	Boeing 777-300
BALL	Balloons
BDOG	British Aerospace Bulldog
BE9L	Beech King Air 90
BE10	Beech King Air 100
BE17	Beech Staggerwing/Traveler
BE18	Beech 18/Expeditor/Navigator
BE19	Beech 19/Musketeer
BE20	Beech C-12 Huron/Commuter
BE23	Beech 23 Sundowner/Musketeer
BE24	Beech 24 Sierra/Musketeer
BE30	Beech Super King Air 300
BE33	Beech 33 Bonanza/Denonair
BE35	Beech 35 Bonanza
BE36	Beech 36 Bonanza
BE40	Beech 400 Beech Jet/Jayhawk
BE50	Beech 50 Seminole/Twin Bonanza
BE55	Beech 55 Baron/Cochise
BE56	Beech56 Turbo Barron
BE58	Beech 58 Baron
BE60	Beech 60 Duke
BE65	Beech 65 Queen Air
BE70	Beech 70 Queen Air
BE76	Beech 76 Duchess
BE77	Beech 77 Skipper
BE80	Beech 80 Zamair/Queenaire
BE88	Beech 88 Queen Air
BE95	Beech 95 Travel Air
BE99	Beech 99 Airliner
BK17	Eurocopter BK 17
BN2P	Britten-Norman Islander/Defender
BN2T	Britten Norman Turbine Islander/Defender
BOLT	Steen Skybolt
C02T	Cessna 402 (Turbine)
C04T	Cessna 404 (Turbine)
C06T	Cessna 206 (Turbine)
C07T	Cessna 207 (Turbine)
C14T	Cessna 414 (Turbine)
C17	Boeing Globemaster 3
C25A	Cessna Citation CJ2
C25B	Cessna Citation CJ3
C30J	Lockheed Martin EC-130 Hercules
C56X	Cessna Citation Excel
C72R	Cessna 172RG Cutlass
C77R	Cessna 177 RG Cardinal
C82R	Cessna 182 RG Skylane
C120	Cessna 120
C130	Lookheed Martin KC-130 Hercules
C140	Cessna 140
C150	Cessna 150
C152	Cessna 152
C170	Cessna 170
C172	Cessna 172 Skyhawk/Hawk XP/Reims Rocket
C175	Cessna 175 Skylark
C177	Cessna 177 Cardinal
C180	Cessna 180 Skywagon

General ICAO ACFT designators DCH3

ICAO designator	ACFT type
C182	Cessna 182 Turbo Skylane
C185	Cessna 185 Skywagon
C190	Cessna 190
C195	Cessna 195
C205	Cessna 205
C206	Cessna 206 Stationair/Turbo Stationair/Turbo Skywagon
C207	Cessna 207 Stationair7/Turbo Stationair 8/Skywagon
C208	Cessna Caravan
C210	Cessna Centurion/Turbo Centurion
C303	Cessna 303 Crusader
C310	Cessna 310/T310
C320	Cessna 320 Skynight
C335	Cessna 335
C336	Cessna 336 Skymaster
C337	Cessna 337 Super Skymaster
C340	Cessna 340
C402	Cessna 402
C404	Cessna 404 Titan
C411	Cessna 411
C414	Cessna 414 Chancellor
C421	Cessna 421 Golden Eagle
C425	Cessna 425 Corsair/Conquest I
C441	Cessna 441 Conquest/Conquest II
C500	Cessna Citation I
C525	Cessna Citation Jet 525
C526	Cessna Citation Jet 526
C550	Cessna Citation II/Bravo
C551	Cessna Citation
C560	Cessna Citation V/Ultra
C650	Cessna Citation 6/7/3
C680	Cessna Citation Sovereign
C750	Cessna Citation 10
CAT	Canadair Vickers Catalina
CH40	Champion Lancer 402
CL41	Canadair CL-41
CL44	Canadair CL-44
CL60	Canadair Challenger 600/601/604
CP10	CAP Aviation 10/10B
CP20	CAP Aviation 20/20L
CP21	CAP Aviation 21
CP22	CAP Aviation 222
CP23	CAP Aviation 230/231
CRJ1	Canadair Regional Jet RJ-100
CRJ2	Canadair Regional Jet RJ-200
CRJ7	Canadair Regional Jet RJ-700
CRJ9	Canadair Regional Jet RJ-900
D11	Jodel 111/112/118/119/121/123/124/125/127/128
D18	Jodel 18/19/20
D140	Jodel D140 Abeille
D150	Jodel D 150 Mascaret
DA40	Diamond DA 40/42 Star
DH8A	De Havilland Dash 8
DH82	De Havilland Tiger Moth
DCH1	De Havilland Chipmunk
DCH2	De Havilland Beaver
DCH3	De Havilland Beaver

ICAO designator	ACFT type
DCH4	De Havilland Caribou
DCH5	De Havilland Buffalo
DCH6	De Havilland Twin Otter
DCH7	De Havilland Dash 7
DO27	Dornier Do 27
DO28	Dornier Do 28
DOVE	De Havilland Devon/Sea Devon/Dove
DR22	Centre Est DR220/221
DR30	Robin DR 300/Petit Prince/Prince/Chevalier/Major
DR40	Robing DR-400 Dauphin/Major 80/Earl/Remo/Cadet/Chevalier
DV20	Diamond DV-20/DA-20/Katana/Eclipse/Falcon/Evolution
E45X	Embraer ERJ-145/EMB-145
E110	Embraer Bandeirantte
E120	Embraer Brasilia
E121	Embraer Xingu
E135	Embraer Legacy
E145	Embraer ERJ-145
E170	Embraer 170/175
E190	Embraer 190/195
E200	Extra 200
E230	Extra 230
E300	Extra 300
E314	Embraer Super Tucano
E400	Extra 400
E500	Extra 500
EC20	Eurocopter EC-120 Colibri
EC25	Eurocopter EC-225 Cougar/Super Puma
EC30	Eurocopter EC130
EC35	Eurocopter EC-635
EC45	Eurocopter EC-145
EC55	Eurocopter EC-155
EN28	Enstrom Falcon/Sentinal/Falcon/Shark
EN48	Enstrom TH-29/490
EUFI	Eurofighter C-16/CE-16/Typhoon
EUPA	Europa Aviation Europa
F406	Cessna 406 Caravan 2
F900	Dassault Falcon 900
FA10	Dassault Falcon 10
FA20	Dassault Falcon 200
FA50	Dassault Falcon 50
FOX	Skyfox CA-25 Gazelle/Impala/Kitfox
G109	Grob G109 Ranger/Vigilant
G115	Grob G115/115A/115B/115C/115D
G120	Grob G120 Snunit
G140	Grob G0140TP
G159	Gulfstream 1/Academe
G160	Gron G160 Ranger
GA-7	Grumman American GA-7 Cougar
GAZL	Eurocopter Gazelle
GC1	Globe GC-1 Swift
GLAS	New Glassair
GLF2	Gulfstream American Gulfstream 2/2B/2SP/2TT
GLF3	Gulfstream American Gulfstream 3
GLF4	Gulfstream American Gulfstream 4
GLF5	Gulfstream American Gulfstream 5/500/550
GLID	Glider/Sailplane

ICAO designator	ACFT type
GYRO	Autogyro/Ultralight/Microlight
H25A	Hawker Siddley HS-125
H25B	British Aerospace Bae 125-700
HAWK	Bae Systems T-45 Hawk/Goshawk
HR10	Robin HR-100
HR20	Robin HR-200
J1	Auster J-1 Kingsland/Workmaster/Alpha
J2	Piper Cub J-2
J3	Piper Cub J-3 Prospector
J4	Piper Cub Coupe
J5	Piper Cub Cruiser
JAB4	Jabiru J200/J250/J400/J450
JABI	Babiru SP/SP-6/SK/ST/LSA
JAGR	British Aerospace Jaguar/Shamsher
JS31	British Aerospace Jetstream 31
JS32	British Aerospace Jetstream Super 31
JS41	British Aerospace Jetstream 41
KREZ	Vari-Eze
L8	Luscombe 8 Observer/Trainer/Silvarie/Master
L10	Lockheed Electra
L11	Luscombe 11 Sedan
L14	Lockheed Super Electra
L18	Lockheed Lodestar
L29	Lockheed Delfin
L29A	Lockheed Jetstar 6/8
LA25	Lake LA-250
LA4	Lake Buccaneer
LJ23	Learjet 23
LJ24	Learjet 24
LJ25	Learjet 25
LJ28	Learjet 28
LJ31	Learjet 31
LJ35	Learjet 35
LJ40	Learjet 40
LJ45	Learjet 45
LJ55	Learjet 55
LJ60	Learjet 60
LNC2	Lancair 200/235/320/360
LNC4	Lancair 4
LYNX	Westland Lynx/Super Lynx
M4	Maule M-4 Bee Dee/Rocket/Strata Rocket/Jetasen
M5	Maule M-5 Lunar Rocket/Strata Rocket
M6	Maule M-6 Super Rocket
M7	Maule M-7 Star Craft/Super Rocket/Orion/Comet
M8	Maule M-8
M9	Maule M-9
M10	Mooney Cadet
M20P	Mooney M-20 Ranger/Chaparral/Master/Statesman
M20T	Mooney M-20 Encore/Bravo
M22	Mooney M-22 Mustang
M24	Mooney M-24 Dromader Super
MD11	McDonnell Douglas MD-11
MD81	McDonnell Douglas MD-81
MD82	McDonnell Douglas MD-82
MD83	McDonnell Douglas MD-83
MD87	McDonnell Douglas MD-87

ICAO designator	ACFT type
MD88	McDonnell Douglas MD-88
MD90	McDonnell Douglas MD-90
ME08	Messerschmitt 108 Taifun
ME09	Messerschmitt 109 Buchon
ME62	Messerschmitt 262
MESS	Miles Messenger
MOSQ	De Havilland Mosquito
P28	Piper Cherokee/Archer II/ Dakota/Warrior
P32	Piper Cherokee 6/Lance/Saratoga
P46	Piper Malibu
P57	Partenavia Fachiro
P68	Partenavia Observer/Victor
P68T	Partenavia Sparticus
PA11	Piper Cub
PA12	Piper Super Cruiser
PA14	Piper Family Cruiser
PA15	Piper Vagabond
PA16	Piper Clipper
PA17	Piper Vagabond Trainer
PA18	Piper Super Cub
PA20	Piper Pacer
PA22	Piper Tri-pacer
PA23	Piper Apache
PA24	Piper Comanche
PA25	Piper Pawnee
PA27	Piper Aztec
PA30	Piper Twin Comanche
PA31	Piper Navajo
PA32	Piper Saratoga/Cherokee 6
PA34	Piper Seneca
PA36	Piper Pawnee Brave
PA38	Piper Tomahawk
PA44	Piper Seminole
PA46	Piper Malibu
PAY4	Piper Cheyenne 400
PC6P	Pilatus PC-6 Porter
PC7	Pilatus PC-7 Astra
PC9	Pilatus PC-9 Hudournik
PC12	Pilatus PC-12 Eagle
PC21	Pilatus PC-21
PPRO	Percival Provost
PREN	Percival Prentice
PUMA	Aerospatiale Puma
PUP	Beagle B-121 Pup
R4	Sikorsky Hoverfly
R22	Robinson R22 Beta/Mariner
R44	Robinson R44 Clipper/Raven/Astro
R100	Robin R-1180 Aiglon
R200	Robin R-2112/R-2100/R-2100
R300	Robin R-300/R-3000/R-3120/R-3140
R721	Boeing 727-100
RALL	Morane-Saulnier Rallye Club/Super Club/Rallye Commodore
RF3	Fournier RF3
RF4	Fournier RF4
RF5	Fournier RF5
RF6	Slingsby T3/T67 Firefly

ICAO designator	ACFT type
RF9	Fournier RF9
RJ70	British Aerospace RJ-70
RJ85	British Aerospace RJ-85
RV3	VANS RV-3
RV4	VANS RV-4
RV6	VANS RV06
RV7	VANS RV-7
RV8	VANS RV-8
RV9	VANS RV-9
RV10	VANS RV-10
S38	Sikorsky S-38
S39	Sikorsky S-39
S51	Sikorsky R-5/S-51/H-5
S52	Sikorsky HO5S/S-52
S55P	Sikorsky HRS/HO4S/HH-19/S-55A
S55T	Sikorsky S-55T
S58P	Sikorsky Seahorse/Seabat/Chocktaw
S58T	Sikorsky S-58DT
S61	Sikorsky HS-9 Sea King/Nuri
S61R	Sikorsky HH-3/CG-3
S62	Sikorsky Seaguard
S64	Sikorsky Skycrane/Tarhe
S76	Sikorsky Eagle/Spirit
S92	Sikorsky Helibus
S108	Piper Voyager/Station Wagon
S330	Schweizer 269D/333/330
S360	Aerospatiale SA-360/SA361 Dauphin
SA37	Schweizer SA-2-37A Condor
SA38	Schweizer Twin Condor
SC7	Shorts SC7 Skyvan/Skyline
SF34	Saab SF340
SH33	Shorts 330 Sherpa
SH36	Shorts360
SHAD	CFM Shadow
SHIP	Airship
SS2P	North American Rockwell Thrush Commander
STAR	Beech Starship
SUCO	Bell SuperCobra
SV4	Stampe SV-4
T6	North American Havard
T134	Tupolev Tu-134
T144	Tupolev Tu-144
T154	Tupolev Tu-154
T160	Tupolev Tu-160
T204	Tupolev Tu-224/234/214/204
T334	Tupolev Tu-334
TAMP	Socata TB-9 Tampico
TB30	Socata TB-30 Epsilon
TB31	Socata TB-31 Omega
TBM7	Socata TBM-700
TOBA	Socata TB-200/TB-10 Tobago
TRIN	Socata TB-20/21 Trinidad
TRIS	Pilatus Britten-Norman BN-2A Mk3 Trislander
TUCA	Shorts Tucano
Y18T	Yakovlev Yak-18T
Y112	Yakovlev Yak-112

ICAO designator	ACFT type
Y130	Yakovlev Yak-130
Y141	Yakovlev Yak-141
YAK3	Yakovlev Yak-3
YAK9	Yakovlev Yak-9
YK11	Yakovlev Yak-11
YK12	Yakovlev Yak-12
YK18	Yakovlev Yak-18/18U/18PS/18PM/18P/18A
YK28	Yakovlev Yak-28
YK38	Yakovlev Yak-38
YK40	Yakovlev Yak-40
YK42	Vakovlev Yak-42/142
YK50	Yakovlev Yak-50
YK52	Yakovlev Yak-52
YK53	Yakovlev Yak-53
YK54	Yakovlev Yak-54
YK55	Yakovlev Yak-55
YK58	Yakovlev Yak-58
Z22	Zlin Z-22 Junak
Z26	Zlin Skydevil/Akrobat/Condor/Universal
Z37P	Zlin Sparka/Cmelak
Z37T	Zlin Agro Turbo
Z42	Zlin Z-42 Firnas
Z43	Zlin Z-43 Safir
Z50	Zlin Z-50

Aircraft Registration and Country Codes

ACFT registration	Country	ICAO code	ACFT registration	Country	ICAO code
3A	Monaco	LN		British Indian Ocean	FJ
3B	Mauritius	FI	C	Canada	CU
3C	Equatorial Guinea	FG	C2	Nauru	AN
3D	Swaziland	FD	C5	Gambia	GB
3X	Guinea	GU	C6	Bahamas	MY
4K	Azerbaijan	UB	C9	Mozambique	FQ
4L	Georgia	UG		Canary Islands	GC
4R	Sri Lanka	VC	CC	Chile	SC
4X	Israel	LL	CN	Morocco	GM
5A	Libya	HL		Cocos Islands	AC
5B	Cyprus	LC		Cook Islands	NC
5H	Tanzania	HT	CP	Bolivia	SL
5N	Nigeria	DN	CRM	Macau	LP
5R	Madagascar	FM	CS	Portugal	LP
5T	Mauritania	GQ	CU	Cuba	MU
5U	Niger	DR	CX	Uruguay	SU
5V	Togo	DX	D	Germany	ED
5W	Samoa	NS	D2	Angola	FN
5X	Uganda	HU	D4	Cape Verde Islands	GV
5Y	Kenya	HK	D6	Comoros Islands	FM
6O	Somalia	HC	DQ	Fiji Islands	NF
6V	Senegal	GO	E3	Eritrea	HH
6Y	Jamaica	MK		East Timor	WP
7O	Yemen	OY	EC	Spain	LE
7P	Lesotho	FX	EI	Ireland	EI
7Q	Malawi	FW	EK	Armenia	UG
7T	Algeria	DA	EL	Liberia	GL
8P	Barbados	TB	EP	Iran	OI
8Q	Maldives	VR	ER	Moldova	LU
8R	Guyana	SY	ES	Estonia	EE
8XR	Rwanda	HR	ET	Ethiopia	HA
9A	Croatia	LD	EW	Belarus	UM
9G	Ghana	DG	EY	Tajikistan	UT
9H	Malta	LM	EZ	Turkmenistan	UT
9J	Zambia	FL	F	France	LF
9K	Kuwait	OK		Faroe Islands	EK
9L	Sierra Leone	GF	FO	French West Indies	
9M	Malaysia	WB		French Guiana	SO
9N	Nepal	VN		French Polynesia	NT
9Q	Zaire		G	Jersey	EG
9U	Burundi	HB	G	United Kingdom	EG
9V	Singapore	WS		Georgia	UG
9Y	Trinidad & Tobago	TT		Greenland	BG
A2	Botswanna	FB		Guam	PT
A3	Tonga	NF	H4	Solomon Islands	AG
A40	Oman	OO	HA	Hungary	LH
A5	Bhutan	VQ		Hawaii	PH
A6	United Arab Emirates	OM	HB	Liechtenstein	LS
A7	Qatar	OT	HB	Switzerland	LS
A9	Bahrain	OB	HC	Ecuador	SE
	Alaska	P	HH	Haiti	MT
	American Samoa	NS	HI	Dominican Republic	MD
AP	Pakistan	OP	HK	Columbia	SK
	Armenia	UG	HL	South Korea	RK
B	China	ZB	HP	Panama	MP
B*	Hong Kong	VH	HR	Honduras	MH
B*	Taiwan	RC	HS	Thailand	VT
	Bahrain	OB	HZ	Saudi Arabia	OE
	Baker Island	PB	I	Italy	LI
	Belarus	UM	J2	Djibouti	HF

ACFT registration	Country	ICAO code	ACFT registration	Country	ICAO code
J3	Grenada	MK	TF	Iceland	BI
J6	St Lucia	TL	TG	Guatemala	MG
J7	Dominica	TD	TI	Costa Rica	MR
J8	St Vincent & Grenadines	TV	TJ	Cameroon	FK
JA	Japan	RJ	TL	Central African Republic	FE
	Johnston Island	PJ	TN	Congo	FC
JU	Mongolia	ZM	TR	Gambon	FO
JY	Jordan	OJ	TS	Tunisia	DT
	Kiribati	NG	TT	Chad	FT
	Line Islands	PL	TU	Ivory Coast	DI
LN	Norway	EN	TY	Benin	DB
LV	Argentina	SA	TZ	Mali	GA
LX	Luxembourg	EL	UK	Uzbekistan	UT
LY	Lithuania	EY	UN	Kazakhstan	UA
LZ	Bulgaria	LB	UR	Ukraine	UK
	Madeira	LP	V2	Antigua	TA
	Mariana Islands	PG	V3	Belize	MZ
	Mayotte/Reunion	FM	V4	St Kitts & Nevis	TK
	Melilla	GE	V5	Namibia	FY
	Midways Islands	PM	V6	Micronesia	PT
N	United States (USA)	K*	V7	Marshall Islands	PK
	New Calendonia	NW	V8	Brunei	WB
	Niue Islands	NI	VH	Australia	Y*
	Northern Mariana Islands	AH	VN	Vietnam	VV
OB	Peru	SP	VP-A	Anguilla	TQ
OD	Lebanon	OL	VP-B	Bermuda	TX
OE	Austria	LO	VP-C	Cayman Islands	MW
OH	Finland	EF	VP-F	Falkland Islands	SF
OK	Czech Republic	LK	VP-G	Gibraltar	LX
OM	Slovak Republic	LZ	VP-L	British Virgin Islands	TU
OO	Belgium	EB	VP-M	Montserrat	TR
OY	Denmark	EK	VQ-H	Ascension Islands	FH
P2	Papua New Guinea	AY	VT	India	VA
P4	Aruba	TN	XA	Mexico	MM
	Palau	PT	XT	Burkina Faso	DF
PH	Netherlands	EH	XU	Cambodia	VD
PJ	Netherlands Antilles	TN	XY	Myanmar	VY
PK	Indonesia	WA	YA	Afganistan	OA
PP	Brazil	SB	YI	Iraq	OR
	Puerto Rico	TJ	YJ	Vanuatu	
PZ	Surinam	SM	YK	Syria	OS
RA	CIS	UE	YL	Latvia	EV
RA	Russia Moscow	UW	YN	Nicaragua	MN
RA	Russia OF	UU	YR	Romania	LR
RA	Russia West	UR	YS	El Salvador	MS
RA	Russia	UL	YU	Serbia & Montenegro	LY
RDPL	Laos	VL		Yugoslavia	LY
RP	Philippines	RP	YV	Venezuela	SV
S2	Bangladesh	VG	Z	Zimbabwe	FV
S5	Slovenia	LJ	Z3	Macedonia	LW
S7	Seychelles	FS	Z3	Macedonia	LW
S9	Sao Tome	FP	ZA	Albania	LA
SE	Sweden	ES	ZK	New Zealand	NZ
	Slovak Republic	LZ	ZP	Paraguay	SG
SP	Poland	EP	ZS	South Africa	FA
	St Pierre et Miquelon	CF			
ST	Sudan	HS			
SU	Egypt	HE			
SX	Greece	LG			
T9	Bosnia & Herzegovina	LQ			
TC	Turkey	LT			

Aircraft Registration and County Codes

Country	ACFT registration	ICAO code	Country	ACFT registration	ICAO code
Afganistan	YA	OA	Djibouti	J2	HF
Alaska		P	Dominica	J7	TD
Albania	ZA	LA	Dominican Republic	HI	MD
Algeria	7T	DA	East Timor		WP
American Samoa		NS	Ecuador	HC	SE
Angola	D2	FN	Egypt	SU	HE
Anguilla	VP-A	TQ	El Salvador	YS	MS
Antigua	V2	TA	Equatorial Guinea	3C	FG
Argentina	LV	SA	Eritrea	E3	HH
Armenia		UG	Estonia	ES	EE
Armenia	EK	UG	Ethiopia	ET	HA
Aruba	P4	TN	Falkland Islands	VP-F	SF
Ascension Islands	VQ-H	FH	Faroe Islands		EK
Australia	VH	Y*	Fiji Islands	DQ	NF
Austria	OE	LO	Finland	OH	EF
Azerbaijan	4K	UB	France	F	LF
Bahamas	C6	MY	French Guiana		SO
Bahrain		OB	French Polynesia		NT
Bahrain	A9	OB	French West Indies	FO	
Baker Island		PB	Gambia	C5	GB
Bangladesh	S2	VG	Gambon	TR	FO
Barbados	8P	TB	Georgia		UG
Belarus		UM	Georgia	4L	UG
Belarus	EW	UM	Germany	D	ED
Belgium	OO	EB	Ghana	9G	DG
Belize	V3	MZ	Gibraltar	VP-G	LX
Benin	TY	DB	Greece	SX	LG
Bermuda	VP-B	TX	Greenland		BG
Bhutan	A5	VQ	Grenada	J3	MK
Bolivia	CP	SL	Guam		PT
Bosnia & Herzegovina	T9	LQ	Guatemala	TG	MG
Botswanna	A2	FB	Guinea	3X	GU
Brazil	PP	SB	Guyana	8R	SY
British Indian Ocean		FJ	Haiti	HH	MT
British Virgin Islands	VP-L	TU	Hawaii		PH
Brunei	V8	WB	Honduras	HR	MH
Bulgaria	LZ	LB	Hong Kong	B*	VH
Burkina Faso	XT	DF	Hungary	HA	LH
Burundi	9U	HB	Iceland	TF	BI
Cambodia	XU	VD	India	VT	VA
Cameroon	TJ	FK	Indonesia	PK	WA
Canada	C	CU	Iran	EP	OI
Canary Islands		GC	Iraq	YI	OR
Cape Verde Islands	D4	GV	Ireland	EI	EI
Cayman Islands	VP-C	MW	Israel	4X	LL
Central African Republic	TL	FE	Italy	I	LI
Chad	TT	FT	Ivory Coast	TU	DI
Chile	CC	SC	Jamaica	6Y	MK
China	B	ZB	Japan	JA	RJ
CIS	RA	UE	Jersey	G	EG
Cocos Islands		AC	Johnston Island		PJ
Columbia	HK	SK	Jordan	JY	OJ
Comoros Islands	D6	FM	Kazakhstan	UN	UA
Congo	TN	FC	Kenya	5Y	HK
Cook Islands		NC	Kiribati		NG
Costa Rica	TI	MR	Kuwait	9K	OK
Croatia	9A	LD	Laos	RDPL	VL
Cuba	CU	MU	Latvia	YL	EV
Cyprus	5B	LC	Lebanon	OD	OL
Czech Republic	OK	LK	Lesotho	7P	FX
Denmark	OY	EK	Liberia	EL	GL

General

Aircraft Registration and County Codes

Country	ACFT registration	ICAO code
Libya	5A	HL
Liechtenstein	HB	LS
Line Islands		PL
Lithuania	LY	EY
Luxembourg	LX	EL
Macau	CRM	LP
Macedonia	Z3	LW
Macedonia	Z3	LW
Madagascar	5R	FM
Madeira		LP
Malawi	7Q	FW
Malaysia	9M	WB
Maldives	8Q	VR
Mali	TZ	GA
Malta	9H	LM
Mariana Islands		PG
Marshall Islands	V7	PK
Mauritania	5T	GQ
Mauritius	3B	FI
Mayotte/Reunion		FM
Melilla		GE
Mexico	XA	MM
Micronesia	V6	PT
Midways Islands		PM
Moldova	ER	LU
Monaco	3A	LN
Mongolia	JU	ZM
Montserrat	VP-M	TR
Morocco	CN	GM
Mozambique	C9	FQ
Myanmar	XY	VY
Namibia	V5	FY
Nauru	C2	AN
Nepal	9N	VN
Netherlands Antilles	PJ	TN
Netherlands	PH	EH
New Caledonia		NW
New Zealand	ZK	NZ
Nicaragua	YN	MN
Niger	5U	DR
Nigeria	5N	DN
Niue Islands		NI
Northern Mariana Islands		AH
Norway	LN	EN
Oman	A40	OO
Pakistan	AP	OP
Palau		PT
Panama	HP	MP
Papua New Guinea	P2	AY
Paraguay	ZP	SG
Peru	OB	SP
Philippines	RP	RP
Poland	SP	EP
Portugal	CS	LP
Puerto Rico		TJ
Qatar	A7	OT
Romania	YR	LR
Russia Moscow	RA	UW
Russia OF	RA	UU
Russia West	RA	UR
Russia	RA	UL

Country	ACFT registration	ICAO code
Rwanda	8XR	HR
Samoa	5W	NS
Sao Tome	S9	FP
Saudi Arabia	HZ	OE
Senegal	6V	GO
Serbia & Montenegro	YU	LY
Seychelles	S7	FS
Sierra Leone	9L	GF
Singapore	9V	WS
Slovak Republic		LZ
Slovak Republic	OM	LZ
Slovenia	S5	LJ
Solomon Islands	H4	AG
Somalia	6O	HC
South Africa	ZS	FA
South Korea	HL	RK
Spain	EC	LE
Sri Lanka	4R	VC
St Kitts & Nevis	V4	TK
St Lucia	J6	TL
St Pierre et Miquelon		CF
St Vincent & Grenadines	J8	TV
Sudan	ST	HS
Surinam	PZ	SM
Swaziland	3D	FD
Sweden	SE	ES
Switzerland	HB	LS
Syria	YK	OS
Taiwan	B*	RC
Tajikistan	EY	UT
Tanzania	5H	HT
Thailand	HS	VT
Togo	5V	DX
Tonga	A3	NF
Trinidad & Tobago	9Y	TT
Tunisia	TS	DT
Turkey	TC	LT
Turkmenistan	EZ	UT
Uganda	5X	HU
Ukraine	UR	UK
United Arab Emirates	A6	OM
United Kingdom	G	EG
United States (USA)	N	K*
Uruguay	CX	SU
Uzbekistan	UK	UT
Vanuatu	YJ	
Venezuela	YV	SV
Vietnam	VN	VV
Yemen	7O	OY
Yugoslavia		LY
Zaire	9Q	
Zambia	9J	FL
Zimbabwe	Z	FV

General — Aircraft Registration and County Codes

ICAO code	ACFT reg	Country	ICAO code	ACFT reg	Country
	9Q	Zaire	GA	TZ	Mali
AC		Cocos Islands	GB	C5	Gambia
AG	H4	Solomon Islands	GC		Canary Islands
AH		Northern Mariana Islands	GE		Melilla
AN	C2	Nauru	GF	9L	Sierra Leone
AY	P2	Papua New Guinea	GL	EL	Liberia
BG		Greenland	GM	CN	Morocco
BI	TF	Iceland	GO	6V	Senegal
CF		St Pierre et Miquelon	GQ	5T	Mauritania
CU	C	Canada	GU	3X	Guinea
DA	7T	Algeria	GV	D4	Cape Verde Islands
DB	TY	Benin	HA	ET	Ethiopia
DF	XT	Burkina Faso	HB	9U	Burundi
DG	9G	Ghana	HC	6O	Somalia
DI	TU	Ivory Coast	HE	SU	Egypt
DN	5N	Nigeria	HF	J2	Djibouti
DR	5U	Niger	HH	E3	Eritrea
DT	TS	Tunisia	HK	5Y	Kenya
DX	5V	Togo	HL	5A	Libya
EB	OO	Belgium	HR	8XR	Rwanda
ED	D	Germany	HS	ST	Sudan
EE	ES	Estonia	HT	5H	Tanzania
EF	OH	Finland	HU	5X	Uganda
EG	G	Jersey	K*	N	United States (USA)
EG	G	United Kingdom	LA	ZA	Albania
EH	PH	Netherlands	LB	LZ	Bulgaria
EI	EI	Ireland	LC	5B	Cyprus
EK		Faroe Islands	LD	9A	Croatia
EK	OY	Denmark	LE	EC	Spain
EL	LX	Luxembourg	LF	F	France
EN	LN	Norway	LG	SX	Greece
EP	SP	Poland	LH	HA	Hungary
ES	SE	Sweden	LI	I	Italy
EV	YL	Latvia	LJ	S5	Slovenia
EY	LY	Lithuania	LK	OK	Czech Republic
FA	ZS	South Africa	LL	4X	Israel
FB	A2	Botswanna	LM	9H	Malta
FC	TN	Congo	LN	3A	Monaco
FD	3D	Swaziland	LO	OE	Austria
FE	TL	Central African Republic	LP		Madeira
FG	3C	Equatorial Guinea	LP	CRM	Macau
FH	VQ-H	Ascension Islands	LP	CS	Portugal
FI	3B	Mauritius	LQ	T9	Bosnia & Herzegovina
FJ		British Indian Ocean	LR	YR	Romania
FK	TJ	Cameroon	LS	HB	Liechtenstein
FL	9J	Zambia	LS	HB	Switzerland
FM		Mayotte/Reunion	LT	TC	Turkey
FM	5R	Madagascar	LU	ER	Moldova
FM	D6	Comoros Islands	LW	Z3	Macedonia
FN	D2	Angola	LW	Z3	Macedonia
	FO	French West Indies	LX	VP-G	Gibraltar
FO	TR	Gambon	LY		Yugoslavia
FP	S9	Sao Tome	LY	YU	Serbia & Montenegro
FQ	C9	Mozambique	LZ		Slovak Republic
FS	S7	Seychelles	LZ	OM	Slovak Republic
FT	TT	Chad	MD	HI	Dominican Republic
FV	Z	Zimbabwe	MG	TG	Guatemala
FW	7Q	Malawi	MH	HR	Honduras
FX	7P	Lesotho	MK	6Y	Jamaica
FY	V5	Namibia	MK	J3	Grenada

General

Aircraft Registration and County Codes

ICAO code	ACFT reg	Country
MM	XA	Mexico
MN	YN	Nicaragua
MP	HP	Panama
MR	TI	Costa Rica
MS	YS	El Salvador
MT	HH	Haiti
MU	CU	Cuba
MW	VP-C	Cayman Islands
MY	C6	Bahamas
MZ	V3	Belize
NC		Cook Islands
NF	A3	Tonga
NF	DQ	Fiji Islands
NG		Kiribati
NI		Niue Islands
NS		American Samoa
NS	5W	Samoa
NT		French Polynesia
NW		New Caledonia
NZ	ZK	New Zealand
OA	YA	Afganistan
OB		Bahrain
OB	A9	Bahrain
OE	HZ	Saudi Arabia
OI	EP	Iran
OJ	JY	Jordan
OK	9K	Kuwait
OL	OD	Lebanon
OM	A6	United Arab Emirates
OO	A40	Oman
OP	AP	Pakistan
OR	YI	Iraq
OS	YK	Syria
OT	A7	Qatar
OY	7O	Yemen
P		Alaska
PB		Baker Island
PG		Mariana Islands
PH		Hawaii
PJ		Johnston Island
PK	V7	Marshall Islands
PL		Line Islands
PM		Midways Islands
PT		Guam
PT		Palau
PT	V6	Micronesia
RC	B*	Taiwan
RJ	JA	Japan
RK	HL	South Korea
RP	RP	Philippines
SA	LV	Argentina
SB	PP	Brazil
SC	CC	Chile
SE	HC	Ecuador
SF	VP-F	Falkland Islands
SG	ZP	Paraguay
SK	HK	Columbia
SL	CP	Bolivia
SM	PZ	Surinam
SO		French Guiana
SP	OB	Peru

ICAO code	ACFT reg	Country
SU	CX	Uruguay
SV	YV	Venezuela
SY	8R	Guyana
TA	V2	Antigua
TB	8P	Barbados
TD	J7	Dominica
TJ		Puerto Rico
TK	V4	St Kitts & Nevis
TL	J6	St Lucia
TN	P4	Aruba
TN	PJ	Netherlands Antilles
TQ	VP-A	Anguilla
TR	VP-M	Montserrat
TT	9Y	Trinidad & Tobago
TU	VP-L	British Virgin Islands
TV	J8	St Vincent & Grenadines
TX	VP-B	Bermuda
UA	UN	Kazakhstan
UB	4K	Azerbaijan
UE	RA	CIS
UG		Armenia
UG		Georgia
UG	4L	Georgia
UG	EK	Armenia
UK	UR	Ukraine
UL	RA	Russia
UM		Belarus
UM	EW	Belarus
UR	RA	Russia West
UT	EY	Tajikistan
UT	EZ	Turkmenistan
UT	UK	Uzbekistan
UU	RA	Russia OF
UW	RA	Russia Moscow
VA	VT	India
VC	4R	Sri Lanka
VD	XU	Cambodia
VG	S2	Bangladesh
VH	B*	Hong Kong
VL	RDPL	Laos
VN	9N	Nepal
VQ	A5	Bhutan
VR	8Q	Maldives
VT	HS	Thailand
VV	VN	Vietnam
VY	XY	Myanmar
WA	PK	Indonesia
WB	9M	Malaysia
WB	V8	Brunei
WP		East Timor
WS	9V	Singapore
Y*	VH	Australia
	YJ	Vanuatu
ZB	B	China
ZM	JU	Mongolia

ICAO and IATA airfield designators

AIRFIELD	ICAO	IATA	AIRFIELD	ICAO	IATA
Aberdeen	EGPD	ABZ	Derby	EGBD	
Aberporth	EGUC		Dishforth	EGXD	
Albourne	EGKD		Doncaster Sheffield	EGCN	
Alderney	EGJA	ACI	Dundee	EGPN	DND
Andrewsfield	EGSL		Dunkeswell	EGTU	
Ascot Racecourse	EGLT		Durham Tees Valley	EGNV	MME
Aylesbury	EGTA		Duxford	EGSU	DUX
Bagby	EGNG		Eaglescott	EGHU	
Ballykelly	EGQB	BOL	Earls Colne	EGSR	
Barkston Heath	EGYE		Eday	EGED	
Barra	EGPR	BRR	Edinburgh	EGPH	EDI
Barrow	EGNL	BWF	Elmsett	EGST	
Beccles	EGSM		Elstree	EGTR	
Bedford	EGSB		Enniskillen	EGAB	ENK
Belfast Aldergrove	EGAA	BFS	Exeter	EGTE	EXT
Belfast City	EGAC	BHD	Fair Isle	EGEF	FIE
Bembridge	EGHJ	BBP	Fairford	EGVA	
Benbecula	EGPL	BEB	Fairoaks	EGTF	
Benson	EGUB	BEX	Farnborough	EGLF	FAB
Beverley	EGNY		Farthing Corner	EGMF	
Bicester	EGDD		Fenland	EGCL	
Biggin Hill	EGKB	BQH	Fetlar		FEA
Birmingham	EGBB	BHX	Fife	EGPJ	
Blackbushe	EGLK	BBS	Filton	EFTG	FZO
Blackpool	EGNH	BLK	Flotta		FLH
Bodmin	EGLA		Forrest Moor	EGXF	
Boscombe Down	EGDM		Fowlmere	EGMA	
Bourn	EGSN		Full Sutton	EGNU	
Bournemouth	EGHH	BOH	Glasgow	EGPF	GLA
Bristol	EGGD	BRS	Glasgow City Heliport	EGEG	
Brize Norton	EGVN	BZZ	Glasgow Prestwick	EFPK	PIK
Brooklands	EGLB		Glenforsa		ULL
Brough	EGNB		Gloucestershire	EGBJ	GLO
Caernarfon	EGCK		Goodwood Racecourse	EGKG	
Cambridge	EGSC	CBG	Great Yarmouth	EGSD	
Campbeltown	EGEC	CAL	Guernsey	EGJB	GCI
Cardiff Heliport	EGFC		Halton	EGWN	
Cardiff	EGFF	CWL	Haverfordwest	EGFE	HAW
Carlisle	EGNC	CAX	Hawarden	EGNR	CEG
Chalgrove	EGLJ		Henlow	EGWE	
Challock	EGKE		Henstridge	EGHS	
Cheltenham Racecourse	EGBC		Hereford	EGVH	
Chichester	EGHR	QUG	Hethel	EGSK	
Chivenor	EGDC		Holyhead	EGCH	
Church Fenton	EGXG		Honington	EGXH	BEQ
Clacton	EGSQ		Hucknall	EGNA	
Colerne	EGUO		Humberside	EGNJ	HUY
Coltishall	EGYC	CLF	Inverness	EGPE	INV
Compton Abbas	EGHA		Ipswich		IPW
Coningsby	EGXC		Islay	EGPI	ILY
Cosford	EGWC		Isle of Man	EGNS	IOM
Cottesmore	EGXJ		Isle of Wight	EGHN	
Coventry	EGBE	CVT	Isleworth	EGLI	
Cranfield	EGTC		Jersey	EGJJ	JER
Cranwell	EGYD		Kemble	EGBP	
Crowfield	EGSO		Kinloss	EGQK	FSS
Culdrose	EGDR		Kirkwall	EGPA	KOI
Cumbernauld	EGPG		Lakenheath	EGUL	LKZ
Deanland	EGKL		Lands End	EGHC	LEQ
Denham	EGLD		Langford Lodge	EGAL	

568

AIRFIELD	ICAO	IATA
Lasham	EGHL	QLA
Lashenden	EGKH	
Leconfield	EGXV	
Lee on Solent	EGHF	
Leeds Bradford	EGNM	LBA
Leeming	EGXE	
Leicester	EGBG	
Lerwick	EGET	LWK
Leuchars	EGQL	ADX
Linton on Ouse	EGXU	HRT
Lisburn	EGQD	
Little Gransden	EGMJ	
Liverpool	EGGP	LPL
Llanbedr	EGOD	
London City	EGLC	LCY
London Gatwick	EGKK	LGW
London Heathrow	EGLL	LHR
London Heliport	EGLE	
London Luton	EGGW	LTN
London Stansted	EGSS	STN
Londonderry	EGAE	LDY
Long Marston	EGBL	
Lossiemouth	EGQS	LMO
Lydd	EGMD	LYX
Lyneham	EGDL	LYE
Manchester Barton	EGCB	
Manchester Woodford	EGCD	WFD
Manchester	EGCC	MAN
Manston	EGMH	MSE
Marham	EGYM	
Marshland	EGSI	
Maypole	EGHB	
Middle Wallop	EGVP	
Mildenhall	EGUN	MHZ
Mount Pleasant	EGYP	
Mountwise	EGDB	
Netheravon	EGDN	
Netherthorpe	EGNF	
Newcastle	EGNT	NCL
Newmarket Racecourse	EGSW	
Newtownards	EDAD	
North Ronaldsay	EGEN	NRL
North Weald	EGSX	
Northampton	EGBK	ORM
Northolt	EGVC	NHT
Northolt	EGWU	NHT
Norwich	EGSH	NWI
Nottingham East Midlands	EGNX	EMA
Nottingham	EGBN	NQT
Oaksey Park	EFTW	
Oban	EGEO	OBN
Odiham	EGVO	ODH
Old Buckenham	EGSV	
Old Sarum	EGLS	
Outer Skerrie		OUK
Oxford	EGTK	OXF
Panshanger	EGLG	
Papa Stour		PSV
Papa Westray	EGEP	PPW
Pembrey	EGFP	
Pembrey	EGOP	

AIRFIELD	ICAO	IATA
Penzance Heliport	EGHK	PZE
Perranporth	EGTP	
Perth	EGPT	PSL
Peterborough Conington	EGSF	
Peterborough Sibson	EGSP	
Peterhead Heliport	EGPS	
Plymouth	EGHD	PLH
Plymouth	EGVE	PLH
Popham	EGHP	
Portland	EGDP	
Portsmouth	EGVF	
Redhill	EGKR	KRH
Reftford	EGNE	
Rochester	EGTO	RCS
Sanday	EGES	NDY
Sandtoft	EGCF	
Scampton	EGXP	
Scatsta	EGPM	SCS
Scilly Isles	EGHE	ISC
Seething	EGSJ	
Shawbury	EGOS	
Sheffield City	EGSY	SZD
Sherburn in Elmet	EGCJ	
Shipdham	EGSA	
Shobdon	EGBS	
Shoreham	EGKA	ESH
Shuttleworth	EGTH	
Silverstone	EGBV	
Skegness	EGNI	
Sleap	EGCV	
Southampton	EGHI	SOU
Southend	EGMC	SEN
Southport Birkdale Sands	EGCO	
St Athan	EGDX	
St Mawgan	EGDG	NQY
Stapleford	EGSG	
Stornoway	EGPO	SYY
Stronsay	EGER	
Strubby Heliport	EGCG	
Sturgate	EGCS	
Sumburgh	EGPB	LSI
Swansea	EGFH	SWS
Tatenhill	EGBM	
Ternhill	EGOE	
Thorne	EGCP	
Thruxton	EGHO	
Tilstock	EGCT	
Tiree	EGPU	TRE
Topcliffe	EGXZ	
Tresco Heliport	EGHT	
Truro	EGHY	
Turweston	EGBT	
Unst	EGPW	UNT
Upavon	EGDJ	UPV
Uxbridge	EGUU	
Valley	EGOV	
Waddington	EGXW	WTN
Warton	EGNO	
Wattisham	EGUW	
Wellesbourne Mountford	EGBW	
Welshpool	EGCW	

569

IATA and ICAO airfield designators

AIRFIELD	ICAO	IATA
West Freugh	EGOY	
West Wales	EGFA	
Westray	EGEW	WRY
Whalsay	EGEH	WHS
White Waltham	EGLM	
Wick	EGPC	WIC
Wickenby	EGNW	
Wittering	EGXT	
Wolverhampton	EGBO	
Woodvale	EGOW	
Wrexham	EGCE	
Wycombe Air Park	EGTB	
Wyton	EGUY	
Yeovil	EGHG	
Yeovilton	EGDY	YEO

IATA	AIRFIELD	ICAO
	Aberporth	EGUC
ABZ	Aberdeen	EGPD
ACI	Alderney	EGJA
ADX	Leuchars	EGQL
	Albourne	EGKD
	Andrewsfield	EGSL
	Ascot Racecourse	EGLT
	Aylesbury	EGTA
	Bagby	EGNG
	Barkston Heath	EGYE
BBP	Bembridge	EGHJ
BBS	Blackbushe	EGLK
BEB	Benbecula	EGPL
	Beccles	EGSM
	Bedford	EGSB
BEQ	Honington	EGXH
	Beverley	EGNY
BEX	Benson	EGUB
BFS	Belfast Aldergrove	EGAA
BHD	Belfast City	EGAC
BHX	Birmingham	EGBB
	Bicester	EGDD
BLK	Blackpool	EGNH
	Bodmin	EGLA
BOH	Bournemouth	EGHH
BOL	Ballykelly	EGQB
	Boscombe Down	EGDM
	Bourn	EGSN
BQH	Biggin Hill	EGKB
	Brooklands	EGLB
	Brough	EGNB
BRR	Barra	EGPR
BRS	Bristol	EGGD
BWF	Barrow	EGNL
BZZ	Brize Norton	EGVN
	Caernarfon	EGCK
CAL	Campbeltown	EGEC
	Cardiff Heliport	EGFC
CAX	Carlisle	EGNC
CBG	Cambridge	EGSC
CEG	Hawarden	EGNR
	Chalgrove	EGLJ
	Challock	EGKE
	Cheltenham Racecourse	EGBC
	Chivenor	EGDC

IATA	AIRFIELD	ICAO
	Church Fenton	EGXG
	Clacton	EGSQ
CLF	Coltishall	EGYC
	Colerne	EGUO
	Compton Abbas	EGHA
	Coningsby	EGXC
	Cosford	EGWC
	Cottesmore	EGXJ
	Cranfield	EGTC
	Cranwell	EGYD
	Crowfield	EGSO
	Culdrose	EGDR
	Cumbernauld	EGPG
CVT	Coventry	EGBE
CWL	Cardiff	EGFF
	Deanland	EGKL
	Denham	EGLD
	Derby	EGBD
	Dishforth	EGXD
DND	Dundee	EGPN
	Doncaster Sheffield	EGCN
	Dunkeswell	EGTU
DUX	Duxford	EGSU
	Eaglescott	EGHU
	Earls Colne	EGSR
	Eday	EGED
EDI	Edinburgh	EGPH
	Elmsett	EGST
	Elstree	EGTR
EMA	Nottingham East Midlands	EGNX
ENK	Enniskillen	EGAB
ESH	Shoreham	EGKA
EXT	Exeter	EGTE
FAB	Farnborough	EGLF
FIE	Fair Isle	EGEF
	Fairford	EGVA
	Fairoaks	EGTF
	Farthing Corner	EGMF
	Fenland	EGCL
FEA	Fetlar	
	Fife	EGPJ
FLH	Flotta	
	Forrest Moor	EGXF
	Fowlmere	EGMA
FSS	Kinloss	EGQK
	Full Sutton	EGNU
FZO	Filton	EFTG
GCI	Guernsey	EGJB
GLA	Glasgow	EGPF
	Glasgow City Heliport	EGEG
GLO	Gloucestershire	EGBJ
	Goodwood Racecourse	EGKG
	Great Yarmouth	EGSD
	Halton	EGWN
HAW	Haverfordwest	EGFE
	Henlow	EGWE
	Henstridge	EGHS
	Hereford	EGVH
	Hethel	EGSK
	Holyhead	EGCH
HRT	Linton on Ouse	EGXU

General — IATA and ICAO airfield designators

IATA	AIRFIELD	ICAO	IATA	AIRFIELD	ICAO
	Hucknall	EGNA	NWI	Norwich	EGSH
HUY	Humberside	EGNJ		Oaksey Park	EFTW
ILY	Islay	EGPI	OBN	Oban	EGEO
INV	Inverness	EGPE	ODH	Odiham	EGVO
IOM	Isle of Man	EGNS		Old Buckenham	EGSV
IPW	Ipswich			Old Sarum	EGLS
ISC	Scilly Isles	EGHE	ORM	Northampton	EGBK
	Isle of Wight	EGHN	OUK	Outer Skerrie	
	Isleworth	EGLI	OXF	Oxford	EGTK
JER	Jersey	EGJJ		Panshanger	EGLG
	Kemble	EGBP		Pembrey	EGFP
KOI	Kirkwall	EGPA		Pembrey	EGOP
KRH	Redhill	EGKR		Perranporth	EGTP
	Langford Lodge	EGAL		Peterborough Conington	EGSF
	Lashenden	EGKH		Peterborough Sibson	EGSP
LBA	Leeds Bradford	EGNM		Peterhead Heliport	EGPS
LCY	London City	EGLC	PIK	Glasgow Prestwick	EGPK
LDY	Londonderry	EGAE	PLH	Plymouth	EGHD
	Leconfield	EGXV	PLH	Plymouth	EGVE
	Lee on Solent	EGHF		Popham	EGHP
	Leeming	EGXE		Portland	EGDP
	Leicester	EGBG		Portsmouth	EGVF
LEQ	Lands End	EGHC	PPW	Papa Westray	EGEP
LGW	London Gatwick	EGKK	PSL	Perth	EGPT
LHR	London Heathrow	EGLL	PSV	Papa Stour	
	Lisburn	EGQD	PZE	Penzance Heliport	EGHK
	Little Gransden	EGMJ	QLA	Lasham	EGHL
LKZ	Lakenheath	EGUL	QUG	Chichester	EGHR
	Llanbedr	EGOD	RCS	Rochester	EGTO
LMO	Lossiemouth	EGQS		Reftford	EGNE
	London Heliport	EGLE		Sandtoft	EGCF
	Long Marston	EGBL		Scampton	EGXP
LPL	Liverpool	EGGP	SCS	Scatsta	EGPM
LSI	Sumburgh	EGPB		Seething	EGSJ
LTN	London Luton	EGGW	SEN	Southend	EGMC
LWK	Lerwick	EGET		Shawbury	EGOS
LYE	Lyneham	EGDL		Sherburn in Elmet	EGCJ
LYX	Lydd	EGMD		Shipdham	EGSA
MAN	Manchester	EGCC		Shobdon	EGBS
	Manchester Barton	EGCB		Shuttleworth	EGTH
	Marham	EGYM		Silverstone	EGBV
	Marshland	EGSI		Skegness	EGNI
	Maypole	EGHB		Sleap	EGCV
MHZ	Mildenhall	EGUN	SOU	Southampton	EGHI
	Middle Wallop	EGVP		Southport Birkdale Sands	EGCO
MME	Durham Tees Valley	EGNV		St Athan	EGDX
	Mount Pleasant	EGYP		Stapleford	EGSG
	Mountwise	EGDB	STN	London Stansted	EGSS
MSE	Manston	EGMH		Stronsay	EGER
NCL	Newcastle	EGNT		Strubby Heliport	EGCG
NDY	Sanday	EGES		Sturgate	EGCS
	Netheravon	EGDN	SWS	Swansea	EGFH
	Netherthorpe	EGNF	SYY	Stornoway	EGPO
	Newmarket Racecourse	EGSW	SZD	Sheffield City	EGSY
	Newtownards	EDAD		Tatenhill	EGBM
NHT	Northolt	EGVC		Ternhill	EGOE
NHT	Northolt	EGWU		Thorne	EGCP
	North Weald	EGSX		Thruxton	EGHO
NQT	Nottingham	EGBN		Tilstock	EGCT
NQY	St Mawgan	EGDG		Topcliffe	EGXZ
NRL	North Ronaldsay	EGEN	TRE	Tiree	EGPU

571

ICAO and IATA airfield designators

IATA	AIRFIELD	ICAO
	Tresco Heliport	EGHT
	Truro	EGHY
	Turweston	EGBT
ULL	Glenforsa	
UNT	Unst	EGPW
UPV	Upavon	EGDJ
	Uxbridge	EGUU
	Valley	EGOV
	Warton	EGNO
	Wattisham	EGUW
	Wellesbourne Mountford	EGBW
	Welshpool	EGCW
	West Freugh	EGOY
	West Wales	EGFA
WFD	Manchester Woodford	EGCD
	White Whaltham	EGLM
WHS	Whalsay	EGEH
WIC	Wick	
WIC	Wick	EGPC
	Wickenby	EGNW
	Wittering	EGXT
	Wolverhampton	EGBO
	Woodvale	EGOW
	Wrexham	EGCE
WRY	Westray	EGEW
WTN	Waddington	EGXW
	Wycombe Air Park	EGTB
	Wyton	EGUY
YEO	Yeovilton	EGDY
	Yeovil	EGHG

ICAO	AIRFIELD	IATA
EGAA	Belfast Aldergrove	BFS
EGAB	Enniskillen	ENK
EGAC	Belfast City	BHD
EGAD	Newtownards	
EGAE	Londonderry	LDY
EGAL	Langford Lodge	
EGBB	Birmingham	BHX
EGBC	Cheltenham Racecourse	
EGBD	Derby	
EGBE	Coventry	CVT
EGBG	Leicester	
EGBJ	Gloucestershire	GLO
EGBK	Northampton	ORM
EGBL	Long Marston	
EGBM	Tatenhill	
EGBN	Nottingham	NQT
EGBO	Wolverhampton	
EGBP	Kemble	
EGBS	Shobdon	
EGBT	Turweston	
EGBV	Silverstone	
EGBW	Wellesbourne Mountford	
EGCB	Manchester Barton	
EGCC	Manchester	MAN
EGCD	Manchester Woodford	WFD
EGCE	Wrexham	
EGCF	Sandtoft	
EGCG	Strubby Heliport	
EGCH	Holyhead	
EGCJ	Sherburn in Elmet	

ICAO	AIRFIELD	IATA
EGCK	Caernarfon	
EGCL	Fenland	
EGCN	Doncaster Sheffield	
EGCO	Southport Birkdale Sands	
EGCP	Thorne	
EGCS	Sturgate	
EGCT	Tilstock	
EGCV	Sleap	
EGCW	Welshpool	
EGDB	Mountwise	
EGDC	Chivenor	
EGDD	Bicester	
EGDG	St Mawgan	NQY
EGDJ	Upavon	UPV
EGDL	Lyneham	LYE
EGDM	Boscombe Down	
EGDN	Netheravon	
EGDP	Portland	
EGDR	Culdrose	
EGDX	St Athan	
EGDY	Yeovilton	YEO
EGEC	Campbeltown	CAL
EGED	Eday	
EGEF	Fair Isle	FIE
EGEG	Glasgow City Heliport	
EGEH	Whalsay	WHS
EGEN	North Ronaldsay	NRL
EGEO	Oban	OBN
EGEP	Papa Westray	PPW
EGER	Stronsay	
EGES	Sanday	NDY
EGET	Lerwick	LWK
EGEW	Westray	WRY
EGFA	West Wales	
EGFC	Cardiff Heliport	
EGFE	Haverfordwest	HAW
EGFF	Cardiff	CWL
EGFH	Swansea	SWS
EGFP	Pembrey	
EGGD	Bristol	BRS
EGGP	Liverpool	LPL
EGGW	London Luton	LTN
EGHA	Compton Abbas	
EGHB	Maypole	
EGHC	Lands End	LEQ
EGHD	Plymouth	PLH
EGHE	Scilly Isles	ISC
EGHF	Lee on Solent	
EGHG	Yeovil	
EGHH	Bournemouth	BOH
EGHI	Southampton	SOU
EGHJ	Bembridge	BBP
EGHK	Penzance Heliport	PZE
EGHL	Lasham	QLA
EGHN	Isle of Wight	
EGHO	Thruxton	
EGHP	Popham	
EGHR	Chichester	QUG
EGHS	Henstridge	
EGHT	Tresco Heliport	
EGHU	Eaglescott	

ICAO	AIRFIELD	IATA	ICAO	AIRFIELD	IATA
EGHY	Truro		EGPB	Sumburgh	LSI
EGJA	Alderney	ACI	EGPC	Wick	WIC
EGJB	Guernsey	GCI	EGPD	Aberdeen	ABZ
EGJJ	Jersey	JER	EGPE	Inverness	INV
EGKA	Shoreham	ESH	EGPF	Glasgow	GLA
EGKB	Biggin Hill	BQH	EGPG	Cumbernauld	
EGKD	Albourne		EGPH	Edinburgh	EDI
EGKE	Challock		EGPI	Islay	ILY
EGKG	Goodwood Racecourse		EGPJ	Fife	
EGKH	Lashenden		EGPK	Glasgow Prestwick	PIK
EGKK	London Gatwick	LGW	EGPL	Benbecula	BEB
EGKL	Deanland		EGPM	Scatsta	SCS
EGKR	Redhill	KRH	EGPN	Dundee	DND
EGLA	Bodmin		EGPO	Stornoway	SYY
EGLB	Brooklands		EGPR	Barra	BRR
EGLC	London City	LCY	EGPS	Peterhead Heliport	
EGLD	Denham		EGPT	Perth	PSL
EGLE	London Heliport		EGPU	Tiree	TRE
EGLF	Farnborough	FAB	EGPW	Unst	UNT
EGLG	Panshanger		EGQB	Ballykelly	BOL
EGLI	Isleworth		EGQD	Lisburn	
EGLJ	Chalgrove		EGQK	Kinloss	FSS
EGLK	Blackbushe	BBS	EGQL	Leuchars	ADX
EGLL	London Heathrow	LHR	EGQS	Lossiemouth	LMO
EGLM	White Whaltham		EGSA	Shipdham	
EGLS	Old Sarum		EGSB	Bedford	
EGLT	Ascot Racecourse		EGSC	Cambridge	CBG
EGMA	Fowlmere		EGSD	Great Yarmouth	
EGMC	Southend	SEN	EGSF	Peterborough Conington	
EGMD	Lydd	LYX	EGSG	Stapleford	
EGMF	Farthing Corner		EGSH	Norwich	NWI
EGMH	Manston	MSE	EGSI	Marshland	
EGMJ	Little Gransden		EGSJ	Seething	
EGNA	Hucknall		EGSK	Hethel	
EGNB	Brough		EGSL	Andrewsfield	
EGNC	Carlisle	CAX	EGSM	Beccles	
EGNE	Reftford		EGSN	Bourn	
EGNF	Netherthorpe		EGSO	Crowfield	
EGNG	Bagby		EGSP	Peterborough Sibson	
EGNH	Blackpool	BLK	EGSQ	Clacton	
EGNI	Skegness		EGSR	Earls Colne	
EGNJ	Humberside	HUY	EGSS	London Stansted	STN
EGNL	Barrow	BWF	EGST	Elmsett	
EGNM	Leeds Bradford	LBA	EGSU	Duxford	DUX
EGNO	Warton		EGSV	Old Buckenham	
EGNR	Hawarden	CEG	EGSW	Newmarket Racecourse	
EGNS	Isle of Man	IOM	EGSX	North Weald	
EGNT	Newcastle	NCL	EGSY	Sheffield City	SZD
EGNU	Full Sutton		EGTA	Aylesbury	
EGNV	Durham Tees Valley	MME	EGTB	Wycombe Air Park	
EGNW	Wickenby		EGTC	Cranfield	
EGNX	Nottingham East Midlands	EMA	EGTE	Exeter	EXT
EGNY	Beverley		EGTF	Fairoaks	
EGOD	Llanbedr		EGTG	Filton	FZO
EGOE	Ternhill		EGTH	Shuttleworth	
EGOP	Pembrey		EGTK	Oxford	OXF
EGOS	Shawbury		EGTO	Rochester	RCS
EGOV	Valley		EGTP	Perranporth	
EGOW	Woodvale		EGTR	Elstree	
EGOY	West Freugh		EGTU	Dunkeswell	
EGPA	Kirkwall	KOI	EGTW	Oaksey Park	

573

ICAO	AIRFIELD	IATA
EGUB	Benson	BEX
EGUC	Aberporth	
EGUL	Lakenheath	LKZ
EGUN	Mildenhall	MHZ
EGUO	Colerne	
EGUU	Uxbridge	
EGUW	Wattisham	
EGUY	Wyton	
EGVA	Fairford	
EGVC	Northolt	NHT
EGVE	Plymouth	PLH
EGVF	Portsmouth	
EGVH	Hereford	
EGVN	Brize Norton	BZZ
EGVO	Odiham	ODH
EGVP	Middle Wallop	
EGWC	Cosford	
EGWE	Henlow	
EGWN	Halton	
EGWU	Northolt	NHT
EGXC	Coningsby	
EGXD	Dishforth	
EGXE	Leeming	
EGXF	Forrest Moor	
EGXG	Church Fenton	
EGXH	Honington	BEQ
EGXJ	Cottesmore	
EGXP	Scampton	
EGXT	Wittering	
EGXU	Linton on Ouse	HRT
EGXV	Leconfield	
EGXW	Waddington	WTN
EGXZ	Topcliffe	
EGYC	Coltishall	CLF
EGYD	Cranwell	
EGYE	Barkston Heath	
EGYM	Marham	
EGYP	Mount Pleasant	
	Fetlar	FEA
	Flotta	FLH
	Glenforsa	ULL
	Ipswich	IPW
	Outer Skerrie	OUK
	Papa Stour	PSV

ICAO document listing

Annexes

1	Personnel Licensing
2	Rules of the Air
3	Meteorological Services for International Air Navigation
4	Aeronautical Charts
5	Units of Measurement to be Used in Air and Ground Operations
6	operation of Aircraft
6 Part 1	International Commercial Air Transport – Aeroplanes
6 Part 2	International General Aviation – Aeroplanes
6 Part 3	International Operations – Helicopters
7	Aircraft Nationality and Registration Marks
8	Airworthiness of Aircraft
9	Facilitation
10	Aeronautical Telecommunications
10 Volume 1	Radio Navigation Aids
10 Volume 2	Communication Procedures including those with PANS status
10 Volume 3 Part I	Digital Data Communication Systems
10 Volume 3 Part II	Voice Communications Systems
10 Volume 4	Surveillance Radar and Collision Avoidance Systems
10 Volume 5	Aeronautical Radio Frequency Spectrum Utilisation
11	Air Traffic Services
12	Search and Rescue
13	Aircraft Accident and Incident Investigation
14 Volume 1	Aerodrome Design and Operations
14 Volume 2	Heliports
15	Aeronautical Information Services
16 Volume 1	Aircraft Noise
16 Volume 2	Aircraft Engine Emissions
17	Security
18	The Safe Transport of Dangerous Goods

Documents

4444	Air Traffic Management
7030	Regional Supplementary Procedures
7100	Tariffs for Airports and Air Navigation Services
7192	Training Manual
7192 Part B5	Integrated Commercial Pilot Course
7192 Part B5 Vol 1	Course details
7192 Part B5 Vol 2	Instructor Briefing Sheets
7192 Part D1	Aircraft Maintenance
7192 Part D3	Flight Operations Officers/Flight Dispatchers
7192 Part E1	Cabin Attendants Safety Training
7192 Part F1	Meteorology for Air Traffic Controllers and Pilots
7300	Convention on International Civil Aviation
7910	Location Indicators
8071	Manual on Testing of Radio Navigation Aids
8071 Volume I	Testing of Ground-based Radio Navigation Systems
8071 Volume II	Testing of Satellite-based Radio Navigation Systems
8071 Volume III	Testing of Surveillance Radar Systems
8168	Aircraft Operations
8168 Volume 1	Flight Procedures
8168 Volume 2	Construction of Visual and Instrument Flight Procedures

General — ICAO document listing

8400	ICAO Abbreviations and Codes
8585	Designators for Aircraft Operating Agencies, Aeronautical Authorities and Services
8643	Aircraft Type Designators
8697	Aeronautical Chart Manual
8984	Manual of Civil Aviation Medicine
9137	Airport Services Manual
9137 Part 1	Rescue and Fire Fighting
9137 Part 2	Pavement Surface Conditions
9137 Part 3	Bird Control and Reduction
9137 Part 5	Removal of Disabled Aircraft
9137 Part 6	Control of Obstacles
9137 Part 7	Airport Emergency Planning
9137 Part 8	Airport Operational Services
9137 Part 9	Airport Maintenance Practices
9150	Stolport Manual
9157	Aerodrome Design Manual
9157 Part 1	Runways
9157 Part 2	Taxiways, Aprons and Holding Bays
9157 Part 3	Pavements
9157 Part 4	Visual Aids
9157 Part 5	Electrical Systems
9184	Airport Planning Manual
9184 Part 1	Master Planning
9184 Part 2	Land Use and Environmental control
9184 Part 3	Guidelines for Consultant/Construction Services
9261	Heliport Manual
9284	Technical Instructions for the Safe Transport of Dangerous Goods by Air
9332	Manual on the ICAO Bird Strike Information System (IBIS)
9365	Manual of All-Weather Operations
9432	Manual of Radiotelephony
9481	Emergency Response Guidance for Aircraft Incidents involving Dangerous Goods
9640	Manual of Aircraft Ground De/Anti-icing Operations
9683	Human Factors Training Manual
9684	Manual of the Secondary Surveillance Radar (SSR) systems
9731	International Aeronautical and Maritime Search and Rescue (IAMSAR) Manual
9731 Volume 1	Organisation and Management
9731 Volume 2	Mission Co-Ordination
9731 Volume 3	Mobile Facilities
9760	Airworthiness Manual

See **www.afeonline.com** for full document listing
For information and to purchase documents contact:
AFE Ltd, 1a Ringway Trading Estate, Shadowmoss Road, Manchester M225LH
Tel: 0161 499 0023   Fax: 0161 499 0298   **www.afeonline.com**

ICAO wake turbulence classifications

Aircraft	Category
Airbus A300 B2	High
Airbus A300 B4	High
Airbus A300-600	High
Airbus A300-600R	High
Airbus A310-200	High
Airbus A310-300	High
Airbus A318-100	Medium
Airbus A319-100	Medium
Airbus A320-100	Medium
Airbus A320-200	Medium
Airbus A321-100	Medium
Airbus A330-200	High
Airbus A330-300	High
Airbus A340-200	High
Airbus A340-300	High
Airbus A340-500	High
Airbus A340-600	High
Airbus A380-800	High
Airbus A380-800F	High
ATR 42	Medium
ATR 72	Medium
BAC 1-11/200	Medium
BAC 1-11/300	Medium
BAC 1-11/400	Medium
BAC 1-11/475	Medium
BAC 1-11/500	Medium
BAe 146-100	Medium
BAe 146-200	Medium
BAe 146-300	Medium
BAe ATP	Medium
BAe 125-700	Medium
Bae 125-800	Medium
B707-120N	Medium
B707-320B	High
B707-320C	High
B707-320/420	High
B717-200	Medium
B727-100	Medium
B727-100C	Medium
B727-200	Medium
B737-100	Medium
B737-200	Medium
B737-300	Medium
B737-200C	Medium
B737-400	Medium
B737-500	Medium
B737-600	Medium
B737-700	Medium
B737-800	Medium
B737-900	Medium
B747-100B SR	High
B747-100	High
B747-100 B	High
B747 SP	High
B747-200B	High
B747-200C	High
B747-200F	High
B747-300	High
B747-400	High
B747-400 Combi	High
B747-400F	High
B757-200	Medium
B757-300	Medium
B767-200	High
B767-200 ER	High
B767-300	High
B767-300 ER	High
B767-400 ER	High
B777-200	High
B777-300	High
B777-200 LR	High
B777-300 ER	High
Canadair CL44	Medium
Caravelle Series 10	Medium
Cessna CIT550	Low
Cessna CIT650	Medium
Concorde	High
DC-3	Medium
DC-4	Medium
DC-8-43	High
DC-8-55	High
DC-8-61/71	High
DC-8-62/72	High
DC-8-63/73	High
DC-9-15	Medium
DC-9-21	Medium
DC-9-32	Medium
DC-9-41	Medium
DC-9-51	Medium
DC-10-10	High
DC-10-15	High
DC-10-30	High
DC-10-40	High
DC-10-30/40	High
DHC 7	Medium
DHC 8-300	Medium
Domier 228-101/201	Low
Domier 228-202	Low
Domier 228-212	Low
Fokker F27 MK 500	Medium
Fokker F28 MK 1000 LTP	Medium
Fokker F28 MK 1000 HTP	Medium
Fokker F50 HTP	Medium
Fokker F50 LTP	Low
Fokker F100	Medium
Gulfstream II	Medium
HS125-1B	Medium
HS125-400	Medium
HS125-600	Medium
HS748	Medium
IL-62	High
IL-62M	High
IL-76T	High
IL-86	High
L-100-20/30	High
L1011-1	High
L-1011-100/200	High
L-1011-500	High
MD-11	Medium
MD-81	Medium
MD-82	Medium
MD-82-88	Medium
MD-83	Medium
MD-87	Medium
Skyvan SC7-3A	Low
SD-330	Medium
SD-360	Medium
TU-134A	Medium
TU-154B	Medium

Aerodrome fire/crash protection categories

CIVIL/ICAO	RAF	MOD	NATO	FAA	USAF
1	Nil	Nil	1	A	1
2	1	1A	2	A	2
3	2 + 3	2A	4	B	2
4	3 + 4	3A	5	B	2
5	4	3A	6	B	2
6	5	3A	7	C	2
7	5 + 6	4A	8	C	2
8	7	5A	9	D	2
9	–	6A	–	D	3
–	–	–	–	E	4
–	–	–	–	–	5

HF Volmet broadcasts

Royal Air Force Volmet

Frequencies: 5450 & 11253

Time past the hour

00	06	12	18	24
30	36	42	48	54
Benson	Belfast Aldergrove	Split	Keflavik	Adana
Coltishall	Birmingham	Budapest	Ascension	Akrotiri
Coningsby	Brize Norton	Bari	Banjul	Al Udeid
Leeming	Cranwell	Gioia Del Colle	Dakar	Amman
Leichars	Nottingham East Midlands	Aviano	Gibraltar	Basrah
Lossiemouth	Kinloss	Rimini	Mombasa	Bahrain
Marham	Lyneham	Ancona	Nairobi	Cairo
Odiham	Manchester	Rome	Rio de Janeiro	Kabul
Shawbury	Northolt	Constanta	Montevideo	Kuwait
Hannover	Prestwick	Bucharest	Brize Norton	Muscat
Geilenkirchen	St Mawgan	Pristina	Lyneham	Al Kharj
Oslo	London Stansted	Skopje	Waddington	Salalah
Trondheim	Waddington		Fujairah	Thumrait

Shannon Volmet

Frequencies: 3413, 5505, 8957 & 13264

Time past the hour

00	05	10	15	20
Brussels	London Heathrow	Copenhagen Kastrup	Madrid Barajas	Rome Fiumiciono
Hamburg	London Gatwick	Stockholm Alarnda	Lisbon	Milan Malpensa
Frankfurt	Shannon	Gottenburg Landvetter	Paris Orly	Zurich
Cologne-Bonn	Prestwick	Bergen	Santa Maria	Geneva
Dusseldorf	Amsterdam	Oslo	Paris Charles de Gaulle	Turin
Munich	Manchester	Helsinki Vantaa	Lyon Satolas	Keflavik
		Dublin		
		Barcelona		

30	35	40	45	50
Frankfurt	London Gatwick	Copenhagen Kastrup	Santa Maria	Zurich
Cologne Bonn	London Heathrow	Stockholm Arlanda	Athens	Geneva
Brussels	Amsterdam	Gotenburg Landvetter	Paris Charles de Gaulle	Rome Fiumicino
Hamburg	Manchester	Bergen	Madrid Barajas	Milan Malpensa
Frankfurt	Shannon	Olso	Lisbon	Zurich
Dusseldorf	Prestwick	Helsinki Vantaa	Paris Orly	Turin
Munich		Dublin	Lyon St Exupery	Keflavik
		Barcelona		

VOR Tacan channel pairings

500	108.10	18X	568	109.45	31Y	636	114.15	88Y
502	108.30	20X	570	109.55	32Y	638	114.25	89Y
504	108.50	22X	572	109.65	33Y	640	114.35	90Y
506	108.70	24X	574	109.75	34Y	642	114.45	9IY
508	108.90	26X	576	109.85	35Y	644	114.55	92Y
510	109.10	28X	578	109.95	36Y	646	114.65	93Y
512	109.30	30X	580	110.05	37Y	648	114.75	94Y
514	109.50	32X	582	110.15	38Y	650	114.85	95Y
516	109.70	34X	584	110.25	39Y	652	114.95	96Y
518	109.90	36X	586	110.35	40Y	654	115.05	97Y
520	110.10	38X	588	110.45	41Y	656	115.15	98Y
522	110.30	40X	590	110.55	42Y	658	115.25	99Y
524	110.50	42X	592	110.65	43Y	660	115.35	100Y
526	110.70	44X	594	110.75	44Y	662	115.45	I01Y
528	110.90	46X	596	110.85	45Y	664	115.55	102Y
530	111.10	48X	598	110.95	46Y	666	115.65	103Y
532	111.30	50X	600	111.05	47Y	668	115.75	104Y
534	111.50	52X	602	111.15	48Y	670	115.85	105Y
536	111.70	54X	604	111.25	49Y	672	115.95	106Y
538	111.90	56X	606	111.35	50Y	674	116.05	107Y
540	108.05	17Y	608	111.45	51Y	676	116.15	108Y
542	108.15	18Y	610	111.55	52Y	678	116.25	109Y
544	108.25	19Y	612	111.65	53Y	680	116.35	110Y
546	108.35	20Y	614	111.75	54Y	682	116.45	111Y
548	108.45	21Y	616	111.85	55Y	684	116.55	112Y
550	108.55	22Y	618	111.95	56Y	686	116.65	113Y
552	108.65	23Y	620	113.35	80Y	688	116.75	114Y
554	108.75	24Y	622	113.45	8IY	690	116.85	115Y
556	108.85	25Y	624	113.55	82Y	692	116.95	116Y
558	108.95	26Y	626	113.65	83Y	694	117.05	117Y
560	109.05	27Y	628	113.75	84Y	696	117.15	118Y
562	109.15	28Y	630	113.85	85Y	698	117.25	119Y
564	109.25	29Y	632	113.95	86Y			
566	109.35	30Y	634	114.05	87Y			

VOR Tacan channel pairings

Channel	Frequency	Channel	Frequency	Channel	Frequency
1X	134.40	30X	109.30	51X	111.40
2X	134.50	30Y	109.35	51Y	111.45
3X	134.60	31X	109.40	52X	111.50
4X	134.70	31Y	109.45	52Y	111.55
5X	134.80	32X	109.50	53X	111.60
6X	134.90	32Y	109.55	53Y	111.65
7X	135.00	33X	109.60	54X	111.70
8X	135.10	33Y	109.65	54Y	111.75
9X	135.20	34X	109.70	55X	111.80
10X	135.30	34Y	109.75	55Y	111.85
11X	135.40	35X	109.80	56X	111.90
12X	135.50	35Y	109.85	56Y	111.95
13X	135.60	36X	109.90	57X	112.00
14X	135.70	36Y	109.95	57Y	112.05
15X	135.80	37X	110.00	58X	112.10
16X	135.90	37Y	110.05	58Y	112.15
17X	108.00	38X	110.10	59X	112.20
17Y	108.05	38Y	110.15	59Y	112.25
18X	108.10	39X	110.20	60X	133.30
18Y	108.15	39Y	110.25	61X	133.40
19X	108.20	40X	110.30	62X	133.50
19Y	108.25	40Y	110.35	63X	133.60
20X	108.30	41X	110.40	64X	133.70
20Y	108.35	41Y	110.45	65X	133.80
21X	108.40	42X	110.50	66x	133.90
21Y	108.45	42Y	110.55	67X	134.00
22X	108.50	43X	110.60	68X	134.10
22Y	108.55	43Y	110.65	69X	134.20
23X	108.60	44X	110.70	70X	112.30
23Y	108.65	44Y	110.75	70Y	112.35
24X	108.70	45X	110.80	71X	112.40
24Y	108.75	45Y	110.85	71Y	112.45
25X	108.80	46X	110.90	72X	112.50
25Y	108.85	46Y	110.95	72Y	112.55
26X	108.90	47X	111.00	73X	112.60
26Y	108.95	47Y	111.05	73Y	112.65
27X	109.00	48X	111.10	74X	112.70
27Y	109.05	48Y	111.15	74Y	112.75
28X	109.10	49X	111.20	75X	112.80
28Y	109.15	49Y	111.25	75Y	112.85
29X	109.20	50X	111.30	76X	112.90
29Y	109.25	50Y	111.35	76Y	112.95

Channel	Frequency	Channel	Frequency	Channel	Frequency
77X	113.00	98Y	115.15	119Y	117.25
77Y	113.05	99X	115.20	120X	117.30
78X	113.10	99Y	115.25	120Y	117.35
78Y	113.15	100X	115.30	121X	117.40
79X	113.20	l00Y	115.35	121Y	117.45
79Y	113.25	101X	115.40	122X	117.50
80X	113.30	l01Y	115.45	122Y	117.55
80Y	113.35	102X	115.50	123X	117.60
81X	113.40	102Y	115.55	123Y	117.65
8lY	113.45	103X	115.60	124X	117.70
82X	113.50	103Y	115.65	124Y	117.75
82Y	113.55	104X	115.70	125X	117.80
83X	113.60	104Y	115.75	125Y	117.85
83Y	113.65	105X	115.80	126X	117.90
84X	113.70	105Y	115.85	126Y	117.95
84Y	113.75	106X	115.90		
85X	113.80	106Y	115.95		
85Y	113.85	107X	116.00		
86X	113.90	107Y	116.05		
86Y	113.95	108X	116.10		
87X	114.00	108Y	116.15		
87Y	114.05	109X	116.20		
88X	114.10	l09Y	116.25		
88Y	114.15	110X	116.30		
89X	114.20	110Y	116.35		
89Y	114.25	111X	116.40		
90X	114.30	111Y	116.45		
90Y	114.35	112X	116.50		
91X	114.40	112Y	116.55		
91Y	114.45	113X	116.60		
92X	114.50	113Y	116.65		
92Y	114.55	114X	116.70		
93X	114.60	114Y	116.75		
93Y	114.65	115X	116.80		
94X	114.70	115Y	116.85		
95X	114.80	116X	116.90		
95Y	114.85	116Y	116.95		
96X	114.90	117X	117.00		
96Y	114.95	117Y	117.05		
97X	115.00	118X	117.10		
97Y	115.05	118Y	117.15		
98X	115.10	119X	117.20		

World times & map

London –	Local time	London –	Local time	London –	Local time
Algiers	GMT	Dublin	GMT	New York	GMT -5
Amsterdam	GMT +1	Dusseldorf	GMT +1	Nice	GMT +1
Ankara	GMT +3	Frankfurt	GMT +1	Oslo	GMT +1
Athens	GMT +2	Geneva	GMT +1	Palma	GMT +1
Auckland	GMT +12	Gibraltar	GMT +1	Paris	GMT +1
Baghdad	GMT +3	Gothenburg	GMT +1	Perth	GMT +8
Bahrain	GMT +3	Helsinki	GMT +2	Prague	GMT +1
Bangkok	GMT +7	Hong Kong	GMT +8	Rio de Janeiro	GMT -3
Barbados	GMT -4	Honolulu	GMT 8	Rome	GMT +1
Barcelona	GMT +1	Istanbul	GMT +3	Salzburg	GMT +1
Beijing	GMT +8	Johannesburg	GMT +2	San Francisco	GMT -8
Beirut	GMT +2	Kano, Nigeria	GMT +1	Singapore	GMT +7
Belgrade	GMT +1	Karachi	GMT +5	Stockholm	GMT +1
Bergen	GMT +1	Kingston	GMT -5	Sydney	GMT +10
Berlin	GMT +1	Kuala Lumpur	GMT +7	Tehran	GMT +3
Bermuda	GMT -4	Kuwait	GMT +8	Tel Aviv	GMT +2
Bombay	GMT +5	Lagos	GMT +1	Tokyo	GMT +9
Brisbane	GMT +10	Lisbon	GMT +1	Toronto	GMT -5
Brussels	GMT +1	Madrid	GMT +1	Tripoli	GMT +2
Budapest	GMT +1	Malta	GMT +1	Turin	GMT +1
Buenos Aires	GMT -3	Mauritius	GMT +4	Vancouver	GMT -8
Cairo	GMT +2	Milan	GMT +1	Venice	GMT +1
Calcutta	GMT +5	Montreal	GMT -5	Vienna	GMT +1
Chicago	GMT -6	Moscow	GMT +3	Warsaw	GMT +1
Dakar	GMT	Munich	GMT +1	Washington	GMT -5
Darwin	GMT +9	Nairobi	GMT +3	Yangon	GMT +6
Delhi	GMT+5	Naples	GMT +1	Zurich	GMT +1
Detroit	GMT +5	Nassau	GMT -5		

GMT is called Greenwich Mean (or Meridian) Time because it is measured from the Greenwich Meridian Line at the Royal Observatory in Greenwich. It is the place from where all time zones are measured.
GMT remains the same all year round.
Spring – clocks forwards
Autumn – clocks backwards
The Greenwich Meridian (Prime Meridian or Longitude Zero degrees) marks the starting point of every time zone in the World.
GMT is Greenwich Mean (or Meridian) Time is the mean (average) time that the earth takes to rotate from noon to noon. GMT is World Time and the basis of every world time zone which sets the time of day and is at the centre of the time zone map. GMT sets current time or official time around the globe. Most time changes are measured by GMT.
GMT has been replaced by atomic time (UTC) it is still widely regarded as the correct time for every international time zone.

World times & map

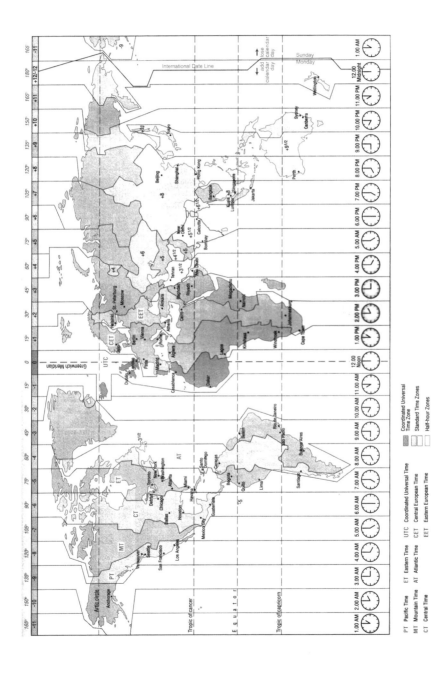

World time map

SR-SS tables

All times UTC

DATE	EGAA Belfast Aldergrove SR/SS	EGBB Birmingham SR/SS	EGFF Cardiff SR/SS	EGPH Edinburgh SR/SS	EGLL London Heathrow SR/SS	EGCC Manchester SR/SS	LFPG Paris SR/SS	EHAM Amsterdam SR/SS	EIDW Dublin SR/SS	ENOS Oslo SR/SS	LEBB Bilbao SR/SS
Jan 1	0846/1609	0818/1604	0819/1615	0844/1550	0806/1603	0825/1600	0745/1602	0733/1554	0841/1616	0802/1423	0745/1646
Jan 15	0837/1630	0811/1623	0812/1633	0834/1611	0800/1621	0817/1620	0739/1616	0727/1613	0833/1635	0807/1449	0742/1700
Feb 1	0805/1711	0742/1701	0745/1710	0800/1655	0732/1658	0747/1700	0722/1646	0706/1641	0811/1707	0734/1530	0720/1727
Feb 15	0736/1740	0715/1728	0719/1735	0729/1726	0707/1723	0719/1727	0701/1709	0642/1708	0745/1734	0659/1607	0715/1738
Mar 1	0703/1808	0645/1754	0649/1800	0654/1756	0637/1748	0647/1754	0632/1733	0612/1740	0713/1803	0617/1645	0650/1800
Mar 15	0629/1835	0612/1819	0618/1824	0618/1825	0606/1812	0614/1821	0604/1755	0540/1759	0640/1829	0535/1720	0630/1815
Apr 1	0556/1900	0542/1842	0548/1846	0543/1851	0536/1834	0542/1844	0528/1821	0501/1828	0558/1901	0443/1802	0553/1840
Apr 15	0521/1928	0510/1906	0517/1910	0507/1920	0507/1857	0509/1910	0459/1842	0430/1852	0525/1926	0401/1836	0530/1854
May 1	0435/2007	0427/1942	0436/1944	0418/2002	0423/1932	0424/1948	0429/1906	0356/1919	0450/1955	0316/1916	0506/1913
May 15	0410/2032	0404/2005	0414/2006	0451/2028	0402/1953	0401/2019	0408/1925	0332/1942	0425/2019	0241/1949	0449/1927
June 1	0353/2051	0349/2023	0400/2023	0333/2050	0348/2011	0345/2030	0350/1946	0311/0304	0403/2043	0208/2025	0434/1944
June 15	0347/2103	0344/2033	0355/2033	0326/2102	0343/2021	0339/2041	0345/1956	2005/2016	0356/2055	0155/2042	0430/1955
July 1	0352/2103	0349/2033	0400/2033	0332/2101	0348/2021	0344/2041	0350/1957	0309/2018	0402/2056	0201/2024	0434/1957
July 15	0407/2051	0403/2023	0413/2024	0348/2049	0401/2011	0359/2030	0401/1950	0322/2009	0416/2047	0222/2024	0444/1951
Aug 1	0441/2017	0427/1959	0443/1954	0425/2012	0430/1942	0431/1958	0422/1929	0346/1946	0441/2020	0259/1947	0559/1936
Aug 15	0506/1946	0456/1924	0505/1927	0452/1940	0452/1915	0455/1929	0441/1906	0408/1920	0505/1953	0332/1911	0514/1919
Sept 1	0533/1913	0520/1853	0527/1857	0519/1905	0515/1845	0520/1856	0506/1833	0436/1843	0535/1914	0413/1822	0533/1847
Sept 15	0558/1838	0543/1820	0549/1825	0546/1828	0537/1813	0544/1822	0526/1803	0459/1811	0559/1841	0446/1739	0549/1823
Oct 1	0637/1745	0619/1731	0623/1737	0628/1732	0611/1725	0621/1731	0550/1729	0525/1734	0627/1801	0523/1651	0608/1756
Oct 15	0704/1711	0644/1659	0647/1707	0657/1657	0635/1655	0647/1659	0611/1700	0549/1702	0652/1728	0557/1609	0624/1731
Nov 1	0732/1642	0709/1632	0712/1640	0726/1626	0700/1628	0714/1630	0638/1629	0619/1628	0725/1652	0640/1522	0644/1707
Nov 15	0800/1617	0734/1610	0736/1620	0755/1600	0724/1607	0740/1608	0700/1609	0644/1605	0751/1628	0715/1449	0704/1647
Dec 1	0824/1602	0757/1556	0758/1607	0821/1543	0745/1555	0804/1553	0723/1555	0710/1548	0818/1610	0752/1422	0724/1637
Dec 15	0841/1558	0813/1553	0813/1604	0839/1538	0801/1552	0820/1550	0738/1552	0726/1543	0835/1606	0814/1412	0738/1636

General — SR-SS tables

Continuously above horizon

	OA Baghdad SR/SS	VIDD Delhi SR/SS	VTBD Bangkok SR/SS	WIII Jakarta SR/SS	YSBK Sydney SR/SS	NZWN Wellington SR/SS	RJTT Tokyo SR/SS	SABA Buenos Aires SR/SS	SVCC Caracas SR/SS	MMMX Mexico City SR/SS	KIAD Washington SR/SS
Jan 1	0407/1406	0144/1206	2342/1101	2242/1111	1848/0910	1652/0755	2151/0740	0845/2312	1045/2218	1310/0009	1229/2158
Jan 15	0407/1417	0145/1217	2346/1109	2249/1115	1900/0909	1707/0752	2151/0752	0858/2310	1049/2226	1310/0019	1227/2212
Feb 1	0400/1434	0140/1230	2346/1118	2255/1118	1917/0901	1728/0739	2143/0809	0915/2302	1051/2233	1311/0028	1216/2231
Feb 15	0349/1446	0131/1241	2342/1122	2258/1116	1931/0849	1745/0723	2131/0823	0928/2251	1048/2236	1303/0034	1202/2247
Mar 1	0331/1500	0117/1251	2335/1126	2259/1112	1943/0833	1801/0703	2113/0837	0941/2232	1043/2239	1256/0041	1143/2302
Mar 15	0315/1510	0054/1300	2326/1129	2258/1107	1955/0814	1817/0640	2054/0849	0954/2214	1035/2239	1253/0045	1122/2317
Apr 1	0251/1523	0043/1309	2314/1130	2255/1058	2008/0751	1835/0613	2030/0903	1007/2115	1025/2239	1229/0049	1055/2333
Apr 15	0233/1533	0026/1317	2305/1131	2254/1052	2018/0733	1849/0551	2011/0915	1017/2132	1018/2239	1217/0055	1034/2347
May 1	0215/1545	0011/1326	2256/1134	2253/1046	2030/0715	1905/0528	1951/0928	1029/2113	1011/2240	1206/0059	1012/0001
May 15	0203/1555	0001/1335	2251/1138	2253/1044	2041/0703	1918/0513	1913/0939	1040/2100	1007/2242	1201/0105	0957/0014
Jun 1	0154/1607	2354/1345	2249/1143	2256/1044	2052/0654	1934/0501	1928/0951	1053/2051	1006/2246	1157/0111	0946/0028
Jun 15	0152/1613	2353/1350	2250/1146	2300/1046	2059/0653	1942/0459	1926/0959	1100/2050	1007/2250	1159/0115	0944/0036
Jul 1	0156/1616	2356/1353	2253/1150	2303/1050	2101/0657	1945/0515	1930/1002	1103/2054	1011/2253	1201/0117	0948/0039
Jul 15	0204/1614	0003/1352	2257/1150	2304/1052	2058/0704	1957/0515	1938/0958	1059/2101	1014/2254	1206/0114	0956/0035
Aug 1	0215/1602	0012/1343	2302/1146	2304/1055	2048/0715	1945/0531	1950/0946	1049/2113	1018/2251	1211/0112	1010/0022
Aug 15	0224/1550	0029/1331	2305/1140	2300/1054	2034/0725	1925/0545	2000/0932	1035/2123	1019/2246	1215/0107	1023/0006
Sep 1	0236/1528	0029/1314	2306/1130	2253/1052	2018/0737	1856/0604	2014/0910	1014/2135	1019/2237	1220/0050	1038/2341
Sep 15	0245/1509	0036/1257	2307/1120	2245/1050	1954/0746	1834/0618	2024/0849	0954/2146	1018/2229	1223/0035	1051/2319
Oct 1	0256/1447	0044/1238	2308/1108	2238/1047	1932/0757	1802/0638	2036/0826	0934/2157	1017/2219	1226/0023	1105/2253
Oct 15	0306/1430	0052/1220	2309/1059	2231/1045	1914/0808	1741/0652	2049/0808	0913/2208	1017/2211	1231/0013	1119/2232
Nov 1	0321/1412	0103/1207	2313/1051	2226/1046	1855/0823	1714/0714	2104/0747	0854/2224	1018/2205	1236/0002	1137/2210
Nov 15	0334/1401	0114/1158	2318/1047	2225/1049	1843/0836	1657/0735	2117/0736	0841/2237	1023/2205	1243/2358	1152/2156
Dec 1	0348/1355	0127/1154	2327/1047	2288/1055	1837/0851	1647/0752	2132/0729	0835/2253	1029/2205	1253/2356	1209/2148
Dec 15	0359/1357	0137/1157	2334/1052	2234/1102	1838/0902	1647/0806	2144/0731	0835/2304	1037/2210	1300/0002	1221/2149

General — SR-SS tables

BGTL	CYUL HKJK Montreal SR/SS	PAED FACT Anchorage SR/SS	GOOY Thule SR/SS	OSDI Nairobi SR/SS	LIII Cape Town SR/SS	LTBA Dakar SR/SS	UUEE Tripoli SR/SS	Roma SR/RR	Istanbul SR/SS	Moscow SR/SS
Jan 1	1233/2124	1914/0054	–/–	0330/1542	0339/1801	0735/1852	0610/1612	0638/1548	0530/1446	0601/1306
Jan 15	1230/2139	1859/0119	–/–	0337/1548	0350/1800	0739/1900	0611/1623	0635/1603	0528/1500	0552/1328
Feb 1	1215/2203	1821/0207	–/–	0341/1553	0407/1752	0735/1909	0603/1639	0624/1624	0517/1521	0525/1403
Feb 15	1158/2221	1741/0249	0549/0932	0342/1552	0419/1740	0735/1914	0551/1652	0607/1642	0501/1538	0458/1431
Mar 1	1134/2242	1705/0325	0339/1138	0341/1549	0433/1723	0727/1918	0536/1705	0545/1700	0440/1556	0422/1504
Mar 15	1108/2301	1619/0401	0159/1312	0338/1545	0445/1706	0718/1921	0518/171	0522/1717	0417/1612	0346/15335
Apr 1	1036/2322	1521/0448	0001/1502	0334/1539	0458/1641	0706/1922	0455/1727	0453/1753	0348/1630	0301/1608
Apr 15	1011/2340	1443/0519	2157/1652	0331/1535	0508/1625	0656/1924	0437/1737	0429/1751	0326/1644	0225/1636
May 1	0945/0000	1350/0606	–/–	0328/1532	0519/1606	0647/1927	0419/1749	0406/1809	0303/1702	0147/1709
May 15	0926/2417	1316/0639	–/–	0327/1531	0531/1553	0642/1931	0409/1759	0350/1823	0247/1716	0118/1736
Jun 1	0912/0034	1237/0720	–/–	0329/1532	0542/1545	0640/1936	0400/1811	0337/1839	0235/1731	0052/1804
Jun 15	0908/0043	1222/0741	–/–	0331/1534	0549/1544	0640/1940	0357/1817	0334/1846	0232/1738	0043/1818
Jul 1	0912/0045	1227/0739	–/–	0335/1538	0552/1548	0644/1943	0402/1820	0339/1849	0237/1741	0048/1819
Jul 15	0922/0039	1253/0718	–/–	0337/1540	0549/1554	0648/1943	0409/1816	0348/1844	0246/1736	/0104/1807
Aug 1	0940/0022	1328/0642	–/–	0337/1541	0539/1606	0652/1940	0420/1806	0403/1829	0300/1722	0133/1738
Aug 15	0956/0002	1407/0600	–/–	0335/1539	0525/1616	0656/1933	0430/1752	0417/1811	0314/1704	0200/1707
Sep 1	1017/2333	1453/0504	2239/1613	0330/1535	0505/1627	0658/1922	0442/1732	0436/1744	0332/1638	0234/1625
Sep 15	1033/2307	1523/0426	0011/1426	0235/1530	0447/1637	0659/1912	0450/1714	0450/1720	0345/1614	0301/1548
Oct 1	1053/2236	1608/0330	0151/1237	0320/1526	0425/1648	0700/1900	0501/1652	0507/1652	0401/1547	0332/1506
Oct 15	1110/2211	1639/0252	0324/1056	0315/1522	0405/1659	0701/1850	0512/1635	0522/1629	0416/1525	0401/1430
Nov 1	1133/2144	1728/0159	–/–	0311/1521	0346/1713	0706/1842	0525/1617	0542/1604	0436/1501	0436/1350
Nov 15	1152/2127	1809/0120	–/–	0312/1522	0334/1726	0711/1838	0537/1606	0600/1549	0453/1446	0506/1322
Dec 1	1213/2115	1846/0052	–/–	0316/1528	0329/1742	0720/1838	0551/1601	0618/1540	0511/1437	0537/1301
Dec 15	1226/2114	1910/0041	–/–	0321/1534	0329/1753	0727/1842	0603/1602	0631/1540	0523/1438	0555/1255

AERAD ENC chart legends

NEW ENC SPECIFICATION

The following pages describes the new ENC specification, and is valid for ENC charts dated 23 JUL 04 or later.

ENC

The Enroute Navigation Charts (ENC) are compiled from official documents and topographical reference charts. The ENCs have been designed primarily for instrument enroute navigation. The information contained is kept to a minimum, consistent with the function of the chart.

Numbering

The ENC:s are normally combined High Level (HL) and Low Level (LL) charts. Due to the complexity of the route structure such a combined chart would be insufficient, the HL charts and the LL charts are produced separately.

Africa	AF	Europe	EU
Asia	AS	North America	NAM
Atlantic	AT	Pacific	PAC
Australia	AUS	Polar area	POLAR
Central America	CAM	South America	SAM

Each series of ENCs within the same area is individually numbered, starting with 1.

ENC FRONT COVER

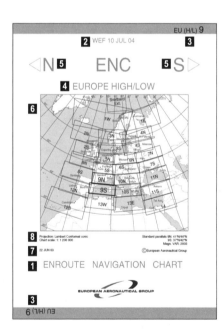

1 Date when the latest obtained amendments are included.

2 Effective date.

3 ENC number.

4 Indicates gegraphical area and combined high level (HL) and low level (LL) chart. Combined HL/LL chart is also indicated by grey stripe along top/bottom of front cover. A blue stripe indicates HL chart only, and no stripe indicates LL chart only.

5 Indication of coverage on each side of chart.

6 Chart index shows the area covered by the ENC and also indicates adjacent ENCs.

7 Chart scale chosen for a chart depends on the extension of the geographical area covered and on the density of the information within that area.

8 Projection used is Lambert conformal conic projection except for polar areas where a polar stereographic projection is used.

General — AERAD ENC chart legends

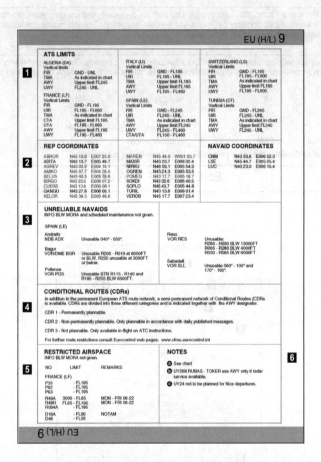

1 ATS limits in addition to those given in the chart. Listed countrywise and in alphabetical order.

2 Rep coordinates and Nav coordinates not presented in the ENC. Listed in alphabetical order. When combined HL and LL chart, Rep coordinates refering to LL routes are printed in blue colour.

3 Description of unreliable Nav aids.

4 Other restrictions, e.g. CDRs in Europe.

5 Restricted airspace specifications for areas indicated in the ENC. Listed countrywise and in alphabetical order.

6 Explanation of notes not clarified in the front cover panel.

General AERAD ENC chart legends

GRATICULE, ISOGONIC LINES

Isogonic lines with value followed by "E" or "W". Spacing normally 1°.
Graticule of meridians and parallels with latitude and longitude values outside and close to chart border. Graduation of ticks spaced at 5 minute intervals. At high LAT and in chart with small scale bigger spacing may be used.

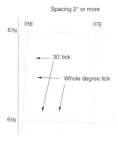

LAT/LONG figures are placed outside and close to the chart border.

In addition, LAT/LONG figures are given at meridians/parallels where graduation ticks are marked.

Scale bar printed in upper right and lower left corner of the ENC, outside the chartframe.

GENERAL INFORMATION

Zurich
LSZH

Civil or combined military/civil aerodromes, with city name and 4-letter ICAO code. Minimum RWY length 5000FT / 1524M.

087→△

En-route holding pattern shown exeptionally only. MT shown when not corresponding with route track figure.

El Volcano
N43 28.0
E003 43.2

Active volcanoes.

591

General — AERAD ENC chart legends

MNM Off Route Altitude (MORA).
Indicated in hundreds of feet for each LAT/LONG square as used in the chart. MORA is based on a tolerance of 1500FT above terrain and obstructions up to 5000FT, and 2000FT above terrain and obstructions over 5000FT, rounded up to the nearest 100FT.

In charts with small scale, northern LAT or condensed areas two or more squares may be used. When MORA includes more than two LAT/LONG squares this will be explained in an index in the chart.

EGYPT
CAIRO FIR CTA HECA

KHARTOUM FIR HSSS
SUDAN

FIR-, UIR-, boundary and when CTA-, UTA-, and OCA-boundary coincides with FIR/UIR boundary. Boundary symbol together with FIR/UIR name, country name and 4-letter ICAO code. In W-Europe, USA and Canada, country name will only be presented when political border coincides with FIR/UIR border. Vertical limits are tabulated on ENC front cover panel.

RVSM
NORWAY
STAVANGER FIR/UIR/UTA ENSV

Reduced Vertical Separation Minima areas (RVSM) indicated along FIR border.

REYKJAVIK FIR/OCA BIRD
ICELAND
NAT MNPS/RVSM

North atlantic Minimum Navigation Performance Specification area (MNPS) indicated along FIR border.

TOKYO OCA

OCA boundary with name. Vertical limits are tabulated on ENC front cover panel.

CTA

CTA border, when not coinciding with FIR or UIR. For vertical limits see ATS LIMITS on ENC

CTA - 120 (D)

CTA boundary with upper limit in hundreds of FT when not in accordance with ATS-limits. Airspace classification within brackets.

TMA (C)

TMA boundary. Airspace classification within brackets.

TMA - 30 (D)

TMA boundary with upper vertical limit in hundreds of FT when not in accordance with ATS-limits. Where TMA is divided into sectors, only outer border and MAX upper limits are indicated in chart. QNH area boundary, with text if required inside the QNH area.

QNH below FL140

QNH boundary, with text if required inside the QNH area.

KOREA
TAGU FIR RKTT

TOKYO FIR RJTG   QNH below FL140
JAPAN

QNH area coinciding with FIR or UIR boundary, with text if required inside the QNH area.

INDIA
MYANMAR

Political border with country name indicated when not consistent with FIR- or CTA-border.

ADIZ

ADIZ-, CDIZ- boundary. Selected boundaries only.

ERRATIC AREA

Boundary of "erratic" area of magnetic compass reliability (polar- and sub-polar area).

DATE LINE

Date line with text.

Overlapping area between ENC charts.

ENC 8S

Adjacent chart.

Complete INFO see ENC 9S

Complete information on another chart.

RESTRICTED AIRSPACE

No information about areas below MORA is published. Ident is given in an abbreviated format. Find official identification by adding the two first letters of the 4-letter ICAO FIR code. Specifications of areas respectively are tabulated on ENC front cover panel.

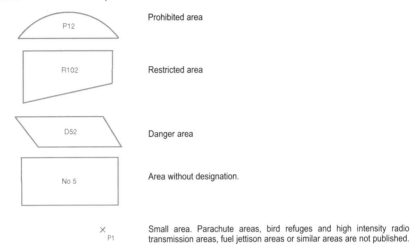

RADIO AIDS

Radio facility normally only given when included in the route structure.

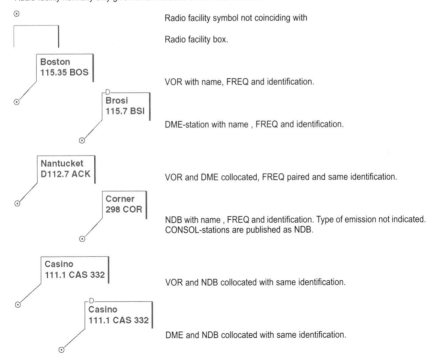

General

AERAD ENC chart legends

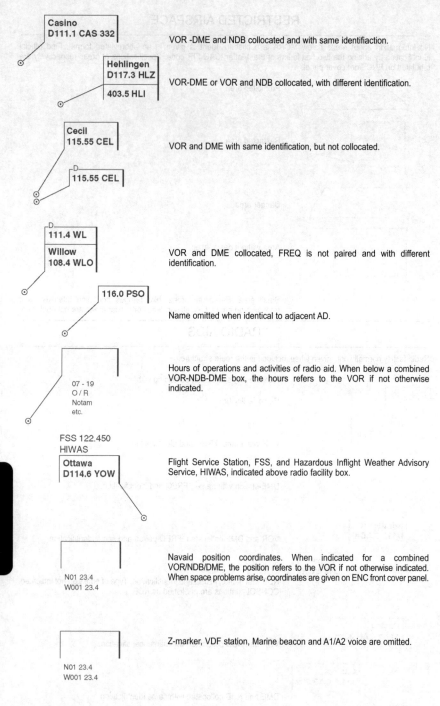

Casino
D111.1 CAS 332 — VOR-DME and NDB collocated and with same identifiaction.

Hehlingen
D117.3 HLZ
403.5 HLI — VOR-DME or VOR and NDB collocated, with different identification.

Cecil
115.55 CEL
115.55 CEL — VOR and DME with same identification, but not collocated.

111.4 WL
Willow
108.4 WLO — VOR and DME collocated, FREQ is not paired and with different identification.

116.0 PSO — Name omitted when identical to adjacent AD.

07 - 19
O / R
Notam
etc. — Hours of operations and activities of radio aid. When below a combined VOR-NDB-DME box, the hours refers to the VOR if not otherwise indicated.

FSS 122.450
HIWAS
Ottawa
D114.6 YOW — Flight Service Station, FSS, and Hazardous Inflight Weather Advisory Service, HIWAS, indicated above radio facility box.

N01 23.4
W001 23.4 — Navaid position coordinates. When indicated for a combined VOR/NDB/DME, the position refers to the VOR if not otherwise indicated. When space problems arise, coordinates are given on ENC front cover panel.

N01 23.4
W001 23.4 — Z-marker, VDF station, Marine beacon and A1/A2 voice are omitted.

594

COMMUNICATION FREQUENCIES

Shannon SOTA Centre
135.800

Amsterdam RAD below FL300
131.380 134.705
131.380 134.705

En-route communication frequencies presented within a box. Placement corresponding to the area where frequencies are to be used.

Maastricht CTL
131.380 134.705
131.380 134.705
131.380 134.705
131.380 134.705
131.380 134.705
131.380 134.705
131.380 134.705

Maastricht CTL/RAD FL300 and above
131.380 134.705
131.380 134.705
131.380 134.705
131.380

Amsterdam RAD below FL300
131.380 134.705
131.380 134.705

Air defence clearance required on 126.125 in Toulouse TMA below FL120.

Other communication information is presented in a similar box.

Enroute WX
Toulouse 128.175

REPORTING POINTS

Compulsory   **Non compulsory**

▲   △   High or High/Low ALT with or without radio aid.

▲   △   Low ALT only with or without radio aid.

Coordinates will always be indicated. However if space not available, coordinates will be presented on ENC front cover panel.

GERON
N01 23.4
W123 45.6

REP with name.

AIREP, section 3.

AIREP, section 3 in one direction only.

AIREP, section 3 for both directions with directional arrows to indicate a certain route.

SWOPA
N42 26.9
E001 02.2

Transfer of control point.

Reporting point for U/A9 only.

Compulsory reporting point for U/A9 and non-compulsory reporting point for UG5.

595

General

AERAD ENC chart legends

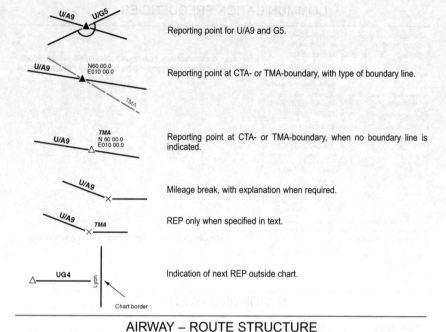

AIRWAY – ROUTE STRUCTURE
TRACK LINE AND DESIGNATION IN AIRSPACE CLASSIFICATION A-E.

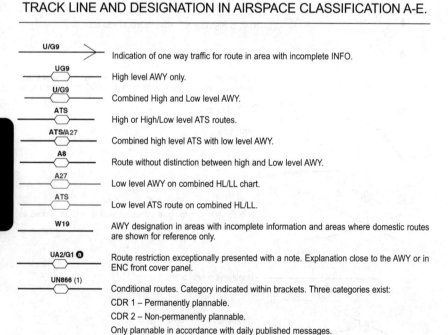

ROUTES WITHIN AIRSPACE CLASSIFICATION F and G

– – – – – – – – – – – – – – Advisory route.

– – – – –〈UJ5F〉– – – – – Advisory route with designator.

DISTANCE

DIST in distance box always between REP.

▲———〈27〉———▲ Distance box for two way AWY.

▲———〈27〉———▲ Distance box for two way LL AWY on combined HL/LL chart.

[30〉 One way AWY DIST box.

[30〉 One way LL AWY DIST box on combined HL/LL chart.

Distance given for combined high level (HL) and low level (LL) chart.

71 / 62 |〈133〉— COP not coinciding at fix, showing DIST to or from VOR.

TRACKS AND RADIALS

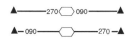

▲——270〈 〉090——▲ Average magnetic track placed close to DIST box, indicating track to REP.

▲—090〈 〉270—▲ Radial, placed close to VOR station and indicating radial from the VOR.

▲-092——270〈 〉090——268-▲ Radial and average magnetic tracks on great circle tracks for long legs exceeding coverage of VOR to VOR or VOR to NDB when track figure differs more than 2°.

▲——029T〈273〉209T——▲ True track (T) when used in "erratic area" of compass reliability.

▲—271〈 〉091—▲〈 〉—▲ Repeated track figures may be excluded when space problems arise.

MNM OBSTACLE CLEARANCE ALT (MOCA)

▲—〈 〉—△—▲
  59   30

MOCA is given in hundreds of FT. Lowest indicated MOCA is 2000FT (20). On exclusive HL chart MOCA for the LL route is omitted. MOCA is based on a tolerance of 1500FT above terrain and obstructions up to 5000FT, and 2000FT above terrain and obstructions over 5000FT, rounded up to the nearest 100FT.

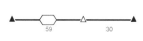

MOCA is valid for a corridor extending 10NM to either side of route centerline and including a 10NM radius beyond the fix or mileage break defining the MOCA segment MNM ENROUTE ALT (MEA) MNM Enroute Altitude (MEA) without direction as published by the authorities. For exclusive HL routes MEA is only shown where it differs from the standard altitude given in ATS LIMITS on ENC front cover panel.

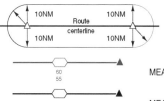

———〈 〉———▲
   60
   55

MEA (55) on LL-route.

———〈 〉———▲
   165
   200

MEA (200) on HL/LL or HL route.

General

AERAD ENC chart legends

MNM FLIGHT ALTITUDE /FLIGHT LEVEL

MNM Flight ALT/FL as published by the authorities with directional arrow.

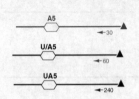

MNM Flight ALT/FL on LL-route on combined HL/LL chart.

MNM Flight ALT/FL on HL/LL-route.

MNM Flight ALT/FL on HL-route.

Where upper limits of AWYs are standardized (countrywise) according to tabulated upper limits in ATS LIMITS on ENC front cover panel, no indication will be in the chart. If not according to standard, the upper limits are shown as below. UNL is not indicated.

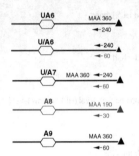

MNM Flight ALT/FL for a HL -route with Maximum Authorized Altitude (MAA) at FL360 without direction.

MNM Flight ALT/FL for a LL- and HL-route with direction.

MNM FL/ALT for LL-route, and HL-route segment from FL240 to FL360.

MNM FL/ALT with direction and MAA on LLroute, on combined HL/LL chart.

MNM FL/ALT with direction and MAA for HLroute, without distinction between high and low airspace.

POSITION LINES

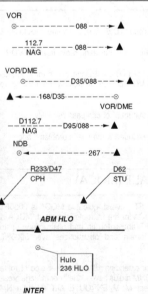

Radial from VOR station.

Radial with FREQ and identification of distant VOR station.

Radial and DIST from VOR/DME station.

Radial and DIST, FREQ and identification of a distant VOR/DME station.

Magnetic bearing to NDB station.

Radial and DIST from a VOR and or a DME station, presented on a "hook" when space required.

Abeam position where "abeam" (ABM) is included in REP.

Abeam position where "abeam" (ABM) is not included in REP.

Minimum Obstacle Clearance Altitude (MOCA)

The MOCA calculation described below is acc to the EAG method which is one of the JAA approved methods.

MOCA is the sum of:

- the maximum terrain or obstacle elevation whichever is highest, plus – 1500 FT for elevation up to and including 5000 FT, or – 2000 FT elevation exceeding 5000 FT rounded up to the next 100 FT.

MOCA in hundreds of feet, is given in the ENC, SID and STAR.

The lowest MOCA to be indicated is 2000 FT (20).

MOCA is valid for a corridor as indicated below.

Straight segment:

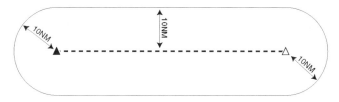

Intersecting segment:

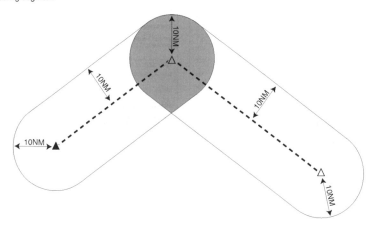

Minimum Off-Route Altitude (MORA)

MORA is the sum of:

- the maximum terrain or obstacle elevation whichever is highest, plus – 1500 FT for elevation up to and including 5000 FT, or – 2000 FT for elevation exceeding 5000 FT rounded up to the next 100 FT.

MORA in hundreds of feet is given in the ENC.

The lowest MORA to be indicated is 2000 FT (20).

MORA is normally valid in an area bounded by every LAT/LONG squares on the ENC. In small scale ENCs or northern LAT areas, one MORA figure may cover two or more LAT/LONG squares.

GEN 2.3 – CAA CHART LEGENDS

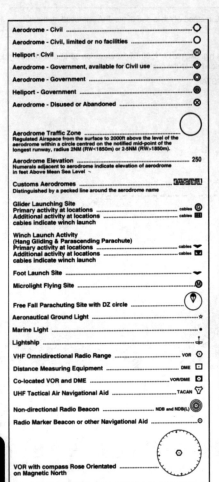

General

CAA chart legends

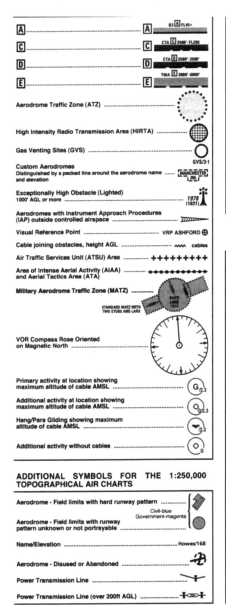

Interception procedures

Interception of Aircraft

The interception of a civilian aircraft by a military one is thankfully a rare occurrence. To reduce the risk to both aircraft in the unlikely event of an interception, a series of internationally agreed signals and procedures have been devised for use in such a situation. A civilian aircraft may be intercepted if it is in or near a dangerous area, or an area sensitive for security reasons. The intercepting aircraft may lead the other away from the area, and might require it to land at an aerodrome. All international flights by a UK-registered aircraft are required to carry details of interception procedures.

If you are intercepted, there are four steps to take at once:

1) Stay calm and comply immediately with the instructions or signals of the intercepting aircraft. This is of the utmost importance for obvious reasons.
2) Notify the ATSU with which you are in contact.
3) Attempt to establish communication with the interceptor on the emergency frequency of 121.5, giving your callsign and flight details.
4) Set the distress code – 7700 Mode C – on your transponder, unless instructed otherwise by an ATSU.

The intercepting aircraft will probably first give instructions by visual signals.

Remember – comply with any instructions **immediately**. You can ask questions later.

INTERCEPTING AIRCRAFT	MEANING	INTERCEPTED AIRCRAFT	MEANING
Takes up position ahead and to the left and rocks wings. After acknowledgement turns slowly on to desired heading.	You have been intercepted. Follow me.	Rocks wings and follows.	Understood will comply.

Note: Conditions may make it necessary for the intercepting aircraft to take up position to the right and to turn to the right.

By night both aircraft flash their navigation lights at irregular intervals, in addition to the above signals.

If the intercepted aircraft is too slow for the intercepting aircraft, the intercepting aircraft will fly a race-track pattern and rock its wings each time it passes the intercepted aircraft.

The intercepting aircraft may give further instructions.

INTERCEPTING AIRCRAFT	MEANING	INTERCEPTED AIRCRAFT	MEANING
An abrupt breakaway climbing turn through more than 90° without crossing ahead of the aircraft.	You may proceed.	Rocks wings.	Understood will comply.

The intercepting aircraft may lead you to an aerodrome for landing, where it will signal as follows.

INTERCEPTING AIRCRAFT	MEANING	INTERCEPTED AIRCRAFT	MEANING
Circles aerodrome, lowering landing gear and overflying runway in direction of landing.	Land at this aerodrome.	Lowers landing gear, overflies runway and proceeds to land.	Understood will comply.

At night the intercepting aircraft will also show steady navigation lights. The intercepted aircraft should do the same and show a steady landing light if possible.

General — Interception procedures

The intercepted aircraft can also display one of three signals.
Note: The CAA advises against using hand signals because they could be misinterpreted!

INTERCEPTED AIRCRAFT	MEANING	INTERCEPTING AIRCRAFT	MEANING
Irregular flashing of all available lights.	In distress.	Makes an abrupt breakaway climbing turn.	Understood.
Regular switching on and off of all available lights.	Cannot comply.	Makes an abrupt breakaway climbing turn.	Understood.
Raises landing gear whilst overflying runway between 1000 and 2000ft above the aerodrome. Continues to circle aerodrome. Also flashes landing light or all available lights by night.	The aerodrome you have designated is inadequate.	Raises landing gear and signals intercepted aircraft to follow it. OR Makes an abrupt breakaway climbing turn.	Understood follow me. Understood you may proceed.

When attempting to establish contact by radio the following phrases should be used (remembering that the pilot of the intercepting aircraft may not speak your language, nor you his).

Phrases used by the intercepting aircraft:

PHRASE	PRONUNCIATION	MEANING
Callsign	KOL SA-IN	What is your callsign?
Follow	FOL LO	Follow me
Descend	DEE SEND	Descend for landing
You Land	YOU LAAND	Land at this aerodrome
Proceed	PRO SEED	You may proceed

If any of the instructions given by the interceptor (either by signals or radio) conflict with those from the ATSU, obey the intercepting aircraft whilst requesting clarification.

Phrases used by the intercepted aircraft:

PHRASE	PRONUNCIATION	MEANING
Callsign	KOL SA-IN	My callsign is…
Wilco	VILL-KO	Understood, will comply
Can not	KANN NOTT	Unable to comply
Repeat	REE PEET	Repeat your instruction
Am lost	AM LOSST	Position unknown
Mayday	MAYDAY	I am in distress
Hijack	HI-JACK	I have been hijacked
Land (place name)	LAAND (place name)	I request to land at (place name)
Descend	DEE SEND	I require descent

RTF Standard words and phrases

Acknowledge	Confirm that you have received and understood this message
Affirm	Yes (shortened from affirmative)
Air Taxi	A helicopter proceeding at slow speed (less than 20 knots), close to the surface in ground effect
Approved	Permission for proposed action is granted
Break	This is a separation between messages
Break Break	Indicates a separation between messages in a busy RTF environment
Cancel	Annul the previously transmitted clearance
Changing to	I intend to call [unit] on [frequency]
Check	Examine a system or procedure
Cleared	Authorised to proceed under the conditions specified
Climb	Climb and maintain
Confirm	I request verification of [message/information]? Did you correctly receive this message?
Confirm squawk [code]	Confirm the code and mode you have set on the transponder
Contact	Establish radio contact with [unit], (they have your details)
Correct	True or Accurate
Correction	An error has been made. The correct message is.....
Descend	Descend and maintain
Disregard	Ignore
Fanstop	I am initiating a practice engine failure after take-off
Freecall	Call [ATSU], (they do not have your details)
Ground Taxi	A helicopter moving under its own power whilst in contact with the surface
Hold Short	Stop before reaching the specified location
Hover	A helicopter holding position whilst airborne. Hover allows spot/axial turns whilst maintaining position
How do you read?	What is the readability of my transmissions?
I Say Again	I will repeat (for clarity or emphasis)
Lift	The manoeuvre where a helicopter becomes airborne into a hover
Maintain	Continue in accordance with condition specified
Monitor	Listen out on [frequency]
Negative	No, or That is not correct, or Permission is not granted
Out	This exchange is ended and no response is expected
Over	This transmission is ended and I expect a response
Pass your message	Proceed with your message
Read Back	Repeat all, or a specified part, of the message; exactly as you received it
Report	Pass requested information
Request	I want to know, or I want to obtain
Roger	I have received all of your last transmission
Say Again	Repeat all, or a specified part, of your last transmission
Securité	May be used by a military ground unit to indicate a message containing information that affects safety (e.g. commencing live firing in a danger area)
Speak Slower	Slow down your rate of speaking
Squawk [code and mode]	Set this mode and code
Squawk Charlie	Set the transponder to mode C
Squawk ident	Operate the ident button on the transponder
Squawk standby	Set the transponder to the standby position
Standby	Wait and I will call you
Taxi	For a helicopter, 'taxi' can mean air taxi or ground taxi
Unable	I cannot comply with your request, instruction or clearance (a reason is normally given)
Wilco	I understand your message and will co-operate with it
Words Twice	As a request – please transmit each word twice As information – I will transmit each word twice

Fluids – Fuel

Fuels

BP	SHELL	EXXON MOBIL	NATO	JOINT SERVICES
AVGAS 80 (a)	AVGAS 80 (a)	AVGAS 80 (a)	–	AVGAS 80 (a)
–	AVGAS 82UL (b)	–	–	–
AVGAS 100 ©	AVGAS 100 ©	AVGAS 100 ©	–	AVGAS 100 ©
AVGAS 100LL (d)	AVGAS 100LL (d)	AVGAS 100LL (d)	–	AVGAS 100LL (d)
JET A-1 with FSII	–	JET A/A-1	F34	AVTUR/FSII
JET A	JET A	–	–	–
JET A-1	JET A-1	JET A-1	F35	AVTUR
JET TS-1	–	JET TS-1 Premium	–	–
–	–	JET TS-1 Regular	–	–
JET RT-1	–	–	–	–
JET TH	–	–	–	–
–	RP-1	–	–	–
–	RP-2	–	–	–
JET FUEL No 3	JET FUEL No 3	–	–	–
–	JP-8 + 100 (e)	–	F37	–
–	JP-4	JP-4	F40	AVTAG/FSII
–	JET B	JET-B	–	–
–	RP-4	–	–	–
–	–	JP-5/JP-8 ST	–	–
–	JP-5	JP-5	F44	AVCAT/FSII
–	–	JP-7	–	–
–	–	QAV-1	Kerosine	–
–	–	High Flash Kerosine	–	–
–	–	T-1	–	–
–	–	T-1S	–	–
–	–	T-2	–	–
–	–	RT	–	–

(a) AVGAS 80 – Dyed red (b) AVGAS 82UL – Dyed purple
© AVGAS 100 – Dyed green (d) AVGAS 100LL – Dyed blue
(e) Using Aeroshell performance additive 101

General — Fluids – Oil

Oils

Fluids Cross Reference

BP	SHELL	EXXON MOBIL	NATO	JOINT SERVICES	US MILITARY
Castrol Aviator S65	Aeroshell 65 Aeroshell W65	–	O-113 (obsolete)	OM-107 (obsolete)	J-1966 SAE Grade 30 J-1899 SAE Grade 30
BP Aviation Oil 80 Castrol Aviator S80	Aeroshell 80	–	–	OM-170 (obsolete)	J-1966 SAE Grade 40
BP Aviation Oil 10 Castrol Aviator S100	Aeroshell 100	Exxon Aviation Oil 100	O-117 (obsolete)	OM-270	J-1966 SAE Grade 50
–	–	Exxon Aviation Oil 120	–	–	J-1966 SAE Grade 60
BP Aero Oil D80 Castrol Aviator AD80	Aeroshell W80	Exxon Aviation Oil EE80	O-123 (obsolete)	OMD-160	J-1899 SAE Grade 40
–	Aeroshell W 15W-50	Exxon Aviation Oil Elite 20W-50	O-162 (obsolete)	OMD-162	J-1899 SAE Grade multigrade
Multigrade Aero Oil D20W/50	–	–	Exxon Aviation Oil EE 20W-50	–	–
Multigrade Aero Oil D25W/50	–	–	Exxon Aviation Oil EE 20W-60	–	–
BP Aero Oil D100 Castrol Aviator AD 100	Aeroshell W100	Exxon Aviation Oil EE100	O-125 (obsolete)	OMD-250	J-1899 SAE Grade 50
BP Aero Oil D120 Castrol Aviator AD120	Aeroshell W120	Exxon Aviation Oil EE120	O-128 (obsolete)	OMD-370	J-1899 SAE Grade 60
Brayco 460 Grade 1010	Aeroshell Turbine Oil 2	AVREX M TURBO 201/1010	O-133	OM-10 (obsolete)	MIL-PRF-6081D
–	Aeroshell Fluid 1	–	O-134	OM-13	–
–	Aeroshell Turbine Oil 3 Aeroshell turbine Oil 3SP	–	O-135	OM-11	–
Brayco 363	Aeroshell Fluid 3	–	O-142	OM-12	MIL-PRF-7870C
Brayco 885	Aeroshell Fluid 12	–	O-147	OX-14	MIL-PRF-6085D
Turbo Oil 2389 Castrol 399	Aeroshell Turbine Oil 308	AVREX S TURBO 256 TURBO 284 (RM 284A)	O-148	OX-9	MIL-PRF-7808L Grade 3
Turbo Oil 274 Castrol 98	Aeroshell Turbine Oil 750	–	O-149	OX-38	–
–	–	–	O-150 (note 4)	–	–
–	Aeroshell Turbine Oil 531	–	O-152	–	MIL-PRF-23699F Grade C/I
–	Aeroshell Fluid 5L-A	–	O-153	OEP-30	MIL-PRF-6086

General — Fluids – Hydraulic

Oil

BP	SHELL	EXXON MOBIL	NATO	JOINT SERVICES	US MILITARY	
–	Aeroshell Turbine Oil 560	Mobil Jet Oil 254, 291	O-154	–	Light Grade	
					MIL-PRF-23699F Grade HTS	
–	Aeroshell Fluid 5M-A	–	O-155	OEP-70	MIL-PRF-6086E Medium Grade	
Turbo Oil 2380/2197 Castrol 5000/5050	Aeroshell Turbine Oil 500 Aeroshell Turbine Oil 529	Mobil Jet Oil II	O-156	OX-27	MIL-PRF-23699F ST MIL-PRF-23699F C/I	
Brayco 855	–	–	O-157	–	MIL-PRF-14107D	
Turbo Oil 25 Castrol 599	Aeroshell Turbine Oil 555	–	O-160	OX-26	DOD-L-85734 (AS)	
–	–	–	note 3	OMD-80	–	
Brayco 300	Aeroshell Fluid 18	–	0-190 (obsolete)	OX-18 (obsolete) OX-24	MIL-PRF-32033	
Brayco 350	–	–	0-192	–	MIL-PRF-3150D	
–	Aeroshell Turbine Oil 390	–	note 3	OX-7	–	

Note 3 – The code used for oils is the NATO designation. When an item has no NATO code the Joint Services Designation is used.

Hydraulic Fluids

BP	SHELL	EXXON MOBIL	NATO	JOINT SERVICES	US MILITARY
BP Aero Hydraulic 1F Brayco Micronic 756	Aeroshell Fluid 41	ESSO INVAROL FJ13	H-515	OM-15	MIL-PRF-5606H DEF STAN 91-48/2
–	Aeroshell Fluid 4	ESSO INVAROL J13	H-520	OM-18	MIL-H-5606A
–	–	–	H-536	OX-50	–
Brayco Micronic 882	Aeroshell Fluid 31	MOBIL AERO HFS	H-537	OX-19	MIL-PRF-83282D
Brayco Micronic 881	Aeroshell Fluid 51	–	H-538	OX-538	MIL-PRF-87257A
Brayco Micronic 883	Aeroshell Fluid 61	–	H-544 Type II	–	MIL-PRF-46170C Type C
–	–	–	H-548	OX-75	–
–	–	–	H-572	OM-65	–
Brayco 717	–	–	H-575	–	MIL-DTL-17111C
OX-20	–	–	H-576	OM-33	–

Fluids Cross Reference

Conversion Factors

To convert	Into	Multiply by
Inches	Centimetres	2.5400
Centimetres	Inches	0.3937
Feet	Metres	0.3048
Metres	Feet	3.2808
Miles	Kilometres	1.6093
Kilometres	Miles	0.6214
Miles	Nautical Miles	0.8684
Nautical Miles	Miles	1.1515
Kilometres	Nautical Miles	0.5396
Nautical Miles	Kilometres	1.8532
Knots	Miles per Hr	1.1515
Miles per Hr	Knots	0.8684
Pints	Litres	0.5682
Litres	Pints	1.7598
Imperial Gallons	Litres	4.5459
Litres	Imperial Gallons	0.2200
Imperial Gallons	US Gallons	1.2009
US Gallons	Imperial Gallons	0.8327
Litres	US Gallons	0.2642
US Gallons	Litres	3.7853
Ounces	Grammes	28.3497
Grammes	Ounces	0.0353
Pounds	Kilogrammes	0.4536
Kilogrammes	Pounds	2.2046
Tons	Tonnes	1.0161
Tonnes	Tons	0.9842
Horsepower	Kilowatts	0.7400
Kilowatts	Horsepower	1.3400
Fahrenheit	Centigrade	$C = 5/9(F-32)$

UK pressure setting regions

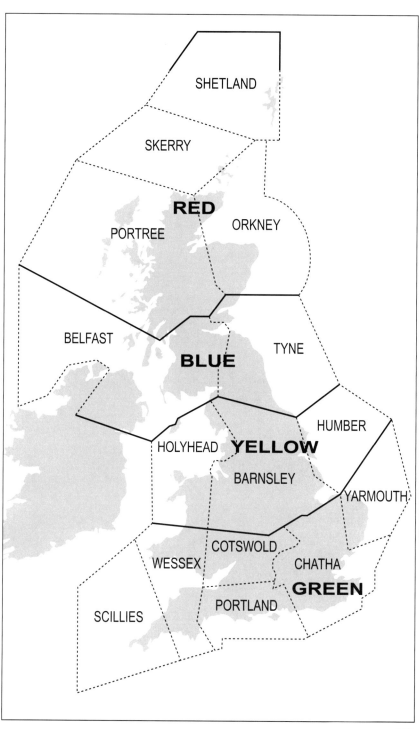

UK call signs

Call Sign	3 letter code	Operator
Atlantic	AAG	Air Atlantique
Foyle	UPA	Air Foyle
Airscan	SCY	Air Scandic
Baby	BMI	Baby BMI
Bond	BND	Bond Offshore Helicopters
Thomson	TOM	Britannia Airways (Budget Air)
Britannia	BAL	Britannia Airways Ltd
Speedbird	BAW	British Airways
Santa	XMS	British Airways Santa
Shuttle	SHT	British Airways Shuttle
Penguin	BAN	British Antarctic Survey
Brintel	BIH	British International
Bee Med	LAJ	British Mediterranean Airlines
Midland	BMA	British Midland Airways
British	BRT	British Regional Airlines
All Weather	AWX	CAA Directorate of Airspace Policy
Exam	EXM	CAA Flight Examiners
Minair	CFU	CAA Flying Unit
Standards	SDS	CAA Training Standards
Channex	EXS	Channel Express
Flyer	CFE	City Flyer Express
Comex	CDE	Comed Group
Typhoon	CBY	RAF Coningsby FTU
Surveyor	SVY	Cooper Aerial Surveys
Colt	COH	RAF Coltishall FTU
Church Fenton	CFN	RAF Church Fenton FTU
Bristow	BHL	Bristow Helicopters Group
Sky News	SKH	Bristish Sky Broadcasting
Gauntlet	BDN	RAF Boscombe Down DERA
Felix	BAE	BAE Systems (Corporate Air Travel)
	BKH	RAF Barkston Heath FTU
Ayline	AUR	Aurigny Air Services
Cottesmore	COT	RAF Cottesmore FTU
Cranwell	CWI	RAF Cranwell FTU
Oxford	CSE	CSE Aviation
World Express	DHK	DHL Air
Eastflight	EZE	Eastern Airways
Easy	EZY	Easyjet Airlines
Echelon	ECY	Euroceltic Airways
Agency	EMN	Examiner Training Agency
Flightline	FLT	Flightline
Fordair	FOB	Ford Motor Company
Geebee Airways	GBL	G B Airways
Goldair	GDA	Gold Air International
Bizjet	HJL	Hamlin Jet
Coastguard		HM Coastguard
Jetplan	JPN	Jeppesen UK
Kinloss	KIN	RAF Kinloss FTU
Javelin	LEE	RAF Leeming FTU
Leuchars	LCS	RAF Leuchars FTU
Linton on Ouse	LOP	RAF Linton on Ouse FTU
Logan	LOG	Logan Air
Lossie	LOS	RAF Lossiemouth FTU
Manx	MNX	Manx Airlines
Marham	MRH	RAF Marham FTU
Marshall	MCE	Marshall Aerospace
Macline	MCH	McAlpine Helicopters
Metman	MET	Meteorological Research Flight
Watchdog	WDG	Ministry of Agriculture Fisheries and Food
Monarch	MON	Monarch Airlines
Kestrel	MYT	MyTravel Airways

General

UK call signs

Call Sign	3 letter code	Operator
Mylite	LIZ	MyTravel Lite
Grid	GRD	National Grid Company
Science	EVM	Natural Environment Research Council
Netax	NEX	Northern Executive Aviation
Woodstock	WDK	Oxford Air Services
Scampton	SMZ	RAF Scampton FTU
Red Devils	DEV	Red Devils Parachute Display Team
Reed Aviation	RAV	Reed Aviation
Rols	BTU	Rolls Royce Bristol Engine Division
Merlin	RRI	Rolls Royce Military Aviation
Rafair	RFR	Royal Air Force
Navy	NVY	Royal Navy
Servisair	SGH	Servisair
Suckling	SAY	Suckling Airways
Top Jet	TCX	Thomas Cook Airlines
Zap	AWC	Titan Airways
Nirto	NTR	TNT International Aviation
Helimed	HLE	UK HEMS
Police	UKP	UK Home Office
Kitty	RRF	UK Royal Positioning Flight (Militrary Aircraft)
Leopard	LPD	UK Royal/HRH Duke of York
Sparrowhawk	KRH	UK Royal/VIP Flight (Civil Aircraft)
Kittyhawk	KRF	UK Royal/VIP Flight (Military Aircraft)
Rainbow	TQF	UK Royal/VIP Helicopter Flight
Anglesey	VYT	RAF Valley FTU
Vickers	VSB	BAE Systems Vickers
Virgin	VIR	Virgin Atlantic
Balloon Virgin		Virgin Balloon Flight
Virgin Express	VEX	Virgin Express
Green Isle	VEI	Virgin Express Ireland
Vulcan	WAD	RAF Waddington FTU
Tarnish	WTN	Warton Military Flight Ops
Westland	WHE	Westland Helicopters
Striker	WIT	RAF Wittering FTU
Avro	WFD	BAE Woodford Flight Test
Shamrock	EIN	Aer Lingus
Aeroflot	AFL	Aeroflot - Rissian Int Airlines
Lifeline	AMQ	Aeromedicare
Air Berlin	BER	Air Berlin
Air Canada	ACA	Air Canada
Air Baltic	BTI	Air Baltic
Air China	CCA	Air China
Air India	AIC	Air India
Jamaica	AJM	Air Jamaica
Air Maurtius	MAU	Air Maurtius
New Zealand	ANZ	Air New Zealand
Ann Air	ANK	All Nipon Airways
American	AAL	American Airlines
Bangladesh	BBC	Biman Bangladesh Airlines
West Indian	BWA	BWIA
Cathay	CPA	Cathay Pacific
Continental	COA	Continental Airlines
Egyptair	MSR	Egyptair
Emirates	UAE	Emirates
Ethiopian	ETH	Ethiopian Airlines
Ghana	GHA	Ghana Airways
Gulf Air	GFA	Gulf Air
Iran Air	IRA	Iran Air
Japanair	JAL	Japan Airlines
Koreanair	KAL	Korean Air
Kuwaiti	KAC	Kuwait Airways

Call Sign	3 letter code	Operator
Lithuania Air	LIL	Lithuanian Airways
Malaysian	MAS	Malaysia Airlines
Cedar Jet	MEA	Middle East Airlines
Pakistan	PIA	Pakistan Int Airlines
Qutari	QTR	Qatar Airways
Brunei	RBA	Royal Brunei
Jordanian	RJA	Royal Jordanian Airlines
Saudia	SVA	Saudi Arabian Airlines
Sudanair	SUD	Sudan Airways
Thai	THA	Thai Airways
		Turkish Airlines
Turkmenistan	TUA	Turkmenistan Airlines
United	UAL	United Airlines
	VLO	Varig
Adria	ADR	Adria Airways
Argentina	ARG	Aeolineas Argentinas
Air Algerie	DAH	Air Algerie
Airfrans	AFR	Air France
Alitalia	AZA	Alitalia
Armenian	RME	Armenian Airlines
Austrian	AUA	Austrian Airlines
Croatia	CTN	Croatia Airlines
CSA Lines	CSA	Czech Airlines
Iberia	IBE	Iberia
Jat	JAT	JAT Airways
Lauda Air	LDA	Lauda Air
Hansaline	CLH	Lufthansa
	LXO	Luxor Air
Malev	MAH	Malev Hungarian Airlines
Olympic	OAL	Olympic Airways
Royal Air Maroc	RAM	Royal Air Maroc
Swiss	SWR	Swiss Airlines
Syrianair	SYR	Syrian Airlines
Presidente	TPE	TAP Air Portugal
Tarom	ROT	TAROM Romanian
Transoviet	TSO	Transaero Airlines
Tunair	TAR	Tunis Air
Uzbek	UZB	Uzbekistan Airways
Yemeni	IYE	Yemenia Airways
Jambo	AFJ	Alliance
Cyprus	CYP	Cyprus Airways
Elal	ELY	El Al Israel Airlines
Finnair	FIN	Finnair
Iceair	ICE	Icelandair
Pollot	LOT	LOT Polish Airlines
Springbok	SAA	South African Airways
Air Malta	AMC	Air Malta
Kenya	KQA	Kenya Airways
KLM	KLM	KLM Royal Dutch Airlines
Qantas	QFA	Qantas
Srilankan	ALK	Sri Lankan Airlines

General UK call signs

3 letter code	Operator	Call Sign
AAG	Air Atlantique	Atlantic
UPA	Air Foyle	Foyle
SCY	Air Scandic	Airscan
BMI	Baby BMI	Baby
BND	Bond Offshore Helicopters	Bond
TOM	Britannia Airways (Budget Air)	Thomson
BAL	Britannia Airways Ltd	Britannia
BAW	British Airways	Speedbird
XMS	British Airways Santa	Santa
SHT	British Airways Shuttle	Shuttle
BAN	British Antarctic Survey	Penguin
BIH	British International	Brintel
LAJ	British Mediterranean Airlines	Bee Med
BMA	British Midland Airways	Midland
BRT	British Regional Airlines	British
AWX	CAA Directorate of Airspace Policy	All Weather
EXM	CAA Flight Examiners	Exam
CFU	CAA Flying Unit	Minair
SDS	CAA Training Standards	Standards
EXS	Channel Express	Channex
CFE	City Flyer Express	Flyer
CDE	Comed Group	Comex
CBY	RAF Coningsby FTU	Typhoon
SVY	Cooper Aerial Surveys	Surveyor
COH	RAF Coltishall FTU	Colt
CFN	RAF Church Fenton FTU	Church Fenton
BHL	Bristow Helicopters Group	Bristow
SKH	Bristish Sky Broadcasting	Sky News
BDN	RAF Boscombe Down DERA	Gauntlet
BAE	BAE Systems (Corporate Air Travel)	Felix
BKH	RAF Barkston Heath FTU	
AUR	Aurigny Air Services	Ayline
COT	RAF Cottesmore FTU	Cottesmore
CWI	RAF Cranwell FTU	Cranwell
CSE	CSE Aviation	Oxford
DHK	DHL Air	World Express
EZE	Eastern Airways	Eastflight
EZY	Easyjet Airlines	Easy
ECY	Euroceltic Airways	Echelon
EMN	Examiner Training Agency	Agency
FLT	Flightline	Flightline
FOB	Ford Motor Company	Fordair
GBL	G B Airways	Geebee Airways
GDA	Gold Air International	Goldair
HJL	Hamlin Jet	Bizjet
	HM Coastguard	Coastguard
JPN	Jeppesen UK	Jetplan
KIN	RAF Kinloss FTU	Kinloss
LEE	RAF Leeming FTU	Javelin
LCS	RAF Leuchars FTU	Leuchars
LOP	RAF Linton on Ouse FTU	Linton on Ouse
LOG	Logan Air	Logan
LOS	RAF Lossiemouth FTU	Lossie
MNX	Manx Airlines	Manx
MRH	RAF Marham FTU	Marham
MCE	Marshall Aerospace	Marshall
MCH	McAlpine Helicopters	Macline
MET	Meteorological Research Flight	Metman
WDG	Ministry of Agriculture Fisheries and Food	Watchdog
MON	Monarch Airlines	Monarch
MYT	MyTravel Airways	Kestrel

3 letter code	Operator	Call Sign
LIZ	MyTravel Lite	Mylite
GRD	National Grid Company	Grid
EVM	Natural Environment Research Council	Science
NEX	Northern Executive Aviation	Netax
WDK	Oxford Air Services	Woodstock
SMZ	RAF Scampton FTU	Scampton
DEV	Red Devils Parachute Display Team	Red Devils
RAV	Reed Aviation	Reed Aviation
BTU	Rolls Royce Bristol Engine Division	Rols
RRI	Rolls Royce Military Aviation	Merlin
RFR	Royal Air Force	Rafair
NVY	Royal Navy	Navy
SGH	Servisair	Servisair
SAY	Suckling Airways	Suckling
TCX	Thomas Cook Airlines	Top Jet
AWC	Titan Airways	Zap
NTR	TNT International Aviation	Nirto
HLE	UK HEMS	Helimed
UKP	UK Home Office	Police
RRF	UK Royal Positioning Flight (Militrary Aircraft)	Kitty
LPD	UK Royal/HRH Duke of York	Leopard
KRH	UK Royal/VIP Flight (Civil Aircraft)	Sparrowhawk
KRF	UK Royal/VIP Flight (Military Aircraft)	Kittyhawk
TQF	UK Royal/VIP Helicopter Flight	Rainbow
VYT	RAF Valley FTU	Anglesey
VSB	BAE Systems Vickers	Vickers
VIR	Virgin Atlantic	Virgin
	Virgin Balloon Flight	Balloon Virgin
VEX	Virgin Express	Virgin Express
VEI	Virgin Express Ireland	Green Isle
WAD	RAF Waddington FTU	Vulcan
WTN	Warton Military Flight Ops	Tarnish
WHE	Westland Helicopters	Westland
WIT	RAF Wittering FTU	Striker
WFD	BAE Woodford Flight Test	Avro
EIN	Aer Lingus	Shamrock
AFL	Aeroflot - Rissian Int Airlines	Aeroflot
AMQ	Aeromedicare	Lifeline
BER	Air Berlin	Air Berlin
ACA	Air Canada	Air Canada
BTI	Air Baltic	Air Baltic
CCA	Air China	Air China
AIC	Air India	Air India
AJM	Air Jamaica	Jamaica
MAU	Air Maurtius	Air Maurtius
ANZ	Air New Zealand	New Zealand
ANK	All Nipon Airways	Ann Air
AAL	American Airlines	American
BBC	Biman Bangladesh Airlines	Bangladesh
BWA	BWIA	West Indian
CPA	Cathay Pacific	Cathay
COA	Continental Airlines	Continental
MSR	Egyptair	Egyptair
UAE	Emirates	Emirates
ETH	Ethiopian Airlines	Ethiopian
GHA	Ghana Airways	Ghana
GFA	Gulf Air	Gulf Air
IRA	Iran Air	Iran Air
JAL	Japan Airlines	Japanair
KAL	Korean Air	Koreanair
KAC	Kuwait Airways	Kuwaiti

General

UK call signs

3 letter code	Operator	Call Sign
LIL	Lithuanian Airways	Lithuania Air
MAS	Malaysia Airlines	Malaysian
MEA	Middle East Airlines	Cedar Jet
PIA	Pakistan Int Airlines	Pakistan
QTR	Qatar Airways	Qutari
RBA	Royal Brunei	Brunei
RJA	Royal Jordanian Airlines	Jordanian
SVA	Saudi Arabian Airlines	Saudia
SUD	Sudan Airways	Sudanair
THA	Thai Airways	Thai
	Turkish Airlines	
TUA	Turkmenistan Airlines	Turkmenistan
UAL	United Airlines	United
VLO	Varig	
ADR	Adria Airways	Adria
ARG	Aeolineas Argentinas	Argentina
DAH	Air Algerie	Air Algerie
AFR	Air France	Airfrans
AZA	Alitalia	Alitalia
RME	Armenian Airlines	Armenian
AUA	Austrian Airlines	Austrian
CTN	Croatia Airlines	Croatia
CSA	Czech Airlines	CSA Lines
IBE	Iberia	Iberia
JAT	JAT Airways	Jat
LDA	Lauda Air	Lauda Air
CLH	Lufthansa	Hansaline
LXO	Luxor Air	
MAH	Malev Hungarian Airlines	Malev
OAL	Olympic Airways	Olympic
RAM	Royal Air Maroc	Royal Air Maroc
SWR	Swiss Airlines	Swiss
SYR	Syrian Airlines	Syrianair
TPE	TAP Air Portugal	Presidente
ROT	TAROM Romanian	Tarom
TSO	Transaero Airlines	Transoviet
TAR	Tunis Air	Tunair
UZB	Uzbekistan Airways	Uzbek
IYE	Yemenia Airways	Yemeni
AFJ	Alliance	Jambo
CYP	Cyprus Airways	Cyprus
ELY	El Al Israel Airlines	Elal
FIN	Finnair	Finnair
ICE	Icelandair	Iceair
LOT	LOT Polish Airlines	Pollot
SAA	South African Airways	Springbok
AMC	Air Malta	Air Malta
KQA	Kenya Airways	Kenya
KLM	KLM Royal Dutch Airlines	KLM
QFA	Qantas	Qantas
ALK	Sri Lankan Airlines	Srilankan

General UK call signs

Operator	Call Sign	3 letter code
Air Atlantique	Atlantic	AAG
Air Foyle	Foyle	UPA
Air Scandic	Airscan	SCY
Baby BMI	Baby	BMI
Bond Offshore Helicopters	Bond	BND
Britannia Airways (Budget Air)	Thomson	TOM
Britannia Airways Ltd	Britannia	BAL
British Airways	Speedbird	BAW
British Airways Santa	Santa	XMS
British Airways Shuttle	Shuttle	SHT
British Antarctic Survey	Penguin	BAN
British International	Brintel	BIH
British Mediterranean Airlines	Bee Med	LAJ
British Midland Airways	Midland	BMA
British Regional Airlines	British	BRT
CAA Directorate of Airspace Policy	All Weather	AWX
CAA Flight Examiners	Exam	EXM
CAA Flying Unit	Minair	CFU
CAA Training Standards	Standards	SDS
Channel Express	Channex	EXS
City Flyer Express	Flyer	CFE
Comed Group	Comex	CDE
RAF Coningsby FTU	Typhoon	CBY
Cooper Aerial Surveys	Surveyor	SVY
RAF Coltishall FTU	Colt	COH
RAF Church Fenton FTU	Church Fenton	CFN
Bristow Helicopters Group	Bristow	BHL
Bristish Sky Broadcasting	Sky News	SKH
RAF Boscombe Down DERA	Gauntlet	BDN
BAE Systems (Corporate Air Travel)	Felix	BAE
RAF Barkston Heath FTU		BKH
Aurigny Air Services	Ayline	AUR
RAF Cottesmore FTU	Cottesmore	COT
RAF Cranwell FTU	Cranwell	CWI
CSE Aviation	Oxford	CSE
DHL Air	World Express	DHK
Eastern Airways	Eastflight	EZE
Easyjet Airlines	Easy	EZY
Euroceltic Airways	Echelon	ECY
Examiner Training Agency	Agency	EMN
Flightline	Flightline	FLT
Ford Motor Company	Fordair	FOB
G B Airways	Geebee Airways	GBL
Gold Air International	Goldair	GDA
Hamlin Jet	Bizjet	HJL
HM Coastguard	Coastguard	
Jeppesen UK	Jetplan	JPN
RAF Kinloss FTU	Kinloss	KIN
RAF Leeming FTU	Javelin	LEE
RAF Leuchars FTU	Leuchars	LCS
RAF Linton on Ouse FTU	Linton on Ouse	LOP
Logan Air	Logan	LOG
RAF Lossiemouth FTU	Lossie	LOS
Manx Airlines	Manx	MNX
RAF Marham FTU	Marham	MRH
Marshall Aerospace	Marshall	MCE
McAlpine Helicopters	Macline	MCH
Meteorological Research Flight	Metman	MET
Ministry of Agriculture Fisheries and Food	Watchdog	WDG
Monarch Airlines	Monarch	MON
MyTravel Airways	Kestrel	MYT

General UK call signs

Operator	Call Sign	3 letter code
MyTravel Lite	Mylite	LIZ
National Grid Company	Grid	GRD
Natural Environment Research Council	Science	EVM
Northern Executive Aviation	Netax	NEX
Oxford Air Services	Woodstock	WDK
RAF Scampton FTU	Scampton	SMZ
Red Devils Parachute Display Team	Red Devils	DEV
Reed Aviation	Reed Aviation	RAV
Rolls Royce Bristol Engine Division	Rols	BTU
Rolls Royce Military Aviation	Merlin	RRI
Royal Air Force	Rafair	RFR
Royal Navy	Navy	NVY
Servisair	Servisair	SGH
Suckling Airways	Suckling	SAY
Thomas Cook Airlines	Top Jet	TCX
Titan Airways	Zap	AWC
TNT International Aviation	Nirto	NTR
UK HEMS	Helimed	HLE
UK Home Office	Police	UKP
UK Royal Positioning Flight (Militrary Aircraft)	Kitty	RRF
UK Royal/HRH Duke of York	Leopard	LPD
UK Royal/VIP Flight (Civil Aircraft)	Sparrowhawk	KRH
UK Royal/VIP Flight (Military Aircraft)	Kittyhawk	KRF
UK Royal/VIP Helicopter Flight	Rainbow	TQF
RAF Valley FTU	Anglesey	VYT
BAE Systems Vickers	Vickers	VSB
Virgin Atlantic	Virgin	VIR
Virgin Balloon Flight	Balloon Virgin	
Virgin Express	Virgin Express	VEX
Virgin Express Ireland	Green Isle	VEI
RAF Waddington FTU	Vulcan	WAD
Warton Military Flight Ops	Tarnish	WTN
Westland Helicopters	Westland	WHE
RAF Wittering FTU	Striker	WIT
BAE Woodford Flight Test	Avro	WFD
Aer Lingus	Shamrock	EIN
Aeroflot - Rissian Int Airlines	Aeroflot	AFL
Aeromedicare	Lifeline	AMQ
Air Berlin	Air Berlin	BER
Air Canada	Air Canada	ACA
Air Baltic	Air Baltic	BTI
Air China	Air China	CCA
Air India	Air India	AIC
Air Jamaica	Jamaica	AJM
Air Maurtius	Air Maurtius	MAU
Air New Zealand	New Zealand	ANZ
All Nipon Airways	Ann Air	ANK
American Airlines	American	AAL
Biman Bangladesh Airlines	Bangladesh	BBC
BWIA	West Indian	BWA
Cathay Pacific	Cathay	CPA
Continental Airlines	Continental	COA
Egyptair	Egyptair	MSR
Emirates	Emirates	UAE
Ethiopian Airlines	Ethiopian	ETH
Ghana Airways	Ghana	GHA
Gulf Air	Gulf Air	GFA
Iran Air	Iran Air	IRA
Japan Airlines	Japanair	JAL
Korean Air	Koreanair	KAL
Kuwait Airways	Kuwaiti	KAC

Operator	Call Sign	3 letter code
Lithuanian Airways	Lithuania Air	LIL
Malaysia Airlines	Malaysian	MAS
Middle East Airlines	Cedar Jet	MEA
Pakistan Int Airlines	Pakistan	PIA
Qatar Airways	Qutari	QTR
Royal Brunei	Brunei	RBA
Royal Jordanian Airlines	Jordanian	RJA
Saudi Arabian Airlines	Saudia	SVA
Sudan Airways	Sudanair	SUD
Thai Airways	Thai	THA
Turkish Airlines		
Turkmenistan Airlines	Turkmenistan	TUA
United Airlines	United	UAL
Varig		VLO
Adria Airways	Adria	ADR
Aeolineas Argentinas	Argentina	ARG
Air Algerie	Air Algerie	DAH
Air France	Airfrans	AFR
Alitalia	Alitalia	AZA
Armenian Airlines	Armenian	RME
Austrian Airlines	Austrian	AUA
Croatia Airlines	Croatia	CTN
Czech Airlines	CSA Lines	CSA
Iberia	Iberia	IBE
JAT Airways	Jat	JAT
Lauda Air	Lauda Air	LDA
Lufthansa	Hansaline	CLH
Luxor Air		LXO
Malev Hungarian Airlines	Malev	MAH
Olympic Airways	Olympic	OAL
Royal Air Maroc	Royal Air Maroc	RAM
Swiss Airlines	Swiss	SWR
Syrian Airlines	Syrianair	SYR
TAP Air Portugal	Presidente	TPE
TAROM Romanian	Tarom	ROT
Transaero Airlines	Transoviet	TSO
Tunis Air	Tunair	TAR
Uzbekistan Airways	Uzbek	UZB
Yemenia Airways	Yemeni	IYE
Alliance	Jambo	AFJ
Cyprus Airways	Cyprus	CYP
El Al Israel Airlines	Elal	ELY
Finnair	Finnair	FIN
Icelandair	Iceair	ICE
LOT Polish Airlines	Pollot	LOT
South African Airways	Springbok	SAA
Air Malta	Air Malta	AMC
Kenya Airways	Kenya	KQA
KLM Royal Dutch Airlines	KLM	KLM
Qantas	Qantas	QFA
Sri Lankan Airlines	Srilankan	ALK

CAA Safety Sense Leaflets

UK AIM
The UK Aeronautical Information Manual

Safety Sense Leaflets

1 Good Airmanship

SAFETYSENSE LEAFLET 1e - GOOD AIRMANSHIP

1 INTRODUCTION	12 PERFORMANCE	23 EN-ROUTE
2 REPORTING	13 FUEL PLANNING	24 DIVERSION
3 STATISTICS	14 DESTINATION	25 LOST
4 REFRESHER TRAINING	15 FLYING ABROAD	26 SPEED CONTROL
5 LIMITATIONS	16 FLIGHT OVER WATER	27 ENVIRONMENTAL
6 PREPARATION	17 PILOT FITNESS	28 WIND & WAKE TURBULENCE
7 UNFAMILIAR AIRCRAFT	18 PRE FLIGHT	
8 WEATHER	19 STARTING ENGINE	29 CIRCUIT PROCEDURE
9 VFR NAVIGATION	20 TAKEOFF	30 LANDING
10 RADIO	21 LOOKOUT	31 SUMMARY
11 WEIGHT & BALANCE	22 AIRSPACE	

1 INTRODUCTION

a   Although this guide is mainly intended for Private Pilots of fixed wing aircraft, much of the advice will be relevant to all pilots, whatever their experience or the type of aircraft they fly. However, there are specific leaflets giving more detailed advice for helicopter (no 17) and balloon (no 16) pilots.

b   Any review of General Aviation Accidents shows that most should not have happened. They are a result of a combination of the following:
- use of incorrect techniques
- lack of preparation before flight
- being out of practice
- lack of appreciation of weather
- overconfidence
- flying illegally or outside licence privileges
- failing to maintain control
- a complacent attitude
- the 'it will be alright' syndrome.

c   Comprehensive Knowledge, careful Preparation and frequent flying Practice are key elements in developing 'Good Airmanship' which is the best insurance against appearing as an accident statistic.

2 KNOWLEDGE - REPORTING

a) "Learn from the mistakes of others; you might not live long enough to make all of them yourself".

b) Share your knowledge and experience with others, preferably by reporting to the CAA* (BMAA, BGA etc) anything from which you think others could learn. Your report could prevent someone else's accident. Photographs often help to illustrate a problem.

c) Improve your knowledge by reading the CAA's GASIL, published every quarter, the Air Accident Investigation Branch's monthly Bulletin*, the General Aviation Safety Council's quarterly Bulletin* and the Confidential Human Factors Incident Reporting Programme's GA Feedback leaflet.

Details of reported light aircraft occurrences are held by the CAA's Safety Investigation & Data Department*, and available for safety purposes.

d) More specific information is available in other Safety Sense Leaflets, in Aeronautical Information Circulars (available by subscription or free from the ais web site)* particularly the pink Safety ones, and in other publications.

3 STATISTICS

a. There is an average of one fatal GA accident a month in the United Kingdom.

b. The main fatal accident causes during the last 20 years have been:
- continued flight into bad weather, including impact with high ground and loss of control in IMC
- loss of control in visual met conditions, including stall/spin
- low aerobatics and low flying
- mid-air collisions (sometimes either pilot knew the other was there)
- runway too short for the aircraft's weight or performance
- colliding with obstacles, perhaps being too low on the approach

c. A high proportion of stall/spin fatal accident pilots were not in good flying practice.

d. Loss of control in flight is the major cause of fatal accidents in gliding and microlighting.

e. The main causes of twin-engined aircraft fatal accidents were:
- pressing on into bad weather (often to aerodromes with limited navigational facilities) resulting in controlled flight into terrain or loss of control IMC
- loss of control VFR particularly following engine failure

4 REFRESHER TRAINING

Revise your basic knowledge and skills by having a regular flight, at least every year, with an instructor which includes:
- steep turns
- slow flight and stalls (clean and with flap) so that you recognise buffet, pitch attitude, control loads etc.

Note: in a level 60° banked turn, the stall speed increases by about 42%, - a 50kt straight & level stall becomes 71kts.

Practise at a safe height,
- if the aircraft is aerobatic or cleared for spinning, practise full spins as well as incipient spin recovery from a safe height. Aim to recover by 3000 feet above ground.
- practise forced landing procedures
- instrument flying and cloud avoidance
- take-offs and landings, including normal, cross-wind, flapless and short
- if you fly a twin, practise engine out procedures and power off stalls. Manufacturers quote a minimum safe speed for flight with one engine inoperative, V_{MCA}. Age and modifications may increase this for your aircraft

5 LIMITATIONS

a. You must know the aircraft's limitations and HEED THEM. If it is placarded 'NO AEROBATICS', it means it!

b. Know your own limitations; if you do not have a valid Instrument or IMC Rating, then you must fly clear of cloud, in sight of the surface and with a flight visibility of 3000 metres. If not in practice, you are not as good as you were!

6 PREPARATION

a. Make sure that your personal paperwork (licence/rating, Certificate of Test/ Experience and medical), is up to date. Also check that the aircraft's documents, including Certificates of Airworthiness/ Permit to Fly, Maintenance Release and Insurance are current.

b. Make sure that the Check List you use conforms to the Flight Manual of that aircraft.

7 UNFAMILIAR AIRCRAFT

a. Before you fly a new aircraft type, ensure any 'Differences Training' is completed.

b. Before you fly either a new aircraft type, one you have not flown for a while or one you do not fly often, study the Pilot's Operating Handbook/Flight Manual and be thoroughly familiar with:-
- airframe and engine limitations
- normal and emergency procedures
- operating, stall and best glide speeds
- weight and balance calculation
- take-off, cruise and landing performance.

c. Familiarise yourself with the external and ground checks, cockpit layout and fuel system, e.g. don't confuse the carb heat control with the mixture control.

d. Even if not legally required, try to have one or more thorough check flight with an instructor, particularly if converting to a tail wheel type. (In the case of a single seat aircraft, make thoroughly pre-briefed exploratory flights.) Include the items in para 4, Refresher Training.

e. If you have not flown the type in the last six months, treat it as 'new'. Many clubs require a check-flight if you have not flown the type in the last 28 days.

8 WEATHER

a. Get an aviation weather forecast, heed what it says and make a carefully reasoned GO/NO–GO decision. Do not let 'Get-there/home–itis' affect your judgement and do not worry about 'disappointing' your passenger(s). Establish clearly in your mind the current en-route conditions, the forecast and the 'escape route' to good weather. Plan an alternative route if you intend to fly over high ground where cloud is likely to lower and thicken.

b. Note the freezing level. Don't forget to check on crosswind at the destination.

c. The various methods of obtaining aviation weather, (including codes) are described in the booklet 'GET MET', available free from the Met Office*. Aerodrome and area forecasts and reports are freely available on the met office web site www.meto.gov.uk .

d. Know the conditions that lead to the formation of carburettor or engine icing and stay alert for this hazard. Do a carb heat check at top of climb and periodically use it in the cruise and with the first indication of a loss of power due to icing; once formed it may take more than 15 seconds of heat to melt the ice. In the circuit, check carb heat during pre-landing checks and use carb heat at low power settings as directed in the Pilot's Operating Handbook/Flight Manual. (See SafetySense Leaflet 14 'Piston Engine Icing'.)

9 VFR NAVIGATION

a. Use appropriate current aeronautical charts. (See SafetySense Leaflet 5 'VFR Navigation'.) Amendments to charts are available on the website www.caa.co.uk/dap

b. Check NOTAMs, Temporary Navigation Warnings, AICs etc for changes issued since your chart was printed or which are of a temporary nature, such as a closed runway, an air display, navaid or ATC frequency change. These are available on the AIS web site at www.ais.org.uk Refer to the GASIL 'changes' sheet.

c. Information on Temporary Restricted or Controlled Airspace, Red Arrows displays and Emergency Restrictions is available on Freephone 0500 354 802, updated daily.

d. Prepare your Route Plan thoroughly, with particular reference to minimum flying altitude and suitable diversions. Familiarise yourself with the geographical features, time points, airspace en-route and frequencies.

e. Note masts and other obstructions in planning your minimum flying altitude; note Maximum Elevation Figures (MEF) printed on the charts.

f. Allow extra height over hilly terrain, particularly in windy conditions, to minimise turbulence and the effects of down draughts.

g. Plan to reach your destination at least one hour before sunset unless qualified and prepared for night flight. Note aerodrome operating hours.

h. In any aircraft, the minimum height over a congested (i.e. built- up) area is not less than 1000ft above the highest object within 600 metres. In any aircraft other than a helicopter, you must not fly over congested areas without sufficient height to safely alight clear of the area in the event of engine failure. This could be higher than 1000ft (note: Permit to Fly aircraft may not be allowed over congested areas).

i. Do not plan to fly below 1000ft agl, (where most military low flying takes place –see SafetySense Leaflet 18 'Military Low Flying), unless necessary. If your engine fails you may need time to select a safe landing field.

j. Know the procedure if you get lost, see para 25.

k. If you use GPS to back up your visual navigation, double check any way-points when working them out and entering them. Progress **must** be monitored by map reading and not by implicitly trusting the GPS. (See SafetySense leaflet 25)

10 RADIO

a. Know what to do in the event of radio failure, including when flying Special VFR in controlled airspace. Know your way round your radio switches.

b. Note all useful radio frequencies, including destination and diversion aerodromes, VOLMET, LARS, Danger Area Crossing Service etc.

c. Note the frequencies and morse ident of radio NAVAIDs for back-up to the visual navigation.

d. Remind yourself about radio procedures, phraseology etc (*See CAP 413 'Radiotelephony Manual' and SafetySense leaflet 22*).

11 WEIGHT AND BALANCE

a. Use the *actual* empty weight and CG from the latest Weight and Balance Schedule of the specific aircraft you are flying. Aircraft get heavier due to extra equipment, coats of paint etc. Use people's actual weights, too.

b. Check that the aircraft maximum weight is complied with. If too heavy, you must reduce the weight by off- loading passengers, baggage or fuel.

c. Check that the CG is within limits for take-off and throughout the flight. If your calculations show that it will not stay within the approved range, including the restricted range for spinning or aerobatics, you must make some changes.

d. Never attempt to fly an aircraft which is outside the permitted weight/CG range and performance limitations. It is extremely dangerous (sudden loss of control likely), as well as illegal, invalidates the C of A and almost certainly your insurance. (*See Safety Sense Leaflet 9 'Weight and Balance'.*)

12 PERFORMANCE

a. Make sure that the runways you are going to operate from are long enough for take-off and landing. Use the Pilot's Operating Handbook/ Flight Manual to calculate the distances that you need. Check for any CAA Supplements that may downgrade the performance.

b. Note that any factors given for elevation, temperature, slope, grass, snow, tail wind etc are all cumulative and must be multiplied, e.g. 1.3 x 1.2 etc.

c. The performance figures given in the Handbook/Manual were obtained by a test pilot on a new aircraft, so in addition to the published factors, apply a safety factor of 1.33 for take-off and 1.43 for landing. These give acceptable safety margins, and will offset an out-of-practice pilot/tired engine. On a few aircraft these may have been included in the manufacturers information as 'factored' data. (*See SafetySense Leaflet 7 'Aeroplane Performance'*)

d. Short wet grass is slippery and may need a factor of up to 1.6!

13 FUEL PLANNING

a. Always plan to land by the time the tank(s) are down to the greater of ¼ tank or 45 minutes cruise flight, but don't rely solely on gauge(s) which may be unreliable. Remember, headwinds may be stronger than forecast and frequent use of carb heat will reduce range.

b. Understand the operation and limitations of the fuel system, gauges, pumps, mixture control, unusable fuel etc and remember to lean the mixture if it is permitted.

c. Don't assume you can achieve the Handbook/ Manual fuel consumption. As a rule of thumb, due to service and wear, expect to use 20% more fuel than the 'book' figures.

14 DESTINATION

a. Check for any special procedures and activities at your destination such as gliding, parachuting, or microlighting. Update the UK Aeronautical Information Publication (UK AIP) or other Flight Guides with NOTAMs from the AIS web site at www.ais.org.uk or their fax pollback numbers

b. If your destination is a strip, remember that the environment may be very different from the licensed aerodrome at which you learnt to fly, or from which you normally operate. There may be hard to see cables or other obstructions on the approach path, or hills, trees and buildings close to the strip giving wind shear and/ or unusual air currents.

c. Before going to a strip, it is suggested that you are checked out by an instructor or someone who knows the strip well. If you can't arrange either, go by road and have a look at the potential problems for different wind/surface conditions. Assess the slope; it may be visually deceptive. (*See SafetySense leaflet 12 'Strip Sense'*).

d. You **must** obtain permission if the destination is "Prior Permission Required (PPR)". If flying non-radio, always phone to find out the procedures.

e. Prepare a Flight Plan for filing on the day if you are going over a sparsely populated area, or more than 10nm from the UK coast. . (*See UK AIP Enroute [ENR] 1.10 and Safety Sense leaflet 20*)

15 FLYING ABROAD

a. Make sure you are conversant with the aeronautical (and customs) regulations, charts (including scale and units, e.g. feet or metres), airspace restrictions etc for each country you are flying over. Their individual AIS web site may help. Remember, an IMC rating is not valid outside the UK.

b. Ensure you know how to find weather forecasts and reports for your return flight.

c. Take the aircraft documents, your licence, and a copy of 'Interception Procedures' (AIP ENR 1.12 and Safety Sense leaflet 11).

d. Before crossing an international FIR boundary you must file a Flight Plan. Check that it has been accepted and the DEParture message sent once you are airborne. (*See SafetySense leaflet 20 'VFR Flight Plans'*)

e. Check the Terrorism Act's restrictions on flights to & from Ireland, Channel Isles and Isle of Man (UK AIP GEN 1.2.1).

f. Ensure you have informed Customs and Immigration if you are returning from an EU country and not using a Customs aerodrome. See AIP GEN 1.2.1.2 (1.2.1.3 covers flight from non-EU countries).

g. In some countries, e.g. Germany and France, it is a legal requirement to have a 760 channel radio which can transmit and receive on frequencies between 118 and 137 MHz.

16 FLIGHT OVER WATER

a. The weather over the sea can often be very different from the land, e.g. sea fog.

b. When flying over water, everyone in a single-engined aircraft should, as a minimum, wear lifejackets. In the event of an emergency there will be neither time nor space to put it on.

c. The water around the UK coast is very cold in winter and cold in summer. Survival time in normal clothing may be as low as 15 minutes (about the time needed to scramble an SAR helicopter but not for it to reach you). A good quality insulated survival suit, with the hood up and well sealed, should provide over 3 hours survival time. In water, the body loses heat 100 times faster than in cold air.

d. In addition, take a life-raft; it's heavy, so re-check weight and balance. A life-raft is much easier to see and will help rescuers find you. It should be properly secured in the aircraft, but easily accessible, you will not have much time.

e. Make sure that lifejackets, survival suits and life-raft have been tested recently by an approved organisation – they **must be serviceable** when needed.

f. You are strongly urged to carry an approved Emergency Locator Transmitter and flares.

g. Remain in contact with an appropriate aeronautical radio station.

h. Know the ditching procedure.

i. Pilots and passengers who regularly fly over water are advised to attend an underwater escape training and Sea Survival Course. (*See SafetySense Leaflet 21 'Ditching'.*)

17 PILOT FITNESS

a. Don't fly when unfit – it is better to cancel a flight than to wreck an aircraft or hurt yourself! (*See Safety Sense leaflet 24 'Pilot Health'*) Are you fit to fly? – Check against the 'I'm Safe" list below.

I Illness (any symptom)
M Medication (your family doctor may not know you are a pilot)
S Stress (upset following an argument?)
A Alcohol/ Drugs
F Fatigue (good night's sleep etc)
E Eating (food keeps blood-sugar correct).

b. Plan to use oxygen when flying above 10,000ft. Use it at lower altitude when flying at night or if you are a smoker (more carbon monoxide in the blood). **Do not smoke** when using oxygen.

c. If you need to wear spectacles or contact lenses for flying, make sure that the required spare pair of glasses is readily accessible.

d. Wear clothes that cover the limbs and will give some protection in the event of fire. Avoid synthetic material which melts into the skin. In winter, take additional warm clothing in case of heater failure or a forced landing.

e. Use the seat belts/harnesses provided for everyone's protection. Wear a helmet in open-cockpit aircraft.

18 PRACTICE- PREFLIGHT INSPECTION

a. Remove tie-downs, control locks, pitot cover and tow bar, then complete a thorough pre-flight inspection. Use the Check List unless you are very familiar with the aircraft.

b. Determine **visually** that you have enough fuel of the right type. If necessary, use a dip-stick to check fuel levels. Personally supervise re- fuelling. Don't let anyone confuse AVGAS and AVTUR. Make sure the filler caps are properly secured. With the fuel selector ON, check fuel drains for water and other contamination. Be aware of the danger of static electricity during re-fuelling.

c. Check engine oil level and if necessary top up with the correct grade; do not over-fill.

d. Remove **all** ice, frost, and snow from the aircraft. Even frost spoils the airflow over aerofoil surfaces resulting in loss of lift and abnormal control effects. Beware of re-freezing. Use only authorised de-icing fluids. (See SafetySense leaflet 3 'Winter Flying').

e. If you find anything with which you are unhappy, seek further advice.

f. Check **visually** that the flying control surfaces move in the correct sense in response to control inputs.

g. Properly secure any baggage so that nothing can foul the flying controls. Beware of loose items, e.g. passengers' cameras

h. The law requires you **must brief** passengers on location and use of doors, emergency exits and equipment, as well as procedures to be followed in the event of an emergency. Personally secure doors and luggage hatches. (See Safety Sense Leaflet 2 'Care of Passengers'.)

i. Confirm all seats are upright for takeoff and properly locked in place.

19 STARTING ENGINE

a. Know where to find and how to use the aircraft's fire extinguisher, as well as the location of any others in the vicinity.

b. **Never** attempt to hand swing a propeller (or allow anyone else to swing your propeller) unless you know the proper, safe procedure, and there is a suitably briefed person at the controls, the brakes are ON and/or the wheels are chocked. Check that the area behind the aircraft is clear.

c. Use a Check List which details the correct sequence for starting the engine. Make sure the brakes are ON (or chocks in place) and that avionics are OFF before starting engine(s).

20 TAKE-OFF

a. Never attempt to take-off unless you are sure the surface and length available are suitable.

b. Visually check the approach and runway are clear before lining up and taking-off.

c. Choose an acceleration check point from which you can stop if the aircraft hasn't achieved a safe speed. If you haven't reached for example 2/3 of your rotate speed by 1/3 of the way along the runway, abandon the take-off!

d. In the event of engine failure after take-off, if the runway remaining is long enough, re-land and if not, **never** attempt to turn back. Use areas ahead of you and go for the best site. It is a question of knowing your aircraft, your level of experience and practice and working out beforehand your best option at the aerodrome in use. (One day, at a safe height, and well away from the circuit, try a 180° turn at idle rpm and see how much height you lose!)

21 LOOK OUT

a. Always keep a good look-out (and listen-out) for other aircraft, particularly over radio beacons and in the vicinity of aerodromes and Visual Reference Points.

b. The most hazardous conflicts are those aircraft with the least relative movement to your own. These are the ones that are difficult to see and the ones you are most likely to hit. Beware of blind spots and move your head or the aircraft to uncover these areas. Scan effectively. (See SafetySense Leaflet 13 'Collision Avoidance'.)

c. Remember the Rules of the Air, which include flying on the right side of line features and giving way to traffic on your right.

d. If the aircraft has strobe lights, use them in the air. Especially in a crowded circuit, use landing lights as well.

e. Spend as little time as possible with your head 'in the office'.

f. If you have a transponder, select and transmit the conspicuity code 7000 with Mode C (altitude reporting) unless another is appropriate or ATC advise.

22 AIRSPACE

a. Do not enter controlled airspace unless **properly authorised**. At times, you might have to orbit and wait for permission. Keep out of Restricted Airspace including Danger Areas. Don't forget the Danger Area Crossing and Information Services.

b. Use the Lower Airspace Radar Service (LARS), available from many aerodromes, particularly on week days. It may prevent you from getting a nasty fright from military or other aircraft. (*See SafetySense Leaflet 8 'Air Traffic Services Outside Controlled Airspace'.*)

c. Radar Advisory Service (RAS) can tell you about conflicting aircraft and offer advice to avoid. Radar Information Service (RIS) can give you details of conflicting aircraft, but you have to decide if avoiding action is necessary. Make sure you know which service you are receiving. **Pilots are always responsible for their own terrain and obstacle clearance.**

d. Allocation of a transponder code does not mean that you are receiving a service.

23 EN-ROUTE

a. If you encounter deteriorating weather, turn back or divert before you are caught in cloud. A 180° turn in cloud will not be as easy as in the skills test!

b. Do not attempt to fly between lowering cloud and rising ground. Many pilots have come to grief because a lowering cloud base has forced them lower and lower into the hills. You MUST avoid 'scud running'.

c. If forced into or above cloud, do not fly below your planned Safety Altitude.

d. Don't overlook en-route checks such as FREDA – fuel, radio, engine, DI and altimeter. 'Engine' should include a carb heat check.

24 DIVERSION

a. Unless you have a valid IMC or Instrument Rating, and are flying a suitably equipped aircraft, you must remain in sight of the surface. Before take-off, make plans for a retreat or diversion to an alternative aerodrome in the event of encountering lowering cloud base or deteriorating visibility,. If cloud base lowers to your calculated minimum flying altitude, or in flight visibility drops to 3 km, carry out these plans *immediately*. Turn back before entering cloud. Don't fly above clouds unless they are widely scattered and you can remain in sight of the surface.

b. Divert to the nearest aerodrome if the periodic fuel check indicates you won't have your planned fuel reserve at destination.

c. An occasional weather check from VOLMET is always worthwhile.

25 LOST

a. If you become unsure of your position, then **tell someone**. Transmit first on your working frequency. If you have lost contact on that frequency or they cannot help you, then change to 121.5 MHz and use Training Fix, PAN or MAYDAY, whichever is appropriate (*See CAP 413 'Radiotelephony Manual'*). If you have a transponder, you may wish to select the emergency code, which is 7700. It will instantly alert a radar controller.

b. Few pilots like to admit a problem on the radio. However, if any 2 of the items below apply to you, you should call for assistance quickly, **'HELP ME'**:

H   High ground/ obstructions – are you near any?
E   Entering controlled airspace – are you close?
L   Limited experience, low time or student pilot (let them know)
P   Position uncertain, get a 'Training Fix' in good time; don't leave it too late
M   MET conditions; is the weather deteriorating?
E   Endurance – fuel remaining; is it getting short?

c. As a last resort, make an early decision to land in a field while you have the fuel and daylight to do so. Choose a field with care by making a careful reconnaissance. Do not take off again without obtaining a weather update or further advice.

26 SPEED CONTROL

a. Good airspeed control can prevent inadvertent stalling or spinning, a major killer in aviation.

b. When landing, aim for the flight handbook speed, or 1.3 times the stall speed with flap, over the threshold and reduce speed in the round-out. If it is turbulent or gusty, add a margin of, say, 5kts or half the gust factor, whichever is the greater. If your speed is high, the landing distance required is likely to be more than you calculated.

c. A spin occurs when an aircraft is 'out of balance' at the stall, so always practise keeping the ball in the centre.

d. If you have not practised slow flight for some time, get an instructor to accompany you while you do so (at a safe altitude).
e. Do not exceed the limiting speeds for your aircraft. That includes maximum manoeuvring speed Va.
f. Do not apply extreme control movements at any time.
g. In aeroplanes with fixed pitch propellers, beware of maximum rpm.

27 ENVIRONMENTAL
a. Many people don't like aircraft noise and several aerodromes are under threat of closure due to this, so it is vital to be a good neighbour.
b. Adhere to noise abatement procedures and do NOT fly over published or briefed noise sensitive areas near aerodromes..
c. Select sites for practice forced landings or aerobatics very carefully. HASELL includes 'LOCATION'.
d. When en-route, fly at a height/ power setting to minimise noise nuisance, in addition to complying with Rule 5 'Low Flying'.
e. When flying a variable pitch propeller aircraft, change pitch slowly to avoid excessive noise. When flying twins, synchronise the engines to avoid 'beats'.
f. Select engine run-up areas to minimise disturbance to people, animals etc.
g. NEVER be tempted to fly low or 'beat up' the countryside.

28 WIND & WAKE TURBULENCE
a. Know the maximum demonstrated cross-wind for the aircraft type you are flying and factor this for your experience and recency.

b. Remember, that was obtained by a test pilot! If the wind approaches what you have decided is your own limit, be ready to divert. You may retain better control on landing by not using full flap.
c. Use the 'Sixth Sense' rule to work out the cross-wind component.
   10° off runway = 1/ 6 of the wind
   20° off runway = 2/ 6 wind
   30° off runway = 3/ 6 wind etc.
d. If there is a cross-wind, the reduced head-wind component will cause the take-off and landing runs to be longer.

e. If another runway which is more into wind is available, use it (after asking Air Traffic Control if there is one). You may have to wait a few minutes to fit in with other traffic.
f. When winds or gusts exceed 66% of the aircraft's stall speed (50% for taildraggers), in general, don't go flying! If you have to, use outside assistance for taxiing such as a wing walker. Taxi very slowly when winds exceed 30% of the stall speed (unless the POH specifies otherwise) , and be VERY careful when the wind is from your rear.
g. On the ground, stay 1000ft clear of the 'blast' end of powerful aircraft.
h. Beware of wake turbulence behind heavier aircraft on take-off, during the approach or on landing. You should remain 8nm, or 4 minutes or more, behind large aircraft. Note that wake turbulence lingers when wind conditions are very light. These very powerful vortices are invisible. Heed Air Traffic warnings. (SafetySense Leaflet 15 'Wake Turbulence'.)

29 CIRCUIT PROCEDURES
a. When joining or re-joining, make your radio call early and keep radio transmissions to the point. If non-radio (or your radio has failed), know the procedures. (See CAP 413 and SafetySense Leaflet 6 'Aerodrome Sense'.)
b. Check that the change from QNH to QFE reduces the altimeter reading by the aerodrome elevation. If landing using QNH, e.g. at a strip, don't forget to add aerodrome elevation to your planned circuit height.
c. Use the correct joining procedures for your destination aerodrome. Unless otherwise published, make a standard join from the overhead (See CAP 413 & poster "Standard Overhead Join"). Check circuit height and direction. Look out for other aviation activity such as gliding, parachuting.
d. Check windsock/signals square or nearby smoke to ensure you land in the right direction. Be very sure of the wind direction and strength before committing yourself to an approach at a non-radio aerodrome.
e. Make radio calls in the circuit at the proper places and listen and look for other circuit traffic. Don't forget pre- landing checks, which are easily forgotten if you make a straight-in approach.

f. Be aware of optical illusions at unfamiliar aerodromes with sloping runway or terrain, or with very long, or very wide runways.
g. Take care at aerodromes where the runways can be confused, e.g. 02 and 20. Make sure you know whether the circuit is left- hand or right- hand, as this will determine the dead side. If in doubt – ASK.
h. In most piston engined aircraft, apply full carb heat BEFORE reducing power. You may decide to cancel carb heat at 200ft or so above the ground.

30 LANDING

a. A good landing is a result of a good approach. If your approach is bad, make an early decision to go- around. Don't try to scrape in.
b. Plan to touch down at the right speed, close to the runway threshold, unless the field length allows otherwise. Use any approach guidance (PAPI/ VASI) to cross-check your descent.

c. Go-around if not solidly 'on' in the first third of the runway, or the first quarter if the runway is wet grass. However, if the runway is very long, plan your landing to minimise runway occupancy – think of the next user.

d. Wait until you are clear of the active runway, then stop to carry out the after landing checks. Double check the lever you intend moving is the flaps and NOT the landing gear.
e. If the clearance between the propeller and the ground is small, or grass is long and hiding obstructions, be especially watchful to prevent taxiing accidents.
f. If you are changing passengers, shut down the engine. Do not do 'running changes'; propellers are very dangerous.
g. Remember, the flight isn't over until the engine(s) are shutdown and all checks completed.
h. 'Book in' and close any Flight Plan.

31 SUMMARY

- Keep in current flying practice, have an annual check-out with particular emphasis on stall recognition and asymmetric practice in twins.
- Get an aviation weather forecast.
- Prepare a thorough Route Plan using the latest charts, check on NOTAMs, Temporary Nav warnings etc.
- Know the aircraft thoroughly.
- Don't over-load the aircraft.
- Make sure the runway is long enough in the conditions.
- Over water in a single- engined aircraft, wear a lifejacket (perhaps also an immersion suit), carry an accessible life-raft.
- Pre-flight properly with special emphasis on fuel/oil contents and flying controls.
- In a single-engined aircraft, bear in mind the consequences of engine failure.
- Maintain a good look- out, scan effectively.
- If the weather deteriorates, or night approaches, make the decision to divert or return early
- Don't end up in weather outside your ability or licence privileges.
- NEVER descend below your Safety Altitude in IMC.
- Request help early if lost or have other problems, e.g. fuel shortage.
- Keep out of controlled airspace unless you have clearance.
- Make regular cruise checks including fuel contents/selection and carb heat.
- Maintain flying speed, avoid inadvertent stall/ spin, don't fly low and slow.
- Don't do anything stupid - become an old pilot, NOT a bold pilot.

Finally -
- Pilots exercising GOOD AIRMANSHIP never sit there 'doing nothing', they always think 15 to 20 miles ahead.

SAFETYSENSE LEAFLET 2b — CARE OF PASSENGERS

1 INTRODUCTION
2 PRE-FLIGHT PREPARATION
3 BEFORE BOARDING
4 ON BOARD BEFORE STARTING ENGINE(S)
5 EMERGENCIES
6 EXTRA PRECAUTIONS OVER WATER
7 PASSENGERS NEW TO FLYING IN LIGHT AIRCRAFT
8 CHILD RESTRAINTS
9 SUMMARY FOR PASSENGERS

1 INTRODUCTION

a) The Commander of an aircraft is responsible for the safety and well-being of his passengers and the law requires a pre-flight safety briefing in any UK registered aircraft. This applies to **ALL** aircraft, including gliders, balloons, microlights and helicopters, as well as 'conventional' aeroplanes.

b) Article 44 of the Air Navigation Order (ANO) 2000 requires the Commander of an aircraft registered in the UK to take all reasonable steps to ensure that before take-off all passengers are familiar with the position and method of use of emergency exits, safety belts and harnesses, lifejackets and other emergency equipment. He/she must also ensure that passengers are instructed on the actions to take in an emergency.

c) Although the guidance in this Leaflet is comprehensive and too long to be used on every flight, it is up to the pilot to decide what is appropriate on each occasion. He/she should use simple language, as some words (e.g. leading edge, trailing edge, port and starboard) may not be understood by all passengers. Remember, three quarters of the UK population have never flown.

d) Passengers in light aircraft may find it helpful to have a pre-flight discussion on the differences from larger aircraft (see para 6).

2 PRE-FLIGHT PREPARATION

The pilot must:

a) Comply with any airworthiness requirements such as having controls removed from passenger seats. Even if not required, consider this if permitted. While not a requirement, it is useful to place sick bags in easily accessible places without making it obvious to the passengers.

b) Ensure luggage is not so heavy that it adversely affects the weight and balance. The same applies to the passengers themselves. A set of scales (checked for accuracy) are useful to have available- many people are unsure of their weight and often under-estimate it. See SafetySense leaflet 9..

June 2005

c) Check that luggage is properly secured and does **not** contain hazardous items, such as:
- flammable liquids and solids, e.g. matches, fire-lighters, paint
- explosives, e.g. fireworks, toy gun caps
- magnetic materials, e.g. loudspeakers
- corrosives, e.g. acids, alkalis, car batteries.
- compressed gases, e.g. camping gas, aqualung cylinders
- active mobile telephones or other electronic devices

d) Advise passengers of the restrictions on smoking in or near the aircraft.

e) Suggest that passengers wear sensible shoes and clothing. Bare limbs are at risk and thin nylon melts if there is a fire. In winter, warm clothing should be available in case of heater failure, diversion or forced landing; walking down a Scottish hillside in winter would be no fun in shirt-sleeves and sandals!

f) Tell passengers it is best not to fly if they are unwell or even recovering from a cold.

g) **NOT** take passengers who are under the influence of alcohol (or anything worse). They could hazard the flight. Drunkenness in an aircraft is an offence under ANO Art 65.

h) Tell passengers not to distract the pilot at critical times when he/she is busy, by asking questions in the middle of radio calls, when carrying out the Vital Actions or by interrupting the pilot's navigation/monitoring of the flight. (Don't be distracted by an airsick or frightened passenger. FLY THE AIRCRAFT.)

3 BEFORE BOARDING

a) Check **personally** that external baggage doors are closed and locked, don't leave it to others.

b) Escort passengers when going to and from the aircraft.

c) Point out that propellers and helicopter rotors are extremely hazardous and should be avoided at all times, even when stationary. Rotating propellers and rotors (particularly helicopter tail rotors) may be hard to see, especially from the side or at night. The hazard may not be noticed if nearby aircraft have engines running.

d) **Always** shut down the engine(s) when passengers are boarding or leaving, avoid 'running changes', or passengers approaching the aircraft while a propeller/rotor is turning, unless they are escorted by properly briefed helpers.

e) Advise passengers that when going to and from a propeller-driven aeroplane, they must approach/depart from *behind* the wing. The only exceptions are a small number of types with pusher propellers or entry doors forward of the wing. With these aeroplanes the engine(s) must always be stopped when passengers are boarding or leaving.

f) Ensure that even if the engine is stopped passengers do not step forward off the wing leading edge towards a propeller.

g) If flying a helicopter, refer to Leaflet No 17 'Helicopter Airmanship', para 4.2, covering safe conduct of passengers with rotors running.

h) Arrange that someone is in charge of children, particularly small ones, both in flight and when going to and from the aircraft. It is safest to hold their hands.

i) Ensure everyone is aware of hazards under the wings of high-winged aircraft, e.g. struts, pitot tubes.

j) Show passengers the location of any steps or handholds, if there are wing walk-ways, show passengers where they can step to prevent holed fabric or dented skin.

k) Help passengers with external door catches and locks. A door, caught by a gust of wind, can injure passengers or pilots and cause damage to the door hinges.

l) For balloons, gliders, microlights etc explain any additional specific instructions.

4 ON BOARD BEFORE STARTING ENGINE(S)

The pilot must brief passengers so that they:
- know how to adjust and lock their seats/seat backs securely in position.
- know how to fasten, unfasten and adjust seat belts/harnesses. Strongly suggest they keep them fastened through the flight in case of unexpected turbulence etc.
- know how to unlock and open doors or canopy noting that some aircraft have a double locking system. Locks and handles should be left alone once the doors are closed. **Personally** supervise the closure and locking of doors etc, don't be rushed.

- do not obstruct the controls with objects such as cameras, handbags, knees or feet.
- do not put metallic or magnetic objects near the compass.
- switch OFF all mobile telephones and electronic devices before flight.
- do not interfere with the controls in flight.
- know how to use the headsets.
- can use the intercom, if fitted, and know how to communicate if there is no intercom.
- know where to find the sick-bags.
- know the emergency procedures detailed below.

5 EMERGENCIES

a) Before flight, the pilot **must** brief passengers on how to brace themselves if a forced landing or ditching appears likely. There are two main reasons for this:
- to reduce injury due to striking objects inside the aircraft.
- to reduce 'flailing' of the body.

b) Passengers in forward facing seats **WITHOUT** a control wheel/ stick in front of them should, if possible, be briefed to adopt the 'brace' position. The upper body should be bent forward as far as possible with the chest close to the thighs and knees and the head touching the back of the seat in front. The hands should be placed one on top of the other on top of the head with the forearms tucked in against the side of the face. Fingers should NOT be interlocked. The lower legs should be inclined aft of the vertical with the feet flat on the floor. The seat belt should be as tight as possible and low on the torso.

c) Check that front seat occupants have got their belt and upper torso restraint as tight as possible prior to impact.

d) Tell passengers to kick or force out a window if the doors or canopy cannot be opened or if the aircraft has overturned.

e) Remind rear seat passengers how to operate the seat-back release on the front seats (thus allowing rear seat passengers to vacate the aircraft).

f) Agree the order in which the aircraft should be evacuated.

g) Remind passengers that harnesses and belts should be as tight as possible and at the last minute headsets removed, unplugged and stowed.

h) Brief passengers to unlock, but not unfasten, the cabin doors/emergency exits just before landing (or ditching).

i) Make it clear that seat belts/harnesses must be kept fastened until the aircraft has stopped, undo belts, open doors and get out fast.

j) Explain that you must not leave a helicopter until the main rotor has stopped.

k) Explain the position, release method and how to use the fire extinguisher as well as the location of the first aid kit.

6 EXTRA PRECAUTIONS OVER WATER

a) Lifejackets

- Before flying over water in a single-engined aircraft, make sure that passengers are **wearing** lifejackets, know how to inflate them and how to use any ancillary items, e.g. light, whistle.
- If the aircraft is twin-engined, point out the location of lifejackets and how to put them on. If one engine stops, consider asking the passengers to put on their lifejackets - it's now a single-engined aircraft!
- Impress on your passengers that lifejackets must NOT be inflated until **outside** the aircraft.

b) Life-rafts

- The life-raft should be secured such that it cannot strike people's heads during deceleration. Make sure it is accessible in an emergency. Assign responsibility for getting the life-raft out – it's too late when the aircraft has sunk. It may be heavy, so a strong passenger should be chosen. Do not tie the life-raft to the aircraft after ditching. Passengers should know how to inflate the life-raft and what emergency equipment it contains, e.g. fluorescein dye, flares.
- Brief passengers to swim away from the aircraft before inflating the life-raft so that it cannot be holed on anything sharp. When inflated, make sure it does not blow away, leaving some or all of the passengers still in the water.

c) **Above all**, impress on your passengers not to panic. There will be a lot of water flying around, perhaps through a broken windscreen, but there is usually at least a couple of minutes to get everybody out.

d) Safety Sense Leaflet No. 21 'Ditching' contains comprehensive advice.

7 PASSENGERS NEW TO FLYING IN LIGHT AIRCRAFT

Those who are more used to package holiday jets may find a light aircraft a very different experience. No one wants an early return with a sick or frightened passenger. Chat to them beforehand about:

a) *The higher noise level*: headsets, ear defenders or cotton wool in the ears may help.

b) *Turbulence* – a light aircraft will be more affected. Don't fight it, relax and go with the motion.

c) *Pressure changes and the ears* – most light aircraft are un-pressurised and climb quite slowly so the ears automatically compensate. Plan to descend at about 300 ft per minute. However, during fast descents, holding the nose and attempting to blow with the mouth closed, will equalise the pressure. Alternatively, follow the practice of some airlines and hand out a few chewy sweets.

d) *Stall and other warning horns*. Mention horns and bells, the sudden unexpected noise on landing may startle nervous passengers.

e) *Lookout* – discuss the usefulness of extra pairs of eyes throughout the flight, particularly when joining the circuit. Agree on how passengers should attract your attention. Explain the blind spots. Tell them that high flying traffic can be ignored.

f) *Motion Sickness* – What to do if feeling unwell, but don't mention the word 'sick'. (Make sure there are sick bags handy.)

g) *Toilets* – The lack of a toilet, even in some larger twin-engined aircraft.

h) *Children* – Special care is needed so that they:

- do not touch the controls, door release etc
- keep their legs clear of the controls when sitting on a booster cushion
- keep quiet when the pilot is talking on the radio or is very busy
- tell the pilot if they see another aircraft (keeping their eyes outside helps prevent air sickness).

It helps if you:

- keep talking to them during the flight pointing out landmarks etc
- avoid turbulent or windy days so that they remember their flight with PLEASURE

8 CHILD RESTRAINTS

a. The ANO and some flight manuals have requirements about safety restraint if children under the age of 2 (or 3 in some cases) are on board. These can be fulfilled as follows.

- For children up to the age of 6 months, approved belt loops as used in commercial airliners must be carried.
- For children between 6 months and 2 years old, either these approved belt loops must be carried, or the child must be strapped into a suitable car-type safety seat as described below.
- For children between 2 and 3 years old, the child must be strapped into either a car-type safety seat as described below, or secured properly by adult seat belts.
- Children 3 years old or more must be restrained using the aircraft seat belts.

b. The safety seats referred to must
- have a well-defined shell
- be designed to allow quick securing and removal from the seat
- have a single point of release for the harness which the child cannot easily release
- secure at least the torso, lap and shoulders
- have straps at least 1" wide

c. The safety seat must be installed so that:-
- it is secured to the aircraft seat in the direction of flight with the aircraft seat belt or harness
- it does not interfere with the aircraft controls or exits
- the lower part of it does not extend unreasonably beyond the aircraft seat
- the aircraft seat belt buckle does not lie on any sub-frame member of the safety seat
- only one set of straps secures the child.

9 SUMMARY FOR PASSENGERS

Have you been told how to use:

- seats/ locking mechanism
- seat belts/ harnesses
- door and emergency exit release
- front seat- back release
- fire extinguisher
- lifejackets and life-raft if carried?

where to find the first aid kit?

and what to do:

- in a forced landing
- in a ditching?

Remember, its a LEGAL requirement for the pilot to tell you.

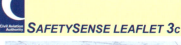

SAFETYSENSE LEAFLET 3c — WINTER FLYING

1 INTRODUCTION
2 AIRCRAFT PREPARATION
3 FLIGHT PREPARATION
4 PRE-FLIGHT
5 DEPARTURE
6 EN ROUTE
7 LANDING
8 AFTER FLIGHT
9 SUMMARY

1 INTRODUCTION

The purpose of this leaflet is to advise pilots/operators of aeroplanes, helicopters and microlights of some of the problems they may encounter while flying in winter.

2 AIRCRAFT PREPARATION

a) During the pleasant days of summer, items of equipment may have 'disappeared'. Make sure the aircraft has serviceable pitot head covers, static vent plugs, control surface locks and, if parked outside, proper tie- downs. Having made sure you have got them – **use them**.

b) Some engines may need the aircraft manufacturer's approved winter cooling restrictor to allow the oil and cylinders to reach and maintain correct operating temperatures. After fitting, keep an eye on the oil temperature/cylinder head temperature, especially if the weather turns warmer.

c) The grade of engine oil may need to be changed when operating in colder conditions. Consult the Manufacturers Manual or Maintenance Organisation.

d) Check that the cabin heater/demister is working properly before you really need to use it. A faulty cabin heater, either combustion or exhaust, can allow exhaust gases, including carbon monoxide, into the cabin. If in doubt, have the heater pressure-tested (see *Airworthiness Notices Nos. 40 and 41*). Carbon monoxide is colourless, odourless, tasteless, insidious in its effects and lethal. One of the first symptoms may be a severe headache, drowsiness or dizziness.

e) 'Spot' type carbon monoxide detectors only have a limited life when unwrapped. Use a 'fresh' one and read the instructions.

f) The pitot-static system should be checked for water which can freeze and block the system. If static drains are fitted, know where they are and how to use them.

g) The battery is worked harder in winter, so make sure it is in good condition and well charged. If you've had to make prolonged attempts to start the engine, when it does start allow plenty of time for the battery to re- charge before using heavy electrical loads. In a single-engined aircraft it's all you are left with if the electrical charging system fails in flight.

h) Some aircraft require the addition of Iso-propyl alcohol in the fuel for operation in low ambient temperatures. (See Flight Manual and, in the case of the Cessna 300/ 400 series, see *Airworthiness Notice No.8.*)

i) Check that all the airframe, propeller and windscreen systems are operating correctly. De- icing systems suffer from neglect and may prove faulty when required. Leaks may have developed in inflatable boots especially on the tailplane (due to stones thrown up by the landing gear/propellers), so check that they ALL inflate properly.

635

j) Make sure engine crankcase oil breather pipes are clear and free from deposits which can freeze, causing a pressure build-up that could force engine oil seals out of their housings.

k) Control cable tensions may need to be adjusted.

3 FLIGHT PREPARATION

a) If you are planning to visit another aerodrome, make sure it is open. Mud, snow, flooding or frozen ruts may have necessitated closure. Remember also that daylight and airport operating hours are much shorter in winter.

b) **Never** fly in icing conditions for which the aircraft is not cleared. Do not be misled into thinking that because an aircraft is fitted with de-icing, or anti-icing, equipment, it is necessarily effective in all conditions. Most general aviation aeroplanes are not cleared for flight in icing conditions, although some protection may be given. Those cleared are generally cleared only for flight in light icing conditions (the equivalent of a build-up of 12 mm (1/2 inch) of ice in 40 nautical miles). General aviation helicopters are *not* cleared. (See Pilots' Operating Handbooks, Flight Manuals, etc.)

c) Continued flight into bad weather is the number one killer in UK general aviation. Get an up to date aviation weather forecast. The current 'GET MET' booklet explains how (copies available from the Met Office).

d) The most likely temperature range for airframe icing is from 0 to −10° C; it rarely occurs at −20° C or colder (see para 6(c) for carburettor icing conditions). Pay attention to any icing warnings. Note the freezing level, it can be surprisingly low even in Spring and Autumn; you may need to descend below it to melt an ice build-up; but beware of high ground. Remember also that altimeters overread in very low air temperatures, by as much as several hundred feet. You can be lower than you think.

e) If you are likely to encounter ice en- route, have you room to descend to warmer air? Will the airspace or performance allow you to climb to cold, clear air? (Note that any ice build up may not melt and will degrade cruise performance). Can you land safely at your destination? If the answers to these questions are NO, don't go.

f) Prepare an accurate route plan with time markers, including an alternative in case you do encounter ice/snow. The countryside looks very different when covered by a blanket of snow and familiar landmarks may have disappeared.

g) Wet snow, slush or mud can seriously lengthen the take-off run or prevent take-off altogether. Check the Flight Manual and Safety Sense Leaflet 7 '*Aeroplane Performance*', and allow a generous safety margin, especially from grass.

h) Have a cloth handy for de-misting the inside of the windows while taxiing.

i) Dress sensibly, (you should spend some time outside whilst pre-flighting the aircraft), and have additional warm clothing available in case of heater failure or a forced landing.

j) Some parts of the UK will be pretty inhospitable in winter (e.g. much of Wales and Scotland) so, if you are in a single-engined aircraft, file a flight plan and carry a few survival items in case of a forced landing, e.g. warm clothing, silvered survival bag, torch/ mirror and whistle for signalling.

k) Be prepared to divert and carry a night-stop kit. Don't put pressure on yourself to get home if the weather deteriorates.

l) Read AICs 61/1999 (Pink 195) 'Risks and Factors Associated with Operations on Runways Contaminated with Snow, Slush or Water' and 106/2004 (Pink 74) 'Frost, Ice and Snow on Aircraft'. These are orientated to larger aircraft but do have useful information for General Aviation.

m) When snow has fallen, check SNOWTAMS in the NOTAM series, if available, to find out if your proposed destination, and alternate(s), are open and which operational areas have been cleared. If there is an eight digit code at the end of a METAR, it shows that winter conditions affect that aerodrome. It may be easiest to telephone them. The first two digits, of the eight digit code, are the runway and the last two the braking action. AIP, GEN para 3.5.10.13, page 3-5-30 gives further details/decode. Know the effect that braking action described as, for example POOR, will have on the landing/abandoned take-off distance you need to have available. Bear in mind the effects of a crosswind combined with an icy runway.

4 PRE-FLIGHT

a) There may be a greater risk of water condensation in aircraft fuel tanks in winter. Drain fluid from *all* water drains (there can be as many as thirteen on some single-engined aircraft). Drain it into a clear container so that you can see any water.

b) When refuelling, ensure the aircraft is properly earthed. The very low humidity on a crisp, cold day can be conducive to a build-up of static electricity.

c) After flying high such that integral wing tank fuel has been 'cold soaked', and the ambient air is humid and cool, frost will form. If it is raining, almost invisible clear ice may form.

d) Tests have shown that frost, ice or snow with the thickness and surface roughness of medium or coarse sandpaper reduces lift by as much as 30% and increases drag by 40%. Even a small area can significantly affect the airflow, particularly on a laminar flow wing.

e) Ensure that the entire aircraft is properly de-iced and check visually that all snow, ice and even frost, which can produce a severe loss of lift, is cleared. This includes difficult-to-see 'T' tails. If water has collected in a spinner or control surface and then frozen, this produces serious out-of-balance forces. There is no such thing as a little ice.

f) The most effective equipment for testing for the presence of frost and ice are your eyes and your hands.

g) The best way to remove snow is by using a broom or brush. Frozen snow, ice and frost can be removed by using approved de-icing fluid in a pressure sprayer similar to a garden sprayer. An alternative is to melt the ice with hot water and then leather the aircraft dry to prevent re-freezing. Make sure that control surface hinges, vents etc are not contaminated. A scraper might damage aircraft skins and transparencies.

h) Do not rely on snow blowing off during the take off run. The 'clean aircraft concept' is the only way to fly safely – there should be nothing on the outside of the aircraft that does not belong there.

i) Check that the pitot heater really is warming the pitot head – but don't burn your hand (use the back of it) or flatten the battery.

j) Beware of wheel fairings jammed full of mud, snow and slush – particularly mud, as it is dense and doesn't melt (on one occasion 41 kg, nearly 100 lb, of mud was removed from the three wheel fairings of a 4 seat tourer). If the fairings are removed, there may be a loss of performance and removal may invalidate the aircraft's C of A. Check that retractable gear mechanisms are not contaminated. Also, remove mud from the under-side and leading edge of wings and tail plane; it seriously affects airflow.

k) Water-soaked engine air intake filters can freeze and block the airflow.

l) If hand-swinging a propeller, perhaps because of a flat battery, move the aircraft to a part of the airfield which isn't slippery. Don't try it unless you've been trained. Use chocks and a qualified person in the cockpit.

m) During the engine run-up, check that use of carburettor heat gives a satisfactory drop in rpm or manifold pressure.

n) Check any de-icing boots, particularly the tailplane, for condition, holes etc. Wiping the boots with approved anti-icing fluid will enhance their resistance to ice build up.

5 DEPARTURE

a) Remember that taxiways and aerodrome obstructions may be hidden by snow, so ask if you are not certain.

b) Check the cabin heater/demister operation as early as possible. Be prepared to use the DV window.

c) Taxi slowly to avoid throwing up snow and slush into wheel wells or onto the aircraft's surfaces. Taxiing slowly is safer in case the tyres slide on an icy surface. Stop well clear of obstructions if there is any doubt about braking effectiveness.

d) Allow gyro instruments extra time to spin-up when they are cold.

e) You may consider using a 'Soft Field' take off technique – if so be sure that you are fully aware of recommended procedures.

f) Ensure that no carburettor ice is present prior to take-off by carrying out a 15 second carb heat check as in Safety Sense leaflet 14, both during power checks and before take-off. Ensure the engine is developing full power before taking off.

6 EN ROUTE

a) After take-off on a slushy or snowy runway, select the gear UP-DOWN-UP. This may loosen accumulated slush before it freezes the gear in the up position.

b) Monitor VOLMET and turn back or divert early if the weather deteriorates. **Don't** wait until you are in a blinding snowstorm or covered in ice.

c) Carburettor icing is one of the worst enemies. The chart shows when it is most likely to occur. (See also *Leaflet No 14 – 'Piston Engine Icing'*.)

d) Carburettor ice forms stealthily, so monitor engine instruments for loss of rpm (fixed pitch propeller) or manifold pressure (constant speed propeller), which may mean carb ice is forming.

e) Apply full carb heat periodically (every 10-15 minutes) and keep it on long enough to be effective. As a guide, carb heat should be applied for a minimum of 15 seconds, or longer if necessary. The engine may run roughly for a short period while the ice melts.

f) Use carb heat as an intermittent ON/ OFF control – either full hot or full cold. Do not use carb heat continuously or at high power settings unless the Handbook/Flight Manual allows it. At low power settings, eg descent, the application of heat **before** reducing power, and its continuous use while power is low, is recommended.

- During a descent, when using small throttle openings, with full carb heat, increase rpm periodically to warm the engine.
- Remember carb heat increases fuel consumption.
- At low rpm, use full heat but if appropriate cancel it prior to touchdown in accordance with Manual/Handbook instructions..

CARB ICING

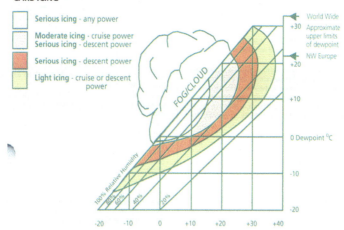

In the absence of dewpoint information assume high humidity when:

- the ground is wet (even dew)
- in precipitation or fog
- just below cloud base

g) If the aircraft has de-icing boots, it's a good idea to cycle the boots from time to time, even when ice is not expected. This prevents the valves in pneumatic systems from sticking.

h) If you are flying just above clouds to stay clear of airframe icing, remember that the cloud tops will quickly rise as you fly:
- across high ground;
- towards a warm, cold or occluded front;
- towards a low pressure area.

If you fly into the top of clouds, the concentration of water droplets is often greatest near the cloud top and ice could build up quickly.

i) Airframe Icing is most frequently encountered within convective clouds, Cumulus or Cumulonimbus (CU/CB) where the build up of ice can be very rapid. In these clouds the icing layer can be several thousand feet thick and a dramatic change of altitude will be required to avoid icing. It is better to avoid flying through these clouds if you can, either by turning back or changing your route.

Icing can also occur in thin layered clouds, especially during the winter. During the autumn, winter and spring an extensive sheet of Stratocumulus (SC) may frequently form just below a temperature inversion, with the temperature in the cloud between 0 to –10° C. Such clouds may only be one to two thousand feet deep but within the cloud layer ice may build up quickly. This icing can be avoided by descending below the cloud, provided there is sufficient height available above the ground, or by climbing above the cloud layer, but remember paragraph h.

j) If you see ice forming anywhere on the aircraft, act promptly to get out of the conditions, don't wait until the aircraft is loaded with ice. Ice forms easiest on thin edges. As the tailplane generally has a smaller leading edge radius than the wing it means that if you can see it on the wing, the tailplane (or propeller blades) will already have a heavier load. Pilots have reported that ice builds up 3 to 6 times faster on the tailplane than the wing and up to double that on a windshield wiper arm. On some aircraft the tailplane cannot be seen from the cockpit. In fact the pencil like OAT probe is often the first place ice forms. If ice does form, keep the speed up; **Don't fly too slowly.** The stall speed will have increased. The Manual/Handbook may give a minimum speed to cope with increased drag and weight due to ice build-up.

k) The stall warning system may be iced up or otherwise affected. It is in any case designed and calibrated to provide indication of wing stall, not the tailplane!

l) If you've got a big build-up of ice, the drag and weight are increasing while the climb performance is decreasing so you can't climb to get above it. High ground may prevent you from descending.

m) Tell ATC so that others can be warned.

n) Most of the time snow, which is already frozen, will not stick to an aircraft, but occasionally wet snow with a high moisture content will stick. Treat it like ice.

o) Freezing rain can occur during the winter months either at or near the ground, or in a layer above the ground. It occurs when warm moist air is moving into a cold region. The invading warm moist air may cause a layer of air, where the temperature is higher than zero° C, to overrun a much colder layer beneath where the temperatures are below zero° C. Under these conditions precipitation forming in the high cloud layers will melt to form rain as it falls through the warm air which will then fall into the sub-freezing layer beneath. This rain will quickly freeze again in the cold air forming a solid layer of clear ice over everything. This clear ice will build up very quickly and be difficult to 'shake off'.

p) Freezing rain is the most severe form of airframe icing. It can be encountered in flight up to altitudes of 10 000 feet, or it may be encountered on the ground or when flying close to the ground. Aircraft parked outside will be quickly coated with a layer of clear ice, and similarly aircraft in flight. If such conditions are encountered in flight near the ground it is best to land as soon as possible, or if the severe icing is encountered at a higher altitude descend, if possible, into a warmer layer below.

q) If you are in trouble, tell someone clearly and in good time and make sure the transponder is ON and set to code 7700. The Emergency Services can receive a transponder return much better than the primary radar return.

r) Ice forming on an aircraft can cause odd vibrations and noises. An aerial iced up may begin to vibrate (and can fall off). Don't panic, remember **AVIATE, NAVIGATE, COMMUNICATE.**

s) Monitor any autopilot, it may have been surreptitiously altering the trim to compensate, possibly, for the effect of an ice build-up.

639

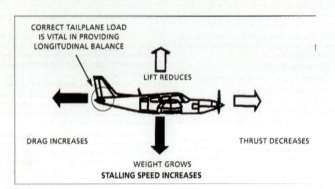

7 LANDING

a) If on arrival you descend with an iced up aeroplane and windshield and cannot see, use the DV window.

b) Most icing accidents occur when the pilot loses control during approach or landing. Even a thin coat of ice on the aircraft justifies a 20% increase in approach speed. It will extend the landing run – perhaps on a slippery runway. The handling may be different, don't make large or abrupt changes in power or flap settings.

c) If you suspect, because of changed stick forces or vibration, that there is ice on the tailplane, a flapless or partial flap landing may be advisable (the handbook/manual gives flapless-approach speeds). This reduces the tailplane load and the likelihood of tailplane stall, which can result in a VERY severe pitch down. Recovery is by REDUCING THE FLAP angle and by pulling hard – over 50 kg (110 lbs) may be necessary.

d) Another unpleasant surprise due to tailplane ice could be when the aircraft is being flown on autopilot, which has been slowly and silently re-trimming nose-up and reaches the limit. When the flaps are lowered, the autopilot could disconnect and it may require 4 strong arms to recover. Again, go for the flap selector.

e) When landing on a very wet or icy runway, particularly in a crosswind, the aircraft may aquaplane or slide and directional control can be lost. In such circumstances an alternate runway or diversion is necessary. Aircraft with castoring nosewheels may be more vulnerable.

f) Remember that ground temperatures fall quickly during the late afternoon on an exposed airfield and by dusk ice may be forming on any wet runways. The ice may form as a clear sheet which is invisible and has a coefficient of friction of zero!

g) Helicopter pilots should beware of 'white-out' due to blowing snow when hovering. (See Safety Sense Leaflet No. 17 'Helicopter Airmanship'.)

8 AFTER FLIGHT

a) Take care when getting out of the aircraft. Jumping from the aircraft walkway onto an icy apron could lead to a painful tumble.

b) If parked outside, use control locks and proper tie-downs to guard against winter gales. Face into the prevailing or forecast wind. Put proper pitot and static covers on – make sure the pitot has cooled down!

c) If it is muddy or slushy, inspect wheel fairings, landing gear bays, flaps and tailplane for loose mud or slush. These are easier to remove when soft than when frozen.

d) Notify Air Traffic if the actual weather was different, or worse, than forecasted. It might be important for other pilots to know.

9 SUMMARY

- Stay out of icing conditions for which the aircraft has NOT been cleared.
- Note freezing level in the aviation weather forecast. Don't go unless the aircraft is equipped for the conditions.
- Have warm clothing available for pre-flight and in case of heater failure or forced landing.
- Mud, snow and slush will lengthen take off and landing runs. Work out your distances.
- Remove all frost, ice and snow from the aircraft – there is no such thing as a little ice.
- Check carefully that all essential electrical services, especially pitot heat, are working properly.
- Check that the heater/demister are effective. Watch out for any signs of carbon monoxide poisoning.
- Be extra vigilant for carb ice.
- If ice does start to form, act promptly, get out of the conditions by descending (beware of high ground), climbing or diverting.
- If you encounter ice, tell ATC so that others can be warned.
- During the approach if you suspect tailplane ice, or suffer a severe pitch down, RETRACT THE FLAPS.
- If you have to land with an iced up aeroplane, add at least 20% to the approach speed.
- Snow covered, icy or muddy runways will make the landing run much longer and crosswinds harder to handle.

THERE IS NO SUCH THING AS A LITTLE ICE

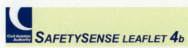

SAFETYSENSE LEAFLET 4b

USE OF MOGAS

1. INTRODUCTION
2. MOGAS SPECIFICATION & SUPPLY
3. OPERATING LIMITATIONS
4. HANDLING AND TESTING OF FUEL
5. PRE-FLIGHT AND MAINTENANCE PRECAUTIONS
6. PRE-TAKE-OFF
7. IN-FLIGHT
8. RECORDING USE OF MOGAS
9. PROBLEMS
10. SUMMARY

1 INTRODUCTION

a. Before an aircraft is granted a certificate of airworthiness or a permit to fly it must be demonstrated that the aircraft, including its engine(s), complies with the applicable airworthiness requirements. The aircraft or engine designer will normally define, by reference to a recognised specification, the fuel or fuels he is using when showing compliance. The evidence of that compliance will normally be based upon testing using the nominated fuel(s) only. Consequently, when the CAA is satisfied that compliance with the requirements has been shown and issues an approval for the aircraft or engine, that approval will be conditional upon the use of fuels conforming with the particular nominated specifications.

b. For many decades the industry-standard fuel for piston engine aircraft has been 100LL Avgas, conforming with specification DEFSTAN 91-90. This specification is comprehensive, and any changes to be made to it are subject to wide consultation and rigorous analysis before adoption. Production of Avgas is subject to stringent quality control and chemical analysis of the product. The delivery of Avgas to aerodromes is subject to procedures to protect the fuel from contamination and to maintain its quality and traceability. The Air Navigation Order (ANO) places obligations on the managers of aviation fuel installations at aerodromes and personnel carrying out refuelling, to apply procedures to maintain the quality of the Avgas. CAP434 provides guidance on these matters.

c. Compared with the Avgas specification the specifications for motor gasolines, (MOGAS), allow greater variability in the composition of the fuels, and proposed changes to the specifications themselves are not subject to the same level of scrutiny. The major oil companies consider that the systems in place for the production and delivery of Avgas are essential for a fuel which is to be used in aircraft and consequently they do not support the use of MOGAS for aviation purposes. Most major aircraft engine manufacturers, (other than a few whose engines have been developed from car engines), have aligned themselves with the oil industry and have refused to obtain approval for the use of MOGAS in their engines. However, some third-parties, mostly in the US, have provided evidence to the various Airworthiness Authorities to justify continued compliance with the airworthiness requirements for some engine and aircraft types when using certain motor gasolines.

d. Information on the types of aircraft which have been approved by the CAA to use leaded or unleaded MOGAS is given in Airworthiness Notices (ANs) 98, 98A, 98B, and 98C. This leaflet provides guidance on the use of MOGAS and does **not** override the ANs.

e. CAA AN 98 permits leaded motor gasoline, (Leaded MOGAS), to be used with certain engine/aircraft combinations **provided** that the fuel is obtained from an aerodrome aviation fuel installation in full compliance with the applicable requirements of the ANO; (equivalent to the storage and quality control procedures applied to Avgas). Therefore the permissions granted under AN 98 **exclude** the use of fuel obtained from a filling station/ garage forecourt.

f. AN 98A provides a partial exemption from the relevant Article of the Air Navigation Order to allow certain light aircraft to use Leaded MOGAS obtained from garage forecourts **subject to the conditions contained in the Notice**.

g. AN 98B provides a partial exemption from the relevant Article of the Air Navigation Order to allow microlight aeroplanes to use Unleaded MOGAS obtained from garage forecourts **subject to the conditions contained in the Notice**.

h. AN 98C provides a partial exemption from the relevant Article of the ANO to allow certain light aircraft to use Unleaded MOGAS obtained from garage forecourts **subject to the conditions contained in the Notice**.

i. It should be noted that the CAA does not accept any responsibility for any infringement of manufacturers warranties, possible accelerated deterioration of engine or airframe components or any other long term damaging effects resulting from the use of MOGAS.

2 MOGAS SPECIFICATION AND SUPPLY

a. *Leaded MOGAS*

The CAA approvals to use Leaded MOGAS apply **only** to motor gasoline conforming with BSI Specification BS4040:1988. If you use a mixture AVGAS/MOGAS with more than 25% MOGAS, it will be assumed that your aircraft is using MOGAS. Note: "Lead Replacement Petrol", "Unleaded 4-Star", and other products intended to replace leaded petrol are **not** equivalent to BS:4040 and their use is **not** approved

b. *Unleaded MOGAS*

Where approval has been given to use Unleaded MOGAS, this **must** conform with BSI Specification BS:7070 or EN228:1995. Beware of other fuels which are widely available (often advertised as having special properties).

c. Aircraft with certificates of airworthiness carrying out flights for the purposes of commercial air transport or aerial work, and multi-engine aircraft with certificates of airworthiness when flying for any purpose are not approved to use fuels obtained from garage forecourts. Such aircraft must obtain their fuels in full compliance with the specific demands of the Air Navigation Order; (Aviation Fuel at Aerodromes - Article 112 of ANO 2000).

d. Because the sampling, analysis and acceptance controls for MOGAS obtained from garage forecourts are less stringent than those for Avgas obtained at aerodromes, it is essential to ensure that the MOGAS is free from water, alcohol and other contamination. The following conditions must be met:

• The engine/airframe combination must be approved to use MOGAS, either by being listed in Schedule 1 of AN98, or by specific approval as specified in the Airworthiness Notices

• The aircraft must be either:
   o a microlight aeroplane,
   o a powered sailplane,
   o a gyroplane, or
   o a single-engine light aircraft with maximum authorised weight below 2730 kg operating on a private flight.

e. If your engine is a two-stroke don't forget to add the correct quantity of oil, or purchase premixed fuel/oil to the correct ratio.

3 OPERATING LIMITATIONS

Motor gasolines have a higher vapour pressure than Avgas and are also subject to seasonal variation. To reduce the likelihood of interruption of fuel flow to the engine due to vapour lock, the following operating limitations are imposed for all flights using MOGAS:

a. Prior to take-off, the temperature of the fuel in the aircraft tank(s) must be **less than 20 ^{0}C**

b. The aircraft must not be flown at altitudes greater than **6000 ft**, unless the CAA has agreed, in writing, to different limitations for that particular aircraft.

4 HANDLING AND TESTING OF FUEL

a. *Fuel supply*

MOGAS is more volatile than AVGAS, especially in winter; (to help coldstarting). Consequently MOGAS is more susceptible to fuel vaporisation at above average ambient temperatures, so beware of hot weather in the Spring and:

- Use freshly obtained fuel from a major supplier with a high turnover. (Note that local Regulations may only allow transportation of limited quantities in your own vehicle).

- Avoid long storage in the aircraft fuel tanks.

- Record the source of supply. (Note that most credit card receipts show the type of fuel, the quantity, and when and where it was purchased. Retention of such receipts is a means of satisfying this requirement).

b. **Testing for alcohol**
The use of fuel containing alcohol is prohibited. Alcohol may be added to motor gasolines by the oil companies in order to improve the octane rating. However, if water accumulates in the fuel tanks, or forms within them due to condensation, the alcohol may migrate from the fuel and combine with the water. This may cause loss of power in two ways:
- Firstly, the aqueous alcohol solution may be drawn into the engine in place of the fuel and cause the engine to stop.
- Secondly, the migration of the alcohol away from the fuel and into the water will lower the octane rating of the fuel. Operation using fuel of insufficient octane rating may damage the engine.
- Also, alcohol is incompatible chemically with certain rubbers and plastics used in "O" rings and seals, and with certain adhesives, sealants, pipelines, gaskets etc.
Because of these potential adverse effects MOGAS must be tested to ensure that no alcohol is present. Testing for alcohol can be carried out as follows:
1) Obtain a clear tube, (like a test tube or fuel drain device), and mark a line on it about 10% from the bottom.
2) Add water to the tube until it comes to the line. Now, fill the tube with your fuel sample until it is near the top.
3) Shake vigorously for 10-15 seconds, let it settle and if the meniscus is on the line, the fuel sample is alcohol free.
4) If it is above the line (because the alcohol has mixed with the water) alcohol is present and the fuel must not be used in an aircraft.

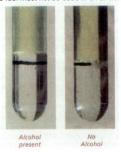

Alcohol present   No Alcohol

c. **Water and other contaminants**
Fuel must be filtered to remove water and any other contaminants. Either use a chamois and funnel or one of the proprietary devices which are available.

d. **Fuel containers**
These must be properly labelled, clean and free from corrosion etc. There is always a risk of fire when refuelling from cans due to static electrical discharge. There have been several cases of fire, including one in the UK.
Plastic fuel containers SHOULD NOT be used. The process of filling, as well as sloshing in the can during transportation, can cause an electro-static charge to build up, which then discharges as the can is brought near to the aircraft filler neck. Use a METAL container and funnel, earth them both.
Make up a proper earthing device from copper braid, heavy duty crocodile clips and a 'ground stake' so that the tank, funnel and fuel container are ALL earthed. Static charges build up most easily in dry air. The driest days in the UK can be clear, crisp days in winter.

e. **Maintenance indoors**
Refuelling and working with fuel, or on fuel systems, in enclosed areas is hazardous because of the accumulation of fuel vapour, which is heavier than air. In an incident overseas, mechanics were killed by an explosion when pouring fuel into a container attached to a wing fuel tank inside a heated hangar on a very cold (and possibly dry) day. Such work must be done outside and there must be effective electrical bonding between the aircraft, fuel source, piping or funnel and the ground.

REMEMBER - PETROL IS DANGEROUS THERE IS ALWAYS THE POSSIBILITY OF FIRE OR EXPLOSION

Note: CAP 434 'Aviation Fuel at Aerodromes' contains further information on the storage and handling of fuels.

5 PRE-FLIGHT & MAINTENANCE PRECAUTIONS

a. Non-metallic parts

Because of the different constituents of MOGAS and AVGAS, non-metallic fuel pipes and seals must be carefully inspected for signs of leakage or deterioration.

b. Water Drains

If the aircraft has been standing overnight or longer, check the drains for water. (This should be normal practice).

c. Fuel temperature

Prior to flight you should make sure that the 20 ^{0}C limitation will not be exceeded, ideally by measuring the fuel temperature; (the top of the tank will be several degrees warmer than the bottom), or by considering;

- The length of time the ambient temperature has been above 20 ^{0}C.
- Whether the aircraft has been, or will be, standing in the sunshine. (In metal aircraft with integral wing fuel tanks, white-painted wings significantly reduce the rate at which fuel temperature increases, compared with dark ones. Even in the UK a fuel temperature rise of 15^{0}C, (from 19^{0}C to 34^{0}C), in 3 hours has been measured in an aircraft with light-coloured wings and integral wing fuel tanks).
- How long it has been since the aircraft was refuelled, noting the method of fuel storage; eg underground tank or small bowser standing in the sunshine.

6 PRE-TAKE-OFF

a. Carburettor icing is more likely when using MOGAS because it has a higher volatility, (and possibly a higher water content), than AVGAS. Pay particular attention to the serviceability of carburettor heating, (if fitted). If carburettor heating is selectable, ensure that a satisfactory RPM drop is obtained when heating is selected on during pre-take-off checks. Note: If there is an increase above the original engine speed afterwards, it shows that ice was already present when heating was selected on.

b. After any prolonged period of 'heat soak' at low fuel flow, (eg during taxying and holding before take-off on a hot day), local hot spots in the engine bay may induce vapour lock in fuel pipes. Before becoming committed to taking off, ensure full power is available and can be maintained. Be particularly alert for the possibility of power loss necessitating abandonment of the take-off.

c. On certain aircraft, the front fuel tank **must** be used for take-off, initial climb and landing. This is because the tank is higher than the engine and provides a positive head of fuel thus reducing the likelihood of vapour-lock. These aircraft are listed in Notices 98, and 98C.

7 IN FLIGHT

a. *Fuel pressure*.

Pay particular attention to the fuel pressure gauge, (if fitted), and be on the alert for any signs of power loss when you switch off the electric fuel pump, (if fitted), after takeoff. For aircraft fitted with electric fuel pumps; in the event of:

- fuel pressure fluctuations,
- loss of fuel pressure,
- engine misfiring when temperature or altitude are high,

switch the pump ON immediately.

b. *Carburettor heating*

Make regular selections of full carburettor heat lasting at least 15 seconds duration; longer if your engine is particularly prone to carburettor icing.

8 RECORDING THE USE OF MOGAS

The airframe log book must be annotated such that the operating hours using MOGAS can be determined. Block records must be transferred at appropriate intervals into the engine log book(s) where applicable.

9 PROBLEMS

If you experience any problems when using MOGAS, do not hesitate to contact the CAA Safety Investigation & Data Department. Please provide as much detail as possible about the circumstances AND the source of the fuel as soon as practical after the incident.

10 SUMMARY

Do not fly using MOGAS if the fuel tank temperature is greater than 20^0C.
Do not fly above 6000 ft using MOGAS.
Only use MOGAS, (leaded or unleaded), if your aircraft/engine combination is approved to do so.
Use Leaded MOGAS conforming with BS:4040, or Unleaded MOGAS conforming with BS:7070 or EN228 as applicable.
Always use fresh fuel from a major supplier with a high turnover, (or fuel from a managed aerodrome installation).
Test for the presence of alcohol.
Filter the fuel to ensure it is free from contaminants and water.
When refuelling use metal containers and earth everything properly.
Certain aircraft must use the front fuel tank during take-off, climb, and landing; (See AN98 and 98C).
In the event of fuel pressure fluctuations or engine misfiring, switch any fuel pump on.
Be aware that carburettor icing is more likely.
Install a placard, visible to the pilot, providing the following information:

1 For *Leaded* MOGAS:

USE OF LEADED MOGAS
(See Airworthiness Notice 98A)
Use freshly obtained fuel conforming with the specification BS:4040
Test the fuel to ensure that it is free from water and alcohol.
Inspect fuel system non-metallic pipes and seals daily for deterioration and leaks.
Verify correct functioning of the carburettor heating system.
Verify take-off power prior to committing to take-off.
Fuel tank temperature not to exceed 20 degrees Celsius.
Maximum operating altitude 6000 ft.
CARBURETTOR ICING AND VAPOUR LOCK ARE MORE LIKELY WITH MOGAS

2 For *Unleaded* MOGAS:

USE OF UNLEADED MOGAS
(See Airworthiness Notice 98C)
Use freshly obtained fuel conforming with the specification EN228 or BS:7070.
Test the fuel to ensure that it is free from water and alcohol.
Inspect fuel system non-metallic pipes and seals daily for deterioration and leaks.
Verify correct functioning of the carburettor heating system.
Verify take-off power prior to committing to take-off.
Fuel tank temperature not to exceed 20 degrees Celsius.
Maximum operating altitude 6000 ft.
CARBURETTOR ICING AND VAPOUR LOCK ARE MORE LIKELY WITH MOGAS

Record the use of MOGAS in the log books.
Report any problems involving MOGAS to the CAA Safety Investigation and Data Department.

SAFETYSENSE LEAFLET 5d   VFR NAVIGATION

1 INTRODUCTION
2 THE CHARTS
3 UP-TO-DATE INFORMATION
4 PLANNING THE ROUTE
5 THE ROUTE PLAN/LOG
6 AIRBORNE
7 UNSURE OF POSITION
8 LOST
9 APPROACHING YOUR DESTINATION
10 POST FLIGHT
11 SUMMARY

1 INTRODUCTION

a) This leaflet contains useful advice for pilots of all aircraft, including balloons, gliders and microlights. It is particularly relevant to aircraft flying in UK airspace. It should be noted that Visual Flight Rules are defined in Rules 24 to 27 of the Rules of the Air. Some pilots seem to think VMC stands for Very Marginal Conditions!

b) This Leaflet should be read in conjunction with other General Aviation Safety Sense Leaflets.

2 THE CHARTS

a) The law requires, and good airmanship demands, that you **must** carry all the charts you need for your flight and for any diversion which may reasonably be expected.

b) The best 'all round' charts for VFR flight within the United Kingdom airspace are the Aeronautical Charts ICAO 1: 500,000. Their scale and degree of topographical, hydrographical, and terrain detail are suited to map reading at the speeds and altitudes commonly flown by general aviation aircraft. The chart shows aeronautical information up to and including flight level 245, and is amended frequently.

c) If you need greater detail, this is provided by the larger scale 1: 250,000 topographical charts, (e.g. major power lines are shown). Remember that Controlled Airspace with a lower limit above an altitude of 5000 ft is not shown. You should always carry a 1: 500,000 as well.

d) Instrument approach and landing charts and aerodrome charts, are published in the UK AIP (AD) for licensed aerodromes. Although not primarily intended for the VFR pilot, these charts contain a wealth of information which can make it easier to recognise, and make a good final approach to the *right* aerodrome whilst keeping you clear of 'instrument' traffic. Also, commercial Flight Guides carry many other aerodrome charts which help in identifying your destination aerodrome.

CAA Safety Sense Leaflet — 5 VFR Navigation

3 UP-TO-DATE INFORMATION

a) Confirm that each of your charts are the latest edition by referring to Green Aeronautical Information Circulars (AICs), or your Chart Agent. Note the date of the 'Validity of Aeronautical Information' at the bottom of each chart and then check the Aeronautical Information Publication (AIP), NOTAMs etc for amendments which affect your area of operation or route (e.g. a changed frequency). Chart **and frequency** updates are published on the CAA web site at www.caa.co.uk, through "airspace policy" and "aeronautical charts"

b) AIS pre-flight bulletin information is available from the AIS website on the internet at www.ais.org.uk or by fax. The system is fully described in AIC 66/2003 (White 84)*. Bulletins available include Aerodrome Information and UK Daily Nav Warnings, which can be accessed individually, in groups, or as complete bulletins. Information for VFR flight in adjacent FIRs is also available, as are the UKAIP, AICs and AIP Supplements.

4 PLANNING THE ROUTE

a) Erase all previous track lines and pencil information from the chart.

b) Draw in your intended route. Does it cross:
- a major hazard;
  Why fly in a straight line over high ground (weather hazards/ few forced landing options) when a slightly longer track could keep you over a friendly valley and well clear of cloud and other weather- related hazards? AIC 6/2003 (Pink 48)* 'Flight Over and in the Vicinity of High Ground' contains useful advice on mountain waves, turbulence etc.
- Controlled Airspace;
- an aerodrome with an active Aerodrome Traffic Zone (ATZ) – *follow rule 39 of the Rules of the Air Regulations in CAP 393);*
- an active aerodrome *without* an ATZ –
- a Prohibited, Restricted, or Danger Area;
- a Military Aerodrome Traffic Zone (MATZ);
- an ATS Advisory Route;
- the extended runway of an aerodrome with an Instrument Approach Procedure (IAP) in the open FIR, indicated by a 'cone' symbol;
- a gliding, parachuting, hang gliding, microlight or paragliding site;
- an area of intense aerial activity (AIAA)
- an air navigation obstruction;
- an high- intensity radio transmission area, a nuclear power station or gaol;
- a bird sanctuary;
- an altimeter setting region (ASR) boundary;

- a NOTAM or mauve AIC restriction, e.g. one which may apply to an air display, a military exercise or Royal low- level corridor and purple airspace (see para 5l).

Do any of these items affect you? If you are not sure, read the chart legend. Many will require a change of route; others will require prior permission, or even a positive ATC clearance to transit at certain altitudes. If your final intended track relies on weather or clearances, make sure you plan an alternate route, complete with timings and fuel.

c) Now study the topography, hydrography and terrain of the en-route area.

d) Where is the high ground? Identify the spot heights and contours and remember that the highest point en-route is often the top of an air navigation obstruction. Current charts show Maximum Elevation Figures (MEF), which are the highest known feature including terrain and obstacles in each quadrant in thousands and hundreds of feet amsl. THESE ARE NOT SAFETY ALTITUDES.

Note: Land- based obstacles up to a height of 299 ft AGL are not normally shown.

e) Do not plan to fly below 1000 ft AGL, this airspace can be heavily used by high speed military aircraft (see *Safety Sense Leaflet No. 18 'Military Low Flying'*). It also reduces options in the event of engine failure.

f) Where are the best line features? If a river, valley, railway, road, ridge or tree line is reasonably close and runs roughly parallel to the direct track, then (airspace constraints permitting and not forgetting the right-hand traffic rule, Rule 19) plan to keep it in sight. A modest increase in track distance is a small price to pay for being sure of your position. Line features at right angles to the route can be useful ETA checks.

g) How can you best pin-point your position? Look for distinctive areas of water; line features which cross one another; prominent obstructions etc. Look again and check that they will not be hidden by hills, ridges, or woods. Is there a similar pin-point nearby which could lead to confusion?

h) Large built-up areas do not make good pin-points. If you overfly them, your minimum permissible height may be dictated by engine-out height limitations (Rule 5.1). Think twice about using active aerodromes as pin-points — the smaller grass ones are often difficult to identify, and some of them will have other aircraft in the area. Do not use aerodromes with a parachuting symbol as a turning point, hard to see free-fall parachutists could be using the area. Disused aerodromes with hard runways can be useful as check points, but in some areas there are many of them, so another feature should be used as a cross-check.

i) As a landmark feature, the hard runway pattern at both active and disused aerodromes is shown on the 1: 250 000 Series, although information for disused aerodromes cannot be guaranteed.

j) The best pin-points have line features which lead you to them. Use these, wherever possible, for turning points and for any airspace entry and exit points. Because others will also choose them, it is a good idea to pass to one side (ideally right) of them. The same applies to the Visual Reference Points (VRPs) marked near Controlled Airspace and busy aerodromes; use them as references, not aiming points, although a published 'Entry Point' is just that. Unprotected Instrument Approach Procedures, indicated by 'cones' ⊶▬, do not mean that the approaches will always be to the runway with the 'cone'.

k) If you are flying to an unfamiliar aerodrome, it will be easier to spot if the sun is to one side or behind you. Arriving into sun will make it harder to see.

l) Taking all these factors — and the weather — into account, decide on your final route, altitudes and diversion aerodromes. Unless everything is 'GO', consider postponing your flight.

5 THE ROUTE PLAN/LOG

a) You should never fly a route without a written route plan, containing, at the very least:
- Magnetic headings, time/ distance marks, minimum safe VFR altitudes, freezing level and planned altitude for each leg, including that to any diversion aerodromes;
- Total distance, time, fuel to destination and diversion aerodromes;
- time available on reserve fuel;
- weather for the Route and Destination/ Diversion aerodromes;
- estimated time of arrival (ETA) at each pin-point and turning point so you can log and compare it with your actual time of arrival.

b) Have you a foolproof system for adjusting ETAs as you pass each check point? Have you marked 'Drift Lines' on the chart? These remove the guess-work if you do get off-track.

c) Have you made best use of 'ETA Check' line features? You should aim to check the ETA at a maximum of 15 minute intervals.

d) Note down your plans for your alternate routings and other en-route contingencies. You may have to remain clear of, or change your route through, Controlled Airspace; in any case you must be ready to pass entry/exit ETAs.

e) Which aerodromes do you plan to use if the weather deteriorates, your radio fails, or some mechanical failure occurs? The nearest aerodrome might not necessarily be the best, but will you have enough fuel, bearing in mind a possible headwind, to get to the one that is?

f) Have you made a note of all the contact frequencies, including parachute drop zone activity information services? Do you know which ones are on the chart? Does the aircraft equipment operate on all the frequencies you may want to use? Do you know how to select 25 kHz channels?

g) Make use of the Lower Airspace Radar Advisory Service (LARS). Brief details, including frequencies, are on the chart. Many units are military aerodromes operating only on weekdays, and a map showing the areas of coverage is in the AIP (ENR 1.6). There is a full explanation in Safety Sense Leaflet 8, 'Air Traffic Services outside Controlled Airspace'.

h) If your route penetrates a MATZ, you should plan to make contact on the MATZ frequency (it's on the chart) at least 15 nm or 5 minutes' flying time from the zone boundary. Have you planned a pin-point to help you? Details on MATZ penetrations are in SafetySense leaflet 26, and in AIC 9/2001 (Yellow 39)* Military Aerodrome Traffic Zones.

5 VFR Navigation

i) If you plan to fly over water more than 20 miles wide or over a sparsely- populated area, file a Flight Plan (see *Safety Sense Leaflet 20 VFR Flight Plans*). A Flight Plan is mandatory if leaving UK airspace. You may need to cancel that Flight Plan on arrival, or if you divert.

j) Does your aircraft have a transponder? Transmit code 7000 with Mode C at all times unless told otherwise. Be prepared to use the emergency codes if needed

k) Are there any activities which could lead to special procedures, e.g. gliding at your destination? Is there a noise sensitive area?

l) Use Freephone **0500 354 802** to check on Red Arrows displays and Emergency Restrictions. The information is updated daily and available from the evening before.

m) Finally, *check for legibility*. Does the route and all other information stand out clearly on the chart and route plan?

n) Don't forget to 'book-out' and it's a sensible idea to start with a clean wind shield.

o) If you are using a GPS to back-up your visual navigation, check and double check that you have programmed it correctly and do not use it unless you are thoroughly conversant with **all** its modes of operation. Read *SafetySense leaflet 25 "Use of GPS"*.

6 AIRBORNE

a) Air traffic are there to help you, but they are not clairvoyant. If you are permitted to do so, set heading from overhead the aerodrome. Check that you really are heading the right way from landmarks or the sun and haven't, for instance, confused zero-three with three-zero.

b) Frequency changes are best made with a landmark in sight ahead. You can then concentrate on the transmission and report your position with confidence.

c) You should try to stay in R/T contact at all times. If you use the Flight Information Service, do remember it is generally a non - radar service. If you lose contact, continue to transmit your position 'blind' at regular intervals to inform others of your presence.

d) Check your DI for precession against the magnetic compass (remember the errors inherent in the compass), try to ensure level, balanced flight when synchronising and double-check using those line features parallel to track. Don't forget a **FREDA** check every 10 minutes:

- Fuel
- Radio
- Engine instruments, carburettor heat
- DI
- Altimeter

e) Call ATC in plenty of time before entering Controlled Airspace, Danger Areas with a crossing service, MATZs and Advisory Routes. If there is any doubt about your clearance, then orbit over a chosen pin-point until clearance is positively obtained, or fly the planned alternative route around it. Many unlawful and hazardous infringements of controlled airspace occur because this advice is not heeded.

f) If you use radio nav-aids to confirm your visual observations don't forget to ident the beacon. Radio aids and GPS are to assist visual navigation, NOT substitute for it.

g) Look ahead constantly for potential weather problems. **If the weather deteriorates, don't press on — turn back or divert. Don't be lulled into a false sense of security by still being able to see blue sky. Stay within your licence privileges and your current capabilities. If necessary, carry out a forced-landing with power (see para 7) (not doing this has killed many people).**

7 UNSURE OF POSITION

a) Immediately you become unsure of your position *note the time* and if you are in touch with an ATC unit, especially a radar unit, request assistance. If not, but you think you are near Controlled Airspace, call the Distress and Diversion Cell on 121.5 MHz. Otherwise, check that the DI and compass are still synchronised. Continue to fly straight and level and on route plan heading. Then calculate a rough distance travelled since you last had a positive pin- point.

b) Now compare the outside with your estimated position. Does the general picture make sense? Look at the terrain e.g. ridges, hill lines, valleys, escarpments. Can you see a distinctive line feature e.g. motorway, dual-carriageway, railway, river? Look at the shape of the woodland. Is there a coastline visible? As a general rule work from the ground to the chart.

c) Keep checking that heading and do not relax your lookout for other aircraft.

d) If you are happy with the general picture, then narrow your sights to more specific features, but remember to up-date your estimated position regularly. Look for unique features such as a lake or reservoir, a TV mast, a road with a river or railway running alongside.

8 LOST

a) If you are still uncertain, then **TELL SOMEONE**. Transmit first on your 'working' frequency and do not mince words. Say you are LOST. If you have lost contact on that frequency, then change to 121.5 MHz and make a PAN call. If you have a transponder, set it to Code 7700. It will flash on a radar screen to alert the controller.

b) If you are lost and any of the items below apply to you, call for assistance – **'HELP ME'**:

H High ground/ obstructions – are you near any?
E Entering controlled airspace – are you close?
L Limited experience, low time or student pilot, let them know
P PAN call in good time – don't leave it too late
M Met conditions – is the weather deteriorating?
E Endurance – fuel remaining; is it getting low?

c) Transmit as much of the following information as you can:

- **PAN PAN/ PAN PAN/ PAN PAN**
- Call sign and aircraft type
- Nature of emergency
- Your intentions
- Your best estimate of position, flight level/ altitude and heading
- Whether you are a
  – student pilot
  – pilot with NO instrument qualification
  – pilot with IMC rating or full Instrument Rating
- Fuel endurance
- Transponder equipped
- Persons on board

d) The Emergency Service may be terrain-limited if you are flying below 3000 ft amsl so, if requested, climb above that altitude, but **do not** agree to climb into IMC unless you are in current practice to fly on instruments, in which case climb above Safety Altitude. Don't forget to cancel your PAN CALL when you are safe.

e) If conditions (weather, terrain, R/T) preclude safe use of the emergency service:
- maintain VFR;
- note your fuel state;
- calculate the time left to look for an area suitable for a precautionary landing.

f) Transmit your intention to make a precautionary landing and carry out the appropriate actions.

g) Give yourself plenty of time to make at least one low pass to check approaches for obstacles (e.g. aerials, cables), the surface and wind direction.

9 APPROACHING YOUR DESTINATION

a) With your destination area in sight, do not put aside your chart until you have positively identified the **correct** aerodrome (and any Visual Reference Points).

b) Select the appropriate radio frequency in plenty of time to obtain landing information. Don't forget a last **FREDA** check.

c) Check the Minimum Safe Altitude and noise sensitive areas. Note the aerodrome elevation, and remember that an ATZ extends to 2000 ft *above aerodrome level*. Check your altimeter setting and confirm that the change from QNH to QFE equals the aerodrome elevation.

d) Have you positively identified the high ground and significant obstructions within the ATZ?

e) Make sure it's the right aerodrome, plenty of pilots have got it wrong.

f) Do not just rely on the compass or DI to establish the circuit pattern. Use line features to help you to line up with the **correct** runway.

g) Unless you know of different procedures, or safety reasons or Controlled Airspace prevent it, join the circuit pattern in the standard overhead manner, as shown on the poster on the CAA web site and in LASORS. *See SafetySense leaflet 6 "Aerodrome sense"*.

CAA Safety Sense Leaflet — 5 VFR Navigation

10 POST FLIGHT

a) Were you satisfied with your navigation, or would more pre-flight preparation have helped? Study your route, tracks made good and actual timings to try to learn from the flight.

b) Are there any hints and tips which might be useful to other pilots flying that route? If so, publish them through club Newsletter, or via the CAA General Aviation Safety Information Leaflet (GASIL).

c) If you think that the chart would benefit from any change, contact the:
VFR Chart Editor
Aeronautical Charts and Data Section
CAA House K6
45– 59 Kingsway
London WC2B 6TE.
Tel: 020 7453 6572
Fax: 020 7453 6565
Helpful advice as well as your Charts can be obtained from Chart Room staff.

*The AICs referred to in this leaflet may have been superseded, please check that you are consulting the latest edition.

11 SUMMARY

- Use up to date charts
- Prepare a thorough written route plan which takes into account other airspace users, high ground etc
- If the weather deteriorates, know your safety altitude and resist any temptation to fly lower
- Plan to fly above 1000 ft agl to keep clear of military traffic
- Get an aviation weather forecast and if it turns out to be worse than predicted **KNOW WHEN TO TURN BACK OR DIVERT**
- Check NOTAMs, Bi-weekly Bulletin for latest airspace/frequency information and Freephone 0500 354802 for Royal Flights/ Red Arrows Displays
- Let someone responsible know where and when you are going, your ETA, or file a Flight Plan
- Check DI against compass at regular intervals as part of your **FREDA** check
- If you encounter bad weather, turn back, divert or land
- Use the Lower Airspace Radar Service (LARS)
- Obtain permission before entering anyone else's airspace
- Know what to do if you become lost or suffer an emergency – set the transponder to 7700
- Check when near your destination that it really is the correct aerodrome
- Fly within your licence privileges and current capability
- Don't forget to look out.

TO FAIL TO PREPARE IS TO PREPARE TO FAIL

SAFETYSENSE LEAFLET 6d - AERODROME SENSE

1 INTRODUCTION
2 BEFORE SETTING OFF
3 ARRIVAL
4 CIRCUIT PATTERN
5 AFTER LANDING
6 AFTER SHUTDOWN
7 REFUELLING
8 DEPARTURE
9 MISCELLANEOUS
10 SUMMARY

1 INTRODUCTION

This leaflet is intended to be a reminder of good sense and consideration for others which is expected of aerodrome users. It will help you to pave the way so that your visit does not cause problems for others and is at the same time pleasant for yourself and your passengers.

2 BEFORE SETTING OFF

a) Look up the aerodrome in the UK AIP (Aerodromes), [which does not include unlicensed or most government aerodromes] or in *Pooley's* or other commercial *Flight Guide*. Check on runway lengths, displaced thresholds, location of general aviation parking areas, runway lighting, local regulations, noise sensitive areas, special activities, warnings, opening hours, fuel availability etc. Also check on ATC procedures and visual reference points to save you a nasty surprise when ATC ask you to 'report when passing X' and you have no idea where X is!

b) Use the UK AIP to find all the frequencies you may need. Check NOTAMs and AIS Information Bulletins for any radio frequency changes, work in progress, change in opening hours etc. These are available on the web site www.ais.org.uk

c) If it is a 25 kHz frequency, make sure you know how to select it on the aircraft equipment.

d) Safety Sense Leaflet 12 gives comprehensive advice when using unlicensed aerodromes and private strips. Leaflet 26 gives additional guidance on the use of military aerodromes.

e) Check whether the aerodrome requires prior permission (PPR). At unlicensed aerodromes and strips this generally needs to be obtained by writing or telephoning before hand. At Licensed Aerodromes permission can normally be obtained by radio. Check on this as well as operating hours. Note that you may not be allowed to land outside promulgated operating hours.

f) Check what air traffic services are available. Air Traffic Controllers will provide instructions within the ATZ, but Flight Information Service Officers may only give instructions on the ground, and Air to Ground Communications Operators can only provide information

g) If you are non-radio or there is no air traffic service at your destination, phone to get the correct procedures, as well as the runway and altimeter setting details. Know the signal square markings.

h) There may be special procedures for helicopters or microlights.

i) Know the procedures in the event of radio failure.

j) Make sure you know about aerodrome lighting and markings. See Rules 42 to 46 of Rules of the Air or CAP 637, available free on the CAA web site www.caa.co.uk .

653

CAA Safety Sense Leaflet 6 Aerodrome Sense

3 ARRIVAL
a) Make sure that you have carried out field approach (including altimeter) checks, and have identified the correct aerodrome.

b) If an Automatic Terminal Information Service (ATIS) is provided, listen early, copy the details, and use the code letter in your initial call

c) Identify the runway in use. Beware of confusing directions by 180 degrees!

d) Check the circuit direction. Make all turns near the aerodrome (especially inside an ATZ) in that direction.

e) Identify the 'dead side' – if there is one!

f) Descend outside the circuit pattern, using the procedure illustrated here (taken from CAP 413 and also on the "standard overhead join" poster available under "Safety", "General aviation", "information", "posters" on the CAA web site) unless another is published.

g) Avoid noise sensitive areas and keep to published circuit height.

h) Consider using your landing lights, especially in poor visibility.

4 CIRCUIT PATTERN
a) Follow the pattern illustrated, if there is no different procedure published.

b) Remember wake turbulence separations if it's an airport with larger aeroplanes or helicopters (see Safety Sense Leaflet No. 15, Wake Vortex),

c) Keep radio calls brief and unambiguous. Know the non-radio procedure, look for light signals.

d) If the controller tells you to "orbit", maintain circuit height while flying turns in the circuit direction or as ordered, through 360 degrees. Allow for wind, aiming to return to the same point over the ground after every orbit.

e) Ensure you have completed your pre-landing checks – it is easy to be distracted at an unfamiliar aerodrome

f) Check you are aiming for the correct runway threshold (displaced?)

g) Be prepared (expect) to go-around, especially on the first approach to an unfamiliar runway. If you have to go-around, remember to side-step to the dead side so that you are flying parallel with the runway while able to see it.

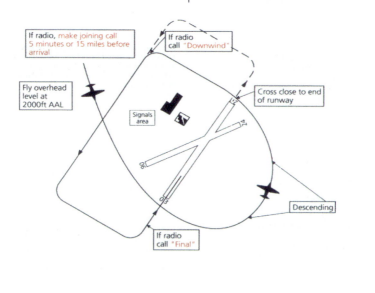

Light Signal	Meaning to Aircraft in flight	Meaning to Aircraft on Aerodrome
STEADY GREEN	Authorised to land if pilot satisfied no collision risk exists	Authorised to take-off if pilot satisfied no collision risk exists
STEADY RED	Do not land. Give way to other aircraft and continue circling	Stop
GREEN FLASHES	Return, wait for permission to land	Authorised to taxi IF pilot satisfied no collision risk exists
RED FLASHES	- Do not land - Aerodrome not available for landing	Taxi clear of landing area in use
WHITE FLASHES	Land at this aerodrome, after receiving continuous green light	Return to starting point on aerodrome

5 AFTER LANDING

a) On an aerodrome without marked runways, turn left after landing (Rule 17)

b) Taxi clear of the runway and stop before doing your after-landing checks. Before raising the flaps, check visually that you are not about to move the undercarriage selector instead!

c) If you are unsure of your route to the parking area, wait clear of the runway and call the tower for assistance via a 'Follow Me' service

d) **Never** cross an active runway without permission from the controller or FISO, or informing the Air/Ground Communications Operator: there may be more than one active runway.

e) Keep a lookout for parallel grass runways, glider strips and tow cables or parachuting areas, and have a good look before crossing any runway. If you are non-radio or the aerodrome has no Air Traffic Service, have an especially good look.

f) Look for any marshaller's signals, but remember you are still responsible for your aircraft's safety. Most common aeroplane marshalling signals are shown in this leaflet. A full list is at Rule 47 of Rules of the Air.

g) When under marshaller's control, reduce speed to a walking pace.

h) If you are flying a helicopter, do **not** land or hover near parked aeroplanes.

CAA Safety Sense Leaflet 6 Aerodrome Sense

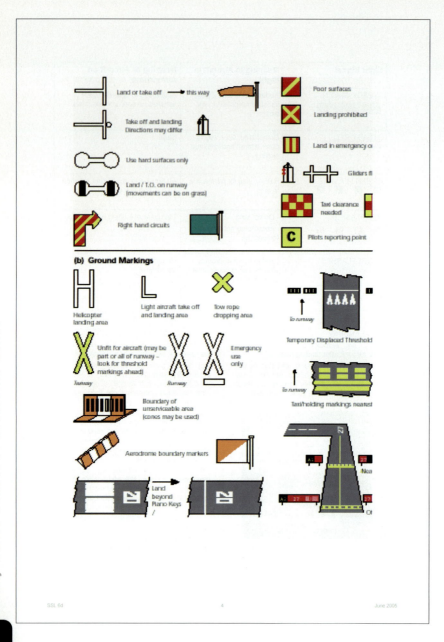

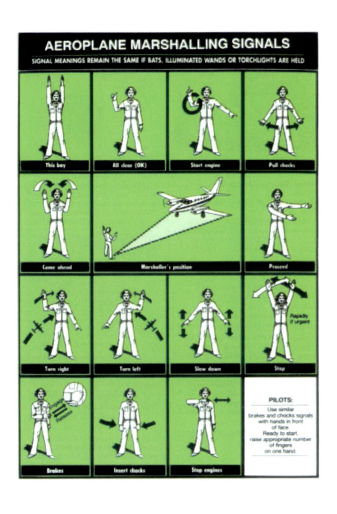

6 AFTER SHUT DOWN

Control locks are in place

a) Before leaving the aircraft ensure:
- it is parked into wind (if you can't get hangarage).
- all the electrics are off and the magnetos are safe,
- control locks are in place (another aircraft's propwash can be more severe than a strong wind),
- the parking brake is on and/or the wheels are chocked,
- pitot covers etc. are in place if you are staying for more than a couple of hours,
- it is locked, unless aerodrome personnel ask otherwise (remove or hide items which might be stolen),

b) Don't leave the aircraft in the way of others and then disappear with the key in your pocket.

c) If you are leaving your aircraft overnight or for a long time, check the weather forecast and, if necessary or in doubt, arrange for the aircraft to be tied down.

d) If you have to walk across a busy apron, keep well clear of aircraft with engines running and keep an eye on taxiing aircraft. Take particular care to escort passengers who may be in a completely strange environment. Local rules often require a high-visibility jacket.

e) Report to the building which shows a black **C** on a yellow background.

f) **Close any Flight Plan**

7 REFUELLING

a) **Always** supervise the refuelling of your aircraft because **you** are responsible for what goes into the tanks and how much. JET A1 and AVGAS mistakes are easily made, and diesel-engined types are becoming ever more popular.

Diesel-engined types

b) Ensure that earthing wires are attached before delivery begins and that the nozzle is earthed. Make sure a fire extinguisher is available.

c) After refuelling, personally check **all** filler caps and access panels for security.

d) Make sure hoses and earthing wires are wound back clear after use.

e) When you sign or pay for the fuel, double-check the invoice for the correct type/grade of fuel and **quantity**. (JET A1 in a piston-engined aircraft has been discovered at this stage.)

8 DEPARTURE

a) Don't forget to pay landing and parking fees.

b) Book out and/or let the 'tower' know your departure intentions (and if going to or from certain UK islands the Terrorism Act applies, see UK AIP GEN 1-2-1). Especially if you are non-radio, get the runway and altimeter setting details. There may be an aerodrome terminal information service (ATIS) available by telephone or on a dedicated frequency.

c) **Always** get the latest weather information, even if staying in the circuit. Allow time to obtain recorded or faxed weather information.

d) **Thoroughly** pre-flight the aircraft, making sure that no damage has occurred nor that birds have built a nest overnight. Don't forget the obvious things like pitot covers, tie-down blocks, external control locks, towing bar or baggage doors.

Check baggage doors

e) **Visually** check that your fuel has not 'disappeared' overnight. Always check fuel drains for water etc especially if parked outside in heavy rain. Water can get into the tank via worn filler cap seals.

f) at some aerodromes you must obtain permission to start engines. Before start-up, ensure that no-one is near the propeller/rotors and that the brakes are on and/or chocks in place, particularly when hand-swinging a propeller.

g) If a marshaller is standing by for start-up, give clear and unmistakable signals.

h) Never start engines in a hangar, nor immediately in front of open doors.

i) Don't use high power settings when another aircraft is parked close behind in your slipstream.

j) Switch on the red anti-collision beacon, prior to starting the engine [Rule 9(2)(b)]. Do not cause dazzle with strobes.

k) Do not taxi on the manoeuvring area without agreement from the 'tower'. If a controller or FISO is on duty, their permission is required.

l) Although aircraft have right of way over vehicles on the manoeuvring area, (except those towing aircraft) if in doubt **STOP** until the way ahead is clear.

m) When taxiing, don't just follow someone else – they might be wrong.

n) If returning to land at the aerodrome don't assume it's the same runway as when you took off – the wind may have swung round.

9 MISCELLANEOUS

a) Don't leave chocks, tie-down ropes or weights where they might be struck by other aircraft.

b) Don't drop litter or allow it to blow out of your aircraft – it could be ingested by the engines of other aircraft.

c) Comply with aerodrome warnings and signs, e.g. CRASH EXIT KEEP CLEAR.

d) Do not smoke or allow others to smoke inside hangars or near aircraft, nor on aprons or manoeuvring areas.

e) Do not taxi onto a Customs area unless you are clearing Customs.

f) If you note any obstructions, debris, pot holes, etc. on the aerodrome – **tell someone in authority at once!**

10 SUMMARY

- Before setting off, obtain aerodrome details including frequencies, reporting points, runway and taxiway layout, operating hours etc.
- If necessary, obtain permission by telephoning destination
- Call the aerodrome in good time and be ready to follow joining procedures/ reporting points.
- If no procedure is published, use the standard joining and circuit pattern.
- On arrival, make sure it is the correct runway – and aerodrome.
- If you are uncertain of your taxiing route, **STOP** and **ASK**.
- Book in and close any Flight Plan.
- Supervise re-fuelling yourself.
- **YOU** are responsible for the passengers' safety until they are in the clubroom/terminal.
- When departing, allow time to obtain weather information, file Flight Plan, book out etc.

ALL AERODROME USERS BRING HAPPINESS

SOME BY ARRIVING

, OTHERS BY DEPARTING !

SAFETYSENSE LEAFLET 7c   AEROPLANE PERFORMANCE

1   INTRODUCTION
2   WHERE TO FIND THE INFORMATION
3   USE OF PERFORMANCE DATA
4   PERFORMANCE PLANNING
5   GENERAL POINTS

6   TAKE OFF – POINTS TO NOTE
7   LANDING – POINTS TO NOTE
8   SAFETY FACTORS
9   ADDITIONAL INFORMATION
10 SUMMARY

1 INTRODUCTION

a) Accidents such as failure to get airborne, collision with obstacles after take-off and over-run on landing, occur frequently to light aeroplanes – (over 20 cases per year). Many have happened at short strips, often when operating out of wind or where there was a slope. Poor surfaces such as long or wet grass, mud or snow, were often contributory factors. These were performance accidents and many, if not all, of these accidents could have been avoided if the pilots had been fully aware of the performance limitations of their aeroplanes.

b) The pilot in command has a legal obligation under Article 38 of the Air Navigation Order, which requires the pilot to check that the aeroplane will have adequate performance for the proposed flight. The purpose of this leaflet is to remind you of the actions needed to ensure that your aeroplane's take-off, climb and landing performance will be adequate. It may not of course, be necessary before every flight. If you are using a 3000 metre runway a cursory check of performance will do, but where is the dividing line – 700, 1000 or 1500 metres? This will be decided by a large number of variables and only by reference to performance data, including climb performance, can the safety, or otherwise, of the particular flight be properly determined.

2 WHERE TO FIND THE INFORMATION

The data needed to predict the performance in the expected conditions may be in any one of the following:
- The UK Flight Manual, or for a few older aeroplanes, the Performance Schedule.
- The Pilot's Operating Handbook or the Owner's Manual. This is applicable to most light aeroplanes and sometimes contains CAA Change Sheets and/or Supplements giving additional performance data which may either supplement or override data in the main document, e.g. a 'fleet downgrade'.
- For some imported aeroplanes, an English language Flight Manual approved by the airworthiness authority in the country of origin, with a UK supplement containing the performance data approved by the CAA.

3 USE OF PERFORMANCE DATA

a) Many light aeroplanes are in performance group E, and certificated with UNFACTORED data, being the performance achieved by the manufacturer using a *new aeroplane and engine(s) in ideal conditions* flown by a highly experienced pilot. The CAA does not verify the Performance Data on all foreign aeroplanes; in some cases a single spot check is made.

b) To ensure a high level of safety on UK **Public Transport** flights, there is a *legal* requirement to apply specified safety factors to un-factored data (the result is called Net Performance Data). It is strongly recommended that those same factors be used for private flights in order to take account of:

- Your lack of practice
- Incorrect speeds/techniques
- Aeroplane and engine wear and tear
- Less than favourable conditions

c) Performance data in manuals for UK manufactured aeroplanes certificated for the purposes of Public Transport may include the Public Transport factors, (i.e. Net Performance) but manuals and handbooks for the smaller aeroplanes often do not. For foreign manufactured aeroplanes the Net Performance may be included as a Supplement. Manuals usually make it clear if factors are included but if in any doubt you should consult the CAA Safety Regulation Group (see para 9e).

d) Any 'Limitations' given in the Certificate of Airworthiness, the Flight Manual, the Performance Schedule or the Owner's Manual/Pilot's Operating Handbook are mandatory on all flights. (Note that there can be a UK Limitation contained in a Supplement which is not referred to in the text of the main document.)

e) If any advice/information given in this leaflet differs from that given in the Flight Manual, (or Pilot's Operating Handbook), then you must always comply with the manual or handbook – these are the authoritative documents.

4 PERFORMANCE PLANNING

A list of variables affecting performance together with Factors for non-Public Transport operations are shown in tabular form at the end of this leaflet. These represent the increase in take-off distance to a height of 50 feet or the increase in landing distance from 50 feet. It is intended that the tabular form will be suitable for attachment to a pilot's clipboard for easy reference. When specific Factors are given in the aeroplane's manual, handbook or supplement, they must be considered the minimum acceptable. The primary source is the Flight Manual or Pilot's Operating Handbook but cross check using this leaflet and use this where other Information is not available.

5 GENERAL POINTS

a) **Aeroplane weight:** use the actual aircraft Basic Empty weight stated on the Weight and Balance Schedule for the *individual* aeroplane you plan to fly. The weight of aeroplanes of a given type can vary considerably dependent upon the level of equipment, by as much as 77 kg (170 lb) – the "invisible passenger", for a well equipped single-engined aeroplane. Do not use the 'example weight' shown in the weight and balance section, it may be a new aeroplane with minimum equipment. Remember, on many aeroplanes it may not be possible to fill all the fuel tanks, all the seats and the baggage area. Safety Sense Leaflet 9, (Weight and Balance) provides further guidance.

b) **Aerodrome elevation:** performance deteriorates with altitude and you should use the pressure altitude at the aerodrome for calculations. (This equates to the height shown on the altimeter on the ground at the aerodrome with the sub- scale set at 1013 mb.)

c) **Slope:** an uphill slope increases the take-off ground run, and a downhill slope increases the landing distance. Any benefit arising from an upslope on landing or a downslope on take-off should be regarded as a 'bonus'. There are a few 'one way strips' where the slope is so great that in most wind conditions it is best to land up the hill and take off downhill.

d) **Temperature:** performance decreases on a hot day. On really hot days many pilots have been surprised by the loss of power in ambient temperatures of 30° C and above. Remember, temperature may be low on a summer morning but very high in the afternoon.

e) **Wind:** even a slight tailwind increases the take-off and landing distances very significantly. Note that if there is a 90° crosswind there is no beneficial headwind component and aircraft controllability may be the problem. Where data allows adjustment for wind, it is recommended that not more than 50% of the headwind component and not less than 150% of the tailwind component of the reported wind be assumed. In some manuals these factors are already included; check the relevant section.

f) **Cloudbase and visibility:** if you have to make a forced landing or fly a low-level circuit and re-land, you **MUST** be able to see obstacles and the ground. Thus, cloudbase and visibility have to be appropriate.

g) **Turbulence and windshear:** will adversely affect the performance, you must be aware of these when working out the distances needed.

h) **Surrounding terrain:** if there are hills or mountains nearby, check that you will have a rate or angle of climb sufficient to out-climb the terrain. This is particularly important if there is any wind, it may cause significant down drafts.

i) **Rain drops, mud, insects and ice:** these have a significant effect on aeroplanes, particularly those with laminar flow aerofoils. Stall speeds are increased and greater distances are required. Note that any ice, snow or frost affects all aerofoils, including the propeller and also increases the aircraft's weight – you must clear it all before flight. (AIC 104/ 98 (Pink 176) – Frost, Ice and Snow on Aircraft, refers.)

j) **Tyre pressure:** low tyre pressure (perhaps hidden by grass or wheel fairings) will increase the take- off run, as will wheel fairings jammed full of mud, grass, slush, etc.

k) **Engine failure:** since an engine failure or power loss (even on some twin-engined aircraft) may result in a forced landing, this must be borne in mind during all stages of the flight.

l) **Performance during aerobatics:** remember that variations in aeroplane weight will directly affect its performance during aerobatics (even, for example, steep turns) and outside air temperature/ altitude will similarly affect engine power available. Hot day aerobatics in a heavier than normal aeroplane require careful planning and thought.

6 TAKE OFF – POINTS TO NOTE

a) **Cross wind:** a cross wind on take off may require use of brakes to keep straight, and will increase the take off distance.

b) **Decision point:** you should work out the runway point at which you can stop the aeroplane in the event of engine or other malfunctions e.g. low engine rpm, loss of ASI, lack of acceleration or dragging brakes. Do NOT mentally programme yourself in a GO-mode to the exclusion of all else.

If the ground is soft or the grass is long and the aeroplane is still on the ground and not accelerating, stick to your decision-point and abandon take off. If the grass is wet or damp, particularly if it is very short, you will need a lot more space to stop.

c) **Twin engines:** on twin engined aircraft, if there is an engine failure after lift off, you may not reach the scheduled single engine rate of climb until:
- the landing gear and flaps have retracted (there may be a temporary degradation as the gear doors open)
- the best single engine climb speed, 'blue line speed', has been achieved.

Under limiting conditions an engine failure shortly after lift off may preclude continued flight and a forced landing will be necessary. Where the performance is marginal, the following points must be considered when deciding the best course of action:

- while flying with asymmetric power it is **vital** that airspeed is maintained comfortably above the minimum control speed, V_{MC}. A forced landing under control is infinitely preferable to the loss of directional control with the aircraft rolling inverted at low altitude. If there are signs you are losing directional control, lower the nose immediately if height permits to regain speed and if all else fails reduce power on the operating engine. (Care must be taken to maintain normal margins above the stall.)
- performance and stall speed margins will be reduced in turns. All manoeuvres must be kept to gentle turns.

KEEP IN ASYMMETRIC PRACTICE

d) **Use of available length:** make use of the full length of the runway, there is no point in turning a good length runway into a short one by doing an 'intersection' take off. On short fields use any 'starter strip'.

e) **Rolling take off:** although turning onto the runway, and applying full power without stopping can reduce the take off run, it should only be used with great care (due to landing gear side loads and directional control) and your propwash must not hazard other aircraft. If you are having to do this sort of thing, then the runway is probably TOO SHORT.

f) **Surface and slope:** grass, soft ground or snow increase rolling resistance and therefore the take- off ground run. When the ground is soft, a heavy aircraft may 'dig in' and never reach take off speed. Keeping the weight off the nosewheel or getting the tail up on a tail wheel

663

aircraft, may help. An uphill slope reduces acceleration.

For surface and slope, remember that the increases shown are the take-off and landing distances to or from a height of 50 feet. The correction to the ground run will usually be proportionally greater.

g) **Flap setting:** use the settings recommended in Pilots Handbook/Flight Manual but check for any Supplement attached to your manual/handbook. The take-off performance shown in the main part of the manual may give some flap settings which are not approved for Public Transport operations by aeroplanes on the UK Civil Aircraft register. Do not use settings which are 'folk-lore'.

h) **Humidity:** high humidity can have an adverse effect on engine performance and this is usually taken into account during certification; however there may be a correction factor applicable to your aeroplane. Check in the manual/handbook.

i) **Abandoned take-off:** Many multi-engined aeroplane manuals include data on rejected take-off distances. Some aircraft quote a minimum engine rpm that should be available during the take off run.

j) **Engine power:** check early in the take off run that engine(s) rpm/manifold pressure are correct. If they are low, abandon take off when there is plenty of room to stop. Brief use of carb heat at the hold should ensure carb ice is not forming.

7 LANDING – POINTS TO NOTE

a) When landing at places where the length is not generous, make sure that you touch down on or very close to your aiming point (beware of displaced thresholds). If you've misjudged it, make an early decision to go around if you have any doubts – don't float half way along the runway before deciding.

b) Landing on a wet surface, or snow, can result in increased ground roll, despite increased rolling resistance. This is because of the amount of braking possible is reduced, due to lack of tyre friction. Very short wet grass with a firm subsoil will be slippery and can give a 60% distance increase (1.6 factor).

c) When landing on grass the pilot cannot see or always know whether the grass is wet or covered in dew.

d) The landing distances quoted in the Pilot's Operating Handbook/Flight Manual assume the correct approach speed and technique is flown, use of higher speed will add significantly to the distance required whilst a lower speed will erode stall margins.

8 SAFETY FACTORS

a) **Take-off**

It is strongly recommended that the appropriate Public Transport factor, or one corresponding to that requirement, should be applied for all flights. For take-off this factor is x 1.33 and applies to all single engined aeroplanes and to multi-engined aeroplanes with limited performance scheduling (Group E). Manuals for aeroplanes in other Performance Groups may give factored data.

Pilots of these latter Performance Group aeroplanes and other complex types are expected to refer to the Flight Manual for specific information on all aspects of performance planning. It is therefore important to check which Performance Group your aeroplane is in.

The table at the end of this leaflet gives guidance for pilots of aeroplanes for which there is only UNFACTORED data. It is taken from AIC 67/2002 (Pink 36).

Don't forget, where several factors are relevant, they must be **multiplied**. The resulting Take-Off Distance Required to a height of 50 feet, (TODR), can become surprisingly high.

For example:

In still air, on a level dry hard runway at sea level with an ambient temperature of 10° C, an aeroplane requires a measured take-off distance to a height of 50 feet of 390m. This should be multiplied by the safety factor of 1.33 giving a TODR of 519m.

The same aeroplane in still air from a dry, short-grass strip (factor of 1.2) with a 2% uphill slope (factor of 1.1), 500 feet above sea-level (factor of 1.05) at 20° C (factor of 1.1), including the safety factor (factor of 1.33) will need TODR of:– 390 x 1.2 x 1.1 x 1.05 x 1.1 x 1.33 = **791m**

You should always ensure that, after applying all the relevant factors, including the safety factor, the TODR does not exceed the take-off distance available. If it does, you **must** offload passengers, fuel or baggage. Better a disappointed passenger than a grieving widow! Do not rely on the 'It will be alright' syndrome.

b) **Climb (and Go-around)**
In order that the aeroplane climb performance does not fall below the prescribed minimum, some manuals/handbooks quote take-off and landing weights that should not be exceeded at specific combinations of altitude and temperature ('WAT' limits). They are calculated using the pressure altitude and temperature at the relevant aerodrome.
Remember rate of climb decreases with altitude – don't allow yourself to get into a situation where the terrain outclimbs your aeroplane!

c) **Landing**
It is recommended that the Public Transport factor should be applied for all flights. For landing, this factor is x 1.43 (so that you should be able to land in 70% of the distance available).
Again when several factors are relevant, they must be multiplied. As with take-off, the total distance required may seem surprisingly high.
You should always ensure that after applying all the relevant factors, including the safety factor, the Landing Distance Required (LDR) from a height of 50 feet does not exceed Landing Distance Available.

9 ADDITIONAL INFORMATION

a) **Engine failure:** bear in mind the glide performance, miles per 1000 ft, of single-engined types and the ability to make a safe forced landing throughout the flight. Where possible, the cruise altitude should be selected accordingly.

b) **Obstacles:** it is essential to be aware of any obstacles likely to impede either the take-off or landing flight path and to ensure there is adequate performance available to clear them by a safe margin. AGA 3 section of the UK AIP includes obstacle data for a number of UK aerodromes. Excessive angles of bank shortly after take off greatly reduce rate of climb.

c) **Aerodrome distances:** for many aerodromes information on available distances is published in the Aerodrome section of the AIP or in one of the Flight Guides. At aerodromes where no published information exists, distances can be paced out. The pace length should be established accurately or assumed to be no more than 0.75 metres (2 ½ft). It is better to measure the length accurately with the aid of a rope of known length.
Slopes can be calculated if surface elevation information is available, if not they should be estimated. For example, an altitude difference of 50 ft on a 750 metre (2,500 ft) strip indicates a 2% slope.
Be sure not to mix metres and feet in your calculation and remember, for instance, that a metre is more than a yard (see Conversion Table below).
Beware of intersection take-offs, displaced runway thresholds or soft ground which may reduce the available runway length to less than the published figures. Check NOTAMs, Local Notices etc.

d) **Runway surface:** operations from strips or aerodromes covered in snow, slush or extensive standing water are inadvisable and should not be attempted without first reading AIC 126/96 (Pink 131), 'Risks and Factors Associated with Operations on Runways Contaminated with Snow, Slush or Water'. A short wait could help in the case of standing water, hail, etc.

e) **Advice:** where doubt exists on the source of data to be used or its application in given circumstances, advice should be sought from the Flight Department, Safety Regulation Group, Civil Aviation Authority, Aviation House, Gatwick Airport South, West Sussex RH6 0YR, Telephone (01293)573113 Fax (01293)573977.

Conversion Table:

1 kg	= 2·205 lb	1 lb	= 0.454 kg
1 inch	= 2.54 cm	1 cm	= 0.394 in
1 foot	= 0.305 m	1 metre	= 3.28 ft
1 Imp gal	= 4.546 litres	1 litre	= 0.22 Imp gal
1 US gal	= 3.785 litres	1 litre	= 0.264 US gal
1 Imp gal	= 1.205 US gal	1 US gal	= 0.83 Imp gal

10 SUMMARY:

	FACTORS MUST BE MULTIPLIED i.e. 1.20 x 1.35			
	TAKE-OFF		LANDING	
CONDITION	INCREASE IN TAKE-OFF DISTANCE TO HEIGHT 50 FEET	FACTOR	INCREASE IN LANDING DISTANCE FROM 50 FEET	FACTOR
A 10% increase in aeroplane weight, e.g. another passenger	20%	1.20	10%	1.10
An increase of 1,000 ft in aerodrome elevation	10%	1.10	5%	1.05
An increase of 10°C in ambient temperature	10%	1.10	5%	1.05
Dry grass* - Up to 20 cm (8 in) (on firm soil)	20%	1.20	15%⁺	1.15
Wet grass* - Up to 20 cm (8 in) (on firm soil)	30%	1.3	35%⁺ Very short grass may be slippery, distances may increase by up to 60%	1.35
Wet paved surface	-	-	15%	1.15
A 2% slope*	Uphill 10%	1.10	Downhill 10%	1.10
A tailwind component of 10% of lift-off speed	20%	1.20	20%	1.20
Soft ground or snow*	25% or more	1.25 +	25%⁺ or more	1.25 +
NOW USE ADDITIONAL SAFETY FACTORS (if data is unfactored)		**1.33**		**1.43**

Notes: 1.   * Effect on Ground Run/ Roll will be greater.

2. ⁺ For a few types of aeroplane e. g. those without brakes, grass surfaces may decrease the landing roll. However, to be on the safe side, assume the INCREASE shown until you are thoroughly conversant with the aeroplane type.

3. Any deviation from normal operating techniques is likely to result in an increased distance.

If the distance required exceeds the distance available, changes will HAVE to be made.

SAFETYSENSE LEAFLET 8d Air Traffic Services outside controlled airspace

1. INTRODUCTION
2. NON-RADAR SERVICES
3. RADAR SERVICES
4. HOW TO OBTAIN A SERVICE
5. LIMITATIONS OF RADAR SERVICE
6. TERRAIN AND OBSTACLE CLEARANCE
7. CHANGING FREQUENCY
8. AVAILABILITY OF SERVICES

1 INTRODUCTION

a) In this leaflet, 'controlled airspace' means airspace of Classes A, B, C, D and E (the UK does not use C at present, although it may be introduced at some future date). The leaflet describes only the types of air traffic service available outside controlled airspace; i.e. airspace Classes F and G – commonly known as the 'open' FIR. These types of service are as follows:

- Flight Information Service (FIS).
- Radar Information Service (RIS).
- Radar Advisory Service (RAS).
- Procedural Service.
- Alerting (i.e. Emergency) Service.

These services apply to the departure, en-route and arrival stages of your flight.

b) If you require FIS, RIS or RAS you must ask an appropriate air traffic service unit (ATSU) to provide the particular service you wish to receive. However, some ATSUs are only staffed by Flight Information Service Officers (FISO) or Air/Ground Operators. Such ATSUs will **ONLY** be able to provide an information service for their local area.

c) If the service you requested cannot be provided, for instance due to workload or equipment problems, another service may be offered.

d) Procedural Service and Alerting Service (known as Emergency Service) are provided automatically by ATSUs, (see also paras 2b and 2c). It is not necessary for you to request these services.

e) It is important that you understand the benefits and limitations of these air traffic services so that you can ask the controller for the best one to suit your needs. (Further details are given in the yellow AIC entitled 'Services to Aircraft Outside Controlled Airspace'.) (Currently 2/2001, Yellow 33.)

f) You do not have to hold an instrument rating to fly in accordance with the Instrument Flight Rules (IFR) outside controlled airspace. There is nothing mysterious about IFR within UK airspace, they are there to ensure that you have adequate clearance from ground obstacles and that you are safely separated in the vertical plane, according to your magnetic track, from other aircraft in flight. The UK regulations for IFR flights outside controlled airspace are as follows:

- Except when necessary for take off and landing, or when authorised by the appropriate authority, you must be flying at least 1000 ft above the highest obstacle within 5 nm of your aircraft. However, you may disregard this if you are flying IFR below 3000 ft and you are **clear of cloud and in sight of the surface.***

- If you are in level flight above 3000 ft amsl you MUST fly at a level appropriate to your magnetic track in accordance with the Quadrantal System.

g) There is often confusion about the IFR in relation to VMC and IMC. VMC and IMC refer to the weather conditions encountered during flight and are terms used to denote actual weather conditions, in relation to the VFR minima.

h) The VFR minima (weather conditions for flying in accordance with Visual Flight Rules) for flight outside controlled airspace are contained in the box below:

	Distance from cloud		Flight Visibility
	Horizontal	Vertical	
at and above FL 100	1500 m	1000 ft	8 km
below FL 100	1500 m	1000 ft	5 km
OR **at or below 3000 ft amsl**	clear of cloud and in sight of the surface		5 km
Aircraft, other than helicopters, flying at 140 kts IAS or less	clear of cloud and in sight of the surface		1500 m
Helicopters	clear of cloud and in sight of the surface		compatible with speed

Note that licence privileges may prevent a pilot from flying in conditions of low flight visibility.

PILOTS MUST BE AWARE OF THE MINIMUM SAFE HEIGHT AND MUST COMPLY WITH RULE 5 OF THE RULES OF THE AIR REGULATIONS (THE LOW FLYING RULES.)

2 NON- RADAR SERVICES

a) **Flight Information Service (FIS):**

This non-radar service provides information to assist with the safe and efficient conduct of your flight. The information available may include:

- Weather.
- Serviceability of navigation and approach aids.
- Conditions at aerodromes.
- Other aircraft reported in your area, which are in contact with or known to the FIS.
- Other information pertinent to flight safety.

Remember that use of FIS is not intended to replace pre-flight planning, nor is it intended to be a comprehensive source of information on the presence of other aircraft. The controller may be able to provide information on aircraft in your vicinity that have contacted him, but it is **most unlikely** that he will be aware of **all** aircraft that may affect your flight, i.e.: warnings of conflicting traffic are far less likely to be given under a FIS than under RAS or RIS. Most ATSUs can provide a FIS within their local areas. Those ATSUs which provide RAS and RIS can normally offer a FIS when conditions prevent them from providing a radar service.

b) **Procedural Service:**

This non-radar service provides separation between participating traffic and is only based on position reports. For example, it is used for IFR traffic which is carrying out pilot-interpreted approaches when radar is not available, or for aircraft flying along Advisory Routes. It may also be used when radar contact is temporarily lost with an aircraft receiving RAS in the vicinity of other participating traffic. However, on the RT, civil ATSUs do not use the term 'Procedural Service'.

c) **Alerting (Emergency) Service:**

When the controller becomes aware, or suspects, that you need Search and Rescue assistance, he will notify the appropriate organisations; this is known as an Alerting Service. It is not a service which you request – it is provided automatically. Remember, the best way of making sure that the controller realises that you have an emergency situation is to make a clear MAYDAY or PAN call, whichever is appropriate (see AIC 117/99 Pink 202 entitled *'Use of the VHF International Aeronautical Emergency Service'*).

3 RADAR SERVICES

a) **Radar Information Service (RIS):**

RIS is a radar service which aims to provide you with information on conflicting traffic, but **no** avoiding action will be offered. Hence you are responsible for maintaining separation from other aircraft. This service is tailor-made for letting you get on with your flying in VMC while the controller provides you with an extra pair of eyes. This is a very useful facility when carrying out general handling, or when flying through busy airspace where repeated avoiding action under RAS may be unnecessary and time-wasting. The controller may provide radar headings for his planning purpose or at the pilot's request. The pilot still remains responsible for separation from other aircraft and may decide not to accept the heading. However, you **must** tell the controller **before** you change level, level- band, or route. RIS may be requested under any flight rules or meteorological conditions, but in IMC it is better to obtain and use RAS (if available).

b) **Radar Advisory Service (RAS):**
This service is only available to flights operating under IFR irrespective of meteorological conditions and aims to provide you with the information and the advisory avoiding action necessary to maintain separation from other aircraft. It is the radar service used by many pilots, particularly when flying in IMC. But remember, if you are:

- not qualified to fly in IMC, or
- qualified but out-of-practice,

you **must NOT** accept an advisory turn or level change which will put you into IMC.

However, if you do not take the controller's advice, or if for any reason you cannot accept heading or level changes, you **must tell the controller**, who may be able to offer alternative avoiding action. You **must** also inform the controller before making any other changes in heading or level, because it may affect your separation from other aircraft. If you request RAS, but the controller is unable to provide that service, you may be offered RIS instead.

4 HOW TO OBTAIN A SERVICE

a) You should contact the appropriate ATSU and ask for the service you require. The controller will tell you whether your request can be met. If you don't specify the type of service, the controller may apply the best available. You can request a change in the type of service at any time. The ATSU should be given the following information:

- call sign and type of aircraft
- estimated position
- heading
- level (or level-band for traffic carrying out general handling)
- intention (next reporting/turning point, destination or general handling area)
- flight rules (IFR/VFR)
- type of service requested (RAS/ RIS).

b) Services are available from civil and military ATSUs, subject to their operating hours and controller workload. But remember that, at weekends, many military ATSUs are closed so you may not be able to obtain a RAS or RIS for some part of your route. In this case you should consider contacting the FIR controller for a FIS, or aerodromes along your route who may be able to provide a more comprehensive FIS for their local area.

NB: Remember, even when only providing a FIS, a controller may wish to identify your aircraft on radar to confirm your position – but that does **not** mean that a radar service will subsequently be provided. Furthermore, just because you have been allocated a transponder code, AND IDENTIFIED, it **does not mean that you are receiving any service.**

5 LIMITATIONS OF RADAR SERVICE

Gliders, microlights, balloons and very slow moving aircraft do not always show on radar. When they do, they are often indistinguishable from the radar returns of birds, road vehicles etc; this is an inherent limitation in radar services. It is important that you are aware of this and **maintain the best possible look-out for other aircraft** even though you are receiving RIS or RAS. When a radar service is adversely affected by other factors, e.g. weather returns on the radar, poor radar performance, high traffic density, controller workload etc it is described as 'limited'. The controller will then give you a specific warning of the situation e.g. 'You are near the base of radar cover'. You should note the warning and conduct your flight accordingly, such as by adjusting your look-out scan.

6 TERRAIN AND OBSTACLE CLEARANCE

Pilots are always responsible for providing their own **terrain and obstacle** clearance whilst flying under VFR. However, ATSUs will only provide RAS above levels/ altitudes which they consider safe.

7 CHANGING FREQUENCY

When you are in sight of your destination or wish to change to another frequency, always tell the FIR/Radar Controller that you are leaving their frequency and your subsequent intentions.

8 AVAILABILITY OF SERVICES

Any ATSU may provide the services described in this Leaflet but you should particularly note the following:

a). Lower Airspace Radar Service:

Although many ATSUs can provide RAS/RIS, those whose location makes them particularly suitable for providing radar service to transit traffic, at and below FL 95, participate in a system called the Lower Airspace Radar Service (LARS). Details are in AIC 2/2001, (Yellow 33) *Services to Aircraft Outside Controlled Airspace* and in the UK AIP ENR 1–6–4, latest chart of coverage shown in 6–1–6–3. The service is mostly available weekdays 0800 to 1700 local.

b) ATCC FIR Service:

The London and Scottish ATCCs try to provide FIS in their FIRs. Details are in the UK AIP ENR 1-1-2-1-2. You should consider a call if you have not obtained any service elsewhere. (Note that all London Flight Information Regions may be operated by one person).

c) Military Aerodrome Traffic Zone (MATZ) Penetration Service:

This is available for aerodromes which have MATZs. The Penetration Service will often include provision of RAS. Details are in the Yellow AIC listed in para 1e, in AIC 9/01 (Yellow39) *'Military Aerodrome Traffic Zones'* and in the UK AIP ENR 2– 2– 3.

d) Danger Area Services:

Nominated Service Units (see UK AIP ENR 5– 1– 3– 1 to –24 and the legend on the CAA 1:500,000 charts) provide (†) Danger Area Crossing Service (DACS) or (§) Danger Area Activity Information Service (DAAIS). MERELY OBTAINING INFORMATION UNDER DAAIS DOES **NOT** GIVE YOU A CLEARANCE TO CROSS AN ACTIVE DANGER AREA. **YOU MUST HAVE A SPECIFIC CLEARANCE.**

d) Areas of Intense Aerial Activity and Aerial Tactics Areas:

Intense civil and/or military activity takes place within these areas which are listed in ENR 5–2. Pilots of non- participating aircraft who are unable to avoid AIAAs/ATAs, must keep a good look out and call the appropriate frequency which is also shown on the 1:500,000 chart.

e) Free-fall Parachute Drop Zones:

Intense free-fall is conducted up to FL 150 at permanent drop zones (see UK AIP ENR 1–1–5–7). Activity information may be available from certain ATSUs but pilots are advised to assume a drop zone (DZ) is active if no information can be obtained. Parachute dropping aircraft and, on occasions, parachutists may be encountered outside the DZ circle shown on the chart and pilots are strongly advised to give a wide berth to all active DZs

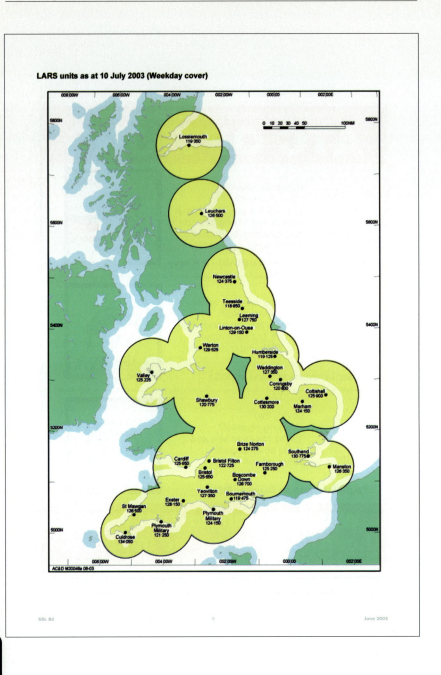

SAFETYSENSE LEAFLET 9   **WEIGHT AND BALANCE**

photo- John Thorpe

1 INTRODUCTION
2 THE LAW AND INSURANCE
3 WEIGHT
4 BALANCE (CENTRE OF GRAVITY)
5 CALCULATION
6 SUMMARY

1 INTRODUCTION

a) The principles of weight and balance should have been understood by all pilots during their initial training. It is clear that, afterwards, some forget, don't bother or are caught in 'traps' There have been several fatal accidents to UK registered general aviation aircraft in which overloading, or out-of-limits centre of gravity (cg), were contributory factors.

b) An overloaded aircraft may fail to become airborne, while out-of-limits centre of gravity seriously affects the stability and controllability. Pilots must appreciate the effects of weight and balance on the performance and handling of aircraft, *particularly* in combination with performance reducing factors, such as long or wet grass, a 'tired' engine(s), severe or un-coordinated manoeuvres, turbulence, high ambient temperatures and emergency situations (see also Safety Sense Leaflet No 7 – *Aeroplane Performance*).

c) This Leaflet is intended to remind pilots of the main points of weight and balance.

2 THE LAW AND INSURANCE

a) Article 43(d) of Air Navigation (No. 2) Order 2000 states that 'the Commander of an aircraft registered in the United Kingdom shall satisfy himself before the aircraft takes off that the load carried by the aircraft is of such weight, and is so distributed and secured, that it may safely be carried on the intended flight'. The CAA has successfully prosecuted pilots who have failed to comply with this Article.

b) In addition ANO Article 8 requires that all aircraft have a valid Certificate of Airworthiness or Permit to Fly. These documents, either directly, or by reference to a Flight Manual/Pilots Operating Handbook which forms part of a C of A, specify the weight and centre of gravity limits within which the aircraft must be operated. If these limitations are not observed, the pilot is failing to comply with a legal condition for the operation of his aircraft, thus insurers could reject any claim in the event of a mishap.

3 WEIGHT

a) The effects of overloading include:

- reduced acceleration and increased take-off speed, requiring a longer take-off run and distance to clear a 50 ft obstacle;
- decreased angle of climb reducing obstacle clearance capability after take-off;
- higher take-off speeds imposing excessive loads on the landing gear, especially if the runway is rough;
- reduced ceiling and rate of climb;
- reduced range;
- impaired manoeuvrability;
- impaired controllability;
- increased stall speeds;
- increased landing speeds, requiring a longer runway;
- reduced braking effectiveness;
- reduced structural strength margins;
- on twin-engined aircraft, failure to climb or maintain height on one engine.

Photo – John Thorpe

b) It **must** be realised, that with many four and six seat aircraft, it is **not possible** to fill all the seats, use the maximum baggage allowance, fill all the fuel tanks **and** remain within the approved weight and centre of gravity limits. You may have to reduce the number of passengers, baggage, or fuel load or possibly a combination of all three. Better that a passenger travels by bus or by train than in an ambulance!

c) The aircraft weight used in the example calculation in the Flight Manual/Pilot's Operating Handbook is for a **new** aircraft usually with little or no equipment. The weight and/or other data used in the example **MUST NOT** be used as the basis for operational weight and balance calculations. Whenever significant equipment is added a new empty weight and cg position must be provided for the Weight and Balance Schedule. This is the **only** valid source of data. You **must** use this actual equipped weight and be sure whether this includes such items as engine oil, fire extinguisher, first aid kit, life jackets, etc. The actual weight of a well equipped single engined aircraft can be as much as 170 lb (77 kg) greater than a basic aircraft – the invisible passenger! Periodic re-weighing of an aircraft is sensible – many owners have been surprised by the increase.

d) Estimating the weight of baggage can result in variations from half to double the correct weight. If there is a remote possibility of being close to the maximum take-off weight, you **must** weigh the baggage. (Pocket-sized spring balances can be obtained from fishing/hardware shops and are a handy standby if 'scales' are not available.) Note that, on some aircraft, if the maximum baggage allowance is used, restrictions are placed on rear seat occupancy. When carrying freight, check for any gross errors in the declared weight. There may also be a weight per unit area limitation on the baggage compartment floor. Make sure the baggage/freight is properly stowed and secured so that it cannot move and does not obstruct exits or emergency equipment.

e) Beware of items such as flammable substances, acids, mercury, magnetic materials, etc which are classified as Dangerous Goods with special controls that apply even in general aviation aircraft. Further assistance is available from Dangerous Goods Office, phone (01293) 573800 fax (01293) 573991.

f) Again, if the aircraft is anywhere near maximum weight, the passengers must be weighed or asked for their weight (even if it means embarrassing your spouse or friends). The risk of embarrassment is a better option than the effect of the aircraft being overweight. Remember, passengers' weight when flying is NOT their stripped weight. Allow for clothes, shoes, wallets and handbags! Check your own weight as equipped for flying and compare it with the weight you admit to.

g) Fuel gauges are often inaccurate and estimates of the weight of part filled fuel tanks should err on the high side for weight (but NOT endurance) purposes. Be careful of mixed units such as litres/lbs/kgs/Imp gallons/US gallons.

h) If a long range or extra tank(s) have been fitted, the extra fuel could add a lot to the weight. Check that the contents marked at the filler cap(s) are the same as in the Pilot's Handbook/Flight Manual or Supplement and are the ones you used for your calculations.

i) See para 4(g) on weight restrictions of Normal and Utility category.

Note:
1 kg =2.205 lb	1 lb = 0.454kg
1 inch =2.54 cm	1 cm = 0.394 inches
1 ft = 0.305 metre	1 m = 3.28 ft
1 Imp gall =4.546 litres	1 litre = 0.22 Imp gall
1 US gall = 3.785 litres	1 litre = 0.264 USG
1 Imp gall =1.205 US gall	1 USG = 0.83 Imp gall

4 BALANCE (CENTRE OF GRAVITY)

a) Balance refers to the location of the centre of gravity (cg) along the longitudinal axis of the aircraft. The cg is the point about which an aircraft would balance if it were possible to suspend it from that point. There are forward and aft limits established during certification flight testing; they are the extreme cg positions at which the longitudinal stability requirements can be met. Operation outside these limits means you would be flying in an area where the aircraft's handling has not been investigated, or is unsatisfactory. The limits for each aircraft are contained in the Pilot's Operating Handbook/Flight Manual, UK Supplement or Weight and cg Schedule referred to in 3(c). The aircraft must not be flown outside these limits.

b) The cg is measured from a datum reference, which varies from one aircraft type to another, check the Handbook/ Flight Manual. The arm is the horizontal distance (defined by the manufacturer) from the reference datum to the item of weight. The moment is the product of the weight of an item multiplied by its arm. Remember the see-saw, where a small weight at a large distance can be balanced by a large weight at a small distance.

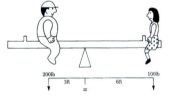

c) Exceeding the forward cg limit usually results in:
- difficulty in rotating to take-off attitude;
- increased stall or minimum flying speed against full up elevator;
- extra tail downforce requires more lift from wing resulting in greater induced drag. This means higher fuel consumption and reduced range;
- inadequate nose up trim in the landing configuration necessitating a pull force throughout the approach making it more difficult to fly a stable approach;
- difficulty in flaring and holding the nose wheel off after touch down. Many modern aircraft have deliberately restricted elevator travel (for stall behaviour reasons). Inability to hold the nose up during a bounce on landing can result in damaged nose landing gear and propeller;

- increased loads on the nose landing gear.

d) Exceeding the aft cg usually results in:
- pitch up at low speed and high power, leading to premature rotation on take-off or to inadvertent stall in the climb or during a go-around;
- on a tail wheel type, difficulty in raising the tail and in maintaining directional control on the ground;
- difficulty in trimming especially at high power;
- longitudinal instability, particularly in turbulence, with the possibility of a reversal of control forces;
- degraded stall qualities to an unknown degree;

- more difficult spin recovery, unexplored spin behaviour, delayed or even inability to recover.

e) Relatively small, but very heavy objects can make a big difference, e.g. a tool box or spare parts. Be careful where you stow them and make sure they cannot move.

f) On many aircraft the cg moves as fuel is used; on some aircraft types it could move the cg forward to beyond the forward limit when flying solo. On other types the cg moves rearward with fuel use, thus, on a loaded aircraft the cg could move to beyond the aft limit. Aft mounted long range tanks have a large effect. Careful cg calculation prior to flight will reveal any likely problems.

g) The following cg terms may be used (mainly on aircraft certificated to US regulations):
Normal category – normal flying, no spinning or aerobatic manoeuvres, bank angle may be restricted to 60°.
Utility category – manoeuvres in which bank angles exceed 60°, spinning (if permitted). No aerobatics.

h) There may be cg or weight restrictions on certain manoeuvres e.g. steep turns, spinning, aerobatics etc, imposed by the Pilot's Operating Handbook, Flight Manual or UK Supplement (e.g.: on the Socata Rallye, the rear seats must be removed to remain within the permitted cg range for spinning or aerobatics).

i) Very light (or heavy) pilots flying solo may need ballast or other measures, particularly in some homebuilt and tandem two seat aircraft.

j) Any ballast (permanent or temporary) **must** be securely fixed.

k) When parachute dropping remember the effect of the movement of parachutists prior to and immediately after dropping.

5 CALCULATION

The Pilot's Operating Handbook or Flight Manual contains a Weight and Balance section, with a worked example. The Limitations Section contains the permitted weight and cg limits. (Check to see if there are any CAA Supplements which further restrict weight or cg range.) The presentation varies from aircraft to aircraft and may be diagrammatic, graphical or tabular. You must be familiar with the method for **your** aircraft. Examples follow:

1 SAMPLE LOADING CALCULATION

	Weight (kg)	Moment* (kg.m)
1 Empty weight (includes unusable fuel, full oil and other fluids) as well as extra equipment and navaids	662	663
2 Fuel 139 litres at 0.72 kg/litre (standard tanks)	100	120
3 Pilot and front passenger	150	140
4 Rear passenger	80	150
5 Baggage or child's seat (54 kg max)	40	100
TOTAL WEIGHT AND MOMENT	1032	1173

* The moments are obtained by applying the known weights to the loading graph in item 2.

2

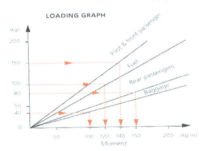
LOADING GRAPH

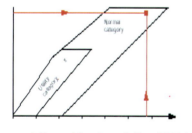

4 WEIGHT LIMITATIONS

Normal category
Maximum weight for take-off   1043 kg
Maximum weight for landing   1043 kg
Maximum weight for baggage
Or optional child seat   54 kg

Utility category
Maximum weight for take-off   907 kg
Maximum weight for landing   907 kg

FUEL CAPACITIES
2 Standard tanks of gallons)   81.5 litres   (21.5 US
Total fuel   163 litres   (43 US gallons)
Total usable fuel   152 litres   (40 US gallons)
Unusable fuel   11 litres   (3 US gallons)

In this example it can be seen that the weight is below the maximum allowed and the cg is within limits

6 SUMMARY

- Obtain **actual** (not 'typical') empty weight and cg of the **individual aircraft** you are operating from the latest Weight and Balance Schedule.
- Check that the aircraft maximum take-off weight is not exceeded. If it is, you **MUST** reduce the weight by off-loading passengers, baggage or fuel.
- Check that the cg is within limits before take-off and will remain within limits throughout the flight. If it does not stay within the approved range, you **MUST** make some changes to one or more of the following:
    - position of baggage or cargo
    - allocation of seats according to
    - passenger weight
    - fuel load and tank position
    - planned type of flight
- Before certain manoeuvres, e.g. spinning or aerobatics, check and if necessary act upon any weight or cg range restrictions.
- **DO NOT** forget the effect of weight changes on runway length requirements and safety factors given in Safety Sense Leaflet No 7 – Aeroplane Performance.
- **NEVER** consider flying an aircraft which is outside the permitted weight and cg range.

Note: Weight is used throughout this leaflet but European Regulations may refer to Mass.

SAFETYSENSE LEAFLET 10b — BIRD AVOIDANCE

1 INTRODUCTION
2 PLANNING THE FLIGHT
3 AT THE AERODROME AND IN FLIGHT
4 AFTER FLIGHT
5 SUMMARY
6 CIVIL AVIATION AUTHORITY

1 INTRODUCTION

Many pilots do not realise that if they collide with a soft feathery bird, the effect of speed turns it into a missile capable of inflicting considerable damage. This has included smashed windshields (killing pilots), blocked engine air intakes, broken pitot heads, damaged brake hoses, holed structures and helicopter tail rotor damage. Out of about 100 incidents *reported* each year by UK general aviation pilots, about 5% result in damage. The advice given in this Leaflet may provide greater awareness of the problem, and perhaps further reduce the number of collisions as well as help pilots to minimise the consequences if a bird strike does occur.

2 PLANNING THE FLIGHT

a. Check aerodrome documentation and NOTAMS (issued by some countries as BIRDTAMS) for information about permanent or seasonal bird problems at both departure and destination aerodromes.

b. Plan to fly as high as possible, only 1% of general aviation bird strikes occur above 2500 ft (although a jet airliner struck a vulture at 37,000 ft off the W. African coast!).

c. Do not fly over bird and wildlife sanctuaries detailed in UK AIP ENR 5-6-1 or marked on aeronautical charts.

d. Avoid flying along rivers or shore lines, especially at low altitude. Birds as well as pilots use these useful navigational features.

e. Note also that inland waters and shallow estuaries, even outside the breeding season, may contain large numbers of gulls, waders and wildfowl which make regular flights around dawn and dusk. In order to minimise the possibility of bird strikes and unnecessary disturbance of birds, DO NOT fly low over such areas. Note: It is an offence to deliberately disturb nesting birds, pilots have been successfully prosecuted for doing so.

f. Avoid off-shore islands, headlands, cliffs, inland waters and shallow estuaries, so as not to disturb nesting colonies.

g. Helicopters cause more disturbance to bird colonies than fixed wing aeroplanes.

h. Bear in mind that birds **do** fly at night.

i. If there are two pilots, discuss emergency procedures before departure, including those if the cockpit communications are lost.

j. Up to 80–90 kts, birds usually have time to get out of your way, but the higher the speed, the greater the chance of a strike.

k. If your flying requires lengthy periods at low level, consider wearing head protection with polycarbonate visor. Pilots' lives have been saved by their helmets, particularly in helicopters. Use goggles and a head protection during air racing.

l. In July and August the risk of a strike is at its greatest because many inexperienced young birds are present. Also, the flying abilities of adults may be impaired as they moult their flight feathers.

m. Birds of Prey have been known to attack aircraft!

3 AT THE AERODROME AND IN FLIGHT

a. In springtime, pre-flight the aircraft thoroughly as birds can build a nest almost overnight. Any signs of grass etc may necessitate further investigation of hard to inspect corners. A nest under the cowling could catch fire, or one in the tail area can restrict the flying controls.

b. As you taxi out, listen for any warnings of bird activity on the ATIS e.g. a mass release of racing pigeons.

c. While you are taxiing, look for birds on the aerodrome. Note that the most frequently struck birds, gulls, have a grey or black back which makes them hard to see on concrete or tarmac runways.

d. In general terms, the slower a bird's wing beat, the bigger the bird and the more hazardous it could be.

e. If birds are observed on the aerodrome, request aerodrome personnel to disperse them before you take-off. This is particularly important for turbo-prop and jet powered aircraft operating at aerodromes mainly used by smaller general aviation aircraft (the birds may have got used to slow aircraft).

f. Never use an aircraft to scare birds away.

g. Some aircraft have windshield heating, remember that its use, in accordance with the Pilots Operating Handbook or Flight Manual, will make the windshield more pliable and better able to withstand bird impact.

h. Use landing lights during take-off, climb, descent, approach and landing. Although there is no conclusive evidence that birds see and avoid aircraft lights, their use will make the aircraft more visible.

i. If you experience a bird strike during the take-off run, provided there is sufficient runway remaining – stop. Taxi off the runway and shut down. Inspect the intake, engine etc for damage/ingestion, or for bird remains blocking cooling or other airflow ducts. Several airline incidents have occurred where turbine engine damage or high vibration developed during subsequent flights because of undetected engine damage. Don't forget to check landing gear and brake hydraulic lines, downlocks, weight switches etc.

j. Where the take-off must be continued, with an engine problem, properly identify the affected engine and execute emergency procedures and tell the aerodrome why you are returning. It is essential to FLY THE AIRCRAFT.

k. If you see bird(s) ahead of you, and it is safe to do so, attempt to pass above them as birds usually break-away downwards when threatened. Be careful when near the ground, and never do anything that will lead to a stall or spin.

l. As you pass through a flock, or feel a strike, FLY THE AIRCRAFT. Maintain the correct speed and use whatever performance remains to reach a safe height.

m. If structural or control system damage is suspected (or the windshield is holed) consider the need for a controllability check before attempting a landing. During such a check at a safe height, do not slow down below threshold speed. Be wary of unseen helicopter tail rotor damage.

n. If the windshield is broken (or cracked), slow the aircraft to reduce wind blast, follow approved procedures (depressurise a pressurised aircraft), use sunglasses or smoke goggles to reduce the effect of wind, precipitation, or debris, but remember to fly the aircraft. Don't be distracted by the blood, feathers, smell and windblast. Small general aviation aeroplane and helicopter windshields are not required to be tested against bird impact and the propeller gives little protection. Gulls, pigeons, lapwings and even swifts can hole light aircraft windshields.

o. If dense bird concentrations are expected, avoid high-speed descent and approach. Halving the speed results in a quarter of the impact energy.

p. If flocks of birds are visible on the approach, go-around early for a second attempt, the approach may then be clear.

4 AFTER FLIGHT

a. After landing, if you have had a bird strike, check the aircraft for damage.

b. Report **all** bird strikes on the Bird Strike Report Form Freepost CA 1282, available on the CAA website at www.caa.co.uk/publications/search.asp via "general aviation" (A copy is at the back of this leaflet). Send to the address below. AIC 81/2002 (Pink 37) refers.

c. If you are unsure of the bird species send the remains (even feathers can be sufficient) for identification (marked "bird remains") to the

Bird Strike Avoidance Team,
Central Science Laboratory,
Sand Hutton,
York YO41 1LZ.

d. Photograph any damage, and send to the same address as the reporting form:

Civil Aviation Authority
Aerodrome Standards Dept
Freepost RCC1456
Crawley RH6 0YR

Fax: 01293 573971

5 SUMMARY

- Check NOTAMS/ATIS for bird activity at departure and destination aerodrome.
- Plan to fly as high as possible, most birds fly below 2500 ft. -
- Avoid bird sanctuaries and coastlines in spring. -
- Pre- flight the aircraft thoroughly, birds nests can be built (or rebuilt) in a few hours. -
- Many hazardous species are coloured such that they merge into the background. -
- If you see hazardous birds on or near runways, get aerodrome personnel to move them BEFORE you take off. -
- The higher the speed, the greater the risk and consequential damage.
- Birds usually escape by diving, so try to fly over them, but do NOT risk a stall or spin. -
- Most general aviation aircraft windshields etc are NOT required to be able to withstand bird strikes. -
- If the windshield is broken, avoid distraction – FLY THE AIRCRAFT. -
- Report ALL bird strikes using the Reporting Form CA 1282. (Photos of damage are helpful.)

If you are NOT CERTAIN of the bird species send remains to the Central Science Laboratory.

BIRDSTRIKE OCCURRENCE FORM - CA 1282 (Amended 02/2003)

To be completed on discovering evidence that a birdstrike has, or may have, occurred.
To be completed for all birdstrikes, whether or not damage has been caused.
Copies of this form should be sent as indicated at Note 1 below.

Aircraft Operator..................................
Aircraft type & series.............................
Aircraft reg. ....
Date (dd/mm/yy) ..../......./.......
Time (local) ....Hrs (24 hr)
Dawn ☐ Day ☐ Dusk ☐ Night ☐
Aerodrome ....
Runway in use ....
Height (agl) ....ft
Speed (IAS)..................kts
Position (if en route)...........................

Phase of Flight
Taxi ☐   Descent ☐
Take-off run ☐ Approach ☐
Climb ☐   Landing roll ☐
En Route ☐   Ground checks ☐

Part(s) of Aircraft   Struck damaged*
                            (describe)
Radome   ☐   ☐
Windshield   ☐   ☐
Nose (if not one of the above) ☐ ☐
Engine nos.   1   ☐   ☐
              2   ☐   ☐
              3   ☐   ☐
              4   ☐   ☐
Propeller   ☐   ☐
Wing/rotor (inc high lift devices) ☐ ☐
Fuselage   ☐   ☐
Landing Gear   ☐   ☐
Tail   ☐   ☐
Lights   ☐   ☐
Other (specify*)   ☐   ☐

Effect on flight
None ☐   Returned ☐
Aborted t/off ☐ Diverted ☐
Other ☐

Other Reports raised
Mandatory Occurrence Report (MOR) ☐
Air Safety Report (ASR) ☐
Other* (specify) ☐

Send to:
Civil Aviation Authority
Aerodrome Standards Dept
FREEPOST RCC1456
Crawley RH6 0YR

Fax No 01293 573971
Web site: www.caa.co.uk

Precipitation:
None ☐ Fog ☐ Rain ☐ Sleet/Snow ☐

Bird Species/description (e.g. Herring gull, Woodpigeon)
....

If you are not *certain* of the bird species, please send a copy of this form and any remains (e.g. a wing, but even the smallest of remains are useful) to: -

BIRDSTRIKE AVOIDANCE TEAM, CENTRAL SCIENCE LABORATORY, SAND HUTTON, YORK YO41 1LZ, UK.

Please mark the container "Bird remains"
This identification service is provided free to UK aerodromes and aircraft operators.

Bird remains sent for identification   Yes ☐ No ☐

Number of birds
          seen   struck* (enter actual number if known)
1   ☐   ☐
2-10   ☐   ☐
11-100   ☐   ☐
100+   ☐   ☐

Pilot warned of birds   Yes ☐ No ☐

Note 1: Copies of this form should be submitted as soon as practicable to the recipients shown below. (It is not necessary to wait for confirmation of bird species.)

Aerodrome ☐
Aircraft Operator ☐
Civil Aviation Authority (address overleaf) ☐
Bird Strike Avoidance Team ☐ (if identification required)

Remarks and other relevant information*:

Reporter Details
Name ....
Employer ....
Tel no .... Date ....

SAFETYSENSE LEAFLET 11 - INTERCEPTION PROCEDURES

1 INTRODUCTION
2 PROCEDURES
3 INTERCEPTING AIRCRAFT SIGNALS AND YOUR RESPONSES
4 SIGNALS INITIATED BY YOUR AIRCRAFT AND RESPONSES BY INTERCEPTING AIRCRAFT
5 COMMUNICATION
6 AFTER FLIGHT

1 INTRODUCTION

a. In order to comply with ICAO standards, the Air Navigation Order includes an item in Schedule 11 – 'Documents to be carried by aircraft registered in the UK'. This requires that on **INTERNATIONAL** flights **ALL** aircraft must carry a copy of 'Signals for Use in the Event of Interception'. These are detailed in the UK AIP ENR 1– 12. This leaflet is intended to expand on the AIP and may be carried by pilots whose international flights require details of the Signals and Procedures.

b. Under Article 9 of the Convention on International Civil Aviation, each contracting state reserves the right, for reasons of military necessity or public safety, to restrict or prohibit the aircraft from other states from flying over certain areas of its territory.

c. The regulations of a state may prescribe the need to investigate the identity of aircraft. Accordingly, it may be necessary to lead an aircraft of another nation, which has been intercepted, away from a particular area (such as a prohibited area) or the aircraft may for security reasons be required to land at a particular aerodrome.

d. In order to reduce the possibility of interception, pilots should adhere to flight plans and ATC procedures, as well as maintaining a listening watch on the appropriate ATC frequency. If details of your flight are in doubt, all possible efforts will be made to identify it through the appropriate Air Traffic Services Units.

e. As interception of civil aircraft can be potentially hazardous, interception procedures will only be used as a last resort. If you are fired upon, there is little advice that can be offered!

f. The word 'interception' does **not** include the intercept and escort service provided on request to an aircraft in distress in accordance with Search and Rescue procedures.

g. Remember, the intercepting aircraft may not be able to fly as slowly as a low speed general aviation aircraft.

684

2 PROCEDURES

If you are intercepted by another aircraft you must immediately:

- a. follow the instruction given by the intercepting aircraft, interpreting and responding to visual signals in accordance with paragraph 3,
- b. notify, if possible, the appropriate Air Traffic Services Unit,
- c. attempt to establish radio communication with the intercepting aircraft or with the appropriate intercept control unit, by making a general call on the emergency frequency 121.50 MHz, giving your identity and the nature of the flight. If no contact has been established and if UHF is fitted, repeat the call on the emergency frequency 243 MHz,
- d. select mode A, code 7700 and Mode C if equipped with a transponder, unless otherwise instructed by the appropriate Air Traffic Services Unit.

3 INTERCEPTING AIRCRAFT SIGNALS AND YOUR RESPONSES

a. **'You have been intercepted, follow me'**
Day - the intercepting aircraft rocks its wings from a position slightly above and ahead of, and normally to the left of your aircraft and, after acknowledgement, makes a slow level turn, normally to the left, onto the desired heading.
Night - same, also flashes navigation lights at irregular intervals
Note 1 – Meteorological conditions or terrain may require the intercepting aircraft to take up a position slightly above and ahead of, and to the right of your aircraft and to make the subsequent turn to the right.
Note 2 – If your aircraft can't keep pace with the intercepting aircraft, he is expected to fly a series of racetrack patterns and to rock his wings each time he passes your aircraft.

Your response to show you have understood and will comply:

Aeroplanes:

Day - rock your wings and follow him.

Night - same and in addition flash navigation lights at irregular intervals.

Helicopters:

Day or Night - rock your helicopter, flash navigation (or landing lights) at irregular intervals and follow.

Note: You must also try to communicate as in Para 5 overleaf.

CAA Safety Sense Leaflet — 11 Interception Procedures

b 'You may proceed'

Intercepting aircraft signals by day or night with an abrupt break away manoeuvre away from your aircraft consisting of a climbing turn of 90° or more without crossing the line of flight of your aircraft.

Your response to show you have understood and will comply:

Aeroplanes:

Day or Night - rock your wings.

Helicopters:

Day or Night - rock your helicopter.

CAA Safety Sense Leaflet — 11 Interception Procedures

c **'Land at this aerodrome'**

Day - the intercepting aircraft signals by circling the aerodrome, lowering his landing gear and over flying runway in direction of landing, or if your aircraft is a helicopter he signals by over flying the helicopter landing area.

Night - same and, in addition, shows steady landing lights.

Your response to show you have understood and will comply:

Aeroplanes:

Day - lower landing gear (if possible), following the intercepting aircraft and, if after over flying the runway you consider landing is safe, proceed to land.

Night - same and, in addition, show steady landing lights (if fitted).

Helicopters:

Day or Night - following the intercepting aircraft and proceeding to land showing a steady landing light (if fitted).

4 SIGNALS INITIATED BY YOUR AIRCRAFT AND RESPONSES

a. **'Aerodrome designated is inadequate for my aeroplane'**,

Day - raise landing gear (if possible), while passing over landing runway at a height exceeding 300 m (1000 ft) but not exceeding 600 m (2000 ft) above the aerodrome level, and continue to circle the aerodrome.

Night - in addition, flash landing lights while passing over landing runway as above. If unable to flash landing lights, flash any other available lights.

The intercepting aircraft responds to show he has understood.

– If it is desired that you follow him to an alternate aerodrome, he will raise his landing gear and use the signals prescribed for intercepting aircraft in paragraph 3a. However, if he has under- stood and decides you may proceed, he will use the manoeuvre prescribed in paragraph 3b.

b **'I cannot comply'**

Day or Night - switch all available lights on and off at regular intervals but in such a manner as to be distinct from flashing lights.

The intercepting aircraft responds to show he has understood by using the manoeuvre at paragraph 3b.

c **'I am in distress'**

Day or Night - flash all available lights at irregular intervals.

The intercepting aircraft responds to show he has understood by using the manoeuvre described in paragraph 3b.

5 COMMUNICATION

a. If radio contact with the intercepting aircraft is established, but communication in a common language is not possible, you should attempt to convey essential information and acknowledgement of instructions by using the following phrases and pronunciations (ICAO Annex 2, Appendix 2 and Attachment A refer):

Phrase	Pronunciation	Meaning
CALL SIGN	KOL SA- IN	My call sign is (call sign)
WILCO	VILL- CO	Understood Will comply
CAN NOT	KANN NOTT	Unable to comply
REPEAT	REE- PEET	Repeat your instruction
AM LOST	AM LOSST	Position unknown
MAYDAY	MAYDAY	I am in distress
HIJACK	HI-JACK	I have been hi-jacked
LAND (place name)	LAAND (place name)	I request to land at (place name)
DESCEND	DEE- SEND	I require descent

b. The following phrases should to be used by the intercepting aircraft in the circumstances prescribed above:

Phrase	Pronunciation	Meaning
CALL SIGN	KOL SA- IN	What is your call sign?
FOLLOW	FOL- LO	Follow me
DESCEND	DEE- SEND	Descend for landing
YOU LAND	YOU LAAND	Land at this aerodrome
PROCEED	PRO- SEED	You may proceed

c. If any instructions received by radio from other sources conflict with those given by the intercepting aircraft's visual signals or radio instructions, you must request immediate clarification while continuing to comply with the visual instructions.

d. Beware of making hand gestures, these could be misinterpreted!

6 AFTER FLIGHT

As interceptions are very rare, others may learn from your experience. Please tell the CAA's Safety Investigation and Data Department.

# SAFETYSENSE LEAFLET 12d   STRIP SENSE

1 INTRODUCTION
2 ASSESSING THE STRIP
3 OPERATIONAL CONSIDERATIONS
4 OVERNIGHT CONSIDERATIONS
5 FLYING CONSIDERATIONS
6 SETTING UP YOUR OWN STRIP
7 MAIN POINTS

1 INTRODUCTION

a. Unlicensed aerodromes and private strips are often used by pilots and private owners. They may be more convenient or cheaper than licensed aerodromes; however they do require special consideration. Approximately one third of GA Reportable Accidents in the UK occur during take off or landing at unlicensed aerodromes. The proportion of flying activity is not known.

b. This Leaflet is intended to start you thinking about the differences and particular needs of such flying, and also to give some guidelines about operating from, or establishing, your own strip. It should be read in conjunction with the relevant parts of SafetySense leaflet 6, Aerodrome Sense

2 ASSESSING THE STRIP

a. It is important to realise that the CAA criteria for the licensing of an aerodrome, e. g. clear approaches without power or other cables, no trees or obstructions close to the runway and so on, are unlikely to have been applied to the strip. Since in almost all cases **Prior Permission is Required (PPR) before landing**, your phone call should also include discussion of any difficulties, obstructions, noise sensitive areas to be avoided and the useable length of the strip.

b. Find out the arrangements for grass cutting. It is no use landing only to find the grass is so long that it prevents you taking off again. As a rule of thumb, the grass length should not be more than 30% of the diameter of the wheel.

c. Use an Ordnance Survey map to find out accurately the elevation above mean sea level of the strip – modern maps are in metres.

d. The orientation of the strip may have been laid out to fit in with the needs of agriculture. Establish the direction of the prevailing winds in the area and note the location of any windsock. Will it be affected by nearby trees or buildings? A well located windsock will give you the ground level wind speed and direction. Beware of strips near the coast; sea breezes can change rapidly from onshore to offshore, morning and evening.

e. Tell the operator of the strip what experience you have, which strips you have used recently, and what aeroplane you intend using. He has probably seen pilots with similar aeroplanes flying into and out of the strip and you can benefit from local knowledge. He does not want an accident any more than you do! Exchange telephone numbers in case of a last minute hitch. If possible visit it by road to see for yourself, but best of all carry out the advice of paras 5a, 5b and 5c.

f. The length of the strip *must* be accurately established. If you pace it out, remember an average pace is *not* one metre, but considerably less (the British army's marching pace is only 30 inches). This may decrease still further after walking several hundred metres. A proper measuring device is better; for example a rope of accurately known length.

g. The strip should be adequately drained or self-draining. Visit it after heavy rain to see whether it remains waterlogged or muddy. Rain after long dry periods may not soak away and can remain hidden by the grass.

h. The surface should be free from ruts and holes and should be properly and regularly rolled. One way of assessing the surface is to drive a car along the strip. If at about 30 mph the ride is comfortable, there should be no problems.

i. If it is a disused wartime airfield, some of the runway may be unusable, while other parts may have a surface in poor condition – including loose gravel and stones. These can be picked up by the propeller wash and can damage windscreens, tail and, of course, the propeller itself. Stone damage can be very expensive.

j. Carefully examine from the ground, air or maps the approaches to the strip and the go-around area, with particular reference to any runway slope, obstructions or hills within 5 km, windshear or turbulence from nearby woods/ buildings and other considerations.

k. Look closely at neighbouring properties; a climb out above the breeding pens or stud farm next door will soon bring an end to everyone's operation.

3 OPERATIONAL CONSIDERATIONS

a. Aeroplane performance *must* be appropriate for the proposed strip. You must be fully familiar with the contents of Safety Sense Leaflet No. 7 (Aeroplane Performance) or AIC 67/2002 (Pink 36) 'Take off, Climb and Landing Performance of Light Aeroplanes'. Remember, the figures shown in the Pilots Operating Handbook are obtained using a new aeroplane, flown by an expert pilot under near ideal conditions, i.e. the best possible results. On the strip, the grass may be different from the 'short, dry, mown grass' of the Handbook. There may be a slight uphill gradient, tall trees or cables at the far end, or a cross wind. Short wet grass should be treated with utmost caution, it can increase landing distances by 60% – it's like an icy surface! Take account of all of these most carefully and then add an additional margin for safety before deciding. (SafetySense Leaflet No. 7, Aeroplane Performance, recommends a 33% safety factor for take-off but 43% for landing.)

b. Your own abilities as a pilot need critical and honest assessment. The ability to land smoothly on a long hard runway is very different from the skills needed for this type of operation.

c. Most importantly the combination of YOU and YOUR aeroplane must be satisfactory. A weakness in either of these could show up in the accident statistics

d. The CAA poster 'AIRSTRIPS, think Hedgerow NOT Heathrow' reminds pilots of the operational considerations, and is available for free download from the CAA web site www.caa.co.uk through "safety", "general aviation" and "information".

e. Some strips are located on hills where, up to a certain wind speed, take offs are downhill and landings uphill. Re- read the above paragraphs, for although such strips are not necessarily dangerous, they should not be attempted unless you are totally confident about paragraphs a, b and c.

f. You must check that the insurance covers operation from an unlicensed aerodrome or a strip. It is important that you give Insurers fullest possible written details before the visit.

g. Find out about the local arrangements for booking in and booking out; usually a Movements Log is provided.

h. Ensure that passengers and spectators are properly briefed about where they may go, where they may stand and what they may or may not touch.

i. Leave details of route, ETA and passengers in the Movements Log **AND** with someone who will react appropriately and alert the Emergency Services if you fail to arrive/return.

j. If you are planning to go abroad direct from the strip, then nominating a 'responsible person' is even more important.. Remember customs and immigration requirements, and those of the Terrorism Act if going to or from Northern Ireland, the Isle of Man, or the Channel Islands. Consult the UK AIP GEN 1.2.1 and SafetySense Leaflet No. 20, 'VFR Flight Plans'.

4 OVERNIGHT CONSIDERATIONS

a. If you intend to leave the aircraft overnight at a strip, it may be necessary for you to arrange your own tie- downs and wheel chocks. Ensure that control locks are in place and the aircraft is properly secured. If the wind is likely to increase, then position your aircraft so as to minimise the possibility of it moving and be prepared to reposition it if the wind direction changes. Covers should be used to keep insects and water out of the pitot tube and static vents.

b. Next morning your pre- flight inspection should be more careful than usual just in case birds or other wildlife have taken up residence; birds can build a nest overnight. Check the pitot head, static and tank vents for insects.

c. If the strip is shared with cows, horses or sheep, then an electric or other suitable fence to separate them from your aeroplane is essential. Cows are very partial to the taste of aeroplane dope and their rough tongues have been known to strip fabric from wings. Metal aeroplanes do not escape their attentions, since they make suitable back-scratchers.

d. Discuss with the strip operator the security of the aeroplane. Vandalism and fuel thefts may be a problem.

5 FLYING CONSIDERATIONS

a. Consider having a familiarisation flight to and from the strip with a pilot who knows the strip and is both current on your aeroplane and operations into grass strips.

b. In any case you must know and fly the correct speeds for your aeroplane and remember the importance of using appropriate techniques, keeping the weight off the nosewheel etc.

c. If the strip is shorter than you are used to or has difficult approaches, you should arrange for a flying instructor to appraise your flying skills and revise and improve short field, soft field, general circuit and airmanship skills. It is not the intention of this leaflet to list the skills – that is the instructor's task. Listen and learn. If an instructor is not available, at least practice your short landings on a long runway before attempting to land at a short strip.

d. Airmanship and look- out must be of the highest order; there is unlikely to be any form of ATC service to advise you of the presence of other aircraft, their position or intentions, so be especially vigilant. Low flying military aircraft may NOT avoid strips.

e. Circuit practice at unlicensed aerodromes could be unpopular with the neighbours and may be in breach of part of Rule 5 of the Rules of the Air if you are within 500 ft of persons, vessels, vehicles or structures. However, if you find a problem with turbulence or crosswind, surface or slope, do not hesitate to go around in accordance with normal aviation practice.

f. Plan your circuit using the best available QNH, for example from a nearby aerodrome. Failing that you could use the most recent 'regional pressure setting (RPS)' but be aware your altimeter will certainly over-read if you use RPS. You should already know the elevation of the strip, so add this figure to the appropriate height that you would use in a normal circuit. Thus, if the strip is 250 ft amsl, downwind will be e.g. 1250 ft QNH.

g. Get into the habit of flying a compact circuit using engine and propeller handling techniques that will minimise noise disturbance. Avoid long flat and noisy approaches, these are not conducive to good neighbourliness nor necessarily the best short landing technique. If your approach is bad, make an early decision to go-around. It is often useful to plan to make a go-around from your first approach (avoiding persons, vessels vehicles and structures by 500 feet).

h. Note carefully the position and height of any obstructions on the approach especially hard- to- see local power and phone cables. Make sure that you can clear them (and any crop) by an adequate margin, and provided that you maintain this clearance, always aim to touch down close to the threshold – not halfway down the strip.

i. Always start your take off run as close as possible to the beginning of the strip, unless there are very good reasons not to do so. Work out an acceleration check point from which you can stop if you haven't reached sufficient speed to make a safe take-off

j. Bear in mind when turning off the strip, Rule 17(7) of the Rules of the Air and other arriving aircraft.

k. When performing power checks or engine runs try to minimise any noise nuisance and ensure that the slipstream is not creating a problem. Unexpected noise etc can terrify livestock; be considerate when choosing the site for engine checks.

l. After take off, reduce power and propeller rpm when it is safe. Climb to at least 500 ft agl before turning.

m. If you are a regular strip user, decide your weather and wind limits and be clear about your Go/ No Go decision process.

6 SETTING UP YOUR OWN STRIP

a. If you are planning to move your aeroplane to a strip, or perhaps start your own, the points below should be considered, in addition to any others in CAP 428 'Safety Standards at Unlicensed Aerodromes'.

b. Remember that Rule 5 of the Rules of the Air includes, amongst other requirements, the prohibition of flights below 1000 feet over 'congested' areas except when aircraft are taking off or landing at a licensed or government aerodrome. It is therefore most important that climb out, approach and circuit paths at an unlicensed aerodrome are clear of 'congested' areas. Such areas are legally defined as 'in relation to a city, town, or settlement, any area which is substantially used for residential, industrial, commercial or recreational purposes'.

c. Talk to nearby aerodrome operators to ensure that you will not conflict with their activities.

d. Look again at the performance of the aeroplane and your abilities. If operating from this strip means that *every* take off and landing, even when the aeroplane is lightly loaded, is 'tight', change to a more suitable aeroplane or strip.

e. Remember that, unless there is 'established use', aircraft operations may be in contravention of local regulations. It may of course be possible to obtain planning permission from the outset for your strip, although this would probably involve you in a great deal of hassle. However, this is much better than having it compulsorily closed by the local council if they decide that your operations are in contravention of Planning Regulations. It is in your interests to establish this from the outset and it is furthermore a good idea to talk to all of the neighbours and the planning authority *before* you do anything.

f. Cutting the grass and generally maintaining the surface has been discussed earlier; however, if you are responsible for the upkeep of the strip it is important to establish who will cut the grass, roll it and how often. This needs to be a regular activity – we all know only too well how much our lawns grow in a week.

g. Beware when mowing. Instances have occurred of pilots following the mown lines instead of the strip direction.

h. Grass seed mixtures which will give reduced rolling resistance and slower growth are available. Consult a seed merchant.

i. In deciding the orientation of the strip/landing run, consider carefully the local wind effects. It may be possible to re- orientate the strip by some 10 or 20 degrees which could reduce the cross wind effect. This is particularly important for some tailwheel types where the maximum crosswind component that can be tolerated may be as little as 10 knots.

j. Remember that whilst taking off down a slope or landing up a slope is acceptable, taking off and landing across the slope is dangerous. Ensure that the orientation of the strip eliminates excessive lateral slope.

k. It is essential to mark any obstacles, potholes or bad ground at this stage and runway markers or even runway numbers will help people to line up and operate more accurately. It is also possible to have local power lines and telephone lines moved by paying the costs.

l. You must decide in advance on your fuel arrangements. If you are intending to store fuel, then you must comply with Article 101 of the Air Navigation Order and CAP 434 'Aviation Fuel at Aerodromes'. It may be possible to obtain relatively small quantities of aviation fuel by sharing the delivery with a nearby aerodrome or strip. It is normally necessary to obtain local council permission to store fuel.

m. Decide on your maintenance arrangements, your engineer may require coaxing/ persuasion to visit your strip at short notice to rectify a defect.

n. If you own or fly a wood or fabric covered aeroplane it should be hangared – ideally all aircraft should be. However, storing it in a farm barn brings its own particular problems – rodents. Mice are nimble creatures, able to climb landing gear legs and set up home in your aeroplane. We heard of a squirrel that got into the wing structure and stored its winter supply of acorns near the wing tip. Over 30 lbs of acorns were removed! A tray of rat poison encircling each wheel should be considered.

Birds also find aircraft irresistible nesting sites; a nest removed in the morning may be substantially rebuilt by late afternoon. Pre-flight checking the aeroplane becomes very important. Insects may take over your aircraft. Given a few days undisturbed progress, a wasps' nest could appear.

o. It is vital to remove all live- stock from the runway prior to take off and prior to landing. Thus, if animals have access to the strip, assistance by a friend or farmhand is essential. Animals are unpredictable.

p. Cows leave other evidence of their presence – cow pats! Not only does this look unsightly on the aeroplane, but a build up of this, and mud, add to the drag and weight of the aeroplane. Mud and animal contaminants may also be corrosive, so regular washing of the aeroplane, especially the underside, becomes a necessity. Check regularly that spats are clear of mud and grass. Temporary removal of the spats must be agreed with a CAA Regional Office.

q. The farmer and/ or his workers may need gentle reminders about the fragile nature of your aeroplane compared with farm machinery, should they need to move it. They may not know about the dangers of propellers/ helicopter rotors.

r. Consider sitting a small hut or caravan on the strip. This will give secure storage for oil, fire extinguishers, fire axe, polish, foot pump and so on. It is suggested that this should have a large letter C painted on it to make it clear that it is a reporting point for pilots and where the Movements Log is kept. A notice board inside is useful to display information such as local instructions, NOTAMs, the engineer's telephone number, accident procedures and any temporary obstructions, soft ground and grass cutting rotas. Make sure there is enough room to park visiting aircraft well clear of the landing area.

s. Get into the habit of checking the strip each day before starting flying. Any ruts, soft ground or other problems should be dealt with or publicised on the notice board so that they can be avoided on take off and landing.

7 MAIN POINTS

DO obtain permission from the owner/operator prior to visiting the strip. Talk to pilots who have used the strip before and can advise you on procedures/obstructions.
DO check that the combination of you **and** your aeroplane **can** safely cope with this strip.
DO always leave details of ETA route, destination and how many are on board in the Movements Log.
DO always nominate a 'responsible person' as described in Safety Sense Leaflet 20 'VFR Flight Plans', who knows how to raise the alarm if you fail to arrive/return.
DO follow the requirements for customs, immigration and the Terrorism Act if flying to or from overseas.
DO talk to neighbouring aerodromes or to the Flight Information Service on the radio.
DO build up a working relationship with your nearest aerodrome. You may need them for fuel, weather information and maintenance.
DO be ready for unexpected effects from trees, barns, windshear, downdraught, etc.
DO work hard at being a good neighbour and improving the Public's perception of General Aviation by minimising noise nuisance.
DO check that the strip **really** is long enough, with a 30% margin for safety.
DO check on the effect of power and other cables.
DO check whether any slope makes it a 'one way' strip.
DO obtain and display a copy of the CAA's AIRSTRIPS poster.
DO NOT 'beat up' the strip or engage in other forms of reckless, illegal and unsociable flying.
DO NOT attempt to take off or land if the grass is long, the ground is muddy or weather is marginal. There will always be a better day to fly or you can always divert into a neighbouring aerodrome.
DO NOT run- up an engine where the noise affects others or slipstream can be a nuisance.
DO NOT attempt to 'scrape' in from a bad approach.

FINALLY, ensure that safety is the first consideration. A safe flight will almost always be an enjoyable and rewarding one.

SAFETYSENSE LEAFLET 13a   **COLLISION AVOIDANCE**

1 INTRODUCTION
2 CAUSES OF MID-AIR COLLISIONS
3 LIMITATIONS OF THE EYE
4 VISUAL SCANNING TECHNIQUE
5 HOW TO SCAN

6 SCAN PATTERNS
7 THE TIME-SHARING PLAN
8 AIRPROX REPORTING
9 OPERATIONAL TECHNIQUES
10 SUMMARY

1 INTRODUCTION

a. 'See-and-avoid' is recognised as the main method that a pilot uses to minimise the risk of collision when flying in visual meteorological conditions. 'See- and- avoid' is directly linked with a pilot's skill at looking outside the cockpit or flight deck and becoming aware of what is happening in his/ her surrounding. Its effectiveness can be greatly improved if the pilot can acquire skills to compensate for the limitations of the human eye. These skills include the application of:

- effective visual scanning
- the ability to listen selectively to radio transmissions from ground stations and other aircraft,
- creating a mental picture of the traffic situation, and
- the development of 'good airmanship'.

b. This Leaflet, based on ICAO Circular 213–AN/ 130, aims to help pilots to make 'look-out' more effective and is mainly for pilots who do most of their flying under visual flight rules (VFR). It should be of interest to all pilots, however, regardless of the type of aircraft they fly and the flight rules under which they operate since no pilot is immune to collisions.

c. A study of over two hundred reports of mid- air collisions in the US and Canada showed that they can occur in all phases of flight and at all altitudes. However, nearly all mid-air collisions occur in daylight and in excellent visual meteorological conditions, mostly at lower altitudes where most VFR flying is carried out. Collisions also can and do occur at higher altitudes. Because of the concentration of aircraft close to aerodromes, most collisions occurred near aerodromes when one or both aircraft were descending or climbing. Although some aircraft were operating as instrument flight rules (IFR) flights, most were VFR.

d. The pilots involved in the collisions ranged in experience from first solo to 15,000 hours, and the types of flight were equally varied. In one case a private pilot flying cross-country, legally VFR, in a single-engine aircraft collided with a turboprop aircraft under IFR control flown by two experienced airline pilots. In another case, a 7000 hour commercial pilot on private business in a twin-engine aircraft overtook a single-engine aircraft on its final approach piloted by a young instructor giving dual instruction to a student pilot. Two commercial pilots, each with well over 1000 hours, collided while ferrying a pair of new single-engine aircraft.

e. Experienced or inexperienced pilots can be involved in a mid-air collision. While a novice pilot has much to think about and so may forget to maintain an adequate look- out, the experienced pilot, having flown many hours of routine flight without spotting any hazardous traffic, may grow complacent and forget to scan.

f. There appears to be little difference in mid-air collision risk between high-wing and low-wing aircraft.

g. If you learn to use your eyes and maintain vigilance, you can reduce the risk of mid-air collisions. Studies show that there are certain definite warning patterns.

2 CAUSES OF MID-AIR COLLISIONS

a. What contributes to mid-air collisions? Undoubtedly, traffic congestion and aircraft speeds are part of the problem. In the head-on situation, for instance, a jet and a light twin-engine aircraft may have a closing speed of about 650 kt. It takes a minimum of 10 seconds for a pilot to spot traffic, identify it, realise it is a collision risk, react, and have the aircraft respond. But two aircraft converging at 650 kt could be less than 10 seconds apart when the pilots are first **able** to see each other! Furthermore, the field of view from the flight deck of a large aircraft can be more restricted than that from the cockpit of a small aircraft.

b. In addition, some air traffic control and radar facilities are overloaded or limited by terrain or weather. Thus they may not be able to offer the service you require.

c. These factors are all contributory causes, but the reason most often noted in the mid-air collision statistics reads 'failure of pilot to see other aircraft in time' — i.e., failure of the see-and-avoid system. In most cases at least one of the pilots involved could have seen the other aircraft in time to avoid the collision if that pilot had been watching properly. Therefore, it could be said that it is really the eye which is the leading contributor to mid-air collisions. Take a look at how its limitations affect you.

3 LIMITATIONS OF THE EYE

a. The human eye is a very complex system. Its function is to receive images and transmit them to the brain for recognition and storage. About 80 per cent of our total information intake is through the eyes, thus the eye is our prime means of identifying what is going on around us.

b. In the air we depend on our eyes to provide most of the basic input necessary for flying the aircraft, e.g. attitude, speed, direction and proximity to opposing traffic. As air traffic density and aircraft closing speeds increase, the problem of mid- air collision increases considerably, and so does the importance of effective scanning. A basic understanding of the eyes' limitations in target detection is one of the best insurances a pilot can have against collision.

c. The eye, and consequently vision, is vulnerable to many things including dust, fatigue, emotion, germs, fallen eyelashes, age, optical illusions, and the effect of alcohol and certain medications. In flight, vision is influenced by atmospheric conditions, glare, lighting, windshield deterioration and distortion, aircraft design, cabin temperature, oxygen supply (particularly at night), acceleration forces and so forth. If you need glasses to correct your vision, make sure that you have regular checks that the prescription is still correct and that you carry any required second pair.

d. Most importantly, the eye is vulnerable to the vagaries of the mind. We can 'see' and identify only what the mind permits us to see. A daydreaming pilot staring out into space is probably the prime candidate for a mid-air collision.

e. One inherent problem with the eye is the time required for accommodation or refocusing. Our eyes automatically accommodate for near and far objects, but the change from something up close, like a dark instrument panel two feet away, to a bright landmark or aircraft a mile or so away, takes one to two seconds. That can be a long time when you consider that you need 10 seconds to avoid a mid-air collision.

f. Another focusing problem usually occurs when there is nothing to specifically focus on, which happens at very high altitudes, as well as at lower levels on vague, colourless days above a haze or cloud layer with no distinct horizon. People experience something known as 'empty-field myopia', i.e. staring but seeing nothing, not even opposing traffic entering their visual field.

g. To accept what we see, we need to receive cues from **both eyes** (binocular vision). If an object is visible to only one eye, but hidden from the other by a windshield post or other obstruction, the total image is not always acceptable to the mind. Therefore, it is essential that pilots move their heads when scanning around obstructions.

h. Another inherent eye problem is the narrow field of vision. Although our eyes accept light rays from an arc of nearly 200°, they are limited to a relatively narrow area (approximately 10–15°) in which they can actually focus on and classify an object. Although movement on the periphery can be perceived, we cannot identify what is happening there, and we tend not to believe what we see out of the corner of our eyes. This, aided by the brain, often leads to 'tunnel vision'.

i. Motion or contrast is needed to attract the eyes' attention, and tunnel vision limitation can be compounded by the fact that at a distance **an aircraft on a collision course will appear to be motionless**. The aircraft will remain in a seemingly stationary position, without appearing to move or to grow in size, for a relatively long time, and then **suddenly** bloom into a huge mass, almost filling up one of the windows. This is known as the 'blossom effect'. It is frightening that a large insect smear or dirty spot on the windshield can hide a converging aircraft until it is too close to be avoided.

j. In addition to its inherent problems, the eye is also severely limited by environment. Optical properties of the atmosphere alter the appearance of aircraft, particularly on hazy days. 'Limited visibility' actually means 'reduced vision'. You may be legally VFR when you have 5 km visibility, but at that distance on a hazy day you may have difficulty in detecting opposing traffic; at that range, even though another aircraft may be visible, a collision may be **unavoidable** because of the high closing speeds involved.

k. Light also affects our visual efficiency. Glare, usually worse on a sunny day over a cloud layer or during flight directly into the sun, makes objects hard to see and scanning uncomfortable. An aircraft that has a high degree of contrast against the background will be easy to see, while one with low contrast at the same distance may be impossible to see. In addition, when the sun is behind you, an opposing aircraft will stand out clearly, but if you are looking into the sun, the glare of the sun will usually prevent you from seeing the other aircraft. A dirty, scratched, opaque or distorted windshield will make matters worse. Keep it clean, and if it has deteriorated, consider fitting a new windshield or using a proprietary re-furbishing kit.

l. Another problem with contrast occurs when trying to sight an aircraft against a cluttered background. If the aircraft is between you and terrain that is varicoloured or heavily dotted with buildings, **it will blend into the background** until the aircraft is quite close.

m. In daylight, the colours and shapes are seen by 'cones' which are light sensitive cells occupying a small central area of the retina of the eye. At night, the cones become inactive, and vision is taken over by 'rods' which make up the rest of the retina, and which provide peripheral vision by day. The problem with rods is that they cannot distinguish colour, they are not as good at distinguishing shapes as cones, and at night there is now an area in the centre of the retina (populated by inactive cones), which cannot see anything.

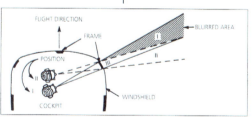

This explains why it is easier to see a faint star by looking away (by about 10 degrees) than straight at it. Rods take 30 minutes in the dark to reach their efficiency. They are insensitive to red light, and that was the reason why WWII night fighter pilots sat around in dim red rooms before jumping into dim red cockpits. Nowadays it is felt more important to interpret a normally lit instrument correctly, than run the risk of misinterpreting a dim red instrument, even though the pilot's outside night vision might be marginally better in the latter case. However, it obviously makes sense for pilots to try and avoid looking at bright lights at night. It is important to maintain a scan at night, but because peripheral rods are being used, it is

CAA Safety Sense Leaflet — 13 Collision Avoidance

better to use a continuous scan which will cause an image (aircraft lights) to move on the retina, rather than trying to focus on one area of sky (because the fine focusing cones are not working). Since the rods are sensitive to movement, they are more likely to be alerted by this technique.

n. Finally, there are the tricks that the mind can play, which can distract the pilot to the point of not seeing anything at all, or cause cockpit myopia — staring at one instrument without even 'seeing' it.

o. It can be realised that visual perception is affected by many factors. Pilots, like others, tend to **overestimate their visual abilities** and to misunderstand their eyes' limitations. Since a major cause of mid-air collisions is the failure to adhere to the practice of see-and-avoid, it can be concluded that the best way to avoid collisions is to learn how to use your eyes for an efficient scan.

4 VISUAL SCANNING TECHNIQUE

a. To avoid collisions you must scan effectively from the moment the aircraft moves until it comes to a stop at the end of the flight. Collision threats are present on the ground, at low altitudes in the vicinity of aerodromes, and at cruising levels.

b. Before take-off, check the runway visually to ensure that there are no aircraft or other objects in the take-off area. Check the approach and circuit to be sure of the position of other aircraft. Assess the traffic situation from radio reports. After take-off, continue to scan to ensure that there will be no obstacles to your safe departure.

c. During the climb and descent beware of the blind spot under the nose – manoeuvre the aircraft so that you can check.

d. During climb or descent, listen to radio exchanges between air traffic and other aircraft and form a mental image of the traffic situation and positions of aircraft on opposing and intersecting headings, anticipating further developments. Scan with particular care in the area of airway (route) intersections and when near a radio beacon or VRP. You should remain constantly alert to all traffic within your normal field of vision, as well as periodically scanning the entire visual field outside the aircraft to ensure detection of conflicting traffic. Remember that the performance capabilities of many aircraft, in both speed and rates of climb/descent, result in high closure rates, limiting the time available for detection, decision, and evasive action.

5 HOW TO SCAN

a. The best way to develop effective scanning is by eliminating bad habits. Naturally, not looking out at all is the poorest scan technique! Glancing out at intervals of five minutes or so is also poor when considering that it takes only seconds for a disaster to happen. Check the next time the aircraft is climbing out or making an approach to see how long you spend without looking outside.

b. Glancing out and 'giving the old once-around' without stopping to focus on anything is practically useless; so is staring out into one spot for long periods of time.

c. There is no one technique that is best for all pilots. The most important thing is for each pilot to develop a scan that is both comfortable and workable.

d. Learn how to scan properly by knowing where and how to concentrate your search on the areas most critical to you at any given time. In the circuit especially, **always** look out before you turn and make sure your path is clear. Look out for traffic making an improper entry into the circuit.

e. During that very critical final approach stage, do not forget to scan all around to avoid tunnel vision. Pilots often fix their eyes on the point of touchdown. You may never arrive at the runway if another pilot is also aiming for the same runway threshold at that time!

f. In normal flight, you can generally avoid the risk of a mid-air collision by scanning an area at least 60° left and right of your flight path. Be aware that constant angle collisions often occur **when the other aircraft initially appears motionless** at about your 10 o'clock or 2 o'clock positions. This does not mean you should forget the rest of the area you can see. You should also scan at least 10° above and below the projected flight path of your aircraft. This will allow you to spot any aircraft that is at an altitude that might prove hazardous to you, whether it is level with you, climbing from below or descending from above.

g. The more you look outside, the less the risk of a collision. Certain techniques may be used to increase the effectiveness of the scan. To be most effective, the gaze should be shifted and refocused at regular intervals. Most pilots do this in the process of scanning the instrument panel but it is also important to focus outside the cockpit or flight deck to set up the visual system for effective target acquisition. Proper scanning requires the constant sharing of attention with other piloting tasks, thus it is easily degraded by such conditions as distraction, fatigue, boredom, illness, anxiety or preoccupation.

h. Effective scanning is accomplished by a series of short, regularly- spaced eye movements that bring successive areas of the sky into the central visual field. Each movement should not exceed 10°, and each area should be observed for at least one second to enable detection. Although horizontal back-and-forth eye movements seem preferred by most pilots, each pilot should develop the scanning pattern that is most comfortable and then keep to it. Peripheral vision can be useful in spotting collision risks. It is essential to remember, however, that if another aircraft appears to have no relative motion, it is likely to be on a collision course with you. If the other aircraft shows no horizontal or vertical motion on the windshield, but is increasing in size, **take immediate evasive action**.

6 SCAN PATTERNS

a. Two scanning patterns described below have proved to be very effective for pilots and involve the 'block' system of scanning. This system is based on the premise that traffic detection can be made only through a series of eye fixations at different points in space. In application, the viewing area (wind- shield) is divided into segments, and the pilot methodically scans for traffic in each block in sequential order.

i. *Side- to- side scanning method*
Start at the far left of your visual area and make a methodical sweep to the right, pausing very briefly in each block of the viewing area to focus your eyes. At the end of the scan, return to and scan the instrument panel and then repeat the external scan.

ii. *Front- to- side scanning method*
Start in the centre block of your visual field (centre of front windshield); move to the left, focusing very briefly in each block, then swing quickly back to the centre block after reaching the last block on the left and repeat the action to the right. Then, after scanning the instrument panel, repeat the external scan.

b. There are other methods of scanning, of course, some of which may be as effective as the two described above. However, unless some series of fixations is made, there is little likelihood that you will be able to detect all targets in your scan area. When the head is in motion, vision is blurred and the mind will not register potential targets.

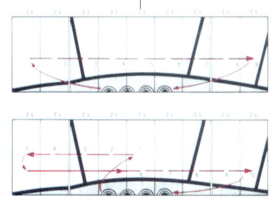

7 THE TIME-SHARING PLAN

a. External scanning is just part of the pilot's total visual work. To achieve maximum efficiency in flight, a pilot also has to establish a good internal scan and learn to give each scan its proper share of time, depending, to some extent, on the work-load inside the cockpit and the density of traffic outside. Generally, the external scan will take considerably longer than the look at the instrument panel.

b. During an experimental scan training course, using military pilots whose experience ranged from 350 hours to over 4000 hours of flight time, it was discovered that the average time they needed to maintain a steady state of flight was three seconds for the instrument panel scan and 18 to 20 seconds for the outside scan.

c. An efficient instrument scan is good practice, even when flying VFR. The ability to scan the panel quickly permits more time to be allotted to exterior scanning, thus improving collision avoidance.

d. Developing an efficient time-sharing plan takes a lot of work and practice, but it is just as important as developing good landing techniques. The best way is to start on the ground, in your own aeroplane or the one you usually fly, and then use your scans in actual practice at every opportunity.

e. During flight, if one crew member is occupied with essential work inside the cockpit, another crew member, if available, must expand his scan to include both his usual sector of observation and that of the other crew member.

8 AIRPROX REPORTING

If you consider that your aircraft has been endangered during flight by the proximity of another aircraft such that a risk of collision existed, report it by radio to the Air Traffic unit with which you are in communication. The call should be prefixed 'AIRPROX'. If this is not possible, immediately after landing (in the UK) telephone or by other means contact any UK ATS unit, but preferably an ATCC. Prompt action is important. Confirm in writing within 7 days using CA 1094 'Airprox Report Form'.

9 OPERATIONAL TECHNIQUES

a. Collision avoidance involves more than proper scanning techniques. You can be the most conscientious scanner in the world and still have an in-flight collision if you neglect other important factors.

- **Check yourself**

Start with yourself - your eyesight, and consequently your safety, depend on your mental and physical condition. If you are preoccupied you should not fly – absent-mindedness and distraction are the main enemies of concentrated attention during flight. Age affects your eyes, so if you are a mature pilot have regular eye checks. If you need glasses to correct your vision, then wear them and ensure that you have the required spare pair with you.

- **Plan ahead**

To minimise the time spent 'head-down' in the cockpit, plan your flight, have charts folded in proper sequence and within reach. Be familiar with headings, frequencies, distances, etc. so that you spend minimum time with your head down in your charts. Pilots should record these things on a flight log before take-off. Lift anything you need to read up to the coaming, rather than look down. Check your maps, NOTAM, etc. in advance for potential hazards such as military low-level routes and other high-density areas. See Safety Sense Leaflet 18, Military Low Flying.

- **Clean windows**

During the pre-flight walk-around, make sure your windshield is clean and in good condition. If possible, keep all windows clear of obstructions such as opaque sun visors and curtains.

- **Night Flying**

Be aware of the limitations of vision at night and give your eyes time to adjust. Avoid blinding others with the careless use on the ground of your strobes or landing lights.

- **Adhere to procedures**

Follow established operating procedures and regulations, such as correct flight levels (quadrantal or semi-circular) and proper circuit practices. You can get into trouble, for instance, by 'sneaking' out of your proper level as cumulus clouds begin to tower higher and higher below you, or by skimming along the tops of clouds without observing proper cloud clearance. Some typical situations involving in-flight mishaps around airports include: entering a right-hand circuit at an airport with left-hand traffic or entering downwind so far ahead of the circuit that you may interfere with traffic taking off and heading out in your direction. Beware of pilots flying large circuits with long final approaches. In most in-flight collisions at least one of the pilots involved was not where he was supposed to be.

- **Avoid crowded airspace**

Avoid crowded airspace, e.g. over a VRP or radio beacon. Aircraft can be training over navigation beacons, even in good weather. If you cannot avoid aerodromes en route, fly over them well above ATZ height and if appropriate give them a call stating your intentions.

- **Compensate for blind spots**

Compensate for your aircraft's design limitations. If you are short, or the aircraft has a high coaming, a suitable cushion can be helpful. All aircraft have blind spots; know where they are in yours. For example, a high-wing aircraft has a wing down in a turn that blocks the view of the area you are turning into, so lift the wing slightly for a good look before turning. One of the most critical potential mid-air collision situations exists when a faster low-wing aircraft is overtaking and descending onto a high-wing aircraft on final approach.

- **Equip to be seen**

Your aircraft lights can help avoid collisions. High intensity strobe lights, which can be installed at relatively low cost, increase your contrast and conspicuity considerably by day and even more by night. In areas of high traffic density, strobe lights are often the first indication another pilot receives of your presence. Transponders, especially with altitude encoding (Mode C) allow radar controllers to identify your aircraft in relation to other traffic and provide you with traffic information. They also indicate your aircraft to commercial aircraft which carry ACAS (aircraft collision avoidance system). If you show mode C, ACAS can guide the commercial aircraft away from you! The carriage of transponders is now mandatory in some airspace, even when operating VFR. If ATC do not allocate you a code, use code 7000 (with Mode C), but switch it off if instructed. Aircraft with one high contrast colour can be seen more easily than those with a pattern or one low contrast colour. Recent tests have shown that matt black (or gloss black) gives greatest contrast. Consider the use of landing lights during the circuit and landing especially on hazy days.

- **Talk and listen**

Use your ears as well as your eyes by taking advantage of all the information that you receive over the radio (but beware, non-radio aircraft may be in the same airspace). Pilots reporting their position to the tower are also reporting to you. Approaching an aerodrome, call the tower when you are 10 km from the airport, or such other distance or time prescribed by the ATS authority, and report your position, height/altitude and intentions. When flying in areas where there are no air traffic services, change to the FIR or nearest aerodrome frequency.

- **Make use of information**

Since detecting a small aircraft at a distance is not the easiest thing to do, make use of any hints you get over the radio. Your job is much easier when you are told that traffic is 'three miles at one o'clock'. Once that particular traffic is sighted, do not forget the rest of the sky. If the traffic seems to be moving on the windshield, you're most probably not on a collision course, so continue your scan but watch the traffic from time to time. If it has little relative motion you should watch it very carefully – he may not have seen you.

- **Use all available eyes**

If you normally fly with another pilot, establish crew procedures which ensure that an effective scan is maintained at all times. Otherwise, use passengers to help in looking for traffic you have been made aware of, while you monitor the movement of other aircraft. Remember, however, that the responsibility for avoiding collisions is yours and you must maintain your vigilance at all times.

- **Scan**

The most important item, of course, is to keep looking out at where you are going and to watch for other traffic. Make use of your scan constantly.

b. Stick to good airmanship; if you keep yourself and your aircraft in good condition, and develop an effective scan time-sharing system, you will have the basic tools for avoiding a mid-air collision.

701

10 SUMMARY

- If you need glasses, carry any required spare pair.
- Clean the windshield and side windows (if either is badly scratched, have a new one fitted).
- If you are short or the aircraft has a high coaming, use a cushion.
- Beware of blind spots, move your head or manoeuvre the aircraft.
- Spend the minimum time with your head down checking the charts (or GPS) changing radio frequencies etc.
- The aircraft with little or no relative motion is the one which is hard to see – and the most hazardous.
- Aircraft below you may blend into the background of buildings etc.
- High intensity strobes can be useful on dull days.
- Use the radio to form a mental picture of what is going on. Don't rely solely on it – someone could be NON- RADIO e.g. a glider.
- Develop and use an effective scan pattern.
- Don't move the eyes continuously, stop and give them a chance to SEE.
- The external scan should take much longer than your instrument scan.
- When you have spotted another aircraft, do not fix on it and forget the rest of the surroundings.
- Use landing lights in the circuit.
- Scan
- Encourage your passengers to assist in the look- out.
- Report any AIRPROX

SAFETYSENSE LEAFLET 14b   PISTON ENGINE ICING

1 INTRODUCTION
2 TYPES OF ICING
3 ATMOSPHERIC CONDITIONS
4 RECOGNITION AND GENERAL PRACTICES
5 PILOT PROCEDURES
6 SUMMARY

1. INTRODUCTION

a. This leaflet is intended to assist pilots of carburetted piston engined aircraft operating below 10,000 feet. Although it may appear to be mainly aimed at aeroplane operations, much of its content applies at least equally to piston-engined helicopters and gyroplanes.

b. Piston engine induction system icing is commonly referred to as carburettor icing, although, as described later, carb icing is only one form. Such icing can occur at any time, **even on warm days, particularly if they are humid**. It can be so severe that unless **correct** action is taken the engine may stop (especially at low power settings during descent, approach or during helicopter autorotation).

c. Every year engine induction system icing is assessed as being a likely contributory factor in several aircraft accidents. Unfortunately the evidence rapidly disappears.

d. Some aircraft/ engine combinations are more prone to icing than others and this should be borne in mind when flying different aircraft types.

e. The aircraft Flight Manual is the primary source of information for individual aircraft. The advice in this leaflet should only be followed where it does not contradict that Flight Manual.

2. TYPES OF ICING

There are three main types of induction system icing:

BUILD-UP OF ICING IN INDUCTION SYSTEM

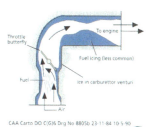

CAA Carto DO C(G)6 Drg No 8805b 23-11-84 10-5-90

a. Carburettor Icing

The most common, earliest to show, and the most serious, is carburettor (carb) icing caused by a combination of the sudden temperature drop due to fuel vaporisation and pressure reduction as the mixture passes through the carburettor venturi and past the throttle valve.

If the temperature drop brings the air below its dew point, condensation results, and if the drop brings the mixture temperature below freezing, the condensed water will form ice on the surfaces of the carburettor. This ice gradually blocks the venturi, which upsets the fuel/ air ratio causing a progressive, smooth loss of power and slowly 'strangles' the engine. Conventional float type carburettors are more prone to icing than pressure jet types.

b. Fuel Icing

Less common is fuel icing which is the result of water, held in suspension in the fuel, precipitating and freezing in the induction piping, especially in the elbows formed by bends.

c. Impact Ice

Ice which builds up on air intakes, filters, alternate air valves etc is called impact ice. It forms on the aircraft in snow, sleet, sub-zero cloud and rain, (if either the rain or the aircraft is below zero°C).

This type of icing can affect fuel injection systems as well as carburettors. In general, impact ice is the main hazard for turbocharged engines.

d. Testing has shown that because of its greater and seasonally variable volatility and higher water content, carb icing is more likely when MOGAS is used.

e. Engines at reduced power settings are more prone to icing because engine induction temperatures are lower. Also, the partially closed butterfly can more easily be restricted by the ice build-up. This is a particular problem if the engine is de-rated as in many piston-engined helicopters and some aeroplanes.

Note: For the sake of simplicity, in the rest of this leaflet, the term Carb Icing includes Induction Icing and Carb Heat includes Alternate Air.

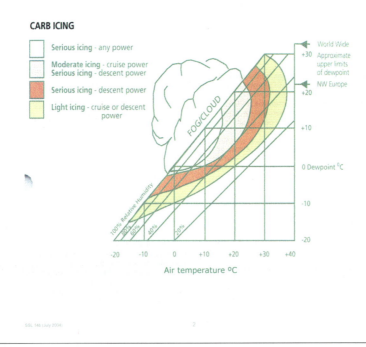

3. ATMOSPHERIC CONDITIONS

a. Carb icing is **not** restricted to cold weather. It will occur on **warm days** if humidity is high, especially at **low power settings**. Flight tests have produced serious icing at descent power with ambient (not surface) temperature above 25° C, even with relative humidity as low as 30%. At cruise power, icing occurred at 20° C when relative humidity was 60% or more. (Cold, clear winter days are less of a hazard than humid summer days because cold air holds less moisture than warm air.) In the United Kingdom and Europe where high humidity is common, pilots must be constantly on the alert for carb icing and take corrective action before an irretrievable situation arises. If the engine fails due to carb icing, it may not re-start (even if it does, the delay could be critical).

b. Carb icing can occur in clear air and is therefore made more dangerous by the lack of any visual warning. In cloud, the icing risk may be higher but the pilot is less likely to be caught unawares.

c. Specific warnings of induction system icing are not normally included in aviation weather forecasts. Pilots must therefore use knowledge and experience. The closer the temperature and dewpoint readings, the greater the relative humidity. However, the humidity reported at an aerodrome may bear little relation to the humidity at flying altitudes. When dewpoint information is not available, assume high humidity particularly when:

- in cloud and fog; these are water droplets and the relative humidity should be assumed to be 100%.
- in clear air where cloud or fog may have just dispersed, or just below the top of a haze layer;
- just below cloud base or between cloud layers (highest liquid water content is at cloud tops);
- in precipitation, especially if persistent;
- if the surface and low level visibility is poor, especially in early morning and late evening, and particularly near a large area of water;
- when the ground is wet (even with dew) and the wind is light.

However, the lack of such indications does not mean low humidity.

d. The chart shows the wide range of ambient conditions in which carb icing is most likely. Particular note should be taken of the much greater risk of serious icing with descent power.

4. RECOGNITION & GENERAL PRACTICES

Paragraphs 4 and 5 are intended as a general guide to assist you to avoid icing, but reference must be made to the relevant sections of the Pilot's Operating Handbook or Flight Manual for specific procedures related to the particular airframe/ engine combinations. **These may vary for a different model of the same aircraft type.**

a. With a fixed pitch propeller, a slight drop in rpm and performance (airspeed and/or altitude) are the most likely indications of the onset of carb icing. This **loss of rpm** can be smooth and gradual and the usual reaction is to open the throttle slightly to compensate. However, this, whilst restoring power, hides the loss. As icing builds up, rough running, vibration, further loss of performance and ultimately engine stoppage may follow. The primary detection instrument is the **rpm gauge** in conjunction with ASI and altimeter.

b. With a constant speed propeller, and in a helicopter, the loss of power would have to be large before a reduction in rpm occurs. Onset of icing is even more insidious, but there will be a **drop in manifold pressure** and in a performance reduction. In this case the primary detection instrument is the **manifold pressure gauge.**

c. An exhaust gas temperature gauge, if fitted, may show a decrease in temperature before any significant decrease in engine and aircraft performance.

d. Carb icing is normally cleared by the pilot selecting an alternative air source which supplies air, (heated in an exhaust heat exchanger) which melts the ice obstruction. This source by-passes the normal intake filter.

e. Engines with fuel injection generally have an alternate air intake located within the engine cowling via a valve downstream from the normal air intake. This alternate air is warmed by engine heat, even though it does not normally pass through a heat exchanger.

f. Always use **full** heat whenever carb heat is applied; partial hot air should only be used if an intake temperature gauge is fitted and only then if specifically recommended in the approved Flight Manual or Pilot's Operating Handbook

g. . Hot air should be selected:
- as a matter of routine, at regular intervals to prevent ice build up,
- whenever a drop in rpm or manifold pressure, or rough engine running, is experienced,
- when carb icing conditions are suspected, and
- when flying in conditions within the high probability ranges indicated in the chart.

But always be aware that hot air, while selected, reduces engine power. This may be critical in certain flight phases.

h. During the cruise, carburettor heat should be applied at regular intervals, to prevent carburettor ice forming. It should be selected for long enough (at the very least 15 seconds but considerably more in certain aircraft) to pre-empt the loss of engine power or restore power to the original level.

i. If icing has caused a loss of power, and the hot air disperses it, re- selection of cold air should produce an increase in rpm or manifold pressure over the earlier reading. This is a useful check to see whether ice is forming, but does not prove that all the ice has melted!. Carry out further checks until there is **no** resultant increase, monitor the engine instruments, and increase the frequencies of the routine checks, as it may re-occur. Lack of carb icing should produce no increase in rpm or manifold pressure beyond that noted prior to the use of hot air.

j. Remember, selection of hot air when ice is present may at first make the situation appear worse, due to an increase in rough running as the ice melts and passes through the engine. If this happens the temptation to return to cold air must be resisted so that the hot air has time to clear the ice. This time may be in the region of 15 seconds, which will, in the event, feel like a very long time!

k. Unless necessary, the continuous use of hot air at high power settings should be avoided. However, carburettor heat should be applied early enough **before** descent to warm the intake, and should remain fully applied during that descent, as the engine is more susceptible to carb icing at low power settings.

5. **PILOT PROCEDURES**

a. *Maintenance*
Periodically check the carb heating system and controls for proper condition and operation. Pay particular attention to seals which may have deteriorated, allowing the hot air to become diluted by cold air.

b. *Start Up*
Start up with the carb heat control in the **COLD** position.

c. *Taxiing*
Generally, use of carb heat is not recommended while taxiing - the air is usually unfiltered when in the HOT position. However, ice may build up at the low taxiing power settings, and if not removed may cause engine failure after take-off. If carburettor heat is needed – USE IT.

d. *Ground Power Checks*
Select carburettor heat fully ON for at least 15 seconds. Check that there is a **significant** power decrease when hot air is selected (typically 75– 100 rpm or 3– 5" of manifold pressure) and that power is regained (but to a level no higher than before) when cold air is re-selected. If the power returns to a higher value, ice was present and further checks should be carried out until the ice has cleared.

e. *Immediately Prior to Take-Off.*
Since icing can occur when taxiing with low power settings, or when the engine is idling, select carb heat ON for a minimum of 15 seconds and then OFF, immediately before take off to clear any build- up. If the aircraft is kept waiting at the holding point in conditions of high humidity, it may be necessary to carry out the run- up drill more than once to clear ice which may have formed.

f. *Take-Off*
Take-off should **only** be commenced when you are sure the engine is developing full power. When at full power and as airspeed is building, you must check that the full throttle rpm and/ or manifold pressure is as expected. **Carburettor heat must NOT be used during take-off** unless specifically authorized in the Flight Manual or Pilots Operating Handbook.

g. *Climb*
Be alert for symptoms of carb icing, especially when visible moisture is present or if conditions are in the high probability ranges in the chart. Be aware if your Flight Manual restricts the use of carb heat at full power.

h. *Cruise*

Avoid clouds as much as possible. (Note that few piston engined aircraft are cleared for flight in airframe icing conditions). Monitor appropriate engine instruments for any changes which could indicate icing. Make a carb heat check (see below) at least every 10 minutes, (more frequently if conditions are conducive to icing). Use full heat and note the warning of para 4 (j), it may take 15 seconds or more to clear the ice and the engine will continue to run roughly as the ice melts and passes through the engine. If the icing is so severe that the engine has died, keep the hot air selected as residual heat in the rapidly cooling exhaust may be effective (opening the throttle fully and closing the mixture control for a while may also help)

i. *Carburettor heat check*

- Note the RPM/ Manifold Pressure (consider slightly increasing power beforehand to prevent a reduction in performance during the check)
- Apply full Carb heat for at least 15 seconds.
- Return Carb heat to Cold. The RPM/ Manifold Pressure will return to approximately the earlier indication if there was no icing. If it is **higher** - icing was present, and may not yet be completely clear, so repeat the check until no increase results.

j. *Descent and Approach*

Carb icing is much more likely at reduced power, so select hot air before, rather than after, power is reduced for the descent, and especially for a practice forced landing or a helicopter autorotation, i.e., before the exhaust starts to cool. (A full carb heat check just before selecting heat for the descent is advisable). Maintain FULL heat during long periods of flight with reduced power settings. At intervals of about 500 ft or more frequently if conditions require, increase power to cruise setting to warm the engine and to provide sufficient heat to melt any ice.

k. *Downwind*

Ensure that the downwind check includes the cruise carburettor heat check at 5(i) above. If you select and leave the heat on, speed or altitude will reduce on the downwind leg unless you have added some power beforehand.

l. *Base Leg and Final Approach*

Unless otherwise stated in the Pilot's Operating Handbook or Flight Manual, the HOT position should be selected before power is reduced and retained to touchdown. On some engine installations, to ensure better engine response and to permit a go-around to be initiated without delay, it may be recommended that the carb heat be returned to COLD at about 200/ 300 ft on finals.

m. *Go-around or Touch and Go*

Ensure the carb heat is COLD, ideally before, or simultaneously as, power is applied for a go-around.

n. *After Landing*

Return to the COLD setting before taxiing, if not already set COLD.

6. SUMMARY

- Icing forms stealthily.
- Some aircraft/ engine combinations are more susceptible than others.
- Icing may occur in warm humid conditions and is a possibility at any time of the year in the UK.
- Mogas makes carb icing more likely.
- Low power settings, such as in a descent or in the circuit, are more prone to give carb icing.
- Use full carb heat frequently when flying in conditions where carb icing is likely. Remember that the RPM gauge is your primary indication for a fixed pitch propeller; manifold pressure for variable pitch.
- Treat the carb heat as an ON/OFF control – either full hot or full cold.
- It takes time for the heat to work and the engine may run roughly while ice is clearing.
- Timely use of appropriate procedures can PREVENT THIS PROBLEM.

FINALLY

In the event of carb heat system failure in flight:

- Avoid likely carb icing conditions.
- Maintain high throttle settings – full throttle if possible.
- Weaken the mixture slightly.
- Land as soon as reasonably possible.

SAFETYSENSE LEAFLET 15b

WAKE VORTEX

photos: Bob Stoyles, Cathay Pacific via 'Crewsnews'

1 INTRODUCTION
2 VORTEX ENCOUNTERS
3 AIR TRAFFIC CONTROL
4 VORTEX AVOIDANCE – APPROACH
5 VORTEX AVOIDANCE – DEPARTURE
6 HELICOPTERS
7 REPORTING
8 FURTHER INFORMATION
9 SUMMARY

1 INTRODUCTION

a. There have been serious and fatal accidents in the UK to light aircraft because pilots were unable to maintain control after being caught in the wake vortex or helicopter downwash generated by heavier aircraft. The hazard to light aircraft is most likely at airports where general aviation mixes with airline traffic.

b. All aircraft generate vortices at the wing tips as a consequence of producing lift. **The heavier the aircraft and the slower it is flying, the stronger the vortex.** Among other factors, the size of the vortex is proportional to the span of the aircraft which generates it, for instance a Boeing 747, with a span of 65 metres trails a vortex from both wingtips each with a diameter of around 65 metres.

c. At low altitudes, vortices generally persist for as long as 80 seconds, but **in very light or calm wind conditions, they can last for up to two and a half minutes.** Once formed, vortices continue to descend until they decay (or reach the ground). Decay is usually sudden and occurs more quickly in windy conditions. Cross-winds can carry a vortex away from the flight path of the aircraft. For each nautical mile behind an aircraft, the vortex the aircraft generates will typically have descended between 100 and 200 ft.

d. Generally, the lighter the aircraft you are flying, the greater the degree of upset if you encounter a wake vortex. Thus, a light aircraft will be vulnerable to the vortices of a similar sized aircraft ahead of it, and microlight aircraft will be even more vulnerable.

e. Aeronautical Information Circular (AIC) 17/1999 (Pink 188) 'Wake Turbulence' provides detailed information including aircraft weight categories and recommended spacings.

f. The AIC provides advice for avoiding vortices in all phases of flight. The simple advice for light aircraft pilots is, 'Avoid crossing below or close behind the flight path of a heavier aircraft'.

g. Jet blast and prop wash may also cause considerable turbulence, but are not covered in this leaflet.

2 VORTEX ENCOUNTERS

a. A light aircraft penetrating a vortex from a larger aircraft on a similar trajectory and axis can experience a severe roll. In the worst cases it may be beyond the power of the ailerons to counteract the roll. Even executive jets have been rolled upside down.

Same Trajectory Encounter

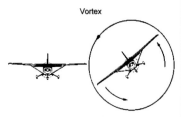

b. If the vortex is entered at right angles to its axis, rapid vertical and pitch displacements with airspeed changes are likely. An oblique entry, the most likely event, will have symptoms of both.

Right Angle Encounter

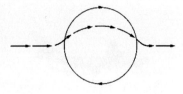

Vertical and Pitch Changes   *Airspeed Change*   *Vertical and Pitch Changes*

c. Although a vortex encounter at altitude is uncomfortable and alarming, it should be recoverable. However, any loose objects in the cockpit may be scattered about. A Piper PA23 Aztec was flying north– south at 1000 ft, 7 ½nm west of Heathrow, underneath the approach path. The Aztec was almost turned on its back by the vortex from a Boeing 757 on the approach which had crossed its track at 2500 ft. The wind at Heathrow was calm.

d. A significant proportion of the wake vortex incidents reported in the UK occur below 200 feet i.e. just before landing where there may not be room to recover. An accident in the UK badly damaged a Robin aircraft, which it appears got too close behind a landing Short SD360. At 100–150 ft the right wing and nose dropped and the aircraft did not respond to control inputs, descended rapidly and hit a hedge. Estimated separation was about 3 nm. **The wind speed was reported as 2kt**. Incidents including fatal accidents have also occurred shortly after take-off, which is when the affected aircraft is most likely to be directly behind a larger aircraft.

e. Close to the ground, vortices generally persist for about 80 seconds where their effect is most hazardous. They tend to move apart at about 5 knots in still air, so a crosswind component of 5 knots can keep the upwind vortex stationary on or near the runway while the downwind vortex moves away at about 10 knots. In crosswinds of more than 5 knots, the area of hazard is not necessarily aligned with the flight path of the aircraft ahead. Take particular care at airfields where intersecting runways are both in use.

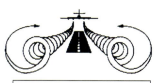

HEADWIND LESS THAN 10 KNOTS

CROSSWIND LESS THAN 5 KNOTS

CROSSWIND OVER 5 KNOTS

At very low altitude the area of hazard is not necessarily aligned with the flightpath of the aircraft ahead

3 AIR TRAFFIC CONTROL

a. At UK airports where there are commercial movements, an ATC service will be provided with the possible exception of some Highlands and Islands aerodromes. The controllers will advise pilots of the recommended interval; e.g. 'Golf November Tango, you are number two to a Boeing 737, the recommended wake vortex spacing is 6 miles, report final'.

b. **For VFR arrivals vortex spacing is the responsibility of the pilot**, however, the recommended distance will be given by ATC **but not by AFISO/ Air Ground Service**. If in doubt, use greater spacing.

c. Read the AIC so that you will be familiar with the weight categories, e.g. 'heavy' includes all wide-bodied airliners. Also become familiar with the spacing minima which ATC will apply.

d. Some large narrow bodied aircraft present a particular hazard to lighter aircraft. Experience has shown that the Boeing 757 creates particularly strong vortices. Caution is recommended for all pilots following such aircraft, particularly on approach. Additional spacing (with the agreement of ATC) is recommended.

4 VORTEX AVOIDANCE APPROACH

a. Since the vortices are invisible, although occasionally the cores can be seen in very humid conditions, they are difficult to avoid unless you have a good 'mental' picture of where they occur.

b. There are two techniques which can be employed:

- Distance can be judged visually by runway length – most major airports have runways between one and 2 nautical miles long (1850 and 3700 metres). Thus, if the recommended spacing is 6 miles, then you need 3 to 6 runway lengths between yourself and the aircraft ahead.

- If the aircraft on the approach ahead of you is much heavier than your own type, try to keep it in sight. In general, vortices drift downwards, **so fly above** and to the upwind side of the lead aircraft's flight path. Obviously as you get closer to the runway lateral displacement has to be reduced, so land beyond the point where the heavier aircraft touched down as generation of vortices ceases when the nosewheel contacts the runway. The heavier the type ahead, the longer the runway is likely to be, so stopping a light aircraft should not be a problem – it may even save you some taxi time! Airliners almost always approach on a 3° glide slope, light aircraft can readily accept steeper angles.

Courtesy- Hargreaves

5 AVOIDANCE – DEPARTURE

a. **Vortices are generated as the aircraft rotates on take off**, so the time interval between departures specified in the AIC starts from rotation. For example, a light aircraft taking off behind a Boeing 737 should allow an interval of at least 2 minutes if commencing take-off from the same point, and at least 3 minutes if taking off from a point part- way up the same runway.

b. Although you may think you can avoid the vortex by lifting off early and climbing above the vortex, most commercial aircraft will climb much more steeply than a light aircraft once they have accelerated. In order to avoid entering the vortex you would need to turn early and fly well clear of the preceeding aircraft's flight path.

6 HELICOPTERS

a. The AIC specifies minimum spacing between light aircraft and the Sikorsky S61N or similar large helicopters (there was a fatal accident to a Piper Warrior in 1992 at Oxford). **It is considered that a helicopter in forward flight generates more intense vortices than a fixed-wing aircraft of a similar weight.** When following a helicopter, pilots of light aircraft should consider allowing a greater spacing than would normally be used behind a fixed-wing aircraft of similar size, perhaps treating each helicopter as being one category higher than that listed in the AIC.

SikorskyS76 – Photo FAA Technical Center, Mr J Sackett

b. Helicopters with rotors turning create a blast of air outwards in all directions, the strongest effect being downwind. This effect is not so significant when the helicopter with rotors turning is on the ground. It is most severe during hovering and hover taxiing, when the rotors are generating enough lift to support the full weight of the helicopter, and this creates the greatest downwash. During an approach it may not be possible to determine which of the stages of flight the helicopter is at, nor the helicopter pilot's immediate intentions. In these circumstances, pilots of light aircraft should aim to keep as far away as possible. In particular, if there is a helicopter on or near the runway, and if runway length permits, consider landing further down the runway to avoid being caught by rotorwash. **If in doubt, make an early decision to go-around.**

7 MICROLIGHTS

a. Microlights and very light aircraft are more susceptible than other GA aircraft to the effects from wake vortex. Control problems have been experienced when one has encountered the vortex of a single engined piston training aeroplane.

b. Pilots of microlights should consider treating every aircraft in front of them as being one category higher than listed in the AIC.

c. Hang gliders and paragliders (including powered parachutes) can expect to be affected even more than microlights. Parachute canopies may collapse, as shown below.

photo courtesy 'Skywings'

8 REPORTING

National Air traffic Services (NATS) maintains a wake vortex database to monitor incident rates. All suspected wake vortex incidents should be **reported immediately to ATC by radio** and followed up after landing using form CA 1695 'Wake Vortex Report Form' (Forms are available from Documedia Limited) If an Occurrence Report (Form CA 1673) is used to report wake turbulence, this will automatically be copied to the database office, a separate Wake Vortex Report need not be sent but detailed information is most useful. Reports should be sent to:

Wake Vortex and Radar Analysis Incidents,
Air Traffic Management Development Centre,
National Air Traffic Services Ltd, Swanwick,

Tel: 01489 615813

9 FURTHER INFORMATION

A graphic 17 minute video, AF 9468 'Wake Turbulence – The Unseen Menace', is available from The British Defence Film Library, Chalfont Grove, Chalfont St Peter, Gerrards Cross, Bucks SL9 8TN. Tel: 01494 878237 Fax: 01494 878007. It provides a useful illustration of the problem to those who fly both small and large aircraft, and also for Air Traffic Services personnel.

10 SUMMARY

- Wake vortices are generally invisible.
- Vortices last longer in calm or light wind conditions and are therefore at their most hazardous then.
- They are most dangerous close to the ground.
- The heavier an aircraft, and the slower it is flying, the stronger its vortex and the greater the risk to following aircraft.
- The lighter the aircraft you are flying, the more vulnerable it is.
- When an aeroplane's nosewheel is on the ground, there are no vortices.
- On departure, use the appropriate time interval when following a heavier aircraft –
    2 minutes if starting the take-off at the same point,
    3 minutes if taking off part-way along the same runway.
- When taking off behind a departing heavier aircraft, note its rotation point so that you can lift- off before that point and climb above the vortex. If you cannot – WAIT.
- On the approach, avoid vortices by flying above and upwind of the lead aircraft's flightpath.
- When following a heavier aircraft which has already landed, note its touchdown point and land beyond it. If there isn't room – GO AROUND.
- Apply the spacing advised by ATC, using runway length as a guide to judging distance.
- When following a large helicopter consider allowing a bigger gap than for the equivalent sized aeroplane.
- Keep well away from helicopters with rotors turning, they may be hovering or hover taxiing – it can be difficult to judge.
- If in doubt – WAIT.
- All encounters should be reported.

Full details are published in AIC 17/1999 (Pink 188), 'Wake Turbulence'

GENERAL AVIATION SAFETY SENSE LEAFLET 16A
BALLOON AIRMANSHIP GUIDE

Photo: Steve Moss

1 INTRODUCTION

a. This guide is intended for pilots of hot air balloons, however much of the advice will apply to any lighter-than-air craft.

b. The growth of hot air ballooning into a significant leisure activity, with many commercial operators, makes the use of safe operating practices vitally important. The objectives are to safeguard persons and property on the ground as well as the balloon and its occupants. The invaluable work of the representative organisation, the British Balloon and Airship Club (BBAC)*, is gratefully acknowledged and all owners, operators and pilots are encouraged to join and participate in its work.

c. The safety record of ballooning is excellent and many of the criteria which apply to the safe conduct of balloon flights apply equally to the safe conduct of any flight. These can be summarised as sound Knowledge, careful Preparation, and the exercise of good Practice. These are detailed below.

2 KNOWLEDGE

a. Reporting

Learn from the mistakes of others; you might not live long enough to make them all yourself. Share your

* See addresses at end of leaflet

knowledge and experience with others by making a report to the BBAC, and to the Safety Data Department* of the Civil Aviation Authority, on any incidents from which you think others might learn. Your report could prevent someone else's accident. Improve your knowledge by reading as many accounts of other people's ballooning problems as you can. The BBAC Pilot's Circular, CAA Balloon Notices, Bulletins issued by the General Aviation Safety Council* and by the Air Accident Investigation Branch* of the Department of Transport, are regular sources of safety information.

b. Statistics

Accurate statistics for ballooning are difficult to obtain. A total of almost 1700 hot air balloons are now registered in the UK. There is no record of the total number of balloon flights, however, in 1996 there were nearly 9000 flights by about 200 commercial balloons, during which over 66,000 passengers were carried. During 1995/96 there were 14 Reportable accidents to balloons; fortunately there were no fatalities, but 11 passengers received injuries and a number of balloons were damaged. The most potentially serious situation which a balloon can encounter, other than mid-air collision, is to be in close proximity to, and up-wind of, over-head electricity cables. In 1995/96 nearly 40% of accidents were due to wire strikes.

c. Training

Keep in regular flying practice; you will handle difficult situations more effectively if you are current. Check that you have sufficient recent flying experience to maintain the validity of the licence, and that your medical certificate is current. Occasionally fly with an experienced pilot/instructor/examiner who can identify any bad flying habits you may have inadvertently acquired.

3 PREPARATION FOR FLIGHT

a. Paperwork

Formal documentation for private flight is relatively simple, but keeping a balloon and Personal Flying Log Book requires a responsible attitude. Commercial operators of balloons for Public Transport must keep a Technical Log and sign for the condition and loading of the balloon before flight. Documents and data, ie Flight Manuals and Log Books, must be kept up to date for all balloons so that the condition and loading of the balloon can be established before the next flight. The Annual or 100 hour inspection must be completed when due and properly certificated before further flight.

b. Balloon Condition

Commercial balloons must be maintained to a high standard and operators are required to provide evidence of this in the Technical Log and Balloon Log Book. Private balloons must be maintained to no less a standard by using similar procedures. You must ensure that all damage or defects, even minor (balloons are not 'self-healing'), are put right in accordance with the Manufacturer's instructions, before further flight. It is the pilot's responsibility to ensure that the balloon is airworthy before flight. **Never use unauthorised or improper parts in a balloon.**

c. Weather

Make maximum use of weather forecasts for ballooning. These are available by phone and fax from a number of commercial sources in addition to the existing Met. Office Services*. The various methods of obtaining aviation weather, (including codes) are available in a small booklet **'GET MET'**. This is available free from CAA Safety Promotion Section. Please send SAE, (address on back page). Avoid personal or commercial pressure to fly and, if in doubt about the conditions, don't. Remember the adage about its better to be on the ground wishing you were flying than the other way round! Know and comply with the Visual Flight Rules as well as with the wind speed limits of the balloon, and what the pilot feels comfortable with.

d. Maps and Charts

The Air Navigation Order (ANO) requires you to carry an up-to-date chart, for obvious and sensible reasons. Balloon pilots probably need more information than most from their maps and charts because their landing area is uncertain until the final stages of the flight. Ordnance Survey maps, marked with sensitive landing and over-flying areas, are needed as well as aeronautical charts. Keep yours up-to-date and don't fly without them.

e. Flight Planning and Navigation

The more time you spend in preparation, the better you will enjoy the flight. Armed with appropriate maps and charts, and in possession of a recent aviation weather forecast, you should be able to make a reasonable estimate of the expected track. Study the available information for the route, including NOTAMs. Check for the proximity of regulated airspace, Danger Areas, sensitive areas and other hazards, including major overhead power lines. Attention to this will avoid last minute unfolding of maps in flight and thumbing through flight guides for the frequency of an airfield you are approaching. Call the AIS **Freephone number, 0500 354802,** for the latest information on Royal Flights, Red Arrows Displays and Airspace Restrictions.

f. Landing Area

Plan to land in an area which provides a choice of suitable sites. Avoid being committed to land in an area which does not offer any alternatives if an initial approach has to be abandoned. **You must not plan to land within the congested area of a city, town or settlement.**

Photo: Steve Moss

g. Radio

Carry a radio if there is a possibility that you will approach controlled airspace or an active Aerodrome Traffic Zone. Ground crews prefer that they, and you, have a well-charged radio and know how to use it. A VHF R/T licence is required if an aeronautical frequency is used. A licence for any R/T equipment must be obtained from the Department of Trade and Industry*. Keep transmissions brief and to the point, and brush-up your radio procedures by reading the Radiotelephony Manual, CAP413 available from Westward Digital Limited* (formerly CAA Printing and Publications).

h. Loading

Make sure that the empty weight of the balloon, including equipment, fuel cylinders and contents, is available and accurate. Use actual weights for passengers and crew and add an allowance for miscellaneous items such as camera bags. (You may need to have 'bathroom scales' available.) Use Flight Manual figures for calculating available lift and do not exceed this figure. Remember, there is significant loss of lifting capability on a hot day. Excessive heating will shorten the life of the balloon.

i. Re-fuelling and Fuel Planning

Use fuel from a reliable source. As Propane is considerably more volatile than petrol and is stored under pressure, treat it with the respect it deserves. Re-fill cylinders in well-ventilated surroundings free of static discharge or other source of combustion. Make sure no-one is smoking. Keep equipment in good condition and observe the Propane Code, available from the BBAC. Propane vapour pressure is reduced by low ambient temperature resulting in loss of burner efficiency, and thus balloon performance. Nitrogen pressurisation has largely replaced warming of cylinders as the preferred method of maintaining pressure on cold days.

j. Safety Equipment

The following safety equipment must be carried:

- alternative method of ignition
- protective gloves
- fire extinguisher
- first aid kit

Have you been appropriately trained within the last 3 years? Is the equipment 'in date'?

The items below are recommended:

- binoculars, to help spot power lines etc., and
- protective helmets for all on board when flying conditions dictate.

k. Ground Crew

Crews should receive training as recommended by the BBAC which is provided in many parts of the country. The training should include inflation, launching and tethering, emergency procedures, refuelling, use of radio, map reading and landowner relations.

l. Flight Over Water

Even though balloons float, life jackets should be carried when the flight is likely to be more than 1 nm from the nearest land, supplemented when water temperatures are low, by a life raft capable of accommodating all occupants. Check that loading figures take this equipment into account, and that these items have recently been tested by an Approved Organisation. (See Safety Sense Leaflet No. 21 'Ditching' for a list.)

m. Tethering, Clearances and Permissions

- Anywhere:
  - In the unlikely event that the top of a tethered balloon is to be more than 60 metres above ground level, a permission in writing must be obtained in good time from the CAA General Aviation Dept.* (ANO Article 86(2)(b)(iii)).

Photo – Courtesy of Aerostat

- Permission must be obtained from the person in charge of any vessel, vehicle or structure within 60 metres of a tethered balloon (ANO Article 86(2)(a)).
- You must obtain the landowners permission.
- The local Police Authority should be notified of any intended flight.

- Controlled Airspace
  - In addition to the above, if the flight is within, or will enter, controlled airspace, a clearance must be obtained from the appropriate Air Traffic Control Unit (ANO Article 86(4) and (5)).

- Equipment
  - Equipment for tethering must be in good condition and provision made for crowd control at public events. Check the Flight Manual for instructions and restrictions which apply to tethered flight, especially wind limits.

n. Night Flying

Free balloons, and tethered balloons above a height of 60 metres, must display the lights specified in the Rules of the Air Regulations (Rules 13 and 14) when flying between half an hour after sunset and half an hour before sunrise. Particular regard must be paid to the 'endangering' Articles of the ANO (Articles 63 and 64) when contemplating a night flight in a balloon, and the pilot's licence privileges and insurance must include night flying before such flights are made.

o. Large Events

Go to the briefing, pay attention to what is said, and comply. There are no prizes for being first off the ground, or for appearing braver than anyone else by setting off in unsuitable weather. Lives could be endangered and the future of the event could be jeopardised by unreasonable behaviour. BBAC agreed practices are now contained in CAP403, the Flying Display Manual, which must be used by organisers when planning an event.

p. Flight Abroad

Make certain that you are aware of the rules and regulations for ballooning, and of NOTAMs, airspace etc, for the country concerned. Take with you all balloon documents, Log Book, Certificates of Registration and Airworthiness, Flight Manual, procedures to be used in the event of interception, and your personal log book and licence. Check insurance cover. You have to file a Flight Plan before flying across an International Boundary.

q. Pilot Fitness

Don't fly if unfit. It is better to cancel a flight than to scrap a balloon. Check the following I'M SAFE check-list:

I – Illness, any symptoms?

M – Medication, does your family doctor know you are a pilot?

S – Stress, any serious personal upsets?

A – Alcohol/drugs.

F – Fatigue, good night's sleep?

E – Eating, recent meals?

4 PRACTICE

a. Selection of Take-off Site

Check that the selected site is sheltered, unobstructed by overhead lines or other hazards, clear of built-up (congested) areas unless an Exemption to Low Flying Rule 5 (1)(a)(ii) is in force (see para 4g) in which case carefully check the conditions under which it is issued. Also, you **MUST** have the landowners permission. Check that the expected track from the site is clear of controlled airspace, Danger Areas and other airspace restrictions as well as built-up areas. Check that there is a good choice of landing sites along the expected track within the planned flight time.

b. Inflation

Brief an adequate number of ground crew and check that they, as well as the pilot and passengers, are wearing gloves and suitable long-sleeved non-synthetic clothing. Check that the balloon is serviceable, the cylinders are re-fuelled and that loading will be within limits. Test burners and check for leaks. Attach the quick-release tether to a vehicle which has an effective hand brake and is in gear. Before starting the cold inflation, attach the flying wires prior to laying out the envelope and the rip line. Carry a lighter, and spare matches or striker. Keep passengers well clear, but paying attention, and move spectators to a safe distance. The fire extinguisher should be readily available.

c. Pre Take-off

Take your time, use a check-list, and do not hesitate to **cancel the flight** if all is not well with the condition of the balloon, its instruments and equipment, with the take-off site or the weather. Test the deflation system, and double check all burner systems for leaks, contents and correct functioning. Leave nothing to chance.

d. Passenger Briefing

Article 39 of the ANO requires that all passengers **MUST** be given a briefing on what they should do in the event of an emergency etc. The briefing must include the following:

- do not hold on to hoses, valves or control lines

- hold on to the internal rope handles or fuel cylinder rims or (except when landing) burner supports

- on landing, normally face backwards hold on tightly and always pay attention to the pilot's instructions, keep arms inside the basket and , **do not leave the basket without the pilot's permission.**

It is recommended that children are not carried on any flight unless they are old enough to understand the briefing, and are tall enough to see over the edge of the basket unaided.

e. Burner Handling and Fuel Management

Test all systems before take-off. Memorise cylinders in use and know the state of the others. Plan cylinder changes in advance, watch the gauges and change cylinders **before** the pressure drops. Check the burner after changing cylinders. Double check all hose connections. Know and practice emergency procedures for pilot-light failure, burner failure and fire in the air.

f. Take-off

Make a final communications check with the ground crew. Agree a contact telephone number and hand over the vehicle keys. Use a take-off technique appropriate to the prevailing wind conditions. Employ both ground crew and a quick-release tether in other than calm conditions to ensure a clean departure. Immediately after take-off, check again that all systems are 'go'.

g. Low Flying Rules

The Rules of the Air Regulations apply to ALL aircraft, which includes balloons. Fly no lower than 1,500 ft above the highest fixed object within 600 metres of the balloon when over a congested area (unless in possession of an Exemption to Rule 5 (1)(a)(ii) for take-off and climb-out). Over open country, remain at least 500 feet clear of any persons vessels, vehicles or structures (Rule 5 (1)(e)) unless taking off or landing in accordance with normal aviation practice (Rule 5 (2)(d)(i)).

h. Avoidance of Obstacles and Overhead Lines

Although binoculars are helpful, if obstacles or power lines are seen at the last moment, make a decision to climb or to land and then **stick to this decision.** For a fast climb, use all burners together, but be careful not to exceed the maximum permitted rate of climb or envelope temperature. To descend, use the deflation valve but be ready to slow the sink rate with the burner when it is safe to do so. Remember, it is easier to maintain or increase the vertical motion of a balloon, either up or

Photo: John Thorpe

down, than to reverse it. From level flight a balloon responds faster when put into a descent than when asked to climb. Do not deliberately fly near power lines (bear in mind the 500 ft rule) and **avoid touching them at any cost.** If contact is inevitable, descend as fast as possible so that the envelope and **not** the basket assembly contacts the wire. Shut down the fuel system and vent fuel lines before contact. If the balloon is caught in the wires, **do not** touch any metal parts. If possible, remain in the basket until the electrical power is switched off. Do not allow ground crew near the balloon until the power is switched off.

i. Controlled Airspace

See ANO Article 86 'Balloon Regulations'. Before embarking on a flight in a direction which could involve approaching controlled airspace, talk to the authority responsible for the airspace. Preferably telephone well in advance, but if an unexpected change of wind during flight is the cause, make early contact by radio to give the controlling authority as much time as possible to consider the situation. Air Traffic often think you are closer to them than is in fact the case. They sometimes forget how big a balloon is and it may not show on radar, so call early! If you cannot make contact and have not previously made arrangements for this situation, you *must* land before entering the airspace.

j. Sensitive Areas and the National Farmers Union (NFU) Code of Practice

While not really a safety matter, it is relevant that the pilot should be fully aware of the NFU Code of Conduct and observe sensitive areas along the line of flight. Don't frighten animals or damage crops, these are the farmer's livelihood. Ask for permission before retrieving the balloon.

k. Landing

Tell the ground crew where you are planning to land. Check that the approach to the selected site is clear and that sufficient fuel remains in the cylinder in use in case it is necessary to make an approach to an alternative site. Brief passengers emphasising that they must **NOT** leave the basket until told to do so. Stow all loose articles. Locate the rip-line and prepare to use it. Turn off fuel just before touch-down. Avoid the risk of setting fire to crops or scorching grass by extinguishing the pilot-light and checking there is no residual flame. Make sure that spectators are not at risk. When you have finished the retrieve make certain gates are left as you found them, and remember to thank the landowner.

Useful Addresses:

- Civil Aviation Authority
  Safety Regulation Group
  Aviation House
  Gatwick Airport South
  West Sussex RH6 0YR
  - General Aviation Dept
    Tel: 01293 573506
    Fax: 01293 573973
  - Aircraft Registration
    Tel: 020 7453 6299
    Fax: 020 7453 6262
  - Licensing Dept:
    Tel: 01293 573700
    Fax: 01293 573996
  - Medical Division
    Tel: 01293 573685
    Fax: 01293 573995
  - Safety Data Dept
    Tel: 01293 573220/1
    Fax: 01293 573972
- Aircraft Accidents Investigation Branch
  Dept. of Transport
  Defence Research Agency
  Farnborough
  Hampshire GU14 6TD
  Tel: 01252 5103000
        01252 512299 (24 hr)
- British Balloon & Airship Club
  6 Langstone Court
  Aylesbury, Bucks HP20 2DF
  Tel: 01296 436930
  Fax: 01844 346309
- General Aviation Safety Council
  Holly Tree Cottage
  Park Corner
  Nettlebed
  Oxon RG9 6DP
  Tel & Fax: 01491 641735
- Meteorological Office
  Central Forecasting Division
  London Road
  Bracknell, Berks RG12 2SZ
  Tel: 01344 856 267
  Fax: 01344 854 920
- Radiocommunications Agency
  Dept. of Trade & Industry
  New Kings Beam House
  22 Upper Ground
  London SE1 9SA
  Attn: RA2/AMACB
        26 N/26.4
  Tel: 020 7211 0122
  Fax: 020 7211 0228

5 MAIN POINTS

- Learn from the mistakes of others
- Keep in current flying practice
- Stick to your limitations and those of the balloon
- Get a proper weather forecast
- Use the latest maps and charts, and check NOTAMs
- Know the balloon, its systems and equipment
- Observe the Propane Code
- Load the balloon correctly
- Have completed first aid and fire training
- Keep spectators at a safe distance.
- Check everything thoroughly before take-off
- Stay out of controlled airspace unless clearance has been obtained to enter it
- Know and observe the Regulations in the Rules of the Air
- Regularly check fuel contents and cylinder in use
- Observe the BBAC/NFU Code of Practice
- Keep well away from power lines etc.

SAFETYSENSE LEAFLET 17c HELICOPTER AIRMANSHIP

1. INTRODUCTION
2. KNOWLEDGE
3. PREPARATION
4. PRACTICE
5. MAIN POINTS

1 INTRODUCTION

a. Although this guide is mainly intended for helicopter pilots, much of the advice will be equally relevant to gyroplane pilots.

b. A review of 42 fatal accidents during a recent 15 year period to helicopters of less than 5700 kg, reveals that most should not have happened. Broadly, they are the result of the following:

- low flying including wire strikes   8
- controlled flight into terrain   8
- loss of control VMC   6
- technical failures   5
- third party into rotors   4
- loss of control IMC/night   4
- collision with ground objects   4
- mid-air collision   1
- unknown   2

c. Comprehensive **knowledge**, careful **preparation** and frequent flying **practice** are the best insurance against becoming an accident statistic. Avoid a complacent 'it will be all right' attitude.

2 KNOWLEDGE

2.1 Reporting

a. Learn from the mistakes of others; you might not live long enough to make **all** of them yourself. Improve your knowledge via other peoples' problems by reading the CAA's GASIL, the Air Accident Investigations Branch's monthly Bulletin and the General Aviation Safety Council's quarterly Flight Safety Bulletin.

b. Share your knowledge and experience with others, preferably by reporting to the Civil Aviation Authority Safety Investigation & Data Department, British Helicopter Advisory Board, the Helicopter Club of Great Britain, or for gyroplanes the Popular Flying Association, anything from which you think others could learn. Your report could prevent someone else's accident. Photographs often help to illustrate a problem.

c. Details of all helicopter occurrences are on the CAA's Safety Investigation & Data Department database.

d. If there is a Manufacturer's Safety Course, improve your knowledge by participating – it could result in cheaper insurance!

2.2 Refresher Training

Revise your basic knowledge and flying skills by having a regular check flight, (at least every 6 months), with an instructor which should include:

- practice engine failure so that in a single- engine helicopter it is a reflex response to lower the collective IMMEDIATELY and to enter autorotation
- in multi-engined helicopters, practice simulated engine-out procedures
- sloping ground take-offs and landings
- appropriate emergency procedures for the type of helicopter, including emergency R/T call, either on the intercom or by a practice PAN call
- AWARENESS of (but not necessarily demonstrated) height-velocity curve, dynamic roll-over, vortex ring, ground resonance and engine icing situations
- awareness of the importance of maintaining rotor rpm, and proficiency at recognising and recovering from low rotor rpm conditions, both with power ON and power OFF
- operation from confined areas.
- assessment of flight visibility
- other flying that you or your instructor feel would be beneficial.

2.3 Limitations

a. You must know the helicopter's limitations and your own – HEED THEM BOTH.

b. Experienced fixed wing pilots, but with low rotorcraft hours, may be confident and relaxed in the air but will not yet have developed the reflex responses, control feel, co-ordination and sensitivity necessary in a helicopter. They may well react incorrectly to a low rotor rpm warning. (See paragraph 2.2 and 4.9b.) A more cautious approach is necessary.

3 PREPARATION

3.1 Paperwork

Make sure that your licence/rating, certificate of experience and medical are up-to-date. Also check that the helicopter's documents, including Certificate of Airworthiness/Permit to Fly, Maintenance Releases and Insurance are current.

3.2 The Helicopter

a. If you do not fly very often, prior to flight study the Pilot's Operating Handbook/Flight Manual etc. so that you are thoroughly familiar with:

- limitations
- normal and emergency procedures
- rotor speeds/power settings
- the height-velocity avoid areas
- weight and balance calculations
- operation of radio and navigation instruments.

b. Sit in the helicopter and re- familiarise yourself with the external and ground checks, cockpit layout, fuel system and position of all controls etc.

c. Carry out refresher training as described in paragraph 2.2 if you have not flown the type in the last six months. (Many commercial operators require a check-flight if their pilots have not flown the type in the last 28 days!)

3.3 Weather

a. Get an aviation weather forecast, heed what it says, and make a carefully reasoned GO/NO GO decision. Do not let 'Get-there/home- itis' influence your judgement. Establish clearly in your mind the current en-route conditions, the forecast and the 'escape route' back to good weather. Take account of the freezing level. Plan a more suitable route if you are likely to fly over high ground which may be cloud covered.

b. The various methods of obtaining aviation weather (including codes), are available in a small booklet 'GET MET'. This is available free from the Met Office.

c. Know the conditions that lead to the formation of piston engine icing (Safety Sense Leaflet 14). Know the Flight Manual/ Pilot's Operating Handbook instructions regarding the use of Carb heat or Engine anti-ice and comply with them. Include Carb Air Temp and OAT in your regular scan of engine instruments.

d. Beware of turbulent and windy conditions, especially if your experience is limited.

e. In wet weather beware of misting of windshield and windows, especially when carrying wet passengers.

3.4 Winter Flying

a. In addition to much of the information in Safety Sense Leaflet 3 'Winter Flying', helicopter pilots should also beware of 'white-out', due to blowing snow, when landing on a snow covered surface.

b. It should also be noted that there are **NO** general aviation helicopters cleared for flight in icing conditions. You must use weather forecasts to avoid snow and icing conditions.

c. Wear warm clothing in case of heater failure or a forced/precautionary landing – you can't put them on in flight!

d. A Canadian gyroplane accident was the result of the pilot's eye balls freezing. He lost control and crashed.

e. If operating from an icy surface, take care to open and close the throttle slowly and lead with the appropriate yaw pedal to avoid the possibility of the helicopter rotating on the spot.

3.5 VFR Navigation

a. Use appropriate current aeronautical charts, ready folded to show the planned track. It may be too late when you are airborne.

b. Check NOTAMs, Temporary Navigation Warnings, AICs etc for changes issued since your chart was printed or which are of a temporary nature, such as an air display, or ATC frequency change. (Internet site www.ais.org.uk)

c. Information on Red Arrows displays and Emergency Restrictions of Flying is available on the AIS website and Freephone **0500 354 802**, updated daily. See AIC 103/ 2002 (Pink 44) entitled 'Helicopter Flights in Restricted (Emergency/ Incident/Accident) Airspace'.

d. Prepare your Route Plan thoroughly with particular reference to Safety Altitude, icing hazards and suitable diversions. Familiarise yourself with geographical features, time points, airspace en-route and the procedures in any helicopter special routes.

e. If you fly a single engined helicopter and your proposed route takes you over a congested area, forest, lake etc. where a forced landing due to engine failure could be hazardous to yourself or those on the ground, plan a different route – so that you can make a safe forced landing.

f. Note congested areas, high ground, masts and other obstructions in planning your safe altitude; note Maximum Elevation Figures (MEF) on charts. Remember you must not fly over some High Security Prisons and other sites in a helicopter, these may not be all shown on your chart, but are listed in the UK AIP ENR 5-1-2.

g. Plan to reach your destination at least one hour before sunset, unless qualified, equipped and prepared for night flying. (Public transport night flying is prohibited in single-engined helicopters.) You may not spot fog or low cloud at night.

h. In order to comply with Rule 5 of the Rules of the Air, 'Low Flying', you must NOT fly:

- within 500 ft of persons, vessels, vehicles and structures, unless taking off or landing in accordance with normal aviation practice or,
- over or within 1000 m of any assembly in the open air of more than 1000 persons at an organised event without complying with the procedures in rule 5 of the Rules of the Air,
- over a congested area, i.e. city, town, or settlement, below 1000 ft above the highest fixed object within 600 m of the helicopter, unless flying on a notified route under 'Special VFR',
- at such a height/speed combination that persons or property on the surface are endangered in the event of an engine failure,
- in the London 'Specified Area', except on the approved routes

Note: If your proposed flight appears to be limited by Rule 5, first check the full terms of the Rule and, if necessary, seek further advice from the CAA's General Aviation Dept*.

i. If you intend to fly below 1000 ft agl (where most military low flying takes place), use Freephone **0800 515544** for the Civil Aircraft Notification Procedure (CANP) or Pipeline and Power line Inspection Procedures (PINS) to let them know where and when you will be operating on relevant activities (see appropriate AICs and Safety Sense Leaflet 18 'Military Low Flying').

j. Know the procedure if you get lost, see paragraph 4.7.

k. Above all, prepare a thorough route plan (Safety Sense Leaflet 5 'VFR Navigation').

l. GPS is a back-up to other methods of navigation NOT a substitute for them. Double check way- point calculation and entry.

3.6 Radio

a. Know what to do in the event of radio failure, including when flying Special VFR in controlled airspace etc.

b. Have all necessary radio frequencies to hand, including those for destination and diversion aerodromes, VOLMET, LARS, Danger Area Crossing Service etc.

c. When using RADIO-NAV to back-up your visual navigation, note the frequencies and Morse idents of radio NAVAIDs.

d. Brush-up periodically on radio procedures, phraseology etc (CAP 413 'Radiotelephony Manual').

3.7 Weight and Balance

a. Use the actual (not typical) empty weight and centre of gravity (cg) from the latest Weight and Balance Schedule of the **actual** helicopter you are operating. Helicopters get heavier due to extra equipment etc. Take account of ground handling equipment, camera installations, etc.

b. Check that the helicopter's maximum/ minimum weights are complied with. If too heavy, you must adjust the weight by off-loading passengers, baggage or fuel.

c. Check that the cg is within limits for take-off and throughout the flight. If it does not stay within the approved range, e.g. after passengers have been unloaded, or with low fuel and two heavy crew in front; then in some helicopters, you may run out of cyclic control for landing. You may have to carry ballast; make sure it is suitable and properly secured.

d. **Never** attempt to fly a helicopter which is outside the permitted weight/ cg range and performance limitations. It is dangerous as well as illegal, invalidates the C of A and almost certainly your insurance.

3.8 Performance

a. Make sure that the sites you intend using are going to be large enough for take-off and landing. Use the Pilot's Operating Handbook/Flight Manual to calculate the space and power required. Calculate your density altitude.

b. Use the recommended take-off and landing profiles. **Minimise** flight in the height-velocity avoid areas.

3.9 Fuel Planning

a. Always plan to land by the time the tank(s) are down to the greater of 1/4 tank or 45 minutes, but don't rely solely on the gauge(s) or low fuel warning. Remember, a headwind may be stronger than forecast, which particularly affects slower flying helicopters. Frequent use of carb heat/ hot air will also increase fuel consumption.

b. Know the hourly fuel consumption of your helicopter. In flight, check that the gauge(s) agree with your calculations.

c. Understand the operation and limitations of the fuel system, gauges, pumps, mixture control (do not lean mixture unless it is permitted), unusable fuel etc.

3.10 Destination

a. Check for any special procedures due to activities at your destination, such as parachuting, gliding, microlighting etc. Use the UK Aeronautical Information Publication (UK AIP) or other Flight Guides including NOTAMs and Temporary Navigation Warnings, etc. to find out where the helicopter operating area is located.

b. If your destination is a private landing site, the surroundings may be very different from the licensed aerodrome at which you learnt to fly, or from which you normally operate. The final approach and take-off area should be at least twice the length of the helicopter including rotor blades. There may be hard-to-see cables or other obstructions in the approach path, or hills, trees and buildings close to the site giving wind shear and/or unusual wind patterns. Read the guidelines published by the British Helicopter Advisory Board (BHAB) on their website www.bhab.org

c. Try to chose a landing site where you can use the recommended profiles, but if that is impossible consider:

- a check out with an instructor or someone who knows the site well, or
- a check from the ground of the potential problems associated with different wind directions, or the reduced climb on a hot day.

Always minimise the time that the helicopter is at greatest risk from engine failure.

d. In a helicopter, you cannot just land anywhere – you need the landowner's (or his Agent's) permission. This also applies at strips and most aerodromes, where Prior Permission is Required (PPR).

3.11 Flying Abroad

a. Make sure you are conversant with the aeronautical rules, charts (including scale and units, e.g. feet or metres), airspace etc for each country you are flying to/over.

b. Take the helicopter's documents which include for example – in some countries the insurance details written in their language, e.g. Spain, your licence and a copy of 'Interception Procedures' (Safety Sense Leaflet 11).

c. Before crossing an International FIR boundary you must file a Flight Plan, check that it has been accepted. (Safety Sense Leaflet 20 'VFR Flight Plans').

d. Don't forget the Prevention of Terrorism restrictions for flights to Ireland, Channel Isles and Isle of Man. (UK AIP GEN 1-2-1 paragraph 5.)

e. Permit to Fly aircraft may need special permission in many countries.

3.12 Over Water

a. Before flying over water, read Safety Sense Leaflet No 21 'Ditching'. Some helicopter manuals/handbooks contain specific advice on ditching including the need to apply full lateral cyclic control as the helicopter contacts the water to stop the main rotor blades.

b. The weather over the sea can often be very different from the land, e.g. sea fog.

c. When flying over water, everyone in a single-engined helicopter should, as a minimum, **wear** a life jacket. In the event of an emergency there will be neither time nor space to put one on.

d. The water around the UK coast is cold even in summer and survival time may be only 15 minutes (about the time needed to scramble an SAR helicopter). A good quality insulated immersion suit, with warm clothing underneath and the hood up and well sealed, should provide over 3 hours survival time. In water, the body loses heat 100 times faster than in cold air.

e. In addition, take a life raft. It's heavy, so re-check weight and balance. A life raft is much easier to see and will help the rescuers find you. It should be properly secure, but easily accessible as a helicopter will sink faster than an aeroplane.

f. Make sure that lifejackets, immersion suits and life raft have been tested recently by an approved organisation – they **must** be serviceable when needed.

g. You are strongly urged to carry a Personal Locator Beacon (PLB) and flares. If flying more than 10 minutes over water, carry an approved ELT.

h. Remain on an appropriate aeronautical radio station.

i. Pilots and passengers who regularly fly over water, are advised to attend an underwater escape training and Sea Survival Course (details in 'Ditching' leaflet).

j. If the helicopter is fitted with flotation equipment, make sure you are familiar with its operation.

k. Minimise over water time in single-engined helicopters. (Public transport helicopters are limited to 10 minutes over water when crossing sea areas around the UK.)

3.13 Night Flying

Night flying is a combination of visual and instrument flight, the ratio depending on the weather and background lighting including moonlight. You must have a Night Rating and you should be in current instrument flying practice (e.g. during the previous 28 days). For night take offs and landings, the site and any relevant obstacles, should be illuminated by external means.

3.14 Pilot Fitness

a. Don't fly when ill or tired – it is better to cancel a flight than to wreck a helicopter or hurt yourself!

Are you fit to fly – 'I'm Safe' checklist

I Illness (any symptom)

M Medication (your family doctor may not know you are a pilot)

S Stress (upset following an argument)

A Alcohol/ Drugs

F Fatigue (good night's sleep etc)

E Eating (food keeps blood-sugar level correct).

b. If you have to wear glasses for flying, make sure that the required spare pair is readily accessible. Sunglasses and a peaked cap may be useful.

c. During hot weather, beware of de-hydration, have water available, the cabin can be like a greenhouse.

d. Wear clothes that cover the limbs and give some protection in the event of fire. Avoid synthetic material which melts into the skin.

4 PRACTICE

4.1 Pre-Flight

a. After removing blade tie-downs, pitot and engine covers, complete a thorough pre-flight inspection, paying particular attention that swashplate, control rods etc. are secure and in good condition – climbing may be necessary. Don't forget any 'telatemps' designed to show overheating. Use the check list.

b. Check the surrounding area for loose objects that could blow about in the rotor wash and that the rotor disc will be well clear of obstacles.

c. Determine visually that you have enough fuel of the right type. Don't let anyone confuse AVGAS and AVTUR. Personally supervise re-fuelling and be aware of the danger of static electricity. If necessary use a dip-stick to check fuel levels. Make sure the filler caps are properly secured and the earthing cable disconnected. With the fuel selector ON, check fuel drains for water and other contamination. Minimise 'Rotors Running' refuelling, which should only be done if approved in the Flight Manual.

d. Check engine and transmission oil levels and, if necessary, top them up. Don't be fooled by a 'tide line' on the sight glass, this has led to failures as there was no oil in the gear box. (See also Airworthiness Notice No 12 Appendix 43 'Helicopter Gearbox Oil Level Sight Glasses'.)

e. Check engine intake(s) for foreign objects, particularly on turbine helicopters.

f. Remove all ice, snow and FROST from the helicopter. Even light frost can disturb the air flow over an aerofoil surface. (Beware of re-freezing.) Only use authorised de-icing fluids on rotor blades, due to the possibility of damaging the bonding of metal fittings and composite rotors.

g. If you find anything which you are not happy about, get further advice.

h. When doing the internal checks, use the check list. Confirm visually that the rotor blades move correctly in response to control inputs.

i. Properly secure any baggage so that nothing can foul the controls. Beware of loose items, e.g. cameras being carried by passengers.

j. Make sure all baggage doors are properly closed and locked.

4.2 Passengers

a. Removal or blanking of dual controls will prevent passenger interference.

b. The law requires that you MUST brief passengers on the location and use of doors, emergency exits and safety harnesses as well as emergency procedures. Personally check that doors and hatches are secure (Safety Sense Leaflet 2 'Care of Passengers').

c. Centralise the controls and switch on the beacon/strobe. Do not start the engine until all ground personnel are well clear of the helicopter and all passengers are seated inside with the doors secure.

d. Do not let passengers step up into the helicopter and then wave to their friend, their hands may be much too close to the rotor disc.

e. If it is necessary for passengers to get in or out with the rotors turning, brief someone to escort passengers to and from the helicopter. Passengers may behave oddly and do silly things in the wind and noise of a running helicopter, childrens' hands should be firmly held. Always approach from the front, wait outside the rotor disc until the pilot has given a 'thumbs up'. NEVER walk uphill away from a helicopter or downhill towards a helicopter, the rotor tip may do more than part your hair!

Note: AIC 'Propeller and Rotor Markings', provides good advice on colours to make rotors much more visible (AIC 83/ 1997, Pink 150).

f. Some passengers may be affected by flicker vertigo, see AIC 73/ 2001 (Pink 23) 'The Effect of Flickering Light on Passengers and Crew'.

4.3 Starting Engine/s

a. Know where to find and how to use the helicopter's fire extinguisher, as well as the location of any others in the vicinity.

b. Use the check-list and closely monitor the appropriate gauge(s).

c. If parked on snow or ice don't forget the possibility of the helicopter yawing. Open and close the throttle carefully

4.4 Take-off

a. Know the helicopter Marshalling Signals.

b. Make sure you know the maximum demonstrated sideways speed for the helicopter type you are flying and factor this for your experience and recency.

c. Ensure skids are not stuck to the ground by mud or ice. This has caused helicopters to roll over on take-off

d. Take particular care if you have to lift off crosswind or downwind, there may only be marginal control if there is a crosswind of 10–12 kts from the critical side. This can also affect hover taxiing.

e. Beware of hovering close to tall buildings and hangars when there is a possibility that the helicopter downwash will not dissipate uniformly and may re-circulate through the top of the rotor disc. This will require more power to hold hover height and produces a dynamic force towards the obstruction. As a rule of thumb, re-circulation can occur when the helicopter is hovering closer than two thirds of the rotor diameter from an obstruction.

f. Before lifting off, always carry out a clearing turn. Consider your options such that engine failure will not be a hazard to persons or property on the ground (see paragraph 4.11 on rotor wash).

g. Lift-off slowly into a low hover and check engine gauges including manifold pressure/rpm and control effectiveness.

h. If you take off into a strong wind and then turn downwind with constant pitch and attitude, the speed 'perceived' from ground reference will appear to increase by an amount equivalent to the wind speed. If you then attempt to reduce 'perceived' speed by increasing the attitude, it can lead to the use of high power, together with a reduced rate of climb and in severe cases a high sink rate. You are now in the classic vortex ring condition, near the trees with low IAS and full power. Now get out of that! (see paragraph 4.13c).

4.5 Look Out

a. Always keep a good look-out (and listen-out), for other aircraft, particularly over and close to radio beacons, Visual Reference Points and in the vicinity of aerodromes. The most hazardous conflicts are those aircraft with the least relative movement to your own. These are the ones that are difficult to see and the ones you are more likely to hit. Beware of blind spots and move your head, or the helicopter, to uncover these areas. Scan effectively. (Safety Sense Leaflet 13 'Collision Avoidance'.)

b. Helicopters are harder to see than aeroplanes so if the fixed wing pilot hasn't seen you, you had better keep an exemplary look out and make sure you've seen him!

c. Remember the Rules of the Air which include flying on the right-hand side of line features (even if your helicopter is flown from the right) and give way to traffic on your right.

d. If the helicopter has strobe lights, use them. If you are in a crowded circuit environment, use landing lights as well.

e. Spend as little time as possible with your head 'in the office'.

4.6 Airspace

a. Do not enter controlled airspace unless properly authorised. You might have to orbit and wait for permission. Keep out of Danger and other Prohibited Areas. If you need to transit, contact the Danger Area Crossing and Information Services.

b. Use the Lower Airspace Radar Service (LARS), which is available from many RAF and civil aerodromes, particularly on week days. It may prevent you from getting a nasty fright from military or other aircraft.

c. A Radar Advisory Service (RAS) will tell you about conflicting aircraft and offer avoidance advice. If you take alternative action or consider no action necessary, then you must tell the controller. A Radar Information Service (RIS) gives you details of conflicting aircraft, but you have to decide if avoiding action is necessary. Make sure you know which service you are receiving. Pilots are always responsible for their own terrain and obstacle clearance. (Safety Sense Leaflet 8 'ATC Services Outside Controlled Airspace'.)

d. Allocation of a transponder code does NOT mean that you are receiving a service.

4.7 En-route Diversion

a. You must not lose sight of the surface unless appropriately qualified, in current practice, and flying a suitably equipped helicopter. Don't fly above clouds unless they are widely scattered and you can remain in sight of the surface.

b. If you encounter deteriorating weather **turn back or divert before you are caught in cloud**. A 180° turn in cloud can easily become a death spiral!

c. Maintain a safe cruising altitude. Many pilots have come to grief because a lowering cloud base has forced them lower and lower into the hills. You MUST avoid 'scud running'.

d. Unless you have an instrument rating, you are strongly advised not to continue if flight visibility is below 3000 metres. In conditions of low visibility or lowering cloud, turn back, divert or make a precautionary landing. Don't PRESS ON – LAND ON!

e. An occasional weather check from VOLMET is always worthwhile.

f. Divert if the periodic cruise check, such as FREDA (fuel, radio, engine, DI, altimeter) indicates you won't have 45 minutes fuel reserve at destination.

4.8 Lost

a. If you are lost (or temporarily unsure of your position) then tell someone. Transmit first on your working frequency. If you have lost contact on that frequency or they cannot help you, then change to 121.5 MHz and make your PAN or MAYDAY call. If you have a transponder, the emergency code is 7700, it will instantly alert a radar controller. Select Mode C, if fitted.

b. If you are lost and any of the items below apply to you, call for assistance – **'HELP ME'**:

H - High ground/ obstructions - are you near any?

E - Entering controlled airspace – are you close?

L - Limited experience, low time or student pilot, let them know –

P - PAN call in good time – don't leave it too late

M - Met conditions – is the weather deteriorating?

E - Endurance – fuel remaining; is it getting low?

c. As a last resort, make an early decision to land while you have the fuel and daylight to do so. Choose a site with care and afterwards use a telephone so that you can advise that you are safe and obtain a weather update or further help.

4.9 Control Considerations

a. Fly at a safe speed in relation to visibility. Stay out of the 'height-velocity avoid curve'.

Above all, maintain rotor speed, needles should be at the top of the green band rather than the bottom.

b. In most helicopters, particularly two bladed teetering rotor types and especially gyroplanes, you MUST avoid any push-over manoeuvre resulting in negative 'g'. This can be one of the causes for the main rotor striking the tail boom with catastrophic results.

c. When flying a helicopter (or gyroplane), with an articulated or teetering head beware of retreating blade stall, especially at or near VNE in turbulent conditions. This may cause pitch up and roll. Recover by reducing speed and pitch.

4.10 Environmental

a. The public don't like helicopter noise. Several aerodromes and landing sites are under threat of closure due to this, so it is vital to be a good neighbour. Read the 'Code of conduct' on the BHAB web site www.bhab.org. Know the noise pattern for your helicopter; most comes from the tail rotor. Often a turn of 90° can direct the noise away from a neighbour. Avoid 'blade slap' on descent by slowing down early with no sudden manoeuvres.

b. Adhere to noise abatement procedures and do NOT fly over noise or other sensitive areas. These are detailed in the UK AIP or other Flight Guides or may be established on a local basis.

c. When en-route, fly at a height/ power setting which will minimise noise nuisance, as well as complying with Rule 5, 'Low Flying' (see paragraph 3.5h).

NEVER be tempted to 'beat up' the countryside.

d. Select sites for practice auto rotations very carefully – HASELL includes 'LOCATION'.

SOME HELICOPTER MARSHALLING SIGNALS

Hover: Arms horizontally sideways, palms downward.

Move Backwards: Arms by sides, palms facing forward, arms swept forward and upward repeatedly to shoulder height.

Stop: Arms repeatedly crossed above the head.

Move Downward: Arms extended horizontally sidways, with palms turned down, beckoning downwards.

Move Sideways: Either arm placed horizontally sideways, then the other arm moved in front of the body to that side, in the direction of the required movement; repeated several times.

Land: Arms placed down and crossed in front of the body.

Move Upward: Arms extended horizontally sideways, with palms up, beckoning upward.

Move Ahead: Arms repeatedly moved upward and backward beckoning onward.

Cut Engine(s): Either arm and hand placed level with the chest, then moved laterally with the palm downwards.

Note: In many cases the speed of arm movement indicates the rate/urgency.

4.11 Wake Turbulence and Rotor Wash

a. Don't operate in conditions worse than those stated in the Pilot's Operating Handbook/Flight Manual. Remember, these were obtained by a test pilot! If in doubt – replan.

b. Stay well clear of the 'blast' end of powerful aircraft.

c. Always be mindful of the effect your own rotor wash can have on parked aeroplanes and other surface objects.

d. Beware of wake turbulence behind heavier aircraft on take-off, during the approach or on landing. You should remain 8 nm, or 4 minutes or more behind large aircraft. Hover- taxiing helicopters, particularly large ones, generate very powerful vortices. (Safety Sense Leaflet 15 'Wake Vortex' provides further guidance.)

e. Note that wake turbulence lingers **when wind conditions are very light**. These very powerful vortices are invisible. Heed Air Traffic warnings.

4.12 Circuit Procedures

a. When joining or re-joining make your radio call early and keep radio transmissions to the point – 'cut the chat'. If non-radio (or your radio has failed), know the procedures.

b. Check that the change from QNH to QFE reduces the altimeter reading by the landing site elevation. If landing using QNH, don't forget to add the site elevation to your planned circuit height.

c. Use the appropriate joining procedures at your destination aerodrome. Check circuit height and look out for other aviation activity, e.g. gliding, parachuting.

d. Check the windsock or nearby smoke to ensure you land into wind. Be very sure of the wind direction and strength before committing yourself to an approach direction.

e. Make radio calls in the circuit at the proper places and listen and look for other traffic. Remember pre-landing checks – easily forgotten if you make a straight-in approach.

f. If you have to fly a fixed wing circuit, maintain your speed, do not slow down or hover thus creating a collision hazard from following traffic.

g. Be aware of optical illusions at unfamiliar landing sites, e.g. those with sloping terrain.

h. Take care at aerodromes where identification of the runways can be confused, e.g. 02 and 20. Make sure you know whether the circuit is left-hand or right-hand, as this will determine the dead side. If in doubt – ASK.

i. In most piston engined helicopters, apply carb heat well BEFORE reducing power. You may decide to return to cold at 200 ft plus above ground.

j. Reduce rate of descent before reducing airspeed.

4.13 Landing

a. Don't land in tall dry grass, the hot exhaust could start a fire.

b. A good landing is a result of a good approach. If your approach is bad, make an early decision and go around.

c. Avoid conditions likely to result in Vortex Ring:

- Power On
- Low IAS (below 35 kts)
- High rate of descent (over 300 ft per min).

See AIC 2/2005 (Pink 78) 'Vortex Ring'.

d. The unplanned downwind approach is particularly hazardous. It can lead to over-pitching, loss of rotor rpm and lift, resulting in a hard contact with the ground. (Correlators are less effective at high power settings, so maintain rotor rpm by leading with the throttle before applying pitch.)

e. If there is a white H marking, you must use that area.

f. If you are loading passengers, have them escorted to/from the helicopter, or else make them wait until the rotors have stopped. They **must** be aware of the danger of the main and tail rotor (see paragraph 4.2e).

g. Remember, the flight isn't over until the engine(s) are shutdown and all checks completed and the rotors have stopped.

h. 'Book in' and close any Flight Plan, if necessary by phoning the local Air Traffic Service Unit. See Safety Sense Leaflet No 20A 'VFR Flight Plans'.

> A helicopter has the unique ability to land almost anywhere. If, despite our advice, you find yourself in a weather, fuel, navigation or other difficulty – simply LAND and sort out the situation.

USEFUL ADDRESSES

- Safety Promotion GAD, Aviation House, Gatwick Airport, West Sussex RH6 0YR.
  Tel: 01293 573225
  Fax: 01293 573973
- Safety Data Department (address as above)
  Tel: 01293 573220/1
  Fax: 01293 573972
- Air Accident Investigation Branch, Berkshire Copse Road, Aldersot, Hants GU11 2HH
  Tel: 01252 512299
  Fax: 01252 376999

  British Helicopter Advisory Board
  Graham Suite, West Entrance,
  Fairoaks Airport, Chobham,
  Woking GU24 8HX
  Tel 01276 856100
  Fax 01276 856126
  www.bhab.org

5 MAIN POINTS

- If the engine fails in a single-engined helicopter, you must have a reflex response to lower the collective **IMMEDIATELY**.
- Keep current. Regular simulated engine-off landing practice with an instructor is recommended.
- Know the helicopter thoroughly.
- Always get an aviation weather forecast, and update it through the day.
- Prepare a thorough Route Plan using latest charts and check on NOTAMs, Temporary Nav warnings etc.
- Keep time over water to a minimum in a single-engined helicopter and wear a lifejacket (and a survival suit), carry a life-raft.
- Pre-flight thoroughly with special emphasis on fuel, engine and transmission oil contents, and flying controls.
- Brief passengers/ground staff about getting in and out of helicopters. Either have passengers escorted or shut down the engine(s).
- Don't over-load the helicopter.
- In a single-engined helicopter, bear in mind the possibility and consequences of engine failure.
- Minimise time in the 'avoid curve'.
- Maintain a good look-out, scan effectively.
- Make regular cruise checks of OAT or carb air temperature and when necessary use carb heat.
- Keep out of controlled airspace unless you have clearance.
- Request help early (or land) if lost or have other problems, e.g. fuel shortage.
- Return or land if the weather deteriorates. Maintain a safe altitude.
- **Maintain rotor rpm.**
- Avoid retreating blade stall in turbulent conditions or near VNE – **SLOW DOWN**.
- Push-over negative 'g' manoeuvres can be catastrophic, particularly in gyroplanes.
- Remain at the controls until the rotors have stopped turning.
- Don't do anything stupid – become an old pilot, NOT a bold pilot.

SAFETYSENSE LEAFLET 18 – MILITARY LOW FLYING

1. INTRODUCTION
2. THE UNITED KINGDOM MILITARY LOW FLYING SYSTEM
3. MILITARY LOW FLYING ACTIVITY
4. CIVIL LOW LEVEL ACTIVITY
5. OTHER AREAS OF INTENSE ACTIVITY
6. REPORTING
7. MAIN POINTS

1 INTRODUCTION

The purpose of this Leaflet, which is based on AIC 41/ 2005 (Yellow 165), is to inform civilian pilots, in the interests of mutual flight safety, about military low flying training operations in the UK. Low flying training is an essential element of an effective air force and regular training in a realistic environment is necessary to maintain operational capabilities. Over the UK, low flying is carried out by the Royal Air Force, the Royal Navy, and the Army Air Corps. A small amount of low flying is also undertaken by other NATO air forces.

2 THE UNITED KINGDOM MILITARY LOW FLYING SYSTEM

a. The United Kingdom Low Flying System (UK LFS) covers the whole of the UK and surrounding over-sea areas, from the surface to 2,000 ft. This permits wide distribution of the activity so as to reduce the impact on the environment. Military pilots must avoid major built-up areas, Controlled Airspace, Aerodrome Traffic Zones (ATZ) and some other sensitive locations. Inevitably, the protection given to these areas creates unavoidable concentrations of military low flying activity where corridors are formed between them. Where necessary, military pilots, except those of helicopters flying below 200 ft Minimum Separation Distance (MSD), (MSD is the authorised minimum separation, in all directions, between an aircraft and the ground, water or any obstacle), follow established uni-directional flows below 2,000 ft to reduce the risk of confliction. These flow arrangements, which apply in daylight hours only, over areas and through 'choke' points, are published on CAA chart UK AIP ENR 6–5–2–1. 'Areas of Intense Aerial Activity, Aerial Tactical Areas and Military Low Flying System'.

b. For administrative purposes, the UK LFS is divided into Low Flying Areas (LFA). Certain LFA, nominated Dedicated User Areas (DUA), are allocated for specific use, e.g. concentrated helicopter training, and are managed under local arrangements. Salisbury Plain and the surrounding area is a DUA. It is used mainly by Army Air Corps helicopters, although other military aircraft may be encountered. Civil pilots should be aware that night exercises are frequently conducted in this area without, or with limited, navigation lights. Details of the Salisbury Plain night training area are in the ENR 1-1-5 section of the UK AIP. Similar night exercises may be conducted in the airspace of Northern Ireland. Details are promulgated by **UK NOTAM** when such exercises are conducted in other areas of the UK LFS.

CAA Safety Sense Leaflet — 18 Military Low Flying

c. In the North of Scotland, the Highlands Restricted Area (HRA) designated EGR 610A, B, C and D is used for special training, often in Instrument Meteorological Conditions (IMC). To ensure safety, entry by civil and non-participating military aircraft is prohibited during the promulgated operating hours – between 15.00 and 23.00 (local time) Monday to Thursday in winter, and 1600– 2200 in summer. Details of the HRA are contained in AIC 40/2005 (Pink 81) and UK AIP ENR 5-12. During operating hours crossing permission for Areas 610C and D may be available from **Tain Range on 122.750 MHz**. If the HRA has not been booked for specific military flying, access to the whole of the HRA airspace can be obtained from the Low Flying Booking Cell, on the Freephone number given in para 4d. A civil pilot will be given clearance to operate in the HRA airspace for up to 3½ hours from the time of the telephone application. The airspace is available for normal use outside the above and during Scottish Bank holidays.

d. UK Danger Areas are regularly used for weapons training. This can lead to an increased amount of low flying in the surrounding airspace. Details of Danger Areas can be found in the UK AIP ENR 5-1.

3 MILITARY LOW FLYING ACTIVITY

a. Military fixed-wing aircraft (except light aircraft and helicopters) are considered to be low flying when less than 2,000 ft MSD. The lowest height at which fixed wing military aircraft normally fly is 250 ft MSD. However, in three specially designated areas, known as Tactical Training Areas (TTA) located in Mid- Wales, in the Borders/SW Scotland and in the North of Scotland, a small number of flights may be authorised to fly down to 100 ft MSD. Military light propeller aircraft and helicopters are considered to be low flying below 500 ft MSD. In practice, most military low flying takes place between 250 ft and 600 ft MSD, decreasing in intensity up to 1,000 ft MSD and reducing further in the 1,000 ft to 2,000 ft height band. However, occasionally military aircraft perform high energy manoeuvres between 250 ft and 2,000 ft during which they rapidly change height, speed and direction.

b. Most low flying training is during weekdays and daylight hours, although it is necessary to carry out some low flying at night and occasionally at weekends. Fast jet aircraft are normally limited to a speed of 450 kts (7½ miles per minute), although speeds of up to 550 kts can be authorised for short periods during simulated attacks and practice interceptions.

c. Low flying takes place in the UK Flight Information Regions (FIR), outside Controlled Airspace, where ground radio and radar coverage is not adequate to provide a radar service. It would be impractical for military jet aircraft to avoid each other by contacting ATC units. With the exception of the HRA, military low flying is only conducted in Visual Meteorological Conditions (VMC), where pilots not only fly with visual reference to the surface, but also apply the see and avoid principle regarding other aircraft.

CAA Safety Sense Leaflet 18 Military Low Flying

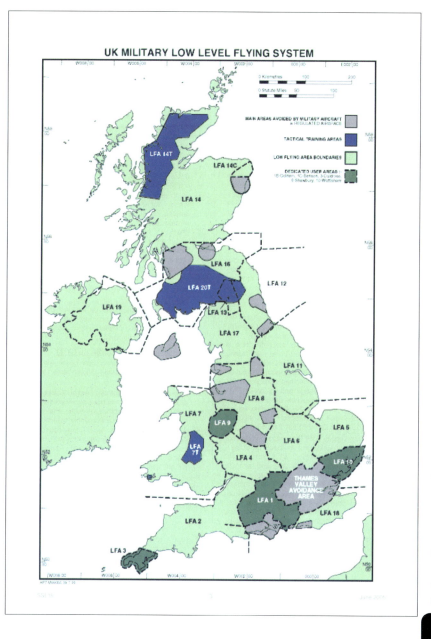

4 CIVIL LOW LEVEL ACTIVITY

a. The Low Flying Booking Cell disseminates the information notified from the Civil Aircraft Notification Procedure (CANP), to all military flying units.

b. Civil pilots engaged in low level aerial work may be subject to aircraft manoeuvring limitations and/or restricted lookout. CANP exists to provide military aircrew with information on aircraft below 1,000 ft agl engaged in crop spraying, photography, surveys or helicopter under slung load work close to a declared site. Military aircraft at speeds in excess of 140 kts will avoid laterally, or by overflying with a separation of **not less than 500 ft** the notified CANP area of operation. However, no provision is made for commercial (public transport) transit flights at low level.

c. **Recreational** activities notified under CANP will not normally be provided with CANP avoidance areas. However, where five or more aircraft (gliders, hang and paragliders, free balloons or microlights) plan to operate at a site which is not normally used, or will be outside the published hours, the Low Flying Booking Cell will issue a warning to military pilots.

d. The Low Flying Booking Cell should be contacted **not less than 4 hours** beforehand, but preferably earlier, to discuss CANP. This minimum period of 4 hours for notification is required so that aircrew can be advised during their flight planning. Notifications with less than 4 hours notice will generally be accepted but as the notice period diminishes, so does the likelihood of the message getting through. A Freephone facility is available on **0800 515544** or Freefax on **0500 300130**. Full information on the use of CANP is published in AIC 43/2005 (Yellow 167) and UK AIP ENR 1-10-13.

e. Pilots should note that information about the Temporary Restricted Airspace associated with Red Arrows displays, **of 6 nm radius**, which may be at country fairs and seaside resorts, is available on **Freephone 0500 354802**. The information, which also includes Temporary Controlled and Restricted Airspace, is updated daily, at about 19.00 hours local. During summer weekends the Red Arrows and other display aircraft may transit at low level between displays and on weekdays may fly contrary to the flow arrows during the run-in to a display. A free sticker is available from Safety Promotion Section, please send a SAE (address at end).

f. Commercial helicopter operators who conduct pipeline inspection flights should refer to AIC 42/2005 (Yellow 166). 'Helicopter Pipeline and Powerline Inspection Procedures'.

g. To reduce the risk of confliction with low flying military aircraft, pilots of civil aircraft on Visual Flight Rules (VFR) flights during the working week are advised to:

- fly above 2,000 ft agl if possible
- avoid particularly, operating in the 250 to 1,000 ft agl height band
- climb above 1,000 ft as soon as possible when departing from aerodromes (or landing sites) in the open FIR, and to remain above 1,000 ft for as long as possible when approaching such aerodromes or sites
- where an ATZ is established, fly circuits and procedures within the ATZ (military pilots are directed to avoid ATZs)

NOTE: at aerodromes without an ATZ, military pilots will apply the see and avoid principles

- keep a good lookout at all times, military aircraft smoke trails can be visible before the camouflaged aircraft can be seen. (*Safety Sense Leaflet 13, 'Collision Avoidance' may be helpful.*)

5 OTHER AREAS OF INTENSE ACTIVITY

In addition to the Military Low Level flying system the following areas should also be noted:

- AIAAs, (Areas of Intense Aerial Activity) airspace within which military or civil aircraft, singly or in combination with others, regularly participate in unusual manoeuvres.
- ATAs, (Aerial Tactics Areas) airspace of defined dimensions designated for air combat training within which high energy manoeuvres are regularly practiced by aircraft formations.

Pilots of non-participating aircraft who are unable to avoid these areas are strongly advised to make use of a radar service and maintain a particularly good lookout.

CAA Safety Sense Leaflet   18 Military Low Flying

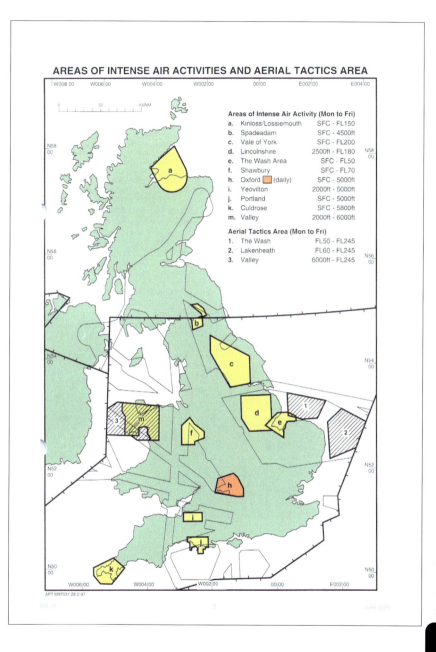

6 REPORTING

Whenever, in the opinion of a pilot (or a controller), the distance between aircraft as well as their relative positions and speed have been such that the safety of the aircraft involved was or may have been compromised the pilot should:

- immediately report by radio to the ATS Unit with which the pilot is in communication, prefixing the message AIRPROX. If this is not possible, immediately after landing in the UK, report by telephone or other means to any UK ATS Unit but preferably to an Area Control Centre

Note: In the event of a an alleged CANP infringement, in order that radar tracing can be implemented as soon as possible, use Freephone 0800 515544

- It is highly important that every AIRPROX is reported immediately to the UK Airprox Board, when the incident occurred in UK airspace, with confirmation in writing, using Airprox Report Form CA1094 (available from address below) within 7 days to:

  Director, UK Airprox Board, Hillingdon House, Uxbridge, Middx, UB10 0RU,

  Tel: 01895 276121 or 276122 (normal hours),

  Fax: 01895 276124,

  Telex: 934725 AFTN: EGGFYTYA.

7 MAIN POINTS

In the airspace used by the military low flying system, as elsewhere in the open FIR, collision avoidance depends on pilots seeing and avoiding other aircraft. Civil pilots can minimise the risk by:

- being aware that military fast jet activity is of a lower intensity on Friday afternoons and, does not normally take place on Saturdays or Sundays. However, there may be a few Hercules flights, some helicopter operations and transits by the Red Arrows and other display aircraft. Public holidays (Vbank holidays in Scotland) are avoided.

- using the Freephone 0500 354802 to find out about Red Arrows Displays etc

- giving at least 4 hours notice to the Low Flying Booking Cell of low level aerial work and other activities notifiable under CANP

- checking NOTAMs etc for details of military exercises, particularly those which include low flying

- flying **above 2000 ft agl** whenever possible

- where possible **avoiding flying below 1000 ft agl**

- climbing above 1000 ft as soon as possible when departing at aerodromes and landing site in the open FIR

- staying above 1000 ft as long as possible when arriving at such aerodromes,

- keeping the circuit inside an ATZ

- **keeping a good look-out at all times.**

SAFETYSENSE LEAFLET 19a   AEROBATICS

1	INTRODUCTION	7	PREPARATION FOR FLIGHT
2	AIRCRAFT SUITABILITY	8	PRE-AEROBATIC VITAL ACTIONS
3	PHYSIOLOGICAL ASPECTS	9	SPORTING AND COMPETITION ASPECTS
4	PERSONAL EQUIPMENT AND CLOTHING	10	AIR DISPLAYS AND PUBLIC EVENTS
5	INSTRUCTION	11	SUMMARY
6	AIRCRAFT CHECKS		

1 INTRODUCTION

a. Aerobatics, whether in a glider or a powered aircraft, provide an opportunity for pilots to learn and participate in a new facet of sporting aviation. It is, however, vital to keep safety in mind, since a reckless or careless attitude can result in serious injury or death. Almost every year accidents occur where the height available proves insufficient to recover from an intentional or, more usually, a badly executed aerobatic manoeuvre.

b. The motivation to acquire aerobatic skills is usually a desire to experience the pleasure of being able to control the aircraft while precisely drawing a defined manoeuvre in the sky. A side benefit is that it also adds to the pilot's ability to cope with unusual attitudes and unexpected upsets, such as wake turbulence, in a safe manner.

c. Only a limited range of aircraft attitudes will have been encountered during a pilot's training towards a private licence. Learning aerobatics will extend the pilot's knowledge of the aircraft's performance envelope, while emphasising the need to co-ordinate use of the flying and engine controls to achieve the desired manoeuvre.

2 AIRCRAFT SUITABILITY

a. The particular aircraft which you propose to use must be cleared for the aerobatic manoeuvres intended, so a review of the Certificate of Airworthiness/Permit to Fly and the Flight Manual/Pilot's Operating Handbook, including all Supplements, is essential before flight. These will detail the permitted speeds (e.g. Va – manoeuvring speed, the maximum speed at which controls can be fully deflected under normal circumstances), as well as the permitted manoeuvres and load factors, which may vary between two outwardly identical aircraft. The aircraft should, ideally, be fitted with a 'g' meter to confirm that it has been flown within its permitted 'g' envelope. In addition, try to avoid sudden large control movements at any speed, especially when reversing direction.

b. On most aircraft the maximum weight and centre of gravity (cg) position permitted for aerobatics is restricted. Fuel and oil system design may also limit manoeuvres which are possible, duration of inverted flight etc.

c. Aircraft with fixed pitch propellers need particular care to ensure rpm limits are not exceeded at high speed.

d. If the aircraft is fitted with a Turn Co-ordinator, be warned that it can give incorrect indications in an inverted spin, whereas a conventional Turn and Slip indicator will always indicate the correct yaw direction.

e. For aerobatics the aircraft must have a full harness, but a lap strap and diagonal is permitted for spinning. Be sure that you understand the restraint system – some aerobatic aircraft have two separate, independent restraints.

f. A serviceable transponder can be used to warn air traffic radar units that you are carrying out aerobatics. Select 7004 (with Mode 'C' if fitted) a few minutes before starting your practice.

g. If there are any aspects concerning the aircraft or its suitability that you do not fully understand, seek advice from a suitable, knowledgeable person.

3 PHYSIOLOGICAL ASPECTS

a. Aerobatic manoeuvres involving changes of aircraft attitude cause marked effects on the balance apparatus of the inner ear. Without good visual cues, erroneous messages from this to the brain can lead to disorientation, so a good horizon and good visibility are essential. Even in perfect conditions, the mismatch between balance cues (which have an in-built time lag) and visual cues (which are instantaneous) can cause motion sickness, although experience and practice will usually overcome this. Non-pilot occupants, or non-aerobatic pilots will be more prone to this feeling and should be warned that it might happen, and that it is a normal physiological response. Any attempt to continue aerobatic flight after the other occupant has gone pale and quiet will inevitably lead to a messy cockpit unless a sick bag is readily available.

b. During aerobatics, 'g' loading causes shifts of blood within the body. Positive 'g' moves blood toward the feet and away from the brain. At about +3½ to +4 'g' a relaxed human being will suffer vision changes, initially loss of colour and peripheral detail (greyout) and then a complete loss of sight (blackout). If the 'g' load increases, loss of consciousness will occur ("g-loc"). Tensing the stomach and leg muscles and grunting will help prevent this sequence (guidance should be sought from a pilot who is familiar with the technique). Other occupants should be advised to carry out the technique when approaching positive 'g' manoeuvres of this magnitude (the natural muscle tensing of the aerobatic passenger may help). They should also keep their head still during application of 'g' to minimise the risk of neck injuries and reduce the likelihood of disorientation. Negative 'g' manoeuvres cause blood to accumulate in the head, and the increased blood pressure can occasionally cause damage. Little can be done to mitigate the effects of negative 'g', which is poorly tolerated and more uncomfortable than positive 'g'. A person's tolerance to 'g' tends to increase with exposure and reduce with age.

c. Because aerobatic flight places extra stresses on the body you should seek medical advice from your AME, (Authorised Medical Examiner), or airsport medical advisor about the wisdom of performing aerobatics if you are over 60 years of age or suffer from any possibly relevant medical condition.

4 PERSONAL EQUIPMENT AND CLOTHING

Whilst there are no requirements to wear or use specific garments or equipment, the following options are strongly recommended.

- Gloves help to protect against fire and abrasion in an accident. They also absorb perspiration, improving grip.
- Overalls made from natural fibres, with zippered pockets, and close fitting ankles, collar and wrists also give protection, as do leather flying boots.
- Particularly when flying open cockpit aeroplanes, a lightweight helmet gives protection while minimising discomfort under increased 'G' loadings.
- Parachutes are useful emergency equipment and in the event of failure to recover from a manoeuvre may be the only alternative to a fatal accident. However, for physical or weight and balance reasons their carriage may not be possible or practicable in some aircraft. Even if their carriage is practicable, the effort required and height lost while exiting the aircraft (and while the canopy opens) must be considered. If worn, the parachute should be comfortable and well fitting with surplus webbing tucked away before flight. It should be maintained in accordance with manufacturer's recommendations. Know, and regularly rehearse, how to use it, and remember the height required to abandon your aircraft when deciding the minimum recovery height for your manoeuvres.
- Don't take any potential loose articles e.g. coins, keys.

5 INSTRUCTION

a. As with any other aspect of aviation, the acquisition of skill and knowledge is most effective and enjoyable with high quality instruction. (There is a formal qualification to enable instructors to teach aerobatics.) Effective pre-flight briefing is essential if full benefit is to be gained from any course of training. Initially, keep the flight lessons as short as possible concentrating on simple, positive 'g' manoeuvres, such as loops and barrel rolls, to start with. Make the post flight analysis and discussion session a worthwhile contribution towards the next training flight.

b. The Aircraft Owners and Pilots Association, (AOPA), have published an aerobatic syllabus and training is available at some flying clubs. General handling, to revise those skills acquired during PPL training is necessary before learning basic aerobatic manoeuvres. Since the PPL syllabus now only includes incipient spinning, you **must** now become familiar with entry to and recovery from a fully developed spin since a poorly executed aerobatic manoeuvre can result in an unintentional spin. Training in recovery from incorrectly executed manoeuvres and unusual attitudes is essential. On completion of the AOPA Aerobatic Course a pilot should be capable of flying a simple sequence of manoeuvres in a safe manner.

c. Know the spin characteristics of the aircraft even though you may have no intention of entering a spin. Know also the different symptoms of erect and inverted spins and the appropriate recovery drills for each type of spin.

d. Ensure you learn the safest way of recovering from each manoeuvre if it goes wrong and be prepared to use it in the future. Continuing to pull is usually less safe than rolling to the nearest horizon.

e. Novices should not attempt new manoeuvres without proper qualified instruction, the result could be an over-stressed aircraft or an accident.

6 AIRCRAFT CHECKS

a. Maintain a close liaison with the person/organisation responsible for maintenance so that the maintenance schedule may be interpreted to its best effect when taking into account the particular needs of an aerobatic aircraft. Homebuilt and other 'Permit' aircraft are not subject to a formal Maintenance Schedule, thus the engineer who maintains the aircraft should be familiar with the type and the critical areas to inspect.

b. The pre-flight inspection needs to be carried out with extra care, since the aircraft will be flown nearer to its performance and structural limits than usual. Ask other owners/users of the specific aircraft type about items which need particular scrutiny.

c. Check that items of cockpit equipment, such as the fire extinguisher, are properly secured and check VERY carefully for any loose objects which might be present. Even the most insignificant item could lodge in such a manner as to restrict control movement. Dust and dirt from the floor, under negative 'g' situations can get in the pilot's eyes.

d. Make sure that there is sufficient fuel for the flight whilst still remaining within the aerobatic weight and cg envelope.

7 PREPARATION FOR FLIGHT

a. Make sure you all are tightly strapped in, yet still able to move the controls to their full travel without difficulty. It is essential that you feel part of the aircraft and not a loose object within it. Tuck away the surplus harness adjustment.

b. Check that the rudder, which on the ground may be restricted by nosewheel steering or braking, does have FULL travel.

8 PRE-AEROBATIC VITAL ACTIONS

a. Weather conditions must be suitable. There must be good visibility, a clear horizon all round and space to remain clear of cloud under VFR.

b. Allow plenty of height from ground to cloud base. Recognition and recovery from an inadvertent spin and the subsequent dive may require many hundreds of feet, (e.g. a Chipmunk requires 250 feet per turn and 1200 feet for the dive recovery). In an aircraft with a low power to weight ratio, remember to allow sufficient height to complete an aerobatic sequence before reaching the base height.

c. Be considerate to those on the ground. Do not always use a particular area for aerobatic practice to the annoyance of those who desire peace and quiet. Avoid also regular VFR routes and areas well known to have frequent traffic, e.g. PPL training areas.

d. The standard HASELL check needs to be carried out with particular vigilance.

- **Height** - depends on experience of pilot, but novices should commence at no less than 5000 ft above ground level and all manoeuvres should be completed by 3000 ft agl.
- **Airframe** - flaps up, brakes off, (in some aircraft brake application restricts rudder movement), wheels up, etc to suit your particular aircraft.
- **Security** - all harnesses fastened, canopy/doors secure and no loose articles.
- **Engine** - all engine instruments reading normally, mixture rich, carb heat check, adequate fuel selected and electric fuel pump on if applicable.
- **Location** - clear of congested areas and outside or below any controlled airspace (unless appropriate permission from the controlling ATC unit has been given). An area offering good forced landing options in the event of engine problems is wise. Note a good landmark to assist orientation.
- **Look-out** - clearing turns in both directions and check above and particularly below.

e. Look-out needs to be comprehensive at all times, checking between manoeuvres and sequences, to avoid any risk of confliction with other aircraft.

9 SPORTING AND COMPETITION ASPECTS

a. Once the basic skills have been mastered, many pilots are quite content with the occasional aerobatic flight in a club aircraft to enhance their pleasures of aviation.

b. However, some pilots enter competitions to measure their ability against others at a similar level of attainment.

c. Competition aerobatics is an international sport under the Federation Aeronautique Internationale. The relevant sporting regulations have been prepared by the International Aerobatics Committee who sanction both World and Continental championships. The Royal Aero Club of the United Kingdom has recognised the British Aerobatic Association* as the sport's representative body to foster its development and to organise national competitions.

d. Contests are held at a number of venues each year at the various levels of pilot skill. These are Beginners, Standard, Intermediate, Advanced and Unlimited. Aircraft performance is a major factor in progression up through the system, however a well flown sequence in a basic aircraft can be just as competitive.

*Address:
BAeA, White Waltham Airfield,
Nr Maidenhead,
Berks.
SL6 3NJ
Phone: (01455) 617211

10 AIR DISPLAYS AND PUBLIC EVENTS

Before a pilot can perform at an Air Display or public events, he/she MUST have a Display Authorisation permitting aerobatics issued by the CAA (see CAP 403 – Flying Displays and Special Events: a guide to safety and administrative arrangements), for details.

11 SUMMARY

- **Get dual instruction before attempting aerobatics.**
- Check that the aircraft is cleared for aerobatics and know both the aircraft and your own limitations.
- Be proficient with recoveries from spinning and unusual attitudes.
- Start with sufficient height to give plenty of margin if things go wrong.
- Maintain a good look-out and monitor your height constantly.
- Do not exceed the 'g' limits, or use large control movements near or above maximum manoeuvring speed Va. Do not exceed Vne, the never exceed speed.
- Do not exceed maximum engine RPM or manifold pressure limitations.
- Ensure you know the escape route for each manoeuvre if it goes wrong – and use it when necessary.
- **Never be tempted to show off with low aerobatics or beat-ups**
- .

SAFETYSENSE LEAFLET 20b

VFR FLIGHT PLANS

1 INTRODUCTION
2 LEGISLATION
3 DEPARTURES FROM AIRPORTS
4 DEPARTURES FROM STRIPS ETC
5 RETURNING TO THE UK
6 COMPLETION OF THE FLIGHT PLAN
7 SOME GENERAL TIPS FOR VFR FLIGHT PLANNING
ANNEX A – COMMON ICAO DESIGNATORS

1 INTRODUCTION

With the removal of barriers in the European Community, it is now convenient for General Aviation pilots to fly both from their local airfield/ airport, as well as their farm strip, direct to the Continent. However, although British Customs & Excise and Immigration have simplified their systems, the French Authorities have not and it is still necessary to land at a French airport with Customs and Immigration facilities in order to enter France. It is not this leaflet's intention to describe the relaxed procedures operating for Customs here in the UK – readers are advised to contact their local Customs and Excise Office to discuss their own individual arrangements.

2 LEGISLATION

a. VFR flight plans (FPLs) **must** be filed for the following flights:

- A flight to or from the United Kingdom which will cross the United Kingdom FIR boundary.
- A flight within Class D control zones/control areas. However, this requirement may be satisfied by passing flight details by Radio Telephony (RT).
- A flight within the Scottish and London Upper Flight Information Regions, (but since this will be above Flight Level 245, it seems

unlikely that many GA pilots will be concerned with this situation).

b. Other requirements exist for flights where an aircraft's maximum take- off weight exceeds 5700 kg (12500 lbs).

c. In addition, it is **advisable** to file a VFR FPL if the flight involves flying over the sea, more than 10 nm from the UK coastline or flying over sparsely populated areas where Search and Rescue operations might be difficult. In addition, a VFR FPL **may** be filed for any flight at the pilot's discretion.

d. The Terrorism Act 2000 applies to flights between the mainland UK and the Republic of Ireland, Northern Ireland, the Isle of Man and the Channel Islands.

e. Some European Countries do not accept aircraft which only have a Permit to Fly, (homebuilt aircraft/microlights etc). It is the responsibility of the pilot/ operator to obtain permission beforehand from the State concerned.

f. In addition, some if not all of the following documents may be required to be carried in the aircraft: Tech. Log; Certificates of Registration, Airworthiness & Maintenance Release; Radio Licence; Interception Procedures (*leaflet 11*); Load Sheet; Pilot's Licence; Insurance Certificates and your passport.

3 DEPARTURES FROM AIRPORTS

a. Assuming that the departure and destination aerodromes are both major airports, the operation of the FPL is as follows. You complete the FPL at the Air Traffic Service Unit (ATSU) of your departure aerodrome and they will file it into the system on your behalf. The effect of this filing will be to inform your destination airfield, together with any of your alternates, that the flight is going to take place.

b. Once you get airborne, the ATSU will file a 'departure (DEP)' message and this will 'activate' the FPL. Thus the destination airfield, knowing your estimated time en-route from the filed FPL, and now knowing your departure time, will have an estimated time of arrival (ETA) at their airport.

c. Once you arrive, they will 'close' the FPL on your behalf, and that marks the end of the operation. If, however, you do not arrive **within 30 minutes** of your ETA, they will institute overdue action and subsequently, Search and Rescue operations may commence. It is therefore essential that if you land at any airfield other than your destination, you **MUST** inform your original destination of this fact, otherwise they will institute overdue and Search and Rescue action, the cost of which may be passed onto you.

d. This has covered the ideal situation where others handle it for you.

4 DEPARTURES FROM STRIPS etc

a. If the aerodrome that you operate from is:
- an airfield or airport with an ATSU, but your operations are outside their normal hours, or
- an airfield without an ATSU, or
- a private strip.

The responsibility for filing, activating and closing a FPL now rests with the pilot.

b. At this stage, it is important to understand the concept of the 'parent ATSU'. The UK is divided into a total of three areas, each of which has a parent ATSU and the map overleaf shows their areas of responsibility and the table beneath shows the telephone and fax numbers of the Flight Briefing Unit that you should telephone or fax when flight planning.

c. To file a FPL, telephone or fax the Flight Briefing Unit at least 60 minutes before the intended flight. A fax is cheaper than a telephone call. Prior to departure, arrange for a **responsible person** on the ground to telephone the Flight Briefing Unit as soon as you are airborne in order to pass a departure time. This has now activated the FPL. This is a very simple procedure and a suitable **responsible person** could be your spouse, relative, friend, fellow pilot or secretary. Passing an airborne time over the RT could lead to a delay if the controller is busy. If it is not possible to file a FPL on the ground, it can be filed while airborne with any ATSU, but normally with the FIR controller responsible for the area in which the aircraft is flying. In such cases the message should begin with the words 'I wish to file an airborne flight plan'. Once again, when this method of filing is used, delays can occur due to controller workload.

5 RETURNING TO THE UK

a. Prior to departure for the return flight to an airfield without an ATSU (when closed for instance) or to a private strip, pilots are responsible for informing a **responsible person** at the destination of the estimated time of arrival. The responsible person is required to notify the parent ATSU if the aircraft fails to arrive within 30 minutes of the ETA. This action will then trigger the parent ATSU into alerting, overdue and Search and Rescue action. Obviously this person **MUST** have the telephone numbers of the appropriate parent ATSU. If the parent ATSU fails to hear anything, it will assume that the flight landed safely i.e. NO NEWS IS GOOD NEWS and no further action is required. If the responsible person does inform the parent ATSU of your non-arrival, then the parent ATSU will go back to the filed FPL to check departure times, routings and so on as part of the Search and Rescue procedures.

b. It can be seen that the **responsible person** is crucial to this operation. If no one is expecting you, no one will be looking for you if you do not arrive. If, in an extreme case, the pilot fails to find a **responsible person** at his destination, then he may contact his parent ATSU prior to departure and request then to act in the capacity of the **responsible person**. Should the pilot follow this course of action, he will be required to contact the parent ATSU within 30 minutes of landing at his destination or diversion airfield, to confirm his arrival. Failure to do this, will automatically result in the parent ATSU initiating alerting action.

CAA Safety Sense Leaflet   20 VFR Flight Plans

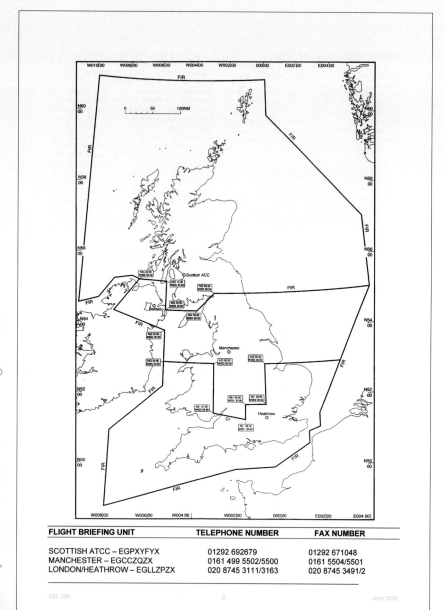

FLIGHT BRIEFING UNIT	TELEPHONE NUMBER	FAX NUMBER
SCOTTISH ATCC – EGPXYFYX	01292 692679	01292 671048
MANCHESTER – EGCCZQZX	0161 499 5502/5500	0161 5504/5501
LONDON/HEATHROW – EGLLZPZX	020 8745 3111/3163	020 8745 3491/2

6 COMPLETION OF THE FLIGHT PLAN

(Note that this is an abbreviated explanation intending to cover simple VFR flights. Full details are obtainable from CAP 694 (The UK Flight plan Guide). An ICAO poster on completing Flight Plans is available from Documedia Solutions at Cheltenham.
Enter all details in block capitals.

Leave the top part of the form blank, i.e. start at item 7.

ITEM 7 AIRCRAFT IDENTIFICATION

INSERT AIRCRAFT REGISTRATION when the radiotelephony call sign will be the aircraft registration (OMIT THE HYPHEN)

ITEM 8 FLIGHT RULES

TYPE OF FLIGHT

INSERT V – VFR to denote the category of flight rules (other letters apply if you plan to fly under IFR)

INSERT G – General Aviation to denote the type of flight

ITEM 9 NUMBER

TYPE OF AIRCRAFT

WAKE TURBULENCE CAT

INSERT Number of aircraft only if more than 1

INSERT AIRCRAFT TYPE DESIGNATOR or ZZZZ if no designator or formation flight comprising more than one type (see item 18 TYP) Note: Aircraft Type Designators for most types are shown in Annex A to this leaflet.

INSERT L – Light (17 000 kg or less)

ITEM 10 EQUIPMENT

INSERT Preceding the oblique stroke one letter as follows:

N – if no COM NAV Approach aid equipment for the route to be flown is carried, or the equipment is unserviceable. OR

S – if standard COM NAV Approach aid equipment for the route to be flown is carried and serviceable. (Standard equipment is considered to be VHF RTF, ADF, VOR and ILS unless another combination is prescribed by the appropriate ATS Authority.) Individual letters apply to each item of navigation equipment.

THEN following the oblique stroke
INSERT one of the following to describe the serviceable SSR equipment carried

N – nil
A –Transponder Mode A 4096 Codes
C – Transponder Mode A 4096 Codes and Mode C

ITEM 13 DEPARTURE AERODROME

TIME

INSERT LOCATION INDICATOR of the departure aerodrome or ZZZZ if no ICAO location indicator assigned (see item 18 – DEP).

INSERT ESTIMATED OFF- BLOCK TIME in Universal Co-ordinated Time (UTC).
Note: Location Indicators are given in UK AIP and most flight guides.

ITEM 15 CRUISING SPEED

LEVEL

ROUTE

CAA Safety Sense Leaflet — 20 VFR Flight Plans

INSERT CRUISING TRUE AIR SPEED for initial or whole cruise as follows:

N (knots) followed by 4 digits (e.g. N0125)
(K = kilometres per hour)
Note: there is no provision for statute mph

INSERT CRUISING LEVEL for initial or whole cruise as follows:

A – Altitude in hundreds of feet (use 3 digits e.g. A025)
F – Flight Level (use 3 digits e.g. F055). OR
VFR – for uncontrolled VFR flights.

INSERT the ROUTE to be flown as follows:

for flights OFF designated routes, list points normally not more than 30 minutes flying time apart and enter DCT (DIRECT) between successive points. Points may be navigation aids, or bearing/distances from these (as "DVR05010", or co-ordinates.

ITEM 16 DESTINATION AERODROME

[ ]

TOTAL EET

[ ]

ALTN AERODROME

[ ]

AERODROME

[ ]

INSERT LOCATION INDICATOR of the designation aerodrome or ZZZZ if no assigned indicator (see item 18 – DEST)

INSERT TOTAL ESTIMATED ELAPSED TIME (EET) en route as a four figure group expressed in hours and minutes.

INSERT LOCATION INDICATOR(S) of no more than two alternate aerodromes or ZZZZ if no assigned indicator(s) (see item 18 ALTN).

ITEM 18 OTHER INFORMATION

INSERT 0 (zero) if no other information OR any other necessary information in the preferred sequence shown hereunder, in the form of the appropriate indicator followed by an oblique stroke and the information to be recorded

EET/ – Significant points or FIR boundary designators and accumulated Estimated Elapsed Times to such points or FIR boundaries, when so prescribed on the basis of regional air navigation agreements or by ATS authority (e.g. EET/EGTT0020 LFF0105 Or EET/ EINN0204)

TYP/ – TYPe(s) of aircraft, preceded by the number(s) of aircraft in a formation flight, if ZZZZ is used in item 9.

DEP/ – Name of DEParture aerodrome if ZZZZ is inserted in item 13.

DEST/ – Name of DESTination aerodrome, if ZZZZ is inserted in item 16.

ALTN/ – Name of ALTerNate aerodrome(s) if ZZZZ is inserted in item 16.

RMK/ – any additional information.

ITEM 19 SUPPLEMENTARY INFORMATION (NOT TO BE TRANSMITTED IN FPL MESSAGES)

ENDURANCE – used a four-figure group to express fuel endurance.

PERSONS ON BOARD –includes passengers and crew, use TBN if number not known at time of filing.

EMERGENCY RADIO –cross out equipment not available.

SURVIVAL EQUIPMENT –cross out equipment not available including S if none carried.

JACKETS – same as above and cross out J if no jackets carried.

DINGHIES – cross out both D and C if no dinghies carried.

REMARKS – enter other remarks regarding survival equipment or cross out N if no remarks.

FILED BY – insert name of the unit, agency or person filing the flight plan.

FLIGHT PLAN — ATS COPY

PRIORITY: ≪≡ FF

ADDRESSEE(S):

FILING TIME → ORIGINATOR ≪≡

SPECIFIC IDENTIFICATION OF ADDRESSEE(S) AND/OR ORIGINATOR

- 3 MESSAGE TYPE: ≪≡ (FPL
- 7 AIRCRAFT IDENTIFICATION: G B O M Z
- 8 FLIGHT RULES: V
- TYPE OF FLIGHT: G ≪≡
- 9 NUMBER: —
- TYPE OF AIRCRAFT: P A 3 8
- WAKE TURBULENCE CAT: / L
- 10 EQUIPMENT: S/C
- 13 DEPARTURE AERODROME: E G K A
- TIME: 1 1 0 0 ≪≡
- 15 CRUISING SPEED: N 0 1 2 0
- LEVEL: V F R
- ROUTE: DCT SAM DCT YVL DCT

≪≡

- 16 DESTINATION AERODROME: Z Z Z Z
- TOTAL EET HR MIN: 0 0 4 5
- ALTN AERODROME: E G T E
- 2ND ALTN AERODROME: E G H H ≪≡
- 18 OTHER INFORMATION: DEST/DUNKESWELL

) ≪≡

SUPPLEMENTARY INFORMATION (NOT TO BE TRANSMITTED IN FPL MESSAGES)

- 19 ENDURANCE HR MIN: –E/ 0 3 0 0
- PERSONS ON BOARD: P/ 0 0 2
- EMERGENCY RADIO UHF: R/ X
- VHF: X
- ELBA: E
- SURVIVAL EQUIPMENT: S
- POLAR: X
- DESERT: X
- MARITIME: M
- JUNGLE: X
- JACKETS: J
- LIGHT: / L
- FLUORES: F
- UHF: X
- VHF: V
- DINGHIES NUMBER: D/ 0 1
- CAPACITY: 0 0 8
- COVER: C
- COLOUR: YELLOW ≪≡
- AIRCRAFT COLOUR AND MARKINGS: A/ BLUE/WHITE
- REMARKS: X / ≪≡
- PILOT IN COMMAND: C/ EDWARDS) ≪≡

FILED BY — SPACE RESERVED FOR ADDITIONAL REQUIREMENTS

CA48/RAF F2919 (REVISED NOVEMBER 1985) Certo D.O. Drg No 8810

7 SOME GENERAL TIPS FOR VFR FLIGHT PLANNING

a. The procedures as outlined above will work when filing FPLs over inhospitable areas or mountainous terrain in the UK. In this case, it can be seen that you will need a **responsible person** at both your departure and destination airfield and both of those will need to have the telephone number of the parent ATSUs in both your departure area and your destination area if they are different.

b. To make the process of filing a FPL over the telephone as speedy as possible, have a copy of your FPL ready filled in, so that you can pass the information quickly in the correct order.

c. Many pilots now file their FPLs by fax. It is suggested that you include a contact telephone number in the remarks section, or better still, **phone the office direct to confirm that the plan has been received**.

d. A test showed that it took well over a minute to fax the top copy of the FPL due to the shaded area, while the non-shaded COM copy took under 15 seconds. Either copy is acceptable for this purpose.

e. If your FPL is for a future date, make sure that the date is entered clearly in the remarks section, item 18 (e.g. RMK/ DATE OF FLIGHT 12 APRIL).

f. It is essential that ATC is advised of cancellations, delays over 30 minutes and changes to FPL details. To prevent a double entry into the computer which would lead to confusion, always cancel the first FPL and resubmit.

g. When departing from smaller airfields, do not assume that the Air Ground Operator or FISO will automatically telephone a departure time to the parent ATSU on your behalf, check with them or, once again, find a **responsible person** to do this for you.

h. All in all, the procedure is intended to simplify VFR FPLs and to move the onus for safe operation on to pilots.

ANNEX A
ICAO TYPE DESIGNATORS

(This list only covers some common light aircraft/ helicopters on the UK Civil Register. The complete list is in ICAO Document 8643, available at most large aerodromes or through the ICAO web site http://www.icao.int/anb/ais/8643/index.cfm

AGUSTA	A109
AGUSTA/ BELL	
206 Jet Ranger, Long Ranger	B06
BEAGLE Pup	PUP
Terrier	AUS6
BEECH (RAYTHEON)	
most as types e.g. 19	BE19
CESSNA (INC REIMS)	
most as numbers e.g.	C152
except some complex e.g.	C82R
DE HAVILLAND as types e.g.	
Tiger Moth	DH82
and Chipmunk	DHC1
DIAMOND DA-20/22 Katana	DV20
EUROPA	EUPA
FOURNIER as types e.g.	RF4
FUJI FA-200	SUBA
GROB most as types	
except complex e.g.	G109
GRUMMAN AMERICAN	
most as type e.g.	AA5
JODEL most as types	
e.g. D-9 and series	D9
LUSCOMBE Silvaire	L8
MOONEY M-20, 201	M20P
231 etc (turbo charged)	M20T
MORANE SAULNIER Rallye	RALL
MUDRY most as number	
e.g. CAP- 10	CP10
PIPER most as type nos e.g.	J3
but most PA28 piston, fixed gear	P28A
PA28 Arrows	P28R
PA28 RT	P28T
PA23 Aztec	PA27
ROBIN DR- 400 series	DR40
ROBINSON as type nos e.g. R- 22	R22
ROCKWELL Commander	
112, 114 etc	AC11
RUTAN Varieze	VEZE
SLINGSBY T67 Firefly	RF6

SAFETYSENSE LEAFLET 21b

DITCHING

Piel Emeraude – Irish Sea 1991

1 INTRODUCTION
2 KNOWLEDGE
3 PREPARATION
4 PRACTICE
5 MAIN POINTS
   SUPPLEMENT A
   SUPPLEMENT B

1 INTRODUCTION

a. Ditching is a deliberate emergency landing on water. It is **NOT** an uncontrolled impact.

b. Available data from both UK and USA indicates that 88% of controlled ditchings are carried out with few injuries to pilots or passengers. There is no statistical survival difference between high wing and low wing aeroplanes. However, despite most ditchings being survivable, approximately 50% of survivors die before help arrives.

c. This leaflet is mainly aimed at private operators of aeroplanes but much of the advice will be equally relevant for helicopters. It includes details of how to improve the chances of survival after a ditching.

d. Details of the UK Search and Rescue System together with appropriate advice, are available in the AIP GEN 3-6.

2 KNOWLEDGE

a. Do you know how far YOUR aircraft can glide per 1000 ft of altitude in still air? It's in the Pilots Operating Handbook or Flight Manual.

b. The main cause of death after ditching is drowning, usually hastened by hypothermia and/or exhaustion. It is essential to consider the reasons for this and how the risks may be minimised.

c. In many cases, the deceased persons did not have lifejackets, either worn or available to them. It is vital TO WEAR a suitable lifejacket whilst flying in a single engined aircraft over water beyond gliding range from land.

d. Selection of the correct lifejacket is most important, since there are many different types available. Some so-called 'lifejackets' are in fact little more than buoyancy aids which are used for leisure boating and have a permanent buoyancy of about 7 kg (15 lbs). This kind of 'lifejacket' will not keep an unconscious person afloat. Worse still, the inherent buoyancy may prevent a person from escaping from an inverted aircraft.

e. A proper lifejacket provides 16 kg (35 lb) of buoyancy which can be enough to keep an unconscious person afloat with the head above water. It is essential to use a lifejacket designed for constant wear since this has the ruggedness and durability to prevent tearing and other damage during normal use.

f. Many automatically inflated lifejackets, used by the sailing community, are activated when a soluble tablet becomes wet. This type is totally unsuitable for general aviation use as they will inflate inside a water-filled cabin, thus seriously hindering escape.

g. Airline lifejackets provided for passengers are unsuitable for GA use, because they are not durable enough for significant constant wear.

h. When being worn, the lifejacket should not become entangled in harness/belt. It should include the following (see supplement B):
- a light activated by pulling a toggle or by immersion in sea water;
- a whistle for attracting attention;
- a crotch strap to stop the lifejacket from riding up over the face;
- a spray hood or plastic face mask which can be pulled over the face and lobes of the jacket. It will reduce heat loss through the head as well as the amount of water flowing across the face;
- high visibility colour with reflective tape.

i. Wearing a suitable lifejacket, is not the end of the story. When not in use, the lifejacket must be properly stored in a dry environment and regularly serviced.

j. A lifejacket should be serviced at least every year (more frequently if required by the manufacturer) by an approved servicing organisation or appropriately licensed engineer. The weight, and thus contents, of the gas cylinder will be checked, and the life-jacket itself examined for damage and leaks; and ancillary equipment inspected for serviceability.

k. Whilst properly fitted lifejackets can prevent people from drowning, none provide any protection against hypothermia

l. Hypothermia is defined as lowering of the 'core' body temperature. In cold water, the skin and peripheral tissues cool very rapidly, but it can be 10 to 15 minutes before the temperature of the heart and brain begin to decrease. Intense shivering occurs in a body's attempt to increase its heat production and counteract the large heat loss. Decreasing consciousness, mental confusion and the loss of the will to live occur when core body temperature falls from the normal 37° C to about 32° C. Heart failure is the usual cause of death when core body temperature falls below 30° C.

m. **The temperature of the sea around British coasts is at its coldest in March, and below 10°C between October and April.** Survival times for individuals in cold water will vary greatly depending on water temperature, individual build, metabolism, fitness and the amount of clothing worn. The graph shows average survival times. Note that without a life-raft or survival suit **there is little difference between survival times in summer and winter**.

LIKELY SURVIVAL TIME FOR RELATIVELY THIN PERSON IN CALM WATERS WITH NO LIFERAFT

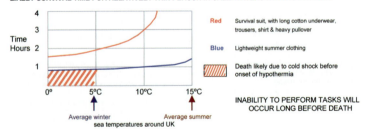

INABILITY TO PERFORM TASKS WILL OCCUR LONG BEFORE DEATH

n. In addition, several other responses to the shock of sudden immersion in cold water can cause death:
- heart failure is possible for those with weak circulatory systems, particularly the elderly;
- hyperventilation can increase the risk of swallowing water;
- cold makes coordinated movement difficult;
- ability to hold ones breath is severely curtailed, perhaps to just a few seconds, thus reducing the chances of successful escape from a submerged aircraft.

The effect of shock and panic can amplify the above effects, so it is important to consider ways of reducing the risk of both cold shock and hypothermia.

o. Clearly, the ideal solution is to get out of the water by using a life-raft.

p. As with lifejackets, an aviation life-raft, with a recognised approval, is the safest option and this must also be regularly serviced and properly stored when not in use. The use of a life-raft, together with other survival tips, are detailed later in this leaflet. However, it is important to know how to use all your survival equipment.

q. A marine life-raft is NOT suitable for aviation use because of a significant difference in the inflation system. Any malfunction of a marine CO 2 cylinder will cause it to vent INTO the life-raft, inflating it, and filling the cockpit possibly causing catastrophic results. Aviation life-raft cylinders are designed to vent to atmosphere in the event of a malfunction. (Just in case, carry a pocket knife or screwdriver.)

r. If, for any reason, a life-raft is not available, the survival time in cold water can be significantly increased by wearing suitable protective clothing.

s. A survival suit specially tailored for general aviation use is most effective, and can prolong life by keeping hypothermia at bay for the longest time. Whilst some pilots may feel that this level of protection is 'over the top' for a cross Channel flight, there have been cases where lives have been saved by the wearing of such clothing. A leak-proof suit, properly worn, can increase survival time from 3 to 10 times depending on the insulating qualities of the clothes worn underneath. Wear several layers of suitable clothing to create layers of air.

t. As with all safety and survival equipment, it should be the correct type, with a recognised approval, be a comfortable fit, properly maintained and serviced, and carefully stored when not in use.

u. If a survival suit is not used, then generally, the more layers of clothes that are worn, the longer will be the survival time. This will vary considerably depending on the type of clothing and the amount being worn. If time permits, put on as much clothing as possible, including headwear, since a very large proportion of body heat escapes through the head. Wet wool retains 50% of its insulating properties, whereas wet cotton retains only 10%. Watersport suits could also be considered.

v. A Personal Locator Beacon (PLB) is a portable radio transmitter which will greatly assist in locating you after ditching. It should be able to float, have a satisfactory power output and provide a continuous signal for 24 hours. The modern generation operate on 406 MHz although older versions operating on 121.5 MHz are still available. Some have GPS and transmit. position information.

755

w. Signals from both types can be received by orbiting INMARSAT or dedicated COSPAS/SARSAT satellites. These relay alerts to the Aeronautical Rescue Co-ordination Centres. The time between activation and alerting the RCC should not exceed 90 minutes in the worst case.

x. Pilots should attempt to transmit an initial distress call on a conventional communications radio BEFORE ditching to alert the RCC. The PLB transmissions can then guide the rescue services to you.

y. Some PLBs are designed to float in the water with the transmitting aerial pointing upwards – the aerial's optimum transmitting position. Most PLBs have a battery test facility. Users **MUST NOT** test the transmit facility – this must only be done by an avionics engineer.

3 PREPARATION

a. Many ditchings and subsequent drownings could have been prevented by careful planning and preparation.

b. Those who frequently fly over-water should consider attending a survival course. Here, in a non-threatening environment, you will be taught the correct operation of lifejackets, methods of getting into life-rafts and the problems which might be encountered after a ditching.

c. Some specialist companies arrange sessions in swimming pools with wave machines whilst others have light aircraft structures which can be used as 'dunkers' to practice underwater escapes.

d. On the day of the flight, obtaining and correctly interpreting the weather forecast is vital. Whilst the weather might be pleasant on one side of the Channel, it may be very different on the other side. It would be no fun to leave English shores in CAVOK, only to struggle against unexpected headwinds, find sea fog or lowering cloudbases resulting from warm air over the cold sea any of which could force you to return.

e. Use forecast wind to ensure that enough fuel is onboard for the flight, **plus any diversions**, which may include a return from overhead the destination or else to a suitable alternative airfield. In many accidents and some ditchings, the reason for engine stoppage has proved to be fuel exhaustion.

f. Thorough pre-flight inspection of the aircraft is essential, including double-checking that fuel and oil levels are satisfactory.

g. A 4-person life-raft can weigh as much as 15 kg (35 lb) and is a significant extra load. Take care to determine the total weight and centre of gravity position and take these into account (*see SafetySense leaflet 9 – Aircraft weight and balance*).

h. Pilots must review any recommended procedures contained in the Aircraft Flight Manual or Pilot's Operating Handbooks for both a power-on and power-off ditching.

i. The law requires that, as commander of the aircraft, you **MUST** brief the passengers on the emergency escape features of the aircraft, operation of the seats, seatbelts etc. On a flight across water in a single engined aircraft, this briefing should be extended to ensure that each passenger knows how to operate the lifejacket they should be wearing. Brief the passengers on the contents and the features found on the lifejacket, including how to inflate it if the bottle fails.

j. Before boarding the aircraft, brief the passengers carefully:

- on the location of the life-raft;
- on the order in which people should vacate the aircraft in the event of a ditching and who will be responsible for taking the life-raft with them;
- that lifejackets must **not** be inflated until clear of the aircraft and that the instructions normally state – 'pull the toggle' to inflate;
- to remove headsets and glasses and to stow glasses on their person prior to touchdown
- tighten seat straps/harnesses prior to touchdown on the water. Rear seat passengers should assume a braced position;
- indicate reference points on the aircraft's internal structure that they should reach for when exiting the aircraft as well as any features which might impede exit.

CAA Safety Sense Leaflet   21 Ditching

k. The life-raft must be SECURED in an accessible position. If flying alone, place the life-raft on the front passenger seat and secure it with the harness. Check it will not interfere with the controls, lookout or exit.

l. Some pilots have a hand-held VHF radio or mobile phone; put them in a sealed plastic bag along with any hand held GPS in order to keep them dry. A waterproof torch or better still a portable waterproof strobe could also be useful.

m. Once airborne, particularly over the sea, it is prudent to fly as high as can be safely and legally flown. This will give better radio reception and more time between the onset of a problem and ditching. Consider a high level longer crossing compared with a short one at low level.

n. Before crossing the coast, carry out a particularly careful cruise check (FREDA check) to ensure that everything is normal.

4 PRACTICE
4.1 Ditching

a. The worst has happened – you are unable to maintain height and a ditching appears likely. If you are flying a twin-engined aircraft and one engine stops, everyone should put on a lifejacket. Make a PAN call.

b. Immediately adjust the airspeed for the best glide speed and taking into account the wind direction either aim towards the nearest coast or towards shipping. Remember, that a medium size vessel is the best choice to ditch near, since a large ship may take many miles to slow down. In any event, avoid landing immediately in front; landing alongside and slightly ahead is better.

c. At this stage, transmit a MAYDAY call, using the frequency you are working on or the emergency frequency of 121.5 MHz. If fitted, immediately select transponder code to 7700, unless you are already using an allocated code. Transmit the best position fix that you can, this may be by means of VOR, DME or GPS or even your estimate in relation to the coastline. **Make this as accurate as you can.**

d. Check immediately for any problem which can be dealt with by vital actions such as: selecting carburettor heat, change of fuel tank, use of the electric fuel pump, etc.
ABOVE ALL, THROUGHOUT, FLY THE AIRCRAFT.

e. Conventional wisdom is that the swell direction is more important than wind direction when planning a ditching. By the time you are down to 2000 ft, the swell should be apparent and your aim should be to touchdown parallel to the line of the swell, attempting, if possible, to land along the crest. The table below describes sea states.

Wind Speed	Appearance of Sea	Effect on Ditching
0– 6 knots (Beaufort 0– 2)	Glassy calm to small ripples	Height very difficult to judge above glassy surface. Ditch parallel to swell
7– 10 knots (Beaufort 3)	Small waves; few if any white caps.	Ditch parallel to swell.
11– 21 knots (Beaufort 4– 5)	Larger waves with many white caps.	Use headwind component but still ditch along general line of swell.
22– 33 knots (Beaufort 6– 7)	Medium to large waves , some foam crests, numerous white caps.	Ditch into wind on crest or downslope of swell.
34 knots and above (Beaufort 8+)	Large waves, streaks of foam, wave crests forming spindrift	Ditch into wind on crest or downslope of swell **Avoid at all costs ditching into face of rising swell**

757

f. If you can see spray and spume on the surface, then the surface wind is strong. In this case it is probably better to plan to land into wind, rather than along the swell. Winds of 35 to 40 kts are generally associated with spray blowing like steam across waves and in these cases the waves could be 10 ft or more in height. Aim for the crest again or, failing that, into the downslope.

g. The force of impact can be high so ditch as slowly as possible whilst maintaining control.

h. Retractable gear aircraft should be ditched with the gear retracted (beware of automatic lowering systems). The Flight Manual/ Pilot's Operating Handbook may provide suitable advice. Consider unlatching the door(s).

i. Hold the aircraft off the water so as to land taildown at the lowest possible forward speed, but do not stall into the water from a height of several feet.

j. There will often be one or two minor touches, 'skips', before the main impact with the water. This main impact will usually result in considerable deceleration with the nose bobbing downwards and water rushing over the cowling and windshield. It may even smash the windshield – leading you to think that the aircraft has submerged.

k. With a high wing aircraft, it may be necessary to wait until the cabin has filled with water before it is possible to open the doors. A determined push or kick on the windows may remove them.

l. The shock of cold water may adversely affect everyone's actions and this is where the importance of the proper pre- flight passenger briefing which emphasised reference points and the agreed order in which to vacate the aircraft becomes apparent. Do **NOT** inflate lifejackets inside the aircraft, inflate them as soon as you are outside. The natural buoyancy of the un-inflated life-raft may make it hard to pull it below the surface to get it out of a sinking aircraft.

m. Consider leaving the master switch and the anti- collision beacon or strobes on. If the aircraft floats for a while or sinks in shallow water, the lights may continue operating and provide a further sign of your position. Exit the aircraft as calmly, but as swiftly as possible. If it is afloat after the passengers are clear, provided you don't put yourself in danger, deploy loose items that could float on the surface and help rescuers spot you, e.g. blankets, overnight bags, seat cushions. The first aid kit and plastic bag with PLB, handheld radio, phone etc. should be taken with you.

4.2 The Life-raft

a. Before inflating the life-raft, it should be tied to someone holding firmly onto the aircraft, so that it doesn't blow away. (It will float even before it is inflated.) Do **NOT** attach it to the sinking aircraft. The lifejacket harness or belt would be a good attachment point. If possible, inflate the life-raft on the downwind side so that it is not blown against the aircraft and damaged. (A pocket knife to cut the cord would be easier than trying to undo a wet knot.) If necessary and you are able to stand on the wing, it may be easier to turn the life-raft upright.

b. Should the life-raft need to be turned upright while you are in the water, get downwind of it and rotate it so that the inflation cylinder is towards you. The weight of the cylinder and the wind will help turn it over. Avoid getting tangled in the attaching cord.

c. If possible, get into the life-raft from the wing, or lower yourself gently into the water to keep your head dry. Remove high heeled shoes and **do not** leap or jump into the life-raft as this may damage it. If you have to enter the water first, hold the bottom of your lifejacket with one hand and place the other hand over your mouth and nose.

d. Climb into the life-raft. If anyone is in the water and injured or cannot climb aboard, position their back towards the entrance. Two people should then hold the person under the armpits, (not by the arms), while any others balance the life-raft by sitting at the far end. Push the person initially down into the water, then give a good pull as the buoyancy from the lifejacket pushes the person back up again. Warn them first!

e. Once everyone is aboard the life-raft, inflate the floor, trail the sea anchor as soon as possible, and erect the canopy to prevent wind chill hypothermia affecting wet bodies. **PROTECTION is the key to survival.** Get all the water out using the bailer and mop up with a sponge or spare item of clothing. If necessary, fully inflate the buoyancy chambers. All should be firm, but not rock hard.

f. Ensure that at least one person is tied to the life-raft just in case a large wave should overturn it; then it should be possible to get back into it and help the others aboard.

g. To avoid vomiting, ensure that everyone takes a sea sickness pill straight away – do not wait for the onset of sickness. The smell inside the life-raft and the loss of visual references will increase the risk of sickness. (Vomiting causes serious fluid loss). The sea sickness pills will normally be found in the equipment pouch inside the life-raft. You can survive around the UK without water for over 4 days. **NEVER** drink sea water.

h. Once the canopy is erected, you will have PROTECTION. Wring out your clothes as much as possible and if you have anything suitable, insulate the floor.

i. Even on a warm day, keep the cover up to provide protection from the sun.

j. Treat any injuries and administer appropriate first aid. It will have been a traumatic experience, some survivors may be suffering from shock, which can affect mental processes.

k. The second element of survival is LOCATION, so switch on your PLB. Rig it as high as possible with the aerial vertical. DO NOT leave the PLB lying on the floor. If the hand-held radio is available in the waterproof bag, now is the time to make sure it is ON and working. Selecting 121.5 MHz will confirm that your PLB is working. This is where a mobile phone or GPS could be useful.

l. Use any other signalling equipment which might be available. However, with pyrotechnics do read the instructions first and check, then check again since some are double-ended. (It would be disastrous if you thought you were about to set off a smoke signal only to discover a white hot magnesium flare burning inside the life-raft!).

m. Take turns to keep watch and only use flares or smoke signals when you are sure somebody will see them, not, for instance, as a search aircraft is flying away from you. Flares should be held at arms length, outside and pointing away from the life-raft as they often drop hot deposits. If you have any gloves or other protection, wear them when using pyrotechnics. Sweep the horizon with the heliograph, (mirror), at any time when the sun is shining. Any marker dye will normally last around 3 hours in the vicinity of the life-raft, so make an intelligent guess as to when to use it – normally once a search aircraft is seen.

4.3 No Life-raft

a. If you do not have a life-raft, but have to survive in the water with only a lifejacket, then this is a life-threatening situation. However, **do NOT give up hope, the will to survive is the most powerful force to prolong life.**

b. The sea is cold, UK waters only reach 15° C even in summer and are below 10° C from October to April. If you are not wearing an immersion suit; then it is **ESSENTIAL** that you and any other survivors immediately adopt the following measures in order to conserve body heat:

- The cold will cause you to lose the use of your hands very quickly, so perform any manual tasks straightaway while you are still able and if possible tie yourselves together.

- Ideally tie the PLB onto the lifejacket. Try to keep the aerial vertical.

- Do NOT swim in an attempt to keep warm. The heat generated due to more blood circulation in the arms, legs and skin will just be transferred to the cold water.

- Generally, don't attempt to swim to the shore unless the distance is say less than 1 km and you are a strong swimmer.

- The main aim is to conserve heat. The most critical areas of the body for heat loss are the head, sides of the chest and the groin region. If the lifejacket has one, cover your head with the spray hood.

c. If there is a group of survivors, tie yourselves together and huddle with the sides of your chests and lower bodies pressed together. If there are children, sandwich them within the middle of the group for extra protection.

d. A lone survivor should adopt the 'HELP' position (this is the Heat Escaping Lessening Posture). The use of this position significantly increases survival times.

- Hold the inner sides of your arms in contact with the side of the chest. Hold your thighs together and raise them slightly to protect the groin region.

e. A single floating person is very difficult to see from the air. When a search aircraft is close enough to be able to see you, signal using your heliograph (mirror). If this is not available, sparkling light reflected by splashing water with your arms, may attract attention.

f. To attract the attention of surface vessels, use the whistle attached to the jacket; shouting is much less effective and more exhausting to the survivor.

4.4 No Lifejacket or Life-raft

a. This is a very life threatening situation, again **DO NOT give up hope**.

b. Use anything from the aircraft such as seat cushions, plastic boxes or pieces of polystyrene that will help you stay afloat.

c. If all else fails an inflated plastic bag or wet shirt are better than nothing.

d. Follow the advice of earlier paragraphs.

4.5 The Rescue

a. If survival equipment is dropped to you, it may consist of two attached packs, get into the raft and investigate the equipment in the other pack.

b. When help arrives, whether it is a boat or helicopter, stop signalling and wait for instructions from the rescuer. DO NOT:

- attempt to stand up
- try doing things on your own initiative.

c. If a helicopter is making the rescue, wait for the winch man to tell you what to do, do not reach out and grab the cable.

d. The winch man will most likely use a strop and carry out a double lift, i.e. go up with the survivor. When the strop is secure, the survivor should put both hands by his side, or better still hold hands behind his back. Many people try to hold on to the cable on the way up. This is unnecessary and could be dangerous as it increases the risk of falling out of the strop. Equally, on approaching the door sill, don't grab at the helicopter or try to help yourself in, the crew are much better at this than you!!

e. Once in the helicopter, your inflated lifejacket is a hazard. You will either be asked to deflate it, or you will be given a new jacket by the crew.

f. In most cases, the rescue services will deflate the life-raft after rescuing you and take it away. It is neither practical nor safe to try to recover it intact and leaving it afloat may result in a false alarm.

g. There is further information on SAR in the UK AIP GEN 3-6 'Search and Rescue'.

5 MAIN POINTS

- Don't panic – Ditchings are SURVIVABLE. The key elements are a good ditching then PROTECTION and LOCATION. Water and particularly food are by comparison minor considerations.
- Correct actions increase your chances of survival and early rescue.
- Always wear a properly maintained constant wear lifejacket when beyond gliding range from land in a single engined aircraft.
- Carry a serviceable aviation life-raft, stowed so that it is accessible, or else wear a survival suit, particularly when the sea temperature is below 10° C.
- Carry a Personal Locator Beacon (and flares).
- In single engined aircraft, route to minimise the time over water or fly high to increase your glide range. Know the range per 1000 ft of altitude.
- Carefully pre-flight the aircraft and make sure there is enough fuel for all contingencies.
- Before take off, brief passengers on ditching procedures and survival equipment.
- Transmit a Mayday preferably on 121.5 MHz; and select 7700 on the transponder.
- Ditch along the crest of the swell, unless there is a very strong wind.
- Touch-down as slowly as possible – but don't stall.
- Inflate lifejackets once clear of the aircraft cabin.
- Get everyone into the life-raft as quickly as possible and get the cover up.
- Switch on the PLB (or hand held radio, mobile phone).
- If in the water with no life-raft, conserve energy and heat by huddling together to reduce the risk of hypothermia. The will to live is the single most important factor in surviving until you are rescued.
- Have the other signalling devices e.g. pyrotechnics, heliograph etc ready for use.
- Let the rescuer take control of the actual rescue.

Supplement A

CAA APPROVED COMPANIES WHICH SERVICE LIFEJACKETS AND LIFERAFTS

*Aviation Engineering & Maintenance Ltd
Stansted Division
Stansted Airport
Stansted
Essex CM24 1RB
Tel: 01279 680030 ext 200
Fax: 01279 680395

Seaweather Aviation Services Ltd
625 Princes Road
Dartford
Kent
DA2 6FF
Tel: 01322 275513
Fax: 01322 292639

Bristow Helicopters Ltd
Safety Equipment Section
Aberdeen Airport
Dyce
Aberdeen AB2 0ES
Tel: 01224 723151
Fax: 01224 770120

** SEMS Aerosafe
13 & 25 Olympic Business Centre
Paycocke Road
Basildon
Essex SS14 3EX
Tel: 01268 534427
Fax: 01268 281009

*FAA Approved
** They also undertake practice evenings in a pool with wave machine and have a rental service.

CAA APPROVED LIFEJACKET AND LIFERAFT MANUFACTURERS

Beaufort Air- Sea Equipment Ltd   0151 652 9151 ext 211
International Safety Products   0151 922 2202
ML Lifeguard Equipment   01824 704314
RFD Ltd   01232 301531 ext 102

COMPANIES KNOWN TO PROVIDE SURVIVAL TRAINING USING A 'DUNKER'

Fleetwood Offshore Survival Centre
Broadwater, Fleetwood
Lancashire FY7 8JZ
Tel: 01253 779123
Fax: 01253 773014

Robert Gordon Institute of Technology
338 King Street
Aberdeen AB24 5BQ
Tel: 01224 619500
Fax: 01224 619519

Humberside Offshore Training Association
Malmo Road
Sutton Fields Industrial Estate
Hull
East Yorks HU7 0YF
Tel: 01482 820567
Fax: 01482 823202

Warsash Maritime Centre
Newtown Road
Warsash
Southampton SO31 9ZL
(using ANDARK facility)
Tel: 01489 576161
Fax: 01489 579388

Supplement B

SUITABLE LIFEJACKETS

- CAA Approved equipment is only required for Public Transport aircraft use and with the exception of North Sea helicopter operations, are NOT intended for constant wear. (Note: when serviced approx. 50% of airline style lifejackets used for GA purposes are found to be defective, versus less than 25% of the constant wear jackets.) Thus, on non- Public Transport flights it is up to you what to wear since not all lifejackets designed for constant wear are CAA Approved. (See Supplement A.)
- There are lifejackets available that are 'Approved' to US or to European Community Standards, some are designed to meet marine criteria.
- It is thus impossible to provide specific details on which are likely to be satisfactory. The subject should be discussed with manufacturers, stockists and maintainers.
- When choosing a lifejacket it will need to be a compromise of:
    o comfort when worn
    o convenience yet avoiding it becoming
    o entangled in seat belt/ harness
    o price
    o durability

SAFETYSENSE LEAFLET 22b - RADIOTELEPHONY

1. INTRODUCTION
2. WIRELESS TELEGRAPHY (WT) ACT
3. FLIGHT RADIOTELEPHONY OPERATORS LICENCE (FRTOL)
4. AIRCRAFT VHF RADIO EQUIPMENT
5. RADIOTELEPHONY (RTF) PHRASEOLOGY
6. MICROPHONE TECHNIQUE
7. AERODROME AERONAUTICAL RADIO STATIONS
8. AIR TRAFFIC CONTROL (ATC)
9. AIR TRAFFIC CONTROL (ATC) SERVICE
10. MILITARY AIR TRAFFIC CONTROL (ATC)
11. RADIO OPERATION
12. EMERGENCY PROCEDURES
13. THE PRACTICAL COMMUNICATIONS TEST FOR THE FRTOL
14. SUMMARY

1 INTRODUCTION

a. Radiotelephony (RTF) is essential for the safe operation of aircraft in a busy environment. RTF enables a pilot to obtain aerodrome information, weather information, and instructions relating to the safe movement of air traffic. Many student pilots find the process of learning to use the radio more daunting than learning to fly.

b. Radio waves are not confined by national boundaries and for this reason radiocommunications are regulated at International, European and national levels. The International Telecommunications Union (ITU) Radio regulations, which are reviewed regularly at World Radio Conferences, form the foundation of international agreements on the use of the radio frequency spectrum. The European Conference of Postal and Telecommunications Administrations (CEPT) committees supported by the European Radiocommunications Office (ERO) provide a forum for the discussion of regulatory issues for posts and telecommunications for the member states, with the UK Office of Communications (Ofcom) representing the interests of the UK with participation from the UK Civil Aviation Authority.

2 WIRELESS TELEGRAPHY (WT) ACT

a. Under the Wireless Telegraphy (WT) Act 1949 it is an offence to install or use radio transmission equipment without a licence. The Office of Communications (Ofcom) is responsible for managing that part of the radio spectrum used for civil purposes in the UK as set out in the Communications Act 2003 and has contracted the Civil Aviation Authority (CAA), Directorate of Airspace Policy (DAP) to administer WT Act radio licences for aircraft, aeronautical ground stations and navigation aids on their behalf.

b. An Aircraft Licence is required for radio equipment installed in an aircraft. A Transportable Licence is required for any handheld VHF radio equipment (even if only for 'back-up' use), with an integral antenna and power supply, for use on multiple aircraft as well as gliders, microlight aircraft, balloons, hand gliders and for other aviation related activities such as parachuting and paragliding.

c. The aircraft radio equipment, whether installed or handheld, is required to have been approved either by the UK CAA under the British Civil Airworthiness Requirements (BCARs) or by the European Aviation Safety Agency (EASA), who are now responsible for all aircraft radio equipment approvals, under the European Technical Standard Order (ETSO) Authorisations process.

d. An Aeronautical (Ground) Station Licence is required for the operation of any radio equipment on the ground; even for handheld VHF radio equipment already covered by a Transportable Licence for use in aircraft.

3 FLIGHT RADIOTELEPHONY OPERATORS LICENCE

a. Under Article 26 of the Air Navigation Order the CAA issues 'appropriate licences' for aircraft station flight radiotelephony operators, generally being a Flight Radiotelephony Operators Licence (FRTOL) issued either as a stand-alone licence or in conjunction with a flight crew licence. Operators of Aeronautical Radio Stations providing Air Traffic Services are also required to be similarly qualified, either holding an Air Traffic Controllers Licence, Flight Information Service Officer's Licence or Radio Station Operator's Certificate of Competence. Glider pilots and student pilots under training are, subject to certain conditions, exempt under ANO Article 26 from the requirement to hold a FRTOL. However, glider pilots without a FRTOL are not permitted to use the radio to communicate with an Air Traffic Control (ATC) unit.

b. The Flight Radiotelephony Operators Licence (FRTOL) entitles the holder to operate the radio equipment in any aircraft. FRTOLs issued prior to April 1998 contain the word 'Restricted', this is often mistaken for a 'VHF Only' limitation which, if applicable, will be endorsed on the reverse of the licence (the frequencies that will be used by a General Aviation aircraft are almost exclusively VHF, the 'VHF Only' limitation will not therefore be likely to cause any difficulties). In the UK the term 'Restricted' referred only to the type of equipment that may be operated (see ANO Schedule 8). Older radio equipment designed for use by specialist radio operators who were 'Unrestricted' is no longer in use, therefore the word 'Restricted' has been eliminated from the FRTOL. The privileges however remain unchanged.

c. When the FRTOL is limited to 'VHF Only', the holder may not use transmitting equipment operating in the HF aeronautical bands below 30 MHz. This limitation may be removed by obtaining a pass in the HF written examination with an RTF Examiner authorised to conduct the HF examination, or by obtaining a pass in the JAA Navigation group examinations at CPL/ATPL level.

d. It is essential that the holder of a FRTOL is familiar with the phraseology and procedures used for aeronautical communication. ATC frequencies are often busy, necessitating the use of concise phrases without ambiguity. **Long winded radio calls waste time and may endanger others.**

e. On 27th March 1977 two heavily laden Boeing 747s collided on the runway at Los Rodeos airport Tenerife in poor visibility, resulting in 575 fatalities. A KLM 747 commenced take-off whilst a Pan Am 747 was still taxiing towards it on the same runway. There was clearly a breakdown in communication; perhaps a misunderstood radio call! The Pan Am aircraft had been asked by the controller, who was unable to see either aircraft due to low cloud, **'Are you CLEAR of the runway'?** The KLM aircraft had already commenced the take-off roll without clearance; could the KLM pilot have mistaken the call to the other aircraft thinking that he was **'CLEAR to Take-Off'?** The answer remains a mystery; the cure is straightforward, use the correct RTF phraseology, which is designed to be unambiguous, acknowledge and read back all clearances and above all, if in doubt **ASK!!**

f. As a direct result of aircraft accidents RTF phraseology has been progressively modified to avoid any possibility of ambiguity or confusion. Specific phrases have well defined meanings and should not be modified by the operator. Some recreational pilots consider that they don't need to know the full vocabulary used for RTF communication, yet when communicating with an Air Traffic Service Unit (ATSU) they may encounter any aspect of it. Every radio user must be fully conversant with the nature of the air traffic service provided, and be able to understand the radio calls they may hear.

4 AIRCRAFT VHF RADIO EQUIPMENT

a. Aircraft VHF radio equipment used for communications operates in the aeronautical mobile band 117.975 MHz to 137.000 MHz with a channel spacing of either 25 kHz or 25 kHz/8.33 kHz. Channel spacing of 25 kHz and 8.33 kHz provide 760 and 2280 frequencies respectively within the band. At present, the mandatory carriage of 8.33 kHz capable VHF radio equipment is only required in certain airspace within Europe which is unlikely to be used by a GA pilot. However, due to the congestion in the VHF aeronautical mobile band, the introduction of 8.33 kHz channel spacing to other airspace or users may be necessary at some time in the future.

b. GA aircraft VHF radio equipment typically has 760 channels spaced at 25 kHz; some older radios may have only 720 channels with an upper limit of 135.975 MHz, these are not allowed for IFR flight, or for VFR flight in certain countries including Germany.

c. Some installed and handheld VHF radio equipment also includes coverage of the aeronautical radio navigation band 108.000 MHz – 117.975 MHz which is used by radio navigation facilities such as VOR and ILS. Air Traffic Information Services (ATIS) broadcasts of aerodrome information to aircraft may be carried on some VORs in addition to ATIS frequency assignments in the aeronautical mobile band.

d. Aircraft VHF radio equipment is fitted with a minimum of controls. Rotary knobs or switches select the operating frequency, allowing it to be adjusted in steps of 1 MHz, 100 kHz, and either 50, 25 or 8.33 kHz. On some equipment an additional switch selection is necessary in order to select 25 kHz resolution. This may take the form of a toggle switch or require a rotary selector knob to be pulled out. Many radios do not display the third decimal of the frequency. This creates the impression that the frequency ends in a (. x2) or a (. x7) rather than (. x25 or .x75). E.g.

121.025	shown as	121.02
121.050	shown as	121.05
121.075	shown as	121.07

This may mislead the user into thinking that a particular frequency cannot be selected (although it is important to remember that ATC will transmit such frequencies using only two digits after the point).

Frequencies spaced at 8.33kHz are at present only used above FL245 in the UK but at lower levels over parts of Europe. 8.33 kHz frequencies are currently designated as "channels" with 6 figures, e.g. 118.033, 118.058 etc. Channels such as 118.055 may be allocated, and these apply to communications on what would be a normal 108.05 frequency, but requiring equipment with a narrow enough bandwidth to use it without causing interference to the neighbouring 8.33 kHz channels.

e. All too often the receiver VOLUME and SQUELCH controls may be incorrectly set. SQUELCH is an electronic switch that mutes the receiver audio output when no signal is received. This facility is designed to reduce operator fatigue, which can result from continuous exposure to background noise. When a continuous radio signal (carrier) is received, it activates or 'lifts' the SQUELCH causing the speaker or headphones to be activated. Where a variable SQUELCH control is fitted, this allows the operator to determine the strength of the received signal required to lift the SQUELCH, which may also be activated by bursts of noise. The correct setting procedure for the SQUELCH control is:

- set the volume control to approximately halfway;
- turn the SQUELCH control up until a hiss appears, this is background 'static' noise;
- turn back the SQUELCH control until the hiss just stops, this occurs quite abruptly;
- leave the SQUELCH control in this position.

Some radios are not fitted with an external SQUELCH control, but incorporate a switch marked TEST. Operating this switch 'lifts' the SQUELCH and allows the volume control to be set at a level where the background hiss is audible, or alternatively where the receiver volume is acceptable.

Note:

- The SQUELCH cannot be set correctly whilst you are receiving a station.
- If the VOLUME control is set excessively high, distortion may occur within the radio making it more difficult to hear stations. Ideally the VOLUME control should not exceed 70% of its rotation.

f. VHF aeronautical radios use amplitude modulation (AM), the same system used by broadcast radio stations in the long and medium wave bands. When two AM stations transmit simultaneously on the same frequency the signals can mix together and may render one or both stations unreadable. If the two transmitters are not exactly on the same frequency, an annoying whistle or 'heterodyne' equal to the difference between the two frequencies will be heard. **Do not transmit at the same time as another station or you may render both signals unreadable. Always listen before speaking and keep transmissions short.**

g. If you experience difficulty contacting another station the following checks should be made:
- The correct frequency is selected *
- Frequencies ending (.025 MHz) and (.250 MHz) are easily transposed.
- The correct radio has been selected on the comms panel e.g. COM 1, or COM 2. (Transmit and receive switching are often independent)
- The ground station is open for watch
- The station is within range (This varies with altitude)
- Volume and Squelch are correctly set

h. Many light aircraft are fitted with a an intercom system which may be integrated into one of the radios or a be a separate unit. Before flight these should be checked and adjusted independently of the radio equipment. It is important to obtain a good balance between intercom volume and radio volume to prevent radio calls being swamped by the intercom. Always seek instruction if you are unfamiliar with a particular radio installation. KNOW THE EQUIPMENT.

i. Most light aircraft are equipped with a **Transponder**. This important aid to flight safety permits an air traffic controller to positively identify an aircraft. The transponder transmits a 4 digit code (SQUAWK), set by the pilot, to the ground station where it is displayed on the radar screen. The code is either issued to the specific aircraft by an air traffic controller or, if no specific code has been issued, one of the special use codes may be selected by the pilot to indicate the type of flight being undertaken by the aircraft. Most transponders incorporate Mode C (Charlie), which transmits and displays the aircraft's level (relative to 1013.2 mb) on the

ATC radar screen when the transponder mode switch is selected to 'Altitude' (ALT). **Adjustment of the altimeter pressure setting has no effect on the Mode C altitude information.** A switch marked 'IDENT' is provided on the transponder, this enables the symbol shown on the radar display to be modified so that the controller can positively identify the aircraft. The IDENT switch should not be operated unless requested by ATC.

j. In recent years an Airborne Collision Avoidance System (ACAS) has been employed in airliners and helicopters in order to provide automatic collision avoidance information. Mode C information from the transponder is important for ACAS to be effective. **Pilots should always fly with their transponder switched on, with ALT selected, unless advised otherwise by ATC.** One of the most commonly known ACAS systems is the Traffic Alert and Collision Avoidance System (TCAS) pronounced 'TEEKAS'.

k. In the absence of a code allocated by ATC, the pilot should set the "conspicuity code" 7000 on the transponder, or in the case of specialist activities the appropriate code.

5 USE OF PHRASEOLOGY

a. The correct radio phraseology to be used in the UK is detailed in CAP413 Radiotelephony Manual. In some cases it may seem very pedantic, however, it must be remembered that it has evolved for a purpose, primarily to avoid ambiguity. Many incorrect phrases are regularly heard.

b. **FINAL** is a position in the circuit pattern between 4 nautical miles and the landing threshold, in circuit parlance it is singular not plural! An Air Traffic Controller Officer hearing a call such as **'ON FINALS'** might easily believe the traffic to be **'LONG FINAL'** (a position between 4 and 8 miles from the landing threshold); in poor visibility, such a mistake could result in the controller giving another aircraft clearance to Take-Off as he believes the landing traffic to be in excess of 4 miles away, when in reality, it may be as little as only half a mile away! There is no official report 'SHORT FINAL' however; the distance from the landing threshold may serve as a more accurate indication of position i.e. 'Half Mile Final'.

c. At Aerodromes with an Aerodrome Flight Information Service (AFIS), the phrase **'at your discretion'** is used to indicate that the Flight Information Service Officer (FISO) is not issuing a clearance. Pilots should NOT respond using the phrase 'at my discretion' but rather reply with their intentions, for example **'landing'**.

d. **'Land at your discretion'** is not a clearance to land. Pilots must exercise their own judgement and comply with the rules of the air. e.g. An aircraft may not land on a runway whilst another aircraft is on that runway unless authorised to do so by an air traffic controller. (Rules of the Air Rule 17).

e. Requests for **'landing instructions'** should not be made.

f. Requests for **'instructions'** should not be made to stations providing a A/GAGCS or FIS. A/GAGCS operators and Flight Information Officers (FISO) are not permitted to give instructions.

g. Public correspondence messages (including air to air conversations) are not permitted on the VHF aeronautical band.

6 MICROPHONE TECHNIQUE

a. Use a headset, it cuts out aircraft noise and avoids the distraction of a handheld microphone.

b. Keep the microphone close to your mouth.

c. Speak directly into the microphone.

d. Don't 'clip' your transmissions - ensure that the transmit button is held firmly pressed BEFORE you speak until AFTER you have finished speaking.

7 AERODROME AERONAUTICAL RADIO STATIONS

a. The nature of the ground radio facilities at an aerodrome is usually dependent upon the number of air traffic movements. Some minor aerodromes have no provision for radio at all, whilst others may have an allocated frequency but the AGCS is seldom manned. The majority of aerodromes have a ground radio station and provide one of three types of air traffic service:
- Air/Ground communication service (AGCS) Callsign **'RADIO'**
- Flight Information Service Callsign **'INFORMATION'**
- Air Traffic Control (ATC) service Callsigns: **'GROUND; TOWER; APPROACH; RADAR; DIRECTOR; DELIVERY'**

Each service employs different procedures and it is important for pilots to be familiar with the differences and the implications for the pilot's actions in response. Small aerodromes may provide an AGCS or aerodrome FIS utilising a single frequency, whereas a busy airport will have an ATC service with separate frequencies for Radar, Approach, Tower, Ground and possibly an Automatic Terminal Information Service (ATIS).

b. **SAFETYCOM (135.475 MHz)** is a common frequency allocated for use by aircraft flying in the vicinity of aerodromes not assigned a discrete frequency. Because there is no frequency assigned for the aerodrome there is no ground radio station. SAFETYCOM is designed to allow pilots to broadcast their intentions to other aircraft that may be operating on or in the vicinity of the aerodrome. Transmissions shall only be made when the aircraft is below 2000 ft aal or below 1000 ft above circuit height within 10 miles of the aerodrome. Calls should be kept concise. Aircraft taxiing, taking-off, landing and flying in the circuit pattern should self announce their position and intentions on the SAFETYCOM frequency to alert other pilots of their presence. Initial calls should be addressed to *'Airfield Name'* with the suffix **'TRAFFIC'**.

e.g. *'WILTON TRAFFIC G-ABCD downwind 24 left to land'*.

The intention of the airborne aircraft is then obvious to a pilot taxiing or waiting to back-track the runway. The pilot of the taxiing aircraft may choose to broadcast his intentions,

e.g. *'WILTON TRAFFIC G-ZZXY holding point 06 awaiting landing traffic'* in order to make his intentions known to the traffic in the circuit. Avoid using the word 'CLEAR', it may be mistaken as a clearance!

Announce your intentions in order to assist other traffic whilst making your presence noticed. Altimeter settings will need to be determined in relation to the aerodrome elevation. The QNH of a neighbouring aerodrome will be approximately correct whereas the regional pressure setting, which has a built in safety margin will result in the aircraft being higher than shown on the altimeter..

c. An **Air/Ground communication service (AGCS) Station** is the simplest form of aeronautical radio communication. The call sign uses the aerodrome name followed by the suffix 'RADIO'. The ground radio operator is not an air traffic controller and **must not give any air traffic instructions or clearances however he may relay instructions and clearances given by a controller** e.g. an airways clearance to departing traffic.

This service provides aerodrome and traffic information only. In some instances the AGCS station may be located in a flying club or building that does not have an unrestricted view of the aerodrome.

In order to operate an AGCS station the operator must be in possession of a Radio Operators Certificate of Competence (CAA Form CA1308), which must be countersigned by the aeronautical radio station licensee.

AGCS operators will NOT use the expression: **'At your discretion'**.

The AGCS operator may pass information to a pilot such as the runway, pressure settings, wind velocity and details of any known reported traffic. Pilots should not request clearances or instructions, as they cannot be given.

Before entering the Aerodrome Traffic Zone (ATZ) during the published hours of operation of an aerodrome with a notified AGCS service, a pilot must obtain **'information'** from the AGCS radio station operator to ensure that the flight can be conducted safely. The AGCS radio station operator may pass messages on behalf of the aerodrome operator but any such message must be passed as information and must include details of the originator of the message.

e.g. *'G- AYZZ Message from the airport manager. You are requested to report to the Control Tower after landing'*

On arrival at an aerodrome with an AGCS service, taxiing and parking are also the responsibility of the pilot. The AGCS operator may not give taxi instructions but, may suggest a suitable parking location if requested by the pilot.

'is there a convenient parking space? G-ZZ'
'G-ZZ there is parking space available next to the blue Cessna'

AN AGCS STATION CANNOT GIVE CLEARANCES OR INSTRUCTIONS TO AN AIRCRAFT.

d. A Flight Information Service Officer is qualified to provide an aerodrome Flight Information Service (FIS) in order to pass:
- 'Instructions' to vehicles and persons on the aerodrome, to aircraft on the ground up to the holding point and, in the case of aircraft landing, after the landing roll is completed;
- 'Information' for the safe conduct of aerodrome traffic on the runway and within an ATZ.

In practice, there is little difference between AFIS and AGCS service, however the FISO is required to undergo training and is tested by the CAA. The FIS call sign uses the suffix **'INFORMATION'** to identify the type of service. Air traffic clearances must not be given, but may be relayed by a FISO.

The service may revert to AGCS if a qualified FISO is not available, it is promulgated by NOTAM, and the AGCS operator is appropriately certificated. The call sign suffix then reverts to **'RADIO'**.

The phrase 'At your discretion' may be used by a FISO and will follow any advisory information. Pilots requesting departure may be advised:
'Take off at your discretion'

The pilot **should** not **respond** by repeating the phrase: **'at my discretion'.** No clearance has been given, there is no requirement to read one back. The pilot should simply respond:
'G –XX Roger' or 'G- XX taking off'

e. Examples of AGCS and FISO RTF phraseology are contained in CAP413 Radiotelephony Manual.

An aerodrome Flight Information Service Officer (FISO) may control aircraft on the ground up to the holding point and after the landing roll is complete. Pilots are reminded that they are responsible at all times for the safety of their aircraft and collision avoidance, LOOKOUT is always paramount.

8 AIR TRAFFIC CONTROL SERVICE

a. Pilots familiar with small aerodromes providing either an AGCS or FIS may find larger aerodromes somewhat daunting. Busy aerodromes will employ separate controllers for Ground, Tower, Approach and possibly Radar. If the purpose of each is fully understood, it will help to eliminate any confusion regarding who to talk to and when.

b. The **GROUND** controller is responsible for all movements on the manoeuvring area; this will include all taxiing aircraft and vehicular traffic equipped with radio. Initial calls will be made to GROUND, including taxi clearance, (start clearance at some aerodromes), departure clearance* if applicable, and, normally, all calls up to the holding point. Landing traffic will normally be instructed to change to GROUND after vacating the runway.

*The departure clearance tells a pilot what he is required to do on departure and will include any frequency changes required, together with routeing instructions and altitude restrictions.
Note: this is NOT a clearance to take- off or to enter an active runway.

c. The **TOWER** controller is responsible for all traffic using the runway and in close proximity to the aerodrome, including the circuit. Normally an aircraft will be instructed to change to TOWER when at the holding point, at which time the pilot should have completed all of his

checks and be ready for departure. The first call will usually be:
'WRAYTON TOWER G-ABCD holding point RW 30 Ready for departure'.
Aircraft remaining in the circuit will remain with **TOWER**, whereas departing aircraft will change to either **APPROACH** or **RADAR**. Pilots arriving at an aerodrome will usually be instructed by APPROACH to contact TOWER at a suitable point in order to obtain circuit joining instructions. After landing, aircraft should vacate the runway, unless otherwise instructed, at the first available taxiway that the aircraft reaches having slowed to taxiing speed and advise the controller:
'Runway Vacated G-XX'
The pilot will normally then be instructed to change to the GROUND frequency. Do not use the phrases:
'Clear the Active' or 'Clear of the Runway'

d. **GROUND** and **TOWER** controllers are located in the glass uppermost part of the ATC Tower; they are invariably located side by side and should have a good view of the aerodrome and circuit.

e. **APPROACH** controllers are usually located in the lower part of the ATC tower and have no visual contact with the aerodrome. Control may be either radar or non-radar. At busy aerodromes RADAR controllers may be used in addition to the APPROACH controller to provide services for traffic transiting the area.

f. It is not uncommon for controllers to conduct more than one function when traffic is light; The **RADAR** and **APPROACH** controllers work in close proximity such that the jobs may be combined. The **GROUND** and **TOWER** controllers are also ideally situated to combine functions. At the very small provincial airports, **TOWER** and **APPROACH** control may also be provided by one controller. Occasionally at smaller airports the service may revert to a **FIS** outside the busy period at weekends, in which case, the service will be apparent from the Callsign Suffix **'INFORMATION'** and no clearances or instructions will be given.

g. **ATIS** uses a dedicated frequency on which a recording of aerodrome information is broadcast continuously. This information is updated at least hourly. Such a facility allows pilots to obtain weather and aerodrome information without having to establish radio contact with the aerodrome, thus considerably reducing the workload of the controller and enabling the pilot to plan ahead. ATIS information is coded using a letter of the alphabet to enable both pilot and controller to ascertain which broadcast the pilot has received.
e.g. *'This is Langford information Delta time zero nine five zero'*
the message concludes:
'on initial contact with Langford advise information Delta received'.
The pilot advises ATC on his initial call that he has received ATIS Delta. Pilots who call ATC without passing the ATIS code may be asked if they have received the latest 'ATIS information'. **To the unwary this may cause confusion!** If a pilot does not report the latest broadcast identification letter the controller will advise the pilot of any updated information.

9 OTHER SERVICES

Lower Airspace Radar Service (LARS) is available to pilots when flying outside controlled airspace below FL95. The General Aviation Safety Sense Leaflet 8 provides details of Air Traffic Services Outside Controlled Airspace.

10 MILITARY ATC

a. Military ATC units often provide a LARS. The terminology used by military controllers differs in some details from that used by civil controllers. Military controllers are not obliged to adhere to civil Rules of the Air when issuing instructions, and it is possible that you may be asked to fly in a manner that might not conform to civil practices or law. It is the pilot's responsibility to advise the controller if he/she is unable to comply with the instruction and why, e.g. being asked to over fly a built up area below 1000 ft or at a height where it is not possible to glide clear, or if altitude changes might place an unqualified pilot in IMC.

b. Military ATC use frequencies in the UHF band (225- 380 MHz) for their primary function of providing services to military aircraft whilst operating VHF frequencies that facilitate communication with civil aircraft is normally a secondary function. When calling a military ATC unit on VHF always allow time for the controller to reply as he may be in communication with a military aircraft on UHF. Often you will hear only one side of the conversation when transmissions are made on both VHF and UHF simultaneously; you hear the VHF transmission from ATC, but not the reply from the aircraft on UHF. Information on operating at and in the vicinity of military aerodromes is contained in SafetySense leaflet 26 "Operations at Military Aerodromes".

769

11 RADIO OPERATION

a. It is not intended to reproduce CAP 413 Radiotelephony Manual, but rather to highlight certain aspects of radio operation.

b. **Radio Check.** Before embarking upon a flight it is essential to know that the radio equipment is working. Listening to other stations will check the radio receiver but in order to check the transmitter, it is necessary to talk to another station and let them confirm that they have received your transmission in an intelligible form. It is also important to be sure that the equipment switches channels and that the channel indicated is the correct one. Where two frequencies are in use at an airfield, radios may be checked by selecting the frequencies alternately. The transmitter may be checked on the initial call for the aerodrome information. When more than one radio is installed, the second radio should be checked on a subsequent call.

c. The golden rule of RTF operation is: **know what you are going to say before you say it**. Whilst this may seem obvious, once the transmit switch is pressed the human brain often forgets the obvious. Secondly, **anticipate what the reply is likely to be**. That way, it will not be a surprise. For example when calling for aerodrome information, the reply will include the QFE, QNH, surface wind and runway (R/W) in use. The pilot may get an idea of the pressure settings in advance by using the altimeter, whilst a good indication of the R/W in use and wind direction can be obtained by observing the windsock and any other traffic. **Always read-back the reply in the same order that it was given – avoid reversing the order.**

d. A **Departure** clearance can often pose problems for the inexperienced; it may be a lengthy clearance, which must be read back to the controller. Prior to departure it is normal to 'Book- Out' with the ATSU, specifying your departure details; flying instructors should allow students to observe and practice this procedure as part of the learning process. If the pilot has any questions about the departure route or the clearance that he is likely to receive, it is a good idea to ask when Booking Out rather than to wait until having to ask on the radio. The departure clearance will normally be a confirmation of the routeing already requested, although occasionally it may involve changes. It should therefore be no surprise when the controller passes a clearance that closely resembles the information passed by the pilot when 'Booking Out'. If a frequency change is required on departure it will be to a published frequency. Know where to look it up and whenever possible, select it on a second radio as a reminder. Invariably, when departing VFR, the first two digits of the transponder code will remain the same for a particular ATSU.

Remember: The departure clearance is NOT a clearance to enter a runway or to take- off!
After take- off you are required to follow the departure clearance, remember the basic rule:
- **Aviate**
- **Navigate** and then
- **Communicate**

When safely airborne and established in the climb you can expect TOWER to instruct you to change frequency:
e.g. *G-ABCD to APPROACH 126.1*
To omit this call could result in uncertainty over your whereabouts!

e. **En-Route** calls usually take the form of position reports. The initial call to an ATSU should begin:
- **Station** being called
- Aircraft **Callsign** in full
- **Request**
e.g. *'WILTON RADAR G- AAXX request Radar Information Service'*

Do not say any more until the ATSU invites you to 'pass your message'. If you are advised to "Standby", do so but **do not acknowledge**. When requested to **"pass your message"** it should consist of:
- **Full call sign** – so that the controller can write it down.
- **Type** – PA28, C172, Robin 400 etc
- **Departure/Destination** –the point of departure and destination; – the controller will write these on a handling slip. **do not include a list of turning points.** If you are returning to the point of departure it is satisfactory to say *Navex from Wilton to Wilton*.
- **Present Position** – should be given relative to a point on a 1:500,000 chart, the controller may not be familiar with small features in the area.
- **Altitude/ Level** – together with the pressure setting this will enable the controller to assess if there is any confliction with other traffic in the vicinity at the same level!
- **Additional details** – What service or information do you require? e.g. Flight Information Service (FIS), regional pressure settings next turning point etc.

Common mistakes are a failure to make any request of the controller, and inadequate or misleading position reporting, leaving the controller unaware of your present position and/ or the next turning point. In order to provide you with a Service, the controller needs to know:

- Who you are
- Where you are and
- What you want

Then **WAIT...**

If you cannot remember what to say, **stop transmitting.** The controller will ask you for anything you miss out!

f. Many pilots will avoid flight through a Control Zone (CTR) by flying a longer route around it. The majority of CTRs in the UK are designated Class D airspace, which permits VFR flight subject to an ATC clearance. In the case of a CTR designated Class A airspace, a Special VFR (SVFR) clearance is required if the aircraft is being flown visually. Requesting a VFR or SVFR clearance is straightforward. The controller will form a mental picture of a pilot from the radio calls made. He is unaware of a pilot's qualifications, experience or status from the aircraft call sign alone. A radio call delivered in a professional manner will be treated accordingly, whereas a poorly structured and hesitant call may lead the controller to be cautious about issuing a clearance that is complex or requires the pilot to fly very accurately. A badly delivered request for a clearance may result in a routing that avoids controlled airspace rather than the route requested. For example

'SOLENT APPROACH G- ABCD request zone transit'

followed by:

'G- ABCD; Cessna 172; Popham to Sandown VFR; 10 miles North of Winchester; Altitude Two Thousand Feet on One Zero Zero Six. Estimate Sierra Alpha Mike, Two Five; request zone transit'.

will probably result in a reply:

'G- CD is cleared to enter the Southampton zone abeam Winchester VFR not below altitude two thousand feet Solent QNH One Zero Zero Nine. Report Sierra Alpha Mike'

Whereas a call:

'SOUTHAMPTON this is G- ABCD Err! a 172 at two thousand feet Err! Point of departure Popham. 4 Persons on board. Err! Can we transit over Southampton to the Isle of Wight Sir? or if not we will go round. Err! we are North of Winchester. Over'

may result in the reply:

'G- ABCD remain outside controlled airspace. Route via Romsey, Totton and Calshot for the IOW., Solent QNH One Zero Zero Nine. Report Romsey'.

g. **Aerodrome Arrival.** Unless you have filed a Flight Plan (CA48) or have telephoned in advance, (essential at PPR aerodromes) VFR flights usually arrive at an aerodrome without prior knowledge of ATC, the FISO or AGCS radio station operator. You may arrive at the same time as other VFR or IFR traffic. If the aerodrome provides a RADAR service it is a good idea to talk to them as soon as you are within range, they may look after you until you are in visual contact with the aerodrome at which point you will be asked to contact **TOWER**. If there is no radar service the initial call will be to **APPROACH** not greater than 25 nm from the aerodrome. Joining procedure will depend upon the type of traffic when you arrive, if there is IFR traffic arriving and departing it is unlikely that you will be able to join overhead. You may be asked to report your position relative to one of the established Visual Reference Points (VRPs). Occasionally, you may be asked to route via a position not obvious to you, **if in doubt ASK.** The change to **TOWER** can occur quite late. On landing you may be asked to vacate the runway at a specific point and change to **GROUND**. Be prepared for references to published ground positions, stand numbers and holding points. In other words, use a plan of the aerodrome! For arrival at a small aerodrome with either **AGCS** or **AFIS**, initial contact should be made within 10 miles of the aerodrome. If unfamiliar with the aerodrome an overhead join is preferred (but not always permitted – see the UK AIP) as it enables orientation with the aerodrome and circuit traffic. Remember you must establish radio contact with the aerodrome **BEFORE** you enter the ATZ. See General Aviation Safety Sense Leaflet 6B, *Aerodrome Sense*.

h. Any pilot arriving at an unfamiliar aerodrome will experience a high workload and may not recognise geographical features. The aircraft has to descend; there are checks to be completed and frequencies to be selected. It is essential to **LOOKOUT, listen out** and keep your wits about you. Be prepared, have a plan and select the required frequencies as far in advance as possible. **Check the Aeronautical Information Publications (AIP) and NOTAMs prior to departure and do not use out of date documents.**

12 EMERGENCY PROCEDURES

a. Fortunately emergencies are rare. However, there have been a number of occasions when a pilot has recognised the need to land as soon as possible, (e.g. no oil pressure but the engine is still running OK) but has not wanted to 'make a fuss about it'. Clearly if a situation arises where there is a possibility of danger or a worsening situation it is in your best interest to make an URGENCY call, that way immediate help, or a priority landing, is available to prevent the situation getting out of hand.

b. **The states of EMERGENCY are**:
- **Distress**. (MAYDAY) A condition of being threatened by serious or imminent danger and of requiring immediate assistance.
- **Urgency**. (PAN PAN) A condition concerning the safety of an aircraft or other vehicle, or some person on board or within sight, *but does not require immediate assistance.*

c. The EMERGENCY MESSAGE advises others:
- **Who you are!**
- **What the problem is,**
- **What you intend to do about it and**
- **Where you are!**

The format is as follows:
- **MAYDAY** (repeated 3 times) or **PAN PAN** (repeated 3 times)
- **STATION** addressed when appropriate
- **CALLSIGN** (once)
- **TYPE** of Aircraft
- **NATURE** of emergency
- **INTENTION** of person in command
- **POSITION – HEIGHT and HDG**
- **Pilot qualification**: e.g. Student pilot, no instrument qualification, IMC rating or full Instrument Rating (IR) (Not required by ICAO).
- **Any other information** – POB, endurance etc

It is probable that in a real emergency you will not wish to be bothered with talking further on the radio. By ending the call: *MAYDAY OUT* you will convey the message that you do not expect a reply.

d. Further attention can be attracted in an emergency by selecting the appropriate code on the transponder:
Emergency   7700
Radio Failure   7600

13 THE PRACTICAL COMMUNICATIONS TEST FOR THE FRTOL

a. Candidates wishing to obtain a FRTOL are required to sit a written examination and a practical communications test with an authorised RTF examiner. The practical test involves the use of an approved RTF simulator; this may provide basic radio facilities, or be a PC based system with a moving map and associated communications equipment. The candidate is briefed to follow a typical light aircraft route from one aerodrome to another passing through a Military Air Traffic Zone (MATZ) and possibly at some stage into or through a CTR. The candidate is required to make all the appropriate radio calls and frequency selections as if he were actually flying the route. The examiner performs the function of an AGCS radio station operator, FISO or controller. Other aircraft may be heard so the candidate is required to listen out. At some stage there will be an emergency involving either the candidate or another aircraft. At all stages of the test the candidate is required to make the appropriate radio calls. There are a number of options available to the candidate and in most cases it is the candidates responsibility to select an appropriate agency with whom to communicate with. The candidate is provided with a route map, a completed navigation flight plan and a list of all communications facilities available to him. The candidate must be familiar with the procedure for obtaining VHF Direction Finding (VDF) bearings from stations equipped with this facility.

b. A typical examination route would be for a C172 aircraft routeing from Shipdham in East Anglia to East Midlands Airport via Huntingdon and Melton Mowbray. The aircraft is equipped with a single channel radio and a transponder with no mode C. The pilot is assumed to be a PPL holder. On this particular route a LARS service is available for most of the route, it would be a shame not to use it. The route passes South of RAF Marham and then through the Combined MATZ (CMATZ) at RAF Wittering and RAF Cottesmore, finally arriving at East Midlands, which is in Class D airspace. Special entry and exit lanes are provided to assist VFR and SVFR traffic.

MAP OF ROUTE (1:500,000 CAA VFR Chart)

c. A typical narrative for the flight could be as follows:
Note: for the sake of clarity numerals are used in this example in preference to spelling out numbers.
Shipdham Radio G-ZAON request radio check 119.55

G-ZAON Shipdham Radio Readability 5

G- ZAON; request airfield information; Taxi VFR to East Midlands

G-ON RW 20; Surface Wind 250/ 07; QNH 1009;

R/ W 20; QNH 1009 G-ON

G-ON Ready for departure.

G-ON traffic is a Cessna 152 on a half-mile FINAL.

ROGER G-ON

G-ON reported traffic, surface wind 260/ 05.

Roger taking off G-ON.

G-ON ROGER

G-ON overhead altitude 2500 ft QNH 1009, changing to Marham 124.15

G-ON ROGER

Marham Approach G- ZAON request FIS

G-ZAON Marham Approach pass your message

G-ZAON Cessna 172 Shipdham to East Midlands 2 miles North of Watton Altitude 2500 ft 1009 Estimating Alconbury at 35 Request Radar Information Service and Chatham Pressure

G-ON Chatham 1005, Squawk 2632

Chatham 1005, Squawk 2632, negative Charlie, G-ON

G-ON identified 8 miles SE of Marham; Flight Information Service; report passing Chatteris

FIS;; WILCO G-ON

G-ON 5 milesNorth of Ely request change to Cambridge 123.6 for VDF

G-ON Squawk 7000 contact Cambridge 123.6

Squawk 7000; Cambridge 123.6 G-ON

Cambridge Homer

G-ZAON request true bearing G-ZAON

G- ZAON Cambridge Homer transmit for bearing.

True Bearing/ True Bearing G- ZAON request true bearing G- ZAON

G-ZAON Cambridge Homer true bearing 355 degrees class Bravo; I say again 355 degrees class Bravo

True bearing 355 degrees class Bravo; changing to Wyton 134.05 G- ZAON

G-ON ROGER

Wyton Approach G- ZAON request FIS.

G-ZAON Wyton approach pass your message.

*G-ZAON
Cessna 172
Shipdham to East Midlands
2 miles South of Chatteris
Altitude 2500 ft 1005
Estimating Alconbury at 35
Request RIS*

G-ON ROGER; Report turning at Alconbury The Wyton Circuit is active with three Vigilants.

WILCO G-ON

*G-ON Overhead Alconbury Altitude 2500ft 1005
Estimating Melton Mowbray at 03*

G-on ROGER Freecall Cottesmore on 130.2

Cottesmore 130.2 G-ON

Cottesmore Approach G- ZAON request MATZ penetration

G-ZAON Cottesmore Approach pass your message

*G- ZAON
Cessna 172
Shipdham to East Midlands
3 miles South of Conington
Altitude 2500 ft Chatham 1005 Estimating Melton Mowbray at 03;
request MATZ penetration;
Radar Information Service and Barnsley pressure*

G-ON Barnsley 1002 Squawk 6554

Barnsley 1002 Squawk 6554 G-ON

G-ON identified; Radar Information Service; maintain 2500 ft
Cottesmore QFE 993 millibars

*Maintain height 2500 ft
QFE 993 millibars;*

Radar Information Service, G-ON

G-ON ROGER, report abeam Oundle

WILCO G-ON

G-ON abeam Oundle

G-ON is cleared to cross the CMATZ at 2500 ft QFE 993 millibars; maintain VFR; report abeam Oakham

Cleared to cross the CMATZ at height 2500ft 993 millibars; Wilco G-ON

G-ON abeam Oakham

G-ON Squawk 7362

Squawk 7362 G-ON

G-ON contact East Midlands Radar 119.65

East Midlands Radar 119.65 G-ON

(If possible listen to East Midlands ATIS 128.225 MHz to obtain airfield information)

East Midlands Radar G- ZAON inbound from Shipdham with Information 'Golf' (The ATIS code)

G-ZAON Stand-by

G-ON expect zone entry via the Shepshed Lane VFR; RW 27 surface wind 270/ 08 QFE 998 millibars, report approaching Shepshed

Route via the Shepshed Lane RW 27 QFE 998 millibars, G-ON

Approaching the Shepshed Lane, G-ON

G- ON Cleared to enter the zone VFR report field in sight

Clear to enter the Zone VFR, WILCO G-ON

G-ON Field in Sight

G-ON contact East Midlands Tower 124.0

East Midlands Tower 124.0 G-ON

East Midlands Tower G-ZAON

G-ZAON join left base RW 27; QFE 998 millibars; No 2 to a Boeing 737 on a 1 mile FINAL.

Join left base RW 27; QFE 998 millibars; No 2. G-ZAON

G-ON report FINAL caution vortex wake the recommended spacing is 6 miles.

WILCO G-ON

G-ON FINAL

G-ON continue approach surface wind 265/ 07

Continue approach G-ON

G-ON Cleared to land RW 27 surface wind 270/07

Cleared to land RW 27 G-ON

G-ON landing time 1417 vacate next left

Vacate next Left, G-ON

G-ON contact East Midlands Ground 121.9

Ground 121.9 G-ON

East Midlands Ground G-ZAON Runway vacated

G-ZAON turn right onto taxi-way Alpha turn left at Alpha 2 for the flying club

Taxi-way Alpha via Alpha 2 for the Flying Club G-ZAON

G-ON report closing down

WILCO G-ON

G-ON Closing Down

Notes:

At some stage in the practical test the candidate will be required to make both Urgency and Emergency calls. They must be made in accordance with CAP413. Failure to make these calls correctly will result in a mandatory failure of the test.

RTF practical tests are conducted using an approved RTF simulator where the candidate must be isolated from the examiner. Only Authorised RTF Examiners may conduct this test. Tests may not be conducted in an aircraft, with the candidate in the same room as the examiner, or without the RTF simulator equipment.

A radiotelephony training record form SRG 1171 is available on the SRG/PLD website to enable candidates to cover all test items with their flight instructor.

Useful References:
CAP 413 Radiotelephony Manual
AIC 19/2004 (White 95) Flight Radiotelephony Operators Licence (VHF and HF) Examinations
Listing of all authorised RTF Examiners can be found on the SRG/PLD website
CAA Flight Safety Poster FSP 4 'Cut the Chat'
The Private Pilot's Licence Course – Air Law and Radiotelephony by Jeremy M Pratt – AFE
The Air Pilots Manual – Volume 7 – by Trevor Thom – Airlife Publishing Ltd
CAA publications can be viewed or downloaded from the CAA web site www.caa.co.uk. Many are available in printed form for purchase from TSO.

14 SUMMARY

- A Wireless Telegraphy (WT) Act Licence is required for aeronautical radio equipment installed or used in aircraft and aeronautical radio stations.
- Aircraft radio equipment must be approved either by the UK CAA or EASA.
- Know how to use the aircraft radio equipment
- Be familiar with CAP413, it is revised from time to time with new phraseology
- Use correct phraseology, it is designed to prevent ambiguity
- Use a headset, speak directly into the microphone positioned close to the mouth
- Listen out before transmitting
- Keep transmissions short
- If uncertain of what to say, STOP TRANSMITTING!
- Know the types of Air Traffic Service provided and the limitations
- Know the Emergency Procedures

775

SAFETYSENSE LEAFLET 23   PILOTS – IT'S YOUR DECISION

1 INTRODUCTION
2 TO GO OR NOT TO GO
3 DIFFERENT RISKS FOR DIFFERENT PEOPLE
4 ONLY HUMAN
5 ONLY A MACHINE
6 HOW DO ACCIDENTS HAPPEN? COMMON SCENARIOS IN THE CAA REVIEW
7 SUMMARY

1 INTRODUCTION

A CAA study examined 166 fatal accidents to UK light aircraft. That review was published as CAP 667 'Review of General Aviation Fatal Accidents 1985 – 1994', and this highlights some of the points made. Most accidents are the result of the pilot's actions. This includes their skill level and, most important of all, **the decisions that they make**. This leaflet details some of the factors that can affect how the pilot's decisions do – or don't – keep the aircraft in one piece and the occupants safe.

2 TO GO OR NOT TO GO

a. Weather

Probably the single most important factor in General Aviation flight safety is the decision of a pilot to begin, or continue, a flight in unsuitable weather conditions. As you might expect, weather was a major factor in fatal accidents: over 80% of Controlled Flight Into Terrain (CFIT) accidents happened when the pilot either continued flying into adverse weather, or did not appreciate the actual effects of the weather conditions. Of those pilots who lost control in Instrument Meteorological Conditions (IMC), only one had an Instrument Rating.

Crosswind landings seldom result in fatalities, but they still feature in many accidents resulting in broken aircraft and painful injuries.

Weather does not stay constant, it doesn't always do what the forecast predicts, and it can deteriorate very fast. Respect the weather, and the implications for flight safety. That doesn't just mean **other** less experienced people who can't fly so well are the ones who should respect the weather; it means **you**.

b. I Can't Turn Back Now!

Any competent pilot knows that weather can, and will, change en-route. If it does, it is essential that the pilot is prepared and willing to divert or turn back if conditions deteriorate. It does not reflect badly on your ability as a pilot if you turn back in poor weather. In fact, **it reflects good judgement and realistic assessment of the situation**. It is also important that diverting is feasible in practical terms. Have you got enough fuel, money to get home, or pay for a hotel? Have you promised to be somewhere important? **Never** put yourself in a position where you would not feel able and willing to turn back if necessary. No Monday job is worth dying for on a Sunday, so carry your driving licence and credit card.

The decision to turn back will be made easier if you have practised, in advance, to fly the relevant manoeuvres **on instruments**, for example: a 180% turn and if necessary climb to a higher Minimum Safe Altitude (MSA).

c. Chain of Events

In aviation accidents, it is common to find a chain of events where one shortcut or poor judgement leads to another. For example, the apparent 'cause' of an accident may be that the pilot has attempted a landing in marginal weather conditions, has not diverted or turned back despite reducing visibility, or has descended below the Minimum Safety Altitude (MSA) to try to establish their position. Consider why they chose to do this – was it really an isolated bad judgement, or could they have been short of fuel due to poor planning and lack of contingency time?

MSA is at least 1000 feet above the Maximum Elevation Figure (MEF) in the relevant chart lat/long square. Remember that good planning, proper use of forecasts, awareness of terrain features en route and relevant safety altitudes, are not just good practice – they save lives.

d. But I've Done it Before!

Why do some highly experienced pilots believe that they can safely fly in marginal conditions, ignore their MSA, or attempt extreme aircraft manoeuvres? One of the reasons could be that either they, or others that they know, have done it before and 'got away with it'. This may well be true, but it certainly does not prove that it is safe. Imagine if your son or daughter tried to convince you that it was quite safe for them to cross a busy road blindfold, because they did it yesterday, and survived? What would you say to them?

e. But I know someone else who does it!

People vary in all kinds of ways, experience, concentration, skill, how they are feeling on a certain day, how much sleep they had, how much sleep they need, the after effects of recent illness, and their personal or domestic circumstances. The fact that someone else, on a particular day, can land in a marginal crosswind does not mean that you can necessarily do the same. The fact that you can do this does not mean that you should encourage someone else to do the same.

Being a competent pilot means correctly assessing your own limitations on a particular occasion. It does not mean pretending that if someone can do it, then everyone can do it every time; or that if someone else is doing it, that necessarily makes it safe or wise.

f. Exercising Sound Judgement

Pilots enjoy a great deal of freedom, despite the unforgiving nature of flying. The reason for this is that the regulatory authorities place a great deal of trust in the pilot to exercise competent judgement concerning flight safety. Qualified pilots are thought to be capable of making responsible decisions about whether it is safe to fly, taking into account their experience level, aircraft type, location, personal physical and emotional state, and prevailing or expected weather conditions. There are two serious threats to the use of this judgement: The pilot may have an excessively optimistic view of the situation or of his own ability; or he may be persuaded by other people to proceed with a flight against his better judgement. How can this happen?

g. But You Promised!

Never promise to fly on a certain day or to be somewhere important, if you can only get there by flying. If it really is important to be there, leave yourself time for alternative surface transport. Tell friends or relations that you may be able to take them flying weather permitting. Better still, keep it as a 'surprise', decide on the day if you feel prepared and fit, the weather is fine, and the aircraft is serviceable, and offer to take them flying. They won't know you had to book the aircraft a month in advance. It is always disappointing to cancel a flight if non-aviator people, especially children, are looking forward to the trip. This is particularly true if the reasons are not easy for them to understand.

h. Peer Pressure

There will always be people who will pressure you in subtle ways to take risks that you don't feel comfortable with. They can be prevalent in clubrooms, asking you if you flew on a certain windy day, and smiling smugly if you say that you cancelled whilst they braved the crosswind, low cloud or lack of horizon. 'You diverted? What an idiot! I'd have carried on and got there... '. Perhaps they would; alternatively they might have carried on and not got there. Perhaps they are just full of bravado and wouldn't have carried on at all. Perhaps they have more experience, a better equipped aircraft, or suicidal tendencies. It doesn't really matter. The fact is that the world of aviation relies on competent and independent pilot judgement, and the pilot is you. If you are swayed by clubhouse buffoons, then you are more afraid of their dubious opinions than of your own death. If this applies to you, you may not have the character that is expected of a pilot licence holder.

Audiences: are you impressing anyone?

In the review of fatal accidents, more than half of the low flying and aerobatic accidents involved an 'audience' – seldom at a formal air show, but more often to impress friends on the ground, at the clubhouse, or even passengers taken for a flight. The temptation to 'show off', to impress those watching, proved fatal in too many cases. (In fact, the 'audience' are not necessarily filled with admiration while watching these antics. They may simply be wondering when the accident will happen, and what this person is doing with a licence.) Before you decide to take such a risk, ask yourself: *would the people who are watching be prepared to risk their lives to impress you? What would you think of them if they were?*

j. Joint Decisions

It is a well known phenomenon that a joint decision made by a group of like minded people is usually more extreme than the decision that any one of them, alone, would have made. Pilots tend to be, by their nature, fairly adventurous individuals who are willing to face a certain amount of risk in order to pursue their activities. Committee decision: 'we'll give it a go!'

3 DIFFERENT PEOPLE - DIFFERENT RISKS

a: Age Groups

The review of fatal accidents suggested that the risks for young pilots were a little different from those of more mature years. Young pilots – especially young male pilots – sometimes took quite unnecessary risks in terms of low flying and aerobatic manoeuvres, often in front of friends or others watching (see 'Audiences' above). Older pilots seem less tempted to perform spectacular or risky manoeuvres, but they may take a different kind of risk. Pilots who fly into terrain, under full control of their aircraft and without any significant technical failures are, on average, older than pilots involved in other kinds of fatal accident. Typically, these pilots continued flying into adverse weather conditions, and / or ignored their MSA (if indeed one had been calculated).

b. Total Experience Level

Pilots involved in the fatal low flying and aerobatics accidents are usually highly experienced. Perhaps they believe that because of their very high hours, they can fly safely in these very unforgiving regimes. Pilots in fatal CFIT accidents are also typically very experienced. Again, they may believe that their long experience might allow them to fly safely in conditions that others are advised to avoid. If this thought ever enters your mind, remember that all of those highly experienced pilots in the fatal accident reports also thought that 'it would be all right'.

Pilots with low flying hours may be vulnerable to a different kind of accident. Those with very low hours feature less in the accident reports than those with 200 – 500 hours. The latter group seem to be more likely to lose control of the aircraft during visual conditions. This is probably not very surprising, given that these pilots are still quite inexperienced, and may be moving for the first time toward some slightly more ambitious flying.

c. Use It or Lose It

Recency may also be a safety issue; the fact that you could do something perfectly six weeks ago does not mean you can immediately do it now. A skill is like a message written in chalk on an outdoor wall – *it gets eroded a little every day*. If the writing is retraced repeatedly it will become more enduring. Even then, it will be eroded eventually if it is not periodically refreshed. Skills are refreshed via practice, annual or recency checks or post qualification training.

4 ONLY HUMAN

a. Trust Me, I'm a Pilot

Despite what some people may think, pilots are only human, and have normal human limitations. The fact that pilots are trained, experienced and competent, does not mean that they will always perform perfectly, that they will never experience an 'off day', overload, illusions or distorted perceptions, or that they will never make a mistake. Everyone recognises that physical parts of the aircraft have a certain expected failure rate, and this is (correctly) seen as a realistic, normal performance level. Human pilots also have a 'realistic' performance failure rate, and it is **not** zero.

b. To Err is Human

One characteristic of human beings is that **we all make mistakes**, no matter how well trained, competent, careful, or skilled we may be. **Nobody** is immune from errors, and the person who imagines that they are infallible is the most dangerous of all. There are two general classes of error:

- 'slips and lapses' include 'finger trouble', errors in data entry or recording (such as writing down the wrong digits), or not noticing that an instrument reading has changed;
- 'mistakes' refer to actions that the pilot makes intentionally, and executes correctly, but they turn out to be a bad plan.

In general, mistakes are more easily reduced by training, but they still can and do happen. The important thing is to recognise and rectify mistakes – and to learn from them. Slips and lapses can happen to anyone and are, if anything, more likely in highly skilled, experienced people.

c. Believing is Seeing

There are well known optical illusions that can affect pilots judgement, e.g. height perception when approaching sloping runways. In other circumstances, there can be a mental distortion that is nothing to do with visual illusions as such, but can be just as dangerous. Human beings are selective about what they 'see'. If a person believes something to be true, then they will tend to 'see' only those cues in the environment that are consistent with that belief, treating these as positive confirmation that the belief is correct, and 'not see', 'blot out' or ignore any evidence to the contrary.

Unfortunately, pilots are no exception to this rule. If a pilot has formed the belief that he is at a certain geographic location, then his mind may try to organise whatever cues are present in a manner that will confirm this belief. This means that conscious cross checking to look for differences to expectation are critically important, and frequently a feature of aviation procedures. This principle can even apply to the expectation that instruments should be showing a certain reading, or hearing an ATC clearance that is expected or usual. It is vital that instruments are actually read and messages are really listened to, with at least some anticipation that they may **not** say what you expected. It is difficult for anyone to accept this about themselves, especially if they are highly technically qualified and experienced. Believe it: if you are human, this **does** apply to you.

d. Stress

Stress is a familiar feeling to most people. When people are stressed, their judgement can be affected, and their thinking may be unclear. They may suffer from 'tunnel' thinking, concentrating on (or over-reacting to) one particular problem to the exclusion of all else. This is dangerous. If there is a problem in flight, *the pilot's first priority must be safe flight*. Attention to a faulty radio, airsick passenger, or navigation problem **must** be a secondary task. If you are feeling stressed before a flight, consider whether you should cancel. If you can foresee a period of high workload during the flight, rehearse it mentally, prepare as much as possible ahead of time and, above all, remember that your first priority at all times is to *fly the aircraft*.

5 ONLY A MACHINE

a. Trust Me, I'm Electronic

Just as human beings can make errors, mechanical and electronic devices can also be faulty. THINK about what your instruments should say – do a mental 'reality check'. Always cross check with a second source (e.g. landmarks in the outside view) if possible. Change – especially movement – attracts attention from our senses, but a static condition, or a very slow rate of change, is more likely to go unnoticed. It is important to check all instruments regularly, never think that your attention will automatically be drawn to a deteriorating situation. If your fuel gauge is stuck on full, the needle will remain steady, although actual fuel levels will be dropping. There will be no rapid movement or change to attract your attention.

b. GPS

GPS is becoming a common accessory for GA pilots. It can be tremendously helpful at times and is probably an overall safety 'plus'. However, a few words of caution *(see SafetySense leaflet 25 – use of GPS)*:

- **Never** use GPS as your primary means of navigation
- **Never** use it to land in poor visibility (and that means you too, helicopter pilots!)
- **Never** spend time head down, fiddling with GPS, and ignoring the outside world.
- **Never** believe GPS data without question. It is NOT infallible and it CAN go wrong.
- **Never** fly in conditions that you would normally avoid, because you believe GPS will reduce the risk and get you there safely.

6 HOW DO ACCIDENTS HAPPEN? COMMON SCENARIOS IN A CAA REVIEW:

a. Controlled Flight Into Terrain (CFIT)

In a CFIT accident the pilot does not lose control, and the aircraft has not failed. They simply fly into the ground, often hills or mountains. The pilots who had fatal CFIT accidents were typically over fifty years old, and very experienced. More than a third were flying in their home base local area, and accidents were not restricted to mountainous regions. Of all CFIT accidents, 82% included unwise reaction to weather conditions (such as continuing to fly into worsening weather) and 64% had not adhered to their MSA (if they had calculated one at all), trying to get 'below the weather', or hoping to confirm their position. More than a third found out too late that they had made an error in navigation.

b. Loss of Control in VMC

Loss of control in visual meteorological conditions (VMC) is almost as common as CFIT. In the accident review, it was noticed that many of these loss of control accidents involved an unfamiliar situation, a distraction or a minor technical failure. The inexperienced pilot was probably coping quite well, until they were overloaded by some unforeseen event. This is probably difficult to avoid, but it is worth rehearsing – even mentally – exactly what you would do if you had a technical failure, or encountered a distraction. Also, remember that if the flight you have planned is going to require 100% of your current skill capacity to cope with it, then you won't have anything left in reserve for unplanned or unusual events that crop up.

c. Low Flying / Aerobatic

Highly experienced young male pilots (often with an informal audience) who fly low and perform aerobatics without adequate height are putting themselves and others at risk. Accidents are not unusual in these circumstances.

d. Loss of Control in IMC

All but one of the pilots killed when they lost control in IMC were flying in instrument conditions without an Instrument Rating. This is extremely unwise to say the least. Possibly they believed that their IMC rating was sufficient for prolonged, intentional flight in instrument conditions. Unfortunately, the IMC rating is not sufficient for such conditions. It should only be regarded as a **minimum skill** to 'get out of trouble' if an unintentional excursion into IMC occurs. Disorientation can affect anyone, particularly those who have not been adequately trained to fly on instruments, **and kept in practice**.

And finally, the bottom line is:

Don't gamble, safe flying is ENJOYABLE flying

7 SUMMARY

Most pilots want to enjoy the freedom to fly when, where and how they want to, whilst maintaining safety for themselves and others. The way to achieve and sustain this situation is to:

- **be realistic about the weather**
- **work out a Minimum Safe Altitude (MSA) and keep to it**
- **use your judgement responsibly, don't be pressurised to fly**
- **know your own limitations**
- **prepare thoroughly**
  * **allow for contingency:**
  * **have enough fuel**
  * **be prepared to divert**
- **rehearse possible 'situations'**
- **use good practice in your planning and flying**
- **don't take unnecessary risks.**

This will avoid the need for additional regulations and restrictions, and give you safe, enjoyable flying.

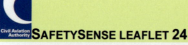

SAFETYSENSE LEAFLET 24 — PILOT HEALTH

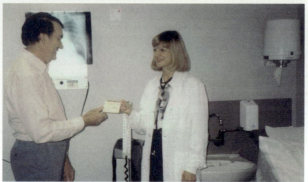

1. INTRODUCTION
2. THE MEDICAL EXAMINATION
3. ENVIRONMENT
4. THE BIOLOGICAL ENGINE
5. HYPOXIA
6. HYPERVENTILATION
7. VISION
8. STRESS AND FATIGUE
9. ILLNESS AND INJURY
10. ALCOHOL
11. EXPANSION OF BODY GASES
12. MEDICATION AND FLYING
13. CARBON MONOXIDE
14. 'I'M SAFE'

1. INTRODUCTION

The CAA requires pilots to hold a medical certificate (for Joint Aviation Authority, JAA, licences) or a declaration of health (for the National Private Pilot Licence, NPPL). The medical assessment is intended to reduce the risk of in-flight incapacitation.

A network of Authorised Medical Examiners (AMEs) across the country are approved by the CAA's Medical Division to undertake the appropriate medical examination for the JAA medical certificate. The AME has received training in aviation medicine and may also be a pilot.

However, pilots who wish to fly only light (up to 2,000 kg) single-engine aircraft (also microlights, gliders or gyroplanes) within the UK and in good weather can obtain the relevant medical documentation for a national PPL without visiting an AME, although they will need to attend their general practitioner. Slightly different requirements apply, depending on which type of flying activity is intended, so it is best to seek advice from your local flying club, or from the NPPL website: www.nppl.uk.com

Advice on health related matters can also be obtained from an airsport medical adviser, an AME, or the CAA Medical Division – see the Medical Division's web site for further details: www.srg.caa.co.uk.

Medical requirements for flying are under regular review and frequently change. Any recent changes will be posted on the Medical Division's website.

2 THE MEDICAL ASSESSMENT

It is particularly important that pilots are aware of their state of health. as what may seem a trivial symptom e.g. mild earache, can assume importance when flying. Whilst a medical assessment by a doctor can be reassuring, such assessments (which, for the NPPL, do not necessarily include a medical examination) are much less useful than the individual's self-determination of fitness, or unfitness. It is primarily the pilot's responsibility to decide if he is fit to fly, and it is also his responsibility to stay on the ground if he suspects he may not be completely well.

3 ENVIRONMENT

The earth's atmosphere consists of a mixture of gases, primarily oxygen and nitrogen, with the former being essential for human life. As an unpressurised aircraft climbs through the atmosphere the cockpit pressure reduces and at 18,000 ft the pilot experiences half the pressure of that at sea level.

4 THE BIOLOGICAL ENGINE

a. The human body converts the substances it absorbs such as food and oxygen into energy by a chemical process, similar to very slow combustion, called 'oxidation'. The body varies its consumption of stored energy sources according to its degree of activity, just like an engine. The intake of food (energy source) is adjusted on a medium to long term basis, whereas oxygen intake can be increased very quickly, in response to a short-term requirement to oxidise more stored nutrients and provide extra energy. When resting we require very little oxygen; under a high physical work load this increases and at maximum effort, oxygen use and energy production can be more than 15 times the resting value.

b. Air is inhaled into the lungs where its oxygen combines with haemoglobin in the red cells of the blood and is then circulated to those tissues where energy is needed. At the cellular level, oxygen combines with food stores to provide energy (with heat as a by-product). All cells need some oxygen to survive and the brain is particularly susceptible to a reduced supply of oxygen. Apart from heat, a main by-product of the oxidation process is carbon dioxide, which is returned to the lungs by the blood and exhaled.

c. Oxygen comprises only one fifth of the air breathed in and its availability for absorption and transport through the body is pressure dependant. Up to about 10,000 ft altitude, the healthy body has compensatory mechanisms to cope with the associated reduction in oxygen availability with increasing altitude without any noticeable detrimental effect. However, if there is an abnormality of the respiratory or cardiovascular system, the individual is likely to be more affected by a reduction in oxygen pressure, and may have symptoms even below 10,000 ft.

d. Reducing the capacity of your oxygen transport system by donating blood may increase your sensitivity to altitude, although this is quickly remedied by the body's reserves. However, a pilot should not fly for at least 24 hours after giving blood.

e. When an individual ascends above 10,000 ft in an aircraft the reduction in oxygen pressure reduces the efficiency of cellular processes, with the brain being the most sensitive of the body's systems. No-one is immune to these effects, which are insidious and often unnoticed by the affected individual. They may lead to hazardous actions, such as forgetting to change fuel tanks or flying off course. The effects become increasingly more serious with increasing altitude and above 18,000 ft, breathing atmospheric air, pilots are likely to eventually lose consciousness. At 25,000 feet this is likely to occur in 2-4 minutes. The mountaineer is able to adapt, to a certain extent, to such altitudes but such adaptation occurs at a rate which is too slow to be of benefit to the aviator used to living near sea level.

5 HYPOXIA

a. When the human body is starved of oxygen at altitude, or is in poor health with regard to its ability to absorb and transport oxygen, its efficiency reduces. When inadequate oxygen is available for normal functioning a condition called 'hypoxia' results. The brain is affected early but symptoms are often unnoticed due to the associated dulling of judgement. The effects are similar to alcohol intoxication. As hypoxia proceeds the individual becomes clumsy, drowsy, develops an inappropriate sense of well being and becomes increasingly error prone. The extent of the symptoms is dependant upon the actual altitude but even short periods above 10,000 ft are likely to produce effects.

b. To prevent hypoxia, flights must be at an altitude less than 10,000 feet, or the aircraft must have a pressurised cabin (as do almost all commercial airliners) or the pilot must utilise an individual oxygen source supplied by a personal mask.

6 HYPERVENTILATION

a. The respiratory system adapts quickly to changes in oxygen demand caused by exercise. However, breathing rapidly does not reduce the effects of hypoxia and can have some disadvantages.

b. The body cells produce carbon dioxide as a by-product of the oxidation process, which is dissolved in the blood and returned to the lungs for exhalation. Increasing the rate and depth of breathing speeds up the removal of carbon dioxide, disturbing the chemical balance in the blood and symptoms similar to hypoxia may result.

c. The most common causes of hyperventilation are stress and anxiety but this can usually be controlled by consciously returning to a normal rate of respiration, and relaxing. Your instructor will give you advice if he notices you are breathing rapidly when under training. If you or a fellow crew member or passenger do experience symptoms which might be attributable to hyperventilation, it is important to first ensure that hypoxia is not the problem.

7 VISION

a. Our sight is something we tend to take for granted. There are, however, two points pilots should be aware of.

b. Firstly, if you use contact lenses or spectacles you should have a spare pair of spectacles immediately available, which can be put on if you become intolerant of your lenses (or lose one, or both, of them) in flight, or you lose or break your spectacles.

c. Secondly, almost all of us will require reading glasses at some point – the lens in each eye stiffens with increasing age and can't adjust for near distances as it can when younger. Generally this process becomes noticeable at about 40 years with the first sign being an inability to read in poor light (because in low lighting conditions the pupil widens and, in photographic parlance, the 'depth of field' reduces and the near point for focussing moves further away from the eye). Unfortunately it will not improve with eye exercises! After your first set of reading glasses you will probably need slightly stronger ones every few years until about age sixty. Do make sure that your reading glasses are suitable for flying. You still need to see clearly into the distance and so you should use bi- focal lenses or the half frame, look-over type so that you can be comfortable looking at a map, your instruments, or at the horizon without having to change or remove your glasses. Full frame near vision spectacles are not acceptable for pilots, because distant vision is adversely affected. 'Varifocal' lenses (those which gradually, rather than abruptly, adjust their refractive power) can be used but make sure you find them suitable for flying before using them in the air, as not everyone can tolerate them (it can make some individuals feel dizzy).

d. There are a number of surgical procedures available which reduce, or even eliminate, the need for spectacles. All involve a reshaping of the clear part at the front of the eye, called the cornea. Although the methods vary, some using lasers, others diamond knives (the older techniques) none offer guaranteed success and all will require a period of grounding with a specialist assessment before being considered fit for flying. The long term effects are not fully known and vision can occasionally be worse after such surgery. Any pilot considering such surgery should look at the CAA Medical Division website (www.srg.caa.co.uk) before submitting to an irreversible procedure.

8 STRESS AND FATIGUE

a. All of us at some time will find our lives affected by stress, fatigue, illness or injury – the important thing is to recognise how these can affect our flying skills and to proceed in a sensible fashion.

b. Stress is considered a modern day ailment, but it is a part of everyday life. It is the reaction to it that may cause a problem. Sleep disturbance, poor appetite and indigestion can all be signs of excess stress, whether at home or at work. Although most consider flying to be a relief from such pressures, it is not sensible to fly when you are experiencing physical symptoms or ruminating over your problems. Any preoccupation can detract from the continuing mental activity needed for safe flying. If you are not feeling 100%, take responsibility for your own flight safety and seek medical advice if you are uncertain of the implications for flying.

c. Short term fatigue is what we experience after strenuous physical or mental exercise. It may be associated with sleepiness and may also be the cause of mistakes and lapses of concentration. Medium to long term fatigue is more often associated with shift work, time zone crossings (which causes 'jet lag') or just regularly cutting back on sleep. It can cause drivers to fall asleep at the wheel or pilots to fall asleep at the controls. The only means of dealing with fatigue is to recognise when it is likely to occur and what can happen as a result. The only means of preventing it is to make sure you get adequate rest before flying.

9 ILLNESS AND INJURY

a. Any illness can be debilitating and recovery can take longer than you think. Most pilots would think that returning to work means they are fit to fly but this is not always the case. As a rule of thumb, any condition requiring medical certification that you are unfit for work should normally require at least an equivalent time back at full employment, without treatment, before flying. Your GP or AME may be able to give you specific guidance if you want to start flying earlier. This particularly applies to some of the modern outpatient surgery or investigations which have been addressed in Aeronautical Information Circular 63/2002 (Pink 35) 'Modern Medical Practice and Flight Safety'. Seek medical advice before flying and ensure you advise your doctor that you are a pilot.

b. If you have an injury ensure you have fully recovered before flying. You do not want to find yourself in severe pain, or with a weak arm or leg when operating an aircraft. Unlike car drivers, pilots do not have the option to stop in a few seconds. Also make sure that you have the full range of movement necessary for flying before returning to the cockpit. A circuit or two with an instructor before going solo can be beneficial if there is any doubt about your fitness after recovery from injury.

10 ALCOHOL

a. The consumption of alcohol produces effects similar to hypoxia. However, breathing oxygen will not reverse the effects. Increasing altitude increases the effects because of the reduced oxygen pressure. It is therefore essential for pilots to separate their flying from alcohol consumption. Since it takes an extended period of time to remove even low levels of alcohol from the blood, pilots should not fly for at least eight hours after consuming modest amounts of alcohol and up to 24 hours (or longer) following a major celebration!

b. Since one of the more subtle effects of alcohol is on the inner ear and can result in an increased susceptibility to disorientation up to three days after taking a large amount of alcohol, pilots should always be careful in the amount of alcohol they consume if they are flying during the next 1-3 days.

11 EXPANSION OF BODY GASES

a. If you take a balloon from sea level to 18,000 ft, its volume will double due to the decrease in pressure (Boyle's Law). Gas in the cavities of your body will do exactly the same thing. Problems can be experienced with air in the sinuses or behind the eardrum (middle ear) as both of these cavities have entrances which can be easily affected by the inflammation from a common cold. The most important point is to avoid flying with a respiratory tract infection (cold). You should know how to 'clear your ears' using the Valsalva technique and if you cannot clear your ears before flight, stay on the ground because you may tear an eardrum, or suffer severe pain in your ears or sinuses on descent (climbing is not usually a problem).

b. It is also possible for the nitrogen gas which is dissolved in our body fluids to come out of solution and form bubbles if exposed to reduced pressure for a prolonged period. This is known as decompression sickness or 'the bends' and is rarely experienced at an altitude below 18,000 ft. However, SCUBA diving exposes the body to increased pressure and dissolves more nitrogen in the body. This may cause decompression sickness during subsequent flying at a very much lower altitude. Most divers are aware of this problem and will not fly, even in a pressurised aircraft, immediately after diving. If you intend to SCUBA dive within 24 hours before flying, seek expert advice about the time interval between the two activities.

12 MEDICATION AND FLYING

a. Doctors can choose from a wide range of medications when treating an illness. There is also a wide range of 'over the counter' treatments which do not require a prescription. Doctors may be unaware of the effects of their prescriptions upon a pilot's flying capability. Some may cause drowsiness, nausea or fatigue and others may reduce resistance to even minor increases in acceleration forces.

b Some quite simple 'over the counter' products carry warnings to avoid operating machinery and they may react with other medication. If the medication you are taking says that driving, or operating machinery may be adversely affected, it is probably unsuitable for use if you are flying. Remember that the underlying condition for which you are taking the medication may preclude flying. Seek specialist advice if you are unsure of whether or not you should be flying, before you take to the air as a pilot.

13 CARBON MONOXIDE

a. An aircraft engine is rather less efficient than your body in that some of its fuel oxidation is incomplete and carbon monoxide rather than dioxide is produced. This would be of academic importance if it were not that many aircraft use their engine exhaust gas heat, through an exchanger, to warm the cabin. Add to that the fact that carbon monoxide bonds very strongly to the blood cells and blocks its oxygen carrying capacity then it becomes necessary to consider the symptoms of carbon monoxide (CO) poisoning.

b. As a gas, CO is colourless, tasteless and lethal! Exposure of pilots to it has been the cause of many fatal accidents. It can usually only be recognised in an aircraft by associated engine exhaust smells. Symptoms are subtle, similar to hypoxia but perhaps with a more obvious headache and it doesn't respond so promptly to oxygen – although using an oxygen mask is likely to restrict further exposure.

c. The best way to deal with CO poisoning is to prevent exposure in the first place but if you do suspect its presence when in flight, increase ventilation, land and try to get an engineer to trace any sources. There are CO monitors on the market and we recommend that one of them be carried. Paper sensors are easily contaminated by other fumes and need to be changed more frequently than their markings would suggest. Electronic detectors often have several functions in addition to a basic warning, but if fitted permanently would constitute a modification, and may place the device outside its operating limits. However, one could be carried as personal equipment.

14 I'M SAFE

a. This acronym gives all pilots a basic checklist for their fitness to fly. The items on that checklist are covered in this leaflet. The bottom line is that a pilot's fitness can change quickly and it is primarily the responsibility of the pilot himself to decide whether or not he is fit to fly.

I llness
M edication
S tress
A lcohol
F atigue
E ating

I'M SAFE
Safety Promotion, GAD, Gatwick

Use this personal checklist before setting off for the airfield, just as you would look at the weather or do a pre-flight check. It is available as a free sticker from Safety Promotion, General Aviation Department, Aviation House, Gatwick Airport South RH6 0YR (please send SAE).

b. If in doubt about any of the items, then take medical advice.

SAFETYSENSE LEAFLET 25 - USE OF GPS

Most illustrations courtesy of Garmin UK and Honeywell

1. INTRODUCTION
2. SYSTEM AND SIGNAL ANOMALIES
3. EQUIPMENT
4. SYSTEM FAMILIARISATION
5. FLIGHT PLANNING
6. PROGRAMMING CROSS-CHECKS
7. THE DATABASE
8. INITIAL STATUS
9. IN-FLIGHT USE
10. DIVERTING FROM THE INTENDED ROUTE
11. INSTRUMENT APPROACHES
12. THE FUTURE
13. SUMMARY

1 INTRODUCTION

a. The most familiar Satellite Navigation (or GNSS) system to most of us in the UK is the US Department of Defence "Navstar" Global Positioning System or GPS. Other systems are available, or in development, but this leaflet is based on the use of the Navstar GPS system.

b. Here you will find background information and guidance for General Aviation pilots in the use of stand-alone GPS equipment (ie. systems not forming part of an integrated Flight Management System).

c. Unless specifically approved for particular purposes, such equipment is only to be used as an aid to other forms of navigation.

2 SYSTEM AND SIGNAL ANOMALIES

a. The GPS system has generally shown exceptional reliability, but it has been known to suffer technical and human failure. Consequently, GPS must not be relied upon as a sole navigation reference in flight critical applications. Common sense dictates that pilots should not only familiarise themselves with the techniques required to use the system properly, but understand how it could go wrong and prepare for the unexpected.

b. AVAILABILITY
The receiver relies on maintaining line of sight between itself and the satellite. It needs to be able to 'see' several satellites (the number depends on the accuracy and integrity required) to provide a fix and, even with 24 satellites in orbit, there are times when insufficient satellites are 'visible' to provide that service.

c. GEOMETRY

Whilst enough satellites may be 'visible' to give a fix, at certain times their angular separation may be small, giving rise to poor accuracy. This reduction in accuracy is called "Dilution of Precision" or "DOP" and may be displayed as a number. A high DOP (more than 6) indicates that GPS position accuracy is significantly degraded and the information should not be used.

d. RAIM

More sophisticated receivers contain a processing algorithm known as Receiver Autonomous Integrity Monitor (RAIM). RAIM compares the information received from a number of satellites and alerts the user to an error. If enough satellites are visible, the function may be able to identify the faulty signal and discard it.

The presence of a working RAIM function only monitors one type of failure and does not guarantee the absence of a position error. RAIM availability at any time and place in the world can be predicted from satellite orbital information by receivers with appropriate software. However, RAIM prediction cannot foresee the failure of a satellite, nor the removal of satellites from service. Neither does it take account of terrain.

e. NOTAMs/NANUs

Notice Advisories to Navstar Users (NANUs) are the means of informing GPS users of planned satellite "outages". NANUs are available in the UK through the NOTAM system, and must be consulted to determine the availability of GPS and RAIM information. Some receivers can be adjusted to manually deselect a particular satellite if it is expected to be out of service.

f. FAILURE / ERROR

The satellite clock (the heart of the system) may drift off time, the satellite may stray from its orbit or its transmitter may simply fail. It can take up to two hours for such failures and errors to be resolved. At such times, unknown position errors have been reported (up to 2 km), despite the presence of RAIM.

g. TERRAIN SHIELDING

At low level, in regions of high terrain or obstacles, satellites can become hidden to the aircraft receiver. This may give rise to unexpected loss of position and/or RAIM.

h. DYNAMIC MASKING

Parts of the aircraft itself may get in the way, for example the outside wing in a turn. If this blanks the signal momentarily, the navigation capability may be degraded or lost, requiring several seconds of straight and level flight to re-establish navigation information.

i. MULTI-PATH REFLECTIONS

The signal may bounce off hills and structures before arriving at the receiver, giving rise to range errors from the satellite. Such errors are generally very small but may appear as a sudden change in position which the receiver interprets as a change in drift and groundspeed. This may lead to distracting messages declaring phenomenal wind shifts, and may be sufficient to destroy the integrity of the navigation information altogether.

j. INTERFERENCE AND JAMMING

The GPS signal received from the satellite is at very low power and is **vulnerable to interference**, either intentionally or otherwise. Sources of unintentional interference include, among others, UHF and microwave television signals, some DME channels, and harmonics from some VHF RT transmissions. It is known that jamming devices are available which can easily disrupt signal coverage across a wide area. Military exercises and trials which include deliberate GPS jamming take place frequently, and are notified. Check NOTAMs for any areas likely to be affected.

k. SUNSPOTS

Because the satellites orbit at very high altitudes, radiation from the sun can affect their transmissions, or even their own navigation system. Particular flares or sunspots cannot be forecast, nor can their effect, but NOTAMs include warnings of possible GPS signal interference when major disturbances are detected.

l. SELECTIVE AVAILABILITY

Finally, the satellites are the property of the US Department of Defence (DoD), which may move satellites around to improve cover over a particular area, thereby reducing the availability over others. Although the signal is promised to remain available for civilian use, the facility exists to insert random errors into the signals to reduce accuracy, or even to switch the whole system off completely.

3 EQUIPMENT

a. CARRIAGE OF EQUIPMENT
The installation or carriage of GPS equipment does not affect the requirement for a primary means of navigation appropriate for the intended route, as detailed in Schedule 5 of the Air Navigation Order.

b. VFR use only
When operating under Visual Flight Rules (VFR) outside controlled airspace, there is no requirement to carry any radio navigation equipment and there is no installation standard for GPS used only as an aid to visual navigation. However, equipment permanently installed (in any way) in an aircraft must be fitted in a manner approved by the CAA. If a hand held unit is carried, care should be taken to ensure that it, the antenna and any leads and fittings for them are secured in such a way that they **cannot interfere** with the normal operation of the aircraft's controls and equipment and do not inhibit the pilots movements or vision in any way. Consideration should also be given to their possible effect on the aircraft occupants if the aircraft comes to a sudden stop.

Equipment permanently installed…

c. IFR certification
If a GPS system has been certified as meeting the "Basic Area Navigation" (BRNAV) requirements this will be stated in a 'Supplement' to the aircraft Flight Manual. Such approval means only that the equipment is regarded as accurate for en-route purposes (within ±5 nautical miles for 95% of the time).
There may be additional approval requirements to operate it in Terminal Areas (including SIDs and STARs) or on an instrument approach. Even systems which are certified for Precision Area Navigation (PRNAV) may not meet the required navigation performance for use on an instrument approach. The use of such equipment for precision navigation will probably also require specific pilot qualification.

4. SYSTEM FAMILIARISATION

a. The individual manufacturers of GPS equipment each provide different functions in the receiver. There may also be major differences between individual receivers from the same manufacturer.

b. Before attempting to use the equipment in the air, pilots should learn about the system in detail, including:
- *Principles of GPS*
- *System Installation & Limitations*
- *Pre-Flight Preparation & Planning*
- *Cross-Checking Data Entry*
- *Use of the System In Flight*
- *Confirmation of Accuracy*
- *Database integrity*
- *Human Error*
- *System Errors & Malfunctions*

More detailed guidance on training is available in other CAA documents.

c. Essential learning, even for VFR use only and preferably with guidance from the manufacturer's representative or an instructor experienced on the individual equipment, should include at least the following:
- Switching on and setting up
- Checking the status of receiver, satellites, battery, and any database used
- Loading waypoints
- Loading a route
- Loading alternate routes
- "Direct" or "GO-TO" functions
- Selecting alternate routes
- What your database contains (and what it doesn't)
- Use of RAIM function if fitted
- Amending RAIM input if fitted
- Regaining the last screen when you have pressed the wrong button!

d. Whether or not you find a suitable instructor, practise using the equipment on the ground before trying it in the air. Then take someone else to fly and navigate for you, while you are becoming totally familiar with the GPS. If you fly a single-seater, ask someone else to fly you in their aircraft while you practise.

e. If the check list supplied with your GPS equipment is complicated, inadequate or non-existent, use part of the learning process to write your own check-list for setting up and use in the air.

f. Although there is no requirement to demonstrate use of the GPS on any UK flight test, it is sensible to use it at least for some of the time when an examiner or instructor is flying with you. You may pick up some useful tips.

5. FLIGHT PLANNING

a. The attention a GPS receiver requires in flight can be minimised with careful planning and preparation before departure, releasing the pilot to other tasks whilst in the air.

b. Most modern units allow the user to enter a series of waypoints as a route or flight plan. Be familiar with how to do this, how to store it, and retrieve it for later use. Doing this significantly reduces the chances of making an error in flight, and allows more time for other things such as lookout or instrument flying.

c. **Plan the flight and prepare a map and log in the normal way.** Then enter the route information from the log, directly into the receiver as a "Flight Plan". This achieves three things;

1 The route information is created visually on a chart, helping to eliminate any gross error.

2 You have a back up should the GPS information become unreliable or unavailable in flight.

3 You are aware of the terrain over which you intend to fly, and can calculate safe altitudes (many databases do not consider terrain).

d. USER WAYPOINTS

i) If the aircraft and GPS receiver are your own, you may want to set it up to your own preferences. For example, you might have a favourite visual navigation route which you follow every time you depart or arrive. Most GPS receivers allow you to set up User Waypoints to guide you along such a route, even if there is an airspace database installed. Keep a record of all loaded User Waypoints for future reference.

ii.) It has been known for one pilot in a group or club to edit the data comprising a stored User Waypoint and leave it with the same name, but in a different position. Deleting or moving existing User Waypoints, or changing their names, should be **expressly prohibited** where the GPS is operated by more than one pilot. Any changes must be agreed by the group.

iii.) This underlines the need to check the position of waypoints in the flight planned route, and any possible alternative or diversion route, before departure. If this is not done, pilots cannot rely on any 'Go Direct' or 'Nearest' function in the air when working with User Waypoints.

iv.) When inserting a User Waypoint, ensure that the latitude and longitude co-ordinates you use are from the correct geodetic datum. The positions of an individual point may be up to a kilometre apart if referred to different datums. Although some receivers have the facility to convert position information between the WGS 84 datum used in GPS equipment and others, these conversions are not always absolutely accurate and can contain errors. Positions may also be in different formats. Many receivers refer to positions as degrees, minutes and decimals of a minute, rather than the degrees minutes and seconds used in documents.

6. PROGRAMMING CROSS-CHECKS

a. Once the route has been entered, 'run' it to make sure you have not missed (or mis-entered) any waypoints. This may be called the 'Simulator' or 'Demo' mode.

b. If you have a map display, it is usually possible to display the route on the screen once it has been entered. Any gross error in the position of a waypoint or turning point should be obvious on the map. If there is no map, or it is too small to be of practical use, **compare the tracks and distances displayed on the GPS with the previously prepared flight log.**

any gross error should be obvious

compare tracks and distances

7 THE DATABASE

a. If you have an aviation database installed, ensure that it is current, and is valid for the area over which you intend to fly. Aerodromes seldom move far, but their serviceability, airspace, frequencies, reporting points and other information change often. An out-of-date database can lead (at best) to embarrassing and expensive error. At worst, it could be catastrophic. **Do not use an out of date database.**

b. Even a current database cannot be automatically assumed to be error free. Instances of database errors have been recorded, and only careful checking against current charts and the AIP may identify these. In addition, NOTAMs must still be consulted before flight.

8 INITIAL STATUS

On start up, check the status of the receiver and its battery. Compare the indicated GPS position with the aircraft's known position. If your aircraft is normally parked in the same place, it helps to enter the coordinates of that position as a User Waypoint. Each time you start up in that position, select 'go direct' to that waypoint. You will then see the current error of the GPS position. You can also compare the relative indicated position of a known database point (such as the Aerodrome Reference Point) with its actual position.

9. IN-FLIGHT USE

a. The GPS system should NEVER be used in isolation. The risk of loss or degradation of the signal, with the attendant possibility of a position error, is genuine. More importantly, the risk of human error in data input and display reading is extremely high and these errors can go unnoticed until it is too late.

b. It is easy to transpose numbers in one's head, and these errors are surprisingly persistent. Do not allow any such errors to lead you into trouble.

c. It may help to go through a three-stage exercise in setting up **any** navigation aid, including GPS;

1 Set it up and satisfy yourself that you have done it correctly.
2 Do something else – even if only for a few seconds.
3 Go back and check again that the set up is still correct.

d. When flying in IMC or above cloud, only use GPS in combination with other radio aids to correlate with dead reckoning of the flight planned route and general situational awareness.

e. If the GPS display agrees with everything else you know, including dead reckoning, the navigation log, map reading and general situational awareness as well as radio navigation, then the GPS display is likely to be providing the most accurate information.

f. The accuracy of GPS will often expose the operational error of other navigation aids. Errors of up to 5° are normal in a VOR display (more than on an ADF), and DME is only accurate to about half a mile. DME indicates slant range but GPS displays horizontal range, giving rise to a further small disparity, which increases as you approach the DME station overhead. Some apparent errors may of course be due to magnetic variation.

g. If flying visually, it is easiest (but not usually particularly accurate) to cross-check your GPS position with a recognisable feature on the ground. You could also compare indications from a radio aid station with the GPS range and bearing to that station. Any difference greater than the normal error associated with the radio aid indicates a problem with one or other aid. If you cannot cross-check with a third system, especially if short of fuel or near controlled airspace, consider asking an ATS radar unit or Distress and Diversion Cell for a position fix.

cross-check your position

h. To avoid becoming totally dependent on the GPS, ask yourself 2 questions regularly throughout the flight;

1 Does the GPS agree with at least one other independent source of navigation information?

2 If the GPS quits completely, *right now*, can I continue safely without it?

If the answer is yes to both questions, you may continue to use the equipment for guidance. However, if the honest answer to either one of the questions is "No", then *you must establish navigation by some other means.*

10. DIVERTING FROM INTENDED ROUTE

a. Re-programming the system in the air is time-consuming, and interferes with other procedures such as lookout. Like any cockpit operation, re-programming should not be undertaken whilst the aircraft is manoeuvring. Unless someone else can fly the aircraft for you, switch operation must be interrupted so that individual selections are interspersed with a thorough lookout (or instrument scan) every few seconds.

b. Anything you can do to reduce this re-programming will help. **Pre-plan likely route changes**, for example around controlled airspace in case you cannot obtain clearance, or around high ground in the event of bad weather. Have a note of the ICAO designators of all suitable diversion aerodromes.

c. Re-programming in the air is also much more likely to produce human errors. If you need to change your planned route, make at least a rough set of mental calculations (and note them down) BEFORE you turn onto the GPS track. Then if your new heading does not agree with your mental calculations, you

will know you have made an error somewhere. Check the new route on a map for terrain and any NOTAMed activity. If your database is not current, you must check for controlled and restricted airspace also.

11. INSTRUMENT APPROACHES

a. Existing instrument approach procedures at some aerodromes are already provided in many receiver databases. These "Overlay" or "Monitored" approaches can present the pilot with a direct comparison with the terrestrial approach aid being used. **If your GPS receiver can do this, you must exercise extreme caution**. VOR and NDB approaches to beacons actually *on* the destination aerodrome usually provide a final approach path or track which is not aligned with the main runway centre-line. Even on a direct approach to a particular runway, pilots should not necessarily expect to be on the extended centreline of the runway.

b. The terrestrial approach procedure may include DME ranges from the threshold, missed approach point (MAP) or some other reference, such as the beacon. The GPS may give distance guidance to a different point, such as the Aerodrome Reference Point. Pilots should be aware of any differences in the distance information given to step-down fixes and/or the MAP, as this has the potential for catastrophic error.

c. **Overlays and Monitored approaches must only be used as supplemental information and the normal equipment for that approach procedure must be used as the primary reference**. Otherwise, disparity between the two displays and the potential for mistakes are just as likely to diminish the safety margins on an instrument approach as enhance them.

d. The safety values in the design criteria of any published approach are applied to known, surveyed obstacles and restrictions to the required flight path. **Disregarding the established approach procedures and published minima, in favour of reliance on** the GPS, is not authorised and is highly dangerous.

11a. USER DEFINED APPROACHES

a. Pilots have been known to produce and follow their own approach procedures using GPS information. This is potentially dangerous. There is no ground based confirmation of position and the risk of mis-entering waypoints is high.

b. Furthermore, when flying towards a waypoint in normal, en-route mode, the course deviation indicator (CDI) normally indicates a track error of 5nm at full-scale deflection (or 1 mile per 'dot'). This is not accurate enough for any final approach, and only changes when either the sensitivity is changed manually or the aircraft is following a published *and correctly activated* GPS approach contained in the database. Changing sensitivity whilst on approach is a hazardous distraction.

c. Unless a *published* approach is activated, the receiver's RAIM function remains in en-route mode (even if the CDI scaling is changed manually) and there may be a position error of up to 2 *nautical miles* before any RAIM alarm is given.

d. User-defined approaches can be dangerous and are not authorised.

12. THE FUTURE

Satellite navigation will one day almost undoubtedly form the basis of our radio navigation but in the mean time, the GPS system is fallible and should be used **with knowledge and caution**, not blind faith.

13. SUMMARY

1. Accuracy is not guaranteed
2. Apparent accuracy does not mean reliability
3. Understand your own equipment.
4. Train before using it.
5. Use standard settings and check lists.
6. Flight plan normally before loading a route.
7. Check the route before flight.
8. Load possible alternative routes.
9. Ensure database is the latest version.
10. Check the status on start-up.
11. Fly and navigate manually, only use the GPS once you have verified its accuracy against something else.
12. Do not rely on GPS for instrument approaches.

--oOo--

Nov 2002

SAFETYSENSE leaflet 26   VISITING MILITARY AERODROMES

Part 1 Visiting during normal operating hours

3   PRE-FLIGHT
4   APPROACHING OR PASSING THE AERODROME
5   INSTRUMENT APPROACHES
6   CIRCUIT PROCEDURES
7   BARRIERS AND CABLES
8   GROUND MOVEMENT
9   DEPARTURE

Part 2 Visiting outside normal hours

10 PRE-FLIGHT
11 APPROACHING THE CLOSED AERODROME
12 CIRCUIT PROCEDURES AT A CLOSED AERODROME

13 SUMMARY

Appendix – Military aerodrome colour codes

1 Introduction

It is Ministry of Defence (MOD) policy to encourage civil use of military aerodromes where this does not conflict with military flying operations. While the same general rules and procedures apply to aircraft at all aerodromes, the specific requirements of military operations mean that the way they are applied often makes them appear quite different to those to which civilian pilots have become accustomed. Military pilots have their own regulations, but pilots of civil aircraft are always subject to the current Air Navigation Order and Rules of the Air Regulations. This leaflet is intended for use by private pilots, although commercial operators may find it useful background. It should be read in conjunction with SafetySense leaflet 6 "Aerodrome Sense".

2 Emergencies

a. Many military aerodromes have long hard surfaced runways. Most have resident fire and rescue services and air traffic controllers who are trained to help pilots of aircraft in distress or in urgent need of assistance, and who have the ability to listen and talk on the emergency frequency 121.5 MHz. These facilities suggest that such aerodromes make excellent diversion destinations for any aircraft with problems. Even if the aerodrome is closed, a long hard runway (or any part of a large flat airfield) is a much more attractive place to land in an emergency than
a farmer's field, and it is even possible that some rescue facilities may still be available.

b. There is a natural reluctance on the part of civilian pilots to make use of military facilities. However, if the pilot is experiencing problems which can be reduced by the use of a military aerodrome, the MOD encourages them to do so by waiving landing fees for any aircraft landing as a result of a diversion for genuine safety reasons. Nevertheless, it should be remembered that military operations normally have total priority. Unless pilots of civilian aircraft make distress ("MAYDAY") or urgency ("PAN, PAN") calls (which again many civilians seem reluctant to do), they are unlikely to be offered the use of these aerodromes.

c. **If you are experiencing problems in the air, do not hesitate to make a PAN call**, especially if there is anything that ATC can do to help. Because of the nature of military operations and the complexity of the aircraft, military air traffic controllers tend to be well practised in emergency procedures. As the saying goes, "an ounce of prevention is worth more than a pound of cure!"

d. Inexperienced pilots may be worried about information being passed to them too quickly for them to absorb. In that case, consider adding the word "TYRO" to your callsign (for example "TYRO Golf Alfa Bravo Charlie Delta") to warn the controller that he needs to give information clearly and more slowly than he would to a qualified military pilot.

Part 1- visiting during normal operating hours

3 Pre-flight

a. As for any flight, the most important part is the planning. Except in an emergency situation, every military aerodrome is **strictly PPR** (prior permission required), well in advance. Some require a minimum of 24 hours notice or longer and permission cannot usually be given instantly over the telephone, so an early request is vital. The published telephone number will normally be to Station Operations (Ops), which may or may not be co-located with Air Traffic Control. In order to consider the request, certain information is usually needed, so be ready to give the following:

- Pilot's name (and those of all passengers) (frequent visitors may require security clearance)
- Aircraft type and registration
- Aerodrome of departure
- Estimated time of arrival at the MATZ (if applicable) or ATZ boundary
- Intended time of departure from the military aerodrome
- Reason for the visit (appointment in nearby town, visit friends etc)
- What the aircraft's insurance covers (temporary £7.5 million Crown Indemnity can be added to the landing fee)
- Fuel type and likely quantity for refuelling if required (AVGAS may not be available)
- Pilot's flying experience and currency, including familiarity with that and other military aerodromes
- Customs, Immigration, and Special Branch clearance requirements (which may not be available)

b. Once permission is granted, Ops will have useful airfield information available. An aerodrome "visiting aircraft brief" may be provided to you either by phone, e-mail, fax or letter if there is enough time. It is expected that you telephone early on the day of arrival, so use that call to obtain more up to date information. Every military aerodrome records regular weather reports and has a dedicated terminal aerodrome forecast (TAF). Although these may not be published by the Met office, the aerodrome Ops will have them available. Ask for and be ready to copy down the latest TAF and METAR. Ops personnel will be aware of local navigation warnings, which they can also pass to you. In addition, they may be able to direct you to an aerodrome Automatic Terminal Information Service (ATIS) giving weather and other essential aerodrome information on a radio frequency and/or a telephone number.

c. Aerodrome and approach charts for military aerodromes are normally not included in the UK AIP, but many are included in commercial guides, and they can be provided by ATC on request or obtained through the internet at www.aidu.mod.uk The "visiting aircraft brief" should be studied in conjunction with the appropriate charts.

d. Even if permission has been granted, always pre-plan a diversion to a suitable alternative aerodrome and carry enough fuel to reach it after allowing for holding time at your intended destination. Emergencies or military operations may require such holding, and prevent you landing even when on final approach. While most military aerodromes have runways long enough to accommodate the majority of light aeroplanes, they may have only one of them. Know your own crosswind limit in the aircraft you will be flying, and do not make an approach if the wind is outside that limit.

4 Approaching or passing the aerodrome

a. Make yourself as obvious as possible to other traffic; consider using the landing light. While it is not mandatory for civilian pilots to recognise a military aerodrome traffic zone (MATZ), if your track passes through or near one (and obviously if you intend landing!) it is **strongly** recommended that you call on the published VHF LARS or zone frequency at least 15 miles or 5 minutes flying time before you expect to enter the MATZ, and comply with requests from ATC. A serviceable transponder, ideally with altitude transmission (mode C), will assist ATC in identifying you but is not essential. Note that, except in a very few cases, the aerodrome traffic zone (ATZ) of a military aerodrome (whether within a MATZ or not) is permanently active, even if the aerodrome is closed, and you must avoid it unless you have permission to enter.

b. Many military aircraft are only equipped with UHF radio equipment, and Air Traffic Control is provided on UHF frequencies which civilian aircraft cannot use. If the controller is talking to an aircraft on UHF, he will not be able to answer a VHF transmission, and indeed may not even hear it if the UHF transmission happens at the same time. When you make the initial call, it is advisable to say on what frequency you are transmitting (e.g. "on 122.1"). Give the controller time to answer, and be prepared to call again if you hear nothing. Once the controller starts talking to you, he may simultaneously transmit on both UHF and VHF frequencies, so listen carefully for your own callsign at all times. You will hear everything he says, whether on the VHF or UHF frequency, but you will not hear UHF transmissions from other aircraft, which may take place while you are transmitting and cause the controller to ask you to repeat your transmission. If you have not received it already, you may be given the "visiting aircraft brief" over the radio, together with pertinent information about the aerodrome facilities.

c. Military procedures use two altimeter settings below transition altitude. Normally, military aircraft will set the Regional Pressure Setting (RPS) on their altimeter when outside the immediate area of the aerodrome and its instrument approach pattern, and controllers may ask you to do the same when receiving a service from them. Otherwise, QFE is the datum, and all heights indicated are above the runway. However, separation from other traffic may dictate that a controller asks you to use a pressure setting which you do not expect.

CAA Safety Sense Leaflet	26 Visiting Military Aerodromes

PAR controller

5 Instrument approaches

a. Expect to set QFE as above. Most aerodromes equipped with radar will provide you with radar assistance until you are visual with the aerodrome, or will direct (vector) you on to a precision or non-precision final approach using that radar. If a surveillance radar approach (SRA) is provided it will usually be more detailed than at a civilian aerodrome, but is still only an aerodrome (non-precision) approach aid.

b. Precision approach radar (PAR) may be provided, which can be likened to a ground controlled ILS. The controller will direct you onto the final approach, and then give heading directions to maintain your flight path on the runway centreline, telling you not to acknowledge such instructions unless requested. Once you reach the glidepath he will tell you regularly whether you are above or below it, but will not give specific rate of descent directions. You must make your own adjustments to follow the glideslope down to your decision height. For any approach, expect the controller to ask you what decision height (minimum descent height for a non-precision approach) you are using; he will pass the procedure minimum with which you must compare your system minimum and add any extra allowance (for example for the IMC rating).

c. On any instrument approach, expect the controller to ask you to "carry out cockpit checks, advise complete" before you turn onto the base leg, and "check gear, acknowledge" during the final approach. Transmit your confirmation when you have completed these checks. On that approach, you may hear the controller talking to the tower controller while he is talking to you. If making an approach towards the runway in use, expect to receive landing clearance or go-around instructions before reaching decision or minimum descent height. Ensure you know the Missed Approach Procedure. If you cannot remember it, ask for "missed approach instructions" well before you reach the final approach.

d. Expect military traffic to be given priority (They usually use a lot of fuel, and often do not carry much spare for diversion – <u>you</u> must!). Consider the aircraft types which normally use the aerodrome. Any major speed difference between these and your own aircraft may result in your being directed perhaps away from the approach to provide separation. The same might apply if there is a major difference in rates of descent; military aircraft often descend quite rapidly, especially above 2000 feet. Be aware that traffic in the visual circuit may pass quite close behind or above you.

e. Once you are able to see the runway, the controller will expect you to land on it if you can, close to the threshold unless there is a cable on the runway (see paragraph 7 below). As always, if you are unable to do so, fly a go-around to join the circuit, manoeuvring onto the dead side as soon as it is safe, or carry out the full Missed Approach Procedure. Beware of jet efflux, and **wake turbulence** from large aeroplanes, or rotor downwash from helicopters which may be using a different but nearby landing area – although the controller will normally be aware of the problem and pass a warning if conditions make such turbulence likely, a lack of warning does not mean a lack of risk. If you are not used to landing on wide runways, beware of the visual illusion which may cause you to round out higher than intended.

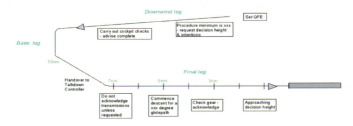

PAR procedure

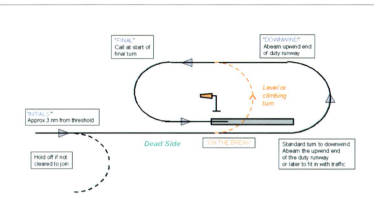

Oval circuit with military join procedures

6 Circuit procedures

a. Many military aerodromes expect visitors to carry out a standard overhead join, as published in the GA safety poster included in LASORS and on the CAA web site. However, depending on the direction of your approach, ATC may direct you to join downwind, or on base leg. Circuit patterns are usually flown at heights which depend on aircraft type. For example, a turboprop trainer may fly the pattern at 1000 feet on QFE, light piston aeroplanes at 800 feet, and, if traffic is mixed, fast jet traffic at 1200 or 1500 feet. The "military standard join" shown above involves approaching parallel with the runway in use from an "initial point" outside the ATZ on the dead side of the runway centreline, at circuit height or lower. A call of "initial" will be made at that "initial point". Some aircraft may approach at high speed for a "run and break", also shown above. Approximately 1-1½ minutes after calling "initials" the aircraft will turn steeply, level or climbing to the circuit height, from the deadside to downwind, calling "on the break" instead of the normal "downwind" call. You may not hear these calls because they will be on a UHF frequency, but ATC may inform you that they have been made. If the aerodrome has "no dead side" (often when helicopters operate together with aeroplanes) the run in may take place over the runway itself. Any non-standard procedures would normally form part of the visiting aircraft brief.

b. Most military circuit patterns are oval. The downwind leg is flown closer than at most civilian aerodromes, because the turn after take-off, and the final turn, both involve continuous 180 degree turns. The "downwind" call is standard, but the call of "final" is given as the aircraft starts its final turn at the end of the downwind leg. It is not easy to fly an accurate military oval circuit, but if you can practise it beforehand, it is very satisfying to be able to fit in. You do not need to change your own normal pattern or radio procedure, but be aware that the controller might be surprised at how late you call "final".

Intentions transmitted by military pilots are slightly different from those found in CAP 413. "Roll" effectively equates to "touch and go". "Land" equates to "full stop". You may also hear "overshoot" which means a pilot will make a low approach to the runway followed by a go-around, while confusingly an instruction to "go-around" is the same as an "orbit" (see next paragraph).

c. **Do not expect ATC to take responsibility for separating aircraft in the visual circuit**. You are expected to fit in with the other traffic, and if that is not possible, go-around. The place to make adjustments is at the turn onto the downwind leg. Do not turn crosswind until it is safe to do so, military aircraft usually climb steeply. Priority is normally given to instrument traffic, and ATC will transmit the position of that instrument traffic with its type. If you have called "downwind" before "instrument traffic at 8 (sometimes 6) miles", unless ATC give you other instructions they will expect you to be able to land and move off the runway before that instrument traffic. If you cannot, or you are told to go-around, the "8(6) mile" call comes before your "downwind" call, go-around at circuit height. This may be referred to, again confusingly, as an "orbit", which involves crossing to the dead side over or just downwind of the threshold at circuit height, rejoining crosswind (ideally over the other threshold), again at circuit height. A call of "instrument traffic at 4(3) miles" is the equivalent of a "final" call; if you have not started the downwind leg you should end up behind the instrument traffic (beware wake turbulence).

d. Once clearance to land is given, the controller will expect you to touch down close to the threshold unless there is a cable on the runway (see paragraph 7 below). Otherwise fly a go-around, manoeuvring onto the dead side as soon as it is safe. Beware of jet efflux, and **wake turbulence** from large aeroplanes, or rotor downwash from helicopters which may be using a different but close landing area – although the controller should be aware of the problem, he may not always have time to remind you.

7 Barriers and cables

a. Several military aerodromes have "arrester cables" which can be laid across the runway to assist fast jet aeroplanes to stop. The mechanism (which may be called RHAG for "rotary hydraulic arrester gear" or a similar sounding acronym) for these cables will normally be permanently fitted on either side of the runway at several hundred metres in from each threshold, one at the threshold or "approach" end, the other (more common) at the "overrun" end (some aerodromes may have more than two). When required, the cables are stretched across the runway between the mechanisms. The position of the cables is marked by yellow discs on vertical boards beside the runway, and often by similar markings on the surface.

b. The vertical position of the cables themselves may be one of three possible. "UP", or possibly "supported", means the inch thick metal cable is raised 3 inches above the runway surface on vertical rubber discs, as shown above. Although certain civil aircraft types may be able to do so, no propeller driven aircraft should attempt to cross a supported cable. "DOWN" or possibly "unsupported" means that the cable is lying on the runway surface, and the supporting discs have been pushed to one side. Crossing even an unsupported cable should be avoided whenever possible, and only attempted in propeller driven aeroplanes at very slow taxying speed. A "DE-RIGGED" cable has been removed from the runway surface completely.

c. Air Traffic Control will pass the state of the cables. Land beyond a rigged approach cable, and aim to turn off before an overrun cable. Similarly, aim to start the take-off run beyond an approach cable and lift off before an overrun cable. This will reduce the available runway length, so adjust your performance calculations to suit. In an emergency, the aerodrome may be able to de-rig a cable for you to make a safe landing, but that may take up to 20 minutes

"UP" cable

RHAG from threshold

d. There may be an "arrester (or jet) barrier" positioned at or beyond the end of the runway. Unlike a cable, this is for emergency use only and does not affect the runway itself. However, an "up" overrun barrier is a 20 foot obstacle affecting the climb after take-off or go-around. If for some reason the approach end barrier is up, it forms a significant obstruction. Propeller driven aeroplanes should not attempt to use an arrester barrier as an aid to stopping in an emergency.

Jet barrier

8 Ground movement

a. Once you have landed, the runway may be required by other aircraft for landing or take-off. You may be asked to vacate the runway quickly ("expedite"). "Expedite" does not mean "rush"! Do not dawdle, but make sure you are totally under control before you make any turns. Pre-flight study will indicate where you may turn off the runway, otherwise check with ATC. Military pilots stop and carry out their after-landing checks when well away from the runway, you should consider doing the same.

b. When taxying, beware of jet efflux or propeller slipstream from larger aircraft, including rotor downwash from helicopters. Several markings around the aerodrome may be different from the ICAO standard ones to which you are used. You should know the taxi route from your briefing, but if in doubt, stop and ask! If the aerodrome uses the military common frequency of 122.1 MHz for ground control, always use the aerodrome callsign when transmitting. Markings may be unfamiliar, if in doubt stop and ask.

c. You will usually be marshalled into your parking position by qualified personnel, rather than choose your own space. Leaflet 6, "Aerodrome Sense", shows the most common signals. You may be offered chocks, but these may be too large for your aircraft, so check them before allowing them to be fitted. Adding weight in the form of fuel may lower wheel spats onto the chocks! If you have asked for fuel, remember that the refuelling personnel will not be familiar with your aircraft. You should supervise the refuelling, paying particular attention to the type of fuel being dispensed. AVGAS and AVTUR (JET A-1) must not be confused! Check that additives are compatible with your aircraft.

d. You should report to Ops to discuss your requirements and future movements. You will probably be required to show your certificate of insurance detailing the level of third party and crown indemnity cover. Ensure you have an appropriate means to pay landing and other fees. Cheques are acceptable, but few military aerodromes have the facilities to accept credit cards. Make sure that if your aircraft has to be moved for operational reasons, the aerodrome authorities are able to either move it or contact you quickly.

9 Departure

a. You should receive a departure briefing from Ops. At that time give them the information you would normally pass on a taxi call, including how you wish to leave the aerodrome area; they will pass this to ATC to reduce radio transmissions. Confirm the frequency to use on start-up. Even if you are departing soon after arrival, a visit to Ops or ATC may provide much useful information and assistance. The staff may be able to help you to file a flight plan, or inform your destination of your intentions. They should certainly be able to update you with TAFs and METARs; NOTAMs should also be available, often already plotted on a map. Check the taxi pattern – find a suitable place for engine and/or navigation equipment checks which will not obstruct the taxiway. If a suitable place for such checks does not exist, consider carrying out whatever checks you can before starting to taxi.

b. When starting engines, you may have the assistance of ground personnel, who will have access to a fire extinguisher and perhaps be able to remove chocks, although 12v ground power is unlikely to be available. Brief them about your intentions, for example if you are delaying taxying to carry out equipment checks, or allowing the engine to warm up. ATC may have asked you to inform them that you are starting engines, but you must always inform your marshaller! A signal to remove chocks is a good way to indicate to him that you are ready to taxi, whether chocks are in place or not.

c. In many cases, aerodrome information is provided in a similar format to an Automatic Terminal Information Service (ATIS). That information will have been displayed in Ops, but may also be available by telephone, or on an ATIS frequency before engine start. When calling for taxi instructions (stating the frequency on which you are calling), add the code letter applicable to the information you have already copied down, and ATC will assume you know it so will not give long instructions. You will be passed the runway in use and QFE when given taxi instructions; the regional pressure setting will be given later if you do not already know it from the aerodrome information.

d. When taxying, again beware of jet efflux, rotor downwash or propeller slipstream. Do not dawdle, but do not rush. Even if you appear to be holding up other traffic, remember safety comes first. You should know the taxi route from your briefing, but if in doubt, stop and ask the controller. You may be given departure instructions (including the regional pressure setting) while taxying, or at the same time as you are given take-off clearance.

e. When ready for departure, look carefully for traffic approaching to land, or taxying onto the runway from the opposite direction. You will probably need greater separation from fast moving traffic than normal. Do not call until you are ready to enter the runway immediately, and do not stay on the runway for longer than necessary. However, essential checks on the runway should not be omitted. Consider the position of any arrester cables and barriers.

f. Once airborne, and at a safe height (500 feet or higher as directed), turn onto your cleared track or heading as advised. When outside the circuit pattern, tell ATC. Your controller may change, either by your changing frequency or by a different voice talking on the same frequency. Once outside the MATZ, you may wish to leave the frequency, although if they can provide a radar service it might be advisable to continue to accept that service for as long as it is offered. If you wish to continue your flight with the aerodrome QNH set, you may need to ask for it before you change frequency.

799

Part 2 – visiting outside normal operating hours

10 Pre-flight

a. A government aerodrome is always PPR (prior permission only). If you wish to use it outside normal operating hours (unless you are making an emergency landing), you must obtain permission during these normal operating hours, as in paragraph 3. Obtain as much of the "visiting aircraft brief" as is relevant. If intending to land later the same day, ask for the TAF and the latest METAR. Check what facilities, if any, will be available (the aerodrome fire and rescue service for example) and how to contact them for assistance if required. Check where you should park, and how anyone who is to meet you can gain access to the apron. Ask how you should pay your fees, and be aware that an aircraft using a military aerodrome outside its normal operating hours may be subject to a surcharge on its landing fees.

b. Find out if any airfield maintenance (grass cutting, runway sweeping etc.) is expected. Check what other activities may take place on the aerodrome (shooting, driving, model flying etc.). Some military aerodromes have gliding clubs operating outside normal hours. Ensure you know how to keep out of their way, and what frequency they operate on.

11 Approaching the closed aerodrome

a. If possible, make use of a Lower Airspace Radar Service (LARS) from a nearby military aerodrome, informing them of your intentions. Except in a very few cases, the ATZ of a military aerodrome is permanently active, even if the aerodrome is closed. If you have permission to land outside operating hours, you will expect to receive no reply when you call on the published VHF zone or approach frequency. However, continue to make 'blind' calls on that frequency. Other civilian aircraft, even flying clubs, may be based at the aerodrome, and will use the frequency when they require it (they may even provide an air/ground communication service). It is also possible that the aerodrome has been re-activated at short notice and the lack of reply is the result of a radio problem!

b. Radio aids to navigation may still be switched on. They can help you find the aerodrome, but do not fly an instrument approach. Any instrument approach to a military aerodrome in IMC must be flown only under Air Traffic Control. In addition, most maintenance is done outside normal operating hours, even if it was not planned when you telephoned, and your instrument indications may not be correct!

12 Circuit procedures at a closed aerodrome

a. Aim to make a standard overhead join unless the "visiting aircraft brief" tells you otherwise. Check the windsock and select the most suitable published runway. Check for obstructions on and close to the runways and taxiways – vehicle drivers and pedestrians will almost certainly not be expecting you. If gliding or powered flying is already taking place, fit in with their established procedures unless it is unsafe to do so, in which case take extreme care.

b. Military aircraft will not be using the aerodrome, so fly your normal circuit pattern with normal calls (on the approach frequency unless advised otherwise) in the correct place. A go-around from the first approach, especially if there is no other flying activity taking place, may act as a warning of your presence to those on the ground. For the same reason, consider using the landing light even in good visibility.

c. The barriers should normally be down and the cables de-rigged. However, that cannot be relied upon; maintenance is a possible reason for them to be up. Look at the position of the barriers during the circuit and initial go-around, and aim to land beyond the approach end cables unless performance limitations apply and you are sure it is safe to do so. Keep a sharp lookout for possible runway intruders, and be ready to go-around. Local people may have become used to having the free run of the aerodrome in the evenings and at weekends.

13 Summary

- Ask for permission well in advance
- Obtain a "visiting airfield brief"
- Check for weather close to arrival time
- Make the radio call early before entering the MATZ
- Be prepared (and pre-planned) to divert at any time
- Listen out carefully
- Priority may be given to military aircraft
- Beware wake turbulence
- Avoid cables and barriers
- Monitor refuelling
- Beware incursions onto the manoeuvring area outside hours

Appendix – Military aerodrome colour codes

In addition to a normal TAF or METAR, military aerodromes may use a colour code, which is a form of shorthand for their crews to reinforce the information in the main message. The meaning of each colour is listed below. PPL holders without instrument qualifications are advised that any code except "blue" or "white" may indicate serious problems, and even "white" is no guarantee that the weather is good, even at the time of the report.

Colour Code	Minimum base of lowest cloud (SCT or more) above aerodrome level	Minimum reported visibility
Blue	2500 feet	8 km
White	1500 feet	5 km
Green	700 feet	3700 m
Yellow	300 feet	1600 m
Amber	200 feet	800 m
Red	below 200 feet (or sky obscured)	Below 800 m
Black	Aerodrome unavailable for reasons other than cloud or visibility	

CAA Safety Sense Leaflet Index

Leaflet 1 Good Airmanship guide
Leaflet 2 Care of Passengers
Leaflet 3 Winter Flying
Leaflet 4 Use of MOGAS
Leaflet 5 VFR Navigation
Leaflet 6 Aerodrome Sense
Leaflet 7 Aeroplane performance
Leaflet 8 Air Traffic Services Outside Controlled Airspace
Leaflet 9 Weight and Balance
Leaflet 10 Bird Avoidance
Leaflet 11 Interception of Civil Aircraft
Leaflet 12 Strip Sense
Leaflet 13 Collision Avoidance
Leaflet 14 Piston Engine Icing
Leaflet 15 Wake Vortex
Leaflet 16 Balloon Airmanship guide
Leaflet 17 Helicopter Airmanship
Leaflet 18 Military Low Flying
Leaflet 19 Aerobatics
Leaflet 20 VFR Flight Plans
Leaflet 21 Ditching
Leaflet 22 Radiotelephony for General Aviation
Leaflet 23 Pilots: Its your decision
Leaflet 24 Pilot Health
Leaflet 25 Use of GPS
Leaflet 26 Visiting Military Aerodromes

UK Aeronautical Information Circulars

UK AIM
The UK Aeronautical Information Manual

AIC 19/2002 (Pink 28) 4 April

LOW ALTITUDE WINDSHEAR

1 Introduction

1.1 This Circular has been produced to provide an understanding of the nature of windshear, and an appreciation of its dangers. Guidance on how best to avoid windshear and how an aircraft may have to be handled during a windshear encounter is also included. It should be noted that research and experiment continues to take place in relation to the phenomenon, nevertheless, windshear encounters still continue to be cited as a primary or contributory cause of accidents and incidents. Although aircraft flying in United Kingdom airspace can experience windshear events, the severity of such encounters and the intensity of the probable causative events are often much less than that experienced elsewhere in the world. Windshear has been the direct cause of accidents; it is not a phenomenon to be treated lightly. Pilots and operators are therefore urged to understand the phenomenon and if planning to fly to destinations or areas where severe weather, turbulence or windshear is known or likely to occur, to obtain appropriate briefings, training and instruction. This circular is written for guidance only, all suggestions regarding flying techniques and similar procedures in this document do not supersede appropriate operations or flight manual instructions.

2 Definitions

2.1 In discussing windshear, it is not easy to find a definition that will satisfy both meteorologist and pilot. **As a consequence, it is possible to find circumstances where an alert has been issued concerning windshear, but where the meteorological understanding of the event differs from that expected by the flight crew.** Thus it is important if operating to areas where windshear can be a regular phenomenon that flight crews fully understand that which is being forecast. At its simplest, windshear can be described as a change in wind direction and/or speed including both downdraughts and updraughts.

2.2 The definitions of windshear used in this circular are:

(a) Windshear:

Variations in the wind vector along the flight path of an aircraft with a pattern, intensity and duration that will displace an aircraft **abruptly** from its intended flight path such that **substantial control input and action is required** to correct it.

(b) Low altitude windshear:

Windshear along the final approach path or along the runway and along the take-off and initial climb-out flight paths.

2.3 Additional qualifying conditions/descriptions related to windshear:

(a) Vertical windshear:

The change of horizontal wind vector with height, as might be determined by two or more anemometers at different heights on a mast.

(b) Horizontal windshear:

The change of the horizontal wind vector with horizontal distance, as might be determined by two or more anemometers mounted at the same height along a runway.

(c) Updraught or downdraught shear:

The changes in the vertical component of wind vector with horizontal distance.

2.4 If the basic windshear definition in paragraph 2.2 (a) is set aside, it can be seen that the additional definitions and qualifying statements allow for changes in wind vector which could include the relatively minor or benign. Notwithstanding these possible (academic) variations in description, this circular is concerned with the basic definition, in particular, the emphasis on **abrupt** displacement of an aircraft from the desired flight path, at low altitude, together with the necessity for **substantial control action** to counteract it. Windshear is therefore highly dynamic which can be extremely uncomfortable and frightening; to think of windshear as an aggravated form of wind gradient is unwise. Windshear can strike suddenly and with devastating effect beyond the recovery powers of both experienced pilots flying the most modern and powerful aircraft. **Thus the first and most vital defence is avoidance; this should be taken to be the recurrent theme of the rest of this circular.**

3 Meteorological Background

3.1 Among the most potent examples of windshear are those associated with thunderstorms, however, windshear can also be experienced in association with other meteorological features such as the passage of a front, a marked temperature inversion, a low level jet (wind maximum) or a turbulent boundary layer. Topography or buildings can create substantial local windshear effects that can be considerably more than might be expected from the average strength of a prevailing wind.

3.2 Thunderstorms

3.2.1 The principles of thunderstorm formation are already covered in AIC 72/2001 (Pink 22). For the purposes of this Circular it is sufficient to remark that thunderstorms are violent events, unpredictable, and associated with turbulence, windshear, lightning and precipitation as separate or joint hazards. Shears will occur, and draughts can strike from all angles. The assessment of the aircraft's actual angle of attack in relation to some of the wind flows will be difficult to judge and in consequence the closeness of the aircraft to the stall will be harder to gauge. In relation to low altitude windshear there are features of thunderstorms which merit further description:

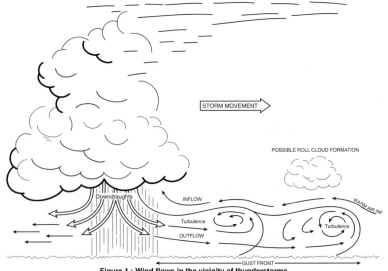

Figure 1 : Wind flows in the vicinity of thunderstorms

(a) **Gust Front:** Some thunderstorms may have a well-defined area of cold air flowing out from a downdraught, but tending to lead the storm along its line of movement. This is labelled a "gust front" (see Figure 1) and may extend some distance (up to 30 km) from the storm centre and affect the area from the surface up to 6000 ft. If the storm is part of an organised line of storms, the gust front may extend an even greater distance from the centre line of the storms. 'Gust fronts' manifest themselves as regions of great turbulence with a potential for vertical shears between out-flowing cold air as it undercuts warmer air. The leading edge of the 'gust front' could be encountered without warning, although roll cloud effects can be associated with it.

(b) **Microburst:** A microburst is a highly concentrated and powerful downdraught of air, typically less that 5 km across, which lasts for about 1 to 5 minutes. (It should be noted that the word 'microburst' is also associated with a phenomenon called 'gap flow microburst', associated with topographical effects and strong winds and is a very strong horizontal shear effect with the energy loss effect on the aircraft as shown in Figure 2). Microbursts associated with thunderstorms have proved to be a most lethal form of windshear giving downdraught speeds of 60 kt or more. As this vertical shaft of air approaches the ground it will 'splay out' and lose its vertical speed component, nevertheless, vertical (downward) components have been recorded as low as 300 ft with surface wind differences of as much as 90 kt. Although these figures are extreme examples they do illustrate that a microburst should not be treated lightly. Microbursts have been well documented in the United States but could easily be found elsewhere in association with thunderstorm activity. Microbursts can be 'wet' or 'dry', ie associated with or without precipitation. The 'dry' microbursts will therefore be difficult to detect on weather radar, but are often associated with high-based cumulus, altocumulus or the cirrus cloud overhang from a cumulonimbus cloud. In each case when precipitation falls from the clouds (indicated by 'virga') it evaporates in the dry air beneath the cloud. This evaporation process requires energy, which further cools the falling air thus enhancing the speed of the downdraught. A 'wet' microburst is associated with intense precipitation that falls in shafts below a cumulonimbus cloud.

3.3 **Frontal Passage**

3.3.1 Fronts vary in strength and normally only well-developed active fronts, with narrow surface frontal zones and with marked temperature differences between the two air masses are likely to carry a windshear risk. Weather charts showing sharp changes in wind direction across the front, temperature differences of 5°C or more, or a speed of frontal advance in the region of 30 kt or more may indicate a potential windshear hazard. Frontal windshear is also a hazard as the following incident illustrates. A twin-jet aircraft was about to land in the UK and was caught by the passage of a cold front. In 10 seconds, the wind changed such that a 10 kt crosswind from the left (with a slight tail wind) changed to an 8 kt crosswind from the right coupled with a 14 kt head wind. A missed approach, from very low level was carried out as directional control became difficult. The actual numerical values of wind speed change in this actual incident may appear less than dramatic to some readers, however, the control difficulties experienced during this incident were considerable. Similar wind **velocity** changes, with an aircraft at critical phases of flight will always be hazardous.

3.4 **Inversions**

3.4.1 A change in wind strength is nearly always present in the boundary layer, ie close to the ground. This is frequently described as 'windshear' but as this normally involves a gradual change in strength with which pilots will be most familiar, it does not fit the definition already given in this circular. A proper windshear hazard can exist, however, when an unexpectedly strong vertical change develops. This is often associated with the following situations:

(a) A low level jet (more accurately referred to as a low level wind maximum) can form just below the top of, or sometimes within, a strong radiation inversion which may develop at night under clear skies. Other low level jets may develop in association with a surface front, particularly ahead of cold fronts.

(b) On occasions, low-level inversions may develop and decouple a relatively strong upper flow from layers of stagnant or slow moving air near the surface. Shear effects may be pronounced across the interface.

3.5 Turbulent Boundary Layer

3.5.1 Within the boundary layer, turbulence can lead to windshear (see definition) in two different situations:

(a) Strong surface winds are generally accompanied by large gusts and lulls (horizontal windshear). Roughly speaking, the stronger the mean wind, the greater is the gust or lull.

(b) Thermal turbulence (updraughts and downdraughts) caused by intense solar heating of the ground. This is more common in hot countries, but can occur in most places given a hot sunny day with little prevailing wind.

3.6 Topographical Windshear

3.6.1 Natural or man-made features can affect the steady state wind flow and cause windshear of varying severity. The strength and direction of the wind relative to the feature (obstacle) are significant and a change in the direction of the wind by relatively few degrees may appreciably alter the effect. The flow of wind across a mountain range is a simple large-scale example, with waves and possibly rotors forming on the lee side of the wind. Wind blowing between two hills or along a valley, or even between two large buildings may be funnelled; change direction and increase in speed, or strong flows may be heavily damped. The shear produced as a result may be considerably more than might be expected from the mean speed as venturi effects may produce additional speed. In any case, the possibility for shear (according to the given definition) is created, with sudden changes in wind vector becoming an appreciable hazard. Larger airport buildings adjacent to busy runways can create hazardous local effects and typical windshear problems, such as loss of airspeed and abrupt crosswind changes, causing upsets to even airliner-sized aircraft. Some locations where topographical windshear hazards exist and published on approach plates are:

(a) Gibraltar: 'Turbulence around the Rock is influenced by both the surface wind and the 1000 ft wind. Generally a 1000 ft wind of less than 15 kt does not produce significant turbulence. However, with a wind direction between 130° and 240° and speed in excess of 15 kt the severity of turbulence increases as the wind speed increases. In some cases the turbulence may make conditions dangerous or hazardous for landing'.

(b) Maderia (Funchal): Considerable information on wind and turbulence reporting is given for this airfield and includes diagrams and tables for the maximum permitted wind strengths beyond which both landing and take-offs are prohibited by the airfield authorities.

(c) Hong Kong (Chep Lap Kok): Again, for this airfield, considerable guidance and information is given on expected turbulence and windshear conditions together with information on the 'Windshear and Turbulence Warning System' used. Of particular note is the use of a possibly non-intuitive phraseology in the warnings and of the use of the term 'microburst' to describe the most powerful horizontal windshear events. 'Microburst' in this context is a truncation of the phrase words 'gap-flow microbursts' which result from the interaction of strong winds and unique topography near the airfield. (ie these are not 'classical' thunderstorm microbursts) – See paragraph 6.

4 The Effects of Windshear on an Aircraft in Flight

4.1 Windshear will affect aircraft in many different ways and during an encounter the situation will be constantly changing, especially during the more dynamic encounters. Conventional thought, in the past, has suggested that such encounters are associated only with thunderstorms; however, there is evidence to show that equally dynamic encounters can be expected with other windshear causes. Particular types of aircraft will vary in their reaction to a given shear; a light high wing piston-engined aircraft may react in a totally different way to a heavy four-engined swept wing jet aircraft. The notes and diagrams that follow describe stylised windshears and the progressive effects that can occur. Windshear can be encountered at any height and the effects will be similar. If the windshear event is at low-level it can be a great hazard and it is this that must be borne in mind in relation to the described effects.

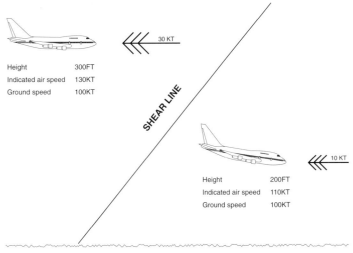

Figure 2 : Effect of loss of headwind - Energy Loss

4.2   To understand the effects of windshear it is important to note the relationship of an aircraft to two reference points. One reference point is the ground below the aircraft and the other is the airmass in which the aircraft is moving. In a windshear encounter it is not only the magnitude of the change of the wind vector, but the rate at which it occurs. For example, an aeroplane at 1000 ft above ground level (agl) may have a head wind component of some 30 kt, but the aerodrome surface report shows this as a 10 kt component close to the runway. That 20 kt difference in wind strength may taper off evenly from 1000 ft to touchdown with no changes in direction; thus its effect and relationship to the aircraft will be that of a reasonable wind gradient with height with which all pilots will be familiar. On the other hand, if the 20 kt difference still exists at 300 ft, it will be obvious that the change, when it occurs, is going to be far more sudden and its effect more marked. Windshear, from the definition, implies a narrow borderline, and the 20 kt of wind strength may be lost over a short vertical distance. If this strength is lost over 100 ft the effect will be as shown in the diagram, with the concomitant loss of aircraft energy.

4.3   The reverse effect will occur when an aeroplane is taking off through such a shear line. In this case however, the gain in energy would be beneficial with aircraft gaining some additional 20 kt of airspeed in a short distance. However, this could potentially produce some handling difficulties, particularly if the aircraft was approaching a flap limiting speed (for example) as the shear line was crossed.

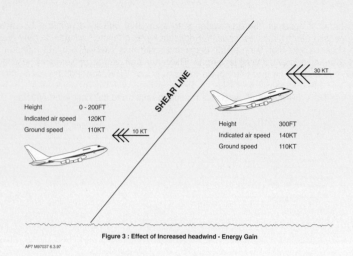

Figure 3 : Effect of Increased headwind - Energy Gain

4.4 In an encounter with a downdraught, the relative airflow across the aircraft's wing will change direction. Normally, an aircraft could be expected to be flying into a head wind and as the downdraught is encountered, this head wind will reduce markedly, thus reducing the energy of the aircraft. In the downdraught itself, the relative airflow across the wing will change as shown in the diagram below. Its effect will be to effectively decrease the angle of attack of the wing, which will be counteracted by the pilot manoeuvring the aircraft in response to the downburst effect, ie increasing the angle of attack. This might be beneficial in terms of increased lift. However, if the wing is already at steep attack angle (eg during an approach), any subsequent changes in airflow may result in the aircraft being very close to the stall.

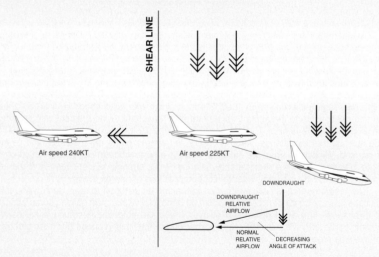

Figure 4 : Effect of downdraught - Energy Loss

4.5 Another version of a windshear event is that occasioned by an encounter with a microburst. Classically, a microburst, as illustrated below, is associated with thunderstorm or similar convective activity associated with areas of heavy rain. However, the term 'microburst' is also associated, in some locations, with very severe windshear warnings, possibly without associated convective activity. (See paragraph 3.6 (c)). **Therefore, if 'microburst' warnings are given, flight crew should be fully aware of which phenomenon they are being warned about and take appropriate action.**

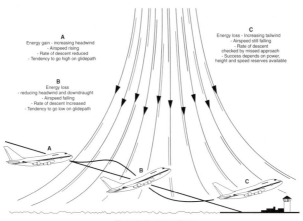

Figure 5 : Microburst effects

4.6 The likely sequence of events during a classic (convective activity) microburst encounter is a combination of the previous events as follows. Approaching the airfield, using the aircraft weather radar, an area of strong radar returns may be observed in the vicinity of the final approach path. This may alert the pilot to the possible dangers, particularly if this is also associated with earlier meteorological forecasts. The recommended course of action would be to initiate a go-around at this early stage in the approach and hold off until the activity moved on. Some airfields may even prohibit approaches during conditions when microburst activity is anticipated. If after the initiation of the go-around, an encounter with the microburst occurred at this stage, the extra energy of the aircraft and possible increased height should ensure that no untoward events occur.

4.7 If, however, the aircraft were to continue its approach into the microburst, the sequence of probable events is as shown in the diagram. A successful go-around will depend, obviously, on the strength of the microburst, its position in relation to the approach path; the aircraft power reserves available and the rate at which they can be increased to give maximum thrust to counteract the energy loss of downdraught and increasing tail wind. The dynamic events will probably be associated with severe buffeting, heavy rain, thick cloud and probably blinding flashes of lightning, and external noise occasioned by heavy rain or hail on the aircraft skin.

4.8 A microburst encounter during take-off could be equally as hazardous notwithstanding the fact that during take-off an aircraft is already developing high thrust and is less constrained by the need to hold a precise glide path. Whilst the initial increase in head wind may improve lift and thus rate of climb, the transition to downdraught and then tail wind may totally negate this increase and the airspeed of the aircraft, possibly still close to the ground may be further reduced. A heavy aircraft, with small reserves of power may not be able to fly through the encounter successfully.

5 Techniques to Counter the Effects of Windshear

5.1 It should now be apparent that windshear can vary enormously in its impact and effect. There is as yet no international agreement on ways of grading windshears, but clearly some shears will be more severe and more dangerous than others. If the definition used by this circular is borne in mind, a windshear encounter is expected to result in **abrupt** changes to the desired flight path requiring **considerable control action** to effect recovery. Thus, if avoidance has failed, any further actions must anticipate the worst effects, bearing in mind that the shear itself will probably be invisible to the crew.

5.2 If the meteorological situation has been carefully studied in advance of flight and updated with latest reports during flight, the possibility of windshear should have already been identified. Strong winds at a destination, particularly where topography is warned as having an effect, or a forecast of thunderstorms, particularly if reinforced, near the destination, with weather radar or visible evidence should trigger a mental 'Windshear Alert'. At this point the prudent decision would be to divert, following company procedures in the case of AOC operators.

5.3 If an approach must be attempted in such conditions, crews should consider a few basic measures to anticipate a windshear encounter. One such measure is to increase the approach speed, however, if a shear event does not materialise, the increased approach speed may then cause problems near the threshold, particularly at airfields with short runways. If an encounter with shear does occur, its overall effect will be to de-stabilise an approach, and any thrust reduction taken to counter an initial extra head wind, may have to be quickly changed to a thrust increase if the head component is removed or reversed. Actions to counter a loss of airspeed close to the ground include:

(a) Briskly increase power (to full go-around if felt necessary);

(b) Raise the nose to check descent;

(c) Co-ordinate power and pitch;

(d) Be prepared to carry out a missed approach rather than risk a landing from a de-stabilised approach.

5.4 The actions required to counter the effect of a downburst or microburst on the approach or during take-off will require more stringent measures. Again, it should be emphasised that if such phenomena are either forecast or suspected, the most sensible course of action would be to delay the take-off or landing, or if airborne to divert to another airfield. In the absence of specific aircraft flight manual or operations manual guidance which must be followed, some **suggested** techniques for dealing with a microburst encounter on an approach are given below:

(a) The presence of thunderstorms should be known and obvious, so that any increase in airspeed caused by the increasing head wind should be seen as the precursor to a downburst or microburst encounter. Any hope, therefore, of a stabilised approach should be abandoned and a missed approach is the recommended action – the technique is to make this as safe as possible;

(b) The initial rise in airspeed and the rise above the approach path should be seen as a bonus and capitalised upon. Without hesitation, the power should be increased to that required for a go-around, whilst being prepared to further increase it to maximum, if necessary, and an appropriate pitch angle, consistent with a missed approach should be selected. Typically, this will be in the region of 15° and this should be held against the buffeting and turbulence that will undoubtedly occur;

(c) The initial 'bonus' of speed and rate of climb may now be rapidly eroded as the downdraught is encountered. Airspeed may now be lost and the aircraft may now begin to descend again despite the high power and pitch angle. It may be impossible to gauge the angle of attack, so there is a possibility that stall warnings may be triggered; in such cases the pitch angle may need to be relaxed slightly;

(d) The point at which the tail wind starts to be encountered may be the most critical. The rate of descent may lessen, but the airspeed may continue to fall and any height loss may now be close to ground obstacle clearance margins. Maximum thrust will by now have been applied and if there is now a risk of hitting the ground or an obstacle it may be necessary to raise the nose until stall warnings start to be triggered and to hold this attitude until the aircraft starts to escape from the effects.

5.5 The likely effects of windshear on take-off have already been discussed in paragraph 4.3. When there is a possibility of shear, without a clear forecast, it may be possible to use a longer runway, preferably pointing away from an area of potential threat. However, this must not be a 'spur of the moment' action and any such decision will need to have been pre-planned taking into account all necessary factors including runway length and obstacle clearances. The high power setting and high pitch angle after rotation may put the aircraft in a reasonable situation should an encounter now occur, however, if it has not been fully thought out there may be unexpected handling and performance difficulties. However, the aircraft will still be low, with a small safety margin. At that point, maximum power should be selected (if it is not already); noise abatement procedures should be ignored and the high pitch attitude (consistent with any stall warnings) should be maintained. Notwithstanding the above, the safest and recommended course of action would be to delay take-off until the possibility of windshear has diminished.

5.6   The vital actions for downburst/microburst encounter in both approach and take-off cases are:

    (a)   Early **recognition** and **committal** to the appropriate action;

    (b)   Follow operations manual or aircraft flight manual techniques;

    (c)   Use maximum power available as soon as possible;

    (d)   Adopt an appropriate pitch angle and try and hold it; do not 'chase' airspeed.

    (e)   Be guided by stall warnings when holding or increasing pitch, easing the back pressure as required to attain or hold a lower pitch attitude if necessary. (In many aircraft types optimum performance is very close to the point of onset of stall warning. It is important, however, not to go beyond the point of onset as it is then not possible for the pilot to know how deeply into the warning the aircraft is).

5.7   It is not possible to be prescriptive in relation to the 'best' technique to use as these will vary between aircraft, and may be documented in appropriate manuals, but could be expected to follow the broad guidelines above. Some of the responses required of the pilot and the attitude and trim forces to be used may sometimes appear to be counter-intuitive. Therefore, the best advice would be to use a windshear training programme, coupled with dedicated simulator exercises to practice the techniques. Any such training should **emphasise** the fact that windshears are to be avoided. The knowledge thus gained and any techniques practised should serve to make the survival of an inadvertent encounter more possible and not encourage pilots to think that windshears can be tackled with impunity.

6   Windshear Warning and Reporting

6.1   Windshear warning is provided in the following ways:

    (a)   Meteorological warning;

    (b)   ATS warning;

    (c)   Pilot warnings;

    (d)   On board pre-encounter warnings.

6.2   Warning of windshear from meteorological sources may start at the pre-flight briefing stage and pinpoint the possibility of frontal or inversion shear. Any forecast of thunderstorm activity should alert pilots to the possibility of downdraught or microburst activity. If the planned destination is one where topographical features are known to cause shear hazards, the direction and strength of forecast winds should be noted carefully and compared with published information. In the United Kingdom, windshear warnings are provided in ATIS broadcasts at London (Heathrow) and Belfast (Aldergrove) if the following conditions exist:

    (a)   The mean surface wind exceeds 20 kt;

(b) The vector difference between the mean surface wind and the gradient wind at about 2000 ft exceeds 40 kt;

(c) Thunderstorms or heavy showers are within about 5 nm of the airfields.

The warnings are broadcast as 'Windshear Forecasts', and if reinforced by pilot (aircraft) reports, the alert becomes 'Windshear forecast and reported'. (See UK AIP GEN 3-5-3 for more information).

6.3 In other parts of the world, windshear warnings can be based not just on meteorological forecasts but on actual observed conditions using, for example, a series of anemometers around an airfield. With such systems, the measured differences in wind velocity between anemometers are used in conjunction with computer programmes and recorded data to produce warnings. Terminal Doppler Weather Radars (TDWR) are also used to measure wind velocities and these can also be configured to produce warnings. **However, a note of caution is necessary.** It will have been noted that no universal standard exists regarding the grading of the severity of windshear, nor is there a universal standard regarding windshear warnings. Furthermore, the term 'microburst' is not always used to describe the classic thunderstorm associated event. 'Microburst' is also used to alert crews to the possibility of a shear event from wind over 30 kt. In this context, it is not a microburst in the sense as shown in Figure 5 (and the associated text). In addition, some locations use a 'Maximum intensity, first encounter' rule, in warnings. This results in warnings as follows:

'Windshear warning, 25 kt, 3 nm final'

This could be interpreted as 'expect an encounter of 25 kt at 3 nm on final approach'. This is incorrect. The correct interpretation (as a result of maximum intensity, first encounter) is:

'Expect encounter(s) with windshear somewhere between 3 nm and touchdown with a maximum intensity of 25 kt'

It should be noted in this context that an aircraft belonging to a UK airline experienced such events with a maximum intensity in the region of 35 (or more) kt ie an encounter with a 'gap-flow microburst' at an approximate height of 150 ft on an approach. From all the available evidence, it would appear that the loss of an aircraft, crew and passengers was very narrowly avoided.

6.4 Pilot reports of windshear encounters are important sources of information to warn other pilots of the danger. The UK AIP (GEN 3-5-17) contains guidance on windshear reporting, which for convenience is repeated below. In this context it should be noted that similar entries in the AIPs of other States may be slightly different and may require the use of different terminology or phraseology.

'Windshear Reporting Criteria

Pilots using navigation systems providing direct wind velocity readout should report the wind and altitude/height above and below the shear layer, and its location. Other pilots should report the loss or gain of airspeed and/or the presence of up or down draughts or a significant change in cross wind effect, the altitude/height and location, their phase of flight and aircraft type. Pilots not able to report windshear in these specific terms should do so in terms of its effect on the aircraft, the altitude/height and location and aircraft type, for example, 'Abrupt windshear at 500 ft QFE on finals, maximum thrust required, B747'. Pilots encountering windshear are requested to make a report even if windshear has previously been forecast or reported'.

6.5 As yet, no perfect on-board system is available for general use and trials continue in this respect. Some aircraft are fitted with predictive windshear warning system, but in most cases the pilot will not receive much advance warning of the presence of windshear. Airborne weather radar may give some clues, however, it must be remembered that most weather radars do not detect turbulence; they only detect precipitation. Warning that an aircraft is about to experience windshear may therefore come from a variety of sources, and in this context it is probably

United Kingdom Aeronautical Information Circular AIC 19/2002 (Pink 28) 4 April

important to ensure that the more sophisticated modern aircraft are configured for a possible encounter in such a way that give the most warning and assistance to the pilot. Guidance for the best configuration to use will come from the manufacturer and it is this that should be used. For less well-appointed aircraft, early clues may come from the airspeed and vertical speed indicators; flight directors may be misleading when a windshear recovery is flown; again manufacturers' guidance should be sought. Finally, although visual clues may have assisted in the early prediction of a windshear event, they will not necessarily be available during an event and its recovery. Similarly, physiological sensations should also be ignored and flight conducted purely by reference to appropriate instruments.

7   Conclusions

7.1   Most pilots will experience changes in wind speed of some form or other; hopefully few will experience windshear as defined in this circular when considerable control inputs will be required to overcome the abrupt changes that the shear encounter has caused. There are no sure and absolute ways of determining the severity of an encounter; therefore the best advice must be to avoid them. If an encounter does occur the generic advice in this circular together with further training and knowledge should help to alleviate the effects.

Recognise	–	that windshear is a hazard
		and
Recognise	–	the signs that may indicate its presence
Avoid	–	windshear by delay or diversion
Prepare	–	for an inadvertent encounter by a 'speed margin' if 'energy loss' is expected
Recover	–	know the techniques recommended for your aircraft and use them without hesitation if windshear is encountered
Report	–	immediately to ATC controlling the airfield at which the incident occurred (see paragraph 6.4) **and** using the Mandatory Occurrence Reporting Scheme, to the Civil Aviation Authority.

AIC 67/2002 (Pink 36) 22 August

Take-off, climb and landing performance of light aeroplanes

1   Introduction

1.1   Accidents, such as failure to get airborne in the distance available, collision with obstacles owing to inadequate climb and over-run on landing, continue to occur fairly frequently to light aeroplanes (ie those below 5700 kg maximum weight). Many such accidents have occurred when operating from short strips, often taking-off or landing out of wind, or with sloping ground. Poor surfaces such as wet grass or ice were also frequent contributory factors. What is not generally realised by many pilots is that these are PERFORMANCE accidents and many, if not all, of these accidents could have been avoided if the pilots had been fully aware of the PERFORMANCE LIMITATIONS of their aeroplanes.

1.2   The pilot-in-command of ANY UK REGISTERED aeroplane has a legal obligation placed on him by Article 43 of the Air Navigation Order 2000 which requires him to check that the aeroplane will have adequate performance for the proposed flight. The purpose of this Circular is to remind pilots of private flights of the actions needed to ensure that the take-off, climb and landing performance will be adequate.

1.3   Aeroplane Performance is subject to many variables including:

   Aeroplane weight;
   Aerodrome altitude;
   Temperature;
   Wind;
   Runway length;
   Slope;
   Surface;
   Flap setting;
   Humidity.

1.3.1   The performance data will usually allow adjustment to be made for these variables. On certification, allowances are made to cater for slight variations in individual pilots' handling of a specific technique.

2   Where to Find the Information

2.1   Performance figures may be given in a variety of publications and it is important for pilots to know where to find the data needed to predict the performance in the expected flight conditions. The appropriate document is specified in the Certificate of Airworthiness and may be any one of the following:

   (a)   The UK Flight Manual;

   (b)   the Owner's Manual or Pilot's Operating Handbook. These documents, which sometimes contain CAA Supplements giving additional performance data which may either supplement or override data in the main document, are the ones applicable to many light aeroplanes;

   (c)   the Performance Schedule (applicable to a few of the older aeroplanes);

   (d)   for some imported aeroplanes, an English language flight manual approved by the Airworthiness Authority in the country of origin, but with a UK Supplement containing the performance data approved by the CAA.

3 Use of Performance Data

3.1 The majority of modern, light aeroplanes were originally classified in Performance Group E for the purposes of public transport. The performance information contained in the Manuals and Handbooks of these aeroplanes is UNFACTORED. This means the data represents the performance achieved by the manufacturer using a new aeroplane in ideal conditions. This level of performance will not be achieved if the flying techniques used by the manufacturer are not followed closely or if the meteorological conditions are not as favourable as those encountered during testing. It is therefore PRUDENT TO ADD SAFETY FACTORS to the data in order to take account of less favourable conditions.

3.2 To ensure a high level of safety on public transport flights, there is a legal requirement to add specified safety factors to the data. It is RECOMMENDED that at least the same factors be used for private flights. When a pilot planning a private flight chooses to accept aerodrome distances or climb performance less than that required for a public transport flight, he should recognise that the level of safety is lowered accordingly.

3.3 Performance data in Manuals for aeroplanes certificated in Performance Groups C, D or F for the purposes of public transport normally include public transport factors. These Manuals usually make it clear if factors are included, but if in any doubt the user should consult the Safety Regulation Group of the CAA.

3.4 It should be remembered that any 'limitations' given in the Certificate of Airworthiness, the Flight Manual, the Performance Schedule or the Owner's Manual/Pilot's Operating Handbook, are MANDATORY ON ALL FLIGHTS.

4 Performance Planning

4.1 A list of variables affecting performance together with guide line factors is summarised in tabular form at the end of this Circular. These represent the increase in take-off distance to a height of 50 ft or the increase in landing distance from 50 ft. It is intended that the tabular form will be suitable for attaching to a pilot's clipboard for easy reference. WHEN SPECIFIC CORRECTIONS ARE GIVEN IN THE AEROPLANE'S MANUAL, HANDBOOK OR SUPPLEMENT, THESE MUST BE CONSIDERED THE MINIMUM ACCEPTABLE.

5 Take-off

5.1 Aeroplane Weight: it is important that the actual weight stated on the weight and balance sheet for the individual aeroplane is used as the basis for calculations. The weight of individual aeroplanes of a given type can vary considerably dependent on the level of equipment. Using the example weight shown in the weight and balance section of the handbook is not satisfactory.

5.1.1 Guide line factor: take-off distance will be increased by 20% for each 10% increase in aeroplane weight (a factor of x 1.20).

5.2 Aerodrome Altitude: aeroplane performance deteriorates with an increase in altitude and the pressure altitude at the aerodrome of departure should be used for calculations. This equates to the height shown on the altimeter on the ground at the aerodrome with the sub-scale set at 1013 mb.

5.2.1 Guide line factor: take-off distance will be increased by 10% for each 1000 ft increase in aerodrome altitude (a factor of x 1.10).

5.3 Temperature: aeroplane performance deteriorates with an increase in ambient temperature.

5.3.1 Guide line factor: take-off distance will be increased by 10% for a 10°C increase in ambient temperature (a factor of x 1.10).

5.4 Wind: a tailwind increases the take-off distance.

5.4.1 Guide line factor: the take-off distance will be increased by 20% for a tailwind component of 10% of the lift-off speed (a factor of x 1.20).

> **Note:** Where the data allows adjustment for wind, it is recommended that not more than 50% of the headwind component and not less than 150% of the tailwind component of the reported wind be assumed. In some Manuals this factoring is already included and it is necessary to check the relevant section.

5.5 Slope: an uphill slope increases the ground run.

5.5.1 Guide line factor: the take-off distance will be increased by 10% for each 2% of uphill slope (a factor of x 1.10) (see also paragraph 8.3).

5.6 Surface: grass, soft ground or snow increase rolling resistance and therefore the ground run.

Guide line factors: for dry grass (under 8 inches), the take-off distance will be increased by 20% (a factor of x 1.20).

For wet grass (under 8 inches), the take-off distance will be increased by 30% (a factor of x 1.30).

> **Note 1:** A take-off should not be attempted if the grass is more than 10 inches high.
> For soft ground or snow the take-off distance will be increased by 25% or more (a factor of at least 1.25).
>
> **Note 2:** For surface and slope factors remember that the increases shown are to the take-off distance to a height of 50 ft. The correction to the ground run will be proportionally greater.

5.7 Flap Setting: read carefully any Supplement attached to your Manual, the take-off performance with or without the use of take-off flap shown in the main part of the Manual may not be approved for use by aeroplanes on the UK register.

5.8 Humidity: high humidity has an adverse affect on performance and this is usually taken into account during certification, however, there may be a correction factor applicable to your aeroplane. Consult the Manual.

5.9 The above factors are cumulative and where several factors are relevant they must be multiplied. The resulting corrected distance required can seem surprisingly high.

5.10 Safety Factors: it is recommended that at least the public transport factors should be applied for all flights. Unless otherwise specified in the aeroplanes manual, handbook or supplement, as factor of 1.33 for take-off is recommended, and should be applied after the application of the corrections for the variables.

5.10.1 Example:

In still air, on a paved level dry runway at sea level with an ambient temperature of 15°C, an aeroplane requires a take-off distance to a height of 50 ft (TODR) of 390 m. This should be multiplied by the safety factor of 1.33 giving a TODR of 519 m.

The same aeroplane in still air on a dry, grass strip (factor x 1.20) with a 2% uphill slope (factor x 1.10), 500 ft above sea-level (factor x 1.05) at 20°C (ie 5°C warmer, therefore factor x 1.05) will have a corrected take-off distance to a height of 50ft of:

390 x 1.20 x 1.10 x 1.05 x 1.05 = 568 m.

The safety factor (x 1.33) should then be applied, giving a TODR of 755m.

5.11 The pilot should always ensure that, after applying all the relevant factors including the safety factor, the take-off distance to a height of 50 ft (TODR) does not exceed the runway length available (or TODA if known).

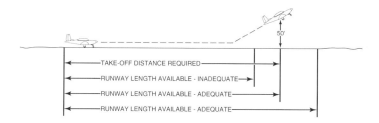

6 Climb

6.1 So that the aeroplane climb performance does not fall below the prescribed minimum, some Manuals give take-off and landing weights that should not be exceeded at specific combinations of altitude and temperature (WAT limits). Unless included in the Limitations section, these weight restrictions are mandatory only for public transport flights. THEY ARE HOWEVER RECOMMENDED FOR PRIVATE FLIGHTS and are calculated using the altitude and temperature at the relevant aerodrome. Where WAT limits are not given the following procedures are recommended:

(a) At the expected take-off and landing weights the aeroplane should be capable of a rate of climb of 700 ft/min if it has a retractable undercarriage, or 500 ft/min if it has a fixed undercarriage. The rates of climb should be assessed at the relevant aerodrome altitude and temperature in the en-route configuration at the en-route climb speed and using maximum continuous power;

(b) For an aeroplane with more than one engine, if conditions are such that during climb to, or descent from, the cruising altitude obstacles cannot be avoided visually, the aeroplane should be able to climb at 150 ft/min with one engine inoperative, at the aerodrome altitude and temperature.

7 Landing

7.1 Aeroplane Weight: See paragraph 5.1.

7.1.1 Guide line factor: landing distance will be increased by 10% for each 10% increase in aeroplane weight (a factor of x 1.10).

7.2 Aerodrome Altitude: aeroplane performance deteriorates with an increase in pressure altitude.

7.2.1 Guide line factor: landing distance will be increased by 5% for each 1000 ft increase in aerodrome pressure altitude (a factor of x 1.05).

7.3 Temperature: aeroplane performance deteriorates with an increase in ambient temperature.

7.3.1 Guide line factor: landing distance will be increased by 5% for a 10°C increase in ambient temperature (a factor of x 1.05).

7.4 Wind: a tailwind increases the landing distance.

7.4.1 Guide line factor: landing distance will be increased by 20% for a tailwind component of 10% of the landing speed (a factor of x 1.20).

> Note: Where the data allows adjustment for wind, it is recommended that not more than 50% of the headwind component and not less than 150% of the tailwind component of the reported wind be assumed. In some Manuals this factoring is already included and it is necessary to check the relevant section.

7.5 Slope: a downhill slope increases the landing distance.

7.5.1 Guide line factor: landing distance will be increased by 10% for each 2% of downhill slope (a factor of x 1.10).

7.6 Surface: grass or snow increase the ground roll, despite increased rolling resistance because brake effectiveness is reduced.

> Guide line factors: for dry grass (under 8 inches) the landing distance will be increased by 20% (a factor of x 1.20).
>
> For wet grass (under 8 inches) the landing distance will be increased by 30% (a factor of x 1.30).
>
> **Note 1**: When the grass is very short, the surface may be slippery and distances may increase by up to 60% (a factor of x 1.60)
>
> For snow, the landing distance will be increased by 25% or more (a factor of at least x 1.25).
>
> **Note 2**: For surface and slope factors, remember the increases shown are to the landing distance from a height of 50 ft. The correction to the ground roll will be greater.

7.7 Safety Factors: it is recommended that the public transport factor should be applied for all flights. For landing, this factor is x 1.43.

7.8 The above factors are cumulative and when several factors are relevant they must be multiplied. As in the take-off case the total distance required may seem surprisingly high.

> For example:
>
> In still air on a paved level, dry runway with an ambient temperature of 15°C an aeroplane requires a landing distance from a height of 50 ft (LDR) of 350 m. This should be multiplied by the safety factor of 1.43 giving a LDR of 501 m.
>
> The same aeroplane landing in still air at a wet grass strip (factor x 1.30) 500 ft above sea-level (factor x 1.025) at 25°C (ie 10°C warmer, therefore factor x 1.05), including the safety factor (factor x 1.43) will require a landing distance of:
>
> 350 x 1.30 x 1.025 x 1.05 x 1.43 = 700 m

7.9 The pilot should always ensure that after applying all the relevant factors including the safety factor the landing distance required from a height of 50 ft (LDR) does not exceed landing distance available.

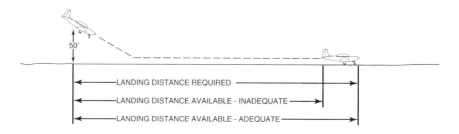

8 Additional Information

8.1 Engine Failure: the possibility of an engine failing during any phase of the flight should also be considered. Considerations should include the one engine inoperative performance of multi-engined types and the glide performance of single engined types. In the latter case, the ability to make a safe forced landing should be borne in mind throughout the flight.

8.2 Obstacles: it is essential to be aware of any obstacles likely to impede either the take-off or landing flight path and to ensure there is adequate performance available to clear them by a safe margin. The AD section of the UK AIP includes obstacle data for a number of UK aerodromes.

8.3 Aerodrome Distances: for many aerodromes, information on available distances is published in some form of aerodrome guide such as the AD section of the UK AIP or commercially available flight guides. At aerodromes where no published information exists, distances should be paced out. The pace length should be established accurately or assumed to be no more than 2.5 ft. Slopes can be calculated if surface elevation information is available, if not they should be estimated. Prior to take-off it might be helpful to taxi the aeroplane from one end of the strip to the other and take an altimeter reading at each end. Most altimeters will show differences down to 20 ft and to find the slope, simply divide altitude difference by strip length and give the result as a percentage. For example an altitude difference of 50 ft on a 2500 ft strip indicates a 2% slope. Be sure not to mix metres and feet in your calculation.

8.4 Operations from strips covered in snow, slush or extensive standing water should not be attempted without first reading AIC 61/1999 (Pink 195).

8.5 Where doubt exists on the source of data to be used or its application in given circumstances, advice should be sought from:

> Flight Test Section
> Flight Department
> Safety Regulation Group
> Civil Aviation Authority
> Aviation House
> South Area
> London Gatwick Airport
> Gatwick
> West Sussex
> RH6 0YR
>
> Tel: 01293-573113.

ANNEXE

TAKE-OFF

Condition	Increase In Take-off Distance To Height 50 ft	Factor
A 10% increase in Aeroplane Weight	20%	1.20
An increase of 1000 ft in Aerodrome Altitude	10%	1.10
An increase of 10°C in Ambient Temperature	10%	1.10
Dry Grass*- Up to 20 cm (8 in) (on firm soil)	20%	1.20
Wet Grass*- Up to 20 cm (8 in) (on firm soil)	30%	1.30
A 2% Uphill Slope*	10%	1.10
A Tailwind Component of 10% of Lift-off Speed	20%	1.20
Soft Ground or Snow*	25% or more	1.25 +

*Effect on Ground Run/Roll will be proportionally greater

LANDING

Condition	Increase In Landing Distance To Height 50 ft	Factor
A 10% increase in Aeroplane Weight	10%	1.10
An increase of 1000 ft in Airfield Altitude	5%	1.05
An increase of 10°C in Ambient Temperature	5%	1.05
A Wet paved runway	15%	1.15
Dry Grass*- Up to 20 cm (8 in) (on firm soil)	15%	1.15
Wet Grass*- Up to 20 cm (8 in) (on firm soil) See Note 3	35%	1.35
A 2% Downhill Slope*	10%	1.10
A Tailwind Component of 10% of Landing Speed	20%	1.20
Snow*	25% or more	1.25 +

*Effect on Ground Run/Roll will be proportionally greater

Note 1: After taking account of the above variables it is recommended that the relevant safety factors 1.33 or take-off; 1.43 for landing are applied.

Note 2: Any deviation from normal operating techniques is likely to result in an increase in the distance required.

Note 3: When the grass is very short, the surface may be slippery and distances may increase by up to 60% (a factor of 1.60).

AIC 6/2003 (Pink 48) 9 January

FLIGHT OVER AND IN THE VICINITY OF HIGH GROUND

1   The aim of this Circular is to remind pilots of the basic theory of airflow over high ground, to describe the effect of the airflow on aircraft in flight and to offer advice on avoiding or minimising the various hazards that may be encountered. This Circular is divided into three parts accordingly :

Part 1 – Meteorology

Part 2 – Flying Aspects

Part 3 – Advice to Pilots

Part 1 – Meteorology

1.1   **Introduction**

1.1.1   The expression 'high ground' is used here to describe mountains, hills, ridges etc which rise to heights in excess of about 500 ft above nearby low lying terrain.

1.1.2   Air flow is more disturbed and turbulent over high ground than over level country and the forced ascent of air over high ground often leads to the formation of cloud on or near the surface, which sometimes extends through a substantial part of the troposphere if the air is moist enough.

1.1.3   Forced ascent also increases instability so that thunderstorms embedded in widespread layer cloud may occur over high ground, even when no convective clouds form over low ground. When the air is generally unstable, cloud development will be greater, icing in the clouds will be more severe and turbulence in the friction layer and in cloud will be intensified over high ground.

1.1.4   The air flowing over high ground may be so dry that, even when it is forced to rise, little or no cloud is formed. The absence of cloud over high ground does not imply the absence of vertical air currents and turbulence.

1.1.5   Strong down currents are caused by the air descending the lee slope and it is, therefore, especially hazardous to fly towards high ground when experiencing a headwind.

1.1.6   On some occasions the disturbance of a transverse airflow by high ground can create an organised flow pattern of waves and large scale eddies in which strong up-draughts and downdraughts and turbulence frequently occur. These organised flow patterns are usually called mountain waves but may also be referred to as lee waves or standing waves and can be associated with relatively low hills and ridges as well as with high mountains.

1.2   **Mountain Waves**

1.2.1   Conditions favourable for the formation of mountain waves are:

(a)   A wind blowing within about 30° of a direction at right angles to a substantial ridge;

(b)   the wind must increase with height with little change in direction (strong waves are often associated with jet streams). A wind speed of more than 15 kt at the crest of the ridge is also usually necessary;

(c)   a marked stable layer (approaching isothermal, or an inversion), with less stable air above and below, between crest level and a few thousand feet above.

1.2.2   Mountain wave systems may extend for many miles downwind of the initiating high ground. Satellite photographs have shown wave clouds extending more than 250 nm from the

824

Pennines and as much as 500 nm downwind of the Andes. However, 50 to 100 nm is a more usual extent of wave systems in most areas. Wave systems may, on occasions, extend well into the stratosphere.

1.2.3 The average wave length of mountain waves in the troposphere is about 5 miles, but much longer waves do occur. A good estimate of the wave length l (nautical miles) can be derived from the mean troposphere wind v (knots) by using the simple formula l= v/7. Disturbances in the stratosphere are often irregular features located very near or just over the initiating mountains. However, when waves to the lee of the high ground are evident, their length is usually greater than in the troposphere; 15 nm is probably a typical wave length but wave lengths of 60 nm have been measured.

1.2.4 The amplitude of waves is much more difficult to determine from meteorological observations. In general, the higher the mountain and the stronger the airflow, the greater is the resulting disturbance; but the most severe conditions occur when the natural frequency of the waves is 'tuned' to the ground profile and conditions for wave motion are only just satisfied. This makes the prediction of wave amplitude uncertain.

1.2.4.1 In the troposphere the double amplitude (peak-to-trough) of waves is commonly 1500 ft with vertical velocities about 1000 ft/min but double amplitudes of about 20,000 ft and vertical velocities over 5000 ft/min have been measured in the USA. Even over the UK vertical velocities up to 2000 ft/min have been recorded, ie well beyond the climbing capability of many light aircraft.

1.2.4.2 Waves in the stratosphere have been measured with double amplitudes up to 1300 ft over the UK and more than 8000 ft over the western USA. Vertical velocities up to 2000 ft/min have been recorded in these waves. In the stratosphere above the Rocky Mountains, disturbances which have been interpreted as rotors, with amplitudes up to 9000 ft, have been observed. A pilot flying in such a disturbance has reported an accelerometer reading of minus 1g.

1.2.4.3 In extensive mountainous areas the wave system generated by one ridge is disturbed by further ridges downwind. Furthermore, the characteristics of an airstream are always changing with time, and occasions when small changes in the airstream give rise to large changes in mountain wave characteristics can be envisaged, but not forecast. Such changes may generate a transient but severe disturbance resulting in violent turbulence (eg due to a wave 'breaking').

1.3 Visual Detection of Mountain Waves

1.3.1 The special clouds which owe their appearance to the nature of wave flow are a valuable indicator to the pilot of the existence of wave formation. Provided there is sufficient moisture available the ascent of air will lead to condensation and formation of characteristic clouds. These clouds form in the crest of standing waves and therefore remain more or less stationary.

1.3.2 They may occur at all heights from the surface to cirrus level and are described briefly in the following paragraphs, to be read in conjunction with the diagram below, which shows the characteristic distribution of clouds and turbulence to the lee of the Sierra Nevada in North America. This is an area in which mountain wave phenomena are exceptionally marked, but the diagram has a fairly general application.

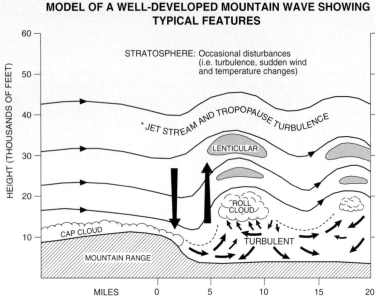

* The tropopause and level of maximum wind are usually located somewhere within this layer

AP7 M97136 2.12.97

(a) Lenticular clouds provide the most unmistakable evidence of the existence of mountain waves. They form within stable layers in the crests of standing waves. Air streams through them, the clouds forming at their up-wind edges and dissipating downwind. They have characteristically smooth lens shaped outlines and may appear at several levels, sometimes resulting in an appearance reminiscent of a stack of inverted saucers. Lenticular clouds usually appear up to a few thousand feet above the mountain crests, but are also seen at any level up to the tropopause and even above. (Mother-of-pearl clouds, seen on rare occasions over mountains, are a form of wave-cloud at an altitude of 80,000 ft or so). Air flow through these clouds is usually smooth unless the edges of the cloud take on a ragged appearance which is an indication of turbulence.

(b) Rotor or roll-clouds appear, at first glance, as harmless bands of ragged cumulus or stratocumulus parallel to and downwind of the ridge. On closer inspection these clouds are seen to be rotating about a horizontal axis. Rotor clouds are produced by local breakdown of the flow into violent turbulence. They occur under the crests of strong waves beneath the stable layers associated with the waves. The strongest rotor normally forms in the first wave downwind of the ridge and is, therefore, usually near or somewhat above the level of the ridge crest, but may occasionally be much deeper (rotor clouds have been reported to extend to 30,000 ft over the Sierra Nevada). There are usually not more than one or two rotor clouds in the lee of the ridge.

(c) Cap clouds form on the ridge crest and strong surface winds which are commonly found sweeping down the lee slope may sometime extend the cap cloud down the slope producing a 'cloud fall' or 'föhn wall'.

1.3.3 Although cloud often provides the most useful visible evidence of disturbances to the airflow, the characteristic cloud types may sometimes be obscured by other cloud systems, particularly frontal cloud. On the other hand, the air may be too dry to form any clouds at all, even in strong wave conditions.

1.4   **Turbulence**

1.4.1   Turbulence at low and medium levels.

1.4.1.1   A strong wind over irregular terrain will produce low level turbulence which increases in depth and intensity with increasing wind speed and terrain irregularity.

1.4.1.2   In a well developed wave system, the rotor zone and the area below are strongly turbulent and reversed flow is often observed at the surface. Strong winds confined to the lower troposphere, with reversed or no flow in the middle and higher troposphere, produce the most turbulent conditions at low levels, sometimes accompanied by 'rotor streaming' comprised of violent rotors which are generated intermittently near lee slopes and move downwind. These low level travelling rotors are distinct from the stationary rotors which form at higher levels in association with strong mountain waves.

1.4.2   **Turbulence in the rotor zone**

1.4.2.1   Rotors lie beneath the crests of lee waves and are often marked by roll clouds. The most powerful rotor lies beneath the first wave crest downstream of the mountains. Rotors give rise to the most severe turbulence to be found in the air flow over high ground. On occasions it may be as violent as that in the worst thunderstorms. Gliders flying in rotor zones in both Europe and the USA have found accelerations of 2g to 4g quite common and 7g has been exceeded in the USA. Several gliders suffered structural damage and one disintegrated.

1.4.3   **Turbulence in waves**

1.4.3.1   Although flight in waves is often remarkably smooth, severe turbulence can occur. The transition from smooth to bumpy flight can be abrupt. Very occasionally violent turbulence may result, sometimes attributed to the wave 'breaking'.

1.4.4   **Turbulence at high levels (near and above the tropopause)**

1.4.4.1 Turbulence near the jet stream

1.4.4.1.1   Turbulence in jet streams is frequently greatly increased in extent and intensity over high ground. Strong vertical wind shears are often concentrated in a few stable layers just above and below the core of the jet stream. Distortion of these layers when the jet stream flows over high ground particularly when mountain waves form, can produce local enhancements of the shears so that the flow in those regions breaks down into turbulence. Usually the cold side of the jet stream is more prone to turbulence, but mountain waves may be more pronounced on the warm side.

1.4.4.2 Turbulence in the stratosphere

1.4.4.2.1   Flight experience has shown that in the stratosphere moderate or severe turbulence is encountered over high ground about four times more frequently than over plains and about seven times more frequently than over the oceans.

1.4.4.2.2   Evidence from research flying over the Rocky Mountains has shown that strong rotors and/or waves may occur well into the stratosphere on days favourable for strong wave formation in the troposphere. The associated severe turbulence can cause serious difficulties to an aircraft flying near its ceiling.

1.5   **Downdraughts**

1.5.1   Whether or not a well developed wave system exists, if the air is stable a strong surface air flow over high ground will produce a substantial and sustained downdraught and/or turbulence on the lee side. Such downdraughts may on occasions be strong enough to defeat the rate of climb capability of some aircraft. In a wave system, a series of downdraughts and updraughts exists, the most powerful being those nearest the high ground.

United Kingdom Aeronautical Information Circular AIC 6/2003 (Pink 48) 9 January

1.6   **Icing**

1.6.1 Adiabatic cooling caused by the forced ascent of air over high ground generally results in a lowering of the freezing level and an increase of liquid water concentration in clouds. Thus airframe icing is likely to be more severe than at the same altitude over lower ground when extensive cloud is present. This hazard is at a maximum a few thousand feet above the freezing level, but in general is unlikely to be serious at altitudes above 20,000 ft, except in cumulonimbus clouds.

Part 2 – Flying Aspects

2.1   The effects of the airflow over high ground on aircraft in flight depend on the magnitude of the disturbance to the airflow, the performance of the aircraft, its altitude and the aircraft's speed and direction in relation to the wave system. A broad distinction may be made between low level hazards (below about 20,000 ft) and high level hazards (above 20,000 ft).

2.2   **Low Altitude Flight**

2.2.1 The main hazards arise from severe turbulence in the rotor zone, from downdraughts and from icing. The presence of roll-clouds in the rotor zone may warn pilots of the region of most severe turbulence, but characteristic cloud formations are not always present or, if they are present, may lose definition in other clouds. Similarly, the updraughts and downdraughts are, in general, not visible. If an aircraft remains for any length of time in a downdraught (eg by flying parallel to the mountains in the descending portion of the wave), which may be remarkably smooth, serious loss of height may occur.

2.2.2 During upwind flight the aircraft's height variations are normally out of phase with the waves; the aircraft is, therefore, liable to be at its lowest height when over the highest ground. The pilot may also find himself being driven down into a roll-cloud over which ample height clearance previously appeared to be available.

2.2.3 Downwind flight may be safer. Height variations are usually in phase with waves, but it must be appreciated that the relative speed of an accidental entry into the rotor zone will be greater than in upwind flight because the rotor zone is stationary with regard to the ground. Thus, the structural loads which may be imposed on the airframe when gusts are encountered are likely to be greater and there will probably be less warning of possible handling difficulties.

2.3   **High Altitude Flight**

2.3.1 The primary danger at high altitude is that of a sudden encounter with localized disturbances (ie turbulence, sudden large wind and temperature changes) at high penetration speeds, this is particularly relevant at cruising levels above FL 300 where the buffet-free margin between the Mach number for 1g buffet and the stall is restricted. In this respect flight downwind is likely to be more critical than flight upwind, especially when the wind is strong. As in the case of low altitude flight the waves are stationary relative to the ground, and the higher relative speed on accidentally encountering a standing wave while flying downwind is likely to place greater loads on the airframe. There will often be no advance warning of the presence of wave activity from preliminary variations in flight instrument readings, or from turbulence. Although downdraughts are present they are unlikely to be hazardous and icing and rotor zone turbulence are unlikely.

2.4   **General**

2.4.1 While flying through strong mountain waves, large fluctuations in wind velocity may be encountered, with associated turbulence; and an aircraft entering a wave system with its auto-pilot (including height and airspeed locks) fully engaged may begin to oscillate in the pitching plane as it attempts to maintain the selected height and airspeed. This oscillation can become unstable and, if unchecked, may put an aircraft into a dangerous flight condition as a result of excessive tailplane deflection. If the aircraft is being flown manually and the pilot chases height or airspeed, a similar result may occur. In either case there is a risk of an upset developing with catastrophic results. This emphasises the importance of the well established technique of flying 'attitude' in these conditions.

United Kingdom Aeronautical Information Circular   AIC 6/2003 (Pink 48) 9 January

Part 3 – Advice to Pilots

3.1   When planning a flight over or close to high ground, pilots should ensure that the possibility of mountain wave conditions is considered in their meteorological briefing, particularly if frontal conditions are present in the area and a jet stream is expected at altitude. Although areas of turbulence associated with mountain waves cannot be forecast with accuracy, Meteorological Offices can help pilots to assess the possibility of mountain waves being encountered and can give advice on the probable height of layers of marked stability. Careful attention should be paid to warnings given in SIGMET messages broadcast during the flight.

3.2   If mountain wave conditions are forecast or known to be present:

(a)   Do not attempt to approach or penetrate rotor clouds or likely rotor zones adjacent to mountain ranges;

(b)   when flying over high ground, maintain a clearance height above the highest ridge at least equal to the height of the ridge above terrain. This should avoid the worst of the lower altitudes hazards;

(c)   choose cruising altitudes well away from the base of layers of marked stability where severe turbulence is most likely to occur (present information suggests that, while there may be more than one stable layer, a margin to 5000 ft on either side of the base of a stable layer, including the tropopause, is advisable);

(d)   be prepared for the occurrence of icing if cloud formations are present.

3.3   When flying in an area in which mountain wave conditions are suspected, always be prepared for turbulence, even in clear air, and take precautions accordingly. These precautions should include:

(a)   Setting up the recommended speed for flight in turbulence;

(b)   re-trimming the aircraft and noting the trim position so that any changes that may occur (due to auto trim action when using the auto-pilot) can be quickly detected;

(c)   ensuring that crew and passengers are securely strapped in and that there are no loose articles;

(d)   following the recommendations on the use of auto-pilot, height and airspeed locks and stability aids (yaw dampers etc) as appropriate;

(e)   do not attempt to chase the gust induced lateral rocking, but aim to keep the aircraft laterally level to within reasonable limits, yaw dampers should remain engaged however;

(f)   try to make all control inputs smoothly and gently.

3.5   **Take-Off and Landing Manoeuvres**

3.5.1   Pilots should be aware of the danger and severity of turbulence which may be encountered in the lee of high ground during take-off and approach to land maoneuvres, when performance margins may be small.

829

AIC 81/2004 (Pink 66) 19 August

THE EFFECT OF THUNDERSTORMS AND ASSOCIATED TURBULENCE ON AIRCRAFT OPERATIONS

1 Introduction

1.1 This Circular has been produced to provide an understanding of the hazard that thunderstorms and their associated effects can pose to all aircraft operations. This Circular has been updated as a consequence of a recent AAIB incident investigation and replaces the guidance originally published in AIC 72/2001 (Pink 22). It has been published for the information and safety of all pilots.

1.2 This Circular has been written with two-pilot operation of larger aircraft in mind; however text that has been highlighted by the use of capital letters is of particular relevance to pilots of all aircraft.

1.3 The overarching advice in this Circular is that flight through thunderstorms should be avoided.

2 Thunderstorm Warnings

2.1 Meteorological Watch Offices (MWO) issue SIGMET (Significant Meteorology) warnings of 'Thunderstorms' when significant cumulonimbus clouds likely to produce thunderstorms are forecast and when these thunderstorms are expected to be difficult to detect visually by a pilot. They could be obscured (OBSC TS), embedded in other clouds (EMBD TS) and could possibly be frequent (FRQ TS) or organised along a line (SQL TS). These warnings include information on the location, movement and development of the thunderstorm areas. As it is expected that all pilots will be aware of the additional phenomena associated with thunderstorms, ie hail, severe icing, and severe turbulence (as expanded on in the Annex to this Circular), these forecast details will not be included in the SIGMET text, although heavy hail (HVYGR) could be included. In addition, aircraft commanders are required to send a Special Aircraft Observation when conditions are encountered likely to affect the safety of aircraft. Such a report could then trigger a SIGMET warning. MWOs do not issue SIGMET warnings in relation to isolated or scattered thunderstorms not embedded in cloud layers or concealed by haze (unless prompted by a Special Aircraft Observation). It should therefore be noted that the absence of a SIGMET warning does not necessarily indicate the absence of thunderstorms.

2.2 Aerodrome Warnings are issued by the Meteorological Office for terminal area operations where there is a forecast likelihood of thunderstorms in the immediate vicinity of an aerodrome. Separate windshear warnings may be issued at some aerodromes (notably London Heathrow and Belfast Aldergrove) where a nearby thunderstorm is the criteria for a windshear warning. Elsewhere, the proximity of a thunderstorm will not necessarily result in such a warning, but the probability of windshear is no less. In relation to windshear hazards at low-level your attention is drawn to AIC 19/2002 (Pink 28).

2.3 Details of the criteria for Special Aircraft Observations and the SIGMET service are given in the UK Aeronautical Information Publication at GEN 3.5.6 and GEN 3.5.8 respectively.

3 Procedures and Flying Techniques

3.1 Notwithstanding the advice that follows, gathered from research and operational experience, the first and most basic advice for all pilots is:

Do not treat thunderstorms lightly and whenever possible AVOID them.

3.2 Thunderstorms should be avoided either visually, by the use of radar, or by other methods. If this cannot be achieved, and in the absence of specific aircraft flight manual or operations manual guidance, the following procedures and techniques are recommended.

    (a) If it is found necessary to penetrate an area of cloud which may contain cumulonimbus clouds:

        (i) ENSURE THAT CREW MEMBERS' SAFETY BELTS OR HARNESSES ARE FIRMLY FASTENED AND SECURE ANY LOOSE ARTICLES BEFOREHAND. Switch on the seat belt notices and make sure that all passengers are securely strapped in and that loose equipment (eg cabin trolleys and galley containers) are firmly secured. Pilots should remember that turbulence is normally worse in the rear of an aircraft than on the flight deck.

        (ii) One pilot should control the aircraft with the other continually monitoring all the flight instruments.

        (iii) SELECT AN ALTITUDE FOR PENETRATION, BEARING IN MIND THE IMPORTANCE OF ENSURING ADEQUATE TERRAIN CLEARANCE IN LIKELY DOWNDRAUGHTS. Investigations have shown that although in some thunderstorms there is very little turbulence at the lower levels, in others there is a great deal; altitude is not necessarily a guide to the degree of turbulence. Increasing height will decrease the buffet margin and up-currents may force the aircraft into buffet owing to an increased angle of attack.

        (iv) SET THE POWER TO GIVE THE RECOMMENDED SPEED FOR FLIGHT IN TURBULENCE, ADJUST THE TRIM AND NOTE ITS POSITION SO THAT ANY EXCESSIVE CHANGES DUE TO AUTOPILOT OR MACH TRIM OPERATION CAN BE QUICKLY ASSESSED. Turbulence speeds quoted in flight or operations manuals provide a single speed or a speed bracket.

        (v) CHECK ALL FLIGHT INSTRUMENTS AND ELECTRICAL SUPPLIES.

        (vi) ENSURE THAT THE PITOT HEATERS ARE SWITCHED ON.

        (vii) CHECK THE OPERATION OF ALL ANTI-ICING AND DE-ICING EQUIPMENT AND OPERATE ALL THESE SYSTEMS IN ACCORDANCE WITH MANUFACTURER'S OR OPERATOR'S INSTRUCTIONS. The operation of leading edge, expanding boot type de-icers should be delayed until some ice has formed, otherwise their effectiveness will be greatly reduced. IN THE ABSENCE OF SPECIFIC INSTRUCTIONS, ENSURE THAT ALL ANTI-ICING SYSTEMS, INCLUDING WINDSCREEN HEATERS, ARE ON.

        (viii) DISREGARD ANY RADIO NAVIGATION INDICATIONS SUBJECT TO INTERFERENCE FROM STATIC, eg ADF.

        (ix) TURN THE COCKPIT LIGHTING FULLY ON AND LOWER THE CREW SEATS AND SUN VISORS TO MINIMISE THE EFFECT OF ANY LIGHTNING FLASHES.

        (x) FOLLOW THE MANUFACTURER'S OR OPERATOR'S RECOMMENDATIONS ON THE USE OF THE FLIGHT DIRECTOR, AUTOPILOT AND MANOMETRIC LOCKS. If these are not stated, height, Mach, rate of climb or descent, and airspeed locks should be disengaged but the yaw damper(s), if fitted, should remain operative. On many aircraft the autopilot, when engaged in a suitable mode (turbulence or basic attitude modes), is likely to produce lower structural loads than would result from manual flight. However, if major trim movements occur due to the autopilot's automatic trim the autopilot should be disengaged. Note that Mach trim operation may also occur on some aircraft but the Mach trim should remain engaged.

(xi) Continue operating, not just monitoring, the weather radar, or other on-board systems, in order to select the safest track for penetration, and to minimise the time of exposure whilst avoiding areas of intense activity.

(xii) Be prepared for turbulence, rail, hail, snow, icing, lightning, static discharge and windshear. In turbine-powered aircraft switch on the continuous ignition system (to reduce the possibility of engine flame-out due to water ingestion) ensuring that limitations on its use, if any, are not exceeded. Also see AIC 29/2004 (Pink 64) – 'Engine Malfunction caused by Lightning Strikes'.

(xiii) AVOID FLYING OVER THE TOP OF A THUNDERSTORM WHENEVER POSSIBLE. Overflying small convective cells close to large storms should also be avoided, particularly if they are on the upwind side of a large storm, because they may grow very quickly. Similarly, do not contemplate flying beneath the cumulonimbus cloud. In addition to the dangers associated with turbulence, rain, hail, snow or lightning, there may well be low cloud base, poor visibility and possibly low-level windshear.

(b) Within the Storm Area:

(i) CONTROL THE AIRCRAFT REGARDLESS OF ALL ELSE.

(ii) CONCENTRATE ON MAINTAINING A CONSTANT PITCH ATTITUDE APPROPRIATE TO CLIMB, CRUISE OR DESCENT, BY REFERENCE TO THE ATTITUDE INDICATORS, CAREFULLY AVOIDING HARSH OR EXCESSIVE CONTROL MOVEMENTS. DO NOT BE MISLED BY CONFLICTING INDICATIONS ON OTHER INSTRUMENTS. DO NOT ALLOW LARGE ATTITUDE EXCURSIONS IN THE ROLLING PLANE TO PERSIST BECAUSE THESE MAY RESULT IN NOSE DOWN PITCH CHANGES.

(iii) MAINTAIN THE ORIGINAL HEADING – IT IS USUALLY THE QUICKEST WAY OUT. DO NOT ATTEMPT ANY TURNS.

(iv) DO NOT CORRECT FOR ALTITUDE GAINED OR LOST THROUGH UP AND DOWN DRAUGHTS UNLESS ABSOLUTELY NECESSARY.

(v) Maintain the trim settings and avoid changing the power setting except when necessary to restore margins from stall warning or high-speed buffet. The target pitch attitude should not be changed unless the mean IAS differs significantly from the recommended penetration speed.

(vi) If trim variations due to the autopilot (auto-trim) are large, the autopilot should be disengaged. Movement of the Mach trim, where it occurs, is however necessary and desirable. Check that the yaw-damper remains engaged.

(vii) If negative 'G' is experienced, temporary warnings (eg low oil pressure) may occur. These should be ignored.

(viii) ON NO ACCOUNT CLIMB IN AN ATTEMPT TO GET OVER THE TOP OF THE STORM.

(c) After a Thunderstorm Encounter – In flight:

(i) If hail has been encountered, considerable damage to the airframe, not visible from the cockpit or cabin, may have occurred. Consideration should therefore be given to diverting to a suitable and nearby aerodrome where the aircraft can be inspected for damage. If this damage has occurred to aerodynamically significant areas, eg a nose radome, the increased drag will affect fuel burn. Thus the aircraft, if continuing to its destination, may burn considerably more fuel than expected or planned. **Actual** fuel usage should now be monitored very closely, bearing in mind that some FMS calculate 'expected overhead destination' fuel, based on data that assumes normal (planned) conditions and normal (ie fully clean) aircraft aerodynamic states.

(ii) If the aircraft has been struck by lightning, treat all magnetic information (eg from direct or remote indicating compasses) with extreme caution. The large electric currents associated with a lightning strike can severely and permanently distort the magnetic field of an aircraft rendering all such information highly inaccurate.

(d) Air Traffic Control Considerations:

(i) Modern ATC radars in general do not display the build up of weather that may constitute a hazard to aircraft and ATC advice on weather avoidance may, therefore, be limited.

(ii) If, as recommended in this Circular, a pilot intends to detour round observed weather when in receipt of an Air Traffic Service that involves ATC responsibility for separation, clearance should first be obtained from ATC so that separation from other aircraft can be maintained. If for any reason the pilot is unable to contact ATC to inform the controller of his/her intended action, any manoeuvre should be limited to the extent necessary to avoid immediate danger and ATC must be informed as soon as possible.

(iii) Because of the constraints on airspeed and flight path and the increased workload of the crew when flying in a Terminal Manoeuvring Area, pilots should consider making a diversion from, or delaying entry to, a Terminal Manoeuvring Area if a storm encounter seems probable.

(e) Take-off and Landing Problems:

(i) The take-off, initial climb, final approach and landing phases of flight present the pilot with additional problems because of the aircraft's proximity to the ground, thus the maintenance of a safe flight path in these phases can be very difficult.

(ii) Some operators give advice on the airspeed adjustments to be made to allow for windshear or turbulence (a speed increase of up to 20 knots according to the type of aircraft and the degree of turbulence may be required). The best advice that can be given to the pilot is that, when there are thunderstorms over or near the aerodrome, he/she should delay take-off or, when approaching to land, hold in an unaffected area or divert to a suitable alternate. For further relevant information see AIC 19/2002 (Pink 28) – 'Low Altitude Windshear'.

(f) Airworthiness and Maintenance Considerations:

(i) Severe weather conditions may cause damage to aircraft and power plant installations, some of which may be invisible to the naked eye. Flight Manuals and Maintenance documents may quantify levels of turbulence which would trigger a maintenance inspection, similar to those that may be applicable to 'heavy landings'. Hail and lightning damage may often be obvious to crews; however, there will be occasions where damage may be restricted to parts of the airframe not normally visible from the ground, or from the cockpit, immediately following a thunderstorm encounter.

(ii) In the event that crews believe that an aircraft has been exposed to hail, lightning, turbulence greater than 'moderate', or a heavy landing, they should record the fact(s) in the technical log on arrival to ensure that an appropriate inspection is completed prior to a subsequent release to service. Operators should ensure that procedures in operation manuals for flight crews and maintenance personnel reflect this advice.

United Kingdom Aeronautical Information Circular AIC 81/2004 (Pink 66) 19 August

(g) Light aircraft operators should ensure that their aircraft are adequately secured on the ground when severe thunderstorm activity is forecast.

4 Concluding Remarks

4.1 DO NOT TAKE-OFF IF A THUNDERSTORM IS OVERHEAD OR APPROACHING.

4.2 AT DESTINATION HOLD CLEAR IF A THUNDERSTORM IS OVERHEAD OR APPROACHING. DIVERT IF NECESSARY.

4.3 AVOID SEVERE THUNDERSTORMS EVEN AT THE COST OF DIVERSION OR AN INTERMEDIATE LANDING. IF AVOIDANCE IS IMPOSSIBLE, THE PROCEDURES RECOMMENDED IN THE FLIGHT OR OPERATIONS MANUAL OR IN THIS CIRCULAR SHOULD BE FOLLOWED.

4.4 Pilots of turbo-jet swept-wing transport aircraft are advised to ensure that they are fully conversant with the control problems that may be met in turbulence with the type of aircraft they fly.

4.5 AFTER AN ENCOUNTER WITH A THUNDERSTORM CONSIDER REPORTING THE EVENT IN THE AIRCRAFT TECHNICAL LOG. THIS WILL ENSURE THAT A FULL AND PROPER INVESTIGATION OF THE AIRCRAFT OCCURS.

ANNEX

1 THUNDERSTORMS, FLIGHT HAZARDS AND WEATHER RADAR

1.1 A thunderstorm cloud, whether of the air mass or frontal type, usually consists of several self-contained cells, each in a different state of development. It must be stressed that the storm clouds are only the visible part of a turbulent system that extends over a much greater area. New and growing cells can be recognised by their cumuliform shape with clear-cut outline and 'cauliflower' top, while the tops of more mature cells will appear less clear-cut and will frequently be surrounded by fibrous cloud. It is important, however, to remember that the development of cells, which can be very rapid, will not always be seen, even in daylight, since other clouds may obscure the view. In frontal or orographic conditions, for instance, where forced ascent of air may give the impetus required for producing vigorous convection currents, extensive layer cloud structures may obscure a view of the development of Cumulonimbus thunderstorm cells or Altocumulus Castellanus; the latter is cumuliform cloud with a base above 8,000 ft and is an indication of middle level instability which often precedes, or is associated with, the development of thunderstorms. Mammatus clouds, udder shaped features seen beneath cumulonimbus clouds, or the associated medium level altocumulus layer clouds (above 8,000 ft), or in association with the high-level cirrus anvil cloud (above 20,000 ft), are an indication of strong vertical winds with associated turbulence.

1.2 The most severe thunderstorms require an increase in the general wind speed and a change in direction with height to maintain a release of energy. With no vertical wind shear, as the cloud grows and the updraught strengthens, precipitation forms in the upper parts of the cloud. As the precipitation falls towards the ground, it exerts a drag on the updraught, which weakens and the cloud decays. However, for a storm that has the downdraught offset from the updraught, particularly where the updraught is not cut off at the surface by the spreading out of the cold downdraught, it can develop into a self-generating system that can last for hours, independently of any surface heating.

1.3 These up and downdraughts are of comparable intensity, often in close proximity to each other and frequently reach speeds in excess of 3,000 ft per minute. Sharp gusts with vertical speeds of 10,000 ft per minute have been measured. The horizontal extent of these vertical draughts may, occasionally, be more than a mile. The top of a developing cell has been observed to rise at more than 5,000 ft per minute. When thunderstorms are associated with frontal conditions, areas of 'line squall' activity can extend for more than 100 miles. The vertical extent of storms will vary considerably but it is not uncommon for them to penetrate the tropopause with cloud tops exceeding 40,000 ft in temperate latitudes and 60,000 ft in sub-tropical and tropical regions. Although an individual cell will usually last for less than an hour, a storm system, with new cells developing and old ones decaying, may persist for several hours.

1.4 Areas in which conditions will be favourable for the development of thunderstorms can usually be forecast successfully several hours in advance but it is not possible at present to determine the precise location and distribution of individual storms. Where up-to-date ground weather radar information is available, however, useful information on the expected movement of an individual storm can be forecast for periods of up to an hour or so ahead.

1.5 As a general rule of thumb, in the UK, the movement of a cumulonimbus cloud is in the direction of the 10,000 ft (700 millibar) wind, though the tendency for large storms to distort wind fields and the development of new cells will cause variations in this general movement.

1.6 All thunderstorms are potentially dangerous. This considered, there are two facts that should be borne in mind. The first is that a severe storm can occur in practically any geographical area in which thunderstorms are known. The second is that no useful correlation exists between the external visual appearance (or the weather radar appearance) of thunderstorm clouds and the turbulence and hail within them.

2 FLIGHT HAZARDS

2.1 Turbulence Associated with Thunderstorms

2.1.1 The air movement in thunderstorms, generally referred to as turbulence and composed of draughts (sustained vertical or sloping currents) and gusts (irregular and local variations), can become violent, dangerous and even destructive, reaching a maximum intensity in developing and mature cells. High rates of roll and large pitching motions have been experienced in these storms, as have large vertical displacements of as much as 5,000 ft. These extreme variations will, of course, only occur in the most severe conditions. Of equal importance is the fact that eddies, which are felt as gusts, can occur some distance outside a thunderstorm cell. The regions around or between adjacent cells are therefore likely to be turbulent – severely so at times – and severe turbulence is often found 15 to 20 miles downwind of a severe storm core. Conditions at or near the surface in the vicinity of thunderstorms are often rough because, during the mature stage of the cells, the outflow from the base is of a turbulent nature and the air is colder than its environment, producing a miniature cold front often accompanied by heavy precipitation and squally conditions. When this is associated with a line of thunderstorms its effects can be felt as much as 40 miles ahead of them. Take-offs and landings in these circumstances are hazardous. Severe turbulence can also be encountered several thousand feet above the tops of active thunderstorm clouds, particularly when the speed of the wind at this level is high (100 kts or more). It is therefore advisable to avoid flying and in particular not to climb, in these areas.

2.1.2 A thunderstorm cell must be well developed before lightning first occurs but it may continue in the decaying cell. Lightning must not, therefore, be regarded as a reliable guide to the degree of turbulence in a cloud.

2.1.3 Accidents involving loss of control of the aircraft have been caused by flying in and around thunderstorms. In some instances there was structural failure that probably occurred during the attempt to regain control.

2.1.4 Stress requirements for modern transport aircraft are set at a level which experience has shown will rarely be reached. Nevertheless, flight research has indicated that, in the extreme conditions that may exist within thunderstorms, abnormal pilot-induced loads are added to already high gust-loads such that stress limits may be exceeded.

2.1.5 In some instances the correct flying technique is difficult to achieve. Indications are that loss of control, which may follow the use of incorrect techniques, is a more serious hazard than the risk of structural failure due directly to an encounter with turbulence. This is because recovery manoeuvres are likely to subject the aircraft to great stresses that may lead to structural failure or serious deformation.

2.2 Thunderstorm Windshear

2.2.1 Accidents have occurred during the take-off, initial climb and final approach phases of flight, which were probably due in part, if not entirely, to the effect of a rapid variation in wind velocity known as windshear. For further information see AIC 19/2002 (Pink 28) – 'Low Altitude Windshear'. Unlike the erratic fluctuations caused by gusts, windshear gives rise to airspeed fluctuations of a more sustained nature and is therefore likely to be more dangerous. Gusts are likely to accompany windshear conditions.

2.2.2 Thunderstorms frequently produce windshear and, although it is hazardous at all levels, it is in the lower levels that windshear may have more drastic consequences. Winds caused by the outflow of cold air from the base of a thunderstorm cell have been known to change in shallow layers of a few hundred feet by as much as 80 kts in speed and 90° or more in direction. Due to the effect of inertia, an aircraft in flight will tend to maintain its ground speed and windshear will therefore produce airspeed variations that can be large enough to be extremely dangerous.

2.3 **Tornadoes**

2.3.1 Tornadoes present a very serious threat to aircraft. A Fokker F-28 flying in cloud at 3,000 ft shortly after take-off from Rotterdam was destroyed by a tornado on 6 October 1981. Tornadoes are generally associated with organised severe local storms. They occur frequently in the United States but can also arise in the UK and Europe although they are less common and seldom as violent. There is evidence that tornado circulation may extend throughout the depth of the storm and constitute a hazard to aircraft at all heights.

2.3.2 The most violent thunderstorms draw air into their cloud bases with great vigour. If the incoming air has any initial rotating motion, it often forms an extremely concentrated vortex from the surface well into the cloud. Meteorologists have estimated that wind velocities in such a vortex can exceed 200 kts. Because pressure inside the vortex is quite low, the strong winds gather dust and debris and the low pressure generates a funnel-shaped cloud extending downward from the cumulonimbus base. If the cloud does not reach the surface, it is a 'funnel cloud'; if it touches the land surface, it is a tornado.

2.3.3 Tornadoes occur with both isolated and squall line thunderstorms. An aircraft entering a tornado vortex is almost certain to suffer structural damage. Since the vortex extends well into the cloud, any pilot flying on instruments in a severe thunderstorm could encounter a hidden vortex.

2.3.4 Families of tornadoes have been observed as appendages of the main cloud extending several miles outward from the area of lightning and precipitation. Thus any cloud connected to a severe thunderstorm carries a threat.

2.4 **Hail**

2.4.1 Notwithstanding all the work that has been done in the field of thunderstorm forecasting, no confirmed or fully reliable method has yet been evolved for recognising a storm that will produce hail. It is safest to assume that hail exists in one part or another of every thunderstorm at some stage in its life. The higher the lapse rate and the greater the moisture content of the air mass, the stronger will be the convective activity which increases the likelihood of the formation of damaging hail. Stability in the upper atmosphere results in the characteristic anvil shape of the spreading-out of the top of the cumulonimbus cloud and strong upper winds will often cause hail to fall from the overhang. Flight beneath the overhang should be avoided.

2.4.2 The maximum size of hailstones which have been found on the ground is around five and a half inches in diameter. It is known that hailstones of four inches in diameter can be encountered at 10,000 ft and damaging hail up to 45,000 ft.

2.4.3 Although hail encounters are usually of short duration, damage to aircraft can be severe. Hail may damage the leading edges and hence reduce the efficiency of the wing. Windscreens or other transparancies may be shattered. In an encounter in the Middle East, hail severely damaged the airframe of a VC 10 that encountered a thunderstorm shortly after take-off. The radome was torn away, denting and damage to the skin occurred in many areas, but there was no evidence of a lightning strike.

2.4.4 Although no fatal accidents to civil aircraft are known to have been attributable entirely to hail damage, hail can be a serious hazard at all altitudes at which civil aircraft operate. Evidence for this comes from a study of military aircraft accidents in the USA, in which aircraft were damaged or destroyed by the combined effect of hail and turbulence and from experience gained through the United States National Severe Storms Project together with individual reports of encounters with hail in normal operations.

2.5 **Rain**

2.5.1 Water ingestion by turbine engines

2.5.1.1   Turbine engines have a limit on the amount of water they can ingest. Updraughts are present in many thunderstorms, particularly those in the developing stages. If the updraught velocity in the thunderstorm approaches or exceeds the terminal velocity of the falling raindrops, very high concentrations of water may occur. It is possible that these concentrations can be in excess of the quantity of water turbine engines are designed to ingest, which could result in flame out and/or structural failure of one or more engines.

2.5.1.2   At the present time, there is no known operational procedure that can completely eliminate the possibility of engine damage/flame out during massive water ingestion but although the exact mechanism of these water induced engine stalls has not been determined, it is believed that thrust changes may have an adverse effect on engine stall margins.

2.5.1.3   To eliminate the risk of engine damage or flame out by heavy rain, it is essential to avoid severe storms. During an unavoidable encounter with extreme precipitation, the best-known recommendation is to follow the severe turbulence penetration procedure contained in the approved aircraft flight manual, with special emphasis on avoiding thrust changes unless excessive airspeed variations occur. Flight research has revealed that water can exist in large quantities at high altitudes even where the ambient temperature is as low as -30° C. Rain, sometimes heavy, may therefore be encountered and give rise to ice accretion and a possibility of the malfunctioning of pressure instruments. Turbine engine igniters must be switched on.

2.5.2   Heavy precipitation, which occurs in cumulonimbus clouds, may often be seen as shafts of rain below the cloud base. Where this precipitation does not reach the surface, the shafts are known as virga. The evaporation cooling associated with virga may intensify existing downdraughts.

2.6   Icing

2.6.1   Flight must not be initiated or continued into areas where the forecast icing conditions will exceed the icing limitations of the aircraft.

2.6.2   Formation of ice on the airframe must always be considered likely when flight takes place through cloud or rain at a temperature below 0° C. The temperature range favourable for ice accretion in thunderstorms is from 0° C down to -45° C, ie where water droplets can exist in a supercooled state. Below about -30° C, however, a large part of the free water content of the atmosphere normally consists of ice particles or crystals and snowflakes and chances of severe icing at these low temperatures are, therefore, greatly reduced. Conversely, because of downdraughts, the freezing level inside thunderstorm clouds must be assumed to drop to the base of the cloud. Airframe icing can therefore be expected everywhere in a thunderstorm cloud.

2.6.3   In piston engines, loss of power can occur over a wide range of temperatures as a result of the formation of ice in the induction system. Proper use of carburettor heat or other induction anti-icing equipment is therefore essential to prevent or minimise the loss of power. Furthermore, in clear air of high humidity (ie of the order of 60% or more), which might exist in areas of thunderstorm activity, carburettor ice can easily form. AIC 145/1997 (Pink 161) – 'Induction System Icing on Piston Engines as Fitted to Aeroplanes, Helicopters and Airships' covers the effect of icing on piston engines in greater detail.

2.6.4   Where turbine engines are concerned, the danger of flame out must be recognised whenever icing conditions are met. Igniters must therefore be switched on and remain on provided they are cleared for continuous operation. In all circumstances operators' or manufacturers' instructions must be strictly followed to achieve maximum protection.

2.6.5   It must be emphasised that, when flying in thunderstorms, anything more than very light ice accretion adds to the problems related to turbulence because of the increased weight of the aircraft, the disturbance of the normal airflow and the reduced effectiveness of the control surfaces.

2.6.6   Experience has shown that, provided the normal precautions are taken (ie using the anti-icing or de-icing equipment correctly) icing conditions need not be a grave hazard if penetration of a thunderstorm area cannot be avoided. However, failure to recognise or anticipate icing conditions, failure to use the equipment properly, equipment unserviceability or extended flight through a storm area will all considerably increase the risks involved.

2.7 Lightning

2.7.1 Lightning can occur both within and away from cumulonimbus clouds, with discharges taking place either within the cloud, between neighbouring clouds, or commonly between a cloud and the ground and less commonly from the top of a cloud upwards. Most recorded lightning strikes have occurred at levels where the temperature is between +10°C and -10°C, ie within about 5,000 ft above or below the freezing level. Some risk also exists outside this band, particularly in the higher levels. Strikes are either electrically positive or negative, although the polarity of the strike is not evident at the time. Positive polarity strikes are likely to be the more severe, (ie cause more damage to the aircraft) and recent investigations have shown that the North Sea is an area prone to a higher than normal frequency of positive strikes, although the overall frequency of strikes per flying hour is similar to that in the rest of Europe. The presence of soft hail has been associated with some positive strikes and may thus be indicative of the conditions conducive to a positive strike. For further information regarding lightning and aircraft engines see AIC 29/2004 (Pink 34) – 'Engine Malfunction Caused by Lightning Strikes'.

2.7.2 The brilliant flash, the smell of burning and the accompanying explosive noise may be alarming and distracting to the pilots of an aircraft struck by lightning. The report on a serious accident, in which a large transport aircraft was destroyed, stated that it was due to a lightning strike causing ignition of vapour in the region of fuel tank vents but fatal accidents due to lightning strikes have fortunately been very few and most aircraft receive only superficial damage when struck.

2.7.3 The effect of lightning strikes upon both direct reading magnetic compasses and magnetically slaved compasses can be severe with deviations of many tens of degrees having been recorded. Magnetic compasses should not be relied upon after an aircraft has been struck and should be checked as soon as possible.

2.8 Static Electricity

2.8.1 This phenomenon will generally first be noticed as noise on the High and Medium frequency radio bands and also, to a lesser extent, on VHF receivers. As the static electricity increases in severity, the noise will increase and in extreme cases a visible discharge, known as St Elmo's fire, will be seen on some parts of the aircraft, particularly around the edges of windscreens. Static electricity is not associated only with thunderstorms but such conditions are particularly favourable to its creation. Although it is not normally dangerous, there have been rare incidents when a static discharge has occurred across a windscreen or plastic panel causing it to break.

2.8.2 An understanding of the effect of static electricity on radio equipment is important. It is detrimental to the performance of MF (eg ADF) and HF equipment but has little or no effect upon VHF and UHF. On HF, static may cause the signal-to-noise ratio to be such that communications are impossible. In these conditions navigational aids such as ADF must be used with extreme caution due to the fluctuating or erroneous indications that may occur.

2.9 Instrumental Errors and Limitations

2.9.1 Altimeters and Vertical Speed Indicators

2.9.1.1 Local pressure variations can occur in or very close to a thunderstorm at all heights and this, together with local gusts, may give rise to errors in the indications of altimeters and vertical speed indicators. There is some doubt as to the magnitude of altitude errors but there is evidence that they can be as much as ± 1,000 ft. It is essential, for ground clearance purposes, that due allowance is made for such errors when flying in or near thunderstorm areas. Near the surface, periods of heavy rain are an indication of the likelihood of pressure variations and gusts.

2.9.2 Airspeed Indicators

2.9.2.1 Despite the precautions taken in the design of pitot heads, there is still a possibility that very heavy rain may cause an airspeed indicator to give a false indication even when the pitot head heaters are used. If the power which gives the safest speed for penetration has been selected before a storm is entered, no action should be taken to correct for violent or short period airspeed indicator oscillations, provided a reasonably level attitude is maintained.

2.9.3 Attitude Indicators

2.9.3.1 Attitude is indicated by instruments presenting pitch and roll information alone or by other more complex flight directors containing attitude indication amongst other elements.

2.9.3.2 The simple artificial horizons fitted to most aircraft, either as the main attitude indicator or as a standby instrument when remote reading indicators are installed, provide indications of pitch angle up to 85° nose up and down and may have complete freedom in the rolling plane. Except in rare circumstances these instruments give an adequate range of indication but may lack referencing, which would enable the pilot to assess attitude accurately at large angles of pitch or be given maximum assistance in recovery from any unusual attitudes.

2.9.3.3 Pitch referencing is also lacking on the attitude indicators of some flight director presentations. Moreover, their range of indication is much less than 85° up and down, in some cases less than 30°. The presentation of information on these earlier instruments does not give an indication to the pilot of the point at which the aircraft's pitch attitude exceeds the limit of indication of the instrument. These instruments therefore give no guidance as to the progress of recovery from attitudes outside their normal range of indication.

2.9.3.4 The wide variety of instruments which may be encountered makes it essential that pilots are fully aware of the limitations of the particular attitude indicator(s) fitted in the aircraft they fly.

2.9.4 Magnetic Compasses

2.9.4.1 Magnetic compasses are likely to be seriously affected by a lightning strike. They should not be relied upon after an aircraft has been struck and should be checked as soon as possible.

2.10 Use of Weather Radar

2.10.1 Pilots should be in no doubt about the function of airborne weather radar. It is provided, principally to enable them to AVOID thunderstorms although they can be of assistance in penetrating areas of storm activity, where avoidance has not been possible. **However, pilots should also be aware of the potential for displayed data to be unreliable when used for calculating the safe vertical clearance for the overflight of active storm cells.**

2.10.2 Pilots should be familiar with the characteristics, and operation of the radar in their aircraft and its limitations. Operators should ensure that their crews are given adequate instructions in relation to the radar equipment fitted to its aircraft, including the operation of the antenna and radar controls and on the adjustment and interpretations of the display.

2.10.3 It should be noted that the subject of airborne weather radar is quite complex and whilst the following notes give a generalised overview, they are no substitute for manufacturers' instructions in relation to specific products.

(a) Most modern airborne weather radars operate in the frequency band of 8-12 GHz (ie wavelengths between 2.5 and 4 cm). This band, sometimes known as the 'X' band was chosen for weather radars as it is highly sensitive to wet precipitation which is a feature of most weather systems that might need to be avoided by pilots. Airborne weather radars do not detect turbulence, although turbulent air, particularly within a thunderstorm, often contains water. In some radars, a change in frequency (a Doppler shift) in the reflected (returned) radar signal caused by moving precipitation is measured and is used to give an indication of likely turbulence.

(b) Although wet precipitation is the most reflective of radar signals, other water products will reflect lesser amounts of incident radar energy. In descending order (ie from most to least reflective) these are: wet hail, rain, hail, ice crystals, wet snow, dry hail and dry snow.

(c) The intensity of the returned radar signal will also be affected by the range of the aircraft from the precipitation, the amplification of signal (gain) being used by the receiver and the aerial tilt setting.

(d) It should be noted that, with weather radars, the significance of radar returns of given intensity usually increases with altitude, but the strength of the echo is not an indication of the strength of any associated turbulence.

(e) Radar return intensities may also be misleading because of attenuation resulting from intervening heavy rain. This may lead to serious underestimation of the severity of the rainfall in a large storm and an incorrect assumption of where the heaviest rainfall is likely to be encountered. The echo from that part of an area of rain furthest from the radar will be relatively weaker, and the actual position of the maximum rainfall at the far edge of the storm area will be further away than indicated on the radar display, sometimes by distances up to several miles. Additionally, a storm cell beyond may be completely masked.

(f) It should also be noted that, notwithstanding recent research and operational experience, it still seems impossible to use radar to detect with certainty areas where large hailstones exist, because clouds containing rain or hail can produce identical radar pictures. Some operators have claimed success in avoiding hail by keeping well clear of cloud echoes that have scalloped edges or pointed or hooked 'fingers' attached. The best advice is to give radar echoes a wide berth, when detouring storms visually.

(g) The high rate of growth of thunderstorms and the danger of flying over or near to the tops both of the main storm and the small convective cells close to it must also be remembered when using weather radar for storm avoidance.

(h) Some guidance on the distances by which thunderstorms should be avoided is given in the table below. It is strongly recommended that the decision to avoid a thunderstorm be taken early.

(i) Where weather information is available from ATC radar, it should be used to supplement the aircraft's weather radar (but see paragraph 3 (c) of this Circular).

3 THUNDERSTORM AVOIDANCE GUIDANCE – WEATHER RADAR

Flight Altitude (ft)	Echo Characteristics			
	Shape	Intensity	Gradient of Intensity*	Rate of Change
0 – 20,000	Avoid by 10 miles echoes with 'hooks', 'fingers', scalloped edges or other protrusions from the main storm return.	Avoid by 10 miles echoes with sharp edges or strong intensities.	Avoid by 10 miles echoes with strong gradients of intensity.	Avoid by 10 miles echoes showing rapid change of shape, height or intensity.
20-25,000	Avoid all echoes by 20 miles.			
25-30,000	Avoid all echoes by 20 miles.			
Above 30,000	Avoid all echoes by 20 miles.			

* Applicable to sets with Iso-Echo or a colour display. Iso-Echo produces a hole in a strong echo when the returned signal is above a pre-set value. Where the return around a hole is narrow, there is a strong gradient of intensity.

3.1 The above avoidance criteria can be simply summarised as: if above 20,000 ft avoid by a minimum of 20 nm; if below avoid by a minimum of 10 nm.

3.2 If storm clouds have to be overflown, always maintain at least 5,000 ft vertical separation from cloud tops. It is possible to estimate this separation (using the principle outlined below), but ATC or Met information on the altitude of the tops may also be available for further guidance:

(a) To ensure that the optimum radar beam is used for this purpose, it will be necessary to adjust the 'gain' control. One particular weather radar manufacturer recommends that with an aircraft in straight and level flight and the aerial tilt set to zero (ie with the center of the weather radar beam (ie along the bore-sight) aligned to the horizontal) the gain should be reduced until the 'radar paint' from the clouds just disappears. The gain should then be increased until a 'solid paint' is produced and the gain left at this setting for the required measurement. The range of the nearest part of this 'paint' should then be recorded.

(b) The beam should now be raised (by adjusting the aerial tilt upwards) until the return 'disappears'. The tilt angle associated with this disappearance should be recorded. Return the tilt to zero in order to continue to monitor the storm and its development and the separation of the aircraft from it. Then either using data provided by the radar manufacturer (as in a 'look-up' table) or by mental arithmetic the approximate height of the cloud top may be obtained. One method of approximation is as follows:

One half of the notional beam-width as quoted by the radar manufacturer (usually in the region of 3° to 4°) should be subtracted from the recorded angle of tilt. Then using the 1:60 rule, this remainder should be applied to the recorded range of the edge of the return to calculate the height (in nautical miles) that the cloud top is above the aircraft.

Example: In an aircraft at 20,000 ft, with a radar whose notional beam width is 4°, a cloud return at 45 nautical miles is made to 'disappear' at an aerial tilt angle of + 3.5°. 3.5 minus 2 (ie fi the notional beam-width) = 1.5, which, when applied to 45 miles using the 1:60 rule, indicates that the cloud tops are 2 nautical miles (or 12,000 ft) above the aircraft. Thus if the cloud is to be overflown with the minimum recommended clearance, of 5,000 ft, a climb to at least 37,000 ft is indicated. If this course of action is followed, do remember that in the finite time it will take the aircraft to climb and to close this distance, the top of the storm cloud itself, if very active, might easily have ascended to a higher altitude.

(c) If the aircraft is not equipped with radar or it is inoperative, avoid by at least 10 miles any storm that by visual inspection is tall, growing rapidly or has an anvil top.

(d) Intermittently monitor long ranges on radar to avoid getting into situations where no alternative remains but to penetrate possibly hazardous areas. Unless otherwise instructed by the radar manufacturer, it is usually necessary to adjust both 'gain' and 'tilt' during this monitoring process to ensure that new weather 'targets' are not missed and that active clouds are continually tracked.

(e) Avoid flying under a cumulonimbus overhang. If such flight cannot be avoided, tilt antenna full up occasionally to determine, if possible, whether precipitation (which may be hail) exists in or is falling from the overhang.

(f) Notwithstanding the principle outlined above, or other guidance provided by radar manufacturers, or by instructors in radar systems, it should always be borne in mind that the result is only an estimate of the height of the storm cloud tops and that the accuracy of the estimate is critically dependant on certain assumptions. These assumptions include radar handling (eg that the beam width in actual use is similar to the quoted notional beam width; and that the tilt control knob has not slipped on the spindle). It should be remembered that weather radars are provided primarily for storm avoidance not penetration or overflight.

4 USE OF INFORMATION FROM A LIGHTNING DISCHARGE MONITOR

4.1 Instruments are available which indicate and record lightning discharges. However, in a similar manner to that of airborne weather radar they should be used for storm avoidance and not penetration. They work on the principle that in mature thunderstorms, air turbulence has changed the normal distribution of charged particles such that large build-ups of electrical charge occur. Lightning dissipates these build-ups. These lightning discharges are detected by the equipment and normally shown on a screen with its centre that of the aircraft. The displayed distance of the discharge from the screen centre is an indication of the strength of the lightning discharge; it is not the actual range of the discharge from the aircraft. The distance calculated uses an algorithm based around the average strength of lightning discharges. Thus, a high power discharge at long range will be displayed at the same distance as a low power discharge at short range.

4.2 Because lightning is more likely to be associated with the most severe turbulence, an area of frequent discharges in a particular direction should be avoided. However, it has been found that the first lightning discharges from recently formed cells (where no discharges have been evident beforehand) may be particularly strong (ie violent). Thus, the lack of an indication of discharge is no guarantee that lightning will not strike. One particular manufacturer recommends that pilots using his equipment should manoeuvre their aircraft such that all discharge clusters are kept at least 25 nautical miles away.

AIC 93/2004 (Pink 68) 16 September

VHF RADIO TELEPHONY EMERGENCY COMMUNICATIONS

1 Need to Communicate at an Early Stage

1.1 Pilots are urged to request assistance from an Air Traffic Service (ATS) unit as soon as there is any doubt about the safe conduct of their flight. The ATS unit will then be better placed to offer guidance and information that will expedite the passage of the aircraft to an aerodrome where it may land safely.

1.2 No ATS unit will know that an aircraft is in difficulty unless this information is communicated in terms that make the situation immediately and clearly apparent.

1.3 The extent to which the ATS unit will be able to offer assistance will depend both on the amount of information that the pilots provide and on it being transmitted at the earliest moment following the realisation that a potentially hazardous situation has arisen, or is in the process of developing.

2 Procedures for use by Pilots

2.1 **Pilots should give thought, as soon as possible after they recognise that a problem has occurred, to declaring to an ATS unit that a hazardous situation has arisen or could arise.**

2.2 **The correct method of communicating this information to ATC is by using the prefix 'MAYDAY, MAYDAY, MAYDAY' or 'PAN PAN, PAN PAN, PAN PAN' as appropriate.** This procedure, which is an international standard, is the single most effective means of alerting the controller to the need to give priority attention to the message that will follow.

2.3 Reluctance by pilots to use the prefix 'MAYDAY' or 'PAN' in situations when either might be appropriate could introduce ambiguity and deny them information that could otherwise improve their situational awareness, thus helping them decide on the best course of action to follow. Air Traffic Controllers will have information available to them that might not be known on the flight deck and will offer this to the crew, but only when they have been told that an emergency exists.

2.4 If pilots do not use the prefix 'MAYDAY' or 'PAN' before transmitting details of technical or procedural difficulties, the controller may ask the pilot whether or not he/she 'wishes to declare an emergency', since the response he/she receives will, thereafter, directly affect the manner in which that aircraft is handled. However, pilots should not rely upon ATC to interpret messages without either prefix as being indications of distress or urgency but, rather, should use 'MAYDAY' or 'PAN' as the means most likely to produce immediate assistance.

2.5 If, subsequent to the transmission of 'MAYDAY, MAYDAY, MAYDAY' or 'PAN PAN, PAN PAN, PAN PAN', the nature of the emergency changes to the extent that the pilots consider the problem appears not to have been as serious as was first thought and they no longer wish to receive priority attention, they may, at their discretion, cancel the emergency condition using procedures specified in the Radiotelephony Manual (CAP 413).

2.6 **The Civil Aviation Authority would prefer that pilots believing themselves to be facing an emergency situation should declare it as early as possible and cancel it later if they decide that the situation allows.**

3 Use of Air Traffic Services

3.1 CAP 413 contains full details of the VHF International Aeronautical Emergency Service and how, when flying in UK airspace, pilots should inform an ATS unit of the emergency (ie use of transponder codes, frequencies, prefixes). CAP 413 also provides examples of distress messages from an aircraft and the response that can be expected from an ATS unit.

3.2 In addition to the services provided on the VHF International Emergency Service frequency of 121.500 MHz, pilots should be aware that all ATC units within the United Kingdom provide an alerting service for aircraft in emergency. **Controllers will offer as much assistance as possible to any aircraft considered to be in an emergency situation. Assistance can include the provision of information on the availability of aerodromes and their associated approach aids, directional guidance, weather information and details of terrain clearance.**

3.3 Where a controller considers that another ATS unit may be able to give more assistance and that in the circumstances it is reasonable to do so, pilots may be asked to change to another frequency. In this event, pilots should ensure that the appropriate prefix is included in the initial message to the new ATS unit whilst the emergency state continues to exist. If communication cannot be established on the new frequency, pilots should revert immediately to the transferring controller.

3.4 The advice given above is intended to emphasise to **all pilots the necessity to make use of Air Traffic Services as soon as it becomes apparent that assistance might be needed.** It does not replace any of the contents of CAP 413.

AIC 106/2004 (Pink 74) 11 November

FROST, ICE AND SNOW ON AIRCRAFT

1 Introduction

1.1 The purpose of this Circular is to emphasise the hazards of aircraft icing to all classes of aeroplanes and helicopters and to remind those concerned of the importance of thorough pre-flight preparation and of the appropriate instructions in relevant publications. The Flight Manual, Operations Manual and the Maintenance Manual contain basic requirements concerning the use and maintenance of anti-icing and de-icing equipment, which must be strictly adhered to.

1.2 All procedures and associated precautions should be checked before each winter, ideally as part of a pre-planned Winter Operations refresher or training programme.

1.3 Nevertheless, icing can occur during flight at any time of the year, and given the global nature of UK air transport operations, frost, ice and snow may be found at times which do not correspond to the European winter period. Therefore the advice, information and instructions contained in this Circular will always be relevant. It should be remembered that any contamination of aircraft aerofoil surfaces will adversely affect performance and handling and even small amounts can have disastrous consequences. It is therefore important that any flight in icing, or potential icing conditions is fully in accordance with the icing clearance of the aircraft.

1.4 The Annex lists some reported incidents and accidents in which icing or other cold weather hazards have been causative factors. These have been included as reminders of the hazards associated with winter period operations and in some cases illustrate the subtle onset of danger.

2 Pre-Flight Preparation

2.1 The aircraft should be free from deposits of frost, ice and snow. When necessary use a de-icing fluid to achieve effective removal of any frost or ice and to provide a measure of protection against any further formation. Only fluids approved for the purpose should be used. The efficiency of the fluid under varying atmospheric conditions is dependant upon the correct mixture strength and methods of application, which must be strictly in accordance with recommended procedures. For example, using fluid diluted with water will effectively remove ice; however, its ability to prevent further ice formation will be significantly reduced. Under certain conditions the fact that the aircraft surfaces have been made wet will actually enhance further accumulation, leading to a dangerous situation if there is a considerable delay between de-icing and take-off. Poor decision-making as to whether or not to de-ice/anti-ice is probably the single biggest cause of icing-related fatal accidents. The importance of making a safe decision cannot be over-emphasised.

2.2 In the absence of adequate advice on approved fluids in the aircraft maintenance manuals, guidance on mixture strength and methods of application should be sought from the aircraft manufacturer and the supplier of the fluid. Ensure, in particular, that aerofoil sections, or all rotor blades, have received similar treatment. After removal of ice, if precipitation is present or conditions are conducive to frost formation, a fluid with anti-icing properties should be applied and the appropriate holdover time guidelines applied. **Note:** Fluids flow off the aircraft before rotation and provide no in-flight ice protection.

2.3 Fit engine blanks and pitot/static covers before de-icing as required. Particular attention should be paid to all leading edges, control surfaces, flaps, slats, their associated mechanisms, hinges and gaps. Excessive use of any non-volatile, viscous de-icing fluids on control surfaces may create out of balance problems by increasing the weight of the surfaces to the aft of the hinge points. Deposits left in operating mechanisms, hinges and gaps may re-freeze during flight and jam controls. Some de-icing fluids can also dilute or wash away essential lubricating greases.

2.4   Ensure that all orifices and guards (eg generator cooling inlets, fuel vents, APU inlets, pressurisation inlets and outlets, static plates, helicopter snow guards, etc) and exposed operating mechanisms (eg nosewheel steering, emergency door and window locks, helicopter rotor heads, etc) are cleared of snow or slush and de-iced when so recommended. As the ingress of moisture, snow, rain and slush to door seals is more likely to occur when the doors are open, the time that they are open should be kept to the minimum practicable and a check made for contamination prior to departure.

2.5   Ensure that the implications of cockpit indications are fully understood. For example, the cockpit indications in relation to an inflatable boot de-icing system may not necessarily confirm the actual operation of the system. Verify, by visual inspection, or by other recommended independent means, the satisfactory operation of any de-icing and/or anti-icing systems.

2.6   Damage to inflatable boots can result in leaks, which will prevent full inflation and significantly reduce their effectiveness and/or allow the ingress of moisture which may subsequently freeze. Their location on leading edges makes them particularly susceptible to damage by erosion, impact or contact with ground equipment. Winter conditions invariably increase the risk of such damage and the importance of regular inspection and functional testing cannot be over-emphasised.

2.7   Ice may block the pitot static system, or melting ice or heavy rain may cause water to enter the pitot static system causing lost or intermittent indications of airspeed. On more integrated fly by wire aircraft, erroneous airspeed indications may cause other system failures also. Make sure you are familiar with how to check pitot heat if it is listed as a crew task in the walk around inspection. Make sure you understand the abnormal checklist procedures associated with loss of air data, locating alternate static sources, etc.

3   Start-up, Taxi, and Take-off Precautions

3.1   Operations to or from runways contaminated with snow, slush or water, by all classes of aeroplane, should be avoided whenever possible. Such operations involve a significant element of risk. The wisest course of action would be to delay departure until conditions improve or, if airborne, and bad conditions are broadcast, divert to a more suitable airfield.

3.2   On some types of engines, icing of engine pressure probes can cause an over-reading in instruments used to indicate engine power delivery. To minimise this possibility and thus of damage to, or flame-out of, the engine, engine anti-icing should be switched on if icing conditions are present or possible. In the absence of Flight Manual, Operations Manual or Pilots Operating Handbook guidance, engine icing can be assumed to be possible if the OAT is less than +10°C and the RVR is less than 1000 m or there is precipitation or standing water. Use carburettor heat and propeller de-icing as appropriate.

3.3   During taxiing in icing conditions, the use of reverse thrust on engines in pods should be avoided, as this can result in ice contamination of leading edges, slats and other flight critical devices. For the same reason, a good distance behind the aircraft ahead should be maintained. In no circumstances should an attempt be made to de-ice an aircraft by positioning it in the jet efflux of the aircraft ahead.

3.4   Before take-off ensure that the wings have remained uncontaminated by ice or snow. Operate the appropriate fuel, propeller, airframe, carburettor, and engine anti-icing or de-icing controls, before, during and after take-off in accordance with the Flight Manual, Operations Manual or Pilots Operating Handbook. Take-off power and aircraft performance should be monitored on more than one instrument.

4 In-Flight Precautions

4.1 The build-up of ice in flight may very rapidly degrade aircraft performance and controllability and pilots should avoid icing conditions for which their aircraft are not approved. Crews should check regularly and thoroughly for the build-up of ice. The instructions in the Flight Manual, Operations Manual or Pilots Operating Handbook, concerning the use of anti-icing and de-icing systems must be followed.

4.2 On some types of aircraft, when ice is present on the tailplane, the action of flap lowering may cause a reduction of longitudinal control. When this happens, the tailplane can stall with a subsequent loss of control from which it may not be possible to recover in the time and height available. Allowing the speed to increase with the flaps extended may also increase this risk of tailplane stall. If longitudinal control difficulties are experienced and it is suspected that ice may have formed on the tailplane, it may be prudent not to lower flaps, or to immediately change the flap setting. It is important that this condition, caused by icing, is not confused with normal pitch changes associated with flap selection or de-selection. Before a decision is taken to carry out a flapless or partial flap landing, the landing performance aspects of such a decision must be assessed. Consideration should be given to diverting to a more suitable aerodrome if runway length and/or condition are limiting factors.

4.3 Operation of anti-icing or de-icing equipment can affect performance and fuel consumption. These effects must be allowed for if flight in icing conditions is planned or possible.

4.4 If pilots or operators have any doubts about the susceptibility of their aircraft to any type of control problem when icing is involved, they should consult the Manufacturer or the Design and Production Standards Division of the Civil Aviation Authority.

5 General Aviation

5.1 This Circular and additional documents listed at paragraph 6 (a), (b), and (c) are applicable to all classes of aeroplane; more specific guidance relating to turbo-prop aeroplanes is at paragraph 6 (d) and that for general aviation is contained in paragraphs 6 (e) and (f).

6 Additional Documents

6.1 Further guidance on specific subjects is contained in the following publications:

(a) AIC 61/1999 (Pink 195) – Risks and Factors Associated with Operations on Runways Affected by Snow, Slush or Water;

(b) CAP 512 – Ground De-icing of Aircraft; www.caa.co.uk > publications > Flight Operations;

(c) AIC 105/2003 (Pink 61) – Recommendations for De-icing/Anti-icing of Aircraft on the Ground;

(d) AIC 98/1999 (Pink 200) – Turbo-prop and Other Propeller Driven Aeroplanes: Icing-Induced Stalls;

(e) AIC 145/1997 (Pink 161) – Induction System Icing on Piston Engines as Fitted to Aeroplanes, Helicopters and Airships;

(f) Winter Flying – General Aviation Safety Sense Leaflet 3C; www.caa.co.uk > publications > General Aviation;

(g) Ice Aware CD-ROM.

Note: A copy of the Ice Aware CD-ROM can be obtained free of charge from the Civil Aviation Authority, Flight Operations Department Admin Section, Tel: 01293-573450 or Fax: 01293-573991.

United Kingdom Aeronautical Information Circular   AIC 106/2004 (Pink 74) 11 November

ANNEX

Frost, Ice and Snow on Aircraft

Incidents and Accidents with Frost, Ice or Snow as a causative factor

1   Frost, Ice and/or snow on aircraft will adversely affect performance and handling and even small amounts have had disastrous consequences. Accidents and incidents have been caused by:

(a) Ice build-up on engine inlet pressure probes causing erroneous indications of engine power;

(b) a thin layer of ice on control surfaces inducing flutter and consequent structural damage;

(c) severe tailplane icing leading to a loss of control on selection of landing flap;

(d) very small deposits of ice on wing leading edges dangerously eroding performance;

(e) windscreens being obscured by snow when operating with an unserviceable heater, leading to a loss of directional control on take-off;

(f) attempting a take-off with wet snow on the wings and tailplane surfaces which had accumulated after earlier de-icing with diluted fluid;

(g) snow/slush on helicopter upper fuselage surfaces entering engine intakes after engine start causing flame-out and engine damage;

(h) engine breather pipes freezing;

(i) inability to open doors after a successful landing. (Although to date such occurrences have not resulted in serious consequences, these conditions could be extremely hazardous in an emergency situation). This problem has been caused by external coverings of ice; ice in locking mechanisms, hinges and seals; and freezing moisture in pressure locking systems;

(j) non-use of engine igniters in potential icing conditions which, in conjunction with other factors, contributed to a double engine failure and consequent forced landing;

(k) very low ambient temperatures at high altitude resulting in apparent fuel freezing leading to subsequent multiple engine rundown, in spite of application of fuel heating systems. (Given a sufficiently long exposure time to low ambient air temperatures, fuel will eventually cool to this temperature which can be well below the freezing point of the fuel). Pilots should therefore be aware of the freezing points of their specified fuels and/or the operational limitations of these fuels and plan accordingly. There are certain aircraft types, including those with piston engines, where the use of special fuel anti-freeze additives are specified as being mandatory in certain conditions;

(l) contamination of retractable landing gear, doors, bays, micro-switches by snow; wet mud or slush. Any contamination should be removed before flight;

(m) carburettor icing. (However, it should be emphasised that this particular problem is not confined to winter operations);

(n) wing upper surface icing due to very low fuel temperatures. Such ice is usually clear and very difficult to detect visually. In addition to any aerodynamic effects caused by this contamination of wing surfaces, there is a potential serious hazard to rear engine aircraft if this ice breaks off. This will often occur during take-off and in such cases ice ingestion and turbine damage have occurred. Typical factors favouring the formation of such ice include:

(i) Low temperature of up-lift fuel close to departure;

(ii) previous long flight times in low ambient temperatures resulting in fuel being cold-soaked to 0°C or below and subsequent cooling of the wing surfaces either by the fuel itself or by conduction from surfaces in contact with the cold fuel. If this is coupled with ambient ground conditions involving high humidity, drizzle, rain or fog in conjunction with temperatures in the range of 0°C to +10°C, ice will form. (However, it should be noted that ice formation has been reported in drizzle and rain even in temperatures between +8°C and +14°C). When carrying out a physical check of the wing upper surface in such conditions it must be borne in mind that this ice may have formed below a layer of slush or snow thus compounding the detection and removal problem.

(o) a twin-engine aeroplane landing in winter conditions experienced a significant wing drop accompanied by a nose-up pitch. Despite application of power and full opposite aileron and rudder the aircraft was slow to recover and the wing tip struck the ground. Control was regained and a safe landing made. Although no ice was seen during a visual check of the wing surfaces prior to landing, the aircraft had been operating all day in icing conditions and prior to this flight had been delayed on the ground in rain conditions for 40 minutes;

(p) a twin-engine aeroplane stalled at an IAS considerably above the basic stall speed and at a much lower than normal angle of attack; the approach to the stall was so insidious that the pilot was unaware that the aircraft had stalled. The pilot did not have the expected visual cues on the rapid accretion of ice and the action of the autopilot in correcting for the aerodynamic effects of the accretion was to actually drive the aircraft further into the stall configuration. Heavy stall buffeting, which was mistaken for propeller icing caused the pilots difficulty in reading instruments. The temperature was much warmer than usual and large water droplets were present;

(q) a twin-engine aeroplane stalled on the approach to an airport, probably as a result of becoming uncontrollable at a speed well above its stalling and minimum control speeds. It was deduced that its handling and flying characteristics had been degraded by ice accumulation;

(r) another twin-engine aeroplane suffered a double engine failure, possibly as a result of ice ingestion. There have been a number of reported flame-outs from this cause, most of which have been suspected as being due to either late or non-selection of engine icing protection systems.

AIC 17/1999 (Pink 188) 25 February

WAKE TURBULENCE

1 Introduction

1.1   Attention is drawn to the dangers associated with turbulence caused by aircraft wake vortices.

1.2   This circular is intended for all airspace users and ATC service providers. This circular states aircraft weight categories used in the UK and in particular gives details of aircraft types in a revised category used at London Heathrow, London Gatwick, London Stansted and Manchester Airports. The circular also re-states the general warning on wake vortex characteristics and illustrates a number of wake vortex avoidance techniques, together with details of current and future research into wake vortex turbulence problems. The suggested wake vortex avoidance techniques are to be considered where appropriate to the aviation community in general and will inevitably be more applicable to operators of smaller aircraft. It is intended, however, that all operators will benefit from this information.

2 Wake Vortex Weight and Spacing Criteria

2.1   The United Kingdom conforms to the ICAO requirements but with certain modifications to the weight and spacing relationship which experience has shown to be advisable for the safety of operations at UK aerodromes.

2.2   Since 1982 the differences between the UK and ICAO criteria have been as follows:

(a) A modification to the Medium and Light weight thresholds and the introduction of a separate category (Small) for spacing purposes (see paragraph 2.4);

(b) some enhancement of the separation minima between certain categories (see paragraph 2.5);

(c) re-classification of the B707, DC8, VC10 and IL62 series of aircraft from the Heavy to the Medium category. This is a special case as experience has shown that these aircraft types conform more to the Medium weight group;

(d) in 1997 a further modification was made for the purposes of spacing in the approach phase at selected airports (London Heathrow, London Gatwick, London Stansted and Manchester), the following aircraft types are classified as Upper Medium for Wake Vortex purposes: B757, B707, DC8, VC10 and IL62. All other 'Medium' aircraft types are classified as Lower Medium at these airports.

2.3   Any differences between ICAO and UK criteria will not affect the composition of flight plans that should be completed in accordance with the Twelfth Edition of PANS RAC. For example, aircraft weight should be entered as H, M or L according to the ICAO Weight Turbulence Categorisation (see paragraph 2.4). It should be noted that aircraft with a maximum take-off weight of 136,000 kg or greater are required to be announced as 'Heavy' on initial contact with an ATC Unit. In the special cases stated in paragraph 2.2(c), this will not be required within the UK as these aircraft types will be considered as Medium (or Upper Medium at selected airports).

851

2.4   Weight parameters (maximum take-off weight in kg)

Category	ICAO and Flight Plan	UK
Heavy (H)	136,000 or greater	136,000 or greater
Medium (M)	7,000 – 136,000	40,000 – 136,000
Small (S) (UK only)	N/A	17,000 – 40,000
Light (L)	7,000 or less	17,000 or less

2.4.1   The wake turbulence group of an aircraft should be indicated on the flight plan (Item 9) as H, M or L **according to the ICAO specification.** For purposes of spacing in the approach and departure phases within the UK, and regardless of the weight category as entered on the flight plan, aircraft 40,000 kg or less and more than 17,000 kg will be treated as Small. Aircraft of 17,000 kg or less MTWA will be treated as Light. Helicopters such as the Sikorsky 61N or larger will be treated as Small.

2.5   Wake Vortex Categories – Fifth Category – Upper Medium

2.5.1   For the purposes of spacing in the approach phase at the selected airports listed in paragraph 2.2(d) the B757, B707, DC8, VC10, IL62 will be treated as Upper Medium. The weight parameters of these aircraft types, and their corresponding wake turbulence separation minima, are listed in the appropriate Manual of Air Traffic Services (Part 2) of that airport and its associated approach/terminal control facility.

2.5.2   Although the UK increased the original ICAO three group scheme to four groups to provide an increased level of protection for certain aircraft types, the data that is available provides **no evidence that a fifth group is required for departing aircraft.**

2.6   Wake Turbulence Spacing Minima – Final Approach

Leading Aircraft	Following Aircraft	Spacing Minima Distance (nm)	
		ICAO	UK
Heavy	Heavy	4	4
Heavy	Medium	5	5
Heavy	Small	N/A	6
Heavy	Light	6	8
Medium	Heavy	3	Note 1
Medium	Medium	3	3 Note 2
Medium	Small	N/A	4
Medium	Light	5	6
Small	Heavy	N/A	Note 1
Small	Medium	N/A	3
Small	Small	N/A	3
Small	Light	N/A	4
Light	Heavy	3	Note 1
Light	Medium	3	Note 1
Light	Small	N/A	Note 1
Light	Light	3	Note 1

Note 1:   Signifies that spacing for wake vortex reasons alone is not necessary

Note 2: Where the leading aircraft is a B757, B707, DC8, IL62 or VC10, the minimum distance shall be increased to 4 miles.

These minima are also to be applied when an aircraft is operating directly behind another aircraft and when crossing behind at the same altitude, or less than 1,000 ft below.

2.7 Wake Turbulence Spacing Minima – Departures

Leading Aircraft	Following Aircraft	Minimum Spacing at the Time Aircraft are Airborne	
Heavy	Medium Small or Light	Departing from the same position	2 Minutes
Medium or Small	Light		2 Minutes
Heavy (Full length Take-off)	Medium Small or Light	Departing from an intermediate point on the same runway	3 Minutes
Medium or Small (Full length Take-off)	Light		3 Minutes

2.7.1 Wake vortex minima on departure are applied by ATC by measuring airborne times between successive aircraft. Take-off clearance may be issued with an allowance for the anticipated take-off run on the runway. This may result in a take-off clearance being issued at less than the prescribed time interval however the **airborne time** will accurately reflect the correct minima.

2.7.2 Pilots do, on occasion, request departure clearance before the minimum time spacing has elapsed. On these occasions ATC should apply the minima as prescribed in the unit instructions (and contained in this AIC) irrespective of request for reduced spacing. It is important for pilots to note that ATC do not have the discretion to reduce spacing minima.

2.8 Wake Turbulence Spacing Minima – Displaced Landing Threshold

2.8.1 A spacing of 2 minutes shall be provided between a Medium, Small or Light aircraft following a Heavy aircraft, and between a Light aircraft following a Medium or Small aircraft when operating on a runway with a displaced threshold when:

(a) A departing Medium, Small or Light aircraft follows a Heavy arrival or a departing Light aircraft follows a Medium or Small arrival;

(b) an arriving Medium, Small or Light aircraft follows a Heavy aircraft departure, or an arriving Light aircraft follows a departing Medium or Small aircraft;

(c) if the projected flight paths are expected to cross. (Fig 4 illustrates vortex generation in the landing and take-off phases of flight).

2.9 Wake Turbulence Spacing Minima – Opposite Direction

2.9.1 A spacing of 2 minutes should be applied between a Medium, Small or Light aircraft and a Heavy aircraft, and between a Medium or Small aircraft and a Light aircraft whenever the heavier aircraft is making a low or missed approach and the lighter aircraft is:

(a) Taking-off on the same runway in the opposite direction;

(b) landing on the same runway in the opposite direction;

(c) landing on a parallel opposite direction runway separated by less than 760 metres.*

* At Aerodromes where a grass strip is in use in addition to the runway(s), the strip will be counted as a runway for the application of vortex wake spacing minima.

2.10 Wake Turbulence Spacing Minima – Crossing and Parallel Runways*

2.10.1 When parallel runways separated by less than 760 metres are in use:

(a) Such runways are considered to be a single runway, for wake turbulence reasons, and the wake vortex minima listed in paragraphs 2.6 and 2.7 apply to landing and departing aircraft respectively.

* At Aerodromes where a grass strip is in use in addition to the runway(s), the strip will be counted as a runway for the application of spacing minima.

2.10.2 The Final Approach Minima listed in paragraph 2.6 will apply to:

(a) Departures from crossing and/or diverging runways if the **projected** flight paths will cross;

(b) departures from parallel runways more than 760 metres apart if the **projected** flight paths will cross.

2.11 Wake Turbulence Spacing Minima – Intermediate Approach

2.11.1 On intermediate approach a minimum wake turbulence spacing of 5 nm will be applied between a Heavy and a Medium, Small or Light aircraft following or crossing behind, if the following or crossing aircraft is at the same level. Controllers and pilots should be aware of the area up to 1,000 ft below and behind a Heavy aircraft, especially at low altitude, where even a momentary wake vortex encounter may be hazardous.

2.12 Application of Wake Turbulence Minima

2.12.1 **Where the separation minima required for IFR purposes is greater than the recommended spacing for wake turbulence, the IFR minima will apply.**

2.12.2 The recommended spacing listed are minima and when applied by ATC may be increased at the discretion of the controller, or at the request of the pilot. It is important to note that a pilot request for increased spacing **must** be made before entering a runway or commencing final approach. Requests made on the runway (or final approach) may result in a departure delay and/or an avoidable missed approach. It is stressed that where an aircraft has lined up on a runway and take-off clearance is issued, the aircraft must commence take-off without delay.

2.13 Probability of Vortex Wake Encounter

2.13.1 It must be emphasised that the spacing minima stated in this circular cannot entirely remove the possibility of a wake turbulence encounter (see paragraph 6 on research). The objectives of the minima are **to reduce the probability of a vortex wake encounter to an acceptably low level, and to minimise the magnitude of the upset when an encounter does occur.**

2.13.2 Care should always be taken when following any substantially heavier aircraft, especially in conditions of light winds. The majority of serious incidents, close to the ground, occur when winds are light.

2.13.3 Particular care should be exercised where the leading aircraft has followed the glide path on final approach from an extended range. Significant wake vortex encounters have been reported where the following aircraft is radar vectored and descended onto final approach behind a significantly larger aircraft. Although the correct spacing minima may be applied by ATC, caution should be exercised by pilots if there is a possibility of a vertical flight profile below that of a larger lead aircraft.

3 Aircraft Wake Vortex Characteristics

3.1 Wake vortices, are present behind every aircraft, including helicopters when in forward flight, but are particularly severe when generated by heavy aircraft. They are most hazardous to aircraft with a small wing span during the take-off, initial climb, final approach and landing phases of flight.

3.2 The characteristics of the wake vortex system generated by an aircraft in flight are determined initially by the aircraft's gross weight, wingspan, aircraft configuration and attitude. Subsequently these characteristics are altered by the interactions between the vortices and the ambient atmosphere. Eventually, after a time varying, according to the circumstances, from a few seconds to a few minutes after the passage of an aircraft, the effects of the vortex become undetectable.

3.3 For practical purposes, the vortex system in the wake of an aircraft may be regarded as made up of two counter-rotating cylindrical air masses trailing aft from the aircraft (Figs 1 and 2). Typically the two vortices are separated by about three quarters of the aircraft's wingspan. In still air they tend to drift slowly downwards and either level off, usually not more than 1,000 ft below the flight path of the aircraft, or, on approaching the ground, move sideways from the track of the generating aircraft at a height roughly equal to half the aircraft's wingspan (see Fig 3).

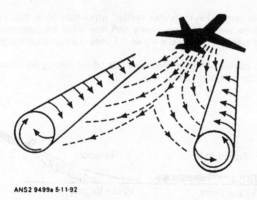

Fig 1   General view of aircraft trailing vortex system.

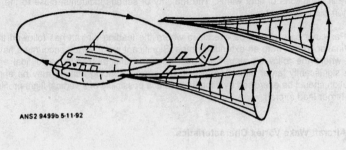

Fig 2   Helicopter Vortices.

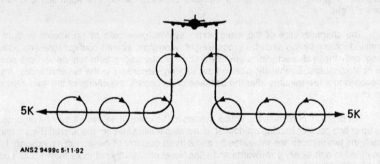

Fig 3   Vortex movement near the ground in still air, viewed from behind the generating aircraft.

3.3.1 The maximum tangential airspeed in the vortex system, which may be as much as 300 ft/sec immediately behind a large aircraft, decays slowly with time. After the passage of the aircraft the tangential airspeed eventually drops sharply as the vortex system disintegrates.

3.4   Wake vortex generation begins when the nosewheel lifts off the runway on take-off and continues until the nosewheel touches down on landing.

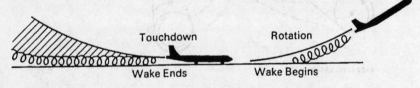

Fig 4   Vortex generation on take-off and landing.

3.4.1 Vortex strength increases with the weight of the generating aircraft. With the aircraft in a given configuration, the vortex strength decreases with increasing aircraft speed; and for a given weight and speed the vortex strength is greatest when the aircraft is in a clean configuration. There is some evidence that for given weight and speed a helicopter produces a stronger vortex than a fixed-wing aircraft.

3.5　In a stable airflow, the wake vortex system described in paragraph 3.3 will drift with the wind. Fig 5 shows the possible effect of a crosswind on the motion of a vortex pair close to the ground.

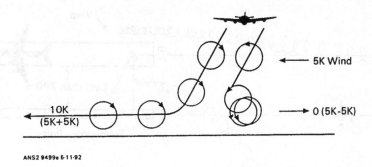

Fig 5　Vortex movement near the ground in a light crosswind, viewed from behind the generating aircraft.

3.5.1　Wind shear causes the two vortices to descend at different rates and, close to the ground, may cause one of the vortices to rise. Atmospheric turbulence and high winds close to the ground hasten the decay and disintegration of vortices. Special attention needs to be given to situations of light wind, when vortices may stay in the approach and touchdown areas of airports or sink to the landing or take-off paths of succeeding aircraft.

4　Wake Vortex Avoidance – Advice to Pilots

4.1　The wake of large aircraft deserves the respect of all pilots. The area up to 1000 ft below and behind such aircraft should be avoided, especially at low altitude where even a momentary wake vortex encounter could be hazardous. When an aircraft is at cruise speed, vortex may persist at considerable distances behind. By far the greater proportion of reported wake turbulence incidents occur in the approach. However, reports of wake vortex encounters do occur in the departure phase of flight. The separation standards listed in paragraph 2, are designed specifically for use in this area. Pilots who find themselves in the position of having to provide their own separation from large aircraft in the approach phase are reminded that the wake vortex spacings listed in paragraph 2 are minima. Where increased spacing is considered necessary, the pilot should inform ATC, where practicable, before joining final approach.

4.2　When the disposition of traffic is such that there appears to be the possibility of a wake vortex encounter, a wake turbulence avoidance manoeuvre of the type listed below may be utilised. Some of the situations represented are more likely to be encountered overseas than in the United Kingdom.

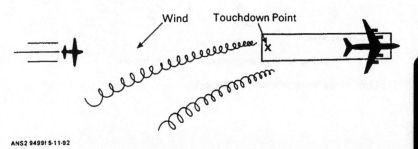

Fig 6　Landing behind a large aircraft on the same runway.

Stay at or above the large aircraft's final approach path. Note its touchdown point and land beyond it.

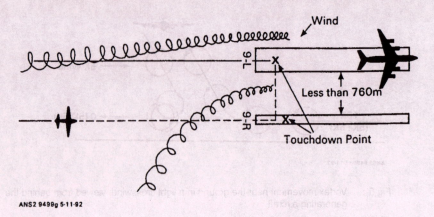

Fig 7   Landing behind a large aircraft on a parallel runway when the parallel runway is closer than 760 metres.

Consider possible drift to the runway. Stay at or above the large aircraft's final approach path and observe its touchdown point.

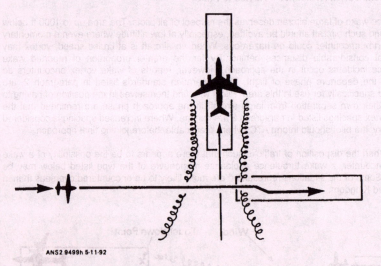

Fig 8   Landing behind a large aircraft – crossing runway.

Cross above the large aircraft's flight path.

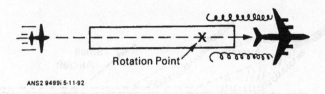

Fig 9   Landing behind a departing large aircraft – same runway.

Note the large aircraft's rotation point and land well before it.

Figs 10 and 11 Landing behind a departing aircraft – crossing runway.

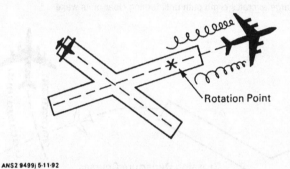

Fig 10   Rotation point beyond the intersection.

Note the large aircraft's rotation point. If it is past the intersection continue the approach and land.

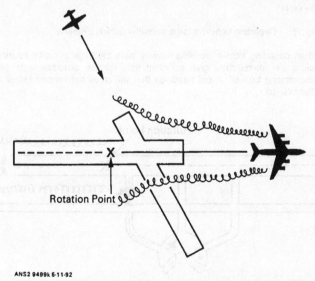

Fig 11   Rotation point prior to intersection.

If the large aircraft rotates prior to the intersection, avoid flight below its flight path. Abandon the approach unless a landing is assured well before reaching the intersection.

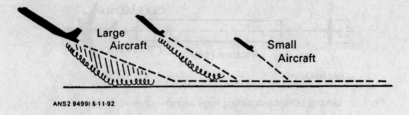

Fig 12   Departing behind a large aircraft – same runway.

Note the large aircraft's rotation point and rotate before it. Climb above and stay upwind of the large aircraft's climb path until turning clear of its wake.

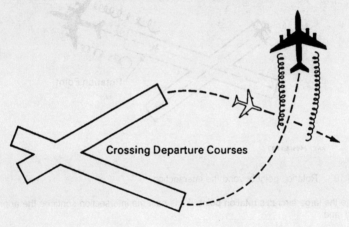

Fig 13   Departing behind a large aircraft – different runway.

When departing from a crossing runway, note the large aircraft's rotation point. If it is before the intersection, give sufficient time for the disturbance to dissipate before commencing take-off. Avoid headings that will cross behind and below a large aircraft after take-off.

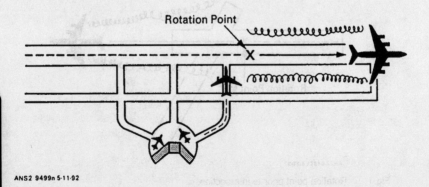

Fig 14   Take-off from an intersection along the same runway.

Be alert to adjacent large aircraft operations, particularly upwind of the runway.

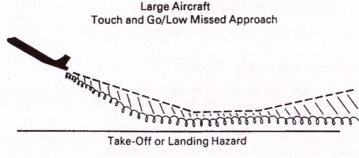

Fig 15   Departing or landing after a large aircraft executing a low missed approach or a touch-and-go landing.

A vortex hazard may exist for about 2 minutes along a runway after a large aircraft has executed a low missed approach or a touch-and-go landing, particularly in light quartering wind conditions.

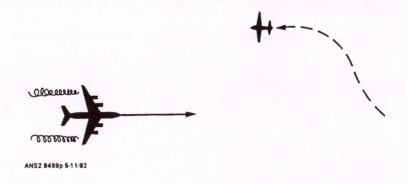

Fig 16   En-route in VMC.

Avoid flight below and behind a large aircraft's flight path. If a large aircraft is observed above on the same track, adjust position laterally, preferably upwind.

5   Helicopters

5.1   In forward flight the downwash from the main rotor(s) of a helicopter is transformed into a pair of trailing vortices similar to the wing-tip vortices of a fixed-wing aircraft (Fig 2). There is some evidence that, per kilogram of gross weight, these vortices are more intense than those of fixed-wing aircraft.

5.2   When hovering, whilst air-taxiing, a helicopter directs downwards a forceful blast of air that rolls outwards in all directions. This can create problems on the apron, in parking areas and to light aircraft movement on taxiway or runways. The risk of damage from this form of turbulence may be reduced if the guidelines listed below are followed:

(a) Whenever possible, ground taxi in a congested parking area rather than air taxi;

(b) if it is necessary to air taxi, ensure that as wide a clearance as possible is maintained from other aircraft or loose ground equipment;

(c) when air taxiing, avoid flying over parked aircraft or vehicles.

6 Research into Wake Vortex Turbulence

6.1 Since 1972, research into the problem of turbulence caused by aircraft wake vortices has included a detailed study of incidents reported by pilots, supplemented by information from Air Traffic Control and the Meteorological Office.

6.2 The reports received to date have been very valuable in obtaining a better understanding of the wake vortex problem. They allow an assessment of the effectiveness of the current standards in providing a satisfactory level of safety. The continued co-operation of pilots and controllers in making these reports is requested. Pilots are reminded that reports are required of incidents occurring behind any class of aircraft and during any phase of flight, ie en-route, climb, descent as well as the approach and take-off phases.

6.3 Pilots of aircraft believed to have created the wake turbulence will be informed by ATC and are requested to complete a report form. The meteorological data is particularly important.

6.4 Wake turbulence report forms are available from the Flight Briefing Unit (FBU) London Heathrow Airport. Verbal reports should be given to other ATC units in the UK and, where available, a report form should be provided. Completed forms should be sent to the address given on the form.

6.5 Reporting procedures for ATSUs are contained in the Manual of Air Traffic Services.

AIC 52/1999 (Pink 193) 6 May

GUIDANCE TO TRAINING CAPTAINS – SIMULATION OF ENGINE FAILURE IN AEROPLANES

1 General

1.1 The initial type rating test for pilots requires the ability to deal with an engine failure during or just after take-off. The majority of engine out training on large aircraft is now carried out in flight simulators. However, there is a continuing need for in-flight training either because simulators are not available or are not suitably qualified. Training Captains conducting this training/testing should have received formal training and be in possession of a TRI qualification. For instruction on aircraft this rating must have included an aircraft upgrade element and also, where appropriate, asymmetric training. This Circular is issued to give guidance to those pilots who may not have had the advantage of formal training and as a reminder to TRI/TREs in order to maintain the highest possible standard of safety.

2 Preparation for Flight

2.1 Thorough briefing of all crew members is essential. Particular emphasis should be placed on minimum heights and control speeds, the conduct of drills, the method of simulating engine failure and all aspects of the prevailing weather conditions. The use and effects of systems particularly relevant to asymmetric flight, such as auto-feathering and rudder boost, should be discussed in detail. Any performance data used should be verifiable.

2.2 Account must be taken of the possible effects on the circuit pattern caused by other traffic, particularly where aircraft with widely varying performance characteristics are using the same runway. Touch-and-go landings demand a particularly high level of crew co-ordination and the briefing should include precise details of the action to be taken by the trainee and Training Captain respectively in relation to the initiation of drills and the setting of throttles, flaps, airbrakes and other controls.

2.3 Practice asymmetric landings in twin-engined aircraft should always be to a full stop. It is inadvisable to touch-and-go because additional factors, particularly resetting displaced rudder trim, could result in a significantly increased ground roll and would be extremely hazardous to the subsequent take-off if overlooked.

2.3.1 Touch-and-go landings provide valuable approach and landing training but are often unrepresentative of normal take-offs (non-standard throttle handling, no V1, ASI bugs not set to take-off speeds etc). For these and other reasons, Training Captains on aeroplanes certificated in Performance Group A are strongly recommended not to conduct simulated engine failure exercises during touch-and-go take-offs. Such training should be confined to standing start take-offs for which performance data has been properly calculated and the crew have been suitably briefed.

2.4 When training is to be conducted away from the aerodrome of departure pre-flight planning should take into account the need for ready access to a diversion aerodrome. This is particularly important where engine shutdown is to be completed in a twin-engined aeroplane.

3 In-Flight Procedures

3.1 Engine failure during take-off below the minimum heights recommended at the Annexe to this Circular should be simulated only by reducing power and never by complete shutdown of the engine. The best method of simulation by power reduction will vary from one class of aeroplane to another. Detailed guidance is given in paragraph 6.

3.2 Immediately before failure is simulated, the Training Captain must position his feet so that he can prevent any application of wrong rudder by the trainee. During and after the simulation he must be particularly vigilant in monitoring airspeed, heading, pitch and roll attitude, rudder position and yaw indication. He must also carefully monitor engine instruments especially on those types of aeroplanes in which a genuine failure of the idling engine would produce an abnormal hazard. He must ensure that any recommended bank angle is correctly applied and after ensuring safe initial rudder application he should monitor the trainee's rudder input by resting his feet lightly on the rudder pedals. He should bring to the trainee's attention any tendency for flight parameters to move significantly from their target values.

3.3 The Training Captain must never allow the trainee to retain control if, due to an incorrect technique such as an exaggerated nose-up attitude, the airspeed is reducing towards minimum control speed. Only at a height which is known to be safe in relation to the control characteristics of the aeroplane should the Training Captain demonstrate, or permit the occurrence of, an actual loss of directional control to the extent that it is necessary to increase airspeed and reduce power in order to regain control.

3.4 When power failure is simulated during take-off, the speed should always be at or above V1 or TOSS and the Training Captain should assume control if there is any indication that action by the trainee is leading to a reduction below these speeds. If either a reduction of power or height loss is necessary for the retention of control the Training Captain must consider whether he simulated failure at too low a speed or whether he took over control too late.

4 Performance Considerations

4.1 Training Captains must only simulate engine failure on take-off in crosswind conditions when they are certain that the speed at which the simulated failure is initiated will, in the prevailing conditions, allow an adequate margin of control.

4.1.1 Certifications under JAR-25 make no allowance for crosswind components in the calculation of VMCG. As general guidance, therefore, Training Captains in aeroplanes so certificated should not simulate engine failure below the greater of:

    (a) V1

    and

    (b) VMCG incremented by 1 kt per kt of crosswind component (to a maximum of 10 kt).

If (b) is the greater value, V1 should **not** be increased but the engine failure initiated at the appropriate speed above V1.

4.1.2 The advice of manufacturers' training departments should be sought before engine failures are simulated in crosswind components greater than 10 kt, and in any case engine failures should never be simulated in crosswind components exceeding 15 kt, or on slippery or contaminated runways. Crosswind conditions make it difficult to monitor the trainee's rudder input and to correct any degree of wrong or inadequate movement.

4.2 Performance Group

4.2.1 A continued take-off following simulated engine failure during the ground run (after V1) should only be practised in aeroplanes certificated in Performance Group A. On no occasion should an engine failure before V1 be followed by a continued take-off. On aeroplanes in other Performance Groups any engine failure on the take-off roll must be followed by a rejected take-off.

5 Actual Engine Shutdown

5.1 For the majority of pilots an actual engine failure or pre-meditated shutdown is a rare event. It is particularly important therefore that operators allocate adequate time for refresher training in the relevant drills and procedures. Ideally this training should take place in conjunction with the biannual base check and Training Captains should emphasise the need to complete drills calmly and methodically and stress the importance of acting without haste. Where appropriate, the autopilot should be engaged. It should be remembered that during certification tests a reasonable allowance is made for pilot reaction time and that incomplete and over-hasty drills are known to have been the cause of a significant number of accidents and incidents. The importance of the methodical completion of drills cannot be over-emphasised.

5.2 Whenever refresher training involving shutdown is carried out in an aircraft in flight, Training Captains must be aware of the time required to restart an engine in the event of an actual failure of a second engine during an engine shutdown demonstration.

5.3 Practice engine shutdown for training purposes should only be carried out if icing conditions can be avoided throughout the exercise.

5.4 Recommended safe heights for practising exercises involving engine shutdown are detailed in the Annexe to this Circular.

6 Recommended Techniques for Simulating Engine Failure on Take-off

6.1 Turbo-jet and Turbo-fan Engines

6.1.1 The Flight Manual VMCG is established on the basis of an instant fuel cut occurring and when the procedure is practised in a flight simulator that method should be adopted. However, when practised in an aircraft, good airmanship requires smooth handling of all controls and the throttle should be closed at a rate commensurate with the engine's deceleration behaviour. It may then be advisable, with certain engines, to position the power lever slightly forward of 'idle' in order to reduce response time if subsequent acceleration of the engine should be required.

6.2 Turbo-prop Engines

6.2.1 The simulation of engine failure by throttling back can introduce particular handling and performance problems. The primary problem arises from the fact that a turbo-prop engine which has been throttled back to flight idle will produce very much more drag than an engine which has failed and auto-feathered. A further problem is that any automatic feathering or drag limiting devices fitted are usually made inoperative when the throttle is closed. Consequently, if an engine which has been throttled back to simulate failure suffers a real failure, it may go to a very high drag 'windmilling' condition, remaining unfeathered unless correct feathering action is taken by the crew. Furthermore, because the engine is in a low power condition, failure may not be noticed until after severe handling difficulties have arisen.

6.2.2 There will also be a reduction in performance which may well lead to decay in airspeed and an inability to maintain adequate clearance over obstacles. Any such loss in airspeed can of course contribute to the loss of directional control.

6.2.3 These potential problems can best be avoided by appropriate methods of simulating engine failure. Advice from engine or aircraft manufacturers specific to type should always be sought and followed, but where this is lacking the following general advice is likely to be appropriate:

> The throttle should be retarded smoothly towards a pre-determined torque setting approximating to zero thrust. This torque setting should be maintained during the remainder of the take-off and initial climb; if it falls due to a suspected malfunction the throttle should be realigned with that of the operative engine.

6.3 Piston Engines

6.3.1 Generally the throttle may initially be moved smoothly to the closed position; the mixture control or Idle Cut-Off should not be used to simulate engine failure. Reference to the engine manufacturer's recommendations should clarify the technique in particular cases. When the trainee has identified the 'failed' engine and completed his 'touch only' feathering drill the throttle should be advanced to the zero thrust position.

6.3.2 Training Captains should be familiar with the advice in AIC 130/1997 (Pink 153) 'Propeller Feathering on Twin Piston Engined Aircraft' before simulating engine failure in aeroplanes in this category.

7 In-flight Precautionary Power Reduction: Propeller-driven Aeroplanes

7.1 An abnormal condition affecting a powerplant of a multi-propeller engined aeroplane may lead to the Commander deciding not to shut the affected engine down but to make a precautionary power reduction. In this event a torque or manifold pressure value equal to or greater than the zero thrust setting for the speed and altitude should be set. Any lower power setting would result in a degradation in both handling and performance which may cause the aeroplane not to meet its certification standard; for example during a go-around or drift down.

7.2 Pilots should be given specific operational guidance in this context during conversion and recurrent training and provided with zero thrust setting data which takes account of variations in speed and altitude; also in temperature and configuration if known to be significant for a particular aeroplane type.

8 Rejected Take-Off (RTO) Training in Aeroplanes

8.1 When a flight simulator is not available for a specific aircraft type, crews must receive RTO training in the aeroplane within the bounds of reasonable safety.

8.2 The content of CAA Form 1179 (application for inclusion of aircraft types in pilot's licence) varies depending on aircraft type; many of these forms contain a test requiring RTO following a simulated engine failure. In the past this has often been accomplished by retarding one of the power levers, after which the candidate's ability to bring the aircraft to a standstill on the runway with adequate directional control is assessed. On a number of occasions aircraft have run off the marked runway areas; this is a particular hazard when the Training Captain is unable to exercise full control over the nosewheel steering if the exercise is mishandled. Therefore the Authority recommends that, where RTO is required to be practised in an aeroplane, the exercise is initiated at a speed no greater than 50% V1 (Performance Group A) or 50% VR (other Performance Groups) by an appropriate emergency call, eg 'Stop Stop', and that asymmetric power reduction should be avoided. This latter method will be accepted by the Authority as the means of satisfying the CAA Form 1179 rejected take-off requirement only for tests conducted in aeroplanes.

8.3 AIC 141/1998 (Pink 182) 'Rejected Take-off (RTO) – UK Registered Aircraft' provides more detailed guidance on RTO considerations.

ANNEXE

RECOMMENDED MINIMUM SAFE HEIGHTS FOR COMPLETE SHUTDOWN OF POWER PLANTS FOR TRAINING PURPOSES

Piston or turbo-prop engined aeroplanes with Maximum Total Weight Authorised (MTWA) not exceeding 5700 kg (exception – Dove – 4000 ft)	3000 ft agl
Four engined aeroplanes (Note 1)	4000 ft agl
(a) Twin piston or turbo-prop engined aeroplanes with MTWA exceeding 5700 kg (exception – Bristol 170 – 6000 ft) (b) Triple turbo jet or turbo-fan engined aeroplanes	5000 ft agl
Twin turbo-jet or turbo-fan engined aeroplanes	8000 ft agl

Note 1: For four-engined Performance Group A aeroplanes (Regulation 7 of the Air Navigation (General) Regulations 1993) this height may be reduced to 1500 ft above ground level provided that the aeroplane's instantaneous weight is such as to permit a gross rate climb of at least 200 ft per minute in the two-engined-out en-route configuration.

Note 2: It is recognised that, for certain aeroplane types, the Authority's Airworthiness Flight Test Schedules require engine shutdown for performance measurement at lower heights than those tabulated above. Such flight tests are conducted by appropriately qualified pilots; the heights in this Annexe are considered to provide minimum safety margins for pilot training and testing.

Note 3: Since different types of aeroplanes have widely differing characteristics, the advice of the Authority should be sought if there is any doubt about the safety of methods and procedures to be adopted for any particular type/marque.

AIC 61/1999 (Pink 195) 3 June

RISKS AND FACTORS ASSOCIATED WITH OPERATIONS ON RUNWAYS AFFECTED BY SNOW, SLUSH OR WATER

1 Introduction

1.1 Operations from contaminated runways, by all classes of aeroplane, should be avoided whenever possible.

1.2 This document has been laid out in a conventional manner with take-off information preceding landing information. It should not be assumed that this indicates that take-offs are considered to be more critical and equal attention should be paid to the later paragraphs dealing with landings on runways affected by snow, slush or water.

1.3 Major UK aerodromes make every effort, within the limits of manpower and equipment available, to keep runways clear of snow, slush and its associated water, but circumstances arise when complete clearance cannot be sustained. In such circumstances, continued operation involves a significant element of risk and the wisest course of action is to delay the departure until conditions improve or, if airborne, divert to another aerodrome.

2 Operational Factors and Reporting Phaseology

2.1 At major UK and Western European aerodromes, when clearing has not been accomplished, the runway surface condition is reported by the following method which is described in the UK AIP Section AD 1.2.2, United Kingdom Snow Plan. The depth of snow or slush is measured by a standard depth gauge, readings being taken at approximately 300 metre intervals, between 5 and 10 metres from the runway centre-line and clear of the effects of rutting. Depth is reported in millimetres for each third of the runway length.

For information on the assessment of wet runway surface conditions, see UK AIP AD 1.1.1.

2.1.1 A subjective assessment is also made of the nature of the surface condition, on the following scale:

(a) Dry Snow (less than 0.35 Specific Gravity);

(b) Wet Snow (0.35 to 0.50 Specific Gravity);

(c) Compacted Snow (over 0.50 Specific Gravity);

(d) Slush (0.50 to 0.80 Specific Gravity);

(e) Standing Water (1.00 Specific Gravity). Note: Where only water is present the reporting scheme of 2.2 will be used.

(Note: Specific Gravity values are stated here to assist in the correlation of conditions with aircraft data, and not necessarily to assist in the determination of conditions as found).

2.2 The presence of water on a runway will be reported to the pilot using the following description:

	For JAR-OPS performance purposes, runways reported as DRY, DAMP or WET should be considered as NOT CONTAMINATED.
DRY	The surface is not affected by water, slush, snow or ice. NOTE: Reports that the runway is dry are not normally passed to pilots. If no runway surface report is passed, the runway can be assumed to be dry.
DAMP	The surface shows a change of colour due to moisture. NOTE: If there is sufficient moisture to produce a surface film or the surface appears reflective, the runway will be reported as WET.
WET	The surface is soaked but no significant patches of standing water are visible. NOTE: Standing water is considered to exist when water on the runway surface is deeper than 3 mm. Patches of standing water covering more than 25% of the assessed area will be reported as WATER PATCHES.
	For JAR-OPS performance purposes, runways reported as WATER PATCHES or FLOODED should be considered as CONTAMINATED.
WATER PATCHES	Significant patches of standing water are visible. NOTE: Water patches will be reported when more than 25% of the assessed area is covered by water more than 3 mm deep.
FLOODED	Extensive patches of standing water are visible. NOTE: Flooded will be reported when more than 50% of the assessed area covered by water more than 3 mm deep.

2.2.1 When reported, the presence or otherwise of surface contaminants on a runway will be assessed over the most significant portion of the runway (ie the area most likely to be used by aircraft taking off and landing). This area may differ slightly from one runway to another but will be approximate to the central two-thirds of the width of the runway extending longitudinally from a point 100 m before the aiming point to 100 m beyond the aiming point of the reciprocal runway.

2.2.2 Reports will be given sequentially for each third of the runway to be used, for example, 'Runway surface is wet, water patches and wet'. Where an aircraft will not make use of all the runway, the third most likely to be used at high speed is the most critical.

2.3 Depths greater than 3 mm of water, slush or wet snow, or 10 mm of dry snow, are likely to have a significant effect on the performance of aeroplanes. The main effects are:

(a) Additional drag – retardation effects on the wheels, spray impingement and increased skin friction;

(b) possibility of power loss or system malfunction due to spray ingestion or impingement;

(c) reduced wheel-braking performance – reduced wheel to runway friction and aquaplaning;

(d) directional control problems;

(e) possibility of structural damage.

2.3.1 A water depth of less than 3 mm is normal during and after heavy rain and in such conditions, no corrections to take-off performance are necessary other than the allowance, where applicable, for the effect of a wet surface. However, on such a runway where the water depth is less than 3 mm and where the performance effect 2.3 (a) is insignificant, isolated patches of standing water or slush of depth in excess of 15 mm located in the latter part of the take-off run may still lead to ingestion and temporary power fluctuations which could impair safety. Some aircraft types are susceptible to power fluctuations at depths greater than 9 mm and AFM limitations should be checked. For depths greater than 9 mm, specific mention will be made in the surface condition report as to location, extent and depth of water patches.

2.4 A continuous depth of water greater than 3 mm on a well constructed runway is unlikely as a result of rain alone, but can occur if very heavy rain combines with a lack of runway camber/crossfall or a crosswind to reduce the rate of water drainage from the runway. In such conditions the water depth is unlikely to persist for more than about 15 minutes after the rain has ceased, and take-off should be delayed accordingly.

2.5 In assessing the performance effect of increased drag (for reasons outlined in paragraph 2.3(a)), the condition of the up-wind two thirds of the take-off runway is most important, ie the area where the aeroplane is travelling at high speed. Small isolated patches of standing water will have a negligible effect on performance, but if extensive areas of standing water, slush or wet snow are present and there is doubt about the depth, take-off should not be attempted.

2.6 It is difficult to measure, or predict, the actual coefficient of friction or value of displacement and impingement drag associated with a contaminated runway. Therefore, it follows that aeroplane performance relative to a particular contaminated runway cannot be scheduled with a high degree of accuracy and hence any 'contaminated runway' data contained in the Flight Manual should be regarded as the best data available.

2.7 The provision of performance information for contaminated runways should not be taken as implying that ground handling characteristics on these surfaces will be as good as can be achieved on dry or wet runways, in particular, in crosswinds and when using reverse thrust. Remember, the use of a contaminated runway should be avoided if at all possible. A short delay in take-off or a short hold before landing can sometimes be sufficient to remove the contaminated runway risk. If necessary a longer delay or diversion to an airport with a more suitable runway should be considered.

3 General Limitations for Take-off

3.1 When operations from contaminated runways are unavoidable the following procedures may assist:

(a) Take-offs should not be attempted in depths of dry snow greater than 60 mm or depths of water, slush or wet snow greater than 15 mm. If the snow is very dry, the depth limit may be increased to 80 mm. In all cases the AFM limits, if more severe should be observed;

(b) ensure that all retardation and anti-skid devices are fully serviceable and check that tyres are in good condition;

(c) consider all aspects when selecting the flap/slat configuration from the range permitted in the Flight Manual. Generally greater increments of flaps/slats will reduce the unstick speed but could, for example, increase the effect of impingement drag for a low wing aircraft. Appropriate field length performance corrections should be made (see paragraph 6);

(d) fuel planning should include a review all aspects of the operation; including whether the carriage of excess fuel is justified;

(e) ensure that de-icing of the airframe and engine intakes, if appropriate, has been properly carried out and that the aircraft is aerodynamically clean at the time of take-off. Necessary de-icing fluids on the aerodynamic surfaces are permitted;

(f) pay meticulous attention to engine and airframe anti-ice drills;

(g) do not attempt a take-off with a tail wind or, if there is any doubt about runway conditions, with a crosswind in excess of the slippery runway crosswind limit. In the absence of a specified limit take-off should not be attempted in crosswinds exceeding 10 kt;

(h) taxi slowly and adopt other taxiing techniques which will avoid snow/slush adherence to the airframe or accumulation around the flap/slat or landing gear areas. Particularly avoid the use of reverse thrust, other than necessary serviceability checks which should be carried out away from contaminated runway areas. Be cautious of making sharp turns on a slippery surface;

(i) use the maximum runway distance available and keep to a minimum the amount of runway used to line up. Any loss should be deducted from the declared distances for the purpose of calculating the RTOW;

(j) power setting procedures appropriate to the runway condition as specified in the AFM should be used. Rapid throttle movements should be avoided and allowances made for take-off distance increases;

(k) normal rotation and take-off safety speeds should be used, (eg where the Flight Manual permits the use of data for overspeed procedures to give improved climb performance, these procedures should not be used). Rotation should be made at the correct speed using normal rate to the normal attitude;

(l) maximum take-off power should be used.

4 The Design and Production Standards Division of the Civil Aviation Authority will advise on the safety of any proposed changes to the above procedures.

5 Aircraft Commanders should also take the following factors into account when deciding whether to attempt a take-off:

(a) The nature of the overrun area and the consequences of an overrun off that particular runway;

(b) weather changes since the last runway surface condition report, particularly precipitation and temperature, the possible effect on stopping or acceleration performance and whether subsequent contaminant depths exceed Flight Manual limits.

6 Take-off Performance

6.1 Aeroplanes in JAR-OPS Performance Classes A and C:

(a) Operators present the limitations and performance corrections in their Operations Manuals in terms of reported precipitation depths. The limiting depth recommended in paragraph 3.1(a) should be used and the Specific Gravities given in the scale in paragraph 2.1 should be assumed;

(b) the take-off weight from a contaminated runway should not exceed the maximum permitted for normal operation when the runway is dry. If a normal take-off weight analysis for the runway results in a weight which is limited by obstacle clearance in the Net Flight Path, it is necessary to assume, for the contaminated runway take-off, that the Take-off Distance Available is equal to the Take-off Distance Required at the dry runway obstacle limited weight;

(c) for aircraft where the AFM does not contain appropriate contaminated runway data, the CAA should be consulted on whether operations can take place or how the data can be obtained.

6.2 Aeroplanes in Performance Class B of MTWA not exceeding 5700 kg:

(a) JAR-OPS requires specific reference to the conditions of the surface of the runway from which the take-off will be made. The ability to give take-off performance information with any confidence depends on knowledge of the slush drag characteristics. For small aeroplanes this knowledge is limited;

(b) In all instances manufacturer's information for operations on a contaminated runway should be sought. In its absence the best information available using conservative assumptions is given in paragraphs 6.2(c) to (e). Before using it to show compliance with JAR-OPS approval should be sought from the appropriate operational authority;

(c) Aerodrome Requirements:

(i) Twin engined Aeroplanes in Class B:

(1) Either a paved runway having an Emergency Distance available not less than 1.5 x Take-Off Distance Required* or 1500 ft whichever is the greater, or a grass runway having an Emergency Distance Available not less than 2.0 x Take-Off Distance Required* or 2000 ft which ever is the greater, should be available;

* As given in the Approved Flight Manual for the conditions of weight, aerodrome altitude, air temperature, runway slope and wind, appropriate to a dry hard-surface runway and including the factors of JAR-OPS 1.530, as appropriate;

(2) So that a check can be made on acceleration, the pilot will need to be able to identify the point on the runway at a distance of 40% of the EDA from the start of take-off. Pilots should request the aerodrome authority to provide, for this purpose, a suitable marker board, flag or readily identifiable object that does not contravene the obstacle requirements for a licensed aerodrome. If the necessary acceleration is not achieved, take-off should be abandoned (see paragraph 6.2(d)).

(ii) Single engined aeroplanes in Class B.

(1) Either a paved runway having an Emergency Distance Available not less than 2.0 x Take-off Distance to 50 ft Height Point* or 1500 ft whichever is the greater, or a grass runway having an Emergency Distance Available not less than 2.66 x Take-off Distance to 50 ft Height Point* or 2000 ft whichever is the greater, should be available;

* As given in the Performance Schedule or 'Owners Manual' for the conditions of weight, aerodrome altitude, air temperature, runway slope and wind, appropriate to a dry hard-surfaced runway, and including the factors of JAR-OPS 1.530, as appropriate;

(2) An acceleration check should be made (as described in paragraph 6.2(c)(i)(2)) and if the necessary acceleration is not achieved take-off should be abandoned.

(d) Flight Procedure:

(i) The aeroplane is accelerated along the runway as for a normal take-off but using a small amount of aft stick to relieve nose gear drag. If a speed of 0.85 V2 has been achieved before reaching the distance marker, the take-off should normally be continued. Rotation should be initiated and lift-off achieved at 0.9 V2. The aeroplane should then be accelerated and climbed away to achieve V2 at 50 ft;

(ii) If a speed of 0.85 V2 has not been achieved on reaching the marker boards the take-off should be abandoned. The throttles should be closed and maximum retardation used, consistent with retaining directional control.

(e) Limitations for Take-Off:

(i) The Procedures recommended in paragraph 3.1(a) to (k) should be followed when operations from contaminated runways are contemplated.

Additionally:

(1) Visibility should be adequate to see the distance marker (recommended in paragraph 6.2(c)(i)(2) and 6.2(c)(ii)(2)) from the start of take-off; and

(2) performance credit for reported headwinds and for downhill runway slopes should not be taken when ascertaining the Take-off Distance Required.

(f) Applicability:

(i) The procedures and practices outlined above in paragraphs 6.2(c), (d) and (e) have been approved for the following aeroplanes:

Class B Aero/Rockwell Commander
Beech 200 and derivatives
BN2 Islander and derivatives
Cessna 401/402/404/414/421/425/441
DH Canada DHC-6 Series 100 and 300
Piper PA23-250
Piper PA31
Short SC 7 Skyvan

(ii) This list is not exhaustive and applicability to other aeroplanes should be checked by reference to the:

Flight Department
Civil Aviation Authority
Safety Regulation Group
Aviation House
South Area
London Gatwick Airport
West Sussex
RH6 0YR

Tel: 01293-573113.

7 Landing

7.1 Attempts to land on heavily contaminated runways involve considerable risk and should be avoided whenever possible. If the destination aerodrome is subject to such conditions, departure should be delayed until conditions improve or an alternate used. It follows that advice in the Flight Manual or Operations Manual concerning landing weights and techniques on very slippery or heavily contaminated runways is there to enable the Commander to make a decision at despatch and, when airborne, as to his best course of action.

7.2 Depths of water or slush, exceeding approximately 3 mm, over a considerable proportion of the length of the runway, can have an adverse effect on landing performance. Under such conditions aquaplaning is likely to occur with its attendant problems of negligible wheel-braking and loss of directional control. Moreover, once aquaplaning is established it may, in certain circumstances, be maintained in much lower depths of water or slush. A landing should only be attempted in these conditions if there is an adequate distance margin over and above the normal Landing Distance Required and when the crosswind component is small. The effect of aquaplaning on the landing roll is comparable with that of landing on an icy surface and guidance is contained in some Flight Manuals on the effect on the basic landing distance of such very slippery conditions.

Appendix 1

Recommended reading for candidates for the UK JAR PPL:

Subject	Reference
Accident and Incident Investigation – civil aviation	23, 162, 875, 881
Acting in a disruptive manner	357, 438
Aerobatic manoeuvres	218, 741, 802
Aerodrome traffic rules	443, 453
Aerodrome traffic zones	199, 207, 210, 230, 265, 296, 454
Aerodrome/heliport	303-309
Aeronautical Chart legends	86-89, 600-601
Aeronautical information services	104-108
AIP Preface	21
Air traffic incidents (including AIRPROX)	253-255
Air traffic services – types of service	117-124
Aircraft equipment	393-410
Airspace restrictions	182-197, 203-210
Altimeter setting Procedures	231-233
ATS airspace classification	203-210
Authority of commander and members of the crew of an aircraft	357
Carburettor icing on piston engines	849, 875, 877
Carriage of dangerous goods	37, 356, 431, 489-495
Carriage of persons	356
Carriage of weapons and of munitions of war	355
Certificate of airworthiness	331-335, 337, 378, 380
Certificate of maintenance review	333, 379
Choice of VFR or IFR	450
Class A Airspace	203
Class B Airspace	204
Class C Airspace	205
Class D Airspace	206
Class E Airspace	208
Class F Airspace	209
Class G Airspace	210
Communication services	122
Definitions for ICAO Terms (UK)	73
Differences from ICAO standards, recommended practices and procedures	42-71
Distress, urgency and safety signals	462
Documents and records, production of	360, 494
Dropping of articles and animals	354
Dropping of persons	354
Endangering safety of an aircraft	357
Entry, transit and departure of aircraft – Documentary requirement	27
Entry, transit and departure of aircraft – Flights to/from EU	25
Entry, transit and departure of aircraft – Flights to/from non-EU countries	26
Entry, transit and departure of aircraft – National regulations	24
Entry, transit and departure of aircraft – Private flights	27
Entry, transit and departure of cargo – Customs and excise requirements	34
Entry, transit and departure of cargo – Hazardous cargo requirements	36
Entry, transit and departure of passengers and crew – Flying Licences and Ratings	36
Entry, transit and departure of passengers and crew – Immigration Requirements	35
Entry, transit and departure of passengers and crew – Public Health	36
Equipment of aircraft	331, 336
Failure of navigation and anti-collision lights	446
Flight crew of aircraft – licences	334, 407-420
Flight crew of aircraft – ratings	336-338, 415-417
Flight planning	235-247
Flying displays	127, 189, 328, 353, 428, 437
Frost, ice and snow on aircraft	846-850
Glider pilot – minimum age	334
Quadrantal rule and semi-circular rule	178, 225, 232, 262, 452
Inspection, overhaul, repair, replacement and modification	330-331
Instruction in flying	44, 305, 325-327, 334-339, 352, 361-363, 407-416, 427
Interception Proceedures	602

Appendix 1

Subject	Reference
Interpretation Differences to ICAO (UK)	73
Launching, picking up and dropping of tow ropes, etc.	454
Lights and pyrotechnic signals for control of aerodrome traffic	121, 459
Log book – flying	343
Low flying	444-445
Lower Airspace Radar Service (LARS)	224-230
Markings for paved runways and taxiways	456
Markings on unpaved manoeuvring areas	457
Marshalling signals	453, 460-462
Measuring system, aircraft markings and holidays	74-75
Meteorological services – automated services	149
Meteorological services – meteorological codes	155
Meteorological services – observations and reports	131-136
Meteorological services – responsible authority	131
Meteorological services – Sigmet	49, 82-84, 145, 149-151, 829-830
Meteorological services – types of service	137
Meteorological services – VOLMET	148
Military Aerodrome Traffic Zones (MATZ)	265-267
Movement of aircraft	45, 117, 198, 362, 381, 453
Nationality and registration marks	330, 389-391
Notification of arrival and departure	449
Operation of radio in aircraft	351
Passenger briefing by commander	350
Penalties	375-376
Practice instrument approaches	446
Pre-flight action by commander of aircraft	350
Radar services and procedures	215-218
Radio and radio navigation equipment	38-41, 413-414
Registration of aircraft	329-330
Reporting hazardous conditions	444
Right-hand traffic rule	449
Rules for avoiding aerial collisions	448
Rules of the air	443-462
Runways affected by snow, slush or water	318-321, 868-874
Search and rescue	163-168
Simulated instrument flight	446
Smoking in aircraft	357
Speed limitation	203-210, 450
Technical log	334, 384
Thunderstorms and associated turbulence	830-843
Towing of gliders	353-354
VFR flight plan	228
VHF international aeronautical emergency service	648-649
Visual flight rules (VFR)	46, 201, 385, 450
Wake turbulence	577, 851
Weather reports and forecasts	141, 229, 448
Weight schedule	337
Windshear	805-815

Appendix 2

Recommended reading for candidates for the UK IMC Rating:

Subject	Reference
AERAD ENC chart legend	589-599
Aerodromes Operating Minima	24, 306-311
Air traffic services – types of services	117-124
Air Traffic Services Outside Controlled Airspace, CAA Safety Sense Leaflet 8	667-672
Aircraft equipment	393-410
Airspace classification	203

Appendix 2

Subject	Reference
Airspace restrictions	113, 195-197
Altimeter Setting Procedures	230-232
Approach Ban	315
ATS routes and upper control areas	176
CAA Chart legend	86-89, 600-601
Carburettor icing	703-708
Choice of VFR or IFR	450
Class A airspace	203, 450,
Class B airspace	204, 297, 298
Class C airspace	189, 205
Class D airspace	206-208
Class E airspace	208
Class F airspace	209
Class G Airspace	182, 210
Communication services	122-128
Distance Measuring Equipment (DME)	124
Flight crew of aircraft – ratings	336, 341 – 342, 416-430, 435-436, 746-752
Flight plan and air traffic control clearance	453
Flight plan	180, 235-239, 248, 451, 453
Flight planning	235-236
Flights crossing ADRs or Published Holding Patterns under IFR	179
Frost, ice and snow on aircraft	846-850
GPS (Use of), CAA Safety Sense Leaflet 25	787-793
High ground, flight over and in the vicinity of	824-829
Holding - En-Route	181
Holding, approach and departure procedures	211
ICAO, Definitions Terms and Differences of Interpretation (UK)	73
Instrument Flight Rules (IFR)	202-203, 452
Instrument Landing System (ILS)	124
Interpretation (of ANO)	377-385
Lower Airspace Radar Service (LARS)	225-227, 289
Meteorological Codes	155-162
Meteorological services	131-139
Minimum height (IFR)	452
Non Directional Beacon (NDB)	124
Obstacles - En-Route	199
Operating minima – aerodromes	24, 306-311
Position Reporting within the London and Scottish FIR/UIR	183-184
Position reports (IFR)	179, 203
Practice instrument approaches	446
Quadrantal rule	452
Radar Advisory Service (RAS)	179, 215
Radar Information Service (RIS)	216
Radar services and procedures	215-217
Radio and radio navigation equipment to be carried	38, 410-414
Radio Failure Procedures	184
Radio navigation aids	61, 96-102, 125-126, 454
Semi-circular rule	452
Simulated instrument flight	446
SSR operating procedure	218-219
Standard Instrument Departure (SID) and Arrival Routes (STAR)	90
Standard Instrument Departure, Standard Instrument Arrival, Standard Terminal Arrival Route and Noise Abatement Charts	113
Thunderstorms	830-843
Transponder operating procedure	218-219
VFR or IFR	450
VHF Direction-Finding (VDF)	124
VHF Omni-directional Radio Range (VOR)	124
VOLMET	148
Weather reports and forecasts	131-139, 155-162, 448

Index

(Former) Pennine Radar Area of Responsibility 230
8.33kHz Channel Spacing in the VHF Radio Communications Band 41

A

Abbreviations used in AIS Publications 76-85
Accelerate-Stop Distance Available' (ASDA) 304
Accuracy of Meteorological Forecasts 141
Addressing Of Flight Plan Messages 248
Addressing of IFR Flight Plans 239
Addressing of VFR Flight Plans 238
Advisory Heights for Surveillance Radar Approach Procedures 214
AERAD ENC Chart Legend 589-599
Aerial Sporting and Recreational Activities 199, 268-286
Aerial Tactics Area 197
Aerodrome Air Navigation Service Charges – Exemptions 173
Aerodrome Air Navigation Service Charges – Minimum charge 173
Aerodrome Air Navigation Service Charges – Rebates 172
Aerodrome Categorisation 321
Aerodrome Fire/Crash Protection Categories 578
Aerodrome Forecasts – Additional Meteorological Services 145
Aerodrome Forecasts – Amended Route/Area Forecast (Advisory Criteria) 144
Aerodrome Forecasts – Cloud 142
Aerodrome Forecasts – Notification Required from Operators 145
Aerodrome Forecasts – Prior Notification for Special Forecasts 145
Aerodrome Forecasts – QFE/QNH 143
Aerodrome Forecasts – Runway Visual Range (RVR) 142
Aerodrome Forecasts – Severe Icing/Turbulence 143
Aerodrome Forecasts – Special Forecasts and Specialised Information 145
Aerodrome Forecasts – Surface Wind 142
Aerodrome Forecasts – TAF Variants/Amendments 143
Aerodrome Forecasts – TREND 143
Aerodrome Forecasts – Visibility 142
Aerodrome Forecasts – Weather 142
Aerodrome Forecasts – (TAF) 141
Aerodrome Obstacle Charts – ICAO Type 'A' 112
Aerodrome Obstacles 198
Aerodrome Operating Minima – Aeroplane Category 307
Aerodrome Operating Minima – Aeroplanes – Multi-engine aeroplanes operating in 310
accordance with Performance Class A
Aerodrome Operating Minima – Aeroplanes – Multi-engine aeroplanes operating in 311
accordance with Performance Class B or C
Aerodrome Operating Minima – Aeroplanes – Multi-engine aeroplanes which do not meet the 311
applicable one engine inoperative climb requirements must be treated as single engine aeroplanes
Aerodrome Operating Minima – Aeroplanes – Single engine aeroplanes 310
Aerodrome Operating Minima – Aeroplanes – Take-off 310
Aerodrome Operating Minima – Aeroplanes 310
Aerodrome Operating Minima – Cloud Ceiling 307
Aerodrome Operating Minima – Commercial Air Transportation 307
Aerodrome Operating Minima – Decision Height (DH) 307
Aerodrome Operating Minima – Final Approach and Take-off area (FATO) 307
Aerodrome Operating Minima – Glossary of Terms 307
Aerodrome Operating Minima – Helicopters – Performance Class 1 operations 313
Aerodrome Operating Minima – Helicopters – Performance Class 2 operations 313
Aerodrome Operating Minima – Helicopters – Performance Class 3 operations 313
Aerodrome Operating Minima – Helicopters 313
Aerodrome Operating Minima (AOM) 307
Aerodrome Operating Minima 306
Aerodrome Traffic Zones 210
Aerodrome/Heliport Charges – Aerodrome Operators 171
Aerodrome/Heliport Charges – Conditions Applicable to the Landing, Parking or Storage of Aircraft ...171
Aerodrome/Heliport Charges – Landing of Aircraft 171

Index

Aerodrome/Heliport Charges 171
Aerodrome/Heliports – Altimeter Error 308
Aerodrome/Heliports – Availability of Ground Services 303
Aerodrome/Heliports – CAT II/III Operations at Aerodromes 304
Aerodrome/Heliports – Closure of Aerodromes 303
Aerodrome/Heliports – Declared Distances 304
Aerodrome/Heliports – Determination of DH/MDH 309
Aerodrome/Heliports – Determination of Take-off Minima 309
Aerodrome/Heliports – Helicopter Performance – Performance Class 1 Operations 307
Aerodrome/Heliports – Helicopter Performance – Performance Class 2 Operations 307
Aerodrome/Heliports – Helicopter Performance – Performance Class 3 Operations 307
Aerodrome/Heliports – Helicopter Performance 307
Aerodrome/Heliports – Heliport – Precision approach – Category I Operations 310
Aerodrome/Heliports – Instrument Approach Procedure (IAP) 308
Aerodrome/Heliports – Landing Distance Available' (LDA) 304
Aerodrome/Heliports – Low Visibility Procedure (LVP) 308
Aerodrome/Heliports – Low Visibility Take-off 308
Aerodrome/Heliports – Minimum Descent Height (MDH) 308
Aerodrome/Heliports – Missed Approach Point (MAPt) 308
Aerodrome/Heliports – Noise Abatement Requirements 303
Aerodrome/Heliports – Obstacle Clearance Height (OCH) 309
Aerodrome/Heliports – Operational Hours 303
Aerodrome/Heliports – Operations Outside Published Operating Hours 304
Aerodrome/Heliports – Prior Permission Requirements 303
Aerodrome/Heliports – Reporting Term – Surface Conditions 306
Aerodrome/Heliports – Runway Friction Assessment 306
Aerodrome/Heliports – Runway Surface Condition Reporting 305
Aerodrome/Heliports – Runway utilisation procedures 305
Aerodrome/Heliports – Runway Visual Range (RVR) 308
Aerodrome/Heliports – Take-off Distance Available' (TODA) 304
Aerodrome/Heliports – Take-off Run Available' (TORA) 304
Aerodrome/Heliports – Use of Government Aerodromes 304
Aerodrome/Heliports – Visual Approach 308
Aerodrome/Heliports – Visual Ground Aids 304
Aerodrome/Heliports – Visual Reference 308
Aerodrome/Heliports 303
Aerodromes without published Instrument Approach Procedures 316
Aeronautical Authority 21
Aeronautical Chart Services Available 110
Aeronautical Charts – Responsible Authority 109
Aeronautical Charts ICAO Scale 1:500,000 – United Kingdom 110-111, 114
Aeronautical Fixed Services 127
Aeronautical Information Circulars (AIC) 105
Aeronautical Information Services 104
Aeronautical Publications 104
AIC 106/2004 Frost, ice and snow on aircraft 846-850
AIC 17/1999 Wake turbulence 851-862
AIC 19/2002 Low altitude windshear 805-815
AIC 52/1999 Guidance to training captains – simulation of engine failure in aeroplanes 863-867
AIC 6/2003 Flight over and in the vicinity of high ground 824-829
AIC 61/1999 Risks and factors associated with operations on runways affected by snow, slush or water 868-874
AIC 67/2002 Take-off, climb and landing performance of light aeroplanes 816-823
AIC 81/2004 The effect of thunderstorms and associated turbulence on aircraft operations 830-843
AIC 93/2004 VHF Radio telephony emergency communications 844-845
AIP Amendment service (AMDT and AMDT AIRAC) 105
AIP structure and amendment interval 21
Air Navigation Obstacle 198
Air Navigation Order General Regulations 2005 463-481
Air Navigation Services Charges 172
Aerodrome Air Navigation Service Charges 172
Air Traffic Incidents 253
Air Traffic Service Advisory Routes 178

Index

Air Traffic Services – Aerodrome Air Traffic Services 117
Air Traffic Services – Aerodrome Distance Measuring Equipment (DME) 124
Air Traffic Services – Aerodrome Flight Information Service (AFIS) 118
Air Traffic Services – AFTN and associated Data Services 123
Air Traffic Services – Approach Control Service (APP) 117
Air Traffic Services – Approach Radar (RAD) 124
Air Traffic Services – Area Air Traffic Services 117
Air Traffic Services – Area of Responsibility 116
Air Traffic Services – Clearance for Immediate Take-Off...119
Air Traffic Services – Communication Services 122
Air Traffic Services – Communications and Navigational Aids at UK Aerodromes 123
Air Traffic Services – Departure Clearances 121
Air Traffic Services – Enquiries and Complaints 123
Air Traffic Services – En-route Telecommunications Services 123
Air Traffic Services – Initial Call 121
Air Traffic Services – Instrument Approaches 118
Air Traffic Services – Instrument Landing System (ILS) and Distance Measuring Equipment (DME) 124
Air Traffic Services – Instrument Landing System (ILS) 124
Air Traffic Services – Lamp, Pyrotechnic and Ground Signals 121
Air Traffic Services – Land after Procedure 119
Air Traffic Services – MF Non-directional Beacon (NDB) 124
Air Traffic Services – Microwave Landing Systems (MLS) 124
Air Traffic Services – Minimum Flight Altitude 121
Air Traffic Services – National Responsibilities 116
Air Traffic Services – Procedures for Arriving VFR Flights 118
Air Traffic Services – Provision of Air Traffic Services (ATS) 117
Air Traffic Services – Radio Navigation Services 123
Air Traffic Services – Responsibility of APP at Aerodromes within Controlled Airspace 117
Air Traffic Services – Responsibility of APP Control at Aerodromes outside Controlled Airspace 117
Air Traffic Services – Runway Utilisation Procedures 119
Air Traffic Services – Runway-in-Use 119
Air Traffic Services – Services and Procedures for Arriving Flights 118
Air Traffic Services – Special Landing Procedures at London Heathrow, London Gatwick, ...119-121
London Stansted and Manchester Airports
Air Traffic Services – Standard Overhead Join 119
Air Traffic Services – The Aeronautical Mobile Service 126
Air Traffic Services – Types of Services 117
Air Traffic Services – VHF Direction-Finding Station (VDF) 124
Air Traffic Services – VHF Omni-directional Radio Range (VOR) and Distance Measuring Equipment (DME) 124
Air Traffic Services – Visual Circuit Reporting Procedures 119
Air Traffic Services Airspace Classification 203
Air Traffic Services 116
AIRAC System 107
Airborne Procedures 177
Aircraft Instruments, Equipment and Flight Documents 37
Aircraft Nationality and Registration Marks 75
Aircraft Registration & Country Codes 562-567
Aircraft Reports – Airframe Icing 147
Aircraft Reports – Routine Aircraft Observations 145
Aircraft Reports – Special Aircraft Observations 145
Aircraft Reports – Turbulence (TURB) 147
Aircraft Reports – Turbulence and Icing Reporting Criteria 147
Aircraft Reports – Windshear Reporting Criteria 147
Aircraft Reports 145
AIRMET Copy Form 153-154
AIRMET Service 150
Airports designated under Prevention of Terrorism Legislation 28
AIRPROX Assessment 254
AIRPROX in Foreign Airspace 254
AIRPROX in UK Airspace 253
AIRPROX Report Forms CA 1094/RAF 765A, CA 1094A and CA 1261 255
AIRPROX Reporting – General 253

Index

AIRPROX Reporting Procedures ... 253
Airspace Classifications ... 203
Airspace other than Controlled Airspace and Advisory Routes – Class G Airspace ... 182
Airspace Restrictions – Danger Area ... 195
Airspace Restrictions – Emergency Restriction of Flying Regulations ... 195
Airspace Restrictions – Gas Venting Operations ... 198
Airspace Restrictions – High Intensity Radio Transmission Area (HIRTA) ... 198
Airspace Restrictions – Laser Sites ... 198
Airspace Restrictions – Prohibited Area ... 195
Airspace Restrictions – Radiosonde Balloon Ascents ... 198
Airspace Restrictions – Restricted Area ... 195
Airspace Restrictions – Restriction of Flying Regulations ... 195
Airspace Restrictions – Small Arms Ranges ... 197
Airspace Restrictions ... 195, 197
Airspace Restrictions, Danger Areas and Hazards to Flight ... 195
Air-to-Air Refuelling Area (AARA) ... 197
Airway Crossings or Penetrations in VMC – Civil Aircraft ... 178
Allocation of Levels ... 178
Alternative Race-track Procedure ... 213
Altimeter Setting Procedures – En-route Outside Controlled Airspace ... 232
Altimeter Setting Procedures – En-route Within Controlled Airspace ... 232
Altimeter Setting Procedures – Flight Planning ... 232
Altimeter Setting Procedures – Missed Approach ... 232
Altimeter Setting Procedures – Selected Transition Altitudes ... 231
Altimeter Setting Procedures – Tables of Cruising Levels ... 233
Altimeter Setting Procedures – Take-off and Climb ... 231
Altimeter Setting Procedures ... 179, 230
Altimeter Setting Regions (ASR) ... 231
AM Radio Stations Broadcasting in the UK ... 531
Annex 1 Personnel Licensing (9th Edition) ... 42
Annex 2 Rules of the Air (9th Edition) (AMDT 37) ... 45
Annex 3 Meteorological Service for International Air Navigation (15th Edition) (AMDT 73) ... 48
Annex 4 Aeronautical Charts (9th Edition) (AMDT 53) ... 49
Annex 5 Units of Measurement to be used in Air and Ground Operations (4th Edition) (AMDT 16) ... 52
Annex 6 Operation of Aircraft Part 1 (International Commercial Air Transport – Aeroplanes) ... 52
(8th Edition) (AMDT 28)
Annex 6 Operation of Aircraft Part II (International General Aviation – Aeroplanes) (6th Edition) (AMDT 23) ... 54
Annex 6 Operation of Aircraft Part III (International Operations – Helicopters) (5th Edition) ... 55
Annex 7 Aircraft Nationality and Registration Marks (5th Edition) (AMDT 5) ... 57
Annex 8 Airworthiness of Aircraft (9th Edition) (AMDT 98) ... 58
Annex 9 Facilitation (12th Edition) (AMDT 19) ... 59
Annex 10 Aeronautical Telecommunications Vol I(Radio Navigation Aids) (5th Edition) (AMDT 79) ... 61
Annex 10 Aeronautical Telecommunications Vol II (Communications ... 61
Procedures including those with PANS status) (6th Edition) (AMDT 79)
Annex 10 Aeronautical Telecommunications Vol III Part 1 ... 62
Digital Data Communications Systems) and Part 2 (Voice Communication Systems) (1st Edition) (AMDT 79)
Annex 10 Aeronautical Telecommunications Vol IV (Surveillance
Radar and Collision Avoidance Systems) (3rd Edition) (AMDT 79) ... 63
Annex 11 Air Traffic Services (Air Traffic Service, Flight Information
Service and Alerting Service) (13th Edition) (AMDT 42) ... 63
Annex 12 Search and Rescue (7th Edition) (AMDT 17) ... 64
Annex 13 Aircraft Accident and Incident Investigation (9th Edition) (AMDT 10) ... 64
Annex 14 Aerodromes: Vol I (Aerodrome Design and Operations) (3rd Edition) (AMDT 6) ... 64
Annex 14 Aerodromes: Vol II (Heliports) (2nd Edition) (AMDT 3) ... 66
Annex 15 Aeronautical Information Services (11th Edition) (AMDT 33) ... 66
Annex 16 Environmental Protection Vol I (Aircraft Noise) (3rd Edition) (AMDT 7) ... 67
Annex 16 Vol II (Aircraft Engine Emissions) (2nd Edition) (AMDT 4) ... 67
Annex 17 Security (6th Edition) (AMDT 10) ... 67
Annex 18 The Safe Transport of Dangerous Goods by Air (3nd Edition) (AMDT 7) ... 67
ANO Article 1 Citation and commencement ... 329
ANO Article 2 Revocation ... 329
ANO Article 3 Aircraft to be registered ... 329

Index

ANO Article 4 Registration of aircraft in the UK 329-330
ANO Article 5 Nationality and registration marks 330
ANO Article 6 Grant of air operators' certificates...331
ANO Article 7 Grant of police air operators' certificates..331
ANO Article 8 Certificate of airworthiness to be in force 331
ANO Article 9 Issue, renewal, etc., of certificates of airworthiness..................................331-332
ANO Article 10 Validity of certificate of airworthiness..332
ANO Article 11 Issue, renewal etc; of permits to fly 332-333
ANO Article 12 Issue of EASA permits fly 333
ANO Article 13 Issue etc; certificates of validation of permits to fly or equivalent document ..333
ANO Article 14 Certificate of maintenance review 333-334
ANO Article 15 Technical log 334
ANO Article 16 Requirement for a certificate of release to service..................................334-335
ANO Article 17 Requirement for a certificate of release to service under Part 145................335
ANO Article 18 Licensing of maintenance engineers...335-336
ANO Article 19 Equipment of aircraft..336
ANO Article 20 Radio equipment of aircraft 336-337
ANO Article 21 Minimum equipment requirements..337
ANO Article 22 Aircraft, engine and propeller log books..337
ANO Article 23 Aircraft weight schedule...337
ANO Article 24 Access and inspection for airworthiness purposes...337
ANO Article 25 Composition of crew of aircraft..338-339
ANO Article 26 Members of flight crew – requirement for licence 339-340
ANO Article 27 Grant, renewal and effect of flight crew licences....................................340-341
ANO Article 28 Maintenance of privileges of aircraft ratings in United Kingdom licences 341
ANO Article 29 Maintenance of privileges of aircraft ratings in JAR-FCL licences, ...342
United Kingdom licences for which there are JAR-FCL equivalents, United Kingdom Basic
Commercial Pilot's licences and United Kingdom Flight Engineer's Licences
ANO Article 30 Maintenance of privileges of aircraft ratings in National Private Pilot's Licences 342
ANO Article 31 Maintenance of privileges of other ratings..342
ANO Article 32 Medical requirements..342
ANO Article 33 Miscellaneous licensing provisions...343
ANO Article 34 Validation of licences 343
ANO Article 35 Personal flying logbook...343
ANO Article 36 Instruction in flying...343-344
ANO Article 37 Glider pilot – minimum age...344
ANO Article 38 Operations Manual..344
ANO Article 39 Police operations manual 344
ANO Article 40 Training Manual 345
ANO Article 41 Flight data monitoring, accident prevention and flight safety programme......345
ANO Article 42 Public transport – operator's responsibilities 345
ANO Article 43 Loading – public transport aircraft and suspended loads........................345-346
ANO Article 44 Public transport – aeroplanes -operating conditions and performance requirements....346
ANO Article 45 Public transport – helicopters – operating conditions and performance requirements 347-348
ANO Article 46 Public transport operations at night or in instrument meteorological 348
conditions aeroplanes with one power unit which are registered elsewhere than in the United Kingdom
ANO Article 47 Public transport aircraft registered in the United Kingdom – aerodrome operating minima ...348-349
ANO Article 48 Public transport aircraft registered elsewhere than in the United Kingdom – 349
aerodrome operating minima
ANO Article 49 Non-public transport aircraft – aerodrome operating minima..................349-350
ANO Article 50 Pilots to remain at controls 350
ANO Article 51 Wearing of survival suits by crew 350
ANO Article 52 Pre-flight action by commander of aircraft..350
ANO Article 53 Passenger briefing by commander 350
ANO Article 54 Public transport of passengers – additional duties of commander.................351
ANO Article 55 Operation of radio in aircraft..351-352
ANO Article 56 Minimum navigation performance...352
ANO Article 57 Height keeping performance – aircraft registered in the United Kingdom......352
ANO Article 58 Height keeping performance – aircraft registered elsewhere than in the United Kingdom 352
ANO Article 59 Area navigation and required navigation performance capabilities – 352-353
aircraft registered in the United Kingdom

Index

ANO Article 60 Area navigation and required navigation performance capabilities – 353
aircraft registered elsewhere than in the United Kingdom
ANO Article 61 Use of airborne collision avoidance system 353
ANO Article 62 Use of flight recording systems and preservation of records 353
ANO Article 63 Towing of gliders 353-354
ANO Article 64 Operation of self-sustaining gliders 354
ANO Article 65 Towing, picking up and raising of persons and articles 354
ANO Article 66 Dropping of articles and animals 354
ANO Article 67 Dropping of persons and grant of parachuting permissions..354-355
ANO Article 68 Grant of aerial application certificates 355
ANO Article 69 Carriage of weapons and of munitions of war 355-356
ANO Article 70 Carriage of dangerous goods 356
ANO Article 71 Method of carriage of persons 356
ANO Article 72 Exits and break-in markings 356-357
ANO Article 73 Endangering safety of an aircraft 357
ANO Article 74 Endangering safety of any person or property 357
ANO Article 75 Drunkenness in aircraft 357
ANO Article 76 Smoking in aircraft 357
ANO Article 77 Authority of commander of an aircraft 357
ANO Article 78 Acting in a disruptive manner 357
ANO Article 79 Stowaways 357
ANO Article 80 Flying displays 357-358
ANO Article 81 Application and interpretation of Part 6..358
ANO Article 82 Fatigue of crew – operator's responsibilities 358-359
ANO Article 83 Fatigue of crew – responsibilities of crew 359
ANO Article 84 Flight times – responsibilities of flight crew 359
ANO Article 85 Protection of aircrew from cosmic radiation 359
ANO Article 86 Documents to be carried 359
ANO Article 87 Keeping and production of records of exposure to cosmic radiation 359
ANO Article 88 Production of documents and records 360
ANO Article 89 Production of air traffic service equipment documents and records..360
ANO Article 90 Power to inspect and copy documents and records 360
ANO Article 91 Preservation of documents, etc 360
ANO Article 92 Revocation, suspension and variation of certificates, licences and other documents 360
ANO Article 93 Revocation, suspension and variation of permissions, etc. granted under article 138 or article 140 361
ANO Article 94 Offences in relation to documents and records 361
ANO Article 95 Rules of the air..362
ANO Article 96 Power to prohibit or restrict flying 362
ANO Article 97 Balloons, kites, airships, gliders and parascending parachutes...362-363
ANO Article 98 Regulation of small aircraft 363
ANO Article 99 Regulation of rockets 363-364
ANO Article 100 Requirement for an air traffic control approval 364
ANO Article 101 Duty of person in charge to satisfy himself as to competence of controllers 364
ANO Article 102 Manual of air traffic services 364
ANO Article 103 Provision of air traffic services 364
ANO Article 104 Making of an air traffic direction in the interests of safety 364
ANO Article 105 Making of a direction for airspace policy purposes 365
ANO Article 106 Use of radio call signs at aerodromes 365
ANO Article 107 Prohibition of unlicensed air traffic controllers and student air traffic controllers 365
ANO Article 108 Grant and renewal of air traffic controller's and student air traffic controller's licences 365
ANO Article 109 Privileges of an air traffic controller licence or a student air traffic controller licence...................366
ANO Article 110 Maintenance of validity of ratings and endorsements..366
ANO Article 111 Obligation to notify rating ceasing to be valid and change of unit 366
ANO Article 112 Requirement for medical certificate 366
ANO Article 113 Appropriate licence...366
ANO Article 114 Incapacity of air traffic controllers 366
ANO Article 115 Fatigue of air traffic controllers – air traffic controller's responsibilities 366
ANO Article 116 Prohibition of drunkenness etc. of controllers 366
ANO Article 117 Failing exams 366
ANO Article 118 Use of simulators 367
ANO Article 119 Approval of courses and persons..367
ANO Article 120 Acting as an air traffic controller and a student air traffic controller 367

883

Index

ANO Article 121 Prohibition of unlicensed flight information service officers...367
ANO Article 122 Licensing of flight information service officers 367
ANO Article 123 Flight information service manual..367
ANO Article 124 Air traffic service equipment 368
ANO Article 125 Air traffic service equipment records 368-369
ANO Article 126 Aerodromes – public transport of passengers and instruction in flying...........................369
ANO Article 127 Use of Government aerodromes 369
ANO Article 128 Licensing of aerodromes 369-370
ANO Article 129 Charges at aerodromes licensed for public use 370
ANO Article 130 Use of aerodromes by aircraft of Contracting States and of the Commonwealth...........370
ANO Article 131 Noise and vibration caused by aircraft on aerodromes..370
ANO Article 132 Aeronautical lights..370-371
ANO Article 133 Lighting of en-route obstacles..371
ANO Article 134 Lighting if wind turbine generators in United Kingdom territorial waters 371
ANO Article 135 Dangerous lights 372
ANO Article 136 Customs and Excise aerodromes 372
ANO Article 137 Aviation fuel at aerodromes 372
ANO Article 138 Restriction with respect to carriage for valuable consideration in aircraft 372
registered elsewhere than the UK
ANO Article 139 Filing and approval of tariffs...373
ANO Article 140 Restriction on aerial photography, aerial survey and aerial work in aircraft 373
registered elsewhere than in the UK
ANO Article 141 Flights over any foreign country 373
ANO Article 142 Mandatory reporting of occurrences..373-374
ANO Article 143 Mandatory reporting of birdstrikes 374
ANO Article 144 Power to prevent aircraft flying 375
ANO Article 145 Right of access to aerodromes and other places 375
ANO Article 146 Obstruction of persons...375
ANO Article 147 Directions 375
ANO Article 148 Penalties 375-376
ANO Article 149 Extra-territorial effect of the Order...376
ANO Article 150 Aircraft in transit over certain United Kingdom territorial waters 376
ANO Article 151 Application of Order to British -controlled aircraft registered elsewhere than in the 376
United Kingdom
ANO Article 152 Application of Order to the Crown and visiting forces, etc..376
ANO Article 153 Exemption from Order 376
ANO Article 154 Appeal to County Court or Sheriff Court...377
ANO Article 155 Interpretation..377-385
ANO Article 156 Meaning of aerodrome traffic zone...385
ANO Article 157 Public transport and aerial work – general rules 385-386
ANO Article 158 Public transport and aerial work – exceptions – flying displays etc 386
ANO Article 159 Public transport and aerial work – exceptions – charity flights..386
ANO Article 160 Public transport and aerial work – exceptions – cost sharing.....................................386-387
ANO Article 161 Public transport and aerial work – exceptions – recovery of direct costs 387
ANO Article 162 Public transport and aerial work – exceptions – jointly owned aircraft............................387
ANO Article 163 Public transport and aerial work – exceptions – parachuting...387
ANO Article 164 Exceptions from application of provisions of the order for certain classes of aircraft 387
ANO Article 165 Approval of persons to furnish reports..387
ANO Article 166 Certificates, authorisations, approvals and permissions 388
ANO Article 167 Competent authority..388
ANO Article 168 Saving 388
ANO General Regulation 4 Particulars and weighing requirements 646-465
ANO General Regulation 5 Aeroplanes to which article 44(5) applies...465
ANO General Regulation 6 Helicopters to which article 45(1) applies..465
ANO General Regulation 7 Weight and performance: general provisions 466
ANO General Regulation 8 Noise and vibration caused by aircraft on aerodromes..................................466
ANO General Regulation 9 Pilots maintenance – prescribed repairs or replacements.......................466-467
ANO General Regulation 10 Aeroplanes flying for the purpose of public transport of passengers – 467
aerodrome facilities for approach to landing and landing
ANO General Regulation 11 Reportable occurrences, time and manner of reporting and 467-468
information to be reported

Index

ANO General Regulation 12 Mandatory reporting of bird strikes – time and manner of reporting 468
and information to be reported
ANO General Regulation 13 Minimum navigation performance and height specifications 468-469
ANO General Regulation 14 Airborne Collision Avoidance System 469
ANO General Regulation 15 Mode S Transponder..469
ANO General Regulation Schedule 2 Aeroplane Performance 1 Weight and performance of 469-471
public transport aeroplanes designated as aeroplanes of performance group A or performance group B
ANO General Regulation Schedule 2 Aeroplane Performance 2 471-473
Weight and performance of public transport aeroplanes designated as aeroplanes of performance group C
ANO General Regulation Schedule 2 Aeroplane Performance 3 473-474
Weight and performance of public transport aeroplanes designated as aeroplanes of performance group D
ANO General Regulation Schedule 2 Aeroplane Performance 4 474-475
Weight and performance of public transport aeroplanes designated as aeroplanes of performance group E
ANO General Regulation Schedule 2 Aeroplane Performance 5 475-476
Weight and performance of public transport aeroplanes designated as aeroplanes of performance group F
ANO General Regulation Schedule 2 Aeroplane Performance 6 476-477
Weight and performance of public transport aeroplanes designated as aeroplanes of performance group X
ANO General Regulation Schedule 2 Aeroplane Performance 6 477-480
Weight and performance of public transport aeroplanes designated as aeroplanes of performance group Z
ANO General Regulation Schedule 2 Aeroplane Performance 469-480
ANO General Regulation Schedule 3 Helicopter Performance 2 480-481
Weight and performance of public transport helicopters carrying out Performance Class 2 operation
ANO General Regulation Schedule 3 Helicopter Performance 3 481
Weight and performance of public transport helicopters carrying out Performance Class 3 operation
ANO General Regulation Schedule 3 Helicopter Performance 1 480
Weight and performance of public transport helicopters carrying out Performance Class 1 operation
ANO General Regulation Schedule 3 Helicopter Performance 480-481
ANO Rules Of The Air Rule 1 Interpretation...444
ANO Rules Of The Air Rule 2 Application of Rules to aircraft...444
ANO Rules Of The Air Rule 3 Misuse of signals and markings...444
ANO Rules Of The Air Rule 4 Reporting hazardous conditions...444
ANO Rules Of The Air Rule 5 Low flying...444-446
ANO Rules Of The Air Rule 6 Simulated Instrument Flight...446
ANO Rules Of The Air Rule 7 Practice instrument approaches...446
ANO Rules Of The Air Rule 8 General...446
ANO Rules Of The Air Rule 9 Display of lights by aircraft...446
ANO Rules Of The Air Rule 10 Failure of navigation and anti-collision lights 446
ANO Rules Of The Air Rule 11 Flying machines 447
ANO Rules Of The Air Rule 12 Gliders...447
ANO Rules Of The Air Rule 13 Free balloons..447
ANO Rules Of The Air Rule 14 Captive balloons and kites 447
ANO Rules Of The Air Rule 15 Airships 448
ANO Rules Of The Air Rule 16 Weather reports and forecasts 448
ANO Rules Of The Air Rule 17 Rules for avoiding aerial collisions...448-449
ANO Rules Of The Air Rule 18 Aerobatic manoeuvres...449
ANO Rules Of The Air Rule 19 Right-hand traffic rule 449
ANO Rules Of The Air Rule 20 Notification of arrival and departure 449
ANO Rules Of The Air Rule 21 Flight in Class A airspace 450
ANO Rules Of The Air Rule 22 Choice of VFR or IFR..450
ANO Rules Of The Air Rule 23 Speed limitation 450
ANO Rules Of The Air Rule 24 Visual flight and reported visibility 450
ANO Rules Of The Air Rule 25 Flight within controlled airspace 450
ANO Rules Of The Air Rule 26 Flight outside controlled airspace..451
ANO Rules Of The Air Rule 27 VFR flight plan and air traffic control clearance...451
ANO Rules Of The Air Rule 28 Instrument Flight Rules 452
ANO Rules Of The Air Rule 29 Minimum height 452
ANO Rules Of The Air Rule 30 Quadrantal rule and semi-circular rule...452
ANO Rules Of The Air Rule 31 Flight plan and air traffic control clearance 453
ANO Rules Of The Air Rule 32 Position reports...453
ANO Rules Of The Air Rule 33 Application of aerodrome traffic rules 453
ANO Rules Of The Air Rule 34 Visual signals..453
ANO Rules Of The Air Rule 35 Movement of aircraft on aerodromes...453
ANO Rules Of The Air Rule 36 Access to and movement of persons and vehicles on the aerodrome 453

ANO Rules Of The Air Rule 37 Right of way on the ground...453-454
ANO Rules Of The Air Rule 38 Launching, picking up and dropping of tow ropes, etc..............................454
ANO Rules Of The Air Rule 39 Flight within aerodrome traffic zones 454
ANO Rules Of The Air Rule 40 Use of radio navigation aids..454
ANO Rules Of The Air Rule 41 General..454-455
ANO Rules Of The Air Rule 42 Signals in the signals area...455-456
ANO Rules Of The Air Rule 43 Markings for paved runways and taxiways 456-457
ANO Rules Of The Air Rule 44 Markings on unpaved manoeuvring areas 457-458
ANO Rules Of The Air Rule 45 Signals visible from the ground..458
ANO Rules Of The Air Rule 46 Lights and pyrotechnic signals for control of aerodrome traffic...................459
ANO Rules Of The Air Rule 47 Marshalling signals (from marshaller to an aircraft) 460-461
ANO Rules Of The Air Rule 48 Marshalling signals (from a pilot of an aircraft to a marshaller) 462
ANO Rules Of The Air Rule 49 Distress, urgency and safety signals 462
ANO Schedule 1 Orders revoked 389-391
ANO Schedule 3 A and B Conditions and categories of certificate of airworthiness 391-393
ANO Schedule 3 Part A A and B Conditions 392-392
ANO Schedule 3 Part B Categories of certificate of airworthiness and purposes for which aircraft may fly...........392
ANO Schedule 4 Aircraft equipment..393-410
ANO Schedule 5 Radio communication and radio navigation equipment to be carried in aircraft......410-414
ANO Schedule 6 Aircraft, engine and propeller log books..414-415
ANO Schedule 7 Areas specified in connection with the carriage of flight navigators as 415-416
members of the flight crews or suitable navigational equipment on public transport aircraft
ANO Schedule 8 Flight crew of aircraft – licences, ratings, qualification and maintenance 416-430
of licence privileges
ANO Schedule 8 Part A Flight crew licences 416-425
ANO Schedule 8 Part B Ratings and qualifications 425-427
ANO Schedule 8 Part C Maintenance of licence privileges...427-430
ANO Schedule 9 Part A Operations Manual...431
ANO Schedule 9 Public Transport – operational requirements 431
ANO Schedule 10 Part B Training Manual...431-432
ANO Schedule 10 Part C Crew training and tests 432-434
ANO Schedule 11 Air traffic controllers – licences, ratings, endorsements and maintenance 434-436
of licence privileges
ANO Schedule 11 Part A Air traffic controller licences 434-435
ANO Schedule 11 Part B Ratings, rating endorsements and licence endorsements 435-436
ANO Schedule 12 Part A Records to be kept in accordance with article 125(1).......................................436
ANO Schedule 12 Part B Records required in accordance with article 125(4)..436
ANO Schedule 12 Air traffic service equipment – records required and matters to which the 436-437
CAA may have regard
ANO Schedule 12 Part C Matters to which the CAA may have regard in granting an approval 436-437
of apparatus under article 125(5)
ANO Schedule 13 Aerodrome manual...437
ANO Schedule 14 Part A Provisions referred to in article 148(5)...437-439
ANO Schedule 14 Part B Provisions referred to in article 148(6)...439-440
ANO Schedule 14 Part C Provisions referred to in article 148(7)...440
ANO Schedule 14 Penalties 437-440
ANO Schedule 15 Parts of straits specified in connection with the flight of aircraft in transit over the 441
United Kingdom territorial waters
ANO The Rules Of The Air...443-462
Approach Ban – All aircraft..315
Approval and Licensing of Aircraft Radio Stations..128
Approval and Licensing of Ground Radio Stations...129
Area Codes..199
Area of Intense Air Activity (AIAA)..197
Areas of Responsibility for Providing Aeronautical Telecommunications...123
Areas with Sensitive Fauna...287
Articles To The Air Navigation Order..325-388
ATC Assistance and Responsibilities..214
ATC Clearance 176
ATC Incidents in Foreign Airspace..255
ATC incidents in UK Airspace – Foreign Pilots/Operators 255
ATS Flight Planning Restrictions...176

Index

ATS Route Designators ... 176
ATS Routes and Upper Control Areas (UTA) ... 176
ATS Routes Description ... 176
ATS Rules and Procedures ... 176
ATS Units Address List ... 122
ATS Units Participating in the Lower Airspace Radar Service ... 225-227
Automatic Query/Response – UK International NOTAM Database ... 106
Availability of the MATZ Penetration Service ... 266
Aviation Organisations ... 515-526

B

Bird Concentration Areas March-July Map ... 299
Bird Concentration Areas October-March Map ... 300
Bird Migration And Areas With Sensitive Fauna ... 287
Bird Migration ... 287
Bird Sanctuaries ... 287
Boscombe Down Advisory Radio Area ... 197
Bristol Filton and Bristol LARS ... 228
Broadcast Text Meteorological Information ... 150

C

CAA Chart Legend ... 86-89, 600-601
CAA Contact Details ... 527-530
CAA Medical Department ... 549-552
CAA Medical Examiners ... 532-548
CAA Regulations 1991 Article 1 Citation and commencement ... 496
CAA Regulations 1991 Article 2 Revocation ... 496
CAA Regulations 1991 Article 3 Interpretation ... 496-497
CAA Regulations 1991 Article 4 Service of documents ... 497
CAA Regulations 1991 Article 5 Publication by the Authority ... 498
CAA Regulations 1991 Article 6 Regulation of the conduct of the Authority ... 498-499
CAA Regulations 1991 Article 7 Reasons for decisions ... 499
CAA Regulations 1991 Article 8 Inspection of aircraft register ... 499
CAA Regulations 1991 Article 9 Dissemination of reports of reportable occurrences ... 499
CAA Regulations 1991 Article 10-14 Substitution of a public use aerodrome licence for an ordinary ... 499-501
aerodrome licence for a public use aerodrome licence
CAA Regulations 1991 Article 15 Regulation of the conduct of the Authority ... 501
CAA Regulations 1991 Article 16 Application for the grant, revocation, suspension or variation of licences ... 501-502
CAA Regulations 1991 Article 17 Revocation, suspension or variation of licences without ... 502
application being made
CAA Regulations 1991 Article 18 Variation of schedules of terms ... 502-503
CAA Regulations 1991 Article 19 Environmental cases ... 503
CAA Regulations 1991 Article 20 Objections and representations ... 503
CAA Regulations 1991 Article 21 Consultation by the Authority ... 503-504
CAA Regulations 1991 Article 22 Furnishing of information by the Authority ... 504
CAA Regulations 1991 Article 23 Preliminary hearings ... 504
CAA Regulations 1991 Article 24 Preliminary hearings of allegations of behaviour damaging to a competitor ... 504-505
CAA Regulations 1991 Article 25 Hearings in connection with licences ... 505
CAA Regulations 1991 Article 26 Procedure at hearings ... 505-506
CAA Regulations 1991 Article 27 Appeals to the Secretary of State ... 506-507
CAA Regulations 1991 Article 28 Appeal from decisions after preliminary hearings of allegations of ... 507
behaviour damaging to a competitor
CAA Regulations 1991 Article 29 Decisions of appeals ... 507-508
CAA Regulations 1991 Article 30 Transfer of licences ... 508
CAA Regulations 1991 Article 31 A Determination by the authority ... 508
CAA Regulations 1991 Article 31 B Representations ... 508-509
CAA Regulations 1991 Article 31 C Hearings in connection with licences ... 509
CAA Regulations 1991 Article 31 D Procedure at hearings ... 509
CAA Regulations 1991 Article 31 E Determination by Authority and Appeal to the Secretary of State ... 509-510
CAA Regulations 1991 Article 31 F Decision by Secretary of State on appeal ... 510
CAA Regulations 1991 Article 31 Surrender of licences ... 508
CAA Regulations 1991 Article 32 Participation in civil proceedings ... 510
CAA Safety Sense Leaflet 1 Good Airmanship guide ... 621-629

Index

CAA Safety Sense Leaflet 2 Care of Passengers 630-634
CAA Safety Sense Leaflet 3 Winter Flying 635-641
CAA Safety Sense Leaflet 4 Use of MOGAS 642-646
CAA Safety Sense Leaflet 5 VFR Navigation 647-652
CAA Safety Sense Leaflet 6 Aerodrome Sense 653-660
CAA Safety Sense Leaflet 7 Aeroplane performance 661-666
CAA safety Sense Leaflet 8 Air Traffic Services Outside Controlled Airspace 667-672
CAA Safety Sense Leaflet 9 Weight and Balance 673-679
CAA Safety Sense Leaflet 10 Bird Avoidance 679-683
CAA Safety Sense Leaflet 11 Interception Procedures 684-688
CAA Safety Sense Leaflet 12 Strip Sense 689-694
CAA Safety Sense Leaflet 13 Collision Avoidance 695-702
CAA Safety Sense Leaflet 14 Piston Engine Icing 703-708
CAA Safety Sense Leaflet 15 Wake Vortex 709-714
CAA Safety Sense Leaflet 16 Balloon Airmanship guide 715-723
CAA Safety Sense Leaflet 17 Helicopter Airmanship 724-734
CAA Safety Sense Leaflet 18 Military Low Flying 735-740
CAA Safety Sense Leaflet 19 Aerobatics 741-745
CAA Safety Sense Leaflet 20 VFR Flight Plans 746-752
CAA Safety Sense Leaflet 21 Ditching 753-762
CAA Safety Sense Leaflet 22 Radiotelephony for General Aviation 763-775
CAA Safety Sense Leaflet 23 Pilots: Its your decision 776-781
CAA Safety Sense Leaflet 24 Pilot Health 782-786
CAA Safety Sense Leaflet 25 Use of GPS 787-793
CAA Safety Sense Leaflet 26 Visiting Military Aerodromes 794-801
CAA Safety Sense Leaflets 619-802
Cable Launching of Gliders, Hang Gliders and Parascending Parachutes 199
Captive and Free Flight Manned Balloon Launch Sites 200
Carriage of Airborne Collision Avoidance Systems (ACAS) in the United Kingdom FIR and UIR 40
Carriage of Radio and Radar Equipment 128
Carriage of Radio and Radio Navigation Equipment 38
Carriage of Radio Communication and Navigation Equipment 178
Carriage of SSR Transponders 39
Charges at London Heathrow, London Gatwick, London Stansted, Aberdeen/Dyce, Edinburgh and Glasgow 172
Aerodromes where National Air Traffic Services Limited (NATS) provides the Navigation Services
Charges For Aerodromes/Heliports and Air Navigation Services 171
Charges for Approach Services provided from an Aerodrome to aircraft which do not land at that Aerodrome 173
Charges for En-route Navigation Services made available by NATS in the Shanwick Oceanic Control Area 173
Charges for the Navigation Services made available by NATS for flights made by Helicopter to a 173
Vessel or Off-shore Installation in the North Sea
Chart Amendment and Revision 109
Civil Aviation Regulations 1991 496-511
Class A – Controlled Airspace 203
Class B – Controlled Airspace 204
Class B Gliding Areas North of 5520N Map 298
Class B Gliding Areas South of 5520N Map 297
Class C – Controlled Airspace 205
Class D – Controlled Airspace 206
Class E – Controlled Airspace 208
Class F – Advisory Airspace 209
Class G Airspace 210
Clearance to enter the Hebrides UTA 180
Common Frequency for Helicopter Departures 126
Common VHF frequency for Use at Aerodromes having no notified Ground Radio Frequency 126
Control Areas (Airways) 176
Conversion Factors 608
Conversion of Reported Meteorological Visibility to RVR – All aircraft 315
Conversion Tables 103
Co-ordination between the Operator and ATS 121
Co-ordination of Civil and Military Aircraft 180
Criteria for Special Meteorological Reports and Forecasts 142
Cruising Levels 179

Index

Customs Contact Points 30
Customs Requirements – Arriving on Flights from another EU Country 34
Customs Requirements – Arriving on Flights from Non-EU Countries 34
Customs Requirements – Customs and Excise Forms 34
Customs Requirements – Departing on Flights to Non-EU Destinations 34
Customs Requirements – Passengers 34
Customs Requirements –Aircrew 34

D

Danger Area – Byelaws 196
Danger Area – Pilotless Target Aircraft 196
Danger Area Activity Information Service 196
Danger Area Crossing Service 196
Dangerous Goods Regulations 2002 Article 1 Citation and commencement 489
Dangerous Goods Regulations 2002 Article 2 Revocation 489
Dangerous Goods Regulations 2002 Article 3 Interpretation 489-490
Dangerous Goods Regulations 2002 Article 4 Requirement for approval of operator 490
Dangerous Goods Regulations 2002 Article 5 Prohibition of carriage of dangerous goods ..490-491
Dangerous Goods Regulations 2002 Article 6 Provision of information by the operator to crew etc .. 491
Dangerous Goods Regulations 2002 Article 7 Acceptance of dangerous goods by the operator 491-492
Dangerous Goods Regulations 2002 Article 8 Method of loading by operator 492
Dangerous Goods Regulations 2002 Article 9 Inspections by the operator for damage, leakage or contamination 492
Dangerous Goods Regulations 2002 489-495
Dangerous Goods Regulations 2002 Article 10 Removal of contamination by the operator 492
Dangerous Goods Regulations 2002 Article 11 Shipper's responsibilities 492-493
Dangerous Goods Regulations 2002 Article 12 Commander's duty to inform air traffic services ... 493
Dangerous Goods Regulations 2002 Article 13 Provision of training 493
Dangerous Goods Regulations 2002 Article 14 Provision of information to passengers 493-494
Dangerous Goods Regulations 2002 Article 15 Provision of information in respect of cargo 494
Dangerous Goods Regulations 2002 Article 16 Keeping of documents and records 494
Dangerous Goods Regulations 2002 Article 17 Production of documents and records 494
Dangerous Goods Regulations 2002 Article 18 Powers in relation to enforcement of the Regulations 494-495
Dangerous Goods Regulations 2002 Article 19 Occurrence reporting 495
Dangerous Goods Regulations 2002 Article 20 Dropping articles for agricultural, horticultural, ...495 forestry or pollution control purposes
Dangerous Goods Regulations 2002 Article 21 Police aircraft 495
Designated Airports – Customs, Immigration and Health 29
Designated Authorities 23
Details of Meteorological Observations and Reports for UK aerodromes 132
Determination of Minima – Additional Cases 316
DIALMET 151
Differences from ICAO Standards, Recommended Practice and Procedures 42-71
Dimensional Units 74
Diversion – Enhanced Non-Standard Flights (ENSFs) – Entry into EG R157 (Hyde Park)/E1G R158 189
(City of London)/EG R169 (Isle of Dogs) Restricted Areas
Diversion – Non Deviating Status (NDS) 190
Diversion – Non-Standard Flights (NSFs) in Controlled Airspace 188
Diversion – Unusual Aerial Activities (UAA) Outside Controlled Airspace 191
Diversion – Unusual Aerial Activities in Controlled Airspace 189
Diversion – VFR Flight in Class C Airspace above FL 245 189
Diversion – VFR Flight in Class C Areas of Delegated ATS 189
Diversion 187
Doc 4444 Procedures for Air Navigation Services Rules of the Air and Air Traffic Services (14th Edition) 67
Doc 8168 Procedures for Air Navigation Services – Aircraft Operations Volume II 68
(Construction of Visual and Instrument Flight Procedures) (4th Edition)
Doc 8168 Procedures for Air Navigation Services Aircraft Operations Vol 1 (Flight Procedures) (4th Edition) 68

E

Eastbound and Westbound Flights 180
Emergency Satellite Voice Calls from Aircraft 127
Emergency Service 127
En-Route Air Navigation Service Charges – Eurocontrol 173
En-Route Air Navigation Service Charges – Methods of Payment 174

En-Route Air Navigation Service Charges – Value Added Tax 174
En-Route Air Navigation Service Charges 173
En-Route High Level Holding 182
En-Route Holding 181
En-Route Obstacles 199
Entry, Transit and Departure of Passengers and Crew 24, 34
European Medium/High Level Spot Wind/Temperature Forecast Chart – Metform 614 139
Evacuation of UK NOTAM Office (NOF) 106

F

Failure of Navigation Equipment – Actions taken by ATC 185
Failure of Navigation Equipment – Instrument Meteorological Conditions (IMC) 185
Failure of Navigation Equipment -Visual Meteorological Conditions (VMC) 185
Failure of Navigation Equipment 184
Failure of Two-way Radio Communications Equipment 184
Flight Level Graph 234
Flight Plan Procedures for Helicopter Operations over Sea Areas around the United Kingdom 264
Flight Planning – Action in the Event of Diversion 236
Flight Planning – Booking Out 235
Flight Planning – Cancelling an IFR Flight Plan in Flight 236
Flight Planning – Persons On Board 236
Flight Planning – Submission Time Parameters 235
Flight Planning – Submitting a Flight Plan Through the Departure Aerodrome ATSU 236
Flight Planning – Types and Categories of Flight Plan 235
Flight Planning – When to file a Flight Plan 235
Flight Planning 235
Flight Plans, ATC clearance and other procedures 180
Flights crossing ADRs or Published Holding Patterns under IFR 179
Flights Crossing Airways in IFR 177
Flights From/To Countries outside the EU – Arrival in the United Kingdom from outside the EU 26
Flights From/To Countries outside the EU – Flights to and from the United Kingdom – 27
Non-scheduled Flights – Commercial
Flights From/To Countries outside the EU – Private Flights 27
Flights From/To Countries outside the EU 26
Flights Joining Airways 177
Fluids 605-607
Flying Licences and Ratings – Experienced Holders of Non-UK Professional Pilot's Licences 36
Flying Licences and Ratings – JAA Licences issued by Joint Aviation Authorities Member States 36
Flying Licences and Ratings – Visiting Pilots – Instrument and Flying Instructor Ratings 36
Flying Licences and Ratings 36
FM Broadcast Interference 214
Foreign NOTAM AFS Addressing 107
Free-Fall Drop Zones 200, 277-282

G

General Conditions under which Aerodromes/Heliports and Associated Facilities are Available for Use ... 303
General Description of CAA Aeronautical Charts 110
Geodetic Reference Data 74
Glider Launching Sites 199, 268-274
Global Upper-air and Temperature Data 140
Government Aerodromes 321
Grouping Of Aerodromes/Heliports 321

H

Hang Gliding and Parascending Sites 199, 275-277
Hazardous Cargo Requirements – Carriage of Dangerous Goods 37
Hazardous Cargo Requirements – Carriage of Munitions of War 36
Hazards and Danger Areas 182
Hazards to Flight 196
Helicopter Routes in the London Control Zone – Scale 1:50,000 113
Helicopters – Airborne Radar Approach (ARA) for Over Water Operations 314
Helicopters – Onshore non-precision approaches 314
Heliports – Categories of precision operation 307

Index

Heliports – Instrument Approaches ... 307
Heliports ... 307
HF VOLMET Broadcasts ... 579-580
HM Coastguard Maritime Rescue Centres ... 170
Holding and Approach to Land Procedures ... 211
Holding, Approach and Departure Procedures ... 211

I

ICAO Aircraft Designators ... 553-561
ICAO and IATA Airfield Designators ... 568-574
ICAO Document Listing ... 575-576
ICAO IAP Charts ... 109
ICAO Wake Turbulence Classification ... 577
IFR Flight – Communications ... 203
IFR Flight – Cruising Levels ... 203
IFR Flight – Position Reports ... 203
IFR Flight Plans – Filing of Flight Plans and Associated Messages ... 239
IFR Flight Plans – Flights Departing from an Aerodrome Within, and then Exiting, the IFPS Zone ... 240
IFR Flight Plans – Flights Entering or Over flying the IFPS Zone ... 239
IFR Flight Plans – Flights Wholly Within the IFPS Zone ... 239
IFR Flight Plans – General Description of IFPS ... 239
IFR Flight Plans – Submission of Flight Plans ... 239
IFR Flight Plans – The Re-Addressing Function ... 240
IFR Flight Plans with portion(s) of flight operated as VFR – Compilation of Flight Plans ... 240
IFR Flight Plans with portion(s) of flight operated as VFR – Compilation of Associated Messages ... 241
IFR Flight Plans with portion(s) of flight operated as VFR – Replacement Flight Plan Procedure ... 241
IFR Flight Plans with portion(s) of flight operated as VFR – Supplementary Flight Plan Information ... 242
IFR Flight Plans with portion(s) of flight operated as VFR ... 240
IFR Flight Plans ... 239
IFR Flight ... 202
Immigration (Carriers' Liability) Act 1987 ... 35
Immigration Requirements – Landing Cards ... 35
Infringements of CANP Airspace ... 246
Instrument Approach and Aerodrome Charts – ICAO ... 111
Instrument Flight Rules – Change from IFR flight to VFR flight ... 202
Instrument Flight Rules – Rules applicable to IFR flights outside Controlled Airspace ... 203
Instrument Flight Rules – Rules applicable to IFR flights within Controlled Airspace ... 202
Instrument Flight Rules ... 202
Integrated Aeronautical Information Package ... 104
Interception Of Civil Aircraft ... 249-252
Interception Procedures ... 602-603
Interference from High Powered Transmitters ... 130
Internet Services ... 151
Investigation of AIRPROX ... 254

K

Keevil Aerodrome ... 200
Kites ... 200

L

Land-Based Air Navigation Obstacles ... 198
Leaving or Joining Advisory Routes ... 179
Licenced Aerodromes ... 321
List of Radio Navigation Aids Decode ... 99-102
List of Radio Navigation Aids Encode ... 96-99
Listening Watch ... 178
Location Indicators Decode ... 93-95
Location Indicators Encode ... 91-93
London ACC –FIS Sectors Map ... 294
Loss of Communication ... 181
Low Level Civil Aircraft Notification Procedures (CANP) – Commercial Aerial Activity ... 244
Low Level Civil Aircraft Notification Procedures (CANP) ... 244
Low Level Cross-Channel Operations – UK/France ... 183

Lower Airspace Radar Service Map...........289
Lower Airspace Radar Service...........224

M

Malfunctions, Maintenance and Test Transmissions...........130
MATZ Participating Aerodromes...........266-267
Measuring System, Aircraft Markings, Holidays...........74
Meteorological charts are available via facsimile from two automated services, Broadcast Fax and METFAX...........149
Meteorological Codes – Aerodrome Forecast (TAF) Codes...........160
Meteorological Codes – Aerodrome Weather Report Codes (Actuals)...........155
Meteorological Codes – Aerodrome, North Sea Helicopter and Shanwick Charges...........174
Meteorological Codes – Air Temperature/Dewpoint...........157
Meteorological Codes – AUTO METAR coding...........159
Meteorological Codes – CAVOK...........157
Meteorological Codes – Cloud...........157
Meteorological Codes – Differences from the MTEAR...........160
Meteorological Codes – Eurocontrol Charges...........174
Meteorological Codes – Examples of METAR...........159
Meteorological Codes – Horizontal Visibility...........156
Meteorological Codes – Identifier...........155
Meteorological Codes – QNH...........157
Meteorological Codes – Reports in Abbreviated Plain Language...........162
Meteorological Codes – RMK...........159
Meteorological Codes – Runway State Group...........158
Meteorological Codes – RVR...........156
Meteorological Codes – Significant Present and Forecast Weather Codes...........157
Meteorological Codes – Supplementary Information...........158
Meteorological Codes – TREND...........159
Meteorological Codes – Weather...........156
Meteorological Codes – Wind...........155
Meteorological Codes...........155
Meteorological Radio Broadcasts (VOLMET)...........148
Meteorological Services – Accuracy of Meteorological Measurement or Observation...........132
Meteorological Services – Aerodrome Warnings...........132
Meteorological Services – Climatological information for certain UK aerodromes...........134-137
Meteorological Services – Forecast Offices providing a service to Civil Aviation...........137
Meteorological Services – Main Forecast Weather Chart and Text...........138
Meteorological Services – Marked Temperature Inversion...........133
Meteorological Services – Observing Systems and Operating Procedures...........131
Meteorological Services – Outlook Chart and Text...........139
Meteorological Services – Pre-flight Briefing...........137
Meteorological Services – Runway Visual Range (RVR)...........133
Meteorological Services – The UK Low Level Forecast (Metform 215)...........138
Meteorological Services – UK Low Level Weather and Spot Wind Forecast Charts – Metform 215/214...........138
Meteorological Services – Weather Forecast Chart Issues...........139
Meteorological Services – Windshear Alerting Service – London Heathrow and Belfast Aldergrove Airports...........133
Meteorological Services...........131
METFAX Services...........150
Microlight Sites...........200, 282-286
Military Aerodrome Traffic Zones Map...........296
Military Aerodromes in UK Territorial Airspace...........316
Military Helicopter Operations in the Salisbury Plain Area...........197
Military Mandatory Radar Service Area (MRSA)...........179
Military Middle Airspace Radar Service Map...........290
Military Middle Airspace Radar Service...........228
Military Personnel – Reporting of AIRPROX...........255
Military Training Area (MTA) and Military Temporary Reserved Airspace (MTRA)...........181
Military Training Area (MTA) or Military Temporary Reserved Airspace (MTRA) – An area of Upper Airspace of defined...........196
Minimum and Maximum Cruising Levels...........178
Miscellaneous AIP Charts...........113
Missed Approach Climb Gradient...........213

Index

Morecambe Bay and Liverpool Bay Gas Fields – Helicopter Support Flights 263
Mountains and Hills with Warning Lights 199
Mu-Meter and Grip Tester Friction Levels 306

N

National Regulations – Aerodrome Operating Minima 24
National Regulations – Arrival and Departure of Civil Aircraft on Flights between Great Britain, Republic of Ireland, Northern Ireland, Isle of Man or the Channel Islands 24
National Regulations – Customs and Excise, Immigration and Health Airports 25
National Regulations – Forced Landings 25
National Regulations – Jet Aircraft 24
National Regulations –Crossing UK Boundaries 24
National Regulations 24
Navigational Assistance 217
Non-Standard Civil Flights and Unusual Aerial Activities in the UK Upper Airspace 180
Northern North Sea and Atlantic Rim Low Level Radar Advisory, Flight Information Service 258-263
and Helicopter Operating Procedures
Northwest Europe Low Level Weather and Spot Wind Forecast Charts – Metforms 415/414 139
NOTAM (Notices to Airmen) 105
NOTAM Construction 106
NOTAM Handling 106

O

Observation Flights Conducted Under the Treaty on Open Skies 194
Obstacle Clearance Heights for Aerodromes with Published Instrument Approach Procedures 212
Obstacle Clearance 211
Off-shore Air Navigation Obstacles 199
Off-Shore Operations – Helicopter Main Routes (HMR) 257
Off-Shore Operations – Helicopter Protected Zone (HPZ) 257
Off-Shore Operations – Out of Hours Helicopter Operations 258
Off-Shore Operations – Southern North Sea Low Level Radar Advisory, Flight Information 255
Service and Helicopter Operating Procedures
Off-Shore Operations – Use of GPS for North Sea Operations 258
Off-Shore Operations 255
Organisation of the UK AIS 104
Other Automated Meteorological Services 149
Other Regulated Airspace Military Aerodrome Traffic Zones 265
Other Temporary Hazards 200
Other Traffic Using ADRs 179
Overseas Aerodromes 316

P

Parachute Flares and Other Illuminants 197
Position Reporting within the London and Scottish FIR/UIR – Climb and Descent 183
Position Reporting within the London and Scottish FIR/UIR – DME Distance Reports to ATC 183
Position Reporting within the London and Scottish FIR/UIR – Minimum Rates of Climb and Descent 183
Position Reporting within the London and Scottish FIR/UIR – Noise Abatement Approach Techniques 184
Position Reporting within the London and Scottish FIR/UIR – Omit Position Report Procedure 183
Position Reporting within the London and Scottish FIR/UIR – Vacating (Leaving) Levels 183
Position Reporting within the London and Scottish FIR/UIR 183
Position Reports 179
Powered Aircraft – Airway Crossings 178
Powered Aircraft – Other penetrations of Airways 178
Powers and Obligations of Captains and Owners or Agents of Aircraft Under the Immigration Act 1971 ... 35
Precision Approach Terrain Charts – ICAO 113
Pre-flight Information Bulletins (PIB) 107
Priority of Messages 129
Priority on ATS Advisory Routes 178
Procedural and Aerodrome Charts 110
Procedure for Glider Operations in Class B Airspace 179
Procedure for Participating Aircraft 178
Procedure Turns 213
Procedures for Military Aircraft 178

Procedures for Penetration of a MATZ by Civil Aircraft	265
Protection of Instrument Approach Procedures at Aerodromes outside Controlled Airspace	214
Public Health Requirements	36
Public Holidays in the UK	75

R

Radar Advisory Service (RAS)	179, 215
Radar Approach Procedures – Obstacle Clearance Heights and Missed Approach Procedures	213
Radar Information Service (RIS)	216
Radar Service Outside Controlled Airspace	215
Radar Services and Procedures	215
Radar Vectoring Area Charts	217
Radar Vectoring Controlled Airspace	217
Radar Vectoring for ILS Approach	217
Radio Communications and Equipment	176
Radio Communications between Aerodrome Fire Services and Aircraft during an Emergency	127
Radio Failure and Loss of Communication Procedures	184
Radio Failure Procedures for Pilots	184
Radio Navigation Aids – Coastal Refraction	125
Radio Navigation Aids – Designated Operational Coverage	125
Radio Navigation Aids – Lack of Failure Warning System	126
Radio Navigation Aids – Limitations of Non-Directional Beacons and Automatic Direction Finding Equipment	125
Radio Navigation Aids – Mountain Effect	125
Radio Navigation Aids – Night Effect	125
Radio Navigation Aids – Static Interference	125
Radio Navigation Aids – Station Interference	125
Radio Telephony Phraseology For Use With SSR	219
Radio Test in Flight Procedure	128, 129
Regional supplementary procedures are applied in accordance with ICAO Doc 7030/4	235
Regional Supplementary Procedures	
Relay of RTF Communications to the Public	127
Release of Racing Pigeons	288
Reporting Points	181
Requests for NOTAM Issue	106
Requests for NOTAM Reception	106
Rescue and Fire Fighting Services and Snow Plan	317
RNAV	37
Royal Flight Callsigns	193
Royal Flights – Establishment of Temporary (Class A/C) Controlled Airspace (CAS-T)	192
Royal Flights – Permanent Control Zones and Areas	192
Royal Flights – Procedures Applicable to Royal Flight CAS-T	193
Royal Flights – Promulgation of Royal Flight information	193
Royal Flights – Special ATC Arrangements for Royal Flights in Fixed-Wing Aircraft	192
Royal Flights – Temporary Control Areas	192
Royal Flights – Temporary Control Zones	192
Royal Flights – Temporary Controlled Airways	192
Royal Flights in Helicopters	193
Royal Flights	192
RPL Submission Procedure	242
RSVM	37, 235
RTF and NDB Frequencies Used on Off-shore Installations	264
RTF Standard Words & Phrases	604

S

SAR Agreements	164
Schedules To The Air Navigation Order	389-441
Scottish ACC – FIS Sectors	295
Search and Rescue – Action at Night	167
Search and Rescue – Action by Survivors	166
Search and Rescue – Aircraft Not Equipped with Radio	166
Search and Rescue – Air-to-Ground Signals	168
Search and Rescue – Alerting	165
Search and Rescue – Callsigns	167

Index

Search and Rescue – Communications 165
Search and Rescue – Crash Landing in Isolated Area 166
Search and Rescue – Difficult Areas for SAR 165
Search and Rescue – Distress Frequencies 164
Search and Rescue – Emergency Locator Transmitters (ELT) 167
Search and Rescue – Emergency Satellite Voice Calls from Aircraft 165
Search and Rescue – Flight in Areas in which Search and Rescue Operations are in Progress 166
Search and Rescue – Ground To Air Emergency Signalling Code 168, 169
Search and Rescue – Procedure for a Pilot-in-Command Requiring SAR Escort Facilities 166
Search and Rescue – Scene of Search Frequencies 164
Search and Rescue – Signals with Surface Craft 167
Search and Rescue – The Rescue Organisation 164
Search and Rescue – Types of Services 163
Search and Rescue – Units with Emergency Facilities on 121.500MHz 166
Search and Rescue Signals 167
Search and Rescue 162
Sectorisation of Visual Manoeuvring (Circling) Area 211
Separation Standards 178
Sigmet Service 149
Snow Plan – Aerodromes participating 321
Snow Plan – Assessment and Notification of Runway Surface and Allied Conditions 319
Snow Plan – Clearance Techniques 318
Snow Plan – Operational Priorities for the Clearance of Movement Areas 318
Snow Plan – Responsibility for Planning and Implementation 318
Snow Plan – Runways Affected by Slush 320
Snow Plan – Runways Affected by Snow and Ice 319
SNOWTAM 107
Special Equipment to be Carried 38
SSR Operating Procedures – Mode A Conspicuity Code 218
SSR Operating Procedures – Mode S Aircraft Identification 219
SSR Operating Procedures – Special Purpose Codes 218
SSR Operating Procedures – Transponder Failure 219
SSR Operating Procedures 218
Standard Instrument Departure (SID) and Arrival Routes (STAR) 90
Standard Instrument Departure, Standard Instrument Arrival, Standard Terminal Arrival Route 113
and Noise Abatement Charts
Submission of Repetitive Flight Plan (RPL) Data to Eurocontrol CFMU – Brussels 242
Summary of Holding and Approach to Land Procedures at Individual Aerodromes Introductory Notes ... 212
Supplements to the AIP (AIP SUP) 105
Supply of UK AIS Documents 107

T

Terrain Clearance – Advice to Pilots 225
Terrain Clearance 181, 217, 225
The Air Navigation (Restriction Of Flying) (City of London) Regulations 2004 488
The Air Navigation (Restriction Of Flying) (Highgrove House) Regulations 1991 486
The Air Navigation (Restriction Of Flying) (Hyde Park) Regulations 2004 487
The Air Navigation (Restriction Of Flying) (Isle of Dogs) Regulations 2004 488
The Air Navigation (Restriction Of Flying) (Nuclear Installations) Regulations 2002 483-484
The Air Navigation (Restriction Of Flying) (Prisons) Regulations 2001 484-485
The Air Navigation (Restriction Of Flying) (Scottish Highlands) Regulations 1981 482
The Air Navigation (Restriction Of Flying) (Specified Areas) Regulations 2005 486
The UK Low Level Spot Wind Forecast (Metform 214) 139
The UK Search and Rescue Region (SRR) 163
The Upper Airspace Control Area and the Hebrides UTA 179
Time System 74
Topographical Air Charts of the United Kingdom – Scale 1:250,000 .... 110-111, 115
Types of Radar Service – Establishing a Service 216
Types of Radar Service 215

U

UK Airspace Restrictions and Hazardous Areas – 1:1,000,000 113
UK Areas of Intense Air Activity, Aerial Tactics Areas and Military Low Flying System – 1:1,000,000 ... 113

UK Call Signs	610-618
UK Definitions for Terms used by ICAO	73
UK Immigration Service Offices	31-33
UK Interpration Differences	72
UK Military Low Flying System	197
UK NOTAM Series – Post-Flight Information	109
UK NOTAM Series – Pre-Flight Briefing	108
UK NOTAM Series	108
UK Parent Unit System	237
UK Pressure Setting Chart	609
UK SSR Code Assignment Plan	219-224
United Kingdom Altimeter Setting and Flight Information Regions Map	292-293
United Kingdom Police Forces	28-29
Unlicensed Aerodromes	321
Upper Airspace Control Area	179
Use of Airborne Collision Avoidance Systems (ACAS) in United Kingdom FIR and UIR	186
Use of Airborne Collision Avoidance Systems (ACAS) in United Kingdom FIR and UIR – Diversion	187
Use of Airborne Collision Avoidance Systems (ACAS) in United Kingdom FIR and UIR – Procedures to be Established	186
Use of Airborne Collision Avoidance Systems (ACAS) in United Kingdom FIR and UIR – TCAS II Operating Characteristics	186
Use of Airborne Collision Avoidance Systems (ACAS) in United Kingdom FIR and UIR – Operation of Aircraft when ACAS II is Unserviceable	186
Use of Airborne Collision Avoidance Systems (ACAS) in United Kingdom FIR and UIR – Operation of TCAS II in RVSM Airspace	186
Use of Airborne Collision Avoidance Systems (ACAS) in United Kingdom FIR and UIR – Guidance for Aircraft Operators and Flight Crews	186
Use of Airborne Collision Avoidance Systems (ACAS) in United Kingdom FIR and UIR – Guidance for Air Traffic Service Providers and for Air Traffic Controllers	187
Use of GPS for North Sea Operations	41
Use Of Radar in Air Traffic Services	215
Use of VHF R/T Channels	126

V

VFR Flight – Radio Communication Failure Procedures	202
VFR Flight Plans – Action When the Destination Aerodrome has no ATSU or AFTN Link	238
VFR Flight Plans – Airborne Time	238
VFR Flight Plans – Submission Time Parameters	238
VFR Flight Plans with portion(s) of flight operated as IFIR	238
VFR Flight Plans	238
Visual Flight Rules – Special VFR Flight	201
Visual Flight Rules – VFR Flight	201
Visual Flight Rules	201
Visual Manoeuvring (Circling) OCHs	213
Visual Manoeuvring (Circling) VM(C) in the Vicinity of the Aerodrome after Completing an Instrument Approach	211-212
VOR TACAN Channel Pairing	581-583

W

Warton Radar Advisory Service Area (RASA) Map	291
Warton Radar Advisory Service Area (RASA)	229
Weather Avoidance	218
When to File a VFR Flight Plan	238
World Time Zones & Map	584-585
World Wide SR-SS Table	586-588